Straight Line

$y = mx + b$ where m = slope = $\Delta y/\Delta x$. Δy is the amount y changes when  changes by Δx.

Quadratic Equation

Roots of $ax^2 + bx + c = 0$ are

$$x_{1,2} = \frac{-b \pm \sqrt{b^2 - 4ac}}{2a}$$

If $(b^2 - 4ac) \geq 0$, roots are real; if $(b^2 - 4ac) < 0$, roots are complex conjugates.

Trigonometric Identities

$\sin(-\alpha) = -\sin\alpha$ $\cos(-\alpha) = \cos\alpha$

$\sin(\omega t \pm 90°) = \pm\cos\omega t$ $\sin(\omega t \pm 180°) = -\sin\omega t$

$\cos(\omega t \pm 90°) = \mp\sin\omega t$ $\cos(\omega t \pm 180°) = -\cos\omega t$

$$\sin^2\omega t = \frac{1}{2}(1 - \cos 2\omega t)$$

Natural Logarithms

Given $e^x = y$. x may be found by noting that $\ln e^x = x$.
For example, if $e^{-5t} = 0.7788$, then

$$\ln e^{-5t} = \ln 0.7788$$
$$-5t = -0.25$$
$$t = 0.05 \text{ s}$$

Calculus

$$\frac{d}{dx}(ax) = a \qquad\qquad \frac{d}{dx}(\sin ax) = a\cos ax$$

$$\frac{d}{dx}(e^{ax}) = ae^{ax} \qquad\qquad \frac{d}{dx}(\cos ax) = -a\sin ax$$

$$\frac{d}{dx}(uv) = u\frac{dv}{dx} + v\frac{du}{dx}$$

$$\int x\,dx = \frac{x^2}{2} \qquad\qquad \int \sin ax\,dx = -\frac{1}{a}\cos ax$$

$$\int \frac{dx}{x} = \ln x \qquad\qquad \int \cos ax\,dx = \frac{1}{a}\sin ax$$

Circuit Analysis
Theory and Practice

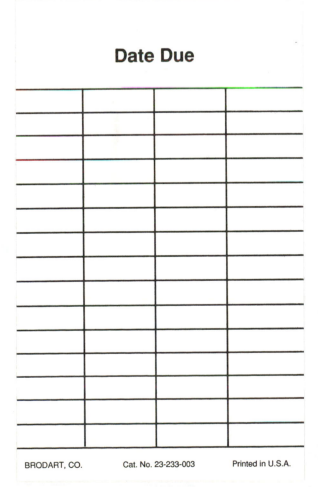

Date Due

SECOND EDITION

Circuit Analysis
Theory and Practice

ALLAN H. ROBBINS
Red River College, Manitoba

WILHELM C. MILLER
Red River College, Manitoba

Delmar
Thomson Learning™

Africa • Australia • Canada • Denmark • Japan • Mexico
New Zealand • Philippines • Puerto Rico • Singapore
Spain • United Kingdom • United States

Notice to the Reader

Delmar Staff

Business Unit Director: Alar Elken
Executive Editor: Sandy Clark
Acquisitions Editor: Gregory L. Clayton
Developmental Editor: Michelle Ruelos Cannistraci
Editorial Assistant: Amy E. Tucker
Executive Marketing Manager: Maura Theriault
Channel Manager: Mona Caron

Marketing Coordinator: Paula Collins
Executive Production Manager: Mary Ellen Black
Production Manager: Larry Main
Senior Project Editor: Christopher Chien
Art Director: Nicole Reamer
Technology Project Manager: Tom Smith

COPYRIGHT © 2000

Delmar is a division of Thomson Learning. The Thomson Learning logo is a registered trademark used herein under license.

Printed in the United States of America

1   2   3   4   5   6   7   8   9   10   XXX   05   04   03   02   01   00

For more information, contact Delmar at 3 Columbia Circle, PO Box 15015, Albany, New York 12212-5015; or find us on the World Wide Web at http://www.delmar.com

For more information, contact:

Asia
Thomson Learning
60 Albert Street, #15-01
Albert Complex
Singapore 189969

Australia/New Zealand
Nelson/Thomson Learning
102 Dodds Street
South Melbourne, Victoria 3205
Australia

Canada
Nelson/Thomson Learning
1120 Birchmont Road
Scarborough, Ontario
Canada M1K 5G4

International Headquarters
Thomson Learning
International Division
290 Harbor Drive, 2nd Floor
Stamford, CT 06902-7477
USA

Japan
Thomson Learning
Palaceside Building 5F
1-1-1 Hitotsubashi, Chiyoda-ku
Tokyo 100 0003
Japan

Latin America
Thomson Learning
Seneca, 53
Colonia Polanco
11560 Mexico D. F. Mexico

South Africa
Thomson Learning
Zonnebloem Building
Constantia Square
526 Sixteenth Road
P.O. Box 2459
Halfway House, 1685
South Africa

Spain
Thomson Learning
Calle Magallanes, 25
28015-Madrid
Espana

UK/Europe/Middle East
Thomson Learning
Berkshire House
168-173 High Holborn
London
WC1V 7AA United Kingdom

Thomas Nelson & Sons Ltd.
Nelson House
Mayfield Road
Walton-on-Thames
KT 12 5PL United Kingdom

Library of Congress Cataloging-in-Publication Data
Robbins, Allan.
   Circuit analysis : theory and practice / Allan H. Robbins, Wilhelm
C. Miller.—2nd ed.
    p.   cm.
   ISBN 0-7668-0626-X
   1. Electric circuit analysis.   I. Miller, Wilhelm (Wilhelm C.)
II. Title.
TK454.R56   1999
621.319'2—dc21
                            99-37574
                            CIP

Contents

Preface   x

PART I
Foundation DC Concepts   1

1   Introduction   2

1.1   Introduction   4
1.2   The SI System of Units   7
1.3   Converting Units   9
1.4   Power of Ten Notation   10
1.5   Prefixes   13
1.6   Significant Digits and Numerical Accuracy   14
1.7   Circuit Diagrams   16
1.8   Circuit Analysis Using Computers   18
Problems   21

2   Voltage and Current   28

2.1   Atomic Theory Review   30
2.2   The Unit of Electrical Charge: The Coulomb   35
2.3   Voltage   36
2.4   Current   38
2.5   Practical DC Voltage Sources   41
2.6   Measuring Voltage and Current   46
2.7   Switches, Fuses, and Circuit Breakers   51
Problems   52

3   Resistance   58

3.1   Resistance of Conductors   60
3.2   Electrical Wire Tables   62
3.3   Resistance of Wires—Circular Mils   65
3.4   Temperature Effects   69
3.5   Types of Resistors   72

3.6   Color Coding of Resistors   76
3.7   Measuring Resistance—The Ohmmeter   78
3.8   Thermistors   81
3.9   Photoconductive Cells   82
3.10   Nonlinear Resistance   83
3.11   Conductance   85
3.12   Superconductors   86
Problems   88

4   Ohm's Law, Power, and Energy   94

4.1   Ohm's Law   96
4.2   Voltage Polarity and Current Direction   101
4.3   Power   104
4.4   Power Direction Convention   107
4.5   Energy   109
4.6   Efficiency   111
4.7   Nonlinear and Dynamic Resistances   114
4.8   Computer-Aided Circuit Analysis   115
Problems   120

PART II
Basic DC Analysis   127

5   Series Circuits   128

5.1   Series Circuits   130
5.2   Kirchhoff's Voltage Law   132
5.3   Resistors in Series   134
5.4   Voltage Sources in Series   137
5.5   Interchanging Series Components   138
5.6   The Voltage Divider Rule   139
5.7   Circuit Ground   142
5.8   Voltage Subscripts   143

5.9 Internal Resistance of Voltage Sources   149
5.10 Voltmeter Design   150
5.11 Ohmmeter Design   155
5.12 Ammeter Loading Effects   158
5.13 Circuit Analysis Using Computers   160
Problems   164

6 Parallel Circuits   176

6.1 Parallel Circuits   178
6.2 Kirchhoff's Current Law   179
6.3 Resistors in Parallel   183
6.4 Voltage Sources in Parallel   189
6.5 Current Divider Rule   190
6.6 Analysis of Parallel Circuits   195
6.7 Ammeter Design   198
6.8 Voltmeter Loading Effects   200
6.9 Circuit Analysis Using Computers   203
Problems   207

7 Series-Parallel Circuits   220

7.1 The Series-Parallel Network   222
7.2 Analysis of Series-Parallel Circuits   223
7.3 Applications of Series-Parallel Circuits   231
7.4 Potentiometers   239
7.5 Loading Effects of Instruments   241
7.6 Circuit Analysis Using Computers   247
Problems   254

8 Methods of Analysis   264

8.1 Constant-Current Sources   266
8.2 Source Conversions   268
8.3 Current Sources in Parallel and Series   272
8.4 Branch-Current Analysis   275
8.5 Mesh (Loop) Analysis   280
8.6 Nodal Analysis   288
8.7 Delta-Wye (Pi-Tee) Conversion   296
8.8 Bridge Networks   303
8.9 Circuit Analysis Using Computers   312
Problems   315

9 Network Theorems   326

9.1 Superposition Theorem   328
9.2 Thévenin's Theorem   332
9.3 Norton's Theorem   341
9.4 Maximum Power Transfer Theorem   352
9.5 Substitution Theorem   357
9.6 Millman's Theorem   359
9.7 Reciprocity Theorem   361
9.8 Circuit Analysis Using Computers   364
Problems   371

PART III
Capacitance and Inductance   383

10 Capacitors and Capacitance   384

10.1 Capacitance   386
10.2 Factors Affecting Capacitance   387
10.3 Electric Fields   390
10.4 Dielectrics   393
10.5 Nonideal Effects   394
10.6 Types of Capacitors   395
10.7 Capacitors in Parallel and Series   400
10.8 Capacitor Current and Voltage   404
10.9 Energy Stored by a Capacitor   407
10.10 Capacitor Failures and Troubleshooting   408
Problems   409

11 Capacitor Charging, Discharging, and Simple Waveshaping Circuits   416

11.1 Introduction   418
11.2 Capacitor Charging Equations   422
11.3 Capacitor with an Initial Voltage   427
11.4 Capacitor Discharging Equations   429
11.5 More Complex Circuits   430
11.6 An RC Timing Application   438
11.7 Pulse Response of RC Circuits   440
11.8 Transient Analysis Using Computers   444
Problems   451

12 Magnetism and Magnetic Circuits   **460**

12.1   The Nature of a Magnetic Field   462
12.2   Electromagnetism   464
12.3   Flux and Flux Density   465
12.4   Magnetic Circuits   467
12.5   Air Gaps, Fringing, and Laminated Cores   468
12.6   Series Elements and Parallel Elements   469
12.7   Magnetic Circuits with DC Excitation   470
12.8   Magnetic Field Intensity and Magnetization Curves   471
12.9   Ampere's Circuital Law   474
12.10 Series Magnetic Circuits: Given Φ, Find NI 475
12.11 Series-Parallel Magnetic Circuits   480
12.12 Series Magnetic Circuits: Given NI, Find Φ 482
12.13 Force Due to an Electromagnet   484
12.14 Properties of Magnetic Materials   485
12.15 Measuring Magnetic Fields   487
Problems   487

13 Inductance and Inductors   **492**

13.1   Electromagnetic Induction   494
13.2   Induced Voltage and Induction   496
13.3   Self-Inductance   499
13.4   Computing Induced Voltage   501
13.5   Inductances in Series and Parallel   503
13.6   Practical Considerations   504
13.7   Inductance and Steady State DC   507
13.8   Energy Stored by an Inductance   509
13.9   Inductor Troubleshooting Hints   510
Problems   511

14 Inductive Transients   **518**

14.1   Introduction   520
14.2   Current Buildup Transients   523
14.3   Interrupting Current in an Inductive Circuit   527
14.4   De-energizing Transients   529
14.5   More Complex Circuits   531
14.6   *RL* Transients Using Computers   537
Problems   542

PART IV
Foundation AC Concepts   **547**

15 AC Fundamentals   **548**

15.1   Introduction   550
15.2   Generating AC Voltages   551
15.3   Voltage and Current Conventions for AC   554
15.4   Frequency, Period, Amplitude, and Peak Value   557
15.5   Angular and Graphic Relationships for Sine Waves   561
15.6   Voltage and Currents as Functions of Time   565
15.7   Introduction to Phasors   570
15.8   AC Waveforms and Average Value   579
15.9   Effective Values   585
15.10 Rate of Change of a Sine Wave (Derivative)   590
15.11 AC Voltage and Current Measurement   590
15.12 Circuit Analysis Using Computers   592
Problems   595

16 *R, L,* and *C* Elements and the Impedance Concept   **604**

16.1   Complex Number Review   606
16.2   Complex Numbers in AC Analysis   611
16.3   *R, L,* and *C* Circuits with Sinusoidal Excitation   617
16.4   Resistance and Sinusoidal AC   617
16.5   Inductance and Sinusoidal AC   619
16.6   Capacitance and Sinusoidal AC   623
16.7   The Impedance Concept   627
16.8   Computer Analysis of AC Circuits   630
Problems   634

17 Power in AC Circuits   **640**

17.1   Introduction   642
17.2   Power to a Resistive Load   643
17.3   Power to an Inductive Load   644
17.4   Power to a Capacitive Load   646
17.5   Power in More Complex Circuits   648
17.6   Apparent Power   650

17.7 The Relationship Between *P, Q,* and *S* 651

17.8 Power Factor 655

17.9 AC Power Measurement 659

17.10 Effective Resistance 662

17.11 Energy Relationships for AC 663

17.12 Circuit Analysis Using Computers 664

Problems 665

PART V
Impedance Networks 671

18 AC Series-Parallel Circuits 672

18.1 Ohm's Law for AC Circuits 674

18.2 AC Series Circuits 681

18.3 Kirchhoff's Voltage Law and the Voltage Divider Rule 689

18.4 AC Parallel Circuits 693

18.5 Kirchhoff's Current Law and the Current Divider Rule 698

18.6 Series-Parallel Circuits 701

18.7 Frequency Effects 704

18.8 Applications 710

18.9 Circuit Analysis Using Computers 714

Problems 720

19 Methods of AC Analysis 736

19.1 Dependent Sources 738

19.2 Source Conversion 739

19.3 Mesh (Loop) Analysis 744

19.4 Nodal Analysis 751

19.5 Delta-to-Wye and Wye-to-Delta Conversions 758

19.6 Bridge Networks 762

19.7 Circuit Analysis Using Computers 768

Problems 771

20 AC Network Theorems 782

20.1 Superposition Theorem—Independent Sources 784

20.2 Superposition Theorem—Dependent Sources 789

20.3 Thévenin's Theorem—Independent Sources 791

20.4 Norton's Theorem—Independent Sources 797

20.5 Thévenin's and Norton's Theorems for Dependent Sources 803

20.6 Maximum Power Transfer Theorem 813

20.7 Circuit Analysis Using Computers 818

Problems 824

21 Resonance 834

21.1 Series Resonance 836

21.2 Quality Factor, *Q* 838

21.3 Impedance of a Series Resonant Circuit 841

21.4 Power, Bandwidth, and Selectivity of a Series Resonant Circuit 842

21.5 Series-to-Parallel *RL* and *RC* Conversion 851

21.6 Parallel Resonance 857

21.7 Circuit Analysis Using Computers 867

Problems 871

22 Filters and the Bode Plot 882

22.1 The Decibel 884

22.2 Multistage Systems 890

22.3 Simple *RC* and *RL* Transfer Functions 893

22.4 The Low-Pass Filter 902

22.5 The High-Pass Filter 909

22.6 The Band-Pass Filter 914

22.7 The Band-Reject Filter 918

22.8 Circuit Analysis Using Computers 920

Problems 924

23 Three-Phase Systems 934

23.1 Three-Phase Voltage Generation 936

23.2 Basic Three-Phase Circuit Connections 937

23.3 Basic Three-Phase Relationships 940

23.4 Examples 948

23.5 Power in a Balanced System 954

23.6 Measuring Power in Three-Phase Circuits 960

23.7 Unbalanced Loads 963

23.8 Power System Loads 967

23.9 Circuit Analysis Using Computers 967

Problems 671

24 **Transformers and Coupled Circuits** **978**

24.1 Introduction 980

24.2 Iron-Core Transformers: The Ideal Model 983

24.3 Reflected Impedance 992

24.4 Transformer Ratings 994

24.5 Transformer Applications 994

24.6 Practical Iron-Core Transformers 1002

24.7 Transformer Tests 1007

24.8 Voltage and Frequency Effects 1009

24.9 Loosely Coupled Circuits 1010

24.10 Magnetically Coupled Circuits with Sinusoidal Excitation 1015

24.11 Coupled Impedance 1017

24.12 Circuit Analysis Using Computers 1019

Problems 1023

25 **Nonsinusoidal Waveforms** **1030**

25.1 Composite Waveforms 1032

25.2 Fourier Series 1034

25.3 Fourier Series of Common Waveforms 1039

25.4 Frequency Spectrum 1046

25.5 Circuit Response to a Nonsinusoidal Waveform 1052

25.6 Circuit Analysis Using Computers 1056

Problems 1059

APPENDIX A
OrCAD-PSpice A/D **1067**

APPENDIX B
Solution of Simultaneous Linear Equations **1076**

APPENDIX C
Maximum Power Transfer Theorem **1083**

APPENDIX D
Answers to Selected Odd-Numbered Problems **1086**

Glossary **1099**

Index **1104**

Preface

Welcome to the second edition of *Circuit Analysis: Theory and Practice.* If you are a student, we hope that this new edition will make your journey into learning circuit theory easier and more rewarding; if you are an instructor, we hope it will better assist you in your role as an educator. Although this book has a new look, it retains all the characteristics and usefulness of the previous version. In addition, it includes new and/or improved features such as an Online Companion™ web resource with RealAudio sound clips, Putting It Into Practice project-like problems that go beyond the usual end-of-chapter exercises, a CD-ROM-based e.resource™ with PowerPoint® Presentations, and more. Additionally, in response to instructor feedback, the end-of-chapter problem sets have been modified and expanded to provide a smoother transition from simple practice exercises to more challenging and in-depth problems.

The Book and Who it is For

Circuit Analysis: Theory and Practice was developed specifically for use in introductory circuit analysis courses. Written primarily as a textbook for electronics students in engineering technology programs, university engineering programs, industrial training programs and the like, it covers fundamentals of dc and ac circuits, methods of analysis, capacitance, inductance, magnetic circuits, basic transients, Fourier analysis, and other topics. When students successfully complete a course using this book, they will have a good working knowledge of basic circuit principles and a demonstrated ability to solve a variety of circuit-related problems.

Text Organization

The book contains 25 chapters and is divided into five main parts: Foundation DC Concepts, Basic DC Analysis, Capacitance and Inductance, Foundation AC Concepts, and Impedance Networks. Chapters 1 through 4 are introductory. They cover the foundation concepts of voltage, current, resistance, Ohm's Law, and power. Chapters 5 through 9 focus on dc analysis methods. Included are Kirchhoff's Laws, series and parallel circuits, mesh and nodal analysis, Y and Δ transformations, source transformations, Thévenin's and Norton's theorems, the maximum power transfer theorem, and so on. Chapters 10 through 14 cover basic concepts of capacitance, magnetism, and inductance, plus magnetic circuits and simple dc transients. Chapters 15 through 17 cover foundation concepts of ac, ac voltage generation, the basic ideas of frequency, period, phase, and so on. Phasors and the impedance concept are introduced and used to solve simple problems. Power in ac circuits is investigated and the concept of power factor and the power triangle are introduced. Chapters 18 through 25 then apply these ideas. Topics include ac versions of earlier dc techniques such as mesh and nodal analysis, Thévenin's theorem, and so on, as well as new ideas such as resonance, filters, Bode techniques, three-phase systems, transformers, and nonsinusoidal waveform analysis.

Several appendices round out the book: Appendix A provides a short tutorial on OrCAD PSpice; Appendix B reviews determinants and the solution of simultaneous equations; Appendix C provides additional material on the maximum power transfer theorem; and finally, Appendix D contains answers to selected odd-numbered end-of-chapter problems.

Features of the Book

New for the Second Edition

- Additional diagrams. The text now includes over 1200 full-color photos and diagrams (many of which incorporate 3-D effects) to illustrate and clarify ideas and to aid visual learners
- More problems, both easy and challenging. We now have over 1600 End-of-Chapter problems, Practice Problems and In-Process Learning Check problems
- Answers to Practice Problems have been moved from the appendix and placed with the practice problems to make it easier for students to verify their work
- OrCAD PSpice® and Electronics Workbench® computer simulation methods have been integrated throughout the text. Problems and examples make use of actual screen captures so that students see in the book exactly what they will see on their own computer screens
- A new feature, Putting It Into Practice, presents students with a group of challenging, project-like problems that require them to reason their way through realistic situations similar to those they experience on the job after graduation
- An Online Companion™ web site has been added. It contains RealAudio sound clips that present a more in-depth discussion of the most difficult topic for each chapter (keyed to the text by an icon)
- An extensive ancillary package has been created to include all solutions, PowerPoint slides, Image Library, Computerized Testbank/Gradebook, and Electronics Workbench® circuit files

Features from the First Edition

- Clearly written, easy-to-understand writing style that emphasizes principles and concepts
- Hundreds of worked-out and clearly illustrated examples to promote student understanding
- In-Process Learning Checks that help identify learning gaps before the student moves on to new material

- Chapter Previews provide a context and a brief overview for the upcoming chapter
- Competency-based objectives define the knowledge or skill that the student is expected to gain from each chapter
- Key terms at the beginning of each chapter identify new terms to be introduced
- Icons and graphics are used to direct the user's attention to focal points of the text
- Answers to odd-numbered problems are provided in an appendix

How to Use This Text

Since the most important attribute of a text is its value to the user, we have created a textbook that not only presents the technical material in an easy-to-read, easy-to-understand style, we have also provided in-text learning features that help the student in other ways. For example, each chapter includes a short vignette that provides insight into the development of the theory, important contributors met along the way, how the material relates to the electronic field in general, and so on. Some of these features are illustrated on the following pages.

Chapter Openers
Each chapter begins with an overview of the chapter, providing a perspective for the following chapter and an answer to "Why am I learning this?".

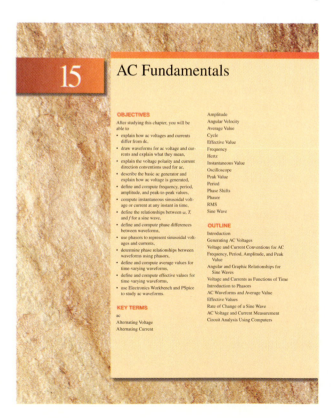

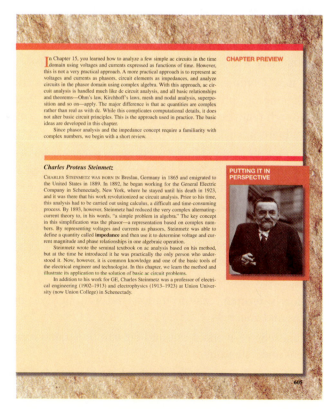

Objectives and Key Terms

Chapter opening Objectives and Key Terms prepare students for recognition of key chapter topics and terms prior to chapter content.

Putting it In Perspective

Short vignettes provide interesting background on the people and events leading to the major contributions in the electrical sciences. While entertaining, they provide insight and add a human element into the study of electric circuits.

Putting it Into Practice

This new feature allows students to develop problem-solving skills that are similar to those used by someone practicing in the electrical/electronics field. Putting it Into Practice offers students a challenging realistic problem to solve utilizing concepts learned in the preceding chapter.

Practice Problems

These problems are placed throughout the textbook (generally at the end of a section) to enable students to practice the skills that were learned in the section. The answers are found immediately after the practice problems so that students do not need to constantly flip through the textbook to see whether they are on the right track.

---

450   **Chapter 11** ■ Capacitor Charging, Discharging, and Simple Waveshaping Circuits

for TR, **0** for TD, **20V** for *V2*, and **0V** for *V1*. (This defines a pulse with a period of 5 s, a width of 1 s, rise and fall times of 1 μs, amplitude of 20 V, and an initial value of 0 V.) Click Apply, then close the Property editor. Double click the capacitor symbol and set *IC* to −10V in the Properties Editor. Set TSTOP to 2s. Place a Voltage Marker as shown, then click Run. You should get the voltage trace of Figure 11–50 on the screen. Add the second axis and the current trace as described in the previous examples. The red current curve should appear.

FIGURE 11–50 Waveforms for the circuit of Figure 11–49 with $V_0 = -10$ V.

Note that voltage starts at −10 V and climbs to 20 V while current starts at $(E - V_0)/R = 30$ V/5 kΩ = 6 mA and decays to zero. When the switch is turned to the discharge position, the current drops from 0 A to −20 V/5 kΩ = −4 mA and then decays to zero while the voltage decays from 20 V to zero. Thus the solution checks.

PUTTING IT INTO PRACTICE

An electronic device employs a timer circuit of the kind shown in Figure 11–32(a), i.e., an *RC* charging circuit and a threshold detector. (Its timing waveforms are thus identical to those of Figure 11–32(b)). The input to the *RC* circuit is a 0 to 5 V ±4% step, $R = 680$ kΩ ±10%, $C = 0.22$ μF ±10%, the threshold detector activates at $v_C = 1.8$ V ±0.05 V and the required delay is 67 ms ±18 ms. You test a number of units as they come off the production line and find that some do not meet the timing spec. Perform a design review and determine the cause. Redesign the timing portion of the circuit in the most economical way possible.

---

332   **Chapter 9** ■ Network Theorems

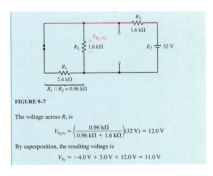

FIGURE 9–7

The voltage across R_2 is

$$V_{R_{2(3)}} = \left(\frac{0.96 \text{ k}\Omega}{0.96 \text{ k}\Omega + 1.6 \text{ k}\Omega}\right)(32 \text{ V}) = 12.0 \text{ V}$$

By superposition, the resulting voltage is

$$V_{R_2} = -4.0 \text{ V} + 3.0 \text{ V} + 12.0 \text{ V} = 11.0 \text{ V}$$

PRACTICE PROBLEMS 1

Use the superposition theorem to determine the voltage across R_1 and R_3 in the circuit of Figure 9–4.

Answers: $V_{R_1} = 27.0$ V, $V_{R_3} = 21.0$ V

IN-PROCESS LEARNING CHECK 1

Use the final results of Example 9–2 and Practice Problem 1 to determine the power dissipated by the resistors in the circuit of Figure 9–4. Verify that the superposition theorem does not apply to power.

(Answers are at the end of the chapter.)

9.2   Thévenin's Theorem

In this section, we will apply one of the most important theorems of electric circuits. **Thévenin's theorem** allows even the most complicated circuit to be reduced to a single voltage source and a single resistance. The importance of such a theorem becomes evident when we try to analyze a circuit as shown in Figure 9–8.

If we wanted to find the current through the variable load resistor when $R_L = 0$, $R_L = 2$ kΩ, and $R_L = 5$ kΩ using existing methods, we would need to analyze the entire circuit three separate times. However, if we could reduce the entire circuit external to the load resistor to a single voltage source in series with a resistor, the solution becomes very easy.

Thévenin's theorem is a circuit analysis technique which reduces any linear bilateral network to an equivalent circuit having only one voltage

PRACTICE
PROBLEMS 8

a. Use OrCAD Capture to input the circuit of Figure 22–35.

b. Use the Probe postprocessor to observe the frequency response from 1 Hz to 100 kHz.

c. From the display, determine the cutoff frequencies and use the cursors to determine the bandwidth.

d. Compare the results to those obtained in Practice Problem 7.

EXAMPLE 22–12 Use Electronics Workbench to obtain the frequency response for the circuit of Figure 22–32. Compare the results to those obtained in Example 22–11.

Solution In order to perform the required measurements, we need to use the function generator and the Bode plotter, both located in the Instruments parts bin. The circuit is constructed as shown in Figure 22–41.

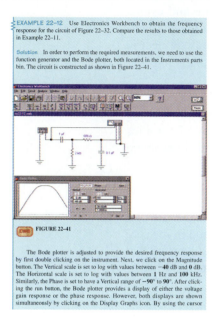

FIGURE 22–41

The Bode plotter is adjusted to provide the desired frequency response by first double clicking on the instrument. Next, we click on the Magnitude button. The Vertical scale is set to log with values between **−40** dB and **0** dB. The Horizontal scale is set to log with values between **1** Hz and **100** kHz. Similarly, the Phase is set to have a Vertical range of **−90°** to **90°**. After clicking the run button, the Bode plotter provides a display of either the voltage gain response or the phase response. However, both displays are shown simultaneously by clicking on the Display Graphs icon. By using the cursor

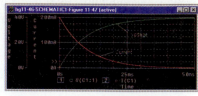

FIGURE 11–47 Waveforms for the circuit of Figure 11–46.

Analysis of Results

Click the Toggle cursor icon, then use the cursor to determine values from the screen. For example, at $t = 5$ ms, you should find $v_C = 15.7$ V and $i_C = 121$ mA. (An analytic solution for this circuit (which is Figure 11–23) may be found in Example 11–10, part (c). It agrees exactly with the PSpice solution.)

As a second example, consider the circuit of Figure 11–21 (shown as Figure 11–48). Create the circuit using the same general procedure as in the previous example, except do not rotate the capacitor. Again, be sure to set V_0 (the initial capacitor voltage) to zero. In the Simulation Profile box, set TSTOP to 50ms. Place differential voltage markers (found on the toolbar at the top of the screen) across C to graph the capacitor voltage. Run the analysis, create a second axis, then add the current plot. You should get the same graph (i.e., Figure 11–47) as you got for the previous example, since its circuit is the Thévenin equivalent of this one.

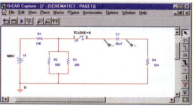

FIGURE 11–48 Differential markers are used to display the voltage across C_1.

As a final example, consider Figure 11–49(a), which shows double switching action.

In-Process Learning Checks

In-Process Learning Checks provide a quick review of material just covered.

Examples

Numerous examples and solutions are included to help clarify topics and to guide the student to solve problems.

Online Companion™ Web Resources, RealAudio Clips

The authors have provided sound clips and have made them available to students via RealAudio on the text's Online Companion™ web site. These sound clips (one per chapter) present more in-depth discussions of the most difficult topic for each chapter, and will tie directly back to the text through a designated icon appearing in the text margins.

Computer Simulation

Two popular computer simulation programs, OrCAD PSpice and Electronics Workbench, are used in the book. Examples provide step-by-step instructions on how to construct circuits, connect meters, and test circuit operation. Results are then validated by comparison to theoretical results. Such simulation packages provide an additional way to enrich and add insight to the study of electrical circuits.

About OrCAD PSpice and MicroSim PSpice

For many years, users of this book have used PSpice from MicroSim Corporation. However, in early 1998, MicroSim was purchased by OrCAD and PSpice has now been integrated into the OrCAD suite of products. Because MicroSim PSpice is no longer available, all PSpice work in this book has been done using the demo version of OrCAD PSpice. However, to ease the migration for users who have not made the change, we have included PSpice Version 8 (i.e., MicroSim) versions of the PSpice material in this text on our web site at www.electronictech.com. For those users wanting a demo disk from OrCAD, please contact their web site at www.orcad.com.

Required Background

Students need a working knowledge of basic algebra and trigonometry and the ability to solve second-order linear equations such as those found in mesh analysis. They should be familiar with the SI metric system and the atomic nature of matter. In terms of higher math, calculus is introduced gradually in later chapters to aid in the development of ideas. (This is in keeping with ABET guidelines, which require the use of some calculus in accredited programs.) However, optional derivations and problems using calculus (which are provided for enrichment purposes) are marked by an ∫ icon and may be omitted in those programs that do not stress the use of calculus.

The Learning Package

The complete ancillary package was developed to achieve two goals:

1. To assist students in learning the essential information needed to prepare for the exciting field of electronics.
2. To assist instructors in planning and implementing their instructional programs for the most efficient use of time and other resources.

The *Circuit Analysis: Theory and Practice* package was created as an integrated whole. Supplements are linked to and integrated with the text to create a comprehensive supplement package that supports students and instructors. The package includes:

Laboratory Manual

Contains instructions for hands-on electronic lab work, plus additional computer simulation labs. It also includes a comprehensive guide to lab equipment and laboratory measurements.
ISBN: 0-7668-0627-8

Instructor's Resource Guide

Contains step-by-step solutions to all end-of-chapter (even and odd) problems, including waveforms, circuit diagrams and more. The Instructor's Resource Guide also includes the e.resource™ CD-ROM in the back of the book. ISBN: 0-7668-0626-8

e.resource™

Available all on one CD-ROM are all the tools and instructional resources that will enrich your classroom. The elements of e.resource link directly to the text and tie together to provide a unified instructional system.

Features contained in the e.resource include:

PowerPoint® Presentation Slides: Provides customizable presentations for classroom use. Slides are prepared for every chapter of the book that helps you present key points and concepts. Graphics from the Image Library or your own images can be imported to create individualized classroom presentations.

Image Library: Includes 200 full-color images from the textbook, providing the instructor with another means of promoting student understanding. The Image Library allows the instructor to display or print images for a classroom presentation.

Computerized Testbank: Over 1000 questions for use in creating tests of varying levels so you can assess student comprehension.

Gradebook: Tracks student performance, prints student progress reports, organizes assignments, and more; simplifies administrative tasks.

Electronics Workbench Circuit Files: 100 circuits taken directly from the textbook. Instructors may copy and distribute these circuit files to students free of charge.

Electronics Technology Homepage: Includes Netscape Navigator so you can link directly to the Delmar Electronics Technology website and to the textbook's Online Companion for additional resources.

Online Companion™

One of the new features of this edition is a companion Internet web site, intended for use by both educators and students. It provides ongoing assistance in the form of additional problems, supplements, circuit schematics, updates on the status of EWB and OrCAD PSpice, and general information, plus a method whereby you can interact with the authors.

Features of the Online Companion include:

- RealAudio Sound Files
- Technology updates

- Internet activities
- Discussion forums
- Comprehensive listing of links to electronics industry and educational sites
- Ask the Authors: Frequently Asked Questions

Please visit our web site at www.electronictech.com for more details.

To the Student

Learning circuit theory should be challenging, interesting, and (hopefully) fun. However, it is also hard work, since the knowledge and skills that you seek can only be gained through practice. We offer a few guidelines.

1. As you go through the material, try to gain an appreciation of where circuit theory comes from—i.e., the basic experimental laws on which it is based. This will help you better understand the foundation ideas on which the theory is built.

2. Learn the terminology and definitions. Important new terms are introduced frequently. Learn what they mean and where they are used.

3. Study each new section carefully and be sure that you understand the basic ideas and how they are put together. Work your way through the examples with your calculator. Try the practice problems, then the end-of-chapter problems. Not every concept will be clear immediately and most likely many will require several readings before you gain an adequate understanding.

4. When you are ready, test your understanding using the In-Process Learning Checks (self-quizzes) located in each chapter.

5. When you have mastered the material, move on to the next block. For those concepts that you are having difficulty with, consult your instructor or some other authoritative source.

Calculators for Circuit Analysis

You will need a good scientific calculator. A good calculator will permit you to more easily master the numerical aspects of problem solving, thereby leaving you more time to concentrate on circuit theory itself. This is especially true for ac, where complex number work dominates. There are some inexpensive calculators on the market that handle complex-number arithmetic almost as easily as real-number arithmetic. Such calculators save an enormous amount of time. You should acquire such a calculator (after consulting with your instructor), and learn to use if proficiently.

Acknowledgements

Many people have contributed to the success of *Circuit Analysis: Theory and Practice*. We begin by expressing our thanks to our students for providing subtle (and sometimes not-so-subtle) feedback. Next, the reviewers and accuracy checkers: no textbook can be successful without the dedication and commitment of such people. We thank the following:

Reviewers

Joe Calabrese, DeVry Institute of Technology, Columbus, OH

Solomon Oldak, DeVry Institute of Technology, Pomona, CA

Walter Bartlett, Durham Technical Community College, Durham, NC

Joseph Booker, DeVry Institute of Technology, Addison, IL

Mohammed Brihoun, DeVry Institute of Technology, Decatur, GA

James Bryant, DeVry Institute of Technology, Columbus, OH

Foster Chin, Tulsa Community College, Tulsa, OK

Jim Davis, Muskingham Area Tech College, Zanesville, OH

Joe Ebden, Seneca College, Ontario, Canada

Ray Fleming, Edison Community College, Pique, OH

Norman Grossman, DeVry Institute of Technology, Long Beach, CA

Robert Hofinger, Purdue University, Indianapolis, IN

William Lin, DeVry Institute of Technology, New Brunswick, NJ

Leei Mao, Greenville Area Technical College, Greenville, SC

Mike Marien, Southern Polytechnic State University, Marietta, GA

Fred Melzer, DeVry Institute of Technology, Alberta, Canada

Vic Quiros, DeVry Institute of Technology, Phoenix, AZ

Stanley Smith, Onondaga Community College, Syracuse, NY

Richard Sturtevant, Springfield Tech Community College, Springfield, MA

Pui Chor Wong, DeVry Institute of Technology, Alberta, Canada

Accuracy Checkers

Nizar Al-holou, University of Detroit Mercy, Detroit, MI

Charles Bray, University of Memphis, Memphis, TN

Yvette Grimmond, Seneca College, Ontario, Canada

Marie Sichler, Red River College, Manitoba, Canada

The following firms and individuals supplied photographs, diagrams and other useful information:

Allen-Bradley
AT & T
AVX Corporation
B+K Precision
Bourns Inc.
Butterworth & Co. Ltd.
Carte International
Condor DC Power Supplies Inc.
Illinois Capacitor Inc.

Interactive Images Technologies
JBL Professional
John Fluke Mfg. Co. Inc.
OrCAD
Siemens Solar Industries
Simpson Electric Company
Tektronix
Transformers Manufacturers Inc.
Vansco Electronics

We express our deep appreciation to the staff at Delmar Publishers for their tireless efforts in putting this book together: To Greg Clayton, our Electronics Editor, for direction and encouragement; Michelle Cannistraci, our Development Editor, for encouragement, advice and making sure we got things done on time; Christopher Chien, Senior Project Editor, for his skill in editing and pulling the final project together; Nicole Reamer, Art Director, for her guidance in preparing the art; Alar Elken, Business Unit Director, and Larry Main, Production Manager, and their staffs for making the project work on such an impossibly short deadline. We also wish to thank Monica Ohlinger of Ohlinger Publishing Services for her work in organizing and administering the project and Ben Shriver for his admirable job of copyediting the manuscript. A special thanks to all of you.

Lastly, we thank our wives and families for their support and perseverance during the preparation of this book.

Allan H. Robbins
Wilhelm C. Miller
August, 1999

About the Authors

Allan H. Robbins graduated from engineering with a Bachelor's degree and a Master's degree in Electrical Engineering. In graduate school, he specialized in circuit theory. Allan, who was formerly head of the Department of Electrical and Computer Technology at Red River College, has been an instructor for over 30 years. In addition to his academic career, he has been a consultant and a small business partner. He began writing as a contributing author for Osborne-McGraw-Hill in the computer field and is also joint author of one other textbook. He has served as Section Chairman for the IEEE and as a member of the board for the Electronics Industry Association of Manitoba.

Wilhelm (Will) C. Miller graduated from Electronic Engineering Technology and obtained his Bachelor of Science degree in Physics and Mathematics from the University of Winnipeg. He worked in the communications field for ten years, including a year's assignment with Saudi PTT in Jeddah, Saudi Arabia. Will has been an instructor in the Electronics and Computer Engineering Technologies for 18 years, having taught at Red River College and College of The Bahamas (Nassau, Bahamas). He is currently vice-president of CTTAM (the Certified Technicians and Technologists Association of Manitoba).

Foundation dc Concepts

PART

I

1 **Introduction**

2 **Voltage and Current**

3 **Resistance**

4 **Ohm's Law, Power, and Energy**

Circuit theory provides the tools and concepts needed to understand and analyze electrical and electronic circuits. The foundations of this theory were laid down nearly two hundred years ago by a number of pioneer researchers. In 1780, Alessandro Volta of Italy developed an electric cell (battery) that provided the first source of what we now call dc voltage. Around the same time, the concept of current was evolved (even though nothing was known about the atomic structure of matter until much later). In 1826, Georg Simon Ohm of Germany brought the two ideas together and experimentally determined the relationship between voltage and current in a resistive circuit. This result, known as Ohm's law, set the stage for the development of modern-day circuit theory.

In Part I, we examine the foundation of this theory. We look at voltage, current, power, energy, and the relationships between them. The ideas developed here are the fundamental ideas upon which circuit theory is built.

Introduction

OBJECTIVES

After studying this chapter, you will be able to

- describe the SI system of measurement,
- convert between various sets of units,
- use power of ten notation to simplify handling of large and small numbers,
- express electrical units using standard prefix notation such as μA, kV, mW, etc.,
- use a sensible number of significant digits in calculations,
- describe what block diagrams are and why they are used,
- convert a simple pictorial circuit to its schematic representation,
- describe generally how computers fit in the electrical circuit analysis picture.

KEY TERMS

Ampere
Block Diagram
Circuit
Conversion Factor
Current
Energy
Joule
Meter
Newton
Pictorial Diagram
Power of Ten Notation
Prefixes
Programming Language
Resistance
Schematic Diagram
Scientific Notation
SI Units
Significant Digits
SPICE
Volt
Watt

OUTLINE

Introduction
The SI System of Units
Converting Units
Power of Ten Notation
Prefixes
Significant Digits and Numerical Accuracy
Circuit Diagrams
Circuit Analysis Using Computers

An electrical circuit is a system of interconnected components such as resistors, capacitors, inductors, voltage sources, and so on. The electrical behavior of these components is described by a few basic experimental laws. These laws and the principles, concepts, mathematical relationships, and methods of analysis that have evolved from them are known as **circuit theory.**

Much of circuit theory deals with problem solving and numerical analysis. When you analyze a problem or design a circuit, for example, you are typically required to compute values for voltage, current, and power. In addition to a numerical value, your answer must include a unit. The system of units used for this purpose is the SI system (Systéme International). The SI system is a unified system of metric measurement; it encompasses not only the familiar MKS (meters, kilograms, seconds) units for length, mass, and time, but also units for electrical and magnetic quantities as well.

Quite frequently, however, the SI units yield numbers that are either too large or too small for convenient use. To handle these, engineering notation and a set of standard prefixes have been developed. Their use in representation and computation is described and illustrated. The question of significant digits is also investigated.

Since circuit theory is somewhat abstract, diagrams are used to help present ideas. We look at several types—schematic, pictorial, and block diagrams—and show how to use them to represent circuits and systems.

We conclude the chapter with a brief look at computer usage in circuit analysis and design. Several popular application packages and programming languages are described. Special emphasis is placed on OrCAD PSpice and Electronics Workbench, the two principal software packages used throughout this book.

Hints on Problem Solving

DURING THE ANALYSIS of electric circuits, you will find yourself solving quite a few problems. An organized approach helps. Listed below are some useful guidelines:

1. Make a sketch (e.g., a circuit diagram), mark on it what you know, then identify what it is that you are trying to determine. Watch for "implied data" such as the phrase "the capacitor is initially uncharged". (As you will find out later, this means that the initial voltage on the capacitor is zero.) Be sure to convert all implied data to explicit data.
2. Think through the problem to identify the principles involved, then look for relationships that tie together the unknown and known quantities.
3. Substitute the known information into the selected equation(s) and solve for the unknown. (For complex problems, the solution may require a series of steps involving several concepts. If you cannot identify the complete set of steps before you start, start anyway. As each piece of the solution emerges, you are one step closer to the answer. You may make false starts. However, even experienced people do not get it right on the first try every time. Note also that there is seldom one "right" way to solve a problem. You may therefore come up with an entirely different correct solution method than the authors do.)
4. Check the answer to see that it is sensible—that is, is it in the "right ballpark"? Does it have the correct sign? Do the units match?

1.1   Introduction

Technology is rapidly changing the way we do things; we now have computers in our homes, electronic control systems in our cars, cellular phones that can be used just about anywhere, robots that assemble products on production lines, and so on.

A first step to understanding these technologies is electric circuit theory. Circuit theory provides you with the knowledge of basic principles that you need to understand the behavior of electric and electronic devices, circuits, and systems. In this book, we develop and explore its basic ideas.

Before We Begin

Before we begin, let us look at a few examples of the technology at work. (As you go through these, you will see devices, components, and ideas that have not yet been discussed. You will learn about these later. For the moment, just concentrate on the general ideas.)

As a first example, consider Figure 1–1, which shows a VCR. Its design is based on electrical, electronic, and magnetic circuit principles. For example, resistors, capacitors, transistors, and integrated circuits are used to control the voltages and currents that operate its motors and amplify the audio and video signals that are the heart of the system. A magnetic circuit (the read/write system) performs the actual tape reads and writes. It creates, shapes, and controls the magnetic field that records audio and video signals on the tape. Another magnetic circuit, the power transformer, transforms the ac voltage from the 120-volt wall outlet voltage to the lower voltages required by the system.

FIGURE 1–1   A VCR is a familiar example of an electrical/electronic system.

Figure 1–2 shows another example. In this case, a designer, using a personal computer, is analyzing the performance of a power transformer. The transformer must meet not only the voltage and current requirements of the application, but safety- and efficiency-related concerns as well. A software application package, programmed with basic electrical and magnetic circuit fundamentals, helps the user perform this task.

Figure 1–3 shows another application, a manufacturing facility where fine pitch surface-mount (SMT) components are placed on printed circuit boards at high speed using laser centering and optical verification. The bottom row of Figure 1–4 shows how small these components are. Computer control provides the high precision needed to accurately position parts as tiny as these.

Before We Move On

Before we move on, we should note that, as diverse as these applications are, they all have one thing in common: all are rooted in the principles of circuit theory.

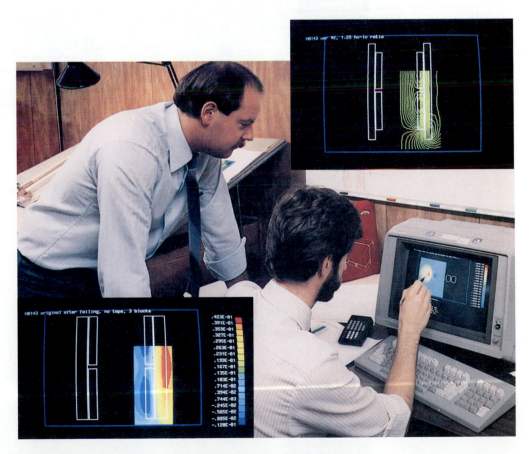

FIGURE 1–2   A transformer designer using a 3-D electromagnetic analysis program to check the design and operation of a power transformer. Upper inset: Magnetic field pattern. *(Courtesy Carte International Inc.)*

FIGURE 1–3 Laser centering and optical verification in a manufacturing process. *(Courtesy Vansco Electronics Ltd.)*

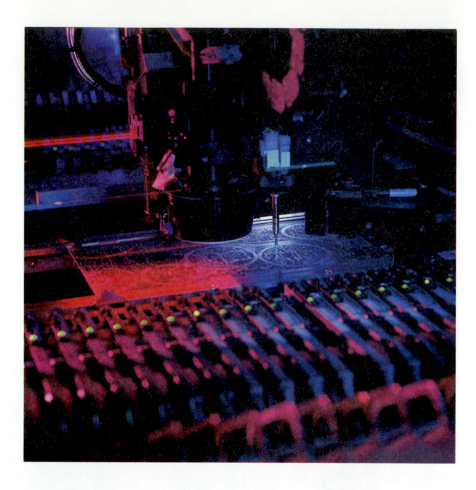

FIGURE 1–4 Some typical electronic components. The small components at the bottom are surface mount parts that are installed by the machine shown in Figure 1–3.

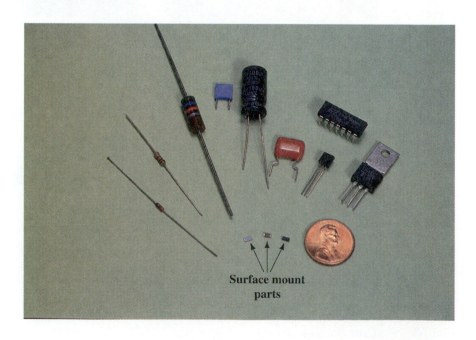

Surface mount parts

1.2   The SI System of Units

The solution of technical problems requires the use of units. At present, two major systems—the English (US Customary) and the metric—are in everyday use. For scientific and technical purposes, however, the English system has been largely superseded. In its place the SI system is used. Table 1–1 shows a few frequently encountered quantities with units expressed in both systems.

The SI system combines the MKS metric units and the electrical units into one unified system: See Tables 1–2 and 1–3. (Do not worry about the electrical units yet. We define them later, starting in Chapter 2.) The units in Table 1–2 are defined units, while the units in Table 1–3 are derived units, obtained by combining units from Table 1–2. Note that some symbols and abbreviations use capital letters while others use lowercase letters.

A few non-SI units are still in use. For example, electric motors are commonly rated in horsepower, and wires are frequently specified in AWG sizes (American Wire Gage, Section 3.2). On occasion, you will need to convert non-SI units to SI units. Table 1–4 may be used for this purpose.

Definition of Units

When the metric system came into being in 1792, the meter was defined as one ten-millionth of the distance from the north pole to the equator and the second as $1/60 \times 1/60 \times 1/24$ of the mean solar day. Later, more accurate definitions based on physical laws of nature were adopted. The meter is now

TABLE 1–1   Common Quantities

1 meter = 100 centimeters = 39.37 inches
1 millimeter = 39.37 mils
1 inch = 2.54 centimeters
1 foot = 0.3048 meter
1 yard = 0.9144 meter
1 mile = 1.609 kilometers
1 kilogram = 1000 grams = 2.2 pounds
1 gallon (US) = 3.785 liters

TABLE 1–2   Some SI Base Units

Quantity	Symbol	Unit	Abbreviation
Length	ℓ	meter	m
Mass	m	kilogram	kg
Time	t	second	s
Electric current	I, i	ampere	A
Temperature	T	kelvin	K

TABLE 1–3   Some SI Derived Units*

Quantity	Symbol	Unit	Abbreviation
Force	F	newton	N
Energy	W	joule	J
Power	P, p	watt	W
Voltage	V, v, E, e	volt	V
Charge	Q, q	coulomb	C
Resistance	R	ohm	Ω
Capacitance	C	farad	F
Inductance	L	henry	H
Frequency	f	hertz	Hz
Magnetic flux	Φ	weber	Wb
Magnetic flux density	B	tesla	T

*Electrical and magnetic quantities will be explained as you progress through the book. As in Table 1–2, the distinction between capitalized and lowercase letters is important.

TABLE 1–4 Conversions

	When You Know	Multiply By	To Find
Length	inches (in)	0.0254	meters (m)
	feet (ft)	0.3048	meters (m)
	miles (mi)	1.609	kilometers (km)
Force	pounds (lb)	4.448	newtons (N)
Power	horsepower (hp)	746	watts (W)
Energy	kilowatthour (kWh)	3.6×10^6	joules* (J)
	foot-pound (ft-lb)	1.356	joules* (J)

Note: 1 joule = 1 newton-meter.

defined as the distance travelled by light in a vacuum in 1/299 792 458 of a second, while the second is defined in terms of the period of a cesium-based atomic clock. The definition of the kilogram is the mass of a specific platinum-iridium cylinder (the international prototype), preserved at the International Bureau of Weights and Measures in France.

Relative Size of the Units*

To gain a feel for the SI units and their relative size, refer to Tables 1–1 and 1–4. Note that 1 meter is equal to 39.37 inches; thus, 1 inch equals 1/39.37 = 0.0254 meter or 2.54 centimeters. A force of one pound is equal to 4.448 newtons; thus, 1 newton is equal to 1/4.448 = 0.225 pound of force, which is about the force required to lift a ¼-pound weight. One joule is the work done in moving a distance of one meter against a force of one newton. This is about equal to the work required to raise a quarter-pound weight one meter. Raising the weight one meter in one second requires about one watt of power.

The watt is also the SI unit for electrical power. A typical electric lamp, for example, dissipates power at the rate of 60 watts, and a toaster at a rate of about 1000 watts.

The link between electrical and mechanical units can be easily established. Consider an electrical generator. Mechanical power input produces electrical power output. If the generator were 100% efficient, then one watt of mechanical power input would yield one watt of electrical power output. This clearly ties the electrical and mechanical systems of units together.

However, just how big is a watt? While the above examples suggest that the watt is quite small, in terms of the rate at which a human can work it is actually quite large. For example, a person can do manual labor at a rate of about 60 watts when averaged over an 8-hour day—just enough to power a standard 60-watt electric lamp continuously over this time! A horse can do considerably better. Based on experiment, Isaac Watt determined that a strong dray horse could average 746 watts. From this, he defined the horsepower (hp) as 1 horsepower = 746 watts. This is the figure that we still use today.

---

*Paraphrased from Edward C. Jordan and Keith Balmain, *Electromagnetic Waves and Radiating Systems*, Second Edition. (Englewood Cliffs, New Jersey: Prentice-Hall, Inc, 1968).

1.3   Converting Units

Often quantities expressed in one unit must be converted to another. For example, suppose you want to determine how many kilometers there are in ten miles. Given that 1 mile is equal to 1.609 kilometers, Table 1–1, you can write 1 mi = 1.609 km, using the abbreviations in Table 1–4. Now multiply both sides by 10. Thus, 10 mi = 16.09 km.

This procedure is quite adequate for simple conversions. However, for complex conversions, it may be difficult to keep track of units. The procedure outlined next helps. It involves writing units into the conversion sequence, cancelling where applicable, then gathering up the remaining units to ensure that the final result has the correct units.

To get at the idea, suppose you want to convert 12 centimeters to inches. From Table 1–1, 2.54 cm = 1 in. Since these are equivalent, you can write

$$\frac{2.54 \text{ cm}}{1 \text{ in}} = 1 \quad \text{or} \quad \frac{1 \text{ in}}{2.54 \text{ cm}} = 1 \tag{1–1}$$

Now multiply 12 cm by the second ratio and note that unwanted units cancel. Thus,

$$12 \text{ cm} \times \frac{1 \text{ in}}{2.54 \text{ cm}} = 4.72 \text{ in}$$

The quantities in equation 1–1 are called **conversion factors.** Conversion factors have a value of 1 and you can multiply by them without changing the value of an expression. When you have a chain of conversions, select factors so that all unwanted units cancel. This provides an automatic check on the final result as illustrated in part (b) of Example 1–1.

EXAMPLE 1–1   Given a speed of 60 miles per hour (mph),

a. convert it to kilometers per hour,

b. convert it to meters per second.

Solution

a. Recall, 1 mi = 1.609 km. Thus,

$$1 = \frac{1.609 \text{ km}}{1 \text{ mi}}$$

Now multiply both sides by 60 mi/h and cancel units:

$$60 \text{ mi/h} = \frac{60 \text{ mi}}{\text{h}} \times \frac{1.609 \text{ km}}{1 \text{ mi}} = 96.54 \text{ km/h}$$

b. Given that 1 mi = 1.609 km, 1 km = 1000 m, 1 h = 60 min, and 1 min = 60 s, choose conversion factors as follows:

$$1 = \frac{1.609 \text{ km}}{1 \text{ mi}}, \quad 1 = \frac{1000 \text{ m}}{1 \text{ km}}, \quad 1 = \frac{1 \text{ h}}{60 \text{ min}}, \quad \text{and } 1 = \frac{1 \text{ min}}{60 \text{ s}}$$

Thus,

$$\frac{60 \text{ mi}}{h} = \frac{60 \text{ mi}}{h} \times \frac{1.609 \text{ km}}{1 \text{ mi}} \times \frac{1000 \text{ m}}{1 \text{ km}} \times \frac{1 \text{ h}}{60 \text{ min}} \times \frac{1 \text{ min}}{60 \text{ s}} = 26.8 \text{ m/s}$$

You can also solve this problem by treating the numerator and denominator separately. For example, you can convert miles to meters and hours to seconds, then divide (see Example 1–2). In the final analysis, both methods are equivalent.

EXAMPLE 1–2   Do Example 1–1(b) by expanding the top and bottom separately.

Solution

$$60 \text{ mi} = 60 \text{ mi} \times \frac{1.609 \text{ km}}{1 \text{ mi}} \times \frac{1000 \text{ m}}{1 \text{ km}} = 96\,540 \text{ m}$$

$$1 \text{ h} = 1 \text{ h} \times \frac{60 \text{ min}}{1 \text{ h}} \times \frac{60 \text{ s}}{1 \text{ min}} = 3600 \text{ s}$$

Thus, velocity = 96 540 m/3600 s = 26.8 m/s as above.

PRACTICE PROBLEMS 1

1. Area = πr^2. Given $r = 8$ inches, determine area in square meters (m²).
2. A car travels 60 feet in 2 seconds. Determine
   a. its speed in meters per second,
   b. its speed in kilometers per hour.

For part (b), use the method of Example 1–1, then check using the method of Example 1–2.

Answers: 1. 0.130 m² 2. a. 9.14 m/s b. 32.9 km/h

1.4   Power of Ten Notation

Electrical values vary tremendously in size. In electronic systems, for example, voltages may range from a few millionths of a volt to several thousand volts, while in power systems, voltages of up to several hundred thousand are common. To handle this large range, the **power of ten notation** (Table 1–5) is used.

To express a number in power of ten notation, move the decimal point to where you want it, then multiply the result by the power of ten needed to restore the number to its original value. Thus, 247 000 = 2.47 × 10⁵. (The number 10 is called the **base,** and its power is called the **exponent.**) An easy way to determine the exponent is to count the number of places (right or left) that you moved the decimal point. Thus,

$$247\,000 = 2\,4\,7\,0\,0\,0 = 2.47 \times 10^5$$
$$5\,4\,3\,2\,1$$

TABLE 1–5 Common Power of Ten Multipliers

$1\,000\,000 = 10^6$	$0.000001 = 10^{-6}$
$100\,000 = 10^5$	$0.00001 = 10^{-5}$
$10\,000 = 10^4$	$0.0001 = 10^{-4}$
$1\,000 = 10^3$	$0.001 = 10^{-3}$
$100 = 10^2$	$0.01 = 10^{-2}$
$10 = 10^1$	$0.1 = 10^{-1}$
$1 = 10^0$	$1 = 10^0$

Similarly, the number 0.003 69 may be expressed as 3.69×10^{-3} as illustrated below.

$$0.003\,69 = 0.0\,0\,3\,6\,9 = 3.69 \times 10^{-3}$$
$$1\,2\,3$$

Multiplication and Division Using Powers of Ten

To multiply numbers in power of ten notation, multiply their base numbers, then add their exponents. Thus,

$$(1.2 \times 10^3)(1.5 \times 10^4) = (1.2)(1.5) \times 10^{(3+4)} = 1.8 \times 10^7$$

For division, subtract the exponents in the denominator from those in the numerator. Thus,

$$\frac{4.5 \times 10^2}{3 \times 10^{-2}} = \frac{4.5}{3} \times 10^{2-(-2)} = 1.5 \times 10^4$$

EXAMPLE 1–3 Convert the following numbers to power of ten notation, then perform the operation indicated:

a. 276×0.009,

b. $98\,200/20$.

Solution

a. $276 \times 0.009 = (2.76 \times 10^2)(9 \times 10^{-3}) = 24.8 \times 10^{-1} = 2.48$

b. $\dfrac{98\,200}{20} = \dfrac{9.82 \times 10^4}{2 \times 10^1} = 4.91 \times 10^3$

Addition and Subtraction Using Powers of Ten

To add or subtract, first adjust all numbers to the same power of ten. It does not matter what exponent you choose, as long as all are the same.

EXAMPLE 1–4   Add 3.25×10^2 and 5×10^3

a. using 10^2 representation,

b. using 10^3 representation.

Solution

a. $5 \times 10^3 = 50 \times 10^2$. Thus, $3.25 \times 10^2 + 50 \times 10^2 = 53.25 \times 10^2$

b. $3.25 \times 10^2 = 0.325 \times 10^3$. Thus, $0.325 \times 10^3 + 5 \times 10^3 = 5.325 \times 10^3$, which is the same as 53.25×10^2

NOTES...

Use common sense when handling numbers. With calculators, for example, it is often easier to work directly with numbers in their original form than to convert them to power of ten notation. (As an example, it is more sensible to multiply 276×0.009 directly than to convert to power of ten notation as we did in Example 1–3(a).) If the final result is needed as a power of ten, you can convert as a last step.

Powers

Raising a number to a power is a form of multiplication (or division if the exponent is negative). For example,

$$(2 \times 10^3)^2 = (2 \times 10^3)(2 \times 10^3) = 4 \times 10^6$$

In general, $(N \times 10^n)^m = N^m \times 10^{nm}$. In this notation, $(2 \times 10^3)^2 = 2^2 \times 10^{3 \times 2} = 4 \times 10^6$ as before.

Integer fractional powers represent roots. Thus, $4^{1/2} = \sqrt{4} = 2$ and $27^{1/3} = \sqrt[3]{27} = 3$.

EXAMPLE 1–5   Expand the following:

a. $(250)^3$ b. $(0.0056)^2$ c. $(141)^{-2}$ d. $(60)^{1/3}$

Solution

a. $(250)^3 = (2.5 \times 10^2)^3 = (2.5)^3 \times 10^{2 \times 3} = 15.625 \times 10^6$

b. $(0.0056)^2 = (5.6 \times 10^{-3})^2 = (5.6)^2 \times 10^{-6} = 31.36 \times 10^{-6}$

c. $(141)^{-2} = (1.41 \times 10^2)^{-2} = (1.41)^{-2} \times (10^2)^{-2} = 0.503 \times 10^{-4}$

d. $(60)^{1/3} = \sqrt[3]{60} = 3.915$

PRACTICE PROBLEMS 2

Determine the following:

a. $(6.9 \times 10^5)(0.392 \times 10^{-2})$

b. $(23.9 \times 10^{11})/(8.15 \times 10^5)$

c. $14.6 \times 10^2 + 11.2 \times 10^1$ (Express in 10^2 and 10^1 notation.)

d. $(29.6)^3$

e. $(0.385)^{-2}$

Answers: a. 2.71×10^3 b. 2.93×10^6 c. $15.7 \times 10^2 = 157 \times 10^1$ d. 25.9×10^3
e. 6.75

1.5   Prefixes

Scientific and Engineering Notation

If power of ten numbers are written with one digit to the left of the decimal place, they are said to be in **scientific notation.** Thus, 2.47×10^5 is in scientific notation, while 24.7×10^4 and 0.247×10^6 are not. However, we are more interested in **engineering notation.** In engineering notation, prefixes are used to represent certain powers of ten; see Table 1–6. Thus, a quantity such as 0.045 A (amperes) can be expressed as 45×10^{-3} A, but it is preferable to express it as 45 mA. Here, we have substituted the prefix *milli* for the multiplier 10^{-3}. It is usual to select a prefix that results in a base number between 0.1 and 999. Thus, 1.5×10^{-5} s would be expressed as 15 μs.

TABLE 1–6   Engineering Prefixes

Power of 10	Prefix	Symbol
10^{12}	tera	T
10^{9}	giga	G
10^{6}	mega	M
10^{3}	kilo	k
10^{-3}	milli	m
10^{-6}	micro	μ
10^{-9}	nano	n
10^{-12}	pico	p

EXAMPLE 1–6   Express the following in engineering notation:

a. 10×10^4 volts   b. 0.1×10^{-3} watts   c. 250×10^{-7} seconds

Solution

a. 10×10^4 V = 100×10^3 V = 100 kilovolts = 100 kV

b. 0.1×10^{-3} W = 0.1 milliwatts = 0.1 mW

c. 250×10^{-7} s = 25×10^{-6} s = 25 microseconds = 25 μs

EXAMPLE 1–7   Convert 0.1 MV to kilovolts (kV).

Solution

$$0.1 \text{ MV} = 0.1 \times 10^6 \text{ V} = (0.1 \times 10^3) \times 10^3 \text{ V} = 100 \text{ kV}$$

Remember that a prefix represents a power of ten and thus the rules for power of ten computation apply. For example, when adding or subtracting, adjust to a common base, as illustrated in Example 1–8.

EXAMPLE 1–8   Compute the sum of 1 ampere (amp) and 100 milliamperes.

Solution   Adjust to a common base, either amps (A) or milliamps (mA). Thus,

$$1 \text{ A} + 100 \text{ mA} = 1 \text{ A} + 100 \times 10^{-3} \text{ A} = 1 \text{ A} + 0.1 \text{ A} = 1.1 \text{ A}$$

Alternatively, 1 A + 100 mA = 1000 mA + 100 mA = 1100 mA.

1. Convert 1800 kV to megavolts (MV).

2. In Chapter 4, we show that voltage is the product of current times resistance—that is, $V = I \times R$, where V is in volts, I is in amperes, and R is in ohms. Given $I = 25$ mA and $R = 4$ kΩ, convert these to power of ten notation, then determine V.

3. If $I_1 = 520$ μA, $I_2 = 0.157$ mA, and $I_3 = 2.75 \times 10^{-4}$ A, what is $I_1 + I_2 + I_3$ in mA?

Answers: 1. 1.8 MV   2. 100 V   3. 0.952 mA

IN-PROCESS
LEARNING
CHECK 1

1. All conversion factors have a value of what?

2. Convert 14 yards to centimeters.

3. What units does the following reduce to?

$$\frac{km}{h} \times \frac{m}{km} \times \frac{h}{min} \times \frac{min}{s}$$

4. Express the following in engineering notation:
   a. 4270 ms   b. 0.001 53 V   c. 12.3×10^{-4} s

5. Express the result of each of the following computations as a number times 10 to the power indicated:
   a. 150×120 as a value times 10^4; as a value times 10^3.
   b. $300 \times 6/0.005$ as a value times 10^4; as a value times 10^5; as a value times 10^6.
   c. $430 + 15$ as a value times 10^2; as a value times 10^1.
   d. $(3 \times 10^{-2})^3$ as a value times 10^{-6}; as a value times 10^{-5}.

6. Express each of the following as indicated.
   a. 752 μA in mA.
   b. 0.98 mV in μV.
   c. 270 μs + 0.13 ms in μs and in ms.

(Answers are at the end of the chapter.)

1.6   Significant Digits and Numerical Accuracy

The number of digits in a number that carry actual information are termed **significant digits.** Thus, if we say a piece of wire is 3.57 meters long, we mean that its length is closer to 3.57 m than it is to 3.56 m or 3.58 m and we have three significant digits. (The number of significant digits includes the first estimated digit.) If we say that it is 3.570 m, we mean that it is closer to 3.570 m than to 3.569 m or 3.571 m and we have four significant digits. When determining significant digits, zeros used to locate the decimal point are not counted. Thus, 0.004 57 has three significant digits; this can be seen if you express it as 4.57×10^{-3}.

Most calculations that you will do in circuit theory will be done using a hand calculator. An error that has become quite common is to show more digits of "accuracy" in an answer than are warranted, simply because the numbers appear on the calculator display. The number of digits that you should show is related to the number of significant digits in the numbers used in the calculation.

To illustrate, suppose you have two numbers, $A = 3.76$ and $B = 3.7$, to be multiplied. Their product is 13.912. If the numbers 3.76 and 3.7 are exact this answer is correct. However, if the numbers have been obtained by measurement where values cannot be determined exactly, they will have some uncertainty and the product must reflect this uncertainty. For example, suppose A and B have an uncertainty of 1 in their first estimated digit—that is, $A = 3.76 \pm 0.01$ and $B = 3.7 \pm 0.1$. This means that A can be as small as 3.75 or as large as 3.77, while B can be as small as 3.6 or as large as 3.8. Thus, their product can be as small as $3.75 \times 3.6 = 13.50$ or as large as $3.77 \times 3.8 = 14.326$. The best that we can say about the product is that it is 14, i.e., that you know it only to the nearest whole number. You cannot even say that it is 14.0 since this implies that you know the answer to the nearest tenth, which, as you can see from the above, you do not.

We can now give a "rule of thumb" for determining significant digits. *The number of significant digits in a result due to multiplication or division is the same as the number of significant digits in the number with the least number of significant digits.* In the previous calculation, for example, 3.7 has two significant digits so that the answer can have only two significant digits as well. This agrees with our earlier observation that the answer is 14, not 14.0 (which has three).

When adding or subtracting, you must also use common sense. For example, suppose two currents are measured as 24.7 A (one place known after the decimal point) and 123 mA (i.e., 0.123 A). Their sum is 24.823 A. However, the right-hand digits 23 in the answer are not significant. They cannot be, since, if you don't know what the second digit after the decimal point is for the first current, it is senseless to claim that you know their sum to the third decimal place! The best that you can say about the sum is that it also has one significant digit after the decimal place, that is,

> 24.7   A   (One place after decimal)
> + 0.123 A
> ─────────────
> 24.823 A → 24.8 A (One place after decimal)

Therefore, when adding numbers, add the given data, then round the result to the last column where all given numbers have significant digits. The process is similar for subtraction.

NOTES...

When working with numbers, you will encounter *exact* numbers and *approximate* numbers. Exact numbers are numbers that we know for certain, while approximate numbers are numbers that have some uncertainty. For example, when we say that there are 60 minutes in one hour, the 60 here is exact. However, if we measure the length of a wire and state it as 60 m, the 60 in this case carries some uncertainty (depending on how good our measurement is), and is thus an approximate number. When an exact number is included in a calculation, there is no limit to how many decimal places you can associate with it—the accuracy of the result is affected only by the approximate numbers involved in the calculation. Many numbers encountered in technical work are approximate, as they have been obtained by measurement.

NOTES...

In this book, given numbers are assumed to be exact unless otherwise noted. Thus, when a value is given as 3 volts, take it to mean exactly 3 volts, not simply that it has one significant figure. Since our numbers are assumed to be exact, all digits are significant, and we use as many digits as are convenient in examples and problems. Final answers are usually rounded to 3 digits.

1. Assume that only the digits shown in $8.75 \times 2.446 \times 9.15$ are significant. Determine their product and show it with the correct number of significant digits.

2. For the numbers of Problem 1, determine

$$\frac{8.75 \times 2.446}{9.15}$$

3. If the numbers in Problems 1 and 2 are exact, what are the answers to eight digits?

4. Three currents are measured as 2.36 A, 11.5 A, and 452 mA. Only the digits shown are significant. What is their sum shown to the correct number of significant digits?

Answers: 1. 196   2. 2.34   3. 195.83288; 2.3390710   4. 14.3 A

1.7   Circuit Diagrams

Electric circuits are constructed using components such as batteries, switches, resistors, capacitors, transistors, interconnecting wires, etc. To represent these circuits on paper, diagrams are used. In this book, we use three types: block diagrams, schematic diagrams, and pictorials.

Block Diagrams

Block diagrams describe a circuit or system in simplified form. The overall problem is broken into blocks, each representing a portion of the system or circuit. Blocks are labelled to indicate what they do or what they contain, then interconnected to show their relationship to each other. General signal flow is usually from left to right and top to bottom. Figure 1–5, for example, represents an audio amplifier. Although you have not covered any of its circuits yet, you should be able to follow the general idea quite easily—sound is picked up by the microphone, converted to an electrical signal, amplified by a pair of amplifiers, then output to the speaker, where it is converted back to sound. A power supply energizes the system. The advantage of a block diagram is that it gives you the overall picture and helps you understand the general nature of a problem. However, it does not provide detail.

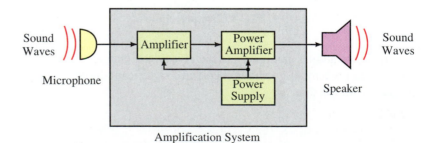

FIGURE 1–5   An example block diagram. Pictured is a simplified representation of an audio amplification system.

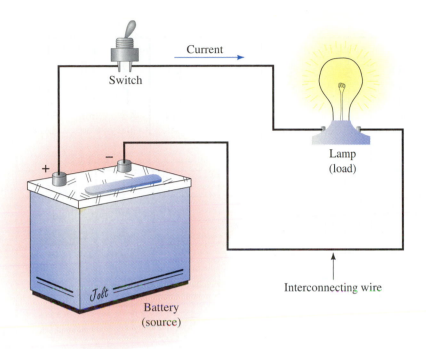

FIGURE 1–6 A pictorial diagram. The battery is referred to as a *source* while the lamp is referred to as a *load*. (The + and − on the battery are discussed in Chapter 2.)

Pictorial Diagrams

Pictorial diagrams are one of the types of diagrams that provide detail. They help you visualize circuits and their operation by showing components as they actually appear. For example, the circuit of Figure 1–6 consists of a battery, a switch, and an electric lamp, all interconnected by wire. Operation is easy to visualize—when the switch is closed, the battery causes current in the circuit, which lights the lamp. The battery is referred to as the source and the lamp as the load.

Schematic Diagrams

While pictorial diagrams help you visualize circuits, they are cumbersome to draw. **Schematic diagrams** get around this by using simplified, standard symbols to represent components; see Table 1–7. (The meaning of these symbols will be made clear as you progress through the book.) In Figure 1–7(a), for example, we have used some of these symbols to create a schematic for the circuit of Figure 1–6. Each component has been replaced by its corresponding circuit symbol.

When choosing symbols, choose those that are appropriate to the occasion. Consider the lamp of Figure 1–7(a). As we will show later, the lamp possesses a property called *resistance* that causes it to resist the passage of charge. When you wish to emphasize this property, use the resistance symbol rather than the lamp symbol, as in Figure 1–7(b).

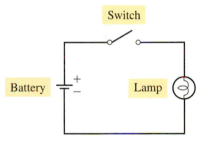

(a) Schematic using lamp symbol

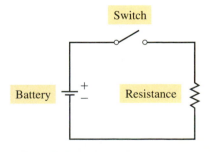

(b) Schematic using resistance symbol

FIGURE 1–7 Schematic representation of Figure 1–6. The lamp has a circuit property called resistance (discussed in Chapter 3).

TABLE 1–7   Schematic Circuit Symbols

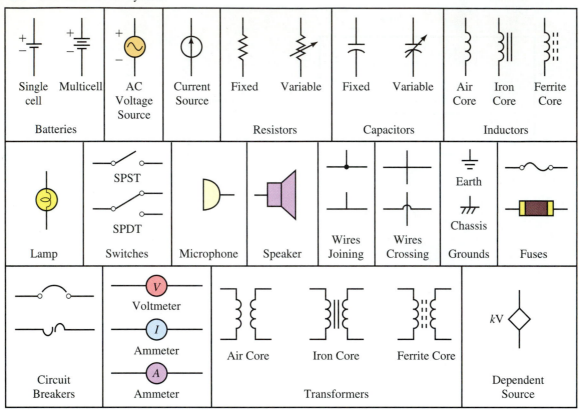

When you draw schematic diagrams, draw them with horizontal and vertical lines joined at right angles as in Figure 1–7. This is standard practice. (At this point you should glance through some later chapters, e.g., Chapter 7, and study additional examples.)

1.8   Circuit Analysis Using Computers

Personal computers are used extensively for analysis and design. Software tools available for such tasks fall into two broad categories: prepackaged application programs (application packages) and programming languages. **Application packages** solve problems without requiring programming on the part of the user, while **programming languages** require the user to write code for each type of problem to be solved.

Circuit Simulation Software

Simulation software is application software; it solves problems by simulating the behavior of electrical and electronic circuits rather than by solving sets of equations. To analyze a circuit, you "build" it on your screen by selecting components (resistors, capacitors, transistors, etc.) from a library of parts, which you then position and interconnect to form the desired circuit. You can

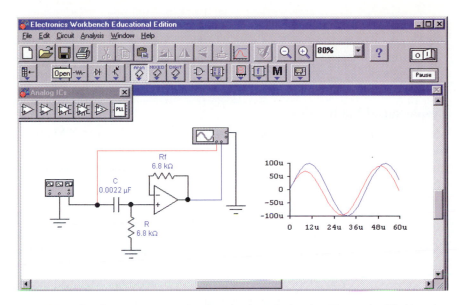

FIGURE 1–8   Computer screen showing circuit analysis using Electronics Workbench.

change component values, connections, and analysis options instantly with the click of a mouse. Figures 1–8 and 1–9 show two examples.

Most simulation packages use a software engine called SPICE, an acronym for *Simulation Program with Integrated Circuit Emphasis*. Popular products are *PSpice, Electronics Workbench*® (EWB) and *Circuit Maker*. In this text, we use Electronics Workbench and OrCAD PSpice, both of which have either evaluation or student versions (see the Preface for more details). Both products have their strong points. Electronics Workbench, for instance, more closely models an actual workbench (complete with realistic meters) than does PSpice and is a bit easier to learn. On the other hand, PSpice has a

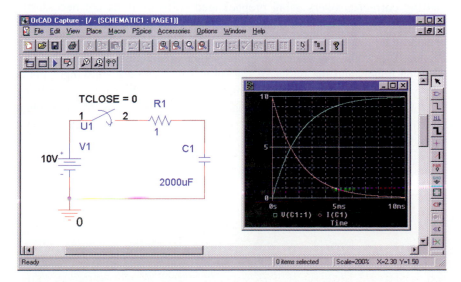

FIGURE 1–9   Computer screen showing circuit analysis using OrCAD PSpice.

more complete analysis capability; for example, it determines and displays important information (such as phase angles in ac analyses and current waveforms in transient analysis) that Electronics Workbench, as of this writing, does not.

Prepackaged Math Software

Math packages also require no programming. A popular product is Mathcad from Mathsoft Inc. With Mathcad, you enter equations in standard mathematical notation. For example, to find the first root of a quadratic equation, you would use

$$x := \frac{-b + \sqrt{b^2 - 4 \cdot a \cdot c}}{2 \cdot a}$$

Mathcad is a great aid for solving simultaneous equations such as those encountered during mesh or nodal analysis (Chapters 8 and 19) and for plotting waveforms. (You simply enter the formula.) In addition, Mathcad incorporates a built-in Electronic Handbook that contains hundreds of useful formulas and circuit diagrams that can save you a great deal of time.

Programming Languages

Many problems can also be solved using programming languages such as BASIC, C, or FORTRAN. To solve a problem using a programming language, you code its solution, step by step. We do not consider programming languages in this book.

A Word of Caution

With the widespread availability of inexpensive software tools, you may wonder why you are asked to solve problems manually throughout this book. The reason is that, as a student, your job is to learn principles and concepts. Getting correct answers using prepackaged software does not necessarily mean that you understand the theory—it may mean only that you know how to enter data. Software tools should always be used wisely. Before you use PSpice, Electronics Workbench, or any other application package, be sure that you understand the basics of the subject that you are studying. This is why you should solve problems manually with your calculator first. Following this, try some of the application packages to explore ideas. Most chapters (starting with Chapter 4) include a selection of worked-out examples and problems to get you started.

1.3   Converting Units

1. Perform the following conversions:
   a. 27 minutes to seconds
   b. 0.8 hours to seconds
   c. 2 h 3 min 47 s to s
   d. 35 horsepower to watts
   e. 1827 W to hp
   f. 23 revolutions to degrees

2. Perform the following conversions:
   a. 27 feet to meters
   b. 2.3 yd to cm
   c. 36°F to degrees C
   d. 18 (US) gallons to liters
   e. 100 sq. ft to m^2
   f. 124 sq. in. to m^2
   g. 47-pound force to newtons

3. Set up conversion factors, compute the following, and express the answer in the units indicated.
   a. The area of a plate 1.2 m by 70 cm in m^2.
   b. The area of a triangle with base 25 cm, height 0.5 m in m^2.
   c. The volume of a box 10 cm by 25 cm by 80 cm in m^3.
   d. The volume of a sphere with 10 in. radius in m^3.

4. An electric fan rotates at 300 revolutions per minute. How many degrees is this per second?

5. If the surface mount robot machine of Figure 1–3 places 15 parts every 12 s, what is its placement rate per hour?

6. If your laser printer can print 8 pages per minute, how many pages can it print in one tenth of an hour?

7. A car gets 27 miles per US gallon. What is this in kilometers per liter?

8. The equatorial radius of the earth is 3963 miles. What is the earth's circumference in kilometers at the equator?

9. A wheel rotates 18° in 0.02 s. How many revolutions per minute is this?

10. The height of horses is sometimes measured in "hands," where 1 hand = 4 inches. How many meters tall is a 16-hand horse? How many centimeters?

11. Suppose $s = vt$ is given, where s is distance travelled, v is velocity, and t is time. If you travel at $v = 60$ mph for 500 seconds, you get upon unthinking substitution $s = vt = (60)(500) = 30,000$ miles. What is wrong with this calculation? What is the correct answer?

12. How long does it take for a pizza cutter traveling at 0.12 m/s to cut diagonally across a 15-in. pizza?

13. Joe S. was asked to convert 2000 yd/h to meters per second. Here is Joe's work: velocity = $2000 \times 0.9144 \times 60/60 = 1828.8$ m/s. Determine conversion factors, write units into the conversion, and find the correct answer.

14. The mean distance from the earth to the moon is 238 857 miles. Radio signals travel at 299 792 458 m/s. How long does it take a radio signal to reach the moon?

15. Your plant manager asks you to investigate two machines. The cost of electricity for operating machine #1 is 43 cents/minute, while that for machine #2 is $200.00 per 8-hour shift. The purchase price and production capacity for both machines are identical. Based on this information, which machine should you purchase and why?

16. Given that 1 hp = 550 ft-lb/s, 1 ft = 0.3048 m, 1 lb = 4.448 N, 1 J = 1 N-m, and 1 W = 1 J/s, show that 1 hp = 746 W.

1.4   Power of Ten Notation

17. Express each of the following in power of ten notation with one nonzero digit to the left of the decimal point:

   a. 8675
   b. 0.008 72
   c. 12.4×10^2
   d. 37.2×10^{-2}
   e. $0.003\ 48 \times 10^5$
   f. $0.000\ 215 \times 10^{-3}$
   g. 14.7×10^0

18. Express the answer for each of the following in power of ten notation with one nonzero digit to the left of the decimal point.

   a. $(17.6)(100)$
   b. $(1400)(27 \times 10^{-3})$
   c. $(0.15 \times 10^6)(14 \times 10^{-4})$
   d. $1 \times 10^{-7} \times 10^{-4} \times 10.65$
   e. $(12.5)(1000)(0.01)$
   f. $(18.4 \times 10^0)(100)(1.5 \times 10^{-5})(0.001)$

19. Repeat the directions in Question 18 for each of the following.

   a. $\dfrac{125}{1000}$
   b. $\dfrac{8 \times 10^4}{(0.001)}$
   c. $\dfrac{3 \times 10^4}{(1.5 \times 10^6)}$
   d. $\dfrac{(16 \times 10^{-7})(21.8 \times 10^6)}{(14.2)(12 \times 10^{-5})}$

20. Determine answers for the following

   a. $123.7 + 0.05 + 1259 \times 10^{-3}$
   b. $72.3 \times 10^{-2} + 1 \times 10^{-3}$
   c. $86.95 \times 10^2 - 383$
   d. $452 \times 10^{-2} + (697)(0.01)$

21. Convert the following to power of 10 notation and, without using your calculator, determine the answers.

   a. $(4 \times 10^3)(0.05)^2$
   b. $(4 \times 10^3)(-0.05)^2$
   c. $\dfrac{(3 \times 2 \times 10)^2}{(2 \times 5 \times 10^{-1})}$
   d. $\dfrac{(30 + 20)^{-2}(2.5 \times 10^6)(6000)}{(1 \times 10^3)(2 \times 10^{-1})^2}$
   e. $\dfrac{(-0.027)^{1/3}(-0.2)^2}{(23 + 1)^0 \times 10^{-3}}$

22. For each of the following, convert the numbers to power of ten notation, then perform the indicated computations. Round your answer to four digits:

a. $(452)(6.73 \times 10^4)$

b. $(0.009\ 85)(4700)$

c. $(0.0892)/(0.000\ 067\ 3)$

d. $12.40 - 236 \times 10^{-2}$

e. $(1.27)^3 + 47.9/(0.8)^2$

f. $(-643 \times 10^{-3})^3$

g. $[(0.0025)^{1/2}][1.6 \times 10^4]$

h. $[(-0.027)^{1/3}]/[1.5 \times 10^{-4}]$

i. $\dfrac{(3.5 \times 10^4)^{-2} \times (0.0045)^2 \times (729)^{1/3}}{[(0.008\ 72) \times (47)^3] - 356}$

23. For the following,

a. convert numbers to power of ten notation, then perform the indicated computation,

b. perform the operation directly on your calculator without conversion. What is your conclusion?

i. 842×0.0014 ii. $\dfrac{0.0352}{0.007\ 91}$

24. Express each of the following in conventional notation:

a. 34.9×10^4

b. 15.1×10^0

c. 234.6×10^{-4}

d. 6.97×10^{-2}

e. $45\ 786.97 \times 10^{-1}$

f. 6.97×10^{-5}

25. One coulomb (Chapter 2) is the amount of charge represented by 6 240 000 000 000 000 000 electrons. Express this quantity in power of ten notation.

26. The mass of an electron is 0.000 000 000 000 000 000 000 000 000 000 899 9 kg. Express as a power of 10 with one non-zero digit to the left of the decimal point.

27. If 6.24×10^{18} electrons pass through a wire in 1 s, how many pass through it during a time interval of 2 hr, 47 min and 10 s?

28. Compute the distance traveled in meters by light in a vacuum in 1.2×10^{-8} second.

29. How long does it take light to travel 3.47×10^5 km in a vacuum?

30. How far in km does light travel in one light-year?

31. While investigating a site for a hydroelectric project, you determine that the flow of water is 3.73×10^4 m³/s. How much is this in liters/hour?

32. The gravitational force between two bodies is $F = 6.6726 \times 10^{-11} \dfrac{m_1 m_2}{r^2}$ N, where masses m_1 and m_2 are in kilograms and the distance r between gravitational centers is in meters. If body 1 is a sphere of radius 5000 miles and density of 25 kg/m³, and body 2 is a sphere of diameter 20 000 km and density of 12 kg/m³, and the distance between centers is 100 000 miles, what is the gravitational force between them?

1.5   Prefixes

33. What is the appropriate prefix and its abbreviation for each of the following multipliers ?

a. 1000

b. 1 000 000

c. 10^9

d. 0.000 001

e. 10^{-3}

f. 10^{-12}

34. Express the following in terms of their abbreviations, e.g., microwatts as μW. Pay particular attention to capitalization (e.g., V, not v, for volts).

    a. milliamperes   b. kilovolts

    c. megawatts   d. microseconds

    e. micrometers   f. milliseconds

    g. nanoamps

35. Express the following in the most sensible engineering notation (e.g., 1270 μs = 1.27 ms).

    a. 0.0015 s   b. 0.000 027 s   c. 0.000 35 ms

36. Convert the following:

    a. 156 mV to volts   b. 0.15 mV to microvolts

    c. 47 kW to watts   d. 0.057 MW to kilowatts

    e. 3.5×10^4 volts to kilovolts   f. 0.000 035 7 amps to microamps

37. Determine the values to be inserted in the blanks.

    a. 150 kV = ___ $\times 10^3$ V = ___ $\times 10^6$ V

    b. 330 μW = ___ $\times 10^{-3}$ W = ___ $\times 10^{-5}$ W

38. Perform the indicated operations and express the answers in the units indicated.

    a. 700 μA $-$ 0.4 mA = ___ μA = ___ mA

    b. 600 MW + 300 $\times 10^4$ W = ___ MW

39. Perform the indicated operations and express the answers in the units indicated.

    a. 330 V + 0.15 kV + 0.2 $\times 10^3$ V = ___ V

    b. 60 W + 100 W + 2700 mW = ___ W

40. The voltage of a high voltage transmission line is 1.15×10^5 V. What is its voltage in kV?

41. You purchase a 1500 W electric heater to heat your room. How many kW is this?

42. While repairing an antique radio, you come across a faulty capacitor designated 39 mmfd. After a bit of research, you find that "mmfd" is an obsolete unit meaning "micromicrofarads". You need a replacement capacitor of equal value. Consulting Table 1–6, what would 39 "micromicrofarads" be equivalent to?

43. A radio signal travels at 299 792.458 km/s and a telephone signal at 150 m/μs. If they originate at the same point, which arrives first at a destination 5000 km away? By how much?

44. a. If 0.045 coulomb of charge (Question 25) passes through a wire in 15 ms, how many electrons is this?

    b. At the rate of 9.36×10^{19} electrons per second, how many coulombs pass a point in a wire in 20 μs?

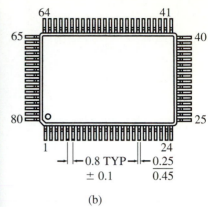

(b)

(a)

FIGURE 1–10

1.6 Significant Digits and Numerical Accuracy

For each of the following, assume that the given digits are significant.

45. Determine the answer to three significant digits:

$$2.35 - 1.47 \times 10^{-6}$$

46. Given $V = IR$. If $I = 2.54$ and $R = 52.71$, determine V to the correct number of significant digits.

47. If $A = 4.05 \pm 0.01$ is divided by $B = 2.80 \pm 0.01$,

   a. What is the smallest that the result can be?

   b. What is the largest that the result can be?

   c. Based on this, give the result A/B to the correct number of significant digits.

48. The large black plastic component soldered onto the printed circuit board of Figure 1–10(a) is an electronic device known as an *integrated circuit*. As indicated in (b), the center-to-center spacing of its leads (commonly called pins) is 0.8 ± 0.1 mm. Pin diameters can vary from 0.25 to 0.45 mm. Considering these uncertainties,

   a. What is the minimum distance between pins due to manufacturing tolerances?

   b. What is the maximum distance?

1.7 Circuit Diagrams

49. Consider the pictorial diagram of Figure 1–11. Using the appropriate symbols from Table 1–7, draw this in schematic form. Hint: In later chapters, there are many schematic circuits containing resistors, inductors, and capacitors. Use these as aids.

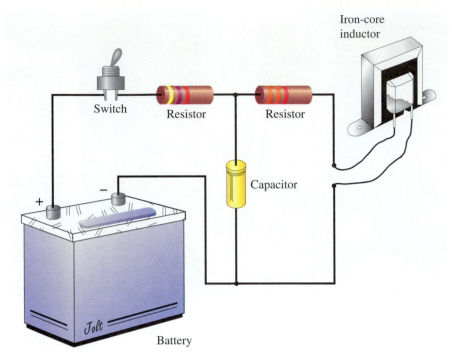

Iron-core
inductor

Switch   Resistor   Resistor

Capacitor

Battery

FIGURE 1–11

50. Draw the schematic diagram for a simple flashlight.

1.8   Circuit Analysis Using Computers

51. Many electronic and computer magazines carry advertisements for computer software tools such as PSpice, SpiceNet, Mathcad, MLAB, Matlab, Maple V, plus others. Investigate a few of these magazines in your school's library; by studying such advertisements, you can gain valuable insight into what modern software packages are able to do.

In-Process Learning Check 1

1. One

2. 1280 cm

3. m/s

4. a. 4.27 s   b. 1.53 mV   c. 1.23 ms

5. a. $1.8 \times 10^4 = 18 \times 10^3$

   b. $36 \times 10^4 = 3.6 \times 10^5 = 0.36 \times 10^6$

   c. $4.45 \times 10^2 = 44.5 \times 10^1$

   d. $27 \times 10^{-6} = 2.7 \times 10^{-5}$

6. a. 0.752 mA   b. 980 μV   c. 400 μs = 0.4 ms

2

Voltage and Current

OBJECTIVES

After studying this chapter, you will be able to

- describe the makeup of an atom,
- explain the relationships between valence shells, free electrons, and conduction,
- describe the fundamental (coulomb) force within an atom, and the energy required to create free electrons,
- describe what ions are and how they are created,
- describe the characteristics of conductors, insulators, and semiconductors,
- describe the coulomb as a measure of charge,
- define voltage,
- describe how a battery "creates" voltage,
- explain current as a movement of charge and how voltage causes current in a conductor,
- describe important battery types and their characteristics,
- describe how to measure voltage and current.

KEY TERMS

Ampere
Atom
Battery
Cell
Circuit Breaker
Conductor
Coulomb
Coulomb's Law
Current
Electric Charge
Electron
Free Electrons
Fuse
Insulator
Ion
Neutron
Polarity
Potential Difference
Proton
Semiconductor
Shell
Switch
Valence
Volt

OUTLINE

Atomic Theory Review
The Unit of Electrical Charge: The Coulomb
Voltage
Current
Practical DC Voltage Sources
Measuring Voltage and Current
Switches, Fuses, and Circuit Breakers

A basic electric circuit consisting of a source of electrical energy, a switch, a load, and interconnecting wire is shown in Figure 2–1. When the switch is closed, current in the circuit causes the light to come on. This circuit is representative of many common circuits found in practice, including those of flashlights and automobile headlight systems. We will use it to help develop an understanding of voltage and current.

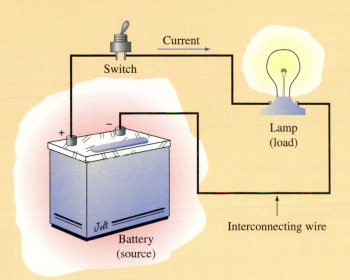

Current

Switch

Lamp
(load)

+   −

Jolt

Battery
(source)

Interconnecting wire

FIGURE 2–1   A basic electric circuit.

Elementary atomic theory shows that the current in Figure 2–1 is actually a flow of charges. The cause of their movement is the "voltage" of the source. While in Figure 2–1 this source is a battery, in practice it may be any one of a number of practical sources including generators, power supplies, solar cells, and so on.

In this chapter we look at the basic ideas of voltage and current. We begin with a discussion of atomic theory. This leads us to free electrons and the idea of current as a movement of charge. The fundamental definitions of voltage and current are then developed. Following this, we look at a number of common voltage sources. The chapter concludes with a discussion of voltmeters and ammeters and the measurement of voltage and current in practice.

PUTTING IT IN
PERSPECTIVE

The Equations of Circuit Theory

IN THIS CHAPTER you meet the first of the equations and formulas that we use to describe the relationships of circuit theory. Remembering formulas is made easier if you clearly understand the principles and concepts on which they are based. As you may recall from high school physics, formulas can come about in only one of three ways, through experiment, by definition, or by mathematical manipulation.

Experimental Formulas

Circuit theory rests on a few basic experimental results. These are results that can be proven in no other way; they are valid solely because experiment has shown them to be true. The most fundamental of these are called "laws." Four examples are Ohm's law, Kirchhoff's current law, Kirchhoff's voltage law, and Faraday's law. (These laws will be met in various chapters throughout the book.) When you see a formula referred to as a law or an experimental result, remember that it is based on experiment and cannot be obtained in any other way.

Defined Formulas

Some formulas are created by definition, i.e., we make them up. For example, there are 60 seconds in a minute because we define the second as 1/60 of a minute. From this we get the formula $t_{sec} = 60 \times t_{min}$.

Derived Formulas

This type of formula or equation is created mathematically by combining or manipulating other formulas. In contrast to the other two types of formulas, the only way that a derived relationship can be obtained is by mathematics.

An awareness of where circuit theory formulas come from is important to you. This awareness not only helps you understand and remember formulas, it helps you understand the very foundations of the theory—the basic experimental premises upon which it rests, the important definitions that have been made, and the methods by which these foundation ideas have been put together. This can help enormously in understanding and remembering concepts.

2.1   Atomic Theory Review

The basic structure of an atom is shown symbolically in Figure 2–2. It consists of a nucleus of protons and neutrons surrounded by a group of orbiting electrons. As you learned in physics, the electrons are negatively charged (−), while the protons are positively charged (+). Each atom (in its normal state) has an equal number of electrons and protons, and since their charges are equal and opposite, they cancel, leaving the atom electrically neutral, i.e., with zero net charge. The nucleus, however, has a net positive charge, since it consists of positively charged protons and uncharged neutrons.

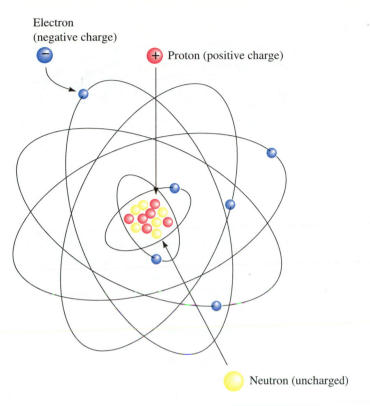

Electron
(negative charge)

+ Proton (positive charge)

Neutron (uncharged)

FIGURE 2–2   Bohr model of the atom. Electrons travel around the nucleus at incredible speeds, making billions of trips in a fraction of a second. The force of attraction between the electrons and the protons in the nucleus keeps them in orbit.

The basic structure of Figure 2–2 applies to all elements, but each element has its own unique combination of electrons, protons, and neutrons. For example, the hydrogen atom, the simplest of all atoms, has one proton and one electron, while the copper atom has 29 electrons, 29 protons, and 35 neutrons. Silicon, which is important because of its use in transistors and other electronic devices, has 14 electrons, 14 protons, and 14 neutrons.

Electrons orbit the nucleus in spherical orbits called **shells,** designated by letters *K, L, M, N,* and so on (Figure 2–3). Only certain numbers of electrons can exist within any given shell. For example, there can be up to 2 electrons in the *K* shell, up to 8 in the *L* shell, up to 18 in the *M* shell, and up to 32 in the *N* shell. The number in any shell depends on the element. For instance, the copper atom, which has 29 electrons, has all three of its inner shells completely filled but its outer shell (shell *N*) has only 1 electron, Figure 2–4. This outermost shell is called its **valence shell,** and the electron in it is called its **valence electron.**

No element can have more than eight valence electrons; when a valence shell has eight electrons, it is filled. As we shall see, the number of valence electrons that an element has directly affects its electrical properties.

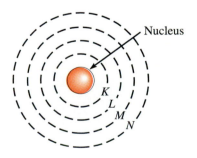

Nucleus

K
L
M
N

FIGURE 2–3   Simplified representation of the atom. Electrons travel in spherical orbits called "shells."

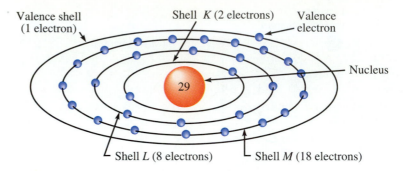

FIGURE 2–4   Copper atom. The valence electron is loosely bound.

Electrical Charge

In the previous paragraphs, we mentioned the word "charge". However, we need to look at its meaning in more detail. First, we should note that electrical charge is an intrinsic property of matter that manifests itself in the form of forces—electrons repel other electrons but attract protons, while protons repel each other but attract electrons. It was through studying these forces that scientists determined that the charge on the electron is negative while that on the proton is positive.

However, the way in which we use the term "charge" extends beyond this. To illustrate, consider again the basic atom of Figure 2–2. It has equal numbers of electrons and protons, and since their charges are equal and opposite, they cancel, leaving the atom as a whole uncharged. However, if the atom acquires additional electrons (leaving it with more electrons than protons), we say that it (the atom) is negatively charged; conversely, if it loses electrons and is left with fewer electrons than protons, we say that it is positively charged. The term "charge" in this sense denotes an imbalance between the number of electrons and protons present in the atom.

Now move up to the macroscopic level. Here, substances in their normal state are also generally uncharged; that is, they have equal numbers of electrons and protons. However, this balance is easily disturbed—electrons can be stripped from their parent atoms by simple actions such as walking across a carpet, sliding off a chair, or spinning clothes in a dryer. (Recall "static cling".) Consider two additional examples from physics. Suppose you rub an ebonite (hard rubber) rod with fur. This action causes a transfer of electrons from the fur to the rod. The rod therefore acquires an excess of electrons and is thus negatively charged. Similarly, when a glass rod is rubbed with silk, electrons are transferred from the glass rod to the silk, leaving the rod with a deficiency and, consequently, a positive charge. Here again, charge refers to an imbalance of electrons and protons.

As the above examples illustrate, "charge" can refer to the charge on an individual electron or to the charge associated with a whole group of electrons. In either case, this charge is denoted by the letter Q, and its unit of measurement in the SI system is the coulomb. (The definition of the coulomb is considered shortly.) In general, the charge Q associated with a group of electrons is equal to the product of the number of electrons times the charge on each individual electron. Since charge manifests itself in the form of forces, charge is defined in terms of these forces. This is discussed next.

Coulomb's Law

The force between charges was studied by the French scientist Charles Coulomb (1736–1806). Coulomb determined experimentally that the force between two charges Q_1 and Q_2 (Figure 2–5) is directly proportional to the product of their charges and inversely proportional to the square of the distance between them. Mathematically, Coulomb's law states

$$F = k\frac{Q_1 Q_2}{r^2} \quad \text{[newtons, N]} \qquad (2\text{–}1)$$

where Q_1 and Q_2 are the charges in coulombs, r is the center-to-center spacing between them in meters, and $k = 9 \times 10^9$. Coulomb's law applies to aggregates of charges as in Figure 2–5(a) and (b), as well as to individual electrons within the atom as in (c).

As Coulomb's law indicates, force decreases inversely as the square of distance; thus, if the distance between two charges is doubled, the force decreases to $(\frac{1}{2})^2 = \frac{1}{4}$ (i.e., one quarter) of its original value. Because of this relationship, electrons in outer orbits are less strongly attracted to the nucleus than those in inner orbits; that is, they are less **tightly bound** to the nucleus than those close by. Valence electrons are the least tightly bound and will, if they acquire sufficient energy, escape from their parent atoms.

Free Electrons

The amount of energy required to escape depends on the number of electrons in the valence shell. If an atom has only a few valence electrons, only a small amount of additional energy is needed. For example, for a metal like copper, valence electrons can gain sufficient energy from heat alone (thermal energy), even at room temperature, to escape from their parent atoms and wander from atom to atom throughout the material as depicted in Figure 2–6. (Note that these electrons do not leave the substance, they simply wander from the valence shell of one atom to the valence shell of another. The material therefore remains electrically neutral.) Such electrons are called **free electrons.** In copper, there are of the order of 10^{23} free electrons per cubic centimeter at room temperature. As we shall see, it is the presence of this large number of free electrons that makes copper such a good conductor of electric current. On the other hand, if the valence shell is full (or nearly full), valence electrons are much more tightly bound. Such materials have few (if any) free electrons.

Ions

As noted earlier, when a previously neutral atom gains or loses an electron, it acquires a net electrical charge. The charged atom is referred to as an **ion.** If the atom loses an electron, it is called a **positive ion;** if it gains an electron, it is called a **negative ion.**

Conductors, Insulators, and Semiconductors

The atomic structure of matter affects how easily charges, i.e., electrons, move through a substance and hence how it is used electrically. Electrically, materials are classified as conductors, insulators, or semiconductors.

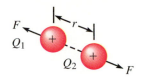

(a) Like charges repel

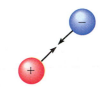

(b) Unlike charges attract

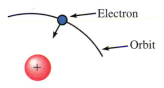

(c) The force of attraction keeps electrons in orbit

FIGURE 2–5 Coulomb law forces.

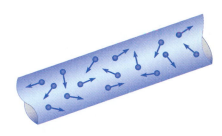

FIGURE 2–6 Random motion of free electrons in a conductor.

Conductors

Materials through which charges move easily are termed **conductors.** The most familiar examples are metals. Good metal conductors have large numbers of free electrons that are able to move about easily. In particular, silver, copper, gold, and aluminum are excellent conductors. Of these, copper is the most widely used. Not only is it an excellent conductor, it is inexpensive and easily formed into wire, making it suitable for a broad spectrum of applications ranging from common house wiring to sophisticated electronic equipment. Aluminum, although it is only about 60% as good a conductor as copper, is also used, mainly in applications where light weight is important, such as in overhead power transmission lines. Silver and gold are too expensive for general use. However, gold, because it oxidizes less than other materials, is used in specialized applications; for example, some critical electrical connectors use it because it makes a more reliable connection than other materials.

Insulators

Materials that do not conduct (e.g., glass, porcelain, plastic, rubber, and so on) are termed **insulators.** The covering on electric lamp cords, for example, is an insulator. It is used to prevent the wires from touching and to protect us from electric shock.

   Insulators do not conduct because they have full or nearly full valence shells and thus their electrons are tightly bound. However, when high enough voltage is applied, the force is so great that electrons are literally torn from their parent atoms, causing the insulation to break down and conduction to occur. In air, you see this as an arc or flashover. In solids, charred insulation usually results.

Semiconductors

Silicon and germanium (plus a few other materials) have half-filled valence shells and are thus neither good conductors nor good insulators. Known as **semiconductors,** they have unique electrical properties that make them important to the electronics industry. The most important material is silicon. It is used to make transistors, diodes, integrated circuits, and other electronic devices. Semiconductors have made possible personal computers, VCRs, portable CD players, calculators, and a host of other electronic products. You will study them in great detail in your electronics courses.

IN-PROCESS LEARNING CHECK 1

1. Describe the basic structure of the atom in terms of its constituent particles: electrons, protons, and neutrons. Why is the nucleus positively charged? Why is the atom as a whole electrically neutral?

2. What are valence shells? What does the valence shell contain?

3. Describe Coulomb's law and use it to help explain why electrons far from the nucleus are loosely bound.

4. What are free electrons? Describe how they are created, using copper as an example. Explain what role thermal energy plays in the process.

5. Briefly distinguish between a normal (i.e., uncharged) atom, a positive ion, and a negative ion.

6. Many atoms in Figure 2–6 have lost electrons and are thus positively charged, yet the substance as a whole is uncharged. Why?

_____

(Answers are at the end of the chapter.)

2.2   The Unit of Electrical Charge: The Coulomb

As noted in the previous section, the unit of electrical charge is the coulomb (C). The **coulomb** is defined as the charge carried by 6.24×10^{18} electrons. Thus, if an electrically neutral (i.e., uncharged) body has 6.24×10^{18} electrons removed, it will be left with a net positive charge of 1 coulomb, i.e., $Q = 1$ C. Conversely, if an uncharged body has 6.24×10^{18} electrons added, it will have a net negative charge of 1 coulomb, i.e., $Q = -1$ C. Usually, however, we are more interested in the charge moving through a wire. In this regard, if 6.24×10^{18} electrons pass through a wire, we say that the charge that passed through the wire is 1 C.

We can now determine the charge on one electron. It is $Q_e = 1/(6.24 \times 10^{18}) = 1.60 \times 10^{-19}$ C.

EXAMPLE 2–1   An initially neutral body has 1.7 μC of negative charge removed. Later, 18.7×10^{11} electrons are added. What is the body's final charge?

Solution   Initially the body is neutral, i.e., $Q_{\text{initial}} = 0$ C. When 1.7 μC of electrons is removed, the body is left with a positive charge of 1.7 μC. Now, 18.7×10^{11} electrons are added back. This is equivalent to

$$18.7 \times 10^{11} \text{ electrons} \times \frac{1 \text{ coulomb}}{6.24 \times 10^{18} \text{ electrons}} = 0.3 \ \mu\text{C}$$

of negative charge. The final charge on the body is therefore $Q_f = 1.7 \ \mu$C $- 0.3 \ \mu$C $= +1.4 \ \mu$C.

To get an idea of how large a coulomb is, we can use Coulomb's law. If two charges of 1 coulomb each were placed one meter apart, the force between them would be

$$F = (9 \times 10^9)\frac{(1 \text{ C})(1 \text{ C})}{(1 \text{ m})^2} = 9 \times 10^9 \text{ N, i.e., about 1 million tons!}$$

PRACTICE PROBLEMS 1

1. Positive charges $Q_1 = 2 \ \mu$C and $Q_2 = 12 \ \mu$C are separated center to center by 10 mm. Compute the force between them. Is it attractive or repulsive?

2. Two equal charges are separated by 1 cm. If the force of repulsion between them is 9.7×10^{-2} N, what is their charge? What may the charges be, both positive, both negative, or one positive and one negative?

3. After 10.61×10^{13} electrons are added to a metal plate, it has a negative charge of 3 μC. What was its initial charge in coulombs?

Answers: 1. 2160 N, repulsive;   2. 32.8 nC, both (+) or both (−);   3. 14 μC (+)

2.3 Voltage

When charges are detached from one body and transferred to another, a *potential difference* or *voltage* results between them. A familiar example is the voltage that develops when you walk across a carpet. Voltages in excess of ten thousand volts can be created in this way. (We will define the volt rigorously very shortly.) This voltage is due entirely to the separation of positive and negative charges.

Figure 2–7 illustrates another example. During electrical storms, electrons in thunderclouds are stripped from their parent atoms by the forces of turbulence and carried to the bottom of the cloud, leaving a deficiency of electrons (positive charge) at the top and an excess (negative charge) at the bottom. The force of repulsion then drives electrons away beneath the cloud, leaving the ground positively charged. Hundreds of millions of volts are created in this way. (This is what causes the air to break down and a lightning discharge to occur.)

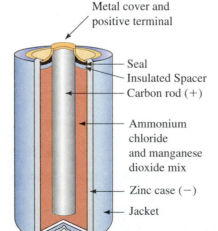

FIGURE 2–7 Voltages created by separation of charges in a thunder cloud. The force of repulsion drives electrons away beneath the cloud, creating a voltage between the cloud and ground as well. If voltage becomes large enough, the air breaks down and a lightning discharge occurs.

Practical Voltage Sources

As the preceding examples show, voltage is created solely by the separation of positive and negative charges. However, static discharges and lightning strikes are not practical sources of electricity. We now look at practical sources. A common example is the battery. In a battery, charges are separated by chemical action. An ordinary flashlight battery (dry cell) illustrates the concept in Figure 2–8. The inner electrode is a carbon rod and the outer electrode is a zinc case. The chemical reaction between the ammonium-chloride/manganese-dioxide paste and the zinc case creates an excess of elec-

Metal cover and positive terminal

Seal
Insulated Spacer
Carbon rod (+)

Ammonium chloride and manganese dioxide mix

Zinc case (−)

Jacket

(a) Basic construction.

(b) C cell, commonly called a flashlight battery.

FIGURE 2–8 Carbon-zinc cell. Voltage is created by the separation of charges due to chemical action. Nominal cell voltage is 1.5 V.

trons; hence, the zinc carries a negative charge. An alternate reaction leaves the carbon rod with a deficiency of electrons, causing it to be positively charged. These separated charges create a voltage (1.5 V in this case) between the two electrodes. The battery is useful as a source since its chemical action creates a continuous supply of energy that is able to do useful work, such as light a lamp or run a motor.

Potential Energy

The concept of voltage is tied into the concept of potential energy. We therefore look briefly at energy.

In mechanics, potential energy is the energy that a body possesses because of its position. For example, a bag of sand hoisted by a rope over a pulley has the potential to do work when it is released. The amount of work that went into giving it this potential energy is equal to the product of force times the distance through which the bag was lifted (i.e., work equals force times distance).

In a similar fashion, work is required to move positive and negative charges apart. This gives them potential energy. To understand why, consider again the cloud of Figure 2–7. Assume the cloud is initially uncharged. Now assume a charge of Q electrons is moved from the top of the cloud to the bottom. The positive charge left at the top of the cloud exerts a force on the electrons that tries to pull them back as they are being moved away. Since the electrons are being moved against this force, work (force times distance) is required. Since the separated charges experience a force to return to the top of the cloud, they have the potential to do work if released, i.e., they possess potential energy.

Definition of Voltage: The Volt

In electrical terms, a difference in potential energy is defined as **voltage.** In general, the amount of energy required to separate charges depends on the voltage developed and the amount of charge moved. By definition, *the voltage between two points is one volt if it requires one joule of energy to move one coulomb of charge from one point to the other.* In equation form,

$$V = \frac{W}{Q} \quad \text{[volts, V]} \tag{2–2}$$

where W is energy in joules, Q is charge in coulombs, and V is the resulting voltage in volts.

Note carefully that voltage is defined between points. For the case of the battery, for example, voltage appears between its terminals. Thus, voltage does not exist at a point by itself; it is always determined with respect to some other point. (For this reason, voltage is also called **potential difference.** We often use the terms interchangeably.) Note also that, although we considered static electricity in developing the energy argument, the same conclusion results regardless of how you separate the charges; this may be by chemical means as in a battery, by mechanical means as in a generator, by photoelectric means as in a solar cell, and so on.

Alternate arrangements of Equation 2–2 are useful:

$$W = QV \quad \text{[joules, J]} \tag{2–3}$$

$$Q = \frac{W}{V} \quad \text{[coulombs, C]} \tag{2–4}$$

EXAMPLE 2–2   If it takes 35 J of energy to move a charge of 5 C from one point to another, what is the voltage between the two points?

Solution

$$V = \frac{W}{Q} = \frac{35\,\text{J}}{5\,\text{C}} = 7\,\text{J/C} = 7\,\text{V}$$

PRACTICE PROBLEMS 2

1. The voltage between two points is 19 V. How much energy is required to move 67×10^{18} electrons from one point to the other?

2. The potential difference between two points is 140 mV. If 280 μJ of work are required to move a charge Q from one point to the other, what is Q?

Answers: 1. 204 J   2. 2 mC

Although Equation 2–2 is the formal definition of voltage, it is a bit abstract. A more satisfying way to look at voltage is to view it as the force or "push" that moves electrons around a circuit. This view is looked at in great detail, starting in Chapter 4 where we consider Ohm's law. For the moment, however, we will stay with Equation 2–2, which is important because it provides the theoretical foundation for many of the important circuit relationships that you will soon encounter.

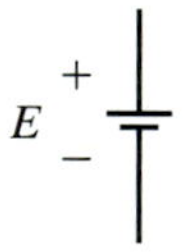

Symbol for DC Voltage Sources

Consider again Figure 2–1. The battery is the source of electrical energy that moves charges around the circuit. This movement of charges, as we will soon see, is called an electric current. Because one of the battery's terminals is always positive and the other is always negative, current is always in the same direction. Such a unidirectional current is called **dc** or **direct current,** and the battery is called a **dc source.** Symbols for dc sources are shown in Figure 2–9. The long bar denotes the positive terminal. On actual batteries, the positive terminal is usually marked POS ($+$) and the negative terminal NEG ($-$).

(a) Symbol for a cell

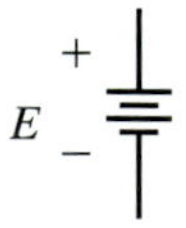

(b) Symbol for a battery

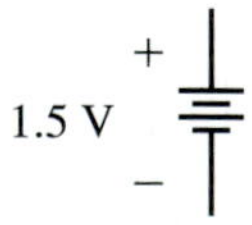

(c) A 1.5 volt battery

FIGURE 2–9   Battery symbol. The long bar denotes the positive terminal and the short bar the negative terminal. Thus, it is not necessary to put $+$ and $-$ signs on the diagram. For simplicity, we use the symbol shown in (a) throughout this book.

2.4   Current

Earlier, you learned that there are large numbers of free electrons in metals like copper. These electrons move randomly throughout the material (Figure 2–6), but their net movement in any given direction is zero.

Assume now that a battery is connected as in Figure 2–10. Since electrons are attracted by the positive pole of the battery and repelled by the neg-

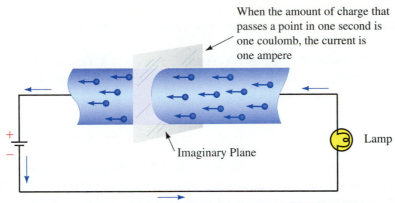

When the amount of charge that passes a point in one second is one coulomb, the current is one ampere

Imaginary Plane

Lamp

Movement of electrons through the wire

FIGURE 2–10 Electron flow in a conductor. Electrons (−) are attracted to the positive (+) pole of the battery. As electrons move around the circuit, they are replenished at the negative pole of the battery. This flow of charge is called an electric current.

ative pole, they move around the circuit, passing through the wire, the lamp, and the battery. This movement of charge is called an **electric current.** The more electrons per second that pass through the circuit, the greater is the current. Thus, current is the *rate of flow* (or *rate of movement*) of charge.

The Ampere

Since charge is measured in coulombs, its rate of flow is coulombs per second. In the SI system, one coulomb per second is defined as one **ampere** (commonly abbreviated A). From this, we get that *one ampere is the current in a circuit when one coulomb of charge passes a given point in one second* (Figure 2–10). The symbol for current is I. Expressed mathematically,

$$I = \frac{Q}{t} \quad \text{[amperes, A]} \qquad (2\text{–}5)$$

where Q is the charge (in coulombs) and t is the time interval (in seconds) over which it is measured. *In Equation 2–5, it is important to note that t does not represent a discrete point in time but is the interval of time during which the transfer of charge occurs.* Alternate forms of Equation 2–5 are

$$Q = It \quad \text{[coulombs, C]} \qquad (2\text{–}6)$$

and

$$t = \frac{Q}{I} \quad \text{[seconds, s]} \qquad (2\text{–}7)$$

EXAMPLE 2–3 If 840 coulombs of charge pass through the imaginary plane of Figure 2–10 during a time interval of 2 minutes, what is the current?

Solution Convert t to seconds. Thus,

$$I = \frac{Q}{t} = \frac{840 \text{ C}}{(2 \times 60)\text{s}} = 7 \text{ C/s} = 7 \text{ A}$$

1. Between $t = 1$ ms and $t = 14$ ms, 8 μC of charge pass through a wire. What is the current?

2. After the switch of Figure 2–1 is closed, current $I = 4$ A. How much charge passes through the lamp between the time the switch is closed and the time that it is opened 3 minutes later?

Answers: 1. 0.615 mA   2. 720 C

Although Equation 2–5 is the theoretical definition of current, we never actually use it to measure current. In practice, we use an instrument called an ammeter (Section 2.6). However, it is an extremely important equation that we will soon use to develop other relationships.

Current Direction

In the early days of electricity, it was believed that current was a movement of positive charge and that these charges moved around the circuit from the positive terminal of the battery to the negative as depicted in Figure 2–11(a). Based on this, all the laws, formulas, and symbols of circuit theory were developed. (We now refer to this direction as the **conventional current direction.**) After the discovery of the atomic nature of matter, it was learned that what actually moves in metallic conductors are electrons and that they move through the circuit as in Figure 2–11(b). This direction is called the **electron flow direction.** However, because the conventional current direction was so well established, most users stayed with it. We do likewise. Thus, *in this book, the conventional direction for current is used.*

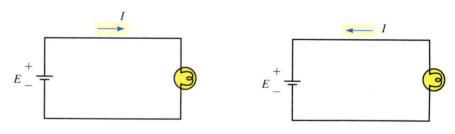

(a) Conventional current direction   (b) Electron flow direction

FIGURE 2–11   Conventional current versus electron flow. In this book, we use conventional current.

Alternating Current (AC)

So far, we have considered only dc. Before we move on, we will briefly mention ac or alternating current. **Alternating current** is current that changes direction cyclically, i.e., charges alternately flow in one direction, then in the other in a circuit. The most common ac source is the commercial ac power system that supplies energy to your home. We mention it here because you will encounter it briefly in Section 2.5. It is covered in detail in Chapter 15.

1. Body *A* has a negative charge of 0.2 μC and body *B* has a charge of 0.37 μC (positive). If 87×10^{12} electrons are transferred from *A* to *B,* what are the charges in coulombs on *A* and on *B* after the transfer?

2. Briefly describe the mechanism of voltage creation using the carbon-zinc cell of Figure 2–8 to illustrate.

3. When the switch in Figure 2–1 is open, the current is zero, yet free electrons in the copper wire are moving about. Describe their motion. Why does their movement not constitute an electric current?

4. If 12.48×10^{20} electrons pass a certain point in a circuit in 2.5 s, what is the current in amperes?

5. For Figure 2–1, assume a 12-V battery. The switch is closed for a short interval, then opened. If *I* = 6 A and the battery expends 230 040 J moving charge through the circuit, how long was the switch closed?

(Answers are at the end of the chapter.)

2.5   Practical DC Voltage Sources

Batteries

Batteries are the most common dc source. They are made in a variety of shapes, sizes, and ratings, from miniaturized button batteries capable of delivering only a few microamps to large automotive batteries capable of delivering hundreds of amps. Common sizes are the AAA, AA, C, and D as illustrated in the various photos of this chapter. All batteries use unlike conductive electrodes immersed in an electrolyte. Chemical interaction between the electrodes and the electrolyte creates the voltage of the battery.

Primary and Secondary Batteries

Batteries eventually become "discharged." Some types of batteries, however, can be "recharged." Such batteries are called **secondary** batteries. Other types, called **primary** batteries, cannot be recharged. A familiar example of a secondary battery is the automobile battery. It can be recharged by passing current through it opposite to its discharge direction. A familiar example of a primary cell is the flashlight battery.

Types of Batteries and Their Applications

The voltage of a battery, its service life, and other characteristics depend on the material from which it is made.

Alkaline

This is one of the most widely used, general-purpose primary cells available. Alkaline batteries are used in flashlights, portable radios, TV remote controllers, cassette players, cameras, toys, and so on. They come in various sizes as depicted in Figure 2–12. Alkaline batteries provide 50% to 100% more total energy for the same size unit than carbon-zinc cells. Their nominal cell voltage is 1.5 V.

FIGURE 2–12   Alkaline batteries. From left to right, a 9-V rectangular battery, an AAA cell, a D cell, an AA cell, and a C cell.

Carbon-Zinc

Also called a **dry cell,** the carbon-zinc battery was for many years the most widely used primary cell, but it is now giving way to other types such as the alkaline battery. Its nominal cell voltage is 1.5 volts.

Lithium

Lithium batteries (Figure 2–13) feature small size and long life (e.g., shelf lives of 10 to 20 years). Applications include watches, pacemakers, cameras,

FIGURE 2–13   An assortment of lithium batteries. The battery on the computer motherboard is for memory backup.

and battery backup of computer memories. Several types of lithium cells are available, with voltages from of 2 V to 3.5 V and current ratings from the microampere to the ampere range.

Nickel-Cadmium

Commonly called "Ni-Cads," these are the most popular, general-purpose rechargeable batteries available. They have long service lives, operate over wide temperature ranges, and are manufactured in many styles and sizes, including C, D, AAA, and AA. Inexpensive chargers make it economically feasible to use nickel-cadmium batteries for home entertainment equipment.

Lead-Acid

This is the familiar automotive battery. Its basic cell voltage is about 2 volts, but typically, six cells are connected internally to provide 12 volts at its terminals. Lead-acid batteries are capable of delivering large current (in excess of 100 A) for short periods as required, for example, to start an automobile.

Battery Capacity

Batteries run down under use. Their **capacity** is specified in **ampere-hours** (Ah). The ampere-hour rating of a battery is equal to the product of its current drain times the length of time that you can expect to draw the specified current before the battery becomes unusable. For example, a battery rated at 200 Ah can theoretically supply 20 A for 10 h, or 5 A for 40 h, etc. The relationship between capacity, life, and current drain is

$$\text{life} = \frac{\text{capacity}}{\text{current drain}} \tag{2–8}$$

The capacity of batteries is not a fixed value as suggested above but is affected by discharge rates, operating schedules, temperature, and other factors. At best, therefore, capacity is an estimate of expected life under certain conditions. Table 2–1 illustrates approximate service capacities for several sizes of carbon-zinc batteries at three values of current drain at 21°C. Under the conditions listed, the AA cell has a capacity of (3 mA)(450 h) = 1350 mAh at a drain of 3 mA, but its capacity decreases to (30 mA)(32 h) = 960 mAh at a drain of 30 mA. Figure 2–14 shows a typical variation of capacity of a Ni-Cad battery with changes in temperature.

Other Characteristics

Because batteries are not perfect, their terminal voltage drops as the amount of current drawn from them increases. (This issue is considered in Chapter 5.) In addition, battery voltage is affected by temperature and other factors that affect their chemical activity. However, these factors are not considered in this book.

TABLE 2–1 Capacity-Current Drain of Selected Carbon-Zinc Cells

Cell	Starting Drain (mA)	Service Life (h)
AA	3.0	450
	15.0	80
	30.0	32
C	5.0	520
	25.0	115
	50.0	53
D	10.0	525
	50.0	125
	100.0	57

Courtesy T. R. Crompton, *Battery Reference Book,* Butterworths & Co. (Publishers) Ltd, 1990.

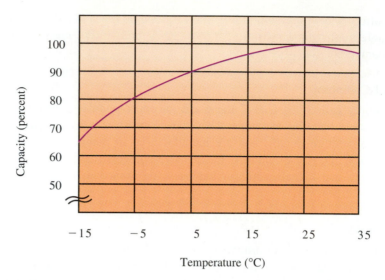

FIGURE 2–14   Typical variation of capacity versus temperature for a Ni-Cad battery.

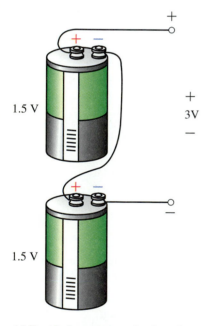

(a) For ideal sources, total voltage is
the sum of the cell voltages

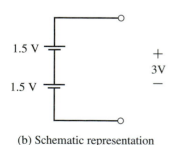

(b) Schematic representation

FIGURE 2–15   Cells connected in
series to increase the available voltage.

EXAMPLE 2–4   Assume the battery of Figure 2–14 has a capacity of 240 Ah at 25°C. What is its capacity at −15°C?

Solution   From the graph, capacity at −15°C is down to 65%. Thus, capacity = 0.65 × 240 = 156 Ah.

Cells in Series and Parallel

Cells may be connected as in Figures 2–15 and 2–16 to increase their voltage and current capabilities. This is discussed in later chapters.

Electronic Power Supplies

Electronic systems such as TV sets, VCRs, computers, and so on, require dc for their operation. Except for portable units which use batteries, they obtain their power from the commercial ac power lines by means of built-in power supplies

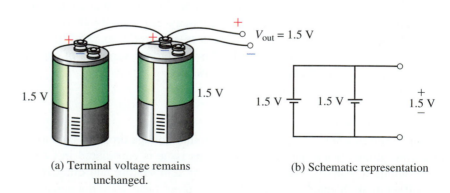

(a) Terminal voltage remains
unchanged.

(b) Schematic representation

FIGURE 2–16   Cells connected in parallel to increase the available current. (Both must have the same voltage.) Do not do this for extended periods of time.

(Figure 2–17). Such supplies convert the incoming ac to the dc voltages required by the equipment. Power supplies are also used in electronic laboratories. These are usually variable to provide the range of voltages needed for prototype development and circuit testing. Figure 2–18 shows a variable supply.

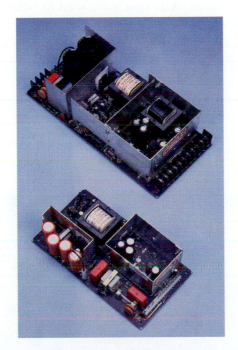

FIGURE 2–17   Fixed power supplies. *(Courtesy of Condor DC Power Supplies Inc.)*

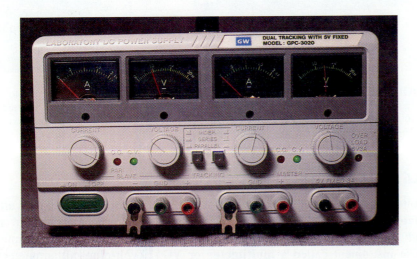

FIGURE 2–18   Variable laboratory power supply.

Solar Cells

Solar cells convert light energy to electrical energy using photovoltaic means. The basic cell consists of two layers of semiconductor material. When light strikes the cell, many electrons gain enough energy to cross from one layer to the other to create a dc voltage.

Solar energy has a number of practical applications. Figure 2–19, for example, shows an array of solar panels supplying power to a commercial ac network. In remote areas, solar panels are used to power communications systems and irrigation pumps. In space, they are used to power satellites. In everyday life, they are used to power hand-held calculators.

DC Generators

Direct current (dc) generators, which convert mechanical energy to electrical energy, are another source of dc. They create voltage by means of a coil of wire rotated through a magnetic field. Their principle of operation is similar to that of ac generators (discussed in Chapter 15).

FIGURE 2–19 Solar panels. Davis California Pacific Gas & Electric PVUSA (Photo-voltaic for Utility Scale Applications). Solar panels produce dc which must be converted to ac before being fed into the ac system. This plant is rated at 174 kilowatts. *(Courtesy Siemens Solar Industries, Camarillo, California)*

2.6 Measuring Voltage and Current

Voltage and current are measured in practice using instruments called **volt-meters** and **ammeters.** While voltmeters and ammeters are available as individual instruments, they are more commonly combined into a multipur-pose instrument called a **multimeter** or **VOM** (volt-ohm-milliammeter). Figure 2–20 shows both digital and analog multimeters. Analog instruments use a needle pointer to indicate measured values, while digital instruments use a numeric readout. Digital instruments are more popular than analog types because they are easier to use.

(a) Analog multimeter.

(b) Hand-held digital multimeter (DMM).
(*Reproduced with permission from the John Fluke Mfg. Co., Inc.*)

FIGURE 2–20   Multimeters. These are multipurpose test instruments that you can use to measure voltage, current and resistance. Some meters use terminal markings of + and −, others use VΩ and COM and so on. Color coded test leads (red and black) are industry standard.

Setting the Multimeter for Voltage and Current Measurement

In what follows, we will concentrate on the digital multimeter (DMM) and leave the analog instruments to your lab course. (It should be noted however that many of the comments below also apply to analog instruments.)

Multimeters typically have a set of terminals marked VΩ, A, and COM as can be seen in Figure 2–20, as well as a function selector switch or set of push buttons that permit you to select functions and ranges. Terminal VΩ is the terminal to use to measure voltage and resistance, while terminal A is used for current measurement. The terminal marked COM is the common terminal for all measurements. (Some multimeters combine the VΩ and A terminals into one terminal marked VΩA.) On many instruments the VΩ terminal is called the + terminal and the COM terminal is called the − terminal, Figure 2–21.

Voltage Select

When set to dc voltage ($\overline{\overline{V}}$), the meter measures the dc voltage between its $V\Omega$ (or $+$) and COM (or $-$) terminals. In Figure 2–21(a), for example, with its leads placed across a 47.2-volt source, the instrument indicates 47.2 V.

Current Select

When set to dc current ($\overline{\overline{A}}$), the multimeter measures the dc current passing through it, i.e., the current entering its A (or $+$) terminal and leaving its COM (or $-$) terminal. In Figure 2–21(b), the meter measures and displays a current of 3.6 A.

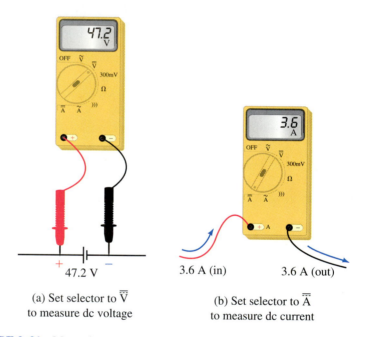

(a) Set selector to $\overline{\overline{V}}$
to measure dc voltage

(b) Set selector to $\overline{\overline{A}}$
to measure dc current

FIGURE 2–21   Measuring voltage and current with a multimeter. By convention, you connect the red lead to the $V\Omega$ ($+$) terminal and the black lead to the COM ($-$) terminal.

How to Measure Voltage

Since voltage is the potential difference between two points, you measure voltage by placing the voltmeter leads *across* the component whose voltage you wish to determine. Thus, to measure the voltage across the lamp of Figure 2–22, connect the leads as shown. If the meter is not autoscale and you have no idea how large the voltage is, set the meter to its highest range, then work your way down to avoid damage to the instrument.

Be sure to note the sign of the measured quantity. (Most digital instruments have an **autopolarity** feature that automatically determines the sign for you.) If the meter is connected as in Figure 2–21(a) with its $+$ lead connected to the $+$ terminal of the battery, the display will show 47.2 as indicated, while if the leads are reversed, the display will show -47.2.

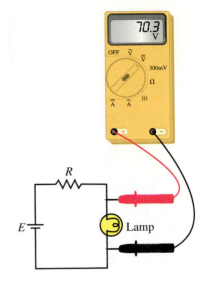

FIGURE 2–22 To measure voltage, place the voltmeter leads across the component whose voltage you wish to determine. If the voltmeter reading is positive, the point where the red lead is connected is positive with respect to the point where the black lead is connected.

How to Measure Current

As indicated by Figure 2–21(b), the current that you wish to measure must pass *through* the meter. Consider Figure 2–23(a). To measure this current, open the circuit as in (b) and insert the ammeter. The sign of the reading will be positive if current enter the A or (+) terminal or negative if it enters the COM (or −) terminal as described in the Practical Note.

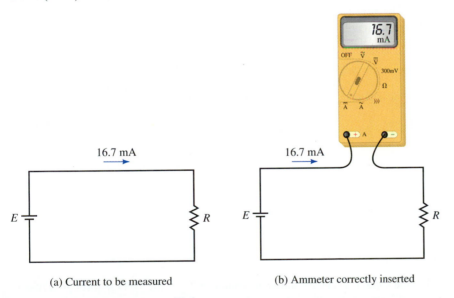

(a) Current to be measured　　　(b) Ammeter correctly inserted

FIGURE 2–23 To measure current, insert the ammeter into the circuit so that the current you wish to measure passes through the instrument. The reading is positive here because current enters the + (A) terminal.

Reading Analog Multimeters

Consider the analog meter of Figure 2–24. Note that it has a selector switch for selecting dc volts, ac volts, dc current, and ohms plus a variety of scales to go with these functions and their ranges. To measure a quantity, set the selector switch to the desired function and range, then read the value from the appropriate scale.

FIGURE 2–24 Analog multimeter. The quantity being measured is indicated on the scale selected by the rotary switch.

EXAMPLE 2–5 The meter of Figure 2–24 is set to the 100 volts dc range. This means that the instrument reads full scale when 100 volts is applied and proportionally less for other voltages. For the case shown, (expanded detail, Figure 2–25), the needle indicates 70 volts.

FIGURE 2–25 Meter indicates 70 volts on the 100-V scale.

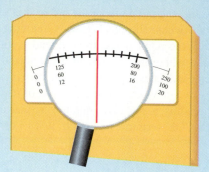

Meter Symbols

In our examples so far, we have shown meters pictorially. Usually, however, they are shown schematically. The schematic symbol for a voltmeter is a circle with the letter *V*, while the symbol for an ammeter is a circle with the letter *I*. The circuits of Figures 2–22 and 2–23 have been redrawn (Figure 2–26) to indicate this.

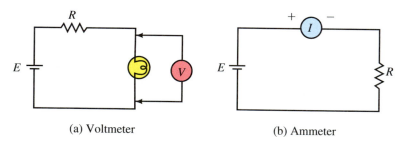

(a) Voltmeter   (b) Ammeter

FIGURE 2–26 Schematic symbols for voltmeter and ammeter.

PRACTICAL NOTES...

1. One sometimes hears statements such as ". . . the voltage through a resistor" or ". . . the current across a resistor." These statements are incorrect. Voltage does not pass through anything; voltage is a potential difference and appears across things. This is why we connect a voltmeter *across* components to measure their voltage. Similarly, current does not appear across anything; current is a flow of charge that passes *through* circuit elements. This is why we put the ammeter in the current path—to measure the current in it. Thus, the correct statements are ". . . voltage across the resistor . . ." and ". . . current through the resistor . . . ."

2. Do *not* connect ammeters directly across a voltage source. Ammeters have nearly zero resistance and damage will probably result.

2.7   Switches, Fuses, and Circuit Breakers

Switches

The most basic switch is a single-pole, single-throw (SPST) switch as shown in Figure 2–27. With the switch open, the current path is broken and the lamp is off; with it closed, the lamp is on. This type of switch is used, for example, for light switches in homes.

Figure 2–28(a) shows a single-pole, double-throw (SPDT) switch. Two of these switches may be used as in (b) for two-way control of a light. This type of arrangement is sometimes used for stairway lights; you can turn the light on or off from either the bottom or the top of the stairs.

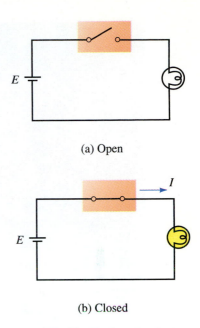

(a) Open

(b) Closed

FIGURE 2–27   Single-pole, single-throw (SPST) switch.

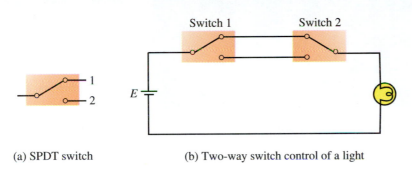

(a) SPDT switch

(b) Two-way switch control of a light

FIGURE 2–28   Single-pole, double-throw (SPDT) switch.

Many other configurations of switches exist in practice. However, we will leave the topic at this point.

Fuses and Circuit Breakers

Fuses and circuit breakers are used to protect equipment or wiring against excessive current. For example, in your home, if you connect too many appliances to an outlet, the fuse or circuit breaker in your electrical panel "blows." This opens the circuit to protect against overloading and possible fire. Fuses and circuit breakers may also be installed in equipment such as your automobile to protect against internal faults. Figure 2–29 shows a variety of fuses and breakers.

Fuses use a metallic element that melts when current exceeds a preset value. Thus, if a fuse is rated at 3 A, it will "blow" if more than 3 amps passes through it. Fuses are made as fast-blow and slow-blow types. Fast-blow fuses are very fast; typically, they blow in a fraction of a second. Slow-blow fuses, on the other hand, react more slowly so that they do not blow on small, momentary overloads.

Circuit breakers work on a different principle. When the current exceeds the rated value of a breaker, the magnetic field produced by the excessive current operates a mechanism that trips open a switch. After the fault or overload condition has been cleared, the breaker can be reset and used again. Since they are mechanical devices, their operation is slower than that of a fuse; thus, they do not "pop" on momentary overloads as, for example, when a motor is started.

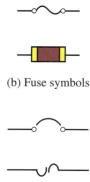

(b) Fuse symbols

(c) Circuit breaker
symbols

(a) A variety of fuses and circuit breakers.

FIGURE 2–29   Fuses and circuit breakers.

PUTTING IT INTO PRACTICE

Your company is considering the purchase of an electrostatic air cleaner system for one of its facilities and your supervisor has asked you to prepare a short presentation for the Board of Directors. Members of the Board understand basic electrical theory but are unfamiliar with the specifics of electrostatic air cleaners. Go to your library (physics books are a good reference) and research and prepare a short description of the electrostatic air cleaner. Include a diagram and a description of how it works.

PROBLEMS

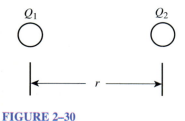

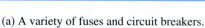

FIGURE 2–30

2.1   Atomic Theory Review

1. How many free electrons are there in the following at room temperature?

   a. 1 cubic meter of copper

   b. a 5 m length of copper wire whose diameter is 0.163 cm

2. Two charges are separated by a certain distance, Figure 2–30. How is the force between them affected if

   a. the magnitudes of both charges are doubled?

   b. the distance between the charges is tripled?

3. Two charges are separated by a certain distance. If the magnitude of one charge is doubled and the other tripled and the distance between them halved, how is the force affected?

4. A certain material has four electrons in its valence shell and a second material has one. Which is the better conductor?

5. a. What makes a material a good conductor? (In your answer, consider valence shells and free electrons.)

   b. Besides being a good conductor, list two other reasons why copper is so widely used.

   c. What makes a material a good insulator?

   d. Normally air is an insulator. However, during lightning discharges, conduction occurs. Briefly discuss the mechanism of charge flow in this discharge.

6. a. Although gold is very expensive, it is sometimes used in electronics as a plating on contacts. Why?

   b. Why is aluminum sometimes used when its conductivity is only about 60% as good as that of copper?

2.2   The Unit of Electrical Charge: The Coulomb

7. What do we mean when we say that a body is "charged"?

8. Compute the force between the following charges and state whether it is attractive or repulsive.

   a. A $+1\ \mu C$ charge and a $+7\ \mu C$ charge, separated 10 mm

   b. $Q_1 = 8\ \mu C$ and $Q_2 = -4\ \mu C$, separated 12 cm

   c. Two electrons separated by 12×10^{-8} m

   d. An electron and a proton separated by 5.3×10^{-11} m

   e. An electron and a neutron separated by 5.7×10^{-11} m

9. The force between a positive charge and a negative charge that are 2 cm apart is 180 N. If $Q_1 = 4\ \mu C$, what is Q_2? Is the force attraction or repulsion?

10. If you could place a charge of 1 C on each of two bodies separated 25 cm center to center, what would be the force between them in newtons? In tons?

11. The force of repulsion between two charges separated by 50 cm is 0.02 N. If $Q_2 = 5Q_1$, determine the charges and their possible signs.

12. How many electrons does a charge of $1.63\ \mu C$ represent?

13. Determine the charge possessed by 19×10^{13} electrons.

14. An electrically neutral metal plate acquires a negative charge of $47\ \mu C$. How many electrons were added to it?

15. A metal plate has 14.6×10^{13} electrons added. Later, $1.3\ \mu C$ of charge is added. If the final charge on the plate is $5.6\ \mu C$, what was its initial charge?

2.3   Voltage

16. Sliding off a chair and touching someone can result in a shock. Explain why.

17. If 360 joules of energy are required to transfer 15 C of charge through the lamp of Figure 2–1, what is the voltage of the battery?

18. If 600 J of energy are required to move 9.36×10^{19} electrons from one point to the other, what is the potential difference between the two points?

19. If 1.2 kJ of energy are required to move 500 mC from one point to another, what is the voltage between the two points?

20. How much energy is required to move 20 mC of charge through the lamp of Figure 2–22?

21. How much energy is gained by a charge of 0.5 μC as it moves through a potential difference of 8.5 kV?

22. If the voltage between two points is 100 V, how much energy is required to move an electron between the two points?

23. Given a voltage of 12 V for the battery in Figure 2–1, how much charge is moved through the lamp if it takes 57 J of energy to move it?

2.4   Current

24. For the circuit of Figure 2–1, if 27 C pass through the lamp in 9 seconds, what is the current in amperes?

25. If 250 μC pass through the ammeter of Figure 2–26(b) in 5 ms, what will the meter read?

26. If the current $I = 4$ A in Figure 2–1, how many coulombs pass through the lamp in 7 ms?

27. How much charge passes through the circuit of Figure 2–23 in 20 ms?

28. How long does it take for 100 μC to pass a point if the current is 25 mA?

29. If 93.6×10^{12} electrons pass through a lamp in 5 ms, what is the current?

30. The charge passing through a wire is given by $q = 10t + 4$, where q is in coulombs and t in seconds,

    a. How much charge has passed at $t = 5$ s?

    b. How much charge has passed at $t = 8$ s?

    c. What is the current in amps?

31. The charge passing through a wire is $q = (80t + 20)$ C. What is the current? Hint: Choose two arbitrary values of time and proceed as in Question 30.

32. How long does it take 312×10^{19} electrons to pass through the circuit of Figure 2–26(b) if the ammeter reads 8 A?

33. If 1353.6 J are required to move 47×10^{19} electrons through the lamp of Figure 2–1 in 1.3 min, what are V and $I?$

2.5   Practical DC Voltage Sources

34. What do we mean by dc? By ac?

35. For the battery of Figure 2–8, chemical action causes 15.6×10^{18} electrons to be transferred from the carbon rod to the zinc can. If 3.85 joules of chemical energy are expended, what is the voltage developed?

36. How do you charge a secondary battery? Make a sketch. Can you charge a primary battery?

37. A battery rated 1400 mAh supplies 28 mA to a load. How long can it be expected to last?

38. What is the approximate service life of the D cell of Table 2–1 at a current drain of 10 mA? At 50 mA? At 100 mA? What conclusion do you draw from these results?

39. The battery of Figure 2–14 is rated at 81 Ah at 5°C. What is the expected life (in hours) at a current draw of 5 A at −15°C?

40. The battery of Figure 2–14 is expected to last 17 h at a current drain of 1.5 A at 25°C. How long do you expect it to last at 5°C at a current drain of 0.8 A?

41. In the engineering workplace, you sometimes have to make estimations based on the information you have available. In this vein, assume you have a battery-operated device that uses the C cell of Table 2–1. If the device draws 10 mA, what is the estimated time (in hours) that you will be able to use it?

2.6   Measuring Voltage and Current

42. The digital voltmeter of Figure 2–31 has autopolarity. For each case, determine its reading.

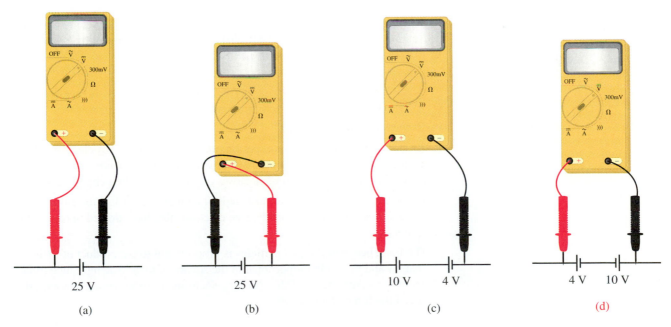

(a)   (b)   (c)   (d)

FIGURE 2–31

43. The current in the circuit of Figure 2–32 is 9.17 mA. Which ammeter correctly indicates the current? (a) Meter 1, (b) Meter 2, (c) both.

44. What is wrong with the statement that the voltage through the lamp of Figure 2–22 is 70.3 V?

45. What is wrong with the metering scheme shown in Figure 2–33? Fix it.

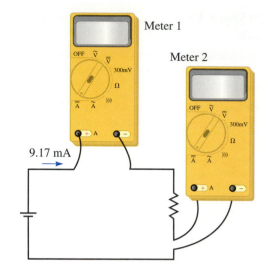

Meter 1

Meter 2

9.17 mA

FIGURE 2–32

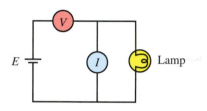

FIGURE 2–33   What is wrong here?

TABLE 2–2

Switch 1	Switch 2	Lamp
Open	Open	Off
Open	Closed	On
Closed	Open	On
Closed	Closed	On

2.7   Switches, Fuses, and Circuit Breakers

46. It is desired to control a light using two switches as indicated in Table 2–2. Draw the required circuit.

47. Fuses have a current rating so that you can select the proper size to protect a circuit against overcurrent. They also have a voltage rating. Why? Hint: Read the section on insulators, i.e., Section 2.1.

ANSWERS TO IN-PROCESS LEARNING CHECKS

In-Process Learning Check 1

1. An atom consists of a nucleus of protons and neutrons orbited by electrons. The nucleus is positive because protons are positive, but the atom is neutral because it contains the same number of electrons as protons, and their charges cancel.

2. The valence shell is the outermost shell. It contains either just the atom's valence electrons or additionally, free electrons that have drifted in from other atoms.

3. The force between charged particles is proportional to the product of their charges and inversely proportional to the square of their spacing. Since force decreases as the square of the spacing, electrons far from the nucleus experience little force of attraction.

4. If a loosely bound electron gains sufficient energy, it may break free from its parent atom and wander throughout the material. Such an electron is called a free electron. For materials like copper, heat (thermal energy) can give an electron enough energy to dislodge it from its parent atom.

5. A normal atom is neutral because it has the same number of electrons as protons and their charges cancel. An atom that has lost an electron is called a positive ion, while an atom that has gained an electron is called a negative ion.

6. The electrons remain in the material.

In-Process Learning Check 2

1. $Q_A = 13.74 \ \mu C$ (pos.) $Q_B = 13.57 \ \mu C$ (neg.)

2. Chemical action creates an excess of electrons at the zinc and a deficiency of electrons at the carbon electrode. Because one pole is positive and the other negative, a voltage exists between them.

3. Motion is random. Since the net movement in all directions is zero, current is zero.

4. 80 A

5. 3195 s

3

Resistance

OBJECTIVES

After studying this chapter, you will be able to

- calculate the resistance of a section of conductor, given its cross-sectional area and length,
- convert between areas measured in square mils, square meters, and circular mils,
- use tables of wire data to obtain the cross-sectional dimensions of various gauges of wire and predict the allowable current for a particular gauge of wire,
- use the temperature coefficient of a material to calculate the change in resistance as the temperature of the sample changes,
- use resistor color codes to determine the resistance and tolerance of a given fixed-composition resistor,
- demonstrate the procedure for using an ohmmeter to determine circuit continuity and to measure the resistance of both an isolated component and one which is located in a circuit,
- develop an understanding of various ohmic devices such as thermistors and photocells,
- develop an understanding of the resistance of nonlinear devices such as varistors and diodes,
- calculate the conductance of any resistive component.

KEY TERMS

Color Codes
Conductance
Diode
Ohmmeter
Open Circuit
Photocell
Resistance
Resistivity
Short Circuit
Superconductance
Temperature Coefficient
Thermistor
Varistor
Wire gauge

OUTLINE

Resistance of Conductors
Electrical Wire Tables
Resistance of Wires—Circular Mils
Temperature Effects
Types of Resistors
Color Coding of Resistors
Measuring Resistance—The Ohmmeter
Thermistors
Photoconductive Cells
Nonlinear Resistance
Conductance
Superconductors

You have been introduced to the concepts of voltage and current in previous chapters and have found that current involves the movement of charge. In a conductor, the charge carriers are the free electrons which are moved due to the voltage of an externally applied source. As these electrons move through the material, they constantly collide with atoms and other electrons within the conductor. In a process similar to friction, the moving electrons give up some of their energy in the form of heat. These collisions represent an opposition to charge movement that is called **resistance.** The greater the opposition (i.e., the greater the resistance), the smaller will be the current for a given applied voltage.

Circuit components (called **resistors**) are specifically designed to possess resistance and are used in almost all electronic and electrical circuits. Although the resistor is the most simple component in any circuit, its effect is very important in determining the operation of a circuit.

Resistance is represented by the symbol R (Figure 3–1) and is measured in units of ohms (after Georg Simon Ohm). The symbol for ohms is the capital Greek letter omega (Ω).

In this chapter, we examine resistance in its various forms. Beginning with metallic conductors, we study the factors which affect resistance in conductors. Following this, we look at commercial resistors, including both fixed and variable types. We then discuss important nonlinear resistance devices and conclude with an overview of superconductivity and its potential impact and use.

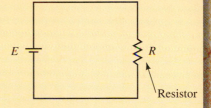

FIGURE 3–1   Basic resistive circuit.

Georg Simon Ohm and Resistance

ONE OF THE FUNDAMENTAL RELATIONSHIPS of circuit theory is that between voltage, current, and resistance. This relationship and the properties of resistance were investigated by the German physicist Georg Simon Ohm (1787–1854) using a circuit similar to that of Figure 3–1. Working with Volta's recently developed battery and wires of different materials, lengths, and thicknesses, Ohm found that current depended on both voltage and resistance. For example, for a fixed resistance, he found that doubling the voltage doubled the current, tripling the voltage tripled the current, and so on. Also, for a fixed voltage, Ohm found that the opposition to current was directly proportional to the length of the wire and inversely proportional to its cross-sectional area. From this, he was able to define the resistance of a wire and show that current was inversely proportional to this resistance; e.g., when he doubled the resistance; he found that the current decreased to half of its former value.

These two results when combined form what is known as Ohm's law. (You will study Ohm's law in great detail in Chapter 4.) Ohm's results are of such fundamental importance that they represent the real beginnings of what we now call electrical circuit analysis.

3.1   Resistance of Conductors

As mentioned in the chapter preview, conductors are materials which permit the flow of charge. However, conductors do not all behave the same way. Rather, we find that the resistance of a material is dependent upon several factors:

• Type of material

• Length of the conductor

• Cross-sectional area

• Temperature

If a certain length of wire is subjected to a current, the moving electrons will collide with other electrons within the material. Differences at the atomic level of various materials cause variation in how the collisions affect resistance. For example, silver has more free electrons than copper, and so the resistance of a silver wire will be less than the resistance of a copper wire having the identical dimensions. We may therefore conclude the following:

The resistance of a conductor is dependent upon the type of material.

If we were to double the length of the wire, we can expect that the number of collisions over the length of the wire would double, thereby causing the resistance to also double. This effect may be summarized as follows:

The resistance of a metallic conductor is directly proportional to the length of the conductor.

A somewhat less intuitive property of a conductor is the effect of cross-sectional area on the resistance. As the cross-sectional area is increased, the moving electrons are able to move more freely through the conductor, just as water moves more freely through a large-diameter pipe than a small-diameter pipe. If the cross-sectional area is doubled, the electrons would be involved in half as many collisions over the length of the wire. We may summarize this effect as follows:

The resistance of a metallic conductor is inversely proportional to the cross-sectional area of the conductor.

The factors governing the resistance of a conductor at a given temperature may be summarized mathematically as follows:

$$R = \frac{\rho \ell}{A} \quad [\text{ohms}, \Omega] \tag{3–1}$$

where

ρ = resistivity, in ohm-meters (Ω-m)

ℓ = length, in meters (m)

A = cross-sectional area, in square meters (m^2).

In the above equation the lowercase Greek letter rho (ρ) is the constant of proportionality and is called the **resistivity** of the material. Resistivity is a physical property of a material and is measured in ohm-meters (Ω-m) in the SI system. Table 3–1 lists the resistivities of various materials at a temperature of 20°C. The effects on resistance due to changes in temperature will be examined in Section 3.4.

TABLE 3–1   Resistivity of Materials, ρ

Material	Resistivity, ρ, at 20°C (Ω-m)
Silver	1.645×10^{-8}
Copper	1.723×10^{-8}
Gold	2.443×10^{-8}
Aluminum	2.825×10^{-8}
Tungsten	5.485×10^{-8}
Iron	12.30×10^{-8}
Lead	22×10^{-8}
Mercury	95.8×10^{-8}
Nichrome	99.72×10^{-8}
Carbon	3500×10^{-8}
Germanium	20–$2300*$
Silicon	$\cong 500*$
Wood	10^{8}–10^{14}
Glass	10^{10}–10^{14}
Mica	10^{11}–10^{15}
Hard rubber	10^{13}–10^{16}
Amber	5×10^{14}
Sulphur	1×10^{15}
Teflon	1×10^{16}

*The resistivities of these materials are dependent upon the impurities within the materials.

Since most conductors are circular, as shown in Figure 3–2, we may deter-mine the cross-sectional area from either the radius or the diameter as follows:

$$A = \pi r^2 = \pi \left(\frac{d}{2}\right)^2 = \frac{\pi d^2}{4} \qquad (3\text{–}2)$$

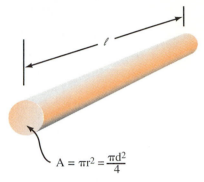

EXAMPLE 3–1 Most homes use solid copper wire having a diameter of 1.63 mm to provide electrical distribution to outlets and light sockets. Deter-mine the resistance of 75 meters of a solid copper wire having the above diameter.

Solution We will first calculate the cross-sectional area of the wire using equation 3–2.

$A = \pi r^2 = \dfrac{\pi d^2}{4}$

FIGURE 3–2 Conductor with a cir-cular cross-section.

$$\begin{aligned}
A &= \frac{\pi d^2}{4} \\
&= \frac{\pi (1.63 \times 10^{-3}\ \text{m})^2}{4} \\
&= 2.09 \times 10^{-6}\ \text{m}^2
\end{aligned}$$

Now, using Table 3–1, the resistance of the length of wire is found as

$$\begin{aligned}
R &= \frac{\rho \ell}{A} \\
&= \frac{(1.723 \times 10^{-8}\ \Omega\text{-m})(75\ \text{m})}{2.09 \times 10^{-6}\ \text{m}^2} \\
&= 0.619\ \Omega
\end{aligned}$$

Find the resistance of a 100-m long tungsten wire which has a circular cross-sec-tion with a diameter of 0.1 mm ($T = 20°C$).

PRACTICE
PROBLEMS 1

Answer: 698 Ω

EXAMPLE 3–2 Bus bars are bare solid conductors (usually rectangular) used to carry large currents within buildings such as power generating sta-tions, telephone exchanges, and large factories. Given a piece of aluminum bus bar as shown in Figure 3–3, determine the resistance between the ends of this bar at a temperature of 20°C.

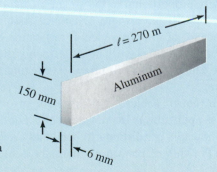

$\ell = 270$ m

Aluminum

150 mm

6 mm

FIGURE 3–3 Conductor with a rectangular cross section.

Solution   The cross-sectional area is

$$A = (150 \text{ mm})(6 \text{ mm})$$
$$= (0.15 \text{ m})(0.006 \text{ m})$$
$$= 0.0009 \text{ m}^2$$
$$= 9.00 \times 10^{-4} \text{ m}^2$$

The resistance between the ends of the bus bar is determined as

$$R = \frac{\rho \ell}{A}$$
$$= \frac{(2.825 \times 10^{-8} \text{ } \Omega\text{-m})(270 \text{ m})}{9.00 \times 10^{-4} \text{ m}^2}$$
$$= 8.48 \times 10^{-3} \text{ } \Omega = 8.48 \text{ m}\Omega$$

IN-PROCESS
LEARNING
CHECK 1

1. Given two lengths of wire having identical dimensions. If one wire is made of copper and the other is made of iron, which wire will have the greater resistance? How much greater will the resistance be?

2. Given two pieces of copper wire which have the same cross-sectional area, determine the relative resistance of the one which is twice as long as the other.

3. Given two pieces of copper wire which have the same length, determine the relative resistance of the one which has twice the diameter of the other.

(Answers are at the end of the chapter.)

3.2   Electrical Wire Tables

Although the SI system is the standard measurement for electrical and other physical quantities, the English system is still used extensively in the United States and to a lesser degree throughout the rest of the English-speaking world. One area which has been slow to convert to the SI system is the designation of cables and wires, where the American Wire Gauge (AWG) is the primary system used to denote wire diameters. In this system, each wire diameter is assigned a gauge number. The higher the AWG number, the smaller the diameter of the cable or wire, e.g., AWG 22 gauge wire is a smaller diameter than AWG 14 gauge. Since cross-sectional area is inversely proportional to the square of the diameter, a given length of 22-gauge wire will have more resistance than an equal length of 14-gauge wire. Because of the difference in resistance, we can intuitively deduce that large-diameter cables will be able to handle more current than smaller-diameter cables. Table 3–2 provides a listing of data for standard bare copper wire.

Even though Table 3–2 provides data for solid conductors up to AWG 4/0, most applications do not use solid conductor sizes beyond AWG 10. Solid conductors are difficult to bend and are easily damaged by mechanical flexing. For this reason, large-diameter cables are nearly always stranded

TABLE 3–2 Standard Solid Copper Wire at 20°C

Size (AWG)	Diameter (inches)	Diameter (mm)	Area (CM)	Area (mm²)	Resistance (Ω/1000 ft)	Current Capacity (A)
56	0.0005	0.012	0.240	0.000122	43 200	
54	0.0006	0.016	0.384	0.000195	27 000	
52	0.0008	0.020	0.608	0.000308	17 000	
50	0.0010	0.025	0.980	0.000497	10 600	
48	0.0013	0.032	1.54	0.000779	6 750	
46	0.0016	0.040	2.46	0.00125	4 210	
45	0.0019	0.047	3.10	0.00157	3 350	
44	0.0020	0.051	4.00	0.00243	2 590	
43	0.0022	0.056	4.84	0.00245	2 140	
42	0.0025	0.064	6.25	0.00317	1 660	
41	0.0028	0.071	7.84	0.00397	1 320	
40	0.0031	0.079	9.61	0.00487	1 080	
39	0.0035	0.089	12.2	0.00621	847	
38	0.0040	0.102	16.0	0.00811	648	
37	0.0045	0.114	20.2	0.0103	521	
36	0.0050	0.127	25.0	0.0127	415	
35	0.0056	0.142	31.4	0.0159	331	
34	0.0063	0.160	39.7	0.0201	261	
33	0.0071	0.180	50.4	0.0255	206	
32	0.0080	0.203	64.0	0.0324	162	
31	0.0089	0.226	79.2	0.0401	131	
30	0.0100	0.254	100	0.0507	104	
29	0.0113	0.287	128	0.0647	81.2	
28	0.0126	0.320	159	0.0804	65.3	
27	0.0142	0.361	202	0.102	51.4	
26	0.0159	0.404	253	0.128	41.0	0.75*
25	0.0179	0.455	320	0.162	32.4	
24	0.0201	0.511	404	0.205	25.7	1.3*
23	0.0226	0.574	511	0.259	20.3	
22	0.0253	0.643	640	0.324	16.2	2.0*
21	0.0285	0.724	812	0.412	12.8	
20	0.0320	0.813	1 020	0.519	10.1	3.0*
19	0.0359	0.912	1 290	0.653	8.05	
18	0.0403	1.02	1 620	0.823	6.39	5.0†
17	0.0453	1.15	2 050	1.04	5.05	
16	0.0508	1.29	2 580	1.31	4.02	10.0†
15	0.0571	1.45	3 260	1.65	3.18	
14	0.0641	1.63	4 110	2.08	2.52	15.0†
13	0.0720	1.83	5 180	2.63	2.00	
12	0.0808	2.05	6 530	3.31	1.59	20.0†
11	0.0907	2.30	8 230	4.17	1.26	
10	0.1019	2.588	10 380	5.261	0.998 8	30.0†
9	0.1144	2.906	13 090	6.632	0.792 5	
8	0.1285	3.264	16 510	8.367	0.628 1	
7	0.1443	3.665	20 820	10.55	0.498 1	
6	0.1620	4.115	26 240	13.30	0.395 2	
5	0.1819	4.620	33 090	16.77	0.313 4	
4	0.2043	5.189	41 740	21.15	0.248 5	
3	0.2294	5.827	52 620	26.67	0.197 1	
2	0.2576	6.543	66 360	33.62	0.156 3	
1	0.2893	7.348	83 690	42.41	0.123 9	
1/0	0.3249	8.252	105 600	53.49	0.098 25	
2/0	0.3648	9.266	133 100	67.43	0.077 93	
3/0	0.4096	10.40	167 800	85.01	0.061 82	
4/0	0.4600	11.68	211 600	107.2	0.049 01	

*This current is suitable for single conductors and surface or loose wiring.

†This current may be accommodated in up to three wires in a sheathed cable. For four to six wires, the current in each wire must be reduced to 80% of the indicated value. For seven to nine wires, the current in each wire must be reduced to 70% of the indicated value.

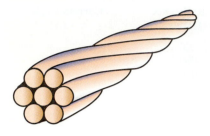

FIGURE 3–4 Stranded wire (7 strands).

rather than solid. Stranded wires and cables use anywhere from seven strands, as shown in Figure 3–4, to in excess of a hundred strands.

As one might expect, stranded wire uses the same AWG notation as solid wire. Consequently, AWG 10 stranded wire will have the same cross-sectional conductor area as AWG 10 solid wire. However, due to the additional space lost between the conductors, the stranded wire will have a larger overall diameter than the solid wire. Also, because the individual strands are coiled as a helix, the overall strand length will be slightly longer than the cable length.

Wire tables similar to Table 3–2 are available for stranded copper cables and for cables constructed of other materials (notably aluminum).

EXAMPLE 3–3   Calculate the resistance of 200 feet of AWG 16 solid copper wire at 20°C.

Solution   From Table 3–2, we see that AWG 16 wire has a resistance of 4.02 Ω per 1000 feet. Since we are given a length of only 200 feet, the resistance will be determined as

$$R = \left(\frac{4.02 \ \Omega}{1000 \ \text{ft}} \right)(200 \ \text{ft}) = 0.804 \ \Omega$$

By examining Table 3–2, several important points may be observed:

- If the wire size increases by three gauge sizes, the cross-sectional area will approximately double. Since resistance is inversely proportional to cross-sectional area, a given length of larger-diameter cable will have a resistance which is approximately half as large as the resistance of a similar length of the smaller-diameter cable.

- If there is a difference of three gauge sizes between cables, then the larger-diameter cable will be able to handle approximately twice as much current as the smaller-diameter cable. The amount of current that a conductor can safely handle is directly proportional to the cross-sectional area.

- If the wire size increases by ten gauge sizes, the cross-sectional area will increase by a factor of about ten. Due to the inverse relationship between resistance and cross-sectional area, the larger-diameter cable will have about one tenth the resistance of a similar length of the smaller-diameter cable.

- For a 10-gauge difference in cable sizes, the larger-diameter cable will have ten times the cross-sectional area of the smaller-diameter cable and so it will be able to handle approximately ten times more current.

EXAMPLE 3–4   If AWG 14 solid copper wire is able to handle 15 A of current, determine the expected current capacity of AWG 24 and AWG 8 copper wire at 20°C.

Solution   Since AWG 24 is ten sizes smaller than AWG 14, the smaller cable will be able to handle about one tenth the capacity of the larger-diameter cable.

AWG 24 will be able to handle approximately 1.5 A of current.

AWG 8 is six sizes larger than AWG 14. Since current capacity doubles for an increase of three sizes, AWG 11 would be able to handle 30 A and AWG 8 will be able to handle 60 A.

PRACTICE PROBLEMS 2

1. From Table 3–2 find the diameters in millimeters and the cross-sectional areas in square millimeters of AWG 19 and AWG 30 solid wire.

2. By using the cross-sectional areas for AWG 19 and AWG 30, approximate the areas that AWG 16 and AWG 40 should have.

3. Compare the actual cross-sectional areas as listed in Table 3–2 to the areas found in Problems 1 and 2 above. (You will find a slight variation between your calculated values and the actual areas. This is because the actual diameters of the wires have been adjusted to provide optimum sizes for manufacturing.)

Answers:

1. $d_{AWG19} = 0.912$ mm $A_{AWG19} = 0.653$ mm^2
   $d_{AWG30} = 0.254$ mm $A_{AWG30} = 0.0507$ mm^2

2. $A_{AWG16} \cong 1.31$ mm^2 $A_{AWG40} \cong 0.0051$ mm^2

3. $A_{AWG16} = 1.31$ mm^2 $A_{AWG40} = 0.00487$ mm^2

IN-PROCESS LEARNING CHECK 2

1. AWG 12-gauge wire is able to safely handle 20 amps of current. How much current should an AWG 2-gauge cable be able to handle?

2. The electrical code actually permits up to 120 A for the above cable. How does the actual value compare to your theoretical value? Why do you think there is a difference?

(Answers are at the end of the chapter.)

3.3   Resistance of Wires—Circular Mils

The American Wire Gauge system for specifying wire diameters was developed using a unit called the **circular mil** (CM), which is defined as the area contained within a circle having a diameter of 1 mil (1 mil = 0.001 inch). A **square mil** is defined as the area contained in a square having side dimensions of 1 mil. By referring to Figure 3–5, it is apparent that the area of a circular mil is smaller than the area of a square mil.

Because not all conductors have circular cross-sections, it is occasionally necessary to convert areas expressed in square mils into circular mils. We will now determine the relationship between the circular mil and the square mil.

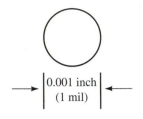

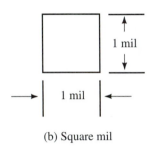

(a) Circular mil

(b) Square mil

FIGURE 3–5

Suppose that a wire has the circular cross section shown in Figure 3–5(a). By applying Equation 3–2, the area, in square mils, of the circular cross section is determined as follows:

$$A = \frac{\pi d^2}{4}$$

$$= \frac{\pi (1 \text{ mil})^2}{4}$$

$$= \frac{\pi}{4} \text{ sq. mil}$$

From the above derivation the following relations must apply:

$$1 \text{ CM} = \frac{\pi}{4} \text{ sq. mil} \qquad (3\text{–}3)$$

$$1 \text{ sq. mil} = \frac{4}{\pi} \text{ CM} \qquad (3\text{–}4)$$

The greatest advantage of using the circular mil to express areas of wires is the simplicity with which calculations may be made. Unlike previous area calculations which involved the use of π, area calculations may be reduced to simply finding the square of the diameter.

If we are given a circular cross section with a diameter, d (in mils) the area of this cross-section is determined as

$$A = \frac{\pi d^2}{4} \quad \text{[square mils]}$$

Using Equation 3–4, we convert the area from square mils to circular mils. Consequently, if the diameter of a circular conductor is given in mils, we determine the area in circular mils as

$$A_{\text{CM}} = d_{\text{mil}}^2 \quad \text{[circular mils, CM]} \qquad (3\text{–}5)$$

EXAMPLE 3–5 Determine the cross-sectional area in circular mils of a wire having the following diameters:

a. 0.0159 inch (AWG 26 wire)

b. 0.500 inch

Solution

a. $d = 0.0159$ inch
$\qquad = (0.0159 \text{ inch})(1000 \text{ mils/inch})$
$\qquad = 15.9 \text{ mils}$

Now, using Equation 3–5, we obtain

$$A_{\text{CM}} = (15.9)^2 = 253 \text{ CM.}$$

From Table 3–2, we see that the above result is precisely the area given for AWG 26 wire.

b. $d = 0.500$ inch
   $= (0.500 \text{ inch})(1000 \text{ mils/inch})$
   $= 500$ mils

$$A_{CM} = (500)^2 = 250\,000 \text{ CM}$$

In Example 3–5(b) we see that the cross-sectional area of a cable may be a large number when it is expressed in circular mils. In order to simplify the units for area, the Roman numeral M is often used to represent 1000. If a wire has a cross-sectional area of 250 000 CM, it is more easily written as 250 MCM.

Clearly, this is a departure from the SI system, where M is used to represent one million. Since there is no simple way to overcome this conflict, the student working with cable areas expressed in MCM will need to remember that the M stands for one thousand and not for one million.

EXAMPLE 3–6

a. Determine the cross-sectional area in square mils and in circular mils of a copper bus bar having cross-sectional dimensions of 0.250 inch × 6.00 inch.

b. If this copper bus bar were to be replaced by AWG 2/0 cables, how many cables would be required?

Solution

a. $A_{\text{sq. mil}} = (250 \text{ mils})(6000 \text{ mils})$
   $\qquad\quad = 1\,500\,000$ sq. mils

The area in circular mils is found by applying Equation 3–4, and this will be

$$A_{CM} = (250 \text{ mils})(6000 \text{ mils})$$

$$= (1\,500\,000 \text{ sq. mils})\left(\frac{4}{\pi} \text{ CM/sq. mil}\right)$$

$$= 1\,910\,000 \text{ CM}$$
$$= 1910 \text{ MCM}$$

b. From Table 3–2, we see that AWG 2/0 cable has a cross-sectional area of 133.1 MCM (133 100 CM), and so the bus bar is equivalent to the following number of cables:

$$n = \frac{1910 \text{ MCM}}{133.1 \text{ MCM}} = 14.4$$

This example illustrates that 15 cables would need to be installed to be equivalent to a single 6-inch by 0.25-inch bus bar. Due to the expense and awkwardness of using this many cables, we see the economy of using solid bus bar. The main disadvantage of using bus bar is that the conductor is not covered with an insulation, and so the bus bar does not offer the same protection as cable. However, since bus bar is generally used in locations where only experienced technicians are permitted access, this disadvantage is a minor one.

TABLE 3–3   Resistivity of Conductors, ρ

Material	Resistivity, ρ, at 20°C (CM-Ω/ft)
Silver	9.90
Copper	10.36
Gold	14.7
Aluminum	17.0
Tungsten	33.0
Iron	74.0
Lead	132.
Mercury	576.
Nichrome	600.

As we have seen in Section 3.1, the resistance of a conductor was determined to be

$$R = \frac{\rho \ell}{A} \quad [\text{ohms, } \Omega] \tag{3–1}$$

Although the original equation used SI units, the equation will also apply if the units are expressed in any other convenient system. If cable length is generally expressed in feet and the area in circular mils, then the resistivity must be expressed in the appropriate units. Table 3–3 gives the resistivities of some conductors represented in circular mil-ohms per foot.

The following example illustrates how Table 3–3 may be used to determine the resistance of a given section of wire.

EXAMPLE 3–7   Determine the resistance of an AWG 16 copper wire at 20°C if the wire has a diameter of 0.0508 inch and a length of 400 feet.

Solution   The diameter in mils is found as

$$d = 0.0508 \text{ inch} = 50.8 \text{ mils}$$

Therefore the cross-sectional area (in circular mils) of AWG 16 is

$$A_{\text{CM}} = 50.8^2 = 2580 \text{ CM}$$

Now, by applying Equation 3–1 and using the appropriate units, we obtain the following:

$$R = \frac{\rho \ell}{A_{\text{CM}}}$$

$$= \frac{\left(10.36 \dfrac{\text{CM-}\Omega}{\text{ft}}\right)(400 \text{ ft})}{2580 \text{ CM}}$$

$$= 1.61 \ \Omega$$

PRACTICE PROBLEMS 3

1. Determine the resistance of 1 mile (5280 feet) of AWG 19 copper wire at 20°C, if the cross-sectional area is 1290 CM.

2. Compare the above result with the value that would be obtained by using the resistance (in ohms per thousand feet) given in Table 3–2.

3. An aluminum conductor having a cross-sectional area of 1843 MCM is used to transmit power from a high-voltage dc (HVDC) generating station to a large urban center. If the city is 900 km from the generating station, determine the resistance of the conductor at a temperature of 20°C. (Use 1 ft ≅ 0.3048 m.)

Answers: 1. 42.4 Ω   2. 42.5 Ω   3. 27.2 Ω

A conductor has a cross-sectional area of 50 square mils. Determine the cross-sectional area in circular mils, square meters, and square millimeters.

IN-PROCESS
LEARNING
CHECK 3

(Answers are at the end of the chapter.)

3.4 Temperature Effects

Section 3.1 indicated that the resistance of a conductor will not be constant at all temperatures. As temperature increases, more electrons will escape their orbits, causing additional collisions within the conductor. For most conducting materials, the increase in the number of collisions translates into a relatively linear increase in resistance, as shown in Figure 3–6.

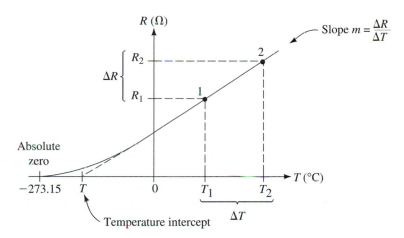

FIGURE 3–6   Temperature effects on the resistance of a conductor.

The rate at which the resistance of a material changes with a variation in temperature is called the **temperature coefficient** of the material and is assigned the Greek letter alpha (α). Some materials have only very slight changes in resistance, while other materials demonstrate dramatic changes in resistance with a change in temperature.

Any material for which resistance increases as temperature increases is said to have a **positive temperature coefficient.**

For semiconductor materials such as carbon, germanium, and silicon, increases in temperature allow electrons to escape their usually stable orbits and become free to move within the material. Although additional collisions do occur within the semiconductor, the effect of the collisions is minimal when compared with the contribution of the extra electrons to the overall flow of charge. As the temperature increases, the number of charge electrons increases, resulting in more current. Therefore, an increase in temperature results in a decrease in resistance. Consequently, these materials are referred to as having **negative temperature coefficients.**

Table 3–4 gives the temperature coefficients, α per degree Celsius, of various materials at 20°C and at 0°C.

TABLE 3–4   Temperature Intercepts and Coefficients for Common Materials

	T (°C)	α (°C)$^{-1}$ at 20°C	α (°C)$^{-1}$ at 0°C
Silver	−243	0.003 8	0.004 12
Copper	−234.5	0.003 93	0.004 27
Aluminum	−236	0.003 91	0.004 24
Tungsten	−202	0.004 50	0.004 95
Iron	−162	0.005 5	0.006 18
Lead	−224	0.004 26	0.004 66
Nichrome	−2270	0.000 44	0.000 44
Brass	−480	0.002 00	0.002 08
Platinum	−310	0.003 03	0.003 23
Carbon		−0.000 5	
Germanium		−0.048	
Silicon		−0.075	

If we consider that Figure 3–6 illustrates how the resistance of copper changes with temperature, we observe an almost linear increase in resistance as the temperature increases. Further, we see that as the temperature is decreased to **absolute zero** ($T = -273.15°C$), the resistance approaches zero.

In Figure 3–6, the point at which the linear portion of the line is extrapolated to cross the abscissa (temperature axis) is referred to as the **temperature intercept** or the **inferred absolute temperature** T of the material.

By examining the straight-line portion of the graph, we see that we have two similar triangles, one with the apex at point 1 and the other with the apex at point 2. The following relationship applies for these similar triangles.

$$\frac{R_2}{T_2 - T} = \frac{R_1}{T_1 - T}$$

This expression may be rewritten to solve for the resistance, R_2 at any temperature, T_2 as follows:

$$R_2 = \frac{T_2 - T}{T_1 - T} R_1 \tag{3–6}$$

An alternate method of determining the resistance, R_2 of a conductor at a temperature, T_2 is to use the temperature coefficient, α of the material. Examining Table 3–4, we see that the temperature coefficient is not a constant for all temperatures, but rather is dependent upon the temperature of the material. The temperature coefficient for any material is defined as

$$\alpha = \frac{m}{R_1} \tag{3–7}$$

The value of α is typically given in chemical handbooks. In the above expression, α is measured in (°C)$^{-1}$, R_1 is the resistance in ohms at a temperature, T_1, and m is the slope of the linear portion of the curve ($m = \Delta R/\Delta T$). It is left as an end-of-chapter problem for the student to use Equations 3–6 and 3–7 to derive the following expression from Figure 3–6.

$$R_2 = R_1[1 + \alpha_1(T_2 - T_1)] \tag{3–8}$$

EXAMPLE 3–8   An aluminum wire has a resistance of 20 Ω at room temperature (20°C). Calculate the resistance of the same wire at temperatures of −40°C, 100°C, and 200°C.

Solution   From Table 3–4, we see that aluminum has a temperature intercept of −236°C.

At **T** = −40°C:
The resistance at −40°C is determined using Equation 3–6.

$$R_{-40°C} = \frac{-40°C - (-236°C)}{20°C - (-236°C)} \, 20 \, \Omega = \frac{196°C}{256°C} \, 20 \, \Omega = 15.3 \, \Omega$$

At **T** = 100°C:

$$R_{100°C} = \frac{100°C - (-236°C)}{20°C - (-236°C)} \, 20 \, \Omega = \frac{336°C}{256°C} \, 20 \, \Omega = 26.3 \, \Omega$$

At **T** = 200°C:

$$R_{200°C} = \frac{200°C - (-236°C)}{20°C - (-236°C)} \, 20 \, \Omega = \frac{436°C}{256°C} \, 20 \, \Omega = 34.1 \, \Omega$$

The above phenomenon indicates that the resistance of conductors changes quite dramatically with changes in temperature. For this reason manufacturers generally specify the range of temperatures over which a conductor may operate safely.

EXAMPLE 3–9   Tungsten wire is used as filaments in incandescent light bulbs. Current in the wire causes the wire to reach extremely high temperatures. Determine the temperature of the filament of a 100-W light bulb if the resistance at room temperature is measured to be 11.7 Ω and when the light is on, the resistance is determined to be 144 Ω.

Solution   If we rewrite Equation 3–6, we are able to solve for the temperature T_2 as follows

$$T_2 = (T_1 - T) \frac{R_2}{R_1} + T$$

$$= [20°C - (-202°C)] \frac{144 \, \Omega}{11.7 \, \Omega} + (-202°C)$$

$$= 2530°C$$

A HVDC (high-voltage dc) transmission line must be able to operate over a wide temperature range. Calculate the resistance of 900 km of 1843 MCM aluminum conductor at temperatures of −40°C and +40°C.

PRACTICE
PROBLEMS 4

Answers: 20.8 Ω; 29.3 Ω

**IN-PROCESS
LEARNING
CHECK 4**

Explain what is meant by the terms *positive temperature coefficient* and *negative temperature coefficient*. To which category does aluminum belong?

(Answers are at the end of the chapter.)

3.5   Types of Resistors

Virtually all electric and electronic circuits involve the control of voltage and/or current. The best way to provide such control is by inserting appropriate values of resistance into the circuit. Although various types and sizes of resistors are used in electrical and electronic applications, all resistors fall into two main categories: fixed resistors and variable resistors.

Fixed Resistors

As the name implies, **fixed resistors** are resistors having resistance values which are essentially constant. There are numerous types of fixed resistors, ranging in size from almost microscopic (as in integrated circuits) to high-power resistors which are capable of dissipating many watts of power. Figure 3–7 illustrates the basic structure of a **molded carbon composition** resistor.

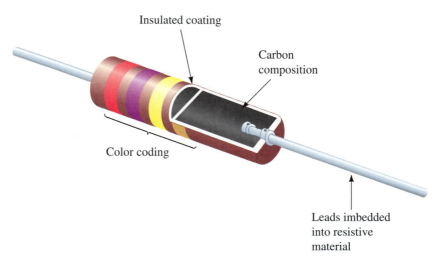

FIGURE 3–7   Structure of a molded carbon composition resistor.

As shown in Figure 3–7, the molded carbon composition resistor consists of a carbon core mixed with an insulating filler. The ratio of carbon to filler determines the resistance value of the component: the higher the proportion of carbon, the lower the resistance. Metal leads are inserted into the carbon core, and then the entire resistor is encapsulated with an insulated coating. Carbon composition resistors are available in resistances from less than 1 Ω to 100 MΩ and typically have power ratings from ⅛ W to 2 W. Figure 3–8 shows various sizes of resistors, with the larger resistors being able to dissipate more power than the smaller resistors.

Although carbon-core resistors have the advantages of being inexpensive and easy to produce, they tend to have wide tolerances and are susceptible to

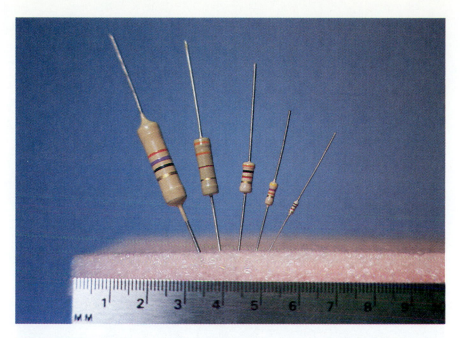

FIGURE 3–8   Actual size of carbon resistors (2 W, 1 W, ½ W, ¼ W, ⅛ W).

large changes in resistance due to temperature variation. As shown in Figure 3–9, the resistance of a carbon composition resistor may change by as much as 5% when temperature is changed by 100°C.

Other types of fixed resistors include **carbon film, metal film, metal oxide, wire-wound,** and **integrated circuit packages.**

If fixed resistors are required in applications where precision is an important factor, then film resistors are usually employed. These resistors consist of either carbon, metal, or metal-oxide film deposited onto a ceramic cylinder. The desired resistance is obtained by removing part of the resistive material, resulting in a helical pattern around the ceramic core. If variation of resistance due to temperature is not a major concern, then low-cost carbon is used. However, if close tolerances are required over a wide temperature range, then the resistors are made of films consisting of alloys such as nickel

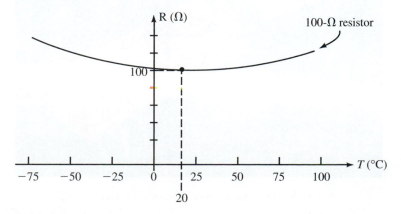

FIGURE 3–9   Variation in resistance of a carbon composition fixed resistor.

chromium, constantum, or manganin, which have very small temperature coefficients.

Occasionally a circuit requires a resistor to be able to dissipate large quantities of heat. In such cases, wire-wound resistors may be used. These resistors are constructed of a metal alloy wound around a hollow porcelain core which is then covered with a thin layer of porcelain to seal it in place. The porcelain is able to quickly dissipate heat generated due to current through the wire. Figure 3–10 shows a few of the various types of power resistors available.

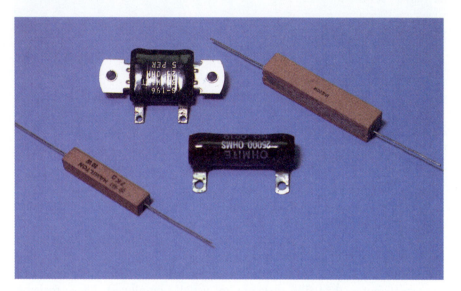

FIGURE 3–10   Power resistors.

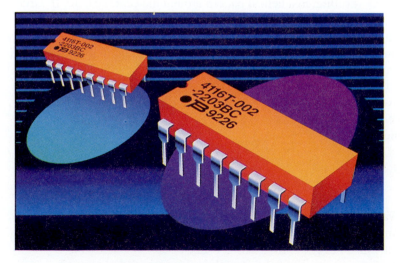

(a) Internal resistor arrangement

(b) Integrated resistor network. *(Courtesy of Bourns, Inc.)*

FIGURE 3–11

In circuits where the dissipation of heat is not a major design consideration, fixed resistances may be constructed in miniature packages (called integrated circuits or ICs) capable of containing many individual resistors. The obvious advantage of such packages is their ability to conserve space on a circuit board. Figure 3–11 illustrates a typical resistor IC package.

Variable Resistors

Variable resistors provide indispensable functions which we use in one form or another almost daily. These components are used to adjust the volume of our radios, set the level of lighting in our homes, and adjust the heat of our stoves and furnaces. Figure 3–12 shows the internal and the external view of typical variable resistors.

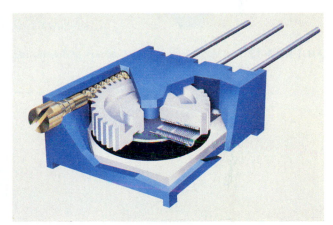

(a) External view of variable resistors.

(b) Internal view of variable resistor.

FIGURE 3–12   Variable resistors. *(Courtesy of Bourns, Inc.)*

In Figure 3–13, we see that variable resistors have three terminals, two of which are fixed to the ends of the resistive material. The central terminal is connected to a wiper which moves over the resistive material when the shaft is rotated with either a knob or a screwdriver. The resistance between the two outermost terminals will remain constant while the resistance between the central terminal and either terminal will change according to the position of the wiper.

If we examine the schematic of a variable resistor as shown in Figure 3–13(b), we see that the following relationship must apply:

$$R_{ac} = R_{ab} + R_{bc} \tag{3–9}$$

Variable resistors are used for two principal functions. **Potentiometers,** shown in Figure 3–13(c), are used to adjust the amount of potential (voltage) provided to a circuit. **Rheostats,** the connections and schematic of which are shown in Figure 3–14, are used to adjust the amount of current within a circuit. Applications of potentiometers and rheostats will be covered in later chapters.

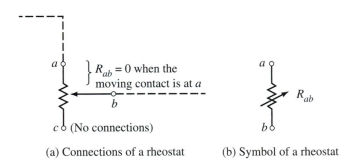

(b) Terminals of a variable resistor

(c) Variable resistor used
as a potentiometer

FIGURE 3–13   (a) Variable resistors. *(Courtesy of Bourns, Inc.)*

(a) Connections of a rheostat   (b) Symbol of a rheostat

FIGURE 3–14

3.6   Color Coding of Resistors

Large resistors such as the wire-wound resistors or the ceramic-encased power resistors have their resistor values and tolerances printed on their cases. Smaller resistors, whether constructed of a molded carbon composition or a metal film, may be too small to have their values printed on the component. Instead, these smaller resistors are usually covered by an epoxy or similar insulating coating over which several colored bands are printed radially as shown in Figure 3–15.

The colored bands provide a quickly recognizable code for determining the value of resistance, the tolerance (in percentage), and occasionally the expected reliability of the resistor. The colored bands are always read from left to right, left being defined as the side of the resistor with the band nearest to it.

The first two bands represent the first and second digits of the resistance value. The third band is called the multiplier band and represents the number of zeros following the first two digits; it is usually given as a power of ten. The fourth band indicates the tolerance of the resistor, and the fifth band (if present) is an indication of the expected reliability of the component. The

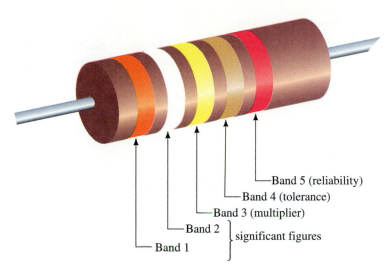

FIGURE 3–15   Resistor color codes.

reliability is a statistical indication of the expected number of components which will no longer have the indicated resistance value after 1000 hours of use. For example, if a particular resistor has a reliability of 1% it is expected that after 1000 hours of use, no more than one resistor in 100 is likely to be outside the specified range of resistance as indicated in the first four bands of the color codes. Table 3–5 shows the colors of the various bands and the corresponding values.

EXAMPLE 3–10   Determine the resistance of a carbon film resistor having the color codes shown in Figure 3–16.

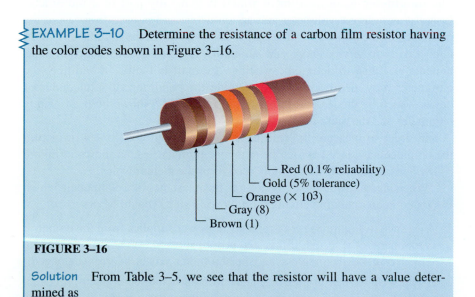

FIGURE 3–16

Solution   From Table 3–5, we see that the resistor will have a value determined as

$$R = 18 \times 10^3 \ \Omega \pm 5\%$$
$$= 18 \ k\Omega \pm 0.9 \ k\Omega \text{ with a reliability of } 0.1\%$$

This specification indicates that the resistance will fall between 17.1 kΩ and 18.9 kΩ. After 1000 hours, we would expect that no more than 1 resistor in 1000 would fall outside the specified range.

TABLE 3–5   Resistor Color Codes

Color	Band 1 Sig. Fig.	Band 2 Sig. Fig.	Band 3 Multiplier	Band 4 Tolerance	Band 5 Reliability
Black		0	$10^0 = 1$		
Brown	1	1	$10^1 = 10$		1%
Red	2	2	$10^2 = 100$		0.1%
Orange	3	3	$10^3 = 1\,000$		0.01%
Yellow	4	4	$10^4 = 10\,000$		0.001%
Green	5	5	$10^5 = 100\,000$		
Blue	6	6	$10^6 = 1\,000\,000$		
Violet	7	7	$10^7 = 10\,000\,000$		
Gray	8	8			
White	9	9			
Gold			0.1	5%	
Silver			0.01	10%	
No color				20%	

PRACTICE PROBLEMS 5

A resistor manufacturer produces carbon composition resistors of 100 MΩ, with a tolerance of ±5%. What will be the color codes on the resistor? (Left to right)

Answer: Brown   Black   Violet   Gold

3.7   Measuring Resistance—The Ohmmeter

The **ohmmeter** is an instrument which is generally part of a multimeter (usually including a voltmeter and an ammeter) and is used to measure the resistance of a component. Although it has limitations, the ohmmeter is used almost daily in service shops and laboratories to measure resistance of components and also to determine whether a circuit is faulty. In addition, the ohmmeter may also be used to determine the condition of semiconductor devices such as diodes and transistors. Figure 3–17 shows both an analog ohmmeter and the more modern digital ohmmeter.

In order to measure the resistance of an isolated component or circuit, the ohmmeter is placed across the component under test, as shown in Figure 3–18. The resistance is then simply read from the meter display.

When using an ohmmeter to measure the resistance of a component which is located in an operating circuit, the following steps should be observed:

1. As shown in Figure 3–19(a), remove all power supplies from the circuit or component to be tested. If this step is not followed, the ohmmeter reading will, at best, be meaningless, and the ohmmeter may be severely damaged.

2. If you wish to measure the resistance of a particular component, it is necessary to isolate the component from the rest of the circuit. This is done by disconnecting at least one terminal of the component from the balance of the circuit as shown in Figure 3–19(b). If this step is not followed, in all likelihood the resistance reading indicated by the ohmmeter will not be the resistance of the desired resistor, but rather the resistance of the combination.

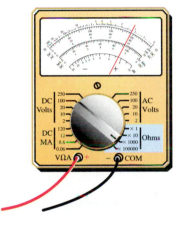

(a) Analog ohmmeter.

(b) Digital ohmmeter. (*Reproduced with permission from the John Fluke Mfg. Co., Inc.*)

FIGURE 3–17

3. As shown in Figure 3–19(b), connect the two probes of the ohmmeter across the component to be measured. The black and red leads of the ohmmeter may be interchanged when measuring resistors. When measuring resistance of other components, however, the measured resistance will be dependent upon the direction of the sensing current. Such devices are covered briefly in a later section of this chapter.

4. Ensure that the ohmmeter is on the correct range to provide the most accurate reading. For example, although a digital multimeter (DMM) can measure a reading for a 1.2-kΩ resistor on the 2-MΩ range, the same ohmmeter will provide additional significant digits (hence more precision) when it is switched to the 2-kΩ range. For analog meters, the best accuracy is obtained when the needle is approximately in the center of the scale.

5. When you are finished, turn the ohmmeter off. Because the ohmmeter uses an internal battery to provide a small sensing current, it is possible to drain the battery if the probes accidently connect together for an extended period.

FIGURE 3–18   Ohmmeter used to measure an isolated component.

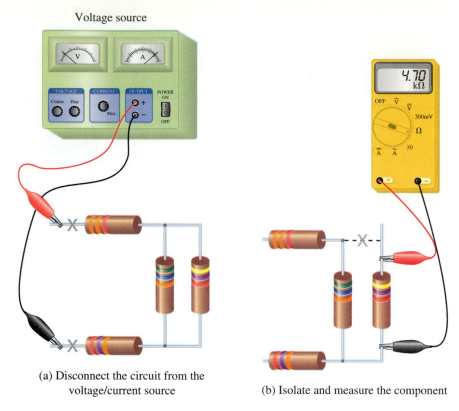

Voltage source

(a) Disconnect the circuit from the voltage/current source

(b) Isolate and measure the component

FIGURE 3–19   Using an ohmmeter to measure resistance in a circuit.

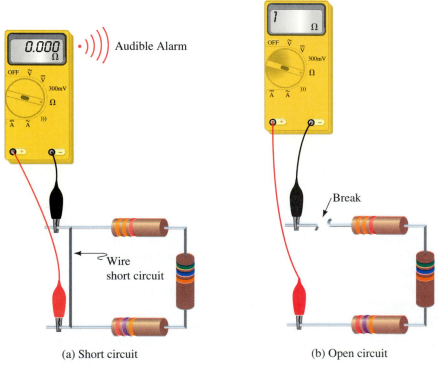

Audible Alarm

Wire short circuit

Break

(a) Short circuit

(b) Open circuit

FIGURE 3–20

In addition to measuring resistance, the ohmmeter may also be used to indicate the continuity of a circuit. Many modern digital ohmmeters have an audible tone which indicates that a circuit is unbroken from one point to another point. As demonstrated in Figure 3–20(a), the audible tone of a digital ohmmeter allows the user to determine continuity without having to look away from the circuit under test.

Ohmmeters are particularly useful instruments in determining whether a given circuit has been short circuited or open circuited.

A **short circuit** occurs when a low-resistance conductor such as a piece of wire or any other conductor is connected between two points in a circuit. Due to the very low resistance of the short circuit, current will bypass the rest of the circuit and go through the short. An ohmmeter will indicate a very low (theoretically zero) resistance when used to measure across a short circuit.

An **open circuit** occurs when a conductor is broken between the points under test. An ohmmeter will indicate infinite resistance when used to measure the resistance of a circuit having an open circuit.

Figure 3–20 illustrates circuits having a short circuit and an open circuit.

PRACTICAL NOTES...

When a digital ohmmeter measures an open circuit, the display on the meter will usually be the digit 1 at the left-hand side, with no following digits. This reading should not be confused with a reading of 1 Ω, 1 kΩ, or 1 MΩ, which would appear on the right-hand side of the display.

An ohmmeter is used to measure across the terminals of a switch.

a. What will the ohmmeter indicate when the switch is closed?

b. What will the ohmmeter indicate when the switch is opened?

Answers: a. 0 Ω (short circuit)
          b. ∞ (open circuit)

PRACTICE
PROBLEMS 6

3.8   Thermistors

In Section 3.4 we saw how resistance changes with changes in temperature. While this effect is generally undesirable in resistors, there are many applications which use electronic components having characteristics which vary according to changes in temperature. Any device or component which causes an electrical change due to a physical change is referred to as a **transducer.**

A **thermistor** is a two-terminal transducer in which resistance changes significantly with changes in temperature (hence a thermistor is a "thermal resistor"). The resistance of thermistors may be changed either by external temperature changes or by changes in temperature caused by current through the component. By applying this principle, thermistors may be used in circuits to control current and to measure or control temperature. Typical appli-

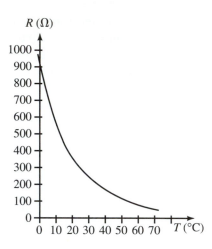

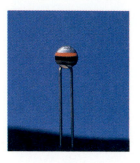

(a) Photograph   (b) Symbol

FIGURE 3–21   Thermistors.

FIGURE 3–22   Thermistor resistance as a function of temperature.

cations include electronic thermometers and thermostatic control circuits for furnaces. Figure 3–21 shows a typical thermistor and its electrical symbol.

Thermistors are constructed of oxides of various materials such as cobalt, manganese, nickel, and strontium. As the temperature of the thermistor is increased, the outermost (valence) electrons in the atoms of the material become more active and break away from the atom. These extra electrons are now free to move within the circuit, thereby causing a reduction in the resistance of the component (negative temperature coefficient). Figure 3–22 shows how resistance of a thermistor varies with temperature effects.

PRACTICE PROBLEMS 7

Referring to Figure 3–22, determine the approximate resistance of a thermistor at each of the following temperatures:

a. 10°C.

b. 30°C.

c. 50°C.

Answers: a. 550 Ω   b. 250 Ω   c. 120 Ω

3.9   Photoconductive Cells

Photoconductive cells or **photocells** are two-terminal transducers which have a resistance determined by the amount of light falling on the cell. Most photocells are constructed of either cadmium sulfide (CdS) or cadmium selenide (CdSe) and are sensitive to light having wavelengths between 4000 Å (blue light) and 10 000 Å (infrared). The angstrom (Å) is a unit commonly used to measure the wavelength of light and has a dimension given as 1 Å = 1×10^{-10} m. Light, which is a form of energy, strikes the material of the photocell and causes the release of valence electrons, thereby reducing the resistance of the component. Figure 3–23 shows the structure, symbol, and resistance characteristics of a typical photocell.

Photocells may be used to measure light intensity and/or to control lighting. They are typically used as part of a security system.

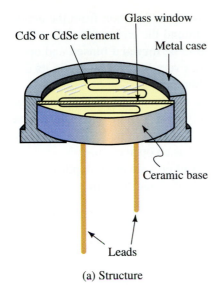

Glass window

CdS or CdSe element

Metal case

Ceramic base

Leads

(a) Structure

(b) Symbol of a photocell

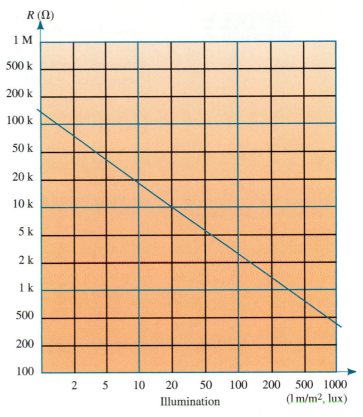

(c) Resistance versus illumination

FIGURE 3–23   Photocell.

3.10   Nonlinear Resistance

Up to this point, the components we have examined have had values of resistance which were essentially constant for a given temperature (or, in the case of a photocell, for a given amount of light). If we were to examine the current versus voltage relationship for these components, we would find that the relationship is linear, as shown in Figure 3–24.

If a device has a linear (straight-line) current-voltage relation then it is referred to as an **ohmic device.** (The linear current-voltage relationship will be will be covered in greater detail in the next chapter.) Often in electronics, we use components which do not have a linear current-voltage relationship; these devices are referred to as **nonohmic devices.** On the other hand, some components, such as the thermistor, can be shown to have both an ohmic region and a nonohmic region. For large current through the thermistor, the component will get hotter. This increase in temperature will result in a decrease of resistance. Consequently, for large currents, the thermistor is a nonohmic device.

We will now briefly examine two common nonohmic devices.

FIGURE 3–24   Linear current-voltage relationship.

Diodes

The diode is a semiconductor device which permits charge to flow in only one direction. Figure 3–25 illustrates the appearance and the symbol of a typical diode.

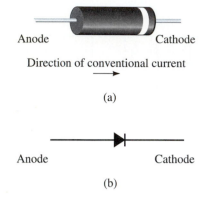

Anode   Cathode

Direction of conventional current

(a)

Anode   Cathode

(b)

FIGURE 3–25 Diode. (a) Typical structure; (b) Symbol.

Conventional current through a diode is in the direction from the anode toward the cathode (the end with the line around the circumference). When current is in this direction, the diode is said to be **forward biased** and operating in its **forward region.** Since a diode has very little resistance in its forward region, it is often approximated as a short circuit.

If the circuit is connected such that the direction of current is from the cathode to the anode (against the arrow in Figure 3–25), the diode is **reverse biased** and operating in its **reverse region.** Due to the high resistance of a reverse-biased diode, it is often approximated as an open circuit.

Although this textbook does not attempt to provide an in-depth study of diode theory, Figure 3–26 shows the basics of diode operation both when forward biased and when reverse biased.

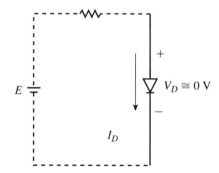

(a) Forward-biased
diode

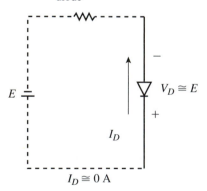

(b) Reverse-biased
diode

FIGURE 3–26 Current-voltage relation for a silicon diode.

PRACTICAL NOTES...

Because an ohmmeter uses an internal voltage source to generate a small sensing current, the instrument may easily be used to determine the terminals (and hence the direction of conventional flow) of a diode. (See Figure 3–27.)

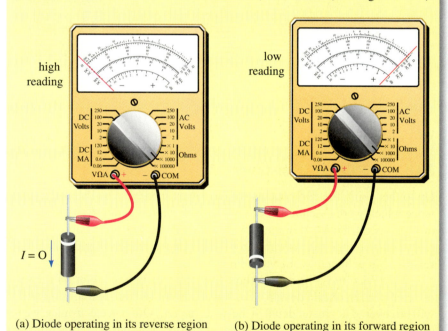

(a) Diode operating in its reverse region   (b) Diode operating in its forward region

FIGURE 3–27 Determining diode terminals with an ohmmeter.

If we measure the resistance of the diode in both directions, we will find that the resistance will be low when the positive terminal of the ohmmeter is connected to the anode of the diode. When the positive terminal is connected to the cathode, virtually no current will occur in the diode and so the indication on the ohmmeter will be a very high resistance (theoretically, $R = \infty \ \Omega$).

Varistors

Varistors, as shown in Figure 3–28, are semiconductor devices which have very high resistances when the voltage across the varistors is below the breakdown value. However, when the voltage across a varistor (either polarity) exceeds the rated value, the resistance of the device suddenly becomes very small, allowing charge to flow. Figure 3–29 shows the current-voltage relation for varistors.

(a) Photograph

(b) Varistor symbols.

FIGURE 3–28   Varistors

Varistors are used in sensitive circuits, such as those in computers, to ensure that if the voltage suddenly exceeds a predetermined value, the varistor will effectively become a short circuit to the unwanted signal, thereby protecting the rest of the circuit from excessive voltage.

3.11   Conductance

Conductance, G, is defined as the measure of a material's ability to allow the flow of charge and is assigned the SI unit the siemens (S). A large conductance indicates that a material is able to conduct current well, whereas a low value of conductance indicates that a material does not readily permit the flow of charge. Mathematically, conductance is defined as the reciprocal of resistance. Thus

$$G = \frac{1}{R} \quad [\text{siemens, S}] \tag{3–10}$$

where R is resistance, in ohms (Ω).

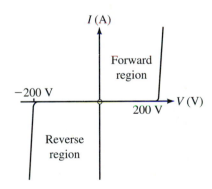

FIGURE 3–29   Current-voltage relation of a 200-V (peak) varistor.

EXAMPLE 3-11   Determine the conductance of the following resistors:

a. 5 Ω
b. 100 kΩ
c. 50 mΩ

Solution

a. $G = \dfrac{1}{5\ \Omega} = 0.2\ \text{S} = 200\ \text{mS}$

b. $G = \dfrac{1}{100\ \text{k}\Omega} = 0.01\ \text{mS} = 10\ \mu\text{S}$

c. $G = \dfrac{1}{50\ \text{m}\Omega} = 20\ \text{S}$

PRACTICE PROBLEMS 8

1. A given cable has a conductance given as 5.0 mS. Determine the value of the resistance, in ohms.

2. If the conductance is doubled, what happens to the resistance?

Answers: 1. 200 Ω   2. It halves.

Although the SI unit of conductance (siemens) is almost universally accepted, older books and data sheets list conductance in the unit given as the mho (ohm spelled backwards) and having an upside-down omega, ℧, as the symbol. In such a case, the following relationship holds:

$$1\ ℧ = 1\ \text{S} \qquad\qquad (3\text{--}11)$$

PRACTICE PROBLEMS 9

A specification sheet for a radar transmitter indicates that one of the components has a conductance of 5 μμ℧.

a. Express the conductance in the proper SI prefix and unit.

b. Determine the resistance of the component, in ohms.

Answers: a. 5 pS   b. $2 \times 10^{11}\ \Omega$

3.12   Superconductors

As you have seen, all power lines and distribution networks have internal resistance which results in energy loss due to heat as charge flows through the conductor. If there was some way of eliminating the resistance of the conductors, electricity could be transmitted farther and more economically. The idea that energy could be transmitted without losses along a "superconductor" transmission line was formerly a distant goal. However, recent discoveries in high-temperature superconductivity promise the almost magical ability to transmit and store energy with no loss in energy.

In 1911, the Dutch physicist Heike Kamerlingh Onnes discovered the phenomenon of superconductivity. Studies of mercury, tin, and lead verified that the resistance of these materials decreases to no more than one ten-billionth of the room temperature resistance when subjected to temperatures of 4.6 K, 3.7 K, and 6 K respectively. Recall that the relationship between kelvins and degrees Celsius is as follows:

$$T_K = T_{(°C)} + 273.15°$$ (3–12)

The temperature at which a material becomes a superconductor is referred to as the **critical temperature,** T_C, of the material. Figure 3–30 shows how the resistance of a sample of mercury changes with temperature. Notice how the resistance suddenly drops to zero at a temperature of 4.6 K.

Experiments with currents in supercooled loops of superconducting wire have determined that the induced currents will remain undiminished for many years within the conductor provided that the temperature is maintained below the critical temperature of the conductor.

A peculiar, seemingly magical property of superconductors occurs when a permanent magnet is placed above the superconductor. The magnet will float above the surface of the conductor as if it is defying the law of gravity, as shown in Figure 3–31.

This principle, which is referred to as the *Meissner effect* (named after Walther Meissner), may be simply stated as follows:

When a superconductor is cooled below its critical temperature, magnetic fields may surround but not enter the superconductor.

The principle of superconductivity is explained in the behavior of electrons within the superconductor. Unlike conductors which have electrons moving randomly through the conductor and colliding with other electrons [Figure 3–32(a)], the electrons in superconductors form pairs which move through the material in a manner similar to a band marching in a parade. The orderly motion of electrons in a superconductor, shown in Figure 3–32(b), results in an ideal conductor, since the electrons no longer collide.

The economy of having a high critical temperature has led to the search for high-temperature superconductors. In recent years, research at the IBM Zurich Research Laboratory in Switzerland and the University of Houston in Texas has yielded superconducting materials which are able to operate at temperatures as high as 98 K (−175°C). While this temperature is still very

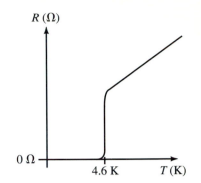

FIGURE 3–30 Critical temperature of mercury.

FIGURE 3–31 The Meissner effect: A magnetic cube hovers above a disk of ceramic superconductor. The disk is kept below its critical temperature in a bath of liquid nitrogen. *(Courtesy of AT&T Bell Laboratories/AT&T Archives)*

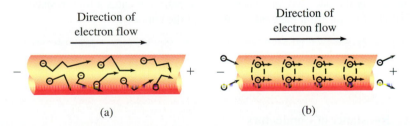

(a)   (b)

FIGURE 3–32 (a) In conductors, electrons are free to move in any direction through the conductor. Energy is lost due to collisions with atoms and other electrons, giving rise to the resistance of the conductor. (b) In superconductors, electrons are bound in pairs and travel through the conductor in step, avoiding all collisions. Since there is no energy loss, the conductor has no resistance.

low, it means that superconductivity can now be achieved by using the readily available liquid nitrogen rather than the much more expensive and rarer liquid helium.

Superconductivity has been found in such seemingly unlikely materials as ceramics consisting of barium, lanthanum, copper, and oxygen. Research is now centered on developing new materials which become superconductors at ever higher temperatures and which are able to overcome the disadvantages of the early ceramic superconductors.

Very expensive, low-temperature superconductivity is currently used in some giant particle accelerators and, to a limited degree, in electronic components (such as superfast Josephson junctions and SQUIDs, i.e., superconducting quantum interference devices, which are used to detect very small magnetic fields). Once research produces commercially viable, high-temperature superconductors, however, the possibilities of the applications will be virtually limitless. High-temperature superconductivity promises to yield improvements in transportation, energy storage and transmission, computers, and medical treatment and research. It is quite possible that high-temperature superconductivity will change electronics as much as the invention of the transistor.

PUTTING IT INTO PRACTICE

You are a troubleshooting specialist working for a small telephone company. One day, word comes in that an entire subdivision is without telephone service. Everyone suspects that a cable was cut by one of several backhoe operators working on a waterline project near the subdivision. However, no one is certain exactly where the cut occurred. You remember that the resistance of a length of wire is determined by several factors, including the length. This gives you an idea for determining the distance between the telephone central office and location of the cut.

First you go to the telephone cable records, which show that the subdivision is served by 26-gauge copper wire. Then, since each customer's telephone is connected to the central office with a pair of wires, you measure the resistance of several loops from the central office. As expected, some of the measurements indicate open circuits. However, several pairs of the wire were shorted by the backhoe, and each of these pairs indicates a total resistance of 338 Ω. How far from the central office did the cut occur?

PROBLEMS

3.1   Resistance of Conductors

1. Determine the resistance, at 20°C, of 100 m of solid aluminum wire having the following radii:

   a. 0.5 mm

   b. 1.0 mm

c. 0.005 mm

d. 0.5 cm

2. Determine the resistance, at 20°C, of 200 feet of iron conductors having the following cross sections:

a. 0.25 inch by 0.25 inch square

b. 0.125 inch diameter round

c. 0.125 inch by 4.0 inch rectangle

3. A 250-foot length of solid copper bus bar, shown in Figure 3–33, is used to connect a voltage source to a distribution panel. If the bar is to have a resistance of 0.02 Ω at 20°C, calculate the required height of the bus bar (in inches).

4. Nichrome wire is used to construct heating elements. Determine the length of 1.0-mm-diameter Nichrome wire needed to produce a heating element which has a resistance of 2.0 Ω at a temperature of 20°C.

5. A copper wire having a diameter of 0.80 mm is measured to have a resistance of 10.3 Ω at 20°C. How long is this wire in meters? How long is the wire in feet?

6. A piece of aluminum wire has a resistance, at 20°C, of 20 Ω. If this wire is melted down and used to produce a second wire having a length four times the original length, what will be the resistance of the new wire at 20°C? (Hint: The volume of the wire has not changed.)

7. Determine the resistivity (in ohm-meters) of a carbon-based graphite cylinder having a length of 6.00 cm, a diameter of 0.50 mm, and a measured resistance of 3.0 Ω at 20°C. How does this value compare with the resistivity given for carbon?

8. A solid circular wire of length 200 m and diameter of 0.4 mm has a resistance measured to be 357 Ω at 20°C. Of what material is the wire constructed?

9. A 2500-m section of alloy wire has a resistance of 32 Ω. If the wire has a diameter of 1.5 mm, determine the resistivity of the material in ohm-meters. Is this alloy a better conductor than copper?

10. A section of iron wire having a diameter of 0.030 inch is measured to have a resistance of 2500 Ω (at a temperature of 20°C).

a. Determine the cross-sectional area in square meters and in square millimeters. (Note: 1 inch = 2.54 cm = 25.4 mm)

b. Calculate the length of the wire in meters.

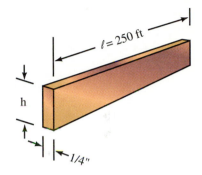

FIGURE 3–33

3.2   Electrical Wire Tables

11. Use Table 3–2 to determine the resistance of 300 feet of AWG 22 and AWG 19 solid copper conductors. Compare the diameters and the resistances of the wires.

12. Use Table 3–2 to find the resistance of 250 m of AWG 8 and AWG 2 solid copper conductors. Compare the diameters and the resistances of the wires.

13. Determine the maximum current which could be handled by AWG 19 wire and by AWG 30 wire.

14. If AWG 8 is rated at a maximum of 40 A, how much current could AWG 2 handle safely?

15. A spool of AWG 36 copper transformer wire is measured to have a resistance of 550 Ω at a temperature of 20°C. How long is this wire in meters?

16. How much current should AWG 36 copper wire be able to handle?

3.3   Resistance of Wires—Circular Mils

17. Determine the area in circular mils of the following conductors ($T = 20$°C):

    a. Circular wire having a diameter of 0.016 inch

    b. Circular wire having a diameter of 2.0 mm

    c. Rectangular bus bar having dimensions 0.25 inch by 6.0 inch

18. Express the cross-sectional areas of the conductors of Problem 17 in square mils and in square millimeters.

19. Calculate the resistance, at 20°C, of 400 feet of copper conductors having the cross-sectional areas given in Problem 17.

20. Determine the diameter in inches and in millimeters of circular cables having cross-sectional areas as given below: (Assume the cables to be solid conductors.)

    a. 250 CM

    b. 1000 CM

    c. 250 MCM

    d. 750 MCM

21. A 200-foot length of solid copper wire is measured to have a resistance of 0.500 Ω.

    a. Determine the cross-sectional area of the wire in both square mils and circular mils.

    b. Determine the diameter of the wire in mils and in inches.

22. Repeat Problem 21 if the wire had been made of Nichrome.

23. A spool of solid copper wire having a diameter of 0.040 inch is measured to have a resistance of 12.5 Ω (at a temperature of 20°C).

    a. Determine the cross-sectional area in both square mils and circular mils.

    b. Calculate the length of the wire in feet.

24. Iron wire having a diameter of 30 mils was occasionally used for telegraph transmission. A technician measures a section of telegraph line to have a resistance of 2500 Ω (at a temperature of 20°C).

    a. Determine the cross-sectional area in both square mils and circular mils.

    b. Calculate the length of the wire in feet and in meters. (Note: 1 ft = 0.3048 m.) Compare your answer to the answer obtained in Problem 10.

3.4   Temperature Effects

25. An aluminum conductor has a resistance of 50 Ω at room temperature. Find the resistance of the same conductor at -30°C, 0°C, and at 200°C.

26. AWG 14 solid copper house wire is designed to operate within a temperature range of -40°C to $+90$°C. Calculate the resistance of 200 circuit feet of wire at both temperatures. Note: A circuit foot is the length of cable needed for a current to travel to and from a load.

27. A given material has a resistance of 20 Ω at room temperature (20°C) and 25 Ω at a temperature of 85°C.

   a. Does the material have a positive or a negative temperature coefficient? Explain briefly.

   b. Determine the value of the temperature coefficient, α, at 20°C.

   c. Assuming the resistance versus temperature function to be linear, determine the expected resistance of the material at 0°C (the freezing point of water) and at 100°C (the boiling point of water).

28. A given material has a resistance of 100 Ω at room temperature (20°C) and 150 Ω at a temperature of −25°C.

   a. Does the material have a positive or a negative temperature coefficient? Explain briefly.

   b. Determine the value of the temperature coefficient, α, at 20°C.

   c. Assuming the resistance versus temperature function to be linear, determine the expected resistance of the material at 0°C (the freezing point of water) and at −40°C.

29. An electric heater is made of Nichrome wire. The wire has a resistance of 15.2 Ω at a temperature of 20°C. Determine the resistance of the Nichrome wire when the temperature of the wire is increased to 260°C.

30. A silicon diode is measured to have a resistance of 500 Ω at 20°C. Determine the resistance of the diode if the temperature of the component is increased with a soldering iron to 30°C. (Assume that the resistance versus temperature function is linear.)

31. An electrical device has a linear temperature response. The device has a resistance of 120 Ω at a temperature of −20°C and a resistance of 190 Ω at a temperature of 120°C.

   a. Calculate the resistance at a temperature of 0°C.

   b. Calculate the resistance at a temperature of 80°C.

   c. Determine the temperature intercept of the material.

32. Derive the expression of Equation 3–8.

3.5   Types of Resistors

33. A 10-kΩ variable resistor has its wiper (movable terminal b) initially at the bottom terminal, c. Determine the resistance R_{ab} between terminals a and b and the resistance R_{bc} between terminals b and c under the following conditions:

   a. The wiper is at c.

   b. The wiper is one-fifth of the way around the resistive surface.

   c. The wiper is four fifths of the way around the resistive surface.

   d. The wiper is at a.

34. The resistance between wiper terminal b and bottom terminal c of a 200-kΩ variable resistor is measured to be 50 kΩ. Determine the resistance which would be measured between the top terminal, a and the wiper terminal, b.

3.6   Color Coding of Resistors

35. Given resistors having the following color codes (as read from left to right), determine the resistance, tolerance and reliability of each component. Express the uncertainty in both percentage and ohms.

a. Brown	Green	Yellow	Silver	
b. Red	Gray	Gold	Gold	Yellow
c. Yellow	Violet	Blue	Gold	
d. Orange	White	Black	Gold	Red

36. Determine the color codes required if you need the following resistors for a project:

   a. 33 kΩ ± 5%, 0.1% reliability

   b. 820 Ω ± 10%

   c. 15 Ω ± 20%

   d. 2.7 MΩ ± 5%

3.7   Measuring Resistance—The Ohmmeter

37. Explain how an ohmmeter may be used to determine whether a light bulb is burned out.

38. If an ohmmeter were placed across the terminal of a switch, what resistance would you expect to measure when the contacts of the switch are closed? What resistance would you expect to measure when the contacts are opened?

39. Explain how you could use an ohmmeter to determine approximately how much wire is left on a spool of AWG 24 copper wire.

40. An analog ohmmeter is used to measure the resistance of a two-terminal component. The ohmmeter indicates a resistance of 1.5 kΩ. When the leads of the ohmmeter are reversed, the meter indicates that the resistance of the component is an open circuit. Is the component faulty? If not, what kind of component is being tested?

3.8   Thermistors

41. A thermistor has the characteristics shown in Figure 3–22.

   a. Determine the resistance of the device at room temperature, 20°C.

   b. Determine the resistance of the device at a temperature of 40°C.

   c. Does the thermistor have a positive or a negative temperature coefficient? Explain.

3.9   Photoconductive Cells

42. For the photocell having the characteristics shown in Figure 3–23(c), determine the resistance

   a. in a dimly lit basement having an illuminance of 10 lux

   b. in a home having an illuminance of 50 lux

   c. in a classroom having an illuminance of 500 lux

3.11   Conductance

43. Calculate the conductance of the following resistances:

    a. $0.25\ \Omega$

    b. $500\ \Omega$

    c. $250\ k\Omega$

    d. $12.5\ M\Omega$

44. Determine the resistance of components having the following conductances:

    a. $62.5\ \mu S$

    b. $2500\ mS$

    c. $5.75\ mS.$

    d. $25.0\ S$

45. Determine the conductance of 1000 m of AWG 30 solid copper wire at a temperature of 20°C.

46. Determine the conductance of 200 feet of aluminum bus bar (at a temperature of 20°C) which has a cross-sectional dimension of 4.0 inches by 0.25 inch. If the temperature were to increase, what would happen to the conductance of the bus bar?

ANSWERS TO IN-PROCESS LEARNING CHECKS

In-Process Learning Check 1

1. Iron wire will have approximately seven times more resistance than copper.

2. The longer wire will have twice the resistance of the shorter wire.

3. The wire having the greater diameter will have one quarter the resistance of the small-diameter wire.

In-Process Learning Check 2

1. 200 A

2. The actual value is less than the theoretical value. Since only the surface of the cable is able to dissipate heat, the current must be decreased to prevent heat build-up.

In-Process Learning Check 3

$A = 63.7\ CM$

$A = 3.23 \times 10^{-8}\ m^2 = 0.0323\ mm^2$

In-Process Learning Check 4

Positive temperature coefficient means that resistance of a material increases as temperature increases. Negative temperature coefficient means that the resistance of a material decreases as the temperature increases. Aluminum has a positive temperature coefficient.

4

Ohm's Law, Power, and Energy

OBJECTIVES

After studying this chapter, you will be able to

- compute voltage, current, and resistance in simple circuits using Ohm's law,
- use the voltage reference convention to determine polarity,
- describe how voltage, current, and power are related in a resistive circuit,
- compute power in dc circuits,
- use the power reference convention to describe the direction of power transfer,
- compute energy used by electrical loads,
- determine energy costs,
- determine the efficiency of machines and systems,
- use OrCAD PSpice and Electronics Workbench to solve Ohm's law problems.

OUTLINE

Ohm's Law
Voltage Polarity and Current Direction
Power
Power Direction Convention
Energy
Efficiency
Nonlinear and Dynamic Resistances
Computer-Aided Circuit Analysis

KEY TERMS

DC Resistance
Dynamic Resistance
Efficiency
Energy
Linear Resistance
Nonlinear Resistance
Ohm
Ohm's Law
Open Circuit
Power
Voltage Reference Convention

CHAPTER PREVIEW

In the previous two chapters, you studied voltage, current, and resistance separately. In this chapter, we consider them together. Beginning with Ohm's law, you will study the relationship between voltage and current in a resistive circuit, reference conventions, power, energy and efficiency. Also in this chapter, we begin our study of computer methods. Two application packages are considered here; they are OrCAD PSpice and Electronics Workbench.

Georg Simon Ohm

PUTTING IT IN PERSPECTIVE

IN CHAPTER 3, WE LOOKED BRIEFLY at Ohm's experiments. We now take a look at Ohm the person.

Georg Simon Ohm was born in Erlangen, Bavaria, on March 16, 1787. His father was a master mechanic who determined that his son should obtain an education in science. Although Ohm became a teacher in a high school, he had aspirations to receive a university appointment. The only way that such an appointment could be realized would be if Ohm could produce important results through scientific research. Since the science of electricity was in its infancy, and because the electric cell had recently been invented by the Italian Conte Alessandro Volta, Ohm decided to study the behavior of current in resistive circuits. Because equipment was expensive and hard to come by, Ohm made much of his own, thanks, in large part, to his father's training. Using this equipment, Ohm determined experimentally that the amount of current transmitted along a wire was directly proportional to its cross-sectional area and inversely proportional to its length. From these results, Ohm was able to define resistance and show that there was a simple relationship between voltage, resistance, and current. This result, now known as Ohm's law, is probably the most fundamental relationship in circuit theory. However, when published in 1827, Ohm's results were met with ridicule. As a result, not only did Ohm miss out on a university appointment, he was forced to resign from his high-school teaching position. While Ohm was living in poverty and shame, his work became known and appreciated outside Germany. In 1842, Ohm was appointed a member of the Royal Society. Finally, in 1849, he was appointed as a professor at the University of Munich, where he was at last recognized for his important contributions.

4.1   Ohm's Law

Consider the circuit of Figure 4–1. Using a circuit similar in concept to this, Ohm determined experimentally that *current in a resistive circuit is directly proportional to its applied voltage and inversely proportional to its resistance.* In equation form, Ohm's law states

$$I = \frac{E}{R} \quad \text{[amps, A]} \tag{4–1}$$

where

E is the voltage in volts,

R is the resistance in ohms,

I is the current in amperes.

From this you can see that the larger the applied voltage, the larger the current, while the larger the resistance, the smaller the current.

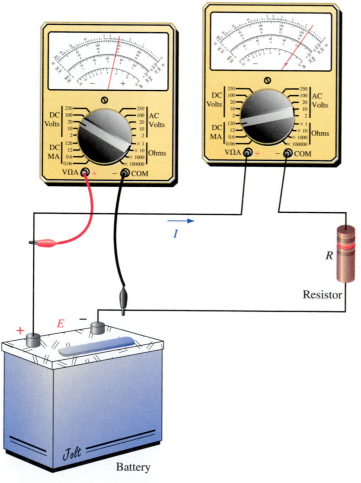

(a) Test circuit

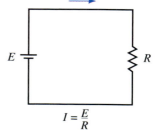

$$I = \frac{E}{R}$$

(b) Schematic, meters not shown

FIGURE 4–1   Circuit for illustrating Ohm's law.

The proportional relationship between voltage and current described by Equation 4–1 may be demonstrated by direct substitution as indicated in Figure 4–2. For a fixed resistance, doubling the voltage as shown in (b) doubles the current, while tripling the voltage as shown in (c) triples the current, and so on.

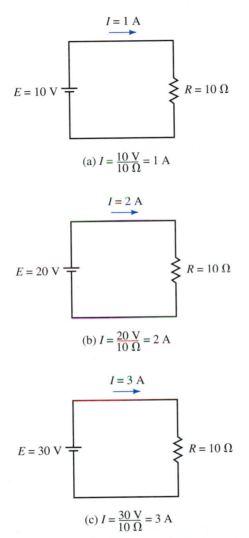

(a) $I = \dfrac{10 \text{ V}}{10 \text{ } \Omega} = 1 \text{ A}$

(b) $I = \dfrac{20 \text{ V}}{10 \text{ } \Omega} = 2 \text{ A}$

(c) $I = \dfrac{30 \text{ V}}{10 \text{ } \Omega} = 3 \text{ A}$

FIGURE 4–2 For a fixed resistance, current is directly proportional to voltage; thus, doubling the voltage as in (b) doubles the current, while tripling the voltage as in (c) triples the current, and so on.

The inverse relationship between resistance and current is demonstrated in Figure 4–3. For a fixed voltage, doubling the resistance as shown in (b) halves the current, while tripling the resistance as shown in (c) reduces the current to one third of its original value, and so on.

Ohm's law may also be expressed in the following forms by rearrangement of Equation 4–1:

$$E = IR \quad \text{[volts, V]} \tag{4–2}$$

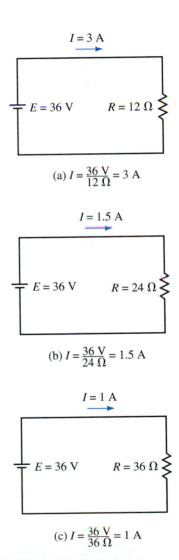

(a) $I = \dfrac{36 \text{ V}}{12 \text{ } \Omega} = 3 \text{ A}$

(b) $I = \dfrac{36 \text{ V}}{24 \text{ } \Omega} = 1.5 \text{ A}$

(c) $I = \dfrac{36 \text{ V}}{36 \text{ } \Omega} = 1 \text{ A}$

FIGURE 4–3 For a fixed voltage, current is inversely proportional to resistance; thus, doubling the resistance as in (b) halves the current, while tripling the resistance as in (c) results in one third the current, and so on.

and

$$R = \frac{E}{I} \quad [\text{ohms, } \Omega] \tag{4–3}$$

When using Ohm's law, be sure to express all quantities in base units of volts, ohms, and amps as in Examples 4–1 to 4–3, or utilize the relationships between prefixes as in Example 4–4.

EXAMPLE 4–1   A 27-Ω resistor is connected to a 12-V battery. What is the current?

Solution   Substituting the resistance and voltage values into Ohm's law yields

$$I = \frac{E}{R} = \frac{12 \text{ V}}{27 \text{ } \Omega} = 0.444 \text{ A}$$

EXAMPLE 4–2   The lamp of Figure 4–4 draws 25 mA when connected to a 6-V battery. What is its resistance?

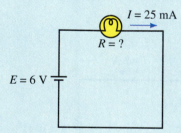

FIGURE 4–4

Solution   Using Equation 4–3,

$$R = \frac{E}{I} = \frac{6 \text{ V}}{25 \times 10^{-3} \text{ A}} = 240 \text{ } \Omega$$

EXAMPLE 4–3   If 125 μA is the current in a resistor with color bands red, red, yellow, what is the voltage across the resistor?

Solution   Using the color code of Chapter 3, $R = 220$ kΩ. From Ohm's law, $E = IR = (125 \times 10^{-6} \text{ A})(220 \times 10^{3} \text{ } \Omega) = 27.5 \text{ V}$.

EXAMPLE 4–4 A resistor with the color code brown, red, yellow is connected to a 30-V source. What is I?

Solution When E is in volts and R in kΩ, the answer comes out directly in mA. From the color code, $R = 120$ kΩ. Thus,

$$I = \frac{E}{R} = \frac{30\ \text{V}}{120\ \text{k}\Omega} = 0.25\ \text{mA}$$

Traditionally, circuits are drawn with the source on the left and the load on the right as indicated in Figures 4–1 to 4–3. However, you will also encounter circuits with other orientations. For these, the same principles apply; as you saw in Figure 4–4, simply draw the current arrow pointing out from the positive end of the source and apply Ohm's law in the usual manner. More examples are shown in Figure 4–5.

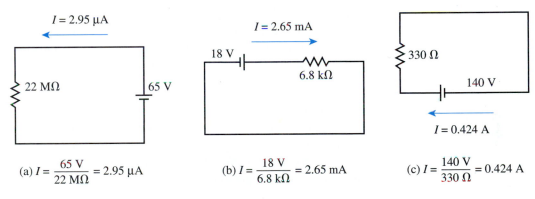

(a) $I = \dfrac{65\ \text{V}}{22\ \text{M}\Omega} = 2.95\ \mu\text{A}$

(b) $I = \dfrac{18\ \text{V}}{6.8\ \text{k}\Omega} = 2.65\ \text{mA}$

(c) $I = \dfrac{140\ \text{V}}{330\ \Omega} = 0.424\ \text{A}$

FIGURE 4–5

1. a. For the circuit of Figure 4–2(a), show that halving the voltage halves the current.

   b. For the circuit of Figure 4–3(a), show that halving the resistance doubles the current.

   c. Are these results consistent with the verbal statement of Ohm's law?

2. For each of the following, draw the circuit with values marked, then solve for the unknown.

   a. A 10 000-milliohm resistor is connected to a 24-V battery. What is the resistor current?

   b. How many volts are required to establish a current of 20 μA in a 100-kΩ resistor?

   c. If 125 V is applied to a resistor and 5 mA results, what is the resistance?

3. For each circuit of Figure 4–6 determine the current, including its direction (i.e., the direction that the current arrow should point).

PRACTICE
PROBLEMS 1

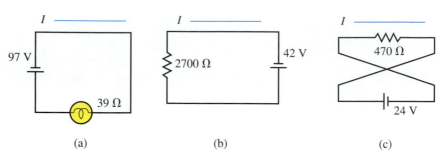

(a)   (b)   (c)

FIGURE 4–6

Answers:

1. a. 5 V/10 Ω = 0.5 A   b. 36 V/6 Ω = 6 A   c. Yes
2. a. 2.4 A   b. 2.0 V   c. 25 kΩ
3. a. 2.49 A, left   b. 15.6 mA, right   c. 51.1 mA, left

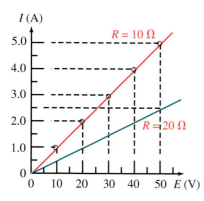

FIGURE 4–7   Graphical representation of Ohm's law. The red plot is for a 10-Ω resistor while the green plot is for a 20-Ω resistor.

Ohm's Law in Graphical Form

The relationship between current and voltage described by Equation 4–1 may be shown graphically as in Figure 4–7. The graphs, which are straight lines, show clearly that the relationship between voltage and current is linear, i.e., that current is directly proportional to voltage.

Open Circuits

Current can only exist where there is a conductive path (e.g., a length of wire). For the circuit of Figure 4–8, I equals zero since there is no conductor between points a and b. We refer to this as an *open circuit*. Since $I = 0$, substitution of this into Equation 4–3 yields

$$R = \frac{E}{I} = \frac{E}{0} \Rightarrow \infty \text{ ohms}$$

Thus, an open circuit has infinite resistance.

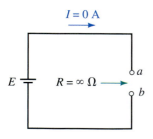

FIGURE 4–8   An open circuit has infinite resistance.

Voltage Symbols

Two different symbols are used to represent voltage. For sources, use uppercase E; for loads (and other components), use uppercase V. This is illustrated in Figure 4–9.

Using the symbol V, Ohm's law may be rewritten in its several forms as

$$I = \frac{V}{R} \quad \text{[amps]} \tag{4–4}$$

$$V = IR \quad \text{[volts]} \tag{4–5}$$

$$R = \frac{V}{I} \quad \text{[ohms]} \tag{4–6}$$

These relationships hold for every resistor in a circuit, no matter how complex the circuit. Since $V = IR$, these voltages are often referred to as *IR drops*.

EXAMPLE 4–5   The current through each resistor of Figure 4–10 is $I = 0.5$ A. Compute V_1 and V_2.

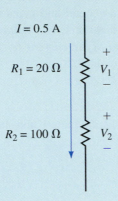

$I = 0.5$ A

$R_1 = 20\ \Omega$ $+$
V_1
$-$

$R_2 = 100\ \Omega$ $+$
V_2
$-$

FIGURE 4–10   Ohm's law applies to each resistor.

Solution $V_1 = IR_1 = (0.5\ \text{A})(20\ \Omega) = 10$ V. Note, I is also the current through R_2. Thus, $V_2 = IR_2 = (0.5\ \text{A})(100\ \Omega) = 50$ V.

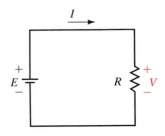

FIGURE 4–9   Symbols used to represent voltages. E is used for source voltages, while V is used for voltages across circuit components such as resistors.

NOTE...

In the interest of brevity (as in Figure 4–10), we sometimes draw only a portion of a circuit with the rest of the circuit implied rather than shown explicitly.

4.2   Voltage Polarity and Current Direction

So far, we have paid little attention to the polarity of voltages across resistors. However, polarity is of extreme importance; fortunately, there is a simple relationship between current direction and voltage polarity. To get at the idea, consider Figure 4–11(a). Here the polarity of V is obvious since the resistor is connected directly to the source. This makes the top end of the resistor positive with respect to the bottom end, and $V = E = 12$ V as indicated by the meters.

FIGURE 4–11   Convention for voltage polarity. Place the plus sign for V at the tail of the current direction arrow.

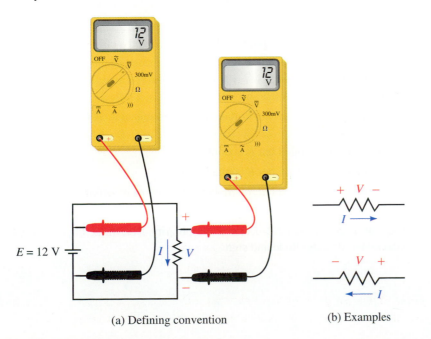

$E = 12$ V

(a) Defining convention

$+$ V $-$
$I \longrightarrow$

$-$ V $+$
$\longleftarrow I$

(b) Examples

Now consider current. The direction of *I* is from top to bottom through the resistor as indicated by the current direction arrow. Examining voltage polarity, we see that the plus sign for *V* is at the tail of this arrow. This observation turns out to be true in general and gives us a convention for marking voltage polarity on circuit diagrams. *For voltage across a resistor, always place the plus sign at the tail of the current reference arrow.* Two additional examples are shown in Figure 4–11(b).

PRACTICE PROBLEMS 2

For each resistor of Figure 4–12, compute *V* and show its polarity.

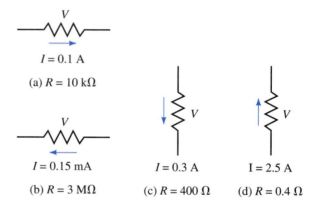

(a) $R = 10$ kΩ, $I = 0.1$ A

(b) $R = 3$ MΩ, $I = 0.15$ mA

(c) $R = 400$ Ω, $I = 0.3$ A

(d) $R = 0.4$ Ω, $I = 2.5$ A

FIGURE 4–12

Answers:
a. 1000 V, + at left   b. 450 V, + at right   c. 120 V, + at top   d. 1 V, + at bottom

IN-PROCESS LEARNING CHECK 1

1. A resistor has color bands brown, black, and red and a current of 25 mA. Determine the voltage across it.

2. For a resistive circuit, what is *I* if *E* = 500 V and *R* is open circuited? Will the current change if the voltage is doubled?

3. A certain resistive circuit has voltage *E* and resistance *R*. If *I* = 2.5 A, what will be the current if:

   a. *E* remains unchanged but *R* is doubled?

   b. *E* remains unchanged but *R* is quadrupled?

   c. *E* remains unchanged but *R* is reduced to 20% of its original value?

   d. *R* is doubled and *E* is quadrupled?

4. The voltmeters of Figure 4–13 have autopolarity. Determine the reading of each meter, its magnitude and sign.

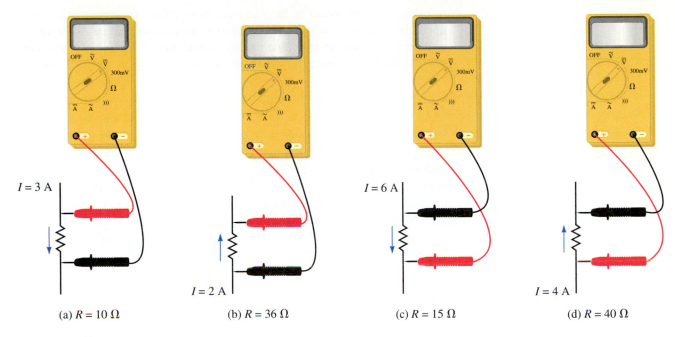

(a) $R = 10\ \Omega$ (b) $R = 36\ \Omega$ (c) $R = 15\ \Omega$ (d) $R = 40\ \Omega$

FIGURE 4–13

(Answers are at the end of the chapter.)

Before We Move On

Before we move on, we will comment on one more aspect of current representation. First, note that to completely specify current, you must include both its value and its direction. (This is why we show current direction reference arrows on circuit diagrams.) Normally we show the current coming out of the plus (+) terminal of the source as in Figure 4–14(a). (Here, $I = E/R = 5$ A in the direction shown. This is the actual direction of the current.) As you can see from this (and all preceding examples in this chapter), determining the actual current direction in single-source networks is easy. However, when analyzing complex circuits (such as those with multiple sources as in later chapters), it is not always easy to tell in advance in what direction all currents will be. As a result, when you solve such problems, you may find that some currents have negative values. What does this mean?

To get at the answer, consider both parts of Figure 4–14. In (a), current is shown in the usual direction, while in (b), it is shown in the opposite direction. To compensate for the reversed direction, we have changed the sign of I. The interpretation placed on this is that a positive current in one direction is the same as a negative current in the opposite direction. Therefore, (a) and (b) are two representations of the same current. Thus, if during the solution of a problem you obtain a positive value for current, this means that its actual direction is the same as the reference arrow; if you obtain a negative value, its direction is opposite to the reference arrow. This is an important idea and one that you will use many times in later chapters. It was

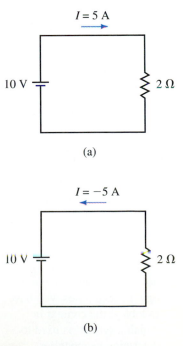

FIGURE 4–14 Two representations of the same current.

introduced at this point to help explain power flow in the electric car of upcoming Example 4–9. However, apart from the electric car example, we will leave its consideration and use to later chapters. That is, we will continue to use the representation of Figure 4–14(a).

4.3   Power

Power is familiar to all of us, at least in a general sort of way. We know, for example, that electric heaters and light bulbs are rated in watts (W) and that motors are rated in horsepower (or watts), both being units of power as discussed in Chapter 1. We also know that the higher the watt rating of a device, the more energy we can get out of it per unit time. Figure 4–15 illustrates the idea. In (a), the greater the power rating of the light, the more light energy that it can produce per second. In (b), the greater the power rating of the heater, the more heat energy it can produce per second. In (c), the larger the power rating of the motor, the more mechanical work that it can do per second.

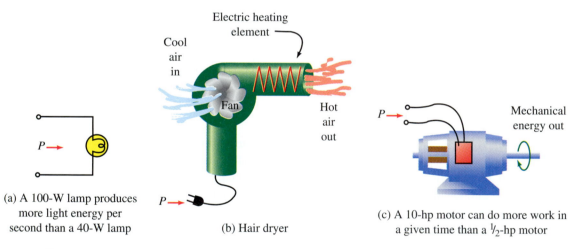

(a) A 100-W lamp produces more light energy per second than a 40-W lamp

(b) Hair dryer

(c) A 10-hp motor can do more work in a given time than a $\frac{1}{2}$-hp motor

FIGURE 4–15   Energy conversion. Power P is a measure of the rate of energy conversion.

NOTES...

1. Time t in Equation 4–7 is a time interval, not an instantaneous point in time.

2. The symbol for energy is W and the abbreviation for watts is W. Multiple use of symbols is common in technology. In such cases, you have to look at the context in which a symbol is used to determine its meaning.

As you can see, power is related to energy, which is the capacity to do work. Formally, **power** is defined as the rate of doing work or, equivalently, as the rate of transfer of energy. The symbol for power is P. By definition,

$$P = \frac{W}{t} \quad \text{[watts, W]} \tag{4–7}$$

where W is the work (or energy) in joules and t is the corresponding time interval of t seconds.

The SI unit of power is the watt. From Equation 4–7, we see that P also has units of joules per second. If you substitute $W = 1$ J and $t = 1$ s you get $P = 1$ J/1 s $= 1$ W. From this, you can see that *one watt equals one joule per second.* Occasionally, you also need power in horsepower. To convert, recall that 1 hp = 746 watts.

Power in Electrical and Electronic Systems

Since our interest is in electrical power, we need expressions for P in terms of electrical quantities. Recall from Chapter 2 that voltage is defined as work per unit charge and current as the rate of transfer of charge, i.e.,

$$V = \frac{W}{Q} \tag{4–8}$$

and

$$I = \frac{Q}{t} \tag{4–9}$$

From Equation 4–8, $W = QV$. Substituting this into Equation 4–7 yields $P = W/t = (QV)/t = V(Q/t)$. Replacing Q/t with I, we get

$$P = VI \quad [\text{watts, W}] \tag{4–10}$$

and, for a source,

$$P = EI \quad [\text{watts, W}] \tag{4–11}$$

Additional relationships are obtained by substituting $V = IR$ and $I = V/R$ into Equation 4–10:

$$P = I^2R \quad [\text{watts, W}] \tag{4–12}$$

and

$$P = \frac{V^2}{R} \quad [\text{watts, W}] \tag{4–13}$$

EXAMPLE 4–6 Compute the power supplied to the electric heater of Figure 4–16 using all three electrical power formulas.

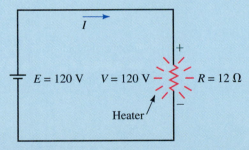

FIGURE 4–16 Power to the load (i.e., the heater) can be computed from any of the power formulas.

Solution $I = V/R = 120 \text{ V}/12 \ \Omega = 10 \text{ A}$. Thus, the power may be calculated as follows:

a. $P = VI = (120 \text{ V})(10 \text{ A}) = 1200 \text{ W}$

b. $P = I^2R = (10 \text{ A})^2(12 \ \Omega) = 1200 \text{ W}$

c. $P = V^2/R = (120 \text{ V})^2/12 \ \Omega = 1200 \text{ W}$

Note that all give the same answer, as they must.

EXAMPLE 4–7 Compute the power to each resistor in Figure 4–17 using Equation 4–13.

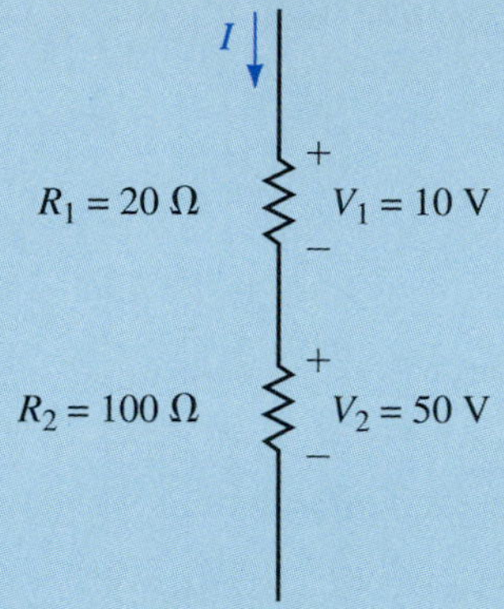

FIGURE 4–17

Solution You must use the appropriate voltage in the power equation. For resistor R_1, use V_1; for resistor R_2, use V_2.

a. $P_1 = V_1^2/R_1 = (10\ \text{V})^2/20\ \Omega = 5\ \text{W}$

b. $P_2 = V_2^2/R_2 = (50\ \text{V})^2/100\ \Omega = 25\ \text{W}$

EXAMPLE 4–8 If the dc motor of Figure 4–15(c) draws 6 A from a 120-V source,

a. Compute its power input in watts.

b. Assuming the motor is 100% efficient (i.e., that all electrical power supplied to it is output as mechanical power), compute its power output in horsepower.

Solution

a. $P_{in} = VI = (120\ \text{V})(6\ \text{A}) = 720\ \text{W}$

b. $P_{out} = P_{in} = 720\ \text{W}$. Converting to horsepower, $P_{out} = (720\ \text{W})/(746\ \text{W/hp}) = 0.965\ \text{hp}$.

PRACTICE PROBLEMS 3

a. Show that $I = \sqrt{\dfrac{P}{R}}$

b. Show that $V = \sqrt{PR}$

c. A 100-Ω resistor dissipates 169 W. What is its current?

d. A 3-Ω resistor dissipates 243 W. What is the voltage across it?

e. For Figure 4–17, $I = 0.5$ A. Use Equations 4–10 and 4–12 to compute power to each resistor. Compare your answers to the answers of Example 4–7.

Answers: c. 1.3 A   d. 27 V   e. $P_1 = 5$ W, $P_2 = 25$ W

Power Rating of Resistors

Resistors must be able to safely dissipate their heat without damage. For this reason, resistors are rated in watts. (For example, composition resistors of the type used in electronics are made with standard ratings of $\frac{1}{8}$, $\frac{1}{4}$, $\frac{1}{2}$, 1, and 2 W as you saw in Figure 3–8.) To provide a safety margin, it is customary to select a resistor that is capable of dissipating two or more times its computed power. By overrating a resistor, it will run a little cooler.

PRACTICAL NOTES...

A properly chosen resistor is able to dissipate its heat safely without becoming excessively hot. However, if through bad design or subsequent component failure its current becomes excessive, it will overheat and damage may result, as shown in Figure 4–18. One of the symptoms of overheating is that the resistor becomes noticeably hotter than other resistors in the circuit. (Be careful, however, as you might get burned if you try to check by touch.) Component failure may also be detected by smell. Burned components have a characteristic odor that you will soon come to recognize. If you detect any of these symptoms, turn the equipment off and look for the source of the problem. Note, however, an overheated component is often the symptom of a problem, rather than its cause.

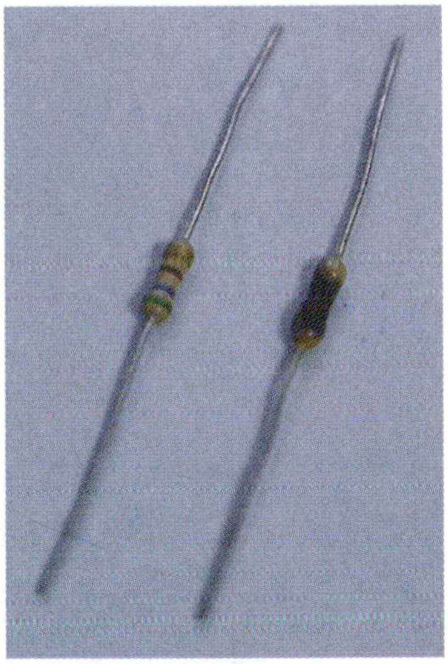

Measuring Power

Power can be measured using a device called a wattmeter. However, since wattmeters are used primarily for ac power measurement, we will hold off their consideration until Chapter 17. (You seldom need a wattmeter for dc circuits since you can determine power directly as the product of voltage times current and V and I are easy to measure.)

FIGURE 4–18   The resistor on the right has been damaged by overheating.

4.4   Power Direction Convention

For circuits with one source and one load, energy flows from the source to the load and the direction of power transfer is obvious. For circuits with multiple sources and loads, however, the direction of energy flow in some parts of the network may not be at all apparent. We therefore need to establish a clearly defined power transfer direction convention.

A resistive load (Figure 4–19) may be used to illustrate the idea. Since the direction of power flow can only be into a resistor, never out of it (since resistors do not produce energy), we define the positive direction of power transfer as from the source to the load as in (a) and indicate this by means of an arrow: $P\rightarrow$. We then adopt the convention that, *for the relative voltage polarities and current and power directions shown in Figure 4–19(a), when power transfer is in the direction of the arrow, it is positive, whereas when it is in the direction opposite to the arrow, it is negative.*

To help interpret the convention, consider Figure 4–19(b), which highlights the source end. From this we see that *power out of a source is positive*

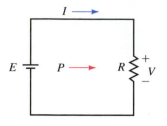

(a) Power transfer

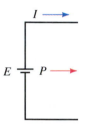

(b) Source end

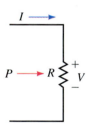

(c) Load end

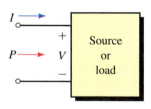

(d) Generalization

FIGURE 4–19   Reference convention for power.

when both the current and power arrows point out from the source, both I and P have positive values, and the source voltage has the polarity indicated.

Now consider Figure 4–19(c), which highlights the load end. Note the relative polarity of the load voltage and the direction of the current and power arrows. From this, we see that *power to a load is positive when both the current and power direction arrows point into the load, both have positive values, and the load voltage has the polarity indicated.*

In Figure 4–19(d), we have generalized the concept. The box may contain either a source or a load. If P has a positive value, power transfer is into the box; if P has a negative value, its direction is out.

EXAMPLE 4–9   Use the above convention to describe power transfer for the electric vehicle of Figure 4–20.

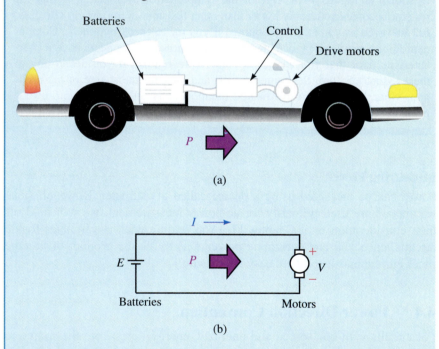

FIGURE 4–20

Solution   During normal operation, the batteries supply power to the motors, and current and power are both positive, Fig. 4–20(b). However, when the vehicle is going downhill, its motors are driven by the weight of the car and they act as generators. Since the motors now act as the source and the batteries as the load, the actual current is opposite in direction to the reference arrow shown and is thus negative (recall Figure 4–14). Thus, $P = VI$ is negative. The interpretation is, therefore, that power transfer is in the direction opposite to the power reference arrow. For example, if $V = 48$ volts and $I = -10$ A, then $P = VI = (48 \text{ V})(-10 \text{ A}) = -480$ W. This is consistent with what is happening, since minus 480 W *into* the motors is the same as plus 480 W *out*. This 480 W of power flows from the motors to the batteries, helping to charge them as the car goes downhill.

4.5   Energy

Earlier (Equation 4–7), we defined power as the rate of doing work. When you transpose this equation, you get the formula for **energy:**

$$W = Pt \qquad \text{(4–14)}$$

If t is measured in seconds, W has units of watt-seconds (i.e., joules, J), while if t is measured in hours, W has units of watthours (Wh). Note that in Equation 4–14, P must be constant over the time interval under consideration. If it is not, apply Equation 4–14 to each interval over which P is constant as described later in this section. (For the more general case, you need calculus.)

The most familiar example of energy usage is the energy that we use in our homes and pay for on our utility bills. This energy is the energy used by the lights and electrical appliances in our homes. For example, if you run a 100-W lamp for 1 hour, the energy consumed is $W = Pt = (100 \text{ W})(1\text{h}) = 100$ Wh, while if you run a 1500-W electric heater for 12 hours, the energy consumed is $W = (1500 \text{ W})(12 \text{ h}) = 18\,000$ Wh.

The last example illustrates that the watthour is too small a unit for practical purposes. For this reason, we use **kilowatthours** (kWh). By definition,

$$\text{energy}_{(kWh)} = \frac{\text{energy}_{(Wh)}}{1000} \qquad \text{(4–15)}$$

Thus, for the above example, $W = 18$ kWh. In most of North America, the kilowatthour (kWh) is the unit used on your utility bill.

For multiple loads, the total energy is the sum of the energy of individual loads.

EXAMPLE 4–10   Determine the total energy used by a 100-W lamp for 12 hours and a 1.5-kW heater for 45 minutes.

Solution   Convert all quantities to the same set of units, e.g., convert 1.5 kW to 1500 W and 45 minutes to 0.75 h. Then,

$$W = (100 \text{ W})(12 \text{ h}) + (1500 \text{ W})(0.75 \text{ h}) = 2325 \text{ Wh} = 2.325 \text{ kWh}$$

Alternatively, convert all power to kilowatts first. Thus,

$$W = (0.1 \text{ kW})(12 \text{ h}) + (1.5 \text{ kW})(0.75 \text{ h}) = 2.325 \text{ kWh}$$

EXAMPLE 4–11   Suppose you use the following electrical appliances: a 1.5-kW heater for $7\frac{1}{2}$ hours; a 3.6-kW broiler for 17 minutes; three 100-W lamps for 4 hours; a 900-W toaster for 6 minutes. At \$0.09 per kilowatthour, how much will this cost you?

Solution   Convert time in minutes to hours. Thus,

$$W = (1500)(7\tfrac{1}{2}) + (3600)\left(\frac{17}{60}\right) + (3)(100)(4) + (900)\left(\frac{6}{60}\right)$$

$$= 13\,560 \text{ Wh} = 13.56 \text{ kWh}$$

$$\text{cost} = (13.56 \text{ kWh})(\$0.09/\text{kWh}) = \$1.22$$

In those areas of the world where the SI system dominates, the **mega-joule** (MJ) is sometimes used instead of the kWh (as the kWh is not an SI unit). The relationship is 1 kWh = 3.6 MJ.

Watthour Meter

In practice, energy is measured by watthour meters, many of which are electromechanical devices that incorporate a small electric motor whose speed is proportional to power to the load. This motor drives a set of dials through a gear train (Figure 4–21). Since the angle through which the dials rotate depends on the speed of rotation (i.e., power consumed) and the length of time that this power flows, the dial position indicates energy used. Note however, that electromechanical devices are starting to give way to electronic meters, which perform this function electronically and display the result on digital readouts.

FIGURE 4–21   Watthour meter. This type of meter uses a gear train to drive the dials. Newer meters are electronic with digital readouts.

Law of Conservation of Energy

Before leaving this section, we consider the law of conservation of energy. It states that energy can neither be created nor destroyed, but is instead converted from one form to another. You saw examples of this above—for example, the conversion of electrical energy into heat energy by a resistor, and the conversion of electrical energy into mechanical energy by a motor. In fact, several types of energy may be produced simultaneously. For example, electrical energy is converted to mechanical energy by a motor, but some heat is also produced. This results in a lowering of efficiency, a topic we consider next.

4.6   Efficiency

Poor efficiency results in wasted energy and higher costs. For example, an inefficient motor costs more to run than an efficient one for the same output. An inefficient piece of electronic gear generates more heat than an efficient one, and this heat must be removed, resulting in increased costs for fans, heat sinks, and the like.

Efficiency can be expressed in terms of either energy or power. Power is generally easier to measure, so we usually use power. The efficiency of a device or system (Figure 4–22) is defined as the ratio of power output P_{out} to power input P_{in}, and it is usually expressed in percent and denoted by the Greek letter η (eta). Thus,

$$\eta = \frac{P_{out}}{P_{in}} \times 100\% \qquad (4\text{--}16)$$

In terms of energy,

$$\eta = \frac{W_{out}}{W_{in}} \times 100\% \qquad (4\text{--}17)$$

Since $P_{in} = P_{out} + P_{losses}$, efficiency can also be expressed as

$$\eta = \frac{P_{out}}{P_{out} + P_{losses}} \times 100\% = \frac{1}{1 + \dfrac{P_{losses}}{P_{out}}} \times 100\% \qquad (4\text{--}18)$$

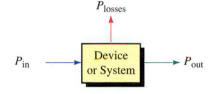

FIGURE 4–22   Input power equals output power plus losses.

The efficiency of equipment and machines varies greatly. Large power transformers, for example, have efficiencies of 98% or better, while many electronic amplifiers have efficiencies lower than 50%. Note that efficiency will always be less than 100%.

EXAMPLE 4–12   A 120-V dc motor draws 12 A and develops an output power of 1.6 hp.

a. What is its efficiency?

b. How much power is wasted?

Solution

a. $P_{in} = EI = (120 \text{ V})(12 \text{ A}) = 1440$ W, and $P_{out} = 1.6$ hp $\times$ 746 W/hp = 1194 W. Thus,

$$\eta = \frac{P_{out}}{P_{in}} = \frac{1194 \text{ W}}{1440 \text{ W}} \times 100 = 82.9\%$$

b. $P_{losses} = P_{in} - P_{out} = 1440 - 1194 = 246$ W

EXAMPLE 4–13   The efficiency of a power amplifier is the ratio of the power delivered to the load (e.g., speakers) to the power drawn from the power supply. Generally, this efficiency is not very high. For example, suppose a power amplifier delivers 400 W to its speaker system. If the power loss is 509 W, what is its efficiency?

Solution

$$P_{in} = P_{out} + P_{losses} = 400 \text{ W} + 509 \text{ W} = 909 \text{ W}$$

$$\eta = \frac{P_{out}}{P_{in}} \times 100\% = \frac{400 \text{ W}}{909 \text{ W}} \times 100\% = 44\%$$

For systems with subsystems or components in cascade (Figure 4–23), overall efficiency is the product of the efficiencies of each individual part, where efficiencies are expressed in decimal form. Thus,

$$\eta_T = \eta_1 \times \eta_2 \times \eta_3 \times \cdots \times \eta_n \qquad (4\text{–}19)$$

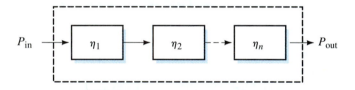

(a) Cascaded system

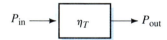

(b) Equivalent of (a)

FIGURE 4–23   For systems in cascade, the resultant efficiency is the product of the efficiencies of the individual stages.

EXAMPLE 4–14

a. For a certain system, $\eta_1 = 95\%$, $\eta_2 = 85\%$, and $\eta_3 = 75\%$. What is η_T?
b. If $\eta_T = 65\%$, $\eta_2 = 80\%$, and $\eta_3 = 90\%$, what is η_1?

Solution

a. Convert all efficiencies to a decimal value, then multiply. Thus, $\eta_T = \eta_1\eta_2\eta_3 = (0.95)(0.85)(0.75) = 0.61$ or 61%.
b. $\eta_1 = \eta_T/(\eta_2\eta_3) = (0.65)/(0.80 \times 0.90) = 0.903$ or 90.3%

EXAMPLE 4–15 A motor drives a pump through a gearbox (Figure 4–24). Power input to the motor is 1200 W. How many horsepower are delivered to the pump?

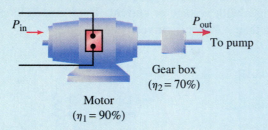

(a) Physical system

(b) Block diagram

FIGURE 4–24 Motor driving pump through a gear box.

Solution The efficiency of the motor-gearbox combination is $\eta_T = (0.90)(0.70) = 0.63$. The output of the gearbox (and hence the input to the pump) is $P_{out} = \eta_T \times P_{in} = (0.63)(1200 \text{ W}) = 756 \text{ W}$. Converting to horsepower, $P_{out} = (756 \text{ W})/(746 \text{ W/hp}) = 1.01 \text{ hp}$.

EXAMPLE 4–16 The motor of Figure 4–24 is operated from 9:00 a.m. to 12:00 noon and from 1:00 p.m. to 5:00 p.m. each day, for 5 days a week, outputting 7 hp to a load. At $0.085/kWh, it costs $22.19 per week for electricity. What is the efficiency of the motor/gearbox combination?

Solution

$$W_{in} = \frac{\$22.19/\text{wk}}{\$0.085/\text{kWh}} = 261.1 \text{ kWh/wk}$$

The motor operates 35 h/wk. Thus,

$$P_{in} = \frac{W_{in}}{t} = \frac{261.1 \text{ kWh/wk}}{35 \text{ h/wk}} = 7460 \text{ W}$$

$$\eta_T = \frac{P_{out}}{P_{in}} = \frac{(7 \text{ hp} \times 746 \text{ W/hp})}{7460 \text{ W}} = 0.7$$

Thus, the motor/gearbox efficiency is 70%.

4.7   Nonlinear and Dynamic Resistances

All resistors considered so far have constant values that do not change with voltage or current. Such resistors are termed **linear** or **ohmic** since their current-voltage (*I-V*) plot is a straight line. However, the resistance of some materials changes with voltage or current. These materials are termed **nonlinear** because their *I-V* plot is curved (Figure 4–25).

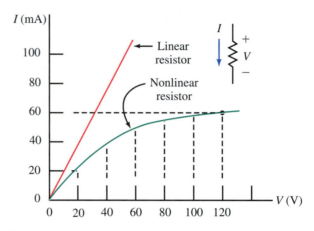

FIGURE 4–25   Linear and nonlinear resistance characteristics.

Since the resistance of all materials changes with temperature, all resistors are to some extent nonlinear, since they all produce heat and this heat changes their resistance. For most resistors, however, this effect is small over their normal operating range, and such resistors are considered to be linear. (The commercial resistors shown in Figure 3–8 and most others that you will encounter in this book are linear.)

Since an *I-V* plot is a graph of Ohm's law, resistance can be computed from the ratio *V/I*. First, consider the linear plot of Figure 4–25. Because the slope is constant, the resistance is constant and you can compute R at any point. For example, at $V = 10$ V, $I = 20$ mA, and $R = 10$ V/20 mA $= 500\ \Omega$. Similarly, at $V = 20$ V, $I = 40$ mA, and $R = 20$ V/40 mA $= 500\ \Omega$, which is the same as before. This is true at all points on this linear curve. The resulting resistance is referred to as dc resistance, R_{dc}. Thus, $R_{dc} = 500\ \Omega$.

An alternate way to compute resistance is illustrated in Figure 4–26. At point 1, $V_1 = I_1 R$. At point 2, $V_2 = I_2 R$. Subtracting voltages and solving for R yields

$$R = \frac{V_2 - V_1}{I_2 - I_1} = \frac{\Delta V}{\Delta I} \quad [\text{ohms, } \Omega] \tag{4–20}$$

where $\Delta V/\Delta I$ is the inverse of the slope of the line. (Here, Δ is the Greek letter delta. It is used to represent a change or increment in value.) To illustrate, if you select ΔV to be 20 V, you find that the corresponding ΔI from Figure 4–26 is 40 mA. Thus, $R = \Delta V/\Delta I = 20$ V/40 mA $= 500\ \Omega$ as before. Resistance calculated as in Figure 4–26 is called **ac** or **dynamic resistance.** For linear resistors, $R_{ac} = R_{dc}$.

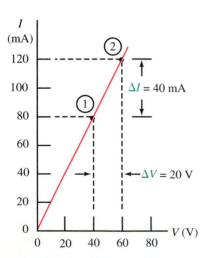

FIGURE 4–26 $R = \Delta V/\Delta I = 20$ V/40 mA $= 500\ \Omega$.

Now consider the nonlinear resistance plot of Figure 4–25. At $V = 20$ V, $I = 20$ mA. Therefore, $R_{dc} = 20$ V/20 mA $= 1.0$ kΩ; at $V = 120$ V, $I = 60$ mA, and $R_{dc} = 120$ V/60 mA $= 2.0$ kΩ. This resistance therefore increases with applied voltage. However, for small variations about a fixed point on the curve, the ac resistance will be constant. This is an important concept in electronics. However, since dynamic resistance is beyond the scope of this book, we must leave it for later courses to explore.

4.8   Computer-Aided Circuit Analysis

We end our introduction to Ohm's law by solving several simple problems using Electronics Workbench and OrCAD PSPice. As noted in Chapter 1, these are application packages that work from a circuit schematic that you build on your screen. (Since the details are different, we will consider the two products separately, using the circuit of Figure 4–27 to get started.) Because this is our first look at circuit simulation, considerable detail is included. (Although the procedures may seem complex, they become quite intuitive with a little practice.) Both packages run under Windows.

ELECTRONICS
WORKBENCH

PSpice

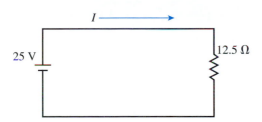

FIGURE 4–27   Simple circuit to illustrate computer analysis.

Electronics Workbench

Figure 4–28 shows an Electronics Workbench (EWB) user interface screen. (This is Version 5, the version current at the time of writing.) Along the top are toolbars with icons and menu items that you can select with your mouse. (Drop-down boxes open to indicate the purpose of your selection.) For example, if you position the mouse pointer over the Basic icon and click the left button, the Basic Parts bin shown in Figure 4–28 opens.

First read the EWB Operational Notes, then build the circuit on the screen as follows:

• Click the New icon to allow creation of a new circuit.

• Click the Sources icon, drag a battery from the parts bin, position it on the screen then release the mouse button.

• Click the Basic icon, drag a resistor from the parts bin, rotate it 90° by clicking the Rotate icon, then position it. (If the resistor won't rotate, it isn't selected—see EWB Operational Notes.)

• To "wire" the circuit, move the mouse pointer to the top battery terminal and when a dot appears, as in Figure 4–29(a), click and drag the wire to the top of the resistor as in (b). Release the mouse button and the wire routes itself neatly as in (c). Similarly, add the bottom wire.

NOTES...

EWB Operational Notes

1. Unless directed otherwise, use the left mouse button for all operations.

2. To drag a component, place the pointer over it, press and hold the left mouse button, drag the mouse across its pad to position the component, then release the button.

3. To select a component on the screen, place the pointer on it and click the left button. The component turns red.

4. To deselect a component, move the cursor to an empty point on the screen and click the left button.

5. To delete a component, select it, then press the Delete key.

6. Some components have default values that you may need to change.

7. The order of wiring a circuit sometimes has an effect. For example, for Figure 4–28, if you connect the ground to the source before you add the wire from the bottom end of the resistor, the connector dot may not be needed.

8. All EWB circuits need a ground.

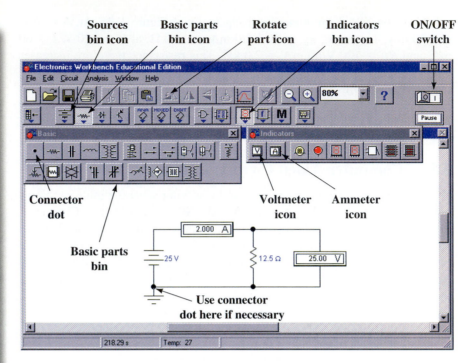

FIGURE 4–28   Electronics Workbench simulation of the circuit of Figure 4–27.

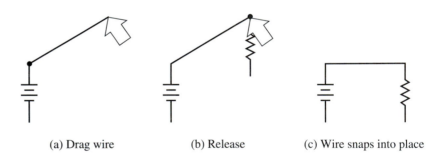

(a) Drag wire   (b) Release   (c) Wire snaps into place

FIGURE 4–29   Routing wires using Electronics Workbench.

- Drag a connector dot from the basic parts bin, position it as shown in Figure 4–28, then add a ground symbol from the sources bin. (Note: Make sure the ground connects properly. See also Note 7.)

- Open the Indicators parts bin and drag an ammeter into position. (EWB automatically rewires the circuit to accommodate the ammeter.)

- Drag a voltmeter from the Indicators parts bin, then wire it into place.

- To change the battery voltage, double click the battery symbol and when the dialog box opens, click the Value tab (if not already selected), type 25, select V, then click OK.

- Similarly, double click the resistor symbol, click the Value tab, type 12.5, select Ω, then click OK.

- Activate the circuit by clicking the ON/OFF power switch at the top right corner of the EWB window.

The voltmeter should show 25 V and the ammeter 2 A as indicated in Figure 4–28.

Repeat the above example, except reverse the source voltage symbol so that the circuit is driven by a −25 V source. Note the voltmeter and ammeter readings and compare to the above solution. Reconcile these results with the voltage polarity and current direction conventions discussed in this chapter and in Chapter 2.

PRACTICE
PROBLEMS 4

OrCAD PSpice

You must specify the type of analysis that you wish to perform. Your choices are DC Sweep (for dc analysis which is what we are doing here), AC Sweep (for AC analysis), Time Domain (for transient analysis), or Bias Point (which we do not consider in this book). When setting up your analysis, there are generally several ways you can proceed, but the following is about the simplest. Lastly, it is assumed that you have loaded PSpice from the demo disk. Now, read the PSpice Operational Notes, then:

- On your Windows screen, click Start, select Programs, OrCAD Demo, then click Capture CIS Demo. This opens the Capture session frame (Appendix A, Figure A–1).

- Click menu item File, select New, and then click Project. The New Project box (Figure A–2) opens. In the Name box, type **Ch 4 PSpice 1,** then click the Analog or Mixed-Signal Circuit Wizard button. Click OK.

- You are prompted to add libraries. Click breakout.olb and click Add, click eval.olb and click Add, and then click Finish. (Various components exist in these libraries. For example, sources are in the SOURCE library, resistors are in the ANALOG library, and so on. You will learn more about where components are located as you go through the PSpice examples.)

- You should be on the Capture schematic editor page. Click anywhere to activate it. You are now ready to build the circuit. Let us start with the voltage source. Click the Place part tool (Figure 4–30) to open the Place Part dialog box. From the Libraries list, select SOURCE, then type **VDC** in the Part box. Click OK and a dc voltage source symbol appears. Position it on the screen as shown in Figure 4–30, click the left button to place it, then press the Esc key to end placement (or click the right button and End Mode as described in Appendix A).

- Click the Place part tool and select ANALOG from the Libraries list, then in the Part box, type **R** and click OK. A resistor appears. Rotate it three times as described in the PSpice Operational Notes box (for reasons described in Appendix A). Place as shown in Figure 4–30 using the left button, and then press Esc.

- To measure current, you need an ammeter. For this, use component IPRINT. Click the Place part tool and select SPECIAL from the Libraries list. In the Part box, type **IPRINT** and then click OK. Place the component and then press Esc.

- Click the Place ground tool, select 0/SOURCE and click OK. Place the ground as in Figure 4–30, and then press Esc.

NOTES...

PSpice Operational Notes

1. Unless directed otherwise, use the left mouse button for all operations.

2. To select a particular component on your schematic, place the pointer on it and click. The selected component changes color.

3. To deselect a component, move the pointer to an empty place on the screen and click.

4. To delete a component from your schematic, select it, then press the Delete key.

5. To drag a component, place the pointer over it, press and hold the left button, drag the component to where you want it, and then release.

6. To change a default value, place the cursor over the numerical value (not the component symbol) and double click. Change the appropriate value.

7. To rotate a component, select it, click the right button, then click Rotate. Alternately, hold the Ctrl key and press the R key. (This is denoted as Ctrl/R).

8. There must be no space between a value and its unit. Thus, use 25V, not 25 V, etc.

9. All PSpice circuits must have a ground.

10. As you build a circuit, you should frequently click the Save Document icon to save your work in case something goes wrong.

Edit Simulation Settings   Voltage marker   Current marker

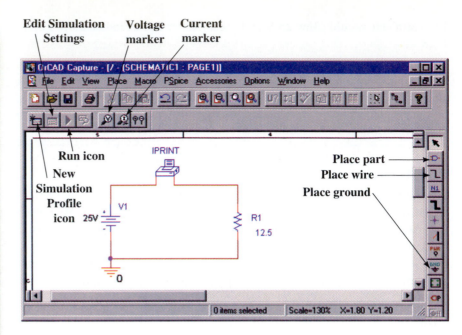

Run icon

New Simulation Profile icon

Place part
Place wire
Place ground

FIGURE 4–30   OrCAD PSpice simulation of Figure 4–27. Note that PSpice uses *V* for sources instead of *E,* although you can change this if you wish.

- To wire the circuit, click the Place wire tool, position the cursor in the little box at the end of the source lead, and click. Then move the cursor to IPRINT and click to place. Wire the rest of the circuit in this manner. Press Esc to end wire placement.

- All components have default values that need to be changed. First, consider the resistor. Its default value is 1kΩ. To change it, double click the 1k value (not the resistor symbol), type **12.5** into the V<u>a</u>lue box, and then click OK. Similarly, double click the IPRINT symbol, click the Parts tab (bottom of screen) in the Properties Editor, then scroll right until you see a cell labeled DC. Type **yes** into the cell and click Apply. Close the editor by clicking the × box in the upper right-hand corner of the Properties Editor window. This returns you to Figure 4–30.

- Click the New Simulation Profile icon (Figure 4–30). Enter a name (e.g., **Figure 4-30**) in the <u>N</u>ame box. Click Create. A Simulations Setting box opens. Click the Analysis tab. From the <u>A</u>nalysis type list, select DC Sweep. From the <u>O</u>ptions list, select Primary Sweep. Under Sweep variable, choose <u>V</u>oltage source. In the <u>N</u>ame box, type the name of your source. (It should be **V1**. Check your schematic to verify.) Click the Value li<u>s</u>t button, type **25V** into the box, and then click OK. (This sets the voltage of your source to 25 V but it does not update the screen display. If you want to update the value on your screen, double click the default value (i.e., 0V, not the battery symbol), type **25V** in the V<u>a</u>lue box, and click OK.) Click the Save Document icon to save your work.

- Click the Run icon, Figure 4–30. When the results screen appears, click View, Output File, then scroll to the bottom of the file where you will find the answers

$$V\_V1 \quad I(V\_PRINT1)$$

$$2.500E+01 \quad 2.000E+00$$

The symbol V_V1 represents the voltage of source V_1 and 2.500E+01 represents its value (in exponent form). Thus, $V_1 = 2.500E+01 = 2.5 \times 10^1 = 25$ (which is the value you entered earlier). Similarly, I(V_PRINT1) represents the current measured by component IPRINT. Thus the current in the circuit is $I = 2.000E+00 = 2.0 \times 10^0 = 2.0$ (which, as you can see, is correct).

- To exit from PSpice, click the $\times$ in the box in the upper right-hand corner of the screen and click yes to save changes.

Plotting Ohm's Law

PSpice can be used to compute and graph results. By varying the source voltage of Figure 4–27 and plotting I for example, you can obtain a plot of Ohm's law. Proceed as follows.

- Follow the steps detailed earlier to call up the Schematic page editor. In the New Project box, type **Ch 4 PSpice 2** into the Name box. Build the circuit of Figure 4–30 again (right up to the point where you have wired the circuit), except omit part IPRINT.

- Click the New Simulation Profile icon and enter Figure 4–32 in the Name box. Click Create. In the Simulations Settings box, click the Analysis tab and from the Analysis type list, select DC Sweep. From the Options list, select Primary Sweep. Under the Sweep Variable, choose Voltage source. In the Name box, type the name of your source. (It should be **V1**.) Under Sweep type, choose Linear. Type **0V** in the Start Value box, type **100V** in the End Value box, and type **5V** in the Increment box, and then click OK. (This will sweep the voltage from 0 to 100 V in 5-V steps.)

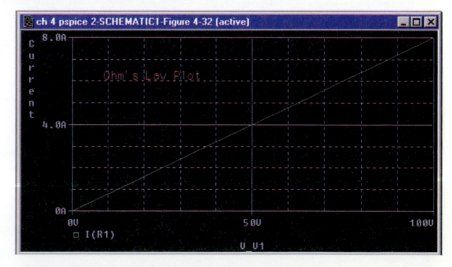

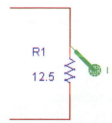

FIGURE 4–31 Using a current marker to display current.

FIGURE 4–32 Ohm's law plot for the circuit of Figure 4–27.

• Click the Current Marker icon and position the marker as shown in Figure 4–31. Click the Save Document icon to save your work. Click the Run icon. The Ohm's law plot of Figure 4–32 appears on your screen. Using the cursor (see Appendix A), read values from the graph and verify with your calculator.

PUTTING IT INTO PRACTICE

You have been assigned to perform a design review and cost analysis of an existing product. The product uses a 12-V ±5% power supply. After a catalog search, you find a new power supply with a specification of 12 V ±2%, which, because of its newer technology, costs only five dollars more than the one you are currently using. Although it provides improved performance for the product, your supervisor won't approve it because of the extra cost. However, the supervisor agrees that if you can bring the cost differential down to $3.00 or less, you can use the new supply. You then look at the schematic and discover that a number of precision 1% resistors are used because of the wide tolerance of the old supply. (As a specific example, a ±1% tolerance 220-kΩ resistor is used instead of a standard 5% tolerance resistor because the loose tolerance of the original power supply results in current through the resistor that is out of spec. Its current must be between 50 mA to 60 mA when the 12-V supply is connected across it. Several other such instances are present in the product, a total of 15 in all.) You begin to wonder whether you could replace the precision resistors (which cost $0.24 each) with standard 5% resistors (which cost $0.03 each) and save enough money to satisfy your supervisor. Perform an analysis to determine if your hunch is correct.

PROBLEMS

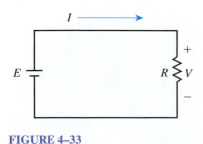

FIGURE 4–33

4.1 Ohm's Law

1. For the circuit of Figure 4–33, determine the current I for each of the following. Express your answer in the most appropriate unit: amps, milliamps, microamps, etc.
   a. $E = 40$ V, $R = 20$ Ω
   b. $E = 35$ mV, $R = 5$ mΩ
   c. $V = 200$ V, $R = 40$ kΩ
   d. $E = 10$ V, $R = 2.5$ MΩ
   e. $E = 7.5$ V, $R = 2.5 \times 10^3$ Ω
   f. $V = 12$ kV, $R = 2$ MΩ

2. Determine R for each of the following. Express your answer in the most appropriate unit: ohms, kilohms, megohms, etc.
   a. $E = 50$ V, $I = 2.5$ A
   b. $E = 37.5$ V, $I = 1$ mA
   c. $E = 2$ kV, $I = 0.1$ kA
   d. $E = 4$ kV, $I = 8 \times 10^{-4}$ A

3. For the circuit of Figure 4–33, compute V for each of the following:
   a. 1 mA, 40 kΩ
   b. 10 μA, 30 kΩ
   c. 10 mA, 4×10^4 Ω
   d. 12 A, 3×10^{-2} Ω

4. A 48-Ω hot water heater is connected to a 120-V source. What is the current drawn?

5. When plugged into a 120-V wall outlet, an electric lamp draws 1.25 A. What is its resistance?

6. What is the potential difference between the two ends of a 20-kΩ resistor when its current is 3×10^{-3} A?

7. How much voltage can be applied to a 560-Ω resistor if its current must not exceed 50 mA?

8. A relay with a coil resistance of 240 Ω requires a minimum of 50 mA to operate. What is the minimum voltage that will cause it to operate?

9. For Figure 4–33, if $E = 30$ V and the conductance of the resistor is 0.2 S, what is I?

10. If $I = 36$ mA when $E = 12$ V, what is I if the 12-V source is

    a. replaced by an 18-V source?

    b. replaced by a 4-V source?

11. Current through a resistor is 15 mA. If the voltage drop across the resistor is 33 V, what is its color code?

12. For the circuit of Figure 4–34,

    a. If $E = 28$ V, what does the meter indicate?

    b. If $E = 312$ V, what does the meter indicate?

13. For the circuit of Figure 4–34, if the resistor is replaced with one with red, red, black color bands, at what voltage do you expect the fuse to blow?

14. A 20-V source is applied to a resistor with color bands brown, black, red, and silver

    a. Compute the nominal current in the circuit.

    b. Compute the minimum and maximum currents based on the tolerance of the resistor.

15. An electromagnet is wound with AWG 30 copper wire. The coil has 800 turns and the average length of each turn is 3 inches. When connected to a 48-V dc source, what is the current

    a. at 20°C?   b. at 40°C?

16. You are to build an electromagnet with 0.643-mm-diameter copper wire. To create the required magnetic field, the current in the coil must be 1.75 A at 20°C. The electromagnet is powered from a 9.6-V dc source. How many meters of wire do you need to wind the coil?

17. A resistive circuit element is made from 100 m of 0.5-mm-diameter aluminum wire. If the current at 20°C is 200 mA, what is the applied voltage?

18. Prepare an Ohm's law graph similar to Figure 4–7 for a 2.5-kΩ and a 5-kΩ resistor. Compute and plot points every 5 V from $E = 0$ V to 25 V. Reading values from the graph, find current at $E = 14$ V.

19. Figure 4–35 represents the I-V graph for the circuit of Figure 4–33. What is R?

20. For a resistive circuit, E is quadrupled and R is halved. If the new current is 24 A, what was the original current?

21. For a resistive circuit, $E = 100$ V. If R is doubled and E is changed so that the new current is double the original current, what is the new value of E?

22. You need to measure the resistance of an electric heater element, but have only a 12-V battery and an ammeter. Describe how you would determine its resistance. Include a sketch.

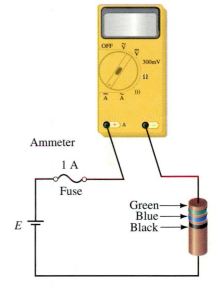

FIGURE 4–34

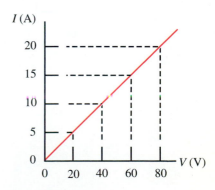

FIGURE 4–35

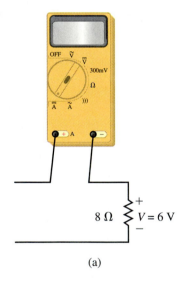

(a) $I = 3$ A

(b) $V = 60$ V

(c) $I = 6$ A

(d) $V = 105$ V

FIGURE 4–36   All resistors are 15 Ω.

23. If 25 m of 0.1-mm-diameter Nichrome wire is connected to a 12-V battery, what is the current at 20°C?

24. If the current is 0.5 A when a length of AWG 40 copper wire is connected to 48 V, what is the length of the wire in meters? Assume 20°C.

4.2   Voltage Polarity and Current Direction

25. For each resistor of Figure 4–36, determine voltage V and its polarity or current I and its direction as applicable.

26. The ammeters of Figure 4–37 have autopolarity. Determine their readings, magnitude, and polarity.

4.3   Power

27. A resistor dissipates 723 joules of energy in 3 minutes and 47 seconds. Calculate the rate at which energy is being transferred to this resistor in joules per second. What is its power dissipation in watts?

28. How long does it take for a 100-W soldering iron to dissipate 1470 J?

29. A resistor draws 3 A from a 12-V battery. How much power does the battery deliver to the resistor?

30. A 120-V electric coffee maker is rated 960 W. Determine its resistance and rated current.

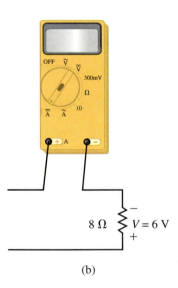

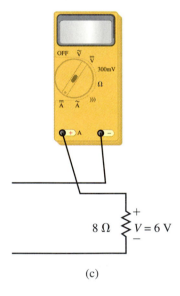

(a)   (b)   (c)

FIGURE 4–37

31. A 1.2-kW electric heater has a resistance of 6 Ω. How much current does it draw?

32. A warning light draws 125 mA when dissipating 15 W. What is its resistance?

33. How many volts must be applied to a 3-Ω resistor to result in a power dissipation of 752 W?

34. What IR drop occurs when 90 W is dissipated by a 10-Ω resistor?

35. A resistor with color bands brown, black, and orange dissipates 0.25 W. Compute its voltage and current.

36. A 2.2-kΩ resistor with a tolerance of $\pm 5\%$ is connected to a 12-V dc source. What is the possible range of power dissipated by the resistor?

37. A portable radio transmitter has an input power of 0.455 kW. How much current does it draw from a 12-V battery?

38. For a resistive circuit, $E = 12$ V.

    a. If the load dissipates 8 W, what is the current in the circuit?

    b. If the load dissipates 36 W, what is the load resistance?

39. A motor delivers 3.56 hp to a load. How many watts is this?

40. The load on a 120-V circuit consists of six 100-W lamps, a 1.2-kW electric heater, and an electric motor drawing 1500 W. If the circuit is fused at 30 A, what happens when a 900-W toaster is plugged in? Justify your answer.

41. A 0.27-kΩ resistor is rated 2 W. Compute the maximum voltage that can be applied and the maximum current that it can carry without exceeding its rating.

42. Determine which, if any, of the following resistors may have been damaged by overheating. Justify your answer.

    a. 560 Ω, ½ W, with 75 V across it.

    b. 3 Ω, 20 W, with 4 A through it.

    c. ¼ W, with 0.25 mA through it and 40 V across it.

43. A 25-Ω resistor is connected to a power supply whose voltage is 100 V $\pm 5\%$. What is the possible range of power dissipated by the resistor?

44. A load resistance made from copper wire is connected to a 24-V dc source. The power dissipated by the load when the wire temperature is 20°C is 192 W. What will be the power dissipated when the temperature of the wire drops to -10°C? (Assume the voltage remains constant.)

4.4 Power Direction Convention

45. Each block of Figure 4–38 may be a source or a load. For each, determine power and its direction.

46. The 12-V battery of Figure 4–39 is being "charged" by a battery charger. The current is 4.5 A as indicated.

    a. What is the direction of current?

    b. What is the direction of power flow?

    c. What is the power to the battery?

4.5 Energy

47. A 40-W night safety light burns for 9 hours.

    a. Determine the energy used in joules.

    b. Determine the energy used in watthours.

    c. At $0.08/kWh, how much does it cost to run this light for 9 hours?

48. An indicator light on a control panel operates continuously, drawing 20 mA from a 120-V supply. At $0.09 per kilowatthour, how much does it cost per year to operate the light?

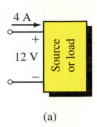

(a)

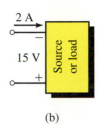

(b)

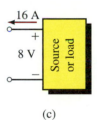

(c)

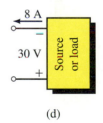

(d)

FIGURE 4–38

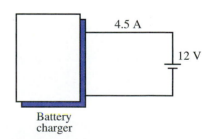

FIGURE 4–39

49. Determine the total cost of using the following at $0.11 per kWh:

    a. a 900-W toaster for 5 minutes,

    b. a 120-V, 8-A heater for 1.7 hours,

    c. an 1100-W dishwasher for 36 minutes,

    d. a 120-V, 288-Ω soldering iron for 24 minutes.

50. An electric device with a cycle time of 1 hour operates at full power (400 W) for 15 minutes, at half power for 30 minutes, then cuts off for the remainder of the hour. The cycle repeats continuously. At $0.10/kWh, determine the yearly cost of operating this device.

51. While the device of Problem 50 is operating, two other loads as follows are also operating:

    a. a 4-kW heater, continuously,

    b. a 3.6-kW heater, on for 12 hours per day.

    Calculate the yearly cost of running all loads.

52. At $0.08 per kilowatthour, it costs $1.20 to run a heater for 50 hours from a 120-V source. How much current does the heater draw?

53. If there are 24 slices in a loaf of bread and you have a two-slice, 1100-W toaster that takes 1 minute and 45 seconds to toast a pair of slices, at $0.13/kWh, how much does it cost to toast a loaf?

4.6   Efficiency

54. The power input to a motor with an efficiency of 85% is 690 W. What is its power output?

    a. in watts   b. in hp

55. The power output of a transformer with $\eta = 97\%$ is 50 kW. What is its power in?

56. For a certain device, $\eta = 94\%$. If losses are 18 W, what are P_{in} and P_{out}?

57. The power input to a device is 1100 W. If the power lost due to various inefficiencies is 190 W, what is the efficiency of the device?

58. A 240-V, 4.5-A water heater produces heat energy at the rate of 3.6 MJ per hour. Compute

    a. the efficiency of the water heater,

    b. the annual cost of operation at $0.09/kWh if the heater is on for 6 h/day.

59. A 120-V dc motor with an efficiency of 89% draws 15 A from the source. What is its output horsepower?

60. A 120-V dc motor develops an output of 3.8 hp. If its efficiency is 87%, how much current does it draw?

61. The power/control system of an electric car consists of a 48-V onboard battery pack, electronic control/drive unit, and motor (Figure 4–40). If 180 A are drawn from the batteries, how many horsepower are delivered to the drive wheels?

62. Show that the efficiency of n devices or systems in cascade is the product of their individual efficiencies, i.e., that $\eta_T = \eta_1 \times \eta_2 \times \cdots \times \eta_n$.

63. A 120-V dc motor drives a pump through a gearbox (Figure 4–24). If the power input to the pump is 1100 W, the gearbox has an efficiency of 75%,

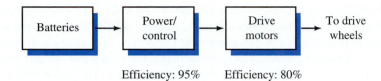

Efficiency: 95%   Efficiency: 80%

FIGURE 4–40

and the power input to the motor is 1600 W, determine the horsepower output of the motor.

64. If the motor of Problem 63 is protected by a 15-A circuit breaker, will it open? Compute the current to find out.

65. If the overall efficiency of a radio transmitting station is 55% and it transmits at 35 kW for 24 h/day, compute the cost of energy used per day at $0.09/kWh.

66. In a factory, two machines, each delivering 27 kW are used on an average for 8.7 h/day, 320 days/year. If the efficiency of the newer machine is 87% and that of the older machine is 72%, compute the difference in cost per year of operating them at $0.10 per kilowatthour.

4.7 Nonlinear and Dynamic Resistances

67. A voltage-dependent resistor has the *I-V* characteristic of Figure 4–41.

   a. At $V = 25$ V, what is *I?* What is R_{dc}?

   b. At $V = 60$ V, what is *I?* What is R_{dc}?

   c. Why are the two values different?

68. For the resistor of Figure 4–41:

   a. Determine $R_{dynamic}$ for *V* between 0 and 40 V.

   b. Determine $R_{dynamic}$ for *V* greater than 40 V.

   c. If *V* changes from 20 V to 30 V, how much does *I* change?

   d. If *V* changes from 50 V to 70 V, how much does *I* change?

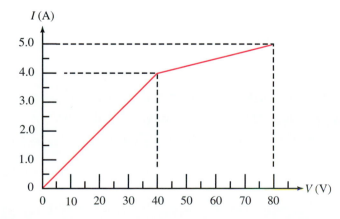

FIGURE 4–41

4.8 Computer-Aided Circuit Analysis

69. **EWB** Set up the circuit of Figure 4–33 and solve for currents for the voltage/resistance pairs of Problem 1a, 1c, 1d, and 1e.

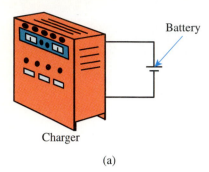

Battery

Charger

(a)

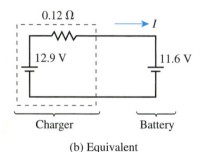

0.12 Ω

→ I

12.9 V

11.6 V

Charger   Battery

(b) Equivalent

FIGURE 4–42

70. **EWB**   A battery charger with a voltage of 12.9 V is used to charge a battery, Figure 4–42(a). The internal resistance of the charger is 0.12 Ω and the voltage of the partially run-down battery is 11.6 V. The equivalent circuit for the charger/battery combination is shown in (b). You reason that, since the two voltages are in opposition, net voltage for the circuit will be 12.9 V − 11.6 V = 1.3 V and, thus, the charging current I will be 1.3 V/0.12 Ω = 10.8 A. Set up the circuit of (b) and use Electronics Workbench to verify your conclusion.

71. **EWB**   Open the Basic parts bin and note the two switches, one a basic switch and the other a time delayed (TD) switch. Drag and place a basic switch on the screen, double click its symbol and when the dialog box opens, select the Value tab, type the letter A then click OK. (This relabels the switch as (A). Press the A key on the keyboard several times and note that the switch opens and closes.) Select a second switch and label it (B), then add a 12-V dc source, so as to set up the two-way light control circuit of Figure 2–28 (page 51). Operate the switches and determine whether you have successfully achieved the desired control.

72. **PSpice**   Repeat Problem 69 using PSpice.

73. **PSpice**   Repeat Problem 70 using PSpice.

74. **PSpice**   Repeat the Ohm's law analysis (Figure 4–32) except use $R = 10$ Ω and sweep the source voltage from −10 V to +10 V in 1-V increments. On this graph, what does the negative value for current mean?

75. **PSpice**   The cursor may be used to read values from PSpice graphs. Get the graph from Problem 74 on the screen and:

a. Click Trace on the menu bar, select Cursor, click Display, then position the cursor on the graph, and click again. The cursor reading is indicated in the box in the lower right hand corner of the screen.

b. The cursor may be positioned using the mouse or the left and right arrow keys. Position the cursor at 2 V and read the current. Verify by Ohm's law. Repeat at a number of other points, both positive and negative.

ANSWERS TO IN-PROCESS LEARNING CHECKS

In-Process Learning Check 1

1. 25 V

2. 0 A; No

3. a. 1.25 A   b. 0.625 A   c. 12.5 A   d. 5 A

4. a. 30 V   b. −72 V   c. −90 V   d. 160 V

Basic dc Analysis

PART II

5 Series Circuits

6 Parallel Circuits

7 Series-Parallel Circuits

8 Methods of Analysis

9 Network Theorems

Part I provided the foundation upon which proficient circuit analysis is built. The terms, units, and definitions used in the previous chapters have developed the necessary vocabulary which is used throughout electrical and electronics technology.

In Part II, we build on this foundation, applying the concepts developed in Part I to initially analyze series dc circuits. Several very important rules, laws, and theorems will be developed in the following chapters, providing additional tools needed to extend circuit theory to the analysis of parallel and series-parallel circuits. Circuit theorems will be used to simplify the most complex circuit into an equivalent circuit which is represented as a single source and a single resistor. Although most circuits are much more complex than the circuits used in this text, all circuits, even the most complex, follow the same laws.

Several different techniques are often available to analyze a given circuit. In order to give a broad overview, this textbook uses the various methods for illustrative purposes. You are encouraged to use different techniques where possible, so that you develop the skills to become proficient in circuit analysis.

5

Series Circuits

OBJECTIVES

After studying this chapter, you will be able to

- determine the total resistance in a series circuit and calculate circuit current,
- use Ohm's law and the voltage divider rule to solve for the voltage across all resistors in the circuit,
- express Kirchhoff's voltage law and use it to analyze a given circuit,
- solve for the power dissipated by any resistor in a series circuit and show that the total power dissipated is exactly equal to the power delivered by the voltage source,
- solve for the voltage between any two points in a series or parallel circuit,
- design a simple voltmeter or ohmmeter given a particular meter movement,
- calculate the loading effect of an ammeter in a circuit,
- use computers to assist in the analysis of simple series circuits.

KEY TERMS

Electric Circuit
Ground
Kirchhoff's Voltage Law

Loading Effect (Ammeter)
Ohmmeter Design
Point Sources
Series Connection
Total Equivalent Resistance
Voltage Divider Rule
Voltage Subscripts
Voltmeter Design

OUTLINE

Series Circuits
Kirchhoff's Voltage Law
Resistors in Series
Voltage Sources in Series
Interchanging Series Components
The Voltage Divider Rule
Circuit Ground
Voltage Subscripts
Internal Resistance of Voltage Sources
Voltmeter Design
Ohmmeter Design
Ammeter Loading Effects
Circuit Analysis Using Computers

In the previous chapter we examined the interrelation of current, voltage, resistance, and power in a single resistor circuit. In this chapter we will expand on these basic concepts to examine the behavior of circuits having several resistors in series.

We will use Ohm's law to derive the voltage divider rule and to verify Kirchhoff's voltage law. A good understanding of these important principles provides an important base upon which further circuit analysis techniques are built. Kirchhoff's voltage law and Kirchhoff's current law, which will be covered in the next chapter, are fundamental in understanding *all* electrical and electronic circuits.

After developing the basic framework of series circuit analysis, we will apply the ideas to analyze and design simple voltmeters and ohmmeters. While meters are usually covered in a separate instruments or measurements course, we examine these circuits merely as an application of the concepts of circuit analysis.

Similarly, we will observe how circuit principles are used to explain the loading effect of an ammeter placed in series with a circuit.

Gustav Robert Kirchhoff

KIRCHHOFF WAS A GERMAN PHYSICIST born on March 12, 1824, in Königsberg, Prussia. His first research was on the conduction of electricity, which led to his presentation of the laws of closed electric circuits in 1845. Kirchhoff's current law and Kirchhoff's voltage law apply to all electrical circuits and therefore are fundamentally important in understanding circuit operation. Kirchhoff was the first to verify that an electrical impulse travelled at the speed of light.

Although these discoveries have immortalized Kirchhoff's name in electrical science, he is better known for his work with R. W. Bunsen in which he made major contributions in the study of spectroscopy and advanced the research into blackbody radiation.

Kirchhoff died in Berlin on October 17, 1887.

Conventional flow

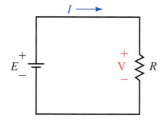

FIGURE 5–1

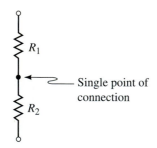

— Single point of connection

FIGURE 5–2   Resistors in series.

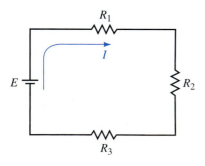

FIGURE 5–3   Series circuit.

5.1   Series Circuits

An **electric circuit** is the combination of any number of sources and loads connected in any manner which allows charge to flow. The electric circuit may be simple, such as a circuit consisting of a battery and a light bulb. Or the circuit may be very complex, such as the circuits contained within a television set, microwave oven, or computer. However, no matter how complicated, each circuit follows fairly simple rules in a predictable manner. Once these rules are understood, any circuit may be analyzed to determine the operation under various conditions.

All electric circuits obtain their energy either from a direct current (dc) source or from an alternating current (ac) source. In the next few chapters, we examine the operation of circuits supplied by dc sources. Although ac circuits have fundamental differences when compared with dc circuits, the laws, theorems, and rules that you learn in dc circuits apply directly to ac circuits as well.

In the previous chapter, you were introduced to a simple dc circuit consisting of a single voltage source (such as a chemical battery) and a single load resistance. The schematic representation of such a simple circuit was covered in Chapter 4 and is shown again in Figure 5–1.

While the circuit of Figure 5–1 is useful in deriving some important concepts, very few practical circuits are this simple. However, we will find that even the most complicated dc circuits can generally be simplified to the circuit shown.

We begin by examining the most simple connection, the **series connection.** In Figure 5–2, we have two resistors, R_1 and R_2, connected at a single point in what is said to be a series connection.

Two elements are said to be in series if they are connected at a single point and if there are no other current-carrying connections at this point.

A **series circuit** is constructed by combining various elements in series, as shown in Figure 5–3. Current will leave the positive terminal of the voltage source, move through the resistors, and return to the negative terminal of the source.

In the circuit of Figure 5–3, we see that the voltage source, E, is in series with R_1, R_1 is in series with R_2, and R_2 is in series with E. By examining this circuit, another important characteristic of a series circuit becomes evident. In an analogy similar to water flowing in a pipe, current entering an element must be the same as the current leaving the element. Now, since current does not leave at any of the connections, we conclude that the following must be true.

The current is the same everywhere in a series circuit.

While the above statement seems self-evident, we will find that this will help to explain many of the other characteristics of a series circuit.

**IN-PROCESS
LEARNING
CHECK 1**

What two conditions determine whether two elements are connected in series?

(Answers are at the end of the chapter.)

The Voltage Polarity Convention Revisited

The +, − sign convention of Figure 5–4(a) has a deeper meaning than so far considered. *Voltage exists between points, and when we place a + at one point and a − at another point, we define this to mean that we are looking at the voltage at the point marked + with respect to the point marked −.* Thus, in Figure 5–4(b), V = 6 volts means that point *a* is 6 V positive with respect to point *b*. Since the red lead of the meter is placed at point *a* and the black at point *b*, the meter will indicate +6 V.

Now consider Figure 5–5(b). [Part (a) has been repeated from Figure 5–4(b) for reference.] Here, we have placed the plus sign at point *b*, meaning that you are looking at the voltage at *b* with respect to *a*. Since point *b* is 6 V negative with respect to point *a*, V will have a value of minus 6 volts, i.e., V = −6 volts. Note also that the meter indicates −6 volts since we have reversed its leads and the red lead is now at point *b* and the black at point *a*. *It is important to realize here that the voltage across R has not changed. What has changed is how we are looking at it and how we have connected the meter to measure it.* Thus, since the actual voltage is the same in both cases, (a) and (b) are equivalent representations.

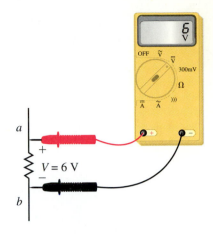

FIGURE 5–4 The +, − symbology means that you are looking at the voltage at the point marked + with respect to the point marked −.

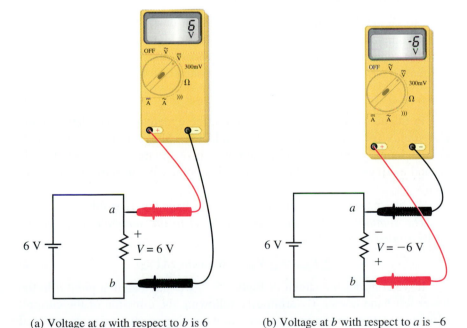

(a) Voltage at *a* with respect to *b* is 6 volts.

(b) Voltage at *b* with respect to *a* is −6 volts.

FIGURE 5–5 Two representations of the same voltage.

EXAMPLE 5–1 Consider Figure 5–6. In (a), I = 3 A, while in (b), I = −3 A. Using the voltage polarity convention, determine the voltages across the two resistors and show that they are equal.

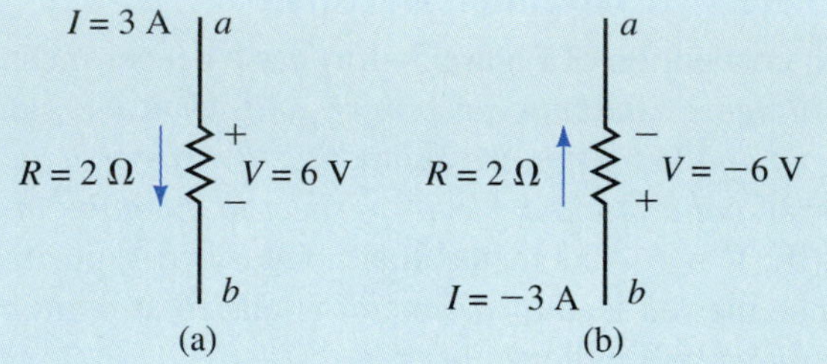

$I = 3\text{ A}$ | a a

$R = 2\ \Omega$ $V = 6\text{ V}$ $R = 2\ \Omega$ $V = -6\text{ V}$

b $I = -3\text{ A}$ | b

(a)   (b)

FIGURE 5–6

Solution In each case, place the plus sign at the tail of the current direction arrow. Then, for (a), you are looking at the polarity of a with respect to b and you get $V = IR = (3\text{ A})(2\ \Omega) = 6\text{ V}$ as expected. Now consider (b). The polarity markings mean that you are looking at the polarity of b with respect to a and you get $V = IR = (-3\text{ A})(2\ \Omega) = -6\text{ V}$. This means that point b is 6 V negative with respect to a, or equivalently, a is 6 V positive with respect to b. Thus, the two voltages are equal.

5.2   Kirchhoff's Voltage Law

Next to Ohm's law, one of the most important laws of electricity is Kirchhoff's voltage law (KVL) which states the following:

The summation of voltage rises and voltage drops around a closed loop is equal to zero. Symbolically, this may be stated as follows:

$$\sum V = 0 \quad \text{for a closed loop} \tag{5–1}$$

In the above symbolic representation, the uppercase Greek letter sigma (Σ) stands for summation and V stands for voltage rises and drops. A **closed loop** is defined as any path which originates at a point, travels around a circuit, and returns to the original point without retracing any segments.

An alternate way of stating Kirchhoff's voltage law is as follows:

The summation of voltage rises is equal to the summation of voltage drops around a closed loop.

$$\sum E_{\text{rises}} = \sum V_{\text{drops}} \quad \text{for a closed loop} \tag{5–2}$$

If we consider the circuit of Figure 5–7, we may begin at point a in the lower left-hand corner. By arbitrarily following the direction of the current, I, we move through the voltage source, which represents a rise in potential from point a to point b. Next, in moving from point b to point c, we pass through resistor R_1, which presents a potential drop of V_1. Continuing through resistors R_2 and R_3, we have additional drops of V_2 and V_3 respectively. By applying Kirchhoff's voltage law around the closed loop, we arrive at the following mathematical statement for the given circuit:

$$E - V_1 - V_2 - V_3 = 0$$

Although we chose to follow the direction of current in writing Kirchhoff's voltage law equation, it would be just as correct to move around the

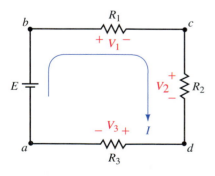

FIGURE 5–7   Kirchhoff's voltage law.

circuit in the opposite direction. In this case the equation would appear as follows:

$$V_3 + V_2 + V_1 - E = 0$$

By simple manipulation, it is quite easy to show that the two equations are identical.

EXAMPLE 5–2   Verify Kirchhoff's voltage law for the circuit of Figure 5–8.

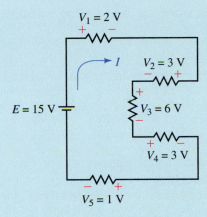

FIGURE 5–8

Solution   If we follow the direction of the current, we write the loop equation as

$$15\,V - 2\,V - 3\,V - 6\,V - 3\,V - 1\,V = 0$$

Verify Kirchhoff's voltage law for the circuit of Figure 5–9.

PRACTICE
PROBLEMS 1

FIGURE 5–9

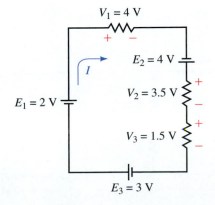

Answer: $2\,V - 4\,V + 4\,V - 3.5\,V - 1.5\,V + 3\,V = 0$

Define Kirchhoff's voltage law.

(Answers are at the end of the chapter.)

5.3   Resistors in Series

Almost all complicated circuits can be simplified. We will now examine how to simplify a circuit consisting of a voltage source in series with several resistors. Consider the circuit shown in Figure 5–10.

Since the circuit is a closed loop, the voltage source will cause a current I in the circuit. This current in turn produces a voltage drop across each resistor, where

$$V_x = IR_x$$

Applying Kirchhoff's voltage law to the closed loop gives

$$E = V_1 + V_2 + \cdots + V_n$$
$$= IR_1 + IR_2 + \cdots + IR_n$$
$$= I(R_1 + R_2 + \cdots + R_n)$$

If we were to replace all the resistors with an equivalent total resistance, R_T, then the circuit would appear as shown in Figure 5–11.

However, applying Ohm's law to the circuit of Figure 5–11 gives

$$E = IR_T \tag{5–3}$$

Since the circuit of Figure 5–11 is equivalent to the circuit of Figure 5–10, we conclude that this can only occur if the total resistance of the n series resistors is given as

$$R_T = R_1 + R_2 + \cdots + R_n \quad [\text{ohms}, \Omega] \tag{5–4}$$

If each of the n resistors has the same value, then the total resistance is determined as

$$R_T = nR \quad [\text{ohms}, \Omega] \tag{5–5}$$

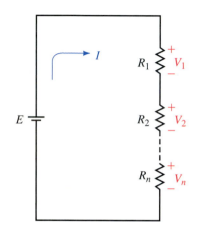

FIGURE 5–10

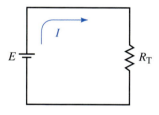

FIGURE 5–11

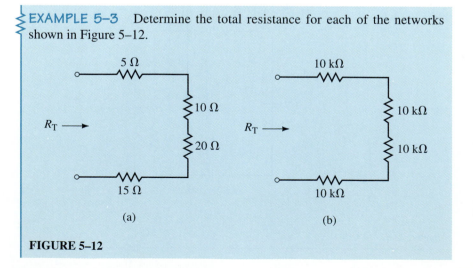

EXAMPLE 5–3   Determine the total resistance for each of the networks shown in Figure 5–12.

FIGURE 5–12

Solution

a. $R_T = 5\ \Omega + 10\ \Omega + 20\ \Omega + 15\ \Omega = 50.0\ \Omega$

b. $R_T = 4(10\ k\Omega) = 40.0\ k\Omega$

Any voltage source connected to the terminals of a network of series resistors will provide the same current as if a single resistance, having a value of R_T, were connected between the open terminals. From Ohm's law we get

$$I = \frac{E}{R_T}\quad \text{[amps, A]} \qquad (5\text{–}6)$$

The power dissipated by each resistor is determined as

$$P_1 = V_1 I = \frac{V_1^2}{R_1} = I^2 R_1 \quad \text{[watts, W]}$$

$$P_2 = V_2 I = \frac{V_2^2}{R_2} = I^2 R_2 \quad \text{[watts, W]} \qquad (5\text{–}7)$$

$$\vdots$$

$$P_n = V_n I = \frac{V_n^2}{R_n} = I^2 R_n \quad \text{[watts, W]}$$

In Chapter 4, we showed that the power delivered by a voltage source to a circuit is given as

$$P_T = EI \quad \text{[watts, W]} \qquad (5\text{–}8)$$

Since energy must be conserved, the power delivered by the voltage source is equal to the total power dissipated by all the resistors. Hence

$$P_T = P_1 + P_2 + \cdots + P_n \quad \text{[watts, W]} \qquad (5\text{–}9)$$

EXAMPLE 5–4

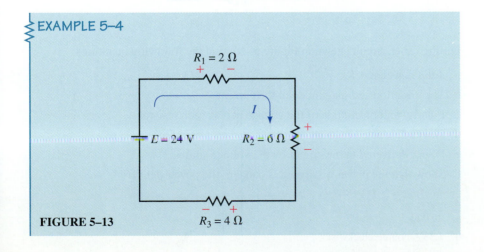

FIGURE 5–13

For the series circuit shown in Figure 5–13, find the following quantities:

a. Total resistance, R_T.

b. Circuit current, I.

c. Voltage across each resistor.

d. Power dissipated by each resistor.

e. Power delivered to the circuit by the voltage source.

f. Verify that the power dissipated by the resistors is equal to the power delivered to the circuit by the voltage source.

Solution

a. $R_T = 2\,\Omega + 6\,\Omega + 4\,\Omega = 12.0\,\Omega$

b. $I = (24\ \text{V})/(12\ \Omega) = 2.00\ \text{A}$

c. $V_1 = (2\ \text{A})(2\ \Omega) = 4.00\ \text{V}$

   $V_2 = (2\ \text{A})(6\ \Omega) = 12.0\ \text{V}$

   $V_3 = (2\ \text{A})(4\ \Omega) = 8.00\ \text{V}$

d. $P_1 = (2\ \text{A})^2(2\ \Omega) = 8.00\ \text{W}$

   $P_2 = (2\ \text{A})^2(6\ \Omega) = 24.0\ \text{W}$

   $P_3 = (2\ \text{A})^2(4\ \Omega) = 16.0\ \text{W}$

e. $P_T = (24\ \text{V})(2\ \text{A}) = 48.0\ \text{W}$

f. $P_T = 8\ \text{W} + 24\ \text{W} + 16\ \text{W} = 48.0\ \text{W}$

PRACTICE PROBLEMS 2

FIGURE 5–14

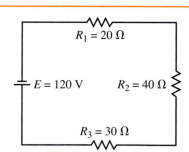

$R_1 = 20\ \Omega$

$E = 120\ \text{V}$ $R_2 = 40\ \Omega$

$R_3 = 30\ \Omega$

For the series circuit shown in Figure 5–14, find the following quantities:

a. Total resistance, R_T.

b. The direction and magnitude of the current, I.

c. Polarity and magnitude of the voltage across each resistor.

d. Power dissipated by each resistor.

e. Power delivered to the circuit by the voltage source.

f. Show that the power dissipated is equal to the power delivered.

Answers:

a. $90.0\ \Omega$

b. $1.33\ \text{A}$ counterclockwise

c. $V_1 = 26.7$ V, $V_2 = 53.3$ V, $V_3 = 40.0$ V

d. $P_1 = 35.6$ V, $P_2 = 71.1$ V, $P_3 = 53.3$ V

e. $P_T = 160.$ W

f. $P_1 + P_2 + P_3 = 160.$ W $= P_T$

Three resistors, R_1, R_2, and R_3, are in series. Determine the value of each resistor if $R_T = 42$ kΩ, $R_2 = 3R_1$, and $R_3 = 2R_2$.

IN-PROCESS LEARNING CHECK 3

(Answers are at the end of the chapter.)

5.4   Voltage Sources in Series

If a circuit has more than one voltage source in series, then the voltage sources may effectively be replaced by a single source having a value that is the sum or difference of the individual sources. Since the sources may have different polarities, it is necessary to consider polarities in determining the resulting magnitude and polarity of the equivalent voltage source.

If the polarities of all the voltage sources are such that the sources appear as voltage rises in given direction, then the resultant source is determined by simple addition, as shown in Figure 5–15.

If the polarities of the voltage sources do not result in voltage rises in the same direction, then we must compare the rises in one direction to the rises in the other direction. The magnitude of the resultant source will be the sum of the rises in one direction minus the sum of the rises in the opposite direction. The polarity of the equivalent voltage source will be the same as the polarity of whichever direction has the greater rise. Consider the voltage sources shown in Figure 5–16.

If the rises in one direction were equal to the rises in the opposite direction, then the resultant voltage source would be equal to zero.

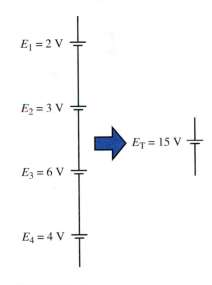

FIGURE 5–15

FIGURE 5–16

A typical lead-acid automobile battery consists of six cells connected in series. If the voltage between the battery terminals is measured to be 13.06 V, what is the average voltage of each cell within the battery?

_____

(Answers are at the end of the chapter.)

5.5   Interchanging Series Components

The order of series components may be changed without affecting the operation of the circuit.

The two circuits in Figure 5–17 are equivalent.

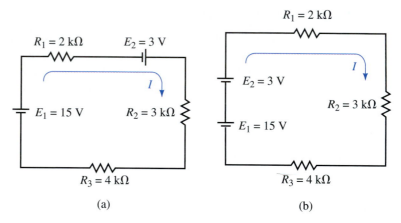

(a)   (b)

FIGURE 5–17

Very often, once the circuits have been redrawn it becomes easier to visualize the circuit operation. Therefore, we will regularly use the technique of interchanging components to simplify circuits before we analyze them.

EXAMPLE 5–5   Simplify the circuit of Figure 5–18 into a single source in series with the four resistors. Determine the direction and magnitude of the current in the resulting circuit.

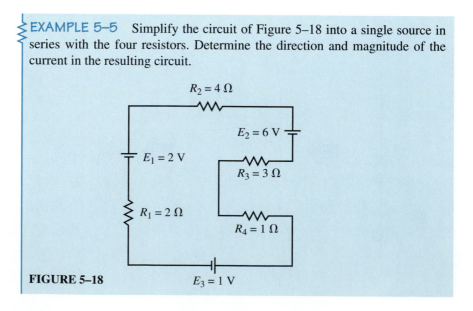

FIGURE 5–18

Solution   We may redraw the circuit by the two steps shown in Figure 5–19. It is necessary to ensure that the voltage sources are correctly moved since it is quite easy to assign the wrong polarity. Perhaps the easiest way is to imagine that we slide the voltage source around the circuit to the new location.

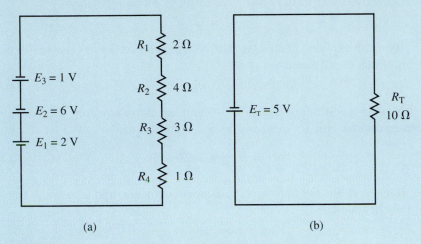

(a)   (b)

FIGURE 5–19

The current in the resulting circuit will be in a counterclockwise direction around the circuit and will have a magnitude determined as

$$I = \frac{E_T}{R_T} = \frac{6\,V + 1\,V - 2\,V}{2\,\Omega + 4\,\Omega + 3\,\Omega + 1\,\Omega} = \frac{5\,V}{10\,\Omega} = 0.500\,A$$

Because the circuits are in fact equivalent, the current direction determined in Figure 5–19 also represents the direction for the current in the circuit of Figure 5–18.

5.6   The Voltage Divider Rule

The voltage dropped across any series resistor is proportional to the magnitude of the resistor. The total voltage dropped across all resistors must equal the applied voltage source(s) by KVL.

Consider the circuit of Figure 5–20.

We see that the total resistance $R_T = 10\,k\Omega$ results in a circuit current of $I = 1\,mA$. From Ohm's law, R_1 has a voltage drop of $V_1 = 2.0\,V$, while R_2, which is four times as large as R_1, has four times as much voltage drop, $V_2 = 8.0\,V$.

We also see that the summation of the voltage drops across the resistors is exactly equal to the voltage rise of the source, namely,

$$E = 10\,V = 2\,V + 8\,V$$

The voltage divider rule allows us to determine the voltage across any series resistance in a single step, without first calculating the current. We

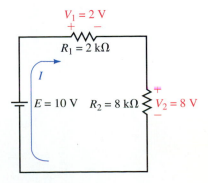

FIGURE 5–20

have seen that for any number of resistors in series the current in the circuit is determined by Ohm's law as

$$I = \frac{E}{R_\text{T}} \quad \text{[Amps, A]} \tag{5–10}$$

where the two resistors in Figure 5–20 result in a total resistance of

$$R_\text{T} = R_1 + R_2$$

By again applying Ohm's law, the voltage drop across any resistor in the series circuit is calculated as

$$V_x = IR_x$$

Now, by substituting Equation 5–4 into the above equation we write the **voltage divider rule** for two resistors as a simple equation:

$$V_x = \frac{R_x}{R_\text{T}} E = \frac{R_x}{R_1 + R_2} E$$

In general, for any number of resistors the voltage drop across any resistor may be found as

$$V_x = \frac{R_x}{R_\text{T}} E \tag{5–11}$$

EXAMPLE 5–6   Use the voltage divider rule to determine the voltage across each of the resistors in the circuit shown in Figure 5–21. Show that the summation of voltage drops is equal to the applied voltage rise in the circuit.

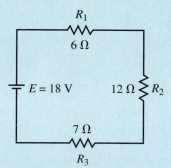

FIGURE 5–21

Solution

$$R_\text{T} = 6\ \Omega + 12\ \Omega + 7\ \Omega = 25.0\ \Omega$$

$$V_1 = \left(\frac{6\ \Omega}{25\ \Omega}\right)(18\ \text{V}) = 4.32\ \text{V}$$

$$V_2 = \left(\frac{12\ \Omega}{25\ \Omega}\right)(18\ \text{V}) = 8.64\ \text{V}$$

$$V_3 = \left(\frac{7\ \Omega}{25\ \Omega}\right)(18\ \text{V}) = 5.04\ \text{V}$$

The total voltage drop is the summation

$$V_\text{T} = 4.32\ \text{V} + 8.64\ \text{V} + 5.04\ \text{V} = 18.0\ \text{V} = E$$

EXAMPLE 5–7 Using the voltage divider rule, determine the voltage across each of the resistors of the circuit shown in Figure 5–22.

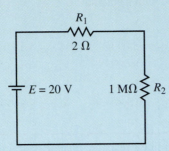

FIGURE 5–22

Solution

$$R_T = 2\ \Omega + 1\ 000\ 000\ \Omega = 1\ 000\ 002\ \Omega$$

$$V_1 = \left(\frac{2\ \Omega}{1\ 000\ 002\ \Omega}\right)(20\ \text{V}) \approx 40\ \mu\text{V}$$

$$V_2 = \left(\frac{1.0\ \text{M}\Omega}{1.000\ 002\ \text{M}\Omega}\right)(20\ \text{V}) = 19.999\ 86\ \text{V}$$

$$\approx 20.0\ \text{V}$$

The previous example illustrates two important points which occur regularly in electronic circuits. If a single series resistance is very large in comparison with the other series resistances, then the voltage across that resistor will be essentially the total applied voltage. On the other hand, if a single resistance is very small in comparison with the other series resistances, then the voltage drop across the small resistor will be essentially zero. As a general rule, if a series resistor is more than 100 times larger than another series resistor, then the effect of the smaller resistor(s) may be effectively neglected.

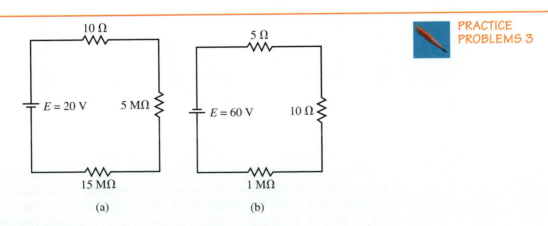

PRACTICE PROBLEMS 3

(a)   (b)

FIGURE 5–23

For the circuits shown in Figure 5–23, determine the approximate voltage drop across each resistor without using a calculator. Compare your approximations to the actual values obtained with a calculator.

Answers:
a. $V_{10\text{-}\Omega} \cong 0$ $V_{5\text{-}M\Omega} \cong 5.00\ \text{V}$ $V_{10\text{-}M\Omega} \cong 15.0\ \text{V}$
    $V_{10\text{-}\Omega} = 10.0\ \mu\text{V}$ $V_{5\text{-}M\Omega} = 5.00\ \text{V}$ $V_{10\text{-}M\Omega} = 15.0\ \text{V}$
b. $V_{5\text{-}\Omega} \cong 0$ $V_{10\text{-}\Omega} \cong 0$ $V_{1\text{-}M\Omega} \cong 60\ \text{V}$
    $V_{5\text{-}\Omega} = 0.300\ \text{mV}$ $V_{10\text{-}\Omega} = 0.600\ \text{mV}$ $V_{1\text{-}M\Omega} = 60.0\ \text{V}$

IN-PROCESS LEARNING CHECK 5

The voltage drops across three resistors are measured to be 10.0 V, 15.0 V, and 25.0 V. If the largest resistor is 47.0 kΩ, determine the sizes of the other two resistors.

(Answers are at the end of the chapter.)

(a) Circuit ground or reference

(b) Chassis ground

FIGURE 5–24

5.7   Circuit Ground

Perhaps one of the most misunderstood concepts in electronics is that of ground. This misunderstanding leads to many problems when circuits are designed and analyzed. The standard symbol for circuit ground is shown in Figure 5–24(a), while the symbol for chassis ground is shown in Figure 5–24(b).

In its most simple definition, **ground** is simply an "arbitrary electrical point of reference" or "common point" in a circuit. Using the ground symbol in this manner usually allows the circuit to be sketched more simply. When the ground symbol is used arbitrarily to designate a point of reference, it would be just as correct to redraw the circuit schematic showing all the ground points connected together or indeed to redraw the circuit using an entirely different point of reference. The circuits shown in Figure 5–25 are exactly equivalent circuits even though the circuits of Figure 5–25(a) and 5–25(c) use different points of reference.

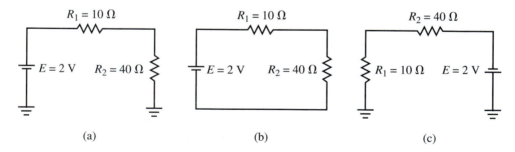

(a)   (b)   (c)

FIGURE 5–25

While the ground symbol is used to designate a common point of reference within a circuit, it usually has a greater meaning to the technologist or engineer. Very often, the metal chassis of an appliance is connected to the circuit ground. Such a connection is referred to as a **chassis ground** and is usually designated as shown in Figure 5–24(b).

In order to help prevent electrocution, the chassis ground is usually further connected to the **earth ground** through a connection provided at the electrical outlet box. In the event of a failure within the circuit, the chassis would redirect current to ground (tripping a breaker or fuse), rather than presenting a hazard to an unsuspecting operator.

As the name implies, the earth ground is a connection which is bonded to the earth, either through water pipes or by a connection to ground rods. Everyone is familiar with the typical 120-Vac electrical outlet shown in Figure 5–26. The rounded terminal of the outlet is always the ground terminal and is used not only in ac circuits by may also be used to provide a common point for dc circuits. When a circuit is bonded to the earth through the ground terminal, then the ground symbol no longer represents an arbitrary connection, but rather represents a very specific type of connection.

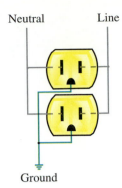

FIGURE 5–26   Ground connection in a typical 120-Vac outlet.

If you measure the resistance between the ground terminal of the 120-Vac plug and the metal chassis of a microwave oven to be zero ohms, what does this tell you about the chassis of the oven?

(Answers are at the end of the chapter.)

**IN-PROCESS
LEARNING
CHECK 6**

5.8   Voltage Subscripts

Double Subscripts

As you have already seen, voltages are always expressed as the potential difference between two points. In a 9-V battery, there is a 9-volt rise in potential from the negative terminal to the positive terminal. Current through a resistor results in a voltage drop across the resistor such that the terminal from which charge leaves is at a lower potential than the terminal into which charge enters. We now examine how voltages within any circuit may be easily described as the voltage between two points. If we wish to express the voltage between two points (say points a and b in a circuit), then we express such a voltage in a subscripted form (e.g., V_{ab}), where the first term in the subscript is the point of interest and the second term is the point of reference.

Consider the series circuit of Figure 5–27.

If we label the points within the circuit a, b, c, and d, we see that point b is at a higher potential than point a by an amount equal to the supply voltage. We may write this mathematically as $V_{ba} = +50$ V. Although the plus sign is redundant, we show it here to indicate that point b is at a higher potential than point a. If we examine the voltage at point a with respect to point b, we see that a is at a lower potential than b. This may be written mathematically as $V_{ab} = -50$ V.

From the above illustration, we make the following general statement:

$$V_{ab} = -V_{ba}$$

for any two points a and b within a circuit.

Current through the circuit results in voltage drops across the resistors as shown in Figure 5–27. If we determine the voltage drops on all resistors

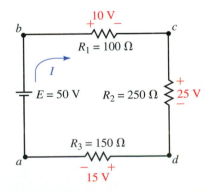

FIGURE 5–27

and show the correct polarities, then we see that the following must also apply:

$$V_{bc} = +10\text{ V} \qquad V_{cb} = -10\text{ V}$$
$$V_{cd} = +25\text{ V} \qquad V_{dc} = -25\text{ V}$$
$$V_{da} = +15\text{ V} \qquad V_{ad} = -15\text{ V}$$

If we wish to determine the voltage between any other two points within the circuit, it is a simple matter of adding all the voltages between the two points, taking into account the polarities of the voltages. The voltage between points b and d would be determined as follows:

$$V_{bd} = V_{bc} + V_{cd} = 10\text{ V} + 25\text{ V} = +35\text{ V}$$

Similarly, the voltage between points b and a could be determined by using the voltage drops of the resistors:

$$V_{ba} = V_{bc} + V_{cd} + V_{da} = 10\text{ V} + 25\text{ V} + 15\text{ V} = +50\text{ V}$$

Notice that the above result is precisely the same as when we determined V_{ba} using only the voltage source. This result indicates that the voltage between two points is not dependent upon the path taken.

EXAMPLE 5–8   For the circuit of Figure 5–28, find the voltages V_{ac}, V_{ad}, V_{cf}, and V_{eb}.

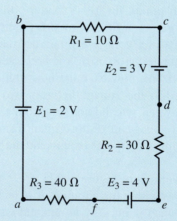

FIGURE 5–28

Solution   First, we determine that the equivalent supply voltage for the circuit is

$$E_T = 3\text{ V} + 4\text{ V} - 2\text{ V} = 5.0\text{ V}$$

with a polarity such that current will move in a counterclockwise direction within the circuit.

Next, we determine the voltages on all resistors by using the voltage divider rule and assigning polarities based on the direction of the current.

$$V_1 = \frac{R_1}{R_T} E_T$$

$$= \left(\frac{10\ \Omega}{10\ \Omega + 30\ \Omega + 40\ \Omega}\right)(5.0\ \text{V}) = 0.625\ \text{V}$$

$$V_2 = \frac{R_2}{R_T} E_T$$

$$= \left(\frac{30\ \Omega}{10\ \Omega + 30\ \Omega + 40\ \Omega}\right)(5.0\ \text{V}) = 1.875\ \text{V}$$

$$V_3 = \frac{R_3}{R_T} E_T$$

$$= \left(\frac{40\ \Omega}{10\ \Omega + 30\ \Omega + 40\ \Omega}\right)(5.0\ \text{V}) = 2.50\ \text{V}$$

The voltages appearing across the resistors are as shown in Figure 5–29.

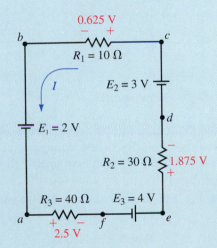

FIGURE 5–29

Finally, we solve for the voltages between the indicated points:

$$V_{ac} = -2.0\ \text{V} - 0.625\ \text{V} = -2.625\ \text{V}$$

$$V_{ad} = -2.0\ \text{V} - 0.625\ \text{V} + 3.0\ \text{V} = +0.375\ \text{V}$$

$$V_{cf} = +3.0\ \text{V} - 1.875\ \text{V} + 4.0\ \text{V} = +5.125\ \text{V}$$

$$V_{eb} = +1.875\ \text{V} - 3.0\ \text{V} + 0.625\ \text{V} = -0.500\ \text{V}$$

Or, selecting the opposite path, we get

$$V_{eb} = +4.0\ \text{V} - 2.5\ \text{V} - 2.0\ \text{V} = -0.500\ \text{V}$$

PRACTICAL NOTES . . .

Since most students initially find it difficult to determine the correct polarity for the voltage between two points, we present a simplified method to correctly determine the polarity and voltage between any two points within a circuit.

1. Determine the circuit current. Calculate the voltage drop across all components.

2. Polarize all resistors based upon the direction of the current. The terminal at which the current enters is assigned to be positive, while the terminal at which the current leaves is assigned to be negative.

3. In order to determine the voltage at point a with respect to point b, start at point a. Refer to Figure 5–30. Now, imagine that you walk around the circuit to the reference point b.

FIGURE 5–30

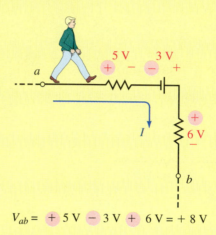

$$V_{ab} = \;+\; 5 \text{ V} \;-\; 3 \text{ V} \;+\; 6 \text{ V} = +8 \text{ V}$$

4. As you "walk" around the circuit, add the voltage drops and rises as you get to them. The assigned polarity of the voltage at any component (whether it is a source or a resistor) is the first polarity that you arrive at for the component.

5. The resulting voltage, V_{ab}, is the numerical sum of all the voltages between a and b.

For Figure 5–30, the voltage V_{ab} will be determined as

$$V_{ab} = +5 \text{ V} - 3 \text{ V} + 6 \text{ V} = 8 \text{ V}$$

PRACTICE PROBLEMS 4

Find the voltage V_{ab} in the circuit of Figure 5–31.

FIGURE 5–31

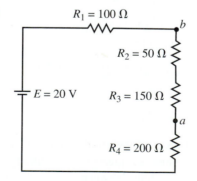

Answer: $V_{ab} = -8.00 \text{ V}$

Single Subscripts

In a circuit which has a reference point (or ground point), most voltages will be expressed with respect to the reference point. In such a case it is no longer necessary to express a voltage using a dual subscript. Rather if we wish to express the voltage at point a with respect to ground, we simply refer to this as V_a. Similarly, the voltage at point b would be referred to as V_b. Therefore, any voltage which has only a single subscript is always referenced to the ground point of the circuit.

EXAMPLE 5–9   For the circuit of Figure 5–32, determine the voltages V_a, V_b, V_c, and V_d.

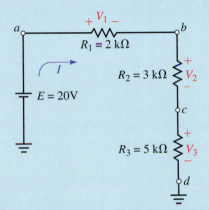

    FIGURE 5–32

Solution   Applying the voltage divider rule, we determine the voltage across each resistor as follows:

$$V_1 = \frac{2\ \text{k}\Omega}{2\ \text{k}\Omega + 3\ \text{k}\Omega + 5\ \text{k}\Omega}(20\ \text{V}) = 4.00\ \text{V}$$

$$V_2 = \frac{3\ \text{k}\Omega}{2\ \text{k}\Omega + 3\ \text{k}\Omega + 5\ \text{k}\Omega}(20\ \text{V}) = 6.00\ \text{V}$$

$$V_3 = \frac{5\ \text{k}\Omega}{2\ \text{k}\Omega + 3\ \text{k}\Omega + 5\ \text{k}\Omega}(20\ \text{V}) = 10.00\ \text{V}$$

Now we solve for the voltage at each of the points as follows:

$$V_a = 4\ \text{V} + 6\ \text{V} + 10\ \text{V} = +20\ \text{V} = E$$
$$V_b = 6\ \text{V} + 10\ \text{V} = +16.0\ \text{V}$$
$$V_c = +10.0\ \text{V}$$
$$V_d = 0\ \text{V}$$

If the voltage at various points in a circuit is known with respect to ground, then the voltage between the points may be easily determined as follows:

$$V_{ab} = V_a - V_b \quad [\text{volts, V}] \qquad \text{(5–12)}$$

EXAMPLE 5–10   For the circuit of Figure 5–33, determine the voltages V_{ab} and V_{cb} given that $V_a = +5$ V, $V_b = +3$ V, and $V_c = -8$ V.

$V_a = +5$ V
$V_b = +3$ V
FIGURE 5–33 $V_c = -8$ V

Solution

$$V_{ab} = +5 \text{ V} - (+3 \text{ V}) = +2 \text{ V}$$
$$V_{cb} = -8 \text{ V} - (+3 \text{ V}) = -11 \text{ V}$$

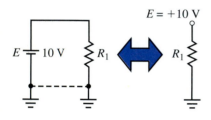

FIGURE 5–34

Point Sources

The idea of voltages with respect to ground is easily extended to include voltage sources. When a voltage source is given with respect to ground, it may be simplified in the circuit as a **point source** as shown in Figure 5–34.

Point sources are often used to simplify the representation of circuits. We need to remember that in all such cases the corresponding points always represent voltages with respect to ground (even if ground is not shown).

EXAMPLE 5–11   Determine the current and direction in the circuit of Figure 5–35.

$E_1 = +5$ V $E_2 = -8$ V

FIGURE 5–35 $R_1 = 52$ kΩ

Solution   The circuit may be redrawn showing the reference point and converting the voltage point sources into the more common schematic representation. The resulting circuit is shown in Figure 5–36.

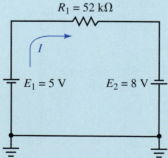

FIGURE 5–36

Now, we easily calculate the current in the circuit as

$$I = \frac{E_\text{T}}{R_1} = \frac{5 \text{ V} + 8 \text{ V}}{52 \text{ k}\Omega} = 0.250 \text{ mA}$$

Voltage measurements are taken at three locations in a circuit. They are $V_a = +5.00$ V, $V_b = -2.50$ V, and $V_c = -5.00$ V. Determine the voltages V_{ab}, V_{ca}, and V_{bc}.

(Answers are at the end of the chapter.)

5.9   Internal Resistance of Voltage Sources

So far we have worked only with ideal voltage sources, which maintain constant voltages regardless of the loads connected across the terminals. Consider a typical lead-acid automobile battery, which has a voltage of approximately 12 V. Similarly, four C-cell batteries, when connected in series, have a combined voltage of 12 V. Why then can we not use four C-cell batteries to operate the car? The answer, in part, is because the lead-acid battery has a much lower internal resistance than the low-energy C-cells. In practice, all voltage sources contain some internal resistance which will reduce the efficiency of the voltage source. We may symbolize any voltage source schematically as an ideal voltage source in series with an internal resistance. Figure 5–37 shows both an ideal voltage source and a practical or actual voltage source.

The voltage which appears between the positive and negative terminals is called the **terminal voltage.** In an ideal voltage source, the terminal voltage will remain constant regardless of the load connected. An ideal voltage source will be able to provide as much current as the circuit demands. However, in a practical voltage source, the terminal voltage is dependent upon the value of the load connected across the voltage source. As expected, the practical voltage source sometimes is not able to provide as much current as the load demands. Rather the current in the circuit is limited by the combination of the internal resistance and the load resistance.

Under a no-load condition ($R_L = \infty\ \Omega$), there is no current in the circuit and so the terminal voltage will be equal to the voltage appearing across the ideal voltage source. If the output terminals are shorted together ($R_L = 0\ \Omega$), the current in the circuit will be a maximum and the terminal voltage will be equal to approximately zero. In such a situation, the voltage dropped across the internal resistance will be equal to the voltage of the ideal source.

The following example helps to illustrate the above principles.

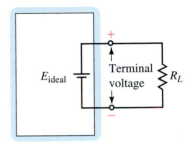

(a) Ideal voltage source

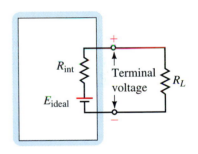

(b) Actual voltage source

FIGURE 5–37

EXAMPLE 5–12   Two batteries having an open-terminal voltage of 12 V are used to provide current to the starter of a car having a resistance of 0.10 Ω. If one battery has an internal resistance of 0.02 Ω and the second battery has an internal resistance of 100 Ω, calculate the current through the load and the resulting terminal voltage for each of the batteries.

Solution   The circuit for each of the batteries is shown in Figure 5–38.

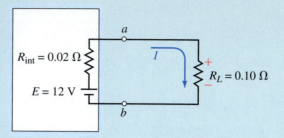

(a) Low internal resistance

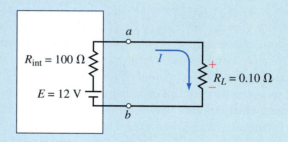

(b) High internal resistance

FIGURE 5–38

$R_{int} = 0.02\ \Omega$:

$$I = \frac{12\ V}{0.02\ \Omega + 0.10\ \Omega} = 100.\ A$$

$$V_{ab} = (100\ A)(0.10\ \Omega) = 10.0\ V$$

$R_{int} = 100\ \Omega$:

$$I = \frac{12\ V}{100\ \Omega + 0.10\ \Omega} = 0.120\ A$$

$$V_{ab} = (0.120\ A)(0.10\ \Omega) = 0.0120\ V$$

This simple example helps to illustrate why a 12-V automobile battery (which is actually 14.4 V) is able to start a car while eight 1.5 V-flashlight batteries connected in series will have virtually no measurable effect when connected to the same circuit.

5.10   Voltmeter Design

We now use the concept of a series circuit to analyze a functional circuit, namely a voltmeter. Although most technologists will seldom need to design a voltmeter, the principles presented here will help you understand the limitations of the instrument and hopefully eliminate some of its mystery.

The typical voltmeter consists of a meter movement in series with a current-limiting resistance. The meter may be either a permanent-magnet moving coil (PMMC) as shown in Figure 5–39, or a digital panel meter (DPM).

Although the digital meter is now more common, we will examine the circuits using the PMMC movement since this type of display is very simple and easily understood with concepts you have learned. The PMMC consists of an electromagnet mounted on a spring. When an external voltage is applied to the terminals of the voltmeter, a small current will occur in the voltmeter. As charge flows through the coils of the electromagnet, a magnetic field is developed. Since this movable coil is located inside a permanent magnet, the magnets will interact causing the coil to deflect proportional to the current within the movement. (In later chapters you will learn more about how magnetic fields may be created by electric currents.)

The amount of current which causes the movement to deflect to its maximum position is referred to as the **full-scale deflection current** and is usually abbreviated as I_{fsd}. The I_{fsd} of an analog meter can be determined from the sensitivity of the meter, S, which is generally printed on the meter face and is given in volts per ohm. The sensitivity is defined to be

$$S = \frac{1}{I_{fsd}}$$

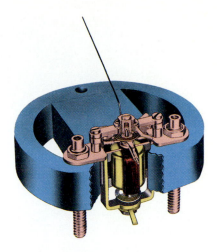

If the current in a circuit is less than I_{fsd}, then the movement will deflect an amount proportional to the current as follows:

$$\text{deflection} = \frac{I}{I_{fsd}} \times 100\% \qquad (5\text{–}13)$$

If excessive current is applied to the meter movement, the fine needle may be bent or the movement itself may be destroyed.

Due to the extreme length of very fine wire in the electromagnet, the PMMC usually has a resistance on the order of several thousand ohms. This resistance, called the **meter resistance,** is abbreviated as R_m. The schematic representation of a typical PMMC movement is shown in Figure 5–40.

By combining the PMMC movement with a single series resistor as shown in Figure 5–41, it is possible to build a simple circuit capable of measuring external voltages.

In the schematic of Figure 5–41, the resistor R_s is used to limit the amount of current so that the meter movement never receives more than I_{fsd}. The value of this resistor is dependent not only on the type of meter movement used, but also on the voltage range of the voltmeter. Clearly we would like to have maximum meter deflection when the meter detects the maximum voltage for the particular range. Once we have decided the voltage range that the meter will be measuring, it is a simple task to scale the meter with numbers and graticules to help in interpreting a measurement. Such a scale is shown in Figure 5–42 for a voltmeter having a 10-V range. In the meter of Figure 5–42, the meter will indicate the maximum voltage (10 V) when the current through the meter movement is equal to I_{fsd}.

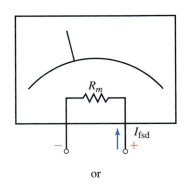

or

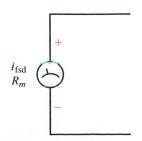

FIGURE 5–40 Schematic representation of a PMMC movement.

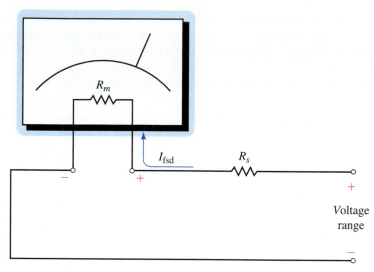

FIGURE 5–41   Simple voltmeter.

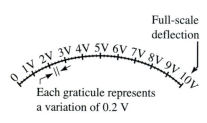

Full-scale deflection

Each graticule represents a variation of 0.2 V

FIGURE 5–42   Typical scaling of a voltmeter.

Using Ohm's law, we determine that the total circuit resistance is determined as

$$R_T = \frac{V_{range}}{I_{fsd}} = V_{range}S$$

Since the meter movement has a resistance of R_m, we may determine the required series resistance as

$$R_s = R_T - R_m$$

or

$$R_s = \frac{V_{range}}{I_{fsd}} - R_m \qquad (5\text{–}14)$$

By adding a selector switch, it is possible to design a multirange voltmeter, shown in Figure 5–43, which is able to select various ranges.

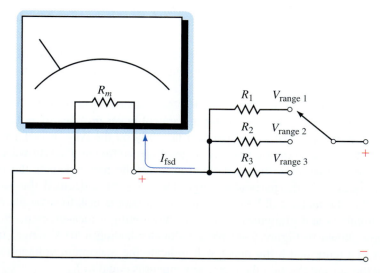

FIGURE 5–43   Multirange voltmeter.

For a multirange voltmeter, the scaling is generally adjusted to allow a simple interpretation of the reading. The following example illustrates how a multirange voltmeter would be designed and how the scale might be indicated on the meter face.

EXAMPLE 5–13 Design a voltmeter which has 20-V, 50-V, and 100-V ranges and uses a meter movement having $I_{fsd} = 1$ mA and $R_m = 2$ kΩ.

Solution **20-V range:** Using Equation 5–14, we determine the required series resistance as

$$R_1 = \frac{20 \text{ V}}{1 \text{ mA}} - 2 \text{ k}\Omega = 18.0 \text{ k}\Omega$$

Similarly, the other two ranges would be designed as follows:

50-V range:

$$R_2 = \frac{50 \text{ V}}{1 \text{ mA}} - 2 \text{ k}\Omega = 48.0 \text{ k}\Omega$$

100-V range:

$$R_3 = \frac{100 \text{ V}}{1 \text{ mA}} - 2 \text{ k}\Omega = 98.0 \text{ k}\Omega$$

The resulting circuit and corresponding scaling is shown in Figure 5–44. Now, by selecting between the various ranges and using the corresponding scale, we are able to use the meter to measure voltages up to 100 V.

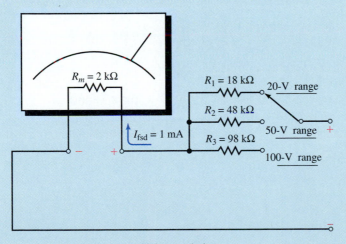

(a) Voltmeter design

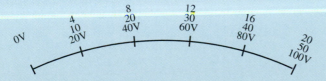

(b) Scaling of the meter face

FIGURE 5–44

EXAMPLE 5–14   If the voltmeter of Example 5–13 is used to measure the voltage across a 40-V voltage source, determine the range(s) which could be used. For the range(s), calculate the current through the meter movement and find the percent of deflection.

Solution   The voltmeter could not be used on its 20-V range, since this would cause the current to exceed I_{fsd}. However, either the 50-V range and the 100-V range could be used.

50-V range: If this range is used, then the current in the circuit is as shown in Figure 5–45.

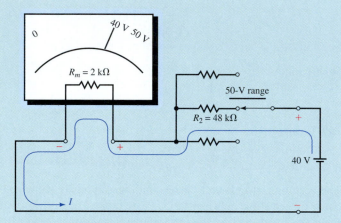

FIGURE 5–45

By applying Ohm's law, we determine the current in the circuit to be

$$I = \frac{40\ \text{V}}{48\ \text{k}\Omega + 2\ \text{k}\Omega} = 0.800\ \text{mA}$$

The deflection of the movement is

$$\text{deflection} = \frac{0.80\ \text{mA}}{1.0\ \text{mA}} \times 100\% = 80\%$$

100-V range: If this range is used, then current in the circuit is as shown in Figure 5–46.

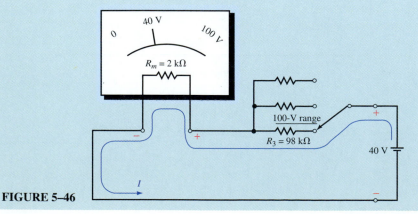

FIGURE 5–46

Applying Ohm's law, we determine the current in the circuit to be

$$I = \frac{40 \text{ V}}{98 \text{ k}\Omega + 2 \text{ k}\Omega} = 0.400 \text{ mA}$$

The deflection when measuring the voltage on this range is

$$\text{deflection} = \frac{0.400 \text{ mA}}{1.00 \text{ mA}} \times 100\% = 40\%$$

Although we are occasionally able to measure voltages on more than one range, it is generally more acceptable to measure on the range which gives the greatest nonmaximum deflection. In this example, the better range would therefore be the 50-V range since it provides the greater deflection.

Design a voltmeter which has a 250-V range and a 1000-V range. Use a meter movement having a sensitivity $S = 5 \text{ k}\Omega/\text{V}$ ($I_{\text{fsd}} = 200 \text{ }\mu\text{A}$) and $R_m = 5 \text{ k}\Omega$ in your design.

PRACTICE
PROBLEMS 5

Answers: $R_1 = 1.25 \text{ M}\Omega$ $R_2 = 5.00 \text{ M}\Omega$

A voltmeter has a sensitivity of 20 kΩ/V. Determine the resistance that you would expect to measure between the terminals when the voltmeter is on its 100-V range.

IN-PROCESS
LEARNING
CHECK 8

(Answers are at the end of the chapter.)

5.11   Ohmmeter Design

In the previous section, we saw how a voltmeter is essentially constructed from a meter movement in series with a resistance. The movement deflects an amount which is proportional to the amount of current passing through it. By employing a similar principle, it is possible to use the same meter movement to construct an ohmmeter.

Unlike the voltmeter, which uses an external voltage to provide the necessary current to cause a deflection within the PMMC movement, an ohmmeter must have an internal voltage source (usually a battery) to provide the required sensing current. The schematic of a simple ohmmeter is shown in Figure 5–47.

In the circuit of Figure 5–47, we can see that no current will be present until an unknown resistance, R_x, is connected across the open terminals of the ohmmeter. The ohmmeter is designed so that maximum current will pass through the meter movement when the resistance connected across the terminals is equal to zero (i.e., a short circuit $R_x = 0$). The scaling of the meter face plate is determined according to the movement deflection for various values of unknown resistance.

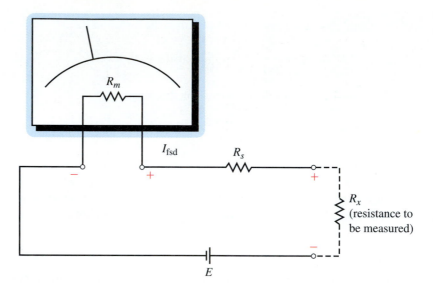

FIGURE 5–47   Simple ohmmeter.

Now, since we want maximum deflection when the terminals are shorted, the value of R_s is calculated in a manner similar to the way the voltmeter was designed, namely

$$R_s = \frac{E}{I_{fsd}} - R_m \qquad \textbf{(5–15)}$$

It is now apparent that when the resistance being measured is a minimum ($R = 0$), current will be maximum. Conversely, when resistance is maximum ($R = \infty$), current will be minimum, namely zero. The scaling for an ohmmeter is shown in Figure 5–48.

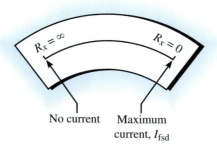

FIGURE 5–48   Scaling of an ohmmeter.

Since current is inversely proportional to the resistance of the circuit, we find that the scale will not be linear. The following example demonstrates this principle.

EXAMPLE 5–15 Design an ohmmeter using a 9-V battery and a PMMC meter movement having $I_{fsd} = 1$ mA and $R_m = 2$ kΩ. Determine the value of R_x when the movement shows 25%, 50%, and 75% deflection.

Solution The value of the series resistance is

$$R_s = \frac{9\text{ V}}{1\text{ mA}} - 2\text{ k}\Omega = 7.00\text{ k}\Omega$$

The resulting circuit is as shown in Figure 5–49(a).

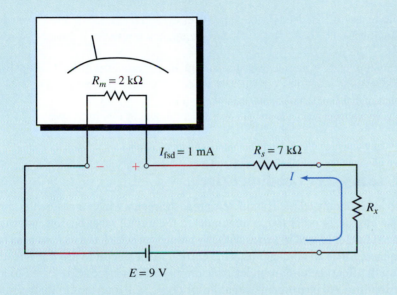

(a) Ohmmeter circuit

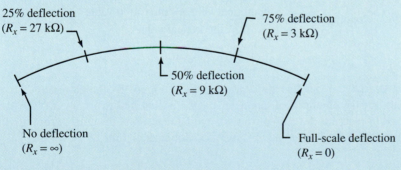

(b) Ohmmeter scaling

FIGURE 5–49

By analyzing the series circuit, we can see that when $R_x = 0\ \Omega$, the current is $I_{fsd} = 1$ mA.

At 25% deflection, the current in the circuit is

$$I = (0.25)(1\text{ mA}) = 0.250\text{ mA}$$

From Ohm's law, the total resistance in the circuit must be

$$R_T = \frac{9\text{ V}}{0.25\text{ mA}} = 36.0\text{ k}\Omega$$

For the given circuit, only the load resistance, R_x, may change. Its value is determined as

$$R_x = R_T - R_s - R_m$$
$$= 36\text{ k}\Omega - 7\text{ k}\Omega - 2\text{ k}\Omega$$
$$= 27.0\text{ k}\Omega$$

Similarly, at 50% deflection, the current in the circuit is $I = 0.500$ mA and the total resistance is $R_T = 18$ kΩ. So, the unknown resistance must be $R_x = 9$ kΩ.

Finally, at 75% deflection, the current in the circuit will be $I = 0.750$ mA, resulting in a total circuit resistance of 12 kΩ. Therefore, for 75% deflection, the unknown resistance is $R_x = 3$ kΩ.

The resulting ohmmeter scale appears as shown in Figure 5–49(b).

5.12 Ammeter Loading Effects

As you have already learned, ammeters are instruments which measure the current in a circuit. In order to use an ammeter, the circuit must be disconnected and the ammeter placed in series with the branch for which the current is to be determined. Since an ammeter uses the current in the circuit to provide a reading, it will affect the circuit under measurement. This effect is referred to as **meter loading.** All instruments, regardless of type, will load the circuit to some degree. The amount of loading is dependent upon both the instrument and the circuit being measured. For any meter, we define the loading effect as follows:

$$\text{loading effect} = \frac{\text{theoretical value} - \text{measured value}}{\text{theoretical value}} \times 100\% \quad \textbf{(5–16)}$$

EXAMPLE 5–16 For the series circuits of Figure 5–50, determine the current in each circuit. If an ammeter having an internal resistance of 250 Ω is used to measure the current in the circuits, determine the current through the ammeter and calculate the loading effect for each circuit.

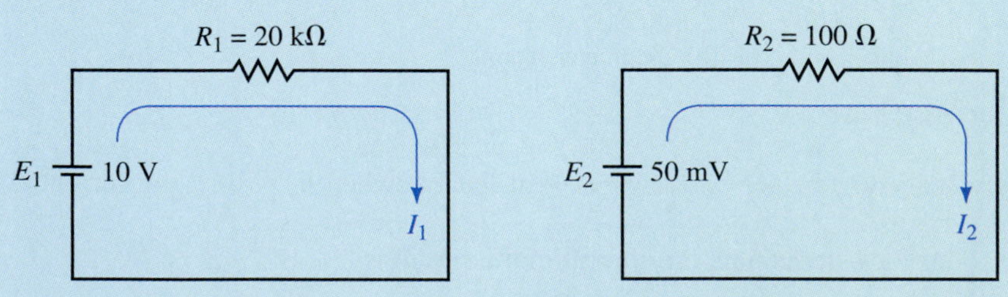

(a) Circuit #1   (b) Circuit #2

FIGURE 5–50

Solution   Circuit No. 1: The current in the circuit is

$$I_1 = \frac{10 \text{ V}}{20 \text{ k}\Omega} = 0.500 \text{ mA}$$

Now, by placing the ammeter into the circuit as shown in Figure 5–51(a) the resistance of the ammeter will slightly affect the operation of the circuit.

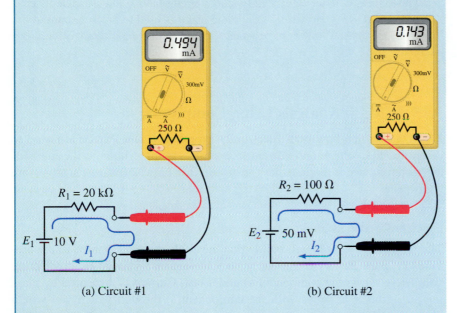

(a) Circuit #1   (b) Circuit #2

FIGURE 5–51

The resulting current in the circuit will be reduced to

$$I_1 = \frac{10 \text{ V}}{20 \text{ k}\Omega + 0.25 \text{ k}\Omega} = 0.494 \text{ mA}$$

Circuit No. 1: We see that by placing the ammeter into circuit No. 1, the resistance of the meter slightly affects the operation of the circuit. Applying Equation 5–16 gives the loading effect as

$$\text{loading effect} = \frac{0.500 \text{ mA} - 0.494 \text{ mA}}{0.500 \text{ mA}} \times 100\%$$

$$= 1.23\%$$

Circuit No. 2: The current in the circuit is also found as

$$I_2 = \frac{50 \text{ mV}}{100 \text{ }\Omega} = 0.500 \text{ mA}$$

Now, by placing the ammeter into the circuit as shown in Figure 5–51(b), the resistance of the ammeter will greatly affect the operation of the circuit.

The resulting current in the circuit will be reduced to

$$I_2 = \frac{50 \text{ mV}}{100 \text{ }\Omega + 250 \text{ }\Omega} = 0.143 \text{ mA}$$

We see that by placing the ammeter into circuit No. 2, the resistance of the meter will adversely load the circuit. The loading effect will be

$$\text{loading effect} = \frac{0.500 \text{ mA} - 0.143 \text{ mA}}{0.500 \text{ mA}} \times 100\%$$

$$= 71.4\%$$

The results of this example indicate that an ammeter, which usually has fairly low resistance, will not significantly load a circuit having a resistance of several thousand ohms. However, if the same meter is used to measure current in a circuit having low values of resistance, then the loading effect will be substantial.

ELECTRONICS WORKBENCH

PSpice

5.13   Circuit Analysis Using Computers

We now examine how both Electronics Workbench and PSpice are used to determine the voltage and current in a series circuit. Although the methods are different, we will find that the results for both software packages are equivalent.

Electronics Workbench

The following example will build upon the skills that you learned in the previous chapter. Just as in the lab, you will measure voltage by connecting voltmeters across the component(s) under test. Current is measured by placing an ammeter in series with the component(s) through which you would like to find the current.

EXAMPLE 5–17   Use Electronics Workbench to solve for the circuit current and the voltage across each of the resistors in Figure 5–52.

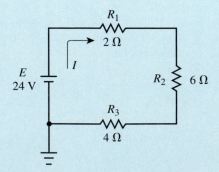

   FIGURE 5–52

Solution   Open Electronics Workbench and construct the above circuit. If necessary, review the steps as outlined in the previous chapter. Remember that your circuit will need to have a circuit ground as found in the Sources parts bin. Once your circuit resembles the circuit shown in Figure 5–52, insert the ammeters and the voltmeters into the circuit as shown in Figure 5–53.

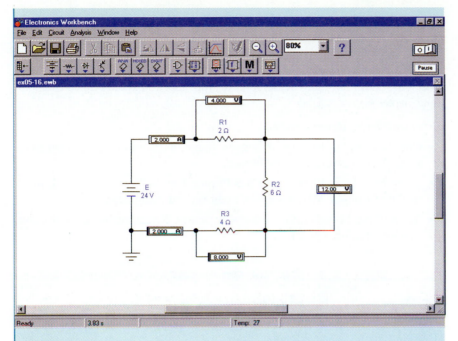

FIGURE 5–53

Notice that an extra ammeter is placed into the circuit. The only reason for this is to show that the current is the same everywhere in a series circuit.

Once all ammeters and voltmeters are inserted with the correct polarities, you may run the simulator by moving the toggle switch to the ON position. Your indicators should show the same readings as the values shown in Figure 5–53. If any of the values indicated by the meters are negative, you will need to disconnect the meter(s) and reverse the terminals by using the Ctrl R function.

Although this example is very simple, it illustrates some very important points that you will find useful when simulating circuit operation.

1. All voltmeters are connected across the components for which we are trying to measure the voltage drop.

2. All ammeters are connected in series with the components through which we are trying to find the current.

3. A ground symbol (or reference point) is required by all circuits that are to be simulated by Electronics Workbench.

PSpice

While PSpice has some differences compared to Electronics Workbench, we find that there are also many similarities. The following example shows how to use PSpice to analyze the previous circuit. In this example, you will use the Voltage Differential tool (one of three markers) to find the voltage across various components in a circuit. If necessary, refer to Appendix A to find the Voltage Differential tool. You may wish to experiment with other markers, namely the Voltage Level marker (which indicates voltage with respect to ground) and the Current Into Pin marker.

EXAMPLE 5–18   Use PSpice to solve for the circuit current and the voltage across each of the resistors in Figure 5–52.

Solution   This example lists some of the more important steps that you will need to follow. For more detail, refer to Appendix A and the PSpice example in Chapter 4.

• Open the CIS Demo software.

• Once you are in the Capture session frame, click on the menu item File, select New, and then click on Project.

• In the New Project box, type **Ch 5 PSpice 1** in the Name text box. Ensure that the Analog or Mixed-Signal Circuit Wizard is activated.

• You will need to add libraries for your project. Select the breakout.olb, and eval.olb libraries. Click Finish.

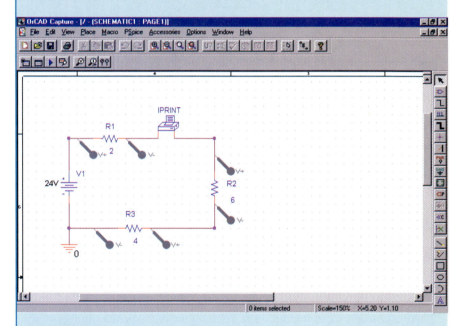

FIGURE 5–54

• You should now be in the Capture schematic editor page. Click anywhere to activate it. Build the circuit as shown in Figure 5–54. Remember to rotate the components to provide for the correct node assignments. Change the component values as required. Remember you will need to change the properties of the IPRINT symbol by going into the Properties Editor and typing **yes** into the DC cell. (Otherwise, you will not get a printout of current.)

• Click the New Simulation Profile icon and enter a name (e.g., **Figure 5–54**) in the Name text box. You will need to enter the appropriate settings for this project in the Simulation Setting box. Click the Analysis tab, and select DC Sweep from the Analysis type list. For this example, we are using a constant circuit voltage. Under Sweep variable, select Voltage source, and type **V1** in the Name text box. In the Sweep type box, select Linear.

Finally, type **24V** in the Start value text box, **24V** in the End value text box, and **1V** in the Increment text box. Click OK and save the document.

- Click on the run icon. You will see a graph of resistor voltages as a function of the source voltage. Since the voltage supply is constant resistor voltages are given for a supply voltage of 24 V only. From the graph, we have the following:

$$V_1 = 4.0 \text{ V}, V_2 = 12.0\text{V and } V_3 = 8.0 \text{ V}.$$

- In order to obtain the results of the IPRINT, we click on View and Output File. From the bottom of the file we have

V_V1	I(V_PRINT1)
2.400E+01	2.000E+00

- For the supply voltage of 24 V, the current is 2.00 A. Clearly, these results are consistent with the theoretical calculations and the results obtained using Electronics Workbench.
- Save your project and exit from PSpice.

PUTTING IT INTO PRACTICE

You are part of a research team in the electrical metering department of a chemical processing plant. As part of your work, you regularly measure voltages between 100 V and 200 V. The only voltmeter available to you today has voltage ranges of 20 V, 50 V and 100 V. Clearly, you cannot safely use the voltmeter to measure the expected voltages. However, after examining the meter, you notice that the meter movement has a resistance of $R_m = 2 \text{ k}\Omega$ and a sensitivity, $S = 20 \text{ k}\Omega/\text{V}$. You realize that the 100-V range can be converted to a 200-V range by adding a series resistor into the circuit. (Naturally, you would take extra precautions when measuring these voltages.) Without changing the internal design of the meter, calculate the value of the series resistor that you would add to the meter. Show the schematic of the design for your modified meter. In your design show the meter and series resistances of the voltmeter as well as the additional resistor that you have added externally.

PROBLEMS

5.1 Series Circuits

1. The voltmeters of Figure 5–55 have autopolarity. Determine the reading of each meter, giving the correct magnitude and sign.

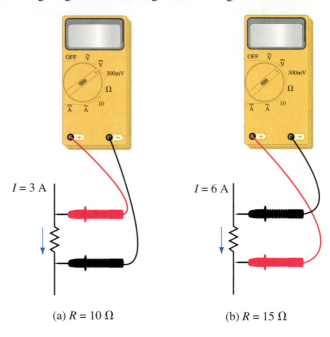

(a) $R = 10 \ \Omega$ (b) $R = 15 \ \Omega$

FIGURE 5–55

2. The voltmeters of Figure 5–56 have autopolarity. Determine the reading of each meter, giving the correct magnitude and sign.

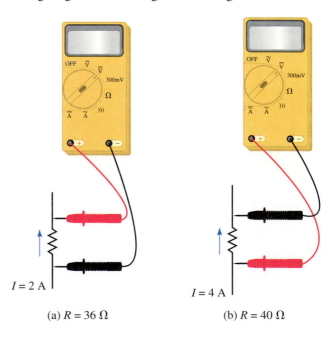

(a) $R = 36 \ \Omega$ (b) $R = 40 \ \Omega$

FIGURE 5–56

3. All resistors in Figure 5–57 are 15 Ω. For each case, determine the magnitude and polarity of voltage V.

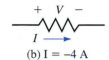

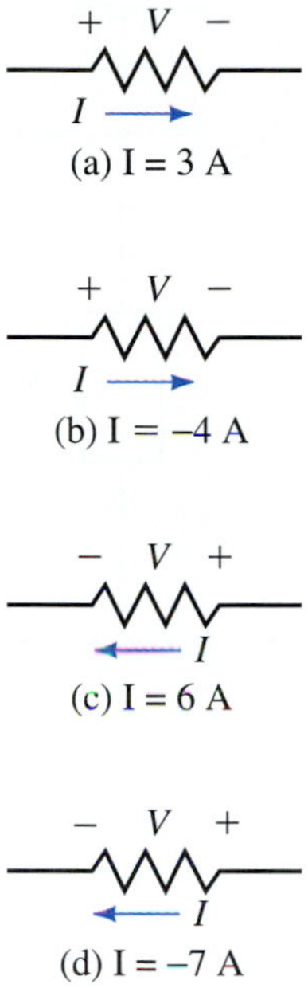

(a) I = 3 A

(b) I = –4 A

(c) I = 6 A

(d) I = –7 A

FIGURE 5–57   All resistors are 15 Ω.

4. The ammeters of Figure 5–58 have autopolarity. Determine their readings, giving the correct magnitude and sign.

5.2   Kirchhoff's Voltage Law

5. Determine the unknown voltages in the networks of Figure 5–59.

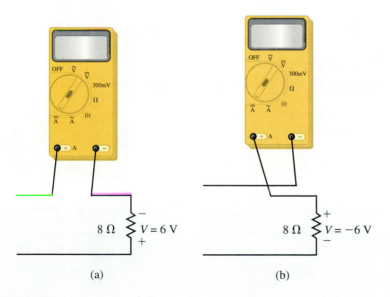

(a)

(b)

FIGURE 5–58

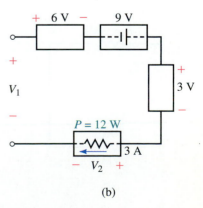

(a)

(b)

FIGURE 5–59

6. Determine the unknown voltages in the networks of Figure 5–60.

7. Solve for the unknown voltages in the circuit of Figure 5–61.

8. Solve for the unknown voltages in the circuit of Figure 5–62.

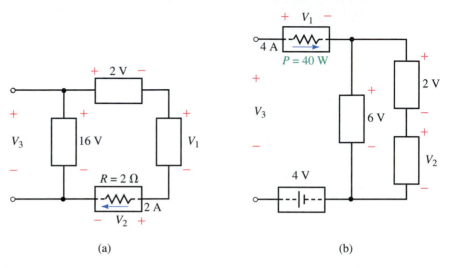

(a)   (b)

FIGURE 5–60

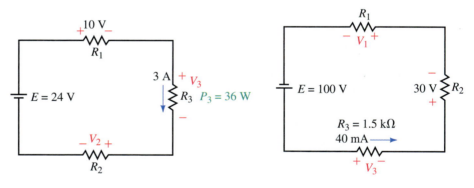

FIGURE 5–61   **FIGURE 5–62**

5.3   Resistors in Series

9. Determine the total resistance of the networks shown in Figure 5–63.

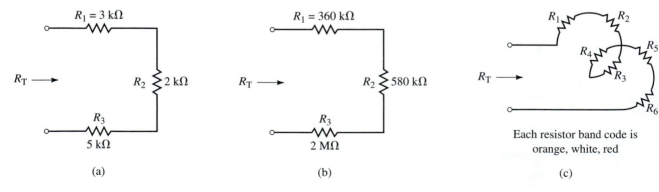

(a)   (b)   (c)

FIGURE 5–63

10. Determine the unknown resistance in each of the networks in Figure 5–64.

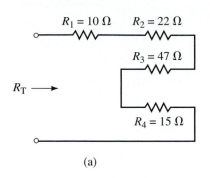

(a)

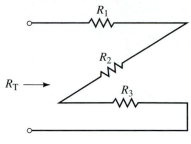

(b) Each resistor band code is
brown, red, orange

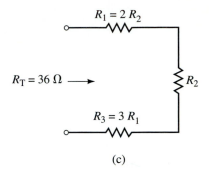

(c)

FIGURE 5–64

11. For the circuits shown in Figure 5–65, determine the total resistance, R_T,
and the current, I.

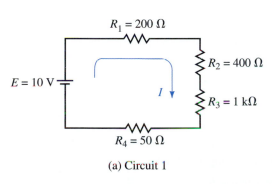

(a) Circuit 1

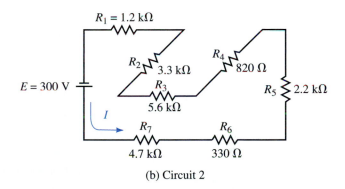

(b) Circuit 2

FIGURE 5–65

12. The circuits of Figure 5–66 have the total resistance, R_T, as shown. For each
of the circuits find the following:
   a. The magnitude of current in the circuit.
   b. The total power delivered by the voltage source.
   c. The direction of current through each resistor in the circuit.
   d. The value of the unknown resistance, R.
   e. The voltage drop across each resistor.
   f. The power dissipated by each resistor. Verify that the summation of pow-
      ers dissipated by the resistors is equal to the power delivered by the volt-
      age source.

13. For the circuit of Figure 5–67, find the following quantities:
   a. The circuit current.
   b. The total resistance of the circuit.
   c. The value of the unknown resistance, R.
   d. The voltage drop across all resistors in the circuit.
   e. The power dissipated by all resistors.

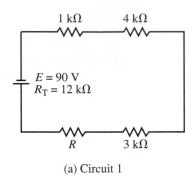

(a) Circuit 1

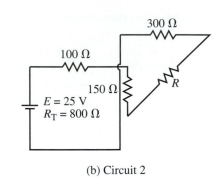

(b) Circuit 2

FIGURE 5–66

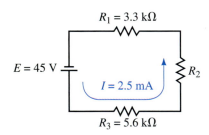

FIGURE 5–67

14. The circuit of Figure 5–68 has a current of 2.5 mA. Find the following quantities:
    a. The total resistance of the circuit.
    b. The value of the unknown resistance, R_2.
    c. The voltage drop across each resistor in the circuit.
    d. The power dissipated by each resistor in the circuit.

15. For the circuit of Figure 5–69, find the following quantities:
    a. The current, I.
    b. The voltage drop across each resistor.
    c. The voltage across the open terminals a and b.

16. Refer to the circuit of Figure 5–70:
    a. Use Kirchhoff's voltage law to find the voltage drops across R_2 and R_3.
    b. Determine the magnitude of the current, I.
    c. Solve for the unknown resistance, R_1.

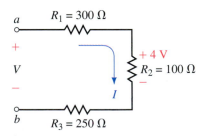

FIGURE 5–69

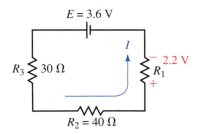

FIGURE 5–70

17. Repeat Problem 16 for the circuit of Figure 5–71.
18. Refer to the circuit of Figure 5–72:
    a. Find R_T.
    b. Solve for the current, I.
    c. Determine the voltage drop across each resistor.
    d. Verify Kirchhoff's voltage law around the closed loop.
    e. Find the power dissipated by each resistor.

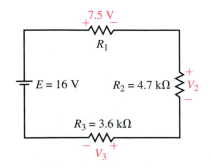

FIGURE 5–68

FIGURE 5–71

f. Determine the minimum power rating of each resistor, if resistors are available with the following power ratings: $\frac{1}{8}$ W, $\frac{1}{4}$ W, $\frac{1}{2}$ W, 1 W, and 2 W.

g. Show that the power delivered by the voltage source is equal to the summation of the powers dissipated by the resistors.

19. Repeat Problem 18 for the circuit of Figure 5–73.

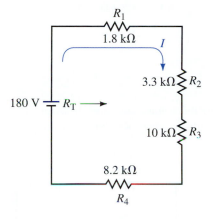

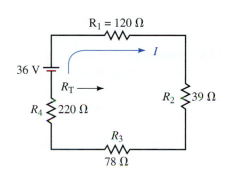

EWB **FIGURE 5–72**   **EWB** **FIGURE 5–73**

20. Refer to the circuit of Figure 5–74.
    a. Calculate the voltage across each resistor.
    b. Determine the values of the resistors R_1 and R_2.
    c. Solve for the power dissipated by each of the resistors.

5.5   Interchanging Series Components

21. Redraw the circuits of Figure 5–75, showing a single voltage source for each circuit. Solve for the current in each circuit.

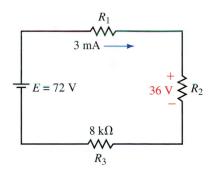

FIGURE 5–74

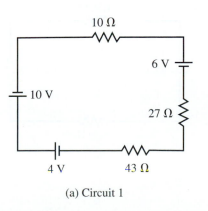

(a) Circuit 1

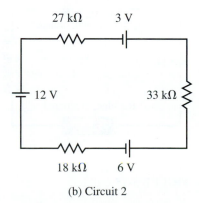

(b) Circuit 2

FIGURE 5–75

22. Use the information given to determine the polarity and magnitude of the unknown voltage source in each of the circuits of Figure 5–76.

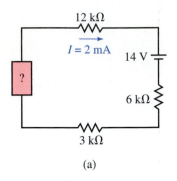

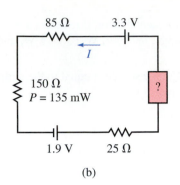

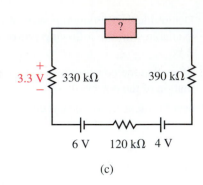

(a)

(b)

(c)

FIGURE 5–76

5.6 Voltage Divider Rule

23. Use the voltage divider rule to determine the voltage across each resistor in the circuits of Figure 5–77. Use your results to verify Kirchhoff's voltage law for each circuit.

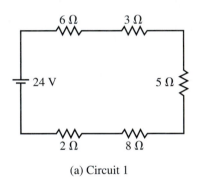

(a) Circuit 1

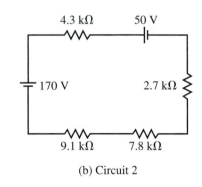

(b) Circuit 2

EWB **FIGURE 5–77**

24. Repeat Problem 23 for the circuits of Figure 5–78.

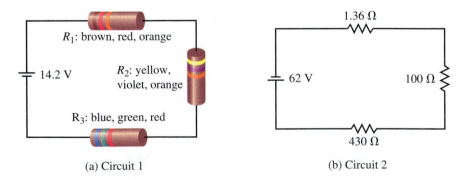

(a) Circuit 1

(b) Circuit 2

FIGURE 5–78

25. Refer to the circuits of Figure 5–79:

    a. Find the values of the unknown resistors.

    b. Calculate the voltage across each resistor.

    c. Determine the power dissipated by each resistor.

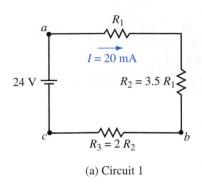

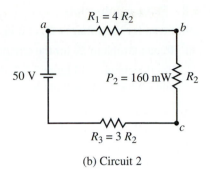

(a) Circuit 1   (b) Circuit 2

26. Refer to the circuits of Figure 5–80:
    a. Find the values of the unknown resistors using the voltage divider rule.
    b. Calculate the voltage across R_1 and R_3.
    c. Determine the power dissipated by each resistor.

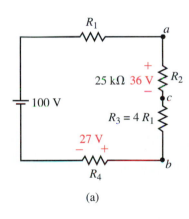

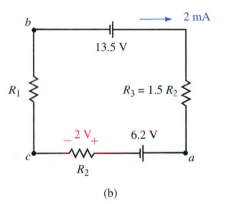

(a)   (b)

27. A string of 24 series light bulbs is connnected to a 120-V supply as shown in Figure 5–81.
    a. Solve for the current in the circuit.
    b. Use the voltage divider rule to find the voltage across each light bulb.
    c. Calculate the power dissipated by each bulb.
    d. If a single light bulb were to become an open circuit, the entire string would stop working. To prevent this from occurring, each light bulb has a small metal strip which shorts the light bulb when the filament fails. If two bulbs in the string were to burn out, repeat Steps (a) through (c).
    e. Based on your calculations of Step (d), what do you think would happen to the life expectancy of the remaining light bulbs if the two faulty bulbs were not replaced?

28. Repeat Problem 27 for a string consisting of 36 light bulbs.

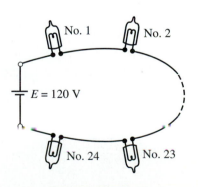

$R = 25 \ \Omega/\text{light bulb}$

5.8   Voltage Subscripts

29. Solve for the voltages V_{ab} and V_{bc} in the circuits of Figure 5–79.

30. Repeat Problem 29 for the circuits of Figure 5–80.

31. For the circuits of Figure 5–82, determine the voltage across each resistor and calculate the voltage V_a.

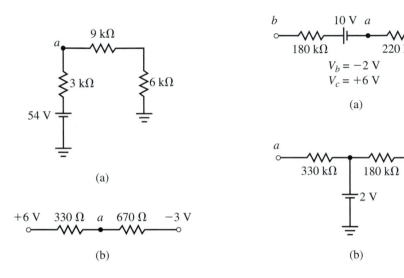

FIGURE 5–82   FIGURE 5–83

32. Given the circuits of Figure 5–83:

    a. Determine the voltage across each resistor.

    b. Find the magnitude and direction of the current in the 180-kΩ resistor.

    c. Solve for the voltage V_a.

5.9   Internal Resistance of Voltage Sources

33. A battery is measured to have an open-terminal voltage of 14.2 V. When this voltage is connected to a 100-Ω load, the voltage measured between the terminals of the battery drops to 6.8 V.

    a. Determine the internal resistance of the battery.

    b. If the 100-Ω load were replaced with a 200-Ω load, what voltage would be measured across the terminals of the battery?

34. The voltage source shown in Figure 5–84 is measured to have an open-circuit voltage of 24 V. When a 10-Ω load is connected across the terminals, the voltage measured with a voltmeter drops to 22.8 V.

    a. Determine the internal resistance of the voltage source.

    b. If the source had only half the resistance determined in (a), what voltage would be measured across the terminals with the 10-Ω resistor connected?

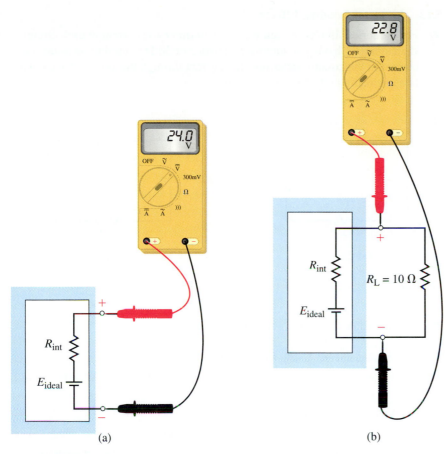

FIGURE 5–84

5.10 Voltmeter Design

35. Given a meter movement having $I_{fsd} = 2$ mA and $R_m = 1$ kΩ, design a voltmeter having the following ranges:

   a. 20-V range.

   b. 100-V range.

   c. 500-V range.

   d. Is it possible to have a 1-V range using the given meter movement? Explain.

36. Repeat Problem 35 using a meter movement having $I_{fsd} = 50$ μA and $R_m = 5$ kΩ.

5.11 Ohmmeter Design

37. Given a meter movement having $I_{fsd} = 2$ mA and $R_m = 1$ kΩ, design an ohmmeter which uses a 9-V battery to provide the sensing current. Determine the values of the unknown resistances which would result in meter deflections of 25%, 50%, 75%, and 100%.

38. Given a meter movement having $I_{fsd} = 50$ μA and $R_m = 5$ kΩ, design an ohmmeter which uses a 6-V battery to provide the sensing current. Determine the percentage meter deflection for a 10-kΩ resistor.

5.12   Ammeter Loading Effects

39. For the series circuits of Figure 5–85, determine the current in each circuit. If an ammeter having an internal resistance of 50 Ω is used to measure the current in the circuits, determine the current through the ammeter and calculate the loading effect for each circuit.

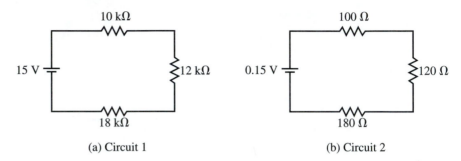

(a) Circuit 1   (b) Circuit 2

FIGURE 5–85

40. Repeat Problem 39 if the ammeter has a resistance of 10 Ω.

5.13   Circuit Analysis Using Computers

41. **EWB**   Refer to the circuits of Figure 5–77. Use Electronics Workbench to find the following:

a. The current in each circuit.

b. The voltage across each resistor in the circuit.

42. **EWB**   Given the circuit of Figure 5–86, use Electronics Workbench to determine the following:

a. The current through the voltage source, *I*.

b. The voltage across each resistor.

c. The voltage between terminals *a* and *b*.

d. The voltage, with respect to ground, at terminal *c*.

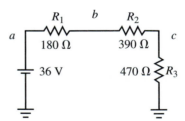

 FIGURE 5–86

43. **PSpice**   Refer to the circuit of Figure 5–73. Use PSpice to find the following:

a. The current in the circuit.

b. The voltage across each resistor in the circuit.

44. **PSpice**   Refer to the circuit of Figure 5–72. Use PSpice to find the following:

    a. The current in the circuit.

    b. The voltage across each resistor in the circuit.

In-Process Learning Check 1

1. Two elements are connected at only one node.

2. No current-carrying element is connected to the common node.

In-Process Learning Check 2

The summation of voltage drops and rises around any closed loop is equal to zero; or the summation of voltage rises is equal to the summation of voltage drops around a closed loop.

In-Process Learning Check 3

$R_1 = 4.2 \text{ k}\Omega$

$R_2 = 12.6 \text{ k}\Omega$

$R_3 = 25.2 \text{ k}\Omega$

In-Process Learning Check 4

$E_{\text{CELL}} = 2.18 \text{ V}$

In-Process Learning Check 5

$R_1 = 18.8 \text{ k}\Omega$

$R_2 = 28.21 \text{ k}\Omega$

In-Process Learning Check 6

The chassis of the oven is grounded when it is connected to the electrical outlet.

In-Process Learning Check 7

$V_{ab} = 7.50 \text{ V}$

$V_{ca} = -10.0 \text{ V}$

$V_{bc} = 2.5 \text{ V}$

In-Process Learning Check 8

$R_{\text{T}} = 2.00 \text{ M}\Omega$

ANSWERS TO IN-PROCESS LEARNING CHECKS

6

Parallel Circuits

OBJECTIVES

After studying this chapter you will be able to

- recognize which elements and branches in a given circuit are connected in parallel and which are connected in series,
- calculate the total resistance and conductance of a network of parallel resistances,
- determine the current in any resistor in a parallel circuit,
- solve for the voltage across any parallel combinations of resistors,
- apply Kirchhoff's current law to solve for unknown currents in a circuit,
- explain why voltage sources of different magnitudes must never be connected in parallel,
- use the current divider rule to solve for the current through any resistor of a parallel combination,
- design a simple ammeter using a permanent-magnet, moving-coil meter movement,
- identify and calculate the loading effects of a voltmeter connected into a circuit,
- use Electronics Workbench to observe loading effects of a voltmeter,
- use PSpice to evaluate voltage and current in a parallel circuit.

KEY TERMS

Ammeter Design
Current Divider Rule
Kirchhoff's Current Law
Loading Effect (Voltmeter)
Nodes
Parallel Circuits
Total Conductance
Total Equivalent Resistance

OUTLINE

Parallel Circuits
Kirchhoff's Current Law
Resistors in Parallel
Voltage Sources in Parallel
Current Divider Rule
Analysis of Parallel Circuits
Ammeter Design
Voltmeter Loading Effects
Circuit Analysis Using Computers

Two fundamental circuits form the basis of all electrical circuits. They are the series circuit and the parallel circuit. The previous chapter examined the principles and rules which applies to series circuits. In this chapter we study the **parallel** (or **shunt**) circuit and examine the rules governing the operation of these circuits.

Figure 6–1 illustrates a simple example of several light bulbs connected in parallel with one another and a battery supplying voltage to the bulbs.

This illustration shows one of the important differences between the series circuit and the parallel circuit. The parallel circuit will continue to operate even though one of the light bulbs may have a defective (open) filament. Only the defective light bulb will no longer glow. If a circuit were made up of several light bulbs in series, however, the defective light bulb would prevent any current in the circuit, and so all the light bulbs would be off.

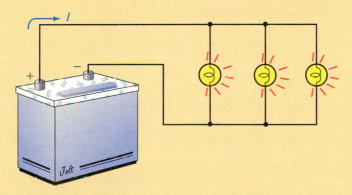

FIGURE 6–1   Simple parallel circuit.

Luigi Galvani and the Discovery of Nerve Excitation

LUIGI GALVANI WAS BORN IN BOLOGNA, Italy, on September 9, 1737.

Galvani's main expertise was in anatomy, a subject in which he was appointed lecturer at the university in Bologna.

Galvani discovered that when the nerves of frogs were connected to sources of electricity, the muscles twitched. Although he was unable to determine where the electrical pulses originated within the animal, Galvani's work was significant and helped to open further discoveries in nerve impulses.

Galvani's name has been adopted for the instrument called the **galvanometer,** which is used for detecting very small currents.

Luigi Galvani died in Bologna on December 4, 1798. Although he made many contributions to science, Galvani died poor and surrounded by controversy.

6.1   Parallel Circuits

The illustration of Figure 6–1 shows that one terminal of each light bulb is connected to the positive terminal of the battery and that the other terminal of the light bulb is connected to the negative terminal of the battery. These points of connection are often referred to as **nodes.**

Elements or branches are said to be in a parallel connection when they have exactly two nodes in common.

Figure 6–2 shows several different ways of sketching parallel elements. The elements between the nodes may be any two-terminal devices such as voltage sources, resistors, light bulbs, and the like.

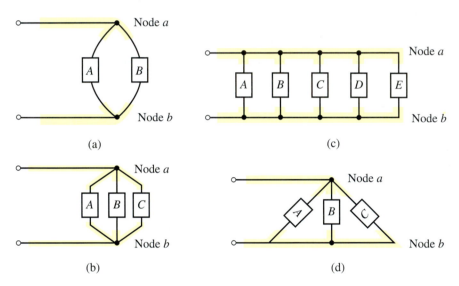

(a)

(b)

(c)

(d)

FIGURE 6–2   Parallel elements.

In the illustrations of Figure 6–2, notice that every element has two terminals and that each of the terminals is connected to one of the two nodes.

Very often, circuits contain a combination of series and parallel components. Although we will study these circuits in greater depth in later chapters, it is important at this point to be able to recognize the various connections in a given network. Consider the networks shown in Figure 6–3.

When analyzing a particular circuit, it is usually easiest to first designate the nodes (we will use lowercase letters) and then to identify the types of connections. Figure 6–4 shows the nodes for the networks of Figure 6–3.

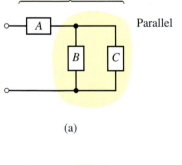

(a)

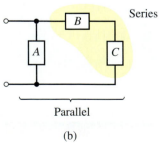

(b)

FIGURE 6–3   Series-parallel combinations.

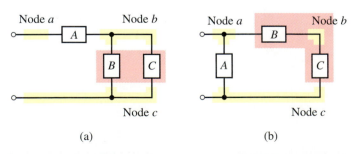

(a)

(b)

FIGURE 6–4

In the circuit of Figure 6–4(a), we see that element *B* is in parallel with element *C* since they each have nodes *b* and *c* in common. This parallel combination is now seen to be in series with element *A*.

In the circuit of Figure 6–4(b), element *B* is in series with element *C* since these elements have a single common node: node *b*. The branch consisting of the series combination of elements *B* and *C* is then determined to be in parallel with element *A*.

6.2   Kirchhoff's Current Law

Recall that Kirchhoff's voltage law was extremely useful in understanding the operation of the series circuit. In a similar manner, Kirchhoff's current law is the underlying principle which is used to explain the operation of a parallel circuit. Kirchhoff's current law states the following:

The summation of currents entering a node is equal to the summation of currents leaving the node.

An analogy which helps us understand the principle of Kirchhoff's current law is the flow of water. When water flows in a closed pipe, the amount of water entering a particular point in the pipe is exactly equal to the amount of water leaving, since there is no loss. In mathematical form, Kirchhoff's current law is stated as follows:

$$\sum I_{\text{entering node}} = \sum I_{\text{leaving node}} \qquad \textbf{(6–1)}$$

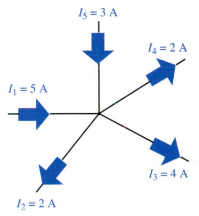

FIGURE 6–5 Kirchhoff's current law.

Figure 6–5 is an illustration of Kirchhoff's current law. Here we see that the node has two currents entering, $I_1 = 5$ A and $I_5 = 3$ A, and three currents leaving, $I_2 = 2$ A, $I_3 = 4$ A, and $I_4 = 2$ A. Now we can see that Equation 6–1 applies in the illustration, namely.

$$\sum I_{\text{in}} = \sum I_{\text{out}}$$
$$5\,\text{A} + 3\,\text{A} = 2\,\text{A} + 4\,\text{A} + 8\,\text{A}$$
$$8\,\text{A} = 8\,\text{A} \quad \text{(checks!)}$$

Verify that Kirchhoff's current law applies at the node shown in Figure 6–6.

PRACTICE
PROBLEMS 1

FIGURE 6–6

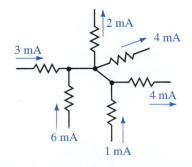

Answer: 3 mA + 6 mA + 1 mA = 2 mA + 4 mA + 4 mA

Quite often, when we analyze a given circuit, we are unsure of the direction of current through a particular element within the circuit. In such cases, we assume a reference direction and base further calculations on this assumption. If our assumption is incorrect, calculations will show that the current has a negative sign. The negative sign simply indicates that the current is in fact opposite to the direction selected as the reference. The following example illustrates this very important concept.

EXAMPLE 6–1   Determine the magnitude and correct direction of the currents I_3 and I_5 for the network of Figure 6–7.

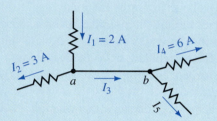

FIGURE 6–7

Solution   Although points a and b are in fact the same node, we treat the points as two separate nodes with 0 Ω resistance between them.

Since Kirchhoff's current law must be valid at point a, we have the following expression for this node:

$$I_1 = I_2 + I_3$$

and so

$$I_3 = I_1 - I_2$$
$$= 2\,\text{A} - 3\,\text{A} = -1\,\text{A}$$

Notice that the reference direction of current I_3 was taken to be from a to b, while the negative sign indicates that the current is in fact from b to a.

Similarly, using Kirchhoff's current law at point b gives

$$I_3 = I_4 + I_5$$

which gives current I_5 as

$$I_5 = I_3 - I_4$$
$$= -1\,\text{A} - 6\,\text{A} = -7\,\text{A}$$

The negative sign indicates that the current I_5 is actually towards node b rather than away from the node. The actual directions and magnitudes of the currents are illustrated in Figure 6–8.

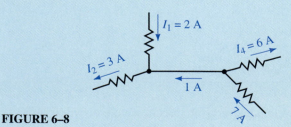

FIGURE 6–8

EXAMPLE 6–2 Find the magnitudes of the unknown currents for the circuit of Figure 6–9.

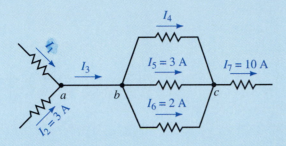

FIGURE 6–9

Solution If we consider point *a*, we see that there are two unknown currents, I_1 and I_3. Since there is no way to solve for these values, we examine the currents at point *b*, where we again have two unknown currents, I_3 and I_4. Finally we observe that at point *c* there is only one unknown, I_4. Using Kirchhoff's current law we solve for the unknown current as follows:

$$I_4 + 3\,\text{A} + 2\,\text{A} = 10\,\text{A}$$

Therefore,

$$I_4 = 10\,\text{A} - 3\,\text{A} - 2\,\text{A} = 5\,\text{A}$$

Now we can see that at point *b* the current entering is

$$I_3 = 5\,\text{A} + 3\,\text{A} + 2\,\text{A} = 10\,\text{A}$$

And finally, by applying Kirchhoff's current law at point *a*, we determine that the current I_1 is

$$I_1 = 10\,\text{A} - 3\,\text{A} = 7\,\text{A}$$

EXAMPLE 6–3 Determine the unknown currents in the network of Figure 6–10.

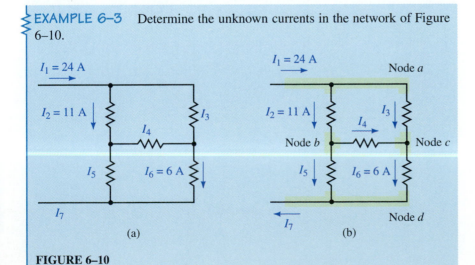

FIGURE 6–10

Solution We first assume reference directions for the unknown currents in the network.

Since we may use the analogy of water moving through conduits, we can easily assign directions for the currents I_3, I_5, and I_7. However, the direction for the current I_4 is not as easily determined, so we arbitrarily assume that its direction is to the right. Figure 6–10(b) shows the various nodes and the assumed current directions.

By examining the network, we see that there is only a single source of current $I_1 = 24$ A. Using the analogy of water pipes, we conclude that the current leaving the network is $I_7 = I_1 = 24$ A.

Now, applying Kirchhoff's current law to node a, we calculate the current I_3 as follows:

$$I_1 = I_2 + I_3$$

Therefore,

$$I_3 = I_1 - I_2 = 24\,\text{A} - 11\,\text{A} = 13\,\text{A}$$

Similarly, at node c, we have

$$I_3 + I_4 = I_6$$

Therefore,

$$I_4 = I_6 - I_3 = 6\,\text{A} - 13\,\text{A} = -7\,\text{A}$$

Although the current I_4 is opposite to the assumed reference direction, we do not change its direction for further calculations. We use the original direction together with the negative sign; otherwise the calculations would be needlessly complicated.

Applying Kirchhoff's current law at node b, we get

$$I_2 = I_4 + I_5$$

which gives

$$I_5 = I_2 - I_4 = 11\,\text{A} - (-7\,\text{A}) = 18\,\text{A}$$

Finally, applying Kirchhoff's current law at node d gives

$$I_5 + I_6 = I_7$$

resulting in

$$I_7 = I_5 + I_6 = 18\,\text{A} + 6\,\text{A} = 24\,\text{A}$$

Determine the unknown currents in the network of Figure 6–11.

FIGURE 6–11

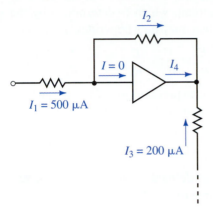

Answers: $I_2 = 500\ \mu A,$ $I_4 = -700\ \mu A$

6.3   Resistors in Parallel

A simple parallel circuit is constructed by combining a voltage source with several resistors as shown in Figure 6–12.

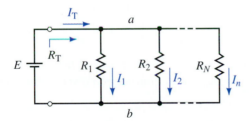

FIGURE 6–12

The voltage source will result in current from the positive terminal of the source toward node *a*. At this point the current will split between the various resistors and then recombine at node *b* before continuing to the negative terminal of the voltage source.

This circuit illustrates a very important concept of parallel circuits. If we were to apply Kirchhoff's voltage law around each closed loop in the parallel circuit of Figure 6–12, we would find that the voltage across all parallel resistors is exactly equal, namely $V_{R_1} = V_{R_2} = V_{R_3} = E$. Therefore, by applying Kirchhoff's voltage law, we make the following statement:

The voltage across all parallel elements in a circuit will be the same.

The above principle allows us to determine the equivalent resistance, R_T, of any number of resistors connected in parallel. The equivalent resistance, R_T, is the effective resistance "seen" by the source and determines the total current, I_T, provided to the circuit. Applying Kirchhoff's current law to the circuit of Figure 6–11, we have the following expression:

$$I_T = I_1 + I_2 + \cdots + I_n$$

However, since Kirchhoff's voltage law also applies to the parallel circuit, the voltage across each resistor must be equal to the supply voltage, E. The total current in the circuit, which is determined by the supply voltage and the equivalent resistance, may now be written as

$$\frac{E}{R_T} = \frac{E}{R_1} + \frac{E}{R_2} + \cdots + \frac{E}{R_n}$$

Simplifying the above expression gives us the general expression for total resistance of a parallel circuit as

$$\frac{1}{R_T} = \frac{1}{R_1} + \frac{1}{R_2} + \cdots + \frac{1}{R_n} \quad \text{(siemens, S)} \qquad \text{(6–2)}$$

Since conductance was defined as the reciprocal of resistance, we may write the above equation in terms of conductance, namely,

$$G_T = G_1 + G_2 + \cdots + G_n \quad \text{(S)} \qquad \text{(6–3)}$$

Whereas series resistors had a total resistance determined by the summation of the particular resistances, we see that any number of parallel resistors have a total conductance determined by the summation of the individual conductances.

The equivalent resistance of n parallel resistors may be determined in one step as follows:

$$R_T = \frac{1}{\dfrac{1}{R_1} + \dfrac{1}{R_2} + \cdots + \dfrac{1}{R_n}} \quad (\Omega) \qquad \text{(6–4)}$$

An important effect of combining parallel resistors is that the resultant resistance will always be smaller than the smallest resistor in the combination.

EXAMPLE 6–4   Solve for the total conductance and total equivalent resistance of the circuit shown in Figure 6–13.

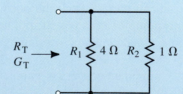

FIGURE 6–13

Solution   The total conductance is

$$G_T = G_1 + G_2 = \frac{1}{4\ \Omega} + \frac{1}{1\ \Omega} = 1.25\ \text{S}$$

The total equivalent resistance of the circuit is

$$R_T = \frac{1}{G_T} = \frac{1}{1.25\ \text{S}} = 0.800\ \Omega$$

Notice that the equivalent resistance of the parallel resistors is indeed less than the value of each resistor.

EXAMPLE 6–5   Determine the conductance and resistance of the network of Figure 6–14.

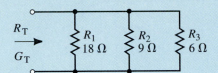

FIGURE 6–14

Solution   The total conductance is

$$G_T = G_1 + G_2 + G_3$$

$$= \frac{1}{18\ \Omega} + \frac{1}{9\ \Omega} + \frac{1}{6\ \Omega}$$

$$= 0.0\overline{5}\ S + 0.1\overline{1}\ S + 0.1\overline{6}\ S$$

$$= 0.3\overline{3}\ S$$

where the overbar indicates that the number under it is repeated infinitely to the right.

The total resistance is

$$R_T = \frac{1}{0.3\overline{3}\ S} = 3.00\ \Omega$$

FIGURE 6–15

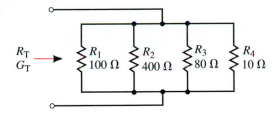

PRACTICE
PROBLEMS 3

For the parallel network of resistors shown in Figure 6–15, find the total conductance, G_T and the total resistance, R_T.

Answers: $G_T = 0.125\ S$
$R_T = 8.00\ \Omega$

---

n Equal Resistors in Parallel

If we have *n* equal resistors in parallel, each resistor, *R,* has the same conductance, *G.* By applying Equation 6–3, the total conductance is found:

$$G_T = nG$$

The total resistance is now easily determined as

$$R_T = \frac{1}{G_T} = \frac{1}{nG} = \frac{R}{n}$$ **(6–5)**

EXAMPLE 6–6 For the networks of Figure 6–16, calculate the total resistance.

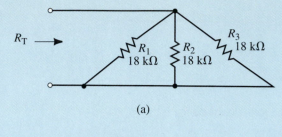

(a)

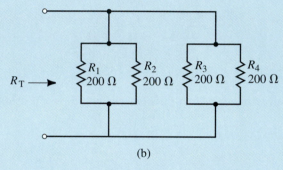

(b)

FIGURE 6–16

Solution

a. $R_T = \dfrac{18\ \text{k}\Omega}{3} = 6\ \text{k}\Omega$

b. $R_T = \dfrac{200\ \Omega}{4} = 50\ \Omega$

Two Resistors in Parallel

Very often circuits have only two resistors in parallel. In such a case, the total resistance of the combination may be determined without the necessity of determining the conductance.

For two resistors, Equation 6–4 is written

$$R_T = \frac{1}{\dfrac{1}{R_1} + \dfrac{1}{R_2}}$$

By cross multiplying the terms in the denominator, the expression becomes

$$R_T = \frac{1}{\dfrac{R_1 + R_2}{R_1 R_2}}$$

Thus, for two resistors in parallel we have the following expression:

$$R_T = \frac{R_1 R_2}{R_1 + R_2} \tag{6–6}$$

For two resistors connected in parallel, the equivalent resistance is found by the product of the two values divided by the sum.

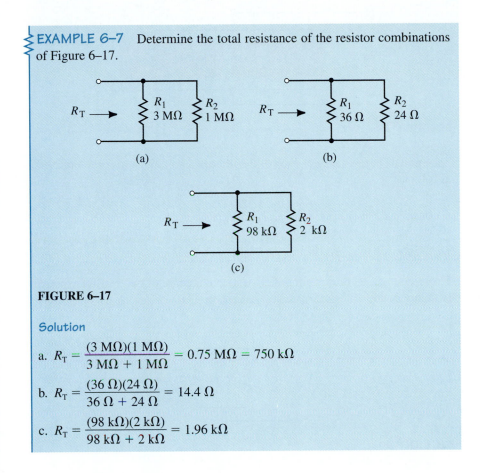

EXAMPLE 6–7   Determine the total resistance of the resistor combinations of Figure 6–17.

(a)

(b)

(c)

FIGURE 6–17

Solution

a. $R_T = \dfrac{(3\ \text{M}\Omega)(1\ \text{M}\Omega)}{3\ \text{M}\Omega + 1\ \text{M}\Omega} = 0.75\ \text{M}\Omega = 750\ \text{k}\Omega$

b. $R_T = \dfrac{(36\ \Omega)(24\ \Omega)}{36\ \Omega + 24\ \Omega} = 14.4\ \Omega$

c. $R_T = \dfrac{(98\ \text{k}\Omega)(2\ \text{k}\Omega)}{98\ \text{k}\Omega + 2\ \text{k}\Omega} = 1.96\ \text{k}\Omega$

Although Equation 6–6 is intended primarily to solve for two resistors in parallel, the approach may also be used to solve for any number of resistors by examining only two resistors at a time.

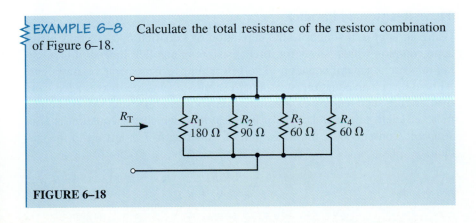

EXAMPLE 6–8   Calculate the total resistance of the resistor combination of Figure 6–18.

FIGURE 6–18

Solution   By grouping the resistors into combinations of two, the circuit may be simplified as shown in Figure 6–19.

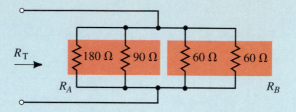

FIGURE 6–19

The equivalent resistance of each of the indicated combinations is determined as follows:

$$R_A = \frac{(180\ \Omega)(90\ \Omega)}{180\ \Omega + 90\ \Omega} = 60\ \Omega$$

$$R_B = \frac{(60\ \Omega)(60\ \Omega)}{60\ \Omega + 60\ \Omega} = 30\ \Omega$$

The circuit can be further simplified as a combination of two resistors shown in Figure 6–20.

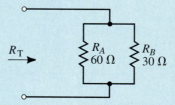

FIGURE 6–20

The resultant equivalent resistance is

$$R_T = \frac{(60\ \Omega)(30\ \Omega)}{60\ \Omega + 30\ \Omega} = 20\ \Omega$$

Three Resistors in Parallel

Using an approach similar to the derivation of Equation 6–6, we may arrive at an equation which solves for three resistors in parallel. Indeed, it is possible to write a general equation to solve for four resistors, five resistors, etc. Although such an equation is certainly useful, students are discouraged from memorizing such lengthy expressions. You will generally find that it is much more efficient to remember the principles upon which the equation is constructed. Consequently, the derivation of Equation 6–7 is left up to the student.

$$R_T = \frac{R_1 R_2 R_3}{R_1 R_2 + R_1 R_3 + R_2 R_3} \tag{6–7}$$

FIGURE 6–21

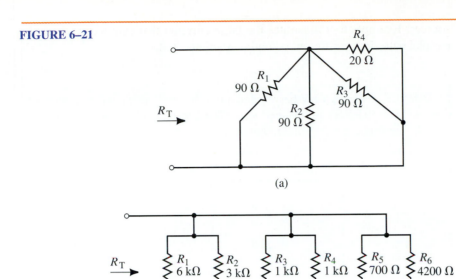

(a)

(b)

Find the total equivalent resistance for each network in Figure 6–21.

Answers: a. 12 Ω   b. 240 Ω

If the circuit of Figure 6–21(a) is connected to a 24-V voltage source, determine the following quantities:

a. The total current provided by the voltage source.

b. The current through each resistor of the network.

c. Verify Kirchhoff's current law at one of the voltage source terminals.

(Answers are at the end of the chapter.)

6.4   Voltage Sources in Parallel

Voltage sources of different potentials should never be connected in parallel, since to do so would contradict Kirchhoff's voltage law. However, when two equal potential sources are connected in parallel, each source will deliver half the required circuit current. For this reason automobile batteries are sometimes connected in parallel to assist in starting a car with a "weak" battery. Figure 6–22 illustrates this principle.

Figure 6–23 shows that if voltage sources of two different potentials are placed in parallel, Kirchhoff's voltage law would be violated around the closed loop. In practice, if voltage sources of different potentials are placed in parallel, the resulting closed loop can have a very large current. The current will occur even though there may not be a load connected across the

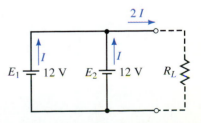

FIGURE 6–22   Voltage sources in parallel.

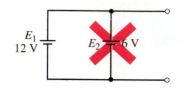

FIGURE 6–23   Voltage sources of different polarities must never be placed in parallel.

sources. Example 6–9 illustrates the large currents that can occur when two parallel batteries of different potential are connected.

EXAMPLE 6–9   A 12-V battery and a 6-V battery (each having an internal resistance of 0.05 Ω) are inadvertently placed in parallel as shown in Figure 6–24. Determine the current through the batteries.

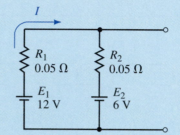

FIGURE 6–24

Solution   From Ohm's law,

$$I = \frac{E_T}{R_T} = \frac{12\ \text{V} - 6\ \text{V}}{0.05\ \Omega + 0.05\ \Omega} = 60\ \text{A}$$

This example illustrates why batteries of different potential must never be connected in parallel. Tremendous currents will occur within the sources resulting in the possibility of a fire or explosion.

6.5   Current Divider Rule

When we examined series circuits we determined that the current in the series circuit was the same everywhere in the circuit, whereas the voltages across the series elements were typically different. The voltage divider rule (VDR) was used to determine the voltage across all resistors within a series network.

In parallel networks, the voltage across all parallel elements is the same. However, the currents through the various elements are typically different. The current divider rule (CDR) is used to determine how current entering a node is split between the various parallel resistors connected to the node.

Consider the network of parallel resistors shown in Figure 6–25.

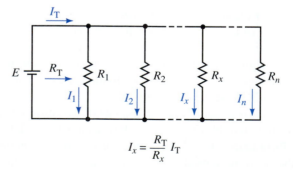

$$I_x = \frac{R_T}{R_x} I_T$$

FIGURE 6–25   Current divider rule.

If this network of resistors is supplied by a voltage source, the total current in the circuit is

$$I_T = \frac{E}{R_T} \qquad \text{(6–8)}$$

Since each of the n parallel resistors has the same voltage, E, across its terminals, the current through any resistor in the network is given as

$$I_x = \frac{E}{R_x} \qquad \text{(6–9)}$$

By rewriting Equation 6–8 as $E = I_T R_T$ and then substituting this into Equation 6–9, we obtain the current divider rule as follows:

$$I_x = \frac{R_T}{R_x} I_T \qquad \text{(6–10)}$$

An alternate way of writing the current divider rule is to express it in terms of conductance. Equation 6–10 may be modified as follows:

$$I_x = \frac{G_x}{G_T} I_T \qquad \text{(6–11)}$$

The current divider rule allows us to calculate the current in any resistor of a parallel network if we know the total current entering the network. Notice the similarity between the voltage divider rule (for series components) and the current divider rule (for parallel components). The main difference is that the current divider rule of Equation 6–11 uses circuit conductance rather than resistance. While this equation is useful, it is generally easier to use resistance to calculate current.

If the network consists of only two parallel resistors, then the current through each resistor may be found in a slightly different way. Recall that for two resistors in parallel, the total parallel resistance is given as

$$R_T = \frac{R_1 R_2}{R_1 + R_2}$$

Now, by substituting this expression for total resistance into Equation 6–10, we obtain

$$I_1 = \frac{I_T R_T}{R_1}$$

$$= \frac{I_T \left(\dfrac{R_1 R_2}{R_1 + R_2} \right)}{R_1}$$

which simplifies to

$$I_1 = \frac{R_2}{R_1 + R_2} I_T \qquad \text{(6–12)}$$

Similarly,

$$I_2 = \frac{R_1}{R_1 + R_2} I_T \qquad \text{(6–13)}$$

Several other important characteristics of parallel networks become evident.

If current enters a parallel network consisting of any number of equal resistors, then the current entering the network will split equally between all of the resistors.

If current enters a parallel network consisting of several values of resistance, then the smallest value of resistor in the network will have the largest amount of current. Inversely, the largest value of resistance will have the smallest amount of current.

This characteristic may be simplified by saying that *most of the current will follow the path of least resistance.*

EXAMPLE 6–10   For the network of Figure 6–26, determine the currents I_1, I_2, and I_3.

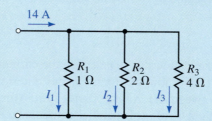

FIGURE 6–26

Solution   First, we calculate the total conductance of the network.

$$G_T = \frac{1}{1\ \Omega} + \frac{1}{2\ \Omega} + \frac{1}{4\ \Omega} = 1.75\ S$$

Now the currents may be evaluated as follows:

$$I_1 = \frac{G_1}{G_T} I_T = \left(\frac{1\ S}{1.75\ S} \right) 14\ A = 8.00\ A$$

$$I_2 = \frac{G_2}{G_T} I_T = \left(\frac{0.5\ S}{1.75\ S} \right) 14\ A = 4.00\ A$$

$$I_3 = \frac{G_3}{G_T} I_T = \left(\frac{0.25\ S}{1.75\ S} \right) 14\ A = 2.00\ A$$

An alternate approach is to use circuit resistance, rather than conductance.

$$R_T = \frac{1}{G_T} = \frac{1}{1.75\ S} = 0.571\ \Omega$$

$$I_1 = \frac{R_T}{R_1} I_T = \left(\frac{0.571\ \Omega}{1\ \Omega} \right) 14\ A = 8.00\ A$$

$$I_2 = \frac{R_T}{R_2} I_T = \left(\frac{0.571\ \Omega}{2\ \Omega} \right) 14\ A = 4.00\ A$$

$$I_3 = \frac{R_T}{R_3} I_T = \left(\frac{0.571\ \Omega}{5\ \Omega} \right) 14\ A = 2.00\ A$$

EXAMPLE 6–11   For the network of Figure 6–27, determine the currents I_1, I_2, and I_3.

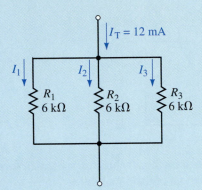

FIGURE 6–27

Solution   Since all the resistors have the same value, the incoming current will split equally between the resistances. Therefore,

$$I_1 = I_2 = I_3 = \frac{12 \text{ mA}}{3} = 4.00 \text{ mA}$$

EXAMPLE 6–12   Determine the currents I_1 and I_2 in the network of Figure 6–28.

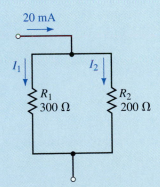

FIGURE 6–28

Solution   Because we have only two resistors in the given network, we use Equations 6–12 and 6–13:

$$I_1 = \frac{R_2}{R_1 + R_2} I_T = \left(\frac{200 \text{ }\Omega}{300 \text{ }\Omega + 200 \text{ }\Omega} \right)(20 \text{ mA}) = 8.00 \text{ mA}$$

$$I_2 = \frac{R_1}{R_1 + R_2} I_T = \left(\frac{300 \text{ }\Omega}{300 \text{ }\Omega + 200 \text{ }\Omega} \right)(20 \text{ mA}) = 12.0 \text{ mA}$$

EXAMPLE 6–13   Determine the resistance R_1 so that current will divide as shown in the network of Figure 6–29.

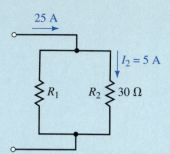

FIGURE 6–29

Solution   There are several methods which may be used to solve this problem. We will examine only two of the possibilities.

Method I: Since we have two resistors in parallel, we may use Equation 6–13 to solve for the unknown resistor:

$$I_2 = \frac{R_1}{R_1 + R_2} I_T$$

$$5\,\text{A} = \left(\frac{R_1}{R_1 + 30\,\Omega}\right)(25\,\text{A})$$

Using algebra, we get

$$(5\,\text{A})R_1 + (5\,\text{A})(30\,\Omega) = (25\,\text{A})R_1$$

$$(20\,\text{A})R_1 = 150\,\text{V}$$

$$R_1 = \frac{150\,\text{V}}{20\,\text{A}} = 7.50\,\Omega$$

Method II: By applying Kirchhoff's current law, we see that the current in R_1 must be

$$I_1 = 25\,\text{A} - 5\,\text{A} = 20\,\text{A}$$

Now, since elements in parallel must have the same voltage across their terminals, the voltage across R_1 must be exactly the same as the voltage across R_2. By Ohm's law, the voltage across R_2 is

$$V_2 = (5\,\text{A})(30\,\Omega) = 150\,\text{V}$$

And so

$$R_1 = \frac{150\,\text{V}}{20\,\text{A}} = 7.50\,\Omega$$

As expected, the results are identical. This example illustrates that there is usually more than one method for solving a given problem. Although the methods are equally correct, we see that the second method in this example is less involved.

Use the current divider rule to calculate the unknown currents for the networks of Figure 6–30.

FIGURE 6–30

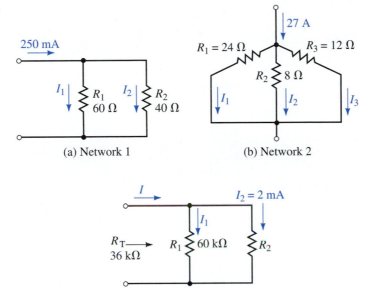

(a) Network 1   (b) Network 2

(c) Network 3

Answers: Network 1: $I_1 = 100$ mA $I_2 = 150$ mA
Network 2: $I_1 = 4.50$ A $I_2 = 13.5$ A $I_3 = 9.00$ A
Network 3: $I_1 = 3.00$ mA $I = 5.00$ mA

Four resistors are connected in parallel. The values of the resistors are 1 Ω, 3 Ω, 4 Ω, and 5 Ω.

a. Using only a pencil and a piece of paper (no calculator), determine the current through each resistor if the current through the 5-Ω resistor is 6 A.

b. Again, without a calculator, solve for the total current applied to the parallel combination.

c. Use a calculator to determine the total parallel resistance of the four resistors. Use the current divider rule and the total current obtained in part (b) to calculate the current through each resistor.

(Answers are at the end of the chapter.)

6.6   Analysis of Parallel Circuits

We will now examine how to use the principles developed in this chapter when analyzing parallel circuits. In the examples to follow, we find that the laws of conservation of energy apply equally well to parallel circuits as to series circuits. Although we choose to analyze circuits a certain way, remember that there is usually more than one way to arrive at the correct answer. As you become more proficient at circuit analysis you will generally use the most efficient method. For now, however, use the method with which you feel most comfortable.

EXAMPLE 6–14

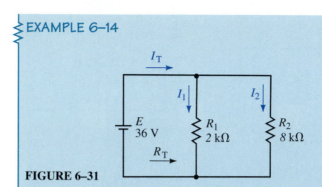

FIGURE 6–31

For the circuit of Figure 6–31, determine the following quantities:

a. R_T

b. I_T

c. Power delivered by the voltage source

d. I_1 and I_2 using the current divider rule

e. Power dissipated by the resistors

Solution

a. $R_T = \dfrac{R_1 R_2}{R_1 + R_2} = \dfrac{(2\text{ k}\Omega)(8\text{ k}\Omega)}{2\text{ k}\Omega + 8\text{ k}\Omega} = 1.6\text{ k}\Omega$

b. $I_T = \dfrac{E}{R_T} = \dfrac{36\text{ V}}{1.6\text{ k}\Omega} = 22.5\text{ mA}$

c. $P_T = EI_T = (36\text{ V})(22.5\text{ mA}) = 810\text{ mW}$

d. $I_2 = \dfrac{R_1}{R_1 + R_2} I_T = \left(\dfrac{2\text{ k}\Omega}{2\text{ k}\Omega + 8\text{ k}\Omega}\right)(22.5\text{ mA}) = 4.5\text{ mA}$

$I_1 = \dfrac{R_2}{R_1 + R_2} I_T = \left(\dfrac{8\text{ k}\Omega}{2\text{ k}\Omega + 8\text{ k}\Omega}\right)(22.5\text{ mA}) = 18.0\text{ mA}$

e. Since we know the voltage across each of the parallel resistors must be 36 V, we use this voltage to determine the power dissipated by each resistor. It would be equally correct to use the current through each resistor to calculate the power. However, it is generally best to use given information rather than calculated values to perform further calculations since it is then less likely that an error is carried through.

$$P_1 = \dfrac{E^2}{R_1} = \dfrac{(36\text{ V})^2}{2\text{ k}\Omega} = 648\text{ mW}$$

$$P_2 = \dfrac{E^2}{R_2} = \dfrac{(36\text{ V})^2}{8\text{ k}\Omega} = 162\text{ mW}$$

Notice that the power delivered by the voltage source is exactly equal to the total power dissipated by the resistors, namely $P_T = P_1 + P_2$.

EXAMPLE 6–15   Refer to the circuit of Figure 6–32:

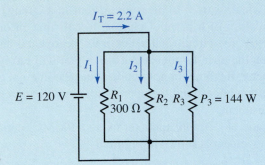

FIGURE 6–32

a. Solve for the total power delivered by the voltage source.

b. Find the currents I_1, I_2, and I_3.

c. Determine the values of the unknown resistors R_2 and R_3.

d. Calculate the power dissipated by each resistor.

e. Verify that the power dissipated is equal to the power delivered by the voltage source.

Solution

a. $P_T = EI_T = (120 \text{ V})(2.2 \text{ A}) = 264 \text{ W}$

b. Since the three resistors of the circuit are in parallel, we know that the voltage across all resistors must be equal to $E = 120$ V.

$$I_1 = \frac{V_1}{R_1} = \frac{120 \text{ V}}{300 \text{ }\Omega} = 0.4 \text{ A}$$

$$I_3 = \frac{P_3}{V_3} = \frac{144 \text{ W}}{120 \text{ V}} = 1.2 \text{ A}$$

Because KCL must be maintained at each node, we determine the current I_2 as

$$I_2 = I_T - I_1 - I_3$$
$$= 2.2 \text{ A} - 0.4 \text{ A} - 1.2 \text{ A} = 0.6 \text{ A}$$

c. $R_2 = \dfrac{V_2}{I_2} = \dfrac{120 \text{ V}}{0.6 \text{ A}} = 200 \text{ }\Omega$

Although we could use the calculated current I_3 to determine the resistance, it is best to use the given data in calculations rather than calculated values.

$$R_3 = \frac{V_3^2}{P_3} = \frac{(120 \text{ V})^2}{144 \text{ W}} = 100 \text{ }\Omega$$

d. $P_1 = \dfrac{V_1^2}{R_1} = \dfrac{(120 \text{ V})^2}{300 \text{ }\Omega} = 48 \text{ W}$

     $P_2 = I_2 E_2 = (0.6 \text{ A})(120 \text{ V}) = 72 \text{ W}$

e. $P_{\text{in}} = P_{\text{out}}$

     $264 \text{ W} = P_1 + P_2 + P_3$

     $264 \text{ W} = 48 \text{ W} + 72 \text{ W} + 144 \text{ W}$

     $264 \text{ W} = 264 \text{ W}$ (checks!)

6.7   Ammeter Design

The ammeter (which is used to measure current in a circuit) is a very good practical application of a parallel circuit. Recall that in order to use the ammeter correctly, the circuit under test must be opened and the ammeter inserted between the open terminals. Although you will rarely be required to design an ammeter, it is important to understand the internal operation of the ammeter. Once you understand the operation, you will appreciate the limitations of the instrument.

The schematic of a simple ammeter using a PMMC (permanent-magnet, moving-coil) movement is shown in Figure 6–33.

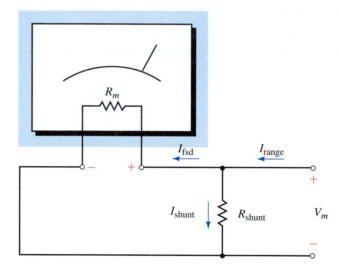

FIGURE 6–33   Simple ammeter.

Since the PMMC movement is able to handle only very small currents, a precisely engineered resistance is placed in parallel with the resistance of the meter movement, R_m. The shunt resistance, R_{shunt}, ensures that the current entering the sensitive meter movement is kept below the I_{fsd} (full-scale deflection current) of the meter movement.

The ammeter is designed to provide full-scale deflection when the current entering the instrument is at the desired range current. The shunt resistance ensures that the excess current bypasses the meter movement. The following example shows how the value of shunt resistance is determined for a given ammeter.

The voltage across the parallel combination of R_m and R_{shunt}, when the meter movement is at its maximum deflection, is determined from Ohm's law as

$$V_m = I_{fsd}R_m$$

From Kirchhoff's current law, the current through the shunt resistor (at full-scale deflection) is

$$I_{shunt} = I_{range} - I_{fsd}$$

And so, from Ohm's law, the shunt resistor has a value determined as

$$R_{\text{shunt}} = \frac{I_{\text{fsd}}R_m}{I_{\text{range}} - I_{\text{fsd}}} = \frac{1}{\left(\dfrac{I_{\text{range}}}{I_{\text{fsd}}} - 1\right)}R_m \qquad (6\text{--}14)$$

Since the above equation is quite complicated, it is generally very difficult to memorize the expression. It is much easier to remember the principles from which the expression is conceived.

EXAMPLE 6–16   Determine the value of shunt resistance required to build a 100-mA ammeter using a meter movement having $I_{\text{fsd}} = 1$ mA and $R_m = 2$ kΩ.

Solution   When the ammeter is detecting the maximum current, the voltage across the meter movement (and the shunt resistance) is

$$V_m = I_{\text{fsd}}R_m = (1 \text{ mA})(2 \text{ k}\Omega) = 2.0 \text{ V}$$

The current in the shunt resistance is

$$I_{\text{shunt}} = I_{\text{range}} - I_{\text{fsd}} = 100 \text{ mA} - 1 \text{ mA} = 99 \text{ mA}$$

And so the shunt resistance must have a value of

$$R_{\text{shunt}} = \frac{2.0 \text{ V}}{99 \text{ mA}} = 20.2 \ \Omega$$

The resulting circuit appears in Figure 6–34.

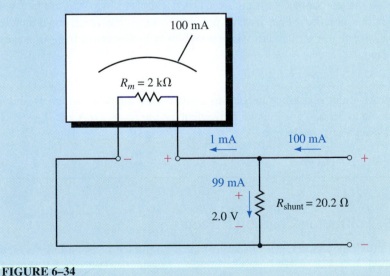

FIGURE 6–34

Figure 6–35 shows the schematic of a simple multirange ammeter. In order to have a multirange ammeter, it is necessary to have a different shunt resistor for each range.

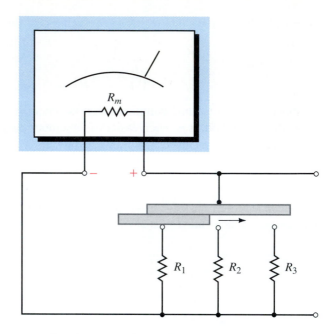

The selector switch must be a "make–before–break" switch to prevent applying excessive current to the meter movement when changing ranges.

FIGURE 6–35 Multirange ammeter.

To prevent damaging the sensitive meter movement, the ammeter of Figure 6–35 uses a "make-before-break" switch. As the name implies, this type of switch makes contact with the new position before breaking contact with the previous position. This feature prevents a large current from damaging the sensitive meter movement while a switch is in transition from one range to the next.

PRACTICE PROBLEMS 6

Design the shunt resistance needed to construct an ammeter which is able to measure up to 50 mA. Use a meter movement having $I_{fsd} = 2$ mA and $R_m = 2$ kΩ. Show a sketch of your design.

Answer: $R_{shunt} = 83.3$ Ω

6.8   Voltmeter Loading Effects

In the previous chapter, we observed that a voltmeter is essentially a meter movement in series with a current-limiting resistance. When a voltmeter is placed across two terminals to provide a voltage reading, the circuit is affected in the same manner as if a resistance were placed across the two terminals. The effect is shown in Figure 6–36.

If the resistance of the voltmeter is very large in comparison with the resistance across which the voltage is to be measured, the meter will indicate essentially the same voltage which was present before the meter was connected. On the other hand, if the meter has an internal resistance which is near in value to the resistance across which the measurement is taken, then the meter will adversely load the circuit, resulting in an erroneous reading. Generally, if the meter resistance is more than ten times larger than the resistance across which the voltage is taken, then the loading effect is considered negligible and may be ignored.

In the circuit of Figure 6–37, there is no current in the circuit since the terminals a and b are open circuited. The voltage appearing between the open terminals must be $V_{ab} = 10$ V. Now, if we place a voltmeter having an internal resistance of 200 kΩ between the terminals, the circuit is closed, resulting in a small current. The complete circuit appears as shown in Figure 6–38.

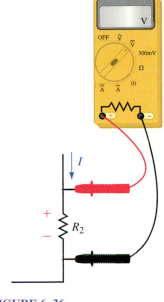

FIGURE 6–36

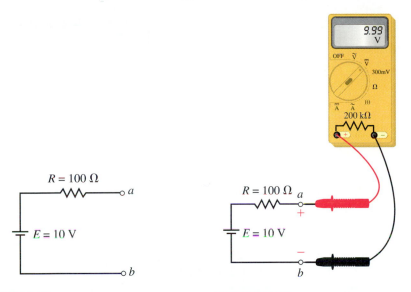

FIGURE 6–37   **FIGURE 6–38**

The reading indicated on the face of the meter is the voltage which occurs across the internal resistance of the meter. Applying Kirchhoff's voltage law to the circuit, this voltage is

$$V_{ab} = \frac{200 \text{ k}\Omega}{200 \text{ k}\Omega + 100 \text{ }\Omega}(10 \text{ V}) = 9.995 \text{ V}$$

Clearly, the reading on the face of the meter is essentially equal to the expected value of 10 V. Recall from the previous chapter that we defined the loading effect of a meter as follows:

$$\text{loading effect} = \frac{\text{actual value} - \text{reading}}{\text{actual value}} \times 100\%$$

For the circuit of Figure 6–38, the voltmeter has a loading effect of

$$\text{loading effect} = \frac{10 \text{ V} - 9.995 \text{ V}}{10 \text{ V}} \times 100\% = 0.05\%$$

This loading error is virtually undetectable for the circuit given. The same would not be true if we had a circuit as shown in Figure 6–39 and used the same voltmeter to provide a reading.

Again, if the circuit were left open circuited, we would expect that $V_{ab} = 10$ V.

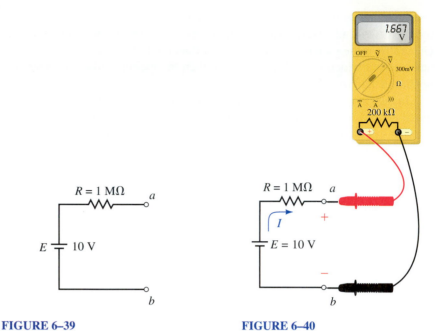

FIGURE 6–39   **FIGURE 6–40**

By connecting the 200-kΩ voltmeter between the terminals, as shown in Figure 6–40, we see that the voltage detected between terminals a and b will no longer be the desired voltage; rather,

$$V_{ab} = \frac{200 \text{ k}\Omega}{200 \text{ k}\Omega + 1 \text{ M}\Omega} (10 \text{ V}) = 1.667 \text{ V}$$

The loading effect of the meter in this circuit is

$$\text{loading effect} = \frac{10 \text{ V} - 1.667 \text{ V}}{10 \text{ V}} \times 100\% = 83.33\%$$

The previous illustration is an example of a problem which can occur when taking measurements in electronic circuits. When an inexperienced technician or technologist obtains an unforeseen result, he or she assumes that something is wrong with either the circuit or the instrument. In fact, both the circuit and the instrument are behaving in a perfectly predictable manner. The tech merely forgot to take into account the meter's loading effect. All instruments have limitations and we must always be aware of these limitations.

EXAMPLE 6–17   A digital voltmeter having an internal resistance of 5 MΩ is used to measure the voltage across terminals a and b in the circuit of Figure 6–40.

a. Determine the reading on the meter.

b. Calculate the loading effect of the meter.

Solution

a. The voltage applied to the meter terminals is

$$V_{ab} = \left(\frac{5 \text{ M}\Omega}{1 \text{ M}\Omega + 5 \text{ M}\Omega} \right)(10 \text{ V}) = 8.33 \text{ V}$$

b. The loading effect is

$$\text{loading error} = \frac{10 \text{ V} - 8.33 \text{ V}}{10 \text{ V}} \times 100\% = 16.7\%$$

All instruments have a loading effect on the circuit in which a measurement is taken. If you were given two voltmeters, one with an internal resistance of 200 kΩ and another with an internal resistance of 1 MΩ, which meter would load a circuit more? Explain.

IN-PROCESS
LEARNING
CHECK 3

(Answers are at the end of the chapter.)

6.9   Computer Analysis

As you have already seen, computer simulation is useful in providing a visualization of the skills you have learned. We will use both Electronics Workbench and PSpice to "measure" voltage and current in parallel circuits. One of the most useful features of Electronics Workbench is the program's ability to accurately simulate the operation of a real circuit. In this section, you will learn how to change the settings of the multimeter to observe meter loading in a circuit.

ELECTRONICS
WORKBENCH

PSpice

Electronics Workbench

EXAMPLE 6–18   Use Electronics Workbench to determine the currents I_T, I_1, and I_2 in the circuit of Figure 6–41. This circuit was analyzed previously in Example 6–14.

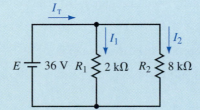

FIGURE 6–41

Solution   After opening the Circuit window:

• Select the components for the circuit from the Parts bin toolbars. You need to select the battery and ground symbol from the Sources toolbar. The resistors are obtained from the Basic toolbar.

• Once the circuit is completely wired, you may select the ammeters from the Indicators toolbar. Make sure that the ammeters are correctly placed into the circuit. Remember that the solid bar on the ammeter is connected to the lower-potential side of the circuit or branch.

• Simulate the circuit by clicking on the power switch. You should see the same results as shown in Figure 6–42.

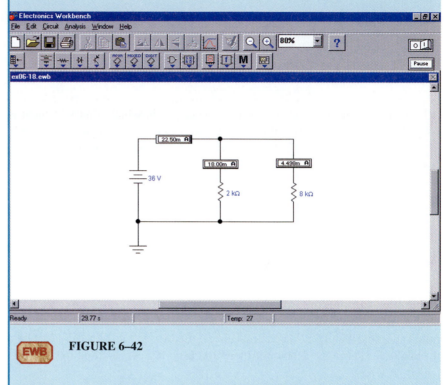

FIGURE 6–42

Notice that these results are consistent with those found in Example 6–14.

EXAMPLE 6–19   Use Electronics Workbench to demonstrate the loading effect of the voltmeter used in Figure 6–40. The voltmeter is to have internal resistance of 200 kΩ.

Solution   After opening the Circuit window:

• Construct the circuit by placing the battery, resistor, and ground as shown in Figure 6–40.
• Select the multimeter from the Instruments toolbar.
• Enlarge the multimeter by double clicking on the symbol.
• Click on the Settings button on the multimeter face.
• Change the voltmeter resistance to 200 kΩ. Accept the new value by clicking on OK.
• Run the simulation by clicking on the power switch. The resulting display is shown in Figure 6–43.

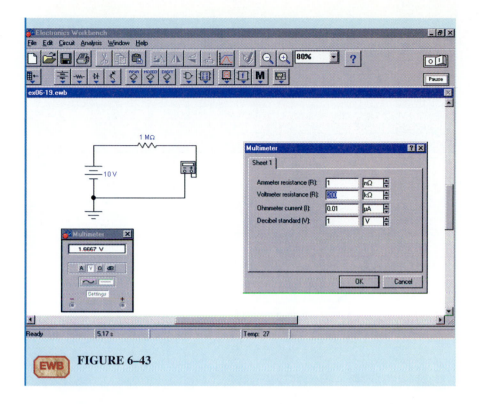

FIGURE 6–43

OrCAD PSpice

In previous PSpice examples, we used the IPRINT part to obtain the current in a circuit. An alternate method of measuring current is to use the Current Into Pin marker.

EXAMPLE 6–20   Use PSpice to determine the currents in the circuit of Figure 6–44.

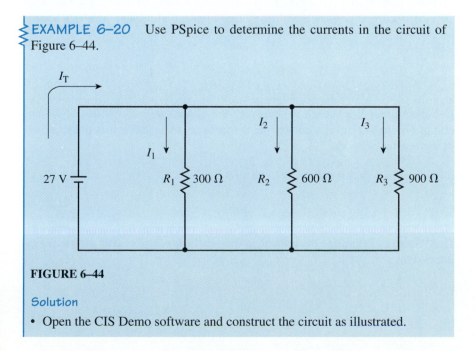

FIGURE 6–44

Solution
• Open the CIS Demo software and construct the circuit as illustrated.

- Place Current Into Pin markers as shown in Figure 6–45. Notice that the marker at the voltage source is placed at the negative terminal, since current enters this terminal.

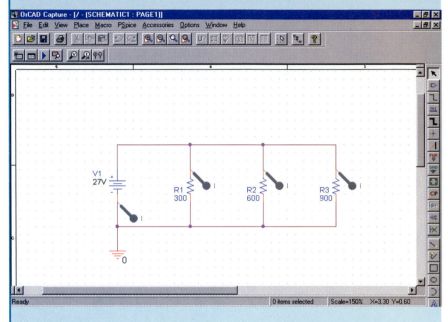

FIGURE 6–45

- Click on the New Simulation Profile and set the simulation so that the voltage source sweeps from 27 V to 27 V in 1-V increments.
- After running the project, you will observe a display of the circuit currents as a function of the source voltage. The currents are $I(\text{R1}) = 90$ mA, $I(\text{R2}) = 45$ mA, $I(\text{R3}) = 30$ mA, and $-I(\text{V1}) = 165$ mA.

PRACTICE PROBLEMS 7

Use Electronics Workbench to determine the current in each resistor of the circuit of Figure 6–21(a) if a 24-V voltage source is connected across the terminals of the resistor network.

Answers: $I_1 = I_2 = I_3 = 0.267$ A, $I_4 = 1.20$ A

PRACTICE PROBLEMS 8

Use PSpice to determine the current in each resistor of the circuit shown in Figure 6–46.

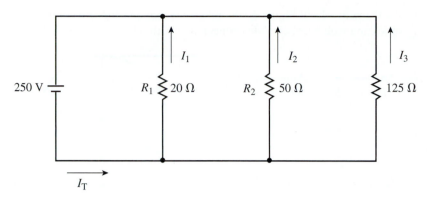

FIGURE 6–46

Answers: $I_1 = 12.5$ A, $I_2 = 5.00$ A, $I_3 = 2.00$ A, $I_T = 19.5$ A

PUTTING IT INTO PRACTICE

You have been hired as a consultant to a heating company. One of your jobs is to determine the number of 1000-W heaters that can be safely handled by an electrical circuit. All of the heaters in any circuit are connected in parallel. Each circuit operates at a voltage of 240 V and is rated for a maximum of 20 A. The normal operating current of the circuit should not exceed 80% of the maximum rated current. How many heaters can be safely installed in each circuit? If a room requires 5000 W of heaters to provide adequate heat during the coldest weather, how many circuits must be installed in this room?

6.1 Parallel Circuits

PROBLEMS

1. Indicate which of the elements in Figure 6–47 are connected in parallel and which elements are connected in series.

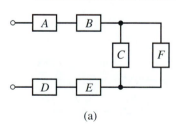

(a)

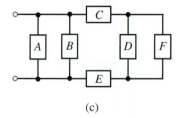

(c)

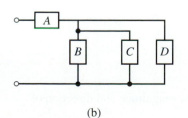

(b)

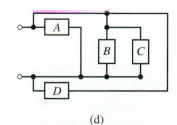

(d)

FIGURE 6–47

2. For the networks of Figure 6–48, indicate which resistors are connected in series and which resistors are connected in parallel.

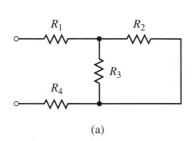

(a)

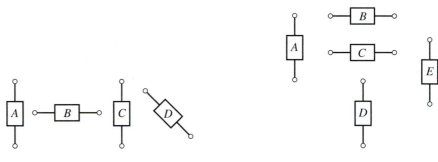

(b)   (c)

FIGURE 6–48

3. Without changing the component positions, show at least one way of connecting all the elements of Figure 6–49 in parallel.

4. Repeat Problem 3 for the elements shown in Figure 6–50.

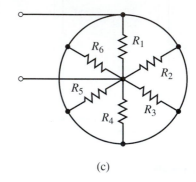

FIGURE 6–49   **FIGURE 6–50**

6.2 Kirchhoff's Current Law

5. Use Kirchhoff's current law to determine the magnitudes and directions of the indicated currents in each of the networks shown in Figure 6–51.

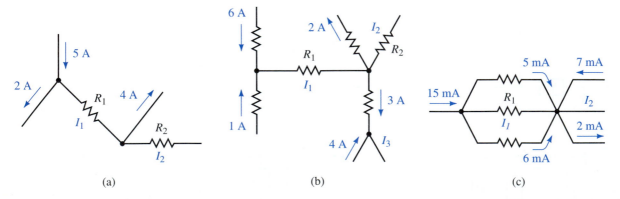

(a)   (b)   (c)

FIGURE 6–51

6. For the circuit of Figure 6–52, determine the magnitude and direction of each of the indicated currents.

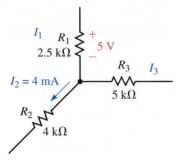

FIGURE 6–52

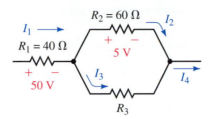

FIGURE 6–53

7. Consider the network of Figure 6–53:
   a. Calculate the currents I_1, I_2, I_3, and I_4.
   b. Determine the value of the resistance R_3.
8. Find each of the unknown currents in the networks of Figure 6–54.

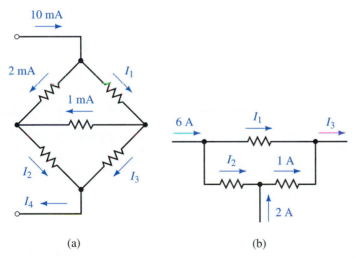

    (a)   (b)

FIGURE 6–54

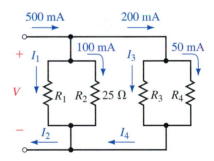

FIGURE 6–55

9. Refer to the network of Figure 6–55:
   a. Use Kirchhoff's current law to solve for the unknown currents, I_1, I_2, I_3, and I_4.
   b. Calculate the voltage, V, across the network.
   c. Determine the values of the unknown resistors, R_1, R_3, and R_4.
10. Refer to the network of Figure 6–56:
    a. Use Kirchhoff's current law to solve for the unknown currents.
    b. Calculate the voltage, V, across the network.
    c. Determine the required value of the voltage source, E. (Hint: Use Kirchhoff's voltage law.)

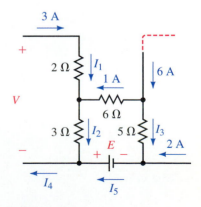

FIGURE 6–56

6.3   Resistors in Parallel

11. Calculate the total conductance and total resistance of each of the networks shown in Figure 6–57.

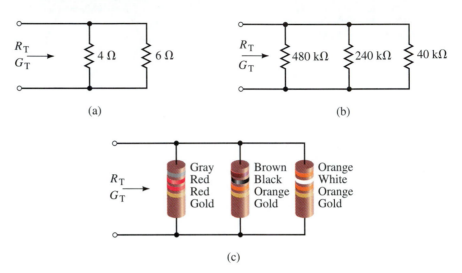

(a)   (b)

(c)

FIGURE 6–57

12. For the networks of Figure 6–58, determine the value of the unknown resistance(s) to result in the given total conductance.

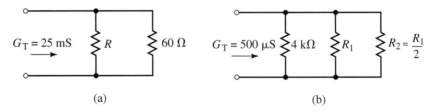

(a)   (b)

FIGURE 6–58

13. For the networks of Figure 6–59, determine the value of the unknown resistance(s) to result in the total resistances given.

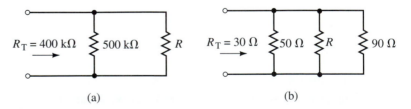

(a)   (b)

FIGURE 6–59

14. Determine the value of each unknown resistor in the network of Figure 6–60, so that the total resistance is 100 kΩ.

15. Refer to the network of Figure 6–61:

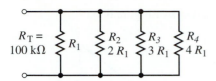

FIGURE 6–60

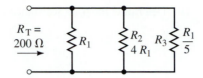

FIGURE 6–61

    a. Calculate the values of R_1, R_2, and R_3 so that the total resistance of the network is 200 Ω.

    b. If R_3 has a current of 2 A, determine the current through each of the other resistors.

    c. How much current must be applied to the entire network?

16. Refer to the network of Figure 6–62:

    a. Calculate the values of R_1, R_2, R_3, and R_4 so that the total resistance of the network is 100 kΩ.

    b. If R_4 has a current of 2 mA, determine the current through each of the other resistors.

    c. How much current must be applied to the entire network?

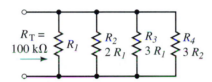

FIGURE 6–62

17. Refer to the network of Figure 6–63:

    a. Find the voltages across R_1 and R_2.

    b. Determine the current I_2.

18. Refer to the network of Figure 6–64:

    a. Find the voltages across R_1, R_2 and R_3.

    b. Calculate the current I_2.

    c. Calculate the current I_3.

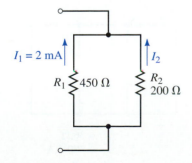

FIGURE 6–63

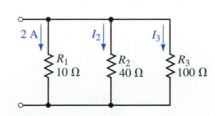

FIGURE 6–64

19. Determine the total resistance of each network of Figure 6–65.

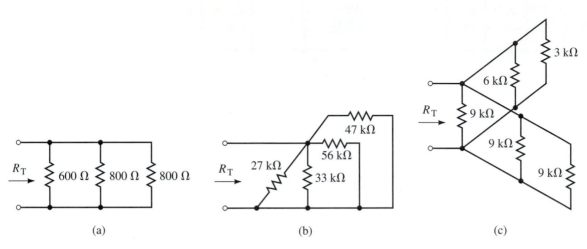

(a)   (b)   (c)

FIGURE 6–65
FIGURE 6–65

20. Determine the total resistance of each network of Figure 6–66.

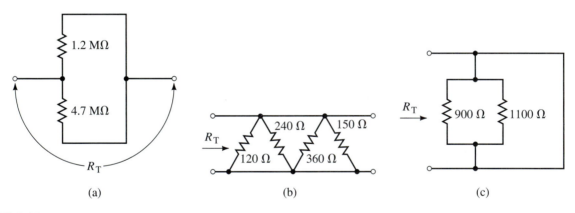

(a)   (b)   (c)

FIGURE 6–66

21. Determine the values of the resistors in the circuit of Figure 6–67, given the indicated conditions.

22. Given the indicated conditions, calculate all currents and determine all resistor values for the circuit of Figure 6–68.

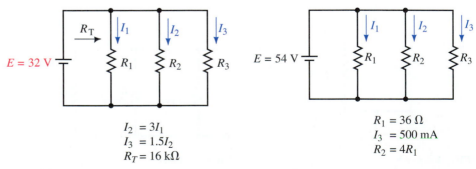

$I_2 = 3I_1$
$I_3 = 1.5I_2$
$R_T = 16\ \text{k}\Omega$

$R_1 = 36\ \Omega$
$I_3 = 500\ \text{mA}$
$R_2 = 4R_1$

FIGURE 6–67   **FIGURE 6–68**

23. Without using a pencil, paper, or a calculator, determine the resistance of each network of Figure 6–69.

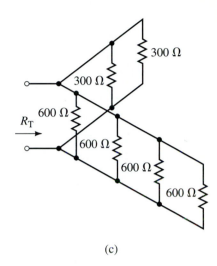

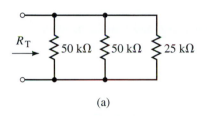

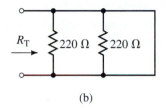

(a)   (b)   (c)

FIGURE 6–69

24. Without using a pencil, paper, or calculator, determine the approximate resistance of the network of Figure 6–70.

25. Without using a pencil, paper, or a calculator, approximate the total resistance of the network of Figure 6–71.

26. Derive Equation 6–7, which is used to calculate the total resistance of three parallel resistors.

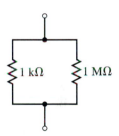

FIGURE 6–70

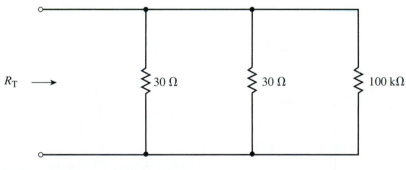

FIGURE 6–71

6.4   Voltage Sources in Parallel

27. Two 20-V batteries are connected in parallel to provide current to a 100-Ω load as shown in Figure 6–72. Determine the current in the load and the current in each battery.

28. Two lead-acid automobile batteries are connected in parallel, as shown in Figure 6–73, to provide additional starting current. One of the batteries is fully charged at 14.2 V and the other battery has discharged to 9 V. If the internal resistance of each battery is 0.01 Ω, determine the current in the batteries. If each battery is intended to provide a maximum current of 150 A, should this method be used to start a car?

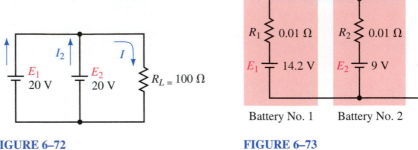

FIGURE 6–72   **FIGURE 6–73**

6.5   Current Divider Rule

29. Use the current divider rule to find the currents I_1 and I_2 in the networks of Figure 6–74.

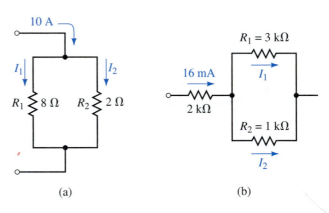

FIGURE 6–74

30. Repeat Problem 29 for the networks of Figure 6–75.

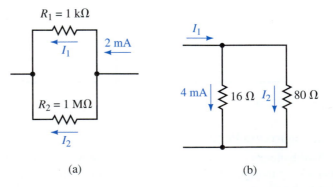

FIGURE 6–75

31. Use the current divider rule to determine all unknown currents for the networks of Figure 6–76.

32. Repeat Problem 31 for the networks of Figure 6–77.

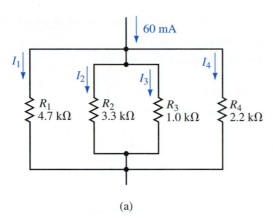

(a)

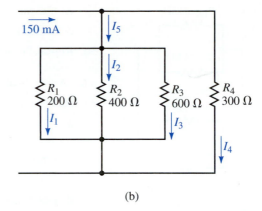

(b)

FIGURE 6–76

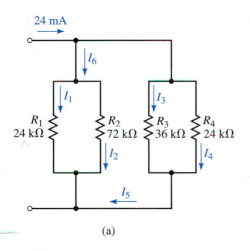

(a)

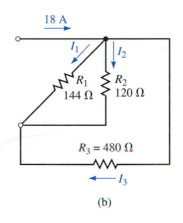

(b)

FIGURE 6–77

33. Use the current divider rule to determine the unknown resistance in the network of Figure 6–78.

34. Use the current divider rule to determine the unknown resistance in the network of Figure 6–79.

35. Refer to the circuit of Figure 6–80:

   a. Determine the equivalent resistance, R_T, of the circuit.

   b. Solve for the current I.

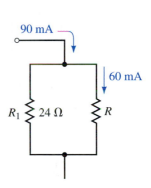

FIGURE 6–78

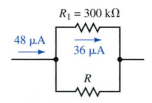

FIGURE 6–79

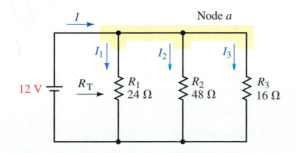

FIGURE 6–80

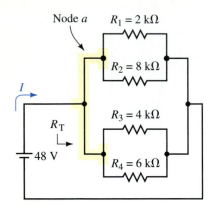

Node *a* $R_1 = 2\ \text{k}\Omega$

$R_2 = 8\ \text{k}\Omega$

$R_3 = 4\ \text{k}\Omega$

R_T

48 V

$R_4 = 6\ \text{k}\Omega$

FIGURE 6–81

c. Use the current divider rule to determine the current in each resistor.

d. Verify Kirchhoff's current law at node *a*.

36. Repeat Problem 35 for the circuit of Figure 6–81.

6.6   Analysis of Parallel Circuits

37. Refer to the circuit of Figure 6–82:

a. Find the total resistance, R_T, and solve for the current, *I,* through the voltage source.

b. Find all of the unknown currents in the circuit.

c. Verify Kirchhoff's current law at node *a*.

d. Determine the power dissipated by each resistor. Verify that the total power dissipated by the resistors is equal to the power delivered by the voltage source.

38. Repeat Problem 37 for the circuit of Figure 6–83.

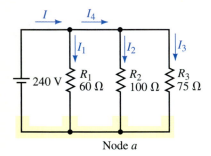

I I_4

I_1 I_2 I_3

240 V R_1 $60\ \Omega$ R_2 $100\ \Omega$ R_3 $75\ \Omega$

Node *a*

FIGURE 6–82

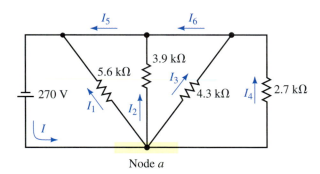

I_5 I_6

3.9 kΩ

5.6 kΩ

I_3

270 V I_1 I_2 4.3 kΩ I_4 2.7 kΩ

I

Node *a*

FIGURE 6–83

39. Refer to the circuit of Figure 6–84:

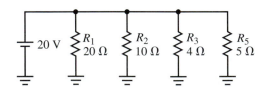

20 V R_1 $20\ \Omega$ R_2 $10\ \Omega$ R_3 $4\ \Omega$ R_5 $5\ \Omega$

FIGURE 6–84

a. Calculate the current through each resistor in the circuit.

b. Determine the total current supplied by the voltage source.

c. Find the power dissipated by each resistor.

40. Refer to the circuit of Figure 6–85.

a. Solve for the indicated currents.

b. Find the power dissipated by each resistor.

c. Verify that the power delivered by the voltage source is equal to the total power dissipated by the resistors.

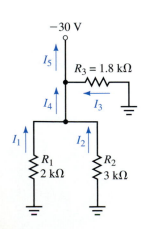

−30 V

I_5 $R_3 = 1.8\ \text{k}\Omega$

I_4 I_3

I_1 I_2

R_1 $2\ \text{k}\Omega$ R_2 $3\ \text{k}\Omega$

FIGURE 6–85

41. Given the circuit of Figure 6–86:

    a. Determine the values of all resistors.

    b. Calculate the currents through R_1, R_2, and R_4.

    c. Find the currents I_1 and I_2.

    d. Find the power dissipated by resistors R_2, R_3, and R_4.

42. A circuit consists of four resistors connected in parallel and connected to a 20-V source as shown in Figure 6–87. Determine the minimum power rating of each resistor if resistors are available with the following power ratings: $\frac{1}{8}$ W, $\frac{1}{4}$ W, $\frac{1}{2}$ W, 1 W, and 2 W.

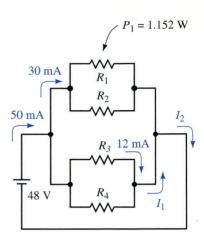

FIGURE 6–86

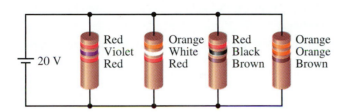

FIGURE 6–87

43. For the circuit of Figure 6–88, determine each of the indicated currents. If the circuit has a 15-A fuse as shown, is the current enough to cause the fuse to open?

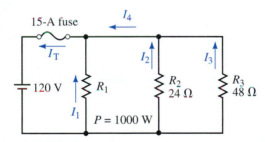

FIGURE 6–88

44. a. For the circuit of Figure 6–88, calculate the value of R_3 which will result in a circuit current of exactly $I_T = 15$ A.

    b. If the value of R_3 is increased above the value found in part (a), what will happen to the circuit current, I_T?

6.7   Ammeter Design

45. A common ammeter uses a meter movement having $I_{fsd} = 50$ μA and $R_m = 5$ kΩ. If the meter has a 10-mA range, determine the value of the shunt resistor needed for this range.

46. Using a meter movement having $I_{fsd} = 50$ μA and $R_m = 2$ kΩ, design an ammeter having a 10-mA range, a 100-mA range, and a 250-mA range. Sketch your design.

47. An ammeter has a 0.5-Ω shunt and a meter movement with $I_{fsd} = 1$ mA and $R_m = 5$ kΩ. Determine the maximum current which can be measured with

this ammeter. If the meter deflects to 62% of its full-scale deflection, how much current is being applied to the ammeter?

48. Using a meter movement having $I_{fsd} = 1$ mA and $R_m = 5$ kΩ, design an ammeter having a 5-mA range, a 20-mA range, and a 100-mA range. Sketch your design.

6.8   Voltmeter Loading Effects

49. A voltmeter having a 1-MΩ internal resistance is used to measure the indicated voltage in the circuit shown in Figure 6–89.

a. Determine the voltage reading which will be indicated by the meter.

b. Calculate the voltmeter's loading effect when used to measure the indicated voltage.

50. Repeat Problem 49 if the 500-kΩ resistor of Figure 6–89 is replaced with a 2-MΩ resistor.

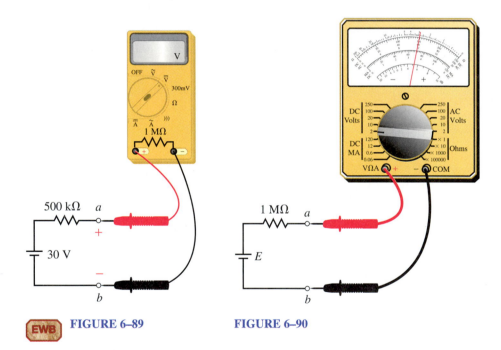

FIGURE 6–89   **FIGURE 6–90**

51. An inexpensive analog voltmeter is used to measure the voltage across terminals a and b of the circuit shown in Figure 6–90. If the voltmeter indicates that the voltage $V_{ab} = 1.2$ V, what is the actual voltage of the source if the resistance of the meter is 50 kΩ?

52. What would be the reading if a digital meter having an internal resistance of 10 MΩ is used instead of the analog meter of Problem 51?

6.9   Computer Analysis

53. **EWB**   Use Electronics Workbench to solve for the current through each resistor in the circuit of Figure 8–82.

54. **EWB**   Use Electronics Workbench to solve for the current through each resistor in the circuit of Figure 8–83.

55. **EWB**   Use Electronics Workbench to simulate a voltmeter with an internal resistance of 1 MΩ used to measure voltage as shown in Figure 6–89.

56. **EWB**   Use Electronics Workbench to simulate a voltmeter with an internal resistance of 500 kΩ used to measure voltage as shown in Figure 6–89.

57. **PSpice**   Use PSpice to solve for the current through each resistor in the circuit of Figure 6–82.

58. **PSpice**   Use PSpice to solve for the current through each resistor in the circuit of Figure 6–83.

In-Process Learning Check 1

a. $I = 2.00$ A

b. $I_1 = I_2 = I_3 = 0.267$ A, $I_4 = 1.200$ A

c. $3(0.267$ A$) + 1.200$ A $= 2.00$ A (as required)

In-Process Learning Check 2

a. $I_{1\Omega} = 30.0$ A, $I_{3\Omega} = 10.0$ A, $I_{4\Omega} = 7.50$ A

b. $I_T = 53.5$ A

c. $R_T = 0.561$ Ω. (The currents are the same as those determined in part a.)

In-Process Learning Check 3

The voltmeter with the smaller internal resistance would load the circuit more, since more of the circuit current would enter the instrument.

7

Series-Parallel Circuits

OBJECTIVES

After studying this chapter, you will be able to

- find the total resistance of a network consisting of resistors connected in various series-parallel configurations,
- solve for the current through any branch or component of a series-parallel circuit,
- determine the difference in potential between any two points in a series-parallel circuit,
- calculate the voltage drop across a resistor connected to a potentiometer,
- analyze how the size of a load resistor connected to a potentiometer affects the output voltage,
- calculate the loading effects of a voltmeter or ammeter when used to measure the voltage or current in any circuit,
- use PSpice to solve for voltages and currents in series-parallel circuits,
- use Electronics Workbench to solve for voltages and currents in series-parallel circuits.

KEY TERMS

Branch Currents
Parallel Branches
Potentiometer Circuits
Series-Parallel Connection
Transistor Bias
Zener Diode

OUTLINE

The Series-Parallel Network
Analysis of Series-Parallel Circuits
Applications of Series-Parallel Circuits
Potentiometers
Loading Effects of Instruments
Circuit Analysis Using Computers

ost circuits encountered in electronics are neither simple series circuits nor simple parallel circuits, but rather a combination of the two. Although series-parallel circuits appear to be more complicated than either of the previous types of circuits analyzed to this point, we find that the same principles apply.

This chapter examines how Kirchhoff's voltage and current laws are applied to the analysis of series-parallel circuits. We will also observe that voltage and current divider rules apply to the more complex circuits. In the analysis of series-parallel circuits, we often simplify the given circuit to enable us to more clearly see how the rules and laws of circuit analysis apply. Students are encouraged to redraw circuits whenever the solution of a problem is not immediately apparent. This technique is used by even the most experienced engineers, technologists, and technicians.

In this chapter, we begin by examining simple resistor circuits. The principles of analysis are then applied to more practical circuits such as those containing zener diodes and transistors. The same principles are then applied to determine the loading effects of voltmeters and ammeters in more complex circuits.

After analyzing a complex circuit, we want to know whether the solutions are in fact correct. As you have already seen, electrical circuits usually may be studied in more than one way to arrive at a solution. Once currents and voltages for a circuit have been found, it is very easy to determine whether the resultant solution verifies the law of conservation of energy, Kirchhoff's current law, and Kirchhoff's voltage law. If there is any discrepancy (other than rounding error), there is an error in the calculation!

Benjamin Franklin

BENJAMIN FRANKLIN WAS BORN IN BOSTON, Massachusetts, in 1706. Although Franklin is best known as a great statesman and diplomat, he also furthered the cause of science with his experiments in electricity. This particularly includes his work with the Leyden jar, which was used to store electric charge. In his famous experiment of 1752, he used a kite to demonstrate that lightning is an electrical event. It was Franklin who postulated that positive and negative electricity are in fact a single "fluid."

Although Franklin's major accomplishments came as a result of his work in achieving independence of the Thirteen Colonies, he was nonetheless a notable scientist.

Benjamin Franklin died in his Philadelphia home on February 12, 1790, at the age of eighty-four.

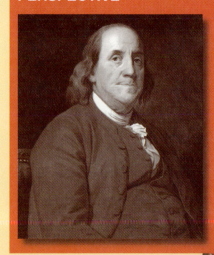

7.1 The Series-Parallel Network

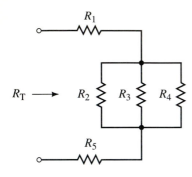

FIGURE 7–1

In electric circuits, we define a **branch** as any portion of a circuit which can be simplified as having two terminals. The components between the two terminals may be any combination of resistors, voltage sources, or other elements. Many complex circuits may be separated into a combination of both series and/or parallel elements, while other circuits consist of even more elaborate combinations which are neither series nor parallel.

In order to analyze a complicated circuit, it is important to be able to recognize which elements are in series and which elements or branches are in parallel. Consider the network of resistors shown in Figure 7–1.

We immediately recognize that the resistors R_2, R_3, and R_4 are in parallel. This parallel combination is in series with the resistors R_1 and R_5. The total resistance may now be written as follows:

$$R_T = R_1 + (R_2\|R_3\|R_4) + R_5$$

EXAMPLE 7–1 For the network of Figure 7–2, determine which resistors and branches are in series and which are in parallel. Write an expression for the total equivalent resistance, R_T.

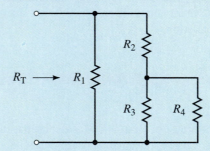

FIGURE 7–2

Solution First, we recognize that the resistors R_3 and R_4 are in parallel: $(R_3\|R_4)$.

Next, we see that this combination is in series with the resistor R_2: $[R_2 + (R_3\|R_4)]$.

Finally, the entire combination is in parallel with the resistor R_1. The total resistance of the circuit may now be written as follows:

$$R_T = R_1 \| [R_2 + (R_3\|R_4)]$$

For the network of Figure 7–3, determine which resistors and branches are in series and which are in parallel. Write an expression for the total resistance, R_T.

FIGURE 7–3

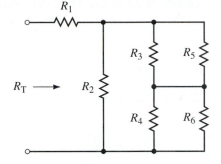

Answer: $R_T = R_1 + R_2 \| [(R_3\|R_5) + (R_4\|R_6)]$

7.2   Analysis of Series-Parallel Circuits

Series-parallel networks are often difficult to analyze because they initially appear confusing. However, the analysis of even the most complex circuit is simplified by following some fairly basic steps. By practicing (not memorizing) the techniques outlined in this section, you will find that most circuits can be reduced to groupings of series and parallel combinations. In analyzing such circuits, it is imperative to remember that the rules for analyzing series and parallel elements still apply.

The same current occurs through all series elements.
The same voltage occurs across all parallel elements.

In addition, remember that Kirchhoff's voltage law and Kirchhoff's current law apply for all circuits regardless of whether the circuits are series, parallel, or series-parallel. The following steps will help to simplify the analysis of series-parallel circuits:

1. Whenever necessary, redraw complicated circuits showing the source connection at the left-hand side. All nodes should be labelled to ensure that the new circuit is equivalent to the original circuit. You will find that as you become more experienced at analyzing circuits, this step will no longer be as important and may therefore be omitted.

2. Examine the circuit to determine the strategy which will work best in analyzing the circuit for the required quantities. You will usually find it best to begin the analysis of the circuit at the components most distant to the source.

3. Simplify recognizable combinations of components wherever possible, redrawing the resulting circuit as often as necessary. Keep the same labels for corresponding nodes.

4. Determine the equivalent circuit resistance, R_T.

5. Solve for the total circuit current. Indicate the directions of all currents and label the correct polarities of the voltage drops on all components.

6. Calculate how currents and voltages split between the elements of the circuit.

7. Since there are usually several possible ways at arriving at solutions, verify the answers by using a different approach. The extra time taken in this step will usually ensure that the correct answer has been found.

EXAMPLE 7–2 Consider the circuit of Figure 7–4.

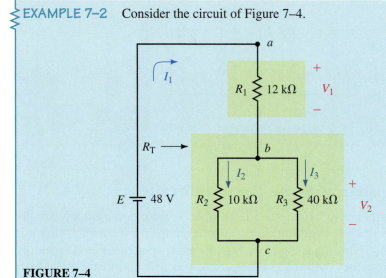

FIGURE 7–4

a. Find R_T.

b. Calculate I_1, I_2, and I_3.

c. Determine the voltages V_1 and V_2.

Solution By examining the circuit of Figure 7–4, we see that resistors R_2 and R_3 are in parallel. This parallel combination is in series with the resistor R_1.

The combination of resistors may be represented by a simple series network shown in Figure 7–5. Notice that the nodes have been labelled using the same notation.

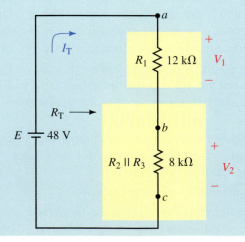

FIGURE 7–5

a. The total resistance of the circuit may be determined from the combination

$$R_T = R_1 + R_2 \| R_3$$

$$R_T = 12 \text{ k}\Omega + \frac{(10 \text{ k}\Omega)(40 \text{ k}\Omega)}{10 \text{ k}\Omega + 40 \text{ k}\Omega}$$

$$= 12 \text{ k}\Omega + 8 \text{ k}\Omega = 20 \text{ k}\Omega$$

b. From Ohm's law, the total current is

$$I_T = I_1 = \frac{48 \text{ V}}{20 \text{ k}\Omega} = 2.4 \text{ mA}$$

The current I_1 will enter node b and then split between the two resistors R_2 and R_3. This current divider may be simplified as shown in the partial circuit of Figure 7–6.

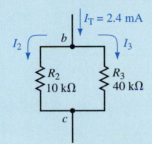

FIGURE 7–6

Applying the current divider rule to these two resistors gives

$$I_2 = \frac{(40 \text{ k}\Omega)(2.4 \text{ mA})}{10 \text{ k}\Omega + 40 \text{ k}\Omega} = 1.92 \text{ mA}$$

$$I_3 = \frac{(10 \text{ k}\Omega)(2.4 \text{ mA})}{10 \text{ k}\Omega + 40 \text{ k}\Omega} = 0.48 \text{ A}$$

c. Using the above currents and Ohm's law, we determine the voltages:

$$V_1 = (2.4 \text{ mA})(12 \text{ k}\Omega) = 28.8 \text{ V}$$

$$V_3 = (0.48 \text{ mA})(40 \text{ k}\Omega) = 19.2 \text{ V} = V_2$$

In order to check the answers, we may simply apply Kirchhoff's voltage law around any closed loop which includes the voltage source:

$$\sum V = E - V_1 - V_3$$
$$= 48 \text{ V} - 28.8 \text{ V} - 19.2 \text{ V}$$
$$= 0 \text{ V (checks!)}$$

The solution may be verified by ensuring that the power delivered by the voltage source is equal to the summation of powers dissipated by the resistors.

Use the results of Example 7–2 to verify that the law of conservation of energy applies to the circuit of Figure 7–4 by showing that the voltage source delivers the same power as the total power dissipated by all resistors.

IN-PROCESS
LEARNING
CHECK 1

(Answers are at the end of the chapter.)

EXAMPLE 7–3   Find the voltage V_{ab} for the circuit of Figure 7–7.

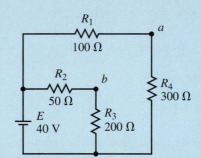

FIGURE 7–7

Solution   We begin by redrawing the circuit in a more simple representation as shown in Figure 7–8.

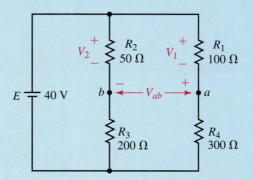

FIGURE 7–8

From Figure 7–8, we see that the original circuit consists of two parallel branches, where each branch is a series combination of two resistors.

If we take a moment to examine the circuit, we see that the voltage V_{ab} may be determined from the combination of voltages across R_1 and R_2. Alternatively, the voltage may be found from the combination of voltages across R_3 and R_4.

As usual, several methods of analysis are possible. Because the two branches are in parallel, the voltage across each branch must be 40 V. Using the voltage divider rules allows us to quickly calculate the voltage across each resistor. Although equally correct, other methods of calculating the voltages would be more lengthy.

$$V_2 = \frac{R_2}{R_2 + R_3}E$$

$$= \left(\frac{50\ \Omega}{50\ \Omega + 200\ \Omega}\right)(40\ \text{V}) = 8.0\ \text{V}$$

$$V_1 = \frac{R_1}{R_1 + R_4}E$$

$$= \left(\frac{100\ \Omega}{100\ \Omega + 300\ \Omega}\right)(40\ \text{V}) = 10.0\ \text{V}$$

As shown in Figure 7–9, we apply Kirchhoff's voltage law to determine the voltage between terminals a and b.

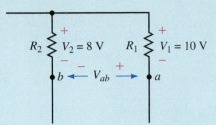

FIGURE 7–9

$$V_{ab} = -10.0 \text{ V} + 8.0 \text{ V} = -2.0 \text{ V}$$

EXAMPLE 7–4   Consider the circuit of Figure 7–10:

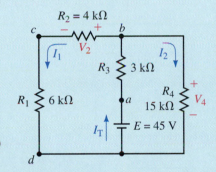

FIGURE 7–10

a. Find the total resistance R_T "seen" by the source E.

b. Calculate I_T, I_1, and I_2.

c. Determine the voltages V_2 and V_4.

Solution   We begin the analysis by redrawing the circuit. Since we generally like to see the source on the left-hand side, one possible way of redrawing the resultant circuit is shown in Figure 7–11. Notice that the polarities of voltages across all resistors have been shown.

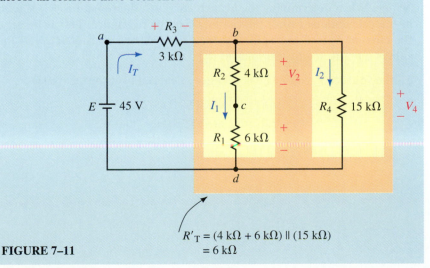

$$R'_T = (4 \text{ k}\Omega + 6 \text{ k}\Omega) \parallel (15 \text{ k}\Omega)$$
$$= 6 \text{ k}\Omega$$

FIGURE 7–11

a. From the redrawn circuit, the total resistance of the circuit is

$$R_T = R_3 + [(R_1 + R_2)\|R_4]$$

$$= 3\text{ k}\Omega + \frac{(4\text{ k}\Omega + 6\text{ k}\Omega)(15\text{ k}\Omega)}{(4\text{ k}\Omega + 6\text{ k}\Omega) + 15\text{ k}\Omega}$$

$$= 3\text{ k}\Omega + 6\text{ k}\Omega = 9.00\text{ k}\Omega$$

b. The current supplied by the voltage source is

$$I_T = \frac{E}{R_T} = \frac{45\text{ V}}{9\text{ k}\Omega} = 5.00\text{ mA}$$

We see that the supply current divides between the parallel branches as shown in Figure 7–12.

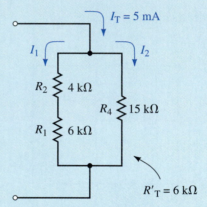

FIGURE 7–12

Applying the current divider rule, we calculate the branch currents as

$$I_1 = I_T \frac{R'_T}{(R_1 + R_2)} = \frac{(5\text{ mA})(6\text{ k}\Omega)}{4\text{ k}\Omega + 6\text{ k}\Omega} = 3.00\text{ mA}$$

$$I_2 = I_T \frac{R'_T}{R_4} = \frac{(5\text{ mA})(6\text{ k}\Omega)}{15\text{ k}\Omega} = 2.00\text{ mA}$$

Notice: When determining the branch currents, the resistance R'_T is used in the calculations rather the the total circuit resistance. This is because the current $I_T = 5$ mA splits between the two branches of R'_T and the split is not affected by the value of R_3.

c. The voltages V_2 and V_4 are now easily calculated by using Ohm's law:

$$V_2 = I_1 R_2 = (3\text{ mA})(4\text{ k}\Omega) = 12.0\text{ V}$$

$$V_4 = I_2 R_4 = (2\text{ mA})(15\text{ k}\Omega) = 30.0\text{ V}$$

EXAMPLE 7–5   For the circuit of Figure 7–13, find the indicated currents and voltages.

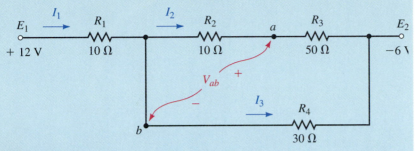

EWB   **FIGURE 7–13**

Solution   Because the above circuit contains voltage point sources, it is easier to analyze if we redraw the circuit to help visualize the operation.

The point sources are voltages with respect to ground, and so we begin by drawing a circuit with the reference point as shown in Figure 7–14.

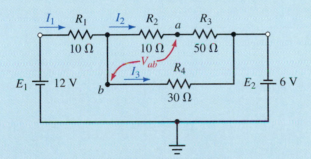

FIGURE 7–14

Now, we can see that the circuit may be further simplified by combining the voltage sources ($E = E_1 + E_2$) and by showing the resistors in a more suitable location. The simplified circuit is shown in Figure 7–15.

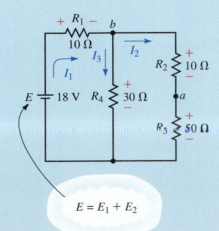

FIGURE 7–15

The total resistance "seen" by the equivalent voltage source is

$$R_T = R_1 + [R_4 \| (R_2 + R_3)]$$

$$= 10\ \Omega + \frac{(30\ \Omega)(10\ \Omega + 50\ \Omega)}{30\ \Omega + (10\ \Omega + 50\ \Omega)} = 30.0\ \Omega$$

And so the total current provided into the circuit is

$$I_1 = \frac{E}{R_T} = \frac{18\ \text{V}}{30\ \Omega} = 0.600\ \text{A}$$

At node *b* this current divides between the two branches as follows:

$$I_3 = \frac{(R_2 + R_3)I_1}{R_4 + R_2 + R_3} = \frac{(60\ \Omega)(0.600\ \text{A})}{30\ \Omega + 10\ \Omega + 50\ \Omega} = 0.400\ \text{A}$$

$$I_2 = \frac{R_4 I_1}{R_4 + R_2 + R_3} = \frac{(30\ \Omega)(0.600\ \text{A})}{30\ \Omega + 10\ \Omega + 50\ \Omega} = 0.200\ \text{A}$$

The voltage V_{ab} has the same magnitude as the voltage across the resistor R_2, but with a negative polarity (since *b* is at a higher potential than *a*):

$$V_{ab} = -I_2 R_2 = -(0.200\ \text{A})(10\ \Omega) = -2.0\ \text{V}$$

PRACTICE PROBLEMS 2

Consider the circuit of Figure 7–16:

FIGURE 7–16

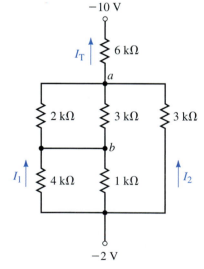

a. Find the total circuit resistance, R_T.

b. Determine the current I_T through the voltage sources.

c. Solve for the currents I_1 and I_2.

d. Calculate the voltage V_{ab}.

Answers: a. $R_T = 7.20\ \text{k}\Omega$ b. $I_T = 1.11\ \text{mA}$ c. $I_1 = 0.133\ \text{mA}$; $I_2 = 0.444\ \text{mA}$
d. $V_{ab} = -0.800\ \text{V}$

7.3   Applications of Series-Parallel Circuits

We now examine how the methods developed in the first two sections of this chapter are applied when analyzing practical circuits. You may find that some of the circuits introduce you to unfamiliar devices. For now, you do not need to know precisely how these devices operate, simply that the voltages and currents in the circuits follow the same rules and laws that you have used up to now.

EXAMPLE 7–6   The circuit of Figure 7–17 is referred to as a *bridge circuit* and is used extensively in electronic and scientific instruments.

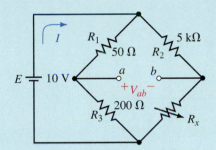

FIGURE 7–17

Calculate the current I and the voltage V_{ab} when

a. $R_x = 0 \; \Omega$ (short circuit)

b. $R_x = 15 \; k\Omega$

c. $R_x = \infty$ (open circuit)

Solution

a. $R_x = 0 \; \Omega$:

The circuit is redrawn as shown in Figure 7–18.

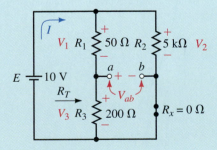

FIGURE 7–18

The voltage source "sees" a total resistance of

$$R_T = (R_1 + R_3)\|R_2 = 250 \; \Omega \| 5000 \; \Omega = 238 \; \Omega$$

resulting in a source current of

$$I = \frac{10 \text{ V}}{238 \; \Omega} = 0.042 \text{ A} = 42.2 \text{ mA}$$

The voltage V_{ab} may be determined by solving for voltage across R_1 and R_2.

The voltage across R_1 will be constant regardless of the value of the variable resistor R_x. Hence

$$V_1 = \left(\frac{50 \ \Omega}{50 \ \Omega + 200 \ \Omega} \right)(10 \ \text{V}) = 2.00 \ \text{V}$$

Now, since the variable resistor is a short circuit, the entire source voltage will appear across the resistor R_2, giving

$$V_2 = 10.0 \ \text{V}$$

And so

$$V_{ab} = -V_1 + V_2 = -2.00 \ \text{V} + 10.0 \ \text{V} = +8.00 \ \text{V}$$

b. $R_x = 15 \ \text{k}\Omega$:

The circuit is redrawn in Figure 7–19.

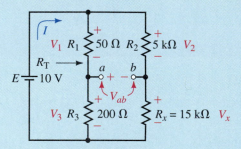

FIGURE 7–19

The voltage source "sees" a circuit resistance of

$$R_T = (R_1 + R_3)\|(R_2 + R_x)$$
$$= 250 \ \Omega \| 20 \ \text{k}\Omega = 247 \ \Omega$$

which results in a source current of

$$I = \frac{10 \ \text{V}}{247 \ \Omega} = 0.0405 \ \text{A} = 40.5 \ \text{mA}$$

The voltages across R_1 and R_2 are

$$V_1 = 2.00 \ \text{V} \quad \text{(as before)}$$

$$V_2 = \frac{R_2}{R_2 + R_x} E$$

$$= \left(\frac{5 \ \text{k}\Omega}{5 \ \text{k}\Omega + 15 \ \text{k}\Omega} \right)(10 \ \text{V}) = 2.50 \ \text{V}$$

Now the voltage between terminals a and b is found as

$$V_{ab} = -V_1 + V_2$$
$$= -2.0 \ \text{V} + 2.5 \ \text{V} = +0.500 \ \text{V}$$

c. $R_x = \infty$:

The circuit is redrawn in Figure 7–20.

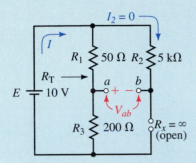

FIGURE 7–20

Because the second branch is an open circuit due to the resistor R_x, the total resistance "seen" by the source is

$$R_T = R_1 + R_3 = 250\ \Omega$$

resulting in a source current of

$$I = \frac{10\ \text{V}}{250\ \Omega} = 0.040\ \text{A} = 40.0\ \text{mA}$$

The voltages across R_1 and R_2 are

$$V_1 = 2.00\ \text{V} \quad \text{(as before)}$$
$$V_2 = 0\ \text{V} \quad \text{(since the branch is open)}$$

And so the resulting voltage between terminals a and b is

$$V_{ab} = -V_1 + V_2$$
$$= -2.0\ \text{V} + 0\ \text{V} = -2.00\ \text{V}$$

The previous example illustrates how voltages and currents within a circuit are affected by changes elsewhere in the circuit. In the example, we saw that the voltage V_{ab} varied from -2 V to $+8$ V, while the total circuit current varied from a minimum value of 40 mA to a maximum value of 42 mA. These changes occurred even though the resistor R_x varied from $0\ \Omega$ to ∞.

A **transistor** is a three-terminal device which may be used to amplify small signals. In order for the transistor to operate as an amplifier, however, certain dc conditions must be met. These conditions set the "bias point" of the transistor. The bias current of a transistor circuit is determined by a dc voltage source and several resistors. Although the operation of the transistor is outside the scope of this textbook, we can analyze the bias circuit of a transistor using elementary circuit theory.

EXAMPLE 7–7   Use the given conditions to determine I_C, I_E, V_{CE}, and V_B for the transistor circuit of Figure 7–21.

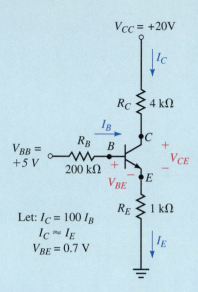

Let: $I_C = 100\,I_B$
$I_C \approx I_E$
$V_{BE} = 0.7$ V

FIGURE 7–21

Solution   In order to simplify the work, the circuit of Figure 7–21 is separated into two circuits: one circuit containing the known voltage V_{BE} and the other containing the unknown voltage V_{CE}.

Since we always start will the given information, we redraw the circuit containing the known voltage V_{BE} as illustrated in Figure 7–22.

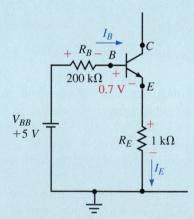

FIGURE 7–22

Although the circuit of Figure 7–22 initially appears to be a series circuit, we see that this cannot be the case, since we are given that $I_E \cong I_C = 100I_B$. We know that the current everywhere in a series circuit must be the same. However, Kirchhoff's voltage law still applies around the closed loop, resulting in the following:

$$V_{BB} = R_B I_B + V_{BE} + R_E I_E$$

The previous expression contains two unknowns, I_B and I_E (V_{BE} is given). From the given information we have the current $I_E \cong 100I_B$, which allows us to write

$$V_{BB} = R_B I_B + V_{BE} + R_E(100I_B)$$

Solving for the unknown current I_B, we have

$$5.0\ \text{V} = (200\ \text{k}\Omega)I_B + 0.7\ \text{V} + (1\ \text{k}\Omega)(100I_B)$$

$$(300\ \text{k}\Omega)I_B = 5.0\ \text{V} - 0.7\ \text{V} = 4.3\ \text{V}$$

$$I_B = \frac{4.3\ \text{V}}{300\ \text{k}\Omega} = 14.3\ \mu\text{A}$$

The current $I_E \cong I_C = 100I_B = 1.43\ \text{mA}$.

As mentioned previously, the circuit can be redrawn as two separate circuits. The circuit containing the unknown voltage V_{CE} is illustrated in Figure 7–23. Notice that the resistor R_E appears in both Figure 7–22 and Figure 7–23.

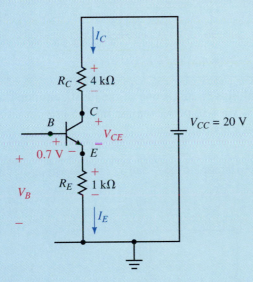

FIGURE 7–23

Applying Kirchhoff's voltage law around the closed loop of Figure 7–23, we have the following:

$$V_{R_C} + V_{CE} + V_{R_E} = V_{CC}$$

The voltage V_{CE} is found as

$$\begin{aligned} V_{CE} &= V_{CC} - V_{R_C} - V_{R_E} \\ &= V_{CC} - R_C I_C - R_E I_E \\ &= 20.0\ \text{V} - (4\ \text{k}\Omega)(1.43\ \text{mA}) - (1\ \text{k}\Omega)(1.43\ \text{mA}) \\ &= 20.0\ \text{V} - 5.73\ \text{V} - 1.43\ \text{V} = 12.8\ \text{V} \end{aligned}$$

Finally, applying Kirchhoff's voltage law from B to ground, we have

$$\begin{aligned} V_B &= V_{BE} + V_{R_E} \\ &= 0.7\ \text{V} + 1.43\ \text{V} \\ &= 2.13\ \text{V} \end{aligned}$$

PRACTICE
PROBLEMS 3

Use the given information to find V_G, I_D, and V_{DS} for the circuit of Figure 7–24.

FIGURE 7–24

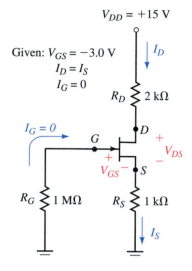

Given: $V_{GS} = -3.0$ V
$I_D = I_S$
$I_G = 0$

Answers: $V_G = 0$, $I_D = 3.00$ mA, $V_{DS} = 6.00$ V

The **universal bias** circuit is one of the most common transistor circuits used in amplifiers. We will now examine how to use circuit analysis principles to analyze this important circuit.

EXAMPLE 7–8 Determine the I_C and V_{CE} for the circuit of Figure 7–25.

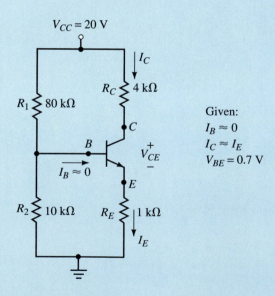

Given:
$I_B \approx 0$
$I_C \approx I_E$
$V_{BE} = 0.7$ V

FIGURE 7–25

Solution If we examine the above circuit, we see that since $I_B \approx 0$, we may assume that R_1 and R_2 are effectively in series. This assumption would be incorrect if the current I_B was not very small compared to the currents through R_1 and R_2. We use the voltage divider rule to solve for the voltage, V_B. (For this reason, the universal bias circuit is often referred to as **voltage divider bias.**)

$$V_B = \frac{R_2}{R_1 + R_2} V_{CC}$$

$$= \left[\frac{10 \text{ k}\Omega}{80 \text{ k}\Omega + 10 \text{ k}\Omega} \right] (20 \text{ V})$$

$$= 2.22 \text{ V}$$

Next, we use the value for V_B and Kirchhoff's voltage law to determine the voltage across R_E.

$$V_{RE} = 2.22 \text{ V} - 0.7 \text{ V} = 1.52 \text{ V}$$

Applying Ohm's law, we now determine the current I_E.

$$I_E = \frac{1.52 \text{ V}}{1 \text{ k}\Omega} = 1.52 \text{ mA} \cong I_C$$

Finally, applying Kirchhoff's voltage law and Ohm's law, we determine V_{CE} as follows:

$$V_{CC} = V_{RC} + V_{CE} + V_{RE}$$
$$V_{CE} = V_{CC} - V_{RC} - V_{RE}$$
$$= 20 \text{ V} - (1.52 \text{ mA})(4 \text{ k}\Omega) - (1.52 \text{ mA})(1 \text{ k}\Omega)$$
$$= 12.4 \text{ V}$$

A **zener diode** is a two-terminal device similar to a varistor (refer to Chapter 3). When the voltage across the zener diode attempts to go above the rated voltage for the device, the zener diode provides a low-resistance path for the extra current. Due to this action, a relatively constant voltage, V_Z, is maintained across the zener diode. This characteristic is referred to as **voltage regulation** and has many applications in electronic and electrical circuits. Once again, although the theory of operation of the zener diode is outside the scope of this textbook, we are able to apply simple circuit theory to examine how the circuit operates.

EXAMPLE 7–9   For the voltage regulator circuit of Figure 7–26, calculate I_Z, I_1, I_2, and P_Z.

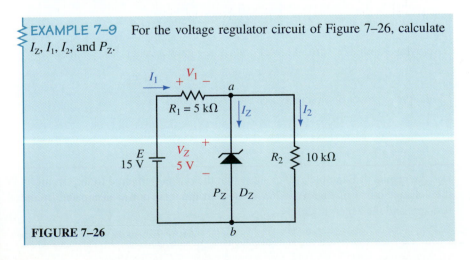

FIGURE 7–26

Solution   If we take a moment to examine the circuit, we see that the zener diode is placed in parallel with the resistor R_2. This parallel combination is in series with the resistor R_1 and the voltage source, E.

In order for the zener diode to operate as a regulator, the voltage across the diode would have to be above the zener voltage without the diode present. If we remove the zener diode the circuit would appear as shown in Figure 7–27.

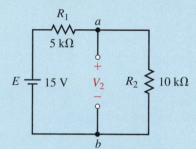

FIGURE 7–27

From Figure 7–27, we may determine the voltage V_2 which would be present without the zener diode in the circuit. Since the circuit is a simple series circuit, the voltage divider rule may be used to determine V_2:

$$V_2 = \frac{R_2}{R_1 + R_2}E = \left(\frac{10\ \text{k}\Omega}{5\ \text{k}\Omega + 10\ \text{k}\Omega}\right)(15\ \text{V}) = 10.0\ \text{V}$$

When the zener diode is placed across the resistor R_2 the device will operate to limit the voltage to $V_Z = 5$ V.

Because the zener diode is operating as a voltage regulator, the voltage across both the diode and the resistor R_2 must be the same, namely 5 V. The parallel combination of D_Z and R_2 is in series with the resistor R_1, and so the voltage across R_1 is easily determined from Kirchhoff's voltage law as

$$V_1 = E - V_Z = 15\ \text{V} - 5\text{V} = 10\ \text{V}$$

Now, from Ohm's law, the currents I_1 and I_2 are easily found to be

$$I_2 = \frac{V_2}{R_2} = \frac{5\ \text{V}}{10\ \text{k}\Omega} = 0.5\ \text{mA}$$

$$I_1 = \frac{V_1}{R_1} = \frac{10\ \text{V}}{5\ \text{k}\Omega} = 2.0\ \text{mA}$$

Applying Kirchoff's current law at node a, we get the zener diode current as

$$I_Z = I_1 - I_2 = 2.0\ \text{mA} - 0.5\ \text{mA} = 1.5\ \text{mA}$$

Finally, the power dissipated by the zener diode must be

$$P_Z = V_Z I_Z = (5\ \text{V})(1.5\ \text{mA}) = 7.5\ \text{mW}$$

IN-PROCESS LEARNING CHECK 2

Use the results of Example 7–9 to show that the power delivered to the circuit by the voltage source of Figure 7–26, is equal to the total power dissipated by the resistors and the zener diode.

(Answers are at the end of the chapter.)

1. Determine I_1, I_z, and I_2 for the circuit of Figure 7–26 if the resistor R_1 is increased to 10 kΩ.

2. Repeat Problem 1 if R_1 is increased to 30 kΩ.

Answers:

1. $I_1 = 1.00$ mA, $I_2 = 0.500$ mA, $I_z = 0.500$ mA

2. $I_1 = I_2 = 0.375$ mA, $I_z = 0$ mA. (The voltage across the zener diode is not sufficient for the device to come on.)

7.4   Potentiometers

As mentioned in Chapter 3, variable resistors may be used as potentiometers as shown in Figure 7–28 to control voltage into another circuit.

The volume control on a receiver or amplifier is an example of a variable resistor used as a potentiometer. When the movable terminal is at the uppermost position, the voltage appearing between terminals b and c is simply calculated by using the voltage divider rule as

$$V_{bc} = \left(\frac{50 \text{ k}\Omega}{50 \text{ k}\Omega + 50 \text{ k}\Omega} \right)(120 \text{ V}) = 60 \text{ V}$$

Alternatively, when the movable terminal is at the lowermost position, the voltage between terminals b and c is $V_{bc} = 0$ V, since the two terminals are effectively shorted and the voltage across a short circuit is always zero.

The circuit of Figure 7–28 represents a potentiometer having an output voltage which is adjustable between 0 and 60 V. This output is referred to as the **unloaded output,** since there is no load resistance connected between the terminals b and c. If a load resistance were connected between these terminals, the output voltage, called the **loaded output,** would no longer be the same. The following example is an illustration of circuit loading.

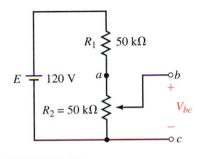

FIGURE 7–28

EXAMPLE 7–10   For the circuit of Figure 7–29, determine the range of the voltage V_{bc} as the potentiometer varies between its minimum and maximum values.

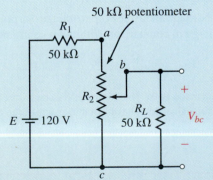

FIGURE 7–29

Solution   The minimum voltage between terminals b and c will occur when the movable contact is at the lowermost contact of the variable resistor. In this position, the voltage $V_{bc} = 0$ V, since the terminals b and c are shorted.

The maximum voltage V_{bc} occurs when the movable contact is at the uppermost contact of the variable resistor. In this position, the circuit may be represented as shown in Figure 7–30.

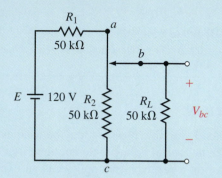

FIGURE 7–30

In Figure 7–30, we see that the resistance R_2 is in parallel with the load resistor R_L. The voltage between terminals b and c is easily determined from the voltage divider rule, as follows:

$$V_{bc} = \frac{R_2 \| R_L}{(R_2 \| R_L) + R_1} E$$

$$= \left(\frac{25\ k\Omega}{25\ k\Omega + 50\ k\Omega} \right)(120\ \text{V}) = 40\ \text{V}$$

We conclude that the voltage at the output of the potentiometer is adjustable from 0 V to 40 V for a load resistance of $R_L = 50\ k\Omega$.

By inspection, we see that an unloaded potentiometer in the circuit of Figure 7–29 would have an output voltage of 0 V to 60 V.

PRACTICE PROBLEMS 5

Refer to the circuit of Figure 7–29.

a. Determine the output voltage range of the potentiometer if the load resistor is $R_L = 5\ k\Omega$.

b. Repeat (a) if the load resistor is $R_L = 500\ k\Omega$.

c. What conclusion may be made about the output voltage of a potentiometer when the load resistance is large in comparison with the potentiometer resistance?

Answers:
a. 0 to 10 V   b. 0 to 57.1 V

c. When R_L is large in comparison to the potentiometer resistance, the output voltage will better approximate the unloaded voltage. (In this example, the unloaded voltage is 0 to 60 V.)

IN-PROCESS LEARNING CHECK 3

A 20-kΩ potentiometer is connected across a voltage source with a 2-kΩ load resistor connected between the wiper (center terminal) and the negative terminal of the voltage source.

a. What percentage of the source voltage will appear across the load when the wiper is one-fourth of the way from the bottom?

b. Determine the percentage of the source voltage which appears across the load when the wiper is one-half and three-fourths of the way from the bottom.

c. Repeat the calculations of (a) and (b) for a load resistor of 200 kΩ.

d. From the above results, what conclusion can you make about the effect of placing a large load across a potentiometer?

(Answers are at the end of the chapter.)

7.5   Loading Effects of Instruments

In Chapters 5 and 6 we examined how ammeters and voltmeters affect the operation of simple series circuits. The degree to which the circuits are affected is called the **loading effect** of the instrument. Recall that, in order for an instrument to provide an accurate indication of how a circuit operates, the loading effect should ideally be zero. In practice, it is impossible for any instrument to have zero loading effect, since all instruments absorb some energy from the circuit under test, thereby affecting circuit operation.

In this section, we will determine how instrument loading affects more complex circuits.

EXAMPLE 7–11   Calculate the loading effects if a digital multimeter, having an internal resistance of 10 MΩ, is used to measure V_1 and V_2 in the circuit of Figure 7–31.

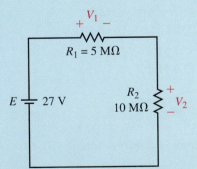

FIGURE 7–31

Solution   In order to determine the loading effect for a particular reading, we need to calculate both the unloaded voltage and the loaded voltage.

For the circuit given in Figure 7–31, the unloaded voltage across each resistor is

$$V_1 = \left(\frac{5 \text{ M}\Omega}{5 \text{ M}\Omega + 10 \text{ M}\Omega} \right)(27 \text{ V}) = 9.0 \text{ V}$$

$$V_2 = \left(\frac{10 \text{ M}\Omega}{5 \text{ M}\Omega + 10 \text{ M}\Omega} \right)(27 \text{ V}) = 18.0 \text{ V}$$

When the voltmeter is used to measure V_1, the result is equivalent to connecting a 10-MΩ resistor across resistor R_1, as shown in Figure 7–32.

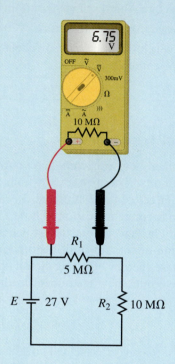

FIGURE 7–32

The voltage appearing across the parallel combination of R_1 and resistance of the voltmeter is calculated as

$$V_1 = \left(\frac{5 \text{ M}\Omega \| 10 \text{ M}\Omega}{(5 \text{ M}\Omega \| 10 \text{ M}\Omega) + 10 \text{ M}\Omega}\right)(27 \text{ V})$$

$$= \left(\frac{3.33 \text{ M}\Omega}{13.3 \text{ M}\Omega}\right)(27 \text{ V})$$

$$= 6.75 \text{ V}$$

Notice that the measured voltage is significantly less than the 9 V that we had expected to measure.

With the voltmeter connected across resistor R_2, the circuit appears as shown in Figure 7–33.

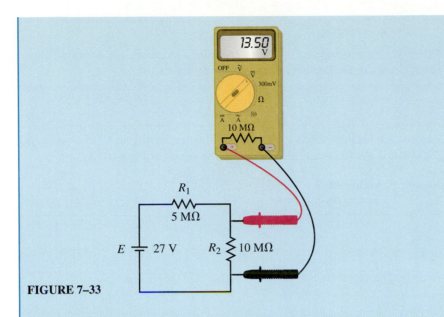

FIGURE 7–33

The voltage appearing across the parallel combination of R_2 and resistance of the voltmeter is calculated as

$$V_2 = \left(\frac{10 \text{ M}\Omega \| 10 \text{ M}\Omega}{5 \text{ M}\Omega + (10 \text{ M}\Omega \| 10 \text{ M}\Omega)} \right)(27 \text{ V})$$

$$= \left(\frac{5.0 \text{ M}\Omega}{10.0 \text{ M}\Omega} \right)(27 \text{ V})$$

$$= 13.5 \text{ V}$$

Again, we notice that the measured voltage is quite a bit less than the 18 V that we had expected.

Now the loading effects are calculated as follows.

When measuring V_1:

$$\text{loading effect} = \frac{9.0 \text{ V} - 6.75 \text{ V}}{9.0 \text{ V}} \times 100\%$$

$$= 25\%$$

When measuring V_2:

$$\text{loading effect} = \frac{18.0 \text{ V} - 13.5 \text{ V}}{18.0 \text{ V}} \times 100\%$$

$$= 25\%$$

This example clearly illustrates a problem that novices often make when they are taking voltage measurements in high-resistance circuits. If the measured voltages $V_1 = 6.75$ V and $V_2 = 13.50$ V are used to verify Kirchhoff's voltage law, the novice would say that this represents a contradiction of the law (since 6.75 V + 13.50 V ≠ 27.0 V). In fact, we see that the circuit is behaving exactly as predicted by circuit theory. The problem occurs when instrument limitations are not considered.

PRACTICE PROBLEMS 6

Calculate the loading effects if an analog voltmeter, having an internal resistance of 200 kΩ, is used to measure V_1 and V_2 in the circuit of Figure 7–31.

Answers: V_1: loading effect = 94.3%
V_2: loading effect = 94.3%

EXAMPLE 7–12   For the circuit of Figure 7–34, calculate the loading effect if a 5.00-Ω ammeter is used to measure the currents I_T, I_1, and I_2.

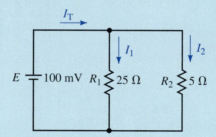

FIGURE 7–34

Solution   We begin by determining the unloaded currents in the circuit. Using Ohm's law, we solve for the currents I_1 and I_2:

$$I_1 = \frac{100 \text{ mV}}{25 \text{ } \Omega} = 4.0 \text{ mA}$$

$$I_2 = \frac{100 \text{ mV}}{5 \text{ } \Omega} = 20.0 \text{ mA}$$

Now, by Kirchhoff's current law,

$$I_T = 4.0 \text{ mA} + 20.0 \text{ mA} = 24.0 \text{ mA}$$

If we were to insert the ammeter into the branch with resistor R_1, the circuit would appear as shown in Figure 7–35.

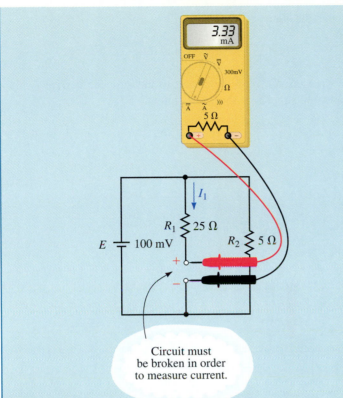

FIGURE 7–35

The current through the ammeter would be

$$I_1 = \frac{100 \text{ mV}}{25 \ \Omega + 5 \ \Omega} = 3.33 \text{ mA}$$

If we were to insert the ammeter into the branch with resistor R_2, the circuit would appear as shown in Figure 7–36.

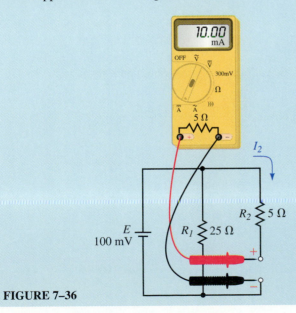

FIGURE 7–36

The current through the ammeter would be

$$I_2 = \frac{100\ \text{mV}}{5\ \Omega + 5\ \Omega} = 10.0\ \text{mA}$$

If the ammeter were inserted into the circuit to measure the current I_T, the equivalent circuit would appear as shown in Figure 7–37.

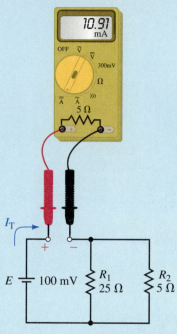

FIGURE 7–37

The total resistance of the circuit would be

$$R_T = 5\ \Omega + 25\ \Omega \| 5\ \Omega = 9.17\ \Omega$$

This would result in a current I_T determined by Ohm's law as

$$I_T = \frac{100\ \text{mV}}{9.17\ \Omega}$$

$$= 10.9\ \text{mA}$$

The loading effects for the various current measurements are as follows: When measuring I_1,

$$\text{loading effect} = \frac{4.0\ \text{mA} - 3.33\ \text{mA}}{4.0\ \text{mA}} \times 100\%$$

$$= 16.7\%$$

When measuring I_2,

$$\text{loading effect} = \frac{20\ \text{mA} - 10\ \text{mA}}{20\ \text{mA}} \times 100\%$$

$$= 50\%$$

When measuring I_T,

$$\text{loading effect} = \frac{24\ \text{mA} - 10.9\ \text{mA}}{24\ \text{mA}} \times 100\%$$

$$= 54.5\%$$

Notice that the loading effect for an ammeter is most pronounced when it is used to measure current in a branch having a resistance in the same order of magnitude as the meter.

You will also notice that if this ammeter were used in a circuit to verify the correctness of Kirchhoff's current law, the loading effect of the meter would produce an apparent contradiction. From KCL,

$$I_T = I_1 + I_2$$

By substituting the measured values of current into the above equation, we have

$$10.91 \text{ mA} = 3.33 \text{ mA} + 10.0 \text{ mA}$$

$$10.91 \text{ mA} \neq 13.33 \text{ mA (contradiction)}$$

This example illustrates that the loading effect of a meter may severely affect the current in a circuit, giving results which seem to contradict the laws of circuit theory. Therefore, wherever an instrument is used to measure a particular quantity, we must always take into account the limitations of the instrument and question the validity of a resulting reading.

Calculate the readings and the loading error if an ammeter having an internal resistance of 1 Ω is used to measure the currents in the circuit of Figure 7–34.

PRACTICE PROBLEMS 7

Answers: $I_{T(RDG)}$ = 19.4 mA; loading error = 19.4%

$\quad\quad\quad I_{1(RDG)}$ = 3.85 mA; loading error = 3.85%

$\quad\quad\quad I_{T(RDG)}$ = 16.7 mA; loading error = 16.7%

7.6   Circuit Analysis Using Computers

Electronics Workbench

The analysis of series-parallel circuits using Electronics Workbench is almost identical to the methods used in analyzing series and parallel circuits in previous chapters. The following example illustrates that Electronics Workbench results in the same solutions as those obtained in Example 7–4.

ELECTRONICS WORKBENCH   **PSpice**

EXAMPLE 7–13   Given the circuit of Figure 7–38, use Electronics Workbench to find the following quantities:

a. Total resistance, R_T

b. Voltages V_2 and V_4

c. Currents I_T, I_1, and I_2

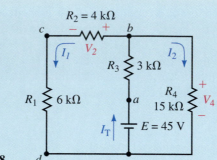

FIGURE 7–38

Solution

a. We begin by constructing the circuit as shown in Figure 7–39. This circuit is identical to that shown in Figure 7–38, except that the voltage source has been omitted and a multimeter (from the Instruments button on the Parts bin toolbar) inserted in its place. The ohmmeter function is then selected and the power switch turned on. The resistance is found to be $R_T = 9.00 \text{ k}\Omega$.

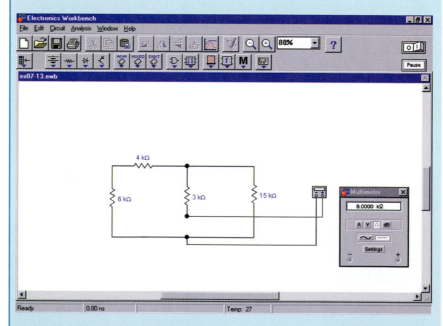

FIGURE 7–39

b. Next, we remove the multimeter and insert the 45-V source. Ammeters and voltmeters are inserted, as shown in Figure 7–40.

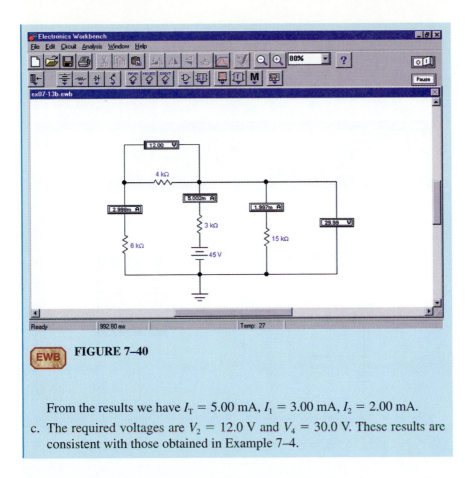

FIGURE 7–40

From the results we have $I_T = 5.00$ mA, $I_1 = 3.00$ mA, $I_2 = 2.00$ mA.

c. The required voltages are $V_2 = 12.0$ V and $V_4 = 30.0$ V. These results are consistent with those obtained in Example 7–4.

The following example uses Electronics Workbench to determine the voltage across a bridge circuit. The example uses a potentiometer to provide a variable resistance in the circuit. Electronics Workbench is able to provide a display of voltage (on a multimeter) as the resistance is changed.

EXAMPLE 7–14 Given the circuit of Figure 7–41, use Electronics Workbench to determine the values of I and V_{ab} when $R_x = 0 \ \Omega$, 15 kΩ, and 50 kΩ.

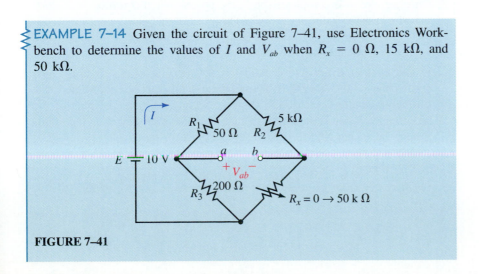

FIGURE 7–41

Solution

1. We begin by constructing the circuit as shown in Figure 7–42. The potentiometer is selected from the Basic button on the Parts bin toolbar. Ensure that the potentiometer is inserted as illustrated.

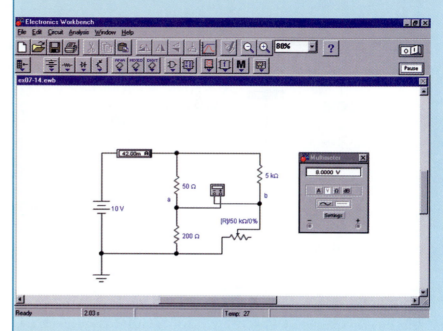

EWB **FIGURE 7–42**

2. Double click on the potentiometer symbol and change its value to 50 kΩ. Notice that the increment value is set for 5%. Adjust this value so that it is at 10%. We will use this in a following step.

3. Once the circuit is completely built, the power switch is turned on. Notice that the resistor value is at 50%. This means that the potentiometer is adjusted so that its value is 25 kΩ The following step will change the value of the potentiometer to 0 Ω.

4. Double click on the potentiometer symbol. The value of the resistor is now easily changed by either Shift **R** (to incrementally increase the value) or **R** (to incrementally decrease the value). As you decrease the value of the potentiometer, you should observe that the voltage displayed on the multimeter is also changing. It takes a few seconds for the display to stabilize. As shown in Figure 7–42, you should observe that $I = 42.0$ mA and $V_{ab} = 8.00$ V when $R_x = 0$ Ω.

5. Finally, after adjusting the value of the potentiometer, we obtain the following readings:

$$R_x = 15 \text{ k}\Omega \ (30\%): \quad I = 40.5 \text{ mA and } V_{ab} = 0.500 \text{ V}$$
$$R_x = 50 \text{ k}\Omega \ (100\%): \quad I = 40.2 \text{ mA and } V_{ab} = -1.09 \text{ V}$$

OrCAD PSpice

PSpice is somewhat different from Electronics Workbench in the way it handles potentiometers. In order to place a variable resistor into a circuit, it is necessary set the parameters of the circuit to sweep through a range of values. The following example illustrates the method used to provide a graphical display of output voltage and source current for a range of resistance. Although the method is different, the results are consistent with those of the previous example.

EXAMPLE 7–15 Use PSpice to provide a graphical display of voltage V_{ab} and current I as R_x is varied from 0 to 50 kΩ in the circuit of Figure 7–41.

Solution OrCAD uses global parameters to represent numeric values by name. This will permit us to set up an analysis that sweeps a variable (in this case a resistor) through a range of values.

- Open the CIS Demo software and move to the Capture schematic as outlined in previous PSpice examples. You may wish to name your project Ch 7 PSpice 1.

- Build the circuit as shown in Figure 7–43. Remember to rotate the components to provide for the correct node assignments. Change all component values (except R4) as required.

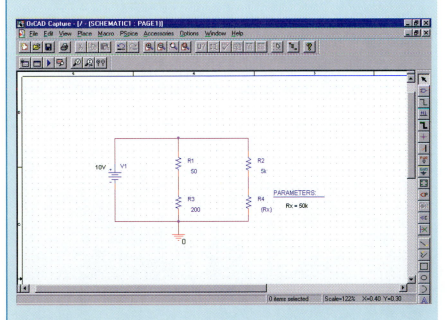

FIGURE 7–43

- Double click on the component value for R4. Enter **{Rx}** in the Value text box of the Display Properties for this resistor. The curly braces tell PSpice to evaluate the parameter and use its value.

- Click on the Place part tool. Select the Special library and click on the PARAM part. Place the PARAM part adjacent to resistor R4.

- Double click on PARAMETERS. Click on New. Type **Rx** in the Property Name text box and click OK. Click in the cell below the Rx column and enter **50k** as the default value for this resistor. Click on Apply. Click on Display and select Name and Value from the Display Format. Exit the Property Editor.

- Click on the New Simulation icon and give the simulation a name such as Fig 7–43.

- In the Simulation Settings, select the Analysis Tab. The Analysis type is DC Sweep. Select Primary Sweep from the Options listing. In the Sweep variable box select Global Parameter. Type **Rx** in the Parameter name text box. Select the Sweep type to be linear and set the limits as follows:

Start value: 100

End value: 50k

Increment: 100.

These settings will change the resistor value from 100 Ω to 50 kΩ in 100-Ω increments. Click OK.

- Click on the Run icon. You will see a blank screen with the abscissa (horizontal axis) showing R_x scaled from 0 to 50 kΩ.

- PSpice is able to plot most circuit variables as a function of R_x. In order to request a plot of V_{ab}, click on Trace and Add Trace. Enter **V(R3:1) − V(R4:1)** in the Trace Expression text box. The voltage V_{ab} is the voltage between node 1 of R_3 and node 1 of R_4. Click OK.

- Finally, to obtain a plot of the circuit current I (current through the voltage source), we need to first add an extra axis. Click on Plot and then click Add Y Axis. To obtain a plot of the current, click on Trace and Add Trace. Select **I(V1).** The resulting display on the monitor is shown in Figure 7–44.

FIGURE 7–44

Notice that the current shown is negative. This is because PSpice sets the reference direction through a voltage source from the positive terminal to the negative terminal. One way of removing the negative sign is to request the current as $-I(V1)$.

Given the circuit of Figure 7–45, use Electronics Workbench to solve for V_{ab}, I, and I_L when $R_L = 100\ \Omega$, $500\ \Omega$, and $1000\ \Omega$.

PRACTICE
PROBLEMS 8

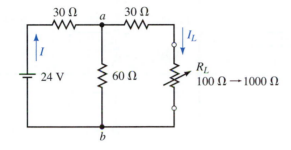

FIGURE 7–45

Answers: $R_L = 100\ \Omega$: $I = 169$ mA, $V_{ab} = 6.93$ V, $I_L = 53.3$ mA
$R_L = 500\ \Omega$: $I = 143$ mA, $V_{ab} = 7.71$ V, $I_L = 14.5$ mA
$R_L = 1000\ \Omega$: $I = 138$ mA, $V_{ab} = 7.85$ V, $I_L = 7.62$ mA

Use PSpice to input file for the circuit of Figure 7–45. The output shall display the source current, I, the load current, I_L, and the voltage V_{ab} as the resistor R_L is varied in 100-Ω increments between 100 Ω and 1000 Ω.

PRACTICE
PROBLEMS 9

PUTTING IT INTO PRACTICE

$\mathbf{V}$ery often manufacturers provide schematics showing the dc voltages that one would expect to measure if the circuit were functioning properly. The following figure shows part of a schematic for an amplifier circuit.

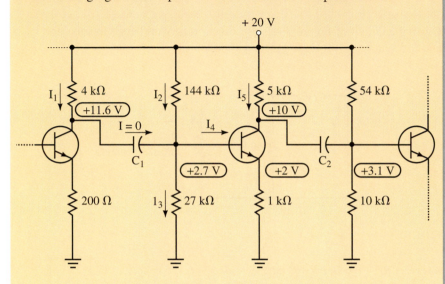

Even though a schematic may include components with which the reader is not familiar, the dc voltages provided on the schematic enable us to determine voltages and current in various parts of the circuit. If the circuit is faulty, measured voltages and currents will be different from the theoretical, allowing the experience troubleshooter to locate the fault.

Examine the circuit shown. Use the voltage information on the schematic to determine the theoretical values of the currents I_1, I_2, I_3, I_4, and I_5. Find the magnitude and correct polarity of the voltage across the device labeled as C_2. (It's a capacitor, and it will be examined in detail in Chapter 10.)

PROBLEMS

7.1   The Series-Parallel Network

1. For the networks of Figure 7–46, determine which resistors and branches are in series and which are in parallel. Write an expression for the total resistance, R_T.

2. For each of the networks of Figure 7–47, write an expression for the total resistance, R_T.

3. Write an expression for both R_{T_1} and R_{T_2} for the networks of Figure 7–48.

4. Write an expression for both R_{T_1} and R_{T_2} for the networks of Figure 7–49.

5. Resistor networks have total resistances as given below. Sketch a circuit which corresponds to each expression.

   a. $R_T = (R_1 \| R_2 \| R_3) + (R_4 \| R_5)$

   b. $R_T = R_1 + (R_2 \| R_3) + [R_4 \| (R_5 + R_6)]$

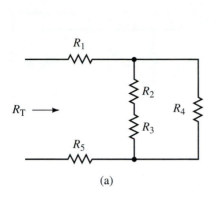

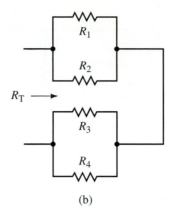

(a)   (b)

FIGURE 7–46

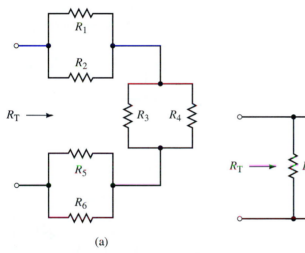

(a)

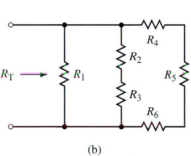

(b)

FIGURE 7–47

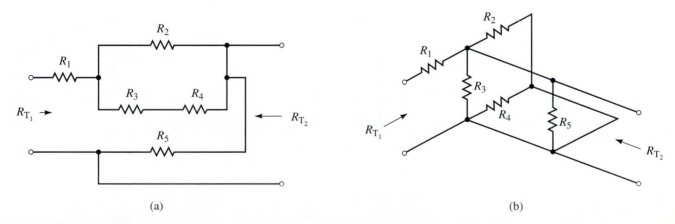

(a)   (b)

FIGURE 7–48

6. Resistor networks have total resistances as given below. Sketch a circuit which corresponds to each expression.

a. $R_T = [(R_1\|R_2) + (R_3\|R_4)]\|R_5$

b. $R_T = (R_1\|R_2) + R_3 + [(R_4 + R_5)\|R_6]$

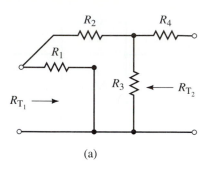

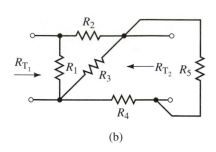

(a)   (b)

FIGURE 7–49

7.2   Analysis of Series-Parallel Circuits

7. Determine the total resistance of each network in Figure 7–50.

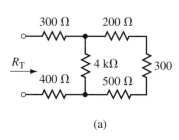

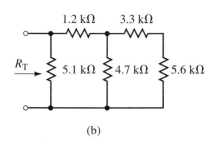

(a)   (b)

FIGURE 7–50

8. Determine the total resistance of each network in Figure 7–51.

All resistors are 1 kΩ.

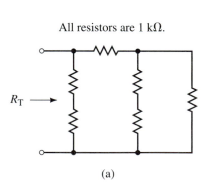

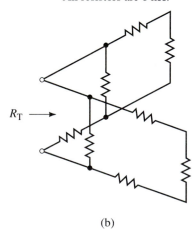

(a)   (b)

FIGURE 7–51

9. Calculate the resistances R_{ab} and R_{cd} in the circuit of Figure 7–52.
10. Calculate the resistances R_{ab} and R_{bc} in the circuit of Figure 7–53.
11. Refer to the circuit of Figure 7–54:
    Find the following quantities:
    a. R_T
    b. I_T, I_1, I_2, I_3, I_4
    c. V_{ab}, V_{bc}.

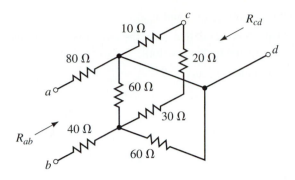

FIGURE 7–52

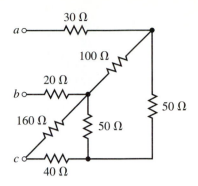

FIGURE 7–53

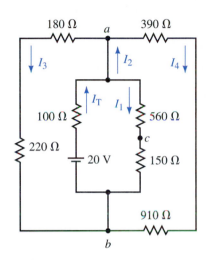

EWB **FIGURE 7–54**

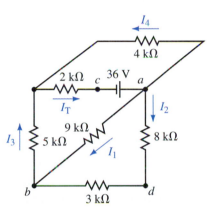

FIGURE 7–55

12. Refer to the circuit of Figure 7–55:

Find the following quantities:

a. R_T (equivalent resistance "seen" by the voltage source).

b. I_T, I_1, I_2, I_3, I_4

c. V_{ab}, V_{bc}, V_{cd}.

13. Refer to the circuit of Figure 7–56:

a. Find the currents I_1, I_2, I_3, I_4, I_5, and I_6.

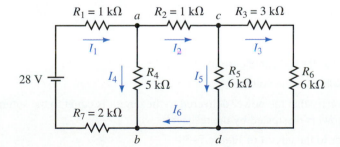

EWB **FIGURE 7–56**

   b. Solve for the voltages V_{ab} and V_{cd}.

   c. Verify that the power delivered to the circuit is equal to the summation of powers dissipated by the resistors.

14. Refer to the circuit of Figure 7–57:

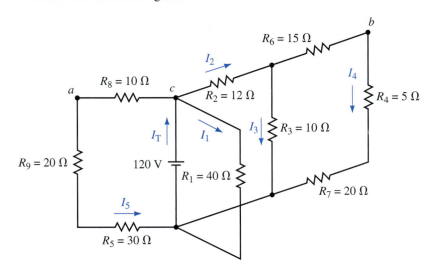

FIGURE 7–57

   a. Find the currents I_1, I_2, I_3, I_4, and I_5.

   b. Solve for the voltages V_{ab} and V_{bc}.

   c. Verify that the power delivered to the circuit is equal to the summation of powers dissipated by the resistors.

15. Refer to the circuits of Figure 7–58:

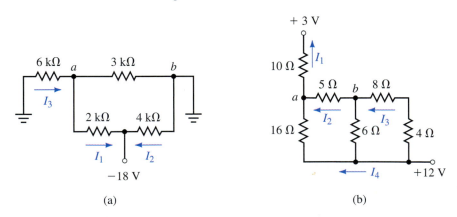

FIGURE 7–58

   a. Find the indicated currents.

   b. Solve for the voltage V_{ab}.

   c. Verify that the power delivered to the circuit is equal to the summation of powers dissipated by the resistors.

16. Refer to the circuit of Figure 7–59:

   a. Solve for the currents I_1, I_2, and I_3 when $R_x = 0\ \Omega$ and when $R_x = 5\ k\Omega$.

b. Calculate the voltage V_{ab} when $R_x = 0\ \Omega$ and when $R_x = 5\ k\Omega$.

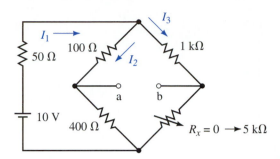

FIGURE 7–59

7.3 Applications of Series-Parallel Circuits

17. Solve for all currents and voltage drops in the circuit of Figure 7–60. Verify that the power delivered by the voltage source is equal to the power dissipated by the resistors and the zener diode.

18. Refer to the circuit of Figure 7–61:

   a. Determine the power dissipated by the 6.2-V zener diode. If the zener diode is rated for a maximum power of ¼ W, is it likely to be destroyed?

   b. Repeat Part (a) if the resistance R_1 is doubled.

19. Given the circuit of Figure 7–62, determine the range of R (maximum and minimum values) which will ensure that the output voltage $V_L = 5.6$ V while the maximum power rating of the zener diode is not exceeded.

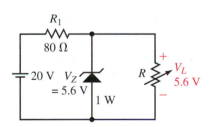

FIGURE 7–62

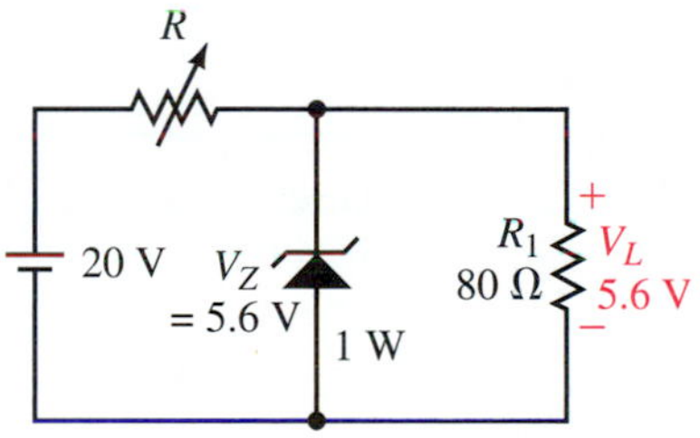

FIGURE 7–63

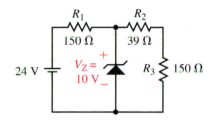

FIGURE 7–60

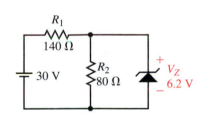

FIGURE 7–61

20. Given the circuit of Figure 7–63, determine the range of R (maximum and minimum values) which will ensure that the output voltage $V_L = 5.6$ V while the maximum power rating of the zener diode is not exceeded.

21. Given the circuit of Figure 7–64, determine V_B, I_C, and V_{CE}.

22. Repeat Problem 21 if R_B is increased to 10 kΩ. (All other quantities remain unchanged.)

23. Consider the circuit of Figure 7–65 and the indicated values:

   a. Determine I_D.

   b. Calculate the required value for R_S.

   c. Solve for V_{DS}.

24. Consider the circuit of Figure 7–66 and the indicated values:

   a. Determine I_D and V_G.

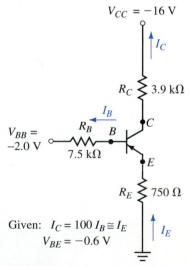

Given: $I_C = 100\ I_B \cong I_E$
$V_{BE} = -0.6$ V

FIGURE 7–64

b. Design the required values for R_S and R_D.

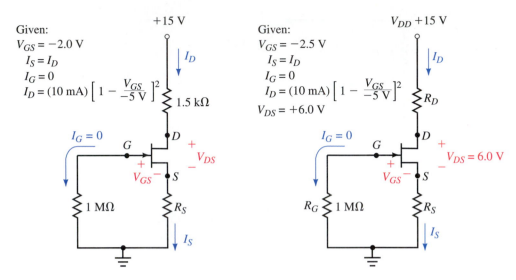

Given:
$V_{GS} = -2.0$ V
$I_S = I_D$
$I_G = 0$
$I_D = (10 \text{ mA}) \left[1 - \dfrac{V_{GS}}{-5 \text{ V}} \right]^2$

$+15$ V

I_D

1.5 kΩ

$I_G = 0$

G

D

$+$

V_{DS}

$−$

$+$

V_{GS} $−$ S

1 MΩ

R_S

I_S

FIGURE 7–65

Given:
$V_{GS} = -2.5$ V
$I_S = I_D$
$I_G = 0$
$I_D = (10 \text{ mA}) \left[1 - \dfrac{V_{GS}}{-5 \text{ V}} \right]^2$
$V_{DS} = +6.0$ V

V_{DD} $+15$ V

I_D

R_D

$I_G = 0$

G

D

$+$

$V_{DS} = 6.0$ V

$−$

$+$

V_{GS} $−$ S

R_G 1 MΩ

R_S

I_S

FIGURE 7–66

25. Calculate I_C and V_{CE} for the circuit of Figure 7–67.

26. Calculate I_C and V_{CE} for the circuit of Figure 7–68.

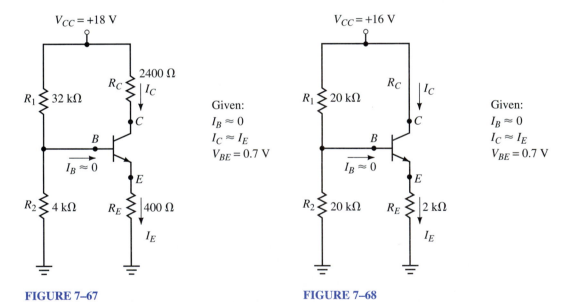

$V_{CC} = +18$ V

2400 Ω
R_C I_C

R_1 32 kΩ

C

B

$I_B \approx 0$

E

R_2 4 kΩ

R_E 400 Ω

I_E

Given:
$I_B \approx 0$
$I_C \approx I_E$
$V_{BE} = 0.7$ V

FIGURE 7–67

$V_{CC} = +16$ V

R_C I_C

R_1 20 kΩ

C

B

$I_B \approx 0$

E

R_2 20 kΩ

R_E 2 kΩ

I_E

Given:
$I_B \approx 0$
$I_C \approx I_E$
$V_{BE} = 0.7$ V

FIGURE 7–68

7.4 Potentiometers

27. Refer to the circuit of Figure 7–69:

a. Determine the range of voltages which will appear across R_L as the potentiometer is varied between its minimum and maximum values.

b. If R_2 is adjusted to be 2.5 kΩ, what will be the voltage V_L? If the load resistor is now removed, what voltage would appear between terminals a and b?

28. Repeat Problem 27 using the load resistor R_L = 30 kΩ.

29. If the potentiometer of Figure 7–70 is adjusted so that R_2 = 200 Ω, determine the voltages V_{ab} and V_{bc}.

30. Calculate the values of R_1 and R_2 required in the potentiometer of Figure 7–70 if the voltage V_L across the 50-Ω load resistor is to be 6.0 V.

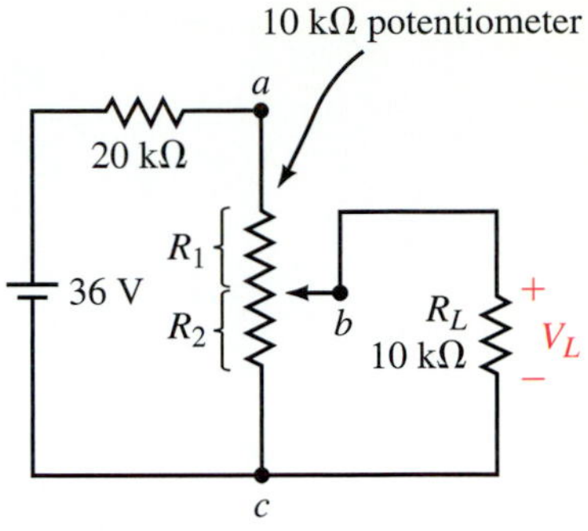

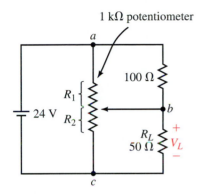

FIGURE 7–70

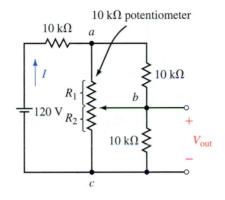

FIGURE 7–71

10 kΩ potentiometer

FIGURE 7–69

31. Refer to the circuit of Figure 7–71:

a. Determine the range of output voltage (minimum to maximum) which can be expected as the potentiometer is adjusted from minimum to maximum.

b. Calculate R_2 when V_{out} = 20 V.

32. In the circuit of Figure 7–71, what value of R_2 results in an output voltage of 40 V?

33. Given the circuit of Figure 7–72, calculate the output voltage V_{out} when R_L = 0 Ω, 250 Ω, and 500 Ω.

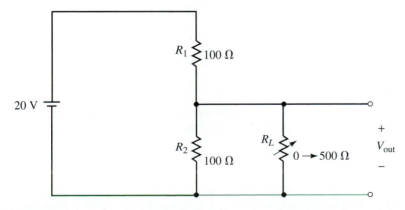

FIGURE 7–72

34. Given the circuit of Figure 7–73, calculate the output voltage V_{out} when R_L = 0 Ω, 500 Ω, and 1000 Ω.

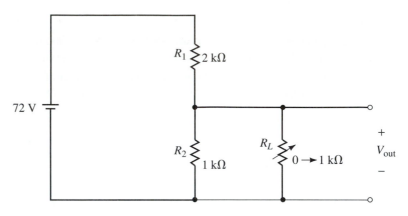

FIGURE 7–73

7.5 Loading Effects of Instruments

35. A voltmeter having a sensitivity $S = 20$ kΩ/V is used on the 10-V range (200-kΩ total internal resistance) to measure voltage across the 750-kΩ resistor of Figure 7–74. The voltage indicated by the meter is 5.00 V.

   a. Determine the value of the supply voltage, E.

   b. What voltage will be present across the 750-kΩ resistor when the voltmeter is removed from the circuit?

   c. Calculate the loading effect of the meter when used as shown.

   d. If the same voltmeter is used to measure the voltage across the 200-kΩ resistor, what voltage would be indicated?

36. The voltmeter of Figure 7–75 has a sensitivity $S = 2$ kΩ/V.

   a. If the meter is used on its 50-V range ($R = 100$ kΩ) to measure the voltage across R_2, what will be the meter reading and the loading error?

   b. If the meter is changed to its 20-V range ($R = 40$ kΩ) determine the reading on this range and the loading error. Will the meter be damaged on this range? Will the loading error be less or more than the error in Part (a)?

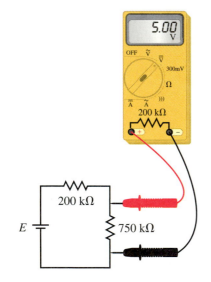

FIGURE 7–74

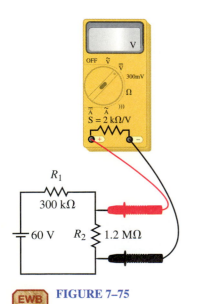

FIGURE 7–75

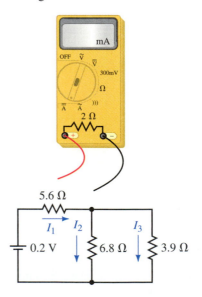

FIGURE 7–76

37. An ammeter is used to measure current in the circuit shown in Figure 7–76.

    a. Explain how to correctly connect the ammeter to measure the current I_1.

    b. Determine the values indicated when the ammeter is used to measure each of the indicated currents in the circuit.

    c. Calculate the loading effect of the meter when measuring each of the currents.

38. Suppose the ammeter in Figure 7–76 has an internal resistance of 0.5 Ω:

    a. Determine the values indicated when the ammeter is used to measure the indicated currents in the circuit.

    b. Calculate the loading effect of the meter when measuring each of the currents.

7.6   Circuit Analysis Using Computers

39. **EWB**   Use Electronics Workbench to solve for V_2, V_4, I_T, I_1, and I_2 in the circuit of Figure 7–10.

40. **EWB**   Use Electronics Workbench to solve for V_{ab}, I_1, I_2, and I_3 in the circuit of Figure 7–13.

41. **EWB**   Use Electronics Workbench to solve for the meter reading in the circuit of Figure 7–75 if the meter is used on its 50-V range.

42. **EWB**   Repeat Problem 41 if the meter is used on its 20-V range.

43. **PSpice**   Use PSpice to solve for V_2, V_4, I_T, I_1, and I_2 in the circuit of Figure 7–10.

44. **PSpice**   Use PSpice to solve for V_{ab}, I_1, I_2, and I_3 in the circuit of Figure 7–13.

45. **PSpice**   Use PSpice to obtain a display of V_{ab} and I_1 in the circuit of Figure 7–59. Let R_x change from 500 Ω to 5 kΩ using 100-Ω increments.

In-Process Learning Check 1

$P_T = 115.2$ mW, $P_1 = 69.1$ mW, $P_2 = 36.9$ mW, $P_3 = 9.2$ mW

$P_1 + P_2 + P_3 = 115.2$ mW as required.

In-Process Learning Check 2

$P_T = 30.0$ mW, $P_{R1} = 20.0$ mW, $P_{R2} = 2.50$ mW, $P_Z = 7.5$ mW

$P_{R1} + P_{R2} + P_Z = 30$ mW as required.

In-Process Learning Check 3

$R_L = 2$ kΩ:

a. $N = \frac{1}{4}$: $V_L = 8.7\%$ of V_{in}

b. $N = \frac{1}{2}$: $V_L = 14.3\%$ of V_{in}

   $N = \frac{3}{4}$: $V_L = 26.1\%$ of V_{in}

$R_L = 200$ kΩ:

c. $N = \frac{1}{4}$: $V_L = 24.5\%$ of V_{in}

   $N = \frac{1}{2}$: $V_L = 48.8\%$ of V_{in}

   $N = \frac{3}{4}$: $V_L = 73.6\%$ of V_{in}

d. If $R_L >> R_1$, the loading effect is minimal.

Methods of Analysis

OBJECTIVES

After studying this chapter you will be able to

- convert a voltage source into an equivalent current source,
- convert a current source into an equivalent voltage source,
- analyze circuits having two or more current sources in parallel,
- write and solve branch equations for a network,
- write and solve mesh equations for a network,
- write and solve nodal equations for a network,
- convert a resistive delta to an equivalent wye circuit or a wye to its equivalent delta circuit and solve the resulting simplified circuit,
- determine the voltage across or current through any portion of a bridge network,
- use PSpice to analyze multiloop circuits;
- use Electronics Workbench to analyze multiloop circuits.

KEY TERMS

Branch-Current Analysis
Bridge Networks
Constant-Current Sources
Delta-Wye Conversions
Linear Bilateral Networks
Mesh Analysis
Nodal Analysis
Wye-Delta Conversions

OUTLINE

Constant-Current Sources
Source Conversions
Current Sources in Parallel and Series
Branch-Current Analysis
Mesh (Loop) Analysis
Nodal Analysis
Delta-Wye (Pi-Tee) Conversion
Bridge Networks
Circuit Analysis Using Computers

The networks you have worked with so far have generally had a single voltage source and could be easily analyzed using techniques such as Kirchhoff's voltage law and Kirchhoff's current law. In this chapter, you will examine circuits which have more than one voltage source or which cannot be easily analyzed using techniques studied in previous chapters.

The methods used in determining the operation of complex networks will include branch-current analysis, mesh (or loop) analysis, and nodal analysis. Although any of the above methods may be used, you will find that certain circuits are more easily analyzed using one particular approach. The advantages of each method will be discussed in the appropriate section.

In using the techniques outlined above, it is assumed that the networks are **linear bilateral networks.** The term **linear** indicates that the components used in the circuit have voltage-current characteristics which follow a straight line. Refer to Figure 8–1.

The term **bilateral** indicates that the components in the network will have characteristics which are independent of the direction of the current through the element or the voltage across the element. A resistor is an example of a linear bilateral component since the voltage across a resistor is directly proportional to the current through it and the operation of the resistor is the same regardless of the direction of the current.

In this chapter you will be introduced to the conversion of a network from a delta (Δ) configuration to an equivalent wye (Y) configuration. Conversely, we will examine the transformation from a Y configuration to an equivalent Δ configuration. You will use these conversions to examine the operation of an unbalanced bridge network.

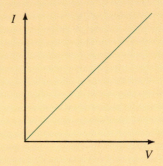

(a) Linear V-I characteristics

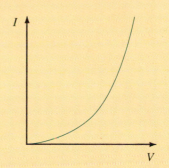

(b) Non-linear V-I characteristics

FIGURE 8–1

Sir Charles Wheatstone

CHARLES WHEATSTONE WAS BORN IN GLOUCESTER, England, on February 6, 1802. Wheatstone's original interest was in the study of acoustics and musical instruments. However, he gained fame and a knighthood as a result of inventing the telegraph and improving the electric generator.

Although he did not invent the bridge circuit, Wheatstone used one for measuring resistance very precisely. He found that when the currents in the Wheatstone bridge are exactly balanced, the unknown resistance can be compared to a known standard.

Sir Charles died in Paris, France, on October 19, 1875.

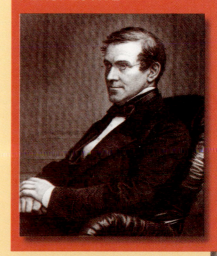

I

FIGURE 8–2 Ideal constant current source.

8.1   Constant-Current Sources

All the circuits presented so far have used voltage sources as the means of providing power. However, the analysis of certain circuits is easier if you work with current rather than with voltage. Unlike a voltage source, a **constant-current source** maintains the same current in its branch of the circuit regardless of how components are connected external to the source. The symbol for a constant-current source is shown in Figure 8–2.

The direction of the current source arrow indicates the direction of conventional current in the branch. In previous chapters you learned that the magnitude and the direction of current through a voltage source varies according to the size of the circuit resistances and how other voltage sources are connected in the circuit. For current sources, the voltage across the current source depends on how the other components are connected.

EXAMPLE 8–1   Refer to the circuit of Figure 8–3:

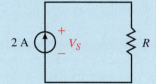

FIGURE 8–3

a. Calculate the voltage V_S across the current source if the resistor is 100 Ω.
b. Calculate the voltage if the resistor is 2 kΩ.

Solution   The current source maintains a constant current of 2 A through the circuit. Therefore,

a. $V_S = V_R = (2\ \text{A})(100\ \Omega) = 200\ \text{V}.$
b. $V_S = V_R = (2\ \text{A})(2\ \text{k}\Omega) = 4000\ \text{V}.$

If the current source is the only source in the circuit, then the polarity of voltage across the source will be as shown in Figure 8–3. This, however, may not be the case if there is more than one source. The following example illustrates this principle.

EXAMPLE 8–2   Determine the voltages V_1, V_2, and V_S and the current I_S for the circuit of Figure 8–4.

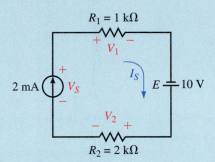

FIGURE 8–4

Solution   Since the given circuit is a series circuit, the current everywhere in the circuit must be the same, namely

$$I_S = 2 \text{ mA}$$

Using Ohm's law,

$$V_1 = (2 \text{ mA})(1 \text{ k}\Omega) = 2.00 \text{ V}$$
$$V_2 = (2 \text{ mA})(2 \text{ k}\Omega) = 4.00 \text{ V}$$

Applying Kirchhoff's voltage law around the closed loop,

$$\sum V = V_S - V_1 - V_2 + E = 0$$
$$V_S = V_1 + V_2 - E$$
$$= 2 \text{ V} + 4 \text{ V} - 10 \text{ V} = -4.00 \text{ V}$$

From the above result, you see that the actual polarity of V_S is opposite to that assumed.

EXAMPLE 8–3   Calculate the currents I_1 and I_2 and the voltage V_S for the circuit of Figure 8–5.

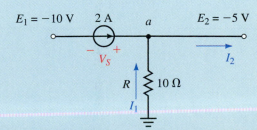

FIGURE 8–5

Solution   Because the 5-V supply is effectively across the load resistor,

$$I_1 = \frac{5 \text{ V}}{10 \text{ }\Omega} = 0.5 \text{ A} \quad \text{(in the direction assumed)}$$

Applying Kirchhoff's current law at point *a*,

$$I_2 = 0.5\,\text{A} + 2.0\,\text{A} = 2.5\,\text{A}$$

From Kirchhoff's voltage law,

$$\sum V = -10\,\text{V} + V_S + 5\,\text{V} = 0\,\text{V}$$
$$V_S = 10\,\text{V} - 5\,\text{V} = +5\,\text{V}$$

By examining the previous examples, the following conclusions may be made regarding current sources:

The constant-current source determines the current in its branch of the circuit.

The magnitude and polarity of voltage appearing across a constant-current source are dependent upon the network in which the source is connected.

8.2   Source Conversions

In the previous section you were introduced to the ideal constant-current source. This is a source which has no internal resistance included as part of the circuit. As you recall, voltage sources always have some series resistance, although in some cases this resistance is so small in comparison with other circuit resistance that it may effectively be ignored when determining the operation of the circuit. Similarly, a constant-current source will always have some shunt (or parallel) resistance. If this resistance is very large in comparison with the other circuit resistance, the internal resistance of the source may once again be ignored. **An ideal current source has an infinite shunt resistance**.

Figure 8–6 shows equivalent voltage and current sources.

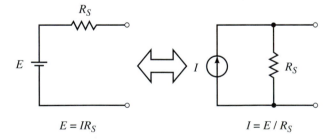

$$E = IR_S \qquad\qquad I = E / R_S$$

FIGURE 8–6

If the internal resistance of a source is considered, the source, whether it is a voltage source or a current source, is easily converted to the other type. The current source of Figure 8–6 is equivalent to the voltage source if

$$I = \frac{E}{R_S} \tag{8–1}$$

and the resistance in both sources is R_S.

Similarly, a current source may be converted to an equivalent voltage source by letting

$$E = IR_S \qquad (8\text{–}2)$$

These results may be easily verified by connecting an external resistance, R_L, across each source. The sources can be equivalent only if the voltage across R_L is the same for both sources. Similarly, the sources are equivalent only if the current through R_L is the same when connected to either source.

Consider the circuit shown in Figure 8–7.
The voltage across the load resistor is given as

$$V_L = \frac{R_L}{R_L + R_S} E \qquad (8\text{–}3)$$

The current through the resistor R_L is given as

$$I_L = \frac{E}{R_L + R_S} \qquad (8\text{–}4)$$

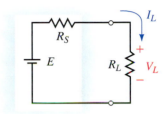

FIGURE 8–7

Next, consider an equivalent current source connected to the same load as shown in Figure 8–8. The current through the resistor R_L is given by

$$I_L = \frac{R_S}{R_S + R_L} I$$

But, when converting the source, we get

$$I = \frac{E}{R_S}$$

And so

$$I_L = \left(\frac{R_S}{R_S + R_L}\right)\left(\frac{E}{R_S}\right)$$

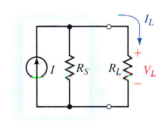

FIGURE 8–8

This result is equivalent to the current obtained in Equation 8–4. The voltage across the resistor is given as

$$V_L = I_L R_L$$
$$= \left(\frac{E}{R_S + R_L}\right) R_L$$

The voltage across the resistor is precisely the same as the result obtained in Equation 8–3. We therefore conclude that the load current and voltage drop are the same whether the source is a voltage source or an equivalent current source.

EXAMPLE 8–4   Convert the voltage source of Figure 8–9(a) into a current source and verify that the current, I_L, through the load is the same for each source.

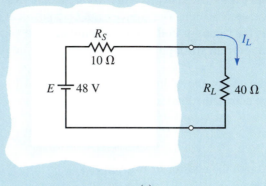

(a)

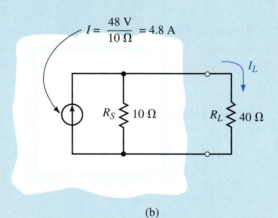

(b)

FIGURE 8–9

Solution   The equivalent current source will have a current magnitude given as

$$I = \frac{48 \text{ V}}{10 \text{ }\Omega} = 4.8 \text{ A}$$

The resulting circuit is shown in Figure 8–9(b).

For the circuit of Figure 8–9(a), the current through the load is found as

$$I_L = \frac{48 \text{ V}}{10 \text{ }\Omega + 40 \text{ }\Omega} = 0.96 \text{ A}$$

For the equivalent circuit of Figure 8–9(b), the current through the load is

$$I_L = \frac{(4.8 \text{ A})(10 \text{ }\Omega)}{10 \text{ }\Omega + 40 \text{ }\Omega} = 0.96 \text{ A}$$

Clearly the results are the same.

EXAMPLE 8–5 Convert the current source of Figure 8–10(a) into a voltage source and verify that the voltage, V_L, across the load is the same for each source.

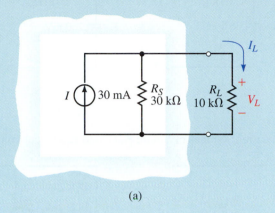

(a)

$$E = (30 \text{ mA})(30 \text{ k}\Omega) = 900 \text{ V}$$

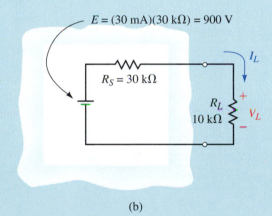

(b)

FIGURE 8–10

Solution The equivalent voltage source will have a magnitude given as

$$E = (30 \text{ mA})(30 \text{ k}\Omega) = 900 \text{ V}$$

The resulting circuit is shown in Figure 8–10(b).

For the circuit of Figure 8–10(a), the voltage across the load is determined as

$$I_L = \frac{(30 \text{ k}\Omega)(30 \text{ mA})}{30 \text{ k}\Omega + 10 \text{ k}\Omega} = 22.5 \text{ mA}$$

$$V_L = I_L R_L = (22.5 \text{ mA})(10 \text{ k}\Omega) = 225 \text{ V}$$

For the equivalent circuit of Figure 8–10(b), the voltage across the load is

$$V_L = \frac{10 \text{ k}\Omega}{10 \text{ k}\Omega + 30 \text{ k}\Omega}(900 \text{ V}) = 225 \text{ V}$$

Once again, we see that the circuits are equivalent.

1. Convert the voltage sources of Figure 8–11 into equivalent current sources.

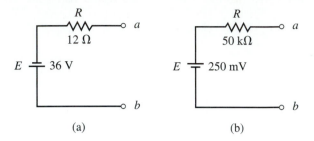

(a)   (b)

FIGURE 8–11

2. Convert the current sources of Figure 8–12 into equivalent voltage sources.

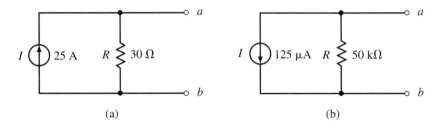

(a)   (b)

FIGURE 8–12

Answers:
1. a. $I = 3.00$ A (downward) in parallel with $R = 12 \, \Omega$
    b. $I = 5.00 \, \mu$A (upward) in parallel with $R = 50 \, k\Omega$
2. a. $E = V_{ab} = 750$ V in series with $R = 30 \, \Omega$
    b. $E = V_{ab} = -6.25$ V in series with $R = 50 \, k\Omega$

8.3   Current Sources in Parallel and Series

When several current sources are placed in parallel, the circuit may be simplified by combining the current sources into a single current source. The magnitude and direction of this resultant source is determined by adding the currents in one direction and then subtracting the currents in the opposite direction.

EXAMPLE 8–6   Simplify the circuit of Figure 8–13 and determine the voltage V_{ab}.

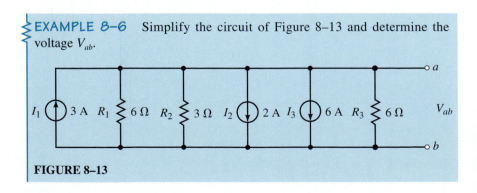

FIGURE 8–13

Solution   Since all of the current sources are in parallel, they can be replaced by a single current source. The equivalent current source will have a direction which is the same as both I_2 and I_3, since the magnitude of current in the downward direction is greater than the current in the upward direction. The equivalent current source has a magnitude of

$$I = 2\,\text{A} + 6\,\text{A} - 3\,\text{A} = 5\,\text{A}$$

as shown in Figure 8–14(a).

The circuit is further simplified by combining the resistors into a single value:

$$R_\text{T} = 6\,\Omega \| 3\,\Omega \| 6\,\Omega = 1.5\,\Omega$$

The equivalent circuit is shown in Figure 8–14(b).

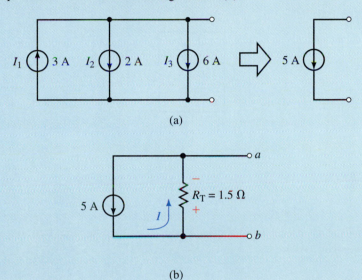

(a)

(b)

FIGURE 8–14

The voltage V_{ab} is found as

$$V_{ab} = -(5\,\text{A})(1.5\,\Omega) = -7.5\,\text{V}$$

EXAMPLE 8–7   Reduce the circuit of Figure 8–15 into a single current source and solve for the current through the resistor R_L.

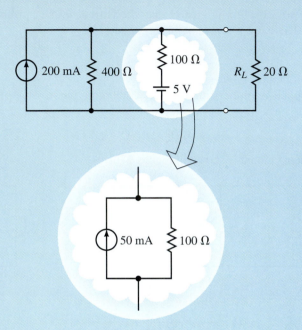

FIGURE 8–15

Solution   The voltage source in this circuit is converted to an equivalent current source as shown. The resulting circuit may then be simpified to a single current source where

$$I_S = 200 \text{ mA} + 50 \text{ mA} = 250 \text{ mA}$$

and

$$R_S = 400 \text{ } \Omega \| 100 \text{ } \Omega = 80 \text{ } \Omega$$

The simplified circuit is shown in Figure 8–16.

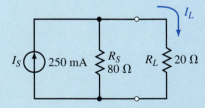

FIGURE 8–16

The current through R_L is now easily calculated as

$$I_L = \left(\frac{80 \text{ } \Omega}{80 \text{ } \Omega + 20 \text{ } \Omega} \right)(250 \text{ mA}) = 200 \text{ mA}$$

Current sources should never be placed in series. If a node is chosen between the current sources, it becomes immediately apparent that the current entering the node is not the same as the current leaving the node. Clearly, this cannot occur since there would then be a violation of Kirchhoff's current law (see Figure 8–17).

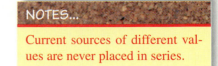

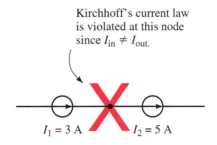

Kirchhoff's current law is violated at this node since $I_{in} \neq I_{out}$.

$I_1 = 3$ A $I_2 = 5$ A

FIGURE 8–17

1. Briefly explain the procedure for converting a voltage source into an equivalent current source.

2. What is the most important rule determining how current sources are connected into a circuit?

(Answers are at the end of the chapter.)

IN-PROCESS
LEARNING
CHECK 1

8.4   Branch-Current Analysis

In previous chapters we used Kirchhoff's circuit law and Kirchhoff's voltage law to solve equations for circuits having a single voltage source. In this section, you will use these powerful tools to analyze circuits having more than one source.

Branch-current analysis allows us to directly calculate the current in each branch of a circuit. Since the method involves the analysis of several simultaneous linear equations, you may find that a review of determinants is in order. Appendix B has been included to provide a review of the mechanics of solving simultaneous linear equations.

When applying branch-current analysis, you will find the technique listed below useful.

1. Arbitrarily assign current directions to each branch in the network. If a particular branch has a current source, then this step is not necessary since you already know the magnitude and direction of the current in this branch.

2. Using the assigned currents, label the polarities of the voltage drops across all resistors in the circuit.

3. Apply Kirchhoff's voltage law around each of the closed loops. Write just enough equations to include all branches in the loop equations. If a branch has only a current source and no series resistance, it is not necessary to include it in the KVL equations.

4. Apply Kirchhoff's current law at enough nodes to ensure that all branch currents have been included. In the event that a branch has only a current source, it will need to be included in this step.

5. Solve the resulting simultaneous linear equations.

EXAMPLE 8–8   Find the current in each branch in the circuit of Figure 8–18.

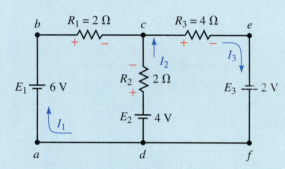

FIGURE 8–18

Solution

Step 1: Assign currents as shown in Figure 8–18.

Step 2: Indicate the polarities of the voltage drops on all resistors in the circuit, using the assumed current directions.

Step 3: Write the Kirchhoff voltage law equations.

Loop *abcda:* $6\ \text{V} - (2\ \Omega)I_1 + (2\ \Omega)I_2 - 4\ \text{V} = 0\ \text{V}$

Notice that the circuit still has one branch which has not been included in the KVL equations, namely the branch *cefd*. This branch would be included if a loop equation for *cefdc* or for *abcefda* were written. There is no reason for choosing one loop over another, since the overall result will remain unchanged even though the intermediate steps will not give the same results.

Loop *cefdc:* $4\ \text{V} - (2\ \Omega)I_2 - (4\ \Omega)I_3 + 2\ \text{V} = 0\ \text{V}$

Now that all branches have been included in the loop equations, there is no need to write any more. Although more loops exist, writing more loop equations would needlessly complicate the calculations.

Step 4: Write the Kirchhoff current law equation(s).

By applying KCL at node *c,* all branch currents in the network are included.

Node *c:* $I_3 = I_1 + I_2$

To simplify the solution of the simultaneous linear equations we write them as follows:

$$2I_1 - 2I_2 + 0I_3 = 2$$
$$0I_1 - 2I_2 - 4I_3 = -6$$
$$1I_1 + 1I_2 - 1I_3 = 0$$

The principles of linear algebra (Appendix B) allow us to solve for the determinant of the denominator as follows:

$$D = \begin{vmatrix} 2 & -2 & 0 \\ 0 & -2 & -4 \\ 1 & 1 & -1 \end{vmatrix}$$

$$= 2 \begin{vmatrix} -2 & -4 \\ 1 & -1 \end{vmatrix} - 0 \begin{vmatrix} -2 & 0 \\ 1 & -1 \end{vmatrix} + 1 \begin{vmatrix} -2 & 0 \\ -2 & -4 \end{vmatrix}$$

$$= 2(2 + 4) - 0 + 1(8) = 20$$

Now, solving for the currents, we have the following:

$$I_1 = \frac{\begin{vmatrix} 2 & -2 & 0 \\ -6 & -2 & -4 \\ 0 & 1 & -1 \end{vmatrix}}{D}$$

$$= \frac{2 \begin{vmatrix} -2 & -4 \\ 1 & -1 \end{vmatrix} - (-6) \begin{vmatrix} -2 & 0 \\ 1 & -1 \end{vmatrix} + 0 \begin{vmatrix} -2 & 0 \\ -2 & -4 \end{vmatrix}}{20}$$

$$= \frac{2(2 + 4) + 6(2) + 0}{20} = \frac{24}{20} = 1.200 \text{ A}$$

$$I_2 = \frac{\begin{vmatrix} 2 & 2 & 0 \\ 0 & -6 & -4 \\ 1 & 0 & -1 \end{vmatrix}}{D}$$

$$= \frac{2 \begin{vmatrix} -6 & -4 \\ 0 & -1 \end{vmatrix} - 0 \begin{vmatrix} 2 & 0 \\ 0 & -1 \end{vmatrix} + 1 \begin{vmatrix} 2 & 0 \\ -6 & -4 \end{vmatrix}}{20}$$

$$= \frac{2(6) + 0 + 1(-8)}{20} = \frac{4}{20} = 0.200 \text{ A}$$

$$I_3 = \frac{\begin{vmatrix} 2 & -2 & 2 \\ 0 & -2 & -6 \\ 1 & 1 & 0 \end{vmatrix}}{D}$$

$$= \frac{2 \begin{vmatrix} -2 & -6 \\ 1 & 0 \end{vmatrix} - 0 \begin{vmatrix} -2 & 2 \\ 1 & 0 \end{vmatrix} + 1 \begin{vmatrix} -2 & 2 \\ -2 & -6 \end{vmatrix}}{20}$$

$$= \frac{2(6) - 0 + 1(12 + 4)}{20} = \frac{28}{20} = 1.400 \text{ A}$$

EXAMPLE 8–9 Find the currents in each branch of the circuit shown in Figure 8–19. Solve for the voltage V_{ab}.

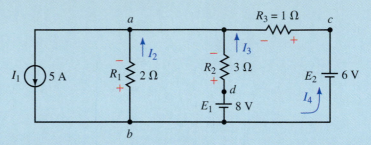

FIGURE 8–19

Solution Notice that although the above circuit has four currents, there are only three **unknown** currents: I_2, I_3, and I_4. The current I_1 is given by the value of the constant-current source. In order to solve this network we will need three linear equations. As before, the equations are determined by Kirchhoff's voltage and current laws.

Step 1: The currents are indicated in the given circuit.

Step 2: The polarities of the voltages across all resistors are shown.

Step 3: Kirchhoff's voltage law is applied at the indicated loops:

Loop *badb:* $\qquad -(2\ \Omega)(I_2) + (3\ \Omega)(I_3) - 8\ \text{V} = 0\ \text{V}$

Loop *bacb:* $\qquad -(2\ \Omega)(I_2) + (1\ \Omega)(I_4) - 6\ \text{V} = 0\ \text{V}$

Step 4: Kirchhoff's current law is applied as follows:

Node *a:* $\qquad\qquad I_2 + I_3 + I_4 = 5\ \text{A}$

Rewriting the linear equations,

$$-2I_2 + 3I_3 + 0I_4 = 8$$
$$-2I_2 + 0I_3 + 1I_4 = 6$$
$$1I_2 + 1I_3 + 1I_4 = 5$$

The determinant of the denominator is evaluated as

$$D = \begin{vmatrix} -2 & 3 & 0 \\ -2 & 0 & 1 \\ 1 & 1 & 1 \end{vmatrix} = 11$$

Now solving for the currents, we have

$$I_2 = \frac{\begin{vmatrix} 8 & 3 & 0 \\ 6 & 0 & 1 \\ 5 & 1 & 1 \end{vmatrix}}{D} = \frac{11}{11} = -1.00\ \text{A}$$

$$I_3 = \frac{\begin{vmatrix} -2 & 8 & 0 \\ -2 & 6 & 1 \\ 1 & 5 & 1 \end{vmatrix}}{D} = \frac{22}{11} = 2.00\ \text{A}$$

$$I_4 = \frac{\begin{vmatrix} -2 & 3 & 8 \\ -2 & 0 & 6 \\ 1 & 1 & 5 \end{vmatrix}}{D} = \frac{44}{11} = 4.00 \text{ A}$$

The current I_2 is negative, which simply means that the actual direction of the current is opposite to the chosen direction.

Although the network may be further analyzed using the assumed current directions, it is easier to understand the circuit operation by showing the actual current directions as in Figure 8–20.

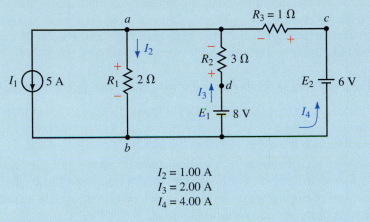

$$I_2 = 1.00 \text{ A}$$
$$I_3 = 2.00 \text{ A}$$
$$I_4 = 4.00 \text{ A}$$

FIGURE 8–20

Using the actual direction for I_2,

$$V_{ab} = +(2 \ \Omega)(1 \text{ A}) = +2.00 \text{ V}$$

Use branch-current analysis to solve for the indicated currents in the circuit of Figure 8–21.

PRACTICE
PROBLEMS 2

FIGURE 8–21

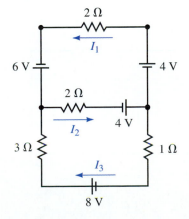

Answers: $I_1 = 3.00$ A, $I_2 = 4.00$ A, $I_3 = 1.00$ A

8.5   Mesh (Loop) Analysis

In the previous section you used Kirchhoff's laws to solve for the current in each branch of a given network. While the methods used were relatively simple, branch-current analysis is awkward to use because it generally involves solving several simultaneous linear equations. It is not difficult to see that the number of equations may be prohibitively large even for a relatively simple circuit.

A better approach and one which is used extensively in analyzing linear bilateral networks is called **mesh** (or **loop**) **analysis.** While the technique is similar to branch-current analysis, the number of simultaneous linear equations tends to be less. The principal difference between mesh analysis and branch-current analysis is that we simply need to apply Kirchhoff's voltage law around closed loops without the need for applying Kirchhoff's current law.

The steps used in solving a circuit using mesh analysis are as follows:

1. Arbitrarily assign a clockwise current to each interior closed loop in the network. Although the assigned current may be in any direction, a clockwise direction is used to make later work simpler.

2. Using the assigned loop currents, indicate the voltage polarities across all resistors in the circuit. For a resistor which is common to two loops, the polarities of the voltage drop due to each loop current should be indicated on the appropriate side of the component.

3. Applying Kirchhoff's voltage law, write the loop equations for each loop in the network. Do not forget that resistors which are common to two loops will have two voltage drops, one due to each loop.

4. Solve the resultant simultaneous linear equations.

5. Branch currents are determined by algebraically combining the loop currents which are common to the branch.

EXAMPLE 8–10   Find the current in each branch for the circuit of Figure 8–22.

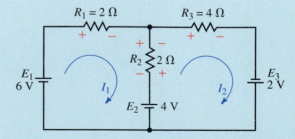

FIGURE 8–22

Solution

Step 1: Loop currents are assigned as shown in Figure 8–22. These currents are designated I_1 and I_2.

Step 2: Voltage polarities are assigned according to the loop currents. Notice that the resistor R_2 has two different voltage polarities due to the different loop currents.

Step 3: The loop equations are written by applying Kirchhoff's voltage law in each of the loops. The equations are as follows:

Loop 1: $\quad 6\text{ V} - (2\text{ }\Omega)I_1 - (2\text{ }\Omega)I_1 + (2\text{ }\Omega)I_2 - 4\text{ V} = 0$

Loop 2: $\quad 4\text{ V} - (2\text{ }\Omega)I_2 + (2\text{ }\Omega)I_1 - (4\text{ }\Omega)I_2 + 2\text{ V} = 0$

Note that the voltage across R_2 due to the currents I_1 and I_2 is indicated as two separated terms, where one term represents a voltage drop in the direction of I_1 and the other term represents a voltage rise in the same direction. The magnitude and polarity of the voltage across R_2 is determined by the actual size and directions of the loop currents. The above loop equations may be simplified as follows:

Loop 1: $\qquad\qquad (4\text{ }\Omega)I_1 - (2\text{ }\Omega)I_2 = 2\text{ V}$

Loop 2: $\qquad\qquad -(2\text{ }\Omega)I_1 + (6\text{ }\Omega)I_2 = 6\text{ V}$

Using determinants, the loop equations are easily solved as

$$I_1 = \frac{\begin{vmatrix} 2 & -2 \\ 6 & 6 \end{vmatrix}}{\begin{vmatrix} 4 & -2 \\ -2 & 6 \end{vmatrix}} = \frac{12 + 12}{24 - 4} = \frac{24}{20} = 1.20\text{ A}$$

and

$$I_2 = \frac{\begin{vmatrix} 4 & 2 \\ -2 & 6 \end{vmatrix}}{\begin{vmatrix} 4 & -2 \\ -2 & 6 \end{vmatrix}} = \frac{24 + 4}{24 - 4} = \frac{28}{20} = 1.40\text{ A}$$

From the above results, we see that the currents through resistors R_1 and R_3 are I_1 and I_2 respectively.

The branch current for R_2 is found by combining the loop currents through this resistor:

$$I_{R_2} = 1.40\text{ A} - 1.20\text{ A} = 0.20\text{ A} \quad \text{(upward)}$$

The results obtained by using mesh analysis are exactly the same as those obtained by branch-current analysis. Whereas branch-current analysis required three equations, this approach requires the solution of only two simultaneous linear equations. Mesh analysis also requires that only Kirchhoff's voltage law be applied and clearly illustrates why mesh analysis is preferred to branch-current analysis.

If the circuit being analyzed contains current sources, the procedure is a bit more complicated. The circuit may be simplified by converting the current source(s) to voltage sources and then solving the resulting network using the procedure shown in the previous example. Alternatively, you may not wish to alter the circuit, in which case the current source will provide one of the loop currents.

EXAMPLE 8–11 Determine the current through the 8-V battery for the circuit shown in Figure 8–23.

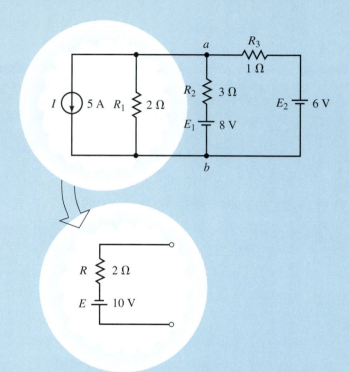

FIGURE 8–23

Solution Convert the current source into an equivalent voltage source. The equivalent circuit may now be analyzed by using the loop currents shown in Figure 8–24.

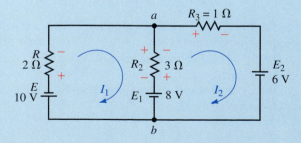

FIGURE 8–24

Loop 1: $-10\text{ V} - (2\text{ }\Omega)I_1 - (3\text{ }\Omega)I_1 + (3\text{ }\Omega)I_2 - 8\text{ V} = 0$

Loop 2: $8\text{ V} - (3\text{ }\Omega)I_2 + (3\text{ }\Omega)I_1 - (1\text{ }\Omega)I_2 - 6\text{ V} = 0$

Rewriting the linear equations, you get the following:

Loop 1: $(5\ \Omega)I_1 - (3\ \Omega)I_2 = -18\ V$

Loop 2: $-(3\ \Omega)I_1 + (4\ \Omega)I_2 = 2\ V$

Solving the equations using determinants, we have the following:

$$I_1 = \frac{\begin{vmatrix} -18 & -3 \\ 2 & 4 \end{vmatrix}}{\begin{vmatrix} 5 & -3 \\ -3 & 4 \end{vmatrix}} = -\frac{66}{11} = -6.00\ A$$

$$I_2 = \frac{\begin{vmatrix} 5 & -18 \\ -3 & 2 \end{vmatrix}}{\begin{vmatrix} 5 & -3 \\ -3 & 4 \end{vmatrix}} = -\frac{44}{11} = -4.00\ A$$

If the assumed direction of current in the 8-V battery is taken to be I_2, then

$$I = I_2 - I_1 = -4.00\ A - (-6.00\ A) = 2.00\ A$$

The direction of the resultant current is the same as I_2 (upward).

The circuit of Figure 8–23 may also be analyzed without converting the current source to a voltage source. Although the approach is generally not used, the following example illustrates the technique.

EXAMPLE 8–12   Determine the current through R_1 for the circuit shown in Figure 8–25.

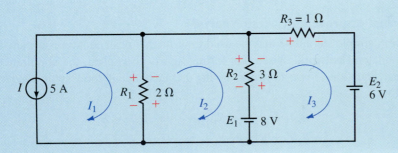

$R_3 = 1\ \Omega$

R_2 ⟩ $3\ \Omega$

R_1 ⟩ $2\ \Omega$

E_1 ⊥ $8\ V$

E_2 $6\ V$

FIGURE 8–25

Solution   By inspection, we see that the loop current $I_1 = -5\ A$. The mesh equations for the other two loops are as follows:

Loop 2: $-(2\ \Omega)I_2 + (2\ \Omega)I_1 - (3\ \Omega)I_2 + (3\ \Omega)I_3 - 8\ V = 0$

Loop 3: $8\ V - (3\ \Omega)I_3 + (3\ \Omega)I_2 - (1\ \Omega)I_3 - 6\ V = 0$

Although it is possible to analyze the circuit by solving three linear equations, it is easier to substitute the known value $I_1 = -5$ V into the mesh equation for loop 2, which may now be written as

Loop 2: $\quad -(2\,\Omega)I_2 - 10\,\text{V} - (3\,\Omega)I_2 + (3\,\Omega)I_3 - 8\,\text{V} = 0$

The loop equations may now be simplified as

Loop 2: $\qquad\qquad (5\,\Omega)I_2 - (3\,\Omega)I_3 = -18\,\text{V}$

Loop 3: $\qquad\qquad -(3\,\Omega)I_2 + (4\,\Omega)I_3 = 2\,\text{V}$

The simultaneous linear equations are solved as follows:

$$I_2 = \frac{\begin{vmatrix} -18 & -3 \\ 2 & 4 \end{vmatrix}}{\begin{vmatrix} 5 & -3 \\ -3 & 4 \end{vmatrix}} = -\frac{66}{11} = -6.00\,\text{A}$$

$$I_3 = \frac{\begin{vmatrix} 5 & -18 \\ -3 & 2 \end{vmatrix}}{\begin{vmatrix} 5 & -3 \\ -3 & 4 \end{vmatrix}} = -\frac{44}{11} = -4.00\,\text{A}$$

The calculated values of the assumed reference currents allow us to determine the actual current through the various resistors as follows:

$$I_{R_1} = I_1 - I_2 = -5\,\text{A} - (-6\,\text{A}) = 1.00\,\text{A} \quad \text{downward}$$
$$I_{R_2} = I_3 - I_2 = -4\,\text{A} - (-6\,\text{A}) = 2.00\,\text{A} \quad \text{upward}$$
$$I_{R_3} = -I_3 = 4.00\,\text{A} \quad \text{left}$$

These results are consistent with those obtained in Example 8–9.

Format Approach for Mesh Analysis

A very simple technique may be used to write the mesh equations for any linear bilateral network. When this format approach is used, the simultaneous linear equations for a network having n independent loops will appear as follows:

$$R_{11}I_1 - R_{12}I_2 - R_{13}I_3 - \cdots - R_{1n}I_n = E_1$$
$$-R_{21}I_1 + R_{22}I_2 - R_{23}I_3 - \cdots - R_{2n}I_n = E_2$$
$$\vdots$$
$$-R_{n1}I_1 - R_{n2}I_2 - R_{n3}I_3 - \cdots + R_{nn}I_n = E_n$$

The terms R_{11}, R_{22}, R_{33}, . . . , R_{nn} represent the total resistance in each loop and are found by simply adding all the resistances in a particular loop. The remaining resistance terms are called the **mutual resistance** terms. These resistances represent resistance which is shared between two loops. For example, the mutual resistance R_{12} is the resistance in loop 1 which is located in the branch between loop 1 and loop 2. If there is no resistance between two loops, this term will be zero.

The terms containing R_{11}, R_{22}, R_{33}, . . . , R_{nn} are positive, and all of the mutual resistance terms are negative. This characteristic occurs because all currents are assumed to be clockwise.

If the linear equations are correctly written, you will find that the coefficients along the principal diagonal (R_{11}, R_{22}, R_{33}, . . . , R_{nn}) will be positive. All other coefficients will be negative. Also, if the equations are correctly written, the terms will be symmetrical about the principal diagonal, e.g., $R_{12} = R_{21}$.

The terms E_1, E_2, E_3, . . . , E_n are the summation of the voltage rises in the direction of the loop currents. If a voltage source appears in the branch shared by two loops, it will be included in the calculation of the voltage rise for each loop.

The method used in applying the format approach of mesh analysis is as follows:

1. Convert current sources into equivalent voltage sources.

2. Assign clockwise currents to each independent closed loop in the network.

3. Write the simultaneous linear equations in the format outlined.

4. Solve the resulting simultaneous linear equations.

EXAMPLE 8–13   Solve for the currents through R_2 and R_3 in the circuit of Figure 8–26.

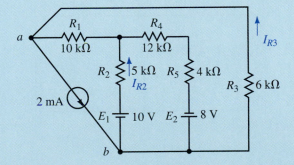

FIGURE 8–26

Solution

Step 1: Although we see that the circuit has a current source, it may not be immediately evident how the source can be converted into an equivalent voltage source. Redrawing the circuit into a more recognizable form, as shown in Figure 8–27, we see that the 2-mA current source is in parallel with a 6-kΩ resistor. The source conversion is also illustrated in Figure 8–27.

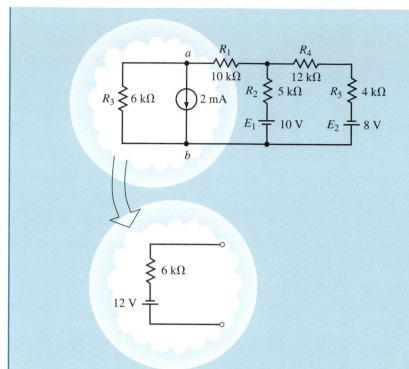

FIGURE 8–27

Step 2: Redrawing the circuit is further simplified by labelling some of the nodes, in this case a and b. After performing a source conversion, we have the two-loop circuit shown in Figure 8–28. The current directions for I_1 and I_2 are also illustrated.

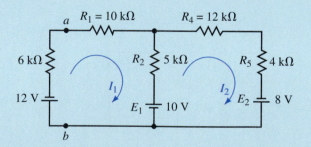

FIGURE 8–28

Step 3: The loop equations are

Loop 1: $(6 \text{ k}\Omega + 10 \text{ k}\Omega + 5 \text{ k}\Omega)I_1 - (5 \text{ k}\Omega)I_2 = -12 \text{ V} - 10 \text{ V}$

Loop 2: $-(5 \text{ k}\Omega)I_1 + (5 \text{ k}\Omega + 12 \text{ k}\Omega + 4 \text{ k}\Omega)I_2 = 10 \text{ V} + 8 \text{ V}$

In loop 1, both voltages are negative since they appear as voltage drops when following the direction of the loop current.

These equations are rewritten as

$$(21 \text{ k}\Omega)I_1 - (5 \text{ k}\Omega)I_2 = -22 \text{ V}$$
$$-(5 \text{ k}\Omega)I_1 + (21 \text{ k}\Omega)I_2 = 18 \text{ V}$$

Step 4: In order to simplify the solution of the previous linear equations, we may eliminate the units ($k\Omega$ and V) from our calculations. By inspection, we see that the units for current must be in milliamps. Using determinants, we solve for the currents I_1 and I_2 as follows:

$$I_1 = \frac{\begin{vmatrix} -22 & -5 \\ 18 & 21 \end{vmatrix}}{\begin{vmatrix} 21 & -5 \\ -5 & 21 \end{vmatrix}} = -\frac{-462 + 90}{441 - 25} = \frac{-372}{416} = -0.894 \text{ mA}$$

$$I_2 = \frac{\begin{vmatrix} 21 & -22 \\ -5 & 18 \end{vmatrix}}{\begin{vmatrix} 21 & -5 \\ -5 & 21 \end{vmatrix}} = \frac{378 + 110}{441 - 25} = \frac{268}{416} = -0.644 \text{ mA}$$

The current through resistor R_2 is easily determined to be

$$I_2 - I_1 = 0.644 \text{ mA} - (-0.894 \text{ mA}) = 1.54 \text{ mA}$$

The current through R_3 is not found as easily. A common mistake is to say that the current in R_3 is the same as the current through the 6-kΩ resistor of the circuit in Figure 8–28. **This is not the case.** Since this resistor was part of the source conversion it is no longer in the same location as in the original circuit.

Although there are several ways of finding the required current, the method used here is the application of Ohm's law. If we examine Figure 8–26, we see that the voltage across R_3 is equal to V_{ab}. From Figure 8–28, we see that we determine V_{ab} by using the calculated value of I_1.

$$V_{ab} = -(6 \text{ k}\Omega)I_1 - 12 \text{ V} = -(6 \text{ k}\Omega)(-0.894 \text{ mA}) - 12 \text{ V} = -6.64 \text{ V}$$

The above calculation indicates that the current through R_3 is upward (since point a is negative with respect to point b). The current has a value of

$$I_{R_3} = \frac{6.64 \text{ V}}{6 \text{ k}\Omega} = 1.11 \text{ mA}$$

Use mesh analysis to find the loop currents in the circuit of Figure 8–29.

PRACTICE PROBLEMS 3

FIGURE 8–29

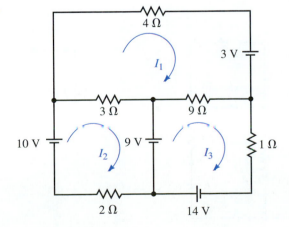

Answers: $I_1 = 3.00$ A, $I_2 = 2.00$ A, $I_3 = 5.00$ A

8.6   Nodal Analysis

In the previous section we applied Kirchhoff's voltage law to arrive at loop currents in a network. In this section we will apply Kirchhoff's current law to determine the potential difference (voltage) at any node with respect to some arbitrary reference point in a network. Once the potentials of all nodes are known, it is a simple matter to determine other quantities such as current and power within the network.

The steps used in solving a circuit using **nodal analysis** are as follows:

1. Arbitrarily assign a reference node within the circuit and indicate this node as **ground**. The reference node is usually located at the bottom of the circuit, although it may be located anywhere.

2. Convert each voltage source in the network to its equivalent current source. This step, although not absolutely necessary, makes further calculations easier to understand.

3. Arbitrarily assign voltages (V_1, V_2, . . . , V_n) to the remaining nodes in the circuit. (Remember that you have already assigned a reference node, so these voltages will all be with respect to the chosen reference.)

4. Arbitrarily assign a current direction to each branch in which there is no current source. Using the assigned current directions, indicate the corresponding polarities of the voltage drops on all resistors.

5. With the exception of the reference node (ground), apply Kirchhoff's current law at each of the nodes. If a circuit has a total of $n + 1$ nodes (including the reference node), there will be n simultaneous linear equations.

6. Rewrite each of the arbitrarily assigned currents in terms of the potential difference across a known resistance.

7. Solve the resulting simultaneous linear equations for the voltages (V_1, V_2, . . . , V_n).

EXAMPLE 8–14   Given the circuit of Figure 8–30, use nodal analysis to solve for the voltage V_{ab}.

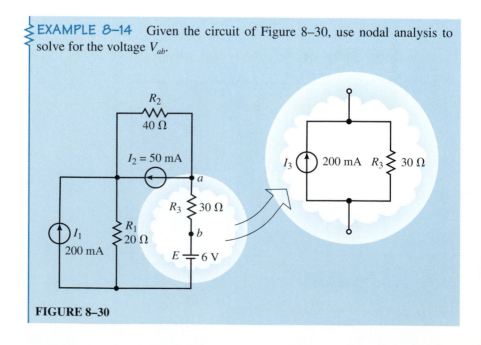

FIGURE 8–30

Solution

Step 1: Select a convenient reference node.

Step 2: Convert the voltage sources into equivalent current sources. The equivalent circuit is shown in Figure 8–31.

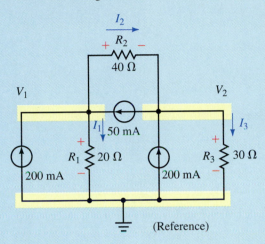

FIGURE 8–31

Steps 3 and 4: Arbitrarily assign node voltages and branch currents. Indicate the voltage polarities across all resistors according to the assumed current directions.

Step 5: We now apply Kirchhoff's current law at the nodes labelled as V_1 and V_2:

Node V_1:
$$\sum I_{\text{entering}} = \sum I_{\text{leaving}}$$
$$200 \text{ mA} + 50 \text{ mA} = I_1 + I_2$$

Node V_2:
$$\sum I_{\text{entering}} = \sum I_{\text{leaving}}$$
$$200 \text{ mA} + I_2 = 50 \text{ mA} + I_3$$

Step 6: The currents are rewritten in terms of the voltages across the resistors as follows:

$$I_1 = \frac{V_1}{20 \ \Omega}$$

$$I_2 = \frac{V_1 - V_2}{40 \ \Omega}$$

$$I_3 = \frac{V_2}{30 \ \Omega}$$

The nodal equations become

$$200 \text{ mA} + 50 \text{ mA} = \frac{V_1}{20 \ \Omega} + \frac{V_1 - V_2}{40 \ \Omega}$$

$$200 \text{ mA} + \frac{V_1 - V_2}{40 \ \Omega} = 50 \text{ mA} + \frac{V_2}{30 \ \Omega}$$

Substituting the voltage expressions into the original nodal equations, we have the following simultaneous linear equations:

$$\left(\frac{1}{20\ \Omega} + \frac{1}{40\ \Omega}\right)V_1 - \left(\frac{1}{40\ \Omega}\right)V_2 = 250\ \text{mA}$$

$$-\left(\frac{1}{40\ \Omega}\right)V_1 + \left(\frac{1}{30\ \Omega} + \frac{1}{40\ \Omega}\right)V_2 = 150\ \text{mA}$$

These may be further simplified as

$$(0.075\ \text{S})V_1 - (0.025\ \text{S})V_2 = 250\ \text{mA}$$

$$-(0.025\ \text{S})V_1 + (0.058\overline{3})V_2 = 150\ \text{mA}$$

Step 7: Use determinants to solve for the nodal voltages as

$$V_1 = \frac{\begin{vmatrix} 0.250 & -0.025 \\ 0.150 & 0.058\overline{3} \end{vmatrix}}{\begin{vmatrix} 0.075 & -0.025 \\ 0.025 & 0.058\overline{3} \end{vmatrix}}$$

$$= \frac{(0.250)(0.058\overline{3}) - (0.150)(-0.025)}{(0.075)(0.058\overline{3}) - (-0.025)(-0.025)}$$

$$= \frac{0.018\overline{3}}{0.00375} = 4.89\ \text{V}$$

and

$$V_2 = \frac{\begin{vmatrix} 0.075 & 0.250 \\ -0.025 & 0.150 \end{vmatrix}}{\begin{vmatrix} 0.075 & 0.025 \\ -0.025 & 0.058\overline{3} \end{vmatrix}}$$

$$= \frac{(0.075)(0.150) - (-0.025)(0.250)}{0.00375}$$

$$= \frac{0.0175}{0.00375} = 4.67\ \text{V}$$

If we go back to the original circuit of Figure 8–30, we see that the voltage V_2 is the same as the voltage V_a, namely

$$V_a = 4.67\ \text{V} = 6.0\ \text{V} + V_{ab}$$

Therefore, the voltage V_{ab} is simply found as

$$V_{ab} = 4.67\ \text{V} - 6.0\ \text{V} = -1.33\ \text{V}$$

EXAMPLE 8–15   Determine the nodal voltages for the circuit shown in Figure 8–32.

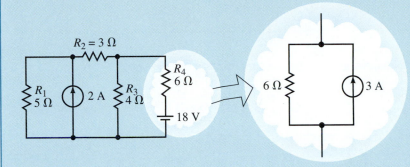

FIGURE 8–32

Solution   By following the steps outlined, the circuit may be redrawn as shown in Figure 8–33.

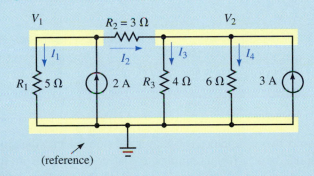

(reference)

FIGURE 8–33

Applying Kirchhoff's current law to the nodes corresponding to V_1 and V_2, the following nodal equations are obtained:

$$\sum I_{\text{leaving}} = \sum I_{\text{entering}}$$

Node V_1: $I_1 + I_2 = 2\,\text{A}$

Node V_2: $I_3 + I_4 = I_2 + 3\,\text{A}$

The currents may once again be written in terms of the voltages across the resistors:

$$I_1 = \frac{V_1}{5\,\Omega}$$

$$I_2 = \frac{V_1 - V_2}{3\,\Omega}$$

$$I_3 = \frac{V_2}{4\,\Omega}$$

$$I_4 = \frac{V_2}{6\,\Omega}$$

The nodal equations become

Node V_1:
$$\frac{V_1}{5\ \Omega} + \frac{(V_1 - V_2)}{3\ \Omega} = 2\ \text{A}$$

Node V_2:
$$\frac{V_2}{4\ \Omega} + \frac{V_2}{6\ \Omega} = \frac{(V_1 - V_2)}{3\ \Omega} + 3\ \text{A}$$

These equations may now be simplified as

Node V_1:
$$\left(\frac{1}{5\ \Omega} + \frac{1}{3\ \Omega}\right)V_1 - \left(\frac{1}{3\ \Omega}\right)V_2 = 2\ \text{A}$$

Node V_2:
$$-\left(\frac{1}{3\ \Omega}\right)V_1 + \left(\frac{1}{4\ \Omega} + \frac{1}{6\ \Omega} + \frac{1}{3\ \Omega}\right)V_2 = 3\ \text{A}$$

The solutions for V_1 and V_2 are found using determinants:

$$V_1 = \frac{\begin{vmatrix} 2 & -0.333 \\ 3 & 0.750 \end{vmatrix}}{\begin{vmatrix} 0.533 & -0.333 \\ -0.333 & 0.750 \end{vmatrix}} = \frac{2.500}{0.289} = 8.65\ \text{V}$$

$$V_2 = \frac{\begin{vmatrix} 0.533 & 2 \\ -0.333 & 3 \end{vmatrix}}{\begin{vmatrix} 0.533 & -0.333 \\ -0.333 & 0.750 \end{vmatrix}} = \frac{2.267}{0.289} = 7.85\ \text{V}$$

In the previous two examples, you may have noticed that the simultaneous linear equations have a format similar to that developed for mesh analysis. When we wrote the nodal equation for node V_1 the coefficient for the variable V_1 was positive, and it had a magnitude given by the summation of the conductance attached to this node. The coefficient for the variable V_2 was negative and had a magnitude given by the mutual conductance between nodes V_1 and V_2.

Format Approach

A simple format approach may be used to write the nodal equations for any network having $n + 1$ nodes. Where one of these nodes is denoted as the reference node, there will be n simultaneous linear equations which will appear as follows:

$$G_{11}V_1 - G_{12}V_2 - G_{13}V_3 - \cdots - R_{1n}V_n = I_1$$
$$-G_{21}V_1 + G_{22}V_2 - G_{23}V_3 - \cdots - R_{2n}V_n = I_2$$
$$\vdots$$
$$-G_{n1}V_1 - G_{n2}V_2 - G_{n3}V_3 - \cdots + R_{nn}V_n = I_n$$

The coefficients (constants) G_{11}, G_{22}, G_{33}, . . . , G_{nn} represent the summation of the conductances attached to the particular node. The remaining coefficients are called the **mutual conductance** terms. For example, the mutual conductance G_{23} is the conductance attached to node V_2, which is common to node V_3. If there is no conductance that is common to two nodes, then this term would be zero. Notice that the terms G_{11}, G_{22}, G_{33}, . . . , G_{nn} are positive and that the mutual conductance terms are negative. Further, if the equations are written correctly, then the terms will be symmetrical about the principal diagonal, e.g., $G_{23} = G_{32}$.

The terms V_1, V_2, . . . , V_n are the unknown node voltages. Each voltage represents the potential difference between the node in question and the reference node.

The terms I_1, I_2, . . . , I_n are the summation of current sources entering the node. If a current source has a current such that it is leaving the node, then the current is simply assigned as negative. If a particular current source is shared between two nodes, then this current must be included in both nodal equations.

The method used in applying the format approach of nodal analysis is as follows:

1. Convert voltage sources into equivalent current sources.
2. Label the reference node as $\perp$. Label the remaining nodes as V_1, V_2, . . . , V_n.
3. Write the linear equation for each node using the format outlined.
4. Solve the resulting simultaneous linear equations for V_1, V_2, . . . , V_n.

The next examples illustrate how the format approach is used to solve circuit problems.

EXAMPLE 8–16 Determine the nodal voltages for the circuit shown in Figure 8–34.

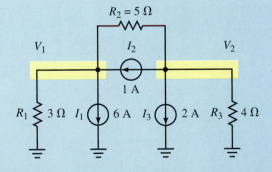

FIGURE 8–34

Solution The circuit has a total of three nodes: the reference node (at a potential of zero volts) and two other nodes, V_1 and V_2.

By applying the format approach for writing the nodal equations, we get two equations:

Node V_1: $\left(\dfrac{1}{3\,\Omega} + \dfrac{1}{5\,\Omega}\right)V_1 - \left(\dfrac{1}{5\,\Omega}\right)V_2 = -6\,\text{A} + 1\,\text{A}$

Node V_2: $-\left(\dfrac{1}{5\,\Omega}\right)V_1 + \left(\dfrac{1}{5\,\Omega} + \dfrac{1}{4\,\Omega}\right)V_2 = -1\,\text{A} - 2\,\text{A}$

On the right-hand sides of the above, those currents that are leaving the nodes are given a negative sign.

These equations may be rewritten as

Node V_1: $(0.533\ \text{S})V_1 - (0.200\ \text{S})V_2 = -5\ \text{A}$

Node V_2: $-(0.200\ \text{S})V_1 + (0.450\ \text{S})V_2 = -3\ \text{A}$

Using determinants to solve these equations, we have

$$V_1 = \frac{\begin{vmatrix} -5 & -0.200 \\ -3 & 0.450 \end{vmatrix}}{\begin{vmatrix} 0.533 & -0.200 \\ -0.200 & 0.450 \end{vmatrix}} = \frac{-2.85}{0.200} = -14.3\ \text{V}$$

$$V_2 = \frac{\begin{vmatrix} 0.533 & -5 \\ -0.200 & -3 \end{vmatrix}}{\begin{vmatrix} 0.533 & -0.200 \\ -0.200 & 0.450 \end{vmatrix}} = \frac{-2.60}{0.200} = -13.0\ \text{V}$$

EXAMPLE 8–17   Use nodal analysis to find the nodal voltages for the circuit of Figure 8–35. Use the answers to solve for the current through R_1.

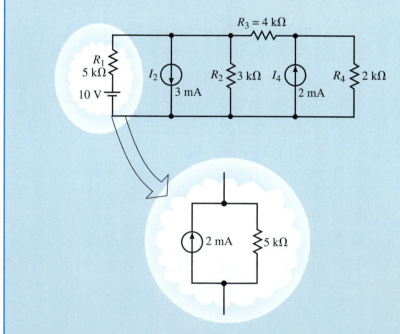

FIGURE 8–35

Solution   In order to apply nodal analysis, we must first convert the voltage source into its equivalent current source. The resulting circuit is shown in Figure 8–36.

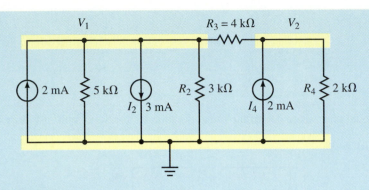

FIGURE 8–36

Labelling the nodes and writing the nodal equations, we obtain the following:

Node V_1: $\left(\dfrac{1}{5\text{ k}\Omega} + \dfrac{1}{3\text{ k}\Omega} + \dfrac{1}{4\text{ k}\Omega}\right)V_1 - \left(\dfrac{1}{4\text{ k}\Omega}\right)V_2 = 2\text{ mA} - 3\text{ mA}$

Node V_2: $-\left(\dfrac{1}{4\text{ k}\Omega}\right)V_1 + \left(\dfrac{1}{4\text{ k}\Omega} + \dfrac{1}{2\text{ k}\Omega}\right)V_2 = 2\text{ mA}$

Because it is inconvenient to use kilohms and milliamps throughout our calculations, we may eliminate these units in our calculations. You have already seen that any voltage obtained by using these quantities will result in the units being "volts." Therefore the nodal equations may be simplified as

Node V_1: $(0.7833)V_1 - (0.2500)V_2 = -1$

Node V_2: $-(0.2500)V_1 + (0.750)V_2 = 2$

The solutions are as follows:

$$V_1 = \frac{\begin{vmatrix} -1 & -0.250 \\ 2 & 0.750 \end{vmatrix}}{\begin{vmatrix} 0.7833 & -0.250 \\ -0.250 & 0.750 \end{vmatrix}} = \frac{-0.250}{0.525} = -0.476\text{ V}$$

$$V_2 = \frac{\begin{vmatrix} 0.7833 & -1 \\ -0.250 & 2 \end{vmatrix}}{\begin{vmatrix} 0.7833 & -0.250 \\ -0.250 & 0.750 \end{vmatrix}} = \frac{1.3167}{0.525} = 2.51\text{ V}$$

Using the values derived for the nodal voltages, it is now possible to solve for any other quantities in the circuit. To determine the current through resistor $R_1 = 5\text{ k}\Omega$, we first reassemble the circuit as it appeared originally. Since the node voltage V_1 is the same in both circuits, we use it in determining the desired current. The resistor may be isolated as shown in Figure 8–37.

FIGURE 8–37

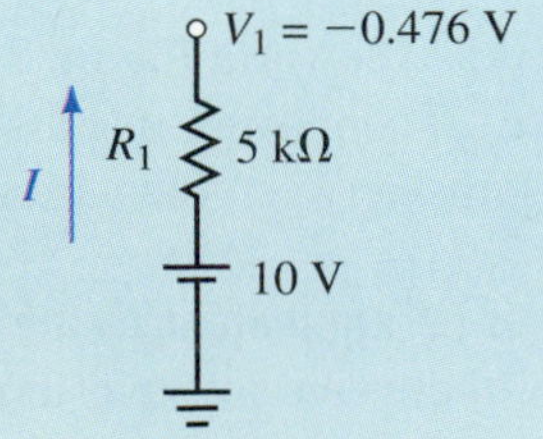

$V_1 = -0.476\text{ V}$

I

R_1 5 kΩ

10 V

The current is easily found as

$$I = 10\,\text{V} - \frac{(-0.476\,\text{V})}{5\,\text{k}\Omega} = 2.10\,\text{mA} \quad \text{(upward)}$$

PRACTICE PROBLEMS 4

Use nodal analysis to determine the node voltages for the circuit of Figure 8–38.

FIGURE 8–38

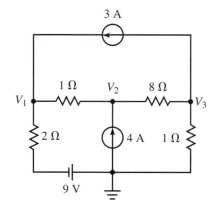

Answers: $V_1 = 3.00\,\text{V}$, $V_2 = 6.00\,\text{V}$, $V_3 = -2.00\,\text{V}$

8.7   Delta-Wye (Pi-Tee) Conversion

Delta-Wye Conversion

You have previously examined resistor networks involving series, parallel, and series-parallel combinations. We will next examine networks which cannot be placed into any of the above categories. While these circuits may be analyzed using techniques developed earlier in this chapter, there is an easier approach. For example, consider the circuit shown in Figure 8–39.

This circuit could be analyzed using mesh analysis. However, you see that the analysis would involve solving four simultaneous linear equations, since there are four separate loops in the circuit. If we were to use nodal analysis, the solution would require determining three node voltages, since there are three nodes in addition to a reference node. Unless a computer is used, both techniques are very time-consuming and prone to error.

As you have already seen, it is occasionally easier to examine a circuit after it has been converted to some equivalent form. We will now develop a technique for converting a circuit from a **delta** (or pi) into an equivalent **wye** (or **tee**) **circuit.** Consider the circuits shown in Figure 8–40. We start by making the assumption that the networks shown in Figure 8–40(a) are equivalent to those shown in Figure 8–40(b). Then, using this assumption, we will determine the mathematical relationships between the various resistors in the equivalent circuits.

The circuit of Figure 8–40(a) can be equivalent to the circuit of Figure 8–40(b) only if the resistance "seen" between any two terminals is exactly

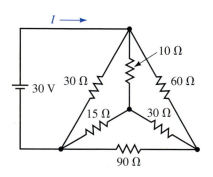

FIGURE 8–39

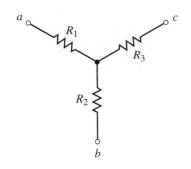

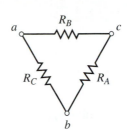

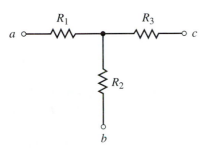

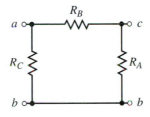

(a) Wye ("Y") or Tee ("T") network   (b) Delta ("Δ") or Pi ("Π") network

FIGURE 8–40

the same. If we were to connect a source between terminals a and b of the "Y," the resistance between the terminals would be

$$R_{ab} = R_1 + R_2 \qquad\qquad (8\text{–}5)$$

But the resistance between terminals a and b of the "Δ" is

$$R_{ab} = R_C \| (R_A + R_B) \qquad\qquad (8\text{–}6)$$

Combining Equations 8–5 and 8–6, we get

$$R_1 + R_2 = \frac{R_C(R_A + R_B)}{R_A + R_B + R_C}$$

$$R_1 + R_2 = \frac{R_A R_C + R_B R_C}{R_A + R_B + R_C} \qquad\qquad (8\text{–}7)$$

Using a similar approach between terminals b and c, we get

$$R_2 + R_3 = \frac{R_A R_B + R_A R_C}{R_A + R_B + R_C} \qquad\qquad (8\text{–}8)$$

and between terminals c and a we get

$$R_1 + R_3 = \frac{R_A R_B + R_B R_C}{R_A + R_B + R_C} \qquad\qquad (8\text{–}9)$$

If Equation 8–8 is subtracted from Equation 8–7, then

$$R_1 + R_2 - (R_2 + R_3) = \frac{R_A R_C + R_B R_C}{R_A + R_B + R_C} - \frac{R_A R_B + R_A R_C}{R_A + R_B + R_C}$$

$$R_1 - R_3 = \frac{R_B R_C - R_A R_B}{R_A + R_B + R_C} \qquad\qquad (8\text{–}10)$$

Adding Equations 8–9 and 8–10, we get

$$R_1 + R_3 + R_1 - R_3 = \frac{R_A R_B + R_B R_C}{R_A + R_B + R_C} + \frac{R_B R_C - R_A R_B}{R_A + R_B + R_C}$$

$$2R_1 = \frac{2R_B R_C}{R_A + R_B + R_C} \tag{8–11}$$

$$R_1 = \frac{R_B R_C}{R_A + R_B + R_C}$$

Using a similar approach, we obtain

$$R_2 = \frac{R_A R_C}{R_A + R_B + R_C} \tag{8–12}$$

$$R_3 = \frac{R_A R_B}{R_A + R_B + R_C} \tag{8–13}$$

Notice that any resistor connected to a point of the "Y" is obtained by finding the product of the resistors connected to the same point in the "Δ" and then dividing by the sum of all the "Δ" resistances.

If all the resistors in a Δ circuit have the same value, R_Δ, then the resulting resistors in the equivalent Y network will also be equal and have a value given as

$$R_Y = \frac{R_\Delta}{3} \tag{8–14}$$

EXAMPLE 8–18   Find the equivalent Y circuit for the Δ circuit shown in Figure 8–41.

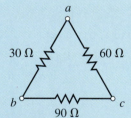

FIGURE 8–41

Solution   From the circuit of Figure 8–41, we see that we have the following resistor values:

$$R_A = 90\ \Omega$$
$$R_B = 60\ \Omega$$
$$R_C = 30\ \Omega$$

Applying Equations 8–11 through 8–13 we have the following equivalent "Y" resistor values:

$$R_1 = \frac{(30\ \Omega)(60\ \Omega)}{30\ \Omega + 60\ \Omega + 90\ \Omega}$$

$$= \frac{1800\ \Omega}{180} = 10\ \Omega$$

$$R_2 = \frac{(30\ \Omega)(90\ \Omega)}{30\ \Omega + 60\ \Omega + 90\ \Omega}$$

$$= \frac{2700\ \Omega}{180} = 15\ \Omega$$

$$R_3 = \frac{(60\ \Omega)(90\ \Omega)}{30\ \Omega + 60\ \Omega + 90\ \Omega}$$

$$= \frac{5400\ \Omega}{180} = 30\ \Omega$$

The resulting circuit is shown in Figure 8–42.

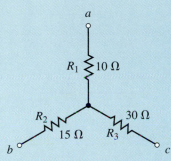

FIGURE 8–42

Wye-Delta Conversion

By using Equations 8–11 to 8–13, it is possible to derive another set of equations which allow the conversion from a "Y" into an equivalent "Δ." Examining Equations 8–11 through 8–13, we see that the following must be true:

$$R_A + R_B + R_C = \frac{R_A R_B}{R_3} = \frac{R_A R_C}{R_2} = \frac{R_B R_C}{R_1}$$

From the above expression we may write the following two equations:

$$R_B = \frac{R_A R_1}{R_2} \qquad \qquad (8\text{–}15)$$

$$R_C = \frac{R_A R_1}{R_3} \qquad \qquad (8\text{–}16)$$

Now, substituting Equations 8–15 and 8–16 into Equation 8–11, we have the following:

$$R_1 = \frac{\left(\dfrac{R_A R_1}{R_2}\right)\left(\dfrac{R_A R_1}{R_3}\right)}{R_A + \left(\dfrac{R_A R_1}{R_2}\right) + \left(\dfrac{R_A R_1}{R_3}\right)}$$

By factoring R_A out of each term in the denominator, we are able to arrive at

$$R_1 = \frac{\left(\dfrac{R_A R_1}{R_2}\right)\left(\dfrac{R_A R_1}{R_3}\right)}{R_A\left[1 + \left(\dfrac{R_1}{R_2}\right) + \left(\dfrac{R_1}{R_3}\right)\right]}$$

$$R_1 = \frac{\left(\dfrac{R_A R_1 R_1}{R_2 R_3}\right)}{\left[1 + \left(\dfrac{R_1}{R_2}\right) + \left(\dfrac{R_1}{R_3}\right)\right]}$$

$$= \frac{\left(\dfrac{R_A R_1 R_1}{R_2 R_3}\right)}{\left(\dfrac{R_1 R_2 + R_1 R_3 + R_2 R_3}{R_2 R_3}\right)}$$

$$= \frac{R_A R_1 R_1}{R_1 R_2 + R_1 R_3 + R_2 R_3}$$

Rewriting the above expression gives

$$R_A = \frac{R_1 R_2 + R_1 R_3 + R_2 R_3}{R_1} \tag{8–17}$$

Similarly,

$$R_B = \frac{R_1 R_2 + R_1 R_3 + R_2 R_3}{R_2} \tag{8–18}$$

and

$$R_C = \frac{R_1 R_2 + R_1 R_3 + R_2 R_3}{R_3} \tag{8–19}$$

In general, we see that the resistor in any side of a "Δ" is found by taking the sum of all two-product combinations of "Y" resistor values and then dividing by the resistance in the "Y" which is located directly opposite to the resistor being calculated.

If the resistors in a Y network are all equal, then the resultant resistors in the equivalent Δ circuit will also be equal and given as

$$R_\Delta = 3R_Y \tag{8–20}$$

EXAMPLE 8–19   Find the Δ network equivalent of the Y network shown in Figure 8–43.

FIGURE 8–43

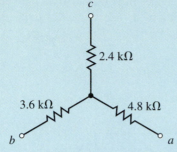

Solution   The equivalent Δ network is shown in Figure 8–44.

FIGURE 8–44

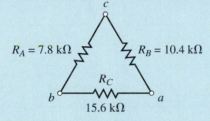

The values of the resistors are determined as follows:

$$R_A = \frac{(4.8 \text{ k}\Omega)(2.4 \text{ k}\Omega) + (4.8 \text{ k}\Omega)(3.6 \text{ k}\Omega) + (2.4 \text{ k}\Omega)(3.6 \text{ k}\Omega)}{4.8 \text{ k}\Omega}$$

$$= 7.8 \text{ k}\Omega$$

$$R_B = \frac{(4.8 \text{ k}\Omega)(2.4 \text{ k}\Omega) + (4.8 \text{ k}\Omega)(3.6 \text{ k}\Omega) + (2.4 \text{ k}\Omega)(3.6 \text{ k}\Omega)}{3.6 \text{ k}\Omega}$$

$$= 10.4 \text{ k}\Omega$$

$$R_C = \frac{(4.8 \text{ k}\Omega)(2.4 \text{ k}\Omega) + (4.8 \text{ k}\Omega)(3.6 \text{ k}\Omega) + (2.4 \text{ k}\Omega)(3.6 \text{ k}\Omega)}{2.4 \text{ k}\Omega}$$

$$= 15.6 \text{ k}\Omega$$

EXAMPLE 8–20   Given the circuit of Figure 8–45, find the total resistance, R_T, and the total current, I.

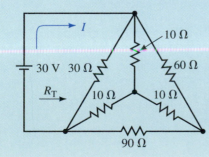

FIGURE 8–45

Solution   As is often the case, the given circuit may be solved in one of two ways. We may convert the "Δ" into its equivalent "Y," and solve the circuit by placing the resultant branches in parallel, or we may convert the "Y" into its equivalent "Δ." We choose to use the latter conversion since the resistors in the "Y" have the same value. The equivalent "Δ" will have all resistors given as

$$R_\Delta = 3(10\ \Omega) = 30\ \Omega$$

The resulting circuit is shown in Figure 8–46(a).

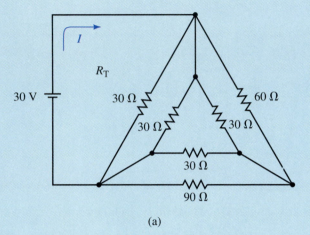

(a)

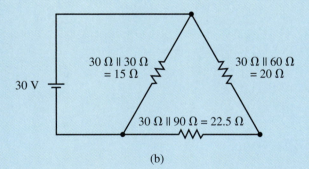

(b)

FIGURE 8–46

We see that the sides of the resulting "Δ" are in parallel, which allows us to simplify the circuit even further as shown in Figure 8–46(b). The total restance of the circuit is now easily determined as

$$R_T = 15\ \Omega \| (20\ \Omega + 22.5\ \Omega)$$
$$= 11.09\ \Omega$$

This results in a circuit current of

$$I = \frac{30\ \text{V}}{11.09\ \Omega} = 2.706\ \text{A}$$

Convert the Δ network of Figure 8–44 into an equivalent Y network. Verify that the result you obtain is the same as that found in Figure 8–43.

Answers: $R_1 = 4.8 \text{ k}\Omega$, $R_2 = 3.6 \text{ k}\Omega$, $R_3 = 2.4 \text{ k}\Omega$

8.8 Bridge Networks

In this section you will be introduced to the **bridge network**. Bridge networks are used in electronic measuring equipment to precisely measure resistance in dc circuits and similar quantities in ac circuits. The bridge circuit was originally used by Sir Charles Wheatstone in the mid-nineteenth century to measure resistance by balancing small currents. The Wheatstone bridge circuit is still used to measure resistance very precisely. The digital bridge, shown in Figure 8–47, is an example of one such instrument.

FIGURE 8–47 Digital bridge used for precisely measuring resistance, inductance, and capacitance.

You will use the techniques developed earlier in the chapter to analyze the operation of these networks. Bridge circuits may be shown in various configurations as seen in Figure 8–48.

Although a bridge circuit may appear in one of three forms, you can see that they are equivalent. There are, however, two different states of bridges: the balanced bridge and the unbalanced bridge.

A **balanced bridge** is one in which the current through the resistance R_5 is equal to zero. In practical circuits, R_5 is generally a variable resistor in

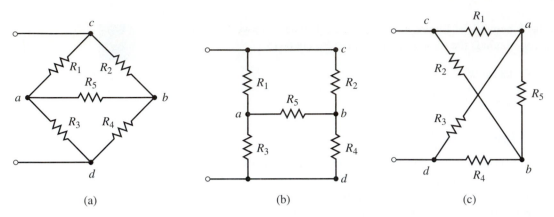

(a)   (b)   (c)

FIGURE 8–48

series with a sensitive galvanometer. When the current through R_5 is zero, then it follows that

$$V_{ab} = (R_5)(0\,\text{A}) = 0\,\text{V}$$

$$I_{R_1} = I_{R_3} = \frac{V_{cd}}{R_1 + R_3}$$

$$I_{R_2} = I_{R_4} = \frac{V_{cd}}{R_2 + R_4}$$

But the voltage V_{ab} is found as

$$V_{ab} = V_{ad} - V_{bd} = 0$$

Therefore, $V_{ad} = V_{bd}$ and

$$R_3 I_{R_3} = R_4 I_{R_4}$$

$$R_3\left(\frac{V_{cd}}{R_1 + R_3}\right) = R_4\left(\frac{V_{cd}}{R_2 + R_4}\right)$$

which simplifies to

$$\frac{R_3}{R_1 + R_3} = \frac{R_4}{R_2 + R_4}$$

Now, if we invert both sides of the equation and simplify, we get the following:

$$\frac{R_1 + R_3}{R_3} = \frac{R_2 + R_4}{R_4}$$

$$\frac{R_1}{R_3} + 1 = \frac{R_2}{R_4} + 1$$

Finally, by subtracting 1 from each side, we obtain the following ratio for a balanced bridge:

$$\frac{R_1}{R_3} = \frac{R_2}{R_4} \qquad\qquad \textbf{(8–21)}$$

From Equation 8–21, we notice that a bridge network is balanced whenever the ratios of the resistors in the two arms are the same.

An **unbalanced bridge** is one in which the current through R_5 is not zero, and so the above ratio does not apply to an unbalanced bridge network. Figure 8–49 illustrates each condition of a bridge network.

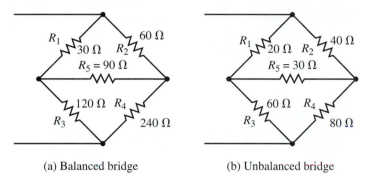

(a) Balanced bridge   (b) Unbalanced bridge

FIGURE 8–49

If a balanced bridge appears as part of a complete circuit, its analysis is very simple since the resistor R_5 may be removed and replaced with either a short (since $V_{R_5} = 0$) or an open (since $I_{R_5} = 0$).

However, if a circuit contains an unbalanced bridge, the analysis is more complicated. In such cases, it is possible to determine currents and voltages by using mesh analysis, nodal analysis, or by using Δ to Y conversion. The following examples illustrate how bridges may be analyzed.

EXAMPLE 8–21   Solve for the currents through R_1 and R_4 in the circuit of Figure 8–50.

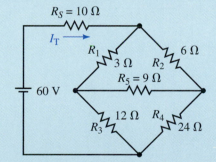

FIGURE 8–50

Solution   We see that the bridge of the above circuit is balanced (since $R_1/R_3 = R_2/R_4$). Because the circuit is balanced, we may remove R_5 and replace it with either a short circuit (since the voltage across a short circuit is zero) or an open circuit (since the current through an open circuit is zero). The remaining circuit is then solved by one of the methods developed in previous chapters. Both methods will be illustrated to show that the results are exactly the same.

Method 1: If R_5 is replaced by an open, the result is the circuit shown in Figure 8–51.

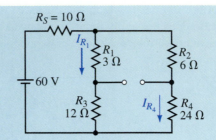

FIGURE 8–51

The total circuit resistance is found as

$$R_T = 10\ \Omega + (3\ \Omega + 12\ \Omega)\|(6\ \Omega + 24\ \Omega)$$
$$= 10\ \Omega + 15\ \Omega\|30\ \Omega$$
$$= 20\ \Omega$$

The circuit current is

$$I_T = \frac{60\ \text{V}}{20\ \Omega} = 3.0\ \text{A}$$

The current in each branch is then found by using the current divider rule:

$$I_{R_1} = \left(\frac{30\ \Omega}{30\ \Omega + 15\ \Omega}\right)(3.0\ \text{A}) = 2.0\ \text{A}$$

$$I_{R_4} = \frac{10\ \Omega}{24\ \Omega + 6\ \Omega}(3.0\ \text{A}) = 1.0\ \text{A}$$

Method 2: If R_5 is replaced with a short circuit, the result is the circuit shown in Figure 8–52.

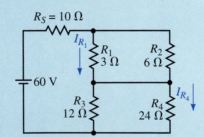

FIGURE 8–52

The total circuit resistance is found as

$$R_T = 10\ \Omega + (3\ \Omega\|6\ \Omega) + (12\ \Omega\|24\ \Omega)$$
$$= 10\ \Omega + 2\ \Omega + 8\ \Omega$$
$$= 20\ \Omega$$

The above result is precisely the same as that found using Method 1. Therefore the circuit current will remain as $I_T = 3.0\ \text{A}$.

The currents through R_1 and R_4 may be found by the current divider rule as

$$I_{R_1} = \left(\frac{6\ \Omega}{6\ \Omega + 3\ \Omega}\right)(3.0\ \text{A}) = 2.0\ \text{A}$$

and

$$I_{R_4} = \left(\frac{12\ \Omega}{12\ \Omega + 24\ \Omega} \right)(3.0\ \text{A}) = 1.0\ \text{A}$$

Clearly, these results are precisely those obtained in Method 1, illustrating that the methods are equivalent. Remember, though, R_5 can be replaced with a short circuit or an open circuit only when the bridge is balanced.

EXAMPLE 8–22   Use mesh analysis to find the currents through R_1 and R_5 in the unbalanced bridge circuit of Figure 8–53.

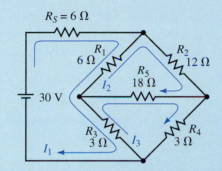

FIGURE 8–53

Solution   After assigning loop currents as shown, we write the loop equations as

Loop 1: $(15\ \Omega)I_1 - (6\ \Omega)I_2 - (3\ \Omega)I_3 = 30\ \text{V}$

Loop 2: $-(6\ \Omega)I_1 + (36\ \Omega)I_2 - (18\ \Omega)I_3 = 0$

Loop 3: $-(3\ \Omega)I_1 - (18\ \Omega)I_2 + (24\ \Omega)I_3 = 0$

The determinant for the denominator will is

$$D = \begin{vmatrix} 15 & -6 & -3 \\ -6 & 36 & -18 \\ 3 & -18 & 24 \end{vmatrix} = 6264$$

Notice that, as expected, the elements in the principal diagonal are positive and that the determinant is symmetrical around the principal diagonal.

The loop currents are now evaluated as

$$I_1 = \frac{\begin{vmatrix} 30 & -6 & -3 \\ 0 & 36 & -18 \\ 0 & -18 & 24 \end{vmatrix}}{D} = \frac{16\ 200}{6264} = 2.586\ \text{A}$$

$$I_2 = \frac{\begin{vmatrix} 15 & 30 & -3 \\ -6 & 0 & -18 \\ -3 & 0 & 24 \end{vmatrix}}{D} = \frac{5940}{6264} = 0.948\ \text{A}$$

$$I_3 = \frac{\begin{vmatrix} 15 & -6 & 30 \\ -6 & 36 & 0 \\ -3 & -18 & 0 \end{vmatrix}}{D} = \frac{6480}{6264} = 1.034\ \text{A}$$

The current through R_1 is found as

$$I_{R_1} = I_1 - I_2 = 2.586\ \text{A} - 0.948\ \text{A} = 1.638\ \text{A}$$

The current through R_5 is found as

$$I_{R_5} = I_3 - I_2 = 1.034\ \text{A} - 0.948\ \text{A}$$
$$= 0.086\ \text{A} \quad \text{to the right}$$

The previous example illustrates that if the bridge is not balanced, there will always be some current through resistor R_5. The unbalanced circuit may also be easily analyzed using nodal analysis, as in the following example.

EXAMPLE 8–23   Determine the node voltages and the voltage V_{R_5} for the circuit of Figure 8–54.

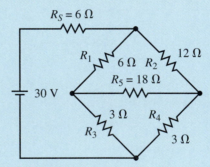

FIGURE 8–54

Solution   By converting the voltage source into an equivalent current source, we obtain the circuit shown in Figure 8–55.

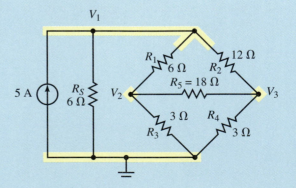

FIGURE 8–55

The nodal equations for the circuit are as follows:

Node 1: $\left(\dfrac{1}{6\,\Omega} + \dfrac{1}{6\,\Omega} + \dfrac{1}{12\,\Omega}\right)V_1 - \left(\dfrac{1}{6\,\Omega}\right)V_2 - \left(\dfrac{1}{12\,\Omega}\right)V_3 = 5\text{ A}$

Node 2: $-\left(\dfrac{1}{6\,\Omega}\right)V_1 + \left(\dfrac{1}{6\,\Omega} + \dfrac{1}{3\,\Omega} + \dfrac{1}{18\,\Omega}\right)V_2 - \left(\dfrac{1}{18\,\Omega}\right)V_3 = 0\text{ A}$

Node 3: $-\left(\dfrac{1}{12\,\Omega}\right)V_1 - \left(\dfrac{1}{18\,\Omega}\right)V_2 + \left(\dfrac{1}{3\,\Omega} + \dfrac{1}{12\,\Omega} + \dfrac{1}{18\,\Omega}\right)V_3 = 0\text{ A}$

The linear equations are

Node 1: $\quad 0.4167V_1 - 0.1667V_2 - 0.0833V_3 = 5\text{ A}$

Node 2: $\quad -0.1667V_1 + 0.5556V_2 - 0.0556V_3 = 0$

Node 3: $\quad -0.0833V_1 - 0.0556V_2 + 0.4722V_3 = 0$

The determinant of the denominator is

$$D = \begin{vmatrix} 0.4167 & -0.1667 & -0.0833 \\ -0.1667 & 0.5556 & -0.0556 \\ -0.0833 & -0.0556 & 0.4722 \end{vmatrix} = 0.08951\text{ A}$$

Again notice that the elements on the principal diagonal are positive and that the determinant is symmetrical about the principal diagonal.

The node voltages are calculated to be

$$V_1 = \dfrac{\begin{vmatrix} 5 & -0.1667 & -0.0833 \\ 0 & 0.5556 & -0.0556 \\ 0 & -0.0556 & 0.4722 \end{vmatrix}}{D} = \dfrac{1.2963}{0.08951} = 14.48\text{ A}$$

$$V_2 = \dfrac{\begin{vmatrix} 0.4167 & 5 & -0.0833 \\ -0.1667 & 0 & -0.0556 \\ -0.0833 & 0 & 0.4722 \end{vmatrix}}{} = \dfrac{0.41667}{0.08951} = 4.66\text{ V}$$

$$V_3 = \dfrac{\begin{vmatrix} 0.4167 & -0.1667 & 5 \\ -0.1667 & 0.0556 & 0 \\ -0.0833 & -0.0556 & 0 \end{vmatrix}}{} = \dfrac{0.2778}{0.08951} = 3.10\text{ V}$$

Using the above results, we find the voltage across R_5:

$$V_{R5} = V_2 - V_3 = 4.655\text{ V} - 3.103\text{ V} = 1.55\text{ V}$$

and the current through R_5 is

$$I_{R5} = \dfrac{1.55\text{ V}}{18\,\Omega} = 0.086\text{ A}\quad\text{to the right}$$

As expected, the results are the same whether we use mesh analysis or nodal analysis. It is therefore a matter of personal preference as to which approach should be used.

A final method for analyzing bridge networks involves the use of Δ to Y conversion. The following example illustrates the method used.

EXAMPLE 8–24 Find the current through R_5 for the circuit shown in Figure 8–56.

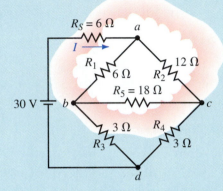

FIGURE 8–56

Solution By inspection we see that this circuit is not balanced, since

$$\frac{R_1}{R_3} \neq \frac{R_2}{R_4}$$

Therefore, the current through R_5 cannot be zero. Notice, also, that the circuit contains two possible Δ configurations. If we choose to convert the top Δ to its equivalent Y, we get the circuit shown in Figure 8–57.

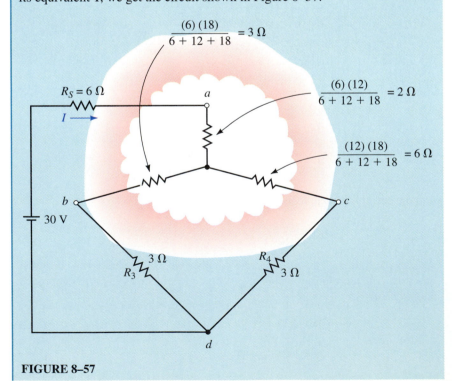

FIGURE 8–57

By combining resistors, it is possible to reduce the complicated circuit to the simple series circuit shown in Figure 8–58.

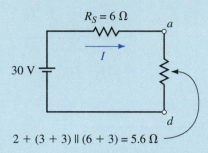

$R_S = 6 \, \Omega$

a

I

30 V

d

$2 + (3 + 3) \parallel (6 + 3) = 5.6 \, \Omega$

FIGURE 8–58

The circuit of Figure 8–58 is easily analyzed to give a total circuit current of

$$I = \frac{30 \text{ V}}{6 \, \Omega + 2 \, \Omega + 3.6 \, \Omega} = 2.59 \text{ A}$$

Using the calculated current, it is possible to work back to the original circuit. The currents in the resistors R_3 and R_4 are found by using the current divider rule for the corresponding resistor branches, as shown in Figure 8–57.

$$I_{R_3} = \frac{(6 \, \Omega + 3 \, \Omega)}{(6 \, \Omega + 3 \, \Omega) + (3 \, \Omega + 3 \, \Omega)} (2.59 \text{ A}) = 1.55 \text{ A}$$

$$I_{R_4} = \frac{(3 \, \Omega + 3 \, \Omega)}{(6 \, \Omega + 3 \, \Omega) + (3 \, \Omega + 3 \, \Omega)} (2.59 \text{ A}) = 1.03 \text{ A}$$

These results are exactly the same as those found in Examples 8–21 and 8–22. Using these currents, it is now possible to determine the voltage V_{bc} as

$$\begin{aligned} V_{bc} &= -(3 \, \Omega) I_{R_4} + (3 \, \Omega) I_{R_3} \\ &= (-3 \, \Omega)(1.034 \text{ A}) + (3 \, \Omega)(3.103 \text{ A}) \\ &= 1.55 \text{ V} \end{aligned}$$

The current through R_5 is determined to be

$$I_{R_5} = \frac{1.55 \text{ V}}{18 \, \Omega} = 0.086 \text{ A} \quad \text{to the right}$$

1. For a balanced bridge, what will be the value of voltage between the midpoints of the arms of the bridge?

2. If a resistor or sensitive galvanometer is placed between the arms of a balanced bridge, what will be the current through the resistor?

3. In order to simplify the analysis of a balanced bridge, how may the resistance R_5, between the arms of the bridge, be replaced?

IN-PROCESS
LEARNING
CHECK 2

(Answers are at the end of the chapter.)

PRACTICE
PROBLEMS 6

FIGURE 8–59

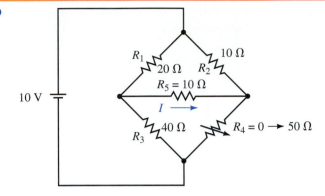

1. For the circuit shown in Figure 8–59, what value of R_4 will ensure that the bridge is balanced?

2. Determine the current I through R_5 in Figure 8–59 when $R_4 = 0\ \Omega$ and when $R_4 = 50\ \Omega$.

Answers: 1. $20\ \Omega$ 2. 286 mA, -52.6 mA

ELECTRONICS
WORKBENCH

PSpice

8.9   Circuit Analysis Using Computers

Electronics Workbench and PSpice are able to analyze a circuit without the need to convert between voltage and current sources or having to write lengthy linear equations. It is possible to have the program output the value of voltage across or current through any element in a given circuit. The following examples were previously analyzed using several other methods throughout this chapter.

EXAMPLE 8–25   Given the circuit of Figure 8–60, use Electronics Workbench to find the voltage V_{ab} and the current through each resistor.

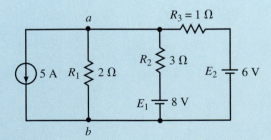

FIGURE 8–60

Solution   The circuit is entered as shown in Figure 8–61. The current source is obtained by clicking on the Sources button in the Parts bin toolbar. As before, it is necessary to include a ground symbol in the schematic although the original circuit of Figure 8–60 did not have one. Make sure that all values are changed from the default values to the required circuit values.

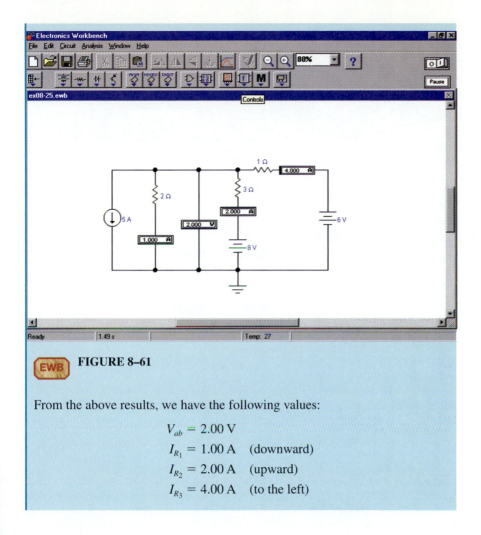

FIGURE 8–61

From the above results, we have the following values:

$$V_{ab} = 2.00 \text{ V}$$
$$I_{R_1} = 1.00 \text{ A} \quad \text{(downward)}$$
$$I_{R_2} = 2.00 \text{ A} \quad \text{(upward)}$$
$$I_{R_3} = 4.00 \text{ A} \quad \text{(to the left)}$$

OrCAD PSpice

EXAMPLE 8–26   Use PSpice to find currents through R_1 and R_5 in the circuit of Figure 8–62.

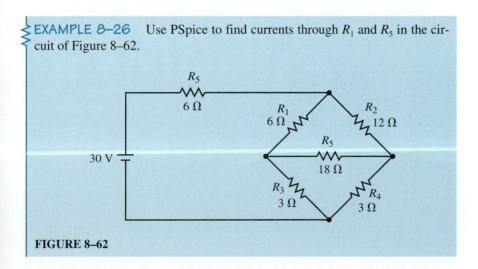

FIGURE 8–62

Solution   The PSpice file is entered as shown in Figure 8–63. Ensure that you enter **Yes** in the DC cell of the Properties Editor for each IPRINT part.

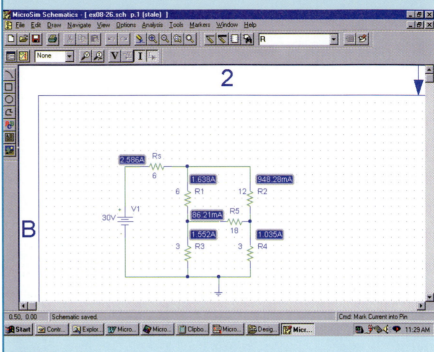

FIGURE 8–63

Once you have selected a New Simulation Profile, click on the Run icon. Select View and Output File to see the results of the simulation. The currents are $I_{R_1} = 1.64$ A and $I_{R_5} = 86.2$ mA. These results are consistent with those obtained in Example 8–22.

PRACTICE PROBLEMS 7

Use Electronics Workbench to determine the currents, I_T, I_{R_1} and I_{R_4} in the circuit of Figure 8–50. Compare your results to those obtained in Example 8–21.

Answers: $I_T = 3.00$ A, $I_{R_1} = 2.00$ A and $I_{R_4} = 1.00$ A

PRACTICE PROBLEMS 8

Use PSpice to input the circuit of Figure 8–50. Determine the value of I_{R_5} when $R_4 = 0$ Ω and when $R_4 = 48$ Ω.

Note: Since PSpice cannot let $R_4 = 0$ Ω, you will need to let it equal some very small value such as 1 μΩ (1e-6).

Answers: $I_{R_5} = 1.08$ A when $R_4 = 0$ Ω and $I_{R_5} = 0.172$ A when $R_4 = 48$ Ω

PRACTICE PROBLEMS 9

Use PSpice to input the circuit of Figure 8–54, so that the output file will provide the currents through R_1, R_2, and R_5. Compare your results to those obtained in Example 8–23.

S train gauges are manufactured from very fine wire mounted on insulated
surfaces which are then glued to large metal structures. These instruments
are used by civil engineers to measure the movement and mass of large
objects such as bridges and buildings. When the very fine wire of a strain
gauge is subjected to stress, its effective length is increased (due to stretching) or decreased (due to compression). This change in length results in a corresponding minute change in resistance. By placing one or more strain
gauges into a bridge circuit, it is possible to detect variation in resistance, ΔR.
This change in resistance can be calibrated to correspond to an applied force.
Consequently, it is possible to use such a bridge as a means of measuring
very large masses. Consider that you have two strain gauges mounted in a
bridge as shown in the accompanying figure.

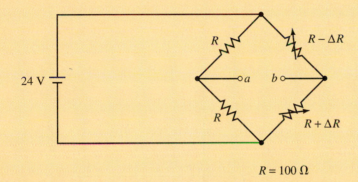

$$R = 100 \ \Omega$$

Strain gauge bridge.

The variable resistors, R_2 and R_4 are strain gauges that are mounted on
opposite sides of a steel girder used to measure very large masses. When a
mass is applied to the girder, the strain gauge on one side of the girder will
compress, reducing the resistance. The strain gauge on the other side of the
girder will stretch, increasing the resistance. When no mass is applied, there
will be neither compression nor stretching and so the bridge will be balanced,
resulting in a voltage $V_{ab} = 0$ V.
    Write an expression for ΔR as a function of V_{ab}. Assume that the scale is
calibrated so that resistance variation of $\Delta R = 0.02 \ \Omega$ corresponds to a mass
of 5000 kg. Determine the measured mass if $V_{ab} = -4.20$ mV.

8.1   Constant-Current Sources   **PROBLEMS**

1. Find the voltage V_S for the circuit shown in Figure 8–64.

2. Find the voltage V_S for the circuit shown in Figure 8–65.

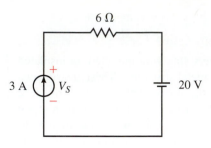

FIGURE 8–64

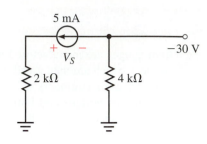

FIGURE 8–65

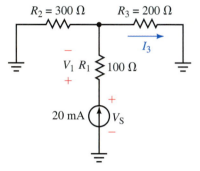

FIGURE 8–66

3. Refer to the circuit of Figure 8–66:
   a. Find the current I_3.
   b. Determine the voltages V_S and V_1.
4. Consider the circuit of Figure 8–67:
   a. Calculate the voltages V_2 and V_S.
   b. Find the currents I and I_3.

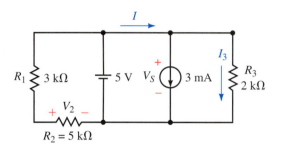

FIGURE 8–67

5. For the circuit of Figure 8–68, find the currents I_1 and I_2.

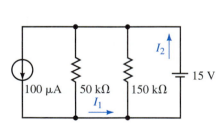

FIGURE 8–68

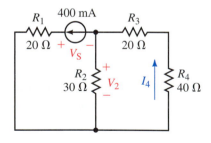

FIGURE 8–69

6. Refer to the circuit of Figure 8–69:
   a. Find the voltages V_S and V_2.
   b. Determine the current I_4.
7. Verify that the power supplied by the sources is equal to the summation of the powers dissipated by the resistors in the circuit of Figure 8–68.

8. Verify that the power supplied by the source in the circuit of Figure 8–69 is equal to the summation of the powers dissipated by the resistors.

8.2 Source Conversions

9. Convert each of the voltage sources of Figure 8–70 into its equivalent current source.

10. Convert each of the current sources of Figure 8–71 into its equivalent voltage source.

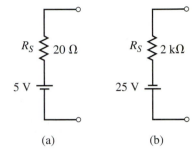

(a)   (b)

FIGURE 8–70

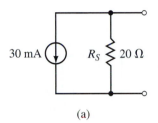

(a)

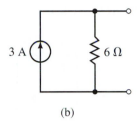

(b)

FIGURE 8–71

11. Refer to the circuit of Figure 8–72:

   a. Solve for the current through the load resistor using the current divider rule.

   b. Convert the current source into its equivalent voltage source and again determine the current through the load.

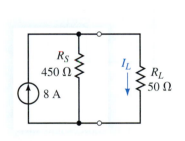

FIGURE 8–72

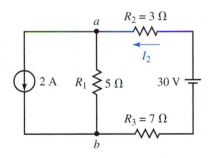

FIGURE 8–73

12. Find V_{ab} and I_2 for the network of Figure 8–73.

13. Refer to the circuit of Figure 8–74:

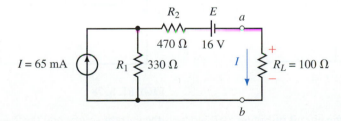

FIGURE 8–74

a. Convert the current source and the 330-Ω resistor into an equivalent voltage source.

b. Solve for the current I through R_L.

c. Determine the voltage V_{ab}.

14. Refer to the circuit of Figure 8–75:

a. Convert the voltage source and the 36-Ω resistor into an equivalent current source.

b. Solve for the current I through R_L.

c. Determine the voltage V_{ab}.

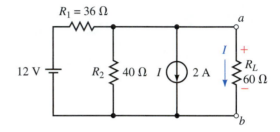

FIGURE 8–75

8.3   Current Sources in Parallel and Series

15. Find the voltage V_2 and the current I_1 for the circuit of Figure 8–76.

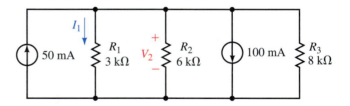

EWB   **FIGURE 8–76**

16. Convert the voltage sources of Figure 8–77 into current sources and solve for the current I_1 and the voltage V_{ab}.

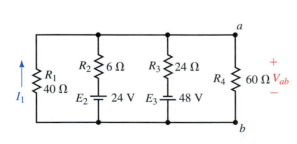

FIGURE 8–77

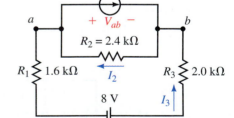

FIGURE 8–78

17. For the circuit of Figure 8–78 convert the current source and the 2.4-kΩ resistor into a voltage source and find the voltage V_{ab} and the current I_3.

18. For the circuit of Figure 8–78, convert the voltage source and the series resistors into an equivalent current source.

   a. Determine the current I_2.

   b. Solve for the voltage V_{ab}.

8.4   Branch-Current Analysis

19. Write the branch-current equations for the circuit shown in Figure 8–79 and solve for the branch currents using determinants.

20. Refer to the circuit of Figure 8–80:

   a. Solve for the current I_1 using branch-current analysis.

   b. Determine the voltage V_{ab}.

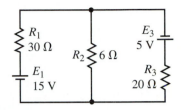

FIGURE 8–79

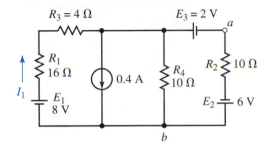

FIGURE 8–80

21. Write the branch-current equations for the circuit shown in Figure 8–81 and solve for the current I_2.

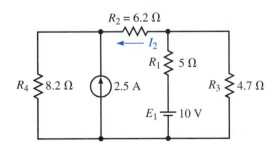

FIGURE 8–81

22. Refer to the circuit shown in Figure 8–82:

   a. Write the branch-current equations.

   b. Solve for the currents I_1 and I_2.

   c. Determine the voltage V_{ab}.

23. Refer to the circuit shown in Figure 8–83:

   a. Write the branch-current equations.

   b. Solve for the current I_2.

   c. Determine the voltage V_{ab}.

24. Refer to the circuit shown in Figure 8–84:

   a. Write the branch-current equations.

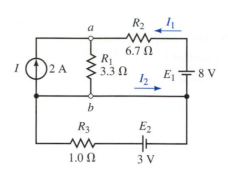

FIGURE 8–82

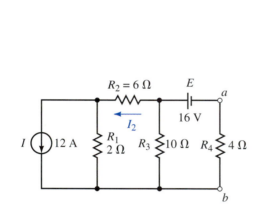

FIGURE 8–83

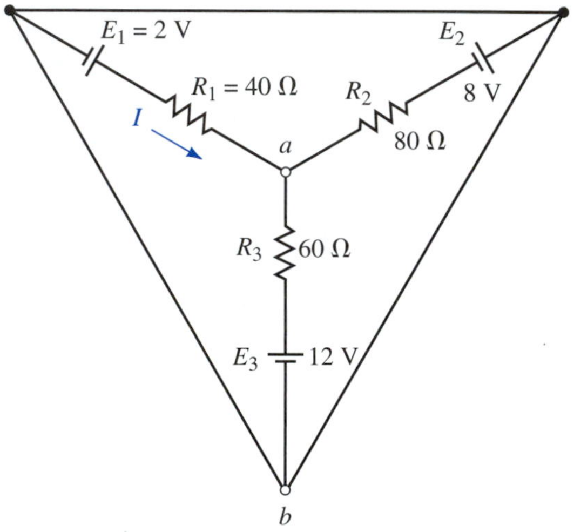

FIGURE 8–84

b. Solve for the current I.

c. Determine the voltage V_{ab}.

8.5   Mesh (Loop) Analysis

25. Write the mesh equations for the circuit shown in Figure 8–79 and solve for the loop currents.

26. Use mesh analysis for the circuit of Figure 8–80 to solve for the current I_1.

27. Use mesh analysis to solve for the current I_2 in the circuit of Figure 8–81.

28. Use mesh analysis to solve for the loop currents in the circuit of Figure 8–83. Use your results to determine I_2 and V_{ab}.

29. Use mesh analysis to solve for the loop currents in the circuit of Figure 8–84. Use your results to determine I and V_{ab}.

30. Using mesh analysis, determine the current through the 6-Ω resistor in the circuit of Figure 8–85.

31. Write the mesh equations for the network in Figure 8–86. Solve for the loop currents using determinants.

32. Repeat Problem 31 for the network in Figure 8–87.

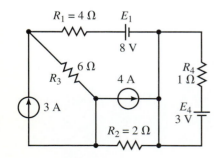

FIGURE 8–85

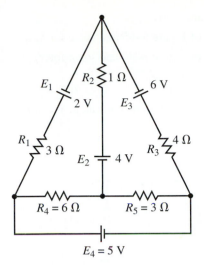

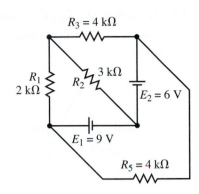

FIGURE 8–86   **FIGURE 8–87**

8.6   Nodal Analysis

33. Write the nodal equations for the circuit of Figure 8–88 and solve for the nodal voltages.

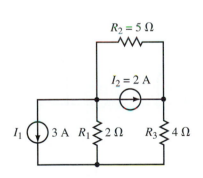

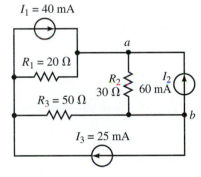

FIGURE 8–88   **FIGURE 8–89**

34. Write the nodal equations for the circuit of Figure 8–89 and determine the voltage V_{ab}.

35. Repeat Problem 33 for the circuit of Figure 8–90.

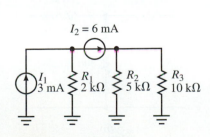

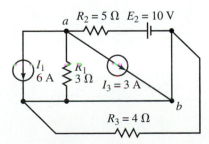

FIGURE 8–90   **FIGURE 8–91**

36. Repeat Problem 34 for the circuit of Figure 8–91.
37. Write the nodal equations for the circuit of Figure 8–85 and solve for $V_{6\,\Omega}$.
38. Write the nodal equations for the circuit of Figure 8–86 and solve for $V_{6\,\Omega}$.

8.7   Delta-Wye (Pi-Tee) Conversion

39. Convert each of the Δ networks of Figure 8–92 into its equivalent Y configuration.

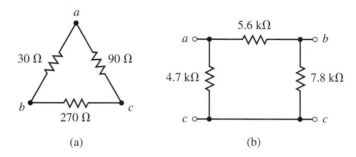

(a)   (b)

FIGURE 8–92

40. Convert each of the Δ networks of Figure 8–93 into its equivalent Y configuration.

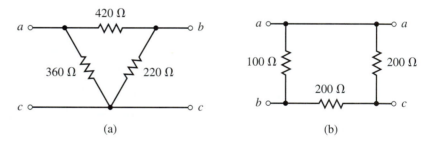

(a)   (b)

FIGURE 8–93

41. Convert each of the Y networks of Figure 8–94 into its equivalent Δ configuration.

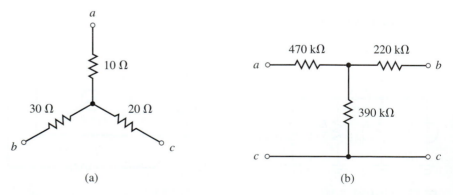

(a)   (b)

FIGURE 8–94

42. Convert each of the Y networks of Figure 8–95 into its equivalent Δ config-
    uration.

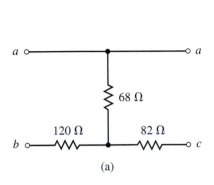

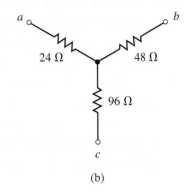

FIGURE 8–95

43. Using Δ-Y or Y-Δ conversion, find the current I for the circuit of Figure 8–96.
44. Using Δ-Y or Y-Δ conversion, find the current I and the voltage V_{ab} for the
    circuit of Figure 8–97.
45. Repeat Problem 43 for the circuit of Figure 8–98.

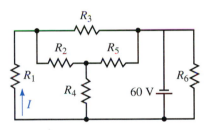

All resistors are 4.5 kΩ

FIGURE 8–96

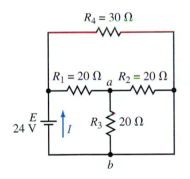

FIGURE 8–97

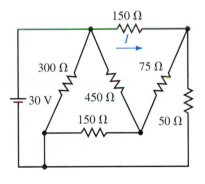

FIGURE 8–98

46. Repeat Problem 44 for the circuit of Figure 8–99.

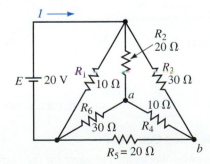

FIGURE 8–99

8.8   Bridge Networks

47. Refer to the bridge circuit of Figure 8–100:
    a. Is the bridge balanced? Explain.
    b. Write the mesh equations.
    c. Calculate the current through R_5.
    d. Determine the voltage across R_5.

48. Consider the bridge circuit of Figure 8–101:
    a. Is the bridge balanced? Explain.
    b. Write the mesh equations.
    c. Determine the current through R_5.
    d. Calculate the voltage across R_5.

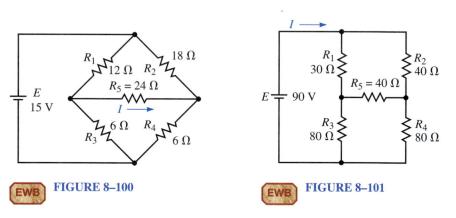

FIGURE 8–100   **FIGURE 8–101**

49. Given the bridge circuit of Figure 8–102, find the current through each resistor.

50. Refer to the bridge circuit of Figure 8–103:
    a. Determine the value of resistance R_x such that the bridge is balanced.
    b. Calculate the current through R_5 when $R_x = 0\ \Omega$ and when $R_x = 10\ \text{k}\Omega$.

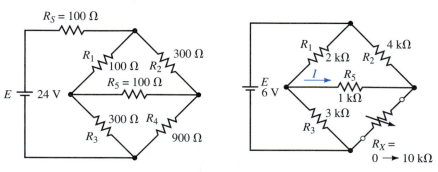

FIGURE 8–102   **FIGURE 8–103**

8.9   Circuit Analysis Using Computers

51. **EWB**   Use Electronics Workbench to solve for the currents through all resistors of the circuit shown in Figure 8–86.

52. **EWB**   Use Electronics Workbench to solve for the voltage across the 5-kΩ resistor in the circuit of Figure 8–90.

53. **PSpice**   Use PSpice to solve for the currents through all resistors in the circuit of Figure 8–96.

54. **PSpice**   Use PSpice to solve for the currents through all resistors in the circuit of Figure 8–97.

In-Process Learning Check 1

1. A voltage source E in series with a resistor R is equivalent to a current source having an ideal current source $I = E/R$ in parallel with the same resistance, R.

2. Current sources are never connected in series.

In-Process Learning Check 2

1. Voltage is zero.

2. Current is zero.

3. R_5 can be replaced with either a short circuit or an open circuit.

ANSWERS TO IN-PROCESS LEARNING CHECKS

9

Network Theorems

OBJECTIVES

After studying this chapter you will be able to

- apply the superposition theorem to determine the current through or voltage across any resistance in a given network,
- state Thévenin's theorem and determine the Thévenin equivalent circuit of any resistive network,
- state Norton's theorem and determine the Norton equivalent circuit of any resistive network,
- determine the required load resistance of any circuit to ensure that the load receives maximum power from the circuit,
- apply Millman's theorem to determine the current through or voltage across any resistor supplied by any number of sources in parallel,
- state the reciprocity theorem and demonstrate that it applies for a given single-source circuit,
- state the substitution theorem and apply the theorem in simplifying the operation of a given circuit.

KEY TERMS

Maximum Power Transfer
Millman's Theorem
Norton's Theorem
Reciprocity Theorem
Substitution Theorem
Superposition Theorem
Thévenin's Theorem

OUTLINE

Superposition Theorem
Thévenin's Theorem
Norton's Theorem
Maximum Power Transfer Theorem
Substitution Theorem
Millman's Theorem
Reciprocity Theorem
Circuit Analysis Using Computers

In this chapter you will learn how to use some basic theorems which will allow the analysis of even the most complex resistive networks. The theorems which are most useful in analyzing networks are the superposition, Thévenin, Norton, and maximum power transfer theorems.

You will also be introduced to other theorems which, while useful for providing a well-rounded appreciation of circuit analysis, have limited use in the analysis of circuits. These theorems, which apply to specific types of circuits, are the substitution, reciprocity, and Millman theorems. Your instructor may choose to omit the latter theorems without any loss in continuity.

André Marie Ampère

ANDRÉ MARIE AMPÈRE WAS BORN in Polémieux, Rhône, near Lyon, France on January 22, 1775. As a youth, Ampère was a brilliant mathematician who was able to master advanced mathematics by the age of twelve. However, the French Revolution, and the ensuing anarchy which swept through France from 1789 to 1799, did not exclude the Ampère family. Ampère's father, who was a prominent merchant and city official in Lyon, was executed under the guillotine in 1793. Young André suffered a nervous breakdown from which he never fully recovered. His suffering was further compounded in 1804, when after only five years of marriage, Ampère's wife died.

Even so, Ampère was able to make profound contributions to the field of mathematics, chemistry, and physics. As a young man, Ampère was appointed as professor of chemistry and physics in Bourg. Napoleon was a great supporter of Ampère's work, though Ampère had a reputation as an "absent-minded professor." Later he moved to Paris, where he taught mathematics.

Ampère showed that two current-carrying wires were attracted to one another when the current in the wires was in the same direction. When the current in the wires was in the opposite direction, the wires repelled. This work set the stage for the discovery of the principles of electric and magnetic field theory. Ampère was the first scientist to use electromagnetic principles to measure current in a wire. In recognition of his contribution to the study of electricity, current is measured in the unit of amperes.

Despite his personal suffering, Ampère remained a popular, friendly human being. He died of pneumonia in Marseille on June 10, 1836 after a brief illness.

PUTTING IT IN PERSPECTIVE

9.1   Superposition Theorem

The **superposition theorem** is a method which allows us to determine the current through or the voltage across any resistor or branch in a network. The advantage of using this approach instead of mesh analysis or nodal analysis is that it is not necessary to use determinants or matrix algebra to analyze a given circuit. The theorem states the following:

The total current through or voltage across a resistor or branch may be determined by summing the effects due to each independent source.

In order to apply the superposition theorem it is necessary to remove all sources other than the one being examined. In order to "zero" a voltage source, we **replace it with a short circuit,** since the voltage across a short circuit is zero volts. A current source is zeroed by **replacing it with an open circuit,** since the current through an open circuit is zero amps.

If we wish to determine the power dissipated by any resistor, we must first find either the voltage across the resistor or the current through the resistor:

$$P = I^2 R = \frac{V^2}{R}$$

EXAMPLE 9–1   Consider the circuit of Figure 9–1:

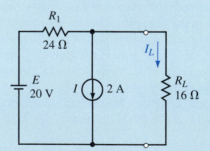

FIGURE 9–1

a. Determine the current in the load resistor, R_L.

b. Verify that the superposition theorem does not apply to power.

Solution

a. We first determine the current through R_L due to the voltage source by removing the current source and replacing in with an open circuit (zero amps) as shown in Figure 9–2.

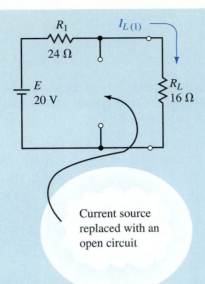

FIGURE 9–2

The resulting current through R_L is determined from Ohm's law as

$$I_{L(1)} = \frac{20 \text{ V}}{16 \text{ }\Omega + 24 \text{ }\Omega} = 0.500 \text{ A}$$

Next, we determine the current through R_L due to the current source by removing the voltage source and replacing it with a short circuit (zero volts) as shown in Figure 9–3.

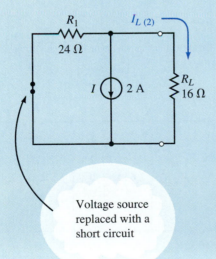

FIGURE 9–3

The resulting current through R_L is found with the current divider rule as

$$I_{L(2)} = -\left(\frac{24 \text{ }\Omega}{24 \text{ }\Omega + 16 \text{ }\Omega}\right)(2 \text{ A}) = -1.20 \text{ A}$$

The resultant current through R_L is found by applying the superposition theorem:

$$I_L = 0.5\,\text{A} - 1.2\,\text{A} = -0.700\,\text{A}$$

The negative sign indicates that the current through R_L is opposite to the assumed reference direction. Consequently, the current through R_L will, in fact, be upward with a magnitude of 0.7 A.

b. If we assume (incorrectly) that the superposition theorem applies for power, we would have the power due the first source given as

$$P_1 = I_{L(1)}^2 R_L = (0.5\,\text{A})^2(16\,\Omega) = 4.0\,\text{W}$$

and the power due the second source as

$$P_2 = I_{L(2)}^2 R_L = (1.2\,\text{A})^2(16\,\Omega) = 23.04\,\text{W}$$

The total power, if superposition applies, would be

$$P_T = P_1 + P_2 = 4.0\,\text{W} + 23.04\,\text{W} = 27.04\,\text{W}$$

Clearly, this result is wrong, since the actual power dissipated by the load resistor is correctly given as

$$P_L = I_L^2 R_L = (0.7\,\text{A})^2(16\,\Omega) = 7.84\,\text{W}$$

The superposition theorem may also be used to determine the voltage across any component or branch within the circuit.

EXAMPLE 9–2   Determine the voltage drop across the resistor R_2 of the circuit shown in Figure 9–4.

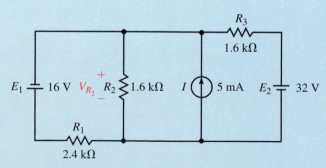

FIGURE 9–4

Solution   Since this circuit has three separate sources, it is necessary to determine the voltage across R_2 due to each individual source.

First, we consider the voltage across R_2 due to the 16-V source as shown in Figure 9–5.

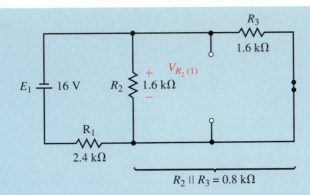

FIGURE 9–5

The voltage across R_2 will be the same as the voltage across the parallel combination of $R_2\|R_3 = 0.8\ \text{k}\Omega$. Therefore,

$$V_{R_2(1)} = -\left(\frac{0.8\ \text{k}\Omega}{0.8\ \text{k}\Omega + 2.4\ \text{k}\Omega}\right)(16\ \text{V}) = -4.00\ \text{V}$$

The negative sign in the above calculation simply indicates that the voltage across the resistor due to the first source is opposite to the assumed reference polarity.

Next, we consider the current source. The resulting circuit is shown in Figure 9–6.

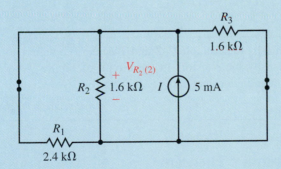

FIGURE 9–6

From this circuit, you can observe that the total resistance "seen" by the current source is

$$R_T = R_1\|R_2\|R_3 = 0.6\ \text{k}\Omega$$

The resulting voltage across R_2 is

$$V_{R_2(2)} = (0.6\ \text{k}\Omega)(5\ \text{mA}) = 3.00\ \text{V}$$

Finally, the voltage due to the 32-V source is found by analyzing the circuit of Figure 9–7.

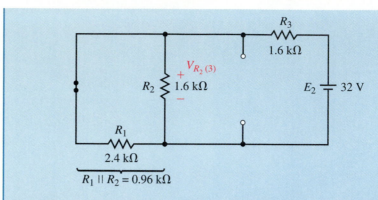

FIGURE 9–7

The voltage across R_2 is

$$V_{R_{2(3)}} = \left(\frac{0.96\text{ k}\Omega}{0.96\text{ k}\Omega + 1.6\text{ k}\Omega}\right)(32\text{ V}) = 12.0\text{ V}$$

By superposition, the resulting voltage is

$$V_{R_2} = -4.0\text{ V} + 3.0\text{ V} + 12.0\text{ V} = 11.0\text{ V}$$

PRACTICE PROBLEMS 1

Use the superposition theorem to determine the voltage across R_1 and R_3 in the circuit of Figure 9–4.

Answers: $V_{R_1} = 27.0\text{ V}$, $V_{R_3} = 21.0\text{ V}$

IN-PROCESS LEARNING CHECK 1

Use the final results of Example 9–2 and Practice Problem 1 to determine the power dissipated by the resistors in the circuit of Figure 9–4. Verify that the superposition theorem does not apply to power.

(Answers are at the end of the chapter.)

9.2   Thévenin's Theorem

In this section, we will apply one of the most important theorems of electric circuits. **Thévenin's theorem** allows even the most complicated circuit to be reduced to a single voltage source and a single resistance. The importance of such a theorem becomes evident when we try to analyze a circuit as shown in Figure 9–8.

If we wanted to find the current through the variable load resistor when $R_L = 0$, $R_L = 2\text{ k}\Omega$, and $R_L = 5\text{ k}\Omega$ using existing methods, we would need to analyze the entire circuit three separate times. However, if we could reduce the entire circuit external to the load resistor to a single voltage source in series with a resistor, the solution becomes very easy.

Thévenin's theorem is a circuit analysis technique which reduces any linear bilateral network to an equivalent circuit having only one voltage

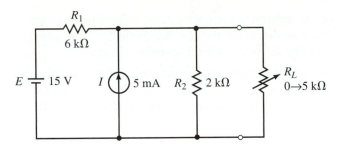

FIGURE 9–8

source and one series resistor. The resulting two-terminal circuit is equivalent to the original circuit when connected to any external branch or component. In summary, Thévenin's theorem is simplified as follows:

Any linear bilateral network may be reduced to a simplified two-terminal circuit consisting of a single voltage source in series with a single resistor as shown in Figure 9–9.

A linear network, remember, is any network that consists of components having a linear (straight-line) relationship between voltage and current. A resistor is a good example of a linear component since the voltage across a resistor increases proportionally to an increase in current through the resistor. Voltage and current sources are also linear components. In the case of a voltage source, the voltage remains constant although current through the source may change.

A bilateral network is any network that operates in the same manner regardless of the direction of current in the network. Again, a resistor is a good example of a bilateral component, since the magnitude of current through the resistor is not dependent upon the polarity of voltage across the component. (A diode is not a bilateral component, since the magnitude of current through the device is dependent upon the polarity of the voltage applied across the diode.)

The following steps provide a technique which converts any circuit into its Thévenin equivalent:

1. Remove the load from the circuit.

2. Label the resulting two terminals. We will label them as *a* and *b,* although any notation may be used.

3. Set all sources in the circuit to zero.

   Voltage sources are set to zero by replacing them with short circuits (zero volts).

   Current sources are set to zero by replacing them with open circuits (zero amps).

4. Determine the Thévenin equivalent resistance, R_{Th}, by calculating the resistance "seen" between terminals *a* and *b.* It may be necessary to redraw the circuit to simplify this step.

5. Replace the sources removed in Step 3, and determine the open-circuit voltage between the terminals. If the circuit has more than one source, it may be necessary to use the superposition theorem. In that case, it will be

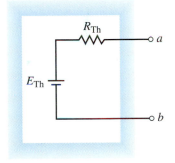

FIGURE 9–9   Thévenin equivalent circuit.

necessary to determine the open-circuit voltage due to each source separately and then determine the combined effect. The resulting open-circuit voltage will be the value of the Thévenin voltage, E_{Th}.

6. Draw the Thévenin equivalent circuit using the resistance determined in Step 4 and the voltage calculated in Step 5. As part of the resulting circuit, include that portion of the network removed in Step 1.

EXAMPLE 9–3   Determine the Thévenin equivalent circuit external to the resistor R_L for the circuit of Figure 9–10. Use the Thévenin equivalent circuit to calculate the current through R_L.

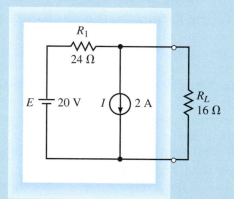

FIGURE 9–10

Solution

Steps 1 and 2: Removing the load resistor from the circuit and labelling the remaining terminals, we obtain the circuit shown in Figure 9–11.

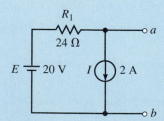

FIGURE 9–11

Step 3: Setting the sources to zero, we have the circuit shown in Figure 9–12.

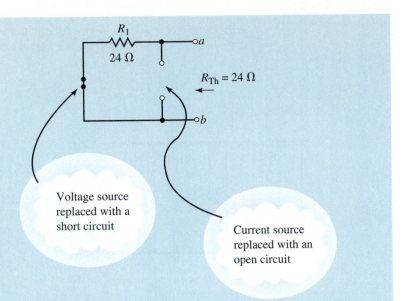

FIGURE 9–12

Step 4: The Thévenin resistance between the terminals is $R_{Th} = 24 \, \Omega$.

Step 5: From Figure 9–11, the open-circuit voltage between terminals a and b is found as

$$V_{ab} = 20 \text{ V} - (24 \, \Omega)(2 \text{ A}) = -28.0 \text{ V}$$

Step 6: The resulting Thévenin equivalent circuit is shown in Figure 9–13.

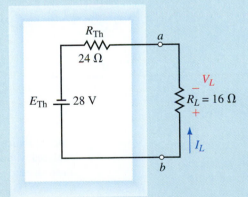

FIGURE 9–13

Using this Thévenin equivalent circuit, we easily find the current through R_L as

$$I_L = \left(\frac{28 \text{ V}}{24 \, \Omega + 16 \, \Omega} \right) = 0.700 \text{ A} \quad \text{(upward)}$$

This result is the same as that obtained by using the superposition theorem in Example 9–1.

EXAMPLE 9–4   Find the Thévenin equivalent circuit of the indicated area in Figure 9–14. Using the equivalent circuit, determine the current through the load resistor when $R_L = 0$, $R_L = 2$ kΩ, and $R_L = 5$ kΩ.

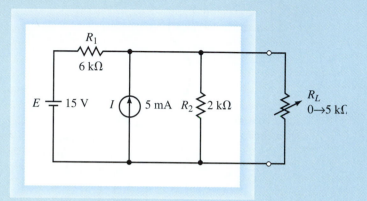

FIGURE 9–14

Solution

Steps 1, 2, and 3: After removing the load, labelling the terminals, and setting the sources to zero, we have the circuit shown in Figure 9–15.

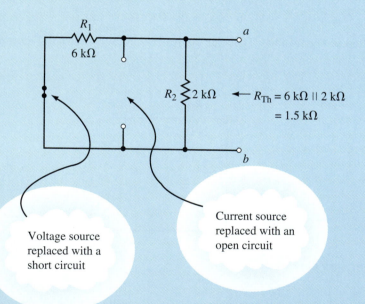

FIGURE 9–15

Step 4: The Thévenin resistance of the circuit is

$$R_{Th} = 6 \text{ k}\Omega \| 2 \text{ k}\Omega = 1.5 \text{ k}\Omega$$

Step 5: Although several methods are possible, we will use the superposition theorem to find the open-circuit voltage V_{ab}. Figure 9–16 shows the circuit for determining the contribution due to the 15-V source.

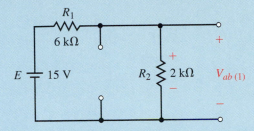

FIGURE 9–16

$$V_{ab(1)} = \left(\frac{2 \text{ k}\Omega}{2 \text{ k}\Omega + 6 \text{ k}\Omega} \right)(15 \text{ V}) = +3.75 \text{ V}$$

Figure 9–17 shows the circuit for determining the contribution due to the 5-mA source.

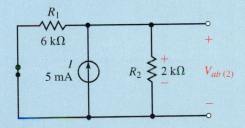

FIGURE 9–17

$$V_{ab(2)} = \left(\frac{(2 \text{ k}\Omega)(6 \text{ k}\Omega)}{2 \text{ k}\Omega + 6 \text{ k}\Omega} \right)(5 \text{ mA}) = +7.5 \text{ V}$$

The Thévenin equivalent voltage is

$$E_{Th} = V_{ab(1)} + V_{ab(2)} = +3.75 \text{ V} + 7.5 \text{ V} = 11.25 \text{ V}$$

Step 6: The resulting Thévenin equivalent circuit is shown in Figure 9–18.

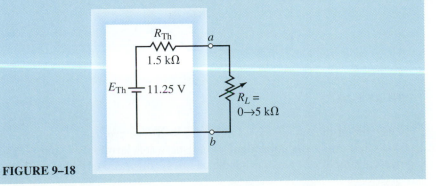

FIGURE 9–18

From this circuit, it is now an easy matter to determine the current for any value of load resistor:

$R_L = 0\ \Omega$: $I_L = \dfrac{11.25\ \text{V}}{1.5\ \text{k}\Omega} = 7.5\ \text{mA}$

$R_L = 2\ k\Omega$: $I_L = \dfrac{11.25\ \text{V}}{1.5\ \text{k}\Omega + 2\ \text{k}\Omega} = 3.21\ \text{mA}$

$R_L = 5\ k\Omega$: $I_L = \dfrac{11.25\ \text{V}}{1.5\ \text{k}\Omega + 5\ \text{k}\Omega} = 1.73\ \text{mA}$

EXAMPLE 9–5   Find the Thévenin equivalent circuit external to R_5 in the circuit in Figure 9–19. Use the equivalent circuit to determine the current through the resistor.

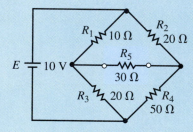

FIGURE 9–19

Solution   Notice that the circuit is an unbalanced bridge circuit. If the techniques of the previous chapter had to be used, we would need to solve either three mesh equations or three nodal equations.

Steps 1 and 2: Removing the resistor R_5 from the circuit and labelling the two terminals a and b, we obtain the circuit shown in Figure 9–20.

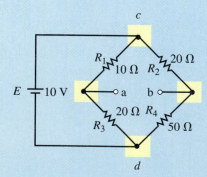

FIGURE 9–20

By examining the circuit shown in Figure 9–20, we see that it is no simple task to determine the equivalent circuit between terminals a and b. The process is simplified by redrawing the circuit as illustrated in Figure 9–21.

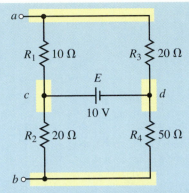

FIGURE 9–21

Notice that the circuit of Figure 9–21 has nodes a and b conveniently shown at the top and bottom of the circuit. Additional nodes (node c and node d) are added to simplify the task of correctly placing resistors between the nodes.

After simplifying a circuit, it is always a good idea to ensure that the resulting circuit is indeed an equivalent circuit. You may verify the equivalence of the two circuits by confirming that each component is connected between the same nodes for each circuit.

Now that we have a circuit which is easier to analyze, we find the Thévenin equivalent of the resultant.

Step 3: Setting the voltage source to zero by replacing it with a short, we obtain the circuit shown in Figure 9–22.

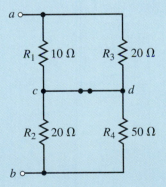

FIGURE 9–22

Step 4: The resulting Thévenin resistance is

$$R_{Th} = 10 \ \Omega \| 20 \ \Omega + 20 \ \Omega \| 50 \ \Omega$$
$$= 6.67 \ \Omega + 14.29 \ \Omega = 20.95 \ \Omega$$

Step 5: The open-circuit voltage between terminals a and b is found by first indicating the loop currents I_1 and I_2 in the circuit of Figure 9–23.

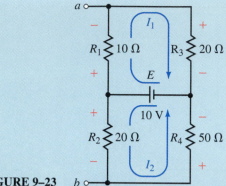

FIGURE 9–23

Because the voltage source, E, provides a constant voltage across the resistor combinations R_1-R_3 and R_2-R_4, we simply use the voltage divider rule to determine the voltage across the various components:

$$V_{ab} = -V_{R_1} + V_{R_2}$$

$$= -\frac{(10\ \Omega)(10\ V)}{30\ \Omega} + \frac{(20\ \Omega)(10\ V)}{70\ \Omega}$$

$$= -0.476\ V$$

Note: The above technique could not be used if the source had some series resistance, since then the voltage provided to resistor combinations R_1-R_3 and R_2-R_4 would no longer be the entire supply voltage but rather would be dependent upon the value of the series resistance of the source.

Step 6: The resulting Thévenin circuit is shown in Figure 9–24.

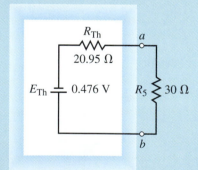

FIGURE 9–24

From the circuit of Figure 9–24, it is now possible to calculate the current through the resistor R_5 as

$$I = \frac{0.476\ V}{20.95\ \Omega + 30\ \Omega} = 9.34\ mA \quad \text{(from } b \text{ to } a\text{)}$$

This example illustrates the importance of labelling the terminals which remain after a component or branch is removed. If we had not labelled the terminals and drawn an equivalent circuit, the current through R_5 would not have been found as easily.

Find the Thévenin equivalent circuit external to resistor R_1 in the circuit of Figure 9–1.

Answer: $R_{\text{Th}} = 16\ \Omega,\ E_{\text{Th}} = 52\ \text{V}$

Use Thévenin's theorem to determine the current through load resistor R_L for the circuit of Figure 9–25.

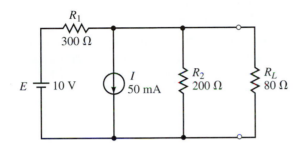

  FIGURE 9–25

Answer: $I_L = 10.0\ \text{mA}$ upward

In the circuit of Figure 9–25, what would the value of R_1 need to be in order that the Thévenin resistance is equal to $R_L = 80\ \Omega$?

(Answers are at the end of the chapter.)

9.3   Norton's Theorem

Norton's theorem is a circuit analysis technique which is similar to Thévenin's theorem. By using this theorem the circuit is reduced to a single current source and one parallel resistor. As with the Thévenin equivalent circuit, the resulting two-terminal circuit is equivalent to the original circuit when connected to any external branch or component. In summary, **Norton's theorem** may be simplified as follows:

   Any linear bilateral network may be reduced to a simplified two-terminal circuit consisting of a single current source and a single shunt resistor as shown in Figure 9–26.

   The following steps provide a technique which allows the conversion of any circuit into its Norton equivalent:

1. Remove the load from the circuit.
2. Label the resulting two terminals. We will label them as *a* and *b*, although any notation may be used.
3. Set all sources to zero. As before, voltage sources are set to zero by replacing them with short circuits and current sources are set to zero by replacing them with open circuits.

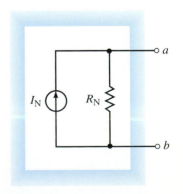

FIGURE 9–26   Norton equivalent circuit.

4. Determine the Norton equivalent resistance, R_N, by calculating the resistance seen between terminals a and b. It may be necessary to redraw the circuit to simplify this step.

5. Replace the sources removed in Step 3, and determine the current which would occur in a short if the short were connected between terminals a and b. If the original circuit has more than one source, it may be necessary to use the superposition theorem. In this case, it will be necessary to determine the short-circuit current due to each source separately and then determine the combined effect. The resulting short-circuit current will be the value of the Norton current I_N.

6. Sketch the Norton equivalent circuit using the resistance determined in Step 4 and the current calculated in Step 5. As part of the resulting circuit, include that portion of the network removed in Step 1.

The Norton equivalent circuit may also be determined directly from the Thévenin equivalent circuit by using the source conversion technique developed in Chapter 8. As a result, the Thévenin and Norton circuits shown in Figure 9–27 are equivalent.

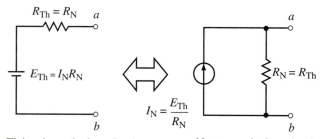

Thévenin equivalent circuit   Norton equivalent circuit

FIGURE 9–27

From Figure 9–27 we see that the relationship between the circuits is as follows:

$$E_{Th} = I_N R_N \tag{9–1}$$

$$I_N = \frac{E_{Th}}{R_{Th}} \tag{9–2}$$

EXAMPLE 9–6   Determine the Norton equivalent circuit external to the resistor R_L for the circuit of Figure 9–28. Use the Norton equivalent circuit to calculate the current through R_L. Compare the results to those obtained using Thévenin's theorem in Example 9–3.

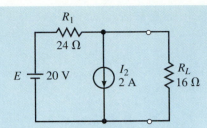

FIGURE 9–28

Solution

Steps 1 and 2: Remove load resistor R_L from the circuit and label the remaining terminals as a and b. The resulting circuit is shown in Figure 9–29.

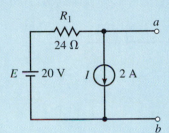

FIGURE 9–29

Step 3: Zero the voltage and current sources as shown in the circuit of Figure 9–30.

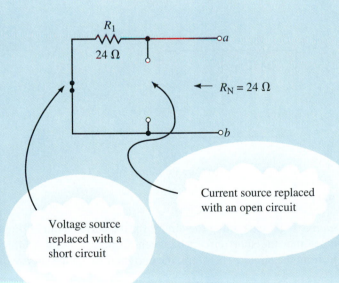

$R_N = 24 \, \Omega$

Current source replaced with an open circuit

Voltage source replaced with a short circuit

FIGURE 9–30

Step 4: The resulting Norton resistance between the terminals is

$$R_N = R_{ab} = 24 \, \Omega$$

Step 5: The short-circuit current is determined by first calculating the current through the short due to each source. The circuit for each calculation is illustrated in Figure 9–31.

Voltage Source, E: The current in the short between terminals *a* and *b* [Figure 9–31(a)] is found from Ohm's law as

$$I_{ab(1)} = \frac{20 \text{ V}}{24 \text{ }\Omega} = 0.833 \text{ A}$$

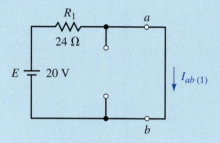

(a) Voltage source

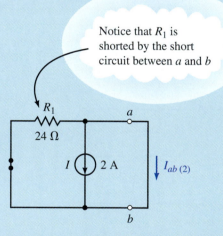

Notice that R_1 is shorted by the short circuit between *a* and *b*

(b) Current source

FIGURE 9–31

Current Source, I: By examining the circuit for the current source [Figure 9–31(b)] we see that the short circuit between terminals *a* and *b* effectively removes R_1 from the circuit. Therefore, the current through the short will be

$$I_{ab(2)} = -2.00 \text{ A}$$

Notice that the current I_{ab} is indicated as being a negative quantity. As we have seen before, this result merely indicates that the actual current is opposite to the assumed reference direction.

Now, applying the superposition theorem, we find the Norton current as

$$I_{\text{N}} = I_{ab(1)} + I_{ab(2)} = 0.833 \text{ A} - 2.0 \text{ A} = -1.167 \text{ A}$$

As before, the negative sign indicates that the short-circuit current is actually from terminal b toward terminal a.

Step 6: The resultant Norton equivalent circuit is shown in Figure 9–32.

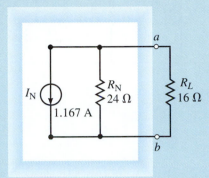

FIGURE 9–32

Now we can easily find the current through load resistor R_L by using the current divider rule:

$$I_L = \left(\frac{24\ \Omega}{24\ \Omega + 16\ \Omega} \right)(1.167\ \text{A}) = 0.700\ \text{A} \quad \text{(upward)}$$

By referring to Example 9–3, we see that the same result was obtained by finding the Thévenin equivalent circuit. An alternate method of finding the Norton equivalent circuit is to convert the Thévenin circuit found in Example 9–3 into its equivalent Norton circuit shown in Figure 9–33.

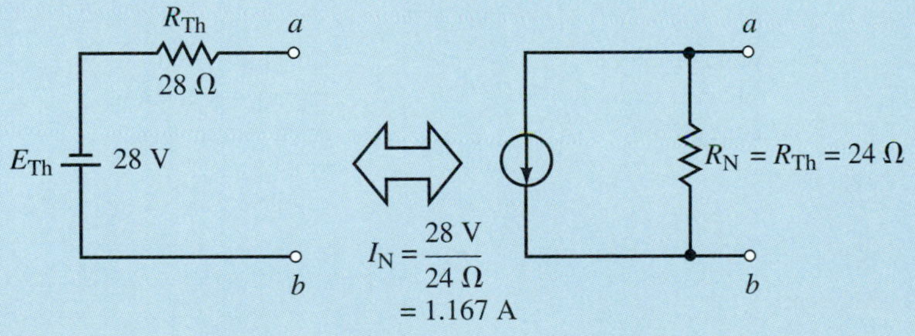

FIGURE 9–33

EXAMPLE 9–7 Find the Norton equivalent of the circuit external to resistor R_L in the circuit in Figure 9–34. Use the equivalent circuit to determine the load current I_L when $R_L = 0$, 2 kΩ, and 5 kΩ.

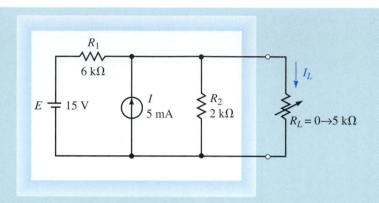

FIGURE 9–34

Solution

Steps 1, 2, and 3: After removing the load resistor, labelling the remaining two terminals *a* and *b,* and setting the sources to zero, we have the circuit of Figure 9–35.

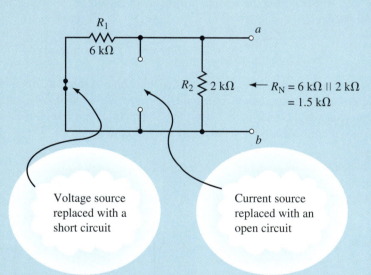

FIGURE 9–35

Step 4: The Norton resistance of the circuit is found as

$$R_N = 6 \text{ k}\Omega \| 2 \text{ k}\Omega = 1.5 \text{ k}\Omega$$

Step 5: The value of the Norton constant-current source is found by determining the current effects due to each independent source acting on a short circuit between terminals *a* and *b.*

Voltage Source, E: Referring to Figure 9–36(a), a short circuit between terminals *a* and *b* eliminates resistor R_2 from the circuit. The short-circuit current due to the voltage source is

$$I_{ab(1)} = \frac{15 \text{ V}}{6 \text{ k}\Omega} = 2.50 \text{ mA}$$

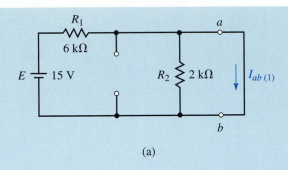

(a)

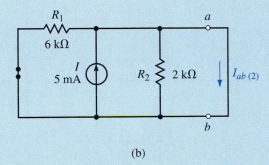

(b)

FIGURE 9–36

Current Source, I: Referring to Figure 9–36(b), the short circuit between terminals a and b eliminates both resistors R_1 and R_2. The short-circuit current due to the current source is therefore

$$I_{ab(2)} = 5.00 \text{ mA}$$

The resultant Norton current is found from superposition as

$$I_N = I_{ab(1)} + I_{ab(2)} = 2.50 \text{ mA} + 5.00 \text{ mA} = 7.50 \text{ mA}$$

Step 6: The Norton equivalent circuit is shown in Figure 9–37.

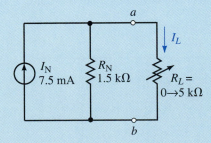

FIGURE 9–37

Let $R_L = 0$: The current I_L must equal the source current, and so

$$I_L = 7.50 \text{ mA}$$

Let $R_L = 2 \text{ k}\Omega$: The current I_L is found from the current divider rule as

$$I_L = \left(\frac{1.5 \text{ k}\Omega}{1.5 \text{ k}\Omega + 2 \text{ k}\Omega} \right)(7.50 \text{ mA}) = 3.21 \text{ mA}$$

Let $R_L = 5\ k\Omega$: Using the current divider rule again, the current I_L is found as

$$I_L = \left(\frac{1.5\ k\Omega}{1.5\ k\Omega + 5\ k\Omega}\right)(7.50\ mA) = 1.73\ mA$$

Comparing the above results to those obtained in Example 9–4, we see that they are precisely the same.

EXAMPLE 9–8   Consider the circuit of Figure 9–38:

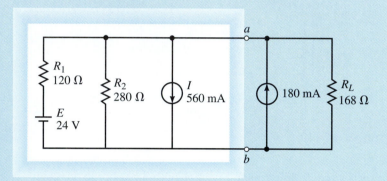

FIGURE 9–38

a. Find the Norton equivalent circuit external to terminals a and b.
b. Determine the current through R_L.

Solution

a. **Steps 1 and 2:** After removing the load (which consists of a current source in parallel with a resistor), we have the circuit of Figure 9–39.

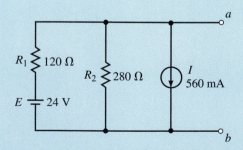

FIGURE 9–39

Step 3: After zeroing the sources, we have the network shown in Figure 9–40.

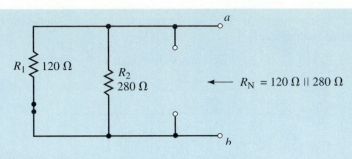

FIGURE 9–40

Step 4: The Norton equivalent resistance is found as

$$R_N = 120\ \Omega \| 280\ \Omega = 84\ \Omega$$

Step 5: In order to determine the Norton current we must again determine the short-circuit current due to each source separately and then combine the results using the superposition theorem.

Voltage Source, E: Referring to Figure 9–41(a), notice that the resistor R_2 is shorted by the short circuit between terminals a and b and so the current in the short circuit is

$$I_{ab(1)} = \frac{24\ \text{V}}{120\ \Omega} = 0.2\ \text{A} = 200\ \text{mA}$$

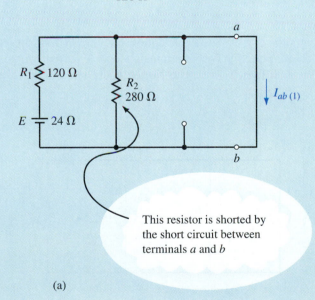

This resistor is shorted by the short circuit between terminals a and b

(a)

FIGURE 9–41

Current Source, I: Referring to Figure 9–41(b), the short circuit between terminals a and b will now eliminate both resistors. The current through the short will simply be the source current. However, since the current will not be from a to b but rather in the opposite direction, we write

$$I_{ab(2)} = -560\ \text{mA}$$

Now the Norton current is found as the summation of the short-circuit currents due to each source:

$$I_\mathrm{N} = I_{ab(1)} + I_{ab(2)} = 200 \text{ mA} + (-560 \text{ mA}) = -360 \text{ mA}$$

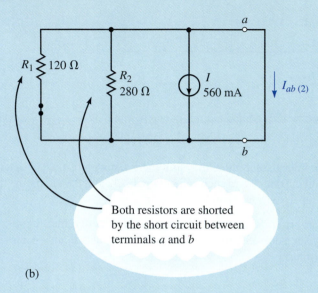

Both resistors are shorted by the short circuit between terminals a and b

(b)

FIGURE 9–41 *continued*

The negative sign in the above calculation for current indicates that if a short circuit were placed between terminals a and b, current would actually be in the direction from b to a. The Norton equivalent circuit is shown in Figure 9–42.

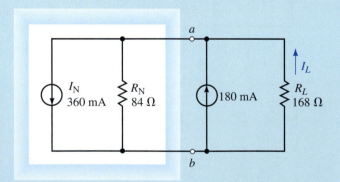

FIGURE 9–42

b. The current through the load resistor is found by applying the current divider rule:

$$I_L = \left(\frac{84 \text{ } \Omega}{84 \text{ } \Omega + 168 \text{ } \Omega}\right)(360 \text{ mA} - 180 \text{ mA}) = 60 \text{ mA} \quad \text{(upward)}$$

Find the Norton equivalent of the circuit shown in Figure 9–43. Use the source conversion technique to determine the Thévenin equivalent of the circuit between points *a* and *b*.

FIGURE 9–43

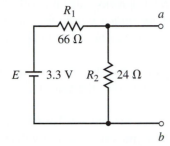

Answers: $R_N = R_{Th} = 17.6\ \Omega$, $I_N = 0.05\ A$, $E_{Th} = 0.88\ V$

Find the Norton equivalent external to R_L in the circuit of Figure 9–44. Solve for the current I_L when $R_L = 0$, 10 kΩ, 50 kΩ, and 100 kΩ.

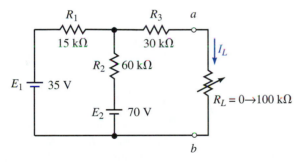

FIGURE 9–44

Answers: $R_N = 42\ k\Omega$, $I_N = 1.00\ mA$
For $R_L = 0$: $I_L = 1.00\ mA$
For $R_L = 10\ k\Omega$: $I_L = 0.808\ mA$
For $R_L = 50\ k\Omega$: $I_L = 0.457\ mA$
For $R_L = 100\ k\Omega$: $I_L = 0.296\ mA$

1. Show the relationship between the Thévenin equivalent circuit and the Norton equivalent circuit. Sketch each circuit.
2. If a Thévenin equivalent circuit has $E_{Th} = 100\ mV$ and $R_{Th} = 500\ \Omega$, draw the corresponding Norton equivalent circuit.
3. If a Norton equivalent circuit has $I_N = 10\ \mu A$ and $R_N = 20\ k\Omega$, draw the corresponding Thévenin equivalent circuit.

 IN-PROCESS LEARNING CHECK 3

(Answers are at the end of the chapter.)

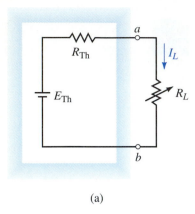

(a)

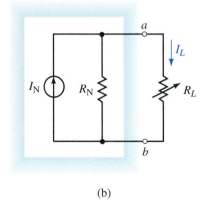

(b)

FIGURE 9–45

9.4   Maximum Power Transfer Theorem

In amplifiers and in most communication circuits such as radio receivers and transmitters, it is often desired that the load receive the maximum amount of power from a source.

The **maximum power transfer theorem** states the following:

A load resistance will receive maximum power from a circuit when the resistance of the load is exactly the same as the Thévenin (Norton) resistance looking back at the circuit.

The proof for the maximum power transfer theorem is determined from the Thévenin equivalent circuit and involves the use of calculus. This theorem is proved in Appendix C.

From Figure 9–45 we see that once the network has been simplified using either Thévenin's or Norton's theorem, maximum power will occur when

$$R_L = R_{Th} = R_N \tag{9–3}$$

Examining the equivalent circuits of Figure 9–45, shows that the following equations determine the power delivered to the load:

$$P_L = \frac{\left(\dfrac{R_L}{R_L + R_{Th}} \times E_{Th}\right)^2}{}$$

which gives

$$P_L = \frac{E_{Th}^2 R_L}{(R_L + R_{Th})^2} \tag{9–4}$$

Similarly,

$$P_L = \left(\frac{I_N R_N}{R_L + R_N}\right)^2 \times R_L \tag{9–5}$$

Under maximum power conditions ($R_L = R_{Th} = R_N$), the above equations may be used to determine the maximum power delivered to the load and may therefore be written as

$$P_{max} = \frac{E_{Th}^2}{4R_{Th}} \tag{9–6}$$

$$P_{max} = \frac{I_N^2 R_N}{4} \tag{9–7}$$

EXAMPLE 9–9   For the circuit of Figure 9–46, sketch graphs of V_L, I_L, and P_L as functions of R_L.

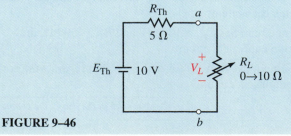

FIGURE 9–46

Solution We may first set up a table of data for various values of resistance, R_L. See Table 9–1. Voltage and current values are determined by using the voltage divider rule and Ohm's law respectively. The power P_L for each value of resistance is determined by finding the product $P_L = V_L I_L$, or by using Equation 9–4.

TABLE 9–1

R_L (Ω)	V_L (V)	I_L (A)	P_L (W)
0	0	2.000	0
1	1.667	1.667	2.778
2	2.857	1.429	4.082
3	3.750	1.250	4.688
4	4.444	1.111	4.938
5	5.000	1.000	5.000
6	5.455	0.909	4.959
7	5.833	0.833	4.861
8	6.154	0.769	4.734
9	6.429	0.714	4.592
10	6.667	0.667	4.444

If the data from Table 9–1 are plotted on a linear graphs, the graphs will appear as shown in Figures 9–47, 9–48, and 9–49.

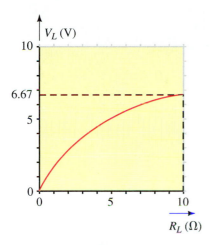

FIGURE 9–47 Voltage versus R_L.

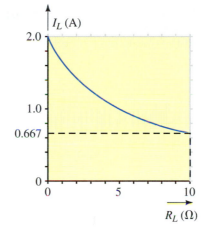

FIGURE 9–48 Current versus R_L.

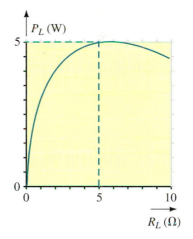

FIGURE 9–49 Power versus R_L.

Notice in the graphs that although voltage across the load increases as R_L increases, the power delivered to the load will be a maximum when $R_L = R_{Th} = 5$ Ω. The reason for this apparent contradiction is because, as R_L increases, the reduction in current more than offsets the corresponding increase in voltage.

EXAMPLE 9–10 Consider the circuit of Figure 9–50:

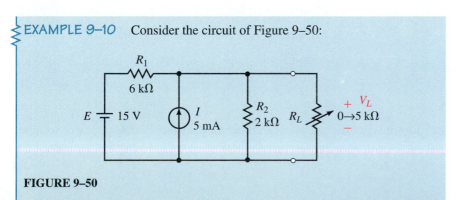

FIGURE 9–50

a. Determine the value of load resistance required to ensure that maximum power is transferred to the load.

b. Find V_L, I_L, and P_L when maximum power is delivered to the load.

Solution

a. In order to determine the conditions for maximum power transfer, it is first necessary to determine the equivalent circuit external to the load. We may determine either the Thévenin equivalent circuit or the Norton equivalent circuit. This circuit was analyzed in Example 9–4 using Thévenin's theorem, and we determined the equivalent circuit to be as shown in Figure 9–51.

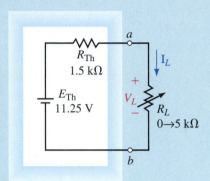

FIGURE 9–51

Maximum power will be transferred to the load when $R_L = 1.5$ kΩ.

b. Letting $R_L = 1.5$ kΩ, we see that half of the Thévenin voltage will appear across the load resistor and half will appear across the Thévenin resistance. So, at maximum power,

$$V_L = \frac{E_{Th}}{2} = \frac{11.25 \text{ V}}{2} = 5.625 \text{ V}$$

$$I_L = \frac{5.625 \text{ V}}{1.5 \text{ k}\Omega} = 3.750 \text{ mA}$$

The power delivered to the load is found as

$$P_L = \frac{V_L^2}{R_L} = \frac{(5.625 \text{ V})^2}{1.5 \text{ k}\Omega} = 21.1 \text{ mW}$$

Or, alternatively using current, we calculate the power as

$$P_L = I_L^2 R_L = (3.75 \text{ mA})^2 (1.5 \text{ k}\Omega) = 21.1 \text{ mW}$$

In solving this problem, we could just as easily have used the Norton equivalent circuit to determine required values.

Recall that efficiency was defined as the ratio of output power to input power:

$$\eta = \frac{P_{out}}{P_{in}}$$

or as a percentage:

$$\eta = \frac{P_{out}}{P_{in}} \times 100\%$$

By using the maximum power transfer theorem, we see that under the condition of maximum power the efficiency of the circuit is

$$\eta = \frac{P_{\text{out}}}{P_{\text{in}}} \times 100\%$$

$$= \frac{\dfrac{E_{\text{Th}}^2}{4R_{\text{Th}}}}{\dfrac{E_{\text{Th}}^2}{2R_{\text{Th}}}} \times 100\% = 0.500 \times 100\% = 50\%$$

(9–8)

For communication circuits and for many amplifier circuits, 50% represents the maximum possible efficiency. At this efficiency level, the voltage presented to the following stage would only be half of the maximum terminal voltage.

In power transmission such as the 115-Vac, 60-Hz power in your home, the condition of maximum power is not a requirement. Under the condition of maximum power transfer, the voltage across the load will be reduced to half of the maximum available terminal voltage. Clearly, if we are working with power supplies, we would like to ensure that efficiency is brought as close to 100% as possible. In such cases, load resistance R_L is kept much larger than the internal resistance of the voltage source (typically $R_L \geq 10R_{\text{int}}$), ensuring that the voltage appearing across the load will be very nearly equal to the maximum terminal voltage of the voltage source.

EXAMPLE 9–11 Refer to the circuit of Figure 9–52, which represents a typical dc power supply.

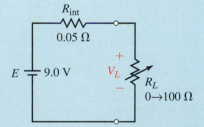

FIGURE 9–52

a. Determine the value of R_L needed for maximum power transfer.

b. Determine terminal voltage V_L and the efficiency when the value of the load resistor is $R_L = 50\ \Omega$.

c. Determine terminal voltage V_L and the efficiency when the value of the load resistor is $R_L = 100\ \Omega$.

Solution

a. For maximum power transfer, the load resistor will be given as $R_L = 0.05\ \Omega$. At this value of load resistance, the efficiency will be only 50%.

b. For $R_L = 50\ \Omega$, the voltage appearing across the output terminals of the voltage source is

$$V_L = \left(\frac{50\ \Omega}{50\ \Omega + 0.05\ \Omega}\right)(9.0\ \text{V}) = 8.99\ \text{V}$$

The efficiency is

$$\eta = \frac{P_{\text{out}}}{P_{\text{in}}} \times 100\%$$

$$= \frac{\dfrac{(8.99\ \text{V})^2}{50\ \Omega}}{\dfrac{(9.0\ \text{V})^2}{50.05\ \Omega}} \times 100\%$$

$$= \frac{1.6168\ \text{W}}{1.6184\ \text{W}} \times 100\% = 99.90\%$$

c. For $R_L = 100\ \Omega$, the voltage appearing across the output terminals of the voltage source is

$$V_L = \left(\frac{100\ \Omega}{100\ \Omega + 0.05\ \Omega}\right)(9.0\ \text{V}) = 8.995\ 50\ \text{V}$$

The efficiency is

$$\eta = \frac{P_{\text{out}}}{P_{\text{in}}} \times 100\%$$

$$= \frac{\dfrac{(8.9955\ \text{V})^2}{100\ \Omega}}{\dfrac{(9.0\ \text{V})^2}{100.05\ \Omega}} \times 100\%$$

$$= \frac{1.6168\ \text{W}}{1.6184\ \text{W}} \times 100\% = 99.95\%$$

From this example, we see that if efficiency is important, as it is in power transmission, then the load resistance should be much larger than the resistance of the source (typically $R_L \geq 10R_{\text{int}}$). If, on the other hand, it is more important to ensure maximum power transfer, then the load resistance should equal the source resistance ($R_L = R_{\text{int}}$).

PRACTICE PROBLEMS 6

Refer to the circuit of Figure 9–44. For what value of R_L will the load receive maximum power? Determine the power when $R_L = R_N$, when $R_L = 25\ \text{k}\Omega$, and when $R_L = 50\ \text{k}\Omega$.

Answers: $R_L = 42\ \text{k}\Omega$: $P_L = 10.5\ \text{mW}$
$R_L = 25\ \text{k}\Omega$: $P_L = 9.82\ \text{mW}$
$R_L = 50\ \text{k}\Omega$: $P_L = 10.42\ \text{mW}$

A Thévenin equivalent circuit consists of $E_{Th} = 10$ V and $R_{Th} = 2$ kΩ. Determine the efficiency of the circuit when

a. $R_L = R_{Th}$

b. $R_L = 0.5R_{Th}$

c. $R_L = 2R_{Th}$

IN-PROCESS
LEARNING
CHECK 4

(Answers are at the end of the chapter.)

1. In what instances is maximum power transfer a desirable characteristic of a circuit?

2. In what instances is maximum power transfer an undesirable characteristic of a circuit?

IN-PROCESS
LEARNING
CHECK 5

(Answers are at the end of the chapter.)

9.5   Substitution Theorem

The **substitution theorem** states the following:

Any branch within a circuit may be replaced by an equivalent branch, provided the replacement branch has the same current through it and voltage across it as the original branch.

This theorem is best illustrated by examining the operation of a circuit. Consider the circuit of Figure 9–53.

The voltage V_{ab} and the current I in the circuit of Figure 9–53 are given as

$$V_{ab} = \left(\frac{6\text{ k}\Omega}{4\text{ k}\Omega + 6\text{ k}\Omega}\right)(10\text{ V}) = +6.0\text{ V}$$

and

$$I = \frac{10\text{ V}}{4\text{ k}\Omega + 6\text{ k}\Omega} = 1\text{ mA}$$

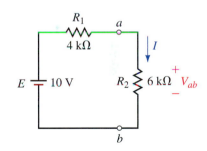

FIGURE 9–53

The resistor R_2 may be replaced with any combination of components, provided that the resulting components maintain the above conditions. We see that the branches of Figure 9–54 are each equivalent to the original branch between terminals a and b of the circuit in Figure 9–53.

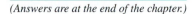

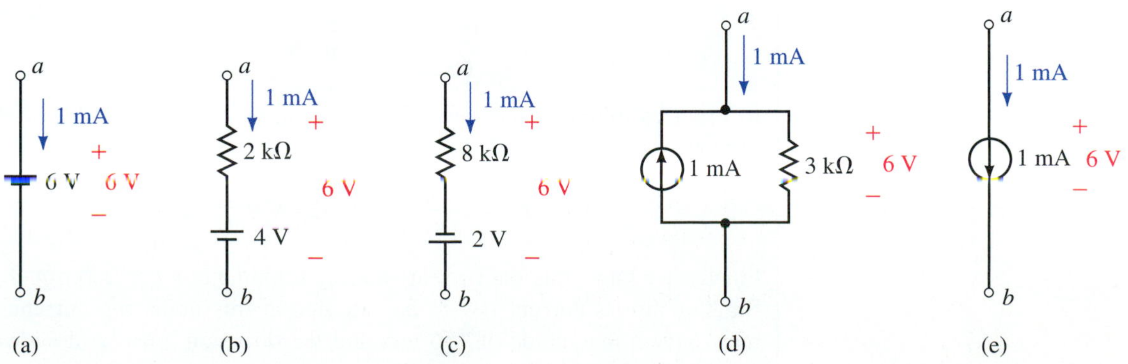

| | | | | |
| (a) | (b) | (c) | (d) | (e) |

FIGURE 9–54

Although each of the branches in Figure 9–54 is different, the current entering or leaving each branch will be same as that in the original branch. Similarly, the voltage across each branch will be the same. If any of these branches is substituted into the original circuit, the balance of the circuit will operate in the same way as the original. It is left as an exercise for the student to verify that each circuit behaves the same as the original.

This theorem allows us to replace any branch within a given circuit with an equivalent branch, thereby simplifying the analysis of the remaining circuit.

EXAMPLE 9–12   If the indicated portion in the circuit of Figure 9–55 is to be replaced with a current source and a 240-Ω shunt resistor, determine the magnitude and direction of the required current source.

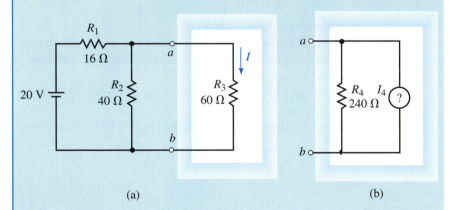

(a)   (b)

FIGURE 9–55

Solution   The voltage across the branch in the original circuit is

$$V_{ab} = \left(\frac{40\ \Omega \| 60\ \Omega}{16\ \Omega + (40\ \Omega \| 60\ \Omega)} \right)(20\ \text{V}) = \left(\frac{24\ \Omega}{16\ \Omega + 24\ \Omega} \right)(20\ \text{V}) = 12.0\ \text{V}$$

which results in a current of

$$I = \frac{12.0\ \text{V}}{60\ \Omega} = 0.200\ \text{A} = 200\ \text{mA}$$

In order to maintain the same terminal voltage, $V_{ab} = 12.0$ V, the current through resistor $R_4 = 240\ \Omega$ must be

$$I_{R_4} = \frac{12.0\ \text{V}}{240\ \Omega} = 0.050\ \text{A} = 50\ \text{mA}$$

Finally, we know that the current entering terminal a is $I = 200$ mA. In order for Kirchhoff's current law to be satisfied at this node, the current source must have a magnitude of 150 mA and the direction must be downward, as shown in Figure 9–56.

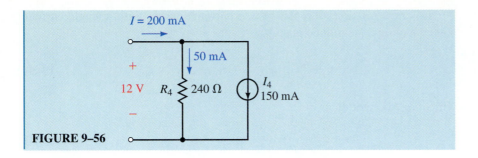

$I = 200$ mA

50 mA

12 V R_4 240 Ω I_4 150 mA

+

−

FIGURE 9–56

9.6   Millman's Theorem

Millman's theorem is used to simplify circuits having several parallel voltage sources as illustrated in Figure 9–57. Although any of the other theorems developed in this chapter will work in this case, Millman's theorem provides a much simpler and more direct equivalent.

In circuits of the type shown in Figure 9–57, the voltage sources may be replaced with a single equivalent source as shown in Figure 9–58.

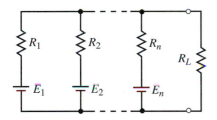

R_1 R_2 R_n

R_L

E_1 E_2 E_n

FIGURE 9–57

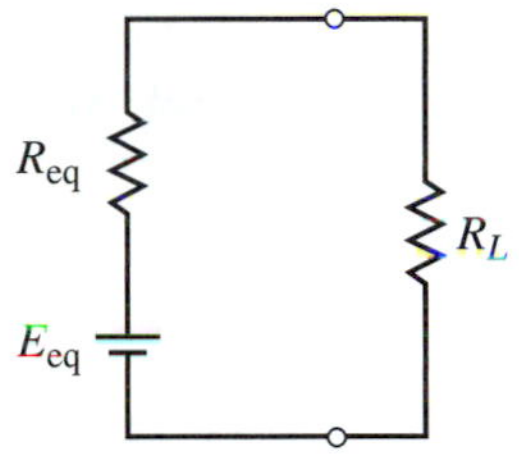

R_{eq}

R_L

E_{eq}

FIGURE 9–58

To find the values of the equivalent voltage source E_{eq} and series resistance R_{eq}, we need to convert each of the voltage sources of Figure 9–57 into its equivalent current source using the technique developed in Chapter 8. The value of each current source would be determined by using Ohm's law (i.e., $I_1 = E_1/R_1$, $I_2 = E_2/R_2$, etc.). After the source conversions are completed, the circuit appears as shown in Figure 9–59.

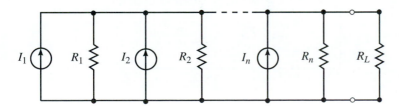

I_1 R_1 I_2 R_2 I_n R_n R_L

FIGURE 9–59

From the circuit of Figure 9–59 we see that all of the current sources have the same direction. Clearly, this will not always be the case since the direction of each current source will be determined by the initial polarity of the corresponding voltage source.

It is now possible to replace the n current sources with a single current source having a magnitude given as

$$I_{eq} = \sum_{x=0}^{n} I_x = I_1 + I_2 + I_3 + \cdots + I_n \qquad (9\text{–}9)$$

which may be written as

$$I_{eq} = \frac{E_1}{R_1} + \frac{E_2}{R_2} + \frac{E_3}{R_3} + \cdots + \frac{E_n}{R_n} \qquad (9\text{–}10)$$

If the direction of any current source is opposite to the direction shown, then the corresponding magnitude would be subtracted, rather than added. From Figure 9–59, we see that removing the current sources results in an equivalent resistance given as

$$R_{eq} = R_1 \| R_2 \| R_3 \| \cdots \| R_n \qquad (9\text{–}11)$$

which may be determined as

$$R_{eq} = \frac{1}{G_{eq}} = \frac{1}{\dfrac{1}{R_1} + \dfrac{1}{R_2} + \dfrac{1}{R_3} + \cdots + \dfrac{1}{R_n}} \qquad (9\text{–}12)$$

The general expression for the equivalent voltage is

$$E_{eq} = I_{eq}R_{eq} = \frac{\dfrac{E_1}{R_1} + \dfrac{E_2}{R_2} + \dfrac{E_3}{R_3} + \cdots + \dfrac{E_n}{R_n}}{\dfrac{1}{R_1} + \dfrac{1}{R_2} + \dfrac{1}{R_3} + \cdots + \dfrac{1}{R_n}} \qquad (9\text{–}13)$$

EXAMPLE 9–13   Use Millman's Theorem to simplify the circuit of Figure 9–60 so that it has only a single source. Use the simplified circuit to find the current in the load resistor, R_L.

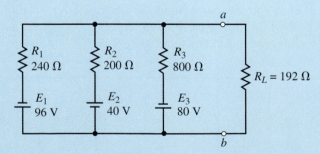

FIGURE 9–60

Solution   From Equation 9–13, we express the equivalent voltage source as

$$V_{ab} = E_{eq} = \frac{\dfrac{-96\ \text{V}}{240\ \Omega} + \dfrac{40\ \text{V}}{200\ \Omega} + \dfrac{-80\ \text{V}}{800\ \Omega}}{\dfrac{1}{240\ \Omega} + \dfrac{1}{200\ \Omega} + \dfrac{1}{800\ \Omega}}$$

$$V_{ab} = \frac{-0.300}{10.42 \text{ mS}} = -28.8 \text{ V}$$

The equivalent resistance is

$$R_{eq} = \frac{1}{\dfrac{1}{240 \ \Omega} + \dfrac{1}{200 \ \Omega} + \dfrac{1}{800 \ \Omega}} = \frac{1}{10.42 \text{ mS}} = 96 \ \Omega$$

The equivalent circuit using Millman's theorem is shown in Figure 9–61. Notice that the equivalent voltage source has a polarity which is opposite to the originally assumed polarity. This is because the voltage sources E_1 and E_3 have magnitudes which overcome the polarity and magnitude of the source E_2.

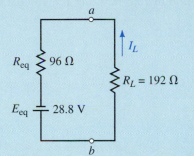

FIGURE 9–61

From the equivalent circuit of Figure 9–61, it is a simple matter to determine the current through the load resistor:

$$I_L = \frac{28.8 \text{ V}}{96 \ \Omega + 192 \ \Omega} = 0.100 \text{ A} = 100 \text{ mA} \quad \text{(upward)}$$

9.7   Reciprocity Theorem

The **reciprocity theorem** is a theorem which can only be used with single-source circuits. This theorem, however, may be applied to either voltage sources or current sources. The theorem states the following:

Voltage Sources

A voltage source causing a current I *in any branch of a circuit may be removed from the original location and placed into that branch having the current* I. *The voltage source in the new location will produce a current in the original source location which is exactly equal to the originally calculated current,* I.

When applying the reciprocity theorem for a voltage source, the following steps must be followed:

1. The voltage source is replaced by a short circuit in the original location.
2. The polarity of the source in the new location is such that the current direction in that branch remains unchanged.

Current Sources

A current source causing a voltage V *at any node of a circuit may be removed from the original location and connected to that node. The current source in the new location will produce a voltage in the original source location which is exactly equal to the originally calculated voltage,* V.

When applying the reciprocity theorem for a current source, the following conditions must be met:

1. The current source is replaced by an open circuit in the original location.

2. The direction of the source in the new location is such that the polarity of the voltage at the node to which the current source is now connected remains unchanged.

The following examples illustrate how the reciprocity theorem is used within a circuit.

EXAMPLE 9–14   Consider the circuit of Figure 9–62:

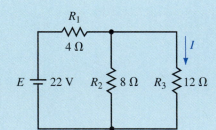

FIGURE 9–62

a. Calculate the current *I*.

b. Remove voltage source *E* and place it into the branch with R_3. Show that the current through the branch which formerly had *E* is now the same as the current *I*.

Solution

a. $V_{12\,\Omega} = \left(\dfrac{8\ \Omega \| 12\ \Omega}{4\ \Omega + (8\ \Omega \| 12\ \Omega)} \right)(22\ \text{V}) = \left(\dfrac{4.8}{8.8} \right)(22\ \text{V}) = 12.0\ \text{V}$

$I = \dfrac{V_{12\,\Omega}}{12\ \Omega} = \dfrac{12.0\ \text{V}}{12\ \Omega} = 1.00\ \text{A}$

b. Now removing the voltage source from its original location and moving it into the branch containing the current *I*, we obtain the circuit shown in Figure 9–63.

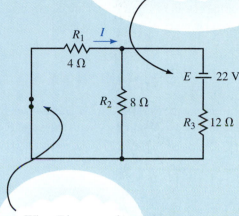

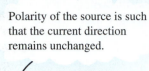

Polarity of the source is such that the current direction remains unchanged.

When E is removed it is replaced by a short circuit.

FIGURE 9–63

For the circuit of Figure 9–63, we determine the current I as follows:

$$V_{4\,\Omega} = \left(\frac{4\ \Omega \| 8\ \Omega}{12\ \Omega + (4\ \Omega \| 8\ \Omega)} \right)(22\ \text{V}) = \left(\frac{2.\overline{6}}{14.\overline{6}} \right)(22\ \text{V}) = 4.00\ \text{V}$$

$$I = \frac{V_{4\,\Omega}}{4\ \Omega} = \frac{4.00\ \text{V}}{4\ \Omega} = 1.00\ \text{A}$$

From this example, we see that the reciprocity theorem does indeed apply.

EXAMPLE 9–15   Consider the circuit shown in Figure 9–64:

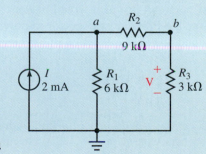

FIGURE 9–64

a. Determine the voltage V across resistor R_3.

b. Remove the current source I and place it between node b and the reference node. Show that the voltage across the former location of the current source (node a) is now the same as the voltage V.

Solution

a. The node voltages for the circuit of Figure 9–64 are determined as follows:

$$R_T = 6 \text{ k}\Omega \| (9 \text{ k}\Omega + 3 \text{ k}\Omega) = 4 \text{ k}\Omega$$

$$V_a = (2 \text{ mA})(4 \text{ k}\Omega) = 8.00 \text{ V}$$

$$V_b = \left(\frac{3 \text{ k}\Omega}{3 \text{ k}\Omega + 9 \text{ k}\Omega}\right)(8.0 \text{ V}) = 2.00 \text{ V}$$

b. After relocating the current source from the original location, and connecting it between node b and ground, we obtain the circuit shown in Figure 9–65.

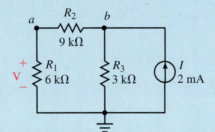

FIGURE 9–65

The resulting node voltages are now found as follows:

$$R_T = 3 \text{ k}\Omega \| (6 \text{ k}\Omega + 9 \text{ k}\Omega) = 2.50 \text{ k}\Omega$$

$$V_b = (2 \text{ mA})(2.5 \text{ k}\Omega) = 5.00 \text{ V}$$

$$V_a = \left(\frac{6 \text{ k}\Omega}{6 \text{ k}\Omega + 9 \text{ k}\Omega}\right)(5.0 \text{ V}) = 2.00 \text{ V}$$

From the above results, we conclude that the reciprocity theorem again applies for the given circuit.

9.8   Circuit Analysis Using Computers

ELECTRONICS
WORKBENCH

PSpice

Electronics Workbench and PSpice are easily used to illustrate the important theorems developed in this chapter. We will use each software package in a somewhat different approach to verify the theorems. Electronics Workbench allows us to "build and test" a circuit just as it would be done in a lab. When using PSpice we will activate the Probe postprocessor to provide a graphical display of voltage, current, and power as a function of load resistance.

Electronics Workbench

EXAMPLE 9–16   Use Electronics Workbench to find both the Thévenin and the Norton equivalent circuits external to the load resistor in the circuit of Figure 9–66:

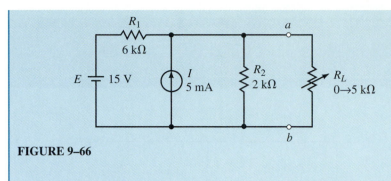

FIGURE 9–66

Solution

1. Using Electronics Workbench, construct the circuit as shown in Figure 9–67.

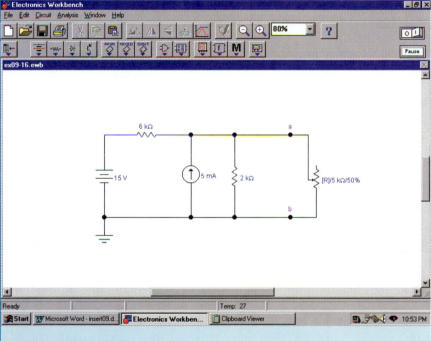

EWB   **FIGURE 9–67**

2. Just as in a lab, we will use a multimeter to find the open circuit (Thévenin) voltage and the short circuit (Norton) current. As well, the multimeter is used to measure the Thévenin (Norton) resistance. The steps in these measurements are essentially the same as those used to theoretically determine the equivalent circuits.

   a. We begin by removing the load resistor, R_L from the circuit (using the Edit menu and the Cut/Paste menu items). The remaining terminals are labeled as a and b.

   b. The Thévenin voltage is measured by simply connecting the multimeter between terminals a and b. After clicking on the power switch, we obtain a reading of $E_{Th} = 11.25$ V as shown in Figure 9–68.

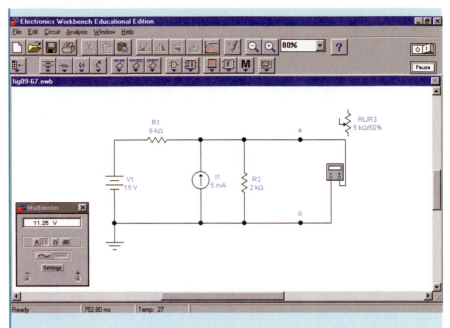

FIGURE 9–68

c. With the multimeter between terminals *a* and *b,* the Norton current is easily measured by switching the multimeter to its ammeter range. After clicking on the power switch, we obtain a reading of $I_N = 7.50$ mA as shown in Figure 9–69.

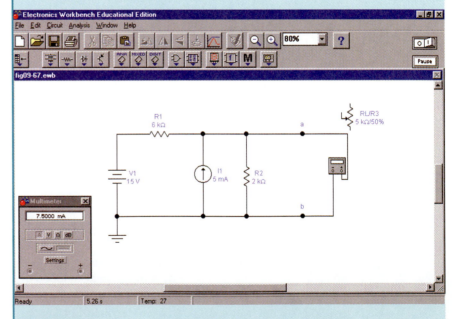

FIGURE 9–69

d. In order to measure the Thévenin resistance, the voltage source is removed and replaced by a short circuit (wire) and the current source is removed and replaced by an open circuit. Now, with the multimeter connected between terminals *a* and *b* it is set to measure resistance (by clicking on the Ω button). After clicking on the power switch, we have the display shown in Figure 9–70. The multimeter provides the Thévenin resistance as $R_{Th} = 1.5 \text{ k}\Omega$.

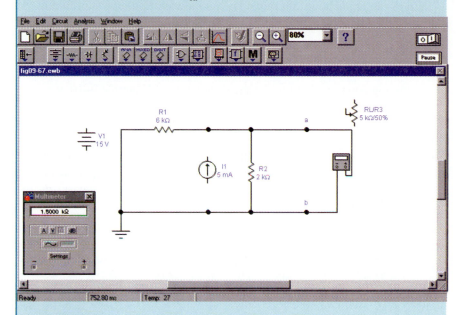

FIGURE 9–70

3. Using the measured results, we are able to sketch both the Thévenin equivalent and the Norton equivalent circuit as illustrated in Figure 9–71.

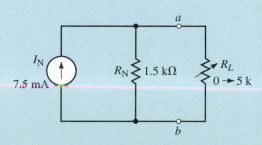

(a) Thévenin equivalent circuit

(b) Norton equivalent circuit

FIGURE 9–71

These results are consistent with those obtained in Example 9–4.

Note: From the previous example we see that it is not necessary to directly measure the Thévenin (Norton) resistance since the value can be easily calculated from the Thévenin voltage and the Norton current. The following equation is an application of Ohm's law and always applies when finding an equivalent circuit.

$$R_{\text{Th}} = R_{\text{N}} = \frac{E_{\text{Th}}}{I_{\text{N}}}$$ (9–14)

Applying equation 9–14 to the measurements of Example 9–16, we get

$$R_{\text{Th}} = R_{\text{N}} = \frac{11.25 \text{ V}}{7.50 \text{ mA}} = 1.50 \text{ k}\Omega$$

Clearly, this is the same result as that obtained when we went through the extra step of removing the voltage and current sources. This approach is the most practical method and is commonly used when actually measuring the Thévenin (Norton) resistance of a circuit.

PRACTICE PROBLEMS 7

Use Electronics Workbench to find both the Thévenin and the Norton equivalent circuits external to the load resistor in the circuit of Figure 9–25.

Answers: $E_{\text{Th}} = 2.00$ V, $I_{\text{N}} = 16.67$ mA, $R_{\text{Th}} = R_{\text{N}} = 120 \ \Omega$

PSpice

As we have already seen, PSpice has an additional postprocessor called PROBE, which is able to provide a graphical display of numerous variables. The following example uses PSpice to illustrate the maximum power transfer theorem.

EXAMPLE 9–17   Use OrCAD PSpice to input the circuit of Figure 9–72 and use the Probe postprocessor to display output voltage, current, and power as a function of load resistance.

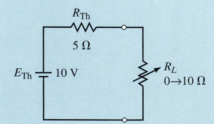

FIGURE 9–72

OrCAD PSpice

Solution   The circuit is constructed as shown in Figure 9–73.

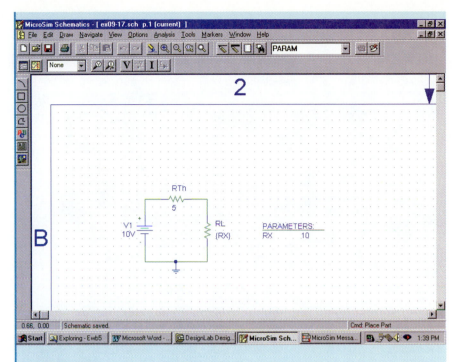

FIGURE 9–73

• Double click on each resistor in the circuit and change the Reference cells to RTH and RL. Click on Apply to accept the changes.

• Double click on the value for RL and enter **{Rx}.** Place the PARAM part adjacent to RL. Use the Property Editor to assign a default value of 10 Ω to Rx. Click on Apply. Have the display show the name and value and then exit the Property Editor.

• Adjust the Simulation Settings to result in a DC sweep of the load resistor from 0.1 Ω to 10 Ω in 0.1 Ω increments. (Refer to Example 7–15 for the complete procedure.)

• Click on the Run icon once the circuit is complete.

• Once the design is simulated, you will see a blank screen with the abscissa (horizontal axis) showing RX scaled from 0 to 10 Ω.

• Since we would like to have a simultaneous display of voltage, current, and power, it is necessary to do the following:

To display V_L: Click Trace and then Add Trace. Select **V(RL:1).** Click OK and the load voltage will appear as a function of load resistance.

To display I_L: First add another axis by clicking on Plot and Add Y Axis. Next, click Trace and then Add Trace. Select **I(RL).** Click OK and the load current will appear as a function of load resistance.

To display P_L: Add another Y axis. Click Trace and Add Trace. Now, since power is not one of the options that can be automatically selected, it is necessary to enter the power into the Trace Expression box. One method of doing this is to enter **I(RL)*V(RL:1)** and then click OK. Adjust the limits of the Y axis by clicking on Plot and Axis Settings. Click on the Y Axis tab and select the User Defined Data Range. Set the limits from 0W to 5W.

The display will appear as shown in Figure 9–74.

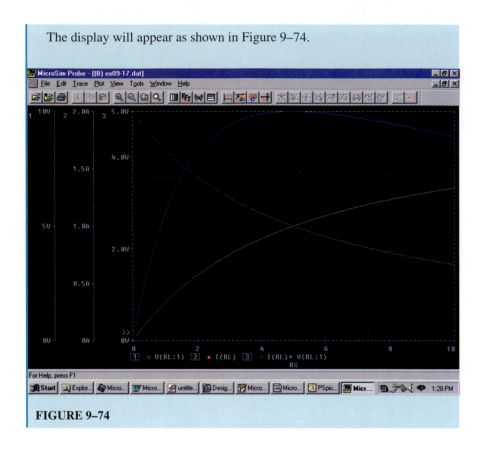

FIGURE 9–74

Use OrCAD PSpice to input the circuit of Figure 9–66. Use the Probe post-processor to obtain voltage, current, and power for the load resistor as it is varied from 0 to 5 kΩ.

PUTTING IT INTO PRACTICE

A simple battery cell (such as a "D" cell) can be represented as a Thévenin equivalent circuit as shown in the accompanying figure.

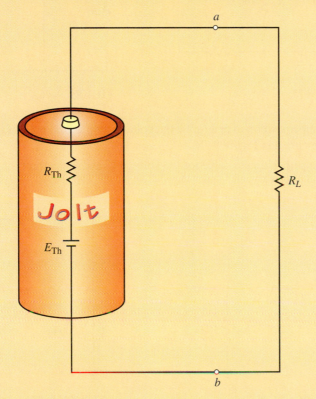

The Thévenin voltage represents the open-circuit (or unloaded) voltage of the battery cell, while the Thévenin resistance is the internal resistance of the battery. When a load resistance is connected across the terminals of the battery, the voltage V_{ab} will decrease due to the voltage drop across the internal resistor. By taking two measurements, it is possible to find the Thévenin equivalent circuit of the battery.

When no load is connected between the terminals of the battery, the terminal voltage is found to be $V_{ab} = 1.493$ V. When a resistance of $R_L = 10.6\ \Omega$ is connected across the terminals, the voltage is measured to be $V_{ab} = 1.430$ V. Determine the Thévenin equivalent circuit of the battery. Use the measurements to determine the efficiency of the battery for the given load.

9.1   Superposition Theorem

PROBLEMS

1. Given the circuit of Figure 9–75, use superposition to calculate the current through each of the resistors.

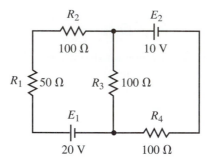

FIGURE 9–75

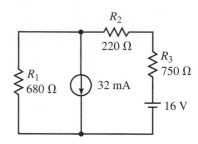

FIGURE 9–76

2. Use superposition to determine the voltage drop across each of the resistors of the circuit in Figure 9–76.

3. Use superposition to solve for the voltage V_a and the current I in the circuit of Figure 9–77.

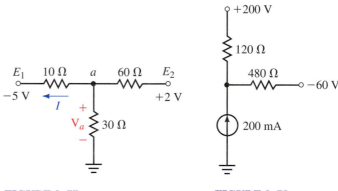

FIGURE 9–77   **FIGURE 9–78**

4. Using superposition, find the current through the 480-Ω resistor in the circuit of Figure 9–78:

5. Given the circuit of Figure 9–79, what must be the value of the unknown voltage source to ensure that the current through the load is $I_L = 5$ mA as shown. Verify the results using superposition.

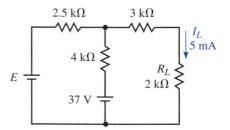

FIGURE 9–79

6. If the load resistor in the circuit of Figure 9–80 is to dissipate 120 W, determine the value of the unknown voltage source. Verify the results using superposition.

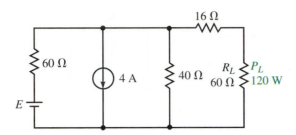

FIGURE 9–80

9.2   Thévenin's Theorem

7. Find the Thévenin equivalent external to R_L in circuit of Figure 9–81. Use the equivalent circuit to find V_{ab}.

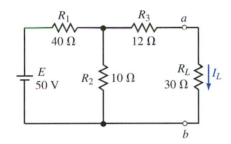

FIGURE 9–81

8. Repeat Problem 7 for the circuit of Figure 9–82.

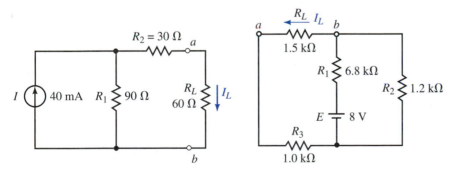

FIGURE 9–82   **FIGURE 9–83**

9. Repeat Problem 7 for the circuit of Figure 9–83.

10. Repeat Problem 7 for the circuit of Figure 9–84.

11. Refer to the circuit of Figure 9–85:

   a. Find the Thévenin equivalent circuit external to R_L.

   b. Use the equivalent circuit to determine V_{ab} when $R_L = 20\ \Omega$ and when $R_L = 50\ \Omega$.

12. Refer to the circuit of Figure 9–86:

   a. Find the Thévenin equivalent circuit external to R_L.

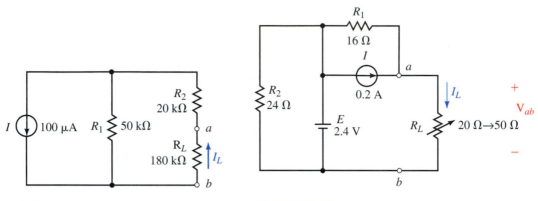

FIGURE 9–84   **FIGURE 9–85**

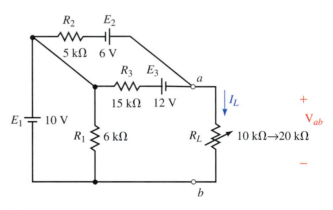

FIGURE 9–86

b. Use the equivalent circuit to determine V_{ab} when $R_L = 10\ k\Omega$ and when $R_L = 20\ k\Omega$.

13. Refer to the circuit of Figure 9–87:

   a. Find the Thévenin equivalent circuit external to the indicated terminals.

   b. Use the Thévenin equivalent circuit to determine the current through the indicated branch.

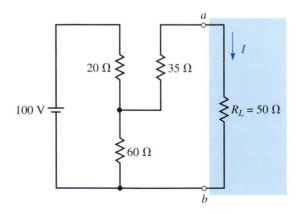

FIGURE 9–87

14. Refer to the circuit of Figure 9–88:

   a. Find the Thévenin equivalent circuit external to R_L.

   b. Use the Thévenin equivalent circuit to find V_L.

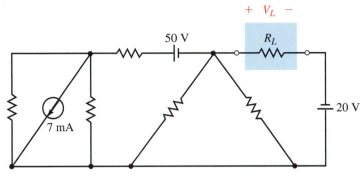

All resistors are 3.3 kΩ.

FIGURE 9–88

15. Refer to the circuit of Figure 9–89:

   a. Find the Thévenin equivalent circuit external to the indicated terminals.

   b. Use the Thévenin equivalent circuit to determine the current through the indicated branch.

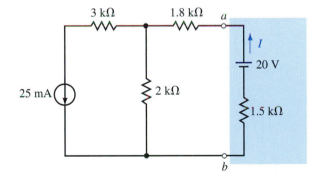

 FIGURE 9–89

16. Refer to the circuit of Figure 9–90:

   a. Find the Thévenin equivalent circuit external to the indicated terminals.

   b. If $R_5 = 1$ kΩ, use the Thévenin equivalent circuit to determine the voltage V_{ab} and the current through this resistor.

17. Refer to the circuit of Figure 9–91:

   a. Find the Thévenin equivalent circuit external to R_L.

   b. Use the Thévenin equivalent circuit to find the current I when $R_L = 0$, 10 kΩ, and 50 kΩ.

18. Refer to the circuit of Figure 9–92:

   a. Find the Thévenin equivalent circuit external to R_L.

   b. Use the Thévenin equivalent circuit to find the power dissipated by R_L.

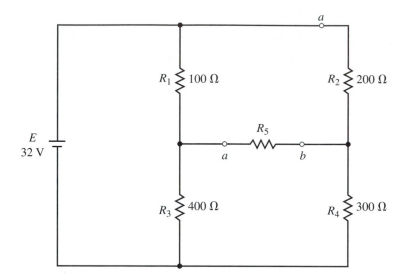

FIGURE 9–90

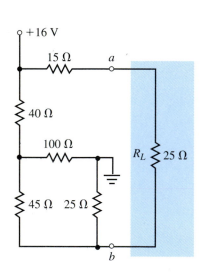

FIGURE 9–92

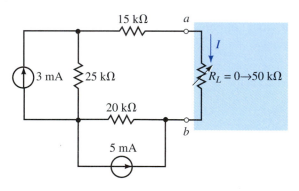

FIGURE 9–91

19. Repeat Problem 17 for the circuit of Figure 9–93.

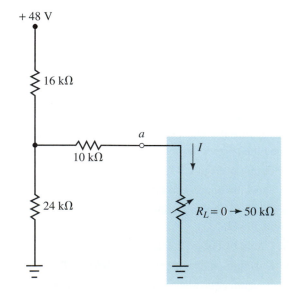

FIGURE 9–93

20. Repeat Problem 17 for the circuit of Figure 9–94.
21. Find the Thévenin equivalent circuit of the network external to the indicated branch as shown in Figure 9–95.

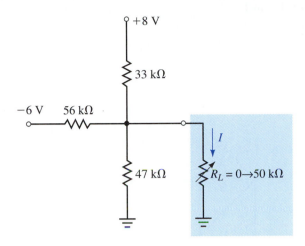

FIGURE 9–94

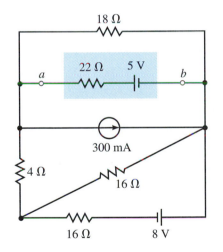

FIGURE 9–95

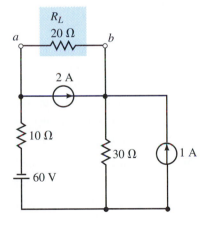

FIGURE 9–96

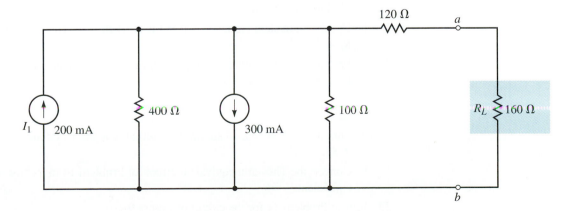

FIGURE 9–97

22. Refer to the circuit of Figure 9–96.

    a. Find the Thévenin equivalent circuit external to the indicated terminals.

    b. Use the Thévenin equivalent circuit to determine the current through the indicated branch.

23. Repeat Problem 22 for the circuit of Figure 9–97.

24. Repeat Problem 22 for the circuit of Figure 9–98.

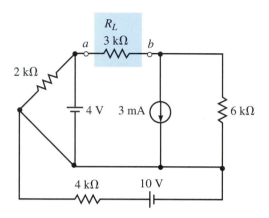

FIGURE 9–98

9.3   Norton's Theorem

25. Find the Norton equivalent circuit external to R_L in the circuit of Figure 9–81. Use the equivalent circuit to find I_L for the circuit.

26. Repeat Problem 25 for the circuit of Figure 9–82.

27. Repeat Problem 25 for the circuit of Figure 9–83.

28. Repeat Problem 25 for the circuit of Figure 9–84.

29. Refer to the circuit of Figure 9–85:

    a. Find the Norton equivalent circuit external to R_L.

    b. Use the equivalent circuit to determine I_L when $R_L = 20 \ \Omega$ and when $R_L = 50 \ \Omega$.

30. Refer to the circuit of Figure 9–86:

    a. Find the Norton equivalent circuit external to R_L.

    b. Use the equivalent circuit to determine I_L when $R_L = 10 \ \mathrm{k}\Omega$ and when $R_L = 20 \ \mathrm{k}\Omega$.

31. a. Find the Norton equivalent circuit external to the indicated terminals of Figure 9–87.

    b. Convert the Thévenin equivalent circuit of Problem 13 to its Norton equivalent.

32. a. Find the Norton equivalent circuit external to R_L in the circuit of Figure 9–88.

    b. Convert the Thévenin equivalent circuit of Problem 14 to its Norton equivalent.

33. Repeat Problem 31 for the circuit of Figure 9–91.

34. Repeat Problem 31 for the circuit of Figure 9–92.

35. Repeat Problem 31 for the circuit of Figure 9–95.

36. Repeat Problem 31 for the circuit of Figure 9–96.

9.4   Maximum Power Transfer Theorem

37. a. For the circuit of Figure 9–91, determine the value of R_L so that maximum power is delivered to the load.

    b. Calculate the value of the maximum power which can be delivered to the load.

    c. Sketch the curve of power versus resistance as R_L is adjusted from $0\ \Omega$ to $50\ k\Omega$ in increments of $5\ k\Omega$.

38. Repeat Problem 37 for the circuit of Figure 9–94.

39. a. For the circuit of Figure 9–99, find the value of R so that $R_L = R_{Th}$.

    b. Calculate the maximum power dissipated by R_L.

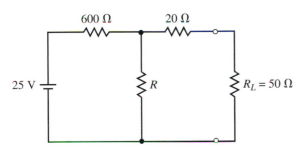

FIGURE 9–99

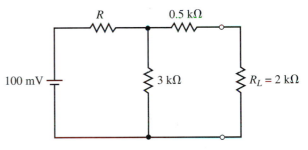

FIGURE 9–100

40. Repeat Problem 39 for the circuit of Figure 9–100.

41. a. For the circuit of Figure 9–101, determine the values of R_1 and R_2 so that the 32-kΩ load receives maximum power.

    b. Calculate the maximum power delivered to R_L.

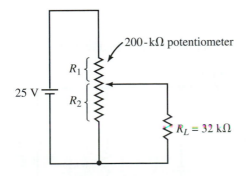

FIGURE 9–101

42. Repeat Problem 41 if the load resistor has a value of $R_L = 25\ k\Omega$.

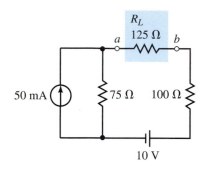

FIGURE 9–102

9.5 Substitution Theorem

43. If the indicated portion of the circuit in Figure 9–102 is to be replaced with a voltage source and a 50-Ω series resistor, determine the magnitude and polarity of the resulting voltage source.

44. If the indicated portion of the circuit in Figure 9–102 is to be replaced with a current source and a 200-Ω shunt resistor, determine the magnitude and direction of the resulting current source.

9.6 Millman's Theorem

45. Use Millman's theorem to find the current through and the power dissipated by R_L in the circuit of Figure 9–103.

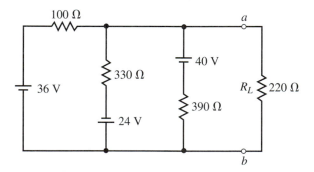

FIGURE 9–103

46. Repeat Problem 45 for the circuit of Figure 9–104.

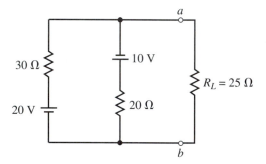

FIGURE 9–104

47. Repeat Problem 45 for the circuit of Figure 9–105.

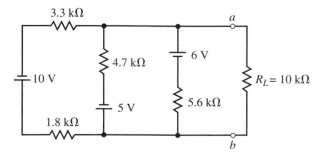

FIGURE 9–105

48. Repeat Problem 45 for the circuit of Figure 9–106.

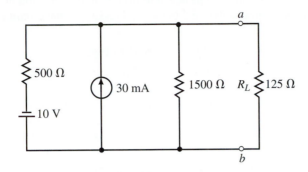

FIGURE 9–106

9.7   Reciprocity Theorem

49. a. Determine the current I in the circuit of Figure 9–107.

   b. Show that reciprocity applies for the given circuit.

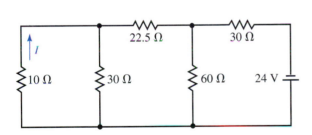

FIGURE 9–107

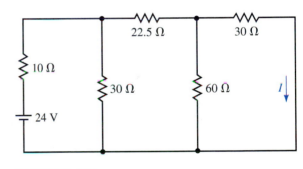

FIGURE 9–108

50. Repeat Problem 49 for the circuit of Figure 9–108.

51. a. Determine the voltage V in the circuit of Figure 9–109.

   b. Show that reciprocity applies for the given circuit.

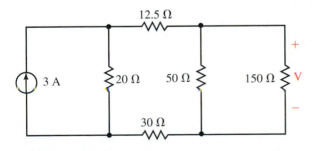

FIGURE 9–109

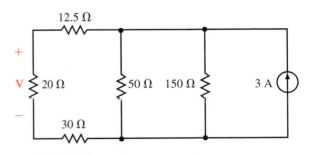

FIGURE 9–110

52. Repeat Problem 51 for the circuit of Figure 9–110.

9.8   Circuit Analysis Using Computers

53. **EWB**   Use Electronics Workbench to find both the Thévenin and the Norton equivalent circuits external to the load resistor in the circuit of Figure 9–81.

54. **EWB**   Repeat Problem 53 for the circuit of Figure 9–82.

55. **PSpice**   Use the schematic editor of PSpice to input the circuit of Figure 9–83 and use the PROBE postprocessor to display output voltage, current, and power as a function of load resistance. Use the cursor in the PROBE postprocessor to determine the value of load resistance for which the load will receive maximum power. Let the load resistance vary from 100 Ω to 4000 Ω in increments of 100 Ω.

56. **PSpice**   Repeat Problem 55 for the circuit of Figure 9–84. Let the load resistance vary from 1 kΩ to 100 kΩ in increments of 1 kΩ.

ANSWERS TO IN-PROCESS LEARNING CHECKS

In-Process Learning Check 1

$P_{R_1} = 304$ mW, $P_{R_2} = 76$ mW, $P_{R_3} = 276$ mW
Assuming that superposition applies for power:
$P_{R_{1(1)}} = 60$ mW, $P_{R_{1(2)}} = 3.75$ mW, $P_{R_{1(3)}} = 60$ mW
But $P_{R_1} = 304$ mW $\neq 123.75$ mW

In-Process Learning Check 2

$R_1 = 133$ Ω

In-Process Learning Check 3

1. Refer to Figure 9–27.
2. $I_N = 200$ μA, $R_N = 500$ Ω
3. $E_{Th} = 0.2$ V, $R_{Th} = 20$ kΩ

In-Process Learning Check 4

a. $\eta = 50\%$.
b. $\eta = 33.3\%$.
c. $\eta = 66.7\%$.

In-Process Learning Check 5

1. In communication circuits and some amplifiers, maximum power transfer is a desirable characteristic.
2. In power transmission and dc voltage sources maximum power transfer is not desirable.

Capacitance and Inductance

R esistance, inductance, and capacitance are the three basic circuit properties that we use to control voltages and currents in electrical and electronic circuits. However, each behaves in a fundamentally different way. Resistance, for example, opposes current, while inductance opposes any change in current, and capacitance opposes any change in voltage. In addition, resistance dissipates energy, while inductance and capacitance both store energy—inductance in its magnetic field and capacitance in its electric field.

Circuit elements that are built to possess capacitance are called capacitors, while elements built to possess inductance are called inductors. In Part III of this book, we explore these elements, their properties, and their behavior in electric circuits.

Note: Because many colleges and universities cover ac principles before transient analysis, the material in Part III has been organized so that Chapters 11 and 14 can be delayed until after the ac topics without loss of continuity.

PART III

10 Capacitors and Capacitance

11 Capacitor Charging, Discharging, and Simple Waveshaping Circuits

12 Magnetism and Magnetic Circuits

13 Inductance and Inductors

14 Inductive Transients

10

Capacitors and Capacitance

OBJECTIVES

After studying this chapter, you will be able to

- describe the basic construction of capacitors,
- explain how capacitors store charge,
- define capacitance,
- describe what factors affect capacitance and in what way,
- describe the electric field of a capacitor,
- compute the breakdown voltages of various materials,
- describe various types of commercial capacitors,
- compute the capacitance of capacitors in series and in parallel combinations,
- compute capacitor voltage and current for simple time-varying waveforms,
- determine stored energy,
- describe capacitor faults and the basic troubleshooting of capacitors.

KEY TERMS

Capacitance
Capacitor
Dielectric
Dielectric Absorption
Dielectric Constant
Electric Field
Electric Field Intensity
Electric Flux
Electric Flux Density
Electrolytic
Farad
Leakage
Permittivity
Voltage Breakdown
Voltage Gradient
Working Voltage

OUTLINE

Capacitance
Factors Affecting Capacitance
Electric Fields
Dielectrics
Nonideal Effects
Types of Capacitors
Capacitors in Parallel and Series
Capacitor Current and Voltage
Energy Stored by a Capacitor
Capacitor Failures and Troubleshooting

A **capacitor** is a circuit component designed to store electrical charge. If you connect a dc voltage source to a capacitor, for example, the capacitor will "charge" to the voltage of the source. If you then disconnect the source, the capacitor will remain charged, i.e., its voltage will remain constant at the value to which it had risen while connected to the source (assuming no leakage). Because of this tendency to hold voltage, *a capacitor opposes changes in voltage*. It is this characteristic that gives capacitors their unique properties.

Capacitors are widely used in electrical and electronic applications. They are used in radio and TV systems, for example, to tune in signals, in cameras to store the charge that fires the photoflash, on pump and refrigeration motors to increase starting torque, in electric power systems to increase operating efficiency, and so on. Photos of some typical capacitors are shown in Figures 10–15 and 10–16.

Capacitance is the electrical property of capacitors: it is a measure of how much charge a capacitor can hold. In this chapter, we look at capacitance and its basic properties. In Chapter 11, we look at capacitors in dc and pulse circuits; in later chapters, we look at capacitors in ac applications.

Michael Faraday and the Field Concept

THE UNIT OF CAPACITANCE, the farad, is named after Michael Faraday (1791–1867). Born in England to a working class family, Faraday received limited education. Nonetheless, he was responsible for many of the fundamental discoveries of electricity and magnetism. Lacking mathematical skills, he used his intuitive ability rather than mathematical models to develop conceptual pictures of basic phenomena. It was his development of the field concept that made it possible to map out the fields that exist around electrical charges.

To get at this idea, recall from Chapter 2 that unlike charges attract and like charges repel, i.e. a force exists between charges. We call the region where this force acts an electric field. To visualize this field, we use Faraday's field concept and draw lines of force (or flux lines) that show at every point in space the magnitude and direction of the force. Now, rather than supposing that one charge exerts a force on another, we instead visualize that the original charges create a field in space and that other charges introduced into this field experience a force due to the field. This concept is helpful in studying certain aspects of capacitors, as you will see in this chapter.

The development of the field concept had a significant impact on science. We now picture several important phenomena in terms of fields, including electric fields, gravitation, and magnetism. When Faraday published his theory in 1844, however, it was not taken seriously, much like Ohm's work two decades earlier. It is also interesting to note that the development of the field concept grew out of Faraday's research into magnetism, not electric charge.

10.1   Capacitance

A capacitor consists of two conductors separated by an insulator. One of its basic forms is the parallel-plate capacitor shown in Figure 10–1. It consists of two metal plates separated by a nonconducting material (i.e., an insulator) called a **dielectric.** The dielectric may be air, oil, mica, plastic, ceramic, or other suitable insulating material.

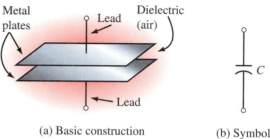

(a) Basic construction   (b) Symbol

FIGURE 10–1   Parallel-plate capacitor.

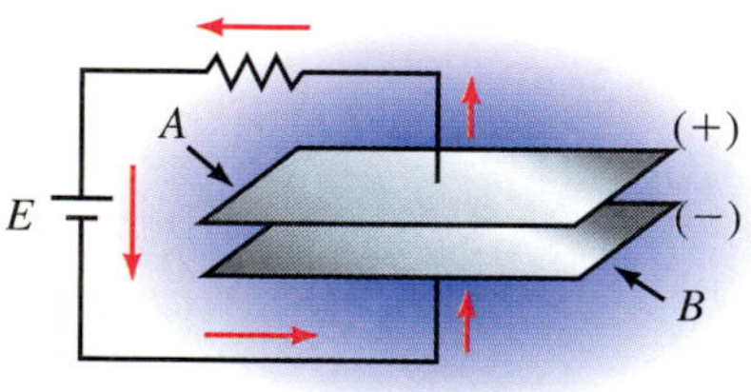

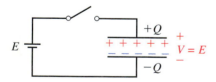

FIGURE 10–2   Capacitor during charging. When the source is connected, electrons are removed from plate A and an equal number deposited on plate B. This leaves the top plate positively charged and the bottom plate negatively charged.

FIGURE 10–3   Capacitor after charging. When the source is disconnected, electrons are trapped on the bottom plate. Thus, charge is stored.

Since the plates of the capacitor are metal, they contain huge numbers of free electrons. In their normal state, however, they are uncharged, that is, there is no excess or deficiency of electrons on either plate. If a dc source is now connected (Figure 10–2), electrons are pulled from the top plate by the positive potential of the battery and the same number deposited on the bottom plate. This leaves the top plate with a deficiency of electrons (i.e., positive charge) and the bottom plate with an excess (i.e., negative charge). In this state, the capacitor is said to be **charged.** If the amount of charge transferred during this process is Q coulombs, we say that the capacitor has a charge of Q.

If we now disconnect the source (Figure 10–3), the excess electrons that were moved to the bottom plate remain trapped as they have no way to return to the top plate. The capacitor therefore remains charged even though no source is present. Because of this, we say that *a capacitor can store charge*. Capacitors with little leakage (Section 10.5) can hold their charge for a considerable time.

Large capacitors charged to high voltages contain a great deal of energy and can give you a bad shock. Always **discharge** capacitors after power has been removed if you intend to handle them. You can do this by shorting a wire across their leads. Electrons then return to the top plate, restoring the charge balance and reducing the capacitor voltage to zero. (However, you also need to be concerned about residual voltage due to dielectric absorption. This is discussed in Section 10.5.)

Definition of Capacitance

The amount of charge Q that a capacitor can store depends on the applied voltage. Experiments show that for a given capacitor, Q is proportional to voltage. Let the constant of proportionality be C. Then,

$$Q = CV \qquad (10\text{–}1)$$

Rearranging terms yields

$$C = \frac{Q}{V} \quad \text{(farads, F)} \tag{10–2}$$

The term C is defined as the capacitance of the capacitor. As indicated, its unit is the **farad.** By definition, *the capacitance of a capacitor is one farad if it stores one coulomb of charge when the voltage across its terminals is one volt.* The farad, however, is a very large unit. Most practical capacitors range in size from picofarads (pF or 10^{-12} F) to microfarads (μF or 10^{-6} F). The larger the value of C, the more charge that the capacitor can hold for a given voltage.

EXAMPLE 10–1

a. How much charge is stored on a 10-μF capacitor when it is connected to a 24-volt source?

b. The charge on a 20-nF capacitor is 1.7 μC. What is its voltage?

Solution

a. From Equation 10–1, $Q = CV$. Thus, $Q = (10 \times 10^{-6}\,\text{F})(24\,\text{V}) = 240\,\mu\text{C}$.

b. Rearranging Equation 10–1, $V = Q/C = (1.7 \times 10^{-6}\,\text{C})/(20 \times 10^{-9}\,\text{F}) = 85\,\text{V}$.

10.2   Factors Affecting Capacitance

Effect of Area

As shown by Equation 10–2, capacitance is directly proportional to charge. This means that the more charge you can put on a capacitor's plates for a given voltage, the greater will be its capacitance. Consider Figure 10–4. The capacitor of (b) has four times the area of (a). Since it has the same number of free electrons per unit area, it has four times the total charge and hence four times the capacitance. This turns out to be true in general, that is, *capacitance is directly proportional to plate area.*

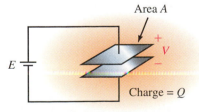

(a) Capacitor with area A
and charge Q

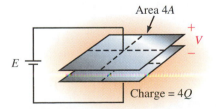

(b) Plates with four times the area
have four times the charge and
therefore, four times the capacitance

FIGURE 10–4   For a fixed separation, capacitance is proportional to plate area.

Effect of Spacing

Now consider Figure 10–5. Since the top plate has a deficiency of electrons and the bottom plate an excess, a force of attraction exists across the gap. For a fixed spacing as in (a), the charges are in equilibrium. Now move the plates closer together as in (b). As spacing decreases, the force of attraction increases, pulling more electrons from within the material of plate *B* to its top surface. This creates a deficiency of electrons in the lower levels of *B*. To replenish these, the source moves additional electrons around the circuit, leaving *A* with an even greater deficiency and *B* with an even greater excess. The charge on the plates therefore increases and hence, according to Equation 10–2, so does the capacitance. We therefore conclude that decreasing spacing increases capacitance, and vice versa. In fact, as we will show later, *capacitance is inversely proportional to plate spacing.*

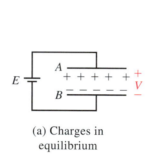

(a) Charges in equilibrium

(b) As spacing decreases, more electrons move from *A* to *B*, increasing the charge on the plates and hence, their capacitance

FIGURE 10–5 Decreasing spacing increases capacitance.

TABLE 10–1 Relative Dielectric Constants (Also Called Relative Permittivities)

Material	ϵ_r (Nominal Values)
Vacuum	1
Air	1.0006
Ceramic	30–7500
Mica	5.5
Mylar	3
Oil	4
Paper (dry)	2.2
Polystyrene	2.6
Teflon	2.1

Effect of Dielectric

Capacitance also depends on the dielectric. Consider Figure 10–6(a), which shows an air-dielectric capacitor. If you substitute different materials for air, the capacitance increases. Table 10–1 shows the factor by which capacitance increases for a number of different materials. For example, if Teflon® is used instead of air, capacitance is increased by a factor of 2.1. This factor is called the **relative dielectric constant** or **relative permittivity** of the mater-

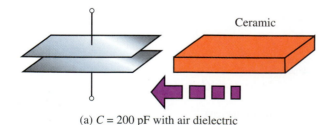

(a) *C* = 200 pF with air dielectric

(b) *C* = 1.5 μF with high permittivity ceramic dielectric

FIGURE 10–6 The factor by which a material increases capacitance is termed its *relative dielectric constant.* The ceramic used here has a value of 7500.

ial. (Permittivity is a measure of how easy it is to establish electric flux in a material.) Note that high-permittivity ceramic increases capacitance by as much as 7500, as indicated in Figure 10–6(b).

Capacitance of a Parallel-Plate Capacitor

From the above observations, we see that capacitance is directly proportional to plate area, inversely proportional to plate separation, and dependent on the dielectric. In equation form,

$$C = \epsilon \frac{A}{d} \quad \text{(F)} \tag{10–3}$$

where area A is in square meters and spacing d is in meters.

Dielectric Constant

The constant ϵ in Equation 10–3 is the **absolute dielectric constant** of the insulating material. Its units are farads per meter (F/m). For air or vacuum, ϵ has a value of $\epsilon_o = 8.85 \times 10^{-12}$ F/m. For other materials, ϵ is expressed as the product of the relative dielectric constant, ϵ_r (shown in Table 10–1), times ϵ_o. That is,

$$\epsilon = \epsilon_r \epsilon_o \tag{10–4}$$

Consider, again, Equation 10–3: $C = \epsilon A/d = \epsilon_r \epsilon_o A/d$. Note that $\epsilon_o A/d$ is the capacitance of a vacuum- (or air-) dielectric capacitor. Denote it by C_o. Then, for any other dielectric,

$$C = \epsilon_r C_o \tag{10–5}$$

EXAMPLE 10–2 Compute the capacitance of a parallel-plate capacitor with plates 10 cm by 20 cm, separation of 5 mm, and

a. an air dielectric,

b. a ceramic dielectric with permittivity of 7500.

Solution Convert all dimensions to meters. Thus, $A = (0.1 \text{ m})(0.2 \text{ m}) = 0.02 \text{ m}^2$, and $d = 5 \times 10^{-3}$ m.

a. For air, $C = \epsilon_o A/d = (8.85 \times 10^{-12})(2 \times 10^{-2})/(5 \times 10^{-3}) = 35.4 \times 10^{-12}$ F = 35.4 pF.

b. For ceramic with $\epsilon_r = 7500$, $C = 7500(35.4 \text{ pF}) = 0.266 \text{ } \mu\text{F}$.

EXAMPLE 10–3 A parallel-plate capacitor with air dielectric has a value of $C = 12$ pF. What is the capacitance of a capacitor that has the following:

a. The same separation and dielectric but five times the plate area?

b. The same dielectric but four times the area and one-fifth the plate spacing?

c. A dry paper dielectric, six times the plate area, and twice the plate spacing?

Solution

a. Since the plate area has increased by a factor of five and everything else remains the same, C increases by a factor of five. Thus, $C = 5(12 \text{ pF}) = 60 \text{ pF}$.

b. With four times the plate area, C increases by a factor of four. With one-fifth the plate spacing, C increases by a factor of five. Thus, $C = (4)(5)(12 \text{ pF}) = 240 \text{ pF}$.

c. Dry paper increases C by a factor of 2.2. The increase in plate area increases C by a factor of six. Doubling the plate spacing reduces C by one-half. Thus, $C = (2.2)(6)(\frac{1}{2})(12 \text{ pF}) = 79.2 \text{ pF}$.

IN-PROCESS LEARNING CHECK 1

1. A capacitor with plates 7.5 cm × 8 cm and plate separation of 0.1 mm has an oil dielectric:

   a. Compute its capacitance;

   b. If the charge on this capacitor is 0.424 μC, what is the voltage across its plates?

2. For a parallel-plate capacitor, if you triple the plate area and halve the plate spacing, how does capacitance change?

3. For the capacitor of Figure 10–6, if you use mica instead of ceramic, what will be the capacitance?

4. What is the dielectric for the capacitor of Figure 10–7(b)?

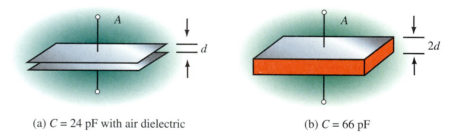

(a) $C = 24$ pF with air dielectric   (b) $C = 66$ pF

FIGURE 10–7

(Answers are at the end of the chapter.)

10.3   Electric Fields

Electric Flux

Electric fields are force fields that exist in the region surrounding charged bodies. Some familiarity with electric fields is necessary to understand dielectrics and their effect on capacitance. We now look briefly at the key ideas.

Consider Figure 10–8(a). As noted in Chapter 2, unlike charges attract and like charges repel, i.e. a force exists between them. The region where this force exists is called an **electric field.** To visualize this field, we use

Faraday's field concept. The direction of the field is defined as the direction of force on a positive charge. It is therefore directed outward from the positive charge and inward toward the negative charge as shown. Field lines never cross, and the density of the lines indicates the strength of the field; i.e., the more dense the lines, the stronger the field. Figure 10–8(b) shows the field of a parallel-plate capacitor. In this case, the field is uniform across the gap, with some fringing near its edges. Electric flux lines are represented by the Greek letter ψ (psi).

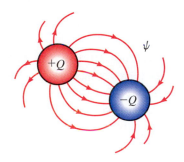
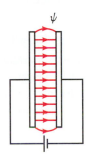

(a) Field about a pair of positive
and negative charges

(b) Field of parallel
plate capacitor

FIGURE 10–8   Some example electric fields.

Electric Field Intensity

The strength of an electric field, also called its **electric field intensity,** is the force per unit charge that the field exerts on a small, positive test charge, Q_t. Let the field strength be denoted by $\mathscr{E}$. Then, by definition,

$$\mathscr{E} = F/Q_t \quad \text{(newtons/coulomb, N/C)} \qquad (10\text{--}6)$$

To illustrate, let us determine the field about a point charge, Q. When the test charge is placed near Q, it experiences a force of $F = kQQ_t/r^2$ (Coulomb's law, Chapter 2). The constant in Coulomb's law is actually equal to $1/4\pi\epsilon$. Thus, $F = QQ_t/4\pi\epsilon r^2$, and from Equation 10–6,

$$\mathscr{E} = \frac{F}{Q_t} = \frac{Q}{4\pi\epsilon r^2} \quad \text{(N/C)} \qquad (10\text{--}7)$$

Electric Flux Density

Because of the presence of ϵ in Equation 10–7, the electric field intensity depends on the medium in which the charge is located. Let us define a new quantity, D, that is independent of the medium. Let

$$D = \epsilon\mathscr{E} \qquad (10\text{--}8)$$

D is known as electric **flux density.** Although not apparent here, D represents the density of flux lines in space, that is,

$$D = \frac{\text{total flux}}{\text{area}} = \frac{\psi}{A} \qquad (10\text{--}9)$$

where ψ is the flux passing through area A.

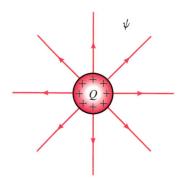

FIGURE 10–9 In the SI system, total flux ψ equals charge Q.

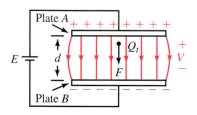

FIGURE 10–10 Work moving test charge Q_t is force times distance.

Electric Flux (Revisited)

Consider Figure 10–9. Flux ψ is due to the charge Q. Although we will not prove it, in the SI system the number of flux lines emanating from a charge Q is equal to the charge itself, that is,

$$\psi = Q \quad \text{(C)} \tag{10–10}$$

An easy way to visualize this is to think of one flux line as emanating from each positive charge on the body as shown in Figure 10–9. Then, as indicated, the total number of lines is equal to the total number of charges.

Field of a Parallel-Plate Capacitor

Now consider a parallel-plate capacitor (Figure 10–10). The field here is created by the charge distributed over its plates. Since plate A has a deficiency of electrons, it looks like a sheet of positive charge, while plate B looks like a sheet of negative charge. A positive test charge Q_t between these sheets is therefore repelled by the positive sheet and attracted by the negative sheet.

Now move the test charge from plate B to plate A. The work W required to move the charge against the force F is force times distance. Thus,

$$W = Fd \quad \text{(J)} \tag{10–11}$$

In Chapter 2, we defined voltage as work divided by charge, i.e., $V = W/Q$. Since the charge here is the test charge, Q_t, the voltage between plates A and B is

$$V = W/Q_t = (Fd)/Q_t \quad \text{(V)} \tag{10–12}$$

Now divide both sides by d. This yields $V/d = F/Q_t$. But $F/Q_t = \mathscr{E}$, from Equation 10–6. Thus,

$$\mathscr{E} = V/d \quad \text{(V/m)} \tag{10–13}$$

Equation 10–13 shows that the electric field strength between capacitor plates is equal to the voltage across the plates divided by the distance between them.

EXAMPLE 10–4   Suppose that the electric field intensity between the plates of a capacitor is 50 000 V/m when 80 V is applied:

a. What is the plate spacing if the dielectric is air? If the dielectric is ceramic?

b. What is $\mathscr{E}$ if the plate spacing is halved?

Solution

a. $\mathscr{E} = V/d$, independent of dielectric. Thus,

$$d = \frac{V}{\mathscr{E}} = \frac{80 \text{ V}}{50 \times 10^3 \text{ V/m}} = 1.6 \times 10^{-3} \text{ m}$$

b. Since $\mathscr{E} = V/d$, $\mathscr{E}$ will double to 100 000 V/m.

1. What happens to the electric field intensity of a capacitor if you do the following:
   a. Double the applied voltage?
   b. Triple the applied voltage and double the plate spacing?

2. If the electric field intensity of a capacitor with polystyrene dielectric and plate size 2 cm by 4 cm is 100 kV/m when 50 V is applied, what is its capacitance?

Answers: 1. a. Doubles   b. Increases by a factor of 1.5   2. 36.8 pF

Capacitance (Revisited)

With the above background, we can examine capacitance a bit more rigorously. Recall, $C = Q/V$. Using the above relationships yields

$$C = \frac{Q}{V} = \frac{\psi}{V} = \frac{AD}{\mathcal{E}d} = \frac{D}{\mathcal{E}}\left(\frac{A}{d}\right) = \in\frac{A}{d}$$

This is the same equation (Equation 10–3) that we developed intuitively in Section 10.2.

10.4   Dielectrics

As you saw in Figure 10–6, a dielectric increases capacitance. We now examine why. Consider Figure 10–11. For a charged capacitor, electron orbits (which are normally circular) become elliptical as electrons are attracted toward the positive (+) plate and repelled from the negative (−) plate. This makes the end of the atom nearest the positive plate appear negative while its other end appears positive. Such atoms are **polarized.** Throughout the bulk of the dielectric, the negative end of a polarized atom is adjacent to the positive end of another atom, and the effects cancel. However, at the surfaces of the dielectric, there are no atoms to cancel, and the net effect is as if a layer of negative charge exists on the surface of the dielectric at the positive plate and a layer of positive charge at the negative plate. This makes the plates appear closer, thus increasing capacitance. Materials for which the effect is largest result in the greatest increase in capacitance.

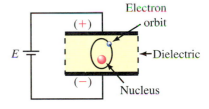

FIGURE 10–11   Effect of the capacitor's electric field on an atom of its dielectric.

TABLE 10–2   Dielectric Strength*

Material	kV/mm
Air	3
Ceramic (high $\in_r$)	3
Mica	40
Mylar	16
Oil	15
Polystyrene	24
Rubber	18
Teflon®	60

*Values depend on the composition of the material. These are the values we use in this book.

Voltage Breakdown

If the voltage of Figure 10–11 is increased beyond a critical value, the force on the electrons is so great that they are literally torn from orbit. This is called **dielectric breakdown** and the electric field intensity (Equation 10–13) at breakdown is called the **dielectric strength** of the material. For air, breakdown occurs when the voltage gradient reaches 3 kV/mm. The breakdown strengths for other materials are shown in Table 10–2. Since the quality of a dielectric depends on many factors, dielectric strength varies from sample to sample. Solid dielectrics are usually damaged by breakdown.

Breakdown is not limited to capacitors; it can occur with any type of electrical apparatus whose insulation is stressed beyond safe limits. (For example, air breaks down and flashovers occur on high-voltage transmission lines when they are struck by lightning.) The shape of conductors also affects breakdown voltage. Breakdown occurs at lower voltages at sharp points than at blunt points. This effect is made use of in lightning arresters.

EXAMPLE 10–5   A capacitor with plate dimensions of 2.5 cm by 2.5 cm and a ceramic dielectric with $\epsilon_r = 7500$ experiences breakdown at 2400 V. What is C?

Solution   From Table 10–2 dielectric strength = 3 kV/mm. Thus, $d = 2400$ V/3000 V/mm = 0.8 mm = 8×10^{-4} m. So

$$C = \epsilon_r \epsilon_o A/d$$
$$= (7500)(8.85 \times 10^{-12})(0.025 \text{ m})^2/(8 \times 10^{-4} \text{ m})$$
$$= 51.9 \text{ nF}$$

PRACTICE PROBLEMS 2

1. At what voltage will breakdown occur for a mylar dielectric capacitor with plate spacing of 0.25 cm?

2. An air-dielectric capacitor breaks down at 500 V. If the plate spacing is doubled and the capacitor is filled with oil, at what voltage will breakdown occur?

Answers: 1. 40 kV   2. 5 kV

Capacitor Voltage Rating

Because of dielectric breakdown, capacitors are rated for maximum operating voltage (called **working voltage**) by their manufacturer (indicated on the capacitor as WVDC or **working voltage dc**). If you operate a capacitor beyond its working voltage, you may damage it.

10.5   Nonideal Effects

So far, we have assumed ideal capacitors. However, real capacitors have nonideal characteristics.

Leakage Current

When a charged capacitor is disconnected from its source, it will eventually discharge. This is because no insulator is perfect and a small amount of charge "leaks" through the dielectric. Similarly, a small leakage current will pass through its dielectric when a capacitor is connected to a source.

The effect of leakage is modeled by a resistor in Figure 10–12. Since leakage is very small, R is very large, typically hundreds of megohms. The larger R is, the longer a capacitor can hold its charge. For most applications, leakage can be neglected.

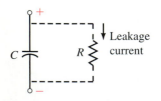

FIGURE 10–12   Leakage current.

Equivalent Series Resistance (ESR)

As a capacitor ages, resistance may develop in its leads as its internal connections begin to fail. This resistance is in series with the capacitor and may eventually cause problems.

Dielectric Absorption

When a capacitor is discharged by temporarily shorting its leads, it should have zero volts when the short is removed. However, atoms sometimes remain partially polarized, and when the short is removed, they cause a residual voltage to appear across the capacitor. This effect is known as **dielectric absorption.** In electronic circuits, the voltage due to dielectric absorption can upset circuit voltage levels; in TV tubes and electrical power apparatus, it can result in large and potentially dangerous voltages.

Temperature Coefficient

Because dielectrics are affected by temperature, capacitance may change with temperature. If capacitance increases with increasing temperature, the capacitor is said to have a **positive temperature coefficient;** if it decreases, the capacitor has a **negative temperature coefficient;** if it remains essentially constant, the capacitor has a **zero temperature coefficient.**

   The temperature coefficient is specified as a change in capacitance in parts per million (ppm) per degree Celsius. Consider a 1-μF capacitor. Since 1 μF = 1 million pF, 1 ppm is 1 pF. Thus, a 1-μF capacitor with a temperature coefficient of 200 ppm/°C could change as much as 200 pF per degree Celsius.

10.6   Types of Capacitors

Since no single capacitor type suits all applications, capacitors are made in a variety of types and sizes. Among these are fixed and variable types with differing dielectrics and recommended areas of application.

Fixed Capacitors

Fixed capacitors are often identified by their dielectric. Common dielectric materials include ceramic, plastic, and mica, plus, for electrolytic capacitors, aluminum and tantalum oxide. Design variations include tubular and interleaved plates. The interleave design (Figure 10–13) uses multiple plates to increase effective plate area. A layer of insulation separates plates, and alternate plates are connected together. The tubular design (Figure 10–14) uses sheets of metal foil separated by an insulator such as plastic film. Fixed capacitors are encapsulated in plastic, epoxy resin, or other insulating material and identified with value, tolerance, and other appropriate data either via body markings or color coding. Electrical characteristics and physical size depend on the dielectric used.

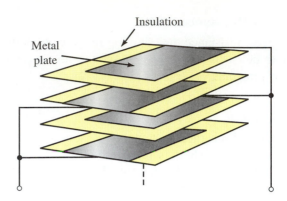

FIGURE 10–13   Stacked capacitor construction. The stack is compressed, leads attached, and the unit coated with epoxy resin or other insulating material.

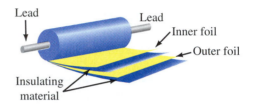

FIGURE 10–14   Tubular capacitor with axial leads.

Ceramic Capacitors

First, consider ceramic. The permittivity of ceramic varies widely (as indicated in Table 10–1). At one end are ceramics with extremely high permittivity. These permit packaging a great deal of capacitance in a small space, but yield capacitors whose characteristics vary widely with temperature and operating voltage. However, they are popular in limited temperature applications where small size and cost are important. At the other end are ceramics with highly stable characteristics. They yield capacitors whose values change little with temperature, voltage, or aging. However, since their dielectric constants are relatively low (typically 30 to 80), these capacitors are physically larger than those made using high-permittivity ceramic. Many surface mount capacitors (considered later in this section) use ceramic dielectrics.

Plastic Film Capacitors

Plastic-film capacitors are of two basic types: film/foil or metalized film. **Film/foil** capacitors use metal foil separated by plastic film as in Figure 10–14, while **metallized-film** capacitors have their foil material vacuum-deposited directly onto plastic film. Film/foil capacitors are generally larger than metallized-foil units, but have better capacitance stability and higher insulation resistance. Typical film materials are polyester, Mylar, polypropylene, and polycarbonate. Figure 10–15 shows a selection of plastic-film capacitors.

Metalized-film capacitors are **self-healing.** Thus, if voltage stress at an imperfection exceeds breakdown, an arc occurs which evaporates the metal-

FIGURE 10–15   Radial lead film capacitors. *(Courtesy Illinois Capacitor Inc.)*

lized area around the fault, isolating the defect. (Film/foil capacitors are not self-healing.)

Mica Capacitors

Mica capacitors are low in cost with low leakage and good stability. Available values range from a few picofarads to about 0.1 μF.

Electrolytic Capacitors

Electrolytic capacitors provide large capacitance (i.e., up to several hundred thousand microfarads) at a relatively low cost. (Their capacitance is large because they have a very thin layer of oxide as their dielectric.) However, their leakage is relatively high and breakdown voltage relatively low. Electrolytics have either aluminum or tantalum as their plate material. Tantalum devices are smaller than aluminum devices, have less leakage, and are more stable.

The basic aluminum electrolytic capacitor construction is similar to that of Figure 10–14, with strips of aluminum foil separated by gauze saturated with an electrolyte. During manufacture, chemical action creates a thin oxide layer that acts as the dielectric. This layer must be maintained during use. For this reason, electrolytic capacitors are polarized (marked with a + and − sign), and the plus (+) terminal must always be kept positive with respect to the minus (−) terminal. Electrolytic capacitors have a **shelf life;** that is, if they are not used for an extended period, they may fail when powered up again. Figure 10–16 shows a selection of aluminum electrolytic devices.

FIGURE 10–16   Radial lead aluminum electrolytic capacitors. *(Courtesy Illinois Capacitor Inc.)*

Tantalum capacitors come in two basic types: wet slug and solid dielectric. Figure 10–17 shows a cutaway view of a solid tantalum unit. The slug, made from powdered tantalum, is highly porous and provides a large internal surface area that is coated with an oxide to form the dielectric. Tantalum capacitors are polarized and must be inserted into a circuit properly.

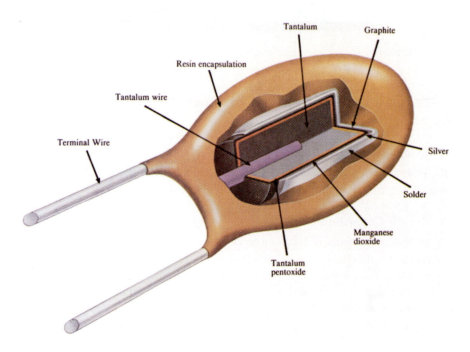

FIGURE 10–17   Cutaway view of a solid tantalum capacitor. *(Courtesy AVX Corporation)*

Surface Mount Capacitors

Many electronic products now use **surface mount devices** (SMDs). (SMDs do not have connection leads, but are soldered directly onto printed circuit boards.) Figure 10–18 shows a surface mount, ceramic chip capacitor. Such devices are extremely small and provide high packaging density.

Variable Capacitors

The most common variable capacitor is that used in radio tuning circuits (Figure 10–19). It has a set of stationary plates and a set of movable plates which are ganged together and mounted on a shaft. As the shaft is rotated, the movable plates mesh with the stationary plates, changing the effective surface area (and hence the capacitance).

Another adjustable type is the **trimmer** or **padder** capacitor, which is used for fine adjustments, usually over a very small range. In contrast to the variable capacitor (which is frequently varied by the user), a trimmer is usually set to its required value, then never touched again.

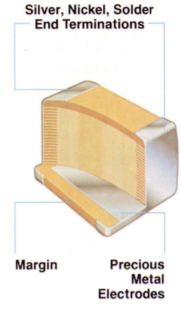

Silver, Nickel, Solder
End Terminations

Margin   Precious
                Metal
                Electrodes

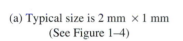

Terminations

(a) Typical size is 2 mm × 1 mm   (b) Cutaway View
    (See Figure 1–4)   (*Courtesy AVX Corporation*)

FIGURE 10–18   Surface mount, ceramic chip capacitor.

(a) Variable capacitor of type used in radios   (b) Symbol

FIGURE 10–19

A 2.5-μF capacitor has a tolerance of $+80\%$ and -20%. Determine what its maximum and minimum values could be.

PRACTICE
PROBLEMS 3

Answer: 4.5 μF and 2 μF

10.7 Capacitors in Parallel and Series

Capacitors in Parallel

For capacitors in parallel, the effective plate area is the sum of the individual plate areas; thus, the total capacitance is the sum of the individual capacitances. This is easily shown. Consider Figure 10–20. The charge on each capacitor is given by Equation 10–1. Thus, $Q_1 = C_1 V$ and $Q_2 = C_2 V$. Since

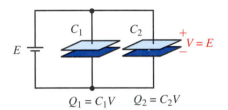

$Q_1 = C_1 V \quad Q_2 = C_2 V$

(a) Parallel capacitors

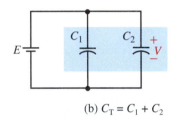

(b) $C_T = C_1 + C_2$

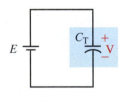

(c) Equivalent

FIGURE 10–20 Capacitors in parallel. Total capacitance is the sum of the individual capacitances.

$Q_T = Q_1 + Q_2$, $Q_T = C_1 V + C_2 V = (C_1 + C_2)V$. But $Q_T = C_T V$. Thus, $C_T = C_1 + C_2$. For more than two capacitors,

$$C_T = C_1 + C_2 + \cdots + C_N \tag{10–14}$$

That is, *the total capacitance of capacitors in parallel is the sum of their individual capacitances.*

EXAMPLE 10–6   A 10-μF, a 15-μF, and a 100-μF capacitor are connected in parallel across a 50-V source. Determine the following:

a. Total capacitance.

b. Total charge stored.

c. Charge on each capacitor.

Solution

a. $C_T = C_1 + C_2 + C_3 = 10\ \mu F + 15\ \mu F + 100\ \mu F = 125\ \mu F$

b. $Q_T = C_T V = (125\ \mu F)(50\ V) = 6.25\ mC$

c. $Q_1 = C_1 V = (10\ \mu F)(50\ V) = 0.5\ mC$

    $Q_2 = C_2 V = (15\ \mu F)(50\ V) = 0.75\ mC$

    $Q_3 = C_3 V = (100\ \mu F)(50\ V) = 5.0\ mC$

Check: $Q_T = Q_1 + Q_2 + Q_3 = (0.5 + 0.75 + 5.0)\ mC = 6.25\ mC$.

---

1. Three capacitors are connected in parallel. If $C_1 = 20\ \mu F$, $C_2 = 10\ \mu F$ and $C_T = 32.2\ \mu F$, what is C_3?

2. Three capacitors are paralleled across an 80-V source, with $Q_T = 0.12$ C. If $C_1 = 200\ \mu F$ and $C_2 = 300\ \mu F$, what is C_3?

3. Three capacitors are paralleled. If the value of the second capacitor is twice that of the first and the value of the third is one quarter that of the second and the total capacitance is 70 μF, what are the values of each capacitor?

Answers: 1. 2.2 μF   2. 1000 μF   3. 20 μF, 40 μF and 10 μF

**PRACTICE
PROBLEMS 4**

Capacitors in Series

For capacitors in series (Figure 10–21), the same charge appears on each. Thus, $Q = C_1 V_1$, $Q = C_2 V_2$, etc. Solving for voltages yields $V_1 = Q/C_1$, $V_2 = Q/C_2$, and so on. Applying KVL, we get $V = V_1 + V_2 + \cdots + V_N$. Therefore,

$$V = \frac{Q}{C_1} + \frac{Q}{C_2} + \cdots + \frac{Q}{C_N} = Q\left(\frac{1}{C_1} + \frac{1}{C_2} + \cdots + \frac{1}{C_N}\right)$$

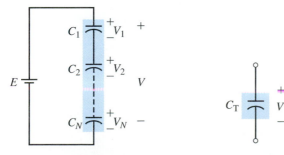

(a) Series connection

(b) Equivalent

FIGURE 10–21   Capacitors in series: $\dfrac{1}{C_T} = \dfrac{1}{C_1} + \dfrac{1}{C_2} + \cdots + \dfrac{1}{C_N}$.

But $V = Q/C_T$. Equating this with the right side and cancelling Q yields

$$\frac{1}{C_T} = \frac{1}{C_1} + \frac{1}{C_2} + \cdots + \frac{1}{C_N}$$

     (10–15)

For two capacitors in series, this reduces to

$$C_T = \frac{C_1 C_2}{C_1 + C_2}$$

     (10–16)

For N equal capacitors in series, Equation 10–15 yields $C_T = C/N$.

NOTES...

1. For capacitors in parallel, total capacitance is always larger than the largest capacitance, while for capacitors in series, total capacitance is always smaller than the smallest capacitance.

2. The formula for capacitors in parallel is similar to the formula for resistors in series, while the formula for capacitors in series is similar to the formula for resistors in parallel.

EXAMPLE 10–7   Refer to Figure 10–22(a):

a. Determine C_T.

b. If 50 V is applied across the capacitors, determine Q.

c. Determine the voltage on each capacitor.

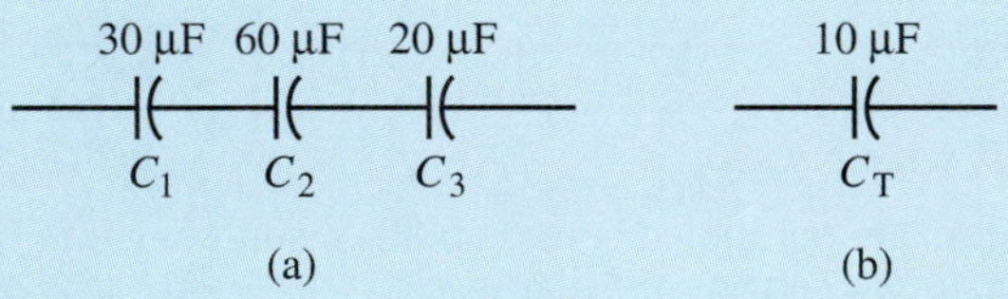

FIGURE 10–22

Solution

a. $\dfrac{1}{C_T} = \dfrac{1}{C_1} + \dfrac{1}{C_2} + \dfrac{1}{C_3} = \dfrac{1}{30\ \mu F} + \dfrac{1}{60\ \mu F} + \dfrac{1}{20\ \mu F}$

$= 0.0333 \times 10^6 + 0.0167 \times 10^6 + 0.05 \times 10^6 = 0.1 \times 10^6$

Therefore as indicated in (b),

$$C_T = \frac{1}{0.1 \times 10^6} = 10\ \mu F$$

b. $Q = C_T V = (10 \times 10^{-6}\ F)(50\ V) = 0.5\ mC$

c. $V_1 = Q/C_1 = (0.5 \times 10^{-3}\ C)/(30 \times 10^{-6}\ F) = 16.7\ V$

     $V_2 = Q/C_2 = (0.5 \times 10^{-3}\ C)/(60 \times 10^{-6}\ F) = 8.3\ V$

     $V_3 = Q/C_3 = (0.5 \times 10^{-3}\ C)/(20 \times 10^{-6}\ F) = 25.0\ V$

Check: $V_1 + V_2 + V_3 = 16.7 + 8.3 + 25 = 50\ V$.

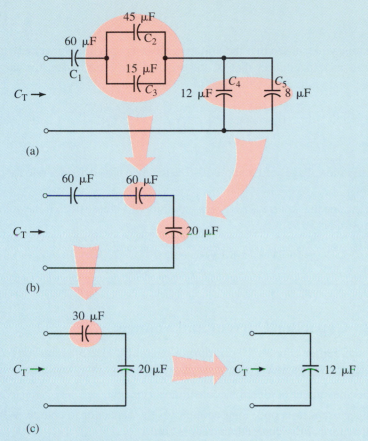

EXAMPLE 10–8   For the circuit of Figure 10–23(a), determine C_T.

(a)

(b)

(c)

FIGURE 10–23   Systematic reduction.

Solution   The problem is easily solved through step-by-step reduction. C_2 and C_3 in parallel yield 45 μF + 15 μF = 60 μF. C_4 and C_5 in parallel total 20 μF. The reduced circuit is shown in (b). The two 60-μF capacitances in series reduce to 30 μF. The series combination of 30 μF and 20 μF can be found from Equation 10–16. Thus,

$$C_T = \frac{30 \ \mu F \times 20 \ \mu F}{30 \ \mu F + 20 \ \mu F} = 12 \ \mu F$$

Alternately, you can reduce (b) directly using Equation 10–15. Try it.

Voltage Divider Rule for Series Capacitors

For capacitors in series (Figure 10–24) a simple voltage divider rule can be developed. Recall, for individual capacitors, $Q_1 = C_1V_1$, $Q_2 = C_2V_2$, etc., and for the complete string, $Q_T = C_TV_T$. As noted earlier, $Q_1 = Q_2 = \cdots = Q_T$. Thus, $C_1V_1 = C_TV_T$. Solving for V_1 yields

$$V_1 = \left(\frac{C_T}{C_1}\right)V_T$$

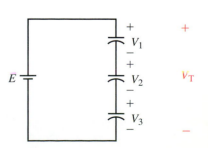

FIGURE 10–24   Capacitive voltage divider.

This type of relationship holds for all capacitors. Thus,

$$V_x = \left(\frac{C_T}{C_x}\right)V_T \tag{10–17}$$

From this, you can see that the voltage across a capacitor is inversely proportional to its capacitance, that is, the smaller the capacitance, the larger the voltage, and vice versa. Other useful variations are

$$V_1 = \left(\frac{C_2}{C_1}\right)V_2, \qquad V_1 = \left(\frac{C_3}{C_1}\right)V_3, \qquad V_2 = \left(\frac{C_3}{C_2}\right)V_3, \qquad \text{etc.}$$

PRACTICE PROBLEMS 5

1. Verify the voltages of Example 10–7 using the voltage divider rule for capacitors.

2. Determine the voltage across each capacitor of Figure 10–23 if the voltage across C_5 is 30 V.

Answers: 1. $V_1 = 16.7$ V $V_2 = 8.3$ V $V_3 = 25.0$ V   2. $V_1 = 10$ V $V_2 = V_3 = 10$ V $V_4 = V_5 = 30$ V

10.8   Capacitor Current and Voltage

As noted earlier (Figure 10–2), during charging, electrons are moved from one plate of a capacitor to the other plate. Several points should be noted.

1. This movement of electrons constitutes a current.

2. This current lasts only long enough for the capacitor to charge. When the capacitor is fully charged, current is zero.

3. Current in the circuit during charging is due solely to the movement of electrons from one plate to the other around the external circuit; no current passes through the dielectric between the plates.

4. As charge is deposited on the plates, the capacitor voltage builds. However, this voltage does not jump to full value immediately since it takes time to move electrons from one plate to the other. (Billions of electrons must be moved.)

5. Since voltage builds up as charging progresses, the difference in voltage between the source and the capacitor decreases and hence the rate of movement of electrons (i.e., the current) decreases as the capacitor approaches full charge.

Figure 10–25 shows what the voltage and current look like during the charging process. As indicated, the current starts out with an initial surge, then decays to zero while the capacitor voltage gradually climbs from zero to full voltage. The charging time typically ranges from nanoseconds to milliseconds, depending on the resistance and capacitance of the circuit. (We study these relationships in detail in Chapter 11.) A similar surge (but in the opposite direction) occurs during discharge.

As Figure 10–25 indicates, current exists only while the capacitor voltage is changing. This observation turns out to be true in general, that is,

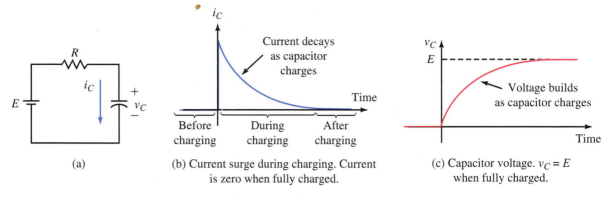

(a)

(b) Current surge during charging. Current is zero when fully charged.

(c) Capacitor voltage. $v_C = E$ when fully charged.

FIGURE 10–25 The capacitor does not charge instantaneously, as a finite amount of time is required to move electrons around the circuit.

current in a capacitor exists only while capacitor voltage is changing. The reason is not hard to understand. As you saw before, a capacitor's dielectric is an insulator and consequently no current can pass through it (assuming zero leakage). The only charges that can move, therefore, are the free electrons that exist on the capacitor's plates. When capacitor voltage is constant, these charges are in equilibrium, no net movement of charge occurs, and the current is thus zero. However, if the source voltage is increased, additional electrons are pulled from the positive plate; inversely, if the source voltage is decreased, excess electrons on the negative plate are returned to the positive plate. Thus, in both cases, capacitor current results when capacitor voltage is changed. As we show next, this current is proportional to the rate of change of voltage. Before we do this, however, we need to look at symbols.

Symbols for Time-Varying Voltages and Currents

Quantities that vary with time are called **instantaneous** quantities. *Standard industry practice requires that we use lowercase letters for time-varying quantities, rather than capital letters as for dc.* Thus, we use v_C and i_C to represent changing capacitor voltage and current rather than V_C and I_C. (Often we drop the subscripts and just use v and i.) Since these quantities are functions of time, they may also be shown as $v_C(t)$ and $i_C(t)$.

Capacitor *v-i* Relationship

The relationship between charge and voltage for a capacitor is given by Equation 10–1. For the time-varying case, it is

$$q = Cv_C \qquad \textbf{(10–18)}$$

But current is the rate of movement of charge. In calculus notation, this is $i_C = dq/dt$. Differentiating Equation 10–18 yields

$$i_C = \frac{dq}{dt} = \frac{d}{dt}(Cv_C) \qquad \textbf{(10–19)}$$

NOTES...

Calculus is introduced at this point to aid in the development of ideas and to help explain concepts. However, not everyone who uses this book requires calculus. Therefore, the material is presented in such a manner that it never relies entirely on mathematics; thus, where calculus is used, intuitive explanations accompany it. However, to provide the enrichment that calculus offers, optional derivations and problems are included, but they are marked with a **J** icon so that they may be omitted if desired.

Since C is constant, we get

$$i_C = C\frac{dv_C}{dt} \quad \text{(A)} \tag{10–20}$$

Equation 10–20 shows that *current through a capacitor is equal to C times the rate of change of voltage across it*. This means that the faster the voltage changes, the larger the current, and vice versa. It also means that if the voltage is constant, the current is zero (as we noted earlier).

Reference conventions for voltage and current are shown in Figure 10–26. As usual, the plus sign goes at the tail of the current arrow. If the voltage is increasing, dv_C/dt is positive and the current is in the direction of the reference arrow; if the voltage is decreasing, dv_C/dt is negative and the current is opposite to the arrow.

The derivative dv_C/dt of Equation 10–20 is the slope of the capacitor voltage versus time curve. When capacitor voltage varies linearly with time (i.e., the relationship is a straight line as in Figure 10–27), Equation 10–20 reduces to

$$i_C = C\frac{\Delta v_C}{\Delta t} = C\frac{\text{rise}}{\text{run}} = C \times \text{slope of the line} \tag{10–21}$$

FIGURE 10–26   The + sign for v_C goes at the tail of the current arrow.

EXAMPLE 10–9   A signal generator applies voltage to a 5-μF capacitor with a waveform as in Figure 10–27(a). The voltage rises linearly from 0 to 10 V in 1 ms, falls linearly to -10 V at $t = 3$ ms, remains constant until $t = 4$ ms, rises to 10 V at $t = 5$ ms, and remains constant thereafter.

a. Determine the slope of v_C in each time interval.

b. Determine the current and sketch its graph.

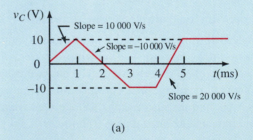

(a)

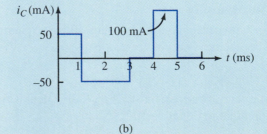

(b)

FIGURE 10–27

Solution

a. We need the slope of v_C during each time interval where slope = rise/run = $\Delta v/\Delta t$.

0 ms to 1 ms: $\Delta v = 10$ V; $\Delta t = 1$ ms; Therefore, slope = 10 V/1 ms = 10 000 V/s.

1 ms to 3 ms: Slope = -20 V/2 ms = $-10 000$ V/s.

3 ms to 4 ms: Slope = 0 V/s.

4 ms to 5 ms: Slope = 20 V/1 ms = 20 000 V/s.

b. $i_C = Cdv_C/dt = C$ times slope. Thus,

0 ms to 1 ms: $i = (5 \times 10^{-6}$ F)(10 000 V/s) = 50 mA.

1 ms to 3 ms: $i = -(5 \times 10^{-6}$ F)(10 000 V/s) = -50 mA.

3 ms to 4 ms: $i = (5 \times 10^{-6}$ F)(0 V/s) = 0 A.

4 ms to 5 ms: $i = (5 \times 10^{-6}$ F)(20 000 V/s) = 100 mA.

The current is plotted in Figure 10–27(b).

EXAMPLE 10–10   The voltage across a 20-μF capacitor is $v_C = 100 \, t \, e^{-t}$ V. Determine current i_C.

Solution   Differentiation by parts using $\dfrac{d(uv)}{dt} = u\dfrac{dv}{dt} + v\dfrac{du}{dt}$ with $u = 100 \, t$ and $v = e^{-t}$ yields

$$i_C = C\frac{d}{dt}(100 \, t \, e^{-t}) = 100 \, C\frac{d}{dt}(t \, e^{-t}) = 100 \, C\left(t\frac{d}{dt}(e^{-t}) + e^{-t}\frac{dt}{dt}\right)$$

$$= 2000 \times 10^{-6}(-t \, e^{-t} + e^{-t}) \text{ A} = 2.0 \, (1 - t)e^{-t} \text{ mA}$$

10.9   Energy Stored by a Capacitor

An ideal capacitor does not dissipate power. When power is transferred to a capacitor, all of it is stored as energy in the capacitor's electric field. When the capacitor is discharged, this stored energy is returned to the circuit.

To determine the stored energy, consider Figure 10–28. Power is given by $p = vi$ watts. Using calculus (see), it can be shown that the stored energy is given by

$$W = \frac{1}{2}CV^2 \quad \text{(J)} \qquad \qquad \text{(10–22)}$$

where V is the voltage across the capacitor. This means that the energy at any time depends on the value of the capacitor's voltage at that time.

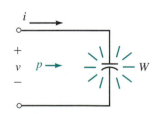

FIGURE 10–28   Storing energy in a capacitor.

Deriving Equation 10–22

Power to the capacitor (Figure 10–28) is given by $p = vi$, where $i = Cdv/dt$. Therefore, $p = Cvdv/dt$. However, $p = dW/dt$. Integrating both sides yields

$$W = \int_0^t pdt = C\int_0^t v\frac{dv}{dt} \, dt = C\int_0^V vdv = \frac{1}{2}CV^2$$

10.10 Capacitor Failures and Troubleshooting

Although capacitors are quite reliable, they may fail because of misapplication, excessive voltage, current, temperature, or simply because they age. They can short internally, leads may become open, dielectrics may become excessively leaky, and they may fail catastrophically due to incorrect use. (If an electrolytic capacitor is connected with its polarity reversed, for example, it may explode.) Capacitors should be used well within their rating limits. Excessive voltage can lead to dielectric puncture creating pinholes that short the plates together. High temperatures may cause an increase in leakage and/or a permanent shift in capacitance. High temperatures may be caused by inadequate heat removal, excessive current, lossy dielectrics, or an operating frequency beyond the capacitor's rated limit.

Basic Testing with an Ohmmeter

Some basic (out-of-circuit) tests can be made with an analog ohmmeter. The ohmmeter can detect opens and shorts and, to a certain extent, leaky dielectrics. First, ensure that the capacitor is discharged, then set the ohmmeter to its highest range and connect it to the capacitor. (For electrolytic devices, ensure that the plus (+) side of the ohmmeter is connected to the plus (+) side of the capacitor.)

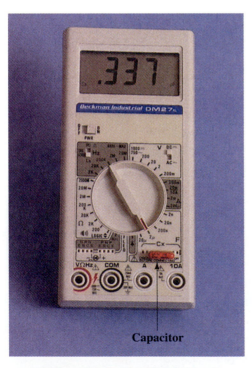

(a) Measuring C with a DMM. (Not all DMMs can measure capacitance)

(b) Capacitor/inductor analyzer. *(Courtesy B + K Precision)*

FIGURE 10–29 Capacitor testing.

Initially, the ohmmeter reading should be low, then for a good capacitor gradually increase to infinity as the capacitor charges through the ohmmeter circuit. (Or at least a very high value, since most good capacitors, except electrolytics, have a resistance of hundreds of megohms.) For small capacitors, however, the time to charge may be too short to yield useful results.

Faulty capacitors respond differently. If a capacitor is shorted, the meter resistance reading will stay low. If it is leaky, the reading will be lower than normal. If it is open circuited, the meter will indicate infinity immediately, without dipping to zero when first connected.

Capacitor Testers

Ohmmeter testing of capacitors has its limitations; other tools may be needed. Figure 10–29 shows two of them. The DMM in (a) can measure capacitance and display it directly on its readout. The LCR (inductance, capacitance, resistance) analyzer in (b) can determine capacitance as well as detect opens and shorts. More sophisticated testers are available that determine capacitance value, leakage at rated voltage, dielectric absorption, and so on.

10.1   Capacitance

PROBLEMS

1. For Figure 10–30, determine the charge on the capacitor, its capacitance, or the voltage across it as applicable for each of the following.
   a. $E = 40$ V, $C = 20$ μF
   b. $V = 500$ V, $Q = 1000$ μC
   c. $V = 200$ V, $C = 500$ nF
   d. $Q = 3 \times 10^{-4}$ C, $C = 10 \times 10^{-6}$ F
   e. $Q = 6$ mC, $C = 40$ μF
   f. $V = 1200$ V, $Q = 1.8$ mC

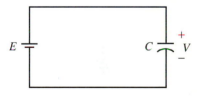

FIGURE 10–30

2. Repeat Question 1 for the following:
   a. $V = 2.5$ kV, $Q = 375$ μC
   b. $V = 1.5$ kV, $C = 0.04 \times 10^{-4}$ F
   c. $V = 150$ V, $Q = 6 \times 10^{-5}$ C
   d. $Q = 10$ μC, $C = 400$ nF
   e. $V = 150$ V, $C = 40 \times 10^{-5}$ F
   f. $Q = 6 \times 10^{-9}$ C, $C = 800$ pF

3. The charge on a 50-μF capacitor is 10×10^{-3} C. What is the potential difference between its terminals?

4. When 10 μC of charge is placed on a capacitor, its voltage is 25 V. What is the capacitance?

5. You charge a 5-μF capacitor to 150 V. Your lab partner then momentarily places a resistor across its terminals and bleeds off enough charge that its voltage falls to 84 V. What is the final charge on the capacitor?

10.2   Factors Affecting Capacitance

6. A capacitor with circular plates 0.1 m in diameter and an air dielectric has 0.1 mm spacing between its plates. What is its capacitance?

7. A parallel-plate capacitor with a mica dielectric has dimensions of 1 cm $\times$ 1.5 cm and separation of 0.1 mm. What is its capacitance?

8. For the capacitor of Problem 7, if the mica is removed, what is its new capacitance?

9. The capacitance of an oil-filled capacitor is 200 pF. If the separation between its plates is 0.1 mm, what is the area of its plates?

10. A 0.01-μF capacitor has ceramic with a dielectric constant of 7500. If the ceramic is removed, the plate separation doubled, and the spacing between plates filled with oil, what is the new value for *C?*

11. A capacitor with a Teflon dielectric has a capacitance of 33 μF. A second capacitor with identical physical dimensions but with a Mylar dielectric carries a charge of 55×10^{-4} C. What is its voltage?

12. The plate area of a capacitor is 4.5 in^2. and the plate separation is 5 mils. If the relative permittivity of the dielectric is 80, what is *C?*

10.3   Electric Fields

13. a. What is the electric field strength $\mathscr{E}$ at a distance of 1 cm from a 100-mC charge in transformer oil?

    b. What is $\mathscr{E}$ at twice the distance?

14. Suppose that 150 V is applied across a 100-pF parallel-plate capacitor whose plates are separated by 1 mm. What is the electric field intensity $\mathscr{E}$ between the plates?

10.4   Dielectrics

15. An air-dielectric capacitor has plate spacing of 1.5 mm. How much voltage can be applied before breakdown occurs?

16. Repeat Problem 15 if the dielectric is mica and the spacing is 2 mils.

17. A mica-dielectric capacitor breaks down when *E* volts is applied. The mica is removed and the spacing between plates doubled. If breakdown now occurs at 500 V, what is *E?*

18. Determine at what voltage the dielectric of a 200 nF Mylar capacitor with a plate area of 0.625 m^2 will break down.

19. Figure 10–31 shows several gaps, including a parallel-plate capacitor, a set of small spherical points, and a pair of sharp points. The spacing is the same for each. As the voltage is increased, which gap breaks down for each case?

20. If you continue to increase the source voltage of Figures 10–31(a), (b), and (c) after a gap breaks down, will the second gap also break down? Justify your answer.

10.5   Nonideal Effects

21. A 25-μF capacitor has a negative temperature coefficient of 175 ppm/°C. By how much and in what direction might it vary if the temperature rises by 50°C? What would be its new value?

22. If a 4.7-μF capacitor changes to 4.8 μF when the temperature rises 40°C, what is its temperature coefficient?

10.7   Capacitors in Parallel and Series

23. What is the equivalent capacitance of 10 μF, 12 μF, 22 μF, and 33 μF connected in parallel?

24. What is the equivalent capacitance of 0.10 μF, 220 nF, and 4.7×10^{-7} F connected in parallel?

25. Repeat Problem 23 if the capacitors are connected in series.

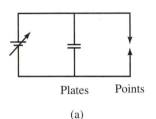

Plates   Points

(a)

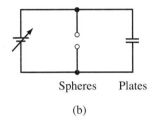

Spheres   Plates

(b)

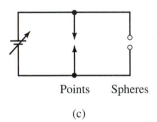

Points   Spheres

(c)

FIGURE 10–31   Source voltage is increased until one of the gaps breaks down. (The source has high internal resistance to limit current following breakdown.)

26. Repeat Problem 24 if the capacitors are connected in series.
27. Determine C_T for each circuit of Figure 10–32.
28. Determine total capacitance looking in at the terminals for each circuit of Figure 10–33.

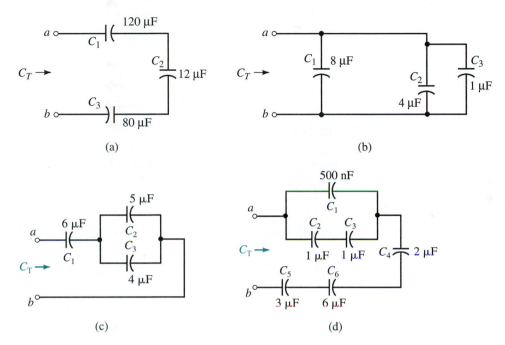

FIGURE 10–32

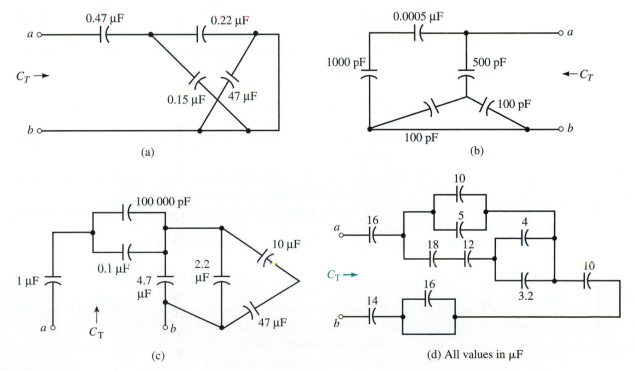

FIGURE 10–33

29. A 30-μF capacitor is connected in parallel with a 60-μF capacitor, and a 10-μF capacitor is connected in series with the parallel combination. What is C_T?

30. For Figure 10–34, determine C_x.

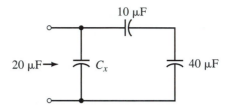

FIGURE 10–34

31. For Figure 10–35, determine C_3 and C_4.

32. For Figure 10–36, determine C_T.

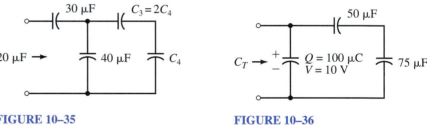

FIGURE 10–35   **FIGURE 10–36**

33. You have capacitors of 22 μF, 47 μF, 2.2 μF and 10 μF. Connecting these any way you want, what is the largest equivalent capacitance you can get? The smallest?

34. A 10-μF and a 4.7-μF capacitor are connnected in parallel. After a third capacitor is added to the circuit, $C_T = 2.695$ μF. What is the value of the third capacitor? How is it connected?

35. Consider capacitors of 1 μF, 1.5 μF, and 10 μF. If $C_T = 10.6$ μF, how are the capacitors connected?

36. For the capacitors of Problem 35, if $C_T = 2.304$ μF, how are the capacitors connected?

37. For Figures 10–32(c) and (d), find the voltage on each capacitor if 100 V is applied to terminals a-b.

38. Use the voltage divider rule to find the voltage across each capacitor of Figure 10–37.

39. Repeat Problem 38 for the circuit of Figure 10–38.

40. For Figure 10–39, $V_x = 50$ V. Determine C_x and C_T.

41. For Figure 10–40, determine C_x.

42. A dc source is connected to terminals a-b of Figure 10–34. If C_x is 12 μF and the voltage across the 40-μF capacitor is 80 V,

   a. What is the source voltage?

   b. What is the total charge on the capacitors?

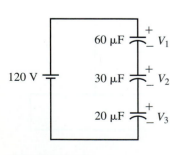

FIGURE 10–37

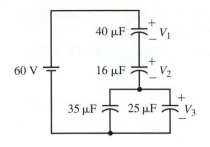

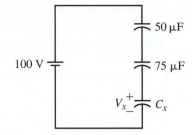

FIGURE 10–38   **FIGURE 10–39**

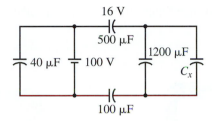

FIGURE 10–40

    c. What is the charge on each individual capacitor?

10.8 Capacitor Current and Voltage

43. The voltage across the capacitor of Figure 10–41(a) is shown in (b). Sketch current i_C scaled with numerical values.

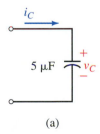

(a)

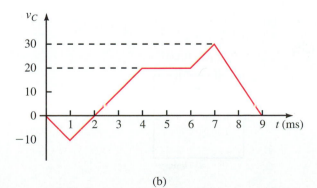

(b)

FIGURE 10–41

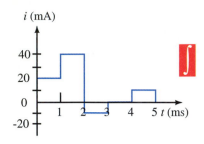

FIGURE 10–42

44. The current through a 1-μF capacitor is shown in Figure 10–42. Sketch voltage v_C scaled with numerical values. Voltage at $t = 0$ s is 0 V.

45. If the voltage across a 4.7-μF capacitor is $v_C = 100e^{-0.05t}$ V, what is i_C?

10.9   Energy Stored by a Capacitor

46. For the circuit of Figure 10–37, determine the energy stored in each capacitor.

47. For Figure 10–41, determine the capacitor's energy at each of the following times: $t = 0$, 1 ms, 4 ms, 5 ms, 7 ms, and 9 ms.

10.10   Capacitor Failures and Troubleshooting

48. For each case shown in Figure 10–43, what is the likely fault?

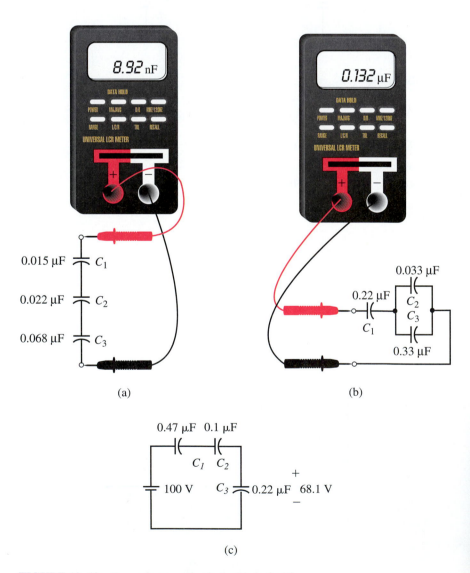

FIGURE 10–43   For each case, what is the likely fault?

In-Process Learning Check 1

1. a. 2.12 nF

   b. 200 V

2. It becomes 6 times larger.

3. 1.1 nF

4. mica

11

Capacitor Charging, Discharging, and Simple Waveshaping Circuits

OBJECTIVES

After studying this chapter, you will be able to

- explain why transients occur in *RC* circuits,
- explain why an uncharged capacitor looks like a short circuit when first energized,
- describe why a capacitor looks like an open circuit to steady state dc,
- describe charging and discharging of simple *RC* circuits with dc excitation,
- determine voltages and currents in simple *RC* circuits during charging and discharging,
- plot voltage and current transients,
- understand the part that time constants play in determining the duration of transients,
- compute time constants,
- describe the use of charging and discharging waveforms in simple timing applications,
- calculate the pulse response of simple *RC* circuits,
- solve simple *RC* transient problems using Electronics Workbench and PSpice.

KEY TERMS

Capacitive Loading
Exponential Functions
Initial Conditions
Pulse
Pulse Width (t_p)
Rise and Fall Times (t_r, t_f)
Step Voltages
Time Constant ($\tau = RC$)
Transient
Transient Duration (5τ)

OUTLINE

Introduction
Capacitor Charging Equations
Capacitor with an Initial Voltage
Capacitor Discharging Equations
More Complex Circuits
An *RC* Timing Application
Pulse Response of *RC* Circuits
Transient Analysis Using Computers

As you saw in Chapter 10, capacitors do not charge or discharge instantaneously. Instead, as illustrated in Figure 10–25, voltages and currents take time to reach their new values. The time taken to reach these new values (i.e., the charge and discharge times) are dependent on the resistance and capacitance of the circuit. During charge, for example, a capacitor charges at a rate determined by its capacitance and the resistance through which it charges, while during discharge, it discharges at a rate determined by its capacitance and the resistance through which it discharges. Since the voltages and currents that exist during these charging and discharging times are transitory in nature, they are called **transients.** Transients do not last very long, typically only a fraction of a second. However, they are important to us for a number of reasons, some of which you will learn in this chapter.

Transients occur in both capacitive and inductive circuits. In capacitive circuits, they occur because capacitor voltage cannot change instantaneously; in inductive circuits, they occur because inductor current cannot change instantaneously. In this chapter, we look at capacitive transients; in Chapter 14, we look at inductive transients. As you will see, many of the basic principles are the same.

Note: Optional problems and derivations using calculus are marked by a ∫ *icon. They may be omitted without loss of continuity by those who do not require calculus.*

Desirable and Undesirable Transients

TRANSIENTS OCCUR IN CAPACITIVE and inductive circuits whenever circuit conditions are changed, for example, by the sudden application of a voltage, the switching in or out of a circuit element, or the malfunctioning of a circuit component. Some transients are desirable and useful; others occur under abnormal conditions and are potentially destructive in nature.

An example of the latter is the transient that results when lightning strikes a power line. Following a strike, the line voltage, which may have been only a few thousand volts before the strike, momentarily rises to many hundreds of thousands of volts or higher, then rapidly decays, while the current, which may have been only a few hundred amps, suddenly rises to many times its normal value. Although these transients do not last very long, they can cause serious damage. While this is a rather severe example of a transient, it nonetheless illustrates that during transient conditions, many of a circuit or system's most difficult problems may arise.

Some transient effects, on the other hand, are useful. For example, many electronic devices and circuits depend on transient effects; these include timers, oscillators, and waveshaping circuits. As you will see in this chapter and in later electronics courses, the charge/discharge characteristic of *RC* circuits is fundamental to their operation.

11.1 Introduction

A basic switched *RC* circuit is shown in Figure 11–1. Most of the key ideas concerning charging and discharging and dc transients in *RC* circuits can be developed from it.

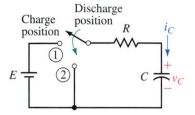

FIGURE 11–1 Circuit for studying capacitor charging and discharging. Transient voltages and currents result when the circuit is switched.

Capacitor Charging

First, assume the capacitor is uncharged and that the switch is open. Now move the switch to the charge position, Figure 11–2(a). At the instant the switch is closed the current jumps to *E/R* amps, then decays to zero, while the voltage, which is zero at the instant the switch is closed, gradually climbs to *E* volts. This is shown in (b) and (c). The shapes of these curves are easily explained.

First, consider voltage. In order to change capacitor voltage, electrons must be moved from one plate to the other. Even for a relatively small capacitor, billions of electrons must be moved. This takes time. Consequently, *capacitor voltage cannot change instantaneously, i.e., it cannot jump*

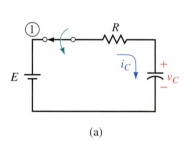

(a)

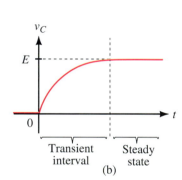

(b)

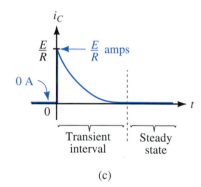

(c)

FIGURE 11–2 Capacitor voltage and current during charging. Time $t = 0$ s is defined as the instant the switch is moved to the charge position. The capacitor is initially uncharged.

abruptly from one value to another. Instead, it climbs gradually and smoothly as illustrated in Figure 11–2(b).

Now consider current. The movement of electrons noted above is a current. As indicated in Figure 11–2(c), this current jumps abruptly from 0 to *E/R* amps, i.e., the current is **discontinuous.** To understand why, consider Figure 11–3(a). Since capacitor voltage cannot change instantaneously, its value just after the switch is closed will be the same as it was just before the switch is closed, namely 0 V. Since the voltage across the capacitor just after the switch is closed is zero (even though there is current through it), *the capacitor looks momentarily like a short circuit.* This is indicated in (b). This is an important observation and is true in general, that is, *an uncharged capacitor looks like a short circuit at the instant of switching.* Applying Ohm's law yields $i_C = E/R$ amps. This agrees with what we indicated in Figure 11–2(c).

Finally, note the trailing end of the current curve Figure 11–2(c). Since the dielectric between the capacitor plates is an insulator, no current can pass through it. This means that the current in the circuit, which is due entirely to the movement of electrons from one plate to the other through the battery, must decay to zero as the capacitor charges.

Steady State Conditions

When the capacitor voltage and current reach their final values and stop changing (Figure 11–2(b) and (c)), the circuit is said to be in **steady state**. Figure 11–4(a) shows the circuit after it has reached steady state. Note that $v_C = E$ and $i_C = 0$. Since the capacitor has voltage across it but no current through it, it looks like an open circuit as indicated in (b). This is also an important observation and one that is true in general, that is, *a capacitor looks like an open circuit to steady state dc.*

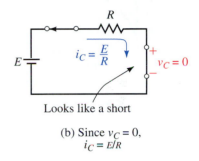

(a) Circuit as it looks just after the switch is moved to the charge position; v_C is still zero

Looks like a short

(b) Since $v_C = 0$, $i_C = E/R$

FIGURE 11–3 An uncharged capacitor initially looks like a short circuit.

(a) $v_C = E$ and $i_C = 0$

(b) Equivalent circuit for the capacitor

FIGURE 11–4   Charging circuit after it has reached steady state. Since the capacitor has voltage across it but no current, it looks like an open circuit in steady state dc.

Capacitor Discharging

Now consider the discharge case, Figures 11–5 and 11–6. First, assume the capacitor is charged to E volts and that the switch is open, Figure 11–5(a). Now close the switch. Since the capacitor has E volts across it just before the switch is closed, and since its voltage cannot change instantaneously, it will still have E volts across it just after as well. This is indicated in (b). The capacitor therefore looks momentarily like a voltage source, (c) and the cur-

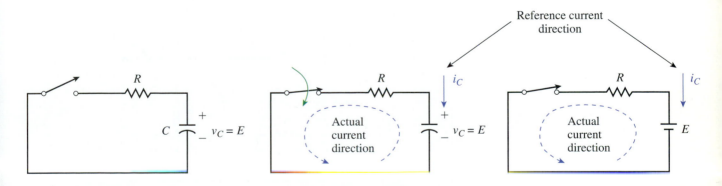

(a) Voltage v_C equals E just before the switch is closed

(b) Immediately after the switch is closed, v_C still equals E

(c) Capacitor therefore momentarily looks like a voltage source. Ohm's law yields $i_C = -E/R$

FIGURE 11–5   A charged capacitor looks like a voltage source at the instant of switching. Current is negative since it is opposite in direction to the current reference arrow.

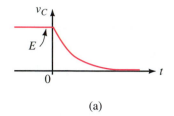

(a)

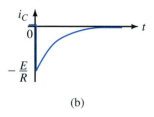

(b)

FIGURE 11–6 Voltage and current during discharge. Time $t = 0$ s is defined as the instant the switch is moved to the discharge position.

rent thus jumps immediately to $-E/R$ amps. (Note that the current is negative since it is opposite in direction to the reference arrow.) The voltage and current then decay to zero as indicated in Figure 11–6.

EXAMPLE 11–1   For Figure 11–1, $E = 40$ V, $R = 10\ \Omega$, and the capacitor is initially uncharged. The switch is moved to the charge position and the capacitor allowed to charge fully. Then the switch is moved to the discharge position and the capacitor allowed to discharge fully. Sketch the voltages and currents and determine the values at switching and in steady state.

Solution   The current and voltage curves are shown in Figure 11–7. Initially, $i = 0$ A since the switch is open. Immediately after it is moved to the charge position, the current jumps to $E/R = 40$ V/10 Ω = 4 A; then it decays to zero. At the same time, v_C starts at 0 V and climbs to 40 V. When the switch is moved to the discharge position, the capacitor looks momentarily like a 40-V source and the current jumps to -40 V/10 Ω = -4 A; then it decays to zero. At the same time, v_C also decays to zero.

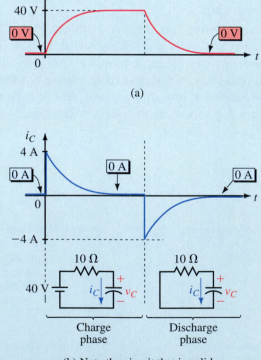

(b) Note the circuit that is valid during each time interval

FIGURE 11–7   A charge/discharge example.

PRACTICAL NOTES...

Some key points to remember are the following:

1. Capacitor voltage cannot change instantaneously, but current can.

2. To determine currents and voltages in a circuit at the instant of switching, replace uncharged capacitors with short circuits and charged capacitors with dc sources equal to their respective voltages at the instant of switching.

3. To determine currents and voltages in a dc circuit after it has reached steady state, replace capacitors with open circuits.

4. For both the charging and discharging cases, show capacitor current i_C such that the plus sign for v_C is at the tail of the current reference arrow. For charging, you will find that current is in the same direction as i_C and hence is positive, while for discharging, current is opposite in direction to i_C and hence is negative.

The Meaning of Time in Transient Analysis

The time t used in transient analysis is measured from the instant of switching. Thus, $t = 0$ in Figure 11–2 is defined as the instant the switch is moved to charge, while in Figure 11–6, it is defined as the instant the switch is moved to discharge. Voltages and currents are then represented in terms of this time as $v_C(t)$ and $i_C(t)$. For example, the voltage across a capacitor at $t = 0$ s is denoted as $v_C(0)$, while the voltage at $t = 10$ ms is denoted as $v_C(10$ ms), and so on.

A problem arises when a quantity is discontinuous as is the current of Figure 11–2(c). Since its value is changing at $t = 0$ s, $i_C(0)$ cannot be defined. To get around this problem, we define two values for 0 s. We define $t = 0^-$ s as $t = 0$ s just prior to switching and $t = 0^+$ s as $t = 0$ s just after switching. In Figure 11–2(c), therefore, $i_C(0^-) = 0$ A while $i_C(0^+) = E/R$ amps. For Figure 11–6, $i_C(0^-) = 0$ A and $i_C(0^+) = -E/R$ amps.

Exponential Functions

As we will soon show, the waveforms of Figures 11–2 and 11–6 are exponential and vary according to e^{-x} or $(1 - e^{-x})$, where e is the base of the natural logarithm. Fortunately, exponential functions are easy to evaluate with modern calculators using their e^x function. You will need to be able to evaluate both e^{-x} and $(1 - e^{-x})$ for any value of x. Table 11–1 shows a tabulation of values for both cases. Note that as x gets larger, e^{-x} gets smaller and approaches zero, while $(1 - e^{-x})$ gets larger and approaches 1. These observations will be important to you in what follows.

TABLE 11–1   Table of Exponentials

x	e^{-x}	$1 - e^{-x}$
0	1	0
1	0.3679	0.6321
2	0.1353	0.8647
3	0.0498	0.9502
4	0.0183	0.9817
5	0.0067	0.9933

PRACTICE PROBLEMS 1

1. Use your calculator and verify the entries in Table 11–1. Be sure to change the sign of x before using the e^x function. Note that $e^{-0} = e^0 = 1$ since any quantity raised to the zeroth power is one.

2. Plot the computed values on graph paper and verify that they yield curves that look like those shown in Figure 11–2(b) and (c).

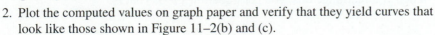

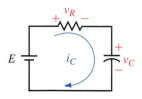

FIGURE 11–8 Circuit for the charging case. Capacitor is initially uncharged.

11.2 Capacitor Charging Equations

We will now develop equations for voltages and current during charging. Consider Figure 11–8. KVL yields

$$v_R + v_C = E \tag{11–1}$$

But $v_R = Ri_C$ and $i_C = C dv_C/dt$ (Equation 10–20). Thus, $v_R = RC dv_C/dt$. Substituting this into Equation 11–1 yields

$$RC\frac{dv_C}{dt} + v_C = E \tag{11–2}$$

Equation 11–2 can be solved for v_C using basic calculus (see). The result is

$$v_C = E(1 - e^{-t/RC}) \tag{11–3}$$

where R is in ohms, C is in farads, t is in seconds and $e^{-t/RC}$ is the exponential function discussed earlier. The product RC has units of seconds. (This is left as an exercise for the student to show.)

∫ Solving Equation 11–2 (Optional Derivation)

First, rearrange Equation 11–2:

$$\frac{dv_C}{dt} = \frac{1}{RC}(E - v_C)$$

Rearrange again:

$$\frac{dv_C}{E - v_C} = \frac{dt}{RC}$$

Now multiply both sides by -1 and integrate.

$$\int_0^{v_C} \frac{dv_C}{v_C - E} = -\frac{1}{RC}\int_0^t dt$$

$$\ln(v_C - E)\Big]_0^{v_C} = -\frac{t}{RC}\Big]_0^t$$

Next, substitute integration limits,

$$\ln(v_C - E) - \ln(-E) = -\frac{t}{RC}$$

$$\ln\left(\frac{v_C - E}{-E}\right) = -\frac{t}{RC}$$

Finally, take the inverse log of both sides. Thus,

$$\frac{v_C - E}{-E} = e^{-t/RC}$$

When you rearrange this, you get Equation 11–3. That is,

$$v_C = E(1 - e^{-t/RC})$$

Now consider the resistor voltage. From Equation 11–1, $v_R = E - v_C$. Substituting v_C from Equation 11–3 yields $v_R = E - E(1 - e^{-t/RC}) = E - E + Ee^{-t/RC}$. After cancellation, you get

$$v_R = Ee^{-t/RC} \qquad (11\text{–}4)$$

Now divide both sides by R. Since $i_C = i_R = v_R/R$, this yields

$$i_C = \frac{E}{R}e^{-t/RC} \qquad (11\text{–}5)$$

The waveforms are shown in Figure 11–9. Values at any time may be determined by substitution.

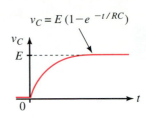
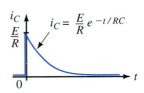
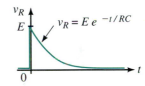

FIGURE 11–9 Curves for the circuit of Figure 11–8.

EXAMPLE 11–2   Suppose $E = 100$ V, $R = 10$ kΩ, and $C = 10$ μF:

a. Determine the expression for v_C.

b. Determine the expression for i_C.

c. Compute the capacitor voltage at $t = 150$ ms.

d. Compute the capacitor current at $t = 150$ ms.

e. Locate the computed points on the curves.

Solution

a. $RC = (10 \times 10^3\ \Omega)(10 \times 10^{-6}\ \text{F}) = 0.1$ s. From Equation 11–3, $v_C = E(1 - e^{-t/RC}) = 100(1 - e^{-t/0.1}) = 100(1 - e^{-10t})$ V.

b. From Equation 11–5, $i_C = (E/R)e^{-t/RC} = (100\ \text{V}/10\ \text{k}\Omega)e^{-10t} = 10e^{-10t}$ mA.

c. At $t = 0.15$ s, $v_C = 100(1 - e^{-10t}) = 100(1 - e^{-10(0.15)}) = 100(1 - e^{-1.5}) = 100(1 - 0.223) = 77.7$ V.

d. $i_C = 10e^{-10t}$ mA $= 10e^{-10(0.15)}$ mA $= 10e^{-1.5}$ mA $= 2.23$ mA.

e. The corresponding points are shown in Figure 11–10.

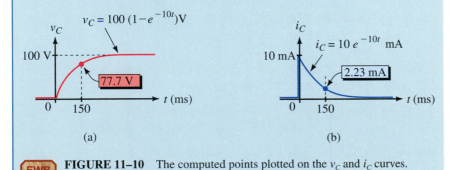

(a)   (b)

EWB   **FIGURE 11–10**   The computed points plotted on the v_C and i_C curves.

In the above example, we expressed voltage as $v_C = 100(1 - e^{-t/0.1})$ and as $100(1 - e^{-10t})$ V. Similarly, current can be expressed as $i_C = 10e^{-t/0.1}$ or as $10e^{-10t}$ mA. Although some authors prefer one notation over the other, both are correct and we will use them interchangeably.

**PRACTICE
PROBLEMS 2**

1. Determine additional voltage and current points for Figure 11–10 by computing values of v_C and i_C at values of time from $t = 0$ s to $t = 500$ ms at 100-ms intervals. Plot the results.

2. The switch of Figure 11–11 is closed at $t = 0$ s. If $E = 80$ V, $R = 4$ kΩ, and $C = 5$ μF, determine expressions for v_C and i_C. Plot the results from $t = 0$ s to $t = 100$ ms at 20-ms intervals. Note that charging takes less time here than for Problem 1.

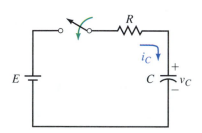

FIGURE 11–11

Answers:

1.

t(ms)	v_C(V)	i_C(mA)
0	0	10
100	63.2	3.68
200	86.5	1.35
300	95.0	0.498
400	98.2	0.183
500	99.3	0.067

2. $80(1 - e^{-50t})$ V $20e^{-50t}$ mA

t(ms)	v_C(V)	i_C(mA)
0	0	20
20	50.6	7.36
40	69.2	2.70
60	76.0	0.996
80	78.6	0.366
100	79.4	0.135

EXAMPLE 11–3   For the circuit of Figure 11–11, $E = 60$ V, $R = 2$ kΩ, and $C = 25$ μF. The switch is closed at $t = 0$ s, opened 40 ms later and left open. Determine equations for capacitor voltage and current and plot.

Solution $RC = (2$ kΩ)(25 μF) $= 50$ ms. As long as the switch is closed (i.e., from $t = 0$ s to 40 ms), the following equations hold:

$$v_C = E(1 - e^{-t/RC}) = 60(1 - e^{-t/50 \text{ ms}}) \text{ V}$$
$$i_C = (E/R)e^{-t/RC} = 30e^{-t/50 \text{ ms}} \text{ mA}$$

Voltage starts at 0 V and rises exponentially. At $t = 40$ ms, the switch is opened, interrupting charging. At this instant, $v_C = 60(1 - e^{-(40/50)}) = 60(1 - e^{-0.8}) = 33.0$ V. Since the switch is left open, the voltage remains constant at 33 V thereafter as indicated in Figure 11–12. (The dotted curve shows how the voltage would have kept rising if the switch had remained closed.)

Now consider current. The current starts at 30 mA and decays to $i_C = 30e^{-(40/50)}$ mA $= 13.5$ mA at $t = 40$ ms. At this point, the switch is opened, and the current drops instantly to zero. (The dotted line shows how the current would have decayed if the switch had not been opened.)

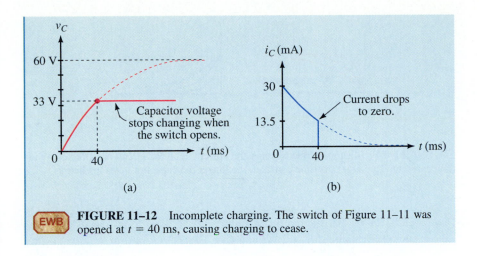

FIGURE 11–12   Incomplete charging. The switch of Figure 11–11 was opened at $t = 40$ ms, causing charging to cease.

The Time Constant

The rate at which a capacitor charges depends on the product of R and C. This product is known as the **time constant** of the circuit and is given the symbol τ (the Greek letter tau). As noted earlier, RC has units of seconds. Thus,

$$\tau = RC \quad \text{(seconds, s)} \qquad \text{(11–6)}$$

Using τ, Equations 11–3 to 11–5 can be written as

$$v_C = E(1 - e^{-t/\tau}) \qquad \text{(11–7)}$$

$$i_C = \frac{E}{R}e^{-t/\tau} \qquad \text{(11–8)}$$

and

$$v_R = Ee^{-t/\tau} \qquad \text{(11–9)}$$

Duration of a Transient

The length of time that a transient lasts depends on the exponential function $e^{-t/\tau}$. As t increases, $e^{-t/\tau}$ decreases, and when it reaches zero, the transient is gone. Theoretically, this takes infinite time. In practice, however, over 99% of the transition takes place during the first five time constants (i.e., transients are within 1% of their final value at $t = 5\tau$). This can be verified by direct substitution. At $t = 5\tau$, $v_C = E(1 - e^{-t/\tau}) = E(1 - e^{-5}) = E(1 - 0.0067) = 0.993E$, meaning that the transient has achieved 99.3% of its final value. Similarly, the current falls to within 1% of its final value in five time constants. Thus, *for all practical purposes, transients can be considered to last for only five time constants* (Figure 11–13). Figure 11–14 summarizes how transient voltages and currents are affected by the time constant of a circuit—the larger the time constant, the longer the duration of the transient.

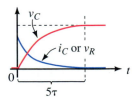

FIGURE 11–13   Transients last five time constants.

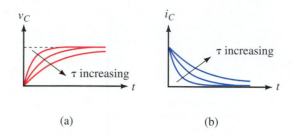

(a)   (b)

FIGURE 11–14   Illustrating how voltage and current in an *RC* circuit are affected by its time constant. The larger the time constant, the longer the capacitor takes to charge.

EXAMPLE 11–4   For the circuit of Figure 11–11, how long will it take for the capacitor to charge if $R = 2$ kΩ and $C = 10$ μF?

Solution $\tau = RC = (2 \text{ k}\Omega)(10 \text{ μF}) = 20$ ms. Therefore, the capacitor charges in $5\,\tau = 100$ ms.

EXAMPLE 11–5   The transient in a circuit with $C = 40$ μF lasts 0.5 s. What is *R?*

Solution $5\,\tau = 0.5$ s. Thus, $\tau = 0.1$ s and $R = \tau/C = 0.1$ s/$(40 \times 10^{-6}$ F$)$ $= 2.5$ kΩ.

Figure 11–15 shows percent capacitor voltage and current plotted versus multiples of time constant. (Points are computed from $v_C = 100(1 - e^{-t/\tau})$ and $i_C = 100e^{-t/\tau}$. For example, at $t = \tau$, $v_C = 100(1 - e^{-t/\tau}) = 100(1 - e^{-\tau/\tau}) = 100(1 - e^{-1}) = 63.2$ V, i.e., 63.2%, and $i_C = 100e^{-\tau/\tau} = 100e^{-1} = 36.8$ A, which is 36.8%, and so on.) These curves, referred to as universal time constant curves, provide an easy method to determine voltages and currents with a minimum of computation.

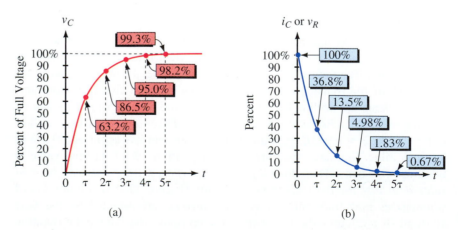

(a)   (b)

FIGURE 11–15   Universal voltage and current curves for *RC* circuit.

EXAMPLE 11–6 Using Figure 11–15, compute v_C and i_C at two time constants into charge for a circuit with $E = 25$ V, $R = 5$ kΩ, and $C = 4$ μF. What is the corresponding value of time?

Solution At $t = 2\,\tau$, v_C equals 86.5% of E or 0.865(25 V) = 21.6 V. Similarly, $i_C = 0.135I_0 = 0.135(E/R) = 0.675$ mA. These values occur at $t = 2\,\tau = 2RC = 40$ ms.

---

1. If the capacitor of Figure 11–16 is uncharged, what is the current immediately after closing the switch?

IN-PROCESS
LEARNING
CHECK 1

FIGURE 11–16

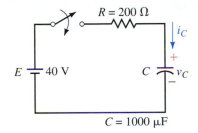

$R = 200\ \Omega$

i_C

$E \doteq 40$ V $C \doteq v_C$

$C = 1000\ \mu F$

2. Given $i_C = 50e^{-20t}$ mA.

   a. What is τ?

   b. Compute the current at $t = 0^+$ s, 25 ms, 50 ms, 75 ms, 100 ms, and 500 ms and sketch it.

3. Given $v_C = 100(1 - e^{-50t})$ V, compute v_C at the same time intervals as in Problem 2 and sketch.

4. For Figure 11–16, determine expressions for v_C and i_C. Compute capacitor voltage and current at $t = 0.6$ s.

5. Refer to Figure 11–10:

   a. What are $v_C(0^-)$ and $v_C(0^+)$?

   b. What are $i_C(0^-)$ and $i_C(0^+)$?

   c. What are the steady state voltage and current?

6. For the circuit of Figure 11–11, the current just after the switch is closed is 2 mA. The transient lasts 40 ms and the capacitor charges to 80 V. Determine E, R, and C.

7. Find capacitor voltage and current for Figure 11–16 at $t = 0.6$ s using the universal time constant curves of Figure 11–15.

(Answers are at the end of the chapter.)

11.3 Capacitor with an Initial Voltage

Suppose a previously charged capacitor has not been discharged and thus still has voltage on it. Let this voltage be denoted as V_0. If the capacitor is now placed in a circuit like that in Figure 11–16, the voltage and current during charging will be affected by the initial voltage. In this case, Equations 11–7 and 11–8 become

$$v_C = E + (V_0 - E)e^{-t/\tau} \tag{11–10}$$

$$i_C = \frac{E - V_0}{R}e^{-t/\tau} \tag{11–11}$$

Note that these revert to their original forms when you set $V_0 = 0$ V.

EXAMPLE 11–7   Suppose the capacitor of Figure 11–16 has 25 volts on it with polarity shown at the time the switch is closed.

a. Determine the expression for v_C.

b. Determine the expression for i_C.

c. Compute v_C and i_C at $t = 0.1$ s.

d. Sketch v_C and i_C.

Solution $\tau = RC = (200 \ \Omega)(1000 \ \mu\text{F}) = 0.2$ s

a. From Equation 11–10,

$$v_C = E + (V_0 - E)e^{-t/\tau} = 40 + (25 - 40)e^{-t/0.2} = 40 - 15e^{-5t} \text{ V}$$

b. From Equation 11–11,

$$i_C = \frac{E - V_0}{R}e^{-t/\tau} = \frac{40 - 25}{200}e^{-5t} = 75e^{-5t} \text{ mA}$$

c. At $t = 0.1$ s,

$$v_C = 40 - 15e^{-5t} = 40 - 15e^{-0.5} = 30.9 \text{ V}$$
$$i_C = 75e^{-5t} \text{ mA} = 75e^{-0.5} \text{ mA} = 45.5 \text{ mA}$$

d. The waveforms are shown in Figure 11–17 with the above points plotted.

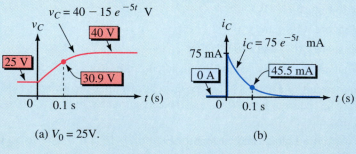

(a) $V_0 = 25$V.   (b)

 FIGURE 11–17   Capacitor with an initial voltage.

PRACTICE PROBLEMS 3

Repeat Example 11–7 for the circuit of Figure 11–16 if $V_0 = -150$ V.

Answers: a. $40 - 190e^{-5t}$ V   b. $0.95e^{-5t}$ A   c. -75.2 V; 0.576 A   d. Curves are similar to Figure 11–17 except that v_C starts at -150 V and rises to 40 V while i_C starts at 0.95 A and decays to zero.

11.4   Capacitor Discharging Equations

To determine the discharge equations, move the switch to the discharge position (Figure 11–18). KVL yields $v_R + v_C = 0$. Substituting $v_R = RCdv_C/dt$ from Section 11.2 yields

$$RC\frac{dv_C}{dt} + v_C = 0 \qquad \text{(11–12)}$$

This can be solved for v_C using basic calculus. The result is

$$v_C = V_0e^{-t/RC} \qquad \text{(11–13)}$$

where V_0 is the voltage on the capacitor at the instant the switch is moved to discharge. Now consider the resistor voltage. Since $v_R + v_C = 0$, $v_R = -v_C$ and

$$v_R = -V_0e^{-t/RC} \qquad \text{(11–14)}$$

Now divide both sides by R. Since $i_C = i_R = v_R/R$,

$$i_C = -\frac{V_0}{R}e^{-t/RC} \qquad \text{(11–15)}$$

Note that this is negative, since, during discharge, the current is opposite in direction to the reference arrow of Figure 11–18. (If you need to refresh your memory, see again Figure 11–5.) Voltage v_C and current i_C are shown in Figure 11–19. As in the charging case, *discharge transients last five time constants.*

In Equations 11–13 to 11–15, V_0 represents the voltage on the capacitor at the instant the switch is moved to the discharge position. If the switch has been in the charge position long enough for the capacitor to fully charge, $V_0 = E$ and Equations 11–13 and 11–15 become $v_C = Ee^{-t/RC}$ and $i_C = -(E/R)e^{-t/RC}$ respectively.

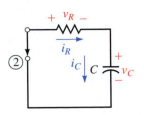

FIGURE 11–18   Discharge case. Initial capacitor voltage is V_0. Note the reference direction for i_C. (To conform to the standard voltage/current reference convention, i_C must be drawn in this direction so that the + sign for v_C is at the tail of the current arrow.) Since the actual current direction is opposite to the reference direction, i_C will be negative. This is indicated in Figure 11–19(b).

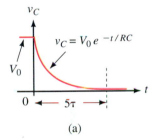

(a)

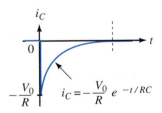

(b) During discharge, i_C is negative as determined in Figure 11–18

FIGURE 11–19   Capacitor voltage and current for the discharge case.

EXAMPLE 11–8   For the circuit of Figure 11–18, assume the capacitor is charged to 100 V before the switch is moved to the discharge position. Suppose $R = 5$ kΩ and $C = 25$ μF. After the switch is moved to discharge,

a. Determine the expression for v_C.

b. Determine the expression for i_C.

c. Compute the voltage and current at 0.375 s.

Solution $RC = (5$ kΩ$)(25$ μF$) = 0.125$ s and $V_0 = 100$ V. Therefore,

a. $v_C = V_0e^{-t/RC} = 100e^{-t/0.125} = 100e^{-8t}$ V.

b. $i_C = -(V_0/R)e^{-t/RC} = -20e^{-8t}$ mA.

c. At $t = 0.375$ s,

$$v_C = 100e^{-8t} = 100e^{-3} = 4.98 \text{ V}$$
$$i_C = -20e^{-8t} \text{ mA} = -20e^{-3} \text{ mA} = -0.996 \text{ mA}$$

The universal time constant curve of Figure 11–15(b) may also be used to solve discharge problems. For example, for the circuit of Example 11–8, at $t = 3\,\tau$, capacitor voltage has fallen to 4.98% of E, which is $(0.0498)(100\ \text{V}) = 4.98\ \text{V}$ and current has decayed to 4.98% of -20 mA which is $(0.0498)(-20\ \text{mA}) = -0.996$ mA. (These agree with Example 11–8 since $3\,\tau$ was also the value of time used there.)

11.5   More Complex Circuits

The charge and discharge equations described previously apply only to circuits of the forms shown in Figures 11–2 and 11–5 respectively. Fortunately, many circuits can be reduced to these forms using standard circuit reduction techniques such as series and parallel combinations, source conversions, Thévenin's theorem, and so on. Once a circuit has been reduced to its series equivalent, you can use any of the equations that we have developed so far.

EXAMPLE 11–9   For the circuit of Figure 11–20(a), determine expressions for v_C and i_C. Capacitors are initially uncharged.

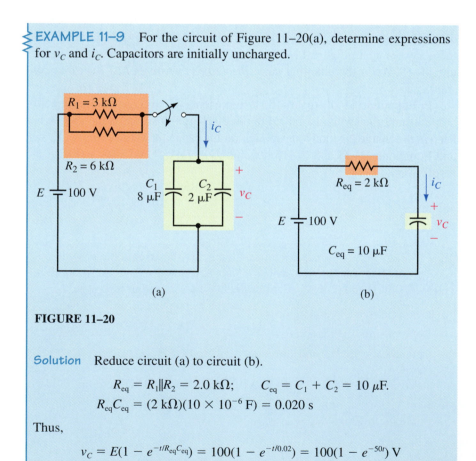

(a)   (b)

FIGURE 11–20

Solution   Reduce circuit (a) to circuit (b).

$$R_{eq} = R_1 \| R_2 = 2.0\ \text{k}\Omega; \qquad C_{eq} = C_1 + C_2 = 10\ \mu\text{F}.$$
$$R_{eq}C_{eq} = (2\ \text{k}\Omega)(10 \times 10^{-6}\ \text{F}) = 0.020\ \text{s}$$

Thus,

$$v_C = E(1 - e^{-t/R_{eq}C_{eq}}) = 100(1 - e^{-t/0.02}) = 100(1 - e^{-50t})\ \text{V}$$

$$i_C = \frac{E}{R_{eq}}e^{-t/R_{eq}C_{eq}} = \frac{100}{2000}e^{-t/0.02} = 50\,e^{-50t}\ \text{mA}$$

EXAMPLE 11–10 The capacitor of Figure 11–21 is initially uncharged. Close the switch at $t = 0$ s.

a. Determine the expression for v_C.

b. Determine the expression for i_C.

c. Determine capacitor current and voltage at $t = 5$ ms.

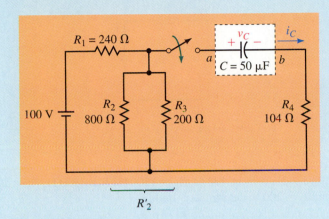

FIGURE 11–21

Solution Reduce the circuit to its series equivalent using Thévenin's theorem:

$$R'_2 = R_2 \| R_3 = 160 \ \Omega$$

From Figure 11–22(a),

$$R_{Th} = R_1 \| R'_2 + R_4 = 240 \| 160 + 104 = 96 + 104 = 200 \ \Omega$$

From Figure 11–22(b),

$$V'_2 = \left(\frac{R'_2}{R_1 + R'_2} \right) E = \left(\frac{160}{240 + 160} \right) \times 100 \text{ V} = 40 \text{ V}$$

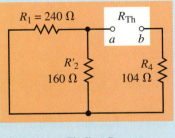

(a) Finding R_{Th}

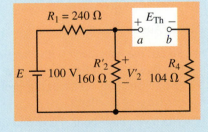

(b) Finding E_{Th}

FIGURE 11–22 Determining the Thévenin equivalent of Figure 11–21 following switch closure.

From KVL, $E_{Th} = V'_2 = 40$ V. The resultant equivalent circuit is shown in Figure 11–23.

$$\tau = R_{Th}C = (200 \ \Omega)(50 \ \mu\text{F}) = 10 \text{ ms}$$

a. $v_C = E_{Th}(1 - e^{-t/\tau}) = 40(1 - e^{-100t})$ V

b. $i_C = \dfrac{E_{Th}}{R_{Th}}e^{-t/\tau} = \dfrac{40}{200}e^{-t/0.01} = 200e^{-100t}$ mA

c. $i_C = 200e^{-100(5\ ms)} = 121$ mA. Similarly, $v_C = 15.7$ V

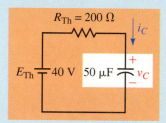

$R_{Th} = 200\ \Omega$

i_C

$E_{Th} = 40$ V   50 μF v_C

FIGURE 11–23   The Thévenin equivalent of Figure 11–21.

PRACTICE PROBLEMS 4

1. For Figure 11–21, if $R_1 = 400\ \Omega$, $R_2 = 1200\ \Omega$, $R_3 = 300\ \Omega$, $R_4 = 50\ \Omega$, $C = 20\ \mu$F, and $E = 200$ V, determine v_C and i_C.

2. Using the values shown in Figure 11–21, determine v_C and i_C if the capacitor has an initial voltage of 60 V.

3. Using the values of Problem 1, determine v_C and i_C if the capacitor has an initial voltage of -50 V.

Answers:

1. $75(1 - e^{-250t})$ V; $0.375e^{-250t}$ A

2. $40 + 20e^{-100t}$ V; $-0.1e^{-100t}$ A

3. $75 - 125e^{-250t}$ V; $0.625e^{-250t}$ A

PRACTICAL NOTES...

Notes About Time References

1. So far, we have dealt with charging and discharging problems separately. For these, we define $t = 0$ s as the instant the switch is moved to the charge position for charging problems and to the discharge position for discharging problems.

2. When you have both charge and discharge cases in the same example, you need to establish clearly what you mean by "time." We use the following procedure:

   a. Define $t = 0$ s as the instant the switch is moved to the first position, then determine corresponding expressions for v_C and i_C. These expressions and the corresponding time scale are valid until the switch is moved to its new position.

   b. When the switch is moved to its new position, shift the time reference and make $t = 0$ s the time at which the switch is moved to its new position, then determine corresponding expressions for v_C and i_C. These new expressions are only valid from the new $t = 0$ s reference point. The old expressions are valid only on the old time scale.

c. We now have two time scales for the same graph. However, we generally only show the first scale explicitly; the second scale is implied rather than shown.

d. Use τ_C to represent the time constant for charging and τ_d to represent the time constant for discharging. Since the equivalent resistance and capacitance for discharging may be different than that for charging, the time constants may be different for the two cases.

EXAMPLE 11–11   The capacitor of Figure 11–24(a) is uncharged. The switch is moved to position 1 for 10 ms, then to position 2, where it remains.

a. Determine v_C during charge.

b. Determine i_C during charge.

c. Determine v_C during discharge.

d. Determine i_C during discharge.

e. Sketch the charge and discharge waveforms.

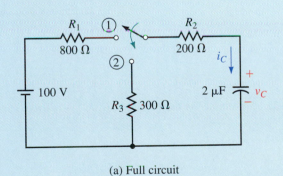

(a) Full circuit

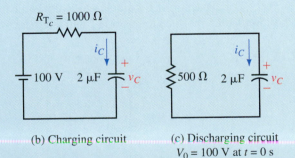

(b) Charging circuit   (c) Discharging circuit
$V_0 = 100$ V at $t = 0$ s

FIGURE 11–24

Solution   Figure 11–24(b) shows the equivalent charging circuit. Here,

$$\tau_c = (R_1 + R_2)C = (1\ k\Omega)(2\ \mu F) = 2.0\ ms.$$

a. $v_C = E(1 - e^{-t/\tau_c}) = 100(1 - e^{-500t})$ V

b. $i_C = \dfrac{E}{R_{T_c}}e^{-t/\tau_c} = \dfrac{100}{1000}e^{-500t} = 100e^{-500t}$ mA

Since $5\tau_C = 10$ ms, charging is complete by the time the switch is moved to discharge. Thus, $V_0 = 100$ V when discharging begins.

c. Figure 11–24(c) shows the equivalent discharge circuit. Note $V_0 = 100$ V.

$$\tau_d = (500\ \Omega)(2\ \mu\text{F}) = 1.0\ \text{ms}$$
$$v_C = V_0 e^{-t/\tau_d} = 100e^{-1000t}\ \text{V}$$

where $t = 0$ s has been redefined for discharge as noted above.

d. $i_C = -\dfrac{V_0}{R_2 + R_3}e^{-t/\tau_d} = -\dfrac{100}{500}e^{-1000t} = -200e^{-1000t}$ mA

e. See Figure 11–25. Note that discharge is more rapid than charge since $\tau_d < \tau_c$.

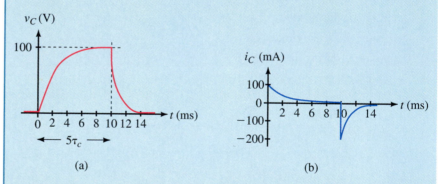

(a)   (b)

FIGURE 11–25   Waveforms for the circuit of Figure 11–24. Note that τ_d is shorter than τ_c.

EXAMPLE 11–12   The capacitor of Figure 11–26 is uncharged. The switch is moved to position 1 for 5 ms, then to position 2 and left there.

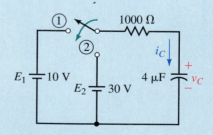

EWB   **FIGURE 11–26**

a. Determine v_C while the switch is in position 1.

b. Determine i_C while the switch is in position 1.

c. Compute v_C and i_C at $t = 5$ ms.

d. Determine v_C while the switch is in position 2.

e. Determine i_C while the switch is in position 2.

f. Sketch the voltage and current waveforms.

g. Determine v_C and i_C at $t = 10$ ms.

Solution

$$\tau_c = \tau_d = RC = (1 \text{ k}\Omega)(4 \text{ } \mu\text{F}) = 4 \text{ ms}$$

a. $v_C = E_1(1 - e^{-t/\tau_c}) = 10(1 - e^{-250t})$ V

b. $i_C = \dfrac{E_1}{R}e^{-t/\tau_c} = \dfrac{10}{1000}e^{-250t} = 10e^{-250t}$ mA

c. At $t = 5$ ms,

$$v_C = 10(1 - e^{-250 \times 0.005}) = 7.14 \text{ V}$$
$$i_C = 10e^{-250 \times 0.005} \text{ mA} = 2.87 \text{ mA}$$

d. In position 2, $E_2 = 30$ V, and $V_0 = 7.14$ V. Use Equation 11–10:

$$v_C = E_2 + (V_0 - E_2)e^{-t/\tau_d} = 30 + (7.14 - 30)e^{-250t}$$
$$= 30 - 22.86e^{-250t} \text{ V}$$

where $t = 0$ s has been redefined for position 2.

e. $i_C = \dfrac{E_2 - V_0}{R}e^{-t/\tau_d} = \dfrac{30 - 7.14}{1000}e^{-250t} = 22.86e^{-250t}$ mA

f. See Figure 11–27.

g. $t = 10$ ms is 5 ms into the new time scale. Thus, $v_C = 30 - 22.86e^{-250(5 \text{ ms})} = 23.5$ V and $i_C = 22.86e^{-250(5 \text{ ms})} = 6.55$ mA. Values are plotted on the graph.

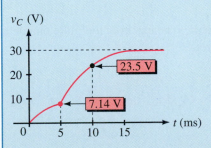
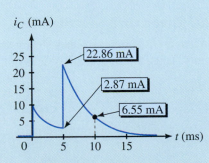

FIGURE 11–27 Capacitor voltage and current for the circuit of Figure 11–26.

EXAMPLE 11–13 In Figure 11–28(a), the capacitor is initially uncharged. The switch is moved to the charge position, then to the discharge position, yielding the current shown in (b). The capacitor discharges in 1.75 ms. Determine the following:

a. E. b. R_1. c. C.

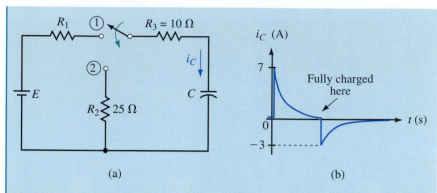

FIGURE 11–28

Solution

a. Since the capacitor charges fully, it has a value of E volts when switched to discharge. The discharge current spike is therefore

$$-\frac{E}{10\ \Omega + 25\ \Omega} = -3\ \text{A}$$

Thus, $E = 105$ V.

b. The charging current spike has a value of

$$\frac{E}{10\ \Omega + R_1} = 7\ \text{A}$$

Since $E = 105$ V, this yields $R_1 = 5\ \Omega$.

c. $5\ \tau_d = 1.75$ ms. Therefore $\tau_d = 350\ \mu$s. But $\tau_d = (R_2 + R_3)C$. Thus, $C = 350\ \mu$s$/35\ \Omega = 10\ \mu$F.

RC Circuits in Steady State DC

When an *RC* circuit reaches steady state dc, its capacitors look like open circuits. Thus, a transient analysis is not needed.

EXAMPLE 11–14   The circuit of Figure 11–29(a) has reached steady state. Determine the capacitor voltages.

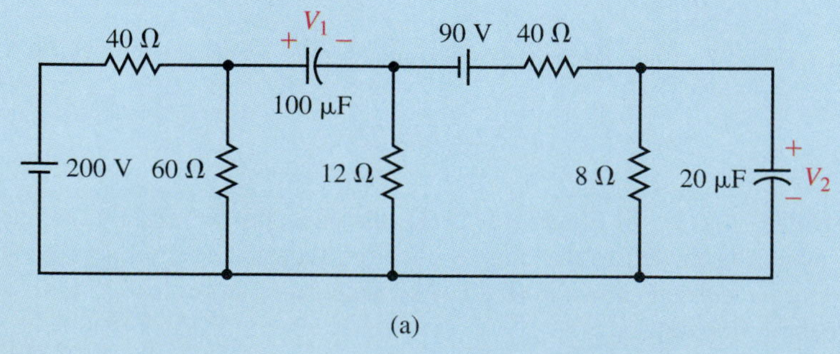

FIGURE 11–29   *Continues*

Solution Replace all capacitors with open circuits. Thus,

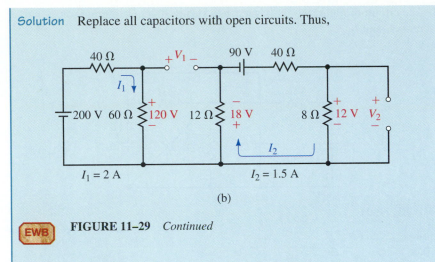

(b)

FIGURE 11–29 *Continued*

$$I_1 = \frac{200 \text{ V}}{40 \text{ }\Omega + 60 \text{ }\Omega} = 2 \text{ A}, \qquad I_2 = \frac{90 \text{ V}}{40 \text{ }\Omega + 8 \text{ }\Omega + 12 \text{ }\Omega} = 1.5 \text{ A}$$

KVL: $V_1 - 120 - 18 = 0$. Therefore, $V_1 = 138$ V. Further,

$$V_2 = (8 \text{ }\Omega)(1.5 \text{ A}) = 12 \text{ V}$$

PRACTICE PROBLEMS 5

1. The capacitor of Figure 11–30(a) is initially unchanged. At $t = 0$ s, the switch is moved to position 1 and 100 ms later, to position 2. Determine v_C and i_C for position 2.

2. Repeat for Figure 11–30(b). Hint: Use Thévenin's theorem.

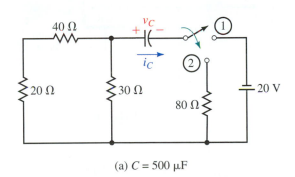

(a) $C = 500 \text{ }\mu\text{F}$

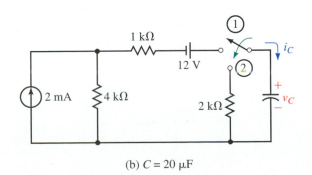

(b) $C = 20 \text{ }\mu\text{F}$

FIGURE 11–30

3. The circuit of Figure 11–31 has reached steady state. Determine source currents I_1 and I_2.

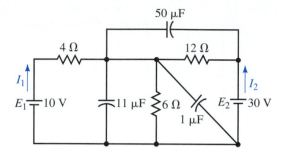

FIGURE 11–31

Answers:

1. $20e^{-20t}$ V; $-0.2e^{-20t}$ A

2. $12.6e^{-25t}$ V; $-6.3e^{-25t}$ mA

3. 0 A;   1.67 A

11.6   An *RC* Timing Application

RC circuits are used to create delays for alarm, motor control, and timing applications. Figure 11–32 shows an alarm application. The alarm unit contains a threshold detector, and when the input to this detector exceeds a preset value, the alarm is turned on.

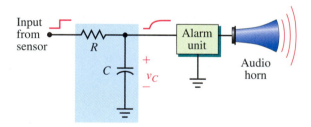

(a) Delay circuit

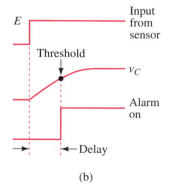

(b)

FIGURE 11–32   Creating a time delay with an *RC* circuit.

EXAMPLE 11–15   The circuit of Figure 11–32 is part of a building security system. When an armed door is opened, you have a specified number of seconds to disarm the system before the alarm goes off. If $E = 20$ V, $C = 40\ \mu$F, the alarm is activated when v_C reaches 16 V, and you want a delay of at least 25 s, what value of R is needed?

Solution $v_C = E(1 - e^{-t/RC})$. After a bit of manipulation, you get

$$e^{-t/RC} = \frac{E - v_C}{E}$$

Taking the natural log of both sides yields

$$-\frac{t}{RC} = \ln\left(\frac{E - v_C}{E}\right)$$

At $t = 25$ s, $v_C = 16$ V. Thus,

$$-\frac{t}{RC} = \ln\left(\frac{20 - 16}{20}\right) = \ln 0.2 = -1.6094$$

Substituting $t = 25$ s and $C = 40\ \mu$F yields

$$R = \frac{t}{1.6094C} = \frac{25\ \text{s}}{1.6094 \times 40 \times 10^{-6}} = 388\ \text{k}\Omega$$

Choose the next higher standard value, namely 390 kΩ.

---

1. Suppose you want to increase the disarm time of Example 11–15 to at least 35 s. Compute the new value of R.

2. If, in Example 11–15, the threshold is 15 V and $R = 1$ MΩ, what is the disarm time?

PRACTICE PROBLEMS 6

Answers:

1. 544 kΩ. Use 560 kΩ.

2. 55.5 s

---

1. Refer to Figure 11–16:

   a. Determine the expression for v_C when $V_0 = 80$ V. Sketch v_C.

   b. Repeat (a) if $V_0 = 40$ V. Why is there no transient?

   c. Repeat (a) if $V_0 = -60$ V.

2. For Part (c) of Question 1, v_C starts at -60 V and climbs to $+40$ V. Determine at what time v_C passes through 0 V, using the technique of Example 11–15.

3. For the circuit of Figure 11–18, suppose $R = 10$ kΩ and $C = 10\ \mu$F:

   a. Determine the expressions for v_C and i_C when $V_0 = 100$ V. Sketch v_C and i_C.

   b. Repeat (a) if $V_0 = -100$ V.

4. Repeat Example 11–12 if voltage source 2 is reversed, i.e., $E_2 = -30$ V.

IN-PROCESS LEARNING CHECK 2

5. The switch of Figure 11–33(a) is closed at $t = 0$ s. The Norton equivalent of the circuit in the box is shown in (b). Determine expressions for v_C and i_C. The capacitor is initially uncharged.

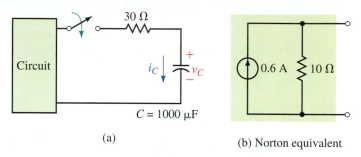

(a)

(b) Norton equivalent

FIGURE 11–33 Hint: Use a source transformation.

(Answers are at the end of the chapter.)

11.7 Pulse Response of *RC* Circuits

In previous sections, we looked at the response of *RC* circuits to switched dc inputs. In this section, we consider the effect that *RC* circuits have on pulse waveforms. Since many electronic devices and systems utilize pulse or rectangular waveforms, including computers, communications systems, and motor control circuits, these are important considerations.

Pulse Basics

A **pulse** is a voltage or current that changes from one level to the other and back again as in Figure 11–34(a) and (b). A **pulse train** is a repetitive stream of pulses, as in (c). If a waveform's high time equals its low time, as in (d), it is called a **square wave.**

The length of each cycle of a pulse train is termed its **period, *T,*** and the number of pulses per second is defined as its **pulse repetition rate** (PRR) or **pulse repetition frequency** (PRF). For example, in (e), there are two complete cycles in one second; therefore, the PRR = 2 pulses/s. With two cycles every second, the time for one cycle is $T = \frac{1}{2}$ s. Note that this is 1/PRR. This is true in general. That is,

$$T = \frac{1}{\text{PRR}} \quad \text{s} \tag{11–16}$$

The width, t_p, of a pulse relative to its period, [Figure 11–34(c)] is its **duty cycle.** Thus,

$$\text{duty cycle} = \frac{t_p}{T} \times 100\% \tag{11–17}$$

A square wave [Figure 11–34(d)] therefore has a 50% duty cycle, while a waveform with $t_p = 1.5$ μs and a period of 10 μs has a duty cycle of 15%.

In practice, waveforms are not ideal, that is, they do not change from low to high or high to low instantaneously. Instead, they have finite **rise** and **fall times.** Rise and fall times are denoted as t_r and t_f and are measured between the 10% and 90% points as indicated in Figure 11–35(a). **Pulse width** is measured at the 50% point. The difference between a real waveform and an ideal waveform is often slight. For example, rise and fall times of real pulses may be only a few nanoseconds and when viewed on an oscilloscope, as in Figure 11–35(b), appear to be ideal. In what follows, we will assume ideal waveforms.

(a) Positive pulse

(b) Negative pulse

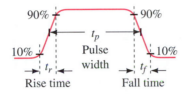

(a) Pulse definitions

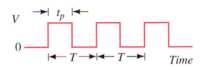

(c) Pulse train. *T* is referred to as the period of the pulse train

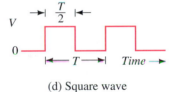

(d) Square wave

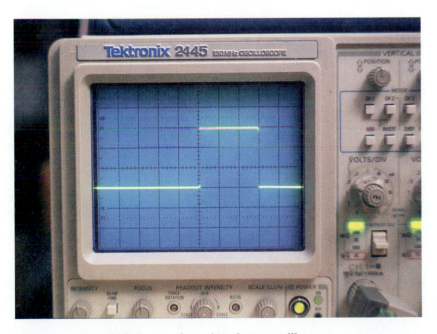

(b) Pulse waveform viewed on an oscilloscope.

FIGURE 11–35   Practical pulse waveforms.

(e) PRR = 2 pulses/s

FIGURE 11–34   Ideal pulses and pulse waveforms.

The Effect of Pulse Width

The width of a pulse relative to a circuit's time constant determines how it is affected by an *RC* circuit. Consider Figure 11–36. In (a), the circuit has been drawn to focus on the voltage across *C;* in (b), it has been drawn to focus on the voltage across *R.* (Otherwise, the circuits are identical.) An easy way to visualize the operation of these circuits is to assume that the pulse is generated by a switch that is moved rapidly back and forth between *V* and common as in (c). This alternately creates a charge and discharge circuit, and thus all of the ideas developed in this chapter apply directly.

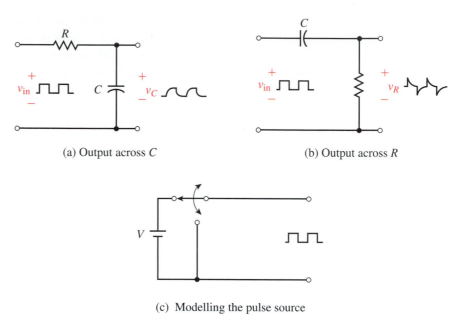

(a) Output across C (b) Output across R

(c) Modelling the pulse source

FIGURE 11–36 *RC* circuits with pulse input.

Pulse Width $t_p \gg 5\,\tau$

First, consider the ouput of circuit (a). When the pulse width and time between pulses are very long compared with the circuit time constant, the capacitor charges and discharges fully, Figure 11–37(b). (This case is similar to what we have already seen in this chapter.) Note, that charging and discharging occur at the transitions of the pulse. The transients therefore increase the rise and fall times of the output. In high-speed circuits, this may be a problem. (You will learn more about this in your digital electronics courses.)

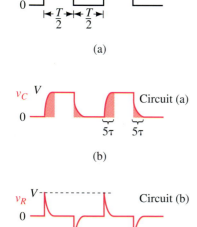

FIGURE 11–37 Pulse width much greater than 5 τ. Note that the shaded areas indicate where the capacitor is charging and discharging. Spikes occur on the input voltage transitions.

> **EXAMPLE 11–16** A square wave is applied to the input of Figure 11–36(a). If $R = 1$ kΩ and $C = 100$ pF, estimate the rise and fall time of the output signal using the universal time constant curve of Figure 11–15(a).
>
> **Solution** Here, $\tau = RC = (1 \times 10^3)(100 \times 10^{-12}) = 100$ ns. From Figure 11–15(a), note that v_C reaches the 10% point at about 0.1 τ, which is $(0.1)(100$ ns$) = 10$ ns. The 90% point is reached at about 2.3 τ, which is $(2.3)(100$ ns$) = 230$ ns. The rise time is therefore approximately 230 ns − 10 ns = 220 ns. The fall time will be the same.

Now consider the circuit in Figure 11–36(b). Here, current i_C will be similar to that of Figure 11–27(b), except that the pulse widths will be narrower. Since voltage $v_R = R\, i_C$, the output will be a series of short, sharp spikes that occur at input transitions as in Figure 11–37(c). Under the conditions here (i.e., pulse width much greater than the circuit time constant), v_R is an approximation to the derivative of v_{in} and the circuit is called a **differentiator circuit.** Such circuits have important practical uses.

Pulse Width $t_p = 5\,\tau$

These waveforms are shown in Figure 11–38. Since the pulse width is 5τ, the capacitor fully charges and discharges during each pulse. Thus waveforms here will be similar to what we have seen previously.

Pulse Width $t_p \ll 5\,\tau$

This case differs from what we have seen so far in this chapter only in that the capacitor does not have time to charge and discharge significantly between pulses. The result is that switching occurs on the early (nearly straight line) part of the charging and discharging curves and thus, v_C is roughly triangular in shape, Figure 11–39(a). As shown below, it has an average value of $V/2$. Under the conditions here, v_C is the approximate integral of v_{in} and the circuit is called an **integrator circuit.**

It should be noted that v_C does not reach the steady state shown in Figure 11–39 immediately. Instead, it works its way up over a period of five time constants (Figure 11–40). To illustrate, assume an input square wave of 5 V with a pulse width of 0.1 s and $\tau = 0.1$ s.

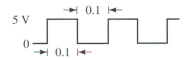

(a) Input waveform

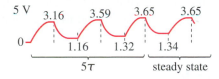

(b) Output voltage v_C

FIGURE 11–40 Circuit takes five time constants to reach a steady state.

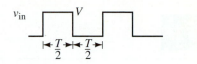

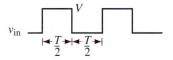

FIGURE 11–38 Pulse width equal to $5\,\tau$. These are the same as Figure 11–37 except that the transients last relatively longer.

Circuit (a)

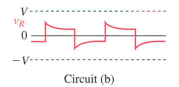

Circuit (b)

FIGURE 11–39 Pulse width much less than $5\,\tau$. The circuit does not have time to charge or discharge substantially.

Pulse 1: $v_C = E(1 - e^{-t/\tau})$. At the end of the first pulse ($t = 0.1$ s), v_C has climbed to $v_C = 5(1 - e^{-0.1/0.1}) = 5(1 - e^{-1}) = 3.16$ V. From the end of pulse 1 to the beginning of pulse 2 (i.e., over an interval of 0.1 s), v_C decays from 3.16 V to $3.16e^{-0.1/0.1} = 3.16e^{-1} = 1.16$ V.

Pulse 2: v_C starts at 1.16 V and 0.1 s later has a value of $v_C = E + (V_0 - E)e^{-t/\tau} = 5 + (1.16 - 5)e^{-0.1/0.1} = 5 - 3.84e^{-1} = 3.59$ V. It then decays to $3.59e^{-1} = 1.32$ V over the next 0.1 s.

Continuing in this manner, the remaining values for Figure 11–40(b) are determined. After $5\,\tau$, v_C cycles between 1.34 and 3.65 V, with an average of $(1.34 + 3.65)/2 = 2.5$ V, or half the input pulse amplitude.

Verify the remaining points of Figure 11–40(b).

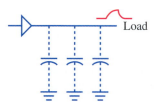

(a) Unloaded driver

Load

(b) Distorted signal

FIGURE 11–41   Distortion caused by capacitive loading.

Capacitive Loading

Capacitance occurs whenever conductors are separated by insulating material. This means that capacitance exists between wires in cables, between traces on printed circuit boards, and so on. In general, this capacitance is undesirable but it cannot be avoided. It is called **stray capacitance.** Fortunately, stray capacitance is often so small that it can be neglected. However, in high-speed circuits, it may cause problems.

To illustrate, consider Figure 11–41. The electronic driver of (a) produces square pulses. However, when it drives a long line as in (b), stray capacitance loads it and increases the signal's rise and fall times (since capacitance takes time to charge and discharge). If the rise and fall times become excessively long, the signal reaching the load may be so degraded that the system malfunctions. (Capacitive loading is a serious issue but we will leave it for future courses to deal with.)

11.8   Transient Analysis Using Computers

ELECTRONICS
WORKBENCH   PSpice

Electronics Workbench and PSpice are well suited for studying transients as they both incorporate easy to use graphing facilities that you can use to plot results directly on the screen. When plotting transients, you must specify the time scale for your plot—i.e., the length of time that you expect the transient to last. A good value to start with is 5τ where τ is the time constant of the circuit. (For complex circuits, if you do not know τ, make an estimate, run a simulation, adjust the time scale, and repeat until you get an acceptable plot.)

Electronics Workbench

Workbench provides two ways to view waveforms—via its oscilloscope or via its analysis graphing facility. Since the analysis graphing facility is easier to use and yields better output, we will begin with it. (In Chapter 14, we introduce the oscilloscope.) As a first example, consider the *RC* charging circuit of Figure 11–42. Determine capacitor voltage at $t = 50$ ms and $t = 150$ ms. (You don't need a switch; you simply tell Workbench to perform a transient analysis.)
Proceed as follows:

• Create the circuit of Figure 11–42 on the screen. (To display node numbering, select Circuit/Schematic Options, enable Show Nodes, then click OK).

• Select Analysis/Transient and in the dialog box, click Initial Conditions Set to Zero. Set End Time (TSTOP) to 0.25, highlight Node 2 (to display the voltage across the capacitor), then click on Add.

• Click Simulate. When the analysis is finished, the graph of Figure 11–43(b) should appear. Click the Toggle Grid icon on the Analysis Graphs menu bar. Expand to full screen.

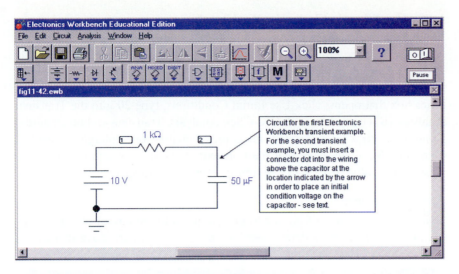

FIGURE 11–42 Electronics Workbench example. The switch is not required since the transient solution is initiated by software.

- Click the Toggle Cursor icon and drag the cursors to the specific time at which you want to read voltages.

Analysis of Results

As indicated in Figure 11–43(a), $v_C = 6.32$ V at $t = 50$ ms and 9.50 V at $t = 150$ ms. (Check by substitution into $v_C = 10(1 - e^{-20t})$. You will find that results agree exactly.)

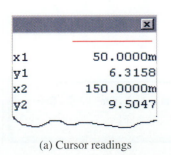

(a) Cursor readings

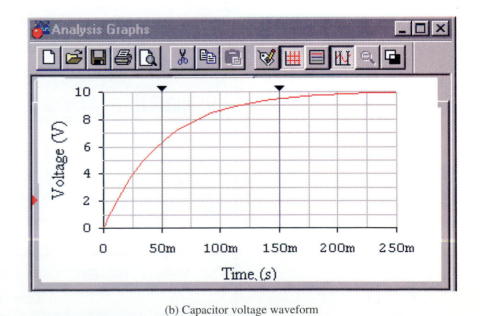

(b) Capacitor voltage waveform

FIGURE 11–43 Solution for the circuit of Figure 11–42. Since $\tau = 50$ ms, run the simulation to at least 250 ms (i.e., 5 τ).

Initial Conditions in Electronics Workbench

Let us change the above problem to include an initial voltage of 20 V on the capacitor. Get the circuit of Fig. 11–42 back on the screen, click the Basic Parts bin icon, drag the connector dot (●) and insert it into the circuit wiring slightly above the capacitor. Double click the dot and in the Connector Properties box that opens, click Use Initial Conditions, type **20** into the Transient Analysis (IC) box, then click OK. Click <u>A</u>nalysis, Transient and under Initial Conditions, click User-Defined, select the capacitor node voltage for display, then click Simulate. Note that the transient starts at 20 V and decays to its steady state value of 10 V in five time constants as expected.

Another Example

Using the clock source from the Sources bin, build the circuit of Figure 11–44. (The clock, with its default settings, produces a square wave that cycles

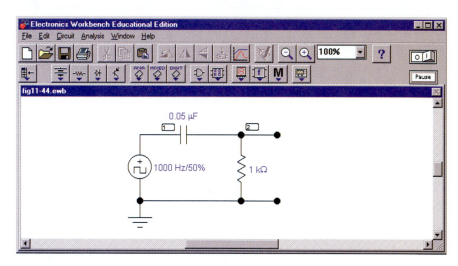

FIGURE 11–44   Applying a square wave source.

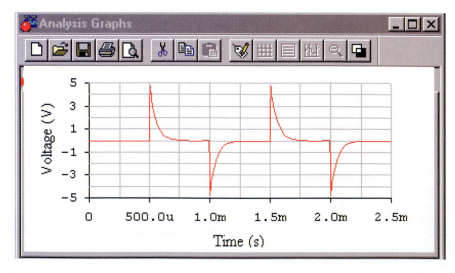

FIGURE 11–45   Output waveform for Figure 11–44. Compare to Figure 11–37(c).

between 0 V and 5 V with a cycle length of $T = 1$ ms.) This means that its *on* time t_p is $T/2 = 500$ μs. Since the time constant of Figure 11–44 is $\tau = RC = 50$ μs, t_p is greater than 5τ and a waveform similar to that of Figure 11–37(c) should result. To verify, follow the procedure of the previous example, except set End Time (TSTOP) to 0.0025 in the Analysis/Transient dialog box. The waveform of Figure 11–45 appears. Note that output spikes occur on the transitions of the input waveform as predicted.

OrCAD PSpice

As a first example, consider Figure 11–2 with $R = 200$ Ω, $C = 50$ μF and $E = 40$ V. Let the capacitor be initially uncharged (i.e., $V_o = 0$ V). First, read the PSpice Operational Notes, then proceed as follows:

- Create the circuit on the screen as in Figure 11–46. (The switch can be found in the EVAL library as part Sw_tClose.) Remember to rotate the capacitor three times as discussed in Appendix A, then set its initial condition (*IC*) to zero. To do this, double click the capacitor symbol, type **0V** into the Property editor cell labeled IC, click Apply, then close the editor. Click the New Simulation Profile icon, enter a name (e.g., **Figure 11–47**) then click Create. In the Simulation Settings box, click the Analysis tab, select Time Domain (Transient) and in Options, select General Settings. Set the duration of the transient (TSTOP) to 50ms (i.e., five time constants). Find the voltage marker on the toolbar and place as shown.

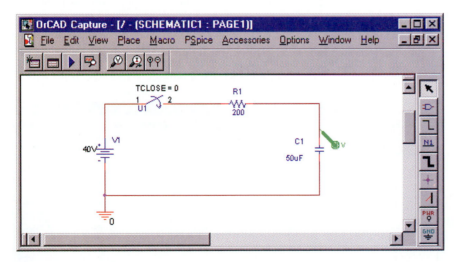

FIGURE 11–46 PSpice example. The voltage marker displays voltage with respect to ground, which, in this case, is the voltage across C_1

- Click the Run icon. When simulation is complete, a trace of capacitor voltage versus time (the green trace of Figure 11–47) appears. Click Plot (on the toolbar) then add Y Axis to create the second axis. Activate the additional toolbar icons described in Operational Note 5, then click the Add Trace icon on the new toolbar. In the dialog box, click I(C1) (assuming your capacitor is designated C_1), then OK. This adds the current trace.

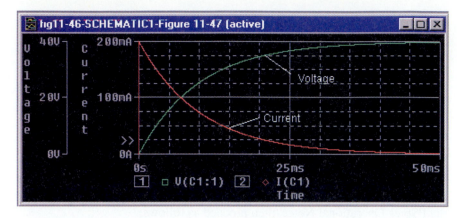

FIGURE 11–47   Waveforms for the circuit of Figure 11–46.

Analysis of Results

Click the Toggle cursor icon, then use the cursor to determine values from the screen. For example, at $t = 5$ ms, you should find $v_C = 15.7$ V and $i_C = 121$ mA. (An analytic solution for this circuit (which is Figure 11–23) may be found in Example 11–10, part (c). It agrees exactly with the PSpice solution.)

As a second example, consider the circuit of Figure 11–21 (shown as Figure 11–48). Create the circuit using the same general procedure as in the previous example, except do not rotate the capacitor. Again, be sure to set V_0 (the initial capacitor voltage) to zero. In the Simulation Profile box, set TSTOP to 50ms. Place differential voltage markers (found on the toolbar at the top of the screen) across C to graph the capacitor voltage. Run the analysis, create a second axis, then add the current plot. You should get the same graph (i.e., Figure 11–47) as you got for the previous example, since its circuit is the Thévenin equivalent of this one.

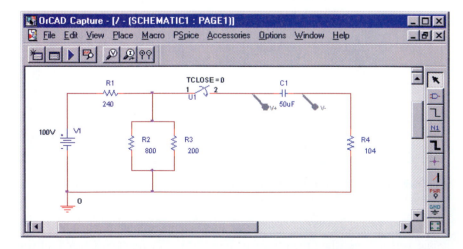

FIGURE 11–48   Differential markers are used to display the voltage across C_1.

As a final example, consider Figure 11–49(a), which shows double switching action.

EXAMPLE 11–17   The capacitor of Figure 11–49(a) has an initial voltage of −10 V. The switch is moved to the charge position for 1 s, then to the discharge position where it remains. Determine curves for v_C and i_C.

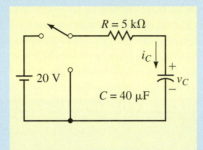

(a) Circuit to be modelled

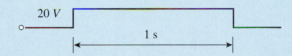

(b) The applied pulse

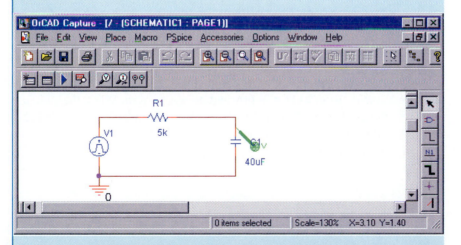

(c) Modelling the switching action using a pulse source

FIGURE 11–49   Creating a charge/discharge waveform using PSpice.

Solution   PSpice has no switch that implements the above switching sequence. However, moving the switch first to charge then to discharge is equivalent to placing 20 V across the *RC* combination for the charge time, then 0 V thereafter as indicated in (b). You can do this with a pulse source (VPULSE) as indicated in (c). (VPULSE is found in the SOURCE library.) To set pulse parameters, double click the VPULSE symbol, then scroll its Property editor as described in the Operational Notes until you find a group of cells labeled PER, PW, etc. Enter **5s** for PER, **1s** for PW, **1us** for TF, **1us**

for TR, **0** for TD, **20V** for *V2*, and **0V** for *V1*. (This defines a pulse with a period of 5 s, a width of 1 s, rise and fall times of 1 μs, amplitude of 20 V, and an initial value of 0 V.) Click Apply, then close the Property editor. Double click the capacitor symbol and set *IC* to $-10V$ in the Properties Editor. Set TSTOP to 2s. Place a Voltage Marker as shown, then click Run. You should get the voltage trace of Figure 11–50 on the screen. Add the second axis and the current trace as described in the previous examples. The red current curve should appear.

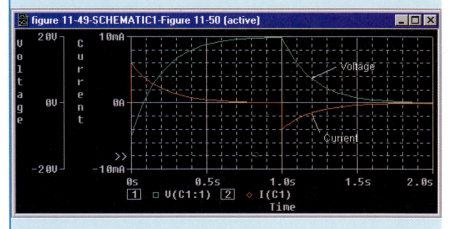

FIGURE 11–50 Waveforms for the circuit of Figure 11–49 with $V_0 = -10$ V.

Note that voltage starts at -10 V and climbs to 20 V while current starts at $(E - V_0)/R = 30$ V/5 k$\Omega = 6$ mA and decays to zero. When the switch is turned to the discharge position, the current drops from 0 A to -20 V/5 k$\Omega = -4$ mA and then decays to zero while the voltage decays from 20 V to zero. Thus the solution checks.

PUTTING IT INTO PRACTICE

An electronic device employs a timer circuit of the kind shown in Figure 11–32(a), i.e., an *RC* charging circuit and a threshold detector. (Its timing waveforms are thus identical to those of Figure 11–32(b)). The input to the *RC* circuit is a 0 to 5 V $\pm 4\%$ step, $R = 680$ kΩ $\pm 10\%$, $C = 0.22$ μF $\pm 10\%$, the threshold detector activates at $v_C = 1.8$ V ± 0.05 V and the required delay is 67 ms ± 18 ms. You test a number of units as they come off the production line and find that some do not meet the timing spec. Perform a design review and determine the cause. Redesign the timing portion of the circuit in the most economical way possible.

11.1   Introduction

1. The capacitor of Figure 11–51 is uncharged.

   a. What are the capacitor voltage and current just after the switch is closed?

   b. What are the capacitor voltage and current after the capacitor is fully charged?

2. Repeat Problem 1 if the 20-V source is replaced by a −60-V source.

3. a. What does an uncharged capacitor look like at the instant of switching?

   b. What does a charged capacitor look like at the instant of switching?

   c. What does a capacitor look like to steady state dc?

   d. What do we mean by $i(0^-)$? By $i(0^+)$?

4. For a charging circuit, $E = 25$ V, $R = 2.2$ kΩ, and the capacitor is initially uncharged. The switch is closed at $t = 0$. What is $i(0^+)$?

5. For a charging circuit, $R = 5.6$ kΩ and $v_C(0^-) = 0$ V. If $i(0^+) = 2.7$ mA, what is E?

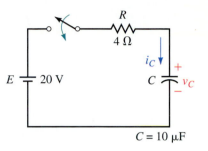

FIGURE 11–51

11.2   Capacitor Charging Equations

6. The switch of Figure 11–51 is closed at $t = 0$ s. The capacitor is initially uncharged.

   a. Determine the equation for charging voltage v_C.

   b. Determine the equation for charging current i_C.

   c. By direct substitution, compute v_C and i_C at $t = 0^+$ s, 40 μs, 80 μs, 120 μs, 160 μs, and 200 μs.

   d. Plot v_C and i_C on graph paper using the results of (c). Hint: See Example 11–2.

7. Repeat Problem 6 if $R = 500$ Ω, $C = 25$ μF, and $E = 45$ V, except compute and plot values at $t = 0^+$ s, 20 ms, 40 ms, 60 ms, 80 ms, and 100 ms.

8. The switch of Figure 11–52 is closed at $t = 0$ s. Determine the equations for capacitor voltage and current. Compute v_C and i_C at $t = 50$ ms.

9. Repeat Problem 8 for the circuit of Figure 11–53.

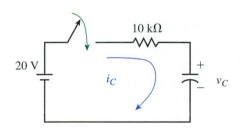

FIGURE 11–52 $V_0 = 0$ V, $C = 10$ μF.

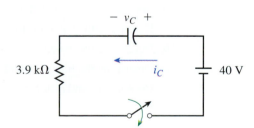

FIGURE 11–53 $C = 10$ μF, $V_0 = 0$ V.

10. The capacitor of Figure 11–2 is uncharged at the instant the switch is closed. If $E = 80$ V, $C = 10$ μF, and $i_C(0^+) = 20$ mA, determine the equations for v_C and i_C.

11. Determine the time constant for the circuit of Figure 11–51. How long (in seconds) will it take for the capacitor to charge?

12. A capacitor takes 200 ms to charge. If $R = 5$ kΩ, what is C?

13. For Figure 11–51, the capacitor voltage with the switch open is 0 V. Close the switch at $t = 0$ and determine capacitor voltage and current at $t = 0^+$, 40 μs, 80 μs, 120 μs, 160 μs, and 200 μs using the universal time constant curves.

14. If $i_C = 25e^{-40t}$ A, what is the time constant τ and how long will the transient last?

15. For Figure 11–2, the current jumps to 3 mA when the switch is closed. The capacitor takes 1 s to charge. If $E = 75$ V, determine R and C.

16. For Figure 11–2, if $v_C = 100(1 - e^{-50t})$ V and $i_C = 25e^{-50t}$ mA, what are E, R, and C?

17. For Figure 11–2, determine E, R, and C if the capacitor takes 5 ms to charge, the current at 1 time constant after the switch is closed is 3.679 mA, and the capacitor charges to 45 volts.

18. For Figure 11–2, $v_C(\tau) = 41.08$ V and $i_C(2\tau) = 219.4$ mA. Determine E and R.

11.3   Capacitor with an Initial Voltage

19. The capacitor of Figure 11–51 has an initial voltage. If $V_0 = 10$ V, what is the current just after the switch is closed?

20. Repeat Problem 19 if $V_0 = -10$ V.

21. For the capacitor of Figure 11–51, $V_0 = 30$ V.
    a. Determine the expression for charging voltage v_C.
    b. Determine the expression for current i_C.
    c. Sketch v_C and i_C.

22. Repeat Problem 21 if $V_0 = -5$ V.

11.4   Capacitor Discharging Equations

23. For the circuit of Figure 11–54, assume the capacitor is charged to 50 V before the switch is closed.
    a. Determine the equation for discharge voltage v_C.
    b. Determine the equation for discharge current i_C.
    c. Determine the time constant of the circuit.
    d. Compute v_C and i_C at $t = 0^+$ s, $t = \tau$, 2τ, 3τ, 4τ, and 5τ.
    e. Plot the results of (d) with the time axis scaled in seconds and time constants.

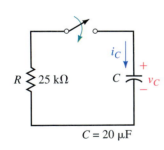

$C = 20$ μF

FIGURE 11–54

24. The initial voltage on the capacitor of Figure 11–54 is 55 V. The switch is closed at $t = 0$. Determine capacitor voltage and current at $t = 0^+$, 0.5 s, 1 s, 1.5 s, 2 s, and 2.5 s using the universal time constant curves.

25. A 4.7-μF capacitor is charged to 43 volts. If a 39-kΘ resistor is then connected across the capacitor, what is its voltage 200 ms after the resistor is connected?

26. The initial voltage on the capacitor of Figure 11–54 is 55 V. The switch is closed at $t = 0$ s and opened 1 s later. Sketch v_C. What is the capacitor's voltage at $t = 3.25$ s?

27. For Figure 11–55, let $E = 200$ V, $R_2 = 1$ kΩ, and $C = 0.5$ μF. After the capacitor has fully charged in position 1, the switch is moved to position 2.

    a. What is the capacitor voltage immediately after the switch is moved to position 2? What is its current?

    b. What is the discharge time constant?

    c. Determine discharge equations for v_C and i_C.

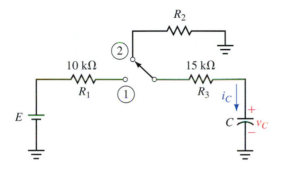

FIGURE 11–55

28. For Figure 11–55, C is fully charged before the switch is moved to discharge. Current just after it is moved is $i_C = -4$ mA and C takes 20 ms to discharge. If $E = 80$ V, what are R_2 and C?

11.5   More Complex Circuits

29. The capacitors of Figure 11–56 are uncharged. The switch is closed at $t = 0$. Determine the equation for v_C. Compute v_C at one time constant using the equation and the universal time constant curve. Compare answers.

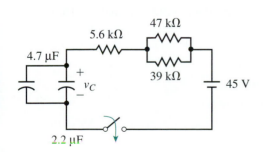

FIGURE 11–56

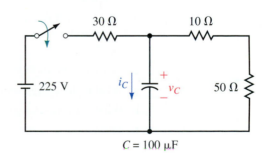

FIGURE 11–57

30. For Figure 11–57, the switch is closed at $t = 0$. Given $V_0 = 0$ V.

    a. Determine the equations for v_C and i_C.

    b. Compute the capacitor voltage at $t = 0^+$, 2, 4, 6, 8, 10, and 12 ms.

c. Repeat (b) for the capacitor current.

d. Why does 225 V/30 Ω also yield $i(0^+)$?

31. Repeat Problem 30, parts (a) to (c) for the circuit of Figure 11–58.

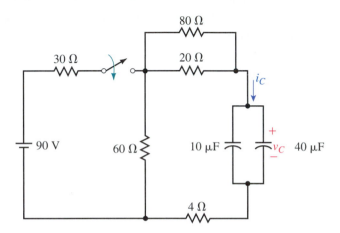

FIGURE 11–58

32. Consider again Figure 11–55. Suppose $E = 80$ V, $R_2 = 25$ kΩ, and $C = 0.5$ μF:

a. What is the charge time constant?

b. What is the discharge time constant?

c. With the capacitor initially discharged, move the switch to position 1 and determine equations for v_C and i_C during charge.

d. Move the switch to the discharge position. How long does it take for the capacitor to discharge?

e. Sketch v_C and i_C from the time the switch is placed in charge to the time that the capacitor is fully discharged. Assume the switch is in the charge position for 80 ms.

33. For the circuit of Figure 11–55, the capacitor is initially uncharged. The switch is first moved to charge, then to discharge, yielding the current shown in Figure 11–59. The capacitor fully charges in 12.5 s. Determine E, R_2, and C.

34. Refer to the circuit of Figure 11–60:

a. What is the charge time constant?

b. What is the discharge time constant?

c. The switch is in position 2 and the capacitor is uncharged. Move the switch to position 1 and determine equations for v_C and i_C.

d. After the capacitor has charged for two time constants, move the switch to position 2 and determine equations for v_C and i_C during discharge.

e. Sketch v_C and i_C.

35. Determine the capacitor voltages and the source current for the circuit of Figure 11–61 after it has reached steady state.

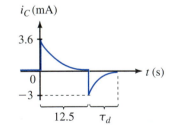

FIGURE 11–59

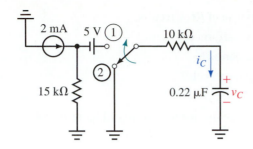

FIGURE 11–60

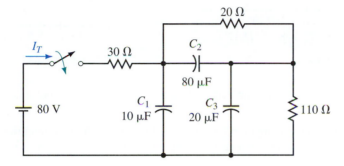

FIGURE 11–61

36. A black box containing dc sources and resistors has open-circuit voltage of 45 volts as in Figure 11–62(a). When the output is shorted as in (b), the short-circuit current is 1.5 mA. A switch and an uncharged 500-μF capacitor are connected as in (c). Determine the capacitor voltage and current 25 s after the switch is closed.

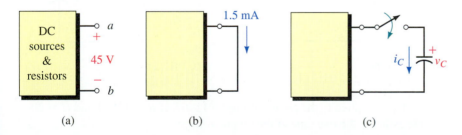

(a)   (b)   (c)

FIGURE 11–62

11.6   An *RC* Timing Application

37. For the alarm circuit of Figure 11–32, if the input from the sensor is 5 V, $R = 750$ kΩ, and the alarm is activated at 15 s when $v_C = 3.8$ V, what is *C*?

38. For the alarm circuit of Figure 11–32, the input from the sensor is 5 V, $C = 47$ μF, and the alarm is activated when $v_C = 4.2$ V. Choose the nearest standard resistor value to achieve a delay of at least 37 s.

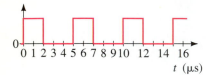

FIGURE 11–63

11.7 Pulse Response of *RC* Circuits

39. Consider the waveform of Figure 11–63.

   a. What is the period?

   b. What is the duty cycle?

   c. What is the PRR?

40. Repeat Problem 39 for the waveform of Figure 11–64.

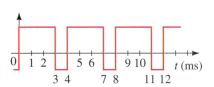

FIGURE 11–64

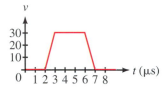

FIGURE 11–65

41. Determine the rise time, fall time, and pulse width for the pulse in Figure 11–65.

42. A single pulse is input to the circuit of Figure 11–66. Assuming that the capacitor is initially uncharged, sketch the output for each set of values below:

   a. $R = 2$ kΩ, $C = 1$ μF.

   b. $R = 2$ kΩ, $C = 0.1$ μF.

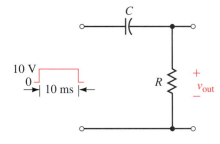

FIGURE 11–66

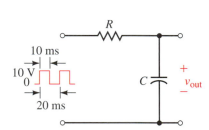

FIGURE 11–67

43. A step is applied to the circuit of Figure 11–67. If $R = 150$ Ω and $C = 20$ pF, estimate the rise time of the output voltage.

44. A pulse train is input to the circuit of Figure 11–67. Assuming that the capacitor is initially uncharged, sketch the output for each set of values below after the circuit has reached steady state:

   a. $R = 2$ kΩ, $C = 0.1$ μF.

   b. $R = 20$ kΩ, $C = 1.0$ μF.

11.8 Transient Analysis Using Computers

45. **EWB** Graph capacitor voltage for the circuit of Figure 11–2 with $E = -25$ V, $R = 40$ Ω, $V_0 = 0$ V, and $C = 400$ μF. Scale values from the plot at $t = 20$ ms using the cursor. Compare to the results you get using Equation 11–3 or the curve of Figure 11–15(a).

46. **EWB**   Obtain a plot of voltage versus time across R for the circuit of Figure 11–68. Assume an initially uncharged capacitor. Use the cursor to read voltage $t = 50$ ms and use Ohm's law to compute current. Compare to the values determined analytically. Repeat if $V_0 = 100$ V.

NOTE...

Using its default setting, Electronics Workbench generates plotting time steps automatically. Sometimes, however, it does not generate enough and you get a jagged curve. To specify more points, under <u>A</u>nalysis/Transient, click the "Generate time steps automatically" box, then click the "Minimum number of time points" button and type in a suitable value (for example, 1000). Experiment until you get a suitably smooth curve.

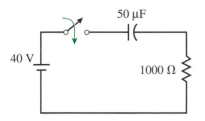

FIGURE 11–68

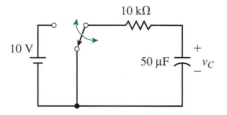

FIGURE 11–69

47. **EWB**   The switch of Figure 11–69 is initially in the discharge position and the capacitor is uncharged. Move the switch to charge for 1 s, then to discharge where it remains. Solve for v_C. (Hint: Use Workbench's time delay (TD) switch. Double click it and set TON to 1 s and TOFF to 0. This will cause the switch to move to the charge position for 1 s, then return to discharge where it stays.) With the cursor, determine the peak voltage. Using Equation 11–3, compute v_C at $t = 1$ s and compare it to the value obtained above.

48. **EWB**   Use Workbench to graph capacitor voltage for the circuit of Problem 34. With the cursor, determine v_C at $t = 10$ ms and 12 ms. Compare to the theoretical answers of 29.3 V and 15.3 V.

49. **PSpice**   Graph capacitor voltage and current for a charging circuit with $E = -25$ V, $R = 40\ \Omega$, $V_0 = 0$ V, and $C = 400\ \mu F$. Scale values from the plot using the cursor. Compare to the results you get using Equations 11–3 and 11–5 or the curves of Figure 11–15.

50. **PSpice**   Repeat the problem of Question 46 using PSpice. Plot both voltage and current.

51. **PSpice**   The switch of Figure 11–70 is closed at $t = 0$ s. Plot voltage and current waveforms. Use the cursor to determine v_C and i_C at $t = 10$ ms.

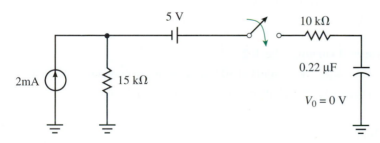

FIGURE 11–70

52. **PSpice**   Redo Example 11–17 with the switch in the charge position for 0.5 s and everything else the same. With your calculator, compute v_C and i_C

at 0.5 s and compare them to PSpice's plot. Repeat for i_C just after moving the switch to the discharge position.

53. **PSpice**   Use PSpice to solve for voltages and currents in the circuit of Figure 11–61. From this, determine the final (steady state) voltages and currents and compare them to the answers of Problem 35.

ANSWERS TO IN-PROCESS LEARNING CHECKS

In-Process Learning Check 1

1. 0.2 A

2. a. 50 ms

   b.

t(ms)	i_C (mA)
0	50
25	30.3
50	18.4
75	11.2
100	6.8
500	0.0

3. a. 20 ms

   b.

t(ms)	v_C (V)
0	0
25	71.3
50	91.8
75	97.7
100	99.3
500	100

4. $40(1 - e^{-5t})$ V; $200e^{-5t}$ mA;   38.0 V;   9.96 mA

5. a. $v_C(0^+) = v_C(0^-) = 0$ b. $i_C(0^-) = 0$; $i_C(0^+) = 10$ mA   c. 100 V 0 A

6. 80 V   40 kΩ   0.2 μF

7. 38.0 V   9.96 mA

In-Process Learning Check 2

1. a. $40 + 40e^{-5t}$ V. v_C starts at 80 V and decays exponentially to 40 V.

   b. There is no transient since initial value = final value.

   c. $40 - 100e^{-5t}$ V. v_C starts at −60 V and climbs exponentially to 40 V.

2. 0.1833 s

3. a. $100e^{-10t}$ V; $-10e^{-10t}$ mA; v_C starts at 100 V and decays to 0 in 0.5 s (i.e., 5 time constants); i_C starts at −10 mA and decays to 0 in 0.5 s

   b. $-100e^{-10t}$ V; $10e^{-10t}$ mA; v_C starts at −100 V and decays to 0 in 0.5 s (i.e., 5 time constants); i_C starts at 10 mA and decays to 0 in 0.5 s

4. a., b., and c.   Same as Example 11–12

   d. $-30 + 37.14e^{-250t}$ V   e. $-37.14e^{-250t}$ mA

   f.

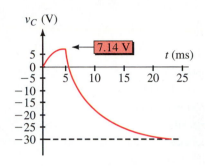

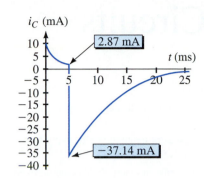

5. $6(1 - e^{-25t})$ V; $150e^{-25t}$ mA

12 Magnetism and Magnetic Circuits

OBJECTIVES

After studying this chapter, you will be able to

- represent magnetic fields using Faraday's flux concept,
- describe magnetic fields quantitatively in terms of flux and flux density,
- explain what magnetic circuits are and why they are used,
- determine magnetic field intensity or magnetic flux density from a *B-H* curve,
- solve series magnetic circuits,
- solve series-parallel magnetic circuits,
- compute the attractive force of an electromagnet,
- explain the domain theory of magnetism,
- describe the demagnetization process.

KEY TERMS

Ampere's Law
Ampere-Turns
Domain Theory
Ferromagnetic
Flux
Flux Density
Fringing
Hall Effect
Hysteresis
Magnetic Circuit

Magnetic Field
Magnetic Field Intensity
Magnetomotive Force
Permeability
Reluctance
Residual Magnetism
Right-Hand Rule
Saturation
Tesla
Weber

OUTLINE

The Nature of a Magnetic Field
Electromagnetism
Flux and Flux Density
Magnetic Circuits
Air Gaps, Fringing, and Laminated Cores
Series Elements and Parallel Elements
Magnetic Circuits with DC Excitation
Magnetic Field Intensity and Magnetization Curves
Ampere's Circuital Law
Series Magnetic Circuits: Given Φ, Find *NI*
Series-Parallel Magnetic Circuits
Series Magnetic Circuits: Given *NI*, Find Φ
Force Due to an Electromagnet
Properties of Magnetic Materials
Measuring Magnetic Fields

Many common devices rely on magnetism. Familiar examples include computer disk drives, tape recorders, VCRs, transformers, motors, generators, and so on. To understand their operation, you need a knowledge of magnetism and magnetic circuit principles. In this chapter, we look at fundamentals of magnetism, relationships between electrical and magnetic quantities, magnetic circuit concepts, and methods of analysis. In Chapter 13, we look at electromagnetic induction and inductance, and in Chapter 24, we apply magnetic principles to the study of transformers.

Magnetism and Electromagnetism

WHILE THE BASIC FACTS about magnetism have been known since ancient times, it was not until the early 1800s that the connection between electricity and magnetism was made and the foundations of modern electromagnetic theory laid down.

In 1819, Hans Christian Oersted, a Danish scientist, demonstrated that electricity and magnetism were related when he showed that a compass needle was deflected by a current-carrying conductor. The following year, Andre Ampere (1775–1836) showed that current-carrying conductors attract or repel each other just like magnets. However, it was Michael Faraday (recall Chapter 10) who developed our present concept of the magnetic field as a collection of flux lines in space that conceptually represent both the intensity and the direction of the field. It was this concept that led to an understanding of magnetism and the development of important practical devices such as the transformer and the electric generator.

In 1873, James Clerk Maxwell (see photo), a Scottish scientist, tied the then known theoretical and experimental concepts together and developed a unified theory of electromagnetism that predicted the existence of radio waves. Some 30 years later, Heinrich Hertz, a German physcist, showed experimentally that such waves existed, thus verifying Maxwell's theories and paving the way for modern radio and television.

12.1   The Nature of a Magnetic Field

Magnetism refers to the force that acts between magnets and magnetic materials. We know, for example, that magnets attract pieces of iron, deflect compass needles, attract or repel other magnets, and so on. This force acts at a distance and without the need for direct physical contact. The region where the force is felt is called the "field of the magnet" or simply, its **magnetic field.** Thus, *a magnetic field is a force field.*

Magnetic Flux

Faraday's flux concept (recall Putting It Into Perspective, Chapter 10) helps us visualize this field. Using Faraday's representation, magnetic fields are shown as lines in space. These lines, called **flux lines** or **lines of force,** show the direction and intensity of the field at all points. This is illustrated in Figure 12–1 for the field of a bar magnet. As indicated, the field is strongest at the **poles** of the magnet (where flux lines are most dense), its direction is from north (N) to south (S) external to the magnet, and flux lines never cross. The symbol for magnetic flux (Figure 12–1) is the Greek letter Φ (phi).

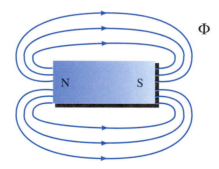

FIGURE 12–1   Field of a bar magnet. Flux is denoted by the Greek letter Φ.

Figure 12–2 shows what happens when two magnets are brought close together. In (a), unlike poles attract, and flux lines pass from one magnet to the other. In (b), like poles repel, and the flux lines are pushed back as indicated by the flattening of the field between the two magnets.

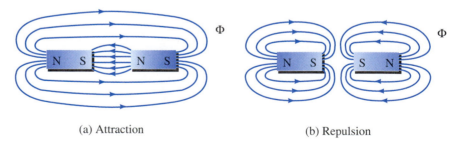

(a) Attraction   (b) Repulsion

FIGURE 12–2   Field patterns due to attraction and repulsion.

Ferromagnetic Materials

Magnetic materials (materials that are attracted by magnets such as iron, nickel, cobalt, and their alloys) are called **ferromagnetic** materials. Ferromagnetic materials provide an easy path for magnetic flux. This is illustrated in Figure 12–3 where the flux lines take the longer (but easier) path through the soft iron, rather than the shorter path that they would normally take (recall Figure 12–1). Note, however, that nonmagnetic materials (plastic, wood, glass, and so on) have no effect on the field.

Figure 12–4 shows an application of these principles. Part (a) shows a simplified representation of a loudspeaker, and part (b) shows expanded

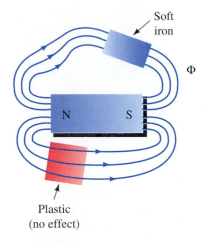

FIGURE 12–3   Magnetic field follows the longer (but easier) path through the iron. The plastic has no effect on the field.

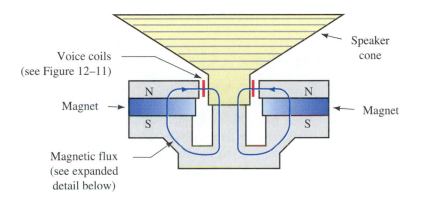

(a) Simplified representation of the magnetic field. Here, the complex field of (b) is represented symbolically by a single line

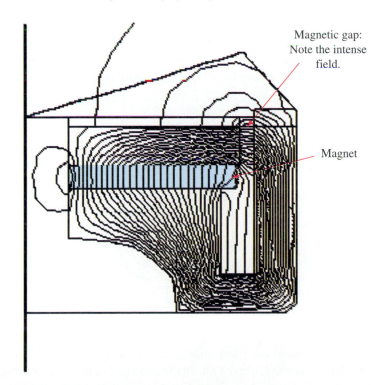

(b) Magnetic field pattern for the loud speaker.
(Courtesy JBL Professional)

FIGURE 12–4   Magnetic circuit of a loudspeaker. The magnetic structure and voice coil are called a "speaker motor". The field is created by the permanent magnet.

details of its magnetic field. (Since the speaker is symmetrical, only half its structure is shown in (b).) The field is created by the permanent magnet, and the iron pole pieces guide the field and concentrate it in the gap where the speaker coil is placed. (For a description of how the speaker works, see Section 12.4.) Within the iron structure, the flux crowds together at sharp interior corners, spreads apart at exterior corners, and is essentially uniform elsewhere. This is characteristic of magnetic fields in iron.

12.2   Electromagnetism

Most applications of magnetism involve magnetic effects due to electric currents. We look first at some basic principles. Consider Figure 12–5. The current, *I*, creates a magnetic field that is concentric about the conductor, uniform along its length, and whose strength is directly proportional to *I*. Note the direction of the field. It may be remembered with the aid of the **right-hand rule.** As indicated in (b), imagine placing your right hand around the conductor with your thumb pointing in the direction of current. Your fingers then point in the direction of the field. If you reverse the direction of the current, the direction of the field reverses. If the conductor is wound into a coil, the fields of its individual turns combine, producing a resultant field as in Figure 12–6. The direction of the coil flux can also be remembered by means of a simple rule: curl the fingers of your right hand around the coil in the direction of the current and your thumb will point in the direction of the field. If the direction of the current is reversed, the field also reverses. Provided no ferromagnetic material is present, the strength of the coil's field is directly proportional to its current.

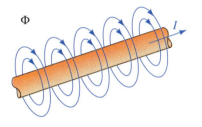

(a) Magnetic field produced by current. Field is proportional to *I*

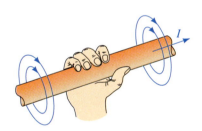

(b) Right-hand rule

FIGURE 12–5   Field about a current-carrying conductor. If the current is reversed, the field reverses direction.

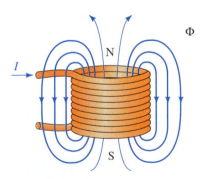

FIGURE 12–6   Field produced by a coil.

If the coil is wound on a ferromagnetic core as in Figure 12–7 (transformers are built this way), almost all flux is confined to the core, although a small amount (called stray or leakage flux) passes through the surrounding air. However, now that ferromagnetic material is present, the core flux is no longer proportional to current. The reason for this is discussed in Section 12.14.

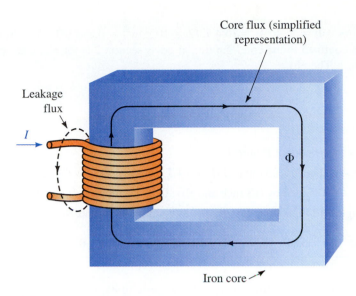

Core flux (simplified
representation)

Leakage
flux

I

Φ

Iron core

FIGURE 12–7   For ferromagnetic materials, most flux is confined to the core.

12.3   Flux and Flux Density

As noted in Figure 12–1, magnetic flux is represented by the symbol Φ. In
the SI system, the unit of flux is the **weber** (Wb), in honor of pioneer
researcher Wilhelm Eduard Weber, 1804–1891. However, we are often more
interested in **flux density** B (i.e., flux per unit area) than in total flux Φ.
Since flux Φ is measured in Wb and area A in m^2, flux density is measured
as Wb/m^2. However, to honor Nikola Tesla (another early researcher, 1856–
1943) the unit of flux density is called the **tesla** (T) where $1\ T = 1\ Wb/m^2$.
Flux density is found by dividing the total flux passing perpendicularly
through an area by the size of the area, Figure 12–8. That is,

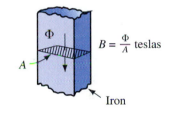

Φ

$B = \dfrac{\Phi}{A}$ teslas

A

Iron

FIGURE 12–8   Concept of flux density. $1\ T = 1\ Wb/m^2$.

$$B = \frac{\Phi}{A}\quad \text{(tesla, T)}\qquad\qquad (12\text{--}1)$$

Thus, if $\Phi = 600\ \mu Wb$ of flux pass perpendicularly through an area $A = 20 \times
10^{-4}\ m^2$, the flux density is $B = (600 \times 10^{-6}\ Wb)/(20 \times 10^{-4}\ m^2) = 0.3$ T.
The greater the flux density, the stronger the field.

EXAMPLE 12–1   For the magnetic core of Figure 12–9, the flux density at
cross section 1 is $B_1 = 0.4$ T. Determine B_2.

FIGURE 12–9

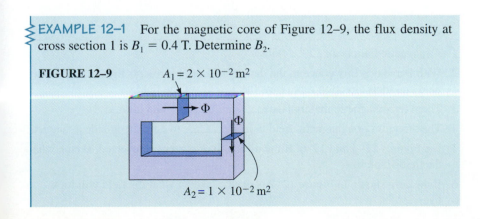

$A_1 = 2 \times 10^{-2}\ m^2$

Φ

Φ

$A_2 = 1 \times 10^{-2}\ m^2$

Solution $\Phi = B_1 \times A_1 = (0.4 \text{ T})(2 \times 10^{-2} \text{ m}^2) = 0.8 \times 10^{-2}$ Wb. Since all flux is confined to the core, the flux at cross section 2 is the same as at cross section 1. Therefore,

$$B_2 = \Phi/A_2 = (0.8 \times 10^{-2} \text{ Wb})/(1 \times 10^{-2} \text{ m}^2) = 0.8 \text{ T}$$

PRACTICE PROBLEMS 1

1. Refer to the core of Figure 12–8:
   a. If A is 2 cm × 2.5 cm and $B = 0.4$ T, compute Φ in webers.
   b. If A is 0.5 inch by 0.8 inch and $B = 0.35$ T, compute Φ in webers.
2. In Figure 12–9, if $\Phi = 100 \times 10^{-4}$ Wb, compute B_1 and B_2.

Answers: 1. a. 2×10^{-4} Wb   b. 90.3 μWb   2. 0.5 T; 1.0 T

To gain a feeling for the size of magnetic units, note that the strength of the earth's field is approximately 50 μT near the earth's surface, the field of a large generator or motor is on the order of 1 or 2 T, and the largest fields yet produced (using superconducting magnets) are on the order of 25 T.

Other systems of units (now largely superseded) are the CGS system and the English systems. In the CGS system, flux is measured in maxwells and flux density in gauss. In the English system, flux is measured in lines and flux density in lines per square inch. Conversion factors are given in Table 12–1. We use only the SI system in this book.

TABLE 12–1 Magnetic Units Conversion Table

System	Flux (Φ)	Flux Density (B)
SI	webers (Wb)	teslas (T) $1 \text{ T} = 1 \text{ Wb/m}^2$
English	lines $1 \text{ Wb} = 10^8$ lines	lines/in^2 $1 \text{ T} = 6.452 \times 10^4$ lines/in^2
CGS	maxwells $1 \text{ Wb} = 10^8$ maxwells	gauss $1 \text{ gauss} = 1 \text{ maxwell/cm}^2$ $1 \text{ T} = 10^4$ gauss

IN-PROCESS LEARNING CHECK 1

1. A magnetic field is a _____ field.
2. With Faraday's flux concept, the density of lines represents the _____ of the field and their direction represents the _____ of the field.
3. Three ferromagnetic materials are _____, _____, and _____.
4. The direction of a magnetic field is from _____ to _____ outside a magnet.
5. For Figures 12–5 and 12–6, if the direction of current is reversed, sketch what the fields look like.
6. If the core shown in Figure 12–7 is plastic, sketch what the field will look like.

7. Flux density *B* is defined as the ratio Φ/A, where *A* is the area (parallel, perpendicular) to Φ.

8. For Figure 12–9, if A_1 is 2 cm $\times$ 2.5 cm, B_1 is 0.5 T, and $B_2 = 0.25$ T, what is A_2?

(Answers are at the end of the chapter.)

12.4 Magnetic Circuits

Most practical applications of magnetism use magnetic structures to guide and shape magnetic fields by providing a well-defined path for flux. Such structures are called **magnetic circuits.** Magnetic circuits are found in motors, generators, computer disk drives, tape recorders, and so on. The speaker of Figure 12–4 illustrates the concept. It uses a powerful magnet to create flux and an iron circuit to guide the flux to the air gap to provide the intense field required by the voice coil. Note how effectively it does its job; almost the entire flux produced by the magnet is confined to the iron path with little leakage into the air.

A second example is shown in Figures 12–10 and 12–11. Tape recorders, VCRs, and computer disk drives all store information magnetically on iron oxide coated surfaces for later retrieval and use. The basic tape recorder scheme is shown symbolically in Figure 12–10. Sound picked up by a microphone is converted to an electrical signal, amplified, and the output applied to the record head. The record head is a small magnetic circuit. Current from the amplifier passes through its coil, creating a magnetic field that magnetizes the moving tape. The magnetized patterns on the tape correspond to the original sound input.

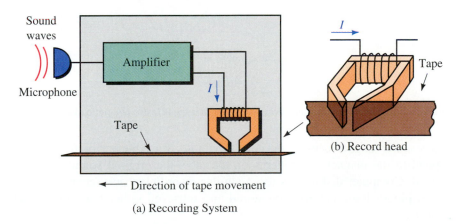

(a) Recording System

FIGURE 12–10   The recording head of a tape recorder is a magnetic circuit.

During playback the magnetized tape is passed by a playback head, as shown in Figure 12–11(a). Voltages induced in the playback coil are amplified and applied to a speaker. The speaker (b) utilizes a flexible cone to reproduce sound. A coil of fine wire attached at the apex of this cone is placed in the field of the speaker air gap. Current from the amplifier passes through this coil, creating a varying field that interacts with the fixed field of

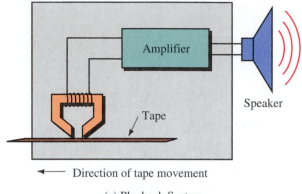

(a) Playback System

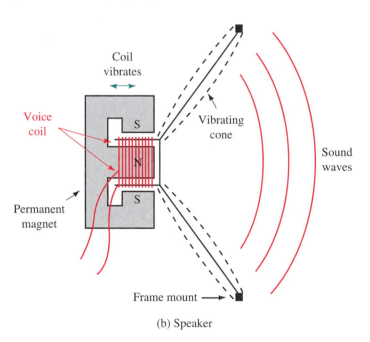

(b) Speaker

FIGURE 12–11   Both the playback system and the speaker use magnetic circuits.

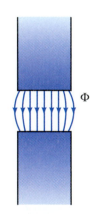

(a) Fringing at gap

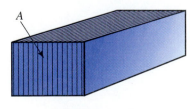

(b) Laminated section.
Effective magnetic area is
less than the physical area

FIGURE 12–12   Fringing and laminations.

the speaker magnet, causing the cone to vibrate. Since these vibrations correspond to the magnetized patterns on the tape, the original sound is reproduced. Computer disk drives use a similar record/playback scheme; in this
case, binary logic patterns are stored and retrieved rather than music and
voice.

12.5   Air Gaps, Fringing, and Laminated Cores

For magnetic circuits with air gaps, **fringing** occurs, causing a decrease in
flux density in the gap as in Figure 12–12(a). For short gaps, fringing can
usually be neglected. Alternatively, correction can be made by increasing
each cross-sectional dimension of the gap by the size of the gap to approximate the decrease in flux density.

EXAMPLE 12–2   A core with cross-sectional dimensions of 2.5 cm by 3 cm has a 0.1-mm gap. If flux density $B = 0.86$ T in the iron, what is the approximate (corrected) flux density in the gap?

Solution

$$\Phi = BA = (0.86 \text{ T})(2.5 \times 10^{-2} \text{ m})(3 \times 10^{-2} \text{ m}) = 0.645 \text{ mWb}$$
$$A_g \simeq (2.51 \times 10^{-2} \text{ m})(3.01 \times 10^{-2} \text{ m}) = 7.555 \times 10^{-4} \text{ m}^2$$

Thus, in the gap

$$B_g \simeq 0.645 \text{ mWb}/7.555 \times 10^{-4} \text{ m}^2 = 0.854 \text{ T}$$

Now consider laminations. Many practical magnetic circuits (such as transformers) use thin sheets of stacked iron or steel as in Figure 12–12(b). Since the core is not a solid block, its effective cross-sectional area (i.e., the actual area of iron) is less than its physical area. A **stacking factor,** defined as the ratio of the actual area of ferrous material to the physical area of the core, permits you to determine the core's effective area.

A laminated section of core has cross-sectional dimensions of 0.03 m by 0.05 m and a stacking factor of 0.9.

a. What is the effective area of the core?

b. Given $\Phi = 1.4 \times 10^{-3}$ Wb, what is the flux density, *B?*

Answers: a. 1.35×10^{-3} m²   b. 1.04 T

12.6   Series Elements and Parallel Elements

Magnetic circuits may have sections of different materials. For example, the circuit of Figure 12–13 has sections of cast iron, sheet steel, and an air gap. For this circuit, flux Φ is the same in all sections. Such a circuit is called a **series magnetic circuit.** Although the flux is the same in all sections, the flux density in each section may vary, depending on its effective cross-sectional area as you saw earlier.

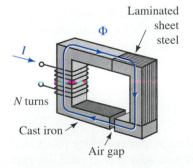

FIGURE 12–13   Series magnetic circuit. Flux Φ is the same throughout.

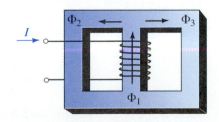

FIGURE 12–14   The sum of the flux entering a junction equals the sum leaving. Here, $\Phi_1 = \Phi_2 + \Phi_3$.

A circuit may also have elements in parallel (Figure 12–14). At each junction, the sum of fluxes entering is equal to the sum leaving. This is the counterpart of Kirchhoff's current law. Thus, for Figure 12–14, if $\Phi_1 = 25\ \mu$Wb and $\Phi_2 = 15\ \mu$Wb, then $\Phi_3 = 10\ \mu$Wb. For cores that are symmetrical about the center leg, $\Phi_2 = \Phi_3$.

IN-PROCESS LEARNING CHECK 2

1. Why is the flux density in each section of Figure 12–13 different?

2. For Figure 12–13, $\Phi = 1.32$ mWb, the cross section of the core is 3 cm by 4 cm, the laminated section has a stacking factor of 0.8, and the gap is 1 mm. Determine the flux density in each section, taking fringing into account.

3. If the core of Figure 12–14 is symmetrical about its center leg, $B_1 = 0.4$ T, and the cross-sectional area of the center leg is 25 cm², what are Φ_2 and Φ_3?

(Answers are at the end of the chapter.)

12.7   Magnetic Circuits with DC Excitation

We now look at the analysis of magnetic circuits with dc excitation. There are two basic problems to consider: (1) given the flux, to determine the current required to produce it and (2) given the current, to compute the flux produced. To help visualize how to solve such problems, we first establish an analogy between magnetic circuits and electric circuits.

MMF: The Source of Magnetic Flux

Current through a coil creates magnetic flux. The greater the current or the greater the number of turns, the greater will be the flux. This flux-producing ability of a coil is called its **magnetomotive force** (mmf). Magnetomotive force is given the symbol $\mathscr{F}$ and is defined as

$$\mathscr{F} = NI \quad \text{(ampere-turns, At)} \tag{12–2}$$

Thus, a coil with 100 turns and 2.5 amps will have an mmf of 250 ampere-turns, while a coil with 500 turns and 4 amps will have an mmf of 2000 ampere-turns.

Reluctance, $\mathscr{R}$: Opposition to Magnetic Flux

Flux in a magnetic circuit also depends on the opposition that the circuit presents to it. Termed **reluctance,** this opposition depends on the dimensions of the core and the material of which it is made. Like the resistance of a wire, reluctance is directly proportional to length and inversely proportional to cross-sectional area. In equation form,

$$\mathscr{R} = \frac{\ell}{\mu A} \quad \text{(At/Wb)} \tag{12–3}$$

where μ is a property of the core material called its **permeability** (discussed in Section 12.8). Permeability is a measure of how easy it is to establish flux in a material. Ferromagnetic materials have high permeability and hence low $\mathscr{R}$, while nonmagnetic materials have low permeability and high $\mathscr{R}$.

Ohm's Law for Magnetic Circuits

The relationship between flux, mmf, and reluctance is

$$\Phi = \mathcal{F}/\mathcal{R} \quad \text{(Wb)} \tag{12–4}$$

This relationship is similar to Ohm's law and is depicted symbolically in Figure 12–15. (Remember however that flux, unlike electric current, does not flow—see note in Section 12.1.)

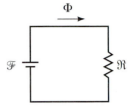

FIGURE 12–15 Electric circuit analogy of a magnetic circuit. $\Phi = \mathcal{F}/\mathcal{R}$.

EXAMPLE 12–3   For Figure 12–16, if the reluctance of the magnetic circuit is $\mathcal{R} = 12 \times 10^4$ At/Wb, what is the flux in the circuit?

FIGURE 12–16

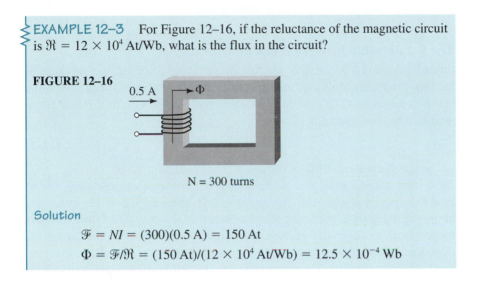

0.5 A

Φ

N = 300 turns

Solution

$$\mathcal{F} = NI = (300)(0.5\,\text{A}) = 150\,\text{At}$$

$$\Phi = \mathcal{F}/\mathcal{R} = (150\,\text{At})/(12 \times 10^4\,\text{At/Wb}) = 12.5 \times 10^{-4}\,\text{Wb}$$

In Example 12–3, we assumed that the reluctance of the core was constant. This is only approximately true under certain conditions. In general, it is not true, since $\mathcal{R}$ is a function of flux density. Thus Equation 12–4 is not really very useful, since for ferromagnetic material, $\mathcal{R}$ depends on flux, the very quantity that you are trying to find. The main use of Equations 12–3 and 12–4 is to provide an analogy between electric and magnetic circuit analysis.

12.8 Magnetic Field Intensity and Magnetization Curves

We now look at a more practical approach to analyzing magnetic circuits. First, we require a quantity called **magnetic field intensity,** H (also known as **magnetizing force**). It is a measure of the mmf per unit length of a circuit.

To get at the idea, suppose you apply the same mmf (say 600 At) to two circuits with different path lengths (Figure 12–17). In (a), you have 600 ampere-turns of mmf to "drive" flux through 0.6 m of core; in (b), you have the same mmf but it is spread across only 0.15 m of path length. Thus the mmf per unit length in the second case is more intense. Based on this idea, one can define magnetic field intensity as the ratio of applied mmf to the length of path that it acts over. Thus,

$$H = \mathcal{F}/\ell = NI/\ell \quad \text{(At/m)} \tag{12–5}$$

For the circuit of Figure 12–17(a), $H = 600\,\text{At}/0.6\,\text{m} = 1000$ At/m, while for the circuit of (b), $H = 600\,\text{At}/0.15 = 4000$ At/m. Thus, in (a) you have

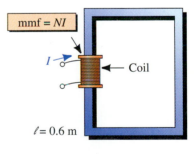

mmf = NI

Coil

$\ell = 0.6$ m

(a) A long path

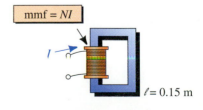

mmf = NI

$\ell = 0.15$ m

(b) A short path

FIGURE 12–17 By definition, H = mmf/length = NI/ℓ.

1000 ampere-turns of "driving force" per meter of length to establish flux in the core, whereas in (b) you have four times as much. (However, you won't get four times as much flux, since the opposition to flux varies with the density of the flux.)

Rearranging Equation 12–5 yields an important result:

$$NI = H\ell \quad \text{(At)} \tag{12–6}$$

In an analogy with electric circuits (Figure 12–18), the NI product is an **mmf source,** while the $H\ell$ product is an **mmf drop.**

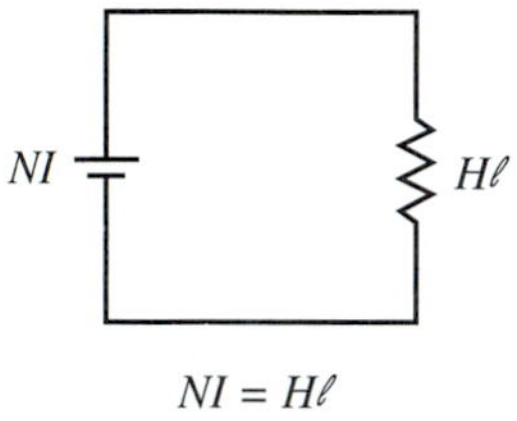

$NI = H\ell$

FIGURE 12–18   Circuit analogy, $H\ell$ model.

The Relationship between B and H

From Equation 12–5, you can see that magnetizing force, H, is a measure of the flux-producing ability of the coil (since it depends on NI). You also know that B is a measure of the resulting flux (since $B = \Phi/A$). Thus, B and H are related. The relationship is

$$B = \mu H \tag{12–7}$$

where μ is the permeability of the core (recall Equation 12–3).

It was stated earlier that permeability is a measure of how easy it is to establish flux in a material. To see why, note from Equation 12–7 that the larger the value of μ, the larger the flux density for a given H. However, H is proportional to current; therefore, the larger the value of μ, the larger the flux density for a given magnetizing current. From this, it follows that the larger the permeability, the more flux you get for a given magnetizing current.

In the SI system, μ has units of webers per ampere-turn-meter. The permeability of free space is $\mu_0 = 4\pi \times 10^{-7}$. For all practical purposes, the permeability of air and other nonmagnetic materials is the same as for a vacuum. Thus, in air gaps,

$$B_g = \mu_0 H_g = 4\pi \times 10^{-7} \times H_g \tag{12–8}$$

Rearranging Equation 12–8 yields

$$H_g = \frac{B_g}{4\pi \times 10^{-7}} = 7.96 \times 10^5 B_g \quad \text{(At/m)} \tag{12–9}$$

PRACTICE PROBLEMS 3

For Figure 12–16, the core cross section is 0.05 m $\times$ 0.08 m. If a gap is cut in the core and H in the gap is 3.6×10^5 At/m, what is the flux Φ in the core? Neglect fringing.

Answer: 1.81 mWb

B-H Curves

For ferromagnetic materials, μ is not constant but varies with flux density and there is no easy way to compute it. In reality, however, it isn't μ that you are interested in: What you really want to know is, given B, what is H, and vice versa. A set of curves, called *B-H* or *magnetization* curves, provides this information. (These curves are obtained experimentally and are available in handbooks. A separate curve is required for each material.) Figure 12–19 shows typical curves for cast iron, cast steel, and sheet steel.

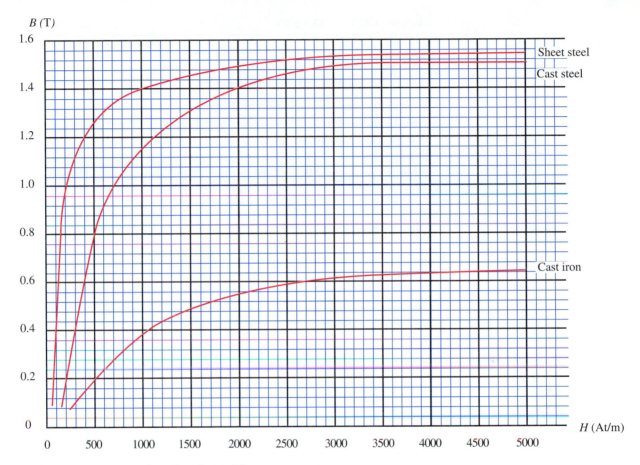

FIGURE 12–19   *B-H* curves for selected materials.

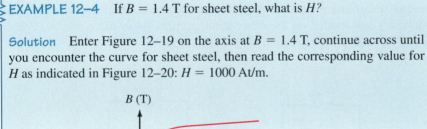

EXAMPLE 12–4   If $B = 1.4$ T for sheet steel, what is H?

Solution   Enter Figure 12–19 on the axis at $B = 1.4$ T, continue across until you encounter the curve for sheet steel, then read the corresponding value for H as indicated in Figure 12–20: $H = 1000$ At/m.

FIGURE 12–20   For sheet steel, $H = 1000$ At/m when $B = 1.4$ T.

The cross section of a sheet steel core is 0.1 m × 0.1 m and its stacking factor is 0.93. If $H = 1500$ At/m, compute flux density B and magnetic flux Φ.

Answer: 1.45 T   13.5 mWb

12.9   Ampere's Circuital Law

One of the key relationships in magnetic circuit theory is **Ampere's circuital law.** Ampere's law was determined experimentally and is a generalization of the relationship $\mathscr{F} = NI = H\ell$ that we developed earlier. Ampere showed that the algebraic sum of mmfs around a closed loop in a magnetic circuit is zero, regardless of the number of sections or coils. That is,

$$\sum_{\circlearrowleft} \mathscr{F} = 0 \qquad\qquad (12\text{–}10)$$

This can be rewrittten as

$$\sum_{\circlearrowleft} NI = \sum_{\circlearrowleft} H\ell \quad \text{At} \qquad\qquad (12\text{–}11)$$

which states that the sum of applied mmfs around a closed loop equals the sum of the mmf drops. The summation is algebraic and terms are additive or subtractive, depending on the direction of flux and how the coils are wound. To illustrate, consider again Figure 12–13. Here,

$$NI - H_{\text{iron}}\ell_{\text{iron}} - H_{\text{steel}}\ell_{\text{steel}} - H_g\ell_g = 0$$

Thus,

$$NI = \underbrace{H_{\text{iron}}\ell_{\text{iron}}}_{\substack{\text{Impressed} \\ \text{mmf}}} + \underbrace{H_{\text{steel}}\ell_{\text{steel}} + H_g\ell_g}_{\text{sum of mmf drops}}$$

which states that the applied mmf NI is equal to the sum of the $H\ell$ drops around the loop. The path to use for the $H\ell$ terms is the mean (average) path.

You now have two magnetic circuit models (Figure 12–21). While the reluctance model (a) is not very useful for solving problems, it helps relate magnetic circuit problems to familiar electrical circuit concepts. The Ampere's law model, on the other hand, permits us to solve practical problems. We look at how to do this in the next section.

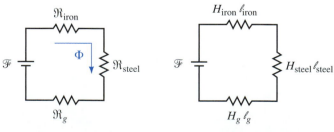

(a) Reluctance model   (b) Ampere's circuital law model

FIGURE 12–21   Two models for the magnetic circuit of Figure 12–13.

1. If the mmf of a 200-turn coil is 700 At, the current in the coil is _____ amps.

2. For Figure 12–17, if $H = 3500$ At/m and $N = 1000$ turns, then for (a), *I* is _____ A, while for (b), *I* is _____ A.

3. For cast iron, if $B = 0.5$ T, then $H = $ _____ At/m.

4. A circuit consists of one coil, a section of iron, a section of steel, and two air gaps (of different sizes). Draw the Ampere's law model.

5. Which is the correct answer for the circuit of Figure 12–22?

   a. Ampere's law around loop 1 yields ($NI = H_1\ell_1 + H_2\ell_2$, or $NI = H_1\ell_1 - H_2\ell_2$).

   b. Ampere's law around loop 2 yields ($0 = H_2\ell_2 + H_3\ell_3$, or $0 = H_2\ell_2 - H_3\ell_3$).

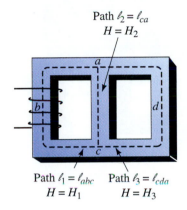

Path $\ell_2 = \ell_{ca}$
$H = H_2$

Path $\ell_1 = \ell_{abc}$ Path $\ell_3 = \ell_{cda}$
$H = H_1$ $H = H_3$

FIGURE 12–22

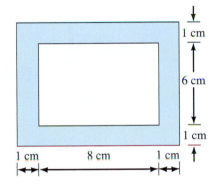

1 cm

6 cm

1 cm

1 cm   8 cm   1 cm

FIGURE 12–23

6. For the circuit of Figure 12–23, the length ℓ to use in Ampere's law is (0.36 m, 0.32 m, 0.28 m). Why? _____

(Answers are at the end of the chapter.)

12.10 Series Magnetic Circuits: Given Φ, Find *NI*

You now have the tools needed to solve basic magnetic circuit problems. We will begin with series circuits where Φ is known and we want to find the excitation to produce it. Problems of this type can be solved using four basic steps:

1. Compute *B* for each section using $B = \Phi/A$.

2. Determine *H* for each magnetic section from the *B-H* curves. Use $H_g = 7.96 \times 10^5 B_g$ for air gaps.

3. Compute *NI* using Ampere's circuital law.

4. Use the computed *NI* to determine coil current or turns as required. (Circuits with more than one coil are handled as in Example 12–6.)

Be sure to use the mean path through the circuit when applying Ampere's law. Unless directed otherwise, neglect fringing.

EXAMPLE 12–5   If the core of Figure 12–24 is cast iron and $\Phi = 0.1 \times 10^{-3}$ Wb, what is the coil current?

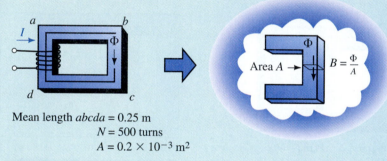

Mean length *abcda* = 0.25 m
N = 500 turns
$A = 0.2 \times 10^{-3}$ m²

FIGURE 12–24

Solution   Following the four steps outlined above:

1. The flux density is

$$B = \frac{\Phi}{A} = \frac{0.1 \times 10^{-3}}{0.2 \times 10^{-3}} = 0.5 \text{ T}$$

2. From the *B-H* curve (cast iron), Figure 12–19, H = 1550 At/m.

3. Apply Ampere's law. There is only one coil and one core section. Length = 0.25 m. Thus,

$$NI = H\ell = 1550 \times 0.25 = 388 \text{ At}$$

4. Divide by N:

$$I = 388/500 = 0.78 \text{ amps}$$

EXAMPLE 12–6   A second coil is added as shown in Figure 12–25. If $\Phi = 0.1 \times 10^{-3}$ Wb as before, but $I_1 = 1.5$ amps, what is I_2?

FIGURE 12–25

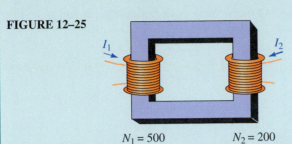

$N_1 = 500$ $N_2 = 200$

Solution   From the previous example, you know that a current of 0.78 amps in coil 1 produces $\Phi = 0.1 \times 10^{-3}$ Wb. But you already have 1.5 amps in coil 1. Thus, coil 2 must be wound in opposition so that its mmf is subtractive. Applying Ampere's law yields $N_1 I_1 - N_2 I_2 = H\ell$. Hence,

$$(500)(1.5\ A) - 200 I_2 = 388\ \text{At}$$

and so $I_2 = 1.8$ amps.

Note.   Since magnetic circuits are nonlinear, you cannot use superposition, that is, you cannot consider each coil of Figure 12–25 by itself, then sum the results. You must consider them simultaneously as we did in Example 12–6.

More Examples

If a magnetic circuit contains an air gap, add another element to the conceptual models (recall Figure 12–21). Since air represents a poor magnetic path, its reluctance will be high compared with that of iron. Recalling our analogy to electric circuits, this suggests that the mmf drop across the gap will be large compared with that of the iron. You can see this in the following example.

EXAMPLE 12–7   The core of Figure 12–24 has a 0.008-m gap cut as shown in Figure 12–26. Determine how much the current must increase to maintain the original core flux. Neglect fringing.

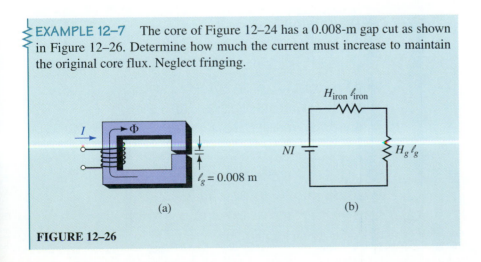

$H_{\text{iron}} \ell_{\text{iron}}$

NI

$H_g \ell_g$

$\ell_g = 0.008$ m

(a)   (b)

FIGURE 12–26

Solution

Iron

$\ell_{iron} = 0.25 - 0.008 = 0.242$ m. Since Φ does not change, B and H will be the same as before. Thus, $B_{iron} = 0.5$ T and $H_{iron} = 1550$ At/m.

Air Gap

B_g is the same as B_{iron}. Thus, $B_g = 0.5$ T and $H_g = 7.96 \times 10^5 B_g = 3.98 \times 10^5$ At/m.

Ampere's Law

$NI = H_{iron}\ell_{iron} + H_g\ell_g = (1550)(0.242) + (3.98 \times 10^5)(0.008) = 375 + 3184 = 3559$ At. Thus, $I = 3559/500 = 7.1$ amps. Note that the current had to increase from 0.78 amp to 7.1 amps in order to maintain the same flux, over a ninefold increase.

EXAMPLE 12–8   The laminated sheet steel section of Figure 12–27 has a stacking factor of 0.9. Compute the current required to establish a flux of $\Phi = 1.4 \times 10^{-4}$ Wb. Neglect fringing.

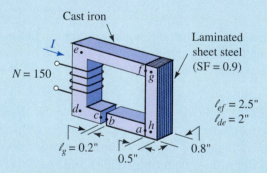

Cross section = 0.5" × 0.8" (all members)
$\Phi = 1.4 \times 10^{-4}$ Wb

FIGURE 12–27

Solution   Convert all dimensions to metric.

Cast Iron

$$\ell_{iron} = \ell_{ab} + \ell_{cdef} = 2.5 + 2 + 2.5 - 0.2 = 6.8 \text{ in} = 0.173 \text{ m}$$
$$A_{iron} = (0.5 \text{ in})(0.8 \text{ in}) = 0.4 \text{ in}^2 = 0.258 \times 10^{-3} \text{ m}^2$$
$$B_{iron} = \Phi/A_{iron} = (1.4 \times 10^{-4})/(0.258 \times 10^{-3}) = 0.54 \text{ T}$$
$$H_{iron} = 1850 \text{ At/m}   \text{(from Figure 12–19)}$$

Sheet Steel

$$\ell_{steel} = \ell_{fg} + \ell_{gh} + \ell_{ha} = 0.25 + 2 + 0.25 = 2.5 \text{ in} = 6.35 \times 10^{-2} \text{ m}$$
$$A_{steel} = (0.9)(0.258 \times 10^{-3}) = 0.232 \times 10^{-3} \text{ m}^2$$
$$B_{steel} = \Phi/A_{steel} = (1.4 \times 10^{-4})/(0.232 \times 10^{-3}) = 0.60 \text{ T}$$
$$H_{steel} = 125 \text{ At/m}   \text{(from Figure 12–19)}$$

Air Gap

$$\ell_g = 0.2 \text{ in} = 5.08 \times 10^{-3} \text{ m}$$
$$B_g = B_{\text{iron}} = 0.54 \text{ T}$$
$$H_g = (7.96 \times 10^5)(0.54) = 4.3 \times 10^5 \text{ At/m}$$

Ampere's Law

$$NI = H_{\text{iron}}\ell_{\text{iron}} + H_{\text{steel}}\ell_{\text{steel}} + H_g\ell_g$$
$$= (1850)(0.173) + (125)(6.35 \times 10^{-2}) + (4.3 \times 10^5)(5.08 \times 10^{-3})$$
$$= 320 + 7.9 + 2184 = 2512 \text{ At}$$
$$I = 2512/N = 2512/150 = 16.7 \text{ amps}$$

EXAMPLE 12–9   Figure 12–28 shows a portion of a solenoid. Flux $\Phi = 4 \times 10^{-4}$ Wb when $I = 2.5$ amps. Find the number of turns on the coil.

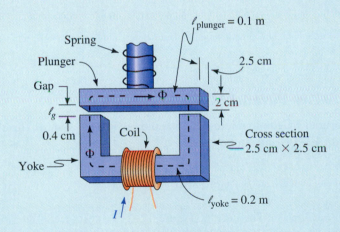

FIGURE 12–28   Solenoid. All parts are cast steel.

Solution

Yoke

$$A_{\text{yoke}} = 2.5 \text{ cm} \times 2.5 \text{ cm} = 6.25 \text{ cm}^2 = 6.25 \times 10^{-4} \text{ m}^2$$
$$B_{\text{yoke}} = \frac{\Phi}{A_{\text{yoke}}} = \frac{4 \times 10^{-4}}{6.25 \times 10^{-4}} = 0.64 \text{ T}$$
$$H_{\text{yoke}} = 410 \text{ At/m}   \text{(from Figure 12–19)}$$

Plunger

$$A_{\text{plunger}} = 2.0 \text{ cm} \times 2.5 \text{ cm} = 5.0 \text{ cm}^2 = 5.0 \times 10^{-4} \text{ m}^2$$
$$B_{\text{plunger}} = \frac{\Phi}{A_{\text{plunger}}} = \frac{4 \times 10^{-4}}{5.0 \times 10^{-4}} = 0.8 \text{ T}$$
$$H_{\text{plunger}} = 500 \text{ At/m}   \text{(from Figure 12–19)}$$

Air Gap

There are two identical gaps. For each,

$$B_g = B_{\text{yoke}} = 0.64 \text{ T}$$

Thus,

$$H_g = (7.96 \times 10^5)(0.64) = 5.09 \times 10^5 \text{ At/m}$$

The results are summarized in Table 12–2.

Ampere's Law

$$NI = H_{\text{yoke}}\ell_{\text{yoke}} + H_{\text{plunger}}\ell_{\text{plunger}} + 2H_g\ell_g = 82 + 50 + 2(2036) = 4204 \text{ At}$$
$$N = 4204/2.5 = 1682 \text{ turns}$$

TABLE 12–2

Material	Section	Length (m)	A (m²)	B (T)	H (At/m)	$H\ell$ (At)
Cast steel	yoke	0.2	6.25×10^{-4}	0.64	410	82
Cast steel	plunger	0.1	5×10^{-4}	0.8	500	50
Air	gap	0.4×10^{-2}	6.25×10^{-4}	0.64	5.09×10^5	2036

12.11 Series-Parallel Magnetic Circuits

Series-parallel magnetic circuits are handled using the sum of fluxes principle (Figure 12–14) and Ampere's law.

EXAMPLE 12–10 The core of Figure 12–29 is cast steel. Determine the current to establish an air-gap flux $\Phi_g = 6 \times 10^{-3}$ Wb. Neglect fringing.

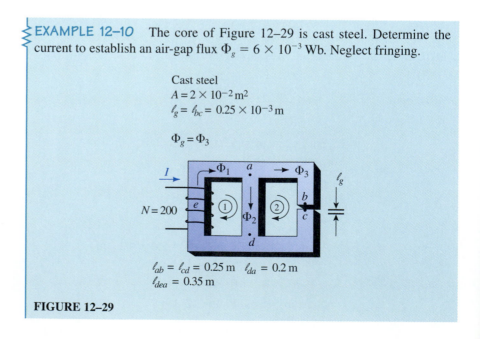

Cast steel
$A = 2 \times 10^{-2} \text{ m}^2$
$\ell_g = \ell_{bc} = 0.25 \times 10^{-3} \text{ m}$

$\Phi_g = \Phi_3$

$N = 200$

$\ell_{ab} = \ell_{cd} = 0.25 \text{ m}$ $\ell_{da} = 0.2 \text{ m}$
$\ell_{dea} = 0.35 \text{ m}$

FIGURE 12–29

Solution   Consider each section in turn.

Air Gap

$$B_g = \Phi_g/A_g = (6 \times 10^{-3})/(2 \times 10^{-2}) = 0.3 \text{ T}$$
$$H_g = (7.96 \times 10^5)(0.3) = 2.388 \times 10^5 \text{ At/m}$$

Sections ab and cd

$$B_{ab} = B_{cd} = B_g = 0.3 \text{ T}$$
$$H_{ab} = H_{cd} = 250 \text{ At/m} \quad \text{(from Figure 12–19)}$$

Ampere's Law (Loop 2)

$\sum_\circlozenge NI = \sum_\circlozenge H\ell$. Since you are going opposite to flux in leg *da*, the corresponding term (i.e., $H_{da}\ell_{da}$) will be subtractive. Also, $NI = 0$ for loop 2. Thus,

$$0 = \sum_\circlozenge {}_{\text{loop2}} H\ell$$
$$0 = H_{ab}\ell_{ab} + H_g\ell_g + H_{cd}\ell_{cd} - H_{da}\ell_{da}$$
$$= (250)(0.25) + (2.388 \times 10^5)(0.25 \times 10^{-3}) + (250)(0.25) - 0.2H_{da}$$
$$= 62.5 + 59.7 + 62.5 - 0.2H_{da} = 184.7 - 0.2H_{da}$$

Thus, $0.2H_{da} = 184.7$ and $H_{da} = 925 \text{ At/m}$. From Figure 12–19, $B_{da} = 1.12 \text{ T}$.

$$\Phi_2 = B_{da}A = 1.12 \times 0.02 = 2.24 \times 10^{-2} \text{ Wb}$$
$$\Phi_1 = \Phi_2 + \Phi_3 = 2.84 \times 10^{-2} \text{ Wb.}$$
$$B_{dea} = \Phi_1/A = (2.84 \times 10^{-2})/0.02 = 1.42 \text{ T}$$
$$H_{dea} = 2125 \text{ At/m} \quad \text{(from Figure 12–19)}$$

Ampere's Law (Loop 1)

$$NI = H_{dea}\ell_{dea} + H_{ad}\ell_{ad} = (2125)(0.35) + 184.7 = 929 \text{ At}$$
$$I = 929/200 = 4.65 \text{ A}$$

The cast-iron core of Figure 12–30 is symmetrical. Determine current *I*. Hint: To find *NI*, you can write Ampere's law around either loop. Be sure to make use of symmetry.

PRACTICE
PROBLEMS 5

FIGURE 12–30 $\Phi_2 = 30 \, \mu\text{Wb}$

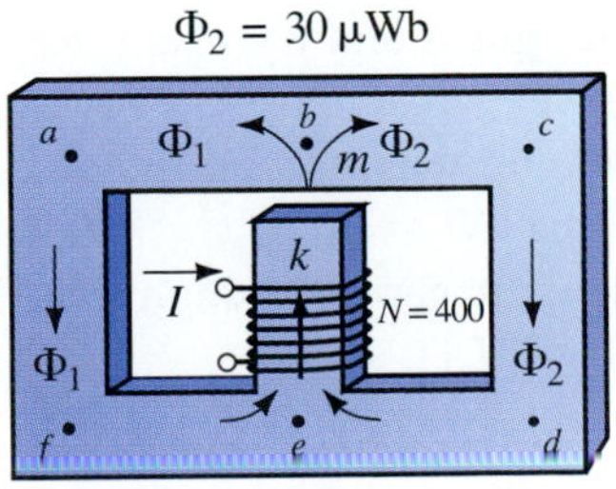

$\ell_{ab} = \ell_{bc} = \ell_{cd} = 4 \text{ cm}$
Gap: $\ell_g = 0.5 \text{ cm}$
$\ell_{ek} = 3 \text{ cm}$

Core dimensions: $1 \text{ cm} \times 1 \text{ cm}$

Answer: 6.5 A

12.12 Series Magnetic Circuits: Given *NI*, Find Φ

In previous problems, you were given the flux and asked to find the current. We now look at the converse problem: given *NI*, find the resultant flux. For the special case of a core of one material and constant cross section (Example 12–11) this is straightforward. For all other cases, trial and error must be used.

EXAMPLE 12–11   For the circuit of Figure 12–31, *NI* = 250 At. Determine Φ.

FIGURE 12–31

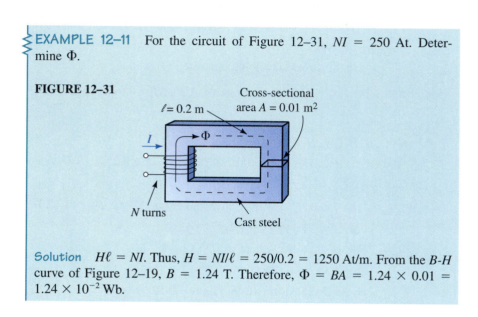

Cross-sectional area $A = 0.01 \text{ m}^2$

$\ell = 0.2$ m

I

Φ

N turns

Cast steel

Solution $H\ell = NI$. Thus, $H = NI/\ell = 250/0.2 = 1250$ At/m. From the *B-H* curve of Figure 12–19, $B = 1.24$ T. Therefore, $\Phi = BA = 1.24 \times 0.01 = 1.24 \times 10^{-2}$ Wb.

For circuits with two or more sections, the process is not so simple. Before you can find *H* in any section, for example, you need to know the flux density. However, in order to determine flux density, you need to know *H*. Thus, neither Φ nor *H* can be found without knowing the other first.

To get around this problem, use trial and error. First, take a guess at the value for flux, compute *NI* using the 4-step procedure of Section 12.10, then compare the computed *NI* against the given *NI*. If they agree, the problem is solved. If they don't, adjust your guess and try again. Repeat the procedure until you are within 5% of the given *NI*.

The problem is how to come up with a good first guess. For circuits of the type of Figure 12–32, note that $NI = H_{\text{steel}}\ell_{\text{steel}} + H_g\ell_g$. As a first guess, assume that the reluctance of the air gap is so high that the full mmf drop appears across the gap. Thus, $NI \simeq H_g\ell_g$, and

$$H_g \simeq NI/\ell_g \tag{12–12}$$

You can now apply Ampere's law to see how close to the given *NI* your trial guess is.

Since we know that some of the mmf drop appears across the steel, we will start at less than 100% for the gap. Common sense and a bit of experience helps. The relative size of the mmf drops also depends on the core material. For cast iron, the percentage drop across the iron is larger than the

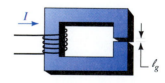

$\mathcal{F} = H_{\text{steel}}\ell_{\text{steel}} + H_g\ell_g$

$\simeq H_g\ell_g$ if $H_g\ell_g \gg H_{\text{steel}}\ell_{\text{steel}}$

$\ell_g = 0.002$ m
$\ell_{\text{steel}} = 0.2$ m

FIGURE 12–32

percentage across a similar piece of sheet steel or cast steel. This is illustrated in the next two examples.

EXAMPLE 12–12   The core of Figure 12–32 is cast steel, $NI = 1100$ At, and the cross-sectional area everywhere is 0.0025 m². Determine the flux in the core.

Solution

Initial Guess
Assume that 90% of the mmf appears across the gap. The applied mmf is 1100 At. Ninety percent of this is 990 At. Thus, $H_g \simeq 0.9NI/\ell = 990/0.002 = 4.95 \times 10^5$ At/m and $B_g = \mu_0 H_g = (4\pi \times 10^{-7})(4.95 \times 10^5) = 0.62$ T.

Trial 1
Since the area of the steel is the same as that of the gap, the flux density is the same, neglecting fringing. Thus, $B_{\text{steel}} = B_g = 0.62$ T. From the *B-H* curve, $H_{\text{steel}} = 400$ At/m. Now apply Ampere's law:

$$NI = H_{\text{steel}}\ell_{\text{steel}} + H_g\ell_g = (400)(0.2) + (4.95 \times 10^5)(0.002)$$
$$= 80 + 990 = 1070 \text{ At}$$

This answer is 2.7% lower than the given *NI* of 1100 At and is therefore acceptable. Thus, $\Phi = BA = 0.62 \times 0.0025 = 1.55 \times 10^{-3}$ Wb.

The initial guess in Example 12–12 yielded an acceptable answer on the first trial. (You are seldom this lucky.)

EXAMPLE 12–13   If the core of Figure 12–32 is cast iron instead of steel, compute Φ.

Solution   Because cast iron has a larger *H* for a given flux density (Figure 12–19), it will have a larger *Hℓ* drop and less will appear across the gap. Assume 75% across the gap.

Initial Guess
$H_g \simeq 0.75 NI/\ell = (0.75)(1100)/0.002 = 4.125 \times 10^5$ At/m.
$B_g = \mu_0 H_g = (4\mu \times 10^{-7})(4.125 \times 10^5) = 0.52$ T.

Trial 1
$B_{\text{iron}} = B_g$. Thus, $B_{\text{iron}} = 0.52$ T. From the *B-H* curve, $H_{\text{iron}} = 1700$ At/m.

Ampere's Law

$$NI = H_{\text{iron}}\ell_{\text{iron}} + H_g\ell_g = (1700)(0.2) + (4.125 \times 10^5)(0.002)$$
$$= 340 + 825 = 1165 \text{ At}   \text{(high by 5.9\%)}$$

Trial 2
Reduce the guess by 5.9% to $B_{\text{iron}} = 0.49$ T. Thus, $H_{\text{iron}} = 1500$ At/m (from the *B-H* curve) and $H_g = 7.96 \times 10^5 B_g = 3.90 \times 10^5$ At/m.

Ampere's Law

$$NI = H_{iron}\ell_{iron} + H_g\ell_g = (1500)(0.2) + (3.90 \times 10^5)(0.002)$$
$$= 300 + 780 = 1080 \text{ At}$$

The error is now 1.82%, which is excellent. Thus, $\Phi = BA = (0.49)(2.5 \times 10^{-3}) = 1.23 \times 10^{-3}$ Wb. If the error had been larger than 5%, a new trial would have been needed.

12.13 Force Due to an Electromagnet

Electromagnets are used in relays, door bells, lifting magnets, and so on. For an electromagnetic relay as in Figure 12–33, it can be shown that the force created by the magnetic field is

$$F = \frac{B_g^2 A_g}{2\mu_0} \tag{12–13}$$

where B_g is flux density in the gap in teslas, A_g is gap area in square meters, and F is force in newtons.

EXAMPLE 12–14   Figure 12–33 shows a typical relay. The force due to the current-carrying coil pulls the pivoted arm against spring tension to close the contacts and energize the load. If the pole face is ¼ inch square and $\Phi = 0.5 \times 10^{-4}$ Wb, what is the pull on the armature in pounds?

FIGURE 12–33   A typical relay.

Solution   Convert to metric units.

$$A_g = (0.25 \text{ in})(0.25 \text{ in}) = 0.0625 \text{ in}^2 = 0.403 \times 10^{-4} \text{ m}^2$$
$$B_g = \Phi/A_g = (0.5 \times 10^{-4})/(0.403 \times 10^{-4}) = 1.24 \text{ T}$$

Thus,

$$F = \frac{B_g^2 A}{2\mu_0} = \frac{(1.24)^2 (0.403 \times 10^{-4})}{2(4\pi \times 10^{-7})} = 24.66 \text{ N} = 5.54 \text{ lb}$$

Figure 12–34 shows how a relay is used in practice. When the switch is closed, the energized coil pulls the armature down. This closes the contacts and energizes the load. When the switch is opened, the spring pulls the contacts open again. Schemes like this use relatively small currents to control large loads. In addition, they permit remote control, as the relay and load may be a considerable distance from the actuating switch.

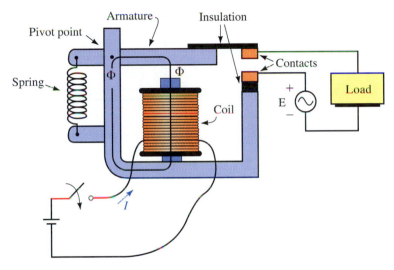

FIGURE 12–34   Controlling a load with a relay.

12.14 Properties of Magnetic Materials

Magnetic properties are related to atomic structure. Each atom of a substance, for example, produces a tiny atomic-level magnetic field because its moving (i.e., orbiting) electrons constitute an atomic-level current and currents create magnetic fields. For nonmagnetic materials, these fields are randomly oriented and cancel. However, for ferromagnetic materials, the fields in small regions, called **domains** (Figure 12–35), do not cancel. (Domains are of microscopic size, but are large enough to hold from 10^{17} to 10^{21} atoms.) If the domain fields in a ferromagnetic material line up, the material is magnetized; if they are randomly oriented, the material is not magnetized.

Magnetizing a Specimen

A nonmagnetized specimen can be magnetized by making its domain fields line up. Figure 12–36 shows how this can be done. As current through the coil is increased, the field strength increases and more and more domains

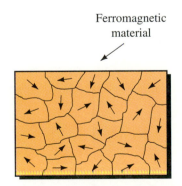

FIGURE 12–35   Random orientation of microscopic fields in a nonmagnetized ferromagnetic material. The small regions are called domains.

align themselves in the direction of the field. If the field is made strong enough, almost all domain fields line up and the material is said to be in **saturation** (the almost flat portion of the *B-H* curve). In saturation, the flux density increases slowly as magnetization intensity increases. This means that once the material is in saturation, you cannot magnetize it much further no matter how hard you try. Path 0-*a* traced from the nonmagnetized state to the saturated state is termed the **dc curve** or **normal magnetization curve.** (This is the *B-H* curve that you used earlier when you solved magnetic circuit problems.)

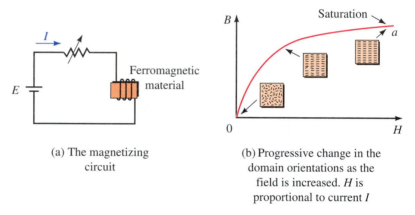

(a) The magnetizing circuit

(b) Progressive change in the domain orientations as the field is increased. *H* is proportional to current *I*

FIGURE 12–36   The magnetization process.

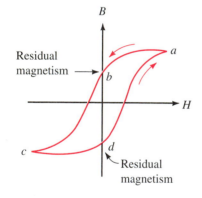

FIGURE 12–37   Hysteresis loop.

Hysteresis

If you now reduce the current to zero, you will find that the material still retains some magnetism, called **residual magnetism** (Figure 12–37, point *b*). If now you reverse the current, the flux reverses and the bottom part of the curve can be traced. By reversing the current again at *d,* the curve can be traced back to point *a*. The result is called a **hysteresis loop.** A major source of uncertainty in magnetic circuit behavior should now be apparent: As you can see, flux density depends not just on current, it also depends on which arm of the curve the sample is magnetized on, i.e., it depends on the circuit's past history. For this reason, *B-H* curves are the average of the two arms of the hysteresis loop, i.e., the dc curve of Figure 12–36.

The Demagnetization Process

As indicated above, simply turning the current off does not demagnetize ferromagnetic material. To demagnetize it, you must successively decrease its hysteresis loop to zero as in Figure 12–38. You can place the specimen inside a coil that is driven by a variable ac source and gradually decrease the coil current to zero, or you can use a fixed ac supply and gradually withdraw the specimen from the field. Such procedures are used by service personnel to "degauss" TV picture tubes.

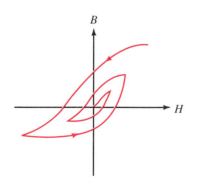

FIGURE 12–38   Demagnetization by successively shrinking the hysteresis loop.

12.15 Measuring Magnetic Fields

One way to measure magnetic field strength is to use the **Hall effect** (after E. H. Hall). The basic idea is illustrated in Figure 12–39. When a strip of semi-conductor material such as indium arsenide is placed in a magnetic field, a small voltage, called the Hall voltage, V_H, appears across opposite edges. For a fixed current I, V_H is proportional to magnetic field strength B. Instruments using this principle are known as **Hall-effect gaussmeters.** To measure a magnetic field with such a meter, insert its probe into the field perpendicular to the field (Figure 12–40). The meter indicates flux density directly.

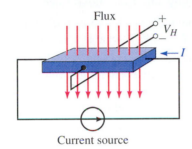

FIGURE 12–39   The Hall effect.

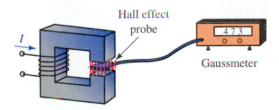

FIGURE 12–40   Magnetic field measurement.

12.3 Flux and Flux Density

PROBLEMS

1. Refer to Figure 12–41:
   a. Which area, A_1 or A_2, do you use to calculate flux density?
   b. If $\Phi = 28$ mWb, what is flux density in teslas?

2. For Figure 12–41, if $\Phi = 250$ μWb, $A_1 = 1.25$ in², and $A_2 = 2.0$ in², what is the flux density in the English system of units?

3. The toroid of Figure 12–42 has a circular cross section and $\Phi = 628$ μWb. If $r_1 = 8$ cm and $r_2 = 12$ cm, what is the flux density in teslas?

4. If r_1 of Figure 12–42 is 3.5 inches and r_2 is 4.5 inches, what is the flux density in the English system of units if $\Phi = 628$ μWb?

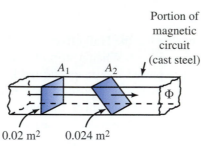

FIGURE 12–41

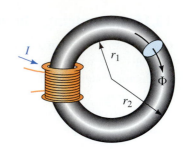

FIGURE 12–42

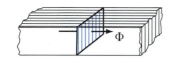

FIGURE 12–43

12.5 Air Gaps, Fringing, and Laminated Cores

5. If the section of core in Figure 12–43 is 0.025 m by 0.04 m, has a stacking factor of 0.85, and $B = 1.45$ T, what is Φ in webers?

6. If the core of Figure 12–43 is 3 cm by 2 cm, has a stacking factor of 0.9, and $B = 8 \times 10^3$ gauss, what is Φ in maxwells?

12.6 Series Elements and Parallel Elements

7. For the iron core of Figure 12–44, flux density $B_2 = 0.6$ T. Compute B_1 and B_3.

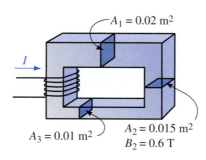

FIGURE 12–44

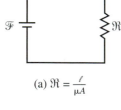

(a) $\Re = \dfrac{\ell}{\mu A}$

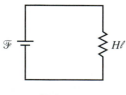

(b) $B = \mu H$

FIGURE 12–46 $\mathscr{F} = NI.$

FIGURE 12–45

8. For the section of iron core of Figure 12–45, if $\Phi_1 = 12$ mWb and $\Phi_3 = 2$ mWb, what is B_2?

9. For the section of iron core of Figure 12–45, if $B_1 = 0.8$ T and $B_2 = 0.6$ T, what is B_3?

12.8 Magnetic Field Intensity and Magnetization Curves

10. Consider again Figure 12–42. If $I = 10$ A, $N = 40$ turns, $r_1 = 5$ cm, and $r_2 = 7$ cm, what is H in ampere-turns per meter?

11. Figure 12–46 shows the two electric circuit equivalents for magnetic circuits. Show that μ in $\Re = \ell/\mu A$ is the same as μ in $B = \mu H$.

12.9 Ampere's Circuital Law

12. Let H_1 and ℓ_1 be the magnetizing force and path length respectively, where flux Φ_1 exists in Figure 12–47 and similarly for Φ_2 and Φ_3. Write Ampere's law around each of the windows.

13. Assume that a coil N_2 carrying current I_2 is added on leg 3 of the core shown in Figure 12–47 and that it produces flux directed upward. Assume, however, that the net flux in leg 3 is still downward. Write the Ampere's law equations for this case.

14. Repeat Problem 13 if the net flux in leg 3 is upward but the directions of Φ_1 and Φ_2 remain as in Figure 12–47.

12.10 Series Magnetic Circuits: Given Φ, Find NI

15. Find the current I in Figure 12–48 if $\Phi = 0.16$ mWb.

16. Let everything be the same as in Problem 15 except that the cast steel portion is replaced with laminated sheet steel with a stacking factor of 0.85.

17. A gap of 0.5 mm is cut in the cast steel portion of the core in Figure 12–48. Find the current for $\Phi = 0.128$ mWb. Neglect fringing.

18. Two gaps, each 1 mm, are cut in the circuit of Figure 12–48, one in the cast steel portion and the other in the cast iron portion. Determine current for $\Phi = 0.128$ mWb. Neglect fringing.

19. The cast iron core of Figure 12–49 measures 1 cm × 1.5 cm, $\ell_g = 0.3$ mm, the air gap flux density is 0.426 T and $N = 600$ turns. The end pieces are half circles. Taking into account fringing, find current I.

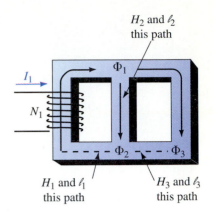

FIGURE 12–47

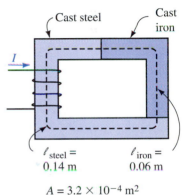

$A = 3.2 \times 10^{-4}$ m²
$N = 300$ turns

FIGURE 12–48

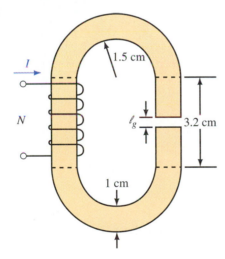

FIGURE 12–49

20. For the circuit of Figure 12–50, $\Phi = 141$ μWb and $N = 400$ turns. The bottom member is sheet steel with a stacking factor of 0.94, while the remainder is cast steel. All pieces are 1 cm × 1 cm. The length of the cast steel path is 16 cm. Find current I.

21. For the circuit of Figure 12–51, $\Phi = 30$ μWb and $N = 2000$ turns. Neglecting fringing, find current I.

22. For the circuit of Figure 12–52, $\Phi = 25{,}000$ lines. The stacking factor for the sheet steel portion is 0.95. Find current I.

23. A second coil of 450 turns with $I_2 = 4$ amps is wound on the cast steel portion of Figure 12–52. Its flux is in opposition to the flux produced by the

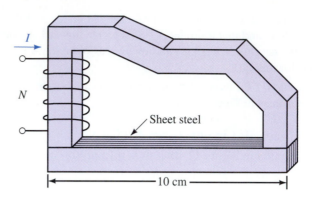

FIGURE 12–50

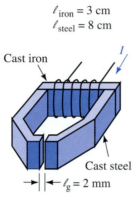

$\ell_{iron} = 3$ cm
$\ell_{steel} = 8$ cm

A (everywhere) = 0.5 cm²

FIGURE 12–51

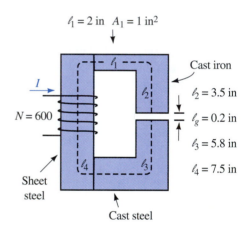

$\ell_1 = 2$ in $A_1 = 1$ in²

$\ell_2 = 3.5$ in
$\ell_g = 0.2$ in
$\ell_3 = 5.8$ in
$\ell_4 = 7.5$ in

Area of all sections (except A_1) = 2 in²

FIGURE 12–52

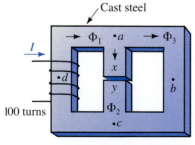

$\ell_g = \ell_{xy} = 0.001$ m
$\ell_{abc} = 0.14$ m
$\ell_{cda} = 0.16$ m
$\ell_{ax} = \ell_{cy} = 0.039$ m
$A = 4$ cm² everywhere

FIGURE 12–53

original coil. The resulting flux is 35 000 lines in the counterclockwise direction. Find the current I_1.

12.11   Series-Parallel Magnetic Circuits

24. For Figure 12–53, if $\Phi_g = 80 \ \mu$Wb, find I.

25. If the circuit of Figure 12–53 has no gap and $\Phi_3 = 0.2$ mWb, find I.

12.12   Series Magnetic Circuits: Given NI, Find Φ

26. A cast steel magnetic circuit with $N = 2500$ turns, $I = 200$ mA, and a cross-sectional area of 0.02 m² has an air gap of 0.00254 m. Assuming 90% of the mmf appears across the gap, estimate the flux in the core.

27. If $NI = 644$ At for the cast steel core of Figure 12–54, find the flux, Φ.

28. A gap $\ell = 0.004$ m is cut in the core of Figure 12–54. Everything else remains the same. Find the flux, Φ.

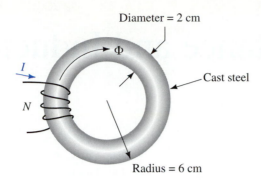

Diameter = 2 cm

Φ

I

Cast steel

N

Radius = 6 cm

FIGURE 12–54

12.13   Force Due to an Electromagnet

29. For the relay of Figure 12–34, if the pole face is 2 cm by 2.5 cm and a force of 2 pounds is required to close the gap, what flux (in webers) is needed?

30. For the solenoid of Figure 12–28, $\Phi = 4 \times 10^{-4}$ Wb. Find the force of attraction on the plunger in newtons and in pounds.

In-Process Learning Check 1

1. Force

2. Strength, direction

3. Iron, nickel, cobalt

4. North, south

5. Same except direction of flux reversed

6. Same as Figure 12–6

7. Perpendicular

8. 10 cm^2

In-Process Learning Check 2

1. While flux is the same throughout, the effective area of each section differs.

2. $B_{iron} = 1.1$ T; $B_{steel} = 1.38$ T; $B_g = 1.04$ T

3. $\Phi_2 = \Phi_3 = 0.5$ mWb

In-Process Learning Check 3

1. 3.5 A

2. a. 2.1 A   b. 0.525 A

3. 1550 At/m

4. Same as Figure 12–21(b) except add $H_{g_2}\ell_{g_2}$.

5. a. $NI = H_1\ell_1 + H_2\ell_2$ b. $0 = H_2\ell_2 - H_3\ell_3$

6. 0.32 m; use the mean path length.

ANSWERS TO IN-PROCESS LEARNING CHECKS

13

Inductance and Inductors

OBJECTIVES

After studying this chapter, you will be able to

- describe what an inductor is and what its effect on circuit operation is,
- explain Faraday's law and Lenz's law,
- compute induced voltage using Faraday's law,
- define inductance,
- compute voltage across an inductance,
- compute inductance for series and parallel configurations,
- compute inductor voltages and currents for steady state dc excitation,
- compute energy stored in an inductance,
- describe common inductor problems and how to test for them.

KEY TERMS

Back Voltage
Choke
Counter EMF
Faraday's Law
Flux Linkage
Henry
Induced Voltage
Inductance
Inductor
L
Lenz's Law
$N\phi$
Stray Inductance

OUTLINE

Electromagnetic Induction
Induced Voltage and Induction
Self-Inductance
Computing Induced Voltage
Inductances in Series and Parallel
Practical Considerations
Inductance and Steady State DC
Energy Stored by an Inductance
Inductor Troubleshooting Hints

In this chapter, we look at self-inductance and inductors. Self-inductance (usually just called inductance) is a circuit property that is due entirely to the magnetic field created by current in a circuit. The effect that **inductance** has on circuit operation is to oppose any change in current—thus, in a sense, inductance can be likened to inertia in a mechanical system.

A circuit element built to possess inductance is called an **inductor.** In its simplest form an inductor is simply a coil of wire, Figure 13–1(a). Ideally, inductors have only inductance. However, since they are made of wire, practical inductors also have some resistance. Initially, however, we assume that this resistance is negligible and treat inductors as ideal (i.e., we assume that they have no property other than inductance). (Coil resistance is considered in Sections 13.6 and 13.7.) In practice, inductors are also referred to as chokes (because they try to limit or "choke" current change), or as reactors (for reasons to be discussed in Chapter 16). In this chapter, we refer to them mainly as inductors.

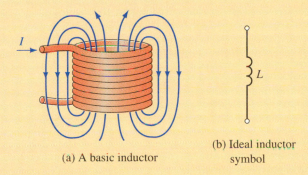

(a) A basic inductor

(b) Ideal inductor symbol

FIGURE 13–1   Inductance is due to the magnetic field created by an electric current.

On circuit diagrams and in equations, inductance is represented by the letter L. Its circuit symbol is a coil as shown in Figure 13–1(b). The unit of inductance is the **henry.**

Inductors are used in many places. In radios, they are part of the tuning circuit that you adjust when you select a station. In fluorescent lamps, they are part of the ballast circuit that limits current when the lamp is turned on; in power systems, they are part of the protection circuitry used to control short-circuit currents during fault conditions.

The Discovery of Electromagnetic Induction

MOST OF OUR IDEAS CONCERNING INDUCTANCE and induced voltages are due to Michael Faraday (recall Chapter 12) and Joseph Henry (1797–1878). Working independently (Faraday in England and Henry—shown at left—in the USA), they discovered, almost simultaneously, the fundamental laws governing electromagnetic induction.

While experimenting with magnetic fields, Faraday developed the transformer. He wound two coils on an iron ring and energized one of them from a battery. As he closed the switch energizing the first coil, Faraday noticed that a momentary voltage was induced in the second coil, and when he opened the switch, he found that a momentary voltage was again induced but with opposite polarity. When the current was steady, no voltage was produced at all.

Faraday explained this effect in terms of his magnetic lines of flux concept. When current was first turned on, he visualized the lines as springing outward into space; when it was turned off, he visualized the lines as collapsing inward. He then visualized that voltage was produced by these lines as they cut across circuit conductors. Companion experiments showed that voltage was also produced when a magnet was passed through a coil or when a conductor was moved through a magnetic field. Again, he visualized these voltages in terms of flux cutting a conductor.

Working independently in the United States, Henry discovered essentially the same results. In fact, Henry's work preceded Faraday's by a few months, but because he did not publish them first, credit was given to Faraday. However, Henry is credited with the discovery of self-induction, and in honor of his work the unit of inductance was named the henry.

NOTES...

Since we work with time-varying flux linkages in this chapter, we use ϕ rather than Φ for flux (as we did in Chapter 12). This is in keeping with the standard practice of using lowercase symbols for time-varying quantities and uppercase symbols for dc quantities.

13.1   Electromagnetic Induction

Since inductance depends on induced voltage, we begin with a review of electromagnetic induction. First, we look at Faraday's and Henry's results. Consider Figure 13–2. In (a), a magnet is moved through a coil of wire, and this action induces a voltage in the coil. When the magnet is thrust into the coil, the meter deflects upscale; when it is withdrawn, the meter deflects downscale, indicating that polarity has changed. The voltage magnitude is proportional to how fast the magnet is moved. In (b), when the conductor is moved through the field, voltage is induced. If the conductor is moved to the right, its far end is positive; if it is moved to the left, the polarity reverses and its far end becomes negative. Again, the voltage magnitude is proportional to how fast the wire is moved. In (c), voltage is induced in coil 2 due to the magnetic field created by the current in coil 1. At the instant the switch is closed, the meter kicks upscale; at the instant it is opened, the meter kicks downscale. In (d) voltage is induced in a coil by its own current. At the instant the switch is closed, the top end of the coil becomes positive, while at the instant it is opened, the polarity reverses and the top end becomes negative.

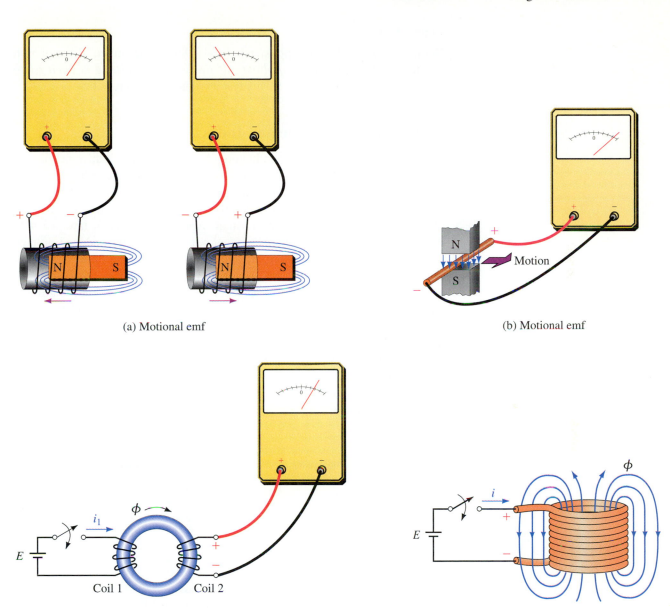

(a) Motional emf

(b) Motional emf

(c) Mutually induced voltage

(d) Self-induced voltage

FIGURE 13–2   Principle of electromagnetic induction. Voltage is induced as long as the flux linkage of the circuit is changing.

Faraday's Law

Based on these observations, Faraday concluded *that voltage is induced in a circuit whenever the flux linking* (i.e., passing through) *the circuit is changing and that the magnitude of the voltage is proportional to the rate of change of the flux linkages.* This result, known as **Faraday's law,** is also sometimes stated in terms of the rate of cutting flux lines. We look at this viewpoint in Chapter 15.

Lenz's Law

Heinrich Lenz (a Russian physicist, 1804–1865) determined a companion result. He showed that *the polarity of the induced voltage is such as to oppose the cause producing it.* This result is known as **Lenz's law.**

13.2   Induced Voltage and Induction

We now turn our attention to inductors. As noted earlier, inductance is due entirely to the magnetic field created by current-carrying conductors. Consider Figure 13–3 (which shows an inductor at three instants of time). In (a) the current is constant, and since the magnetic field is due to this current, the magnetic field is also constant. Applying Faraday's law, we note that, because the flux linking the coil is not changing, the induced voltage is zero. Now consider (b). Here, the current (and hence the field) is increasing. According to Faraday's law, a voltage is induced that is proportional to how fast the field is changing and according to Lenz's law, the polarity of this voltage must be such as to oppose the increase in current. Thus, the polarity of the voltage is as shown. Note that the faster the current increases, the larger the opposing voltage. Now consider (c). Since the current is decreasing, Lenz's law shows that the polarity of the induced voltage reverses, that is, the collapsing field produces a voltage that tries to keep the current going. Again, the faster the rate of change of current, the larger is this voltage.

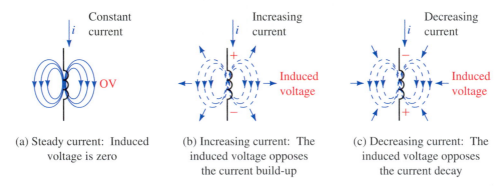

(a) Steady current: Induced
voltage is zero

(b) Increasing current: The
induced voltage opposes
the current build-up

(c) Decreasing current: The
induced voltage opposes
the current decay

FIGURE 13–3   Self-induced voltage due to a coil's own current. The induced voltage opposes the current change. Note carefully the polarities in (b) and (c).

Counter EMF

Because the induced voltage in Figure 13–3 tries to counter (i.e., opposes) changes in current, it is called a **counter emf** or **back voltage.** Note carefully, however, that this voltage does not oppose current, it opposes only changes in current. It also does not prevent the current from changing; it only prevents it from changing abruptly. The result is that current in an inductor changes gradually and smoothly from one value to another as indicated in Figure 13–4(b). The effect of inductance is thus similar to the effect of inertia in a mechanical system. The flywheel used on an engine, for example, prevents abrupt changes in engine speed but does not prevent the engine from gradually changing from one speed to another.

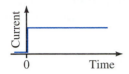

(a) Current cannot jump from one value to another like this

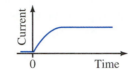

(b) Current must change smoothly with no abrupt jumps

FIGURE 13–4 Current in inductance.

Iron-Core and Air-Core Inductors

As Faraday discovered, the voltage induced in a coil depends on flux linkages, and flux linkages depend on core materials. Coils with ferromagnetic cores (called **iron-core coils**) have their flux almost entirely confined to their cores, while coils wound on nonferromagnetic materials do not. (The latter are sometimes called **air-core coils** because all nonmagnetic core materials have the same permeability as air and thus behave magnetically the same as air.)

First, consider the iron-core case, Figure 13–5. Ideally, all flux lines are confined to the core and hence pass through (link) all turns of the winding. The product of flux times the number of turns that it passes through is defined as the **flux linkage** of the coil. For Figure 13–5, ϕ lines pass through N turns yielding a flux linkage of $N\phi$. By Faraday's law, the induced voltage is proportional to the rate of change of $N\phi$. In the SI system, the constant of proportionality is one and Faraday's law for this case may therefore be stated as

$$e = N \times \text{the rate of change of } \phi \qquad (13–1)$$

In calculus notation,

$$e = N\frac{d\phi}{dt} \quad \text{(volts, V)} \qquad (13–2)$$

where ϕ is in webers, t in seconds, and e in volts. Thus if the flux changes at the rate of 1 Wb/s in a 1 turn coil, the voltage induced is 1 volt.

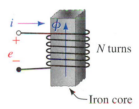

FIGURE 13–5 When flux ϕ passes through all N turns, the flux linking the coil is $N\phi$.

NOTES...

Equation 13–2 is sometimes shown with a minus sign. However, the minus sign is unnecessary. In circuit theory, we use Equation 13–2 to determine the magnitude of the induced voltage and Lenz's law to determine its polarity.

EXAMPLE 13–1 If the flux through a 200-turn coil changes steadily from 1 Wb to 4 Wb in one second, what is the voltage induced?

Solution The flux changes by 3 Wb in one second. Thus, its rate of change is 3 Wb/s.

$$e = N \times \text{rate of change of flux}$$
$$= (200 \text{ turns})(3 \text{ Wb/s}) = 600 \text{ volts}$$

Now consider an air-core inductor (Figure 13–6). Since not all flux lines pass through all windings, it is difficult to determine flux linkages as above. However, (since no ferromagnetic material is present) flux is directly proportional to current. In this case, then, since induced voltage is proportional to the rate of change of flux, and since flux is proportional to current, induced voltage will be proportional to the rate of change of current. Let the constant of proportionality be *L*. Thus,

$$e = L \times \text{rate of change of current} \qquad \textbf{(13–3)}$$

In calculus notation, this can be written as

$$e = L\frac{di}{dt} \quad \text{(volts, V)} \qquad \textbf{(13–4)}$$

L is called the **self-inductance** of the coil, and in the SI system its unit is the henry. (This is discussed in more detail in Section 13.3.)

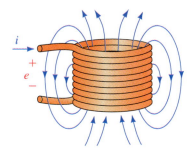

FIGURE 13–6   The flux linking the coil is proportional to current. Flux linkage is *LI*.

We now have two equations for coil voltage. Equation 13–4 is the more useful form for this chapter, while Equation 13–2 is the more useful form for the circuits of Chapter 24. We look at Equation 13–4 in the next section.

IN-PROCESS LEARNING CHECK 1

1. Which of the current graphs shown in Figure 13–7 cannot be the current in an inductor? Why?

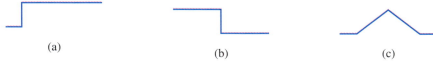

(a)   (b)   (c)

FIGURE 13–7

2. Compute the flux linkage for the coil of Figure 13–5, given $\phi = 500$ mWb and $N = 1200$ turns.

3. If the flux ϕ of Question 2 changes steadily from 500 mWb to 525 mWb in 1 s, what is the voltage induced in the coil?

4. If the flux ϕ of Question 2 changes steadily from 500 mWb to 475 mWb in 100 ms, what is the voltage induced?

(Answers are at the end of the chapter.)

13.3 Self-Inductance

In the preceding section, we showed that the voltage induced in a coil is $e = L\,di/dt$; where L is the self-inductance of the coil (usually referred to simply as inductance) and di/dt is the rate of change of its current. In the SI system, L is measured in henries. As can be seen from Equation 13–4, it is the ratio of voltage induced in a coil to the rate of change of current producing it. From this, we get the definition of the henry. By definition, *the inductance of a coil is one henry if the voltage created by its changing current is one volt when its current changes at the rate of one ampere per second.*

In practice, the voltage across an inductance is denoted by v_L rather than by e. Thus,

$$v_L = L\frac{di}{dt} \quad \text{(V)} \qquad \textbf{(13–5)}$$

Voltage and current references are as shown in Figure 13–8.

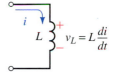

EXAMPLE 13–2 If the current through a 5-mH inductance changes at the rate of 1000 A/s, what is the voltage induced?

Solution

$$v_L = L \times \text{rate of change of current}$$
$$= (5 \times 10^{-3}\ \text{H})(1000\ \text{A/s}) = 5\ \text{volts}$$

FIGURE 13–8 Voltage-current reference convention. As usual, the plus sign for voltage goes at the tail of the current arrow.

1. The voltage across an inductance is 250 V when its current changes at the rate of 10 mA/μs. What is L?

2. If the voltage across a 2-mH inductance is 50 volts, how fast is the current changing?

PRACTICE PROBLEMS 1

Answers: 1. 25 mH   2. 25×10^3 A/s

Inductance Formulas

Inductance for some simple shapes can be determined using the principles of Chapter 12. For example, the approximate inductance of the coil of Figure 13–9 can be shown to be

$$L = \frac{\mu N^2 A}{\ell} \quad \text{(H)} \qquad \textbf{(13–6)}$$

where ℓ is in meters, A is in square meters, N is the number of turns, and μ is the permeability of the core. (Details can be found in many physics books.)

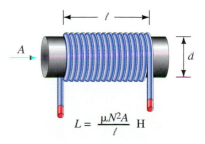

$$L = \frac{\mu N^2 A}{\ell}\ \text{H}$$

FIGURE 13–9 Inductance formula for a single-layer coil.

EXAMPLE 13–3   A 0.15-m-long air-core coil has a radius of 0.006 m and 120 turns. Compute its inductance.

Solution

$$A = \pi r^2 = 1.131 \times 10^{-4} \text{ m}^2$$

$$\mu = \mu_0 = 4\pi \times 10^{-7}$$

Thus,

$$L = 4\pi \times 10^{-7} (120)^2 (1.131 \times 10^{-4})/0.15 = 13.6 \ \mu\text{H}$$

The accuracy of Equation 13–6 breaks down for small ℓ/d ratios. (If ℓ/d is greater than 10, the error is less than 4%.) Improved formulas may be found in design handbooks, such as the *Radio Amateur's Handbook* published by the American Radio Relay League (ARRL).

To provide greater inductance in smaller spaces, iron cores are sometimes used. Unless the core flux is kept below saturation, however, permeability varies and inductance is not constant. To get constant inductance an air gap may be used (Figure 13–10). If the gap is wide enough to dominate, coil inductance is approximately

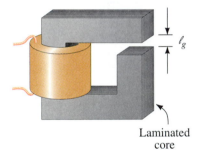

FIGURE 13–10   Iron-core coil with an air gap to control saturation.

$$L = \frac{\mu_0 N^2 A_g}{\ell_g} \quad \text{(H)} \tag{13–7}$$

where μ_0 is the permeability of air, A_g is the area of the air gap, and ℓ_g is its length. (See end-of-chapter Problem 11.) Another way to increase inductance is to use a ferrite core (Section 13.6).

EXAMPLE 13–4   The inductor of Figure 13–10 has 1000 turns, a 5-mm gap, and a cross-sectional area at the gap of $5 \times 10^{-4} \text{ m}^2$. What is its inductance?

Solution

$$L = (4\pi \times 10^{-7})(1000)^2(5 \times 10^{-4})/(5 \times 10^{-3}) = 0.126 \text{ H}$$

PRACTICAL NOTES...

1. Since inductance is due to a conductor's magnetic field, it depends on the same factors that the magnetic field depends on. The stronger the field for a given current, the greater the inductance. Thus, a coil of many turns will have more inductance than a coil of a few turns (L is proportional to N^2) and a coil wound on a magnetic core will have greater inductance than a coil wound on a nonmagnetic form.

2. However, if a coil is wound on a magnetic core, the core's permeability μ may change with flux density. Since flux density depends on current, L becomes a function of current. For example, the inductor of Figure 13–11

has a nonlinear inductance due to core saturation. All inductors encountered in this book are assumed to be linear, i.e., of constant value.

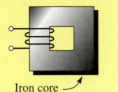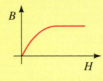

Iron core

FIGURE 13–11   This coil does not have a fixed inductance because its flux is not proportional to its current.

1. The voltage across an inductance whose current changes uniformly by 10 mA in 4 μs is 70 volts. What is its inductance?

2. If you triple the number of turns in the inductor of Figure 13–10, but everything else remains the same, by what factor does the inductance increase?

IN-PROCESS LEARNING CHECK 2

(Answers are at the end of the chapter.)

13.4   Computing Induced Voltage

Earlier, we determined that the voltage across an inductance is given by $v_L = L\,di/dt$, where the voltage and current references are shown in Figure 13–8. Note that the polarity of v_L depends on whether the current is increasing or decreasing. For example, if the current is increasing, di/dt is positive and so v_L is positive, while if the current is decreasing, di/dt is negative and v_L is negative.

To compute voltage, we need to determine di/dt. In general, this requires calculus. However, since di/dt is slope, you can determine voltage without calculus for currents that can be described by straight lines, as in Figure 13–12. For any Δt segment, slope $= \Delta i/\Delta t$, where Δi is the amount that the current changes during time interval Δt.

EXAMPLE 13–5   Figure 13–12 is the current through a 10-mH inductance. Determine voltage v_L and sketch it.

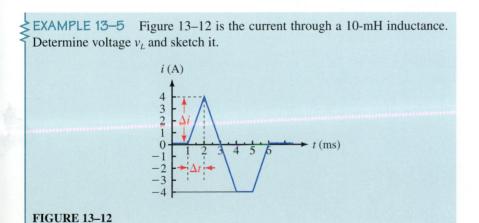

FIGURE 13–12

Solution   Break the problem into intervals over which the slope is constant, determine the slope for each segment, then compute voltage using $v_L = L \times$ slope for that interval:

0 to 1 ms: Slope = 0. Thus, $v_L = 0$ V.

1 ms to 2 ms: Slope = $\Delta i/\Delta t = 4$ A/(1 × 10^{-3} s) = 4 × 10^3 A/s.
                    Thus, $v_L = L\Delta i/\Delta t = (0.010$ H)(4 × 10^3 A/s) = 40 V.

2 ms to 4 ms: Slope = $\Delta i/\Delta t = -8$ A/(2 × 10^{-3} s) = −4 × 10^3 A/s.
                    Thus, $v_L = L\Delta i/\Delta t = (0.010$ H)(−4 × 10^3 A/s) = −40 V.

4 ms to 5 ms: Slope = 0. Thus, $v_L = 0$ V.

5 ms to 6 ms: Same slope as from 1 ms to 2 ms. Thus, $v_L = 40$ V.

The voltage waveform is shown in Figure 13–13.

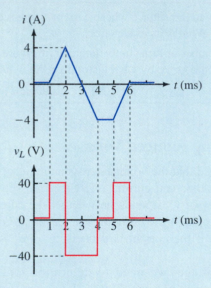

FIGURE 13–13

For currents that are not linear functions of time, you need to use calculus as illustrated in the following example.

EXAMPLE 13–6   What is the equation for the voltage across a 12.5 H inductance whose current is $i = te^{-t}$ amps?

Solution   Differentiate by parts using

$$\frac{d(uv)}{dt} = u\frac{dv}{dt} + v\frac{du}{dt} \text{ with } u = t \text{ and } v = e^{-t}$$

Thus,

$$v_L = L\frac{di}{dt} = L\frac{d}{dt}(te^{-t}) = L[t(-e^{-t}) + e^{-t}] = 12.5e^{-t}(1 - t) \text{ volts}$$

1. Figure 13–14 shows the current through a 5-H inductance. Determine voltage v_L and sketch it.

PRACTICE
PROBLEMS 2

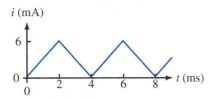

FIGURE 13–14

2. If the current of Figure 13–12 is applied to an unknown inductance and the voltage from 1 ms to 2 ms is 28 volts, what is L?

3. [J] The current in a 4-H inductance is $i = t^2 e^{-5t}$ A. What is voltage v_L?

Answers:
1. v_L is a square wave. Between 0 and 2 ms, its value is 15 V; between 2 ms and 4 ms, its value is -15 V, etc.
2. 7 mH   3. $4e^{-5t}(2t - 5t^2)$ V

1. An inductance L_1 of 50 mH is in series with an inductance L_2 of 35 mH. If the voltage across L_1 at some instant is 125 volts, what is the voltage across L_2 at that instant? Hint: Since the same current passes through both inductances, the rate of change of current is the same for both.

2. Current through a 5-H inductance changes linearly from 10 A to 12 A in 0.5 s. Suppose now the current changes linearly from 2 mA to 6 mA in 1 ms. Although the currents are significantly different, the induced voltage is the same in both cases. Why? Compute the voltage.

IN-PROCESS
LEARNING
CHECK 3

(Answers are at the end of the chapter.)

13.5 Inductances in Series and Parallel

For inductances in series or parallel, the equivalent inductance is found by using the same rules that you used for resistance. For the series case (Figure 13–15) the total inductance is the sum of the individual inductances:

$$L_T = L_1 + L_2 + L_3 + \cdots + L_N \qquad (13–8)$$

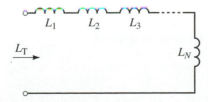

FIGURE 13–15 $L_T = L_1 + L_2 + \cdots + L_N$

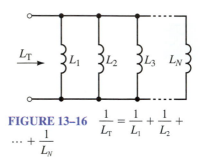

FIGURE 13–16 $\dfrac{1}{L_T} = \dfrac{1}{L_1} + \dfrac{1}{L_2} + \cdots + \dfrac{1}{L_N}$

For the parallel case (Figure 13–16),

$$\frac{1}{L_T} = \frac{1}{L_1} + \frac{1}{L_2} + \frac{1}{L_3} + \cdots + \frac{1}{L_N} \qquad (13\text{–}9)$$

For two inductances, Equation 13–9 reduces to

$$L_T = \frac{L_1 L_2}{L_1 + L_2} \qquad (13\text{–}10)$$

EXAMPLE 13–7   Find L_T for the circuit of Figure 13–17.

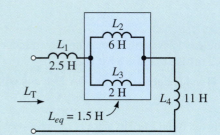

FIGURE 13–17

Solution   The parallel combination of L_2 and L_3 is

$$L_{eq} = \frac{L_2 L_3}{L_2 + L_3} = \frac{6 \times 2}{6 + 2} = 1.5 \text{ H}$$

This is in series with L_1 and L_4. Thus, $L_T = 2.5 + 1.5 + 11 = 15$ H.

PRACTICE PROBLEMS 3

1. For Figure 13–18, $L_T = 2.25$ H. Determine L_x.

FIGURE 13–18

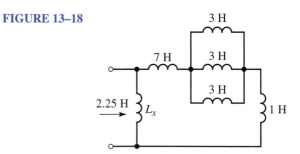

2. For Figure 13–15, the current is the same in each inductance and $v_1 = L_1 di/dt$, $v_2 = L_2 di/dt$, and so on. Apply KVL and show that $L_T = L_1 + L_2 + L_3 + \cdots + L_N$.

Answer: 1. 3 H

13.6   Practical Considerations

Core Types

The type of core used in an inductor depends to a great extent on its intended use and frequency range. (Although you have not studied frequency yet, you can get a feel for frequency by noting that the electrical power system operates at low frequency [60 cycles per second, called 60 hertz], while radio and TV systems operate at high frequency [hundreds of megahertz].) Inductors used in audio or power supply applications generally have iron cores (because they need large inductance values), while inductors for radio-frequency circuits generally use air or ferrite cores. (Ferrite is a mixture of iron oxide in a ceramic binder. It has characteristics that make it suitable for high-frequency work.) Iron cannot be used, however, since it has large power losses at high frequencies (for reasons discussed in Chapter 17).

Variable Inductors

Inductors can be made so that their inductance is variable. In one approach, inductance is varied by changing coil spacing with a screwdriver adjustment. In another approach (Figure 13–19), a threaded ferrite slug is screwed in or out of the coil to vary its inductance.

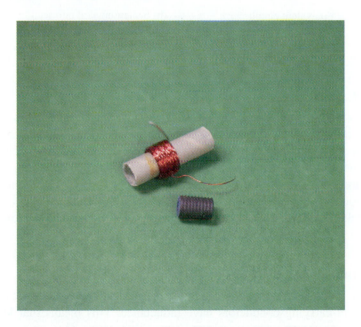

FIGURE 13–19   A variable inductor with its ferrite core removed for viewing.

Circuit Symbols

Figure 13–20 shows inductor symbols. Iron-cores are identified by double solid lines, while dashed lines denote a ferrite core. (Air-core inductors have no core symbol.) An arrow indicates a variable inductor.

Coil Resistance

Ideally, inductors have only inductance. However, since inductors are made of imperfect conductors (e.g., copper wire), they also have resistance. (We can view this resistance as being in series with the coil's inductance as indicated in Figure 13–21(a). Also shown is stray capacitance, considered next.) Although coil resistance is generally small, it cannot always be ignored and thus, must sometimes be included in the analysis of a circuit.

(a) Iron-core

(b) Ferrite-core

(c) Variable

(d) Air-core

FIGURE 13–20   Circuit symbols for inductors.

In Section 13.7, we show how this resistance is taken into account in dc analysis; in later chapters, you will learn how to take it into account in ac analysis.

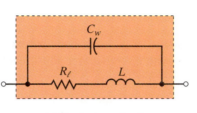

(a) Equivalent circuit

(b) Separating coil into sections helps reduce stray capacitance

FIGURE 13–21   A ferrite-core choke.

Stray Capacitance

Because the turns of an inductor are separated from each other by insulation, a small amount of capacitance exists from winding to winding. This capacitance is called **stray** or **parasitic capacitance.** Although this capacitance is distributed from turn to turn, its effect can be approximated by lumping it as in Figure 13–21(a). The effect of stray capacitance depends on frequency. At low frequencies, it can usually be neglected; at high frequencies, it may have to be taken into account. Some coils are wound in multiple sections as in Figure 13–21(b) to reduce stray capacitance.

Stray Inductance

Because inductance is due entirely to the magnetic effects of electric current, all current-carrying conductors have inductance. This means that leads on circuit components such as resistors, capacitors, transistors, and so on, all have inductance, as do traces on printed circuit boards and wires in cables. We call this inductance **"stray inductance."** Fortunately, in many cases, the stray inductance is so small that it can be neglected.

PRACTICAL NOTES...

Although stray inductance is small, it is not always negligible. In general, stray inductance will not be a problem for short wires at low to moderate frequencies. However, even a short piece of wire can be a problem at high frequencies, or a long piece of wire at low frequencies. For example, the inductance of just a few centimeters of conductor in a high-speed logic system may be nonnegligible because the current through it changes at such a high rate.

13.7 Inductance and Steady State DC

We now look at inductive circuits with constant dc current. Consider Figure 13–22. The voltage across an ideal inductance with constant dc current is zero because the rate of change of current is zero. This is indicated in (a). Since the inductor has current through it but no voltage across it, it looks like a short circuit, (b). This is true in general, that is, *an ideal inductor looks like a short circuit in steady state dc*. (This should not be surprising since it is just a piece of wire to dc.) For a nonideal inductor, its dc equivalent is its coil resistance (Figure 13–23). For steady state dc, problems can be solved using simple dc analysis techniques.

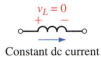

Constant dc current

(a) Since the field is constant, induced voltage is zero

(b) Equivalent of inductor to dc is a short circuit

FIGURE 13–22 Inductance looks like a short circuit to steady state dc.

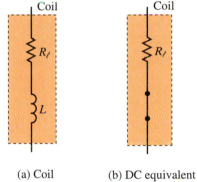

(a) Coil   (b) DC equivalent

FIGURE 13–23 Steady state dc equivalent of a coil with winding resistance.

EXAMPLE 13–8 In Figure 13–24(a), the coil resistance is 14.4 Ω. What is the steady state current I?

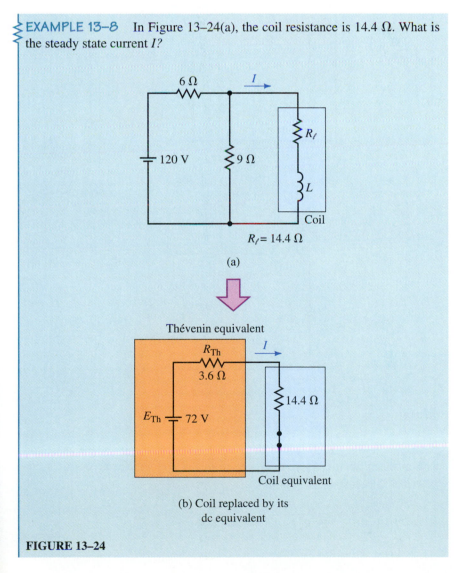

(a)

Thévenin equivalent

(b) Coil replaced by its dc equivalent

FIGURE 13–24

Solution   Reduce the circuit as in (b).

$$E_{Th} = (9/15)(120) = 72 \text{ V}$$
$$R_{Th} = 6\Omega \| 9\Omega = 3.6 \ \Omega$$

Now replace the coil by its dc equivalent circuit as in (b). Thus,

$$I = E_{Th}/R_T = 72/(3.6 + 14.4) = 4 \text{ A}$$

EXAMPLE 13–9   The resistance of coil 1 in Figure 13–25(a) is 30 Ω and that of coil 2 is 15 Ω. Find the voltage across the capacitor assuming steady state dc.

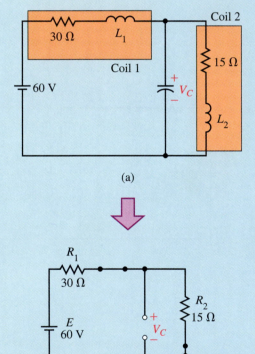

(a)

(b) dc equivalent

FIGURE 13–25

Solution   Replace each coil inductance with a short circuit and the capacitor with an open circuit. As you can see from (b), the voltage across C is the same as the voltage across R_2. Thus,

$$V_C = \frac{R_2}{R_1 + R_2} E = \left(\frac{15 \ \Omega}{45 \ \Omega}\right)(60 \text{ V}) = 20 \text{ V}$$

For Figure 13–26, find I, V_{C_1}, and V_{C_2} in the steady state.

**PRACTICE
PROBLEMS 4**

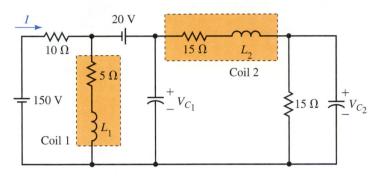

EWB **FIGURE 13–26**

Answers: 10.7 A; 63 V; 31.5 V

13.8   Energy Stored by an Inductance

When power flows into an inductor, energy is stored in its magnetic field. When the field collapses, this energy is returned to the circuit. For an ideal inductor, $R_\ell = 0$ ohm and hence no power is dissipated; thus, an ideal inductor has zero power loss.

To determine the energy stored by an ideal inductor, consider Figure 13–27. Power to the inductor is given by $p = v_L i$ watts, where $v_L = L\,di/dt$. By summing this power (see next **J**), the energy is found to be

$$W = \frac{1}{2}Li^2 \quad \text{(J)} \tag{13–11}$$

where i is the instantaneous value of current. When current reaches its steady state value I, $W = \frac{1}{2}LI^2$ J. This energy remains stored in the field as long as the current continues. When the current goes to zero, the field collapses and the energy is returned to the circuit.

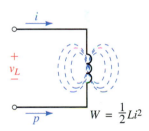

FIGURE 13–27   Energy is stored in the magnetic field of an inductor.

EXAMPLE 13–10   The coil of Figure 13–28(a) has a resistance of 15 Ω. When the current reaches its steady state value, the energy stored is 12 J. What is the inductance of the coil?

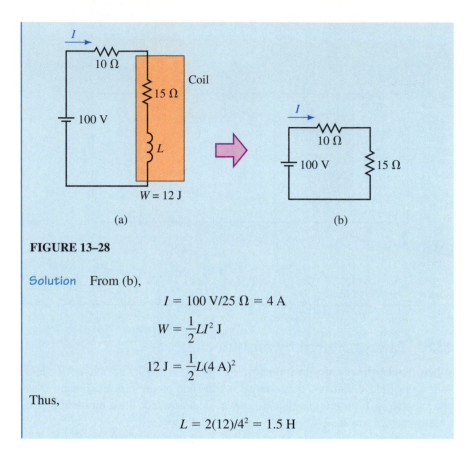

FIGURE 13–28

Solution   From (b),

$$I = 100\ \text{V}/25\ \Omega = 4\ \text{A}$$

$$W = \frac{1}{2}LI^2\ \text{J}$$

$$12\ \text{J} = \frac{1}{2}L(4\ \text{A})^2$$

Thus,

$$L = 2(12)/4^2 = 1.5\ \text{H}$$

Deriving Equation 13–11

The power to the inductor in Figure 13–27 is given by $p = v_L i$, where $v_L = L\,di/dt$. Therefore, $p = Li\,di/dt$. However, $p = dW/dt$. Integrating yields

$$W = \int_0^t p\,dt = \int_0^t Li\frac{di}{dt}dt = L\int_0^i i\,di = \frac{1}{2}Li^2$$

13.9   Inductor Troubleshooting Hints

Inductors may fail by either opening or shorting. Failures may be caused by misuse, defects in manufacturing, or faulty installation.

Open Coil

Opens can be the result of poor solder joints or broken connections. First, make a visual inspection. If nothing wrong is found, disconnect the inductor and check it with an ohmmeter. An open-circuited coil has infinite resistance.

Shorts

Shorts can occur between windings or between the coil and its core (for an iron-core unit). A short may result in excessive current and overheating. Again, check visually. Look for burned insulation, discolored components, an acrid odor, and other evidence of overheating. An ohmmeter can be used to check for shorts between windings and the core. However, checking coil resistance for shorted turns is often of little value, especially if only a few turns are shorted. This is because the shorting of a few windings will not change the overall resistance enough to be measurable. Sometimes the only conclusive test is to substitute a known good inductor for the suspected one.

Unless otherwise indicated, assume ideal inductors and coils.

PROBLEMS

13.2 Induced Voltage and Inductance

1. If the flux linking a 75-turn coil changes at the rate of 3 Wb/s, what is the voltage across the coil?

2. If 80 volts is induced when the flux linking a coil changes at a uniform rate from 3.5 mWb to 4.5 mWb in 0.5 ms, how many turns does the coil have?

3. Flux changing at a uniform rate for 1 ms induces 60 V in a coil. What is the induced voltage if the same flux change takes place in 0.01 s?

13.3 Self-Inductance

4. The current in a 0.4-H inductor is changing at the rate of 200 A/s. What is the voltage across it?

5. The current in a 75-mH inductor changes uniformly by 200 μA in 0.1 ms. What is the voltage across it?

6. The voltage across an inductance is 25 volts when the current changes at 5 A/s. What is L?

7. The voltage induced when current changes uniformly from 3 amps to 5 amps in a 10-H inductor is 180 volts. How long did it take for the current to change from 3 to 5 amps?

8. Current changing at a uniform rate for 1 ms induces 45 V in a coil. What is the induced voltage if the same current change takes place in 100 μs?

9. Compute the inductance of the air-core coil of Figure 13–29, given $\ell = 20$ cm, $N = 200$ turns, and $d = 2$ cm.

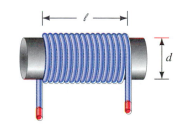

FIGURE 13–29

10. The iron-core inductor of Figure 13–30 has 2000 turns, a cross-section of 1.5 × 1.2 inches, and an air gap of 0.2 inch. Compute its inductance.

11. The iron-core inductor of Figure 13–30 has a high-permeability core. Therefore, by Ampere's law, $NI \simeq H_g \ell_g$. Because the air gap dominates, saturation does not occur and the core flux is proportional to the current, i.e., the flux linkage equals LI. In addition, since all flux passes through the coil, the flux linkage equals $N\Phi$. By equating the two values of flux linkage and using ideas from Chapter 12, show that the inductance of the coil is

$$L = \frac{\mu_0 N^2 A_g}{\ell_g}$$

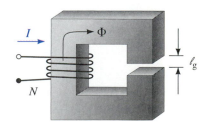

FIGURE 13–30

13.4 Computing Induced Voltage

12. Figure 13–31 shows the current in a 0.75-H inductor. Determine v_L and plot its waveform.

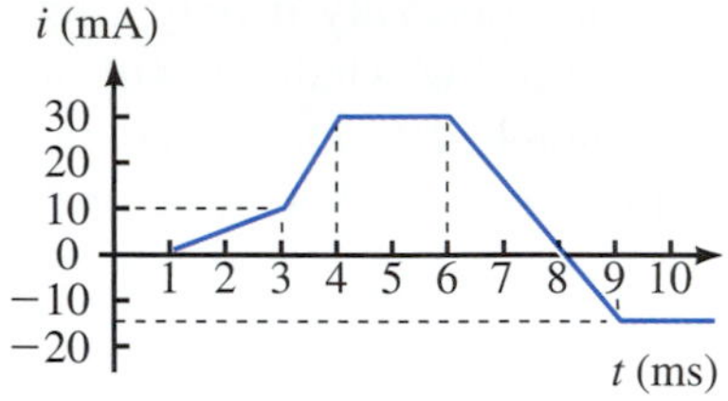

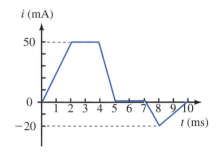

FIGURE 13–31

FIGURE 13–32

13. Figure 13–32 shows the current in a coil. If the voltage from 0 to 2 ms is 100 volts, what is *L?*

14. Why is Figure 13–33 not a valid inductor current? Sketch the voltage across *L* to show why. Pay particular attention to $t = 10$ ms.

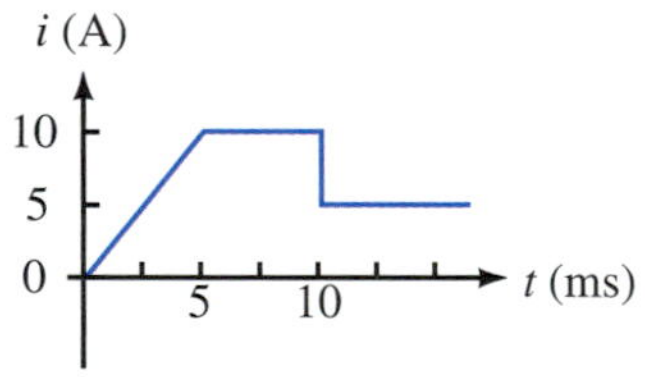

FIGURE 13–33

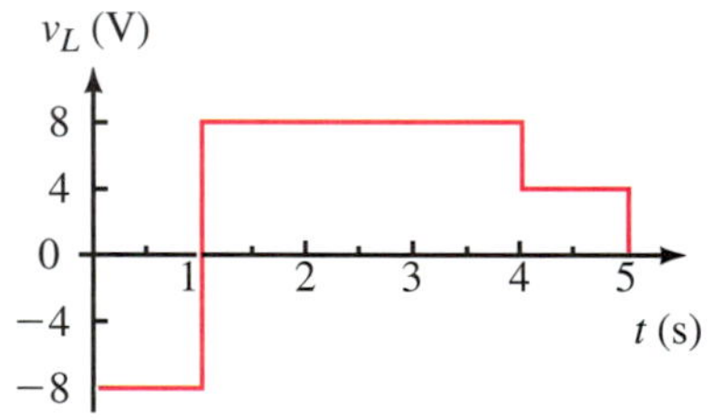

FIGURE 13–34

15. Figure 13–34 shows the graph of the voltage across an inductance. The current changes from 4 A to 5 A during the time interval from 4 s to 5 s.

    a. What is *L?*

    b. Determine the current waveform and plot it.

    c. What is the current at $t = 10$ s?

16. If the current in a 25-H inductance is $i_L = 20e^{-12t}$ mA, what is v_L?

13.5   Inductances in Series and Parallel

17. What is the equivalent inductance of 12 mH, 14 mH, 22 mH, and 36 mH connected in series?

18. What is the equivalent inductance of 0.010 H, 22 mH, 86×10^{-3} H, and 12000 μH connected in series?

19. Repeat Problem 17 if the inductances are connected in parallel.

20. Repeat Problem 18 if the inductances are connected in parallel.

21. Determine L_T for the circuits of Figure 13–35.

22. Determine L_T for the circuits of Figure 13–36.

23. A 30-μH inductance is connected in series with a 60-μH inductance, and a 10-μH inductance is connected in parallel with the series combination. What is L_T?

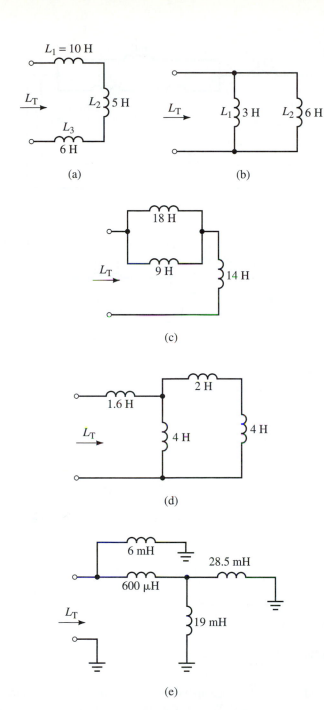

FIGURE 13–35

24. For Figure 13–37, determine L_x.

25. For the circuits of Figure 13–38, determine L_3 and L_4.

26. You have inductances of 24 mH, 36 mH, 22 mH and 10 mH. Connecting these any way you want, what is the largest equivalent inductance you can get? The smallest?

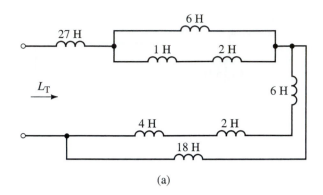

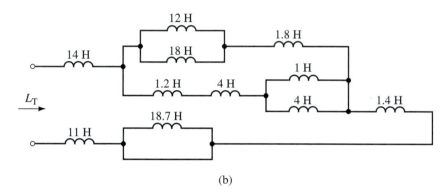

FIGURE 13–36

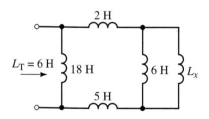

FIGURE 13–37

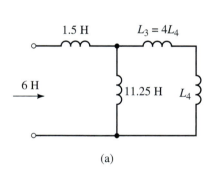

(a)

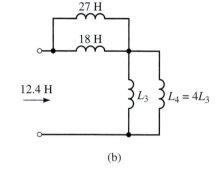

(b)

FIGURE 13–38

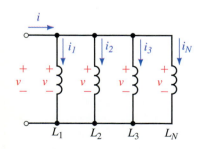

FIGURE 13–39

27. A 6-H and a 4-H inductance are connected in parallel. After a third inductance is added, $L_T = 4$ H. What is the value of the third inductance and how was it connected?

28. Inductances of 2 H, 4 H, and 9 H are connected in a circuit. If $L_T = 3.6$ H, how are the inductors connected?

29. Inductances of 8 H, 12 H, and 1.2 H are connected in a circuit. If $L_T = 6$ H, how are the inductors connected?

30. For inductors in parallel (Figure 13–39), the same voltage appears across each. Thus, $v = L_1 di_1/dt$, $v = L_2 di_2/dt$, etc. Apply KCL and show that $1/L_T = 1/L_1 + 1/L_2 + \cdots + 1/L_N$.

31. By combining elements, reduce each of the circuits of Figure 13–40 to their simplest form.

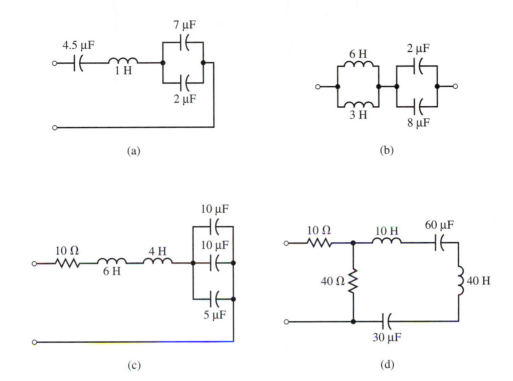

(a) (b)

(c) (d)

FIGURE 13–40

13.7   Inductance and Steady State DC

32. For each of the circuits of Figure 13–41, the voltages and currents have reached their final (steady state) values. Solve for the quantities indicated.

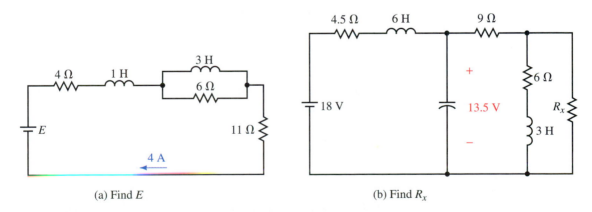

(a) Find E (b) Find R_x

FIGURE 13–41

13.8   Energy Stored by an Inductance

33. Find the energy stored in the inductor of Figure 13–42.

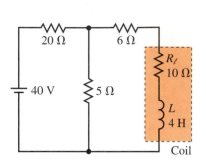

FIGURE 13–42

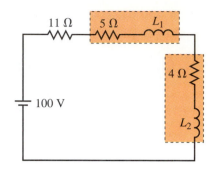

FIGURE 13–43

34. In Figure 13–43, $L_1 = 2L_2$. The total energy stored is $W_T = 75$ J. Find L_1 and L_2.

13.9   Inductor Troubleshooting Hints

35. In Figure 13–44, an inductance meter measures 7 H. What is the likely fault?

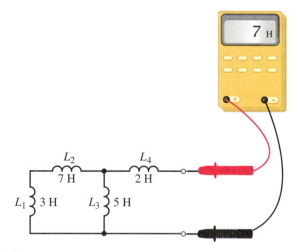

FIGURE 13–44

36. Referring to Figure 13–45, an inductance meter measures $L_T = 8$ mH. What is the likely fault?

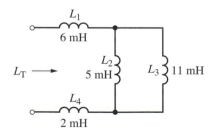

FIGURE 13–45

In-Process Learning Check 1

1. Both a and b. Current cannot change instantaneously.

2. 600 Wb-turns

3. 30 V

4. -300 V

In-Process Learning Check 2

1. 28 mH

2. 9 times

In-Process Learning Check 3

1. 87.5 V

2. Rate of change of current is the same; 20 V

14

Inductive Transients

OBJECTIVES

After studying this chapter, you will be able to

- explain why transients occur in *RL* circuits,
- explain why an inductor looks like an open circuit when first energized,
- compute time constants for *RL* circuits,
- compute voltage and current transients in *RL* circuits during the current buildup phase,
- compute voltage and current transients in *RL* circuits during the current decay phase,
- solve moderately complex *RL* transient problems using circuit simplification techniques,
- solve *RL* transient problems using Electronics Workbench and PSpice.

KEY TERMS

Continuity of Current
De-energizing Transient
Energizing Transient
Initial Condition Circuit
RL Transients
Time Constant

OUTLINE

Introduction
Current Buildup Transients
Interrupting Current in an Inductive Circuit
De-energizing Transients
More Complex Circuits
RL Transients Using Computers

In Chapter 11, you learned that transients occur in capacitive circuits because capacitor voltage cannot change instantaneously. In this chapter, you will learn that transients occur in inductive circuits because inductor current cannot change instantaneously. Although the details differ, you will find that many of the basic ideas are the same.

Inductive transients result when circuits containing inductance are disturbed. More so than capacitive transients, inductive transients are potentially destructive and dangerous. For example, when you break the current in an inductive circuit, an extremely large and damaging voltage may result.

In this chapter, we study basic *RL* transients. We look at transients during current buildup and decay and learn how to calculate the voltages and currents that result.

Inductance, the Dual of Capacitance

INDUCTANCE IS THE DUAL of capacitance. This means that the effect that inductance has on circuit operation is identical with that of capacitance if you interchange the term current for voltage, open circuit for short circuit, and so on. For example, for simple dc transients, current in an *RL* circuit has the same form as voltage in an *RC* circuit: they both rise to their final value exponentially according to $1 - e^{-t/\tau}$. Similarly, voltage across inductance decays in the same manner as current through capacitance, i.e., according to $e^{-t/\tau}$.

Duality applies to steady state and initial condition representations as well. To steady state dc, for example, a capacitor looks like an open circuit, while an inductor looks like a short circuit. Similarly, the dual of a capacitor that looks like a short circuit at the instant of switching is an inductor that looks like an open circuit. Finally, the dual of a capacitor that has an initial condition of V_0 volts is an inductance with an initial condition of I_0 amps.

The principle of duality is helpful in circuit analysis as it lets you transfer the principles and concepts learned in one area directly into another. You will find, for example, that many of the ideas learned in Chapter 11 reappear here in their dual form.

14.1 Introduction

As you saw in Chapter 11, when a circuit containing capacitance is disturbed, voltages and currents do not change to their new values immediately, but instead pass through a transitional phase as the circuit capacitance charges or discharges. The voltages and currents during this transitional interval are called **transients.** In a dual fashion, transients occur when circuits containing inductances are disturbed. In this case, however, transients occur because current in inductance cannot change instantaneously.

To get at the idea, consider Figure 14–1. In (a), we see a purely resistive circuit. At the instant the switch is closed, current jumps from 0 to E/R as required by Ohm's law. Thus, no transient (i.e., transitional phase) occurs because current reaches its final value immediately. Now consider (b). Here, we have added inductance. At the instant the switch is closed, a counter emf appears across the inductance. This voltage attempts to stop the current from changing and consequently slows its rise. Current thus does not jump to E/R immediately as in (a), but instead climbs gradually and smoothly as in (b). The larger the inductance, the longer the transition takes.

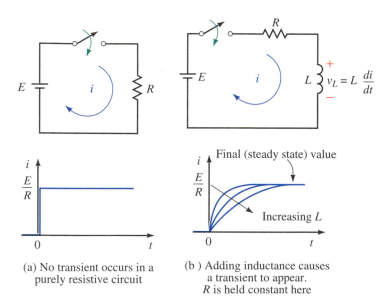

(a) No transient occurs in a purely resistive circuit

(b) Adding inductance causes a transient to appear. R is held constant here

FIGURE 14–1 Transient due to inductance. Adding inductance to a resistive circuit slows the current rise and fall, thus creating a transient.

Continuity of Current

As Figure 14–1(b) illustates, *current through an inductance cannot change instantaneously, i.e., it cannot jump abruptly from one value to another, but must be continuous at all values of time.* This observation is known as the statement of **continuity of current for inductance.** You will find this statement of great value when analyzing circuits containing inductance. We will use it many times in what follows.

Inductor Voltage

Now consider inductor voltage. When the switch is open as in Figure 14–2(a), the current in the circuit and voltage across L are both zero. Now close the switch. Immediately after the switch is closed, the current is still zero, (since it cannot change instantaneously). Since $v_R = Ri$, the voltage across R is also zero and thus the full source voltage appears across L as shown in (b). The inductor voltage therefore jumps from 0 V just before the switch is closed to E volts just after. It then decays to zero, since, as we saw in Chapter 13, the voltage across inductance is zero for steady state dc. This is indicated in (c).

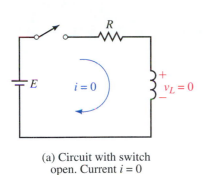

(a) Circuit with switch open. Current $i = 0$

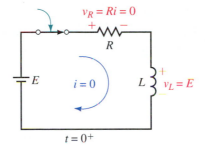

(b) Circuit just after the switch has been closed. Current is still equal to zero. Thus, $v_L = E$

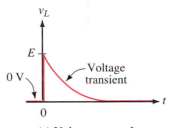

(c) Voltage across L

FIGURE 14–2   Voltage across L.

Open-Circuit Equivalent of an Inductance

Consider again Figure 14–2(b). Note that just after the switch is closed, the inductor has voltage across it but no current through it. It therefore momentarily looks like an open circuit. This is indicated in Figure 14–3. This observation is true in general, that is, *an inductor with zero initial current looks like an open circuit at the instant of switching*. (Later, we extend this statement to include inductors with nonzero initial currents.)

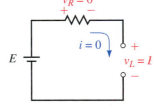

FIGURE 14–3   Inductor with zero initial current looks like an open circuit at the instant the switch is closed.

Initial Condition Circuits

Voltages and currents in circuits immediately after switching must sometimes be calculated. These can be determined with the aid of the open-circuit equivalent. By replacing inductances with open circuits, you can see what a circuit looks like just after switching. Such a circuit is called an **initial condition circuit.**

EXAMPLE 14–1   A coil and two resistors are connected to a 20-V source as in Figure 14–4(a). Determine source current i and inductor voltage v_L at the instant the switch is closed.

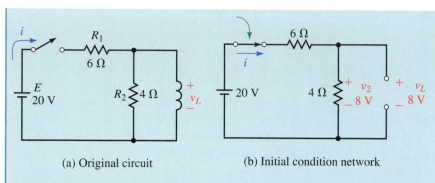

(a) Original circuit (b) Initial condition network

FIGURE 14–4

Solution Replace the inductance with an open circuit. This yields the network shown in (b). Thus $i = E/R_T = 20\text{ V}/10\ \Omega = 2$ A and the voltage across R_2 is $v_2 = (2\text{ A})(4\ \Omega) = 8$ V. Since $v_L = v_2$, $v_L = 8$ volts as well.

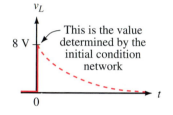

FIGURE 14–5 The initial condition network yields only the value at $t = 0^+$ s.

Initial condition networks yield voltages and currents only at the instant of switching, i.e., at $t = 0^+$ s. Thus, the value of 8 V calculated in Example 14–1 is only a momentary value as illustrated in Figure 14–5. Sometimes such an initial value is all that you need. In other cases, you need the complete solution. This is considered next, in Section 14.2.

PRACTICE PROBLEMS 1

Determine all voltages and currents in the circuit of Figure 14–6 immediately after the switch is closed and in steady state.

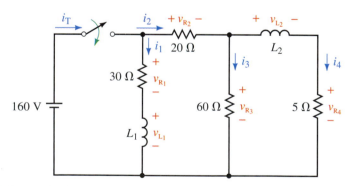

FIGURE 14–6

Answers: Initial: $v_{R_1} = 0$ V; $v_{R_2} = 40$ V; $v_{R_3} = 120$ V; $v_{R_4} = 0$ V; $v_{L_1} = 160$ V; $v_{L_2} = 120$ V; $i_T = 2$ A; $i_1 = 0$ A; $i_2 = 2$ A; $i_3 = 2$ A; $i_4 = 0$ A.

Steady State: $v_{R_1} = 160$ V; $v_{R_2} = 130$ V; $v_{R_3} = v_{R_4} = 30$ V; $v_{L_1} = v_{L_2} = 0$ V; $i_T = 11.83$ A; $i_1 = 5.33$ A; $i_2 = 6.5$ A; $i_3 = 0.5$ A; $i_4 = 6.0$ A

14.2   Current Buildup Transients

Current

We will now develop equations to describe voltages and current during energization. Consider Figure 14–7. KVL yields

$$v_L + v_R = E \qquad (14\text{–}1)$$

Substituting $v_L = Ldi/dt$ and $v_R = Ri$ into Equation 14–1 yields

$$L\frac{di}{dt} + Ri = E \qquad (14\text{–}2)$$

Equation 14–2 can be solved using basic calculus in a manner similar to what we did for *RC* circuits in Chapter 11. The result is

$$i = \frac{E}{R}(1 - e^{-Rt/L}) \quad \text{(A)} \qquad (14\text{–}3)$$

where R is in ohms, L is in henries, and t is in seconds. Equation 14–3 describes current buildup. Values of current at any point in time can be found by direct substitution as we illustrate next. Note that E/R is the final (steady state) current.

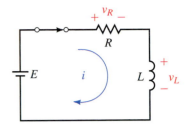

FIGURE 14–7   KVL yields $v_L + v_R = E$.

EXAMPLE 14–2   For the circuit of Figure 14–7, suppose $E = 50$ V, $R = 10\ \Omega$, and $L = 2$ H:

a. Determine the expression for i.

b. Compute and tabulate values of i at $t = 0^+$, 0.2, 0.4, 0.6, 0.8, and 1.0 s.

c. Using these values, plot the current.

d. What is the steady state current?

Solution

a. Substituting the values into Equation 14–3 yields

$$i = \frac{E}{R}(1 - e^{-Rt/L}) = \frac{50\ \text{V}}{10\ \Omega}(1 - e^{-10t/2}) = 5\,(1 - e^{-5t})\ \text{amps}$$

b. At $t = 0^+$ s, $i = 5(1 - e^{-5t}) = 5(1 - e^0) = 5(1 - 1) = 0$ A.
   At $t = 0.2$ s, $i = 5(1 - e^{-5(0.2)}) = 5(1 - e^{-1}) = 3.16$ A.
   At $t = 0.4$ s, $i = 5(1 - e^{-5(0.4)}) = 5(1 - e^{-2}) = 4.32$ A.
   Continuing in this manner, you get Table 14–1.

c. Values are plotted in Figure 14–8. Note that this curve looks exactly like the curves we determined intuitively in Figure 14–1(b).

d. Steady state current is $E/R = 50$ V/10 Ω = 5 A. This agrees with the curve of Figure 14–8.

TABLE 14–1

Time	Current
0	0
0.2	3.16
0.4	4.32
0.6	4.75
0.8	4.91
1.0	4.97

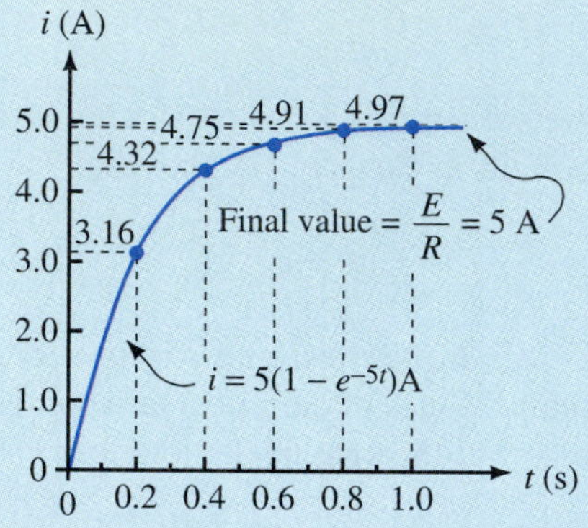

FIGURE 14–8   Current buildup transient.

Circuit Voltages

With i known, circuit voltages can be determined. Consider voltage v_R. Since $v_R = Ri$, when you multiply R times Equation 14–3, you get

$$v_R = E(1 - e^{-Rt/L}) \quad \text{(V)} \tag{14–4}$$

Note that v_R has exactly the same shape as the current. Now consider v_L. Voltage v_L can be found by subtracting v_R from E as per Equation 14–1:

$$v_L = E - v_R = E - E(1 - e^{-Rt/L}) = E - E + Ee^{-Rt/L}$$

Thus,

$$v_L = Ee^{-Rt/L} \tag{14–5}$$

An examination of Equation 14–5 shows that v_L has an initial value of E at $t = 0^+$ s and then decays exponentially to zero. This agrees with our earlier observation in Figure 14–2(c).

EXAMPLE 14–3   Repeat Example 14–2 for voltage v_L.

Solution

a. From equation 14–5,

$$v_L = Ee^{-Rt/L} = 50e^{-5t} \text{ volts}$$

b. At $t = 0^+$ s, $v_L = 50e^{-5t} = 50e^0 = 50(1) = 50$ V.
At $t = 0.2$ s, $v_L = 50e^{-5(0.2)} = 50e^{-1} = 18.4$ V.
At $t = 0.4$ s, $v_L = 50e^{-5(0.4)} = 50e^{-2} = 6.77$ V.

Continuing in this manner, you get Table 14–2.

c. The waveform is shown in Figure 14–9.

d. Steady state voltage is 0 V, as you can see in Figure 14–9.

TABLE 14–2

Time (s)	Voltage (V)
0	50.0
0.2	18.4
0.4	6.77
0.6	2.49
0.8	0.916
1.0	0.337

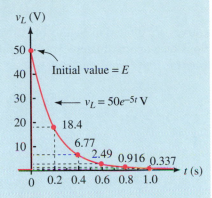

FIGURE 14–9 Inductor voltage transient.

For the circuit of Figure 14–7, with $E = 80$ V, $R = 5$ kΩ, and $L = 2.5$ mH:

a. Determine expressions for i, v_L, and v_R.

b. Compute and tabulate values at $t = 0^+$, 0.5, 1.0, 1.5, 2.0, and 2.5 μs.

c. At each point in time, does $v_L + v_R = E$?

d. Plot i, v_L, and v_R using the values computed in (b).

PRACTICE PROBLEMS 2

Answers: a. $i = 16(1 - e^{-2 \times 10^6 t})$ mA; $v_L = 80e^{-2 \times 10^6 t}$ V; $v_R = 80(1 - e^{-2 \times 10^6 t})$

b.

t (μs)	v_L (V)	i_L (mA)	v_R (V)
0	80	0	0
0.5	29.4	10.1	50.6
1.0	10.8	13.8	69.2
1.5	3.98	15.2	76.0
2.0	1.47	15.7	78.5
2.5	0.539	15.9	79.5

c. Yes

d. i and v_R have the shape shown in Figure 14–8, while v_L has the shape shown in Figure 14–9, with values according to the table shown in b.

Time Constant

In Equations 14–3 to 14–5 L/R is the time constant of the circuit.

$$\tau = \frac{L}{R} \quad \text{(s)} \tag{14–6}$$

Note that τ has units of seconds. (This is left as an exercise for the student.) Equations 14–3, 14–4, and 14–5 may now be written as

$$i = \frac{E}{R}(1 - e^{-t/\tau}) \quad (A) \tag{14-7}$$

$$v_L = Ee^{-t/\tau} \quad (V) \tag{14-8}$$

$$v_R = E(1 - e^{-t/\tau}) \quad (V) \tag{14-9}$$

Curves are plotted in Figure 14–10 versus time constant. As expected, transitions take approximately 5τ; thus, *for all practical purposes, inductive transients last five time constants.*

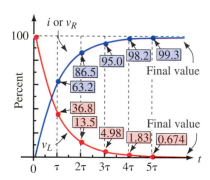

FIGURE 14–10 Universal time constant curves for the *RL* circuit.

EXAMPLE 14–4 In a circuit where $L = 2$ mH, transients last 50 μs. What is R?

Solution Transients last five time constants. Thus, $\tau = 50\ \mu s/5 = 10\ \mu s$. Now $\tau = L/R$. Therefore, $R = L/\tau = 2$ mH/10 μs $= 200\ \Omega$.

EXAMPLE 14–5 For an *RL* circuit, $i = 40(1 - e^{-5t})$ A and $v_L = 100e^{-5t}$ V.

a. What are E and τ?
b. What is R?
c. Determine L.

Solution

a. From Equation 14–8, $v_L = Ee^{-t/\tau} = 100e^{-5t}$. Therefore, $E = 100$ V and
$$\tau = \frac{1}{5} = 0.2s.$$

b. From Equation 14–7,

$$i = \frac{E}{R}(1 - e^{-t/\tau}) = 40(1 - e^{-5t}).$$

Therefore, $E/R = 40$ A and $R = E/40$ A $= 100$ V/40 A $= 2.5\ \Omega$.

c. $\tau = L/R$. Therefore, $L = R\tau = (2.5)(0.2) = 0.5$ H.

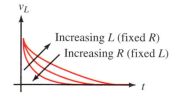

FIGURE 14–11 Effect of R and L on transient duration.

It is sometimes easier to solve problems using the time constant curves than it is to solve the equation. (Be sure to convert curve percentages to a decimal value first, e.g., 63.2% to 0.632.) To illustrate, consider the problem of Examples 14–2 and 14–3. From Figure 14–10 at $t = \tau = 0.2$ s, $i = 0.632E/R$ and $v_L = 0.368E$. Thus, $i = 0.632(5$ A$) = 3.16$ A and $v_L = 0.368(50$ V$) = 18.4$ V as we found earlier.

The effect of inductance and resistance on transient duration is shown in Figure 14–11. The larger the inductance, the longer the transient for a given resistance. Resistance has the opposite effect: for a fixed inductance, the larger the resistance, the shorter the transient. [This is not hard to understand. As R increases, the circuit looks more and more resistive. If you get to a point where inductance is negligible compared with resistance, the circuit looks purely resistive, as in Figure 14–1(a), and no transient occurs.]

IN-PROCESS
LEARNING
CHECK 1

1. For the circuit of Figure 14–12, the switch is closed at $t = 0$ s.

    a. Determine expressions for v_L and i.

    b. Compute v_L and i at $t = 0^+$, 10 μs, 20 μs, 30 μs, 40 μs, and 50 μs.

    c. Plot curves for v_L and i.

FIGURE 14–12

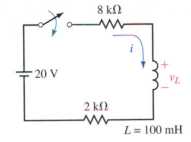

2. For the circuit of Figure 14–7, $E = 85$ V, $R = 50$ Ω, and $L = 0.5$ H. Use the universal time constant curves to determine v_L and i at $t = 20$ ms.

3. For a certain RL circuit, transients last 25 s. If $L = 10$ H and steady state current is 2 A, what is E?

4. An RL circuit has $E = 50$ V and $R = 10$ Ω. The switch is closed at $t = 0$ s. What is the current at the end of 1.5 time constants?

(Answers are at the end of the chapter.)

14.3   Interrupting Current in an Inductive Circuit

We now look at what happens when inductor current is interrupted. Consider Figure 14–13. At the instant the switch is opened, the field begins to collapse, which induces a voltage in the coil. If inductance is large and current

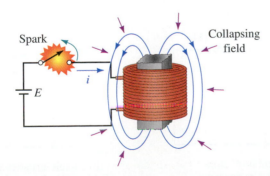

FIGURE 14–13   The sudden collapse of the magnetic field when the switch is opened causes a large induced voltage across the coil. (Several thousand volts may result.) The switch arcs over due to this voltage.

is high, a great deal of energy is released in a very short time, creating a huge voltage that may damage equipment and create a shock hazard. (This induced voltage is referred to as an **inductive kick.**) For example, abruptly breaking the current through a large inductor (such as a motor or generator field coil) can create voltage spikes up to several thousand volts, a value large enough to draw long arcs as indicated in Figure 14–13. Even moderate sized inductances in electronic systems can create enough voltage to cause damage if protective circuitry is not used.

The dynamics of the switch flashover are not hard to understand. When the field collapses, the voltage across the coil rises rapidly. Part of this voltage appears across the switch. As the switch voltage rises, it quickly exceeds the breakdown strength of air, causing a flashover between its contacts. Once struck, the arc is easily maintained, as it creates ionized gases that provide a relatively low resistance path for conduction. As the contacts spread apart, the arc elongates and eventually disappears as the coil energy is dissipated and coil voltage drops below that required to sustain the arc.

There are several important points to note here:

1. Flashovers, as in Figure 14–13, are generally undesirable. However, they can be controlled through proper engineering design. (One way is to use a discharge resistor, as in the next example; another way is to use a diode, as you will see in your electronics course.)

2. On the other hand, the large voltages created by breaking inductive currents have their uses. One is in the ignition system of automobiles, where current in the primary winding of a transformer coil is interrupted at the appropriate time by a control circuit to create the spark needed to fire the engine.

3. It is not possible to rigorously analyze the circuit of Figure 14–13 because the resistance of the arc changes as the switch opens. However, the main ideas can be established by studying circuits using fixed resistors as we see next.

The Basic Ideas

We begin with the circuit of Figure 14–14. Assume the switch is closed and the circuit is in steady state. Since the inductance looks like a short circuit [Figure 14–15(a)], its current is $i_L = 120 \text{ V}/30 \ \Omega = 4 \text{ A}$.

The intuitive explanation here has a sound mathematical basis. Recall, back emf (induced voltage) across a coil is given by

$$v_L = L\frac{di}{dt} \approx L\frac{\Delta i}{\Delta t}$$

where Δi is the change in current and Δt is the time interval over which the change takes place. When you open the switch, current begins to drop immediately toward zero. Since Δi is finite, and $\Delta t \to 0$, the voltage across L rises to a very large value, causing a flashover to occur. After the flashover, current has a path through which to decay and thus Δt, although small, no longer approaches zero. The result is a large but finite voltage spike across L.

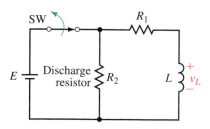

FIGURE 14–14 Discharge resistor R_2 helps limit the size of the induced voltage.

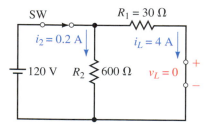

(a) Circuit just before the switch is opened

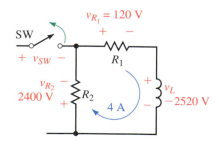

(b) Circuit just after SW is opened. Since coil voltage polarity is opposite to that shown, v_L is negative

 FIGURE 14–15 Circuit of Figure 14–14 immediately before and after the switch is opened. Coil voltage changes abruptly from 0 V to −2520 V for this example.

Now open the switch. Just prior to opening the switch, $i_L = 4$ A; therefore, just after opening the switch, it must still be 4 A. As indicated in (b), this 4 A passes through resistances R_1 and R_2, creating voltages $v_{R_1} = 4$ A $\times$ 30 $\Omega = 120$ V and $v_{R_2} = 4$ A $\times$ 600 $\Omega = 2400$ V with the polarity shown. From KVL, $v_L + v_{R_1} + v_{R_2} = 0$. Therefore at the instant the switch is opened,

$$v_L = -(v_{R_1} + v_{R_2}) = -2520 \text{ volts}$$

appears across the coil, yielding a negative voltage spike as in Figure 14–16. Note that this spike is more than 20 times larger than the source voltage. As we see in the next section, the size of this spike depends on the ratio of R_2 to R_1; the larger the ratio, the larger the voltage.

Consider again Figure 14–15. Note that current i_2 changes abruptly from 0.2 A just prior to switching to -4 A just after. This is permissible, however, since i_2 does not pass through the inductor and only currents through inductance cannot change abruptly.

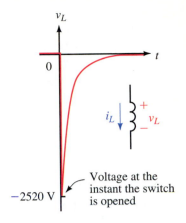

FIGURE 14–16   Voltage spike for the circuit of Figure 14–14. This voltage is more than 20 times larger than the source voltage.

Figure 14–16 shows the voltage across the coil of Figure 14–14. Make a similar sketch for the voltage across the switch and across resistor R_2. Hint: Use KVL to find v_{SW} and v_{R_2}.

PRACTICE
PROBLEMS 3

Answer: v_{SW}: With the switch closed, $v_{SW} = 0$ V; When the switch is opened, v_{SW} jumps to 2520 V, then decays to 120 V. v_{R_2}: Identical to Figure 14–16 except that v_{R_2} begins at -2400 V instead of -2520 V.

Inductor Equivalent at Switching

Figure 14–17 shows the current through L of Figure 14–15. Because the current is the same immediately after switching as it is immediately before, it is constant over the interval from $t = 0^-$ s to $t = 0^+$ s. Since this is true in general, we see that *an inductance with an initial current looks like a current source at the instant of switching.* Its value is the value of the current at switching. This is shown in Figure 14–18.

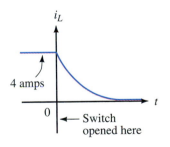

FIGURE 14–17   Inductor current for the circuit of Figure 14–15.

(a) Current at switching   (b) Current source equivalent

FIGURE 14–18   An inductor carrying current looks like a current source at the instant of switching.

14.4   De-energizing Transients

We now look at equations for the voltages and currents described in the previous section. As we go through this material, you should focus on the basic principles involved, rather than just the resulting equations. You will probably forget formulas, but if you understand principles, you should be able to reason your way through many problems using just the basic current and voltage relationships.

Consider Figure 14–19(a). Let the initial current in the inductor be denoted as I_0 amps. Now open the switch as in (b). KVL yields $v_L + v_{R_1} + v_{R_2} = 0$. Substituting $v_L = L di/dt$, $v_{R_1} = R_1 i$, and $v_{R_2} = R_2 i$ yields $L di/dt + (R_1 + R_2)i = 0$. Now using calculus, it can be shown that

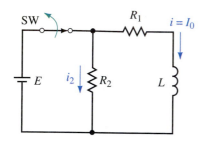

(a) Before the switch
is opened

$$i = I_0 e^{-t/\tau'} \quad \text{(A)} \tag{14–10}$$

where

$$\tau' = \frac{L}{R_T} = \frac{L}{R_1 + R_2} \quad \text{(s)} \tag{14–11}$$

is the time constant of the discharge circuit. If the circuit is in steady state before the switch is opened, initial current $I_0 = E/R_1$ and Equation 14–10 becomes

$$i = \frac{E}{R_1} e^{-t/\tau'} \quad \text{(A)} \tag{14–12}$$

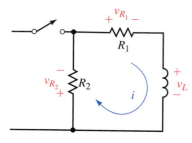

(b) Decay circuit

FIGURE 14–19   Circuit for studying decay transients.

EXAMPLE 14–6   For Figure 14–19(a), assume the current has reached steady state with the switch closed. Suppose that $E = 120$ V, $R_1 = 30\ \Omega$, $R_2 = 600\ \Omega$, and $L = 126$ mH:

a. Determine I_0.

b. Determine the decay time constant.

c. Determine the equation for the current decay.

d. Compute the current i at $t = 0^+$ s and $t = 0.5$ ms.

Solution

a. Consider Figure 14–19(a). Since the circuit is in a steady state, the inductor looks like a short circuit to dc. Thus, $I_0 = E/R_1 = 4$ A.

b. Consider Figure 14–19(b). $\tau' = L/(R_1 + R_2) = 126$ mH/630 Ω = 0.2 ms.

c. $i = I_0 e^{-t/\tau'} = 4e^{-t/0.2\ \text{ms}}$ A.

d. At $t = 0^+$ s, $i = 4e^{-0} = 4$ A.
    At $t = 0.5$ ms, $i = 4e^{-0.5\ \text{ms}/0.2\ \text{ms}} = 4e^{-2.5} = 0.328$ A.

Now consider voltage v_L. It can be shown to be

$$v_L = V_0 e^{-t/\tau'} \tag{14–13}$$

where V_0 is the voltage across L just after the switch is opened. Letting $i = I_0$ in Figure 14–19(b), you can see that $V_0 = -I_0(R_1 + R_2) = -I_0 R_T$. Thus Equation 14–13 can be written as

$$v_L = -I_0 R_T e^{-t/\tau'} \tag{14–14}$$

Finally, if the current has reached steady state before the switch is opened, $I_0 = E/R_1$, and Equation 14–14 becomes

$$v_L = -E\left(1 + \frac{R_2}{R_1}\right)e^{-t/\tau'} \tag{14–15}$$

Note that v_L starts at V_0 volts (which is negative) and decays to zero as shown in Figure 14–20.

Now consider the resistor voltages. Each is the product of resistance times current (Equation 14–10). Thus,

$$v_{R_1} = R_1 I_0 e^{-t/\tau'} \tag{14–16}$$

and

$$v_{R_2} = R_2 I_0 e^{-t/\tau'} \tag{14–17}$$

If current has reached steady state before switching, these become

$$v_{R_1} = E e^{-t/\tau'} \tag{14–18}$$

and

$$v_{R_2} = \frac{R_2}{R_1} E e^{-t/\tau'} \tag{14–19}$$

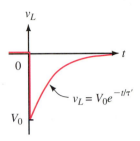

FIGURE 14–20 Inductor voltage during decay phase. V_0 is negative.

Substituting the values of Example 14–6 into these equations, we get for the circuit of Figure 14–19 $v_L = -2520e^{-t/0.2\text{ ms}}$ V, $v_{R_1} = 120e^{-t/0.2\text{ ms}}$ V and $v_{R_2} = 2400e^{-t/0.2\text{ ms}}$ V. These can also be written as $v_L = -2520e^{-5000t}$ V and so on if desired.

Decay problems can also be solved using the decay portion of the universal time constant curves shown in Figure 14–10.

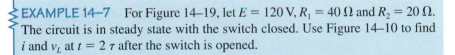

EXAMPLE 14–7 For Figure 14–19, let $E = 120$ V, $R_1 = 40\ \Omega$ and $R_2 = 20\ \Omega$. The circuit is in steady state with the switch closed. Use Figure 14–10 to find i and v_L at $t = 2\ \tau$ after the switch is opened.

Solution $I_0 = E/R_1 = 3$ A. At $t = 2\ \tau$, current will have decayed to 13.5%. Therefore, $i = 0.135 I_0 = 0.405$ A and $v_L = -(R_1 + R_2)i = -(60\ \Omega)(0.405\text{ A}) = -24.3$ V. (Alternately, $V_0 = -(3\text{ A})(60\ \Omega) = -180$ V. At $t = 2\ \tau$, this has decayed to 13.5%. Therefore, $v_L = 0.135(-180\text{ V}) = -24.3$ V as above.)

14.5 More Complex Circuits

The equations developed so far apply only to circuits of the forms of Figures 14–7 or 14–19. Fortunately, many circuits can be reduced to these forms using circuit reduction techniques such as series and parallel combinations, source conversions, Thévenin's theorem, and so on.

EXAMPLE 14–8   Determine i_L for the circuit of Figure 14–21(a) if $L = 5$ H.

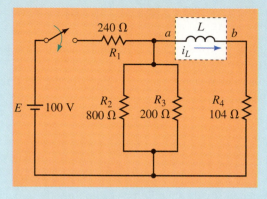

(a) Circuit

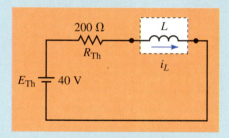

(b) Thévenin equivalent

FIGURE 14–21

Solution   The circuit can be reduced to its Thévenin equivalent (b) as you saw in Chapter 11 (Section 11.5). For this circuit, $\tau = L/R_{Th} = 5$ H/200 Ω = 25 ms. Now apply Equation 14–7. Thus,

$$i_L = \frac{E_{Th}}{R_{Th}}(1 - e^{-t/\tau}) = \frac{40}{200}(1 - e^{-t/25\ ms}) = 0.2\,(1 - e^{-40t}) \quad \text{(A)}$$

EXAMPLE 14–9   For the circuit of Example 14–8, at what time does current reach 0.12 amps?

Solution

$$i_L = 0.2(1 - e^{-40t}) \quad \text{(A)}$$

Thus,

$$0.12 = 0.2(1 - e^{-40t}) \quad \text{(Figure 14–22)}$$
$$0.6 = 1 - e^{-40t}$$
$$e^{-40t} = 0.4$$

Taking the natural log of both sides,

$$\ln e^{-40t} = \ln 0.4$$

$$-40t = -0.916$$

$$t = 22.9 \text{ ms}$$

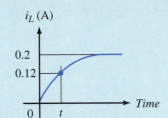

FIGURE 14–22

PRACTICE
PROBLEMS 4

1. For the circuit of Figure 14–21, let $E = 120$ V, $R_1 = 600$ Ω, $R_2 = 3$ kΩ, $R_3 = 2$ kΩ, $R_4 = 100$ Ω, and $L = 0.25$ H:

   a. Determine i_L and sketch it.

   b. Determine v_L and sketch it.

2. Let everything be as in Problem 1 except L. If $i_L = 0.12$ A at $t = 20$ ms, what is L?

Answers:

1. a. $160(1 - e^{-2000t})$ mA   b. $80e^{-2000t}$ V. i_L climbs from 0 to 160 mA with the waveshape of Figure 14–1(b), reaching steady state in 2.5 ms. v_L looks like Figure 14–2(c). It starts at 80 V and decays to 0 V in 2.5 ms.

2. 7.21 H

A Note About Time Scales

Until now, we have considered energization and de-energization phases separately. When both occur in the same problem, we must clearly define what we mean by time. One way to handle this problem (as we did with *RC* circuits) is to define $t = 0$ s as the beginning of the first phase and solve for voltages and currents in the usual manner, then shift the time axis to the beginning of the second phase, redefine $t = 0$ s and then solve the second part. This is illustrated in Example 14–10. Note that only the first time scale is shown explicitly on the graph.

EXAMPLE 14–10   Refer to the circuit of Figure 14–23:

a. Close the switch at $t = 0$ and determine equations for i_L and v_L.

b. At $t = 300$ ms, open the switch and determine equations for i_L and v_L during the decay phase.

c. Determine voltage and current at $t = 100$ ms and at $t = 350$ ms.

d. Sketch i_L and v_L. Mark the points from (c) on the sketch.

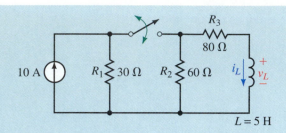

 FIGURE 14–23

Solution

a. Convert the circuit to the left of L to its Thévenin equivalent. As indicated in Figure 14–24(a), $R_{Th} = 60 \| 30 + 80 = 100\ \Omega$. From (b), $E_{Th} = V_2$, where

$$V_2 = (10\ \text{A})(20\ \Omega) = 200\ \text{V}$$

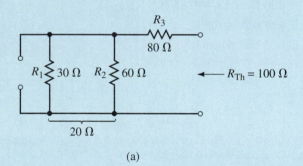

(a)

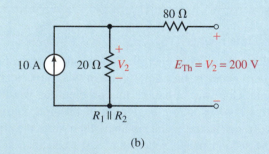

(b)

FIGURE 14–24

The Thévenin equivalent circuit is shown in Figure 14–25(a). $\tau = L/R_{Th} = 50$ ms. Thus during current buildup,

$$i_L = \frac{E_{Th}}{R_{Th}}(1 - e^{-t/\tau}) = \frac{200}{100}(1 - e^{-t/50\ \text{ms}}) = 2\,(1 - e^{-20t})\quad \text{A}$$

$$v_L = E_{Th}e^{-t/\tau} = 200\,e^{-20t}\quad \text{V}$$

b. Current build-up is sketched in Figure 14–25(b). Since $5\tau = 250$ ms, current is in steady state when the switch is opened at 300 ms. Thus $I_0 =$

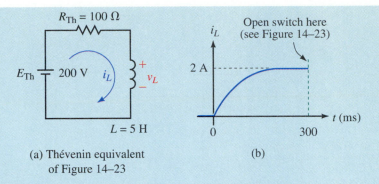

(a) Thévenin equivalent
   of Figure 14–23

(b)

FIGURE 14–25   Circuit and current during the buildup phase.

2 A. When the switch is opened, current decays to zero through a resistance of $60 + 80 = 140 \ \Omega$ as shown in Figure 14–26. Thus, $\tau' = 5H/140 \ \Omega = 35.7$ ms. If $t = 0$ s is redefined as the instant the switch is opened, the equation for the decay is

$$i_L = I_0 e^{-t/\tau'} = 2e^{-t/35.7 \text{ ms}} = 2e^{-28t} \quad \text{A}$$

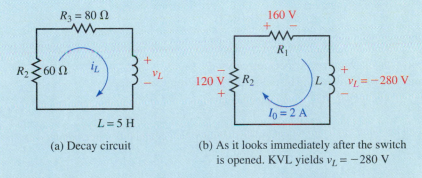

(a) Decay circuit

(b) As it looks immediately after the switch
   is opened. KVL yields $v_L = -280$ V

FIGURE 14–26   The circuit of Figure 14–23 as it looks during the decay phase.

Now consider voltage. As indicated in Figure 14–26(b), the voltage across L just after the switch is open is $V_0 = -280$ V. Thus

$$v_L = V_0 e^{-t/\tau'} = -280e^{-28t} \quad \text{V}$$

c. You can use the universal time constant curves at $t = 100$ ms since 100 ms represents 2τ. At 2τ, current has reached 86.5% of its final value. Thus, $i_L = 0.865(2 \text{ A}) = 1.73$ A. Voltage has fallen to 13.5 %. Thus $v_L = 0.135(200 \text{ V}) = 27.0$ V. Now consider $t = 350$ ms. Note that this is 50 ms into the decay portion of the curve. However, since 50 ms is not a multiple of τ', it is difficult to use the curves. Therefore, use the equations. Thus,

$$i_L = 2 \text{ A } e^{-28(50 \text{ ms})} = 2 \text{ A } e^{-1.4} = 0.493 \text{ A}$$
$$v_L = (-280 \text{ V})e^{-28(50 \text{ ms})} = (-280 \text{ V})e^{-1.4} = -69.0 \text{ V}$$

d. The above points are plotted on the waveforms of Figure 14–27.

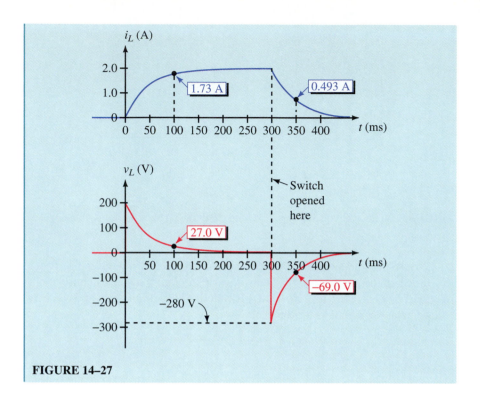

FIGURE 14–27

The basic principles that we have developed in this chapter permit us to solve problems that do not correspond exactly to the circuits of Figure 14–7 and 14–19. This is illustrated in the following example.

EXAMPLE 14–11   The circuit of Figure 14–28(a) is in steady state with the switch open. At $t = 0$ s, the switch is closed.

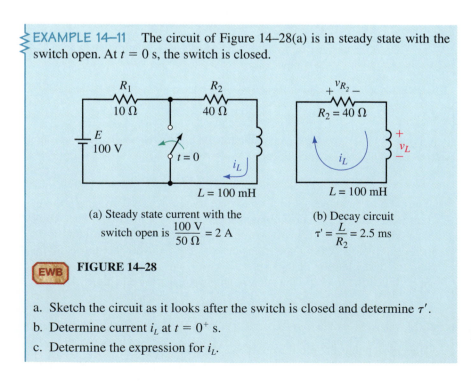

(a) Steady state current with the switch open is $\dfrac{100\ V}{50\ \Omega} = 2\ A$

(b) Decay circuit $\tau' = \dfrac{L}{R_2} = 2.5\ ms$

EWB   **FIGURE 14–28**

a. Sketch the circuit as it looks after the switch is closed and determine τ'.

b. Determine current i_L at $t = 0^+$ s.

c. Determine the expression for i_L.

d. Determine v_L at $t = 0^+$ s.

e. Determine the expression for v_L.

f. How long does the transient last?

g. Sketch i_L and v_L.

Solution

a. When you close the switch, you short out E and R_1, leaving the decay circuit of (b). Thus $\tau' = L/R_2 = 100$ mH/40 Ω = 2.5 ms.

b. In steady state with the switch open, $i_L = I_0 = 100$ V/50 Ω = 2 A. This is the current just before the switch is closed. Therefore, just after the switch is closed, i_L will still be 2 A.

c. i_L decays from 2 A to 0. From Equation 14–10, $i_L = I_0 e^{-t/\tau'} = 2e^{-t/2.5\text{ ms}} = 2e^{-400t}$ A.

d. KVL yields $v_L = -v_{R_2} = -R_2 I_0 = -(40\ \Omega)\ (2\text{A}) = -80$ V. Thus, $V_0 = -80$ V.

e. v_L decays from −80 V to 0. Thus, $v_L = V_0 e^{-t/\tau'} = -80e^{-400t}$ V.

f. Transients last $5\tau' = 5(2.5\text{ ms}) = 12.5$ ms.

FIGURE 14–29

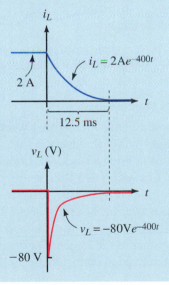

14.6 *RL* Transients Using Computers

Electronics Workbench

Workbench can easily plot voltage, but it has no simple way to plot current. If you want to determine current, use Ohm's law and the applicable voltage waveform. To illustrate, consider Figure 14–21(a). Since the inductor current passes through R_4, we can use the voltage across R_4 to monitor current. Suppose we want to know at what time i_L reaches 0.12 A. (This corresponds to 0.12 A × 104 Ω = 12.48 V across R_4.) Create the circuit as in Figure 14–30.

Select Analysis/Transient, click Initial Conditions Set to Zero, set End Time (TSTOP) to 0.1, select Node 3, click Add, then Simulate. You should get the waveform shown on the screen of Figure 14–30. Expand to full size, then use the cursor to determine the time at which voltage equals 12.48 V. Figure 14–31 shows the result. Rounded to 3 figures, the answer is 22.9 ms (which agrees with the answer we obtained earlier in Example 14–9).

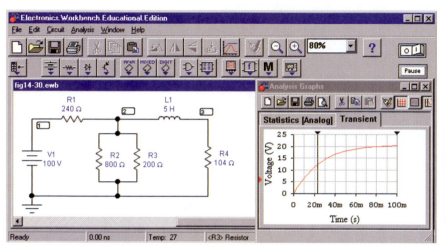

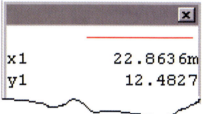

FIGURE 14–30 Electronics Workbench representation of Figure 14–21. No switch is required as the transient solution is initiated by software.

FIGURE 14–31 Values scaled from the waveform of Figure 14–30.

Using Electronics Workbench's Oscilloscope

Waveforms may also be observed using Electronics Workbench's oscilloscope. Close the Analysis Graphs window, click the Instruments Parts bin, position the scope and connect as in Figure 14–32(a). Double click the scope icon, set Time base to 5 ms/div, Channel A to 5 V/div and Y position to −1 (to better fit the trace to the screen.) Select Analysis/Analysis Options, click on the Instruments tab, select Pause after each screen, select Initial Conditions Set to Zero, click OK, then activate the circuit by clicking the ON/OFF power switch in the upper right-hand corner of the screen. The trace shown in Figure 14–32(b) should appear. Click the Expand button on the oscilloscope and drag

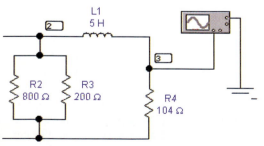

(a) Connecting the oscilloscope

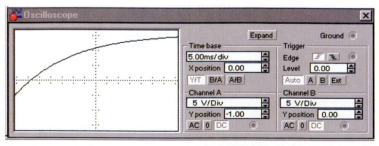

(b) Expanded oscilloscope detail

FIGURE 14–32 Using the Electronics Workbench oscilloscope.

the cursor until V_A reads 12.48 volts (or as close as you can get). The corresponding value of time should be 22.9 ms as determined previously.

OrCAD PSpice

RL transients are handled much like *RC* transients. As a first example, consider Figure 14–33. The circuit is in steady state with the switch closed. At $t = 0$, the switch is opened. Use PSpice to plot inductor voltage and current, then use the cursor to determine values at $t = 100$ ms. Verify manually. (Since the process is similar to that of Chapter 11, abbreviated instructions only are given.)

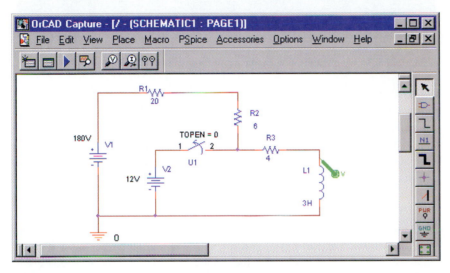

FIGURE 14–33 PSpice can easily determine both voltage and current transients. The voltage marker displays voltage across the inductance. The current display is added after the simulation is run.

Preliminary: First, determine the initial current in the inductor, i.e., the current I_0 that exists at the instant the switch is opened. This is done by noting that the inductor looks like a short circuit to steady state dc. Thus, $I_0 = $ 12 V/ 4Ω = 3 A. Next, note that after the switch is opened, current builds up through R_1, R_2, and R_3 in series. Thus, the time constant of the circuit is $\tau = $ $L_1/R_T = 3$ H/30 Ω = 0.1 s. Now proceed as follows:

• Build the circuit on the screen. Double click the inductor symbol and using the procedure from Chapter 11, set its initial condition IC to 3A. Click the New Profile icon and name the file fig14-33. In the Simulations Settings box, select transient analysis and set TSTOP to 0.5 (five time constants). Click OK.

• Click the Run icon. When simulation is complete, a trace of capacitor voltage versus time appears. Create a second $\underline{Y}$ Axis, then add the current trace I(L1). You should now have the curves of Figure 14–34 on the screen. (The Y-axes can be labeled if desired as described in Appendix A.)

Results: Consider Figure 14–33. With the switch open, steady state current is 180 V/30 Ω = 6 A and the initial current is 3 A. Thus, the current should start at 3 A and rise to 6 A in 5 time constants. (It does.) Inductor voltage

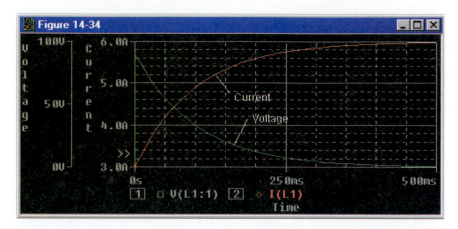

FIGURE 14–34 Inductor voltage and current for the circuit of Figure 14–33.

should start at 180 V − (3 A)(30 Ω) = 90 V and decay to 0 V in 5 time constants. (It does). Thus, the solution checks. Now, with the cursor, scale voltage and current values at t = 100 ms. You should get 33.1 V for v_L and 4.9 A for i_L. (To check, note that the equations for inductor voltage and current are $v_L = 90\, e^{-10\,t}$ V and $i_L = 6 - 3\, e^{-10\,t}$ A respectively. Substitute t = 100 ms into these and verify results.)

EXAMPLE 14–12 Consider the circuit of Figure 14–23, Example 14–10. The switch is closed at t = 0 and opened 300 ms later. Prepare a PSpice analysis of this problem and determine v_L and i_L at t = 100 ms and at t = 350 ms.

Solution PSpice doesn't have a switch that both opens and closes. However, you can simulate such a switch by using two switches as in Figure 14–35. Begin by creating the circuit on the screen using IDC for the current source. Now double click TOPEN of switch U2 and set it to 300ms, then double click the inductor symbol and set IC to 0A. Click the New Profile icon

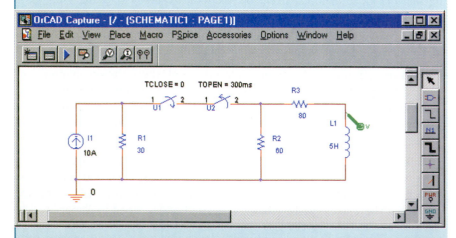

FIGURE 14–35 Simulating the circuit of Example 14–10. Two switches are used to model the closing and opening of the switch of Figure 14–23.

and name the file fig14-35. In the Simulation Settings box, select transient analysis then type in a value of 0.5 for TSTOP. Run the simulation, create a second Y-axis, then add the current trace I(L1). You should now have the curves of Figure 14–36 on the screen. (Compare to Figure 14–27.) Using the cursor, read values at $t = 100$ ms and 350 ms. You should get approximately 27 V and 1.73 mA at $t = 100$ ms and -69 V and 490 mA at $t = 350$ ms. Note how well these agree with the results of Example 14–10.

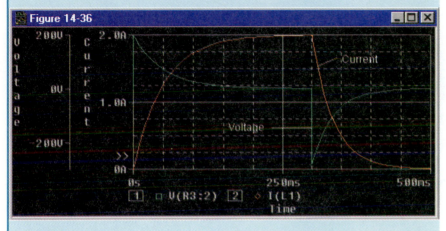

FIGURE 14–36   Inductor voltage and current for the circuit of Figure 14–35.

PUTTING IT INTO PRACTICE

The first sample of a new product that your company has designed has an indicator light that fails. (Symptom: When you turn a new unit on, the indicator light comes on as it should. However, when you turn the power off and back on, the lamp does not come on again.) You have been asked to investigate the problem and design a fix. You acquire a copy of the schematic and study the portion of the circuit where the indicator lamp is located. As shown in the accompanying figure, the lamp is used to indicate the status of the coil; the light is to be on when the coil is energized and off when it is not. Immediately, you see the problem, solder in one component and the problem is fixed. Write a short note to your supervisor outlining the nature of the problem, explaining why the lamp burned out and why your design modification fixed the problem. Note also that your modification did not result in any substantial increase in power consumption (i.e., you did not use a resistor). Note: This problem requires a diode. If you have not had an introduction to electronics, you will need to obtain a basic electronics book and read about it.

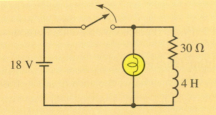

PROBLEMS

14.1   Introduction

1. a. What does an inductor carrying no current look like at the instant of switching?

   b. For each circuit of Figure 14–37, determine i_S and v_L immediately after the switch is closed.

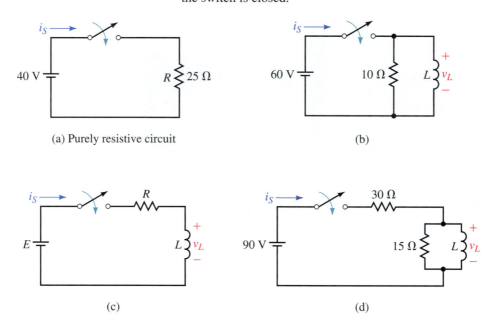

(a) Purely resistive circuit

(b)

(c)

(d)

FIGURE 14–37   The value of L does not affect the solution.

2. Determine all voltages and currents in Figure 14–38 immediately after the switch is closed.

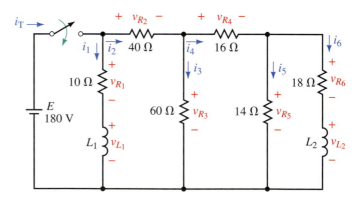

FIGURE 14–38

3. Repeat Problem 2 if L_1 is replaced with an uncharged capacitor.

14.2   Current Buildup Transients

4. a. If $i_L = 8(1 - e^{-500t})$ A, what is the current at $t = 6$ ms?

   b. If $v_L = 125e^{-500t}$ V, what is the voltage v_L at $t = 5$ ms?

5. The switch of Figure 14–39 is closed at $t = 0$ s.
   a. What is the time constant of the circuit?
   b. How long is it until current reaches its steady value?
   c. Determine the equations for i_L and v_L.
   d. Compute values for i_L and v_L at intervals of one time constant from $t = 0$ to $5\,\tau$.
   e. Sketch i_L and v_L. Label the axis in τ and in seconds.

6. Close the switch at $t = 0$ s and determine equations for i_L and v_L for the circuit of Figure 14–40. Compute i_L and v_L at $t = 1.8$ ms.

7. Repeat Problem 5 for the circuit of Figure 14–41 with $L = 4$ H.

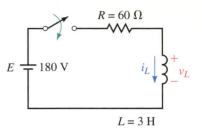

FIGURE 14–39

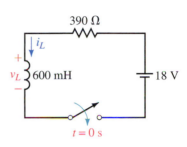

FIGURE 14–40

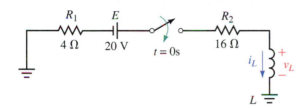

FIGURE 14–41

8. For the circuit of Figure 14–39, determine inductor voltage and current at $t = 50$ ms using the universal time constant curve of Figure 14–10.

9. Close the switch at $t = 0$ s and determine equations for i_L and v_L for the circuit of Figure 14–42. Compute i_L and v_L at $t = 3.4$ ms.

10. Using Figure 14–10, find v_L at one time constant for the circuit of Figure 14–42.

11. For the circuit of Figure 14–1(b), the voltage across the inductance at the instant the switch is closed is 80 V, the final steady state current is 4 A, and the transient lasts 0.5 s. Determine E, R, and L.

12. For an RL circuit, $i_L = 20(1 - e^{-t/\tau})$ mA and $v_L = 40e^{-t/\tau}$ V. If the transient lasts 0.625 ms, what are E, R, and L?

13. For Figure 14–1(b), if $v_L = 40e^{-2000t}$ V and the steady state current is 10 mA, what are E, R, and L?

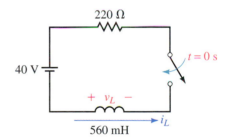

FIGURE 14–42

14.4 De-energizing Transients

14. For Figure 14–43, $E = 80$ V, $R_1 = 200\ \Omega$, $R_2 = 300\ \Omega$, and $L = 0.5$ H.
    a. When the switch is closed, how long does it take for i_L to reach steady state?
    b. When the switch is opened, how long does it take for i_L to reach steady state?
    c. After the circuit has reached steady state with the switch closed, it is opened. Determine equations for i_L and v_L.

15. For Figure 14–43, $R_1 = 20\ \Omega$, $R_2 = 230\ \Omega$, and $L = 0.5$ H, and the inductor current has reached a steady value of 5 A with the switch closed. At $t = 0$ s, the switch is opened.
    a. What is the decay time constant?

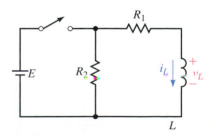

FIGURE 14–43

b. Determine equations for i_L and v_L.

c. Compute values for i_L and v_L at intervals of one time constant from $t = 0$ to 5τ.

d. Sketch i_L and v_L. Label the axis in τ and in seconds.

16. Using the values from Problem 15, determine inductor voltage and current at $t = 3\tau$ using the universal time constant curves shown in Figure 14–10.

17. Given $v_L = -2700\ Ve^{-100t}$. Using the universal time constant curve, find v_L at $t = 20$ ms.

18. For Figure 14–43, the inductor voltage at the instant the switch is closed is 150 V and $i_L = 0$ A. After the circuit has reached steady state, the switch is opened. At the instant the switch is opened, $i_L = 3$ A and v_L jumps to -750 V. The decay transient lasts 5 ms. Determine E, R_1, R_2, and L.

19. For Figure 14–43, $L = 20$ H. The current during buildup and decay is shown in Figure 14–44. Determine R_1 and R_2.

20. For Figure 14–43, when the switch is moved to energization, $i_L = 2$ A $(1 - e^{-10t})$. Now open the switch after the circuit has reached steady state and redefine $t = 0$ s as the instant the switch is opened. For this case, $v_L = -400\ Ve^{-25t}$. Determine E, R_1, R_2, and L.

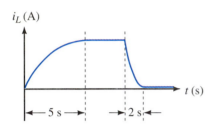

i_L (A)

t (s)

5 s   2 s

FIGURE 14–44

14.5 More Complex Circuits

21. For the coil of Figure 14–45 $R_\ell = 1.7\ \Omega$ and $L = 150$ mH. Determine coil current at $t = 18.4$ ms.

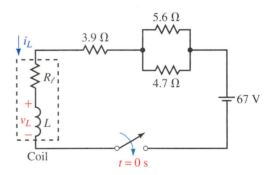

FIGURE 14–45

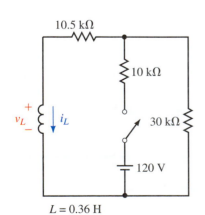

10.5 kΩ

10 kΩ

v_L i_L 30 kΩ

120 V

$L = 0.36$ H

FIGURE 14–46

22. Refer to Figure 14–46:

a. What is the energizing circuit time constant?

b. Close the switch and determine the equation for i_L and v_L during current buildup.

c. What is the voltage across the inductor and the current through it at $t = 20\ \mu s$?

23. For Figure 14–46, the circuit has reached steady state with the switch closed. Now open the switch.

a. Determine the de-energizing circuit time constant.

b. Determine the equations for i_L and v_L.

c. Find the voltage across the inductor and current through it at $t = 17.8\ \mu s$ using the equations determined above.

24. Repeat Part (c) of Problem 23 using the universal time constant curves shown in Figure 14–10.

25. a. Repeat Problem 22, Parts (a) and (b) for the circuit of Figure 14–47.
    b. What are i_L and v_L at $t = 25$ ms?

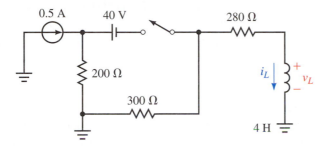

FIGURE 14–47

26. Repeat Problem 23 for the circuit of Figure 14–47, except find v_L and i_L at $t = 13.8$ ms.

27. An unknown circuit containing dc sources and resistors has an open-circuit voltage of 45 volts. When its output terminals are shorted, the short-circuit current is 0.15 A. A switch, resistor, and inductance are connected (Figure 14–48). Determine the inductor current and voltage 2.5 ms after the switch is closed.

28. The circuit of Figure 14–49 is in steady state with the switch in position 1. At $t = 0$, it is moved to position 2, where it remains for 1.0 s. It is then moved to position 3, where it remains. Sketch curves for i_L and v_L from $t = 0^-$ until the circuit reaches steady state in position 3. Compute the inductor voltage and current at $t = 0.1$ s and at $t = 1.1$ s.

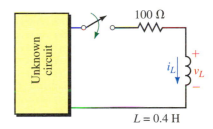

FIGURE 14–48

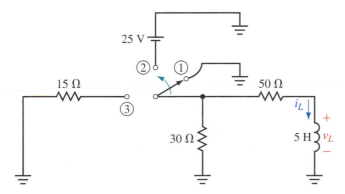

FIGURE 14–49

14.6   *RL* Transients Using Computers

29. **EWB** or **PSpice**   The switch of Figure 14–46 is closed at $t = 0$ and remains closed. Graph the voltage across L and find v_L at 20 μs using the cursor.

30. **EWB** or **PSpice**   For the circuit of Figure 14–47, close the switch at $t = 0$ and find v_L at $t = 10$ ms. (For PSpice, use current source IDC.)

31. **EWB** or **PSpice**   For Figure 14–6, let $L_1 = 30$ mH and $L_2 = 90$ mH. Close the switch at $t = 0$ and find the current in the 30 Ω resistor at $t = 2$ ms. (Answer: 4.61 A) [Hint for Workbench users: Redraw the circuit with L_1 and the 30 Ω resistor interchanged. Finally, use Ohm's law.]

32. **EWB** or **PSpice**   For Figure 14–41, let $L = 4$ H. Solve for v_L and, using the cursor, measure values at $t = 200$ ms and 500 ms. For PSpice users, also find current at these times.

33. **PSpice**   We solved the circuit of Figure 14–21(a) by reducing it to its Thévenin equivalent. Using PSpice, analyze the circuit in its original form and plot the inductor current. Check a few points on the curve by computing values according to the solution of Example 14–8 and compare to values obtained from screen.

34. **PSpice**   The circuit of Figure 14–46 is in steady state with the switch open. At $t = 0$, the switch is closed. It remains closed for 150 μs and is then opened and left open. Compute and plot i_L and v_L. With the cursor, determine values at $t = 60$ μs and at $t = 165$ μs.

ANSWERS TO IN-PROCESS LEARNING CHECKS

In-Process Learning Check 1

1. a. $20e^{-100\,000t}$ V; $2(1 - e^{-100\,000t})$ mA

   b.

$t(\mu s)$	v_L (V)	i_L (mA)
0	20	0
10	7.36	1.26
20	2.71	1.73
30	0.996	1.90
40	0.366	1.96
50	0.135	1.99

   c.

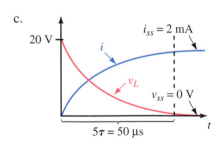

2. 11.5 V;   1.47 A

3. 4 V

4. 3.88 A

Foundation AC Concepts

PART IV

15 AC Fundamentals

16 *R, L,* and *C* Elements and the Impedance Concept

17 Power in AC Circuits

In previous chapters, we concentrated mostly on dc. We now turn our attention to ac (alternating current).

AC is important to us for a number of reasons. Firstly, it is the basis of the electrical power system that supplies our homes and businesses with electrical energy. AC is used instead of dc because it has several important advantages, the chief one being that ac power can be transmitted easily and efficiently over long distances. However, the importance of ac extends far beyond its use in the electrical power industry. The study of electronics, for example, deals to a large extent with ac, whether it be in the field of audio systems, communications systems, control systems, or any number of other areas. In fact, nearly every electrical and electronic device that we use in our daily lives operates from or involves the use of ac in some way.

We begin Part IV of this book with a look at fundamental ac concepts. We examine ways to generate ac voltages, methods used to represent ac voltages and currents, relationships between ac quantities in resistive, inductive, and capacitive circuits, and, finally, the meaning and representation of power in ac systems. This sets the stage for succeeding chapters which deal with ac circuit analysis techniques, including ac versions of the various methods that you have used for dc circuits throughout previous chapters of this book.

15

AC Fundamentals

OBJECTIVES

After studying this chapter, you will be able to

- explain how ac voltages and currents differ from dc,
- draw waveforms for ac voltage and currents and explain what they mean,
- explain the voltage polarity and current direction conventions used for ac,
- describe the basic ac generator and explain how ac voltage is generated,
- define and compute frequency, period, amplitude, and peak-to-peak values,
- compute instantaneous sinusoidal voltage or current at any instant in time,
- define the relationships between ω, T, and f for a sine wave,
- define and compute phase differences between waveforms,
- use phasors to represent sinusoidal voltages and currents,
- determine phase relationships between waveforms using phasors,
- define and compute average values for time-varying waveforms,
- define and compute effective values for time-varying waveforms,
- use Electronics Workbench and PSpice to study ac waveforms.

KEY TERMS

ac
Alternating Voltage
Alternating Current
Amplitude
Angular Velocity
Average Value
Cycle
Effective Value
Frequency
Hertz
Instantaneous Value
Oscilloscope
Peak Value
Period
Phase Shifts
Phasor
RMS
Sine Wave

OUTLINE

Introduction
Generating AC Voltages
Voltage and Current Conventions for AC
Frequency, Period, Amplitude, and Peak Value
Angular and Graphic Relationships for Sine Waves
Voltage and Currents as Functions of Time
Introduction to Phasors
AC Waveforms and Average Value
Effective Values
Rate of Change of a Sine Wave
AC Voltage and Current Measurement
Circuit Analysis Using Computers

Alternating currents (ac) are currents that alternate in direction (usually many times per second), passing first in one direction, then in the other through a circuit. Such currents are produced by voltage sources whose polarities alternate between positive and negative (rather than being fixed as with dc sources). By convention, alternating currents are called *ac currents* and alternating voltages are called *ac voltages*.

The variation of an ac voltage or current versus time is called its waveform. Since waveforms vary with time, they are designated by lowercase letters $v(t)$, $i(t)$, $e(t)$, and so on, rather than by uppercase letters V, I, and E as for dc. Often we drop the functional notation and simply use v, i, and e.

While many waveforms are important to us, the most fundamental is the sine wave (also called sinusoidal ac). In fact, the sine wave is of such importance that many people associate the term ac with sinusoidal, even though ac refers to any quantity that alternates with time.

In this chapter, we look at basic ac principles, including the generation of ac voltages and ways to represent and manipulate ac quantities. These ideas are then used throughout the remainder of the book to develop methods of analysis for ac circuits.

Thomas Alva Edison

NOWADAYS WE TAKE IT FOR GRANTED that our electrical power systems are ac. (This is driven home every time you see a piece of equipment rated "60 hertz ac", for example.) However, this was not always the case. In the late 1800s, a fierce battle—the so-called "war of the currents"—raged in the emerging electrical power industry. The forces favoring the use of dc were led by Thomas Alva Edison, and those favoring the use of ac were led by George Westinghouse (Chapter 24) and Nikola Tesla (Chapter 23).

Edison, a prolific inventor who gave us the electric light, the phonograph, and many other great inventions as well, fought vigorously for dc. He had spent a considerable amount of time and money on the development of dc power and had a lot at stake, in terms of both money and prestige. So unscrupulous was Edison in this battle that he first persuaded the state of New York to adopt ac for its newly devised electric chair, and then pointed at it with horror as an example of how deadly ac was. Ultimately, however, the combination of ac's advantages over dc and the stout opposition of Tesla and Westinghouse won the day for ac.

Edison was born in 1847 in Milan, Ohio. Most of his work was done at two sites in New Jersey—first at a laboratory in Menlo Park, and later at a much larger laboratory in West Orange, where his staff at one time numbered around 5,000. He received patents as inventor or co-inventor on an astonishing 1,093 inventions, making him probably the greatest inventor of all time.

Thomas Edison died at the age of 84 on October 18, 1831.

15.1   Introduction

Previously you learned that dc sources have fixed polarities and constant magnitudes and thus produce currents with constant value and unchanging direction, as illustrated in Figure 15–1. In contrast, the voltages of ac sources alternate in polarity and vary in magnitude and thus produce currents that vary in magnitude and alternate in direction.

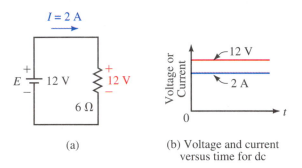

(a)

(b) Voltage and current versus time for dc

FIGURE 15–1   In a dc circuit, voltage polarities and current directions do not change.

Sinusoidal AC Voltage

To illustrate, consider the voltage at the wall outlet in your home. Called a **sine wave** or **sinusoidal ac waveform** (for reasons discussed in Section 15.5), this voltage has the shape shown in Figure 15–2. Starting at zero, the voltage increases to a positive maximum, decreases to zero, changes polarity, increases to a negative maximum, then returns again to zero. One complete variation is referred to as a **cycle.** Since the waveform repeats itself at regular intervals as in (b), it is called a **periodic** waveform.

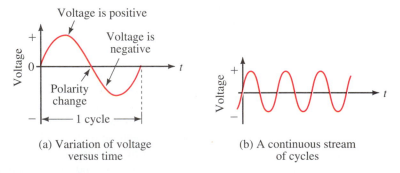

(a) Variation of voltage versus time

(b) A continuous stream of cycles

FIGURE 15–2   Sinusoidal ac waveforms. Values above the axis are positive while values below are negative.

Symbol for an AC Voltage Source

The symbol for a sinusoidal voltage source is shown in Figure 15–3. Note that a lowercase e is used to represent voltage rather than E, since it is a function of time. Polarity marks are also shown although, since the polarity of the source varies, their meaning has yet to be established.

FIGURE 15–3   Symbol for a sinusoidal voltage source. Lowercase letter e is used to indicate that the voltage varies with time.

Sinusoidal AC Current

Figure 15–4 shows a resistor connected to an ac source. During the first half-cycle, the source voltage is positive; therefore, the current is in the clockwise direction. During the second half-cycle, the voltage polarity reverses; therefore, the current is in the counterclockwise direction. Since current is proportional to voltage, its shape is also sinusoidal (Figure 15–5).

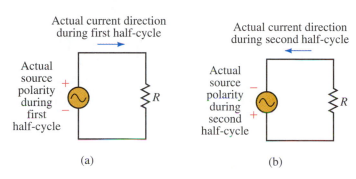

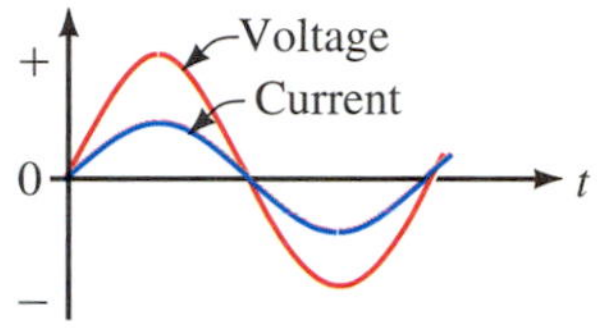

FIGURE 15–5 Current has the same wave shape as voltage.

(a)
(b)

FIGURE 15–4 Current direction reverses when the source polarity reverses.

15.2 Generating AC Voltages

One way to generate an ac voltage is to rotate a coil of wire at constant angular velocity in a fixed magnetic field, Figure 15–6. (Slip rings and brushes connect the coil to the load.) The magnitude of the resulting voltage is proportional to the rate at which flux lines are cut (Faraday's law, Chapter 13), and its polarity is dependent on the direction the coil sides move through the field. Since the rate of cutting flux varies with time, the resulting voltage will also vary with time. For example in (a), since the coil sides are moving parallel to the field, no flux lines are being cut and the induced voltage at this instant (and hence the current) is zero. (This is defined as the 0° position of the coil.) As the coil rotates from the 0° position, coil sides AA' and BB' cut across flux lines; hence, voltage builds, reaching a peak when flux is cut at the maximum rate in the 90° position as in (b). Note the polarity of the voltage and the direction of current. As the coil rotates further, voltage decreases, reaching zero at the 180° position when the coil sides again move parallel to the field as in (c). At this point, the coil has gone through a half-revolution.

During the second half-revolution, coil sides cut flux in directions opposite to that which they did in the first half revolution; hence, the polarity of the induced voltage reverses. As indicated in (d), voltage reaches a peak at the 270° point, and, since the polarity of the voltage has changed, so has the direction of current. When the coil reaches the 360° position, voltage is again zero and the cycle starts over. Figure 15–7 shows one cycle of the resulting waveform. Since the coil rotates continuously, the voltage produced will be a repetitive, periodic waveform as you saw in Figure 15–2(b).

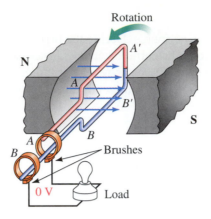

(a) 0° Position: Coil sides move parallel to flux lines. Since no flux is being cut, induced voltage is zero.

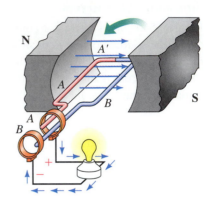

(b) 90° Position: Coil end A is positive with respect to B. Current direction is out of slip ring A.

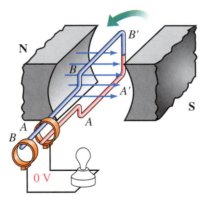

(c) 180° Position: Coil again cutting no flux. Induced voltage is zero.

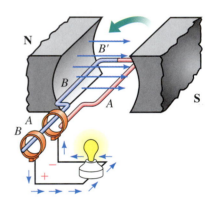

(d) 270° Position: Voltage polarity has reversed, therefore, current direction reverses.

FIGURE 15–6   Generating an ac voltage. The 0° position of the coil is defined as in (a) where the coil sides move parallel to the flux lines.

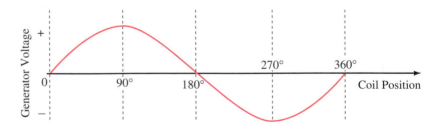

FIGURE 15–7   Coil voltage versus angular position.

PRACTICAL NOTES...

In practice, the coil of Figure 15–6 consists of many turns wound on an iron core. The coil, core, and slip rings rotate as a unit.

   In Figure 15–6, the magnetic field is fixed and the coil rotates. While small generators are built this way, large ac generators usually have the opposite construction, that is, their coils are fixed and the magnetic field is rotated instead. In addition, large ac generators are usually made as three-phase machines with three sets of coils instead of one. This is covered in Chapter 23. However, although its details are oversimplified, the generator of Figure 15–6 gives a true picture of the voltage produced by a real ac generator.

Time Scales

The horizontal axis of Figure 15–7 is scaled in degrees. Often we need it scaled in time. The length of time required to generate one cycle depends on the velocity of rotation. To illustrate, assume that the coil rotates at 600 rpm (revolutions per minute). Six hundred revolutions in one minute equals 600 rev/60 s = 10 revolutions in one second. At ten revolutions per second, the time for one revolution is one tenth of a second, i.e., 100 ms. Since one cycle is 100 ms, a half-cycle is 50 ms, a quarter-cycle is 25 ms, and so on. Figure 15–8 shows the waveform rescaled in time.

Instantaneous Value

As Figure 15–8 shows, the coil voltage changes from instant to instant. The value of voltage at any point on the waveform is referred to as its **instantaneous value.** This is illustrated in Figure 15–9. Figure 15–9(a) shows a photograph of an actual waveform, and (b) shows it redrawn, with values scaled from the photo. For this example, the voltage has a peak value of 40 volts and a cycle time of 6 ms. From the graph, we see that at $t = 0$ ms, the voltage is zero. At $t = 0.5$ ms, it is 20 V. At $t = 2$ ms, it is 35 V. At $t = 3.5$ ms, it is -20 V, and so on.

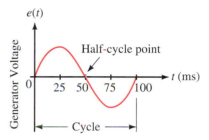

FIGURE 15–8   Cycle scaled in time. At 600 rpm, the cycle length is 100 ms.

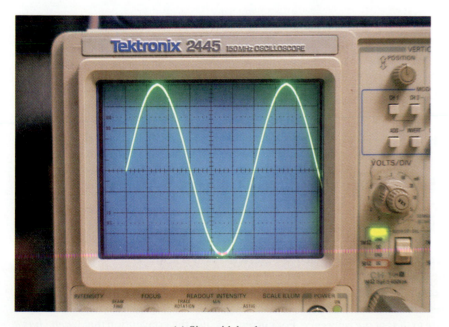

(a) Sinusoidal voltage

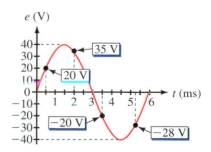

(b) Values scaled from the photograph

FIGURE 15–9   Instantaneous values.

Electronic Signal Generators

AC waveforms may also be created electronically using signal generators. In fact, with signal generators, you are not limited to sinusoidal ac. The general-purpose lab signal generator of Figure 15–10, for example, can produce a variety of variable-frequency waveforms, including sinusoidal, square wave, triangular, and so on. Waveforms such as these are commonly used to test electronic gear.

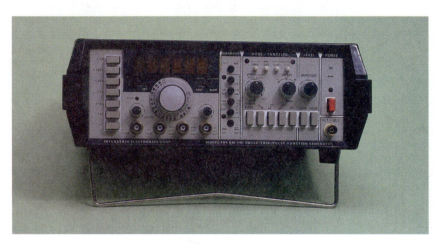

(a) A typical signal generator

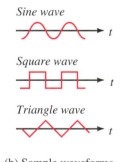

(b) Sample waveforms

FIGURE 15–10   Electronic signal generators produce waveforms of different shapes.

15.3   Voltage and Current Conventions for AC

In Section 15.1, we looked briefly at voltage polarities and current directions. At that time, we used separate diagrams for each half-cycle (Figure 15–4). However, this is unnecessary; one diagram and one set of references is all that is required. This is illustrated in Figure 15–11. First, we assign reference polarities for the source and a reference direction for the current. We then use the convention that, *when e has a positive value, its actual polarity is the same as the reference polarity, and when e has a negative value, its actual polarity is opposite to that of the reference.* For current, we use the convention that *when i has a positive value, its actual direction is the same as the reference arrow, and when i has a negative value, its actual direction is opposite to that of the reference.*

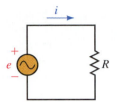

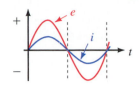

(a) References for voltage and current.

(b) During the first half-cycle, voltage polarity and current direction are as shown in (a). Therefore, e and i are positive. During the second half-cycle, voltage polarity and current direction are opposite to that shown in (a). Therefore, e and i are negative.

FIGURE 15–11   AC voltage and current reference conventions.

To illustrate, consider Figure 15–12. At time t_1, e has a value of 10 volts. This means that at this instant, the voltage of the source is 10 V and its top end is positive with respect to its bottom end. This is indicated in (b). With a voltage of 10 V and a resistance of 5 Ω, the instantaneous value of current is $i = e/R = 10 \text{ V}/5 \text{ } \Omega = 2$ A. Since i is positive, the current is in the direction of the reference arrow.

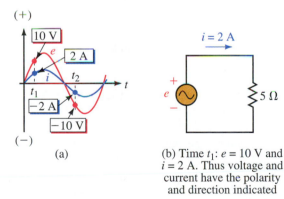

(a)

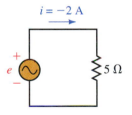

(b) Time t_1: $e = 10$ V and $i = 2$ A. Thus voltage and current have the polarity and direction indicated

(c) Time t_2: $e = -10$ V and $i = -2$ A. Thus, voltage polarity is opposite to that indicated and current direction is opposite to the arrow direction.

FIGURE 15–12   Illustrating the ac voltage and current convention.

Now consider time t_2. Here, $e = -10$ V. This means that source voltage is again 10 V, but now its top end is negative with respect to its bottom end. Again applying Ohm's law, you get $i = e/R = -10 \text{ V}/5 \text{ } \Omega = -2$ A. Since i is negative, current is actually opposite in direction to the reference arrow. This is indicated in (c).

The above concept is valid for any ac signal, regardless of waveshape.

EXAMPLE 15–1   Figure 15–13(b) shows one cycle of a triangular voltage wave. Determine the current and its direction at $t = 0, 1, 2, 3, 4, 5, 6, 7, 8, 9, 10, 11,$ and $12 \ \mu s$ and sketch.

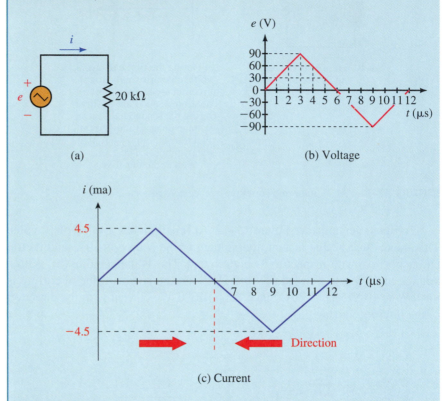

(a)

(b) Voltage

(c) Current

FIGURE 15–13

Solution   Apply Ohm's law at each point in time. At $t = 0 \ \mu s$, $e = 0 \ V$, so $i = e/R = 0 \ V/20 \ k\Omega = 0 \ mA$. At $t = 1 \ \mu s$, $e = 30 \ V$. Thus, $i = e/R = 30 \ V/20 \ k\Omega = 1.5 \ mA$. At $t = 2 \ \mu s$, $e = 60 \ V$. Thus, $i = e/R = 60 \ V/20 \ k\Omega = 3 \ mA$. Continuing in this manner, you get the values shown in Table 15–1. The waveform is plotted as Figure 15–13(c).

TABLE 15–1   Values for Example 15–1

$t \ (\mu s)$	$e \ (V)$	$i \ (mA)$
0	0	0
1	30	1.5
2	60	3.0
3	90	4.5
4	60	3.0
5	30	1.5
6	0	0
7	−30	−1.5
8	−60	−3.0
9	−90	−4.5
10	−60	−3.0
11	−30	−1.5
12	0	0

1. Let the source voltage of Figure 15–11 be the waveform of Figure 15–9. If $R = 2.5$ kΩ, determine the current at $t = 0, 0.5, 1, 1.5, 3, 4.5,$ and 5.25 ms.

2. For Figure 15–13, if $R = 180$ Ω, determine the current at $t = 1.5, 3, 7.5,$ and 9 μs.

Answers:

1. 0, 8, 14, 16, 0, −16, −11.2 (all mA)

2. 0.25, 0.5, −0.25, −0.5 (all A)

15.4 Frequency, Period, Amplitude, and Peak Value

Periodic waveforms (i.e., waveforms that repeat at regular intervals), regardless of their waveshape, may be described by a group of attributes such as frequency, period, amplitude, peak value, and so on.

Frequency

The number of cycles per second of a waveform is defined as its **frequency.** In Figure 15–14(a), one cycle occurs in one second; thus its frequency is one cycle per second. Similarly, the frequency of (b) is two cycles per second and that of (c) is 60 cycles per second. Frequency is denoted by the lower-case letter f. In the SI system, its unit is the **hertz** (Hz, named in honor of pioneer researcher Heinrich Hertz, 1857–1894). By definition,

$$1 \text{ Hz} = 1 \text{ cycle per second} \qquad (15\text{–}1)$$

Thus, the examples depicted in Figure 15–14 represent 1 Hz, 2 Hz, and 60 Hz respectively.

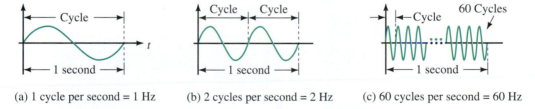

(a) 1 cycle per second = 1 Hz   (b) 2 cycles per second = 2 Hz   (c) 60 cycles per second = 60 Hz

FIGURE 15–14   Frequency is measured in hertz (Hz).

The range of frequencies is immense. Power line frequencies, for example, are 60 Hz in North America and 50 Hz in many other parts of the world. Audible sound frequencies range from about 20 Hz to about 20 kHz. The standard AM radio band occupies from 550 kHz to 1.6 MHz, while the FM band extends from 88 MHz to 108 MHz. TV transmissions occupy several bands in the 54-MHz to 890-MHz range. Above 300 GHz are optical and X-ray frequencies.

Period

The **period,** T, of a waveform, (Figure 15–15) is the duration of one cycle. It is the inverse of frequency. To illustrate, consider again Figure 15–14. In (a), the frequency is 1 cycle per second; thus, the duration of each cycle is $T = 1$ s.

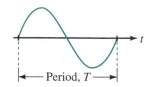

FIGURE 15–15 Period *T* is the duration of one cycle, measured in seconds.

In (b), the frequency is two cycles per second; thus, the duration of each cycle is $T = \frac{1}{2}$ s, and so on. In general,

$$T = \frac{1}{f} \quad (s) \tag{15–2}$$

and

$$f = \frac{1}{T} \quad (Hz) \tag{15–3}$$

Note that these definitions are independent of wave shape.

EXAMPLE 15–2

a. What is the period of a 50-Hz voltage?
b. What is the period of a 1-MHz current?

Solution

$$\text{(a)} \quad T = \frac{1}{f} = \frac{1}{50 \text{ Hz}} = 20 \text{ ms}$$

$$\text{(b)} \quad T = \frac{1}{f} = \frac{1}{1 \times 10^6 \text{ Hz}} = 1 \text{ } \mu s$$

EXAMPLE 15–3   Figure 15–16 shows an oscilloscope trace of a square wave. Each horizontal division represents 50 μs. Determine the frequency.

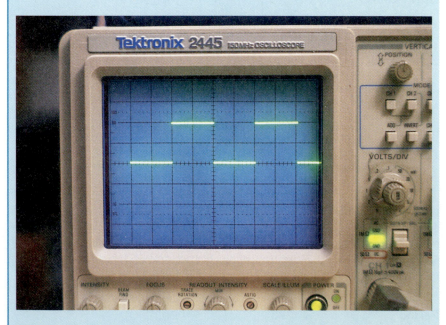

FIGURE 15–16   The concepts of frequency and period apply to nonsinusoidal waveforms.

Solution   Since the wave repeats itself every 200 μs, its period is 200 μs and

$$f = \frac{1}{200 \times 10^{-6} \text{ s}} = 5 \text{ kHz}$$

The period of a waveform can be measured between any two corresponding points (Figure 15–17). Often it is measured between zero points because they are easy to establish on an oscilloscope trace.

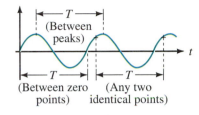

EXAMPLE 15–4   Determine the period and frequency of the waveform of Figure 15–18.

FIGURE 15–17   Period may be measured between any two corresponding points.

FIGURE 15–18

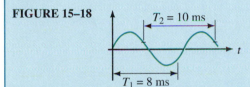

Solution   Time interval T_1 does not represent a period as it is not measured between corresponding points. Interval T_2, however, is. Thus, $T = 10$ ms and

$$f = \frac{1}{T} = \frac{1}{10 \times 10^{-3} \text{ s}} = 100 \text{ Hz}$$

Amplitude and Peak-to-Peak Value

The **amplitude** of a sine wave is the distance from its average to its peak. Thus, the amplitude of the voltage in Figures 15–19(a) and (b) is E_m.

Peak-to-peak voltage is also indicated in Figure 15–19(a). It is measured between minimum and maximum peaks. Peak-to-peak voltages are denoted $E_{p\text{-}p}$ or $V_{p\text{-}p}$ in this book. (Some authors use $V_{pk\text{-}pk}$ or the like.) Similarly, peak-to-peak currents are denoted as $I_{p\text{-}p}$. To illustrate, consider again Figure 15–9. The amplitude of this voltage is $E_m = 40$ V, and its peak-to-peak voltage is $E_{p\text{-}p} = 80$ V.

Peak Value

The **peak value** of a voltage or current is its maximum value with respect to zero. Consider Figure 15–19(b). Here, a sine wave rides on top of a dc value, yielding a peak that is the sum of the dc voltage and the ac waveform amplitude. For the case indicated, the peak voltage is $E + E_m$.

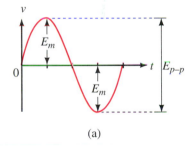

(a)

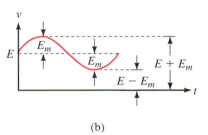

(b)

FIGURE 15–19   Definitions.

1. What is the period of the commercial ac power system voltage in North America?

2. If you double the rotational speed of an ac generator, what happens to the frequency and period of the waveform?

IN-PROCESS
LEARNING
CHECK 1

3. If the generator of Figure 15–6 rotates at 3000 rpm, what is the period and frequency of the resulting voltage? Sketch four cycles and scale the horizontal axis in units of time.

4. For the waveform of Figure 15–9, list all values of time at which $e = 20$ V and $e = -35$ V. Hint: Sine waves are symmetrical.

5. Which of the waveform pairs of Figure 15–20 are valid combinations? Why?

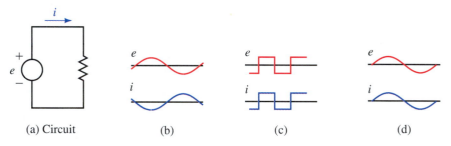

(a) Circuit (b) (c) (d)

FIGURE 15–20   Which waveform pairs are valid?

6. For the waveform in Figure 15–21, determine the frequency.

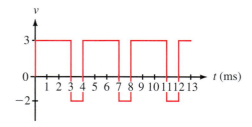

FIGURE 15–21

7. Two waveforms have periods of $T_1 = 10$ ms and $T_2 = 30$ ms respectively. Which has the higher frequency? Compute the frequencies of both waveforms.

8. Two sources have frequencies f_1 and f_2 respectively. If $f_2 = 20f_1$, and T_2 is 1 μs, what is f_1? What is f_2?

9. Consider Figure 15–22. What is the frequency of the waveform?

FIGURE 15–22

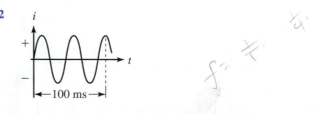

10. For Figure 15–11, if $f = 20$ Hz, what is the current direction at $t = 12$ ms, 37 ms, and 60 ms? Hint: Sketch the waveform and scale the horizontal axis in ms. The answers should be apparent.

11. A 10-Hz sinusoidal current has a value of 5 amps at $t = 25$ ms. What is its value at $t = 75$ ms? See hint in Problem 10.

(Answers are at the end of the chapter.)

15.5   Angular and Graphic Relationships for Sine Waves

The Basic Sine Wave Equation

Consider again the generator of Figure 15–6, reoriented and redrawn in end view as Figure 15–23. The voltage produced by this generator is

$$e = E_m \sin \alpha \quad \text{(V)} \qquad \textbf{(15–4)}$$

where E_m is the maximum coil voltage and α is the instantaneous angular position of the coil. (For a given generator and rotational velocity, E_m is constant.) Note that $\alpha = 0°$ represents the horizontal position of the coil and that one complete cycle corresponds to 360°. Equation 15–4 states that the voltage at any point on the sine wave may be found by multiplying E_m times the sine of the angle at that point.

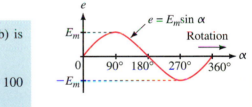

(a) End view showing coil position

EXAMPLE 15–5   If the amplitude of the waveform of Figure 15–23(b) is $E_m = 100$ V, determine the coil voltage at 30° and 330°.

Solution   At $\alpha = 30°$, $e = E_m \sin \alpha = 100 \sin 30° = 50$ V. At 330°, $e = 100 \sin 330° = -50$ V. These are shown on the graph of Figure 15–24.

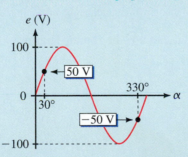

FIGURE 15–24

(b) Voltage waveform

FIGURE 15–23   Coil voltage versus angular position.

Table 15–2 is a tabulation of voltage versus angle computed from $e = 100 \sin \alpha$. Use your calculator to verify each value, then plot the result on graph paper. The resulting waveshape should look like Figure 15–24.

PRACTICE
PROBLEMS 2

TABLE 15–2   Data for Plotting
$e = 100 \sin \alpha$

Angle α	Voltage e
0	0
30	50
60	86.6
90	100
120	86.6
150	50
180	0
210	−50
240	−86.6
270	−100
300	−86.6
330	−50
360	0

Angular Velocity, ω

The rate at which the generator coil rotates is called its **angular velocity.** If the coil rotates through an angle of 30° in one second, for example, its angular velocity is 30° per second. Angular velocity is denoted by the Greek letter ω (omega). For the case cited, $\omega = 30°/s$. (Normally angular velocity is expressed in radians per second instead of degrees per second. We will make this change shortly.) When you know the angular velocity of a coil and the length of time that it has rotated, you can compute the angle through which it has turned. For example, a coil rotating at 30°/s rotates through an angle of 30° in one second, 60° in two seconds, 90° in three seconds, and so on. In general,

$$\alpha = \omega t \tag{15–5}$$

Expressions for t and ω can now be found. They are

$$t = \frac{\alpha}{\omega} \quad (s) \tag{15–6}$$

$$\omega = \frac{\alpha}{t} \tag{15–7}$$

EXAMPLE 15–6   If the coil of Figure 15–23 rotates at $\omega = 300°/s$, how long does it take to complete one revolution?

Solution   One revolution is 360°. Thus,

$$t = \frac{\alpha}{\omega} = \frac{360 \text{ degrees}}{300 \dfrac{\text{degrees}}{\text{s}}} = 1.2 \text{ s}$$

Since this is one period, we should use the symbol T. Thus, $T = 1.2$ s, as in Figure 15–25.

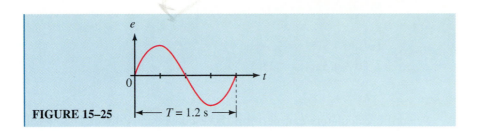

FIGURE 15–25 $T = 1.2$ s

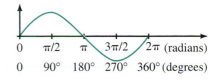
PRACTICE
PROBLEMS 3

If the coil of Figure 15–23 rotates at 3600 rpm, determine its angular velocity, ω, in degrees per second.

Answer: 21 600 deg/s

Radian Measure

In practice, ω is usually expressed in radians per second, where radians and degrees are related by the identity

$$2\pi \text{ radians} = 360° \qquad (15\text{–}8)$$

One radian therefore equals $360°/2\pi = 57.296°$. A full circle, as shown in Figure 15–26(a), can be designated as either 360° or 2π radians. Likewise, the cycle length of a sinusoid, shown in Figure 15–26(b), can be stated as either 360° or 2π radians; a half-cycle as 180° or π radians, and so on.

To convert from degrees to radians, multiply by $\pi/180$, while to convert from radians to degrees, multiply by $180/\pi$.

$$\alpha_{\text{radians}} = \frac{\pi}{180°} \times \alpha_{\text{degrees}} \qquad (15\text{–}9)$$

$$\alpha_{\text{degrees}} = \frac{180°}{\pi} \times \alpha_{\text{radians}} \qquad (15\text{–}10)$$

Table 15–3 shows selected angles in both measures.

(a) 360° = 2π radians

TABLE 15–3   Selected Angles
in Degrees and Radians

Degrees	Radians
30	$\pi/6$
45	$\pi/4$
60	$\pi/3$
90	$\pi/2$
180	π
270	$3\pi/2$
360	2π

(b) Cycle length scaled in degrees
and radians

FIGURE 15–26   Radian measure.

EXAMPLE 15–7

a. Convert 315° to radians.

b. Convert $5\pi/4$ radians to degrees.

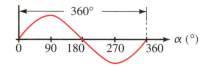

(a) Degrees

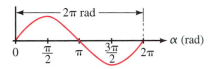

(b) Radians

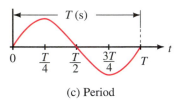

(c) Period

FIGURE 15–27   Comparison of various horizontal scales. Cycle length may be scaled in degrees, radians or period. Each of these is independent of frequency.

Solution

a. $\alpha_{\text{radians}} = (\pi/180°)(315°) = 5.5$ rad

b. $\alpha_{\text{degrees}} = (180°/\pi)(5\pi/4) = 225°$

Scientific calculators can perform these conversions directly. You will find this more convenient than using the above formulas.

Graphing Sine Waves

A sinusoidal waveform can be graphed with its horizontal axis scaled in degrees, radians, or time. When scaled in degrees or radians, one cycle is always 360° or 2π radians (Figure 15–27); when scaled in time, it is frequency dependent, since the length of a cycle depends on the coil's velocity of rotation. However, if scaled in terms of period T instead of in seconds, the waveform is also frequency independent, since one cycle is always T, as shown in Figure 15–27(c).

When graphing a sine wave, you don't actually need many points to get a good sketch: Values every 45° (one eighth of a cycle) are generally adequate. Table 15–4 shows corresponding values for $\sin \alpha$ at this spacing.

TABLE 15–4   Values for Rapid Sketching

α (deg)	α (rad)	t (T)	Value of $\sin \alpha$
0	0	0	0.0
45	$\pi/4$	$T/8$	0.707
90	$\pi/2$	$T/4$	1.0
135	$3\pi/4$	$3T/8$	0.707
180	π	$T/2$	0.0
225	$5\pi/4$	$5T/8$	−0.707
270	$3\pi/2$	$3T/4$	−1.0
315	$7\pi/4$	$7T/8$	−0.707
360	2π	T	0.0

EXAMPLE 15–8   Sketch the waveform for a 25-kHz sinusoidal current that has an amplitude of 4 mA. Scale the axis in seconds.

Solution   The easiest approach is to use $T = 1/f$, then scale the graph accordingly. For this waveform, $T = 1/25$ kHz = 40 μs. Thus,

1. Mark the end of the cycle as 40 μs, the half-cycle point as 20 μs, the quarter-cycle point as 10 μs, and so on (Figure 15–28).

2. The peak value (i.e., 4 mA) occurs at the quarter-cycle point, which is 10 μs on the waveform. Likewise, −4 mA occurs at 30 μs. Now sketch.

3. Values at other time points can be determined easily. For example, the value at 5 μs can be calculated by noting that 5 μs is one eighth of a cycle, or 45°. Thus, $i = 4 \sin 45°$ mA = 2.83 mA. Alternately, from Table 15–4,

at $T/8$, $i = (4\text{ mA})(0.707) = 2.83$ mA. As many points as you need can be computed and plotted in this manner.

4. Values at particular angles can also be located easily. For instance, if you want a value at 30°, the required value is $i = 4 \sin 30°$ mA $= 2.0$ mA. To locate this point, note that 30° is one twelfth of a cycle or $T/12 = (40\ \mu\text{s})/12 = 3.33\ \mu\text{s}$. The point is shown on Figure 15–28.

FIGURE 15–28

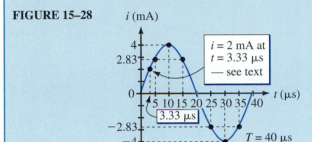

15.6 Voltages and Currents as Functions of Time

Relationship between ω, T, and f

Earlier you learned that one cycle of sine wave may be represented as either $\alpha = 2\pi$ rads or $t = T$ s, Figure 15–27. Substituting these into $\alpha = \omega t$ (Equation 15–5), you get $2\pi = \omega T$. Transposing yields

$$\omega T = 2\pi \quad \text{(rad)} \qquad \text{(15–11)}$$

Thus,

$$\omega = \frac{2\pi}{T} \quad \text{(rad/s)} \qquad \text{(15–12)}$$

Recall, $f = 1/T$ Hz. Substituting this into Equation 15–12 you get

$$\omega = 2\pi f \quad \text{(rad/s)} \qquad \text{(15–13)}$$

EXAMPLE 15–9   In some parts of the world, the power system frequency is 60 Hz; in other parts, it is 50 Hz. Determine ω for each.

Solution   For 60 Hz, $\omega = 2\pi f = 2\pi(60) = 377$ rad/s. For 50 Hz, $\omega = 2\pi f = 2\pi(50) = 314.2$ rad/s.

---

1. If $\omega = 240$ rad/s, what are T and f? How many cycles occur in 27 s?

2. If 56 000 cycles occur in 3.5 s, what is ω?

PRACTICE PROBLEMS 4

Answers:
1. 26.18 ms, 38.2 Hz, 1031 cycles
2. 100.5×10^3 rad/s

Sinusoidal Voltages and Currents as Functions of Time

Recall from Equation 15–4, $e = E_m \sin \alpha$, and from Equation 15–5, $\alpha = \omega t$. Combining these equations yields

$$e = E_m \sin \omega t \qquad \text{(15–14a)}$$

Similarly,

$$v = V_m \sin \omega t \qquad \text{(15–14b)}$$

$$i = I_m \sin \omega t \qquad \text{(15–14c)}$$

EXAMPLE 15–10   A 100-Hz sinusoidal voltage source has an amplitude of 150 volts. Write the equation for e as a function of time.

Solution $\omega = 2\pi f = 2\pi(100) = 628$ rad/s and $E_m = 150$ V. Thus, $e = E_m \sin \omega t = 150 \sin 628t$ V.

Equations 15–14 may be used to compute voltages or currents at any instant in time. Usually, ω is in radians per second, and thus ωt is in radians. You can work directly in radians or you can convert to degrees. For example, suppose you want to know the voltage at $t = 1.25$ ms for $e = 150 \sin 628t$ V.

Working in Rads.   With your calculator in the RAD mode, $e = 150 \sin(628)(1.25 \times 10^{-3}) = 150 \sin 0.785$ rad $= 106$ V.

Working in Degree.   0.785 rad $= 45°$. Thus, $e = 150 \sin 45° = 106$ V as before.

EXAMPLE 15–11   For $v = 170 \sin 2450t$, determine v at $t = 3.65$ ms and show the point on the v waveform.

Solution $\omega = 2450$ rad/s. Therefore $\omega t = (2450)(3.65 \times 10^{-3}) = 8.943$ rad $= 512.4°$. Thus, $v = 170 \sin 512.4° = 78.8$ V. Alternatively, $v = 170 \sin 8.943$ rad $= 78.8$ V. The point is plotted on the waveform in Figure 15–29.

FIGURE 15–29

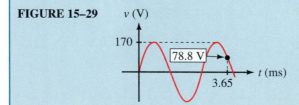

A sinusoidal current has a peak amplitude of 10 amps and a period of 120 ms.

PRACTICE PROBLEMS 5

a. Determine its equation as a function of time using Equation 15–14c.

b. Using this equation, compute a table of values at 10-ms intervals and plot one cycle of the waveform scaled in seconds.

c. Sketch one cycle of the waveform using the procedure of Example 15–8. (Note how much less work this is.)

Answers:

a. $i = 10 \sin 52.36t$ A

c. Mark the end of the cycle as 120 ms, ½ cycle as 60 ms, ¼ cycle as 30 ms, etc. Draw the sine wave so that it is zero at $t = 0$, 10 A at 30 ms, 0 A at 60 ms, -10 A at 90 ms and ends at $t = 120$ ms. (See Figure 15–30.)

Determining when a Particular Value Occurs

Sometimes you need to know when a particular value of voltage or current occurs. Given $v = V_m \sin \alpha$. Rewrite this as $\sin \alpha = v/V_m$. Then,

$$\alpha = \sin^{-1}\frac{v}{V_m} \qquad (15\text{–}15)$$

Compute the angle α at which the desired value occurs using the inverse sine function of your calculator, then determine the time from

$$t = \alpha/\omega$$

EXAMPLE 15–12   A sinusoidal current has an amplitude of 10 A and a period of 0.120 s. Determine the times at which

a. $i = 5.0$ A,

b. $i = -5$ A.

Solution

a. Consider Figure 15–30. As you can see, there are two points on the waveform where $i = 5$ A. Let these be denoted t_1 and t_2 respectively. First, determine ω:

$$\omega = \frac{2\pi}{T} = \frac{2\pi}{0.120 \text{ s}} = 52.36 \text{ rad/s}$$

Let $i = 10 \sin \alpha$ A. Now, find the angle α_1 at which $i = 5$ A:

$$\alpha_1 = \sin^{-1}\frac{i}{I_m} = \sin^{-1}\frac{5 \text{ A}}{10 \text{ A}} = \sin^{-1}0.5 = 30° = 0.5236 \text{ rad}$$

Thus, $t_1 = \alpha_1/\omega = (0.5236 \text{ rad})/(52.36 \text{ rad/s}) = 0.01$ s $= 10$ ms. This is indicated in Figure 15–30. Now consider t_2. Note that t_2 is the same distance back from the half-cycle point as t_1 is in from the beginning of the cycle. Thus, $t_2 = 60$ ms $- 10$ ms $= 50$ ms.

b. Similarly, t_3 (the first point at which $i = -5$ A occurs) is 10 ms past midpoint, while t_4 is 10 ms back from the end of the cycle. Thus, $t_3 = 70$ ms and $t_4 = 110$ ms.

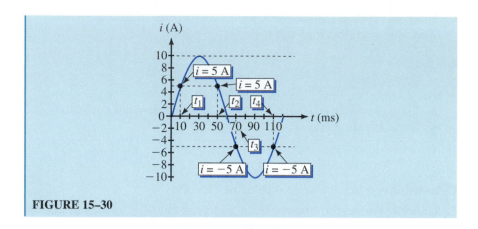

FIGURE 15–30

PRACTICE
PROBLEMS 6

Given $v = 10 \sin 52.36t$, determine both occurrences of $v = -8.66$ V.

Answer: 80 ms   100 ms

Voltages and Currents with Phase Shifts

If a sine wave does not pass through zero at $t = 0$ s as in Figure 15–30, it has
a **phase shift.** Waveforms may be shifted to the left or to the right (see Figure 15–31). For a waveform shifted left as in (a),

$$v = V_m \sin(\omega t + \theta) \tag{15–16a}$$

while, for a waveform shifted right as in (b),

$$v = V_m \sin(\omega t - \theta) \tag{15–16b}$$

(a) $v = V_m \sin(\omega t + \theta)$ (b) $v = V_m \sin(\omega t - \theta)$

FIGURE 15–31   Waveforms with phase shifts. Angle θ is normally measured in degrees, yielding mixed angular units. (See note.)

NOTES...

When applying equations 15–16(a) and (b), it is customary to express ωt in radians and θ in degrees, yielding mixed angular units (as indicated in the following examples). Although this is acceptable when the equations are written in symbolic form, you must convert both angles to the same unit before you make numerical computations.

EXAMPLE 15–13   Demonstrate that $v = 20 \sin(\omega t - 60°)$, where $\omega = \pi/6$ rad/s (i.e, $= 30°$/s), yields the shifted waveform shown in Figure 15–32.

Solution
1. Since ωt and $60°$ are both angles, $(\omega t - 60°)$ is also an angle. Let us define it as x. Then $v = 20 \sin x$, which means that the shifted wave is also sinusoidal.
2. Consider $v = \sin(\omega t - 60°)$. At $t = 0$ s, $v = 20 \sin(0 - 60°) = 20 \sin(-60°) = -17.3$ V as indicated in Figure 15–32.

3. Since $\omega = 30°/s$, it takes 2 s for ωt to reach 60°. Thus, at $t = 2$ s, $v = 20 \sin(60° - 60°) = 0$ V, and the waveform passes through zero at $t = 2$ s as indicated.

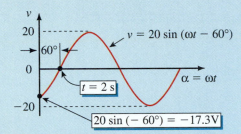

FIGURE 15–32

Summary: Since $v = 20 \sin(\omega t - 60°)$ is a sine wave and since it passes through zero at $t = 2$ s, where $\omega t = 60°$, it represents the shifted wave shown in Figure 15–32.

EXAMPLE 15–14

a. Determine the equation for the waveform of Figure 15–33(a), given $f = 60$ Hz. Compute current at $t = 4$ ms.

b. Repeat (a) for Figure 15–33(b).

Solution

a. $I_m = 2$ A and $\omega = 2\pi(60) = 377$ rad/s. This waveform corresponds to Figure 15–31(b). Therefore,

$$i = I_m \sin(\omega t - \theta) = 2 \sin(377t - 120°)\ \text{A}$$

At $t = 4$ ms, current is

$$i = 2 \sin(377 \times 4\ \text{ms} - 120°) = 2 \sin(1.508\ \text{rad} - 120°)$$
$$= 2 \sin(86.4° - 120°) = 2 \sin(-33.64°) = -1.11\ \text{A}.$$

b. This waveform matches Figure 15–31(a) if you extend the waveform back 90° from its peak as in (c). Thus,

$$i = 2 \sin(377t + 40°)\ \text{A}$$

At $t = 4$ ms, current is

$$i = 2 \sin(377 \times 4\ \text{ms} + 40°) = 2 \sin(126.4°)$$
$$= 1.61\ \text{A}.$$

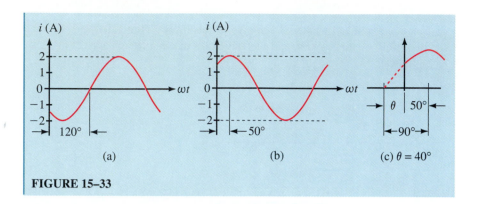

FIGURE 15–33

PRACTICE PROBLEMS 7

1. Given $i = 2 \sin(377t + 60°)$, compute the current at $t = 3$ ms.

2. Sketch each of the following:

   a. $v = 10 \sin(\omega t + 20°)$ V.   b. $i = 80 \sin(\omega t - 50°)$ A.

   c. $i = 50 \sin(\omega t + 90°)$ A.   d. $v = 5 \sin(\omega t + 180°)$ V.

3. Given $i = 2 \sin(377t + 60°)$, determine at what time $i = 1.8$ A.

Answers:

1. 1.64 A

2. a. Same as Figure 15–31(a) with $V_m = 10$ V, $\theta = 20°$.

   b. Same as Figure 15–31(b) with $I_m = 80$ A, $\theta = 50°$.

   c. Same as Figure 15–39(b) except use $I_m = 50$ A instead of V_m.

   d. A negative sine wave with magnitude of 5 V.

3. 0.193 ms

Probably the easiest way to deal with shifted waveforms is to use phasors. We introduce the idea next.

15.7   Introduction to Phasors

A **phasor** is a rotating line whose projection on a vertical axis can be used to represent sinusoidally varying quantities. To get at the idea, consider the red line of length V_m shown in Figure 15–34(a). (It is the phasor.) The vertical

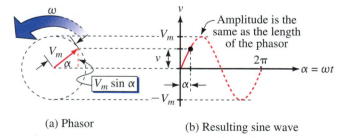

(a) Phasor   (b) Resulting sine wave

FIGURE 15–34   As the phasor rotates about the origin, its vertical projection creates a sine wave. (Figure 15–35 illustrates the process.)

projection of this line (indicated in dotted red) is $V_m \sin \alpha$. Now, assume that the phasor rotates at angular velocity of ω rad/s in the counterclockwise direction. Then, $\alpha = \omega t$, and its vertical projection is $V_m \sin \omega t$. If we designate this projection (height) as v, we get $v = V_m \sin \omega t$, which is the familiar sinusoidal voltage equation.

If you plot a graph of v versus α, you get the sine wave of Figure 15–34(b). Figure 15–35 illustrates the process. It shows snapshots of the phasor and the evolving waveform at various instants of time for a phasor of magnitude $V_m = 100$ V rotating at $\omega = 30°/s$. For example, consider $t = 0, 1, 2$, and 3s:

1. At $t = 0$ s, $\alpha = 0$, the phasor is at its 0° position, and its vertical projection is $v = V_m \sin \omega t = 100 \sin 0° = 0$ V. The point is at the origin.

2. At $t = 1$ s, the phasor has rotated 30° and its vertical projection is $v = 100 \sin 30° = 50$ V. This point is plotted at $\alpha = 30°$ on the horizontal axis.

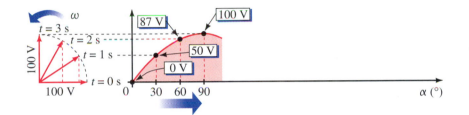

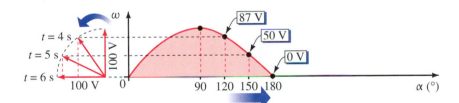

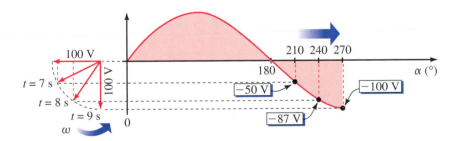

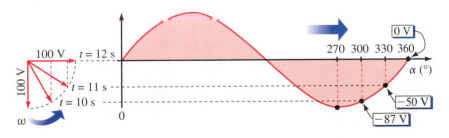

FIGURE 15–35 Evolution of the sine wave of Figure 15–34.

NOTES...

1. Although we have indicated phasor rotation in Figure 15–35 by a series of "snapshots," this is too cumbersome; in practice, we show only the phasor at its $t = 0$ s (reference) position and imply rotation rather than show it explicitly.

2. Although we are using maximum values (E_m and I_m) here, phasors are normally drawn in terms of effective values (considered in Section 15.9). For the moment, we will continue to use maximum values. We make the change in Chapter 16.

3. At $t = 2$ s, $\alpha = 60°$ and $v = 100 \sin 60° = 87$ V, which is plotted at $\alpha = 60°$ on the horizontal axis. Similarly, at $t = 3$ s, $\alpha = 90°$, and $v = 100$ V. Continuing in this manner, the complete waveform is evolved.

From the foregoing, we conclude that *a sinusoidal waveform can be created by plotting the vertical projection of a phasor that rotates in the counterclockwise direction at constant angular velocity* ω. *If the phasor has a length of* V_m, *the waveform represents voltage; if the phasor has a length of* I_m, it represents current. Note carefully: **Phasors apply only to sinusoidal waveforms.**

EXAMPLE 15–15   Draw the phasor and waveform for current $i = 25 \sin \omega t$ mA for $f = 100$ Hz.

Solution   The phasor has a length of 25 mA and is drawn at its $t = 0$ position, which is zero degrees as indicated in Figure 15–36. Since $f = 100$ Hz, the period is $T = 1/f = 10$ ms.

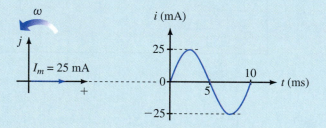

FIGURE 15–36   The reference position of the phasor is its $t = 0$ position.

Shifted Sine Waves

Phasors may be used to represent shifted waveforms, $v = V_m \sin(\omega t \pm \theta)$ or $i = I_m \sin(\omega t \pm \theta)$ as indicated in Figure 15–37. Angle θ is the position of the phasor at $t = 0$ s.

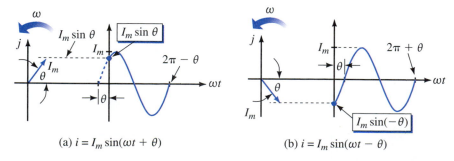

(a) $i = I_m \sin(\omega t + \theta)$ (b) $i = I_m \sin(\omega t - \theta)$

FIGURE 15–37   Phasors for shifted waveforms. Angle θ is the position of the phasor at $t = 0$ s.

EXAMPLE 15–16   Consider $v = 20 \sin(\omega t - 60°)$, where $\omega = \pi/6$ rad/s (i.e., 30°/s). Show that the phasor of Figure 15–38(a) represents this waveform.

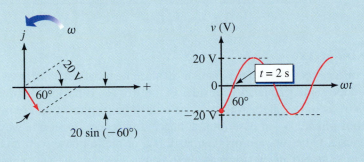

(a) Phasor   (b) $v = 20 \sin(\omega t - 60°)$, $\omega = 30°/s$

FIGURE 15–38

Solution   The phasor has length 20 V and at time $t = 0$ is at $-60°$ as indicated in (a). Now, as the phasor rotates, it generates a sinusoidal waveform, oscillating between ± 20 V as indicated in (b). Note that the zero crossover point occurs at $t = 2$ s, since it takes 2 seconds for the phasor to rotate from $-60°$ to $0°$ at 30 degrees per second. Now compare the waveform of (b) to the waveform of Figure 15–32, Example 15–13. They are identical. Thus, the phasor of (a) represents the shifted waveform $v = 20 \sin(\omega t - 60°)$.

EXAMPLE 15–17   With the aid of a phasor, sketch the waveform for $v = V_m \sin(\omega t + 90°)$.

Solution   Place the phasor at 90° as in Figure 15–39(a). Note that the resultant waveform (b) is a cosine waveform, i.e., $v = V_m \cos \omega t$. From this, we conclude that

$$\sin(\omega t + 90°) = \cos \omega t$$

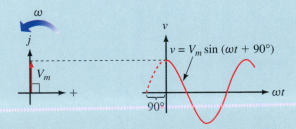

(a) Phasor at 90° position   (b) Waveform can also be
                                       described as a cosine wave

FIGURE 15–39   Demonstrating that $\sin(\omega t + 90°) = \cos \omega t$.

**PRACTICE
PROBLEMS 8**

With the aid of phasors, show that

a. $\sin(\omega t - 90°) = -\cos \omega t$,

b. $\sin(\omega t \pm 180°) = -\sin \omega t$,

Phase Difference

Phase difference refers to the angular displacement between different wave-forms of the same frequency. Consider Figure 15–40. If the angular displace-ment is 0° as in (a), the waveforms are said to be **in phase;** otherwise, they are **out of phase.** When describing a phase difference, select one waveform as reference. Other waveforms then lead, lag, or are in phase with this refer-ence. For example, in (b), for reasons to be discussed in the next paragraph, the current waveform is said to lead the voltage waveform, while in (c) the current waveform is said to lag.

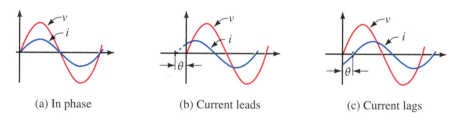

(a) In phase   (b) Current leads   (c) Current lags

FIGURE 15–40   Illustrating phase difference. In these examples, voltage is taken as ref-erence.

The terms **lead** and **lag** can be understood in terms of phasors. If you observe phasors rotating as in Figure 15–41(a), the one that you see passing first is leading and the other is lagging. By definition, *the waveform gener-ated by the leading phasor leads the waveform generated by the lagging phasor and vice versa.* In Figure 15–41, phasor I_m leads phasor V_m; thus cur-rent $i(t)$ leads voltage $v(t)$.

NOTES...

If you have trouble determining which waveform leads and which lags when you are solving a problem, make a quick sketch of their phasors, and the answer will be apparent. Note also that the terms *lead* and *lag* are relative. In Figure 15–41, we said that cur-rent leads voltage; you can just as correctly say that voltage lags current.

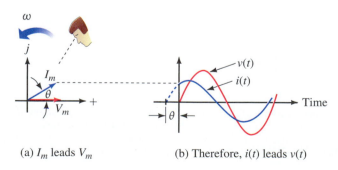

(a) I_m leads V_m (b) Therefore, $i(t)$ leads $v(t)$

FIGURE 15–41   Defining lead and lag.

EXAMPLE 15–18   Voltage and current are out of phase by 40°, and voltage lags. Using current as the reference, sketch the phasor diagram and the corresponding waveforms.

Solution   Since current is the reference, place its phasor in the 0° position and the voltage phasor at −40°. Figure 15–42 shows the phasors and corresponding waveforms.

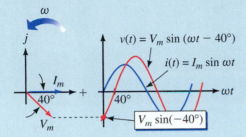

FIGURE 15–42

EXAMPLE 15–19   Given $v = 20 \sin(\omega t + 30°)$ and $i = 18 \sin(\omega t - 40°)$, draw the phasor diagram, determine phase relationships, and sketch the waveforms.

Solution   The phasors are shown in Figure 15–43(a). From these, you can see that v leads i by 70°. The waveforms are shown in (b).

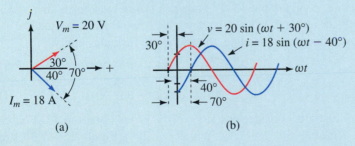

FIGURE 15–43

EXAMPLE 15–20   Figure 15–44 shows a pair of waveforms v_1 and v_2 on an oscilloscope. Each major vertical division represents 20 V and each major division on the horizontal (time) scale represents 20 μs. Voltage v_1 leads. Prepare a phasor diagram using v_1 as reference. Determine equations for both voltages.

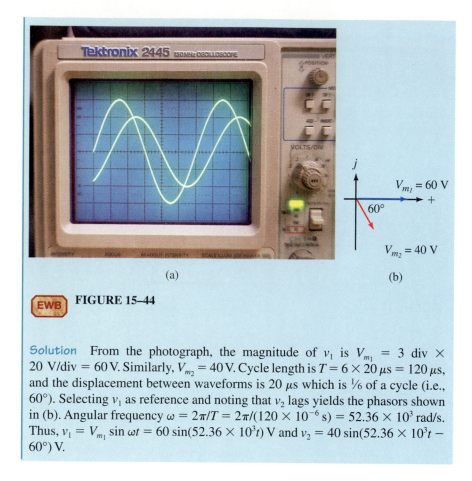

(a)   (b)

EWB   **FIGURE 15–44**

Solution   From the photograph, the magnitude of v_1 is $V_{m_1} = 3$ div $\times$ 20 V/div $= 60$ V. Similarly, $V_{m_2} = 40$ V. Cycle length is $T = 6 \times 20 \,\mu s = 120 \,\mu s$, and the displacement between waveforms is 20 μs which is $\frac{1}{6}$ of a cycle (i.e., 60°). Selecting v_1 as reference and noting that v_2 lags yields the phasors shown in (b). Angular frequency $\omega = 2\pi/T = 2\pi/(120 \times 10^{-6} \text{ s}) = 52.36 \times 10^3$ rad/s. Thus, $v_1 = V_{m_1} \sin \omega t = 60 \sin(52.36 \times 10^3 t)$ V and $v_2 = 40 \sin(52.36 \times 10^3 t - 60°)$ V.

Sometimes voltages and currents are expressed in terms of cos ωt rather than sin ωt. As Example 15–17 shows, a cosine wave is a sine wave shifted by $+90°$, or alternatively, a sine wave is a cosine wave shifted by $-90°$. For sines or cosines with an angle, the following formulas apply.

$$\cos(\omega t + \theta) = \sin(\omega t + \theta + 90°) \tag{15–17a}$$

$$\sin(\omega t + \theta) = \cos(\omega t + \theta - 90°) \tag{15–17b}$$

To illustrate, consider $\cos(\omega t + 30°)$. From Equation 15–17a, $\cos(\omega t + 30°) = \sin(\omega t + 30° + 90°) = \sin(\omega t + 120°)$. Figure 15–45 illustrates this relationship graphically. The red phasor in (a) generates cos ωt as was shown

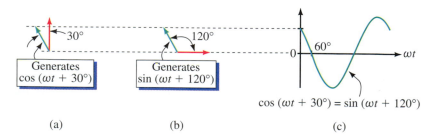

(a)   (b)   (c)

FIGURE 15–45   Using phasors to show that $\cos(\omega t + 30°) = \sin(\omega t + 120°)$.

in Example 15–17. Therefore, the green phasor generates a waveform that leads it by 30°, namely $\cos(\omega t + 30°)$. For (b), the red phasor generates sin ωt, and the green phasor generates a waveform that leads it by 120°, i.e., $\sin(\omega t + 120°)$. Since the green phasor is the same in both cases, you can see that $\cos(\omega t + 30°) = \sin(\omega t + 120°)$. Note that this process is easier than trying to remember equations 15–17(a) and (b).

EXAMPLE 15–21   Determine the phase angle between $v = 30 \cos(\omega t + 20°)$ and $i = 25 \sin(\omega t + 70°)$.

Solution $i = 25 \sin(\omega t + 70°)$ may be represented by a phasor at 70°, and $v = 30 \cos(\omega t + 20°)$ by a phasor at $(90° + 20°) = 110°$, Figure 15–46(a). Thus, v leads i by 40°. Waveforms are shown in (b).

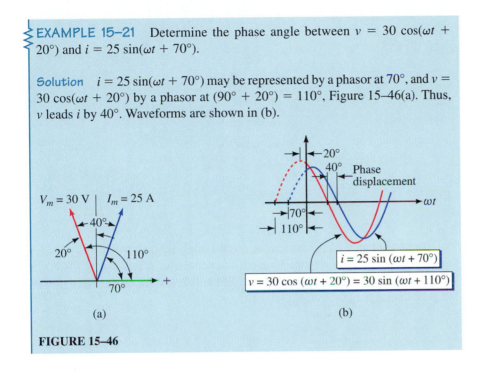

(a)   (b)

FIGURE 15–46

Sometimes you encounter negative waveforms such as $i = -I_m \sin \omega t$. To see how to handle these, refer back to Figure 15–36, which shows the waveform and phasor for $i = I_m \sin \omega t$. If you multiply this waveform by -1, you get the inverted waveform $-I_m \sin \omega t$ of Figure 15–47(a) with corresponding phasor (b). Note that the phasor is the same as the original phasor except that it is rotated by 180°. This is always true—thus, if you multiply a waveform by -1, the phasor for the new waveform is 180° rotated from the original phasor, regardless of the angle of the original phasor.

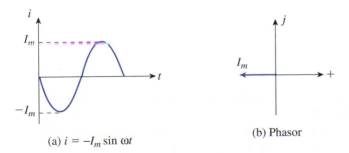

(a) $i = -I_m \sin \omega t$

(b) Phasor

FIGURE 15–47   The phasor for a negative sine wave is at 180°.

EXAMPLE 15–22   Find the phase relationship between $i = -4 \sin(\omega t + 50°)$ and $v = 120 \sin(\omega t - 60°)$.

Solution $i = -4 \sin(\omega t + 50°)$ is represented by a phasor at $(50° - 180°) = -130°$ and $v = 120 \sin(\omega t - 60°)$ by a phasor at $-60°$, Figure 15–48. The phase difference is 70° and voltage leads. Note also that i can be written as $i = 4 \sin(\omega t - 130°)$.

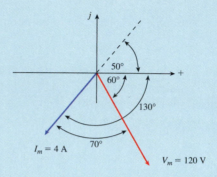

FIGURE 15–48

The importance of phasors to ac circuit analysis cannot be overstated—you will find that they are one of your main tools for representing ideas and for solving problems in later chapters. We will leave them for the moment, but pick them up again in Chapter 16.

**IN-PROCESS
LEARNING
CHECK 2**

1. If $i = 15 \sin \alpha$ mA, compute the current at $\alpha = 0°, 45°, 90°, 135°, 180°, 225°, 270°, 315°,$ and $360°$.

2. Convert the following angles to radians:
   a. 20°   b. 50°
   c. 120°   d. 250°

3. If a coil rotates at $\omega = \pi/60$ radians per millisecond, how many degrees does it rotate through in 10 ms? In 40 ms? In 150 ms?

4. A current has an amplitude of 50 mA and $\omega = 0.2\pi$ rad/s. Sketch the waveform with the horizontal axis scaled in
   a. degrees   b. radians   c. seconds

5. If 2400 cycles of a waveform occur in 10 ms, what is ω in radians per second?

6. A sinusoidal current has a period of 40 ms and an amplitude of 8 A. Write its equation in the form of $i = I_m \sin \omega t$, with numerical values for I_m and ω.

7. A current $i = I_m \sin \omega t$ has a period of 90 ms. If $i = 3$ A at $t = 7.5$ ms, what is its equation?

8. Write equations for each of the waveforms in Figure 15–49 with the phase angle θ expressed in degrees.

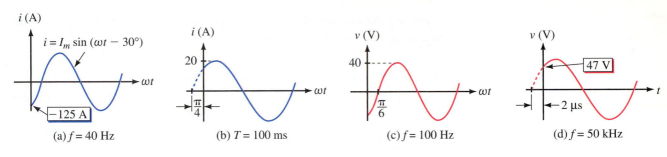

FIGURE 15–49

9. Given $i = 10 \sin \omega t$, where $f = 50$ Hz, find all occurrences of
   a. $i = 8$ A between $t = 0$ and $t = 40$ ms
   b. $i = -5$ A between $t = 0$ and $t = 40$ ms
10. Sketch the following waveforms with the horizontal axis scaled in degrees:
    a. $v_1 = 80 \sin(\omega t + 45°)$ V   b. $v_2 = 40 \sin(\omega t - 80°)$ V
    c. $i_1 = 10 \cos \omega t$ mA   d. $i_2 = 5 \cos(\omega t - 20°)$ mA
11. Given $\omega = \pi/3$ rad/s, determine when voltage first crosses through 0 for
    a. $v_1 = 80 \sin(\omega t + 45°)$ V   b. $v_2 = 40 \sin(\omega t - 80°)$ V
12. Consider the voltages of Question 10:
    a. Sketch phasors for v_1 and v_2.
    b. What is the phase difference between v_1 and v_2?
    c. Determine which voltage leads and which lags.
13. Repeat Question 12 for the currents of Question 10.

(Answers are at the end of the chapter.)

15.8 AC Waveforms and Average Value

While we can describe ac quantities in terms of frequency, period, instantaneous value, etc., we do not yet have any way to give a meaningful value to an ac current or voltage in the same sense that we can say of a car battery that it has a voltage of 12 volts. This is because ac quantities constantly change and thus there is no one single numerical value that truly represents a waveform over its complete cycle. For this reason, ac quantities are generally described by a group of characteristics, including instantaneous, peak, average, and effective values. The first two of these we have already seen. In this section, we look at average values; in Section 15.9, we consider effective values.

Average Values

Many quantities are measured by their average, for instance, test and examination scores. To find the average of a set of marks for example, you add them, then divide by the number of items summed. For waveforms, the process is conceptually the same. For example, to find the average of a waveform, you can sum the instantaneous values over a full cycle, then divide by the number of points used. The trouble with this approach is that waveforms do not consist of discrete values.

Average in Terms of the Area Under a Curve

An approach more suitable for use with waveforms is to find the area under the curve, then divide by the baseline of the curve. To get at the idea, we can use an analogy. Consider again the technique of computing the average for a set of numbers. Assume that you earn marks of 80, 60, 60, 95, and 75 on a group of tests. Your average mark is therefore

$$average = (80 + 60 + 60 + 95 + 75)/5 = 74$$

An alternate way to view these marks is graphically as in Figure 15–50. The area under this curve can be computed as

$$area = (80 \times 1) + (60 \times 2) + (95 \times 1) + (75 \times 1)$$

Now divide this by the length of the base, namely 5. Thus,

$$\frac{(80 \times 1) + (60 \times 2) + (95 \times 1) + (75 \times 1)}{5} = 74$$

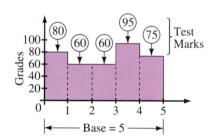

FIGURE 15–50 Determining average by area.

which is exactly the answer obtained above. That is,

$$average = \frac{area\ under\ curve}{length\ of\ base} \qquad (15\text{–}18)$$

This result is true in general. Thus, *to find the average value of a waveform, divide the area under the waveform by the length of its base. Areas above the axis are counted as positive, while areas below the axis are counted as negative.* This approach is valid regardless of waveshape.

Average values are also called **dc values,** because dc meters indicate average values rather than instantaneous values. Thus, if you measure a non-dc quantity with a dc meter, the meter will read the average of the waveform, i.e., the value calculated according to Equation 15–18.

EXAMPLE 15–23

a. Compute the average for the current waveform of Figure 15–51.

b. If the negative portion of Figure 15–51 is −3 A instead of −1.5 A, what is the average?

c. If the current is measured by a dc ammeter, what will the ammeter indicate?

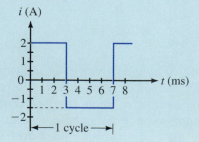

FIGURE 15–51

Solution

a. The waveform repeats itself after 7 ms. Thus, $T = 7$ ms and the average is

$$I_{avg} = \frac{(2\,A \times 3\,ms) - (1.5\,A \times 4\,ms)}{7\,ms} = \frac{6-6}{7} = 0\,A$$

b. $I_{avg} = \frac{(2\,A \times 3\,ms) - (3\,A \times 4\,ms)}{7\,ms} = \frac{-6\,A}{7} = -0.857\,A$

c. A dc ammeter measuring (a) will indicate zero, while for (b) it will indicate -0.857 A.

EXAMPLE 15–24 Compute the average value for the waveforms of Figures 15–52(a) and (c). Sketch the averages for each.

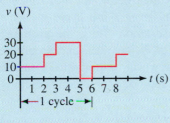

(a)

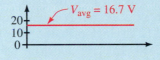

(b)

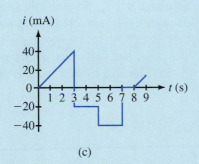

(c)

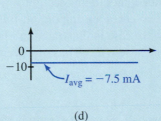

(d)

FIGURE 15–52

Solution For the waveform of (a), $T = 6$ s. Thus,

$$V_{avg} = \frac{(10\,V \times 2\,s) + (20\,V \times 1\,s) + (30\,V \times 2\,s) + (0\,V \times 1\,s)}{6\,s} = \frac{100\,V\text{-}s}{6\,s} = 16.7\,V$$

The average is shown as (b). A dc voltmeter would indicate 16.7 V. For the waveform of (c), $T = 8$ s and

$$I_{avg} = \frac{\frac{1}{2}(40\,mA \times 3\,s) - (20\,mA \times 2\,s) - (40\,mA \times 2\,s)}{8\,s} = \frac{-60}{8}\,mA = -7.5\,mA$$

In this case, a dc ammeter would indicate -7.5 mA.

Determine the averages for Figures 15–53(a) and (b).

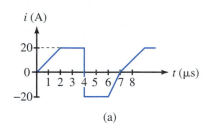

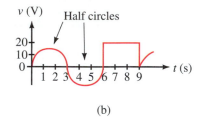

(a)   (b)

FIGURE 15–53

Answers: a. 1.43 A   b. 6.67 V

Sine Wave Averages

Because a sine wave is symmetrical, its area below the horizontal axis is the same as its area above the axis; thus, over a full cycle its net area is zero, independent of frequency and phase angle. Thus, the average of sin ωt, $\sin(\omega t \pm \theta)$, sin $2\omega t$, cos ωt, $\cos(\omega t \pm \theta)$, cos $2\omega t$, and so on are each zero. The average of half a sine wave, however, is not zero. Consider Figure 15–54. The area under the half-cycle may be found using calculus as

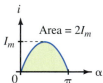

FIGURE 15–54   Area under a half-cycle.

$$\text{area} = \int_0^\pi I_m \sin\alpha \, d\alpha = -I_m \cos \alpha \Big|_0^\pi = 2I_m \qquad (15–19)$$

Similarly, the area under a half-cycle of voltage is $2V_m$. (If you haven't studied calculus, you can approximate this area using numerical methods as described later in this section.)

Two cases are important; full-wave average and half-wave average. The full-wave case is illustrated in Figure 15–55. The area from 0 to 2π is $2(2I_m)$ and the base is 2π. Thus, the average is

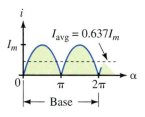

FIGURE 15–55   Full-wave average.

$$I_{\text{avg}} = \frac{2(2I_m)}{2\pi} = \frac{2I_m}{\pi} = 0.637I_m$$

For the half-wave case (Figure 15–56),

$$I_{\text{avg}} = \frac{2I_m}{2\pi} = \frac{I_m}{\pi} = 0.318I_m$$

The corresponding expressions for voltage are

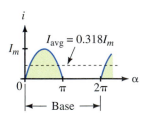

FIGURE 15–56   Half-wave average.

$$V_{\text{avg}} = 0.637V_m \quad \text{(full-wave)}$$
$$V_{\text{avg}} = 0.318V_m \quad \text{(half-wave)}$$

Numerical Methods

If the area under a curve cannot be computed exactly, it can be approximated. One method is to approximate the curve by straight line segments as in Figure 15–57. (If the straight lines closely fit the curve, the accuracy is very good.) Each element of area is a trapezoid (b) whose area is its average height times its base. Thus, $A_1 = \frac{1}{2}(y_0 + y_1)\Delta x$, $A_2 = \frac{1}{2}(y_1 + y_2)\Delta x$, etc. Summing areas and combining terms yields

$$\text{area} = \left(\frac{y_0}{2} + y_1 + y_2 + \cdots + y_{k-1} + \frac{y_k}{2}\right)\Delta x \qquad (15\text{–}20)$$

This result is known as the **trapezoidal rule.** Example 15–25 illustrates its use.

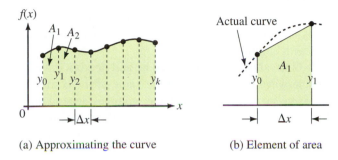

(a) Approximating the curve   (b) Element of area

FIGURE 15–57   Calculating area using the trapezoidal rule.

EXAMPLE 15–25   Approximate the area under $y = \sin(\omega t - 30°)$, Figure 15–58. Use an increment size of $\pi/6$ rad, i.e., 30°.

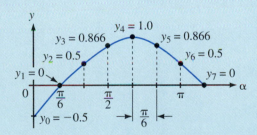

FIGURE 15–58

Solution   Points on the curve $\sin(\omega t - 30°)$ have been computed by calculator and plotted as Figure 15–58. Substituting these values into Equation 15–20 yields

$$\text{area} = \left(\frac{1}{2}(-0.5) + 0 + 0.5 + 0.866 + 1.0 + 0.866 + 0.5 + \frac{1}{2}(0)\right)\left(\frac{\pi}{6}\right) = 1.823$$

The exact area (found using calculus) is 1.866; thus, the above value is in error by 2.3%.

1. Repeat Example 15–25 using an increment size of $\pi/12$ rad. What is the percent error?

PRACTICE
PROBLEMS 10

2. Approximate the area under $v = 50 \sin(\omega t + 30°)$ from $\omega t = 0°$ to $\omega t = 210°$. Use an increment size of $\pi/12$ rad.

Answers:
1. 1.855; 0.59%
2. 67.9 (exact 68.3; error = 0.6%)

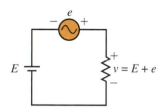

FIGURE 15–59

Superimposed AC and DC

Sometimes ac and dc are used in the same circuit. For example, amplifiers are powered by dc but the signals they amplify are ac. Figure 15–59 shows a simple circuit with combined ac and dc.

Figure 15–60(c) shows superimposed ac and dc. Since we know that the average of a sine wave is zero, the average value of the combined waveform will be its dc component, E. However, peak voltages depend on both components as illustrated in (c). Note for the case illustrated that although the waveform varies sinusoidally, it does not alternate in polarity since it never changes polarity to become negative.

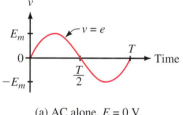
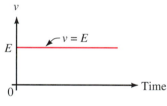
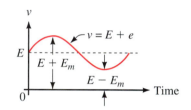

(a) AC alone. $E = 0$ V.
$V_{avg} = 0$ V

(b) DC alone. $e = 0$ V.
$V_{avg} = E$

(c) Superimposed ac and dc.
$V_{avg} = E$

FIGURE 15–60 Superimposed dc and ac.

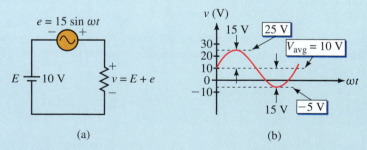

EXAMPLE 15–26   Draw the voltage waveform for the circuit of Figure 15–61(a). Determine average, peak, and minimum voltages.

(a)   (b)

FIGURE 15–61 $v = 10 + 15 \sin \omega t$.

Solution   The waveform consists of a 10-V dc value with 15 V ac riding on top of it. The average is the dc value, $V_{avg} = 10$ V. The peak voltage is $10 + 15 = 25$ V, while the minimum voltage is $10 - 15 = -5$ V. This waveform alternates in polarity, although not symmetrically (as is the case when there is no dc component).

Repeat Example 15–26 if the dc source of Figure 15–61 is $E = -5$ V.

PRACTICE
PROBLEMS 11

Answers: $V_{avg} = -5$ V; positive peak $= 10$ V; negative peak $= -20$ V

15.9   Effective Values

While instantaneous, peak, and average values provide useful information about a waveform, none of them truly represents the ability of the waveform to do useful work. In this section, we look at a representation that does. It is called the waveform's **effective value.** The concept of effective value is an important one; in practice, most ac voltages and currents are expressed as effective values. Effective values are also called **rms values** for reasons discussed shortly.

What Is an Effective Value?

An effective value is an equivalent dc value: it tells you how many volts or amps of dc that a time-varying waveform is equal to in terms of its ability to produce average power. Effective values depend on the waveform. A familiar example of such a value is the value of the voltage at the wall outlet in your home. In North America its value is 120 Vac. This means that the sinusoidal voltage at the wall outlets of your home is capable of producing the same average power as 120 volts of steady dc.

Effective Values for Sine Waves

The effective value of a waveform can be determined using the circuits of Figure 15–62. Consider a sinusoidally varying current, $i(t)$. By definition, the effective value of i is that value of dc current that produces the same average power. Consider (b). Let the dc source be adjusted until its average power is the same as the average power in (a). The resulting dc current is then the effective value of the current of (a). To determine this value, determine the average power for both cases, then equate them.

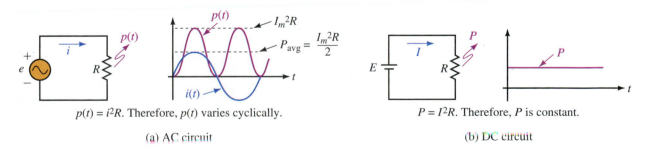

$p(t) = i^2R$. Therefore, $p(t)$ varies cyclically.

(a) AC circuit

$P = I^2R$. Therefore, P is constant.

(b) DC circuit

FIGURE 15–62   Determining the effective value of sinusoidal ac.

First, consider the dc case. Since current is constant, power is constant, and average power is

$$P_{avg} = P = I^2R \qquad (15–21)$$

Now consider the ac case. Power to the resistor at any value of time is $p(t) = i^2R$, where i is the instantaneous value of current. A sketch of $p(t)$ is shown in Figure 15–62(a), obtained by squaring values of current at various points along the axis, then multiplying by R. Average power is the average of $p(t)$. Since $i = I_m \sin \omega t$,

$$
\begin{aligned}
p(t) &= i^2R \\
&= (I_m \sin \omega t)^2 R = I_m^2 R \sin^2 \omega t \\
&= I_m^2 R \left[\frac{1}{2}(1 - \cos 2\omega t) \right]
\end{aligned}
\tag{15–22}
$$

where we have used the trigonometric identity $\sin^2 \omega t = \frac{1}{2}(1 - \cos 2\omega t)$, from the mathematics tables inside the front cover to expand $\sin^2 \omega t$. Thus,

$$
p(t) = \frac{I_m^2 R}{2} - \frac{I_m^2 R}{2} \cos 2\omega t
\tag{15–23}
$$

To get the average of $p(t)$, note that the average of $\cos 2\omega t$ is zero and thus the last term of Equation 15–23 drops off leaving

$$
P_{\text{avg}} = \text{average of } p(t) = \frac{I_m^2 R}{2}
\tag{15–24}
$$

Now equate Equations 15–21 and 15–24, then cancel R.

$$
I^2 = \frac{I_m^2}{2}
$$

Now take the square root of both sides. Thus,

$$
I = \sqrt{\frac{I_m^2}{2}} = \frac{I_m}{\sqrt{2}} = 0.707 I_m
$$

Current I is the value that we are looking for; it is the effective value of current i. To emphasize that it is an effective value, we will initially use subscripted notation I_{eff}. Thus,

$$
I_{\text{eff}} = \frac{I_m}{\sqrt{2}} = 0.707 I_m
\tag{15–25}
$$

Effective values for voltage are found in the same way:

$$
E_{\text{eff}} = \frac{E_m}{\sqrt{2}} = 0.707 E_m
\tag{15–26a}
$$

$$
V_{\text{eff}} = \frac{V_m}{\sqrt{2}} = 0.707 V_m
\tag{15–26b}
$$

As you can see, *effective values for sinusoidal waveforms depend only on magnitude.*

NOTES...

Because ac currents alternate in direction, you might expect average power to be zero, with power during the negative half-cycle being equal and opposite to power during the positive half-cycle and hence cancelling. However, as Equation 15–22 shows, current is squared, and hence power is never negative. This is consistent with the idea that insofar as power dissipation is concerned, the direction of current through a resistor does not matter (Figure 15–63).

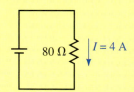

(a) $P = (4)^2(80) = 1280$ W

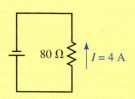

(b) $P = (4)^2(80) = 1280$ W

FIGURE 15–63 Since power depends only on current magnitude, it is the same for both directions.

EXAMPLE 15–27 Determine the effective values of

a. $i = 10 \sin \omega t$ A,

b. $i = 50 \sin(\omega t + 20°)$ mA,

c. $v = 100 \cos 2\omega t$ V

Solution Since effective values depend only on magnitude,

a. $I_{\text{eff}} = (0.707)(10\text{ A}) = 7.07\text{ A}$,

b. $I_{\text{eff}} = (0.707)(50\text{ mA}) = 35.35\text{ mA}$,

c. $V_{\text{eff}} = (0.707)(100\text{ V}) = 70.7\text{ V}$.

To obtain peak values from effective values, rewrite Equations 15–25 and 15–26. Thus,

$$I_m = \sqrt{2}I_{\text{eff}} = 1.414I_{\text{eff}} \qquad \textbf{(15–27)}$$

$$E_m = \sqrt{2}E_{\text{eff}} = 1.414V_{\text{eff}} \qquad \textbf{(15–28a)}$$

$$V_m = \sqrt{2}V_{\text{eff}} = 1.414V_{\text{eff}} \qquad \textbf{(15–28b)}$$

It is important to note that these relationships hold only for sinusoidal waveforms. However, the concept of effective value applies to all waveforms, as we soon see.

Consider again the ac voltage at the wall outlet in your home. Since $E_{\text{eff}} = 120\text{ V}$, $E_m = (\sqrt{2})(120\text{ V}) = 170\text{ V}$. This means that a sinusoidal voltage alternating between $\pm 170\text{ V}$ produces the same average power as 120 V of steady dc (Figure 15–64).

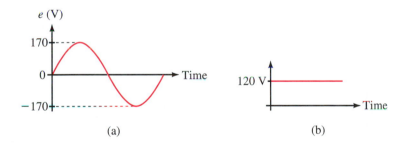

(a)   (b)

FIGURE 15–64 120 V of steady dc is capable of producing the same average power as sinusoidal ac with $E_m = 170\text{ V}$.

General Equation for Effective Values

The $\sqrt{2}$ relationship holds only for sinusoidal waveforms. For other waveforms, you need a more general formula. Using calculus, it can be shown that for any waveform

$$I_{\text{eff}} = \sqrt{\frac{1}{T}\int_0^T i^2 dt} \qquad \textbf{(15–29)}$$

with a similar equation for voltage. This equation can be used to compute effective values for any waveform, including sinusoidal. In addition, it leads to a graphic approach to finding effective values. In Equation 15–29, the integral of i^2 represents the area under the i^2 waveform. Thus,

$$I_{\text{eff}} = \sqrt{\frac{\text{area under the } i^2 \text{ curve}}{\text{base}}} \qquad \textbf{(15–30)}$$

To compute effective values using this equation, do the following:

Step 1: Square the current (or voltage) curve.

Step 2: Find the area under the squared curve.

Step 3: Divide the area by the length of the curve.

Step 4: Find the square root of the value from Step 3.

This process is easily carried out for rectangular-shaped waveforms since the area under their squared curves is easy to compute. For other waveforms, you have to use calculus or approximate the area using numerical methods. For the special case of superimposed ac and dc (Figure 15–60), Equation 15–29 leads to the following formula:

$$I_{eff} = \sqrt{I_{dc}^{\,2} + I_{ac}^{\,2}} \qquad\qquad \textbf{(15–31)}$$

where I_{dc} is the dc current value, I_{ac} is the effective value of the ac component, and I_{eff} is the effective value of the combined ac and dc currents. Equations 15–30 and 15–31 also hold for voltage when V is substituted for I.

RMS Values

Consider again Equation 15–30. To use this equation, we compute the root of the mean square to obtain the effective value. For this reason, effective values are called **root mean square** or **rms** values and **the terms *effective* and *rms* are synonymous.** Since, in practice, ac quantities are almost always expressed as rms values, we shall assume from here on that, unless otherwise noted, *all ac voltages and currents are rms values.*

EXAMPLE 15–28 One cycle of a voltage waveform is shown in Figure 15–65(a). Determine its effective value.

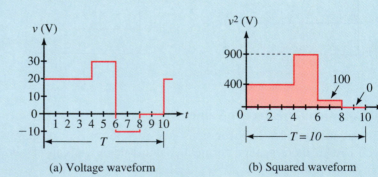

(a) Voltage waveform   (b) Squared waveform

FIGURE 15–65

Solution Square the voltage waveform and plot it as in (b). Apply Equation 15–30:

$$V_{eff} = \sqrt{\dfrac{(400 \times 4) + (900 \times 2) + (100 \times 2) + (0 \times 2)}{10}}$$

$$= \sqrt{\dfrac{3600}{10}} = 19.0 \text{ V}$$

The waveform of Figure 15–65(a) has the same effective value as 19.0 V of steady dc.

EXAMPLE 15–29 Determine the effective value of the waveform of Figure 15–66(a).

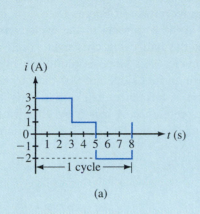

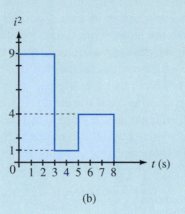

(a)   (b)

FIGURE 15–66

Solution Square the curve, then apply Equation 15–30. Thus,

$$I_{eff} = \sqrt{\frac{(9 \times 3) + (1 \times 2) + (4 \times 3)}{8}}$$

$$= \sqrt{\frac{41}{8}} = 2.26 \text{ A}$$

EXAMPLE 15–30 Compute the effective value of the waveform of Figure 15–61(b).

Solution Use Equation 15–31 (with I replaced by V). First, compute the rms value of the ac component. $V_{ac} = 0.707 \times 15 = 10.61$ V. Now substitute this into Equation 15–31. Thus,

$$V_{rms} = \sqrt{V_{dc}^2 + V_{ac}^2} = \sqrt{(10)^2 + (10.61)^2} = 14.6 \text{ V}$$

1. Determine the effective value of the current of Figure 15–51.
2. Repeat for the voltage graphed in Figure 15–52(a).

PRACTICE PROBLEMS 12

Answers:
1. 1.73 A
2. 20 V

One Final Note

The subscripts *eff* and *rms* are not used in practice. Once the concept is familiar, we drop them.

15.10 Rate of Change of a Sine Wave (Derivative)

Several important circuit effects depend on the rate of change of sinusoidal quantities. The rate of change of a quantity is the slope (i.e., derivative) of its waveform versus time. Consider the waveform of Figure 15–67. As indicated, the slope is maximum positive at the beginning of the cycle, zero at both its peaks, maximum negative at the half-cycle crossover point, and maximum positive at the end of the cycle. This slope is plotted in Figure 15–68. Note that the original waveform and its slope are 90° out of phase. Thus, if sinusoidal waveform A is taken as reference, its slope B leads it by 90°, whereas, if the slope B is taken as reference, A lags it by 90°. Thus, if A is a sine wave, B is a cosine wave, and so on. (This result is important to us in Chapter 16.)

NOTES...

⫿ The Derivative of a Sine Wave

The result developed intuitively here can be proven easily using calculus. To illustrate, consider the waveform $\sin \omega t$ shown in Figure 15–67. The slope of this function is its derivative. Thus,

$$\text{Slope} = \frac{d}{dt}\sin \omega t = \omega \cos \omega t$$

Therefore, the slope of a sine wave is a cosine wave as depicted in Figure 15–68.

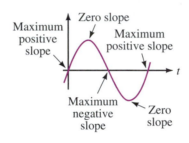

FIGURE 15–67 Slope at various places for a sine wave.

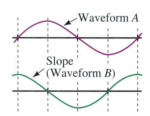

FIGURE 15–68 Showing the 90° phase shift.

15.11 AC Voltage and Current Measurement

Two of the most important instruments for measuring ac quantities are the multimeter and the oscilloscope. Multimeters read voltage, current, and sometimes frequency. Oscilloscopes show waveshape and period and permit determination of frequency, phase difference, and so on.

Meters for Voltage and Current Measurement

There are two basic classes of ac meters: one measures rms correctly for sinusoidal waveforms only (called "average responding" instruments); the other measures rms correctly regardless of waveform (called "true rms" meters). Most common meters are average responding meters.

Average Responding Meters

Average responding meters use a rectifier circuit to convert incoming ac to dc. They then respond to the average value of the rectified input, which, as shown in Figure 15–55, is $0.637V_m$ for a "full-wave" rectified sine wave. However, the rms value of a sine wave is $0.707V_m$. Thus, the scale of such a meter is modified by the factor $0.707V_m/0.637V_m = 1.11$ so that it indicates rms values directly. Other meters use a "half-wave" circuit, which yields the waveform of Figure 15–56 for a sine wave input. In this case, its average is $0.318V_m$, yielding a scale factor of $0.707V_m/0.318V_m = 2.22$. Figure 15–69 shows a typical average responding DMM.

FIGURE 15–69   A DMM. While all DMMs measure voltage, current, and resistance, this one also measures frequency.

True RMS Measurement

To measure the rms value of a nonsinusoidal waveform, you need a true rms meter. A true rms meter indicates true rms voltages and currents regardless of waveform. For example, for the waveform of Figure 15–64(a), any ac meter will correctly read 120 V (since it is a sine wave). For the waveform of 15–61(b), a true rms meter will correctly read 14.6 V (the rms value that we calculated earlier, in Example 15–30) but an average responding meter will yield only a meaningless value. True rms instruments are more expensive than standard meters.

Oscilloscopes

Oscilloscopes (frequently referred to as scopes, Figure 15–70) are used for time domain measurement, i.e., waveshape, frequency, period, phase difference, and so on. Usually, you scale values from the screen, although some higher-priced models can compute and display them for you on a digital readout.

Oscilloscopes measure voltage. To measure current, you need a current-to-voltage converter. One type of converter is a clip-on device, known as a **current gun** that clamps over the current-carrying conductor and monitors its magnetic field. (It works only with ac.) The varying magnetic field induces a voltage which is then displayed on the screen. With such a device, you can monitor current waveshapes and make current-related measurements. Alternately, you can place a small resistor in the current path, mea-

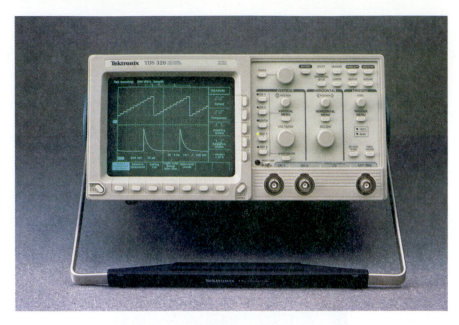

FIGURE 15–70 An oscilloscope may be used for waveform analysis.

sure voltage across it with the oscilloscope, then use Ohm's law to determine the current.

A Final Note

AC meters measure voltage and current only over a limited frequency range, typically from 50 Hz to a few kHz, although others are available that work up to the 100-kHz range. Note, however, that accuracy may be affected by frequency. (Check the manual.) Oscilloscopes, on the other hand, can measure very high frequencies; even moderately priced oscilloscopes work at frequencies up to hundreds of MHz.

15.12 Circuit Analysis Using Computers

**ELECTRONICS
WORKBENCH**

PSpice

Electronics Workbench and PSpice both provide a convenient way to study the phase relationships of this chapter, as they both incorporate easy-to-use graphing facilities. You simply set up sources with the desired magnitude and phase values and instruct the software to compute and plot the results. To illustrate, let us graph $e_1 = 100 \sin \omega t$ V and $e_2 = 80 \sin(\omega t + 60°)$ V. Use a frequency of 500 Hz.

Electronics Workbench

For Electronics Workbench, you must specify rms rather than peak values. Thus, to plot $e_1 = 100 \sin \omega t$ V, key in 70.71 V and 0 degrees. Similarly, for e_2, use 56.57 V and 60°. Here are the steps: Create the circuit of Figure 15–71 on the screen; Double click Source 1 and enter 70.71 V, 0 deg, and 500 Hz in the dialog box; Similarly, set Source 2 to 56.57 V, 60 deg, and

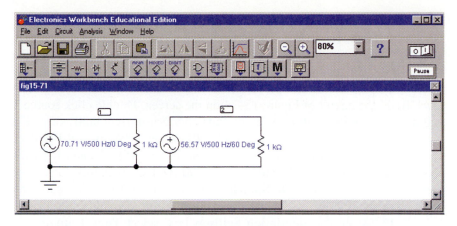

FIGURE 15–71 Studying phase relationships using Electronics Workbench.

500 Hz; Click **Analysis, Transient,** set **TSTOP** to 0.002 (to run the solution out to 2 ms so that you display a full cycle) and **TMAX** to 2e-06 (to avoid getting a choppy waveform); Highlight Node 1 (to display e_1) and click **Add;** Repeat for Node 2 (to display e_2); Click the **Simulate** icon; Following simulation, graphs e_1 and e_2 (Figure 15–72) appear.

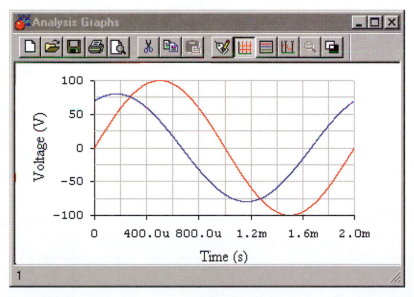

FIGURE 15–72

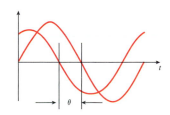

FIGURE 15–73

You can verify the angle between the waveforms using cursors. First, note that the period $T = 2$ ms $= 2000$ μs. (This corresponds to 360°.) Expand the graph to full screen, click the **Grid** icon, then the **Cursors** icon. Using the cursors, measure the time between crossover points, as indicated in Figure 15–73. You should get 333 μs. This yields an angular displacement of

$$\theta = \frac{333 \ \mu s}{2000 \ \mu s} \times 360° = 60°$$

OrCAD PSpice

For this problem, you need a sinusoidal time-varying ac voltage source. Use **VSIN** (it is found in the **SOURCE** library.) For VSIN, you must specify the magnitude, phase, and frequency of the source, as well as its offset (since we do not want an offset in this problem, we will set it to zero.) Proceed as follows. Build the circuit of Figure 15–74 on the screen. Double click source 1 and in the Properties editor, select the **Parts** tab. Scroll right until you find a list of source properties: then in cell **VAMPL,** enter 100V; in cell **PHASE,** enter 0deg; in cell **FREQ,** enter 500Hz; and in cell **VOFF,** enter 0V (this sets source $e_i = 100 \sin \omega t$ V with $\omega = 2\pi(500$ Hz$)$). Click **Apply,** then close the Properties editor window. Similarly, set up source 2 (making sure you use a phase angle of 60°). Click the **New Simulation** icon and enter a name (e.g., **fig15-74**). In the **Simulation Settings** box, select **Time Domain** and **General Settings.** Set **TSTOP** to 2ms (to display a full cycle). Select voltage markers from the toolbar and place as shown. (This causes PSpice to automatically create the photos.) Run the simulation. When the simulation is complete, the waveforms of Figure 15–75 should appear.

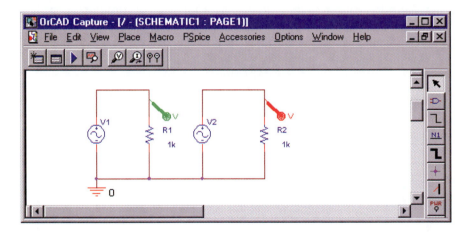

FIGURE 15–74   Studying phase relationships using OrCAD PSpice.

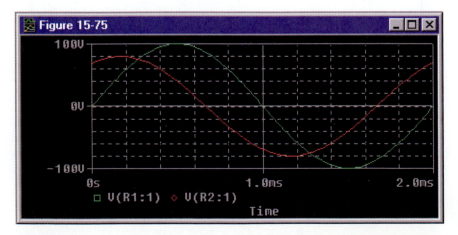

FIGURE 15–75

You can verify the angle between the waveforms using cursors. First, note that the period $T = 2$ ms $= 2000$ μs. (This corresponds to 360°.) Now using the cursors (see Appendix A if you need help), measure the time between crossover points as indicated in Figure 15–73. You should get 333 μs. This yields an angular displacement of

$$\theta = \frac{333 \ \mu s}{2000 \ \mu s} \times 360° = 60°$$

which agrees with the given sources.

15.1   Introduction

PROBLEMS

1. What do we mean by "ac voltage"? By "ac current"?

15.2   Generating AC Voltages

2. The waveform of Figure 15–8 is created by a 600-rpm generator. If the speed of the generator changes so that its cycle time is 50 ms, what is its new speed?

3. a. What do we mean by instantaneous value?

   b. For Figure 15–76, determine instantaneous voltages at $t = 0, 1, 2, 3, 4, 5, 6, 7,$ and 8 ms.

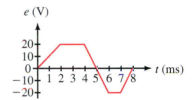

FIGURE 15–76

15.3   Voltage and Current Conventions for ac

4. For Figure 15–77, what is I when the switch is in position 1? When in position 2? Include sign.

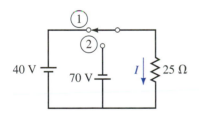

FIGURE 15–77

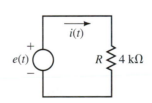

FIGURE 15–78

5. The source of Figure 15–78 has the waveform of Figure 15–76. Determine the current at $t = 0, 1, 2, 3, 4, 5, 6, 7,$ and 8 ms. Include sign.

15.4   Frequency, Period, Amplitude, and Peak Value

6. For each of the following, determine the period:

   a. $f = 100$ Hz   b. $f = 40$ kHz   c. $f = 200$ MHz

7. For each of the following, determine the frequency:

   a. $T = 0.5$ s   b. $T = 100$ ms   c. $5T = 80$ μs

8. For a triangular wave, $f = 1.25$ MHz. What is its period? How long does it take to go through 8×10^7 cycles?

9. Determine the period and frequency for the waveform of Figure 15–79.

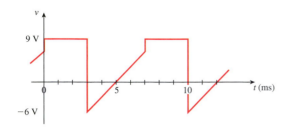

FIGURE 15–79

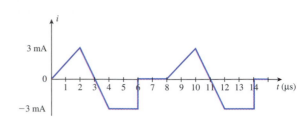

FIGURE 15–80

10. Determine the period and frequency for the waveform of Figure 15–80. How many cycles are shown?

11. What is the peak-to-peak voltage for Figure 15–79? What is the peak-to-peak current of Figure 15–80?

12. For a certain waveform, $625T = 12.5$ ms. What is the waveform's period and frequency?

13. A square wave with a frequency of 847 Hz goes through how many cycles in 2 minutes and 57 seconds?

14. For the waveform of Figure 15–81, determine

    a. period   b. frequency   c. peak-to-peak value

FIGURE 15–81

15. Two waveforms have periods of T_1 and T_2 respectively. If $T_1 = 0.25\ T_2$ and $f_1 = 10$ kHz, what are T_1, T_2, and f_2?

16. Two waveforms have frequencies f_1 and f_2 respectively. If $T_1 = 4\ T_2$ and waveform 1 is as shown in Figure 15–79, what is f_2?

15.5   Angular and Graphic Relationships for Sine Waves

17. Given voltage $v = V_m \sin \alpha$. If $V_m = 240$ V, what is v at $\alpha = 37°$?

18. For the sinusoidal waveform of Figure 15–82,

    a. Determine the equation for i.

    b. Determine current at all points marked.

19. A sinusoidal voltage has a value of 50 V at $\alpha = 150°$. What is V_m?

20. Convert the following angles from radians to degrees:

    a. $\pi/12$ b. $\pi/1.5$ c. $3\pi/2$

    d. 1.43   e. 17   f. 32π

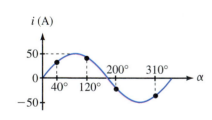

FIGURE 15–82

21. Convert the following angles from degrees to radians:

    a. 10°   b. 25°   c. 80°

    d. 150°   e. 350°   f. 620°

22. A 50-kHz sine wave has an amplitude of 150 V. Sketch the waveform with its axis scaled in microseconds.

23. If the period of the waveform in Figure 15–82 is 180 ms, compute current at $t = 30, 75, 140$, and 315 ms.

24. A sinusoidal waveform has a period of 60 μs and $V_m = 80$ V. Sketch the waveform. What is its voltage at 4 μs?

25. A 20-kHz sine wave has a value of 50 volts at $t = 5$ μs. Determine V_m and sketch the waveform.

26. For the waveform of Figure 15–83, determine v_2.

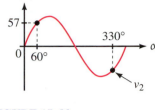

FIGURE 15–83

15.6   Voltages and Currents as Functions of Time

27. Calculate ω in radians per second for each of the following:
   a. $T = 100$ ns   b. $f = 30$ Hz   c. 100 cycles in 4 s
   d. period = 20 ms   e. 5 periods in 20 ms

28. For each of the following values of ω, compute f and T:
   a. 100 rad/s   b. 40 rad in 20 ms   c. 34×10^3 rad/s

29. Determine equations for sine waves with the following:
   a. $V_m = 170$ V, $f = 60$ Hz   b. $I_m = 40$ μA, $T = 10$ ms
   c. $T = 120$ μs, $v = 10$ V at $t = 12$ μs

30. Determine f, T, and amplitude for each of the following:
   a. $v = 75 \sin 200\pi t$ b. $i = 8 \sin 300t$

31. A sine wave has a peak-to-peak voltage of 40 V and $T = 50$ ms. Determine its equation.

32. Sketch the following waveforms with the horizontal axis scaled in degrees, radians, and seconds:
   a. $v = 100 \sin 200\pi t$ V
   b. $i = 90 \sin \omega t$ mA, $T = 80$ μs

33. Given $i = 47 \sin 8260t$ mA, determine current at $t = 0$ s, 80 μs, 410 μs, and 1200 μs.

34. Given $v = 100 \sin \alpha$. Sketch one cycle.
   a. Determine at which two angles $v = 86.6$ V.
   b. If $\omega = 100\pi/60$ rad/s, at which times do these occur?

35. Write equations for the waveforms of Figure 15–84. Express the phase angle in degrees.

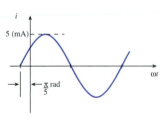

(a) $\omega = 1000$ rad/s

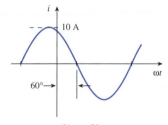

(b) $t = 50$ ms

(c) $f = 900$ Hz

FIGURE 15–84

36. Sketch the following waveforms with the horizontal axis scaled in degrees and seconds:
   a. $v = 100 \sin(232.7t + 40°)$ V
   b. $i = 20 \sin(\omega t - 60°)$ mA, $f = 200$ Hz

37. Given $v = 5 \sin(\omega t + 45°)$. If $\omega = 20\pi$ rad/s, what is v at $t = 20, 75$, and 90 ms?

38. Repeat Problem 35 for the waveforms of Figure 15–85.

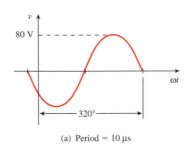

(a) Period = 10 μs

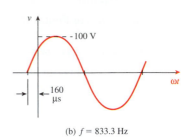

(b) f = 833.3 Hz

FIGURE 15–85

39. Determine the equation for the waveform shown in Figure 15–86.

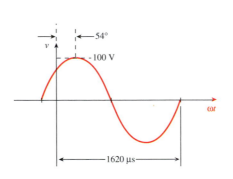

FIGURE 15–86

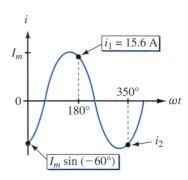

FIGURE 15–87

40. For the waveform of Figure 15–87, determine i_2.

41. Given $v = 30 \sin(\omega t - 45°)$ where $\omega = 40\pi$ rad/s. Sketch the waveform. At what time does v reach 0 V? At what time does it reach 23 V and -23 V?

15.7 Introduction to Phasors

42. For each of the phasors of Figure 15–88, determine the equation for $v(t)$ or $i(t)$ as applicable, and sketch the waveform.

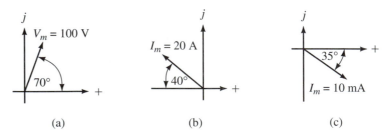

(a)   (b)   (c)

FIGURE 15–88

43. With the aid of phasors, sketch the waveforms for each of the following pairs and determine the phase difference and which waveform leads:

a. $v = 100 \sin \omega t$
   $i = 80 \sin(\omega t + 20°)$

b. $v_1 = 200 \sin(\omega t - 30°)$
   $v_2 = 150 \sin(\omega t - 30°)$

c. $i_1 = 40 \sin(\omega t + 30°)$
   $i_2 = 50 \sin(\omega t - 20°)$

d. $v = 100 \sin(\omega t + 140°)$
   $i = 80 \sin(\omega t - 160°)$

44. Repeat Problem 43 for the following.

    a. $i = 40 \sin(\omega t + 80°)$
        $v = -30 \sin(\omega t - 70°)$

    b. $v = 20 \cos(\omega t + 10°)$
        $i = 15 \sin(\omega t - 10°)$

    c. $v = 20 \cos(\omega t + 10°)$
        $i = 15 \sin(\omega t + 120°)$

    d. $v = 80 \cos(\omega t + 30°)$
        $i = 10 \cos(\omega t - 15°)$

45. For the waveforms in Figure 15–89, determine the phase differences. Which waveform leads?

46. Draw phasors for the waveforms of Figure 15–89.

15.8 AC Waveforms and Average Value

47. What is the average value of each of the following over an integral number of cycles?

    a. $i = 5 \sin \omega t$
        b. $i = 40 \cos \omega t$

    c. $v = 400 \sin(\omega t + 30°)$
        d. $v = 20 \cos 2\omega t$

48. Using Equation 15–20, compute the area under the half-cycle of Figure 15–54 using increments of $\pi/12$ rad.

49. Compute I_{avg} for the waveforms of Figure 15–90.

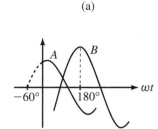

(a)

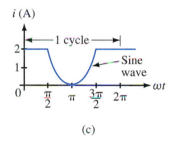

(b)

FIGURE 15–89

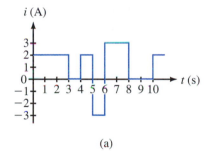

(a)

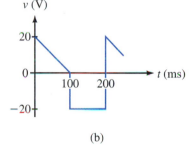

(b)

(c)

FIGURE 15–90

50. For the waveform of Figure 15–91, compute I_m.

51. For the circuit of Figure 15–92, $e = 25 \sin \omega t$ V and period $T = 120$ ms.

    a. Sketch voltage $v(t)$ with the axis scaled in milliseconds.

    b. Determine the peak and minimum voltages.

    c. Compute v at $t = 10, 20, 70,$ and 100 ms.

    d. Determine V_{avg}.

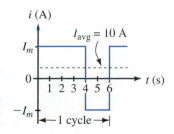

FIGURE 15–91

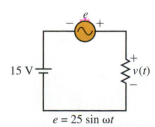

FIGURE 15–92

52. Using numerical methods for the curved part of the waveform (with increment size $\Delta t = 0.25$ s), determine the area and the average value for the waveform of Figure 15–93.

53. ▮ Using calculus, find the average value for Figure 15–93.

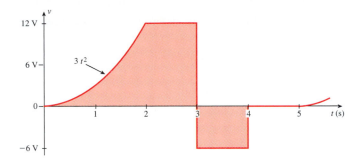

FIGURE 15–93

15.9   Effective Values

54. Determine the effective values of each of the following:

    a. $v = 100 \sin \omega t$ V   b. $i = 8 \sin 377t$ A

    c. $v = 40 \sin(\omega t + 40°)$ V   d. $i = 120 \cos \omega t$ mA

55. Determine the rms values of each for the following.

    a. A 12 V battery   b. $-24 \sin(\omega t + 73°)$ mA

    c. $10 + 24 \sin \omega t$ V   d. $45 - 27 \cos 2 \omega t$ V

56. For a sine wave, $V_{\text{eff}} = 9$ V. What is its amplitude?

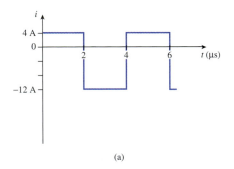

(a)

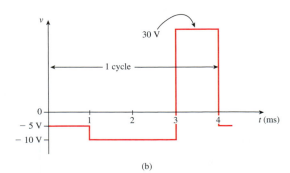

(b)

FIGURE 15–94

57. Determine the root mean square values for
   a. $i = 3 + \sqrt{2}(4) \sin(\omega t + 44°)$ mA
   b. Voltage v of Figure 15–92 with $e = 25 \sin \omega t$ V

58. Compute the rms values for Figures 15–90(a), and 15–91. For Figure 15–91, $I_m = 30$ A.

59. Compute the rms values for the waveforms of Figure 15–94.

60. Compute the effective value for Figure 15–95.

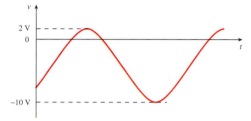

FIGURE 15–95

61. Determine the rms value of the waveform of Figure 15–96. Why is it the same as that of a 24-V battery?

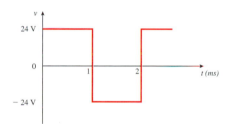

FIGURE 15–96

62. Compute the rms value of the waveform of Figure 15–52(c). To handle the triangular portion, use Equation 15–20. Use a time interval $\Delta t = 1$s.

63. [J] Repeat Problem 62, using calculus to handle the triangular portion.

15.11   AC Voltage and Current Measurement

64. Determine the reading of an average responding AC meter for each of the following cases. (Note: Meaningless is a valid answer if applicable.) Assume the frequency is within the range of the instrument.
   a. $v = 153 \sin \omega t$ V   b. $v = \sqrt{2}(120) \sin(\omega t + 30°)$ V
   c. The waveform of Figure 15–61   d. $v = 597 \cos \omega t$ V

65. Repeat Problem 64 using a true rms meter.

15.12 Circuit Analysis Using Computers

Use Electronics Workbench or PSpice for the following.

66. **EWB   PSpice**   Plot the waveform of Problem 37 and, using the cursor, determine voltage at the times indicated. Don't forget to convert the frequency to Hz.

67. **EWB**   **PSpice**   Plot the waveform of Problem 41. Using the cursor, determine the time at which v reaches 0 V. Don't forget to convert the frequency to Hz.

68. **EWB**   **PSpice**   Assume the equations of Problem 43 all represent voltages. For each case, plot the waveforms, then use the cursor to determine the phase difference between waveforms.

ANSWERS TO IN-PROCESS LEARNING CHECKS

In-Process Learning Check 1

1. 16.7 ms

2. Frequency doubles, period halves

3. 50 Hz; 20 ms

4. 20 V; 0.5 ms and 2.5 ms; -35 V: 4 ms and 5 ms

5. (c) and (d); Since current is directly proportional to the voltage, it will have the same waveshape.

6. 250 Hz

7. $f_1 = 100$ Hz; $f_2 = 33.3$ Hz

8. 50 kHz and 1 MHz

9. 22.5 Hz

10. At 12 ms, direction →; at 37 ms, direction ←; at 60 ms, →

11. At 75 ms, $i = -5$ A

In-Process Learning Check 2

1.

α (deg)	0	45	90	135	180	225	270	315	360
i (mA)	0	10.6	15	10.6	0	-10.6	-15	-10.6	0

2. a. 0.349   b. 0.873
   c. 2.09   d. 4.36

3. 30°; 120°; 450°

4. Same as Figure 15–27 with $T = 10$ s and amplitude = 50 mA.

5. 1.508×10^6 rad/s

6. $i = 8 \sin 157t$ A

7. $i = 6 \sin 69.81t$ A

8. a. $i = 250 \sin(251t - 30°)$ A
   b. $i = 20 \sin(62.8t + 45°)$ A
   c. $v = 40 \sin(628t - 30°)$ V
   d. $v = 80 \sin(314 \times 10^3 t + 36°)$ V

9. a. 2.95 ms; 7.05 ms; 22.95 ms; 27.05 ms
   b. 11.67 ms; 18.33 ms; 31.67 ms; 38.33 ms

10.

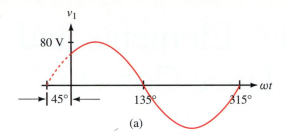

(a)

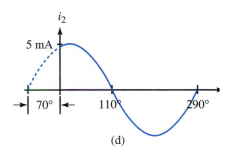

(b)

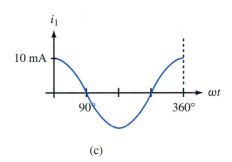

(c)

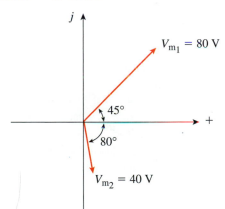

(d)

11. a. 2.25 s   b. 1.33 s

12. a.   b. 125°   c. v_1 leads

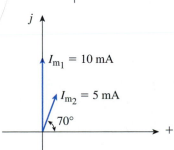

13. a.   b. 20°   c. i_1 leads

16

R, L, and *C* Elements and the Impedance Concept

OBJECTIVES

After studying this chapter, you will be able to

- express complex numbers in rectangular and polar forms,
- represent ac voltage and current phasors as complex numbers,
- represent voltage and current sources in transformed form,
- add and subtract currents and voltages using phasors,
- compute inductive and capacitive reactance,
- determine voltages and currents in simple ac circuits,
- explain the impedance concept,
- determine impedance for *R, L,* and *C* circuit elements,
- determine voltages and currents in simple ac circuits using the impedance concept,
- use Electronics Workbench and PSpice to solve simple ac circuit problems.

KEY TERMS

Capacitive Reactance
Complex Number
Impedance
Inductive Reactance
$j = \sqrt{-1}$
Phasor Domain
Polar Form
Rectangular Form
Time Domain

OUTLINE

Complex Number Review
Complex Numbers in AC Analysis
R, L, and *C* Circuits with Sinusoidal Excitation
Resistance and Sinusoidal AC
Inductance and Sinusoidal AC
Capacitance and Sinusoidal AC
The Impedance Concept
Computer Analysis of AC Circuits

In Chapter 15, you learned how to analyze a few simple ac circuits in the time domain using voltages and currents expressed as functions of time. However, this is not a very practical approach. A more practical approach is to represent ac voltages and currents as phasors, circuit elements as impedances, and analyze circuits in the phasor domain using complex algebra. With this approach, ac circuit analysis is handled much like dc circuit analysis, and all basic relationships and theorems—Ohm's law, Kirchhoff's laws, mesh and nodal analysis, superposition and so on—apply. The major difference is that ac quantities are complex rather than real as with dc. While this complicates computational details, it does not alter basic circuit principles. This is the approach used in practice. The basic ideas are developed in this chapter.

Since phasor analysis and the impedance concept require a familiarity with complex numbers, we begin with a short review.

Charles Proteus Steinmetz

CHARLES STEINMETZ WAS BORN IN Breslau, Germany in 1865 and emigrated to the United States in 1889. In 1892, he began working for the General Electric Company in Schenectady, New York, where he stayed until his death in 1923, and it was there that his work revolutionized ac circuit analysis. Prior to his time, this analysis had to be carried out using calculus, a difficult and time-consuming process. By 1893, however, Steinmetz had reduced the very complex alternating-current theory to, in his words, "a simple problem in algebra." The key concept in this simplification was the phasor—a representation based on complex numbers. By representing voltages and currents as phasors, Steinmetz was able to define a quantity called **impedance** and then use it to determine voltage and current magnitude and phase relationships in one algebraic operation.

Steinmetz wrote the seminal textbook on ac analysis based on his method, but at the time he introduced it he was practically the only person who understood it. Now, however, it is common knowledge and one of the basic tools of the electrical engineer and technologist. In this chapter, we learn the method and illustrate its application to the solution of basic ac circuit problems.

In addition to his work for GE, Charles Steinmetz was a professor of electrical engineering (1902–1913) and electrophysics (1913–1923) at Union University (now Union College) in Schenectady.

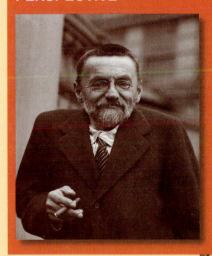

16.1 Complex Number Review

A **complex number** is a number of the form **C** = *a* + *jb,* where *a* and *b* are real numbers and $j = \sqrt{-1}$. The number *a* is called the **real** part of **C** and *b* is called its **imaginary** part. (In circuit theory, *j* is used to denote the imaginary component rather than *i* to avoid confusion with current *i*.)

Geometrical Representation

Complex numbers may be represented geometrically, either in rectangular form or in polar form as points on a two-dimensional plane called the **complex plane** (Figure 16–1). The complex number **C** = 6 + *j*8, for example, represents a point whose coordinate on the real axis is 6 and whose coordinate on the imaginary axis is 8. This form of representation is called the **rectangular form.**

Complex numbers may also be represented in **polar form** by magnitude and angle. Thus, **C** = 10∠53.13° (Figure 16–2) is a complex number with magnitude 10 and angle 53.13°. This magnitude and angle representation is just an alternate way of specifying the location of the point represented by **C** = *a* + *jb.*

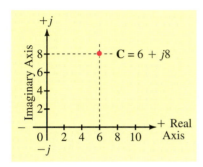

FIGURE 16–1 A complex number in rectangular form.

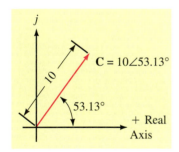

FIGURE 16–2 A complex number in polar form.

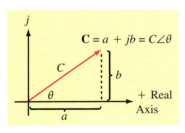

FIGURE 16–3 Polar and rectangular equivalence.

Conversion between Rectangular and Polar Forms

To convert between forms, note from Figure 16–3 that

$$\mathbf{C} = a + jb \quad \text{(rectangular form)} \tag{16–1}$$

$$\mathbf{C} = C\angle\theta \quad \text{(polar form)} \tag{16–2}$$

where *C* is the magnitude of **C.** From the geometry of the triangle,

$$a = C \cos\theta \tag{16–3a}$$

$$b = C \sin\theta \tag{16–3b}$$

where

$$C = \sqrt{a^2 + b^2} \tag{16–4a}$$

and

$$\theta = \tan^{-1}\frac{b}{a} \tag{16–4b}$$

Equations 16–3 and 16–4 permit conversion between forms. When using Equation 16–4b, however, be careful when the number to be converted is in the second or third quadrant, as the angle obtained is the supplementary angle rather than the actual angle in these two quadrants. This is illustrated in Example 16–1 for the complex number **W.**

EXAMPLE 16–1 Determine rectangular and polar forms for the complex numbers **C, D, V,** and **W** of Figure 16–4(a)

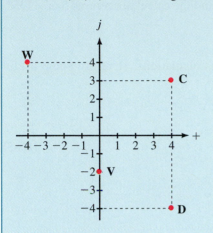

(a) Complex numbers

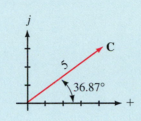

(b) In polar form, **C** = 5∠36.87°

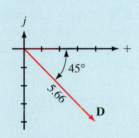

(c) In polar form, **D** = 5.66∠−45°

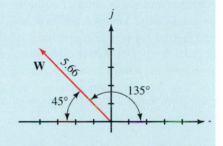

(d) In polar form, **W** = 5.66∠135°

FIGURE 16–4

Solution

Point C: Real part = 4; imaginary part = 3. Thus, **C** = 4 + j3. In polar form, $C = \sqrt{4^2 + 3^2} = 5$ and $\theta_C = \tan^{-1}(3/4) = 36.87°$. Thus, **C** = 5∠36.87° as indicated in (b).

Point D: In rectangular form, **D** = 4 − j4. Thus, $D = \sqrt{4^2 + 4^2} = 5.66$ and $\theta_D = \tan^{-1}(-4/4) = -45°$. Therefore, **D** = 5.66∠−45°, as shown in (c).

Point V: In rectangular form, **V** = −j2. In polar form, **V** = 2∠−90°.

Point W: In rectangular form, **W** = −4 + j4. Thus, $W = \sqrt{4^2 + 4^2} = 5.66$ and $\tan^{-1}(-4/4) = -45°$. Inspection of Figure 16–4(d) shows, however, that this 45° angle is the supplementary angle. The actual angle (measured from the positive horizontal axis) is 135°. Thus, **W** = 5.66∠135°.

In practice (because of the large amount of complex number work that you will do), a more efficient conversion process is needed than that described previously. As discussed later in this section, inexpensive calculators are available that perform such conversions directly—you simply enter the complex number components and press the conversion key. With these, the problem of determining angles for numbers such as **W** in Example 16–1 does not occur; you just enter $-4 + j4$ and the calculator returns $5.66\angle 135°$.

Powers of *j*

Powers of *j* are frequently required in calculations. Here are some useful powers:

$$j^2 = (\sqrt{-1})(\sqrt{-1}) = -1$$
$$j^3 = j^2 j = -j$$
$$j^4 = j^2 j^2 = (-1)(-1) = 1 \qquad \text{(16–5)}$$
$$(-j)j = 1$$
$$\frac{1}{j} = \frac{1}{j} \times \frac{j}{j} = \frac{j}{j^2} = -j$$

Addition and Subtraction of Complex Numbers

Addition and subtraction of complex numbers can be performed analytically or graphically. Analytic addition and subtraction is most easily illustrated in rectangular form, while graphical addition and subtraction is best illustrated in polar form. For analytic addition, add real and imaginary parts separately. Similarly for subtraction. For graphical addition, add vectorially as in Figure 16–5(a); for subtraction, change the sign of the subtrahend, then add, as in Figure 16–5(b).

EXAMPLE 16–2   Given **A** $= 2 + j1$ and **B** $= 1 + j3$. Determine their sum and difference analytically and graphically.

Solution

$$\mathbf{A} + \mathbf{B} = (2 + j1) + (1 + j3) = (2 + 1) + j(1 + 3) = 3 + j4.$$
$$\mathbf{A} - \mathbf{B} = (2 + j1) - (1 + j3) = (2 - 1) + j(1 - 3) = 1 - j2.$$

Graphical addition and subtraction are shown in Figure 16–5.

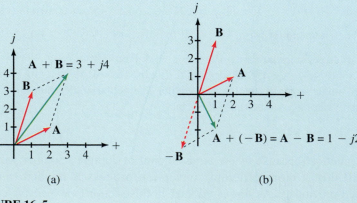

(a)   (b)

FIGURE 16–5

Multiplication and Division of Complex Numbers

These operations are usually performed in polar form. For multiplication, multiply magnitudes and add angles algebraically. For division, divide the magnitude of the denominator into the magnitude of the numerator, then subtract algebraically the angle of the denominator from that of the numerator. Thus, given $\mathbf{A} = A\angle\theta_A$ and $\mathbf{B} = B\angle\theta_B$,

$$\mathbf{A} \cdot \mathbf{B} = AB\underline{/\theta_A + \theta_B} \qquad (16\text{--}6)$$

$$\mathbf{A/B} = A/B\underline{/\theta_A - \theta_B} \qquad (16\text{--}7)$$

EXAMPLE 16–3 Given $\mathbf{A} = 3\angle35°$ and $\mathbf{B} = 2\angle-20°$, determine the product $\mathbf{A} \cdot \mathbf{B}$ and the quotient $\mathbf{A/B}$.

Solution

$$\mathbf{A} \cdot \mathbf{B} = (3\angle35°)(2\angle-20°) = (3)(2)\underline{/35° - 20°} = 6\angle15°$$

$$\frac{\mathbf{A}}{\mathbf{B}} = \frac{(3\angle35°)}{(2\angle-20°)} = \frac{3}{2}\underline{/35° - (-20°)} = 1.5\angle55°$$

EXAMPLE 16–4 For computations involving purely real, purely imaginary, or small integer numbers, it is sometimes easier to multiply directly in rectangular form than it is to convert to polar. Compute the following directly:

a. $(-j3)(2 + j4)$.

b. $(2 + j3)(1 + j5)$.

Solution

a. $(-j3)(2 + j4) = (-j3)(2) + (-j3)(j4) = -j6 - j^2 12 = 12 - j6$

b. $(2 + j3)(1 + j5) = (2)(1) + (2)(j5) + (j3)(1) + (j3)(j5)$
$$= 2 + j10 + j3 + j^2 15 = 2 + j13 - 15 = -13 + j13$$

1. Polar numbers with the same angle can be added or subtracted directly without conversion to rectangular form. For example, the sum of $6 \angle36.87°$ and $4 \angle36.87°$ is $10 \angle36.87°$, while the difference is $6 \angle36.87° - 4 \angle36.87° = 2 \angle36.87°$. By means of sketches, indicate why this procedure is valid.

2. To compare methods of multiplication with small integer values, convert the numbers of Example 16–4 to polar form, multiply them, then convert the answers back to rectangular form.

PRACTICE PROBLEMS 1

Answers:

1. Since the numbers have the same angle, their sum also has the same angle and thus, their magnitudes simply add (or subtract).

Reciprocals

The reciprocal of a complex number $\mathbf{C} = C\angle\theta$ is

$$\frac{1}{\mathbf{C}\angle\theta} = \frac{1}{C}\angle{-\theta} \qquad (16\text{--}8)$$

Thus,

$$\frac{1}{20\angle30°} = 0.05\angle{-30°}$$

Complex Conjugates

The **conjugate** of a complex number (denoted by an asterisk *) is a complex number with the same real part but the opposite imaginary part. Thus, the conjugate of $\mathbf{C} = C\angle\theta = a + jb$ is $\mathbf{C^*} = C\angle{-\theta} = a - jb$. For example, if $\mathbf{C} = 3 + j4 = 5\angle53.13°$, then $\mathbf{C^*} = 3 - j4 = 5\angle{-53.13°}$.

Calculators for AC Analysis

The analysis of ac circuits involves a considerable amount of complex number arithmetic; thus, you will need a calculator that can work easily with complex numbers. There are several inexpensive calculators on the market that are suitable for this purpose in that they can perform all required calculations (addition, subtraction, multiplication, and division) in either rectangular or polar form without the need for conversion. This is important, because it saves you a great deal of time and cuts down on errors. To illustrate, consider Example 16–5. Using a calculator with only basic complex number conversion capabilities requires that you convert between forms as illustrated. On the other hand, a calculator with more sophisticated complex number capabilities (such as that shown in Figure 16–6) allows you to perform the calculation without going through all the intermediate conversion steps. You need to acquire an appropriate calculator and learn to use it proficiently.

FIGURE 16–6   This calculator displays complex numbers in standard mathematical notation.

EXAMPLE 16–5   The following illustrates the type of calculations that you will encounter. Use your calculator to reduce the following to rectangular form:

$$(6 + j5) + \frac{(3 - j4)(10\angle40°)}{6 + 30\angle53.13°}$$

Solution   Using a calculator with basic capabilities requires a number of intermediate steps, some of which are shown below.

$$\text{answer} = (6 + j5) + \frac{(5\angle{-53.13})(10\angle40)}{6 + (18 + j24)}$$

$$= (6 + j5) + \frac{(5\angle{-53.13})(10\angle40)}{24 + j24}$$

$$= (6 + j5) + \frac{(5\angle-53.13)(10\angle40)}{33.94\angle45}$$

$$= (6 + j5) + 1.473\angle-58.13 = (6 + j5) + (0.778 - j1.251)$$

$$= 6.778 + j3.749$$

Using a calculator such as that shown in Figure 16–6 saves steps, as it multiplies $(3 - j4)(10\angle40°)$ and adds $6 + 30\angle53.13°$, etc., directly, without your having to convert forms.

16.2 Complex Numbers in AC Analysis

Representing AC Voltages and Currents by Complex Numbers

As you learned in Chapter 15, ac voltages and currents can be represented as phasors. Since phasors have magnitude and angle, they can be viewed as complex numbers. To get at the idea, consider the voltage source of Figure 16–7(a). Its phasor equivalent (b) has magnitude E_m and angle θ. It therefore can be viewed as the complex number

$$\mathbf{E} = E_m\angle\theta \qquad (16–9)$$

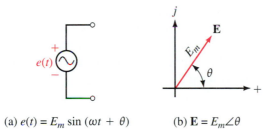

(a) $e(t) = E_m \sin(\omega t + \theta)$ (b) $\mathbf{E} = E_m\angle\theta$

FIGURE 16–7 Representation of a sinusoidal source voltage as a complex number.

From this point of view, the sinusoidal voltage $e(t) = 200 \sin(\omega t + 40°)$ of Figure 16–8(a) and (b) can be represented by its phasor equivalent, $\mathbf{E} = 200 \text{ V}\angle40°$, as in (c).

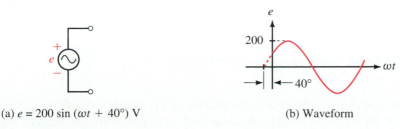

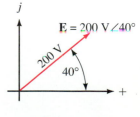

(a) $e = 200 \sin(\omega t + 40°)$ V   (b) Waveform   (c) Phasor equivalent

FIGURE 16–8 Transforming $e = 200 \sin(\omega t + 40°)$ V to $\mathbf{E} = 200 \text{ V}\angle40°$.

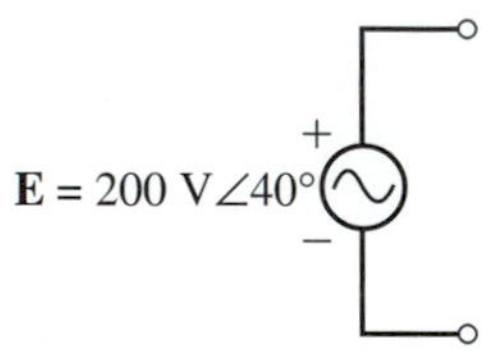

$\mathbf{E} = 200 \text{ V}\angle 40°$

Transformed source

FIGURE 16–9 Direct transformation of the source.

We can take advantage of this equivalence. *Rather than show a source as a time-varying voltage e(t) that we subsequently convert to a phasor, we can represent the source by its phasor equivalent right from the start.* This viewpoint is illustrated in Figure 16–9. Since $\mathbf{E} = 200 \text{ V}\angle 40°$, this representation retains all the original information of Figure 16–8 with the exception of the sinusoidal time variation. However, since the sinusoidal waveform and time variation is implicit in the definition of a phasor, you can easily restore this information if it is needed.

The idea illustrated in Figure 16–9 is of fundamental importance to circuit theory. *By replacing the time function e(t) with its phasor equivalent* $\mathbf{E}$, *we have transformed the source from the time domain to the phasor domain.* The value of this approach is illustrated next.

Before we move on, we should note that both Kirchhoff's voltage law and Kirchhoff's current law apply in the time domain (i.e., when voltages and currents are expressed as functions of time) and in the phasor domain (i.e., when voltages and currents are represented as phasors). For example, $e = v_1 + v_2$ in the time domain can be transformed to $\mathbf{E} = \mathbf{V}_1 + \mathbf{V}_2$ in the phasor domain and vice versa. Similarly for currents.

Summing AC Voltages and Currents

Sinusoidal quantites must sometimes be added or subtracted as in Figure 16–10. Here, we want the sum of e_1 and e_2, where $e_1 = 10 \sin \omega t$ and $e_2 = 15 \sin(\omega t + 60°)$. The sum of e_1 and e_2 can be found by adding waveforms point by point as in (b). For example, at $\omega t = 0°$, $e_1 = 10 \sin 0° = 0$ and $e_2 = 15 \sin(0° + 60°) = 13$ V, and their sum is 13 V. Similarly, at $\omega t = 90°$, $e_1 = 10 \sin 90° = 10$ V and $e_2 = 15 \sin(90° + 60°) = 15 \sin 150° = 7.5$, and their sum is 17.5 V. Continuing in this manner, the sum of $e_1 + e_2$ (the green waveform) is obtained.

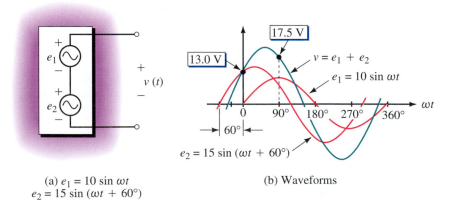

(a) $e_1 = 10 \sin \omega t$
$e_2 = 15 \sin(\omega t + 60°)$

(b) Waveforms

FIGURE 16–10 Summing waveforms point by point.

As you can see, the process is tedious and provides no analytic expression for the resulting voltage. A better way is to transform the sources and use complex numbers to perform the addition. This is shown in Figure 16–11.

Here, we have replaced voltages e_1 and e_2 with their phasor equivalents, $\mathbf{E}_1$ and $\mathbf{E}_2$, and v with its phasor equivalent, $\mathbf{V}$. Since $v = e_1 + e_2$, replacing v, e_1, and e_2 with their phasor equivalents yields $\mathbf{V} = \mathbf{E}_1 + \mathbf{E}_2$. Now $\mathbf{V}$ can be found by adding $\mathbf{E}_1$ and $\mathbf{E}_2$ as complex numbers. Once $\mathbf{V}$ is known, its corresponding time equation and companion waveform can be determined.

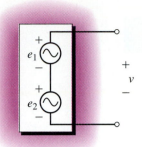

(a) Original network.
$v(t) = e_1(t) + e_2(t)$

EXAMPLE 16–6 Given $e_1 = 10 \sin \omega t$ V and $e_2 = 15 \sin(\omega t + 60°)$ V as before, determine v and sketch it.

Solution $e_1 = 10 \sin \omega t$ V. Thus, $\mathbf{E}_1 = 10 \text{ V}\angle 0°$.

$e_2 = 15 \sin(\omega t + 60°)$ V. Thus, $\mathbf{E}_2 = 15 \text{ V}\angle 60°$.

Transformed sources are shown in Figure 16–12(a) and phasors in (b).

$$\mathbf{V} = \mathbf{E}_1 + \mathbf{E}_2 = 10\angle 0° + 15\angle 60° = (10 + j0) + (7.5 + j13)$$
$$= (17.5 + j13) = 21.8 \text{ V}\angle 36.6°$$

Thus, $v = 21.8 \sin(\omega t + 36.6°)$ V

Waveforms are shown in (c).

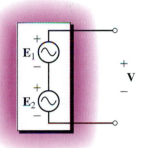

(b) Transformed network.
$\mathbf{V} = \mathbf{E}_1 + \mathbf{E}_2$

FIGURE 16–11 Transformed circuit. This is one of the key ideas of sinusoidal circuit analysis.

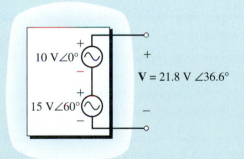

(a) Phasor summation

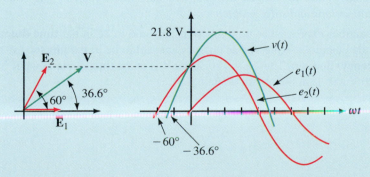

(b) Phasors   (c) Waveforms

FIGURE 16–12

PRACTICE
PROBLEMS 2

Verify by direct substitution that $v = 21.8 \sin(\omega t + 36.6°)$ V, as in Figure 16–12, is the sum of e_1 and e_2. To do this, compute e_1 and e_2 at a point, add them, then compare the sum to $21.8 \sin(\omega t + 36.6°)$ V computed at the same point. Perform this computation at $\omega t = 30°$ intervals over the complete cycle to satisfy yourself that the result is true everywhere. (For example, at $\omega t = 0°$, $v = 21.8 \sin(\omega t + 36.6°) = 21.8 \sin(36.6°) = 13$ V, as we saw earlier in Figure 16–10.)

Answer: Here are the points on the graph at 30° intervals:

ωt	0°	30°	60°	90°	120°	150°	180°	210°	240°	270°	300°	330°	360°
v	13	20	21.7	17.5	8.66	−2.5	−13	−20	−21.7	−17.5	−8.66	2.5	13

IMPORTANT NOTES...

1. To this point, we have used peak values such as V_m and I_m to represent the magnitudes of phasor voltages and currents, as this has been most convenient for our purposes. In practice, however, rms values are used instead. Accordingly, we will now change to rms. Thus, from here on, the phasor $\mathbf{V} = 120$ V $\angle 0°$ will be taken to mean a voltage of 120 volts rms at an angle of 0°. If you need to convert this to a time function, first multiply the rms value by $\sqrt{2}$, then follow the usual procedure. Thus, $v = \sqrt{2}\,(120) \sin \omega t = 170 \sin \omega t$.

2. To add or subtract sinusoidal voltages or currents, follow the three steps outlined in Example 16–6. That is,

   - convert sine waves to phasors and express them in complex number form,

   - add or subtract the complex numbers,

   - convert back to time functions if desired.

3. Although we use phasors to represent sinusoidal waveforms, it should be noted that sine waves and phasors are not the same thing. Sinusoidal voltages and currents are real—they are the actual quantities that you measure with meters and whose waveforms you see on oscilloscopes. *Phasors, on the other hand, are mathematical abstractions that we use to help visualize relationships and solve problems.*

4. Quantities expressed as time functions are said to be in the **time domain,** while quantities expressed as phasors are said to be in the **phasor (or frequency) domain.** Thus, $e = 170 \sin \omega t$ V is in the time domain, while $\mathbf{V} = 120$ V $\angle 0°$ is in the phasor domain.

EXAMPLE 16–7 Express the voltages and currents of Figure 16–13 in both the time and the phasor domains.

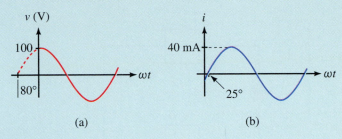

FIGURE 16–13

Solution

a. Time domain: $v = 100 \sin(\omega t + 80°)$ volts.
    Phasor domain: $\mathbf{V} = (0.707)(100 \text{ V} \angle 80°) = 70.7 \text{ V} \angle 80°$.

b. Time domain: $i = 40 \sin(\omega t - 25°)$ mA.
    Phasor domain: $\mathbf{I} = (0.707)(40 \text{ mA} \angle -25°) = 28.3 \text{ mA} \angle -25°$.

EXAMPLE 16–8 If $i_1 = 14.14 \sin(\omega t - 55°)$ A and $i_2 = 4 \sin(\omega t + 15°)$ A, determine their sum, i. Work with rms values.

Solution

$$\mathbf{I}_1 = (0.707)(14.14 \text{ A})\angle -55° = 10 \text{ A} \angle -55°$$

$$\mathbf{I}_2 = (0.707)(4 \text{ A})\angle 15° = 2.828 \text{ A} \angle 15°$$

$$\mathbf{I} = \mathbf{I}_1 + \mathbf{I}_2 = 10 \text{ A} \angle -55° + 2.828 \text{ A} \angle 15°$$

$$= (5.74 \text{ A} - j8.19 \text{ A}) + (2.73 \text{ A} + j0.732 \text{ A})$$

$$= 8.47 \text{ A} - j7.46 \text{ A} = 11.3 \text{ A} \angle -41.4$$

$$i(t) = \sqrt{2}(11.3) \sin(\omega t - 41.4°) = 16 \sin(\omega t - 41.4°) \text{ A}$$

While it may seem silly to convert peak values to rms and then convert rms back to peak as we did here, we did it for a reason. The reason is that very soon, we will stop working in the time domain entirely and work only with phasors. At that point, the solution will be complete when we have the answer in the form $\mathbf{I} = 11.3 \angle -41.4°$. (To help focus on rms, voltages and currents in the next two examples (and in other examples to come) are expressed as an rms value times $\sqrt{2}$.)

EXAMPLE 16–9   For Figure 16–14, $v_1 = \sqrt{2}(16) \sin \omega t$ V, $v_2 = \sqrt{2}(24) \sin(\omega t + 90°)$ and $v_3 = \sqrt{2}(15) \sin(\omega t - 90°)$ V. Determine source voltage e.

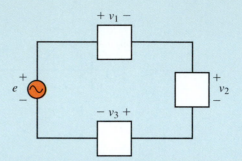

FIGURE 16–14

Solution   The answer can be obtained by KVL. First, convert to phasors. Thus, $\mathbf{V}_1 = 16$ V∠0°, $\mathbf{V}_2 = 24$ V∠90°, and $\mathbf{V}_3 = 15$ V∠−90°. KVL yields $\mathbf{E} = \mathbf{V}_1 + \mathbf{V}_2 + \mathbf{V}_3 = 16$ V∠0° + 24 V∠90° + 15 V∠−90° = 18.4 V∠29.4°. Converting back to a function of time yields $e = \sqrt{2}(18.4) \sin(\omega t + 29.4°)$ V.

EXAMPLE 16–10   For Figure 16–15, $i_1 = \sqrt{2}(23) \sin \omega t$ mA, $i_2 = \sqrt{2}(0.29) \sin(\omega t + 63°)$ A and $i_3 = \sqrt{2}(127) \times 10^{-3} \sin(\omega t - 72°)$ A. Determine current i_T.

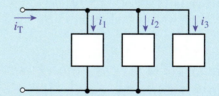

FIGURE 16–15

Solution   Convert to phasors. Thus, $\mathbf{I}_1 = 23$ mA∠0°, $\mathbf{I}_2 = 0.29$ A∠63°, and $\mathbf{I}_3 = 127 \times 10^{-3}$ A∠−72°. KCL yields $\mathbf{I}_T = \mathbf{I}_1 + \mathbf{I}_2 + \mathbf{I}_3 = 23$ mA∠0° + 290 mA∠63° + 127 mA∠−72° = 238 mA∠35.4°. Converting back to a function of time yields $i_T = \sqrt{2}(238) \sin(\omega t + 35.4°)$ mA.

PRACTICE PROBLEMS 3

1. Convert the following to time functions. Values are rms.
   a. $\mathbf{E} = 500$ mV∠−20°   b. $\mathbf{I} = 80$ A∠40°

2. For the circuit of Figure 16–16, determine voltage e_1.

FIGURE 16–16 $e_2 = 141.4 \sin(\omega t + 30°)$ V

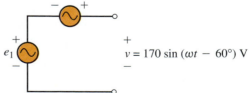

$v = 170 \sin(\omega t - 60°)$ V

Answers:
1. a. $e = 707 \sin(\omega t - 20°)$ mV   b. $i = 113 \sin(\omega t + 40°)$ A
2. $e_1 = 221 \sin(\omega t - 99.8°)$ V

1. Convert the following to polar form:

    a. $j6$ b. $-j4$ c. $3 + j3$ d. $4 - j6$

    e. $-5 + j8$ f. $1 - j2$ g. $-2 - j3$

2. Convert the following to rectangular form:

    a. $4\angle90°$ b. $3\angle0°$ c. $2\angle-90°$ d. $5\angle40°$

    e. $6\angle120°$ f. $2.5\angle-20°$ g. $1.75\angle-160°$

3. If $-\mathbf{C} = 12\angle-140°$, what is $\mathbf{C}$?

4. Given: $\mathbf{C}_1 = 36 + j4$ and $\mathbf{C}_2 = 52 - j11$. Determine $\mathbf{C}_1 + \mathbf{C}_2$ and $\mathbf{C}_1 - \mathbf{C}_2$. Express in rectangular form.

5. Given: $\mathbf{C}_1 = 24\angle25°$ and $\mathbf{C}_2 = 12\angle-125°$. Determine $\mathbf{C}_1 \cdot \mathbf{C}_2$ and $\mathbf{C}_1/\mathbf{C}_2$.

6. Compute the following and express answers in rectangular form:

    a. $\dfrac{6 + j4}{10\angle20°} + (14 + j2)$ b. $(1 + j6) + \left[2 + \dfrac{(12\angle0°)(14 + j2)}{6 - (10\angle20°)(2\angle-10°)}\right]$

7. For Figure 16–17, determine i_T where $i_1 = 10 \sin \omega t$, $i_2 = 20 \sin(\omega t - 90°)$, and $i_3 = 5 \sin(\omega t + 90°)$.

FIGURE 16–17 $i_T = i_1 + i_2 + i_3$

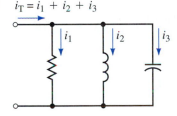

(Answers are at the end of the chapter.)

16.3   *R, L,* and *C* Circuits with Sinusoidal Excitation

R, L, and *C* circuit elements each have quite different electrical properties. Resistance, for example, opposes current, while inductance opposes changes in current, and capacitance opposes changes in voltage. These differences result in quite different voltage-current relationships as you saw earlier. We now investigate these relationships for the case of sinusoidal ac. Sine waves have several important characteristics that you will discover from this investigation:

1. When a circuit consisting of linear circuit elements *R, L,* and *C* is connected to a sinusoidal source, all currents and voltages in the circuit will be sinusoidal.

2. These sine waves have the same frequency as the source and differ from it only in terms of their magnitudes and phase angles.

16.4   **Resistance and Sinusoidal AC**

We begin with a purely resistive circuit. Here, Ohm's law applies and thus, current is directly proportional to voltage. Current variations therefore follow voltage variations, reaching their peak when voltage reaches its peak, changing direction when voltage changes polarity, and so on (Figure 16–18). From this, we conclude that *for a purely resistive circuit, current and voltage are in phase.* Since voltage and current waveforms coincide, their phasors also coincide (Figure 16–19).

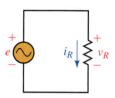

(a) Source voltage is a sine wave. Therefore, v_R is a sine wave.

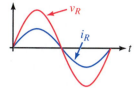

(b) $i_R = v_R/R$. Therefore i_R is a sine wave.

FIGURE 16–18   Ohm's law applies to resistors. Note that current and voltage are in phase.

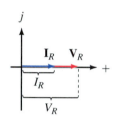

FIGURE 16–19   For a resistor, voltage and current phasors are in phase.

The relationship illustrated in Figure 16–18 may be stated mathematically as

$$i_R = \frac{v_R}{R} = \frac{V_m \sin \omega t}{R} = \frac{V_m}{R} \sin \omega t = I_m \sin \omega t \qquad (16\text{–}10)$$

where

$$I_m = V_m/R \qquad (16\text{–}11)$$

Transposing,

$$V_m = I_m R \qquad (16\text{–}12)$$

The in-phase relationship is true regardless of reference. Thus, if $v_R = V_m \sin(\omega t + \theta)$, then $i_R = I_m \sin(\omega t + \theta)$.

EXAMPLE 16–11   For the circuit of Figure 16–18(a), if $R = 5 \; \Omega$ and $i_R = 12 \sin(\omega t - 18)$ A, determine v_R.

Solution $v_R = Ri_R = 5 \times 12 \sin(\omega t - 18°) = 60 \sin(\omega t - 18°)$ V. The waveforms are shown in Figure 16–20.

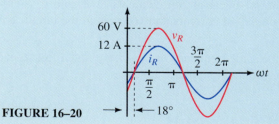

FIGURE 16–20

1. If $v_R = 150 \cos \omega t$ V and $R = 25$ kΩ, determine i_R and sketch both wave-forms.

2. If $v_R = 100 \sin(\omega t + 30°)$ V and $R = 0.2$ MΩ, determine i_R and sketch both waveforms.

Answers:

1. $i_R = 6 \cos \omega t$ mA. v_R and i_R are in phase.

2. $i_R = 0.5 \sin(\omega t + 30°)$ mA. v_R and i_R are in phase.

16.5   Inductance and Sinusoidal AC

Phase Lag in an Inductive Circuit

As you saw in Chapter 13, for an ideal inductor, voltage v_L is proportional to the rate of change of current. Because of this, voltage and current are not in phase as they are for a resistive circuit. This can be shown with a bit of calculus. From Figure 16–21, $v_L = L di_L/dt$. For a sine wave of current, you get when you differentiate

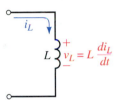

FIGURE 16–21   Voltage v_L is proportional to the rate of change of current i_L.

$$v_L = L\frac{di_L}{dt} = L\frac{d}{dt}(I_m \sin \omega t) = \omega L I_m \cos \omega t = V_m \cos \omega t$$

Utilizing the trigonometric identity $\cos \omega t = \sin(\omega t + 90°)$, you can write this as

$$v_L = V_m \sin(\omega t + 90°) \qquad \text{(16–13)}$$

where

$$V_m = \omega L I_m \qquad \text{(16–14)}$$

Voltage and current waveforms are shown in Figure 16–22, and phasors in Figure 16–23. As you can see, *for a purely inductive circuit, current lags voltage by 90°* (i.e., ¼ cycle). Alternatively you can say that voltage leads current by 90°.

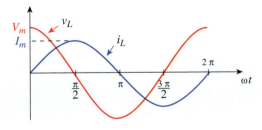

FIGURE 16–22   For inductance, current lags voltage by 90°. Here i_L is reference.

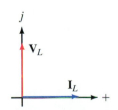

FIGURE 16–23   Phasors for the waveforms of Fig. 16–22 showing the 90° lag of current.

Although we have shown that current lags voltage by 90° for the case of Figure 16–22, this relationship is true in general, that is, current always lags voltage by 90° regardless of the choice of reference. This is illustrated in Figure 16–24. Here, $\mathbf{V}_L$ is at 0° and $\mathbf{I}_L$ at −90°. Thus, voltage v_L will be a sine wave and current i_L a negative cosine wave, i.e., $i_L = -I_m \cos \omega t$. Since

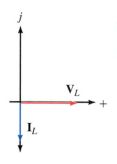

(a) Current $\mathbf{I}_L$ always lags
voltage $\mathbf{V}_L$ by 90°

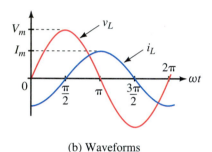

(b) Waveforms

FIGURE 16–24 Phasors and waveforms when $\mathbf{V}_L$ is used as reference.

i_L is a negative cosine wave, it can also be expressed as $i_L = I_m \sin(\omega t - 90°)$. The waveforms are shown in (b).

Since current always lags voltage by 90° for a pure inductance, you can, if you know the phase of the voltage, determine the phase of the current, and vice versa. Thus, if v_L is known, i_L must lag it by 90°, while if i_L is known, v_L must lead it by 90°.

Inductive Reactance

From Equation 16–14, we see that the ratio V_m to I_m is

$$\frac{V_m}{I_m} = \omega L \qquad (16\text{–}15)$$

This ratio is defined as **inductive reactance** and is given the symbol X_L. Since the ratio of volts to amps is ohms, reactance has units of ohms. Thus,

$$X_L = \frac{V_m}{I_m} \quad (\Omega) \qquad (16\text{–}16)$$

Combining Equations 16–15 and 16–16 yields

$$X_L = \omega L \quad (\Omega) \qquad (16\text{–}17)$$

where ω is in radians per second and L is in henries. *Reactance X_L represents the opposition that inductance presents to current for the sinusoidal ac case.*

We now have everything that we need to solve simple inductive circuits with sinusoidal excitation, that is, we know that current lags voltage by 90° and that their amplitudes are related by

$$I_m = \frac{V_m}{X_L} \qquad (16\text{–}18)$$

and

$$V_m = I_m X_L \qquad (16\text{–}19)$$

EXAMPLE 16–12   The voltage across a 0.2-H inductance is $v_L = 100 \sin (400t + 70°)$ V. Determine i_L and sketch it.

Solution $\omega = 400$ rad/s. Therefore, $X_L = \omega L = (400)(0.2) = 80 \ \Omega$.

$$I_m = \frac{V_m}{X_L} = \frac{100 \text{ V}}{80 \ \Omega} = 1.25 \text{ A}$$

The current lags the voltage by 90°. Therefore $i_L = 1.25 \sin(400t - 20°)$ A as indicated in Figure 16–25.

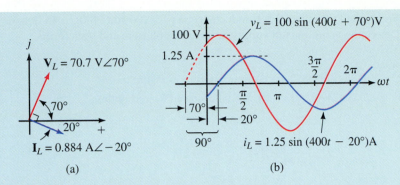

(a)   (b)

FIGURE 16–25   With voltage $\mathbf{V}_L$ at 70°, current $\mathbf{I}_L$ will be 90° later at −20°.

NOTE...

Remember to show phasors as rms values from now on.

EXAMPLE 16–13   The current through a 0.01-H inductance is $i_L = 20 \sin(\omega t - 50°)$ A and $f = 60$ Hz. Determine v_L.

Solution

$$\omega = 2\pi f = 2\pi(60) = 377 \text{ rad/s}$$
$$X_L = \omega L = (377)(0.01) = 3.77 \ \Omega$$
$$V_m = I_m X_L = (20 \text{ A})(3.77 \ \Omega) = 75.4 \text{ V}$$

Voltage leads current by 90°. Thus, $v_L = 75.4 \sin(377t + 40°)$ V as shown in Figure 16–26.

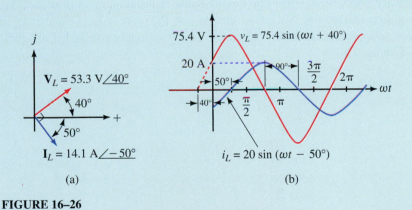

(a)   (b)

FIGURE 16–26

PRACTICE PROBLEMS 5

1. Two inductances are connected in series (Figure 16–27). If $e = 100 \sin \omega t$ and $f = 10$ kHz, determine the current. Sketch voltage and current waveforms.

FIGURE 16–27

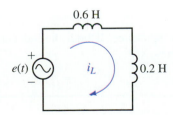

2. The current through a 0.5-H inductance is $i_L = 100 \sin(2400t + 45°)$ mA. Determine v_L and sketch voltage and current phasors and waveforms.

Answers:

1. $i_L = 1.99 \sin(\omega t - 90°)$ mA. Waveforms same as Figure 16–24.

2. $v_L = 120 \sin(2400t + 135°$ V). See following art for waveforms.

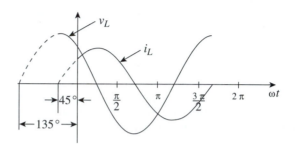

Variation of Inductive Reactance with Frequency

Since $X_L = \omega L = 2\pi f L$, inductive reactance is directly proportional to frequency (Figure 16–28). Thus, if frequency is doubled, reactance doubles, while if frequency is halved, reactance halves, and so on. In addition, X_L is directly proportional to inductance. Thus, if inductance is doubled, X_L is doubled, and so on. Note also that at $f = 0$, $X_L = 0$ Ω. This means that inductance looks like a short circuit to dc. (We already concluded this earlier in Chapter 13.)

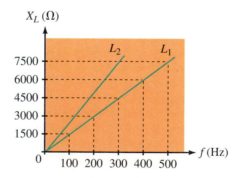

FIGURE 16–28   Variation of X_L with frequency. Note that $L_2 > L_1$.

PRACTICE PROBLEMS 6

A circuit has 50 ohms inductive reactance. If both the inductance and the frequency are doubled, what is the new X_L?

Answer: 200 Ω

16.6   Capacitance and Sinusoidal AC

Phase Lead in a Capacitive Circuit

For capacitance, current is proportional to the rate of change of voltage, i.e., $i_C = C \, dv_C/dt$ [Figure 16–29(a)]. Thus if v_C is a sine wave, you get upon substitution

$$i_C = C\frac{dv_C}{dt} = C\frac{d}{dt}(V_m \sin \omega t) = \omega C V_m \cos \omega t = I_m \cos \omega t$$

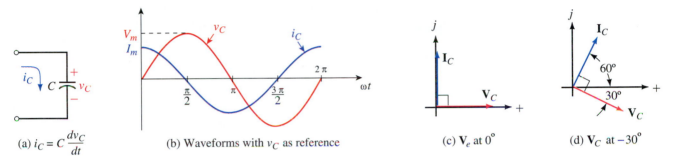

(a) $i_C = C\dfrac{dv_C}{dt}$

(b) Waveforms with v_C as reference

(c) $\mathbf{V}_e$ at $0°$

(d) $\mathbf{V}_C$ at $-30°$

FIGURE 16–29   For capacitance, current always leads voltage by 90°.

Using the appropriate trigonometric identity, this can be written as

$$i_C = I_m \sin(\omega t + 90°) \qquad\qquad \textbf{(16–20)}$$

where

$$I_m = \omega C V_m \qquad\qquad \textbf{(16–21)}$$

Waveforms are shown in Figure 16–29(b) and phasors in (c). As indicated, *for a purely capacitive circuit, current leads voltage by 90°,* or alternatively, voltage lags current by 90°. This relationship is true regardless of reference. Thus, if the voltage is known, the current must lead by 90° while if the current is known, the voltage must lag by 90°. For example, if $\mathbf{I}_C$ is at 60° as in (d), $\mathbf{V}_C$ must be at $-30°$.

1. The current source of Figure 16–30(a) is a sine wave. Sketch phasors and capacitor voltage v_C.

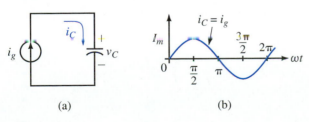

(a)

(b)

FIGURE 16–30

2. Refer to the circuit of Figure 16–31(a):

   a. Sketch the phasors.

   b. Sketch capacitor current i_C.

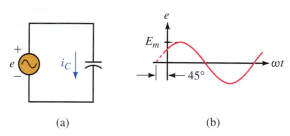

(a)   (b)

FIGURE 16–31

Answers:

1. $\mathbf{I}_C$ is at 0°; $\mathbf{V}_C$ is at −90°; v_C is a negative cosine wave.

2. a. $\mathbf{V}_C$ is at 45° and $\mathbf{I}_C$ is at 135°.

    b. Waveforms are the same as for Problem 2, Practice Problem 5, except that voltage and current waveforms are interchanged.

Capacitive Reactance

Now consider the relationship between maximum capacitor voltage and current magnitudes. As we saw in Equation 16–21, they are related by $I_m = \omega C V_m$. Rearranging, we get $V_m/I_m = 1/\omega C$. The ratio of V_m to I_m is defined as **capacitive reactance** and is given the symbol X_C. That is,

$$X_C = \frac{V_m}{I_m} \quad (\Omega)$$

Since $V_m/I_m = 1/\omega C$, we also get

$$X_C = \frac{1}{\omega C} \quad (\Omega) \tag{16–22}$$

where ω is in radians per second and C is in farads. *Reactance X_C represents the opposition that capacitance presents to current for the sinusoidal ac case.*

We now have everything that we need to solve simple capacitive circuits with sinusoidal excitation, i.e., we know that current leads voltage by 90° and that

$$I_m = \frac{V_m}{X_C} \tag{16–23}$$

and

$$V_m = I_m X_C \tag{16–24}$$

EXAMPLE 16–14   The voltage across a 10-μF capacitance is $v_C = 100$ $\sin(\omega t - 40°)$ V and $f = 1000$ Hz. Determine i_C and sketch its waveform.

Solution

$$\omega = 2\pi f = 2\pi(1000 \text{ Hz}) = 6283 \text{ rad/s}$$

$$X_C = \frac{1}{\omega C} = \frac{1}{(6283)(10 \times 10^{-6})} = 15.92 \text{ }\Omega$$

$$I_m = \frac{V_m}{X_C} = \frac{100 \text{ V}}{15.92 \text{ }\Omega} = 6.28 \text{ A}$$

Since current leads voltage by 90°, $i_C = 6.28 \sin(6283t + 50°)$ A as indicated in Figure 16–32.

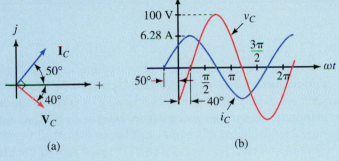

(a)   (b)

FIGURE 16–32   Phasors are not to scale with waveform.

EXAMPLE 16–15   The current through a 0.1-μF capacitance is $i_C = 5$ $\sin(1000t + 120°)$ mA. Determine v_C.

Solution

$$X_C = \frac{1}{\omega C} = \frac{1}{(1000 \text{ rad/s})(0.1 \times 10^{-6} \text{ } F)} = 10 \text{ k}\Omega$$

Thus, $V_m = I_m X_C = (5 \text{ mA})(10 \text{ k}\Omega) = 50$ V. Since voltage lags current by 90°, $v_C = 50 \sin(1000t + 30°)$ V. Waveforms and phasors are shown in Figure 16–33.

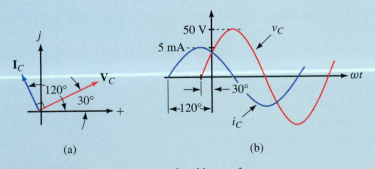

(a)   (b)

FIGURE 16–33   Phasors are not to scale with waveform.

Two capacitances are connected in parallel (Figure 16–34). If $e = 100 \sin \omega t$ V and $f = 10$ Hz, determine the source current. Sketch current and voltage phasors and waveforms.

FIGURE 16–34

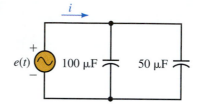

Answer: $i = 0.942 \sin(62.8t + 90°) = 0.942 \cos 62.8t$ A
See Figure 16–29(b) and (c).

Variation of Capacitive Reactance with Frequency

Since $X_C = 1/\omega C = 1/2\pi fC$, the opposition that capacitance presents varies inversely with frequency. This means that the higher the frequency, the lower the reactance, and vice versa (Figure 16–35). At $f = 0$ (i.e., dc), capacitive reactance is infinite. This means that a capacitance looks like an open circuit to dc. (We already concluded this earlier in Chapter 10.) Note that X_C is also inversely proportional to capacitance. Thus, if capacitance is doubled, X_C is halved, and so on.

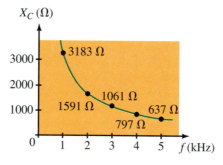

FIGURE 16–35 X_C varies inversely with frequency. Values shown are for $C = 0.05$ μF.

1. For a pure resistance, $v_R = 100 \sin(\omega t + 30°)$ V. If $R = 2$ Ω, what is the expression for i_R?

2. For a pure inductance, $v_L = 100 \sin(\omega t + 30°)$ V. If $X_L = 2$ Ω, what is the expression for i_L?

3. For a pure capacitance, $v_C = 100 \sin(\omega t + 30°)$ V. If $X_C = 2$ Ω, what is the expression for i_C?

4. If $f = 100$ Hz and $X_L = 400$ Ω, what is L?

5. If $f = 100$ Hz and $X_C = 400$ Ω, what is C?

6. For each of the phasor sets of Figure 16–36, identify whether the circuit is resistive, inductive, or capacitive. Justify your answers.

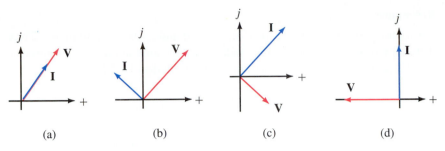

FIGURE 16–36

(Answers are at the end of the chapter.)

16.7 The Impedance Concept

In Sections 16.5 and 16.6, we handled magnitude and phase analysis separately. However, this is not the way it is done in practice. In practice, we represent circuit elements by their impedance, and determine magnitude and phase relationships in one step. Before we do this, however, we need to learn how to represent circuit elements as impedances.

Impedance

The opposition that a circuit element presents to current in the phasor domain is defined as its **impedance.** The impedance of the element of Figure 16–37, for example, is the ratio of its voltage phasor to its current phasor. Impedance is denoted by the boldface, uppercase letter **Z.** Thus,

$$\mathbf{Z} = \frac{\mathbf{V}}{\mathbf{I}} \quad (\text{ohms}) \tag{16–25}$$

(This equation is sometimes referred to as Ohm's law for ac circuits.)

Since phasor voltages and currents are complex, **Z** is also complex. That is,

$$\mathbf{Z} = \frac{\mathbf{V}}{\mathbf{I}} = \frac{V}{I}\angle\theta \tag{16–26}$$

where V and I are the rms magnitudes of **V** and **I** respectively, and θ is the angle between them. From Equation 16–26,

$$\mathbf{Z} = Z\angle\theta \tag{16–27}$$

where $Z = V/I$. Since $V = 0.707V_m$ and $I = 0.707I_m$, Z can also be expressed as V_m/I_m. Once the impedance of a circuit is known, the current and voltage can be determined using

$$\mathbf{I} = \frac{\mathbf{V}}{\mathbf{Z}} \tag{16–28}$$

and

$$\mathbf{V} = \mathbf{IZ} \tag{16–29}$$

Let us now determine impedance for the basic circuit elements R, L, and C.

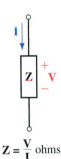

$$Z = \frac{V}{I} \text{ ohms}$$

FIGURE 16–37 Impedance concept.

NOTES...

Although **Z** is a complex number, it is not a phasor since it does not represent a sinusoidally varying quantity.

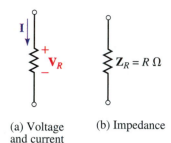

(a) Voltage
and current

(b) Impedance

FIGURE 16–38 Impedance of a pure resistance.

Resistance

For a pure resistance (Figure 16–38), voltage and current are in phase. Thus, if voltage has an angle θ, current will have the same angle. For example, if $\mathbf{V}_R = V_R\angle\theta$, then $\mathbf{I} = I\angle\theta$. Substituting into Equation 16–25 yields:

$$\mathbf{Z}_R = \frac{\mathbf{V}_R}{\mathbf{I}} = \frac{V_R\angle\theta}{I\angle\theta} = \frac{V_R}{I}\angle 0° = R$$

Thus the impedance of a resistor is just its resistance. That is,

$$\mathbf{Z}_R = R \tag{16–30}$$

This agrees with what we know about resistive circuits, i.e., that the ratio of voltage to current is R, and that the angle between them is $0°$.

Inductance

For a pure inductance, current lags voltage by $90°$. Assuming a $0°$ angle for voltage (we can assume any reference we want because we are interested only in the angle between $\mathbf{V}_L$ and $\mathbf{I}$), we can write $\mathbf{V}_L = V_L\angle 0°$ and $\mathbf{I} = I\angle -90°$. The impedance of a pure inductance (Figure 16–39) is therefore

$$\mathbf{Z}_L = \frac{\mathbf{V}_L}{\mathbf{I}} = \frac{V_L\angle 0°}{I\angle -90°} = \frac{V_L}{I}\angle 90° = \omega L\angle 90° = j\omega L$$

where we have used the fact that $V_L/I_L = \omega L$. Thus,

$$\mathbf{Z}_L = j\omega L = jX_L \tag{16–31}$$

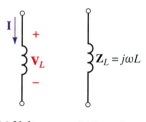

(a) Voltage
and current

(b) Impedance

FIGURE 16–39 Impedance of a pure inductance.

since ωL is equal to X_L.

EXAMPLE 16–16 Consider again Example 16–12. Given $v_L = 100 \sin(400t + 70°)$ and $L = 0.2$ H, determine i_L using the impedance concept.

Solution See Figure 16–40.

$$\mathbf{V}_L = 70.7 \text{ V}\angle 70° \quad \text{and} \quad \omega = 400 \text{ rad/s}$$
$$\mathbf{Z}_L = j\omega L = j(400)(0.2) = j80 \ \Omega$$
$$\mathbf{I}_L = \frac{\mathbf{V}_L}{\mathbf{Z}_L} = \frac{70.7\angle 70°}{j80} = \frac{70.7\angle 70°}{80\angle 90°} = 0.884 \text{ A}\angle -20°$$

In the time domain, $i_L = \sqrt{2}(0.884) \sin(400t - 20°) = 1.25 \sin(400t - 20°)$ A, which agrees with our previous solution.

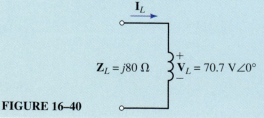

FIGURE 16–40

Capacitance

For a pure capacitance, current leads voltage by 90°. Its impedance (Figure 16–41) is therefore

$$\mathbf{Z}_C = \frac{\mathbf{V}_C}{\mathbf{I}} = \frac{V_C \angle 0°}{I \angle 90°} = \frac{V_C}{I} \angle -90° = \frac{1}{\omega C} \angle -90° = -j\frac{1}{\omega C} \quad \text{(ohms)}$$

Thus,

$$\mathbf{Z}_C = -j\frac{1}{\omega C} = -jX_C \quad \text{(ohms)} \tag{16–32}$$

since $1/\omega C$ is equal to X_C.

FIGURE 16–41   Impedance of a pure capacitance.

EXAMPLE 16–17   Given $v_C = 100 \sin(\omega t - 40°)$, $f = 1000$ Hz, and $C = 10\,\mu F$, determine i_C in Figure 16–42.

Solution

$$\omega = 2\pi f = 2\pi(1000\ \text{Hz}) = 6283\ \text{rads/s}$$

$$\mathbf{V}_C = 70.7\ \text{V} \angle -40°$$

$$\mathbf{Z}_C = -j\frac{1}{\omega C} = -j\left(\frac{1}{6283 \times 10 \times 10^{-6}}\right) = -j15.92\ \Omega.$$

$$\mathbf{I}_C = \frac{\mathbf{V}_C}{\mathbf{Z}_C} = \frac{70.7\angle -40°}{-j15.92} = \frac{70.7\angle -40°}{15.92\angle -90°} = 4.442\ \text{A} \angle 50°$$

In the time domain, $i_C = \sqrt{2}(4.442) \sin(6283t + 50°) = 6.28 \sin(6283t + 50°)$ A, which agrees with our previous solution, in Example 16–14.

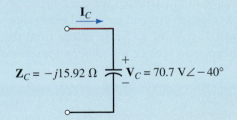

FIGURE 16–42

---

1. If $\mathbf{I}_L = 5\ \text{mA} \angle -60°$, $L = 2$ mH, and $f = 10$ kHz, what is $\mathbf{V}_L$?
2. A capacitor has a reactance of 50 Ω at 1200 Hz. If $v_C = 80 \sin 800t$ V, what is i_C?

PRACTICE PROBLEMS 9

Answers:
1. 628 mV$\angle 30°$

2. 0.170 $\sin(800t + 90°)$ A

A Final Note

The real power of the impedance method becomes apparent when you consider complex circuits with elements in series, parallel, and so on. This we do later, beginning in Chapter 18. Before we do this, however, there are some ideas on power that you need to know. These are considered in Chapter 17.

16.8   Computer Analysis of AC Circuits

ELECTRONICS WORKBENCH   **PSpice**

In Chapter 15, you saw how to represent sinusoidal waveforms using PSpice and Electronics Workbench. Let us now apply these tools to the ideas of this chapter. To illustrate, recall that in Example 16–6, we summed voltages $e_1 = 10 \sin \omega t$ V and $e_2 = 15 \sin(\omega t + 60°)$ V using phasor methods to obtain their sum $v = 21.8 \sin(\omega t + 36.6°)$ V. We will now verify this summation numerically. Since the process is independent of frequency, let us choose $f = 500$ Hz. This yields a period of 2 ms; thus, ½ cycle is 1 ms, ¼ cycle (90°) is 500 μs, 45° is 250 μs, etc.

Electronics Workbench

As noted in Chapter 15, Electronics Workbench requires that you enter rms values for waveforms. Thus, to represent $10 \sin \omega t$ V, enter $(0.7071)(10$ V$) = 7.071$ V as its magnitude, and to represent $15 \sin(\omega t + 60°)$ V, enter 10.607 V. You must also enter its angle. Procedure: Create the circuit of Figure 16–43(a) on the screen. Set Source 1 to **7.071 V, 0 deg,** and **500 Hz** using the procedure of Chapter 15. Similarly, set Source 2 to **10.607 V, 60 deg,** and **500 Hz.** Click Analysis, Transient, set TSTOP to 0.002 (to run the solution out to 2 ms so that you display a full cycle) and TMAX to 2e-06 (to avoid getting a choppy waveform.) Highlight Node 1 (to display e_1) and click Add. Repeat for Node 2 (to display v) Click Simulate. Following simulation, graphs e_1 (the red curve, left-hand scale) and v (the blue curve, right-hand scale) appear. Expand the Analysis Graph window to full screen, then click the cursor icon and drag a cursor to 500 μs or as close as you can get it and read values. (You should get about 10 V for e_1 and 17.5 V for v.) Scale values from the graph at 200-μs increments and tabulate. Now replace the sources with a single source of 21.8

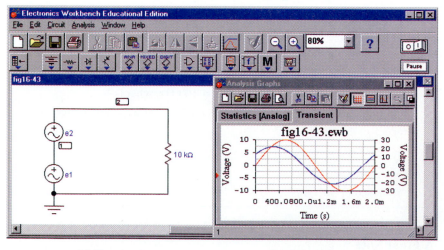

(a) The circuit   (b) Waveforms e_1 and v

FIGURE 16–43   Summing sinusoidal waveforms with Electronics Workbench.

$\sin(\omega t + 36.6°)$ V. (Don't forget to convert to rms). Run a simulation. The resulting graph should be identical to v (the blue curve) of Figure 16–43(b).

OrCAD PSpice

With PSpice, you can plot all three waveforms of Fig. 16–10 simultaneously. Procedure: Create the circuit of Fig. 16–44 using source VSIN. Double click Source 1 and select Parts in the Property Editor. Set VOFF to **0V,** VAMPL to **10V,** FREQ to **500Hz,** and PHASE to **0deg.** Click Apply, then close the Property Editor. Set up source 2 similarly. Now place markers as shown (Note 2) so that PSpice will automatically create the plots. Click the New Simulation Profile icon, choose Transient, set TSTOP to **2ms** (to display a full cycle), set <u>M</u>aximum Step Size to **1us** (to yield a smooth plot), then click OK. Run the simulation and the waveforms of Figure 16–45 should

NOTES...

1. Make sure that the polarities of the sources are as indicated. (You will have to rotate V2 three times to get it into the position shown.)

2. To display V2, use differential markers (indicated as + and − on the toolbar).

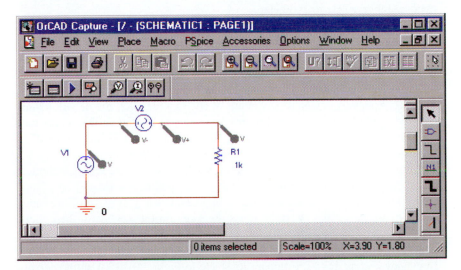

FIGURE 16–44 Summing sinusoidal waveforms with PSpice. Source 2 is displayed using a differential marker.

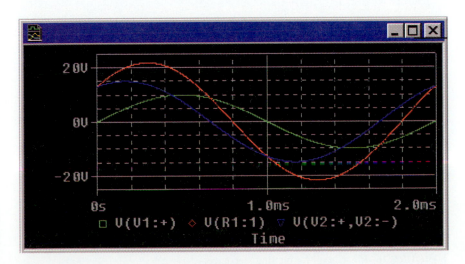

FIGURE 16–45 PSpice waveforms. Compare to Figure 16–10.

appear. Using the cursor, scale voltages at 500 μs. You should get 10 V for e_1, 7.5 V for e_2 and 17.5 V for v. Read values at 200-μs intervals and tabulate. Now replace the sources of Figure 16–44 with a single source of 21.8 sin(ωt + 36.6°) V, run a simulation, scale values, and compare results. They should agree.

Another Example

PSpice makes it easy to study the response of circuits over a range of frequencies. This is illustrated in Example 16–18.

EXAMPLE 16–18   Compute and plot the reactance of a 12-μF capacitor over the range 10 Hz to 1000 Hz.

Solution   PSpice has no command to compute reactance; however, we can calculate voltage and current over the desired frequency range, then plot their ratio. This gives reactance. Procedure: Create the circuit of Figure 16–46 on the screen. (Use source VAC here as it is the source to use for phasor analyses—see Appendix A). Note its default of 0V. Double click the default value (not the symbol) and in the dialog box, enter **120V,** then click OK. Click the New Simulations Profile icon, enter **fig 16-46** for a name and then in the dialog box that opens, select AC Sweep/Noise. For the Start Frequency, key in **10Hz;** for the End Frequency, key **1kHz;** set AC Sweep type to Logarithmic, select Decade and type **100** into the Pts/Decade (points per decade) box. Run the simultation and a set of empty axes appears. Click Trace, Add Trace and in the dialog box, click **V1(C1),** press the / key on the keyboard, then click **I(C1)** to yield the ratio V1(C1)/I(C1) (which is the capacitor's reactance). Click OK and PSpice will compute and plot the capacitor's reactance versus frequency, Figure 16–47. Compare its shape to Figure 16–35. Use the cursor to scale some values off the screen and verify each point using $X_C = 1/\omega C$.

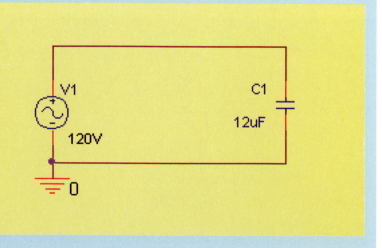

FIGURE 16–46   Circuit for plotting reactance.

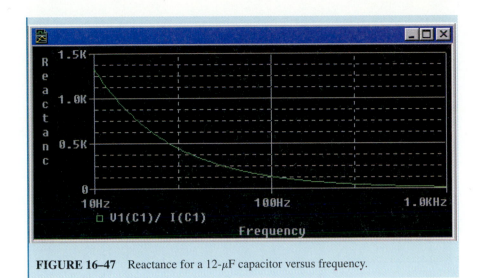

FIGURE 16–47   Reactance for a 12-μF capacitor versus frequency.

Phasor Analysis

As a last example, we will show how to use PSpice to perform phasor analysis—i.e., to solve problems with voltages and currents expressed in phasor form. To illustrate, consider again Example 16–17. Recall, $\mathbf{V}_C = 70.7$ V$\angle-40°$, $C = 10\ \mu$F, and $f = 1000$ Hz. Procedure: Create the circuit on the screen (Figure 16–48) using source VAC and component IPRINT (Note 1). Double click the VAC symbol and in the Property Editor, set ACMAG to **70.7V** and ACPHASE to **−40deg.** (See Note 2). Double click IPRINT and in the Property Editor, type **yes** into cells AC, MAG, and PHASE. Click Apply and close the editor. Click the New Simulation Profile icon, select AC Sweep/Noise, Linear, set both Start Frequency and End Frequency to

1. Component IPRINT is a software ammeter, found in the SPECIAL parts library. In this example, we configure it to display ac current in magnitude and phase angle format. Make sure that it is connected as shown in Figure 16–48, since if it is reversed, the phase angle of the measured current will be in error by 180°.

2. If you want to display the phase of the source voltage on the schematic as in Figure 16–48, double click the source symbol and in the Property Editor, click ACPHASE, Display, then select Value Only.

3. The results displayed by IPRINT are expressed in exponential format. Thus, frequency (Figure 16–49) is shown as 1.000E+03, which is $1.000 \times 10^3 = 1000$ Hz, etc.

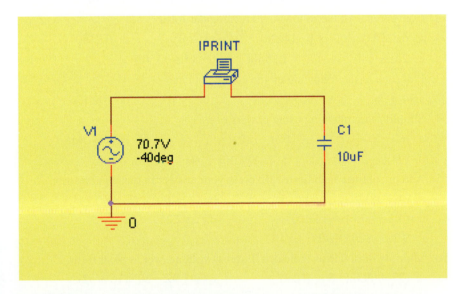

FIGURE 16–48   Phasor analysis using PSpice. Component IPRINT is a software ammeter.

1000Hz and Total Points to **1**. Run the simulation. When the simulation window opens, click View, Output File, then scroll until you find the answers (Figure 16–49 and Note 3). The first number is the frequency (1000 Hz), the second number (IM) is the magnitude of the current (4.442 A), and the third (IP) is its phase (50 degrees). Thus, $\mathbf{I}_C = 4.442 \text{ A}\angle 50°$ as we determined earlier in Example 16–17.

```
FREQ            IM(V_PRINT1)   IP(V_PRINT1)

1.000E+03      4.442E+00       5.000E+01
```

FIGURE 16–49   Current for the circuit of Figure 16–48. $\mathbf{I} = 4.442 \text{ A} \angle 50°$.

PRACTICE PROBLEMS 10

Modify Example 16–18 to plot both capacitor current and reactance on the same graph. You will need to add a second Y-axis for the capacitor current. (See Appendix A if you need help.)

PROBLEMS

16.1   Complex Number Review

1. Convert each of the following to polar form:
   a. $5 + j12$ b. $9 - j6$ c. $-8 + j15$ d. $-10 - j4$

2. Convert each of the following to rectangular form:
   a. $6\angle 30°$ b. $14\angle 90°$ c. $16\angle 0°$ d. $6\angle 150°$
   e. $20\angle -140°$ f. $-12\angle 30°$ g. $-15\angle -150°$

3. Plot each of the following on the complex plane:
   a. $4 + j6$ b. $j4$ c. $6\angle -90°$ d. $10\angle 135°$

4. Simplify the following using powers of j:
   a. $j(1 - j1)$ b. $(-j)(2 + j5)$ c. $j[j(1 + j6)]$
   d. $(j4)(-j2 + 4)$ e. $(2 + j3)(3 - j4)$

5. Add or subtract as indicated. Express your answer in rectangular form.
   a. $(4 + j8) + (3 - j2)$ b. $(4 + j8) - (3 - j2)$
   c. $(4.1 - j7.6) + 12\angle 20°$ d. $2.9\angle 25° - 7.3\angle -5°$
   e. $9.2\angle -120° - (2.6 + j4.1)$

6. Multiply or divide as indicated. Express your answer in polar form.
   a. $(37 + j9.8)(3.6 - j12.3)$ b. $(41.9\angle -80°)(16 + j2)$
   c. $\dfrac{42 + j18.6}{19.1 - j4.8}$ d. $\dfrac{42.6 + j187.5}{11.2\angle 38°}$

7. Reduce each of the following to polar form:

   a. $15 - j6 - \left[\dfrac{18\angle 40° + (12 + j8)}{11 + j11} \right]$

   b. $\dfrac{21\angle 20° - j41}{36\angle 0° + (1 + j12) - 11\angle 40°}$

   c. $\dfrac{18\angle 40° - 18\angle -40°}{7 + j12} - \dfrac{16 + j17 + 21\angle -60°}{4}$

16.2 Complex Numbers in AC Analysis

8. In the manner of Figure 16–9, represent each of the following as trans-formed sources.

   a. $e = 100 \sin(\omega t + 30°)$ V   b. $e = 15 \sin(\omega t - 20°)$ V

   c. $e = 50 \sin(\omega t + 90°)$ V   d. $e = 50 \cos \omega t$ V

   e. $e = 40 \sin(\omega t + 120°)$ V   f. $e = 80 \sin(\omega t - 70°)$ V

9. Determine the sinusoidal equivalent for each of the transformed sources of Figure 16–50.

10. Given: $e_1 = 10 \sin(\omega t + 30°)$ V and $e_2 = 15 \sin(\omega t - 20°)$ V. Determine their sum $v = e_1 + e_2$ in the manner of Example 16–6, i.e.,

    a. Convert e_1 and e_2 to phasor form.

    b. Determine $\mathbf{V} = \mathbf{E}_1 + \mathbf{E}_2$.

    c. Convert $\mathbf{V}$ to the time domain.

    d. Sketch e_1, e_2, and v as per Figure 16–12.

11. Repeat Problem 10 for $v = e_1 - e_2$.

Note: For the remaining problems and throughout the remainder of the book, express phasor quantities as rms values rather than as peak values.

12. Express the voltages and currents of Figure 16–51 as time domain and pha-sor domain quantities.

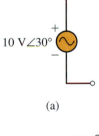

10 V∠30°

(a)

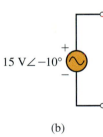

15 V∠−10°

(b)

FIGURE 16–50

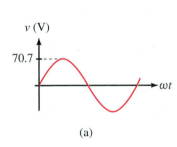

(a)

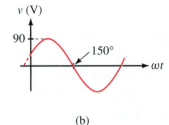

(b)

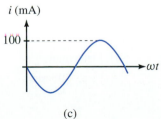

(c)

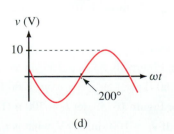

(d)

FIGURE 16–51

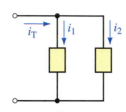

FIGURE 16–52

13. For Figure 16–52, $i_1 = 25 \sin(\omega t + 36°)$ mA and $i_2 = 40 \cos(\omega t - 10°)$ mA.
    a. Determine phasors $\mathbf{I}_1$, $\mathbf{I}_2$ and $\mathbf{I}_T$.
    b. Determine the equation for i_T in the time domain.

14. For Figure 16–52, $i_T = 50 \sin(\omega t + 60°)$ A and $i_2 = 20 \sin(\omega t - 30°)$ A.
    a. Determine phasors $\mathbf{I}_T$ and $\mathbf{I}_2$.
    b. Determine $\mathbf{I}_1$.
    c. From (b), determine the equation for i_1.

15. For Figure 16–17, $i_1 = 7 \sin \omega t$ mA, $i_2 = 4 \sin(\omega t - 90°)$ mA, and $i_3 = 6 \sin(\omega t + 90°)$ mA.
    a. Determine phasors $\mathbf{I}_1$, $\mathbf{I}_2$, $\mathbf{I}_3$ and $\mathbf{I}_T$.
    b. Determine the equation for i_T in the time domain.

16. For Figure 16–17, $i_T = 38.08 \sin(\omega t - 21.8°)$ A, $i_1 = 35.36 \sin \omega t$ A, and $i_2 = 28.28 \sin(\omega t - 90°)$ A. Determine the equation for i_3.

16.4 to 16.6

17. For Figure 16–18(a), $R = 12$ Ω. For each of the following, determine the current or voltage and sketch.
    a. $v = 120 \sin \omega t$ V, $i = $ _____
    b. $v = 120 \sin(\omega t + 27°)$ V, $i = $ _____
    c. $i = 17 \sin(\omega t - 56°)$ mA, $v = $ _____
    d. $i = -17 \cos(\omega t - 67°)$ μA, $v = $ _____

18. Given $v = 120 \sin(\omega t + 52°)$ V and $i = 15 \sin(\omega t + 52°)$ mA, what is R?

19. Two resistors $R_1 = 10$ kΩ and $R_2 = 12.5$ kΩ are in series. If $i = 14.7 \sin(\omega t + 39°)$ mA,
    a. What are v_{R_1} and v_{R_2}?
    b. Compute $v_T = v_{R_1} + v_{R_2}$ and compare to v_T calculated from $v_T = i R_T$.

20. The voltage across a certain component is $v = 120 \sin(\omega t + 55°)$ V and its current is $-18 \cos(\omega t + 145°)$ mA. Show that the component is a resistor and determine its value.

21. For Figure 16–53, $V_m = 10$ V and $I_m = 5$ A. For each of the following, determine the missing quantity:
    a. $v_L = 10 \sin(\omega t + 60°)$ V, $i_L = $ _____
    b. $v_L = 10 \sin(\omega t - 15°)$ V, $i_L = $ _____
    c. $i_L = 5 \cos(\omega t - 60°)$ A, $v_L = $ _____
    d. $i_L = 5 \sin(\omega t + 10°)$ A, $v_L = $ _____

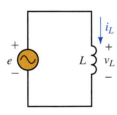

FIGURE 16–53

22. What is the reactance of a 0.5-H inductor at
    a. 60 Hz   b. 1000 Hz   c. 500 rad/s

23. For Figure 16–53, $e = 100 \sin \omega t$ and $L = 0.5$ H. Determine i_L at
    a. 60 Hz   b. 1000 Hz   c. 500 rad/s

24. For Figure 16–53, let $L = 200$ mH.
    a. If $v_L = 100 \sin 377t$ V, what is i_L?
    b. If $i_L = 10 \sin(2\pi \times 400t - 60°)$ mA, what is v_L?

25. For Figure 16–53, if
   a. $v_L = 40 \sin(\omega t + 30°)$ V, $i_L = 364 \sin(\omega t - 60°)$ mA, and $L = 2$ mH, what is f?
   b. $i_L = 250 \sin(\omega t + 40°)$ μA, $v_L = 40 \sin(\omega t + \theta)$ V, and $f = 500$ kHz, what are L and θ?

26. Repeat Problem 21 if the given voltages and currents are for a capacitor instead of an inductor.

27. What is the reactance of a 5-μF capacitor at
   a. 60 Hz   b. 1000 Hz   c. 500 rad/s

28. For Figure 16–54, $e = 100 \sin \omega t$ and $C = 5$ μF. Determine i_C at
   a. 60 Hz   b. 1000 Hz   c. 500 rad/s

29. For Figure 16–54, let $C = 50$ μF.
   a. If $v_C = 100 \sin 377t$ V, what is i_C?
   b. If $i_C = 10 \sin(2\pi \times 400t - 60°)$ mA, what is v_C?

30. For Figure 16–54, if
   a. $v_C = 362 \sin(\omega t - 33°)$ V, $i_C = 94 \sin(\omega t + 57°)$ mA, and $C = 2.2$ μF, what is f?
   b. $i_C = 350 \sin(\omega t + 40°)$ mA, $v_C = 3.6 \sin(\omega t + \theta)$ V, and $f = 12$ kHz, what are C and θ?

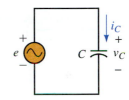

FIGURE 16–54

16.7 The Impedance Concept

31. Determine the impedance of each circuit element of Figure 16–55.

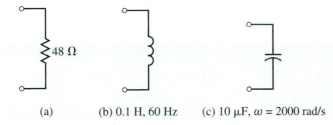

(a)   (b) 0.1 H, 60 Hz   (c) 10 μF, $\omega = 2000$ rad/s

FIGURE 16–55

32. If $\mathbf{E} = 100 \text{ V} \angle 0°$ is applied across each of the circuit elements of Figure 16–56:
   a. Determine each current in phasor form.
   b. Express each current in time domain form.

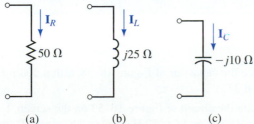

(a)   (b)   (c)

FIGURE 16–56

33. If the current through each circuit element of Figure 16–56 is 0.5 A∠0°:

    a. Determine each voltage in phasor form.

    b. Express each voltage in time domain form.

34. For each of the following, determine the impedance of the circuit element and state whether it is resistive, inductive, or capacitive.

    a. $\mathbf{V} = 240\ \text{V}\angle -30°$, $\mathbf{I} = 4\ \text{A}\angle -30°$.

    b. $\mathbf{V} = 40\ \text{V}\angle 30°$, $\mathbf{I} = 4\ \text{A}\angle -60°$.

    c. $\mathbf{V} = 60\ \text{V}\angle -30°$, $\mathbf{I} = 4\ \text{A}\angle 60°$.

    d. $\mathbf{V} = 140\ \text{V}\angle -30°$, $\mathbf{I} = 14\ \text{mA}\angle -120°$.

35. For each circuit of Figure 16–57, determine the unknown.

36. a. If $\mathbf{V}_L = 120\ \text{V}\ \angle 67°$, $L = 600\ \mu\text{H}$, and $f = 10\ \text{kHz}$, what is $\mathbf{I}_L$?

    b. If $\mathbf{I}_L = 48\ \text{mA}\ \angle -43°$, $L = 550\ \text{mH}$, and $f = 700\ \text{Hz}$, what is $\mathbf{V}_L$?

    c. If $\mathbf{V}_C = 50\ \text{V}\ \angle -36°$, $C = 390\ \text{pF}$, and $f = 470\ \text{kHz}$, what is $\mathbf{I}_C$?

    d. If $\mathbf{I}_C = 95\ \text{mA}\ \angle 87°$, $C = 6.5\ \text{nF}$, and $f = 1.2\ \text{MHz}$, what is $\mathbf{V}_C$?

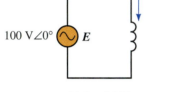

$\mathbf{I}_L = 2\ \text{A}\angle -90°$

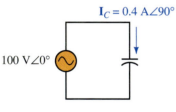

$100\ \text{V}\angle 0°$ *E*

(a) $L = 0.2$ H.
Determine *f*.

$\mathbf{I}_C = 0.4\ \text{A}\angle 90°$

$100\ \text{V}\angle 0°$

(b) $f = 100$ Hz.
Determine *C*.

FIGURE 16–57

16.8 Computer Analysis of AC Circuits

The version of Electronics Workbench current at the time of writing of this book is unable to measure phase angles. Thus, in the problems that follow, we ask only for magnitudes of voltages and currents.

37. **EWB**   Create the circuit of Figure 16–58 on the screen. (Use the ac source from the Sources Parts bin and the ammeter from the Indicators Parts bin.) Double click the ammeter symbol and set Mode to AC. Click the ON/OFF switch at the top right hand corner of the screen to energize the circuit. Compare the measured reading against the theoretical value.

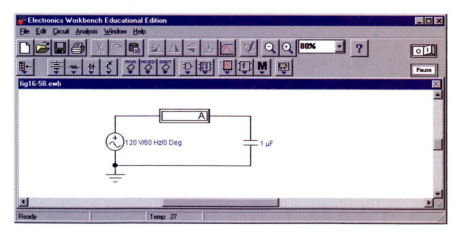

FIGURE 16–58

38. **EWB**   Replace the capacitor of Figure 16–58 with a 200-mH inductor and repeat Problem 37.

39. **PSpice**   Create the circuit of Figure 16–53 on the screen. Use a source of $100\ \text{V}\angle 0°$, $L = 0.2$ H, and $f = 50$ Hz. Solve for current $\mathbf{I}_L$ (magnitude and angle). See note below.

40. **PSpice**   Plot the reactance of a 2.39-H inductor versus frequency from 1 Hz to 500 Hz and compare to Figure 16–28. Change the x-axis scale to linear.

41. **PSpice**   For the circuit of Problem 39, plot current magnitude versus frequency from $f = 1$ Hz to $f = 20$ Hz. Measure the current at 10 Hz and verify with your calculator.

Note: PSpice does not permit source/inductor loops. To get around this, add a very small resistor in series, for example, $R = 0.00001\ \Omega$.

In-Process Learning Check 1

1. a. $6\angle 90°$
   b. $4\angle -90°$
   c. $4.24\angle 45°$
   d. $7.21\angle -56.3°$
   e. $9.43\angle 122.0°$
   f. $2.24\angle -63.4°$
   g. $3.61\angle -123.7°$

2. a. $j4$
   b. $3 + j0$
   c. $-j2$
   d. $3.83 + j3.21$
   e. $-3 + j5.20$
   f. $2.35 - j0.855$
   g. $-1.64 - j0.599$

3. $12\angle 40°$

4. $88 - j7; -16 + j15$

5. $288\angle -100°; 2\angle 150°$

6. a. $14.70 + j2.17$ b. $-8.94 + j7.28$

7. $18.0 \sin(\omega t - 56.3°)$

In-Process Learning Check 2

1. $50 \sin(\omega t + 30°)$ A
2. $50 \sin(\omega t - 60°)$ A
3. $50 \sin(\omega t + 120°)$ A
4. 0.637 H
5. $3.98\ \mu F$
6. a. Voltage and current are in phase. Therefore, R
   b. Current leads by 90°. Therefore, C
   c. Current leads by 90°. Therefore, C
   d. Current lags by 90°. Therefore, L

17

Power in AC Circuits

OBJECTIVES

After studying this chapter, you will be able to

- explain what is meant by active, reactive, and apparent power,
- compute the active power to a load,
- compute the reactive power to a load,
- compute the apparent power to a load,
- construct and use the power triangle to analyze power to complex loads,
- compute power factor,
- explain why equipment is rated in VA instead of watts,
- measure power in single-phase circuits,
- describe why effective resistance differs from geometric resistance,
- describe energy relations in ac circuits,
- use PSpice to study instantaneous power.

KEY TERMS

Active Power
Apparent Power
Average Power
Effective Resistance
F_p
Instantaneous Power

Power Factor Correction
Power Factor
Power Triangle
Q
Reactive Power
S
Skin Effect
VA
VAR
Wattless Power
Wattmeter

OUTLINE

Introduction
Power to a Resistive Load
Power to an Inductive Load
Power to a Capacitive Load
Power in More Complex Circuits
Apparent Power
The Relationship Between P, Q, and S
Power Factor
AC Power Measurement
Effective Resistance
Energy Relationships for AC
Circuit Analysis Using Computers

In Chapter 4, you studied power in dc circuits. In this chapter, we turn our attention to power in ac circuits. In ac circuits, there are additional considerations that are not present with dc. In dc circuits, for example, the only power relationship you encounter is $P = VI$ watts. This is referred to as *real power* or *active power* and is the power that does useful work such as light a lamp, power a heater, run an electric motor, and so on.

In ac circuits, you also encounter this type of power. For ac circuits that contain reactive elements however, (i.e., inductance or capacitance), a second component of power also exists. This component, termed *reactive power*, represents energy that oscillates back and forth throughout the system. For example, during the buildup of current in an inductance, energy flows from the power source to the inductance to create its magnetic field. When the magnetic field collapses, this energy is returned to the circuit. This movement of energy in and out of the inductance constitutes a flow of power. However, since it flows first in one direction, then in the other, it contributes nothing to the average flow of power from the source to the load. For this reason, reactive power is sometimes referred to as *wattless power*. (A similar situation exists regarding power flow to and from the electric field of a capacitor.)

For a circuit that contains resistive as well as reactive elements, some energy is dissipated while the remainder is shuttled back and forth as described above; thus, both active and reactive components of power are present. This combination of real and reactive power is termed *apparent power*.

In this chapter, we look at all three components of power. New ideas that emerge include the concept of power factor, the power triangle, the measurement of power in ac circuits, and the concept of effective resistance.

Henry Cavendish

CAVENDISH, AN ENGLISH CHEMIST and physicist born in 1731, is included here not for what he did for the emerging electrical field, but for what he didn't do. A brilliant man, Cavendish was 50 years ahead of his time, and his experiments in electricity preceded and anticipated almost all the major discoveries that came about over the next half century (e.g., he discovered Coulomb's law before Coulomb did). However, Cavendish was interested in research and knowledge purely for its own sake and never bothered to publish most of what he learned, in effect depriving the world of his findings and holding back the development of the field of electricity by many years. Cavendish's work lay unknown for nearly a century before another great scientist, James Clerk Maxwell, had it published. Nowadays, Cavendish is better known for his work in the gravitational field than for his work in the electrical field. One of the amazing things he did was to determine the mass of the earth using the rather primitive technology of his day.

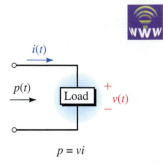

$$p = vi$$

FIGURE 17–1 Voltage, current, and power references. When *p* is positive, power is in the direction of the reference arrow.

17.1 Introduction

At any given instant, the power to a load is equal to the product of voltage times current (Figure 17–1). This means that if voltage and current vary with time, so will power. This time-varying power is referred to as **instantaneous power** and is given the symbol $p(t)$ or just p. Thus,

$$p = vi \quad \text{(watts)} \tag{17–1}$$

Now consider the case of sinusoidal ac. Since voltage and current are positive at various times during their cycle and negative at others, instantaneous power may also be positive at some times and negative at others. This is illustrated in Figure 17–2, where we have multiplied voltage times current point by point to get the power waveform. For example, from $t = 0$ s to $t = t_1$, v and i are both positive; therefore, power is positive. At $t = t_1$, $v = 0$ V and thus $p = 0$ W. From t_1 to t_2, i is positive and v is negative; therefore, p is negative. From t_2 to t_3, both v and i are negative; therefore power is positive, and so on. As discussed in Chapter 4, a positive value for p means that power transfer is in the direction of the reference arrow, while a negative value means that it is in the opposite direction. Thus, during positive parts of the power cycle, power flows from the source to the load, while during negative parts, it flows out of the load back into the circuit.

The waveform of Figure 17–2 is the actual power waveform. We will now show that the key aspects of power flow embodied in this waveform can be described in terms of active power, reactive power, and apparent power.

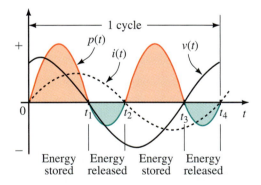

FIGURE 17–2 Instantaneous power in an ac circuit. Positive *p* represents power to the load; negative *p* represents power returned from the load.

Active Power

Since p represents the power flowing to the load, its average will be the average power to the load. Denote this average by the letter P. If P is positive, then, on average, more power flows to the load than is returned from it. (If P is zero, all power sent to the load is returned.) Thus, if P has a positive value, it represents the power that is really dissipated by the load. For this reason, P is called **real power.** In modern terminology, real power is also called **active power.** Thus, *active power is the average value of the instantaneous power, and the terms real power, active power, and average power mean the same* *thing.* (We usually refer to it simply as power.) In this book, we use the terms interchangeably.

Reactive Power

Consider again Figure 17–2. During the intervals that *p* is negative, power is being returned from the load. (This can only happen if the load contains reactive elements: *L* or *C*.) The portion of power that flows into the load then back out is called **reactive power.** Since it first flows one way then the other, *its average value is zero;* thus, reactive power contributes nothing to the average power to the load.

Although reactive power does no useful work, it cannot be ignored. Extra current is required to create reactive power, and this current must be supplied by the source; this also means that conductors, circuit breakers, switches, transformers, and other equipment must be made physically larger to handle the extra current. This increases the cost of a system.

At this point, it should be noted that real power and reactive power do not exist as separate entities. Rather, they are components of the power waveform shown in Figure 17–2. However, as you will see, we are able to conceptually separate them for purposes of analysis.

17.2   Power to a Resistive Load

First consider power to a purely resistive load (Figure 17–3). Here, current is in phase with voltage. Assume $i = I_m\sin \omega t$ and $v = V_m\sin \omega t$. Then,

$$p = vi = (V_m\sin \omega t)(I_m\sin \omega t) = V_mI_m\sin^2\omega t$$

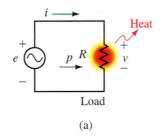

(a)

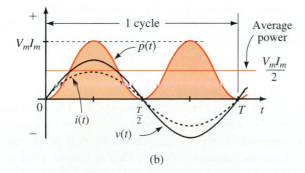

(b)

FIGURE 17–3   Power to a purely resistive load. The peak value of *p* is V_mI_m.

Therefore,

$$p = \frac{V_m I_m}{2}(1 - \cos 2\,\omega t) \qquad \textbf{(17–2)}$$

where we have used the trigonometric relationship $\sin^2 \omega t = \frac{1}{2}(1 - \cos 2\,\omega t)$ from inside the front cover of the book.

A sketch of p versus time is shown in (b). Note that p is always positive (except where it is momentarily zero). This means that power flows only from the source to the load. Since none is ever returned, all power delivered by the source is absorbed by the load. We therefore conclude that *power to a pure resistance consists of active power only*. Note also that the frequency of the power waveform is double that of the voltage and current waveforms. (This is confirmed by the 2ω in Equation 17–2.)

Average Power

Inspection of the power waveform of Figure 17–3 shows that its average value lies half way between zero and its peak value of $V_m I_m$. That is,

$$P = V_m I_m / 2$$

(You can also get the same result by averaging Equation 17–2 as we did in Chapter 15.) Since V (the magnitude of the rms value of voltage) is $V_m/\sqrt{2}$ and I (the magnitude of the rms value of current) is $I_m/\sqrt{2}$, this can be written as $P = VI$. Thus, average power to a purely resistive load is

$$P = VI \quad \text{(watts)} \qquad \textbf{(17–3)}$$

Alternate forms are obtained by substituting $V = IR$ and $I = V/R$ into Equation 17–3. They are

$$P = I^2 R \quad \text{(watts)} \qquad \textbf{(17–4)}$$

$$= V^2/R \quad \text{(watts)} \qquad \textbf{(17–5)}$$

Thus the active power relationships for resistive circuits are the same for ac as for dc.

17.3   Power to an Inductive Load

For a purely inductive load as in Figure 17–4(a), current lags voltage by 90°. If we select current as reference, $i = I_m \sin \omega t$ and $v = V_m \sin(\omega t + 90°)$. A sketch of p versus time (obtained by multiplying v times i) then looks as shown in (b). Note that during the first quarter-cycle, p is positive and hence power flows to the inductance, while during the second quarter-cycle, p is negative and all power transferred to the inductance during the first quarter-cycle flows back out. Similarly for the third and fourth quarter-cycles. Thus, *the average power to an inductance over a full cycle is zero, i.e., there are no power losses associated with a pure inductance.* Consequently, $P_L = 0$ W and the only power flowing in the circuit is reactive power. This is true in general, that is, *the power that flows into and out of a pure inductance is reactive power only.*

To determine this power, consider again equation 17–1.

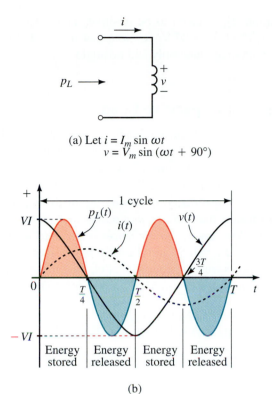

(a) Let $i = I_m \sin \omega t$
$v = V_m \sin(\omega t + 90°)$

(b)

FIGURE 17–4 Power to a purely inductive load. Energy stored during each quarter-cycle is returned during the next quarter-cycle. Average power is zero.

With $v = V_m \sin(\omega t + 90°)$ and $i = I_m \sin \omega t$, $p_L = vi$ becomes

$$p_L = V_m I_m \sin(\omega t + 90°)\sin \omega t$$

After some trigonometric manipulation, this reduces to

$$p_L = VI \sin 2 \omega t \qquad \text{(17–6)}$$

where V and I are the magnitudes of the rms values of the voltage and current respectively.

The product VI in Equation 17–6 is defined as **reactive power** and is given the symbol Q_L. Because it represents "power" that alternately flows into, then out of the inductance, Q_L contributes nothing to the average power to the load and, as noted earlier, is sometimes referred to as wattless power. As you will soon see, however, reactive power is of major concern in the operation of electrical power systems.

Since Q_L is the product of voltage times current, its unit is the volt-amp (VA). To indicate that Q_l represents reactive volt-amps, an "R" is appended to yield a new unit, the *VAR* (*volt-amps reactive*). Thus,

$$Q_L = VI \quad \text{(VAR)} \qquad \text{(17–7)}$$

Substituting $V = IX_L$ and $I = V/X_L$ yields the following alternate forms:

$$Q_L = I^2 X_L = \frac{V^2}{X_L} \quad \text{(VAR)} \qquad \text{(17–8)}$$

By convention, Q_L is taken to be positive. Thus, if $I = 4$ A and $X_L = 2\ \Omega$, $Q_L = (4\ \text{A})^2(2\ \Omega) = +32$ VAR. Note that the VAR (like the watt) is a scalar quantity with magnitude only and no angle.

17.4 Power to a Capacitive Load

For a purely capacitive load, current leads voltage by 90°. Taking current as reference, $i = I_m \sin \omega t$ and $v = V_m \sin(\omega t - 90°)$. Multiplication of v times i yields the power curve of Figure 17–5. Note that negative and positive loops of the power wave are identical; thus, over a cycle, the power returned to the circuit by the capacitance is exactly equal to that delivered to it by the source. This means that *the average power to a capacitance over a full cycle is zero, i.e., there are no power losses associated with a pure capacitance.* Consequently, $P_C = 0$ W and the only power flowing in the circuit is reactive power. This is true in general, that is, *the power that flows into and out of a pure capacitance is reactive power only.* This reactive power is given by

$$p_C = vi = V_m I_m \sin \omega t \sin(\omega t - 90°)$$

which reduces to

$$p_C = -VI \sin 2\omega t \qquad\qquad (17\text{–}9)$$

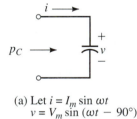

(a) Let $i = I_m \sin \omega t$
$v = V_m \sin (\omega t - 90°)$

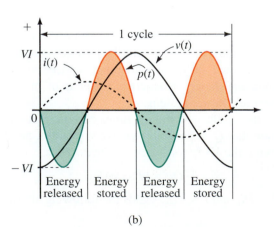

(b)

FIGURE 17–5   Power to a purely capacitive load. Average power is zero.

where V and I are the magnitudes of the rms values of the voltage and current respectively. Now define the product VI as Q_C. This product represents reactive power. That is,

$$Q_C = VI \quad \text{(VAR)} \qquad (17\text{–}10)$$

Since $V = IX_C$ and $I = V/X_C$, Q_C can also be expressed as

$$Q_C = I^2 X_C = \frac{V^2}{X_C} \quad \text{(VAR)} \qquad (17\text{–}11)$$

By convention, reactive power to capacitance is defined as negative. Thus, if $I = 4$ A and $X_C = 2\ \Omega$, then $I^2 X_C = (4\ \text{A})^2 (2\ \Omega) = 32$ VAR. We can either explicitly show the minus sign as $Q_C = -32$ VAR or imply it by stating that Q represents capacitive vars, i.e. $Q_C = 32$ VAR (cap.).

EXAMPLE 17–1   For each circuit of Figure 17–6, determine real and reactive power.

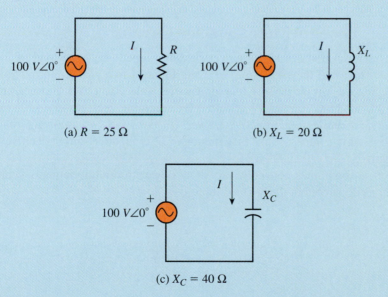

(a) $R = 25\ \Omega$

(b) $X_L = 20\ \Omega$

(c) $X_C = 40\ \Omega$

FIGURE 17–6

Solution   Only voltage and current magnitudes are needed.

a. $I = 100\ \text{V}/25\ \Omega = 4$ A. $P = VI = (100\ \text{V})(4\ \text{A}) = 400$ W. $Q = 0$ VAR

b. $I = 100\ \text{V}/20\ \Omega = 5$ A. $Q = VI = (100\ \text{V})(5\ \text{A}) = 500$ VAR (ind.). $P = 0$ W

c. $I = 100\ \text{V}/40\ \Omega = 2.5$ A. $Q = VI = (100\ \text{V})(2.5\ \text{A}) = 250$ VAR (cap.). $P = 0$ W

The answer for (c) can also be expressed as $Q = -250$ VAR.

1. If the power at some instant in Figure 17–1 is $p = -27$ W, in what direction is the power at that instant?

2. For a purely resistive load, v and i are in phase. Given $v = 10 \sin \omega t$ V and $i = 5 \sin \omega t$ A. Using graph paper, carefully plot v and i at 30° intervals. Now multiply the values of v and i at these points and plot the power. (The result should look like Figure 17–3(b).

   a. From the graph, determine the peak power and average power.

   b. Compute power using $P = VI$ and compare to the average value determined in (a).

3. Repeat Example 17–1 using equations 17–4, 17–5, 17–8, and 17–11.

Answers:
1. From the load to the source.

2. a. 50 W; 25 W

   b. Same

17.5   Power in More Complex Circuits

The relationships described above were developed using the load of Figure 17–1. However, they hold true for every element in a circuit, no matter how complex the circuit or how its elements are interconnected. Further, in any circuit, total real power P_T is found by summing real power to all circuit elements, while total reactive power Q_T is found by summing reactive power, taking into account that inductive Q is positive and capacitive Q is negative.

It is sometimes convenient to show power to circuit elements symbolically as illustrated in the next example.

EXAMPLE 17–2   For the *RL* circuit of Figure 17–7(a), $I = 5$ A. Determine P and Q.

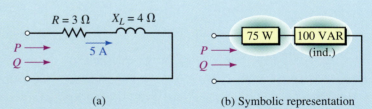

(a)   (b) Symbolic representation

FIGURE 17–7   From the terminals, P and Q are the same for both (a) and (b).

Solution

$$P = I^2R = (5 \text{ A})^2(3 \text{ } \Omega) = 75 \text{ W}$$

$$Q = Q_L = I^2X_L = (5 \text{ A})^2(4 \text{ } \Omega) = 100 \text{ VAR (ind.)}$$

These can be represented symbolically as in Figure 17–7(b).

EXAMPLE 17–3   For the *RC* circuit of Figure 17–8(a), determine *P* and *Q*.

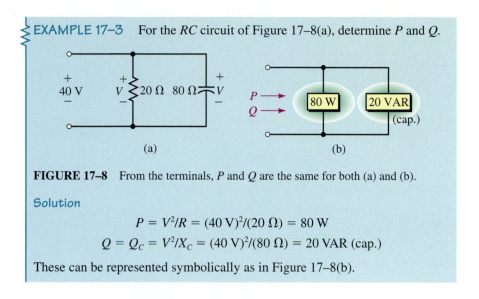

(a)   (b)

FIGURE 17–8   From the terminals, *P* and *Q* are the same for both (a) and (b).

Solution

$$P = V^2/R = (40\text{ V})^2/(20\text{ }\Omega) = 80\text{ W}$$

$$Q = Q_C = V^2/X_C = (40\text{ V})^2/(80\text{ }\Omega) = 20\text{ VAR (cap.)}$$

These can be represented symbolically as in Figure 17–8(b).

In terms of determining total *P* and *Q*, it does not matter how the circuit or system is connected or what electrical elements it contains. Elements can be connected in series, in parallel, or in series-parallel, for example, and the system can contain electric motors and the like, and total *P* is still found by summing the power to individual elements, while total *Q* is found by algebraically summing their reactive powers.

EXAMPLE 17–4

a. For Figure 17–9(a), compute P_T and Q_T.
b. Reduce the circuit to its simplest form.

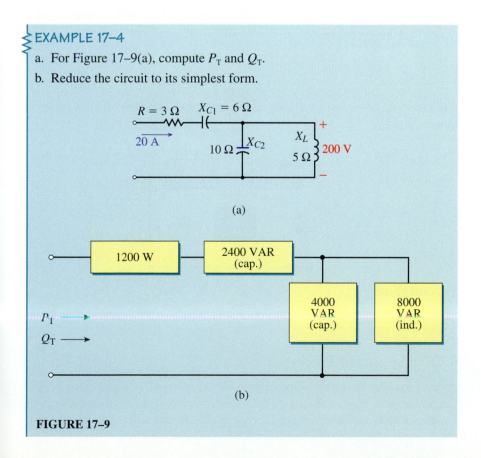

FIGURE 17–9

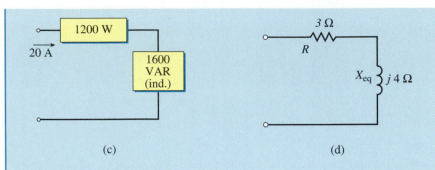

(c)   (d)

FIGURE 17–9   Continued.

Solution

a. $P = I^2R = (20 \text{ A})^2(3 \text{ }\Omega) = 1200 \text{ W}$

   $Q_{C_1} = I^2X_{C_1} = (20 \text{ A})^2(6 \text{ }\Omega) = 2400 \text{ VAR (cap.)}$

   $Q_{C_2} = \dfrac{V_2^2}{X_{C_2}} = \dfrac{(200 \text{ V})^2}{(10 \text{ }\Omega)} = 4000 \text{ VAR (cap.)}$

   $Q_L = \dfrac{V_2^2}{X_L} = \dfrac{(200 \text{ V})^2}{5 \text{ }\Omega} = 8000 \text{ VAR (ind.)}$

These are represented symbolically in part (b). $P_T = 1200 \text{ W}$ and $Q_T =$ $-2400 \text{ VAR} - 4000 \text{ VAR} + 8000 \text{ VAR} = 1600 \text{ VAR}$. Thus, the load is net inductive as shown in (c).

b. $Q_T = I^2X_{eq}$. Thus, $X_{eq} = Q_T/I^2 = (1600 \text{ VAR})/(20 \text{ A})^2 = 4 \text{ }\Omega$. Circuit resistance remains unchanged. Thus, the equivalent is as shown in (d).

PRACTICE PROBLEMS 2

For the circuit of Figure 17–10, $P_T = 1.9 \text{ kW}$ and $Q_T = 900 \text{ VAR (ind.)}$. Determine P_2 and Q_2.

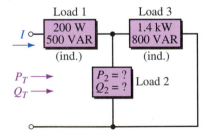

FIGURE 17–10

Answer: 300 W   400 VAR (cap.)

17.6   Apparent Power

When a load has voltage V across it and current I through it as in Figure 17–11, the power that appears to flow to it is VI. However, if the load contains both resistance and reactance, this product represents neither real power nor reactive power. Since it appears to represent power, it is called **apparent power.** Apparent power is given the symbol *S* and has units of volt-amperes (VA). Thus,

$$S = VI \quad \text{(VA)} \tag{17–12}$$

where *V* and *I* are the magnitudes of the rms voltage and current respectively. Since $V = IZ$ and $I = V/Z$, *S* can also be written as

$$S = I^2Z = V^2/Z \quad \text{(VA)} \tag{17–13}$$

For small equipment (such as found in electronics), VA is a convenient unit. However, for heavy power apparatus (Figure 17–12), it is too small and kVA (kilovolt-amps) is frequently used, where

$$S = \frac{VI}{1000} \quad \text{(kVA)} \tag{17–14}$$

In addition to its VA rating, it is common practice to rate electrical apparatus in terms of its operating voltage. Once you know these two, it is easy to determine rated current. For example, a piece of equipment rated at 250 kVA, 4.16 kV has a rated current of $I = S/V = (250 \times 10^3 \text{ VA})/(4.16 \times 10^3 \text{ V}) = 60.1 \text{ A}$.

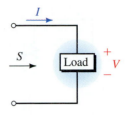

FIGURE 17–11 Apparent power $S = VI$.

FIGURE 17–12 Power apparatus is rated in apparent power. The transformer shown is a 167-kVA unit. (*Courtesy Carte International Ltd.*)

17.7   The Relationship Between *P, Q,* and *S*

Until now, we have treated real, reactive, and apparent power separately. However, they are related by a very simple relationship through the power triangle.

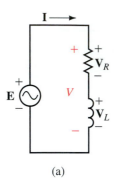

(a)

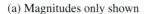

(b)

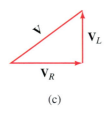

(c)

FIGURE 17–13 Steps in the development of the power triangle.

The Power Triangle

Consider the series circuit of Figure 17–13(a). Let the current through the circuit be $\mathbf{I} = I\angle 0°$, with phasor representation (b). The voltages across the resistor and inductance are $\mathbf{V}_R$ and $\mathbf{V}_L$ respectively. As noted in Chapter 16, $\mathbf{V}_R$ is in phase with $\mathbf{I}$, while $\mathbf{V}_L$ leads it by 90°. Kirchhoff's voltage law applies for ac voltages in phasor form. Thus, $\mathbf{V} = \mathbf{V}_R + \mathbf{V}_L$ as indicated in (c).

The voltage triangle of (c) may be redrawn as in Figure 17–14(a) with magnitudes of V_R and V_L replaced by IR and IX_L respectively. Now multiply all quantities by I. This yields sides of I^2R, I^2X_L, and hypotenuse VI as indicated in (b). Note that these represent P, Q, and S respectively as indicated in (c). This is called the **power triangle.** From the geometry of this triangle, you can see that

$$S = \sqrt{P^2 + Q_L^2} \qquad (17\text{--}15)$$

Alternatively, the relationship between P, Q, and S may be expressed as a complex number:

$$\mathbf{S} = P + jQ_L \qquad (17\text{--}16a)$$

or

$$\mathbf{S} = S\angle\theta \qquad (17\text{--}16b)$$

If the circuit is capacitive instead of inductive, Equation 17–16a becomes

$$\mathbf{S} = P - jQ_C \qquad (17\text{--}17)$$

The power triangle in this case has a negative imaginary part as indicated in Figure 17–15.

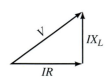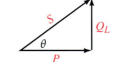

(a) Magnitudes only shown   (b) Multiplied by I (c) Resultant power triangle

FIGURE 17–14 Steps in the development of the power triangle (continued).

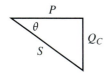

FIGURE 17–15 Power triangle for capacitive case.

The power relationships may be written in generalized forms as

$$\mathbf{S} = \mathbf{P} + \mathbf{Q} \qquad (17\text{--}18)$$

and

$$\mathbf{S} = \mathbf{V}\mathbf{I}^* \qquad (17\text{--}19)$$

where $\mathbf{P} = P\angle 0°$, $\mathbf{Q}_L = jQ_L$, $\mathbf{Q}_C = -jQ_C$, and $\mathbf{I}^*$ is the conjugate of current $\mathbf{I}$. These relationships hold true for all networks regardless of what they contain or how they are configured.

When solving problems involving power, remember that P values can be added to get P_T, and Q values to get Q_T (where Q is positive for inductive elements and negative for capacitive). However, apparent power values cannot be added to get S_T, i.e., $S_T \neq S_1 + S_2 + \cdots + S_N$. Instead, determine P_T and Q_T, then use the power triangle to obtain S_T.

EXAMPLE 17–5 The *P* and *Q* values for a circuit are shown in Figure 17–16(a).

a. Determine the power triangle.

b. Determine the magnitude of the current supplied by the source.

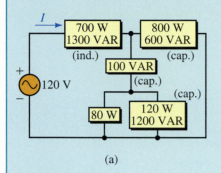

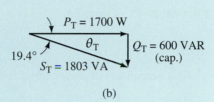

(a)   (b)

FIGURE 17–16

Solution

a. $P_T = 700 + 800 + 80 + 120 = 1700$ W

   $Q_T = 1300 - 600 - 100 - 1200 = -600$ VAR $= 600$ VAR (cap.)

   $\mathbf{S}_T = P_T + jQ_T = 1700 - j600 = 1803\angle -19.4°$ VA

   The power triangle is as shown. The load is net capacitive.

b. $I = S_T/E = 1803$ VA/120 V $= 15.0$ A

EXAMPLE 17–6 A generator supplies power to an electric heater, an inductive element, and a capacitor as in Figure 17–17(a).

a. Find *P* and *Q* for each load.

b. Find total active and reactive power supplied by the generator.

c. Draw the power triangle for the combined loads and determine total apparent power.

d. Find the current supplied by the generator.

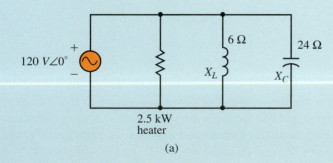

(a)

FIGURE 17–17

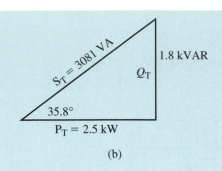

(b)

FIGURE 17–17   Continued.

Solution

a. The components of power are as follows:

Heater: $P_H = 2.5 \text{ kW}$ $Q_H = 0 \text{ VAR}$

Inductor: $P_L = 0 \text{ W}$ $Q_L = \dfrac{V^2}{X_L} = \dfrac{(120 \text{ V})^2}{6 \, \Omega} = 2.4 \text{ kVAR (ind.)}$

Capacitor: $P_C = 0 \text{ W}$ $Q_C = \dfrac{V^2}{X_C} = \dfrac{(120 \text{ V})^2}{24 \, \Omega} = 600 \text{ VAR (cap.)}$

b. $P_T = 2.5 \text{ kW} + 0 \text{ W} + 0 \text{ W} = 2.5 \text{ kW}$

$Q_T = 0 \text{ VAR} + 2.4 \text{ kVAR} - 600 \text{ VAR} = 1.8 \text{ kVAR (ind.)}$

c. The power triangle is sketched as Figure 17–7(b). Both the hypotenuse and the angle can be obtained easily using rectangular to polar conversion. $\mathbf{S}_T = P_T + jQ_T = 2500 + j1800 = 3081\angle 35.8°$. Thus, apparent power is $S_T = 3081 \text{ VA}$.

d. $I = \dfrac{S_T}{E} = \dfrac{3081 \text{ VA}}{120 \text{ V}} = 25.7 \text{ A}$

Active and Reactive Power Equations

An examination of the power triangle of Figures 17–14 and 17–15 shows that P and Q may be expressed respectively as

$$P = VI \cos \theta = S \cos \theta \quad \text{(W)} \tag{17–20}$$

and

$$Q = VI \sin \theta = S \sin \theta \quad \text{(VAR)} \tag{17–21}$$

where V and I are the magnitudes of the rms values of the voltage and current respectively and θ is the angle between them. P is always positive, while Q is positive for inductive circuits and negative for capacitive circuits. Thus, if $V = 120$ volts, $I = 50$ A, and $\theta = 30°$, $P = (120)(50)\cos 30° = 5196$ W and $Q = (120)(50)\sin 30° = 3000$ VAR.

A 208-V generator supplies power to a group of three loads. Load 1 has an apparent power of 500 VA with $\theta = 36.87°$ (i.e., it is net inductive). Load 2 has an apparent power of 1000 VA and is net capacitive with a power triangle angle of $-53.13°$. Load 3 is purely resistive with power $P_3 = 200$ W. Determine the power triangle for the combined loads and the generator current.

PRACTICE PROBLEMS 3

Answers: $S_T = 1300$ VA, $\theta_T = -22.6°$, $I = 6.25$ A

17.8   Power Factor

The quantity $\cos \theta$ in Equation 17–20 is defined as **power factor** and is given the symbol F_p. Thus,

$$F_p = \cos \theta \qquad (17–22)$$

From Equation 17–20, we see that F_p may be computed as the ratio of real power to apparent power. Thus,

$$\cos \theta = P/S \qquad (17–23)$$

Power factor is expressed as a number or as a percent. From Equation 17–23, it is apparent that power factor cannot exceed 1.0 (or 100% if expressed in percent).

The **power factor angle** θ is of interest. It can be found as

$$\theta = \cos^{-1}(P/S) \qquad (17–24)$$

Angle θ is the angle between voltage and current. For a pure resistance, therefore, $\theta = 0°$. For a pure inductance, $\theta = 90°$; for a pure capacitance, $\theta = -90°$. For a circuit containing both resistance and inductance, θ will be somewhere between $0°$ and $90°$; for a circuit containing both resistance and capacitance, θ will be somewhere between $0°$ and $-90°$.

Unity, Lagging, and Leading Power Factor

As indicated by Equation 17–23, a load's power factor shows how much of its apparent power is actually real power. For example, for a purely resistive circuit, $\theta = 0°$ and $F_p = \cos 0° = 1.0$. Therefore, $P = VI$ (watts) and all the load's apparent power is real power. This case ($F_p = 1$) is referred to as *unity* power factor.

For a load containing only resistance and inductance, the load current lags voltage. The power factor in this case is described as *lagging*. On the other hand, for a load containing only resistance and capacitance, current leads voltage and the power factor is described as *leading*. Thus, *an inductive circuit has a lagging power factor, while a capacitive circuit has a leading power factor.*

A load with a very poor power factor can draw excessive current. This is discussed next.

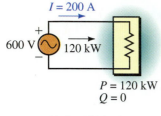

$P = 120$ kW
$Q = 0$

(a) $S = 120$ kVA

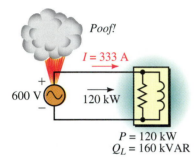

$P = 120$ kW
$Q_L = 160$ kVAR

(b) $S = \sqrt{(120)^2 + (160)^2} = 200$ kVA
The generator is overloaded.

FIGURE 17–18 Illustrating why electrical apparatus is rated in VA instead of watts. Both loads dissipate 120 kW, but the current rating of generator (b) is exceeded because of the power factor of its load.

Why Equipment Is Rated in VA

We now examine why electrical apparatus is rated in VA instead of watts. Consider Figure 17–18. Assume that the generator is rated at 600 V, 120 kVA. This means that it is capable of supplying $I = 120$ kVA/600 V = 200 A. In (a), the generator is supplying a purely resistive load with 120 kW. Since $S = P$ for a purely resistive load, $S = 120$ kVA and the generator is supplying its rated kVA. In (b), the generator is supplying a load with $P = 120$ kW as before, but $Q = 160$ kVAR. Its apparent power is therefore $S = 200$ kVA, which means that the generator current is $I = 200$ kVA/600 V = 333.3 A. Even though it is supplying the same power as in (a), the generator is now greatly overloaded, and damage may result as indicated in (b).

This example illustrates clearly that rating a load or device in terms of power is a poor choice, as its current-carrying capability can be greatly exceeded (even though its power rating is not). Thus, *the size of electrical apparatus (generators, interconnecting wires, transformers, etc.) required to supply a load is governed, not by the load's power requirements, but rather by its VA requirements.*

Power Factor Correction

The problem shown in Figure 17–18 can be alleviated by cancelling some or all of the reactive component of power by adding reactance of the opposite type to the circuit. This is referred to as **power factor correction.** If you completely cancel the reactive component, the power factor angle is 0° and $F_p = 1$. This is referred to as **unity power factor correction.**

Residential customers are charged solely on the basis of energy used. This is because all residential power factors are essentially the same, and the

EXAMPLE 17–7   For the circuit of Figure 17–18(b), a capacitance with $Q_C = 160$ kVAR is added in parallel with the load as in Figure 17–19(a). Determine generator current I.

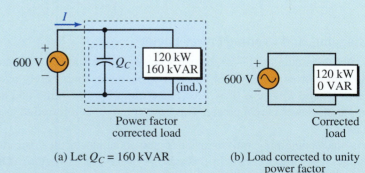

(a) Let $Q_C = 160$ kVAR

(b) Load corrected to unity power factor

FIGURE 17–19   Power factor correction. The parallel capacitor greatly reduces source current.

Solution $Q_T = 160$ kVAR − 160 kVAR = 0. Therefore, $\mathbf{S}_T = 120$ kW + $j0$ kVAR. Thus, $S_T = 120$ kVA, and $I = 120$ kVA/600 V = 200 A. Thus, the generator is no longer overloaded.

power factor effect is simply built into the tariff. Industrial customers, on the other hand, have widely different power factors, and the electrical utility may have to take the power factors of these customers into account.

To illustrate, assume that the loads of Figures 17–18(a) and (b) are two small industrial plants. If the utility based its charge solely on power, both customers would pay the same amount. However, it costs the utility more to supply customer (b) since larger conductors, larger transformers, larger switchgear, and so on are required to handle the larger current. For this reason, industrial customers may pay a penalty if their power factor drops below a prescribed value.

EXAMPLE 17–8   An industrial client is charged a penalty if the plant power factor drops below 0.85. The equivalent plant loads are as shown in Figure 17–20. The frequency is 60 Hz.

a. Determine P_T and Q_T.

b. Determine what value of capacitance (in microfarads) is required to bring the power factor up to 0.85.

c. Determine generator current before and after correction.

Solution

a. The components of power are as follows:

   Lights: $P = 12$ kW, $Q = 0$ kVAR

   Furnace: $P = I^2R = (150)^2(2.4) = 54$ kW

   $\qquad Q = I^2X = (150)^2(3.2) = 72$ kVAR (ind.)

   Motor: $\theta_m = \cos^{-1}(0.8) = 36.9°$. Thus, from the motor power triangle,

   $\qquad Q_m = P_m \tan \theta_m = 80 \tan 36.9° = 60$ kVAR (ind.)

   Total: $P_T = 12$ kW $+ 54$ kW $+ 80$ kW $= 146$ kW

   $\qquad Q_T = 0 + 72$ kVAR $+ 60$ kVAR $= 132$ kVAR

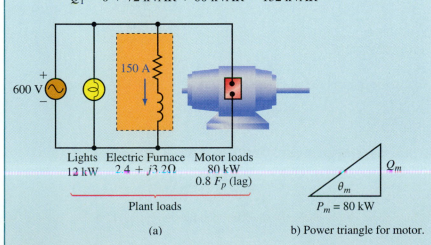

Lights   Electric Furnace   Motor loads
12 kW $2.4 + j3.2\,\Omega$ 80 kW
                              $0.8\ F_p$ (lag)

Plant loads

(a)

b) Power triangle for motor.

FIGURE 17–20

b. The power triangle for the plant is shown in Figure 17–21(a). However, we must correct the power factor to 0.85. Thus we need $\theta' = \cos^{-1}(0.85) = 31.8°$, where θ' is the power factor angle of the corrected load as indicated in Figure 17–21(b). The maximum reactive power that we can tolerate is thus $Q_T' = P_T \tan \theta' = 146 \tan 31.8° = 90.5$ kVAR.

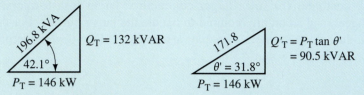

(a) Power triangle for
the plant

(b) Power triangle after correction

FIGURE 17–21   Initial and final power triangles. Note that P_T does not change when we correct the power factor.

Now consider Figure 17–22. $Q_T' = Q_C + 132$ kVAR, where $Q_T' = 90.5$ kVAR. Therefore, $Q_C = -41.5$ kVAR $= 41.5$ kVAR (cap.). But $Q_C = V^2/X_C$. Therefore, $X_C = V^2/Q_C = (600)^2/41.5$ kVAR $= 8.67$ Ω. But $X_C = 1/\omega C$. Thus a capacitor of

$$C = \frac{1}{\omega X_C} = \frac{1}{(2\pi)(60)(8.67)} = 306 \text{ μF}$$

will provide the required correction.

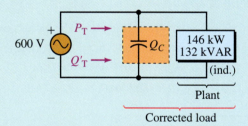

FIGURE 17–22

c. For the original circuit Figure 17–21(a), $S_T = 196.8$ kVA. Thus,

$$I = \frac{S_T}{E} = \frac{196.8 \text{ kVA}}{600 \text{ V}} = 328 \text{ A}$$

For the corrected circuit 17–21(b), $S_T' = 171.8$ kVA and

$$I = \frac{171.8 \text{ kVA}}{600 \text{ V}} = 286 \text{ A}$$

Thus, power factor correction has dropped the current by 42 A.

1. Repeat Example 17–8 except correct the power factor to unity.
2. Due to plant expansion, 102 kW of purely resistive load is added to the plant of Figure 17–20. Determine whether power factor correction is needed to correct the expanded plant to 0.85 F_p, or better.

Answers: 1. 973 μF, 243 A. Other answers remain unchanged.

2. $F_p = 0.88$. No correction needed.

In practice, almost all loads (industrial, residential, and commercial) are inductive due to the presence of motors, fluorescent lamp ballasts, and the like. Consequently, you will likely never run into capacitive loads that need power factor correcting.

1. Sketch the power triangle for Figure 17–9(c). Using this triangle, determine the magnitude of the applied voltage.
2. For Figure 17–10, assume a source of $E = 240$ volts, $P_2 = 300$ W, and $Q_2 = 400$ VAR (cap.). What is the magnitude of the source current I?
3. What is the power factor of each of the circuits of Figure 17–7, 17–8, and 17–9? Indicate whether they are leading or lagging.
4. Consider the circuit of Figure 17–18(b). If $P = 100$ kW and $Q_L = 80$ kVAR, is the source overloaded, assuming it is capable of handling a 120-kVA load?

(Answers are at the end of the chapter.)

17.9 AC Power Measurement

To measure power in an ac circuit, you need a wattmeter (since the product of voltage times current is not sufficient to determine ac power). Figure 17–23 shows such a meter. It is a digital device that monitors voltage and current and from these, computes and displays power. (You may also encounter older electrodynamometer type wattmeters, i.e., electromechanical analog instruments that use a pivoted pointer to indicate the power reading on a scale, much like the analog meters of Chapter 2. Although their details differ dramatically from the electronic types, the manner in which they are connected in a circuit to measure power is the same. Thus, the measurement techniques described below apply to them as well.)

To help understand power measurement, consider Figure 17–1. Instantaneous load power is the product of load voltage times load current, and average power is the average of this product. One way to implement power measurement is therefore to create a meter with a current sensing circuit, a voltage sensing circuit, a multiplier circuit, and an averaging circuit. Figure 17–24 shows a simplified symbolic representation of such an instrument. Current is passed through its current coil (CC) to create a magnetic field proportional to the current, and a sensor circuit connected across the load voltage reacts with this field to produce an output voltage proportional to the product of instantaneous voltage and current (i.e., proportional to

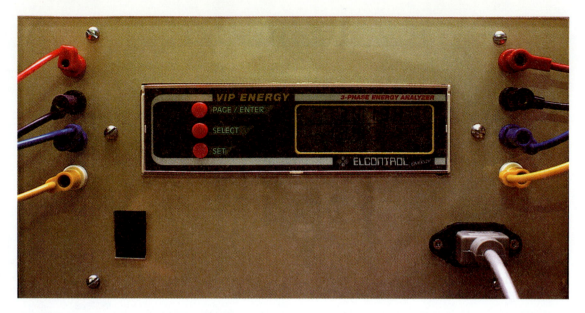

FIGURE 17–23   Multifunction power/energy meter. It can measure active power (W), reactive power (VARs), apparent power (VA), power factor, energy, and more.

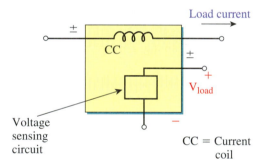

FIGURE 17–24   Conceptual representation of a wattmeter.

instantaneous power). An averaging circuit averages this voltage and drives a display to indicate average power. (The scheme used by the meter of Figure 17–23 is actually considerably more sophisticated than this because it measures many things besides power—e.g., it measures, VARs, VA, energy, etc. However, the basic idea is conceptually correct.)

Figure 17–25 shows how to connect a wattmeter (whether electronic or electromechanical) into a circuit. Load current passes through its current coil circuit, and load voltage is impressed across its voltage sensing circuit. With this connection, the wattmeter computes and displays the product of the magnitude of the load voltage, the magnitude of the load current, and the cosine of the angle between them, i.e. $V_{load} \cdot I_{load} \cdot \cos \theta_{load}$. Thus, it measures load power. Note the $\pm$ marking on the terminals. You usually connect the meter so that load current enters the $\pm$ current terminal and the higher potential end of the load is connected to the $\pm$ voltage terminal. On many meters, the $\pm$ voltage terminal is internally connected so that only three terminals are brought out as in Figure 17–26.

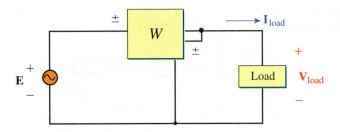

FIGURE 17–25   Connection of wattmeter.

When power is to be measured in a low power factor circuit, a **low power factor wattmeter** must be used. This is because, for low power factor loads, currents can be very high, even though the power is low. Thus, you can easily exceed the current rating of a standard wattmeter and damage it, even though the power indication on the meter is small.

EXAMPLE 17–9   For the circuit of Figure 17–25, what does the wattmeter indicate if

a. $\mathbf{V}_{load} = 100\ \text{V}\angle 0°$ and $\mathbf{I}_{load} = 15\ \text{A}\angle 60°$,

b. $\mathbf{V}_{load} = 100\ \text{V}\angle 10°$ and $\mathbf{I}_{load} = 15\ \text{A}\angle 30°$?

Solution

a. $\theta_{load} = 60°$. Thus, $P = (100)(15)\cos 60° = 750\ \text{W}$,

b. $\theta_{load} = 10° - 30° = -20°$. Thus, $P = (100)(15)\cos(-20°) = 1410\ \text{W}$.

Note: For (b), since $\cos(-20°) = \cos(+20°)$, it does not matter whether we include the minus sign.

EXAMPLE 17–10   For Figure 17–26, determine the wattmeter reading.

FIGURE 17–26   This wattmeter has its voltage side ± terminals connected internally.

Solution   A wattmeter reads only active power. Thus, it indicates 600 W.

It should be noted that the wattmeter reads power only for circuit elements on the load side of the meter. In addition, if the load consists of several elements, it reads the sum of the powers.

PRACTICE PROBLEMS 5

Determine the wattmeter reading for Figure 17–27.

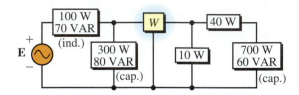

FIGURE 17–27

Answer: 750 W

17.10 Effective Resistance

Up to now, we have assumed that resistance is constant, independent of frequency. However, this is not entirely true. For a number of reasons, the resistance of a circuit to ac is greater than its resistance to dc. While this effect is small at low frequencies, it is very pronounced at high frequencies. AC resistance is known as **effective resistance.**

Before looking at why ac resistance is greater than dc resistance, we need to reexamine the concept of resistance itself. Recall from Chapter 3 that resistance was originally defined as opposition to current, that is, $R = V/I$. (This is ohmic resistance.) Building on this, you learned in Chapter 4 that $P = I^2R$. It is this latter viewpoint that allows us to give meaning to ac resistance. That is, we define ac or effective resistance as

$$R_{\text{eff}} = \frac{P}{I^2} \quad (\Omega) \qquad (17\text{–}25)$$

where P is dissipated power (as determined by a wattmeter). From this, you can see that anything that affects dissipated power affects resistance. For dc and low-frequency ac, both definitions for R, i.e., $R = V/I$ and $R = P/I^2$ yield the same value. However, as frequency increases, other factors cause an increase in resistance. We will now consider some of these.

Eddy Currents and Hysteresis

The magnetic field surrounding a coil or other circuit carrying ac current varies with time and thus induces voltages in nearby conductive material such as metal equipment cabinets, transformer cores, and so on. The resulting currents (called **eddy currents** because they flow in circular patterns like eddies in a brook) are unwanted and create power losses called **eddy current losses.** Since additional power must be supplied to make up for these losses, P in Equation 17–25 increases, increasing the effective resistance of the coil.

If ferromagnetic material is also present, an additional power loss occurs due to hysteresis effects caused by the magnetic field alternately magnetizing the material in one direction, then the other. Hysteresis and eddy current losses are important even at low frequencies, such as the 60-Hz power system frequency. This is discussed in Chapter 24.

Skin Effect

Magnetically induced voltages created inside a conductor by its own changing magnetic field force electrons to the periphery of the conductor (Figure 17–28), resulting in a nonuniform distribution of current, with current density greatest near the periphery and smallest in the center. This phenomenon is known as **skin effect.** Because the center of the wire carries little current, its cross-sectional area has effectively been reduced, thus increasing resistance. While skin effect is generally negligible at power line frequencies (except for conductors larger than several hundred thousand circular mils), it is so pronounced at microwave frequencies that the center of a wire carries almost no current. For this reason, hollow conductors are often used instead of solid wires, as shown in Figure 17–28(c).

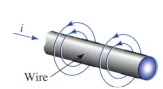

Wire

(a) Varying magnetic field creates forces on electrons within the conductor

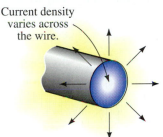

Current density varies across the wire.

(b) This force pushes electrons outward, leaving few free electrons in the center of the conductor

(c) At high frequencies, the effect is so pronounced that hollow conductors may be used

FIGURE 17–28   Skin effect in ac circuits.

Radiation Resistance

At high frequencies some of the energy supplied to a circuit escapes as radiated energy. For example, a radio transmitter supplies power to an antenna, where it is converted into radio waves and radiated into space. The resistance effect here is known as **radiation resistance.** This resistance is much higher than simple dc resistance. For example, a TV transmitting antenna may have a resistance of a fraction of an ohm to dc but several hundred ohms effective resistance at its operating frequency.

> **FINAL NOTES…**
>
> 1. The resistance measured by an ohmmeter is dc resistance.
>
> 2. Many of the effects noted above will be treated in detail in your various electronics courses. We will not pursue them further here.

17.11 Energy Relationships for AC

Recall, power and energy are related by the equation $p = dw/dt$. Thus, energy can be found by integration as

$$W = \int p \, dt = \int vi \, dt \qquad \textbf{(17–26)}$$

Inductance

For an inductance, $v = L \, di/dt$. Substituting this into Equation 17–26, cancelling dt, and rearranging terms yields

$$W_L = \int \left(L \frac{di}{dt} \right) i \, dt = L \int i \, di \qquad \textbf{(17–27)}$$

Recall from Figure 17–4(b), energy flows into an inductor during time interval 0 to $T/4$ and is released during time interval $T/4$ to $T/2$. The process then repeats itself. The energy stored (and subsequently released) can thus be found by integrating power from $t = 0$ to $t = T/4$. Current at $t = 0$ is 0 and current at $t = T/4$ is I_m. Using these as our limits of integration, we find

$$W_L = L \int_0^{I_m} i\, di = \frac{1}{2}LI_m^2 = LI^2 \quad \text{(J)} \tag{17–28}$$

where we have used $I = I_m/\sqrt{2}$ to express energy in terms of effective current.

Capacitance

For a capacitance, $i = C\,dv/dt$. Substituting this into Equation 17–26 yields

$$W_C = \int v \left(C \frac{dv}{dt} \right) dt = C \int v\, dv \tag{17–29}$$

Consider Figure 17–5(b). Energy stored can be found by integrating power from $T/4$ to $T/2$. The corresponding limits for voltage are 0 to V_m. Thus,

$$W_C = C \int_0^{V_m} v\, dv = \frac{1}{2}CV_m^2 = CV^2 \quad \text{(J)} \tag{17–30}$$

where we have used $V = V_m/\sqrt{2}$. You will use these relationships later.

17.12 Circuit Analysis Using Computers

The time-varying relationships between voltage, current and power described earlier in this chapter can be investigated easily using PSpice. To illustrate, consider the circuit of Figure 17–3 with $v = 1.2 \sin \omega t$, $R = 0.8\ \Omega$ and $f = 1000$ Hz. Create the circuit on the screen, including voltage and current markers as in Figure 17–29, then set VSIN parameters to VOFF = **0V,** VAMPL = **1.2V** and FREQ = **1000Hz** as you did in Chapter 16. Click the New Simulation Profile, type **fig17-29** for a name, choose Transient, set TSTOP to **1ms**, then click OK. Run the simulation and the voltage and current waveforms of Figure 17–30 should appear. To plot power (i.e., the product of vi), click <u>T</u>race, then <u>A</u>dd Trace, and when the dialog box opens, use the asterisk to create the product

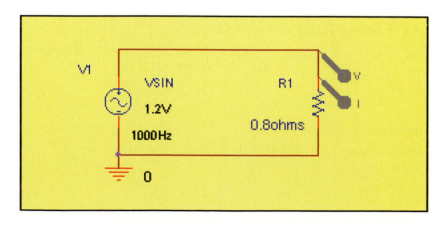

FIGURE 17–29 Using PSpice to investigate instantaneous power in an ac circuit.

V(R1:1)*I(R1), then click OK. The blue power curve should now appear. Compare to Figure 17–3. Note that all curves agree exactly.

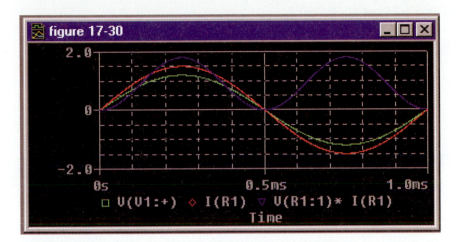

FIGURE 17–30 Voltage, current, and power waveforms for Figure 17–29.

17.1–17.5

1. Note that the power curve of Figure 17–4 is sometimes positive and sometimes negative. What is the significance of this? Between $t = T/4$ and $t = T/2$, what is the direction of power flow?

2. What is real power? What is reactive power? Which power, real or reactive, has an average value of zero?

3. A pair of electric heating elements is shown in Figure 17–31.
   a. Determine the active and reactive power to each.
   b. Determine the active and reactive power delivered by the source.

4. For the circuit of Figure 17–32, determine the active and reactive power to the inductor.

5. If the inductor of Figure 17–32 is replaced by a 40-μF capacitor and source frequency is 60 Hz, what is Q_C?

6. Find R and X_L for Figure 17–33.

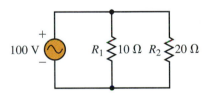

FIGURE 17–31

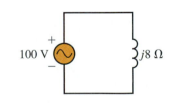

FIGURE 17–32

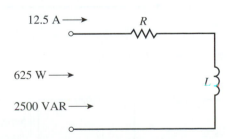

FIGURE 17–33

7. For the circuit of Figure 17–34, $f = 100$ Hz. Find
   a. R b. X_C c. C

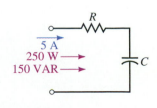

FIGURE 17–34

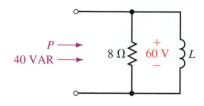

FIGURE 17–35

8. For the circuit of Figure 17–35, $f = 10$ Hz. Find
   a. P b. X_L c. L
9. For Figure 17–36, find X_C.

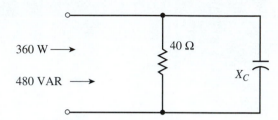

FIGURE 17–36

10. For Figure 17–37, $X_C = 42.5$ Ω. Find R, P, and Q.
11. Find the total average power and the total reactive power supplied by the source for Figure 17–38.

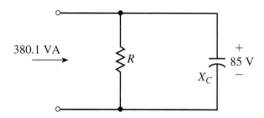

FIGURE 17–37

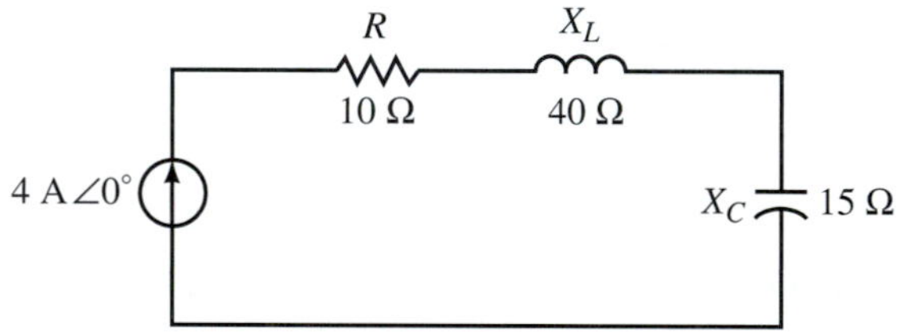

FIGURE 17–38

12. If the source of Figure 17–38 is reversed, what is P_T and Q_T? What conclusion can you draw from this?
13. Refer to Figure 17–39. Find P_2 and Q_3. Is the element in Load 3 inductive or capacitive?

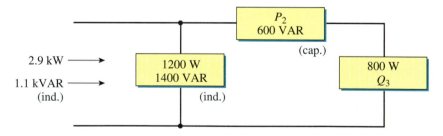

FIGURE 17–39

14. For Figure 17–40, determine P_T and Q_T.

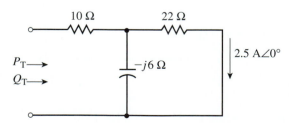

FIGURE 17–40

15. For Figure 17–41, $\omega = 10$ rad/s. Determine
   a. R_T b. R_2 c. X_C d. L_{eq}

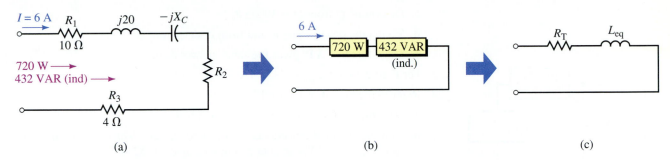

FIGURE 17–41

16. For Figure 17–42, determine the total P_T and Q_T.

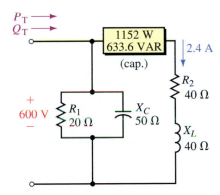

FIGURE 17–42

17.7   The Relationship Between P, Q, and S

17. For the circuit of Figure 17–7, draw the power triangle and determine the apparent power.

18. Repeat Problem 17 for Figure 17–8.

19. Ignoring the wattmeter of Figure 17–27, determine the power triangle for the circuit as seen by the source.

20. For the circuit of Figure 17–43, what is the source current?

21. For Figure 17–44, the generator supplies 30 A. What is R?

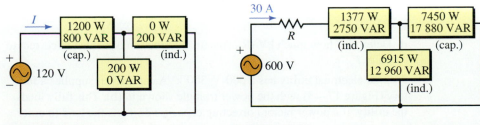

FIGURE 17–43   FIGURE 17–44

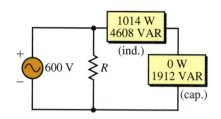

FIGURE 17–45

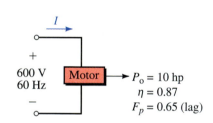

FIGURE 17–47

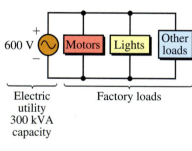

(a)

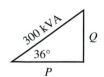

(b) Factory power triangle

FIGURE 17–48

22. Suppose $\mathbf{V} = 100 \text{ V} \angle 60°$ and $\mathbf{I} = 10 \text{A} \angle 40°$:
   a. What is θ, the angle between $\mathbf{V}$ and $\mathbf{I}$?
   b. Determine P from $P = VI \cos \theta$.
   c. Determine Q from $Q = VI \sin \theta$.
   d. Sketch the power triangle and from it, determine $\mathbf{S}$.
   e. Show that $\mathbf{S} = \mathbf{VI}^*$ gives the same answer as (d).

23. For Figure 17–45, $S_{gen} = 4835$ VA. What is R?

24. Refer to the circuit of Figure 17–16:
   a. Determine the apparent power for each box.
   b. Sum the apparent powers that you just computed. Why does the sum not equal $S_T = 1803$ VA as obtained in Example 17–5?

17.8   Power Factor

25. Refer to the circuit of Figure 17–46:
   a. Determine P_T, Q_T, and S_T.
   b. Determine whether the fuse will blow.

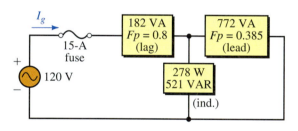

FIGURE 17–46

26. A motor with an efficiency of 87% supplies 10 hp to a load (Figure 17–47). Its power factor is 0.65 (lag).
   a. What is the power input to the motor?
   b. What is the reactive power to the motor?
   c. Draw the motor power triangle. What is the apparent power to the motor?

27. To correct the circuit power factor of Figure 17–47 to unity, a power factor correction capacitor is added.
   a. Show where the capacitor is connected.
   b. Determine its value in microfarads.

28. Consider Figure 17–20. The motor is replaced with a new unit requiring $\mathbf{S}_m = (120 + j35)$ kVA. Everything else remains the same. Find the following:
   a. P_T b. Q_T c. S_T
   d. Determine how much kVAR capacitive correction is needed to correct to unity F_p.

29. A small electrical utility has a 600-V, 300-kVA capacity. It supplies a factory (Figure 17–48) with the power triangle shown in (b). This fully loads the utility. If a power factor correcting capacitor corrects the load to unity

power factor, how much more power (at unity power factor) can the utility sell to other customers?

17.9   AC Power Measurement

30. a. Why does the wattmeter of Figure 17–49 indicate only 1200 watts?

   b. Where would the wattmeter have to be placed to measure power delivered by the source? Sketch the modified circuit.

   c. What would the wattmeter indicate in (b)?

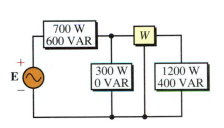

FIGURE 17–49

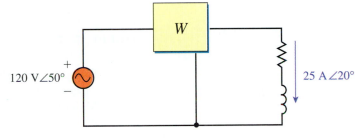

FIGURE 17–50

31. Determine the wattmeter reading for Figure 17–50.

32. Determine the wattmeter reading for Figure 17–51.

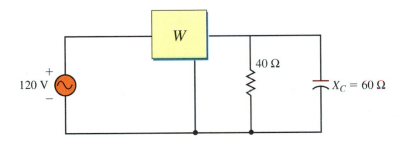

FIGURE 17–51

17.10   Effective Resistance

33. Measurements on an iron-core solenoid coil yield the following values: $V = 80$ V, $I = 400$ mA, $P = 25.6$ W, and $R = 140$ Ω. (The last measurement was taken with an ohmmeter.) What is the ac resistance of the solenoid coil?

17.12   Circuit Analysis Using Computers

34. **PSpice**   An inductance $L = 1$ mH has current $i = 4 \sin (2\pi \times 1000)t$. Use PSpice to investigate the power waveform and compare to Figure 17–4. Use current source ISIN (see Note).

35. **PSpice**   A 10-μF capacitor has voltage $v = 10 \sin(\omega t - 90°)$ V. Use PSpice to investigate the power waveform and compare to Figure 17–5. Use voltage source VSIN with $f = 1000$ Hz.

NOTE...

PSpice represents current into devices. Thus, when you double click a current source symbol (ISIN, IPWL, etc.) and specify a current waveform, you are specifying the current *into* the source.

36. **PSpice**   The voltage waveform of Figure 17–52 is applied to a 200-μF capacitor.

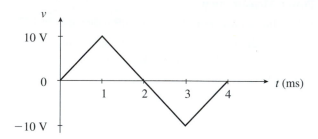

FIGURE 17–52

a. Using the principles of Chapter 10, determine the current through the capacitor and sketch. (Also sketch the voltage waveform on your graph.) Multiply the two waveforms to obtain a plot of $p(t)$. Compute power at its max and min points.

b. Use PSpice to verify the results. Use voltage source VPWL. You have to describe the waveform to the source. It has a value of 0 V at $t = 0$, 10 V at $t = 1$ ms, -10 V at $t = 3$ ms, and 0 V at $t = 4$ ms. To set these, double click the source symbol and enter values via the Property Editor as follows: **0** for T1, **0V** for V1, **1ms** for T2, **10V** for V2, etc. Run the simulation and plot voltage, current, and power using the procedure we used to create Figure 17–30. Results should agree with those of (a).

37. **PSpice**   Repeat Question 36 for a current waveform identical to Figure 17–52 except that it oscillates between 2 A and -2 A applied to a 2-mH inductor. Use current source IPWL (see Note).

ANSWERS TO IN-PROCESS LEARNING CHECKS

In-Process Learning Check 1

1. 100 V

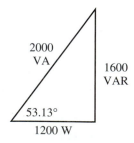

FIGURE 17–53

2. 8.76 A

3. Fig. 17–7: 0.6 (lag); Fig. 17–8: 0.97 (lead); Fig. 17–9: 0.6 (lag)

4. Yes. ($S = 128$ kVA)

Impedance Networks

PART V

18 AC Series-Parallel Circuits

19 Methods of AC Analysis

20 AC Network Theorems

21 Resonance

22 Filters and the Bode Plot

23 Three-Phase Systems

24 Transformers

25 Nonsinusoidal Waveforms

As you have already observed, the impedance of an inductor or a capacitor is dependent upon the frequency of the signal applied to the element. When capacitors and inductors are combined with resistors and voltage or current sources, the circuit will behave in a predictable manner for all frequencies.

The final part of this textbook examines how circuits consisting of various combinations of impedances and sources behave under specific conditions. In particular, we find that all the laws, rules, and theorems developed previously apply to even the most complicated impedance network.

Ohm's law and Kirchhoff's voltage and current laws are easily modified to give the framework for developing methods of network analysis. Just as in dc circuits, Thévenin's and Norton's theorems will allow us to simplify a complicated circuit to a single source and corresponding impedance.

The theorems and methods of analysis are applied to numerous types of circuits which are commonly encountered throughout electrical and electronics technology. Resonant circuits and filter circuits are commonly used to restrict the range of output frequencies for a given range of input frequencies.

The study of three-phase systems and transformers is particularly useful for anyone interested in commercial power distribution. These topics deal with practical applications and the drawbacks of using various types of circuits.

Finally, we examine how a circuit reacts to nonsinusoidal alternating voltages. This topic involves complex signals which are processed by impedance networks resulting in outputs which are often dramatically different from the input.

18

AC Series-Parallel Circuits

OBJECTIVES

After studying this chapter, you will be able to

- apply Ohm's law to analyze simple series circuits,

- apply the voltage divider rule to determine the voltage across any element in a series circuit,

- apply Kirchhoff's voltage law to verify that the summation of voltages around a closed loop is equal to zero,

- apply Kirchhoff's current law to verify that the summation of currents entering a node is equal to the summation of currents leaving the same node,

- determine unknown voltage, current, and power for any series/parallel circuit,

- determine the series or parallel equivalent of any network consisting of a combination of resistors, inductors, and capacitors.

KEY TERMS

AC Parallel Circuits
AC Series Circuits
Current Divider Rule

Frequency Effects
Impedance
Kirchhoff's Current Law
Kirchhoff's Voltage Law
Voltage Divider Rule

OUTLINE

Ohm's Law for AC Circuits
AC Series Circuits
Kirchhoff's Voltage Law and the Voltage Divider Rule
AC Parallel Circuits
Kirchhoff's Current Law and the Current Divider Rule
Series-Parallel Circuits
Frequency Effects
Applications
Circuit Analysis Using Computers

In this chapter we examine how simple circuits containing resistors, inductors, and capacitors behave when subjected to sinusoidal voltages and currents. Principally, we find that the rules and laws which were developed for dc circuits will apply equally well for ac circuits. The major difference between solving dc and ac circuits is that analysis of ac circuits requires using vector algebra.

In order to proceed successfully, it is suggested that the student spend time reviewing the important topics covered in dc analysis. These include Ohm's law, the voltage divider rule, Kirchhoff's voltage law, Kirchhoff's current law, and the current divider rule.

You will also find that a brief review of vector algebra will make your understanding of this chapter more productive. In particular, you should be able to add and subtract any number of vector quantities.

Heinrich Rudolph Hertz

HEINRICH HERTZ WAS BORN IN HAMBURG, Germany, on February 22, 1857. He is known mainly for his research into the transmission of electromagnetic waves.

Hertz began his career as an assistant to Hermann von Helmholtz in the Berlin Institute physics laboratory. In 1885, he was appointed Professor of Physics at Karlsruhe Polytechnic, where he did much to verify James Clerk Maxwell's theories of electromagnetic waves.

In one of his experiments, Hertz discharged an induction coil with a rectangular loop of wire having a very small gap. When the coil discharged, a spark jumped across the gap. He then placed a second, identical coil close to the first, but with no electrical connection. When the spark jumped across the gap of the first coil, a smaller spark was also induced across the second coil. Today, more elaborate antennas use similar principles to transmit radio signals over vast distances. Through further research, Hertz was able to prove that electromagnetic waves have many of the characteristics of light: they have the same speed as light; they travel in straight lines; they can be reflected and refracted; and they can be polarized.

Hertz's experiments ultimately led to the development of radio communication by such electrical engineers as Guglielmo Marconi and Reginald Fessenden.

Heinrich Hertz died at the age of 36 on January 1, 1894.

PUTTING IT IN PERSPECTIVE

18.1   Ohm's Law for AC Circuits

This section is a brief review of the relationship between voltage and current for resistors, inductors, and capacitors. Unlike Chapter 16, all phasors are given as rms rather than as peak values. As you saw in Chapter 17, this approach simplifies the calculation of power.

Resistors

In Chapter 16, we saw that when a resistor is subjected to a sinusoidal voltage as shown in Figure 18–1, the resulting current is also sinusoidal and in phase with the voltage.

The sinusoidal voltage $v = V_m\sin(\omega t + \theta)$ may be written in phasor form as $\mathbf{V} = V\angle\theta$. Whereas the sinusoidal expression gives the instantaneous value of voltage for a waveform having an amplitude of V_m (volts peak), the phasor form has a magnitude which is the effective (or rms) value. The relationship between the magnitude of the phasor and the peak of the sinusoidal voltage is given as

$$V = \frac{V_m}{\sqrt{2}}$$

Because the resistance vector may be expressed as $\mathbf{Z}_R = R\angle 0°$, we evaluate the current phasor as follows:

$$\mathbf{I} = \frac{\mathbf{V}}{\mathbf{Z}_R} = \frac{V\angle\theta}{R\angle 0°} = \frac{V}{R}\angle\theta = I\angle\theta$$

> **NOTES...**
>
> Although currents and voltages may be shown in either time domain (as sinusoidal quantities) or in phasor domain (as vectors), resistance and reactance are never shown as sinusoidal quantities. The reason for this is that whereas currents and voltages vary as functions of time, resistance and reactance do not.

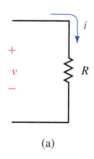

(a)

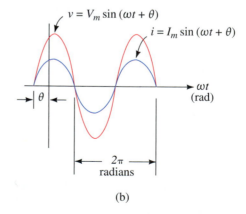

(b)

FIGURE 18–1   Sinusoidal voltage and current for a resistor.

If we wish to convert the current from phasor form to its sinusoidal equivalent in the time domain, we would have $i = I_m \sin(\omega t + \theta)$. Again, the relationship between the magnitude of the phasor and the peak value of the sinusoidal equivalent is given as

$$I = \frac{I_m}{\sqrt{2}}$$

The voltage and current phasors may be shown on a phasor diagram as in Figure 18–2.

Because one phasor is a current and the other is a voltage, the relative lengths of these phasors are purely arbitrary. Regardless of the angle θ, we see that the voltage across and the current through a resistor will always be in phase.

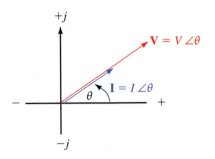

FIGURE 18–2   Voltage and current phasors for a resistor.

EXAMPLE 18–1   Refer to the resistor shown in Figure 18–3:

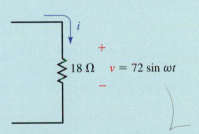

FIGURE 18–3

a. Find the sinusoidal current i using phasors.

b. Sketch the sinusoidal waveforms for v and i.

c. Sketch the phasor diagram of **V** and **I**.

Solution:

a. The phasor form of the voltage is determined as follows:

$$v = 72 \sin \omega t \Leftrightarrow \mathbf{V} = 50.9 \text{ V} \angle 0°$$

From Ohm's law, the current phasor is determined to be

$$\mathbf{I} = \frac{\mathbf{V}}{\mathbf{Z}_R} = \frac{50.9 \text{ V} \angle 0°}{18 \text{ } \Omega \angle 0°} = 2.83 \text{ A} \angle 0°$$

which results in the sinusoidal current waveform having an amplitude of

$$I_m = (\sqrt{2})(2.83 \text{ A}) = 4.0 \text{ A}$$

Therefore, the current i will be written as

$$i = 4 \sin \omega t$$

b. The voltage and current waveforms are shown in Figure 18–4.

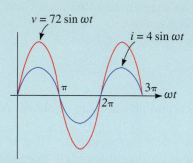

FIGURE 18–4

c. Figure 18–5 shows the voltage and current phasors.

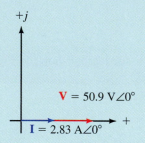

FIGURE 18–5

EXAMPLE 18–2   Refer to the resistor of Figure 18–6:

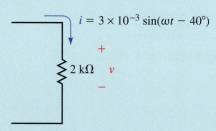

FIGURE 18–6

a. Use phasor algebra to find the sinusoidal voltage, v.
b. Sketch the sinusoidal waveforms for v and i.
c. Sketch a phasor diagram showing **V** and **I.**

Solution

a. The sinusoidal current has a phasor form as follows:

$$i = 3 \times 10^{-3}\sin(\omega t - 40°) \Leftrightarrow \mathbf{I} = 2.12 \text{ mA}\angle -40°$$

From Ohm's law, the voltage across the 2-kΩ resistor is determined as the phasor product

$$\mathbf{V} = \mathbf{I}\mathbf{Z}_R$$
$$= (2.12 \text{ mA}\angle{-40°})(2 \text{ k}\Omega\angle{0°})$$
$$= 4.24 \text{ V}\angle{-40°}$$

The amplitude of the sinusoidal voltage is

$$V_m = (\sqrt{2})(4.24 \text{ V}) = 6.0 \text{ V}$$

The voltage may now be written as

$$v = 6.0 \sin(\omega t - 40°)$$

b. Figure 18–7 shows the sinusoidal waveforms for v and i.

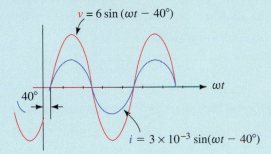

FIGURE 18–7

c. The corresponding phasors for the voltage and current are shown in Figure 18–8.

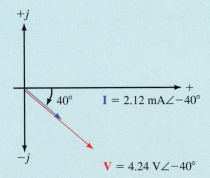

FIGURE 18–8

Inductors

When an inductor is subjected to a sinusoidal current, a sinusoidal voltage is induced across the inductor such that the voltage across the inductor leads the current waveform by exactly 90°. If we know the reactance of an inductor, then from Ohm's law the current in the inductor may be expressed in phasor form as

$$\mathbf{I} = \frac{\mathbf{V}}{\mathbf{Z}_L} = \frac{V\angle\theta}{X_L\angle{90°}} = \frac{V}{X_L}\angle(\theta - 90°)$$

In vector form, the reactance of the inductor is given as

$$\mathbf{Z}_L = X_L \angle 90°$$

where $X_L = \omega L = 2\pi f L$.

EXAMPLE 18–3   Consider the inductor shown in Figure 18–9:

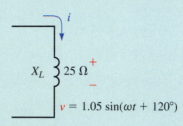

FIGURE 18–9

a. Determine the sinusoidal expression for the current i using phasors.
b. Sketch the sinusoidal waveforms for v and i.
c. Sketch the phasor diagram showing $\mathbf{V}$ and $\mathbf{I}$.

Solution:

a. The phasor form of the voltage is determined as follows:

$$v = 1.05 \sin(\omega t + 120°) \Leftrightarrow \mathbf{V} = 0.742\ \text{V} \angle 120°$$

From Ohm's law, the current phasor is determined to be

$$\mathbf{I} = \frac{\mathbf{V}}{\mathbf{Z}_L} = \frac{0.742\ \text{V} \angle 120°}{25\ \Omega \angle 90°} = 29.7\ \text{mA} \angle 30°$$

The amplitude of the sinusoidal current is

$$I_m = (\sqrt{2})(29.7\ \text{mA}) = 42\ \text{mA}$$

The current i is now written as

$$i = 0.042 \sin(\omega t + 30°)$$

b. Figure 18–10 shows the sinusoidal waveforms of the voltage and current.

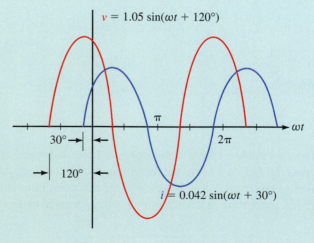

FIGURE 18–10   Sinusoidal voltage and current for an inductor.

c. The voltage and current phasors are shown in Figure 18–11.

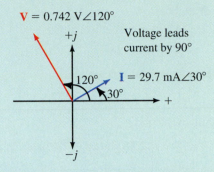

V = 0.742 V∠120°

+j

Voltage leads
current by 90°

120°

I = 29.7 mA∠30°

30°

+

−j

FIGURE 18–11   Voltage and current phasors for an inductor.

Capacitors

When a capacitor is subjected to a sinusoidal voltage, a sinusoidal current results. The current through the capacitor leads the voltage by exactly 90°. If we know the reactance of a capacitor, then from Ohm's law the current in the capacitor expressed in phasor form is

$$\mathbf{I} = \frac{\mathbf{V}}{\mathbf{Z}_C} = \frac{V\angle\theta}{X_L\angle-90°} = \frac{V}{X_L}\angle(\theta + 90°)$$

In vector form, the reactance of the capacitor is given as

$$\mathbf{Z}_C = X_C\angle-90°$$

where

$$X_C = \frac{1}{\omega C} = \frac{1}{2\pi f C}$$

EXAMPLE 18–4   Consider the capacitor of Figure 18–12.

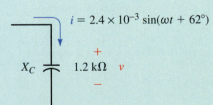

$i = 2.4 \times 10^{-3}\sin(\omega t + 62°)$

+

X_C 1.2 kΩ v

−

FIGURE 18–12

a. Find the voltage v across the capacitor.

b. Sketch the sinusoidal waveforms for v and i.

c. Sketch the phasor diagram showing **V** and **I**.

Solution

a. Converting the sinusoidal current into its equivalent phasor form gives

$$i = 2.4 \times 10^{-3}\sin(\omega t + 62°) \Leftrightarrow \mathbf{I} = 1.70\text{ mA}\angle62°$$

From Ohm's law, the phasor voltage across the capacitor must be

$$\mathbf{V} = \mathbf{IZ}_C$$
$$= (1.70 \text{ mA}\angle 62°)(1.2 \text{ k}\Omega\angle{-90°})$$
$$= 2.04 \text{ V}\angle{-28°}$$

The amplitude of the sinusoidal voltage is

$$V_m = (\sqrt{2})(2.04 \text{ V}) = 2.88 \text{ V}$$

The voltage v is now written as

$$v = 2.88 \sin(\omega t - 28°)$$

b. Figure 18–13 shows the waveforms for v and i.

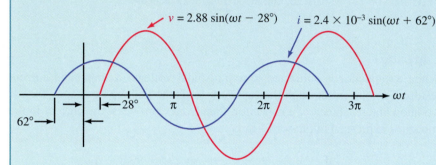

FIGURE 18–13   Sinusoidal voltage and current for a capacitor.

c. The corresponding phasor diagram for **V** and **I** is shown in Figure 18–14.

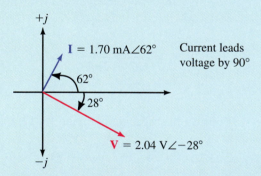

FIGURE 18–14   Voltage and current phasors for a capacitor.

The relationships between voltage and current, as illustrated in the previous three examples, will always hold for resistors, inductors, and capacitors.

1. What is the phase relationship between current and voltage for a resistor?
2. What is the phase relationship between current and voltage for a capacitor?
3. What is the phase relationship between current and voltage for an inductor?

(Answers are at the end of the chapter.)

A voltage source, $\mathbf{E} = 10\ \text{V}\angle 30°$, is applied to an inductive impedance of 50 Ω.

a. Solve for the phasor current, **I.**

b. Sketch the phasor diagram for **E** and **I.**

c. Write the sinusoidal expressions for e and i.

d. Sketch the sinusoidal expressions for e and i.

Answers:

a. $\mathbf{I} = 0.2\ \text{A}\angle -60°$

c. $e = 14.1 \sin(\omega t + 30°)$

$\quad i = 0.283 \sin(\omega t - 60°)$

A voltage source, $\mathbf{E} = 10\ \text{V}\angle 30°$, is applied to a capacitive impedance of 20 Ω.

a. Solve for the phasor current, **I.**

b. Sketch the phasor diagram for **E** and **I.**

c. Write the sinusoidal expressions for e and i.

d. Sketch the sinusoidal expressions for e and i.

Answers:

a. $\mathbf{I} = 0.5\ \text{A}\angle 120°$

c. $e = 14.1 \sin(\omega t + 30°)$

$\quad i = 0.707 \sin(\omega t + 120°)$

18.2   AC Series Circuits

When we examined dc circuits we saw that the current everywhere in a series circuit is always constant. This same applies when we have series elements with an ac source. Further, we had seen that the total resistance of a dc series circuit consisting of n resistors was determined as the summation

$$R_T = R_1 + R_2 + \cdots + R_n$$

When working with ac circuits we no longer work with only resistance but also with capacitive and inductive reactance. *Impedance is a term used to collectively determine how the resistance, capacitance, and inductance "impede" the current in a circuit.* The symbol for impedance is the letter Z and the unit is the ohm (Ω). Because impedance may be made up of any combination of resistances and reactances, it is written as a vector quantity **Z,** where

$$\mathbf{Z} = Z\angle\theta \quad (\Omega)$$

Each impedance may be represented as a vector on the complex plane, such that the length of the vector is representative of the magnitude of the impedance. The diagram showing one or more impedances is referred to as an **impedance diagram.**

Resistive impedance $\mathbf{Z}_R$ is a vector having a magnitude of R along the positive real axis. Inductive reactance $\mathbf{Z}_L$ is a vector having a magnitude of

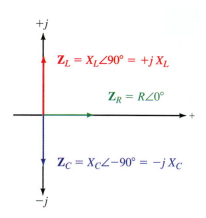

FIGURE 18–15

X_L along the positive imaginary axis, while the capacitive reactance $\mathbf{Z}_C$ is a vector having a magnitude of X_C along the negative imaginary axis. Mathematically, each of the vector impedances is written as follows:

$$\mathbf{Z}_R = R\angle 0° = R + j0 = R$$
$$\mathbf{Z}_L = X_L\angle 90° = 0 + jX_L = jX_L$$
$$\mathbf{Z}_C = X_C\angle 90° = 0 - jX_C = -jX_C$$

An impedance diagram showing each of the above impedances is shown in Figure 18–15.

All impedance vectors will appear in either the first or the fourth quadrants, since the resistive impedance vector is always positive.

For a series ac circuit consisting of n impedances, as shown in Figure 18–16, the total impedance of the circuit is found as the vector sum

$$\mathbf{Z}_T = \mathbf{Z}_1 + \mathbf{Z}_2 + \cdots + \mathbf{Z}_n \qquad\qquad \textbf{(18–1)}$$

Consider the branch of Figure 18–17.

By applying Equation 18–1, we may determine the total impedance of the circuit as

$$\mathbf{Z}_T = (3\ \Omega + j0) + (0 + j4\ \Omega) = 3\ \Omega + j4\ \Omega$$
$$= 5\ \Omega\angle 53.13°$$

The above quantities are shown on an impedance diagram as in Figure 18–18.

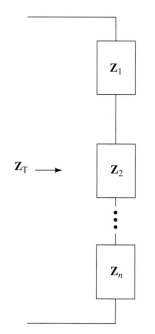

FIGURE 18–16

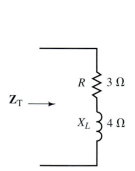

FIGURE 18–17

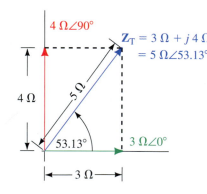

FIGURE 18–18

From Figure 18–18 we see that the total impedance of the series elements consists of a real component and an imaginary component. The corresponding total impedance vector may be written in either polar or rectangular form.

The rectangular form of an impedance is written as

$$\mathbf{Z} = R \pm jX$$

If we are given the polar form of the impedance, then we may determine the equivalent rectangular expression from

$$R = Z\cos\theta \qquad\qquad \textbf{(18–2)}$$

and

$$X = Z\sin\theta \qquad\qquad \textbf{(18–3)}$$

In the rectangular representation for impedance, the resistance term, *R,* is the total of all resistance looking into the network. The reactance term, *X,* is the difference between the total capacitive and inductive reactances. The sign for the imaginary term will be positive if the inductive reactance is greater than the capacitive reactance. In such a case, the impedance vector will appear in the first quadrant of the impedance diagram and is referred to as being an **inductive** impedance. If the capacitive reactance is larger, then the sign for the imaginary term will be negative. In such a case, the impedance vector will appear in the fourth quadrant of the impedance diagram and the impedance is said to be **capacitive.**

The polar form of any impedance will be written in the form

$$\mathbf{Z} = Z\angle\theta$$

The value *Z* is the magnitude (in ohms) of the impedance vector **Z** and is determined as follows:

$$Z = \sqrt{R^2 + X^2} \quad (\Omega) \qquad (18\text{–}4)$$

The corresponding angle of the impedance vector is determined as

$$\theta = \pm\tan^{-1}\left(\frac{X}{R}\right) \qquad (18\text{–}5)$$

Whenever a capacitor and an inductor having equal reactances are placed in series, as shown in Figure 18–19, the equivalent circuit of the two components is a short circuit since the inductive reactance will be exactly balanced by the capacitive reactance.

Any ac circuit having a total impedance with only a real component, is referred to as a **resistive** circuit. In such a case, the impedance vector $\mathbf{Z_T}$ will be located along the positive real axis of the impedance diagram and the angle of the vector will be 0°. The condition under which series reactances are equal is referred to as "series resonance" and is examined in greater detail in a later chapter.

If the impedance **Z** is written in polar form, then the angle θ will be positive for an inductive impedance and negative for a capacitive impedance. In the event that the circuit is purely reactive, the resulting angle θ will be either +90° (inductive) or −90° (capacitive). If we reexamine the impedance diagram of Figure 18–18, we conclude that the original circuit is inductive.

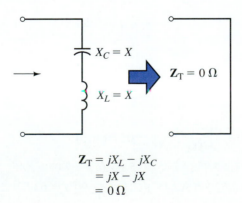

FIGURE 18–19

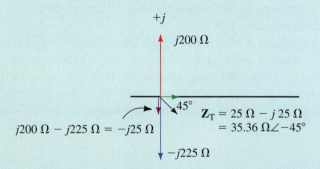

EXAMPLE 18–5   Consider the network of Figure 18–20.

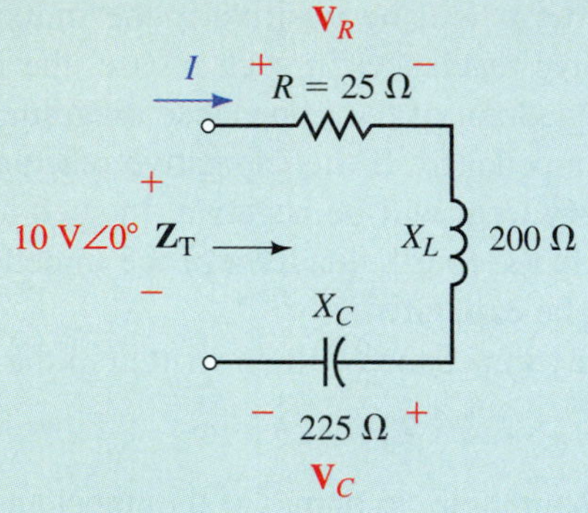

FIGURE 18–20

a. Find $\mathbf{Z}_T$.

b. Sketch the impedance diagram for the network and indicate whether the total impedance of the circuit is inductive, capacitive, or resistive.

c. Use Ohm's law to determine $\mathbf{I}$, $\mathbf{V}_R$, and $\mathbf{V}_C$.

Solution

a. The total impedance is the vector sum

$$\mathbf{Z}_T = 25\ \Omega + j200\ \Omega + (-j225\ \Omega)$$
$$= 25\ \Omega - j25\ \Omega$$
$$= 35.36\ \Omega\angle{-45°}$$

b. The corresponding impedance diagram is shown in Figure 18–21.

FIGURE 18–21

Because the total impedance has a negative reactance term ($-j25\ \Omega$), $\mathbf{Z}_T$ is capacitive.

c.
$$\mathbf{I} = \frac{10\ \text{V}\angle{0°}}{35.36\ \Omega\angle{-45°}} = 0.283\ \text{A}\angle{45°}$$

$$\mathbf{V}_R = (282.8\ \text{mA}\angle{45°})(25\ \Omega\angle{0°}) = 7.07\ \text{V}\angle{45°}$$

$$\mathbf{V}_C = (282.8\ \text{mA}\angle{45°})(225\ \Omega\angle{-90°}) = 63.6\ \text{V}\angle{-45°}$$

   Notice that the magnitude of the voltage across the capacitor is many times larger than the source voltage applied to the circuit. This example illustrates that the voltages across reactive elements must be calculated to ensure that maximum ratings for the components are not exceeded.

EXAMPLE 18–6   Determine the impedance $\mathbf{Z}$ which must be within the indicated block of Figure 18–22 if the total impedance of the network is $13 \ \Omega\angle 22.62°$.

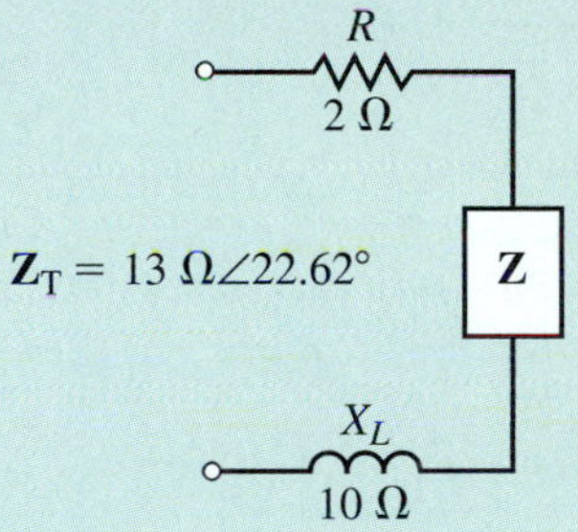

FIGURE 18–22

Solution   Converting the total impedance from polar to rectangular form, we get

$$\mathbf{Z}_T = 13 \ \Omega\angle 22.62° \Leftrightarrow 12 \ \Omega + j5 \ \Omega$$

   Now, we know that the total impedance is determined from the summation of the individual impedance vectors, namely

$$\mathbf{Z}_T = 2 \ \Omega + j10 \ \Omega + \mathbf{Z} = 12 \ \Omega + j5 \ \Omega$$

Therefore, the impedance $\mathbf{Z}$ is found as

$$\mathbf{Z} = 12 \ \Omega + j5 \ \Omega - (2 \ \Omega + j10 \ \Omega)$$
$$= 10 \ \Omega - j5 \ \Omega$$
$$= 11.18 \ \Omega\angle -26.57°$$

   In its most simple form, the impedance $\mathbf{Z}$ will consist of a series combination of a 10-Ω resistor and a capacitor having a reactance of 5 Ω. Figure 18–23 shows the elements which may be contained within $\mathbf{Z}$ to satisfy the given conditions.

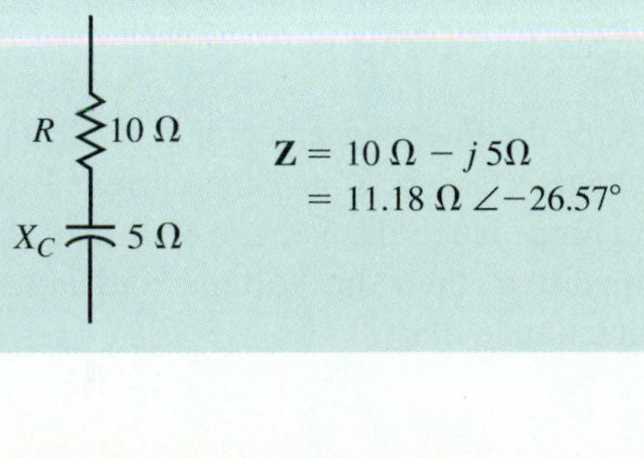

$$\mathbf{Z} = 10 \ \Omega - j \ 5\Omega$$
$$= 11.18 \ \Omega \ \angle -26.57°$$

FIGURE 18–23

EXAMPLE 18–7   Find the total impedance for the network of Figure 18–24. Sketch the impedance diagram showing $\mathbf{Z}_1$, $\mathbf{Z}_2$, and $\mathbf{Z}_T$.

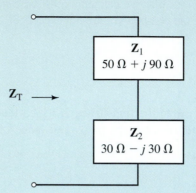

FIGURE 18–24

Solution:

$$\mathbf{Z}_T = \mathbf{Z}_1 + \mathbf{Z}_2$$
$$= (50\ \Omega + j90\ \Omega) + (30\ \Omega - j30\ \Omega)$$
$$= (80\ \Omega + j60\ \Omega) = 100\ \Omega\angle 36.87°$$

The polar forms of the vectors $\mathbf{Z}_1$ and $\mathbf{Z}_2$ are as follows:

$$\mathbf{Z}_1 = 50\ \Omega + j90\ \Omega = 102.96\ \Omega\angle 60.95°$$
$$\mathbf{Z}_2 = 30\ \Omega - j30\ \Omega = 42.43\ \Omega\angle -45°$$

The resulting impedance diagram is shown in Figure 18–25.

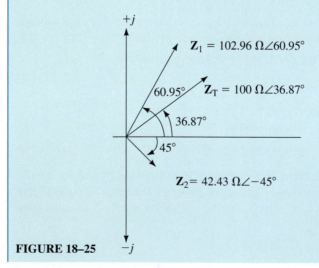

FIGURE 18–25

The phase angle θ for the impedance vector $\mathbf{Z} = Z\angle\theta$ provides the phase angle between the voltage $\mathbf{V}$ across $\mathbf{Z}$ and the current $\mathbf{I}$ through the impedance. For an inductive impedance the voltage will lead the current by θ. If the impedance is capacitive, then the voltage will lag the current by an amount equal to the magnitude of θ.

The phase angle θ is also useful for determining the average power dissipated by the circuit. In the simple series circuit shown in Figure 18–26, we know that only the resistor will dissipate power.

The average power dissipated by the resistor may be determined as follows:

$$P = V_R I = \frac{V_R^2}{R} = I^2 R \qquad \textbf{(18–6)}$$

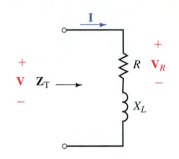

FIGURE 18–26

Notice that Equation 18–6 uses only the **magnitudes** of the voltage, current, and impedance vectors. *Power is never determined by using phasor products.*

Ohm's law provides the magnitude of the current phasor as

$$I = \frac{V}{Z}$$

Substituting this expression into Equation 18–6, we obtain the expression for power as

$$P = \frac{V^2}{Z^2} R = \frac{V^2}{Z}\left(\frac{R}{Z}\right) \qquad \textbf{(18–7)}$$

From the impedance diagram of Figure 18–27, we see that

$$\cos \theta = \frac{R}{Z}$$

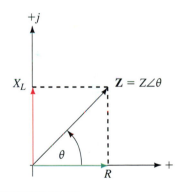

FIGURE 18–27

The previous chapter had defined the power factor as $F_p = \cos \theta$, where θ is the angle between the voltage and current phasors. We now see that for a series circuit, the power factor of the circuit can be determined from the magnitudes of resistance and total impedance.

$$F_p = \cos \theta = \frac{R}{Z} \qquad \textbf{(18–8)}$$

The power factor, F_p, is said to be **leading** if the current leads the voltage (capacitive circuit) and **lagging** if the current lags the voltage (inductive circuit).

Now substituting the expression for the power factor into Equation 18–7, we express power delivered to the circuit as

$$P = VI \cos \theta$$

Since $V = IZ$, power may be expressed as

$$P = VI \cos \theta = I^2 Z \cos \theta = \frac{V^2}{Z} \cos \theta \qquad \textbf{(18–9)}$$

EXAMPLE 18–8   Refer to the circuit of Figure 18–28.

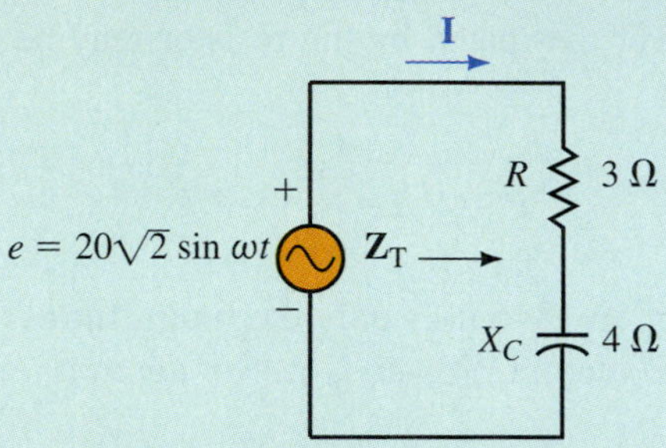

$$\mathbf{I}$$

$e = 20\sqrt{2}\sin \omega t$ $\mathbf{Z_T} \longrightarrow$

R $3\,\Omega$

$+$

$-$

X_C $4\,\Omega$

FIGURE 18–28

a. Find the impedance $\mathbf{Z_T}$.

b. Calculate the power factor of the circuit.

c. Determine $\mathbf{I}$.

d. Sketch the phasor diagram for $\mathbf{E}$ and $\mathbf{I}$.

e. Find the average power delivered to the circuit by the voltage source.

f. Calculate the average power dissipated by both the resistor and the capacitor.

Solution

a. $\mathbf{Z_T} = 3\,\Omega - j4\,\Omega = 5\,\Omega\angle{-53.13°}$

b. $F_p = \cos\theta = 3\,\Omega\backslash5\,\Omega = 0.6$ (leading)

c. The phasor form of the applied voltage is

$$\mathbf{E} = \frac{(\sqrt{2})(20\ \text{V})}{\sqrt{2}}\angle{0°} = 20\ \text{V}\angle{0°}$$

which gives a current of

$$\mathbf{I} = \frac{20\ \text{V}\angle{0°}}{5\,\Omega\angle{-53.13°}} = 4.0\ \text{A}\angle{53.13°}$$

d. The phasor diagram is shown in Figure 18–29.

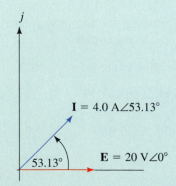

j

$\mathbf{I} = 4.0\ \text{A}\angle{53.13°}$

$53.13°$ $\mathbf{E} = 20\ \text{V}\angle{0°}$

FIGURE 18–29

From this phasor diagram, we see that the current phasor for the capacitive circuit leads the voltage phasor by 53.13°.

e. The average power delivered to the circuit by the voltage source is

$$P = (20 \text{ V})(4 \text{ A}) \cos 53.13° = 48.0 \text{ W}$$

f. The average power dissipated by the resistor and capacitor will be

$$P_R = (4 \text{ A})^2 (3 \text{ Ω}) \cos 0° = 48 \text{ W}$$
$$P_C = (4 \text{ A})^2 (4 \text{ Ω}) \cos 90° = 0 \text{ W} \quad \text{(as expected!)}$$

Notice that the power factor used in determining the power dissipated by each of the elements is the power factor for that element and not the total power factor for the circuit.

As expected, the summation of powers dissipated by the resistor and capacitor is equal to the total power delivered by the voltage source.

A circuit consists of a voltage source $\mathbf{E} = 50 \text{ V}\angle 25°$ in series with $L = 20$ mH, $C = 50$ μF, and $R = 25$ Ω. The circuit operates at an angular frequency of 2 krad/s.

PRACTICE PROBLEMS 3

a. Determine the current phasor, **I.**

b. Solve for the power factor of the circuit.

c. Calculate the average power dissipated by the circuit and verify that this is equal to the average power delivered by the source.

d. Use Ohm's law to find $\mathbf{V}_R$, $\mathbf{V}_L$, and $\mathbf{V}_C$.

Answers:
a. $\mathbf{I} = 1.28 \text{ A}\angle -25.19°$
b. $F_p = 0.6402$
c. $P = 41.0 \text{ W}$
d. $\mathbf{V}_R = 32.0 \text{ V}\angle -25.19°$
   $\mathbf{V}_C = 12.8 \text{ V}\angle -115.19°$
   $\mathbf{V}_L = 51.2 \text{ V}\angle 64.81°$

18.3 Kirchhoff's Voltage Law and the Voltage Divider Rule

When a voltage is applied to impedances in series, as shown in Figure 18–30, Ohm's law may be used to determine the voltage across any impedance as

$$\mathbf{V}_x = \mathbf{IZ}_x$$

The current in the circuit is

$$\mathbf{I} = \frac{\mathbf{E}}{\mathbf{Z}_T}$$

Now, by substitution we arrive at the voltage divider rule for any series combination of elements as

$$\mathbf{V}_x = \frac{\mathbf{Z}_x}{\mathbf{Z}_T} \mathbf{E} \qquad (18\text{--}10)$$

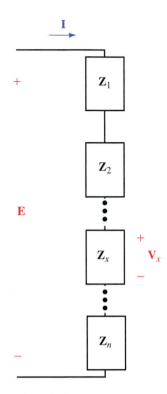

FIGURE 18–30

Equation 18–10 is very similar to the equation for the voltage divider rule in dc circuits. The fundamental differences in solving ac circuits are that we use impedances rather than resistances and that the voltages found are phasors. Because the voltage divider rule involves solving products and quotients of phasors, we generally use the polar form rather than the rectangular form of phasors.

Kirchhoff's voltage law must apply for all circuits whether they are dc or ac circuits. However, because ac circuits have voltages expressed in either sinusoidal or phasor form, Kirchhoff's voltage law for ac circuits may be stated as follows:

The phasor sum of voltage drops and voltage rises around a closed loop is equal to zero.

When adding phasor voltages, we find that the summation is generally done more easily in rectangular form rather than the polar form.

EXAMPLE 18–9   Consider the circuit of Figure 18–31.

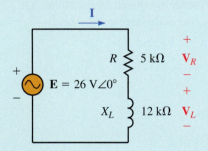

FIGURE 18–31

a. Find $\mathbf{Z}_T$.

b. Determine the voltages $\mathbf{V}_R$ and $\mathbf{V}_L$ using the voltage divider rule.

c. Verify Kirchhoff's voltage law around the closed loop.

Solution

a. $\mathbf{Z}_T = 5 \text{ k}\Omega + j12 \text{ k}\Omega = 13 \text{ k}\Omega\angle67.38°$

b. $\mathbf{V}_R = \left(\dfrac{5 \text{ k}\Omega\angle0°}{13 \text{ k}\Omega\angle67.38°}\right)(26 \text{ V}\angle0°) = 10 \text{ V}\angle{-67.38°}$

$\mathbf{V}_L = \left(\dfrac{12 \text{ k}\Omega\angle90°}{13 \text{ k}\Omega\angle67.38°}\right)(26 \text{ V}\angle0°) = 24 \text{ V}\angle22.62°$

c. Kirchhoff's voltage law around the closed loop will give

$$26 \text{ V}\angle0° - 10 \text{ V}\angle{-67.38°} - 24 \text{ V}\angle22.62° = 0$$
$$(26 + j0) - (3.846 - j9.231) - (22.154 + j9.231) = 0$$
$$(26 - 3.846 - 22.154) + j(0 + 9.231 - 9.231) = 0$$
$$0 + j0 = 0$$

EXAMPLE 18–10   Consider the circuit of Figure 18–32:

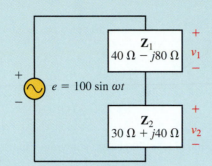

FIGURE 18–32

a. Calculate the sinusoidal voltages v_1 and v_2 using phasors and the voltage divider rule.

b. Sketch the phasor diagram showing $\mathbf{E}$, $\mathbf{V}_1$, and $\mathbf{V}_2$.

c. Sketch the sinusoidal waveforms of e, v_1, and v_2.

Solution

a. The phasor form of the voltage source is determined as

$$e = 100 \sin \omega t \Leftrightarrow \mathbf{E} = 70.71 \angle \text{V } 0°$$

Applying VDR, we get

$$\mathbf{V}_1 = \left(\frac{40\ \Omega - j80\ \Omega}{(40\ \Omega - j80\ \Omega) + (30\ \Omega + j40\ \Omega)} \right)(70.71\ \text{V} \angle 0°)$$

$$= \left(\frac{89.44\ \Omega \angle -63.43°}{80.62\ \Omega \angle -29.74°} \right)(70.71\ \text{V} \angle 0°)$$

$$= 78.4\ \text{V} \angle -33.69°$$

and

$$\mathbf{V}_2 = \left(\frac{30\ \Omega + j40\ \Omega}{(40\ \Omega - j80\ \Omega) + (30\ \Omega + j40\ \Omega)} \right)(70.71\ \text{V} \angle 0°)$$

$$= \left(\frac{50.00\ \Omega \angle 53.13°}{80.62\ \Omega \angle -29.74°} \right)(70.71\ \text{V} \angle 0°)$$

$$= 43.9\ \text{V} \angle 82.87°$$

The sinusoidal voltages are determined to be

$$v_1 = (\sqrt{2})(78.4)\sin(\omega t - 33.69°)$$

$$= 111 \sin(\omega t - 33.69°)$$

and

$$v_2 = (\sqrt{2})(43.9)\sin(\omega t + 82.87°)$$

$$= 62.0 \sin(\omega t + 82.87°)$$

b. The phasor diagram is shown in Figure 18–33.

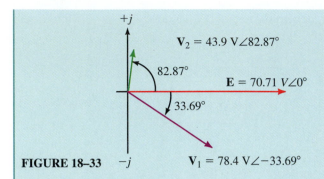

FIGURE 18–33

c. The corresponding sinusoidal voltages are shown in Figure 18–34.

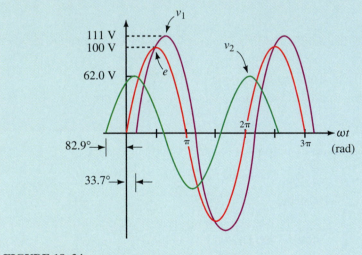

FIGURE 18–34

1. Express Kirchhoff's voltage law as it applies to ac circuits.
2. What is the fundamental difference between how Kirchhoff's voltage law is used in ac circuits as compared with dc circuits?

(Answers are at the end of the chapter.)

PRACTICE
PROBLEMS 4

A circuit consists of a voltage source $\mathbf{E} = 50 \text{ V} \angle 25°$ in series with $L = 20$ mH, $C = 50\ \mu$F, and $R = 25\ \Omega$. The circuit operates at an angular frequency of 2 krad/s.

a. Use the voltage divider rule to determine the voltage across each element in the circuit.

b. Verify that Kirchhoff's voltage law applies for the circuit.

Answers:
a. $\mathbf{V}_L = 51.2 \text{ V} \angle 64.81°$, $\mathbf{V}_C = 12.8 \text{ V} \angle -115.19°$

   $\mathbf{V}_R = 32.0 \text{ V} \angle -25.19°$

b. $51.2 \text{ V} \angle 64.81° + 12.8 \text{ V} \angle -115.19° + 32.0 \text{ V} \angle -25.19° = 50 \text{ V} \angle 25°$

18.4   AC Parallel Circuits

The **admittance Y** of any impedance is defined as a vector quantity which is the reciprocal of the impedance **Z.**

Mathematically, admittance is expressed as

$$\mathbf{Y}_T = \frac{1}{\mathbf{Z}_T} = \frac{1}{Z_T\angle\theta} = \left(\frac{1}{Z_T}\right)\angle-\theta = Y_T\angle-\theta \quad \text{(S)} \qquad \textbf{(18–11)}$$

where the unit of admittance is the siemens (S).

In particular, we have seen that the admittance of a resistor R is called **conductance** and is given the symbol $\mathbf{Y}_R$. If we consider resistance as a vector quantity, then the corresponding vector form of the conductance is

$$\mathbf{Y}_R = \frac{1}{R\angle 0°} = \frac{1}{R}\angle 0° = G\angle 0° = G + j0 \quad \text{(S)} \qquad \textbf{(18–12)}$$

If we determine the admittance of a purely reactive component X, the resultant admittance is called the **susceptance** of the component and is assigned the symbol B. The unit for susceptance is siemens (S). In order to distinguish between inductive susceptance and capacitive susceptance, we use the subscripts L and C respectively. The vector forms of reactive admittance are given as follows:

$$\mathbf{Y}_L = \frac{1}{X_L\angle 90°} = \frac{1}{X_L}\angle-90° = B_L\angle-90° = 0 - jB_L \quad \text{(S)} \qquad \textbf{(18–13)}$$

$$\mathbf{Y}_C = \frac{1}{X_C\angle-90°} = \frac{1}{X_C}\angle 90° = B_C\angle 90° = 0 + jB_C \quad \text{(S)} \qquad \textbf{(18–14)}$$

In a manner similar to impedances, admittances may be represented on the complex plane in an **admittance diagram** as shown in Figure 18–35.

The lengths of the various vectors are proportional to the magnitudes of the corresponding admittances. The resistive admittance vector **G** is shown on the positive real axis, whereas the inductive and capacitive admittance vectors $\mathbf{Y}_L$ and $\mathbf{Y}_C$ are shown on the negative and positive imaginary axes respectively.

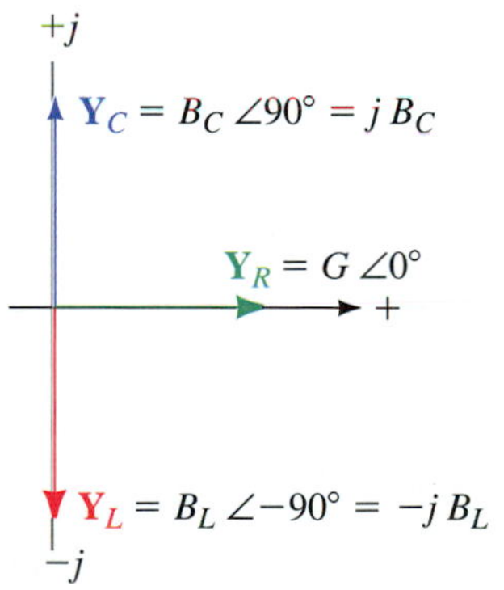

FIGURE 18–35 Admittance diagram showing conductance ($\mathbf{Y}_R$) and susceptance ($\mathbf{Y}_L$ and $\mathbf{Y}_C$).

EXAMPLE 18–11   Determine the admittances of the following impedances. Sketch the corresponding admittance diagram.
a. $R = 10\ \Omega$ b. $X_L = 20\ \Omega$ c. $X_C = 40\ \Omega$

Solutions

a. $\mathbf{Y}_R = \dfrac{1}{R} = \dfrac{1}{10\ \Omega\angle 0°} = 100\ \text{mS}\angle 0°$

b. $\mathbf{Y}_L = \dfrac{1}{X_L} = \dfrac{1}{20\ \Omega\angle 90°} = 50\ \text{mS}\angle-90°$

c. $\mathbf{Y}_C = \dfrac{1}{X_C} = \dfrac{1}{40\ \Omega\angle-90°} = 25\ \text{mS}\angle 90°$

The admittance diagram is shown in Figure 18–36.

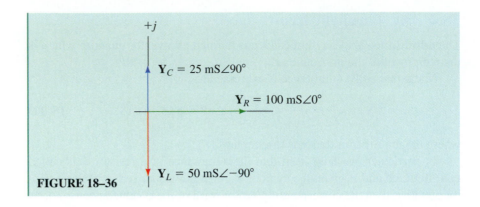

FIGURE 18–36

For any network of n admittances as shown in Figure 18–37, the total admittance is the vector sum of the admittances of the network. Mathematically, the total admittance of a network is given as

$$\mathbf{Y_T} = \mathbf{Y}_1 + \mathbf{Y}_2 + \cdots + \mathbf{Y}_n \quad (S) \tag{18–15}$$

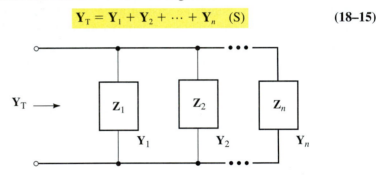

FIGURE 18–37

The resultant impedance of a parallel network of n impedances is determined to be

$$\mathbf{Z_T} = \frac{1}{\mathbf{Y_T}} = \frac{1}{\mathbf{Y}_1 + \mathbf{Y}_2 + \cdots + \mathbf{Y}_n}$$

$$\mathbf{Z_T} = \frac{1}{\dfrac{1}{\mathbf{Z}_1} + \dfrac{1}{\mathbf{Z}_2} + \cdots + \dfrac{1}{\mathbf{Z}_n}} \quad (\Omega) \tag{18–16}$$

EXAMPLE 18–12   Find the equivalent admittance and impedance of the network of Figure 18–38. Sketch the admittance diagram.

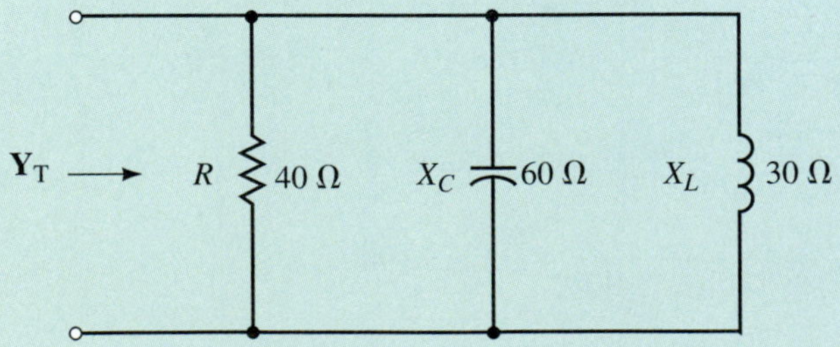

$\mathbf{Y_T} \longrightarrow$ $R \lessgtr 40\ \Omega$ $X_C \eqarrow 60\ \Omega$ $X_L \gtrless 30\ \Omega$

FIGURE 18–38

Solution   The admittances of the various parallel elements are

$$\mathbf{Y}_1 = \frac{1}{40\ \Omega\angle 0°} = 25.0\ \text{mS}\angle 0° = 25.0\ \text{mS} + j0$$

$$\mathbf{Y}_2 = \frac{1}{60\ \Omega\angle -90°} = 16.\overline{6}\ \text{mS}\angle 90° = 0 + j16.\overline{6}\ \text{mS}$$

$$\mathbf{Y}_3 = \frac{1}{30\ \Omega\angle 90°} = 33.\overline{3}\ \text{mS}\angle -90° = 0 - j33.\overline{3}\ \text{mS}$$

The total admittance is determined as

$$\mathbf{Y}_T = \mathbf{Y}_1 + \mathbf{Y}_2 + \mathbf{Y}_3$$
$$= 25.0\ \text{mS} + j16.\overline{6}\ \text{mS} + (-j33.\overline{3}\ \text{mS})$$
$$= 25.0\ \text{mS} - j16.\overline{6}\ \text{mS}$$
$$= 30.0\ \text{mS}\angle -33.69°$$

This results in a total impedance for the network of

$$\mathbf{Z}_T = \frac{1}{\mathbf{Y}_T}$$

$$= \frac{1}{30.0\ \text{mS}\angle -33.69°}$$
$$= 33.3\ \Omega\angle 33.69°$$

The admittance diagram is shown in Figure 18–39.

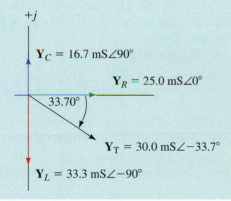

FIGURE 18–39

Two Impedances in Parallel

By applying Equation 18–14 for two impedances, we determine the equivalent impedance of two impedances as

$$\mathbf{Z}_T = \frac{\mathbf{Z}_1\mathbf{Z}_2}{\mathbf{Z}_1 + \mathbf{Z}_2} \quad (\Omega) \qquad\qquad (18\text{–}17)$$

From the above expression, we see that for two impedances in parallel, the equivalent impedance is determined as the product of the impedances over the sum. Although the expression for two impedances is very similar to the expression for two resistors in parallel, the difference is that the calculation of impedance involves the use of complex algebra.

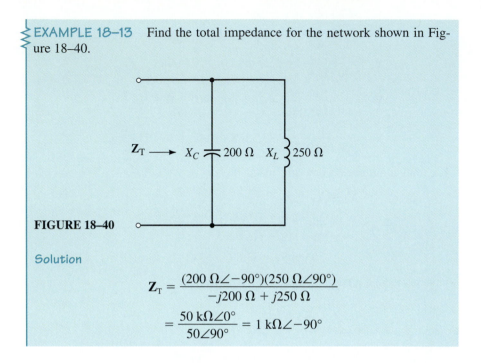

EXAMPLE 18–13   Find the total impedance for the network shown in Figure 18–40.

FIGURE 18–40

Solution

$$\mathbf{Z}_T = \frac{(200 \ \Omega \angle -90°)(250 \ \Omega \angle 90°)}{-j200 \ \Omega + j250 \ \Omega}$$

$$= \frac{50 \ k\Omega \angle 0°}{50 \angle 90°} = 1 \ k\Omega \angle -90°$$

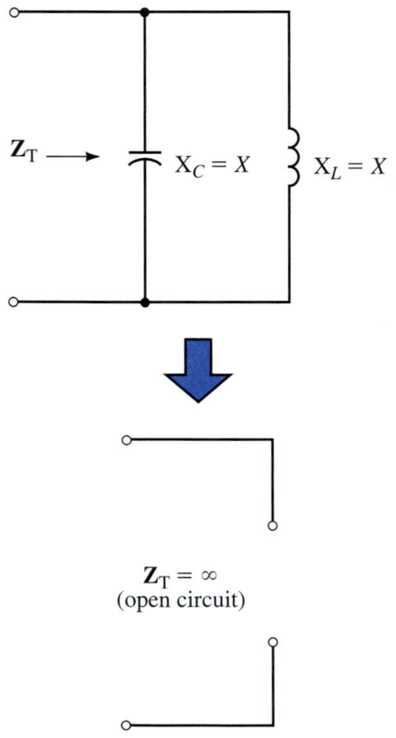

The previous example illustrates that unlike total parallel resistance, the total impedance of a combination of parallel reactances may be much larger that either of the individual impedances. Indeed, if we are given a parallel combination of equal inductive and capacitive reactances, the total impedance of the combination is equal to infinity (namely an open circuit). Consider the network of Figure 18–41.

The total impedance $\mathbf{Z}_T$ is found as

$$\mathbf{Z}_T = \frac{(X_L \angle 90°)(X_C \angle -90°)}{jX_L - jX_C} = \frac{X^2 \angle 0°}{0 \angle 0°} = \infty \angle 0°$$

Because the denominator of the above expression is equal to zero, the magnitude of the total impedance will be undefined ($Z = \infty$). The magnitude is undefined and the algebra yields a phase angle $\theta = 0°$, which indicates that the vector lies on the positive real axis of the impedance diagram.

Whenever a capacitor and an inductor having equal reactances are placed in parallel, the equivalent circuit of the two components is an open circuit.

The principle of equal parallel reactances will be studied in a later chapter dealing with "resonance."

Three Impedances in Parallel

Equation 18–16 may be solved for three impedances to give the equivalent impedance as

$$\mathbf{Z}_T = \frac{\mathbf{Z}_1 \mathbf{Z}_2 \mathbf{Z}_3}{\mathbf{Z}_1 \mathbf{Z}_2 + \mathbf{Z}_1 \mathbf{Z}_3 + \mathbf{Z}_2 \mathbf{Z}_3} \quad (\Omega) \qquad \textbf{(18–18)}$$

although this is less useful than the general equation.

$\mathbf{Z}_T = \infty$
(open circuit)

FIGURE 18–41

EXAMPLE 18–14   Find the equivalent impedance of the network of Figure 18–42.

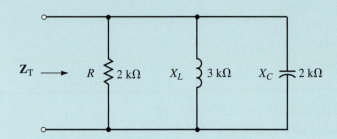

FIGURE 18–42

Solution

$$\mathbf{Z}_T = \frac{(2\,k\Omega\angle 0°)(3\,k\Omega\angle 90°)(2\,k\Omega\angle -90°)}{(2\,k\Omega\angle 0°)(3\,k\Omega\angle 90°) + (2\,k\Omega\angle 0°)(2\,k\Omega\angle -90°) + (3\,k\Omega\angle 90°)(2\,k\Omega\angle -90°)}$$

$$= \frac{12 \times 10^9\ \Omega\angle 0°}{6 \times 10^6\angle 90° + 4 \times 10^6\angle -90° + 6 \times 10^6\angle 0°}$$

$$= \frac{12 \times 10^9\ \Omega\angle 0°}{6 \times 10^6 + j2 \times 10^6} = \frac{12 \times 10^9\ \Omega\angle 0°}{6.325 \times 10^6\angle 18.43°}$$

$$= 1.90\,k\Omega\angle -18.43°$$

And so the equivalent impedance of the network is

$$\mathbf{Z}_T = 1.80\,k\Omega - j0.6\,k\Omega$$

A circuit consists of a current source, $i = 0.030 \sin 500t$, in parallel with $L = 20\,mH$, $C = 50\,\mu F$, and $R = 25\,\Omega$.

a. Determine the voltage **V** across the circuit.

b. Solve for the power factor of the circuit.

c. Calculate the average power dissipated by the circuit and verify that this is equal to the power delivered by the source.

d. Use Ohm's law to find the phasor quantities, $\mathbf{I}_R$, $\mathbf{I}_L$, and $\mathbf{I}_C$.

Answers:
a. $\mathbf{V} = 0.250\,V\angle 61.93°$

b. $F_p = 0.4705$

c. $P_R = 41.0\,W = P_T$

d. $\mathbf{I}_R = 9.98\,mA\angle 61.98°$

   $\mathbf{I}_C = 6.24\,mA\angle 151.93°$

   $\mathbf{I}_L = 25.0\,mA\angle -28.07°$

A circuit consists of a 2.5-A_{rms} current source connected in parallel with a resistor, an inductor, and a capacitor. The resistor has a value of 10 Ω and dissipates 40 W of power.

a. Calculate the values of X_L and X_C if $X_L = 3X_C$.

b. Determine the magnitudes of current through the inductor and the capacitor.

Answers:

a. $X_L = 80\ \Omega, X_C = 26.7\ \Omega$

b. $I_L = 0.25$ mA, $I_C = 0.75$ mA

18.5   Kirchhoff's Current Law and the Current Divider Rule

The current divider rule for ac circuits has the same form as for dc circuits with the notable exception that currents are expressed as phasors. For a parallel network as shown in Figure 18–43, the current in any branch of the network may be determined using either admittance or impedance.

$$\mathbf{I}_x = \frac{\mathbf{Y}_x}{\mathbf{Y}_T}\mathbf{I} \quad \text{or} \quad \mathbf{I}_x = \frac{\mathbf{Z}_T}{\mathbf{Z}_x}\mathbf{I} \qquad (18\text{–}19)$$

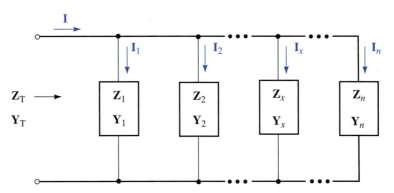

FIGURE 18–43

For two branches in parallel the current in either branch is determined from the impedances as

$$\mathbf{I}_1 = \frac{\mathbf{Z}_2}{\mathbf{Z}_1 + \mathbf{Z}_2}\mathbf{I} \qquad (18\text{–}20)$$

Also, as one would expect, Kirchhoff's current law must apply to any node within an ac circuit. For such circuits, KCL may be stated as follows:

The summation of current phasors entering and leaving a node is equal to zero.

EXAMPLE 18–15   Calculate the current in each of the branches in the network of Figure 18–44.

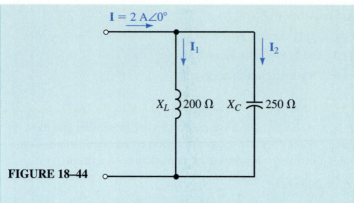

FIGURE 18–44

Solution

$$\mathbf{I}_1 = \left(\frac{250\ \Omega \angle -90°}{j200\ \Omega - j250\ \Omega} \right)(2\ \text{A} \angle 0°)$$

$$= \left(\frac{250\ \Omega \angle -90°}{50\ \Omega \angle -90°} \right)(2\ \text{A} \angle 0°) = 10\ \text{A} \angle 0°$$

and

$$\mathbf{I}_2 = \left(\frac{200\ \Omega \angle 90°}{j200\ \Omega - j250\ \Omega} \right)(2\ \text{A} \angle 0°)$$

$$= \left(\frac{200\ \Omega \angle 90°}{50\ \Omega \angle -90°} \right)(2\ \text{A} \angle 0°) = 8\ \text{A} \angle 180°$$

The above results illustrate that the currents in parallel reactive components may be significantly larger than the applied current. If the current through the component exceeds the maximum current rating of the element, severe damage may occur.

EXAMPLE 18–16   Refer to the circuit of Figure 18–45:

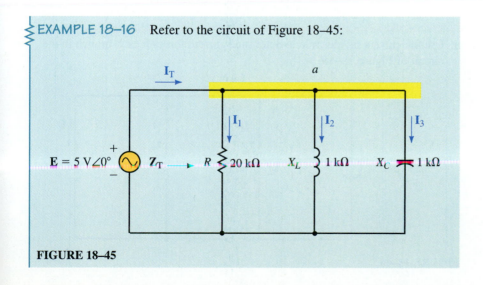

FIGURE 18–45

a. Find the total impedance, $\mathbf{Z}_T$.

b. Determine the supply current, $\mathbf{I}_T$.

c. Calculate $\mathbf{I}_1$, $\mathbf{I}_2$, and $\mathbf{I}_3$ using the current divider rule.

d. Verify Kirchhoff's current law at node a.

Solution

a. Because the inductive and capacitive reactances are in parallel and have the same value, we may replace the combination by an open circuit. Consequently, only the resistor R needs to be considered. As a result

$$\mathbf{Z}_T = 20 \text{ k}\Omega\angle 0°$$

b. $$\mathbf{I}_T = \frac{5 \text{ V}\angle 0°}{20 \text{ k}\Omega\angle 0°} = 250 \text{ }\mu\text{A}\angle 0°$$

c. $$\mathbf{I}_1 = \left(\frac{20 \text{ k}\Omega\angle 0°}{20 \text{ k}\Omega\angle 0°}\right)(250 \text{ }\mu\text{A}\angle 0°) = 250 \text{ }\mu\text{A}\angle 0°$$

$$\mathbf{I}_2 = \left(\frac{20 \text{ k}\Omega\angle 0°}{1 \text{ k}\Omega\angle 90°}\right)(250 \text{ }\mu\text{A}\angle 0°) = 5.0 \text{ mA}\angle -90°$$

$$\mathbf{I}_3 = \left(\frac{20 \text{ k}\Omega\angle 0°}{1 \text{ k}\Omega\angle -90°}\right)(250 \text{ }\mu\text{A}\angle 0°) = 5.0 \text{ mA}\angle 90°$$

d. Notice that the currents through the inductor and capacitor are 180° out of phase. By adding the current phasors in rectangular form, we have

$$I_T = 250 \text{ }\mu\text{A} - j5.0 \text{ A} + j5.0 \text{ A} = 250 \text{ }\mu\text{A} + j0 = 250 \text{ }\mu\text{A}\angle 0°$$

The above result verifies Kirchhoff's current law at the node.

IN-PROCESS LEARNING CHECK 3

1. Express Kirchhoff's current law as it applies to ac circuits.

2. What is the fundamental difference between how Kirchhoff's current law is applied to ac circuits as compared with dc circuits?

(Answers are at the end of the chapter.)

PRACTICE PROBLEMS 7

a. Use the current divider rule to determine current through each branch in the circuit of Figure 18–46.

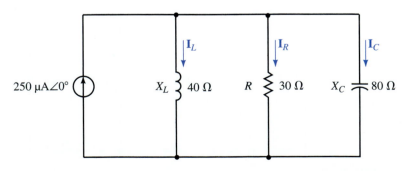

FIGURE 18–46

b. Verify that Kirchhoff's current law applies to the circuit of Figure 18–46.

Answers:

a. $\mathbf{I}_L = 176\ \mu A\angle{-69.44°}$

  $\mathbf{I}_R = 234\ \mu A\angle{20.56°}$

  $\mathbf{I}_C = 86.8\ \mu A\angle{110.56°}$

b. $\Sigma\mathbf{I}_{out} = \Sigma\mathbf{I}_{in} = 250\ \mu A$

18.6 Series-Parallel Circuits

We may now apply the analysis techniques of series and parallel circuits in solving more complicated circuits. As in dc circuits, the analysis of such circuits is simplified by starting with easily recognized combinations. If necessary, the original circuit may be redrawn to make further simplification more apparent. Regardless of the complexity of the circuits, we find that the fundamental rules and laws of circuit analysis must apply in all cases.

Consider the network of Figure 18–47.

We see that the impedances $\mathbf{Z}_2$ and $\mathbf{Z}_3$ are in series. The branch containing this combination is then seen to be in parallel with the impedance $\mathbf{Z}_1$.

The total impedance of the network is expressed as

$$\mathbf{Z}_T = \mathbf{Z}_1 \| (\mathbf{Z}_2 + \mathbf{Z}_3)$$

Solving for $\mathbf{Z}_T$ gives the following:

$$\mathbf{Z}_T = (2\ \Omega - j8\ \Omega)\|(2\ \Omega - j5\ \Omega + 6\ \Omega + j7\ \Omega)$$
$$= (2\ \Omega - j8\ \Omega) \| (8\ \Omega + j2\ \Omega)$$
$$= \frac{(2\ \Omega - j8\ \Omega)(8\ \Omega + j2\ \Omega)}{2\ \Omega - j8\ \Omega + 8\ \Omega + j2\ \Omega)}$$
$$= \frac{(8.246\ \Omega\angle{-75.96°})(8.246\ \Omega\angle{14.04°})}{11.66\ \Omega\angle{-30.96°}}$$
$$= 5.832\ \Omega\angle{-30.96°} = 5.0\ \Omega - j3.0\ \Omega$$

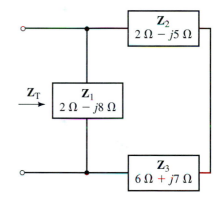

FIGURE 18–47

EXAMPLE 18–17 Determine the total impedance of the network of Figure 18–48. Express the impedance in both polar form and rectangular form.

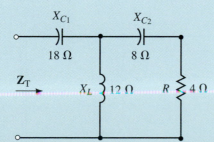

FIGURE 18–48

Solution After redrawing and labelling the given circuit, we have the circuit shown in Figure 18–49.

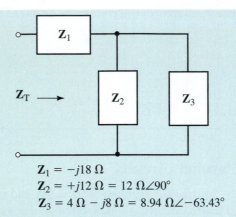

$$Z_1 = -j18 \ \Omega$$
$$Z_2 = +j12 \ \Omega = 12 \ \Omega\angle 90°$$
$$Z_3 = 4 \ \Omega - j8 \ \Omega = 8.94 \ \Omega\angle -63.43°$$

FIGURE 18–49

The total impedance is given as

$$Z_T = Z_1 + Z_2 \parallel Z_3$$

where

$$Z_1 = -j18 \ \Omega = 18\Omega\angle -90°$$
$$Z_2 = j12 \ \Omega = 12\Omega\angle 90°$$
$$Z_3 = 4 \ \Omega - j8 \ \Omega = 8.94 \ \Omega\angle -63.43°$$

We determine the total impedance as

$$Z_T = -j18 \ \Omega + \left[\frac{(12 \ \Omega\angle 90°)(8.94 \ \Omega\angle -63.43°)}{j12 \ \Omega + 4 \ \Omega - j8 \ \Omega} \right]$$

$$= -j18 \ \Omega + \left(\frac{107.3 \ \Omega\angle 26.57°}{5.66\angle 45°} \right)$$

$$= -j18 \ \Omega + 19.0 \ \Omega\angle -18.43°$$

$$= -j18 \ \Omega + 18 \ \Omega - j6 \ \Omega$$

$$= 18 \ \Omega - j24 \ \Omega = 30 \ \Omega\angle -53.13°$$

EXAMPLE 18–18 Consider the circuit of Figure 18–50:

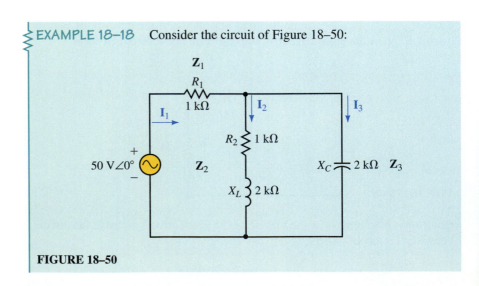

FIGURE 18–50

a. Find $\mathbf{Z}_T$.

b. Determine the currents $\mathbf{I}_1$, $\mathbf{I}_2$, and $\mathbf{I}_3$.

c. Calculate the total power provided by the voltage source.

d. Determine the average powers P_1, P_2, and P_3 dissipated by each of the impedances. Verify that the average power delivered to the circuit is the same as the power dissipated by the impedances.

Solution

a. The total impedance is determined by the combination

$$\mathbf{Z}_T = \mathbf{Z}_1 + \mathbf{Z}_2 \parallel \mathbf{Z}_3$$

For the parallel combination we have

$$\mathbf{Z}_2 \parallel \mathbf{Z}_3 = \frac{(1 \text{ k}\Omega + j2 \text{ k}\Omega)(-j2 \text{ k}\Omega)}{1 \text{ k}\Omega + j2 \text{ k}\Omega - j2 \text{ k}\Omega}$$

$$= \frac{(2.236 \text{ k}\Omega\angle 63.43°)(2 \text{ k}\Omega\angle -90°)}{1 \text{ k}\Omega\angle 0°}$$

$$= 4.472 \text{ k}\Omega\angle -26.57° = 4.0 \text{ k}\Omega - j2.0 \text{ k}\Omega$$

And so the total impedance is

$$\mathbf{Z}_T = 5 \text{ k}\Omega - j2 \text{ k}\Omega = 5.385 \text{ k}\Omega\angle -21.80°$$

$$\mathbf{I}_1 = \frac{50 \text{ V}\angle 0°}{5.385 \text{ k}\Omega\angle -21.80°}$$

$$= 9.285 \text{ mA}\angle 21.80°$$

Applying the current divider rule, we get

$$\mathbf{I}_2 = \frac{(2 \text{ k}\Omega\angle -90°)(9.285 \text{ mA}\angle 21.80°)}{1 \text{ k}\Omega + j2 \text{ k}\Omega - j2 \text{ k}\Omega}$$

$$= 18.57 \text{ mA}\angle -68.20°$$

and

$$\mathbf{I}_3 = \frac{(1 \text{ k}\Omega + j2 \text{ k}\Omega)(9.285 \text{ mA}\angle 21.80°)}{1 \text{ k}\Omega + j2 \text{ k}\Omega - j2 \text{ k}\Omega}$$

$$= \frac{(2.236 \text{ k}\Omega\angle 63.43°)(9.285 \text{ mA}\angle 21.80°)}{1 \text{ k}\Omega\angle 0°}$$

$$= 20.761 \text{ mA}\angle 85.23°$$

c. $$P_T = (50 \text{ V})(9.285 \text{ mA})\cos 21.80°$$

$$= 431.0 \text{ mW}$$

d. Because only the resistors will dissipate power, we may use $P = I^2R$:

$$P_1 = (9.285 \text{ mA})^2(1 \text{ k}\Omega) = 86.2 \text{ mW}$$

$$P_2 = (18.57 \text{ mA})^2(1 \text{ k}\Omega) = 344.8 \text{ mW}$$

Alternatively, the power dissipated by $\mathbf{Z}_2$ may have been determined as $P = I^2Z \cos \theta$:

$$P_2 = (18.57 \text{ mA})^2(2.236 \text{ k}\Omega)\cos 63.43° = 344.8 \text{ mW}$$

> Since $\mathbf{Z}_3$ is purely capacitive, it will not dissipate any power:
>
> $$P_3 = 0$$
>
> By combining these powers, the total power dissipated is found:
>
> $$P_T = 86.2 \text{ mW} + 344.8 \text{ mW} + 0 = 431.0 \text{ mW} \quad \text{(checks!)}$$

PRACTICE
PROBLEMS 8

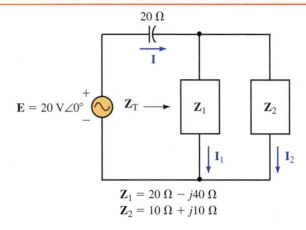

$$\mathbf{Z}_1 = 20 \ \Omega - j40 \ \Omega$$
$$\mathbf{Z}_2 = 10 \ \Omega + j10 \ \Omega$$

FIGURE 18–51

Refer to the circuit of Figure 18–51:

a. Calculate the total impedance, $\mathbf{Z}_T$.

b. Find the current $\mathbf{I}$.

c. Use the current divider rule to find $\mathbf{I}_1$ and $\mathbf{I}_2$.

d. Determine the power factor for each impedance, $\mathbf{Z}_1$ and $\mathbf{Z}_2$.

e. Determine the power factor for the circuit.

f. Verify that the total power dissipated by impedances $\mathbf{Z}_1$ and $\mathbf{Z}_2$ is equal to the power delivered by the voltage source.

Answers:
a. $\mathbf{Z}_T = 18.9 \ \Omega \angle -45°$

b. $\mathbf{I} = 1.06 \text{ A} \angle 45°$

c. $\mathbf{I}_1 = 0.354 \text{ A} \angle 135°$, $\mathbf{I}_2 = 1.12 \text{ A} \angle 26.57°$

d. $F_{P(1)} = 0.4472$ leading, $F_{P(2)} = 0.7071$ lagging

e. $F_P = 0.7071$ leading

f. $P_T = 15.0 \text{ W}$, $P_1 = 2.50 \text{ W}$, $P_2 = 12.5 \text{ W}$

   $P_1 + P_2 = 15.0 \text{ W} = P_T$

18.7   Frequency Effects

As we have already seen, the reactance of inductors and capacitors depends on frequency. Consequently, the total impedance of any network having reactive elements is also frequency dependent. Any such circuit would need to be analyzed separately at each frequency of interest. We will examine several fairly simple combinations of resistors, capacitors, and inductors to see how the various circuits operate at different frequencies. Some of the more important combinations will be examined in greater detail in later chapters which deal with resonance and filters.

RC Circuits

As the name implies, *RC* circuits consist of a resistor and a capacitor. The components of an *RC* circuit may be connected either in series or in parallel as shown in Figure 18–52.

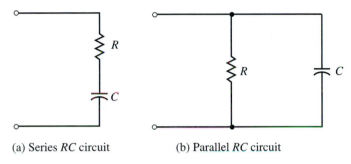

(a) Series *RC* circuit   (b) Parallel *RC* circuit

FIGURE 18–52

Consider the *RC* series circuit of Figure 18–53.
Recall that the capacitive reactance, X_C, is given as

$$X_C = \frac{1}{\omega C} = \frac{1}{2\pi f C}$$

The total impedance of the circuit is a vector quantity expressed as

$$\mathbf{Z}_T = R - j\frac{1}{\omega C} = R + \frac{1}{j\omega C}$$

$$\mathbf{Z}_T = \frac{1 + j\omega RC}{j\omega C} \qquad (18\text{–}21)$$

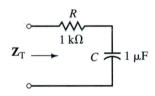

FIGURE 18–53

If we define the **cutoff** or **corner frequency** for an *RC* circuit as

$$\omega_c = \frac{1}{RC} = \frac{1}{\tau} \quad \text{(rad/s)} \qquad (18\text{–}22)$$

or equivalently as

$$f_c = \frac{1}{2\pi RC} \quad \text{(Hz)} \qquad (18\text{–}23)$$

then several important points become evident.
For $\omega \le \omega_c/10$ (or $f \le f_c/10$) Equation 18–21 can be expressed as

$$\mathbf{Z}_T \simeq \frac{1 + j0}{j\omega C} = \frac{1}{j\omega C}$$

and for $\omega \ge 10\omega_c$, the expression of (18–21) can be simplified as

$$\mathbf{Z}_T \simeq \frac{0 + j\omega RC}{j\omega C} = R$$

TABLE 18–1

Angular Frequency, ω (Rad/s)	X_C (Ω)	Z_T (Ω)
0	∞	∞
1	1 M	1 M
10	100 k	100 k
100	10 k	10.05 k
200	5 k	5.099 k
500	2 k	2.236 k
1000	1 k	1.414 k
2000	500	1118
5000	200	1019
10 k	100	1005
100 k	10	1000

Solving for the magnitude of the impedance at several angular frequencies, we have the results shown in Table 18–1.

If the magnitude of the impedance $\mathbf{Z}_T$ is plotted as a function of angular frequency ω, we get the graph of Figure 18–54. Notice that the abscissa and ordinate of the graph are not scaled linearly, but rather logarithmically. This allows for the the display of results over a wide range of frequencies.

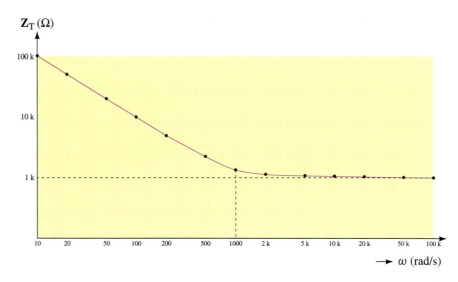

FIGURE 18–54   Impedance versus angular frequency for the network of Figure 18–53.

The graph illustrates that the reactance of a capacitor is very high (effectively an open circuit) at low frequencies. Consequently, the total impedance of the series circuit will also be very high at low frequencies. Secondly, we notice that as the frequency increases, the reactance decreases. Therefore, as the frequency gets higher, the capacitive reactance has a diminished effect in the circuit. At very high frequencies (typically for $\omega \geq 10\omega_c$), the impedance of the circuit will effectively be $R = 1\text{ k}\Omega$.

Consider the parallel RC circuit of Figure 18–55. The total impedance, $\mathbf{Z}_T$, of the circuit is determined as

$$\mathbf{Z}_T = \frac{\mathbf{Z}_R \mathbf{Z}_C}{\mathbf{Z}_R + \mathbf{Z}_C}$$

$$= \frac{R\left(\dfrac{1}{j\omega C}\right)}{R + \dfrac{1}{j\omega C}}$$

$$= \frac{\dfrac{R}{j\omega C}}{\dfrac{1 + j\omega RC}{j\omega C}}$$

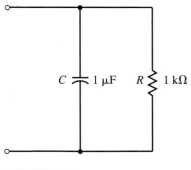

FIGURE 18–55

which may be simplified as

$$\mathbf{Z}_T = \frac{R}{1 + j\omega RC} \qquad (18\text{–}24)$$

As before, the cutoff frequency is given by Equation 18–22. Now, by examining the expression of (18–24) for $\omega \leq \omega_c/10$, we have the following result:

$$\mathbf{Z}_T \simeq \frac{R}{1 + j0} = R$$

For $\omega \geq 10\omega_c$, we have

$$\mathbf{Z}_T \simeq \frac{R}{0 + j\omega RC} = \frac{1}{j\omega C}$$

If we solve for the impedance of the circuit in Figure 18–55 at various angular frequencies, we obtain the results of Table 18–2.

Plotting the magnitude of the impedance $\mathbf{Z}_T$ as a function of angular frequency ω, we get the graph of Figure 18–56. Notice that the abscissa and ordinate of the graph are again scaled logarithmically, allowing for the display of results over a wide range of frequencies.

TABLE 18–2

Angular Frequency, ω (Rad/s)	X_C (Ω)	Z_T (Ω)
0	∞	1000
1	1 M	1000
10	100 k	1000
100	10 k	995
200	5 k	981
500	2 k	894
1 k	1 k	707
2 k	500	447
5 k	200	196
10 k	100	99.5
100 k	10	10

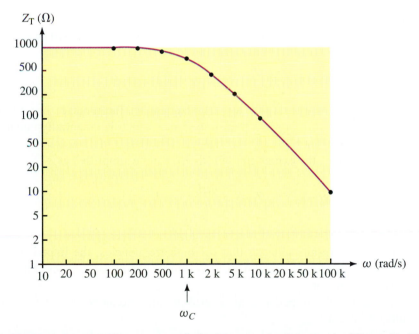

FIGURE 18–56 Impedance versus angular frequency for the network of Figure 18–55.

The results indicate that at dc ($f = 0$ Hz) the capacitor, which behaves as an open circuit, will result in a circuit impedance of $R = 1$ kΩ. As the frequency increases, the capacitor reactance approaches 0 Ω, resulting in a corresponding decrease in circuit impedance.

RL Circuits

RL circuits may be analyzed in a manner similar to the analysis of *RC* circuits. Consider the parallel *RL* circuit of Figure 18–57.

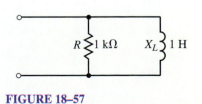

FIGURE 18–57

The total impedance of the parallel circuit is found as follows:

$$\mathbf{Z}_T = \frac{\mathbf{Z}_R\mathbf{Z}_L}{\mathbf{Z}_R + \mathbf{Z}_L}$$

$$= \frac{R(j\omega L)}{R + j\omega L} \qquad (18\text{--}25)$$

$$\mathbf{Z}_T = \frac{j\omega L}{1 + j\omega\dfrac{L}{R}}$$

If we define the *cutoff* or *corner frequency* for an *RL* circuit as

$$\omega_c = \frac{R}{L} = \frac{1}{\tau} \quad (\text{rad/s}) \qquad (18\text{--}26)$$

or equivalently as

$$f_c = \frac{R}{2\pi L} \quad (\text{Hz}) \qquad (18\text{--}27)$$

then several important points become evident.

For $\omega \le \omega_c/10$ (or $f \le f_c/10$) Equation 18–25 can be expressed as

$$\mathbf{Z}_T \simeq \frac{j\omega L}{1 + j0} = j\omega L$$

The above result indicates that for low frequencies, the inductor has a very small reactance, resulting in a total impedance which is essentially equal to the inductive reactance.

For $\omega \ge 10\omega_c$, the expression of (18–25) can be simplified as

$$\mathbf{Z}_T \simeq \frac{j\omega L}{0 + j\omega\dfrac{L}{R}} = R$$

The above results indicate that for high frequencies, the impedance of the circuit is essentially equal to the resistance, due to the very high impedance of the inductor.

Evaluating the impedance at several angular frequencies, we have the results of Table 18–3.

When the magnitude of the impedance $\mathbf{Z}_T$ is plotted as a function of angular frequency ω, we get the graph of Figure 18–58.

RLC Circuits

When numerous capacitive and inductive components are combined with resistors in series-parallel circuits, the total impedance $\mathbf{Z}_T$ of the circuit may rise and fall several times over the full range of frequencies. The analysis of such complex circuits is outside the scope of this textbook. However, for illustrative purposes we examine the simple series *RLC* circuit of Figure 18–59.

TABLE 18–3

Angular Frequency, ω (Rad/s)	X_L (Ω)	Z_T (Ω)
0	0	0
1	1	1
10	10	10
100	100	99.5
200	200	196
500	500	447
1 k	1 k	707
2 k	2 k	894
5 k	5 k	981
10 k	10 k	995
100 k	100 k	1000

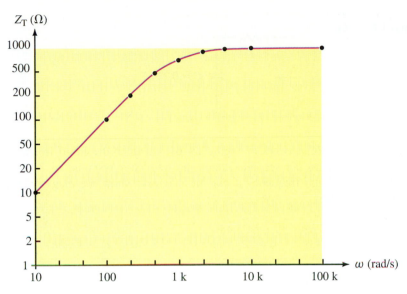

FIGURE 18–58   Impedance versus angular frequency for the network of Figure 18–57.

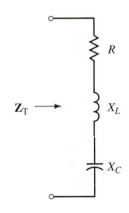

FIGURE 18–59

The impedance $\mathbf{Z}_T$ at any frequency will be determined as

$$\mathbf{Z}_T = R + jX_L - jX_C$$
$$= R + j(X_L - X_C)$$

At very low frequencies, the inductor will appear as a very low impedance (effectively a short circuit), while the capacitor will appear as a very high impedance (effectively an open circuit). Because the capacitive reactance will be much larger than the inductive reactance, the circuit will have a very large capacitive reactance. This results in a very high circuit impedance, $\mathbf{Z}_T$.

As the frequency increases, the inductive reactance increases, while the capacitive reactance decreases. At some frequency, f_0, the inductor and the capacitor will have the same magnitude of reactance. At this frequency, the reactances cancel, resulting in a circuit impedance which is equal to the resistance value.

As the frequency increases still further, the inductive reactance becomes larger than the capacitive reactance. The circuit becomes inductive and the magnitude of the total impedance of the circuit again rises. Figure 18–60 shows how the impedance of a series *RLC* circuit varies with frequency.

The complete analysis of the series *RLC* circuit and the parallel *RLC* circuit is left until we examine the principle of resonance in a later chapter.

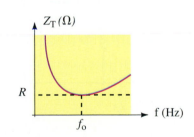

FIGURE 18–60

1. For a series network consisting of a resistor and a capacitor, what will be the impedance of the network at a frequency of 0 Hz (dc)? What will be the impedance of the network as the frequency approaches infinity?

2. For a parallel network consisting of a resistor and an inductor, what will be the impedance of the network at a frequency of 0 Hz (dc)? What will be the impedance of the network as the frequency approaches infinity?

(Answers are at the end of the chapter.)

PRACTICE
PROBLEMS 9

FIGURE 18–61

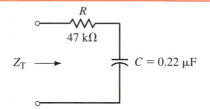

Given the series *RC* network of Figure 18–61, calculate the cutoff frequency in hertz and in radians per second. Sketch the frequency response of Z_T (magnitude) versus angular frequency ω for the network. Show the magnitude Z_T at $\omega_c/10$, ω_c, and $10\omega_c$.

Answers: $\omega_c = 96.7$ rad/s $f_c = 15.4$ Hz
At $0.1\,\omega_c$: $Z_T = 472$ kΩ   At ω_c: $Z_T = 66.5$ kΩ   At $10\,\omega_c$: $Z_T = 47.2$ kΩ

18.8   Applications

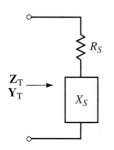

FIGURE 18–62

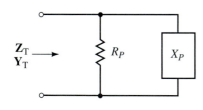

FIGURE 18–63

As we have seen, we may determine the impedance of any ac circuit as a vector $\mathbf{Z} = R \pm jX$. This means that any ac circuit may now be simplified as a series circuit having a resistance and a reactance, as shown in Figure 18–62.

Additionally, an ac circuit may be represented as an equivalent parallel circuit consisting of a single resistor and a single reactance as shown in Figure 18–63. *Any equivalent circuit will be valid only at the given frequency of operation.*

We will now examine the technique used to convert any series impedance into its parallel equivalent. Suppose that the two circuits of Figure 18–62 and Figure 18–63 are exactly equivalent at some frequency. These circuits can be equivalent only if both circuits have the same total impedance, $\mathbf{Z}_T$, and the same total admittance, $\mathbf{Y}_T$.

From the circuit of Figure 18–62, the total impedance is written as

$$\mathbf{Z}_T = R_S \pm jX_S$$

Therefore, the total admittance of the circuit is

$$\mathbf{Y}_T = \frac{1}{\mathbf{Z}_T} = \frac{1}{R_S \pm jX_S}$$

Multiplying the numerator and denominator by the complex conjugate, we obtain the following:

$$\mathbf{Y}_T = \frac{R_S \mp jX_S}{}$$

$$= \frac{R_S \mp jX_S}{R_S^2 + X_S^2}$$ (18–28)

$$\mathbf{Y}_T = \frac{R_S}{R_S^2 + X_S^2} \mp j\frac{X_S}{R_S^2 + X_S^2}$$

Now, from the circuit of Figure 18–63, the total admittance of the parallel circuit may be found from the parallel combination of R_P and X_P as

$$\mathbf{Y}_T = \frac{1}{R_P} + \frac{1}{\pm jX_P}$$

which gives

$$\mathbf{Y}_T = \frac{1}{R_P} \mp j\frac{1}{X_P} \qquad (18\text{–}29)$$

Two vectors can only be equal if both the real components are equal and the imaginary components are equal. Therefore the circuits of Figure 18–62 and Figure 18–63 can only be equivalent if the following conditions are met:

$$R_P = \frac{R_S^2 + X_S^2}{R_S} \qquad (18\text{–}30)$$

and

$$X_P = \frac{R_S^2 + X_S^2}{X_S} \qquad (18\text{–}31)$$

In a similar manner, we have the following conversion from a parallel circuit to an equivalent series circuit:

$$R_S = \frac{R_P X_P^2}{R_P^2 + X_P^2} \qquad (18\text{–}32)$$

and

$$X_S = \frac{R_P^2 X_P}{R_P^2 + X_P^2} \qquad (18\text{–}33)$$

EXAMPLE 18–19   A circuit has a total impedance of $\mathbf{Z}_T = 10\ \Omega + j50\ \Omega$. Sketch the equivalent series and parallel circuits.

Solution   The series circuit will be an inductive circuit having $R_S = 10\ \Omega$ and $X_{LS} = 50\ \Omega$.
  The equivalent parallel circuit will also be an inductive circuit having the following values:

$$R_P = \frac{(10\ \Omega)^2 + (50\ \Omega)^2}{10\ \Omega} = 260\ \Omega$$

$$X_{LP} = \frac{(10\ \Omega)^2 + (50\ \Omega)^2}{50\ \Omega} = 52\ \Omega$$

The equivalent series and parallel circuits are shown in Figure 18–64.

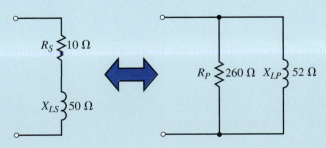

FIGURE 18–64

EXAMPLE 18–20 A circuit has a total admittance of $\mathbf{Y}_T = 0.559 \text{ mS} \angle 63.43°$. Sketch the equivalent series and parallel circuits.

Solution Because the admittance is written in polar form, we first convert to the rectangular form of the admittance.

$$G_P = (0.559 \text{ mS}) \cos 63.43° = 0.250 \text{ mS} \quad \Leftrightarrow \quad R_P = 4.0 \text{ k}\Omega$$

$$B_{CP} = (0.559 \text{ mS}) \sin 63.43° = 0.500 \text{ mS} \quad \Leftrightarrow \quad X_{CP} = 2.0 \text{ k}\Omega$$

The equivalent series circuit is found as

$$R_S = \frac{(4 \text{ k}\Omega)(2 \text{ k}\Omega)^2}{(4 \text{ k}\Omega)^2 + (2 \text{ k}\Omega)^2} = 0.8 \text{ k}\Omega$$

and

$$X_{CS} = \frac{(4 \text{ k}\Omega)^2(2 \text{ k}\Omega)}{(4 \text{ k}\Omega)^2 + (2 \text{ k}\Omega)^2} = 1.6 \text{ k}\Omega$$

The equivalent circuits are shown in Figure 18–65.

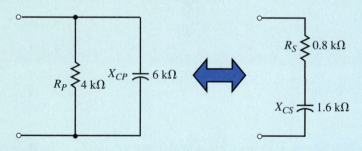

FIGURE 18–65

EXAMPLE 18–21 Refer to the circuit of Figure 18–66.

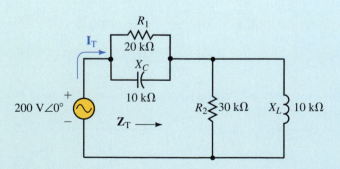

FIGURE 18–66

a. Find $\mathbf{Z}_T$.

b. Sketch the equivalent series circuit.

c. Determine $\mathbf{I}_T$.

Solution

a. The circuit consists of two parallel networks in series. We apply Equations 18–32 and 18–33 to arrive at equivalent series elements for each of the parallel networks as follows:

$$R_{S_1} = \frac{(20 \text{ k}\Omega)(10 \text{ k}\Omega)^2}{(20 \text{ k}\Omega)^2 + (10 \text{ k}\Omega)^2} = 4 \text{ k}\Omega$$

$$X_{CS} = \frac{(20 \text{ k}\Omega)^2(10 \text{ k}\Omega)}{(20 \text{ k}\Omega)^2 + (10 \text{ k}\Omega)^2} = 8 \text{ k}\Omega$$

and

$$R_{S_2} = \frac{(30 \text{ k}\Omega)(10 \text{ k}\Omega)^2}{(30 \text{ k}\Omega)^2 + (10 \text{ k}\Omega)^2} = 3 \text{ k}\Omega$$

$$X_{LS} = \frac{(30 \text{ k}\Omega)^2(10 \text{ k}\Omega)}{(30 \text{ k}\Omega)^2 + (10 \text{ k}\Omega)^2} = 9 \text{ k}\Omega$$

The equivalent circuits are shown in Figure 18–67.

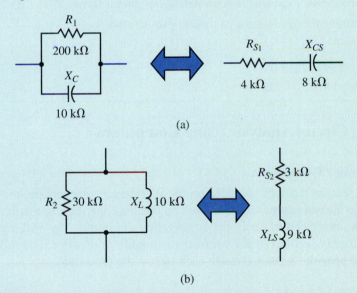

(a)

(b)

FIGURE 18–67

The total impedance of the circuit is found to be

$$\mathbf{Z}_T = (4 \text{ k}\Omega - j8 \text{ k}\Omega) + (3 \text{ k}\Omega + j9 \text{ k}\Omega) = 7 \text{ k}\Omega + j1 \text{ k}\Omega = 7.071 \text{ k}\Omega\angle 8.13°$$

b. Figure 18–68 shows the equivalent series circuit.

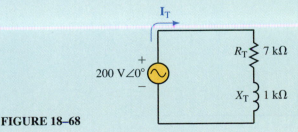

FIGURE 18–68

c.
$$\mathbf{I}_T = \frac{200 \text{ V}\angle 0°}{7.071 \text{ k}\Omega\angle 8.13°} = 28.3 \text{ mA}\angle -8.13°$$

IN-PROCESS LEARNING CHECK 5

An inductor of 10 mH has a series resistance of 5 Ω.

a. Determine the parallel equivalent of the inductor at a frequency of 1 kHz. Sketch the equivalent showing the values of L_P (in henries) and R_P.

b. Determine the parallel equivalent of the inductor at a frequency of 1 MHz. Sketch the equivalent showing the values of L_P (in henries) and R_P.

c. If the frequency were increased still further, predict what would happen to the values of L_P and R_P

(Answers are at the end of the chapter.)

PRACTICE PROBLEMS 10

A network has an impedance of $\mathbf{Z}_T = 50 \text{ k}\Omega\angle 75°$ at a frequency of 5 kHz.

a. Determine the most simple equivalent series circuit (L and R).

b. Determine the most simple equivalent parallel circuit.

Answers:
a. $R_S = 12.9 \text{ k}\Omega$, $L_S = 1.54 \text{ H}$
b. $R_P = 193 \text{ k}\Omega$, $L_P = 1.65 \text{ H}$

18.9   Circuit Analysis Using Computers

ELECTRONICS WORKBENCH

PSpice

Electronics Workbench

In this section we will use Electronics Workbench to simulate how sinusoidal ac measurements are taken with an oscilloscope. The "measurements" are then interpreted to verify the ac operation of circuits. You will use some of the display features of the software to simplify your work. The following example provides a guide through each step of the procedure.

EXAMPLE 18–22   Given the circuit of Figure 18–69.

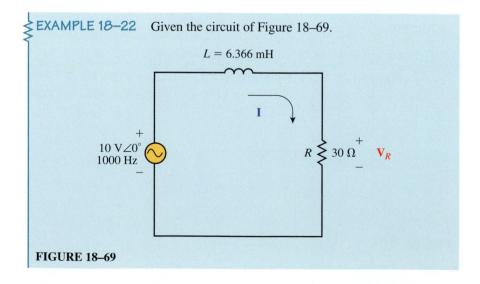

FIGURE 18–69

a. Determine the current **I** and the voltage $\mathbf{V}_R$.

b. Use Electronics Workbench to display the resistor voltage v_R and the source voltage e. Use the results to verify the results of part (a).

Solution

a. $X_L = 2\pi(1000 \text{ Hz})(6.366 \times 10^{-3} \text{ H}) = 40 \text{ }\Omega$

   $\mathbf{Z} = 30 \text{ }\Omega + j40 \text{ }\Omega = 50 \text{ }\Omega\angle 53.13°$

   $\mathbf{I} = \dfrac{10 \text{ V}\angle 0°}{50 \text{ }\Omega\angle 53.13°} = 0.200 \text{ A}\angle -53.13°$

   $\mathbf{V}_R = (0.200 \text{ A}\angle -53.13°)(30 \text{ }\Omega) = 6.00 \text{ V}\angle -53.13°$

b. Use the schematic editor to input the circuit shown in Figure 18–70.

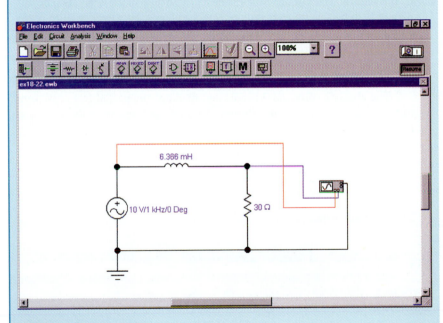

EWB   **FIGURE 18–70**

The ac voltage source is obtained from the Sources parts bin. The properties of the voltage source are changed by double clicking on the symbol and then selecting the Value tab. Change the values as follows:

Voltage (V):	**10 V**
Frequency:	**1 kHz**
Phase:	**0 Deg**

The oscilloscope is selected from the Instruments parts bin and the settings are changed as follows:

Time base:	**0.2 ms/div**
Channel A:	**5 V/Div**
Channel B:	**5 V/Div**

At this point, the circuit could be simulated and the display will resemble actual lab results. However, we can refine the display to provide information that is more useful. First, click on <u>A</u>nalysis/Options. Next click on the Instruments tab and change the following settings for the oscilloscope:

✓Pause after each screen

•Minimum number of time points   **1000**

After returning to the main window, click on the Power switch. It is necessary to press the Pause/Resume button once, since the display on the oscilloscope will not immediately show the circuit steady state values, but rather begins from $t = 0$. Consequently the display will show the transient (which in this case will last approximately 5 $L/R = 1$ ms).

Now, we can obtain a more detailed display by clicking on the Display Graphs button. We will use cursors and a grid to help in the analysis. The display will show several cycles of selected voltages. However, we are really only interested in viewing one complete cycle. This is done as follows:

Click on the Properties button. Select the General tab. We enable the cursors by selecting • All traces and ✓Cursors On from the Cursors box. Next, we enable the grid display by selecting <u>G</u>rid On from the Grid box.

The abscissa (time axis) is adjusted to show one period from $t = 4$ ms to $t = 5$ ms by selecting the Bottom Axis tab and adjusting the values as follows:

Range	Mi<u>n</u>imum	**0.004**
	Ma<u>x</u>imum	**0.005**
Divisions	Numbe<u>r</u>	**20**

The resulting display is shown in Figure 18–71.

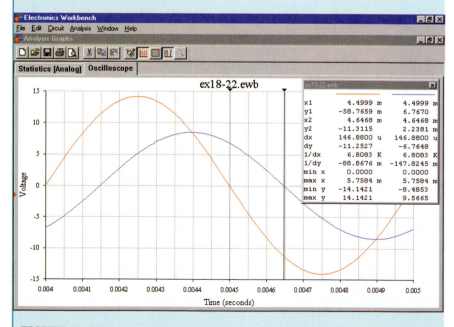

FIGURE 18–71

After positioning the cursors and using the display window, we are able to obtain various measurements for the circuit. The phase angle of the resistor voltage with respect to the source voltage is found by using the difference between the cursors in the bottom axis, $dx = 146.88 \ \mu s$. Now we have

$$\theta = \frac{146.88 \ \mu s}{1000 \ \mu s} \times 360° = 52.88°$$

The amplitude of the resistor voltage is 8.49 V, which results in an rms value of 6.00 V. As a result of the measurements we have

$$\mathbf{V}_R = 6.00 \ \text{V}\angle{-52.88°}$$

and

$$\mathbf{I} = \frac{6 \ \text{V}\angle{-52.88°}}{30 \ \Omega} = 0.200 \ \text{A}\angle{-52.88°}$$

These values correspond very closely to the theoretical results calculated in part (a) of the example.

OrCAD PSpice

In the following example we will use the Probe postprocessor of PSpice to show how the impedance of an RC circuit changes as a function of frequency. The Probe output will provide a graphical result that is very similar to the frequency responses determined in previous sections of this chapter.

EXAMPLE 18–23 Refer to the network of Figure 18–72. Use the OrCAD Capture CIS Demo to input the circuit. Run the Probe postprocessor to provide a graphical display of network impedance as a function of frequency from 50 Hz to 500 Hz.

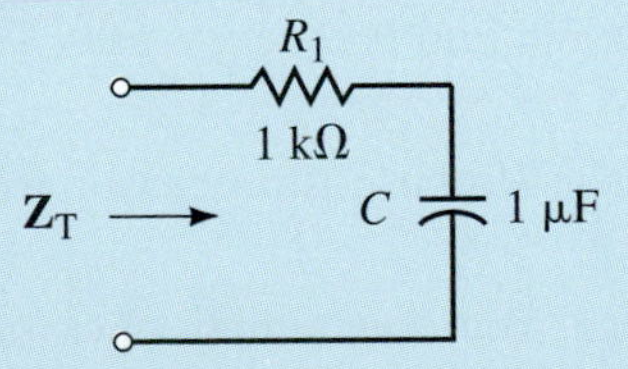

FIGURE 18–72

Solution Since PSpice is unable to analyze an incomplete circuit, it is necessary to provide a voltage source (and ground) for the circuit of Figure 18–72. The input impedance is not dependent on the actual voltage used, and so we may use any ac voltage source. In this example we arbitrarily select a voltage of 10 V.

- Open the CIS Demo software.
- Open a new project and call it Ch 18 PSpice 1. Ensure that the Analog or Mixed-Signal Circuit Wizard is activated.

- Enter the circuit as show in Figure 18–73. Simply click on the voltage value and change its value to **10V** from the default value of **0V.** (There must be no spaces between the magnitude and the units.

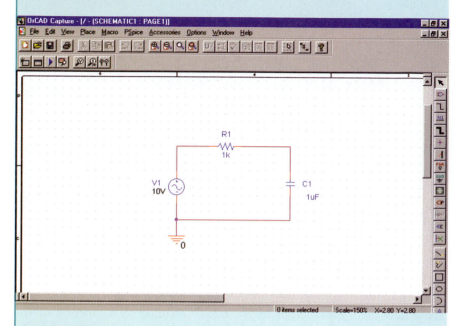

FIGURE 18–73

- Click on PSpice, New Simulation Profile and give a name such as Example 18-23 to the new simulation. The Simulation Settings dialog box will open.

- Click on the Analysis tab and select AC Sweep/Noise as the Analysis type. Select General Settings from the Options box.

- Select Linear AC Sweep Type (Quite often, logarithmic frequency sweeps such as Decade, are used). Enter the following values into the appropriate dialog boxes. Start Frequency; **50,** End Frequency: **500,** Total Points: **1001.** Click OK.

- Click on PSpice and Run. The Probe postprocessor will appear on the screen.

- Click on Trace and Add Trace. You may simply click on the appropriate values from the list of variables and use the division symbol to result in impedance. Enter the following expression into the Trace Expression box:

    V(R1:1)/I(R1)

Notice that the above expression is nothing other than an application of Ohm's Law. The resulting display is shown in Figure 18–74.

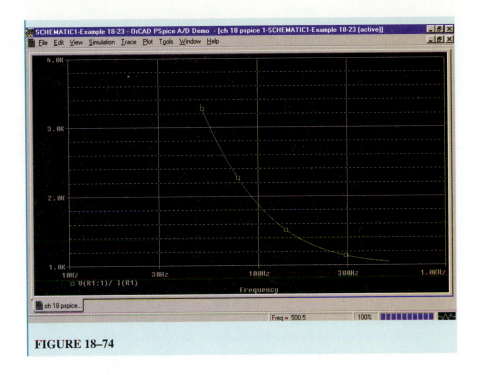

FIGURE 18–74

**PRACTICE
PROBLEMS 11**

EWB   Given the circuit of Figure 18–75.

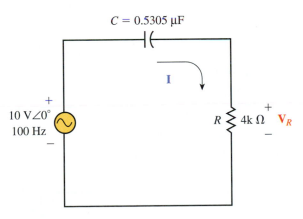

FIGURE 18–75

a. Determine the current **I** and the voltage $\mathbf{V}_R$.

b. Use Electronics Workbench to display the resistor voltage v_R and the source voltage, e. Use the results to verify the results of part (a).

Answers:

a. $\mathbf{I} = 2.00\ \text{mA}\angle 36.87°$, $\mathbf{V}_R = 8.00\ \text{V}\angle 36.87°$

b. $v_R = 11.3\ \sin(\omega t + 36.87°)$, $e = 14.1\ \sin\omega t$

PRACTICE PROBLEMS 12

PSpice   Use OrCAD PSpice to input the circuit of Figure 18–55. Use the Probe postprocessor to obtain a graphical display of the network impedance as a function of frequency from 100 Hz to 2000 Hz.

PUTTING IT INTO PRACTICE

You are working in a small industrial plant where several small motors are powered by a 60-Hz ac line voltage of 120 Vac. Your supervisor tells you that one of the 2-Hp motors which was recently installed draws too much current when the motor is under full load. You take a current reading and find that the current is 14.4 A. After doing some calculations, you determine that even if the motor is under full load, it shouldn't require that much current.

However, you have an idea. You remember that a motor can be represented as a resistor in series with an inductor. If you could reduce the effect of the inductive reactance of the motor by placing a capacitor across the motor, you should be able to reduce the current since the capacitive reactance will cancel the inductive reactance.

While keeping the motor under load, you place a capacitor into the circuit. Just as you suspected, the current goes down. After using several different values, you observe that the current goes to a minimum of 12.4 A. It is at this value that you have determined that the reactive impedances are exactly balanced. Sketch the complete circuit and determine the value of capacitance that was added into the circuit. Use the information to determine the value of the motor's inductance. (Assume that the motor has an efficiency of 100%.)

PROBLEMS

18.1   Ohm's Law for AC Circuits

1. For the resistor shown in Figure 18–76:
   a. Find the sinusoidal current i using phasors.
   b. Sketch the sinusoidal waveforms for v and i.
   c. Sketch the phasor diagram for **V** and **I**.

FIGURE 18–76   **FIGURE 18–77**

2. Repeat Problem 1 for the resistor of Figure 18–77.
3. Repeat Problem 1 for the resistor of Figure 18–78.

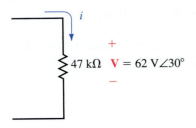

FIGURE 18–78

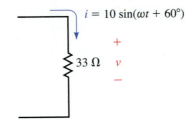

FIGURE 18–79

4. Repeat Problem 1 for the resistor of Figure 18–79.

5. For the component shown in Figure 18–80:

   a. Find the sinusoidal voltage v using phasors.

   b. Sketch the sinusoidal waveforms for v and i.

   c. Sketch the phasor diagram for **V** and **I.**

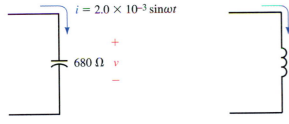

FIGURE 18–80

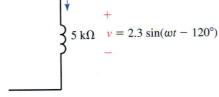

FIGURE 18–81

6. Repeat Problem 5 for the component shown in Figure 18–81.

7. Repeat Problem 5 for the component shown in Figure 18–82.

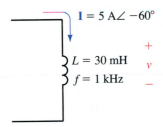

FIGURE 18–82

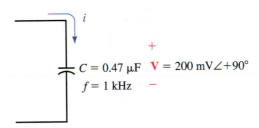

FIGURE 18–83

8. Repeat Problem 5 for the component shown in Figure 18–83.

9. Repeat Problem 5 for the component shown in Figure 18–84.

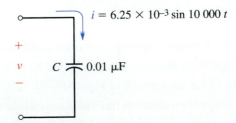

FIGURE 18–84

10. Repeat Problem 5 for the component shown in Figure 18–85.

FIGURE 18–85

$i = 22.5 \times 10^{-3} \cos(20\ 000\ t + 20°)$

$v = 8 \cos(200t - 30°)$

FIGURE 18–86

11. Repeat Problem 5 for the component shown in Figure 18–86.
12. Repeat Problem 5 for the component shown in Figure 18–87.

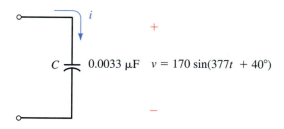

$v = 170 \sin(377t + 40°)$

FIGURE 18–87

18.2 AC Series Circuits

13. Find the total impedance of each of the networks shown in Figure 18–88.

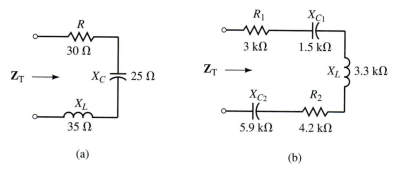

(a)   (b)

FIGURE 18–88

14. Repeat Problem 13 for the networks of Figure 18–89.
15. Refer to the network of Figure 18–90.
    a. Determine the series impedance **Z** which will result in the given total impedance, $\mathbf{Z}_T$. Express your answer in rectangular and polar form.
    b. Sketch an impedance diagram showing $\mathbf{Z}_T$ and **Z.**
16. Repeat Problem 15 for the network of Figure 18–91.
17. A circuit consisting of two elements has a total impedance of $\mathbf{Z}_T = 2\ \text{k}\Omega\angle 15°$ at a frequency of 18 kHz. Determine the values in ohms, henries, or farads of the unknown elements.

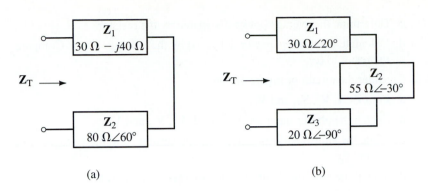

(a)   (b)

FIGURE 18–89

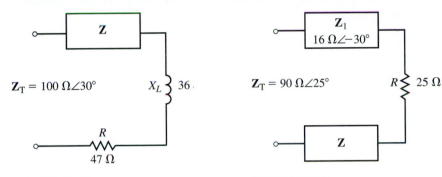

FIGURE 18–90   **FIGURE 18–91**

18. A network has a total impedance of $\mathbf{Z}_T = 24.0 \text{ k}\Omega\angle-30°$ at a frequency of 2 kHz. If the network consists of two series elements, determine the values in ohms, henries, or farads of the unknown elements.

19. Given that the network of Figure 18–92 is to operate at a frequency of 1 kHz, what series components R and L (in henries) or C (in farads) must be in the indicated block to result in a total circuit impedance of $\mathbf{Z}_T = 50 \ \Omega\angle60°$?

20. Repeat Problem 19 for a frequency of 2 kHz.

21. Refer to the circuit of Figure 18–93.

   a. Find $\mathbf{Z}_T, \mathbf{I}, \mathbf{V}_R, \mathbf{V}_L,$ and $\mathbf{V}_C$.

   b. Sketch the phasor diagram showing $\mathbf{I}, \mathbf{V}_R, \mathbf{V}_L,$ and $\mathbf{V}_C$.

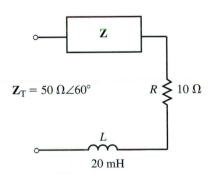

FIGURE 18–92

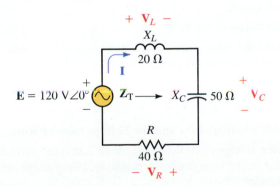

FIGURE 18–93

c. Determine the average power dissipated by the resistor.

d. Calculate the average power delivered by the voltage source. Compare the result to (c).

22. Consider the circuit of Figure 18–94.

   a. Find $\mathbf{Z}_T$, $\mathbf{I}$, $\mathbf{V}_R$, $\mathbf{V}_L$, and $\mathbf{V}_C$.

   b. Sketch the phasor diagram showing $\mathbf{I}$, $\mathbf{V}_R$, $\mathbf{V}_L$, and $\mathbf{V}_C$.

   c. Write the sinusoidal expressions for the current i and the voltages e, v_R, v_C, and v_L.

   d. Sketch the sinusoidal current and voltages found in (c).

   e. Determine the average power dissipated by the resistor.

   f. Calculate the average power delivered by the voltage source. Compare the result to (e).

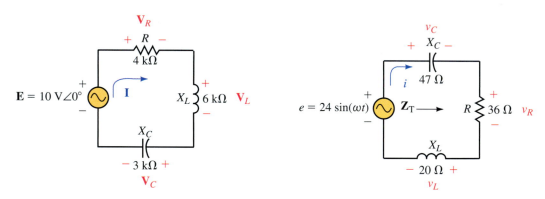

FIGURE 18–94   **FIGURE 18–95**

23. Refer to the circuit of Figure 18–95.

   a. Determine the circuit impedance, $\mathbf{Z}_T$.

   b. Use phasors to solve for i, v_R, v_C, and v_L.

   c. Sketch the phasor diagram showing $\mathbf{I}$, $\mathbf{V}_R$, $\mathbf{V}_L$, and $\mathbf{V}_C$.

   d. Sketch the sinusoidal expressions for the current and voltages found in (b).

   e. Determine the average power dissipated by the resistor.

   f. Calculate the average power delivered by the voltage source. Compare the result to (e).

24. Refer to the circuit of Figure 18–96.

   a. Determine the value of the capacitor reactance, X_C, needed so that the resistor in the circuit dissipates a power of 200 mW.

   b. Using the value of X_C from (a), determine the sinusoidal expression for the current i in the circuit.

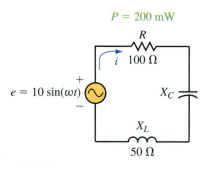

FIGURE 18–96

18.3   Kirchhoff's Voltage Law and the Voltage Divider Rule

25. a. Suppose a voltage of $10\ \text{V}\angle 0°$ is applied across the network in Figure 18–88a. Use the voltage divider rule to find the voltage appearing across each element.

   b. Verify Kirchhoff's voltage law for each network.

26. a. Suppose a voltage of 240 V∠30° is applied across the network in Figure 18–89a. Use the voltage divider rule to find the voltage appearing across each impedance.

    b. Verify Kirchhoff's voltage law for each network.

27. Given the circuit of Figure 18–97:

    a. Find the voltages $\mathbf{V}_C$ and $\mathbf{V}_L$.

    b. Determine the value of R.

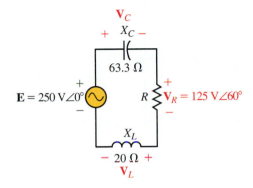

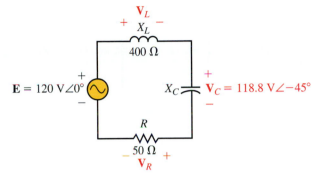

FIGURE 18–97   **FIGURE 18–98**

28. Refer to the circuit of Figure 18–98.

    a. Find the voltages $\mathbf{V}_R$ and $\mathbf{V}_L$.

    b. Determine the value of X_C.

29. Refer to the circuit of Figure 18–99:

    a. Find the voltage across $\mathbf{X}_C$.

    b. Use Kirchhoff's voltage law to find the voltage across the unknown impedance.

    c. Calculate the value of the unknown impedance $\mathbf{Z}$.

    d. Determine the average power dissipated by the circuit.

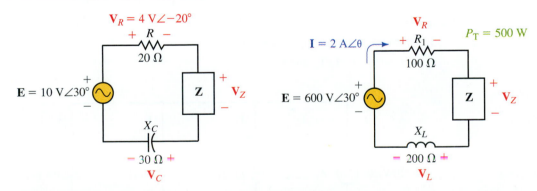

FIGURE 18–99   **FIGURE 18–100**

30. Given that the circuit of Figure 18–100 has a current with a magnitude of 2.0 A and dissipates a total power of 500 W:

    a. Calculate the value of the unknown impedance $\mathbf{Z}$. (Hint: Two solutions are possible.)

b. Calculate the phase angle θ of the current **I**.

c. Find the voltages $\mathbf{V}_R$, $\mathbf{V}_L$, and $\mathbf{V}_Z$.

18.4   AC Parallel Circuits

31. Determine the input impedance, $\mathbf{Z}_T$, for each of the networks of Figure 18–101.

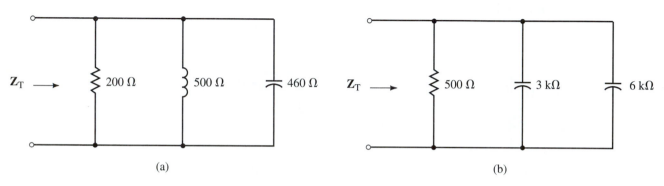

(a)   (b)

FIGURE 18–101

32. Repeat Problem 31 for Figure 18–102.

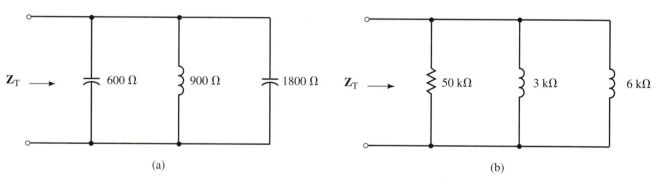

(a)   (b)

FIGURE 18–102

33. Given the circuit of Figure 18–103.

   a. Find $\mathbf{Z}_T$, $\mathbf{I}_T$, $\mathbf{I}_1$, $\mathbf{I}_2$, and $\mathbf{I}_3$.

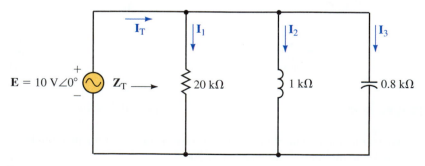

FIGURE 18–103

b. Sketch the admittance diagram showing each of the admittances.

c. Sketch the phasor diagram showing $\mathbf{E}$, $\mathbf{I}_T$, $\mathbf{I}_1$, $\mathbf{I}_2$, and $\mathbf{I}_3$.

d. Determine the average power dissipated by the resistor.

e. Find the power factor of the circuit and calculate the average power delivered by the voltage source. Compare the answer with the result obtained in (d).

34. Refer to the circuit of Figure 18–104.

a. Find $\mathbf{Z}_T$, $\mathbf{I}_T$, $\mathbf{I}_1$, $\mathbf{I}_2$, and $\mathbf{I}_3$.

b. Sketch the admittance diagram for each of the admittances.

c. Sketch the phasor diagram showing $\mathbf{E}$, $\mathbf{I}_T$, $\mathbf{I}_1$, $\mathbf{I}_2$, and $\mathbf{I}_3$.

d. Determine the expressions for the sinusoidal currents i_T, i_1, i_2, and i_3.

e. Sketch the sinusoidal voltage e and current i_T.

f. Determine the average power dissipated by the resistor.

g. Find the power factor of the circuit and calculate the average power delivered by the voltage source. Compare the answer with the result obtained in (f).

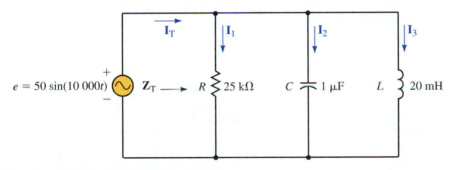

e = 50 sin(10 000t)

FIGURE 18–104

35. Refer to the network of Figure 18–105.

a. Determine $\mathbf{Z}_T$.

b. Given the indicated current, use Ohm's law to find the voltage, $\mathbf{V}$, across the network.

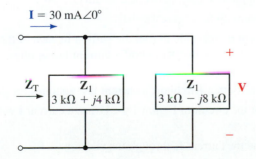

$\mathbf{I} = 30\ \text{mA}\angle 0°$

FIGURE 18–105

36. Consider the network of Figure 18–106.

    a. Determine $\mathbf{Z}_T$.

    b. Given the indicated current, use Ohm's law to find the voltage, **V**, across the network.

    c. Solve for $\mathbf{I}_2$ and **I**.

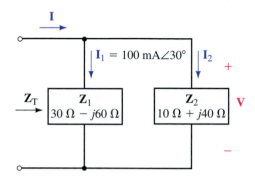

FIGURE 18–106

37. Determine the impedance, $\mathbf{Z}_2$, which will result in the total impedance shown in Figure 18–107.

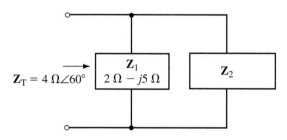

FIGURE 18–107

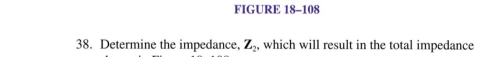

FIGURE 18–108

38. Determine the impedance, $\mathbf{Z}_2$, which will result in the total impedance shown in Figure 18–108.

18.5 Kirchhoff's Current Law and the Current Divider Rule

39. Solve for the current in each element of the networks in Figure 18–101 if the current applied to each network is 10 mA∠−30°.

40. Repeat Problem 39 for Figure 18–102.

41. Use the current divider rule to find the current in each of the elements in Figure 18–109. Verify that Kirchhoff's current law applies.

42. Given that $\mathbf{I}_L = 4$ A∠30° in the circuit of Figure 18–110, find the currents **I**, $\mathbf{I}_C$, and $\mathbf{I}_R$. Verify that Kirchhoff's current law appplies for this circuit.

43. Suppose that the circuit of Figure 18–111 has a current **I** with a magnitude of 8 A:

    a. Determine the current $\mathbf{I}_R$ through the resistor.

    b. Calculate the value of resistance, R.

    c. What is the phase angle of the current **I**?

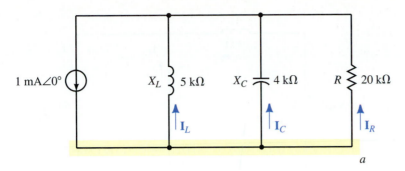

FIGURE 18–109

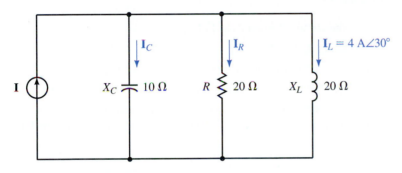

FIGURE 18–110

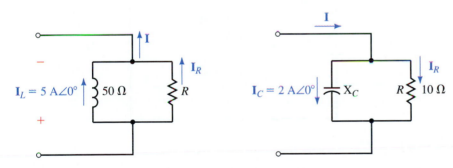

FIGURE 18–111   **FIGURE 18–112**

44. Assume that the circuit of Figure 18–112 has a current **I** with a magnitude of 3 A:

   a. Determine the current $\mathbf{I}_R$ through the resistor.

   b. Calculate the value of capacitive reactance X_C.

   c. What is the phase angle of the current **I?**

18.6 Series-Parallel Circuits

45. Refer to the circuit of Figure 18–113.

   a. Find $\mathbf{Z}_T$, $\mathbf{I}_L$, $\mathbf{I}_C$, and $\mathbf{I}_R$.

   b. Sketch the phasor diagram showing **E**, $\mathbf{I}_L$, $\mathbf{I}_C$, and $\mathbf{I}_R$.

   c. Calculate the average power dissipated by the resistor.

   d. Use the circuit power factor to calculate the average power delivered by the voltage source. Compare the answer with the results obtained in (c).

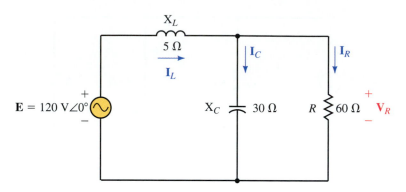

FIGURE 18–113

46. Refer to the circuit of Figure 18–114.
    a. Find $\mathbf{Z}_T$, $\mathbf{I}_1$, $\mathbf{I}_2$, and $\mathbf{I}_3$.
    b. Sketch the phasor diagram showing $\mathbf{E}$, $\mathbf{I}_1$, $\mathbf{I}_2$, and $\mathbf{I}_3$.
    c. Calculate the average power dissipated by each of the resistors.
    d. Use the circuit power factor to calculate the average power delivered by the voltage source. Compare the answer with the results obtained in (c).

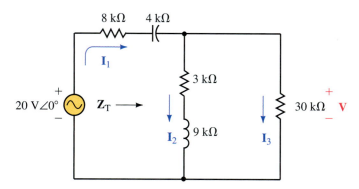

FIGURE 18–114

47. Refer to the circuit of Figure 18–115.
    a. Find $\mathbf{Z}_T$, $\mathbf{I}_T$, $\mathbf{I}_1$, and $\mathbf{I}_2$.
    b. Determine the voltage $\mathbf{V}_{ab}$.

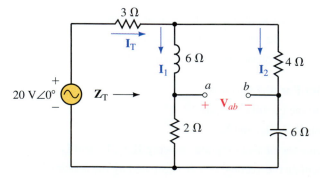

FIGURE 18–115

48. Consider the circuit of Figure 18–116.
    a. Find $\mathbf{Z}_T$, $\mathbf{I}_T$, $\mathbf{I}_1$, and $\mathbf{I}_2$.
    b. Determine the voltage **V**.

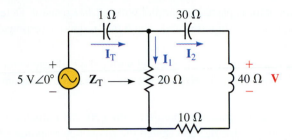

FIGURE 18–116

49. Refer to the circuit of Figure 18–117:
    a. Find $\mathbf{Z}_T$, $\mathbf{I}_1$, $\mathbf{I}_2$, and $\mathbf{I}_3$.
    b. Determine the voltage **V**.

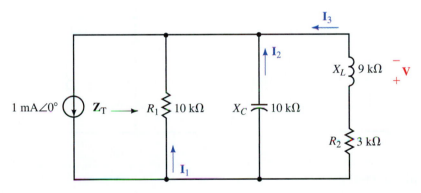

FIGURE 18–117

50. Refer to the circuit of Figure 18–118:
    a. Find $\mathbf{Z}_T$, $\mathbf{I}_1$, $\mathbf{I}_2$, and $\mathbf{I}_3$.
    b. Determine the voltage **V**.

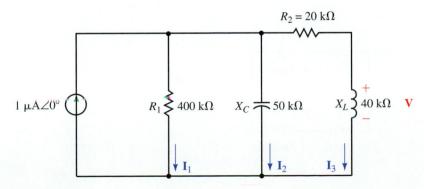

FIGURE 18–118

18.7   Frequency Effects

51. A 50-kΩ resistor is placed in series with a 0.01-μF capacitor. Determine the cutoff frequency ω_C (in rad/s) and sketch the frequency response (Z_T vs. ω) of the network.

52. A 2-mH inductor is placed in parallel with a 2-kΩ resistor. Determine the cutoff frequency ω_C (in rad/s) and sketch the frequency response (Z_T vs. ω) of the network.

53. A 100-kΩ resistor is placed in parallel with a 0.47-μF capacitor. Determine the cutoff frequency f_C (in Hz) and sketch the frequency response (Z_T vs. f) of the network.

54. A 2.7-kΩ resistor is placed in parallel with a 20-mH inductor. Determine the cutoff frequency f_C (in Hz) and sketch the frequency response (Z_T vs. f) of the network.

18.8   Applications

55. Convert each each of the networks of Figure 18–119 into an equivalent series network consisting of two elements.

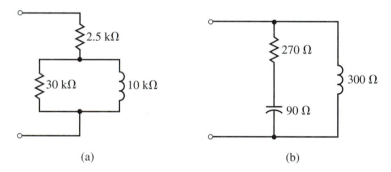

(a)   (b)

FIGURE 18–119

56. Convert each each of the networks of Figure 18–119 into an equivalent parallel network consisting of two elements.

57. Show that the networks of Figure 18–120 have the same input impedance at frequencies of 1 krad/s and 10 krad/s. (It can be shown that these networks are equivalent at all frequencies.)

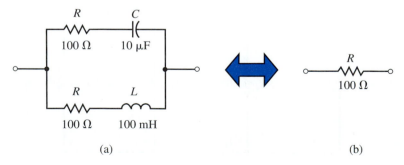

(a)   (b)

FIGURE 18–120

58. Show that the networks of Figure 18–121 have the same input impedance at frequencies of 5 rad/s and 10 rad/s.

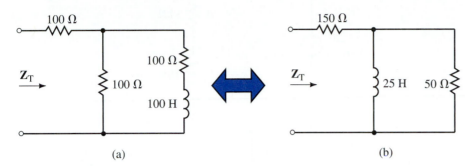

(a)   (b)

FIGURE 18–121

18.9   Circuit Analysis Using Computers

59. **EWB**   Given the circuit of Figure 18–122:

   a. Use Electronics Workbench to simultaneously display the capacitor voltage v_C and e. Record your results and determine the phasor voltage $\mathbf{V}_C$.

   b. Interchange the positions of the resistor and the capacitor relative to the ground. Use Electronics Workbench to simultaneously display the resistor voltage v_R and e. Record your results and determine the phasor voltage $\mathbf{V}_R$.

   c. Compare your results to those obtained in Example 18–8.

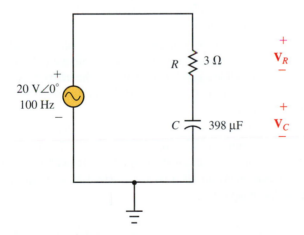

EWB   **FIGURE 18–122**

60. **EWB**   Given the circuit of Figure 18–123:

   a. Use Electronics Workbench to display the inductor voltage v_L and e. Record your results and determine the phasor voltage $\mathbf{V}_L$.

   b. Interchange the positions of the resistor and the inductor relative to the ground. Use Electronics Workbench to display the inductor voltage v_R and e. Record your results and determine the phasor voltage $\mathbf{V}_R$.

   c. Compare your results to those obtained in Example 18–9.

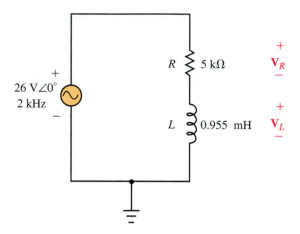

  FIGURE 18–123

61. **PSpice**   A 50-kΩ resistor is placed in series with a 0.01-μF capacitor. Use OrCAD PSpice to input these components into a circuit. Run the Probe postprocessor to provide a graphical display of the network impedance as a function of frequency from 50 Hz to 500 Hz. Let the frequency sweep logarithmically in octaves.

62. **PSpice**   A 2-mH inductor is placed in parallel with a 2-kΩ resistor. Use OrCAD PSpice to input these components into a circuit. Run the Probe postprocessor to provide a graphical display of the network impedance as a function of frequency from 50 kHz to 500 kHz. Let the frequency sweep logarithmically in octaves.

63. **PSpice**   A 100-kΩ resistor is placed in parallel with a 0.47-μF capacitor. Use OrCAD PSpice to input these components into a circuit. Run the Probe postprocessor to provide a graphical display of the network impedance as a function of frequency from 0.1 Hz to 10 Hz. Let the frequency sweep logarithmically in octaves.

64. **PSpice**   A 2.7-kΩ resistor is placed in parallel with a 20-mH inductor. Use OrCAD PSpice to input these components into a circuit. Run the Probe postprocessor to provide a graphical display of the network impedance as a function of frequency from 100 kHz to 1 MHz. Let the frequency sweep logarithmically in octaves.

ANSWERS TO IN-PROCESS LEARNING CHECKS

In-Process Learning Check 1

1. Current and voltage are in phase.

2. Current leads voltage by 90°.

3. Voltage leads current by 90°.

In-Process Learning Check 2

1. The phasor sum of voltage drops and rises around a closed loop is equal to zero.

2. All voltages must be expressed as phasors, i.e., $\mathbf{V} = V\angle\theta = V\cos\theta + jV\sin\theta$.

In-Process Learning Check 3

1. The phasor sum of currents entering a node is equal to the phasor sum of currents leaving the same node.

2. All currents must be expressed as phasors, i.e., $\mathbf{I} = I\angle\theta = I\cos\theta + jI\sin\theta$.

In-Process Learning Check 4

1. At $f = 0$ Hz, $Z = \infty$ (open circuit)

2. As $f \rightarrow \infty$, $Z = R$

In-Process Learning Check 5

1. $R_P = 795\ \Omega$, $L_P = 10.1$ mH

2. $R_P = 790\ \text{M}\Omega$, $L_P = 10.0$ mH

19

Methods of AC Analysis

OBJECTIVES

After studying this chapter, you will be able to

- convert an ac voltage source into its equivalent current source, and conversely, convert a current source into an equivalent voltage source,
- solve for the current or voltage in a circuit having either a dependent current source or a dependent voltage source,
- set up simultaneous linear equations to solve an ac circuit using mesh analysis,
- use complex determinants to find the solutions for a given set of linear equations,
- set up simultaneous linear equations to solve an ac circuit using nodal analysis,
- perform delta-to-wye and wye-to-delta conversions for circuits having reactive elements,
- solve for the balanced condition in a given ac bridge circuit. In particular, you will examine the Maxwell, Hay, and Schering bridges,
- use Electronics Workbench to analyze bridge circuits,
- use PSpice to calculate current and voltage in an ac circuit.

KEY TERMS

Balanced Bridges
Controlling Elements
Delta-Wye Conversion
Dependent Sources
Hay Bridge
Maxwell Bridge
Mesh Analysis
Nodal Analysis
Schering Bridge
Source Conversion

OUTLINE

Dependent Sources
Source Conversion
Mesh (Loop) Analysis
Nodal Analysis
Delta-to-Wye and Wye-to-Delta Conversions
Bridge Networks
Circuit Analysis Using Computers

To this point, we have examined only circuits having a single ac source. In this chapter, we continue our study by analyzing multisource circuits and bridge networks. You will find that most of the techniques used in analyzing ac circuits parallel those of dc circuit analysis. Consequently, a review of Chapter 8 will help you understand the topics of this chapter.

Near the end of the chapter, we examine how computer techniques are used to analyze even the most complex ac circuits. It must be emphasized that although computer techniques are much simpler than using a pencil and calculator, there is virtually no knowledge to be gained by mindlessly entering data into a computer. We use the computer merely as a tool to verify our results and to provide a greater dimension to the analysis of circuits.

Hermann Ludwig Ferdinand von Helmholtz

HERMANN HELMHOLTZ WAS BORN IN POTSDAM (near Berlin, Germany) on August 31, 1821. Helmholtz was a leading scientist of the nineteenth century, whose legacy includes contributions in the fields of acoustics, chemistry, mathematics, magnetism, electricity, mechanics, optics, and physiology.

Helmholtz graduated from the Medical Institute in Berlin in 1843 and practiced medicine for five years as a surgeon in the Prussian army. From 1849 to 1871, he served as a professor of physiology at universities in Königsberg, Bonn, and Heidelberg. In 1871, Helmholtz was appointed Professor of Physics at the University of Berlin.

Helmholtz' greatest contributions were as a mathematical physicist, where his work in theoretical and practical physics led to the proof of the Law of Conservation of Energy in his paper *"Über die Erhaltung der Kraft,"* published in 1847. He showed that mechanics, heat, light, electricity, and magnetism were simply manifestations of the same force. His work led to the understanding of electrodynamics (the motion of charge in conductors), and his theory of the electromagnetic properties of light set the groundwork for later scientists to understand how radio waves are propagated.

For his work, the German emperor Kaiser Wilhelm I made Helmholtz a noble in 1883. This great scientist died on September 8, 1894, at the age of 73.

19.1 Dependent Sources

The voltage and current sources we have worked with up to now have been **independent sources,** meaning that the voltage or current of the supply was not in any way dependent upon any voltage or current elsewhere in the circuit. In many amplifier circuits, particularly those involving transistors, it is possible to explain the operation of the circuits by replacing the device with an equivalent electronic model. These models often use voltage and current sources which have values dependent upon some internal voltage or current. Such sources are called **dependent sources.** Figure 19–1 compares the symbols for both independent and dependent sources.

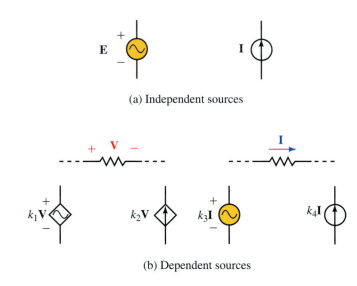

(a) Independent sources

(b) Dependent sources

FIGURE 19–1

Although the diamond is the accepted symbol for representing dependent sources, many articles and textbooks still use a circle. In this textbook we use both forms of the dependent source to familiarize the student with the various notations. The dependent source has a magnitude and phase angle determined by voltage or current at some internal element multiplied by a constant, k. The magnitude of the constant is determined by parameters within the particular model, and the units of the constant correspond to the required quantities in the equation.

EXAMPLE 19–1   Refer to the resistor shown in Figure 19–2. Determine the voltage $\mathbf{V}_R$ across the resistor given that the controlling voltage has the following values:

a. $\mathbf{V} = 0 \text{ V}$.
b. $\mathbf{V} = 5 \text{ V}\angle 30°$.
c. $\mathbf{V} = 3 \text{ V}\angle -150°$.

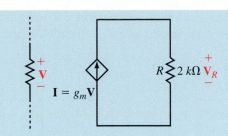

FIGURE 19–2   Given: $g_m = 4$ mS

Solution   Notice that the dependent source of this example has a constant, g_m, called the **transconductance.** Here, $g_m = 4$ mS.

a. $\mathbf{I} = (4 \text{ mS})(0 \text{ V}) = 0$
   $\mathbf{V}_R = 0$ V

b. $\mathbf{I} = (4 \text{ mS})(5 \text{ V}\angle 30°) = 20 \text{ mA}\angle 30°$
   $\mathbf{V}_R = (20 \text{ mA}\angle 30°)(2 \text{ k}\Omega) = 40 \text{ V}\angle 30°$

c. $\mathbf{I} = (4 \text{ mS})(3 \text{ V}\angle -150°) = 12 \text{ mA}\angle -150°$
   $\mathbf{V}_R = (12 \text{ mA}\angle -150°)(2 \text{ k}\Omega) = 24 \text{ V}\angle -150°$

The circuit of Figure 19–3 represents a simplified model of a transistor amplifier.

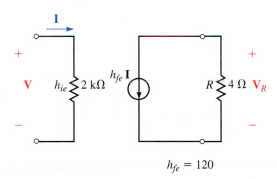

$h_{fe} = 120$

FIGURE 19–3

Determine the voltage $\mathbf{V}_R$ for each of the following applied voltages:

a. $\mathbf{V} = 10$ mV$\angle 0°$.

b. $\mathbf{V} = 2$ mV$\angle 180°$.

c. $\mathbf{V} = 0.03$ V$\angle 90°$.

Answers: a. 2.4 mV$\angle 180°$ b. 0.48 mV$\angle 0°$ c. 7.2 mV$\angle -90°$

19.2   Source Conversion

When working with dc circuits, the analysis of a circuit is often simplified by replacing the source (whether a voltage source or a current source) with its equivalent. The conversion of any ac source is similar to the method used in dc circuit analysis.

A voltage source **E** in series with an impedance **Z** is equivalent to a current source **I** having the same impedance **Z** in parallel. Figure 19–4 shows the equivalent sources.

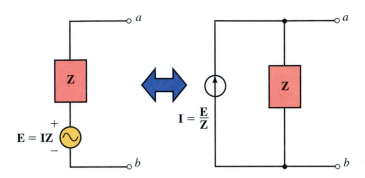

FIGURE 19–4

From Ohm's law, we perform the source conversion as follows:

$$\mathbf{I} = \frac{\mathbf{E}}{\mathbf{Z}}$$

and

$$\mathbf{E} = \mathbf{IZ}$$

It is important to realize that the two circuits of Figure 19–4 are equivalent between points *a* and *b*. This means that any network connected to points *a* and *b* will behave exactly the same regardless of which type of source is used. However, the voltages or currents within the sources will seldom be the same. In order to determine the current through or the voltage across the source impedance, the circuit must be returned to its original state.

EXAMPLE 19–2   Convert the voltage source of Figure 19–5 into an equivalent current source.

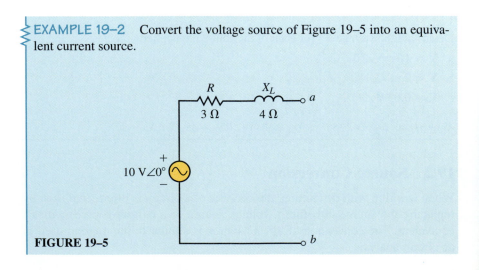

FIGURE 19–5

Solution

$$\mathbf{Z_T} = 3\ \Omega + j4\ \Omega = 5\ \Omega\angle 53.13°$$

$$\mathbf{I} = \frac{10\ \text{V}\angle 0°}{5\ \Omega\angle 53.13°} = 2\ \text{A}\angle -53.13°$$

The equivalent current source is shown in Figure 19–6.

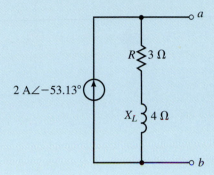

FIGURE 19–6

EXAMPLE 19–3   Convert the current source of Figure 19–7 into an equivalent voltage source.

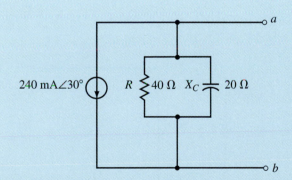

FIGURE 19–7

Solution   The impedance of the parallel combination is determined to be

$$\mathbf{Z} = \frac{(40\ \Omega\angle 0°)(20\ \Omega\angle -90°)}{40\ \Omega - j20\ \Omega}$$

$$= \frac{800\ \Omega\angle -90°}{44.72\angle -26.57°}$$

$$= 17.89\ \Omega\angle -63.43° = 8\ \Omega - j16\ \Omega$$

and so

$$\mathbf{E} = (240\ \text{mA}\angle 30°)(17.89\ \Omega\angle -63.43°)$$

$$= 4.29\ \text{V}\angle -33.43°$$

The resulting equivalent circuit is shown in Figure 19–8.

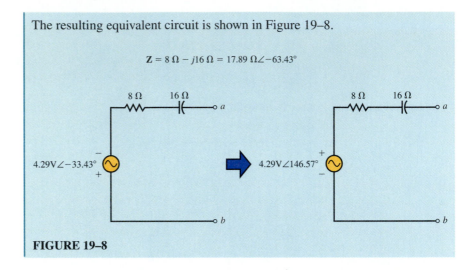

$$\mathbf{Z} = 8\,\Omega - j16\,\Omega = 17.89\,\Omega\angle{-63.43°}$$

FIGURE 19–8

It is possible to use the same procedure to convert a dependent source into its equivalent provided that the controlling element is external to the circuit in which the source appears. *If the controlling element is in the same circuit as the dependent source, this procedure cannot be used.*

EXAMPLE 19–4   Convert the current source of Figure 19–9 into an equivalent voltage source.

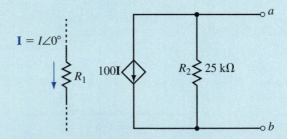

FIGURE 19–9

Solution   In the circuit of Figure 19–9 the controlling element, R_1, is in a separate circuit. Therefore, the current source is converted into an equivalent voltage source as follows:

$$\mathbf{E} = (100\mathbf{I}_1)(\mathbf{Z})$$
$$= (100\mathbf{I}_1)(25\text{ k}\Omega\angle 0°)$$
$$= (2.5 \times 10^6\ \Omega)\mathbf{I}_1$$

The resulting voltage source is shown in Figure 19–10. Notice that the equivalent voltage source is dependent on the current, **I**, just as the original current source.

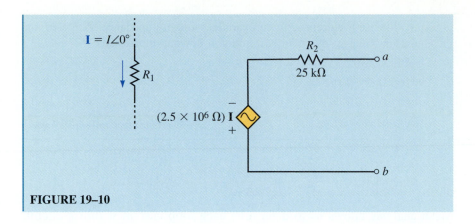

FIGURE 19–10

Convert the voltage sources of Figure 19–11 into equivalent current sources.

PRACTICE
PROBLEMS 2

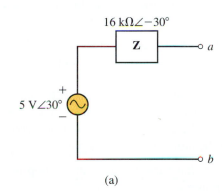

(a)

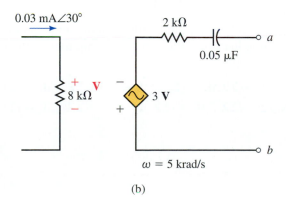

$\omega = 5$ krad/s

(b)

FIGURE 19–11

Answers:
a. **I** = 0.3125 mA∠60° (from *b* to *a*) in parallel with **Z** = 16 kΩ∠−30°
b. **I** = 0.161 mA∠93.43° (from *a* to *b*) in parallel with **Z** = 2 kΩ − *j*4 kΩ

IN-PROCESS
LEARNING
CHECK 1

Given a 40-mA∠0° current source in parallel with an impedance, **Z**. Determine the equivalent voltage source for each of the following impedances:

a. **Z** = 25 kΩ∠30°.

b. **Z** = 100 Ω ∠−90°.

c. **Z** = 20 kΩ − j16 kΩ.

(Answers are at the end of the chapter.)

19.3   Mesh (Loop) Analysis

Mesh analysis allows us to determine each loop current within a circuit, regardless of the number of sources within the circuit. The following steps provide a format which simplifies the process of using mesh analysis:

1. Convert all sinusoidal expressions into equivalent phasor notation. Where necessary, convert current sources into equivalent voltage sources.

2. Redraw the given circuit, simplifying the given impedances wherever possible and labelling the impedances ($\mathbf{Z}_1$, $\mathbf{Z}_2$, etc.).

3. Arbitrarily assign clockwise loop currents to each interior closed loop within a circuit. Show the polarities of all impedances using the assumed current directions. If an impedance is common to two loops, it may be thought to have two simultaneous currents. Although in fact two currents will not occur simultaneously, this maneuver makes the algebraic calculations fairly simple. The actual current through a common impedance is the vector sum of the individual loop currents.

4. Apply Kirchhoff's voltage law to each closed loop in the circuit, writing each equation as follows:

$$\Sigma \, (\mathbf{ZI}) = \Sigma \, \mathbf{E}$$

If the current directions are originally assigned in a clockwise direction, then the resulting linear equations may be simplified to the following format:

Loop 1: $+(\Sigma\mathbf{Z}_1)\mathbf{I}_1 - (\Sigma\mathbf{Z}_{1-2})\mathbf{I}_2 - \cdots - (\Sigma\mathbf{Z}_{1-n})\mathbf{I}_n = (\Sigma\mathbf{E}_1)$

Loop 2: $-(\Sigma\mathbf{Z}_{2-1})\mathbf{I}_1 + (\Sigma\mathbf{Z}_2)\mathbf{I}_2 - \cdots - (\Sigma\mathbf{Z}_{2-n})\mathbf{I}_n = (\Sigma\mathbf{E}_2)$

     .   .
     .   .
     .   .

Loop n: $-(\Sigma\mathbf{Z}_{n-1})\mathbf{I}_1 - (\Sigma\mathbf{Z}_{n-2})\mathbf{I}_2 - \cdots + (\Sigma\mathbf{Z}_n)\mathbf{I}_n = (\Sigma\mathbf{E}_n)$

In the above format, $\Sigma\mathbf{Z}_x$ is the summation of all impedances around loop x. The sign in front of all loop impedances will be positive.

$\Sigma\mathbf{Z}_{x-y}$ is the summation of impedances which are common between loop x and loop y. If there is no common impedance between two loops, this term is simply set to zero. All common impedance terms in the linear equations are given negative signs.

$\Sigma\mathbf{E}_x$ is the summation of voltage rises in the direction of the assumed current $\mathbf{I}_x$. If a voltage source has a polarity such that it appears as a voltage drop in the assumed current direction, then the voltage is given a negative sign.

5. Solve the resulting simultaneous linear equations using substitution or determinants. If necessary, refer to Appendix B for a review of solving simultaneous linear equations.

EXAMPLE 19–5   Solve for the loop equations in the circuit of Figure 19–12.

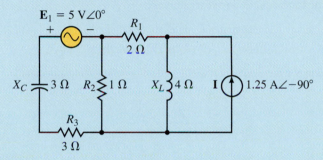

FIGURE 19–12

Solution

Step 1: The current source is first converted into an equivalent voltage source as shown in Figure 19–13.

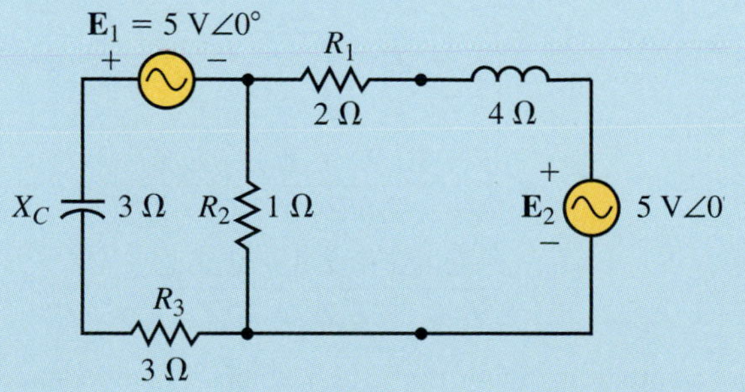

$$\mathbf{E}_2 = (1.25\ \text{A}\angle{-90°})\,(4\ \Omega\angle 90°)$$
$$= 5\ \text{V}\angle 0°$$

FIGURE 19–13

Steps 2 and 3: Next the circuit is redrawn as shown in Figure 19–14. The impedances have been simplified and the loop currents are drawn in a clockwise direction.

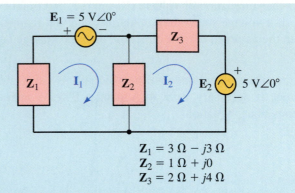

$$\mathbf{Z}_1 = 3\,\Omega - j3\,\Omega$$
$$\mathbf{Z}_2 = 1\,\Omega + j0$$
$$\mathbf{Z}_3 = 2\,\Omega + j4\,\Omega$$

FIGURE 19–14

Step 4: The loop equations are written as

$$\text{Loop 1: } (\mathbf{Z}_1 + \mathbf{Z}_2)\mathbf{I}_1 - (\mathbf{Z}_2)\mathbf{I}_2 = -\mathbf{E}_1$$
$$\text{Loop 2: } -(\mathbf{Z}_2)\mathbf{I}_1 + (\mathbf{Z}_2 + \mathbf{Z}_3)\mathbf{I}_2 = -\mathbf{E}_2$$

The solution for the currents may be found by using determinants:

$$\mathbf{I}_1 = \frac{\begin{vmatrix} -\mathbf{E}_1 & -\mathbf{Z}_2 \\ -\mathbf{E}_2 & \mathbf{Z}_2 + \mathbf{Z}_3 \end{vmatrix}}{\begin{vmatrix} \mathbf{Z}_1 + \mathbf{Z}_2 & -\mathbf{Z}_2 \\ -\mathbf{Z}_2 & \mathbf{Z}_2 + \mathbf{Z}_3 \end{vmatrix}}$$

$$= \frac{(-\mathbf{E}_1)(\mathbf{Z}_2 + \mathbf{Z}_3) - \mathbf{E}_2\mathbf{Z}_2}{(\mathbf{Z}_1 + \mathbf{Z}_2)(\mathbf{Z}_2 + \mathbf{Z}_3) - \mathbf{Z}_2\mathbf{Z}_2}$$

$$= \frac{(-\mathbf{E}_1)(\mathbf{Z}_2 + \mathbf{Z}_3) - \mathbf{E}_2\mathbf{Z}_2}{\mathbf{Z}_1\mathbf{Z}_2 + \mathbf{Z}_1\mathbf{Z}_3 + \mathbf{Z}_2\mathbf{Z}_3}$$

and

$$\mathbf{I}_2 = \frac{\begin{vmatrix} \mathbf{Z}_1 + \mathbf{Z}_2 & -\mathbf{E}_1 \\ -\mathbf{Z}_2 & -\mathbf{E}_2 \end{vmatrix}}{\begin{vmatrix} \mathbf{Z}_1 + \mathbf{Z}_2 & -\mathbf{Z}_2 \\ -\mathbf{Z}_2 & \mathbf{Z}_2 + \mathbf{Z}_3 \end{vmatrix}}$$

$$= \frac{-\mathbf{E}_2(\mathbf{Z}_1 + \mathbf{Z}_2) - \mathbf{E}_1\mathbf{Z}_2}{(\mathbf{Z}_1 + \mathbf{Z}_2)(\mathbf{Z}_2 + \mathbf{Z}_3) - \mathbf{Z}_2\mathbf{Z}_2}$$

$$= \frac{-\mathbf{E}_2(\mathbf{Z}_1 + \mathbf{Z}_2) - \mathbf{E}_1\mathbf{Z}_2}{\mathbf{Z}_1\mathbf{Z}_2 + \mathbf{Z}_1\mathbf{Z}_3 + \mathbf{Z}_2\mathbf{Z}_3}$$

Solving these equations using the actual values of impedances and voltages, we have

$$\mathbf{I}_1 = \frac{-(5)(3 + j4) - (5)(1)}{(3 - j3)(1) + (3 - j3)(2 + j4) + (1)(2 + j4)}$$

$$= \frac{(-15 - j20) - 5}{(3 - j3) + (6 + j6 - j^2 12) + (2 + j4)}$$

$$= \frac{-20 - j20}{23 + j7}$$

$$= \frac{28.28\angle -135°}{24.04\angle 16.93°}$$

$$= 1.18\ \text{A}\angle -151.93°$$

and

$$\mathbf{I}_2 = \frac{-(5)(4-j3)-(5)(1)}{23+j7}$$

$$= \frac{(-20+j15)-(5)}{23+j7}$$

$$= \frac{-25+j15}{23+j7}$$

$$= \frac{29.15\angle 149.04°}{24.04\angle 16.93°}$$

$$= 1.21\ \text{A}\angle 132.11°$$

EXAMPLE 19–6   Given the circuit of Figure 19–15, write the loop equations and solve for the loop currents. Determine the voltage, **V.**

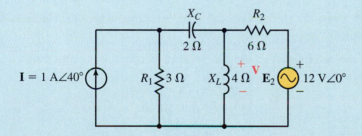

FIGURE 19–15

Solution

Step 1: Converting the current source into an equivalent voltage source gives us the circuit of of Figure 19–16.

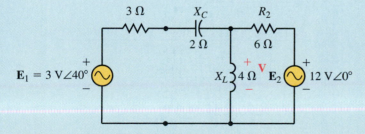

FIGURE 19–16

Steps 2 and 3: After simplifying the impedances and assigning clockwise loop currents, we have the circuit of Figure 19–17.

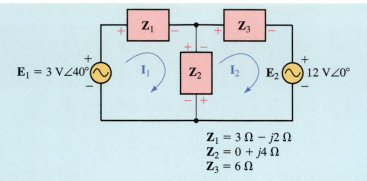

$$\mathbf{Z}_1 = 3\ \Omega - j2\ \Omega$$
$$\mathbf{Z}_2 = 0 + j4\ \Omega$$
$$\mathbf{Z}_3 = 6\ \Omega$$

FIGURE 19–17

Step 4: The loop equations for the circuit of Figure 19–17 are as follows:

$$\text{Loop 1: } (\mathbf{Z}_1 + \mathbf{Z}_2)\mathbf{I}_1 - (\mathbf{Z}_2)\mathbf{I}_2 = \mathbf{E}_1$$
$$\text{Loop 2: } -(\mathbf{Z}_2)\mathbf{I}_1 + (\mathbf{Z}_2 + \mathbf{Z}_3)\mathbf{I}_2 = -\mathbf{E}_2$$

which, after substituting the impedance values into the expressions, become

$$\text{Loop 1: } (3\ \Omega + j2\ \Omega) - (j4\ \Omega) = 3\ \text{V}\angle 40°$$
$$\text{Loop 2: } -(j4\ \Omega) + (6\ \Omega + j4\ \Omega) = -12\ \text{V}\angle 0°$$

Step 5: Solve for the currents using determinants, where the elements of the determinants are expressed as vectors.

Since the determinant of the denominator is common to both terms we find this value first.

$$\mathbf{D} = \begin{vmatrix} 3 + j2 & -j4 \\ -j4 & 6 + j4 \end{vmatrix}$$
$$= (3 + j2)(6 + j4) - (-j4)(-j4)$$
$$= 18 + j12 + j12 - 8 + 16$$
$$= 26 + j24 = 35.38\angle 42.71°$$

Now, the currents are found as

$$\mathbf{I}_1 = \frac{\begin{vmatrix} 3\angle 40° & -j4 \\ -12\angle 0° & 6 + j4 \end{vmatrix}}{\mathbf{D}}$$
$$= \frac{(3\angle 40°)(7.211\angle 33.69°) - (12\angle 0°)(4\angle 90°)}{35.38\angle 42.71°}$$
$$= \frac{21.633\angle 73.69° - 48\angle 90°}{35.38\angle 42.71°}$$
$$= \frac{27.91\angle -77.43°}{35.38\angle 42.71°}$$
$$= 0.7887\ \text{A}\angle -120.14°$$

and

$$I_2 = \frac{\begin{vmatrix} 3 + j2 & 3\angle 40° \\ -j4 & -12\angle 0° \end{vmatrix}}{\mathbf{D}}$$

$$I_2 = \frac{(3.606\angle 33.69°)(-12\angle 0°) + (3\angle 40°)(4\angle 90°)}{35.38\angle 42.71°}$$

$$= \frac{-43.27\angle 33.69° + 12\angle 130°}{35.38\angle 42.71°}$$

$$= \frac{46.15\angle -161.29°}{35.38\angle 42.71°}$$

$$= 1.304\text{ A}\angle 156.00°$$

The current through the 4-Ω inductive reactance is

$$\mathbf{I} = \mathbf{I}_1 - \mathbf{I}_2$$
$$= (0.7887\text{ A}\angle -120.14°) - (1.304\text{ A}\angle 156.00°)$$
$$= (-0.3960\text{ A} - j0.6821\text{ A}) - (-1.1913\text{ A} + j0.5304\text{ A})$$
$$= 0.795\text{ A} - j1.213\text{ A} = 1.45\text{ A}\angle -56.75°$$

The voltage is now easily found from Ohm's law as

$$\mathbf{V} = \mathbf{I}\mathbf{Z}_L$$
$$= (1.45\text{ A}\angle -56.75°)(4\text{ }\Omega\angle 90°) = 5.80\text{ V}\angle 33.25°$$

EXAMPLE 19–7 Given the circuit of of Figure 19–18, write the loop equations and show the determinant of the coefficients for the loop equations. Do not solve them.

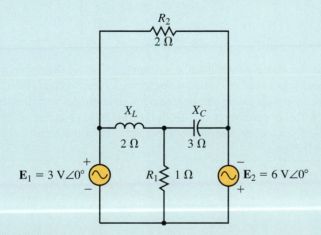

FIGURE 19–18

Solution The circuit is redrawn in Figure 19–19, showing loop currents and impedances together with the appropriate voltage polarities.

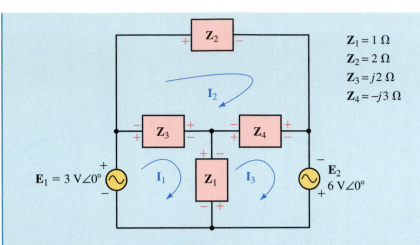

FIGURE 19–19

The loop equations may now be written as

$$\text{Loop 1:} \ (\mathbf{Z}_1 + \mathbf{Z}_3)\mathbf{I}_1 - (\mathbf{Z}_3)\mathbf{I}_2 - (\mathbf{Z}_1)\mathbf{I}_3 = \mathbf{E}_1$$
$$\text{Loop 2:} \ -(\mathbf{Z}_3)\mathbf{I}_1 + (\mathbf{Z}_2 + \mathbf{Z}_3 + \mathbf{Z}_4)\mathbf{I}_2 - (\mathbf{Z}_4)\mathbf{I}_3 = 0$$
$$\text{Loop 3:} \ -(\mathbf{Z}_1)\mathbf{I}_1 - (\mathbf{Z}_4)\mathbf{I}_2 + (\mathbf{Z}_1 + \mathbf{Z}_4)\mathbf{I}_3 = \mathbf{E}_2$$

Using the given impedance values, we have

$$\text{Loop 1:} \ (1\,\Omega + j2\,\Omega)\mathbf{I}_1 - (j2\,\Omega)\mathbf{I}_2 - (1\,\Omega)\mathbf{I}_3 = 3\,\text{V}$$
$$\text{Loop 2:} \ -(j2\,\Omega)\mathbf{I}_1 + (2\,\Omega - j1\,\Omega)\mathbf{I}_2 - (-j3\,\Omega)\mathbf{I}_3 = 0$$
$$\text{Loop 3:} \ -(1\,\Omega)\mathbf{I}_1 - (-j3\,\Omega)\mathbf{I}_2 + (1\,\Omega - j3\,\Omega)\mathbf{I}_3 = 6\,\text{V}$$

Notice that in the above equations, the phase angles ($\theta = 0°$) for the voltages have been omitted. This is because $3\,\text{V}\angle 0° = 3\,\text{V} + j0\,\text{V} = 3\,\text{V}$.

The determinant for the coefficients of the loop equations is written as

$$\mathbf{D} = \begin{vmatrix} 1 + j2 & -j2 & -1 \\ -j2 & 2 - j1 & j3 \\ -1 & j3 & 1 - j3 \end{vmatrix}$$

Notice that the above determinant is symmetrical about the principle diagonal. The coefficients in the loop equations were given the required signs (positive for the loop impedance and negative for all common impedances). However, since the coefficients in the determinant may contain imaginary numbers, it is no longer possible to generalize about the signs ($+/-$) of the various coefficients.

<hr>

PRACTICE PROBLEMS 3

Given the circuit of Figure 19–20, write the mesh equations and solve for the loop currents. Use the results to determine the current **I**.

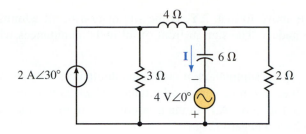

FIGURE 19–20

Answers: $\mathbf{I}_1 = 1.19\ A\angle 1.58°$, $\mathbf{I}_2 = 1.28\ A\angle -46.50°$, $\mathbf{I} = 1.01\ A\angle 72.15°$

Briefly list the steps followed in using mesh analysis to solve for the loop currents of a circuit.

IN-PROCESS LEARNING CHECK 2

(Answers are at the end of the chapter.)

19.4   Nodal Analysis

Nodal analysis allows us to calculate all node voltages with respect to an arbitrary reference point in a circuit. The following steps provide a simple format to apply nodal analysis.

1. Convert all sinusoidal expressions into equivalent phasor notation. If necessary, convert voltage sources into equivalent current sources.

2. Redraw the given circuit, simplifying the given impedances wherever possible and relabelling the impedances as admittances ($\mathbf{Y}_1$, $\mathbf{Y}_2$, etc.).

3. Select and label an appropriate reference node. Arbitrarily assign subscripted voltages ($\mathbf{V}_1$, $\mathbf{V}_2$, etc.) to each of the remaining *n* nodes within the circuit.

4. Indicate assumed current directions through all admittances in the circuit. If an admittance is common to two nodes, it is considered in each of the two node equations.

5. Apply Kirchhoff's current law to each of the *n* nodes in the circuit, writing each equation as follows:

$$\Sigma(\mathbf{YV}) = \Sigma\mathbf{I}_{\text{sources}}$$

The resulting linear equations may be simplified to the following format:

Node 1: $+(\Sigma\mathbf{Y}_1)\mathbf{V}_1 - (\Sigma\mathbf{Y}_{1-2})\mathbf{V}_2 - \cdots - (\Sigma\mathbf{Y}_{1-n})\mathbf{V}_n = (\Sigma\mathbf{I}_1)$

Node 2: $-(\Sigma\mathbf{Y}_{2-1})\mathbf{I}_1 + (\Sigma\mathbf{Y}_2)\mathbf{I}_2 - \cdots - (\Sigma\mathbf{Y}_{2-n})\mathbf{I}_n = (\Sigma\mathbf{I}_2)$

.   .

.   .

.   .

Node *n*: $-(\Sigma\mathbf{Y}_{n-1})\mathbf{I}_1 - (\Sigma\mathbf{Y}_{n-2})\mathbf{I}_2 - \cdots + (\Sigma\mathbf{Y}_n)\mathbf{I}_n = (\Sigma\mathbf{I}_n)$

In the above format, $\Sigma \mathbf{Y}_x$ is the summation of all admittances connected to node x. The sign in front of all node admittances will be positive.

$\Sigma \mathbf{Y}_{x-y}$ is the summation of common admittances between node x and node y. If there are no common admittances between two nodes, this term is simply set to zero. All common admittance terms in the linear equations will have a negative sign.

$\Sigma \mathbf{I}_x$ is the summation of current sources entering node x. If a current source leaves the node, the current is given a negative sign.

6. Solve the resulting simultaneous linear equations using substitution or determinants.

EXAMPLE 19–8   Given the circuit of Figure 19–21, write the nodal equations and solve for the node voltages.

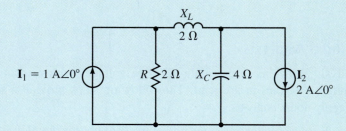

FIGURE 19–21

Solution   The circuit is redrawn in Figure 19–22, showing the nodes and a simplified representation of the admittances.

The nodal equations are written as

$$\text{Node 1: } (\mathbf{Y}_1 + \mathbf{Y}_2)\mathbf{V}_1 - (\mathbf{Y}_2)\mathbf{V}_2 = \mathbf{I}_1$$
$$\text{Node 2: } -(\mathbf{Y}_2)\mathbf{V}_1 + (\mathbf{Y}_2 + \mathbf{Y}_3)\mathbf{V}_2 = -\mathbf{I}_2$$

Using determinants, the following expressions for nodal voltages are obtained:

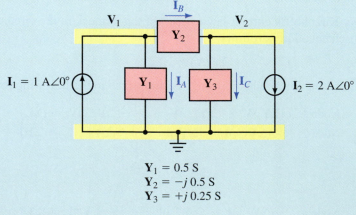

$$\mathbf{Y}_1 = 0.5 \text{ S}$$
$$\mathbf{Y}_2 = -j\,0.5 \text{ S}$$
$$\mathbf{Y}_3 = +j\,0.25 \text{ S}$$

FIGURE 19–22

$$V_1 = \frac{\begin{vmatrix} \mathbf{I}_1 & -\mathbf{Y}_2 \\ -\mathbf{I}_2 & \mathbf{Y}_2 + \mathbf{Y}_3 \end{vmatrix}}{\begin{vmatrix} \mathbf{Y}_1 + \mathbf{Y}_2 & -\mathbf{Y}_2 \\ -\mathbf{Y}_2 & \mathbf{Y}_2 + \mathbf{Y}_3 \end{vmatrix}}$$

$$= \frac{\mathbf{I}_1(\mathbf{Y}_2 + \mathbf{Y}_3) - \mathbf{I}_2\mathbf{Y}_2}{\mathbf{Y}_1\mathbf{Y}_2 + \mathbf{Y}_1\mathbf{Y}_3 + \mathbf{Y}_2\mathbf{Y}_3}$$

and

$$V_2 = \frac{\begin{vmatrix} \mathbf{Y}_1 + \mathbf{Y}_2 & \mathbf{I}_1 \\ -\mathbf{Y}_2 & -\mathbf{I}_2 \end{vmatrix}}{\begin{vmatrix} \mathbf{Y}_1 + \mathbf{Y}_2 & -\mathbf{Y}_2 \\ -\mathbf{Y}_2 & \mathbf{Y}_2 + \mathbf{Y}_3 \end{vmatrix}}$$

$$= \frac{-\mathbf{I}_2(\mathbf{Y}_1 + \mathbf{Y}_2) + \mathbf{I}_1\mathbf{Y}_2}{\mathbf{Y}_1\mathbf{Y}_2 + \mathbf{Y}_1\mathbf{Y}_3 + \mathbf{Y}_2\mathbf{Y}_3}$$

Substituting the appropriate values into the above expressions we find the nodal voltages as

$$V_1 = \frac{1(-j0.5 + j0.25) - (2)(-j0.5)}{(0.5)(-j0.5) + (0.5)(j0.25) + (-j0.5)(j0.25)}$$

$$= \frac{j0.75}{0.125 - j0.125}$$

$$= \frac{0.75\angle 90°}{0.1768\angle -45°}$$

$$= 4.243 \text{ V}\angle 135°$$

and

$$V_2 = \frac{-2(0.5 - j0.5) + (1)(-j0.5)}{(0.5)(-j0.5) + (0.5)(j0.25) + (-j0.5)(j0.25)}$$

$$= \frac{-1 + j0.5}{0.125 - j0.125}$$

$$= \frac{1.118\angle 153.43°}{0.1768\angle -45°}$$

$$= 6.324 \text{ V}\angle 198.43°$$

$$= 6.324 \text{ V}\angle -161.57°$$

EXAMPLE 19–9   Use nodal analysis to determine the voltage **V** for the circuit of Figure 19–23. Compare the results to those obtained when the circuit was analyzed using mesh analysis in Example 19–6.

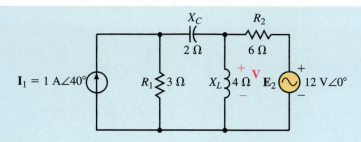

FIGURE 19–23

Solution

Step 1: Convert the voltage source into an equivalent current source as illustrated in Figure 19–24.

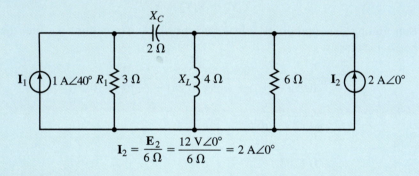

$$I_2 = \frac{E_2}{6\ \Omega} = \frac{12\ V\angle 0°}{6\ \Omega} = 2\ A\angle 0°$$

FIGURE 19–24

Steps 2, 3, and 4: The reference node is selected to be at the bottom of the circuit and the admittances are simplified as shown in Figure 19–25.

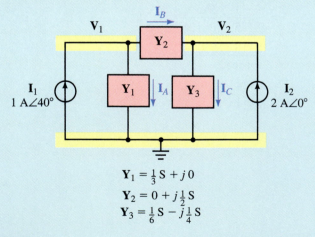

$$\mathbf{Y}_1 = \tfrac{1}{3}\ S + j\,0$$
$$\mathbf{Y}_2 = 0 + j\tfrac{1}{2}\ S$$
$$\mathbf{Y}_3 = \tfrac{1}{6}\ S - j\tfrac{1}{4}\ S$$

FIGURE 19–25

Step 5: Applying Kirchhoff's current law to each node, we have the following:

$$\text{Node 1: } \mathbf{I}_A + \mathbf{I}_B = \mathbf{I}_1$$
$$\mathbf{Y}_1\mathbf{V}_1 + \mathbf{Y}_2(\mathbf{V}_1 - \mathbf{V}_2) = \mathbf{I}_1$$
$$(\mathbf{Y}_1 + \mathbf{Y}_2)\mathbf{V}_1 - \mathbf{Y}_2\mathbf{V}_2 = \mathbf{I}_1$$
$$\text{Node 2: } \mathbf{I}_C = \mathbf{I}_B + \mathbf{I}_2$$
$$\mathbf{Y}_3\mathbf{V}_2 = \mathbf{Y}_2(\mathbf{V}_1 - \mathbf{V}_2) + \mathbf{I}_2$$
$$-\mathbf{Y}_2\mathbf{V}_1 + (\mathbf{Y}_2^2 + \mathbf{Y}_3)\mathbf{V}_2 = \mathbf{I}_2$$

After substituting the appropriate values into the above equations, we have the following simultaneous linear equations:

$$(0.3333 \text{ S} + j0.5 \text{ S})\mathbf{V}_1 - (j0.5 \text{ S})\mathbf{V}_2 = 1 \text{ A}\angle 40°$$
$$-(j0.5 \text{ S})\mathbf{V}_1 + (0.16667 \text{ S} + j0.25 \text{ S})\mathbf{V}_2 = 2 \text{ A}\angle 0°$$

Step 6: We begin by solving for the determinant for the denominator:

$$\mathbf{D} = \begin{vmatrix} 0.3333 + j0.5 & -j0.5 \\ -j0.5 & 0.1667 + j0.25 \end{vmatrix}$$
$$= (0.3333 + j0.5)(0.1667 + j0.25) - (j0.5)(j0.5)$$
$$= 0.0556 + j0.0833 + j0.0833 - 0.125 + 0.25$$
$$= 0.1806 + j0.1667 = 0.2457\angle 42.71°$$

Next we solve the node voltages as

$$\mathbf{V}_1 = \frac{\begin{vmatrix} 1\angle 40° & -j0.5 \\ 2\angle 0° & 0.1667 + j0.25 \end{vmatrix}}{\mathbf{D}}$$

$$= \frac{(1\angle 40°)(0.3005\angle 56.31°) - (2\angle 0°)(0.5\angle -90°)}{}$$

$$= \frac{0.3005\angle 96.31° - 1.0\angle -90°}{0.2457\angle 42.71°}$$

$$= \frac{1.299\angle 91.46°}{0.2457\angle 42.71°} = 5.29 \text{ V}\angle 48.75°$$

and

$$\mathbf{V}_2 = \frac{\begin{vmatrix} 0.3333 + j0.5 & 1\angle 40° \\ -j0.5 & 2\angle 0° \end{vmatrix}}{\mathbf{D}}$$

$$= \frac{(0.6009\angle 56.31°)(2\angle 0°) - (0.5\angle -90°)(1\angle 40°)}{}$$

$$= \frac{1.2019\angle 56.31° - 0.5\angle -50°}{0.2457\angle 42.71°}$$

$$= \frac{1.426\angle 75.98°}{0.2457\angle 42.71°} = 5.80 \text{ V}\angle 33.27°$$

Examining the circuit of Figure 19–23, we see that the voltage $\mathbf{V}$ is the same as the node voltage $\mathbf{V}_2$. Therefore $\mathbf{V} = 5.80 \text{ V}\angle 33.27°$, which is the same result obtained in Example 19–6. (The slight difference in phase angle is the result of rounding error.)

EXAMPLE 19–10   Given the circuit of Figure 19–26, write the nodal equations expressing all coefficients in rectangular form. Do not solve the equations.

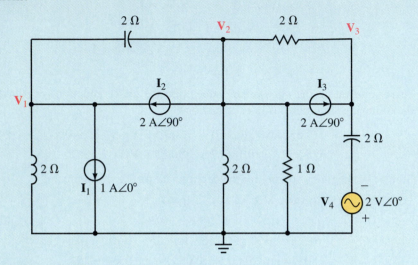

FIGURE 19–26

Solution   As in the previous example, we need to first convert the voltage source into its equivalent current source. The current source will be a phasor **I**$_4$, where

$$\mathbf{I}_4 = \frac{\mathbf{V}_4}{\mathbf{X}_C} = \frac{2\,\text{V}\angle 0°}{2\,\Omega\angle -90°} = 1.0\,\text{A}\angle 90°$$

Figure 19–27 shows the circuit as it appears after the source conversion. Notice that the direction of the current source is downward to correspond with the polarity of the voltage source **V**$_4$.

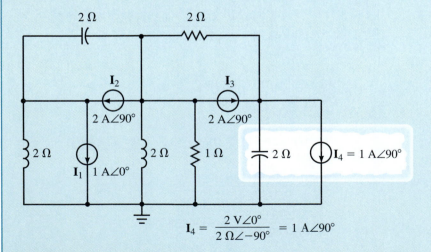

$$\mathbf{I}_4 = \frac{2\,\text{V}\angle 0°}{2\,\Omega\angle -90°} = 1\,\text{A}\angle 90°$$

FIGURE 19–27

Now, by labelling the nodes and admittances, the circuit may be simplified as shown in Figure 19–28.

$\mathbf{Y}_1 = 0 - j\frac{1}{2}\,\text{S}$ $\mathbf{Y}_3 = \frac{1}{2}\,\text{S} + j\,0$ $\mathbf{Y}_5 = 1\,\text{S} - j\frac{1}{2}\,\text{S}$

$\mathbf{Y}_2 = 0 + j\frac{1}{2}\,\text{S}$ $\mathbf{Y}_4 = 0 + j\frac{1}{2}\,\text{S}$

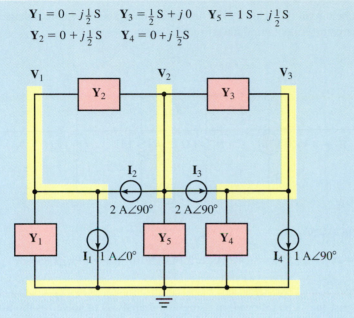

FIGURE 19–28

The admittances of Figure 19–28 are as follows:

$$\mathbf{Y}_1 = 0 - j0.5\,\text{S}$$
$$\mathbf{Y}_2 = 0 + j0.5\,\text{S}$$
$$\mathbf{Y}_3 = 0.5\,\text{S} + j0$$
$$\mathbf{Y}_4 = 0 + j0.5\,\text{S}$$
$$\mathbf{Y}_5 = 1.0\,\text{S} - j0.5\,\text{S}$$

Using the assigned admittances, the nodal equations are written as follows:

Node 1: $(\mathbf{Y}_1 + \mathbf{Y}_2)\mathbf{V}_1 - (\mathbf{Y}_2)\mathbf{V}_2 - (0)\mathbf{V}_3 = -\mathbf{I}_1 + \mathbf{I}_2$

Node 2: $-(\mathbf{Y}_2)\mathbf{V}_1 + (\mathbf{Y}_2 + \mathbf{Y}_3 + \mathbf{Y}_5)\mathbf{V}_2 - (\mathbf{Y}_3)\mathbf{V}_3 = -\mathbf{I}_2 - \mathbf{I}_3$

Node 3: $-(0)\mathbf{V}_1 - (\mathbf{Y}_3)\mathbf{V}_2 + (\mathbf{Y}_3 + \mathbf{Y}_4)\mathbf{V}_3 = \mathbf{I}_3 - \mathbf{I}_4$

By substituting the rectangular form of the admittances and current into the above linear equations, the equations are rewritten as

Node 1: $(-j0.5 + j0.5)\mathbf{V}_1 - (j0.5)\mathbf{V}_2 - (0)\mathbf{V}_3 = -1 + j2$

Node 2: $-(j0.5)\mathbf{V}_1 + (j0.5 + 0.5 + 1 - j0.5)\mathbf{V}_2 - (0.5)\mathbf{V}_3 = -j2 - j2$

Node 3: $-(0)\mathbf{V}_1 - (0.5)\mathbf{V}_2 + (0.5 + j0.5)\mathbf{V}_3 = j2 - j1$

Finally, the nodal equations are simplified as follows:

Node 1: $(0)\mathbf{V}_1 - (j0.5)\mathbf{V}_2 - (0)\mathbf{V}_3 = -1 + j2$

Node 2: $-(j0.5)\mathbf{V}_1 + (1.5)\mathbf{V}_2 - (0.5)\mathbf{V}_3 = -j4$

Node 3: $(0)\mathbf{V}_1 - (0.5)\mathbf{V}_2 + (0.5 + j0.5)\mathbf{V}_3 = j1$

PRACTICE PROBLEMS 4

Given the circuit of Figure 19–29, use nodal analysis to find the voltages V_1 and V_2. Use your results to find the current I.

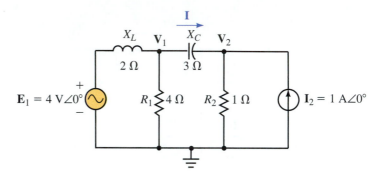

FIGURE 19–29

Answers: $V_1 = 4.22 \text{ V}\angle{-56.89°}$, $V_2 = 2.19 \text{ V}\angle 1.01°$, $I = 1.19 \text{ A}\angle 1.85°$

IN-PROCESS LEARNING CHECK 3

Briefly list the steps followed in using nodal analysis to solve for the node voltages of a circuit.

(Answers are at the end of the chapter.)

19.5   Delta-to-Wye and Wye-to-Delta Conversions

In Chapter 8, we derived the relationships showing the equivalence of "delta" (or "pi") connected resistance to a "wye" (or "tee") configuration.

In a similar manner, impedances connected in a Δ configuration are equivalent to a unique Y configuration. Figure 19–30 shows the equivalent circuits.

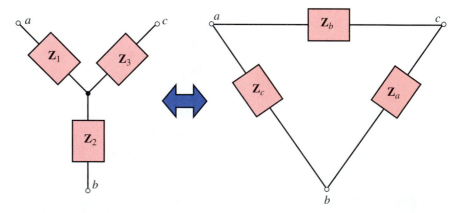

FIGURE 19–30   Delta-wye equivalence.

A Δ configuration is converted to a Y equivalent by using the following:

$$\mathbf{Z}_1 = \frac{\mathbf{Z}_b \mathbf{Z}_c}{\mathbf{Z}_a + \mathbf{Z}_b + \mathbf{Z}_c} \qquad \text{(19–1)}$$

$$\mathbf{Z}_2 = \frac{\mathbf{Z}_a \mathbf{Z}_c}{\mathbf{Z}_a + \mathbf{Z}_b + \mathbf{Z}_c} \qquad (19\text{--}2)$$

$$\mathbf{Z}_3 = \frac{\mathbf{Z}_a \mathbf{Z}_b}{\mathbf{Z}_a + \mathbf{Z}_b + \mathbf{Z}_c} \qquad (19\text{--}3)$$

The above conversion indicates that the impedance in any arm of a Y circuit is determined by taking the product of the two adjacent Δ impedances at this arm and dividing by the summation of the Δ impedances.

If the impedances of the sides of the Δ network are all equal (magnitude and phase angle), the equivalent Y network will have identical impedances, where each impedance is determined as

$$\mathbf{Z}_Y = \frac{\mathbf{Z}_\Delta}{3} \qquad (19\text{--}4)$$

A Y configuration is converted to a Δ equivalent by using the following:

$$\mathbf{Z}_a = \frac{\mathbf{Z}_1 \mathbf{Z}_2 + \mathbf{Z}_1 \mathbf{Z}_3 + \mathbf{Z}_2 \mathbf{Z}_3}{\mathbf{Z}_1} \qquad (19\text{--}5)$$

$$\mathbf{Z}_b = \frac{\mathbf{Z}_1 \mathbf{Z}_2 + \mathbf{Z}_1 \mathbf{Z}_3 + \mathbf{Z}_2 \mathbf{Z}_3}{\mathbf{Z}_2} \qquad (19\text{--}6)$$

$$\mathbf{Z}_c = \frac{\mathbf{Z}_1 \mathbf{Z}_2 + \mathbf{Z}_1 \mathbf{Z}_3 + \mathbf{Z}_2 \mathbf{Z}_3}{\mathbf{Z}_3} \qquad (19\text{--}7)$$

Any impedance in a "Δ" is determined by summing the possible two-impedance product combinations of the "Y" and then dividing by the impedance found in the opposite branch of the "Y."

If the arms of a "Y" have identical impedances, the equivalent "Δ" will have impedances given as

$$\mathbf{Z}_\Delta = 3\mathbf{Z}_Y \qquad (19\text{--}8)$$

EXAMPLE 19–11   Determine the Y equivalent of the Δ network shown in Figure 19–31.

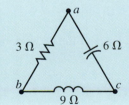

FIGURE 19–31

Solution

$$\mathbf{Z}_1 = \frac{(3\ \Omega)(-j6\ \Omega)}{3\ \Omega - j6\ \Omega + j9\ \Omega} = \frac{-j18\ \Omega}{3 + j3} = \frac{18\ \Omega\angle-90°}{4.242\angle45°}$$

$$= 4.242\ \Omega\angle-135°$$

$$= -3.0\ \Omega - j3.0\ \Omega$$

$$\mathbf{Z}_2 = \frac{(3\ \Omega)(j9\ \Omega)}{3\ \Omega - j6\ \Omega + j9\ \Omega} = \frac{j27\ \Omega}{3 + j3} = \frac{27\ \Omega\angle 90°}{4.242\angle 45°}$$

$$= 6.364\ \Omega\angle 45°$$

$$= 4.5\ \Omega + j4.5\ \Omega$$

$$\mathbf{Z}_3 = \frac{(j9\ \Omega)(-j6\ \Omega)}{3\ \Omega - j6\ \Omega + j9\ \Omega} = \frac{54\ \Omega}{3 + j3} = \frac{54\ \Omega\angle 0°}{4.242\angle 45°}$$

$$= 12.73\ \Omega\angle -45°$$

$$= 9.0\ \Omega - j9.0\ \Omega$$

In the above solution, we see that the given Δ network has an equivalent Y network with one arm having a negative resistance. This result indicates that although the Δ circuit has an equivalent Y circuit, the Y circuit cannot actually be constructed from real components since *negative resistors* do not exist (although some active components may demonstrate negative resistance characteristics). If the given conversion is used to simplify a circuit we would treat the impedance $\mathbf{Z}_1 = -3\ \Omega - j3\ \Omega$ as if the resistance actually were a negative value. Figure 19–32 shows the equivalent Y circuit.

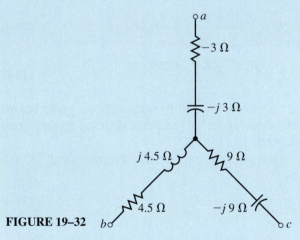

FIGURE 19–32

It is left to the student to show that the Y of Figure 19–32 is equivalent to the Δ of Figure 19–31.

EXAMPLE 19–12   Find the total impedance of the network in Figure 19–33.

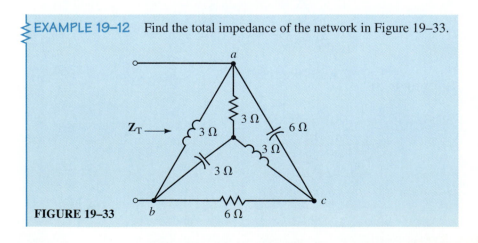

FIGURE 19–33

Solution If we take a moment to examine this network, we see that the circuit contains both a Δ and a Y. In calculating the total impedance, the solution is easier when we convert the Y to a Δ.

The conversion is shown in Figure 19–34.

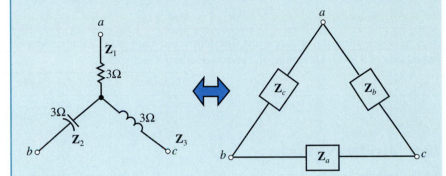

FIGURE 19–34

$$\mathbf{Z}_a = \frac{\mathbf{Z}_1\mathbf{Z}_2 + \mathbf{Z}_1\mathbf{Z}_3 + \mathbf{Z}_2\mathbf{Z}_3}{\mathbf{Z}_1}$$

$$= \frac{(3\ \Omega)(j3\ \Omega) + (3\ \Omega)(-j3\ \Omega) + (j3\ \Omega)(-j3\ \Omega)}{3\ \Omega}$$

$$= \frac{-j^2 9\ \Omega}{3} = 3\ \Omega$$

$$\mathbf{Z}_b = \frac{\mathbf{Z}_1\mathbf{Z}_2 + \mathbf{Z}_1\mathbf{Z}_3 + \mathbf{Z}_2\mathbf{Z}_3}{\mathbf{Z}_2} = \frac{9\ \Omega}{-j3} = j3\ \Omega$$

$$\mathbf{Z}_c = \frac{\mathbf{Z}_1\mathbf{Z}_2 + \mathbf{Z}_1\mathbf{Z}_3 + \mathbf{Z}_2\mathbf{Z}_3}{\mathbf{Z}_3} = \frac{9\ \Omega}{j3} = -j3\ \Omega$$

Now, substituting the equivalent Δ into the original network, we have the revised network of Figure 19–35.

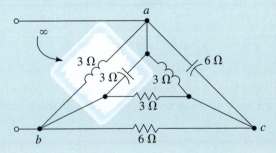

FIGURE 19–35

The network of Figure 19–35 shows that the corresponding sides of the Δ are parallel. Because the inductor and the capacitor in the left side of the Δ have the same values, we may replace the parallel combination of these two components with an open circuit. The resulting impedance of the network is now easily determined as

$$\mathbf{Z}_T = 3\ \Omega \| 6\ \Omega + (j3\ \Omega) \| (-j6\ \Omega) = 2\ \Omega + j6\ \Omega$$

A Y network consists of a 60-Ω capacitor, a 180-Ω inductor, and a 540-Ω resistor. Determine the corresponding Δ network.

Answer: $\mathbf{Z}_a = -1080\ \Omega + j180\ \Omega$, $\mathbf{Z}_b = 20\ \Omega + j120\ \Omega$, $\mathbf{Z}_c = 360\ \Omega - j60\ \Omega$

A Δ network consists of a resistor, inductor, and capacitor, each having an impedance of 150 Ω. Determine the corresponding Y network.

(Answers are at the end of the chapter.)

19.6   Bridge Networks

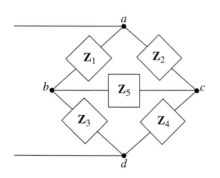

FIGURE 19–36

Bridge circuits, similar to the network of Figure 19–36, are used extensively in electronics to measure the values of unknown components.

Recall from Chapter 8 that any bridge circuit is said to be balanced when the current through the branch between the two arms is zero. In a practical circuit, component values of very precise resistors are adjusted until the current through the central element (usually a sensitive galvanometer) is exactly equal to zero. For ac circuits, the condition of a balanced bridge occurs when the impedance vectors of the various arms satisfy the following condition:

$$\frac{\mathbf{Z}_1}{\mathbf{Z}_3} = \frac{\mathbf{Z}_2}{\mathbf{Z}_4} \tag{19–9}$$

When a balanced bridge occurs in a circuit, the equivalent impedance of the bridge network is easily determined by removing the central impedance and replacing it by either an open or a short circuit. The resulting impedance of the bridge circuit is then found as either of the following:

$$\mathbf{Z}_{\mathrm{T}} = \mathbf{Z}_1 \| \mathbf{Z}_2 + \mathbf{Z}_3 \| \mathbf{Z}_4$$

or

$$\mathbf{Z}_{\mathrm{T}} = (\mathbf{Z}_1 + \mathbf{Z}_3) \| (\mathbf{Z}_2 + \mathbf{Z}_4)$$

If, on the other hand, the bridge is not balanced, then the total impedance must be determined by performing a Δ-to-Y conversion. Alternatively, the circuit may be analyzed by using either mesh analysis or nodal analysis.

EXAMPLE 19–13   Given that the circuit of Figure 19–37 is a balanced bridge.

a. Calculate the unknown impedance, Z_x.

b. Determine the values of L_x and R_x if the circuit operates at a frequency of 1 kHz.

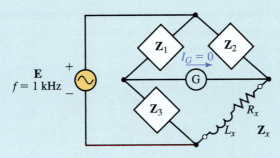

$$Z_1 = 30 \text{ k}\Omega\angle-20°$$
$$Z_2 = 10 \text{ k}\Omega\angle0°$$
$$Z_3 = 100 \ \Omega\angle0°$$

FIGURE 19–37

Solution

a. The expression for the unknown impedance is determined from Equation 19–9 as

$$Z_x = \frac{Z_2 Z_3}{Z_1}$$

$$= \frac{(10 \text{ k}\Omega)(100 \ \Omega)}{30 \text{ k}\Omega\angle-20°}$$

$$= 33.3 \ \Omega\angle20°$$

$$= 31.3 + j11.4 \ \Omega$$

b. From the above result, we have

$$R_x = 31.3 \ \Omega$$

and

$$L_x = \frac{X_L}{2\pi f} = \frac{11.4 \ \Omega}{2\pi(1000 \text{ Hz})} = 1.81 \text{ mH}$$

We will now consider various forms of bridge circuits which are used in electronic circuits to determine the values of unknown inductors and capacitors. As in resistor bridges, the circuits use variable resistors together with very sensitive galvanometer movements to ensure a balanced condition for the bridge. However, rather than using a dc source to provide current in the circuit, the bridge circuits use ac sources operating at a known frequency

(usually 1 kHz). Once the bridge is balanced, the value of unknown inductance or capacitance may be easily determined by obtaining the reading
directly from the instrument. Most instruments using bridge circuitry will
incorporate several different bridges to enable the measurement of various
types of unknown impedances.

Maxwell Bridge

The **Maxwell bridge,** shown in Figure 19–38, is used to determine the
inductance and series resistance of an inductor having a relatively large
series resistance (in comparison to $X_L = \omega L$).

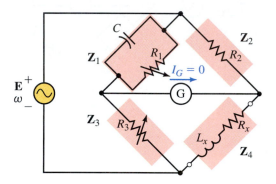

FIGURE 19–38   Maxwell bridge.

Resistors R_1 and R_3 are adjusted to provide the balanced condition (when
the current through the galvanometer is zero: $I_G = 0$).

When the bridge is balanced, we know that the following condition must
apply:

$$\frac{\mathbf{Z}_1}{\mathbf{Z}_2} = \frac{\mathbf{Z}_3}{\mathbf{Z}_4}$$

If we write the impedances using the rectangular forms, we obtain

$$\frac{\left[\dfrac{(R_1)\left(-j\dfrac{1}{\omega C}\right)}{R_1 - j\dfrac{1}{\omega C}}\right]}{R_2} = \frac{R_3}{R_x + j\omega L_x}$$

$$\frac{\left(-j\dfrac{R_1}{\omega C}\right)}{\left(\dfrac{\omega R_1 C - j1}{\omega C}\right)} = \frac{R_2 R_3}{R_x + j\omega L_x}$$

$$\frac{-jR_1}{\omega C R_1} - j = \frac{R_2 R_3}{R_x + j\omega L_x}$$

$$(-jR_1)(R_x + j\omega L_x) = R_2 R_3(\omega C R_1 - j)$$

$$\omega L_x R_1 - jR_1 R_x = \omega R_1 R_2 R_3 C - jR_2 R_3$$

Now, since two complex numbers can be equal only if their real parts are equal and if their imaginary parts are equal, we must have the following:

$$\omega L_x R_1 = \omega R_1 R_2 R_3 C$$

and

$$R_1 R_x = R_2 R_3$$

Simplifying these expressions, we get the following equations for a Maxwell bridge:

$$L_x = R_2 R_3 C \qquad\qquad \textbf{(19–10)}$$

and

$$R_x = \frac{R_2 R_3}{R_1} \qquad\qquad \textbf{(19–11)}$$

EXAMPLE 19–14

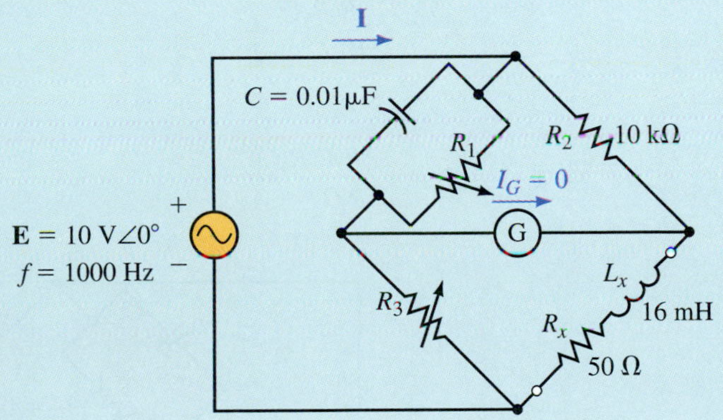

FIGURE 19–39

a. Determine the values of R_1 and R_3 so that the bridge of Figure 19–39 is balanced.

b. Calculate the current **I** when the bridge is balanced.

Solution

a. Rewriting Equations 19–10 and 19–11 and solving for the unknowns, we have

$$R_3 = \frac{L_x}{R_2 C} = \frac{16 \text{ mH}}{(10 \text{ k}\Omega)(0.01 \ \mu\text{F})} = 160 \ \Omega$$

and

$$R_1 = \frac{R_2 R_3}{R_x} = \frac{(10 \text{ k}\Omega)(160 \ \Omega)}{50 \ \Omega} = 32 \text{ k}\Omega$$

b. The total impedance is found as

$$\mathbf{Z}_T = (\mathbf{Z}_C \| \mathbf{R}_1 \| \mathbf{R}_2) + [\mathbf{R}_3 \| (\mathbf{R}_x + \mathbf{Z}_{Lx})]$$

$$\mathbf{Z}_T = (-j15.915 \text{ k}\Omega) \| 32 \text{ k}\Omega \| 10 \text{ k}\Omega + [160 \,\Omega \| (50 \,\Omega + j100.5 \,\Omega)]$$

$$= 6.87 \text{ k}\Omega \angle -25.6° + 77.2 \,\Omega \angle 38.0°$$

$$= 6.91 \text{ k}\Omega \angle -25.0°$$

The resulting circuit current is

$$\mathbf{I} = \frac{10 \text{ V} \angle 0°}{6.91 \text{ k}\Omega \angle -25°} = 1.45 \text{ mA} \angle 25.0°$$

Hay Bridge

In order to measure the inductance and series resistance of an inductor having a small series resistance, a **Hay bridge** is generally used. The Hay bridge is shown in Figure 19–40.

By applying a method similar to that used to determine the values of the unknown inductance and resistance of the Maxwell bridge, it may be shown that the following equations for the Hay bridge apply:

$$L_x = \frac{R_2 R_3 C}{\omega^2 R_1^2 C^2 + 1} \tag{19–12}$$

and

$$R_x = \frac{\omega^2 R_1 R_2 R_3 C^2}{\omega^2 R_1^2 C^2 + 1} \tag{19–13}$$

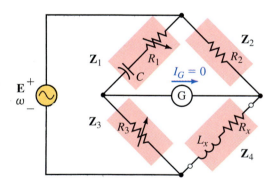

FIGURE 19–40   Hay bridge.

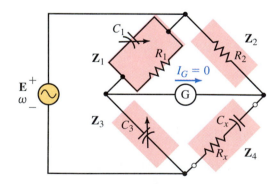

FIGURE 19–41   Schering bridge.

Schering Bridge

The **Schering bridge,** shown in Figure 19–41, is a circuit used to determine the value of unknown capacitance.

By solving for the balanced bridge condition, we have the following equations for the unknown quantities of the circuit:

$$C_x = \frac{R_1 C_3}{R_2} \tag{19–14}$$

$$R_x = \frac{C_1 R_2}{C_3} \qquad\qquad \textbf{(19–15)}$$

EXAMPLE 19–15 Determine the values of C_1 and C_3 which will result in a balanced bridge for the circuit of Figure 19–42.

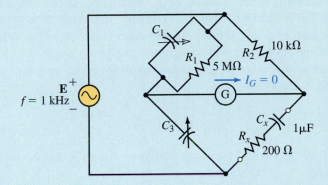

EWB **FIGURE 19–42**

Solution Rewriting Equations 19–14 and 19–15, we solve for the unknown capacitances as

$$C_3 = \frac{R_2 C_x}{R_1} = \frac{(10 \text{ k}\Omega)(1 \ \mu\text{F})}{5 \text{ M}\Omega} = 0.002 \ \mu\text{F}$$

and

$$C_1 = \frac{C_3 R_x}{R_2} = \frac{(0.002 \ \mu\text{F})(200 \ \Omega)}{10 \text{ k}\Omega} = 40 \text{ pF}$$

Determine the values of R_1 and R_3 so that the bridge of Figure 19–43 is balanced.

PRACTICE PROBLEMS 6

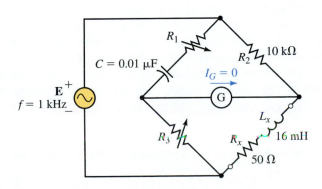

FIGURE 19–43

Answers: $R_1 = 7916 \ \Omega, R_3 = 199.6 \ \Omega$

ELECTRONICS WORKBENCH

PSpice

19.7 Circuit Analysis Using Computers

In some of the examples in this chapter, we analyzed circuits that resulted in as many as three simultaneous linear equations. You have no doubt wondered if there is a less complicated way to solve these circuits without the need for using complex algebra. Computer programs are particularly useful for solving such ac circuits. Both Electronics Workbench and PSpice have individual strengths in the solution of ac circuits. As in previous examples, Electronics Workbench provides an excellent simulation of how measurements are taken in a lab. PSpice, on the other hand, provides voltage and current readings, complete with magnitude and phase angle. The following examples show how these programs are useful for examining the circuits in this chapter.

EXAMPLE 19–16   Use Electronics Workbench to show that the bridge circuit of Figure 19–44 is balanced.

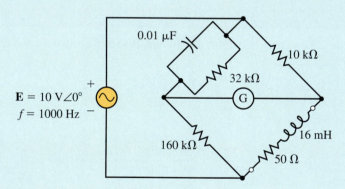

FIGURE 19–44

Solution   Recall that a bridge circuit is balanced when the current through the branch between the two arms of the bridge is equal to zero. In this example, we will use a multimeter set on its ac ammeter range to verify the condition of the circuit. The ammeter is selected by clicking on **A** and it is set to its ac range by clicking on the sinusoidal button. Figure 19–45 shows the circuit connections and the ammeter reading. The results correspond to the conditions that were previously analyzed in Example 19–14. (Note: When using Electronics Workbench, the ammeter may not show exactly zero current in the balanced condition. This is due to the way the program does the calculations. Any current less than 5 μA is considered to be effectively zero.)

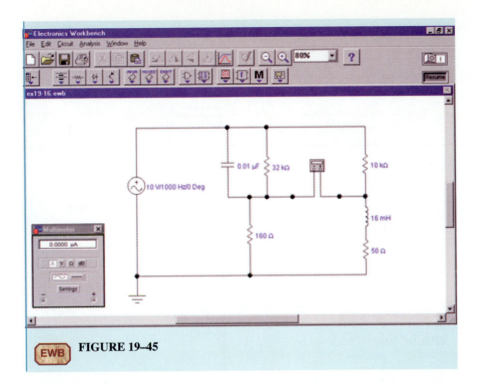

FIGURE 19–45

Use Electronics Workbench to verify that the results obtained in Example 19–15 result in a balanced bridge circuit. (Assume that the bridge is balanced if the galvanometer current is less than 5 μA.)

PRACTICE
PROBLEMS 7

OrCAD PSpice

EXAMPLE 19–17   Use OrCAD Capture CIS to input the circuit of Figure 19–15. Assume that the circuit operates at a frequency of $\omega = 50$ rad/s ($f = 7.958$ Hz). Use PSpice to obtain a printout showing the currents through X_C, R_2, and X_L. Compare the results to those obtained in Example 19–6.

Solution   Since the reactive components in Figure 19–15 were given as impedance, it is necessary to first determine the corresponding values in henries and farads.

$$L = \frac{4\ \Omega}{50\ \text{rad/s}} = 80\ \text{mH}$$

and

$$C = \frac{1}{(2\ \Omega)(50\ \text{rad/s})} = 10\ \text{mF}$$

Now we are ready to use OrCAD Capture to input the circuit as shown in Figure 19–46. The basic steps are reviewed for you. Use the ac current source,

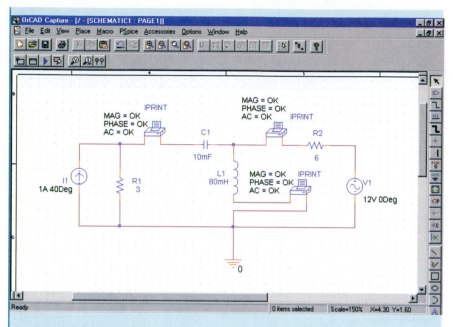

FIGURE 19–46

ISRC from the SOURCE library and place one IPRINT part from the SPE-CIAL library. The resistor, inductor, and capacitor are selected from the ANALOG library and the ground symbol is selected by using the Place ground tool.

Change the value of the current source by double clicking on the part and moving the horizontal scroll bar until you find the field titled AC. Type **1A 40Deg** into this field. A space must be placed between the magnitude and phase angle. Click on Apply. In order for these values to be displayed on the schematic, you must click on the Display button and then <u>V</u>alue Only. Click on OK to return to the properties editor and then close the editor by clicking on X.

The IPRINT part is similar to an ammeter and provides a printout of the current magnitude and phase angle. The properties of the IPRINT part are changed by double clicking on the part and scrolling across to show the appropriate fields. Type **OK** in the AC, MAG, and PHASE fields. In order to display the selected fields on the schematic, you must click on the Display button and then select Name and Value after changing each field. Since we need to measure three currents in the circuit, we could follow this procedure two more times. However, an easier method is to click on the IPRINT part and copy the part by using <Ctrl><C> and <Ctrl><V>. Each IPRINT will then have the same properties.

Once the rest of the circuit is completed and wired, click on the New Simulation Profile tool. Give the simulation a name (such as **ac Branch Currents**). Click on the Analysis tab and select AC Sweep/ Noise as the analysis type. Type the following values:

Start Frequency: **7.958Hz**

End Frequency: **7.958Hz**

Total Points: **1**

Since we do not need the Probe postprocessor to run, it is disabled by selecting the Probe Window tab (from the Simulation Settings dialog box). Click on Display Probe window and exit the simulation settings by clicking on OK.

Click on the Run tool. Once PSpice has successfully run, click on the View menu and select the Output File menu item. Scroll through the file until the currents are shown as follows:

```
FREQ          IM(V_PRINT1) IP(V_PRINT1)
  7.958E+00     7.887E-01  -1.201E+02
FREQ          IM(V_PRINT2) IP(V_PRINT2)
  7.958E+00     1.304E+00   1.560E+02
FREQ          IM(V_PRINT3) IP(V_PRINT3)
  7.958E+00     1.450E+00  -5.673E+01
```

The above printout provides: $\mathbf{I}_1 = 0.7887\ A\angle-120.1°$, $\mathbf{I}_2 = 1.304\ A\angle156.0°$, and $\mathbf{I}_3 = 1.450\ A\angle-56.73°$. These results are consistent with those calculated in Example 19–6.

Use OrCAD PSpice to evaluate the node voltages for the circuit of Figure 19–23. Assume that the circuit operates at an angular frequency of $\omega = 1000$ rad/s ($f = 159.15$ Hz).

PRACTICE PROBLEMS 8

PUTTING IT INTO PRACTICE

The Schering bridge of Figure 19–74 (p. 779) is balanced. In this chapter, you have learned several methods that allow you to find the current anywhere in a circuit. Using any method, determine the current through the galvanometer if the value of $C_x = 0.07\ \mu F$ (All other values remain unchanged.) Repeat the calculations for a value of $C_x = 0.09\ \mu F$. Can you make a general statement for current through the galvanometer if C_x is smaller that the value required to balance the bridge? What general statement can be made if the value of C_x is larger than the value in the balanced bridge?

19.1 Dependent Sources

PROBLEMS

1. Refer to the circuit of Figure 19–47.

    Find **V** when the controlling current **I** is the following:

    a. $20\ \mu A\angle0°$

    b. $50\ \mu A\angle-180°$

    c. $60\ \mu A\angle60°$

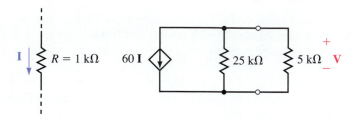

FIGURE 19–47

2. Refer to the circuit of Figure 19–48.

    Find **I** when the controlling voltage, **V**, is the following:

    a. 30 mV∠0°

    b. 60 mV∠−180°

    c. 100 mV∠−30°

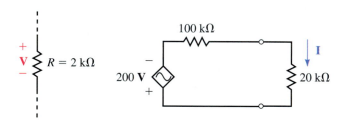

FIGURE 19–48

3. Repeat Problem 1 for the circuit of Figure 19–49.

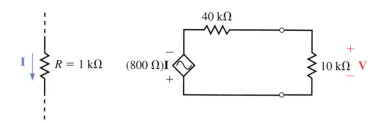

FIGURE 19–49

4. Repeat Problem 2 for the circuit of Figure 19–50.

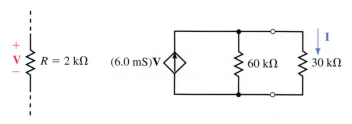

FIGURE 19–50

5. Find the output voltage, **V**_{out}, for the circuit of Figure 19–51.

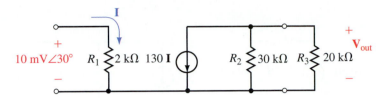

FIGURE 19–51

6. Repeat Problem 5 for the circuit of Figure 19–52.

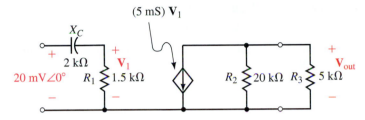

FIGURE 19–52

19.2   Source Conversion

7. Given the circuits of Figure 19–53, convert each of the current sources into an equivalent voltage source. Use the resulting circuit to find V_L.

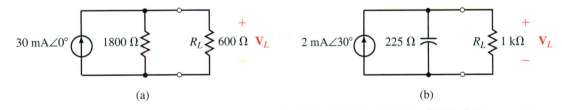

(a)   (b)

FIGURE 19–53

8. Convert each voltage source of Figure 19–54 into an equivalent current source.

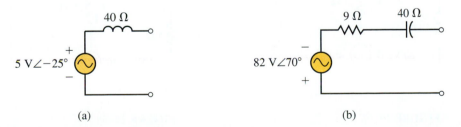

(a)   (b)

FIGURE 19–54

9. Refer to the circuit of Figure 19–55.

   a. Solve for the voltage, **V.**

b. Convert the current source into an equivalent voltage source and again solve for **V.** Compare to the result obtained in (a).

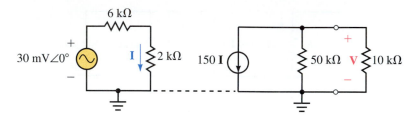

FIGURE 19–55

10. Refer to the circuit of Figure 19–56.

    a. Solve for the voltage, $\mathbf{V}_L$.

    b. Convert the current source into an equivalent voltage source and again solve for $\mathbf{V}_L$.

    c. If $\mathbf{I} = 5\ \mu\text{A}\angle90°$, what is $\mathbf{V}_L$?

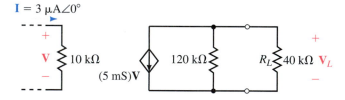

FIGURE 19–56

19.3   Mesh (Loop) Analysis

11. Consider the circuit of Figure 19–57.

    a. Write the mesh equations for the circuit.

    b. Solve for the loop currents.

    c. Determine the current **I** through the 4-Ω resistor.

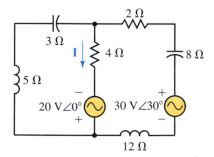

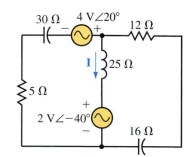

FIGURE 19–57   **FIGURE 19–58**

12. Refer to the circuit of Figure 19–58.

    a. Write the mesh equations for the circuit.

    b. Solve for the loop currents.

    c. Determine the current through the 25-Ω inductor.

13. Refer to the circuit of Figure 19–59.

    a. Simplify the circuit and write the mesh equations.

    b. Solve for the loop currents.

    c. Determine the voltage **V** across the 15-Ω capacitor.

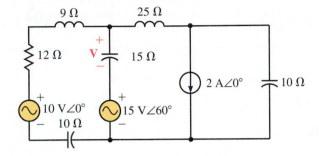

FIGURE 19–59

14. Consider the circuit of Figure 19–60.

    a. Simplify the circuit and write the mesh equations.

    b. Solve for the loop currents.

    c. Determine the voltage **V** across the 2-Ω resistor.

15. Use mesh analysis to find the current **I** and the voltage **V** in the circuit of Figure 19–61.

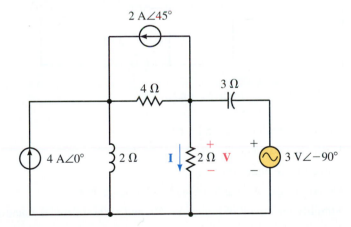

FIGURE 19–60

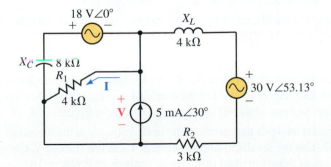

FIGURE 19–61

16. Repeat Problem 15 for the circuit of Figure 19–62.

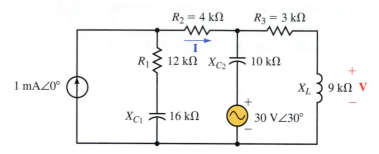

FIGURE 19–62

19.4 Nodal Analysis

17. Consider the circuit of Figure 19–63.

a. Write the nodal equations.

b. Solve for the node voltages.

c. Determine the current **I** through the 4-Ω capacitor.

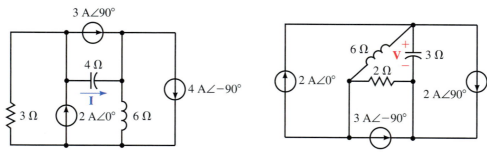

FIGURE 19–63   **FIGURE 19–64**

18. Refer to the circuit of Figure 19–64.

a. Write the nodal equations.

b. Solve for the node voltages.

c. Determine the voltage **V** across the 3-Ω capacitor.

19. a. Simplify the circuit of Figure 19–59, and write the nodal equations.

b. Solve for the node voltages.

c. Determine the voltage across the 15-Ω capacitor.

20. a. Simplify the circuit of Figure 19–60, and write the nodal equations.

b. Solve for the node voltages.

c. Determine the current through the 2-Ω resistor.

21. Use nodal analysis to determine the node voltages in the circuit of Figure 19–61. Use the results to find the current **I** and the voltage **V**. Compare your answers to those obtained using mesh analysis in Problem 15.

22. Use nodal analysis to determine the node voltages in the circuit of Figure 19–62. Use the results to find the current **I** and the voltage **V**. Compare your answers to those obtained using mesh analysis in Problem 16.

19.5   Delta-to-Wye and Wye-to-Delta Conversions

23. Convert each of the Δ networks of Figure 19–65 into an equivalent Y network.

24. Convert each of the Y networks of Figure 19–66 into an equivalent Δ network.

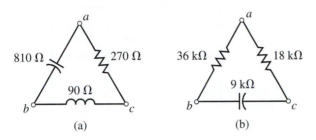

(a)   (b)

FIGURE 19–65

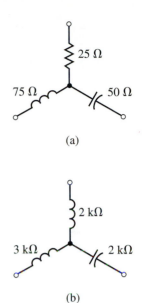

(a)

(b)

FIGURE 19–66

25. Using Δ→Y or Y→Δ conversion, calculate **I** for the circuit of Figure 19–67.

26. Using Δ→Y or Y→Δ conversion, calculate **I** for the circuit of Figure 19–68.

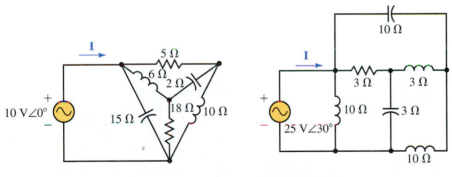

FIGURE 19–67   **FIGURE 19–68**

27. Refer to the circuit of Figure 19–69:
    a. Determine the equivalent impedance, $\mathbf{Z}_T$, of the circuit.
    b. Find the currents **I** and $\mathbf{I}_1$.

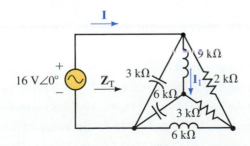

FIGURE 19–69

28. Refer to the circuit of Figure 19–70:
    a. Determine the equivalent impedance, $\mathbf{Z}_T$, of the circuit.
    b. Find the voltages $\mathbf{V}$ and $\mathbf{V}_1$.

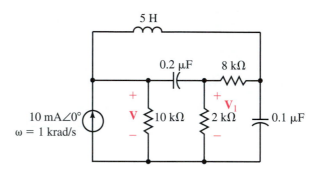

FIGURE 19–70

19.6   Bridge Networks

29. Given that the bridge circuit of Figure 19–71 is balanced:
    a. Determine the value of the unknown impedance.
    b. Solve for the current **I.**

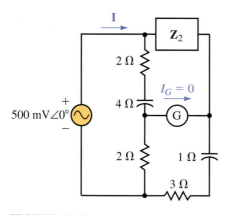

FIGURE 19–71   **FIGURE 19–72**

30. Given that the bridge circuit of Figure 19–72 is balanced:
    a. Determine the value of the unknown impedance.
    b. Solve for the current **I.**
31. Show that the bridge circuit of Figure 19–73 is balanced.
32. Show that the bridge circuit of Figure 19–74 is balanced.
33. Derive Equations 19–14 and 19–15 for the balanced Schering bridge.
34. Derive Equations 19–12 and 19–13 for the balanced Hay bridge.

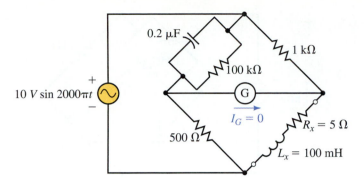

FIGURE 19–73

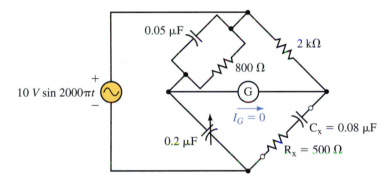

FIGURE 19–74

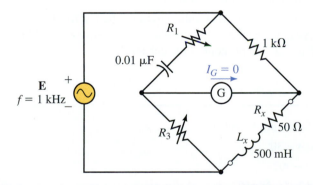

FIGURE 19–75

35. Determine the values of the unknown resistors which will result in a balanced bridge for the circuit of Figure 19–75.

36. Determine the values of the unknown capacitors which will result in a balanced bridge for the circuit of Figure 19–76.

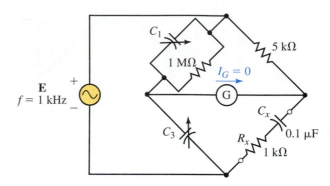

FIGURE 19–76

19.7 Circuit Analysis Using Computers

37. **EWB** Use Electronics Workbench to show that the bridge circuit of Figure 19–73 is balanced. (Assume that the bridge is balanced if the galvanometer current is less than 5 μA.)

38. **EWB** Repeat Problem 37 for the bridge circuit of Figure 19–74.

39. **PSpice** Use the OrCAD Capture to input the file for the circuit of Figure 19–21. Assume that the circuit operates at a frequency $\omega = 2$ krad/s. Use IPRINT and VPRINT to obtain a printout of the node voltages and the current through each element of the circuit.

40. **PSpice** Use the OrCAD Capture to input the file for the circuit of Figure 19–29. Assume that the circuit operates at a frequency $\omega = 1$ krad/s. Use IPRINT and VPRINT to obtain a printout of the node voltages and the current through each element of the circuit.

41. **PSpice** Use the OrCAD Capture to input the file for the circuit of Figure 19–68. Assume that the circuit operates at a frequency $\omega = 20$ rad/s. Use IPRINT to obtain a printout of the current **I.**

42. **PSpice** Use the OrCAD Capture to input the file for the circuit of Figure 19–69. Assume that the circuit operates at a frequency $\omega = 3$ krad/s. Use IPRINT to obtain a printout of the current **I.**

ANSWERS TO IN-PROCESS LEARNING CHECKS

In-Process Learning Check 1

a. **E** = 1000 V∠30°

b. **E** = 4 V∠−90°

c. **E** = 1024 V∠−38.66°

In-Process Learning Check 2

1. Convert current sources to voltage sources.

2. Redraw the circuit.

3. Assign a clockwise current to each loop.

4. Write loop equations using Kirchhoff's voltage law.

5. Solve the resulting simultaneous linear equations to find the loop currents.

In-Process Learning Check 3

1. Convert voltage sources to current sources.

2. Redraw the circuit.

3. Label all nodes, including the reference node.

4. Write nodal equations using Kirchhoff's current law.

5. Solve the resulting simultaneous linear equations to find the node voltages.

In-Process Learning Check 4

$\mathbf{Z}_1 = 150\ \Omega\angle 90°$ $\qquad$ $\mathbf{Z}_2 = 150\ \Omega\angle -90°$ $\qquad$ $\mathbf{Z}_3 = 150\ \Omega\angle 0°$

20

AC Network Theorems

OBJECTIVES

After studying this chapter you will be able to

- apply the superposition theorem to determine the voltage across or current through any component in a given circuit,
- determine the Thévenin equivalent of circuits having independent and/or dependent sources,
- determine the Norton equivalent of circuits having independent and/or dependent sources,
- apply the maximum power transfer theorem to determine the load impedance for which maximum power is transferred to the load from a given circuit,
- use PSpice to find the Thévenin and Norton equivalents of circuits having either independent or dependent sources,
- use Electronics Workbench to verify the operation of ac circuits.

KEY TERMS

Absolute Maximum Power
Maximum Power Transfer
Norton's Theorem
Relative Maximum Power
Superposition Theorem

OUTLINE

Superposition Theorem—Independent Sources

Superposition Theorem—Dependent Sources

Thévenin's Theorem—Independent Sources

Norton's Theorem—Independent Sources

Thévenin's and Norton's Theorems for Dependent Sources

Maximum Power Transfer Theorem

Circuit Analysis Using Computers

In this chapter we apply the superposition, Thévenin, Norton, and maximum power transfer theorems in the analysis of ac circuits. Although the Millman and reciprocity theorems apply to ac circuits as well as to dc circuits, they are omitted since the applications are virtually identical with those used in analyzing dc circuits.

Many of the techniques used in this chapter are similar to those used in Chapter 9, and as a result, most students will find a brief review of dc theorems useful.

This chapter examines the application of the network theorems by considering both independent and dependent sources. In order to show the distinctions between the methods used in analyzing the various types of sources, the sections are labelled according to the types of sources involved.

An understanding of dependent sources is particularly useful when working with transistor circuits and operational amplifiers. Sections 20.2 and 20.5 are intended to provide the background for analyzing the operation of feedback amplifiers. Your instructor may find that these topics are best left until you cover this topic in a course dealing with such amplifiers. Consequently, the omission of sections 20.2 and 20.5 will not in any way detract from the continuity of the important ideas presented in this chapter.

William Bradford Shockley

SHOCKLEY WAS BORN the son of a mining engineer in London, England on February 13, 1910. After graduating from the California Institute of Technology and Massachusetts Institute of Technology, Shockley joined Bell Telephone Laboratories.

With his co-workers, John Bardeen and Walter Brattain, Shockley developed an improved solid-state rectifier using a germanium crystal which had been injected with minute amounts of impurities. Unlike vacuum tubes, the resulting diodes were able to operate at much lower voltages without the need for inefficient heater elements.

In 1948, Shockley combined three layers of germanium to produce a device which was able to not only rectify a signal but to amplify it. Thus was developed the first transistor. Since its humble beginning, the transistor has been improved and decreased in size to the point where now a circuit containing thousands of transistors can easily fit into an area not much bigger than the head of a pin.

The advent of the transistor has permitted the construction of elaborate spacecraft, unprecedented communication, and new forms of energy generation.

Shockley, Bardeen, and Brattain received the 1956 Nobel Prize in Physics for their discovery of the transistor.

20.1   Superposition Theorem—Independent Sources

The superposition theorem states the following:

The voltage across (or current through) an element is determined by summing the voltage (or current) due to each independent source.

In order to apply this theorem, all sources other than the one being considered are eliminated. As in dc circuits, this is done by replacing current sources with open circuits and by replacing voltage sources with short circuits. The process is repeated until the effects due to all sources have been determined.

Although we generally work with circuits having all sources at the same frequency, occasionally a circuit may operate at more than one frequency at a time. This is particularly true in diode and transistor circuits which use a dc source to set a "bias" (or operating) point and an ac source to provide the signal to be conditioned or amplified. In such cases, the resulting voltages or currents are still determined by applying the superposition theorem. The topic of how to solve circuits operating at several different frequencies simultaneously is covered in Chapter 25.

> **NOTES...**
>
> As in dc circuits, the superposition theorem can be applied only to voltage and current; it cannot be used to solve for the total power dissipated by an element. This is because power is not a linear quantity, but rather follows a square-law relationship ($P = V^2/R = I^2R$).

EXAMPLE 20–1 Determine the current **I** in Figure 20–1 by using the superposition theorem.

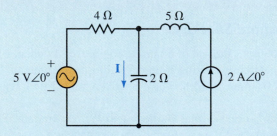

FIGURE 20–1

Solution

Current due to the 5 V∠0° voltage source: Eliminating the current source, we obtain the circuit shown in Figure 20–2.

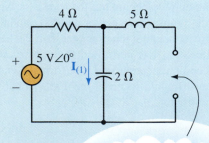

Current source is replaced with an open circuit.

FIGURE 20–2

Applying Ohm's law, we have

$$\mathbf{I}_{(1)} = \frac{5\,\text{V}\angle 0°}{4 - j2\,\Omega} = \frac{5\,\text{V}\angle 0°}{4.472\,\Omega\angle -26.57°}$$

$$= 1.118\,\text{A}\angle 26.57°$$

Current due to the 2 A $\angle 0°$ current source: Eliminating the voltage source, we obtain the circuit shown in Figure 20–3.

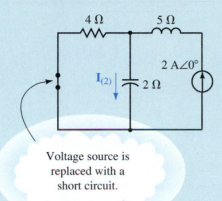

Voltage source is replaced with a short circuit.

FIGURE 20–3

The current $\mathbf{I}_{(2)}$ due to this source is determined by applying the current divider rule:

$$\mathbf{I}_{(2)} = (2\,\text{A}\angle 0°)\,\frac{4\,\Omega\angle 0°}{4\,\Omega - j2\,\Omega}$$

$$= \frac{8\,\text{V}\angle 0°}{4.472\,\Omega\angle -26.57°}$$

$$= 1.789\,\text{A}\angle 26.57°$$

The total current is determined as the summation of currents $\mathbf{I}_{(1)}$ and $\mathbf{I}_{(2)}$:

$$\mathbf{I} = \mathbf{I}_{(1)} + \mathbf{I}_{(2)}$$

$$= 1.118\,\text{A}\angle 26.57° + 1.789\,\text{A}\angle 26.57°$$

$$= (1.0\,\text{A} + j0.5\,\text{A}) + (1.6\,\text{A} + j0.8\,\text{A})$$

$$= 2.6 + j1.3\,\text{A}$$

$$= 2.91\,\text{A}\angle 26.57°$$

EXAMPLE 20–2 Consider the circuit of Figure 20–4:

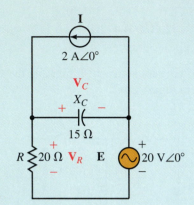

FIGURE 20–4

Find the following:

a. $\mathbf{V}_R$ and $\mathbf{V}_C$ using the superposition theorem.
b. Power dissipated by the circuit.
c. Power delivered to the circuit by each of the sources.

Solution

a. The superposition theorem may be employed as follows:

Voltages due to the current source: Eliminating the voltage source, we obtain the circuit shown in Figure 20–5.

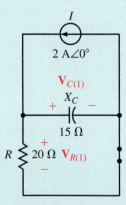

FIGURE 20–5

The impedance "seen" by the current source will be the parallel combination of $\mathbf{R} \| \mathbf{Z}_C$.

$$\mathbf{Z}_1 = \frac{(20\ \Omega)(-j15\ \Omega)}{20\ \Omega - j15\ \Omega} = \frac{300\ \Omega\angle -90°}{25\ \Omega\angle -36.87°} = 12\ \Omega\angle -53.13°$$

The voltage $\mathbf{V}_{R(1)}$ is the same as the voltage across the capacitor, $\mathbf{V}_{C(1)}$. Hence,

$$\mathbf{V}_{R(1)} = \mathbf{V}_{C(1)}$$
$$= (2\ A\angle 0°)(12\ \Omega\angle -53.13°)$$
$$= 24\ V\angle -53.13°$$

Voltages due to the voltage source: Eliminating the current source, we have the circuit shown in Figure 20–6.

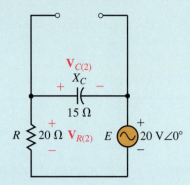

FIGURE 20–6

The voltages $\mathbf{V}_{R(2)}$ and $\mathbf{V}_{C(2)}$ are determined by applying the voltage divider rule,

$$\mathbf{V}_{R(2)} = \frac{20\ \Omega\angle 0°}{20\ \Omega - j15\ \Omega}(20\ \text{V}\angle 0°)$$

$$= \frac{400\ \text{V}\angle 0°}{25\ \angle{-36.87°}} = 16\ \text{V}\angle{+36.87°}$$

and

$$\mathbf{V}_{C(2)} = \frac{-15\ \Omega\angle{-90°}}{20\ \Omega - j15\ \Omega}(20\ \text{V}\angle 0°)$$

$$= \frac{300\ \text{V}\angle 90°}{25\ \angle{-36.87°}} = 12\ \text{V}\angle 126.87°$$

Notice that $\mathbf{V}_{C(2)}$ is assigned to be negative relative to the originally assumed polarity. The negative sign is eliminated from the calculation by adding (or subtracting) 180° from the corresponding calculation.

By applying superposition, we get

$$\mathbf{V}_R = \mathbf{V}_{R(1)} + \mathbf{V}_{R(2)}$$
$$= 24\ \text{V}\angle{-53.13°} + 16\ \text{V}\angle 36.87°$$
$$= (14.4\ \text{V} - j19.2\ \text{V}) + (12.8\ \text{V} + j9.6\ \text{V})$$
$$= 27.2\ \text{V} - j9.6\ \text{V}$$
$$= 28.84\ \text{V}\angle{-19.44°}$$

and

$$\mathbf{V}_C = \mathbf{V}_{C(1)} + \mathbf{V}_{C(2)}$$
$$= 24\ \text{V}\angle{-53.13°} + 12\ \text{V}\angle 126.87°$$
$$= (14.4\ \text{V} - j19.2\ \text{V}) + (-7.2\ \text{V} + j9.6\ \text{V})$$
$$= 7.2\ \text{V} - j9.6\ \text{V}$$
$$= 12\ \text{V}\angle{-53.13°}$$

b. Since only the resistor will dissipate power, the total power dissipated by the circuit is found as

$$P_\text{T} = \frac{(28.84\ \text{V})^2}{20\ \Omega} = 41.60\ \text{W}$$

c. The power delivered to the circuit by the current source is

$$P_1 = V_1 I \cos \theta_1$$

where $\mathbf{V}_1 = \mathbf{V}_C = 12\,\text{V}\angle{-53.13°}$ is the voltage across the current source and θ_1 is the phase angle between $\mathbf{V}_1$ and $\mathbf{I}$.

The power delivered by the current source is

$$P_1 = (12\,\text{V})(2\,\text{A}) \cos 53.13° = 14.4\,\text{W}$$

The power delivered to the circuit by the voltage source is similarly determined as

$$P_2 = E I_2 \cos \theta_2$$

where $\mathbf{I}_2$ is the current through the voltage source and θ_2 is the phase angle between $\mathbf{E}$ and $\mathbf{I}_2$.

$$P_2 = (20\,\text{V})\left(\frac{28.84\,\text{V}}{20\,\Omega}\right) \cos 19.44° = 27.2\,\text{W}$$

As expected, the total power delivered to the circuit must be the summation

$$P_T = P_1 + P_2 = 41.6\,\text{W}$$

PRACTICE PROBLEMS 1

Use superposition to find **V** and **I** for the circuit of Figure 20–7.

FIGURE 20–7

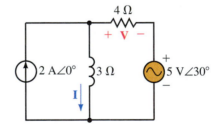

Answers: $\mathbf{I} = 2.52\,\text{A}\angle{-25.41°}$, $\mathbf{V} = 4.45\,\text{V}\angle{104.18°}$

IN-PROCESS LEARNING CHECK 1

A 20-Ω resistor is in a circuit having three sinusoidal sources. After analyzing the circuit, it is found that the current through the resistor due to each of the sources is as follows:

$$I_1 = 1.5\,\text{A}\angle 20°$$
$$I_2 = 1.0\,\text{A}\angle 110°$$
$$I_3 = 2.0\,\text{A}\angle 0°$$

a. Use superposition to calculate the resultant current through the resistor.

b. Calculate the power dissipated by the resistor.

c. Show that the power dissipated by the resistor cannot be found by applying superposition, namely, $P_T \ne I_1^2 R + I_2^2 R + I_3^2 R$.

(Answers are at the end of the chapter.)

20.2   Superposition Theorem—Dependent Sources

Chapter 19 introduced the concept of dependent sources. We now examine ac circuits which are powered by dependent sources. In order to analyze circuits having dependent sources, it is first necessary to determine whether the dependent source is conditional upon a controlling element in its own circuit or whether the controlling element is located in some other circuit.

   If the controlling element is external to the circuit under consideration, the method of analysis is the same as for an independent source. However, if the controlling element is in the same circuit, the analysis follows a slightly different stategy. The next two examples show the techniques used to analyze circuits having dependent sources.

EXAMPLE 20–3   Consider the circuit of Figure 20–8.

a. Determine the general expression for **V** in terms of **I**.
b. Calculate **V** if $\mathbf{I} = 1.0\,\text{A}\angle 0°$.
c. Calculate **V** if $\mathbf{I} = 0.3\,\text{A}\angle 90°$.

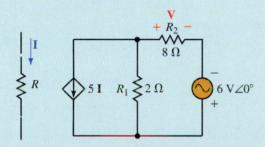

FIGURE 20–8

Solution

a. Since the current source in the circuit is dependent on current through an element which is located outside of the circuit of interest, the circuit may be analyzed in the same manner as for independent sources.

Voltage due to the voltage source: Eliminating the current source, we obtain the circuit shown in Figure 20–9.

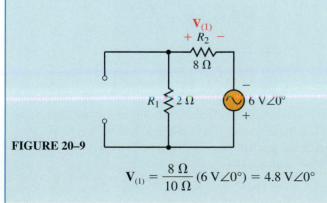

FIGURE 20–9

$$\mathbf{V}_{(1)} = \frac{8\,\Omega}{10\,\Omega}(6\,\text{V}\angle 0°) = 4.8\,\text{V}\angle 0°$$

Voltage due to the current source: Eliminating the voltage source, we have the circuit shown in Figure 20–10.

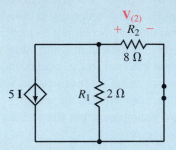

FIGURE 20–10

$$\mathbf{Z}_\mathrm{T} = 2\ \Omega \| 8\ \Omega = 1.6\ \Omega\angle 0°$$

$$\mathbf{V}_{(2)} = \mathbf{V}_{\mathrm{Z_T}} = -(5\mathbf{I})(1.6\ \Omega\angle 0°) = -8.0\ \Omega\mathbf{I}$$

From superposition, the general expression for voltage is determined to be

$$\mathbf{V} = \mathbf{V}_{(1)} + \mathbf{V}_{(2)}$$
$$= 4.8\ \mathrm{V}\angle 0° - 8.0\ \Omega\mathbf{I}$$

b.   If $\mathbf{I} = 1.0\ \mathrm{A}\angle 0°$,

$$\mathbf{V} = 4.8\ \mathrm{V}\angle 0° - (8.0\ \Omega)(1.0\ \mathrm{A}\angle 0°) = -3.2\ \mathrm{V}$$
$$= 3.2\ \mathrm{V}\angle 180°$$

c.   If $\mathbf{I} = 0.3\ \mathrm{A}\angle 90°$,

$$\mathbf{V} = 4.8\ \mathrm{V}\angle 0° - (8.0\ \Omega)(0.3\ \mathrm{A}\angle 90°) = 4.8\ \mathrm{V} - j2.4\ \mathrm{V}$$
$$= 5.367\ \mathrm{V}\angle -26.57°$$

EXAMPLE 20–4   Given the circuit of Figure 20–11, calculate the voltage across the 40-Ω resistor.

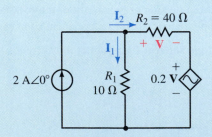

FIGURE 20–11

Solution   In the circuit of Figure 20–11, the dependent source is controlled by an element located in the circuit. Unlike the sources in the previous examples, the dependent source cannot be eliminated from the circuit since doing so would contradict Kirchhoff's voltage law and/or Kirchhoff's current law.

The circuit must be analyzed by considering all effects simultaneously. Applying Kirchhoff's current law, we have

$$\mathbf{I}_1 + \mathbf{I}_2 = 2\ \mathrm{A}\angle 0°$$

From Kirchhoff's voltage law, we have

$$(10 \ \Omega) \ \mathbf{I}_1 = \mathbf{V} + 0.2 \ \mathbf{V} = 1.2 \ \mathbf{V}$$
$$\mathbf{I}_1 = 0.12 \ \mathbf{V}$$

and,

$$\mathbf{I}_2 = \frac{\mathbf{V}}{40 \ \Omega} = 0.025 \ \mathbf{V}$$

Combining the above expressions, we have

$$0.12 \ \mathbf{V} + 0.025 \ \mathbf{V} = 2.0 \ \text{A} \angle 0°$$
$$0.145 \ \mathbf{V} = 2.0 \ \text{A} \angle 0°$$
$$\mathbf{V} = 13.79 \ \text{V} \angle 0°$$

Determine the voltage **V** in the circuit of Figure 20–12.

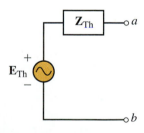

PRACTICE PROBLEMS 2

FIGURE 20–12

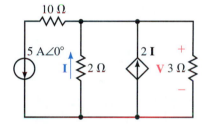

Answer: **V** = 2.73 V∠180°

20.3 Thévenin's Theorem—Independent Sources

Thévenin's theorem is a method which converts any linear bilateral ac circuit into a single ac voltage source in series with an equivalent impedance as shown in Figure 20–13.

The resulting two-terminal network will be equivalent when it is connected to any external branch or component. If the original circuit contains reactive elements, the Thévenin equivalent circuit will be valid only at the frequency at which the reactances were determined. The following method may be used to determine the Thévenin equivalent of an ac circuit having either independent sources or sources which are dependent upon voltage or current in some other circuit. The outlined method may not be used in circuits having dependent sources controlled by voltage or current in the same circuit.

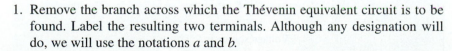

FIGURE 20–13 Thévenin equivalent circuit.

1. Remove the branch across which the Thévenin equivalent circuit is to be found. Label the resulting two terminals. Although any designation will do, we will use the notations *a* and *b*.

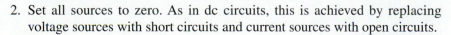

2. Set all sources to zero. As in dc circuits, this is achieved by replacing voltage sources with short circuits and current sources with open circuits.

3. Determine the Thévenin equivalent impedance, $\mathbf{Z}_{Th}$ by calculating the impedance seen between the open terminals a and b. Occasionally it may be necessary to redraw the circuit to simplify this process.

4. Replace the sources removed in Step 3 and determine the open-circuit voltage across the terminals a and b. If any of the sources are expressed in sinusoidal form, it is first necessary to convert these sources into an equivalent phasor form. For circuits having more than one source, it may be necessary to apply the superposition theorem to calculate the open-circuit voltage. Since all voltages will be phasors, the resultant is found by using vector algebra. The open-circuit voltage is the Thévenin voltage, $\mathbf{E}_{Th}$.

5. Sketch the resulting Thévenin equivalent circuit by including that portion of the circuit removed in Step 1.

EXAMPLE 20–5 Find the Thévenin equivalent circuit external to $\mathbf{Z}_L$ for the circuit of Figure 20–14.

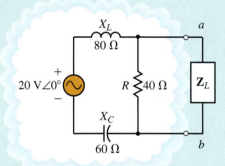

FIGURE 20–14

Solution

Steps 1 and 2: Removing the load impedance $\mathbf{Z}_L$ and setting the voltage source to zero, we have the circuit of Figure 20–15.

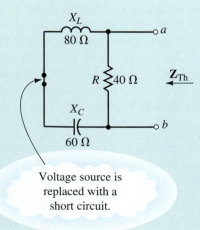

Voltage source is replaced with a short circuit.

FIGURE 20–15

Step 3: The Thévenin impedance between terminals a and b is found as

$$\mathbf{Z}_{\text{Th}} = \mathbf{R}\|(\mathbf{Z}_L + \mathbf{Z}_C)$$

$$= \frac{(40\ \Omega\angle0°)(20\ \Omega\angle90°)}{40\ \Omega + j20\ \Omega}$$

$$= \frac{800\ \Omega\angle90°}{44.72\ \Omega\angle26.57°}$$

$$= 17.89\ \Omega\angle63.43°$$

$$= 8\ \Omega + j16\ \Omega$$

Step 4: The Thévenin voltage is found by using the voltage divider rule as shown in the circuit of Figure 20–16.

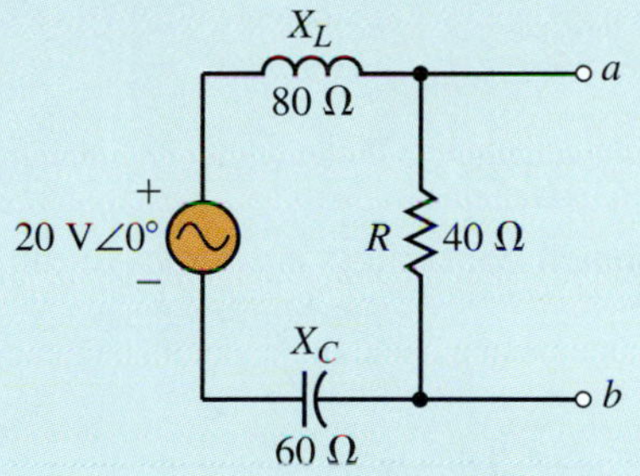

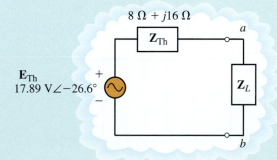

FIGURE 20–16

$$\mathbf{E}_{\text{Th}} = \mathbf{V}_{ab} = \frac{40\ \Omega\angle0°}{40\ \Omega + j80\ \Omega - j60\ \Omega}(20\ \text{V}\angle0°)$$

$$= \frac{800\ \text{V}\angle0°}{44.72\ \Omega\angle26.57°}$$

$$= 17.89\ \text{V}\angle-26.57°$$

Step 5: The resultant Thévenin equivalent circuit is shown in Figure 20–17.

FIGURE 20–17

EXAMPLE 20–6   Determine the Thévenin equivalent circuit external to $\mathbf{Z}_L$ in the circuit in Figure 20–18.

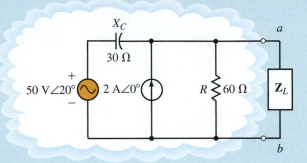

FIGURE 20–18

Solution

Step 1: Removing the branch containing $\mathbf{Z}_L$, we have the circuit of Figure 20–19.

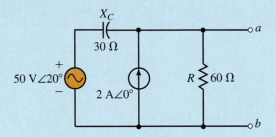

FIGURE 20–19

Step 2: After setting the voltage and current sources to zero, we have the circuit of Figure 20–20.

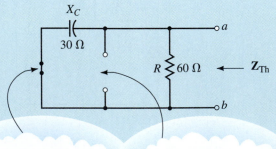

Voltage source replaced with a short circuit.

Current source replaced with an open circuit.

FIGURE 20–20

Step 3: The Thévenin impedance is determined as

$$\mathbf{Z}_{Th} = \mathbf{Z}_C \| \mathbf{Z}_R$$

$$= \frac{(30\ \Omega\angle-90°)(60\ \Omega\angle0°)}{60\ \Omega - j30\ \Omega}$$

$$= \frac{1800\ \Omega\angle-90°}{67.08\ \Omega\angle-26.57°}$$

$$= 26.83\ \Omega\angle-63.43°$$

Step 4: Because the given network consists of two independent sources, we consider the individual effects of each upon the open-circuit voltage. The total effect is then easily determined by applying the superposition theorem. Reinserting only the voltage source into the original circuit, as shown in Figure 20–21, allows us to find the open-circuit voltage, $\mathbf{V}_{ab(1)}$, by applying the voltage divider rule:

$$\mathbf{V}_{ab(1)} = \frac{60\ \Omega}{60\ \Omega - j30\ \Omega}(50\ \text{V}\angle20°)$$

$$= \frac{3000\ \text{V}\angle20°}{67.08\angle-26.57°}$$

$$= 44.72\ \text{V}\angle46.57°$$

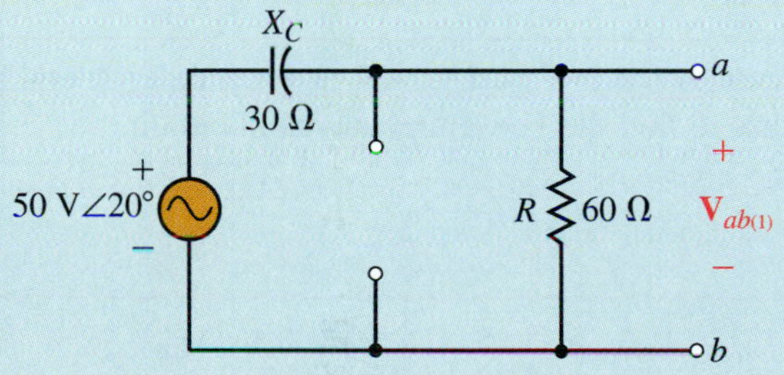

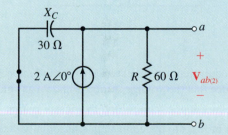

FIGURE 20–21

Now, considering only the current source as shown in Figure 20–22, we determine $\mathbf{V}_{ab(2)}$ by Ohm's law:

FIGURE 20–22

$$\mathbf{V}_{ab(2)} = \frac{(2\ \text{A}\angle0°)(30\ \Omega\angle-90°)(60\ \Omega\angle0°)}{60\ \Omega - j30\ \Omega}$$

$$= (2\ \text{A}\angle0°)(26.83\ \Omega\angle-63.43°)$$

$$= 53.67\ \text{V}\angle-63.43°$$

From the superposition theorem, the Thévenin voltage is determined as

$$\mathbf{E}_{Th} = \mathbf{V}_{ab(1)} + \mathbf{V}_{ab(2)}$$

$$= 44.72 \text{ V}\angle 46.57° + 53.67 \text{ V}\angle{-63.43°}$$

$$= (30.74 \text{ V} + j32.48 \text{ V}) + (24.00 \text{ V} - j48.00 \text{ V})$$

$$= (54.74 \text{ V} - j15.52 \text{ V}) = 56.90 \text{ V}\angle{-15.83°}$$

Step 5: The resulting Thévenin equivalent circuit is shown in Figure 20–23.

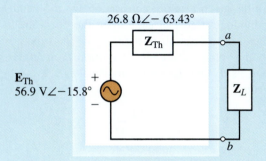

FIGURE 20–23

Refer to the circuit shown in Figure 20–24 of Practice Problem 3. List the steps that you would use to find the Thévenin equivalent circuit.

(Answers are at the end of the chapter.)

PRACTICE
PROBLEMS 3

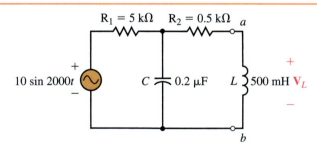

FIGURE 20–24

a. Find the Thévenin equivalent circuit external to the inductor in the circuit in Figure 20–24. (Notice that the voltage source is shown as sinusoidal.)

b. Use the Thévenin equivalent circuit to find the phasor output voltage, $\mathbf{V}_L$.

c. Convert the answer of (b) into the equivalent sinusoidal voltage.

Answers:

a. $\mathbf{Z}_{Th} = 1.5 \text{ k}\Omega - j2.0 \text{ k}\Omega = 2.5 \text{ k}\Omega\angle{-53.13°}$, $\mathbf{E}_{Th} = 3.16 \text{ V}\angle{-63.43°}$

b. $\mathbf{V}_L = 1.75 \text{ V}\angle 60.26°$

c. $v_L = 2.48 \sin(2000t + 60.26°)$

20.4   Norton's Theorem—Independent Sources

Norton's theorem converts any linear bilateral network into an equivalent circuit consisting of a single current source and a parallel impedance as shown in Figure 20–25.

Although Norton's equivalent circuit may be determined by first finding the Thévenin equivalent circuit and then performing a source conversion, we generally use the more direct method outlined below. The steps to find the Norton equivalent cirucuit are as follows:

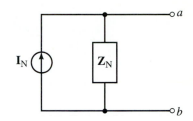

FIGURE 20–25   Norton equivalent circuit.

1. Remove the branch across which the Norton equivalent circuit is to be found. Label the resulting two terminals *a* and *b*.

2. Set all sources to zero.

3. Determine the Norton equivalent impedance, $\mathbf{Z}_N$, by calculating the impedance seen between the open terminals *a* and *b*.

   NOTE: Since the previous steps are identical with those followed for finding the Thévenin equivalent circuit, we conclude that the Norton impedance must be the same as the Thévenin impedance.

4. Replace the sources removed in Step 3 and determine the current that would occur between terminals *a* and *b* if these terminals were shorted. Any voltages and currents that are given in sinusoidal notation must first be expressed in equivalent phasor notation. If the circuit has more than one source it may be necessary to apply the superposition theorem to calculate the total short-circuit current. Since all currents will be in phasor form, any addition must be done using vector algebra. The resulting current is the Norton current, I_N.

5. Sketch the resulting Norton equivalent circuit by inserting that portion of the circuit removed in Step 1.

As mentioned previously, it is possible to find the Norton equivalent circuit from the Thévenin equivalent by simply performing a source conversion. We have already determined that both the Thévenin and Norton impedances are determined in the same way. Consequently, the impedances must be equivalent, and so we have

$$\mathbf{Z}_N = \mathbf{Z}_{Th} \qquad (20\text{–}1)$$

Now, applying Ohm's law, we determine the Norton current source from the Thévenin voltage and impedance, namely,

$$\mathbf{I}_N = \frac{\mathbf{E}_{Th}}{\mathbf{Z}_{Th}} \qquad (20\text{–}2)$$

Figure 20–26 shows the equivalent circuits.

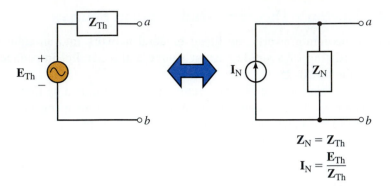

$$\mathbf{Z}_N = \mathbf{Z}_{Th}$$

$$\mathbf{I}_N = \frac{\mathbf{E}_{Th}}{\mathbf{Z}_{Th}}$$

FIGURE 20–26

EXAMPLE 20–7   Given the circuit of Figure 20–27, find the Norton equivalent.

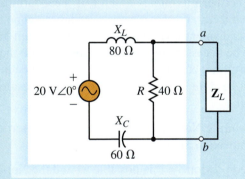

FIGURE 20–27

Solution

Steps 1 and 2: By removing the load impedance, $\mathbf{Z}_L$, and setting the voltage source to zero, we have the network of Figure 20–28.

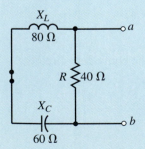

FIGURE 20–28

Step 3: The Norton impedance may now be determined by evaluating the impedance between terminals a and b. Hence, we have

$$\mathbf{Z}_N = \frac{(40\ \Omega\angle 0°)(20\ \Omega\angle 90°)}{40\ \Omega + j20\ \Omega}$$

$$= \frac{800\ \Omega\angle 90°}{44.72\angle 26.57°}$$

$$= 17.89\ \Omega\angle 63.43°$$

$$= 8\ \Omega + j16\ \Omega$$

Step 4: Reinserting the voltage source, as in Figure 20–29, we find the Norton current by calculating the current between the shorted terminals, a and b.

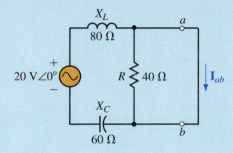

FIGURE 20–29

Because the resistor $R = 40\ \Omega$ is shorted, the current is determined by the impedances X_L and X_C as

$$\mathbf{I}_N = \mathbf{I}_{ab} = \frac{20\ V\angle 0°}{j80\ \Omega - j60\ \Omega}$$

$$= \frac{20\ V\angle 0°}{20\ \Omega\angle -90°}$$

$$= 1.00\ A\angle -90°$$

Step 5: The resultant Norton equivalent circuit is shown in Figure 20–30.

$$\mathbf{Z}_N = 8\ \Omega + j16\ \Omega = 17.89\ \Omega\angle 63.43°$$

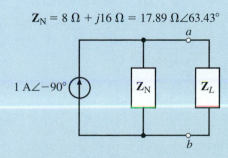

FIGURE 20–30

EXAMPLE 20–8   Find the Norton equivalent circuit external to R_L in the circuit of Figure 20–31. Use the equivalent circuit to calculate the current $\mathbf{I}_L$ when $R_L = 0\ \Omega$, 400 Ω, and 2 kΩ.

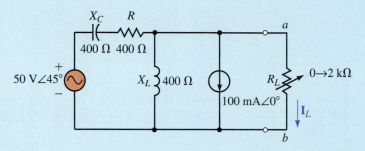

FIGURE 20–31

Solution

Steps 1 and 2: Removing the load resistor and setting the sources to zero, we obtain the network shown in Figure 20–32.

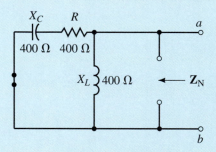

FIGURE 20–32

Step 3: The Norton impedance is determined as

$$\mathbf{Z}_{\mathrm{N}} = \frac{(400\ \Omega\angle 90°)(400\ \Omega - j400\ \Omega)}{j400\ \Omega + 400\ \Omega - j400\ \Omega}$$

$$= \frac{(400\ \Omega\angle 90°)(565.69\ \Omega\angle -45°)}{400\ \Omega\angle 0°}$$

$$= 565.69\ \Omega\angle +45°$$

Step 4: Because the network consists of two sources, we determine the effects due to each source separately and then apply superposition to evaluate the Norton current source.

Reinserting the voltage source into the original network, we see from Figure 20–33 that the short-circuit current between the terminals a and b is easily found by using Ohm's law.

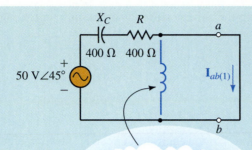

FIGURE 20–33

$$\mathbf{I}_{ab(1)} = \frac{50\ \text{V}\angle 45°}{400\ \Omega - j400\ \Omega}$$

$$= \frac{50\ \text{V}\angle 45°}{565.69\ \Omega\angle -45°}$$

$$= 88.4\ \text{mA}\angle 90°$$

Since short circuiting the current source effectively removes all impedances, as illustrated in Figure 20–34, the short-circuit current between the terminals a and b is given as follows:

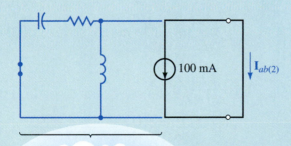

These components are short circuited.

FIGURE 20–34

$$\mathbf{I}_{ab(2)} = -100\ \text{mA}\angle 0°$$

$$= 100\ \text{mA}\angle 180°$$

Now applying the superposition theorem, the Norton current is determined as the summation

$$\mathbf{I}_N = \mathbf{I}_{ab(1)} + \mathbf{I}_{ab(2)}$$
$$= 88.4 \text{ mA}\angle 90° + 100 \text{ mA}\angle 180°$$
$$= -100 \text{ mA} + j88.4 \text{ mA}$$
$$= 133.5 \text{ mA}\angle 138.52°$$

Step 5: The resulting Norton equivalent circuit is shown in Figure 20–35.

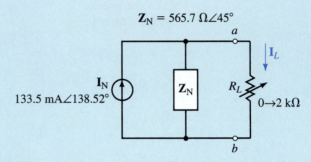

FIGURE 20–35

From the above circuit, we express the current through the load, $\mathbf{I}_L$, as

$$\mathbf{I}_L = \frac{\mathbf{Z}_N}{R_L + \mathbf{Z}_N}\mathbf{I}_N$$

$R_L = 0 \ \Omega$:

$$\mathbf{I}_L = \mathbf{I}_N = 133.5 \text{ mA}\angle 138.52°$$

$R_L = 400 \ \Omega$:

$$\mathbf{I}_L = \frac{\mathbf{Z}_N}{R_L + \mathbf{Z}_N}\mathbf{I}_N$$
$$= \frac{(565.7 \ \Omega\angle 45°)(133.5 \text{ mA}\angle 138.52°)}{400 \ \Omega + 400 \ \Omega + j400 \ \Omega}$$
$$= \frac{75.24 \text{ V}\angle 183.52°}{894.43 \ \Omega\angle 26.57°}$$
$$= 84.12 \text{ mA}\angle 156.95°$$

$R_L = 2 \ \text{k}\Omega$:

$$\mathbf{I}_L = \frac{\mathbf{Z}_N}{R_L + \mathbf{Z}_N}\mathbf{I}_N$$
$$= \frac{(565.7 \ \Omega\angle 45°)(133.5 \text{ mA}\angle 138.52°)}{2000 \ \Omega + 400 \ \Omega + j400 \ \Omega}$$
$$= \frac{75.24 \text{ V}\angle 183.52°}{2433.1 \ \Omega\angle 9.46°}$$
$$= 30.92 \text{ mA}\angle 174.06°$$

Refer to the circuit shown in Figure 20–36 (Practice Problem 4). List the steps that you would use to find the Norton equivalent circuit.

(Answers are at the end of the chapter.)

Find the Norton equivalent circuit external to R_L in the circuit of Figure 20–36. Use the equivalent circuit to find the current I_L.

PRACTICE
PROBLEMS 4

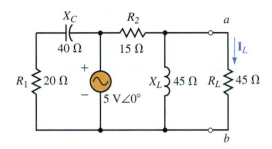

FIGURE 20–36

Answers: $\mathbf{Z}_N = 13.5\ \Omega + j4.5\ \Omega = 14.23\ \Omega\angle 18.43°$, $\mathbf{I}_N = 0.333\ A\angle 0°$
$\mathbf{I}_L = 0.0808\ A\angle 14.03°$

20.5 Thévenin's and Norton's Theorem for Dependent Sources

If a circuit contains a dependent source which is controlled by an element outside the circuit of interest, the methods outlined in Sections 20.2 and 20.3 are used to find either the Thévenin or Norton equivalent circuit.

EXAMPLE 20–9 Given the circuit of Figure 20–37, find the Thévenin equivalent circuit external to R_L. If the voltage applied to the resistor R_1 is 10 mV, use the Thévenin equivalent circuit to calculate the minimum and maximum voltage across R_L.

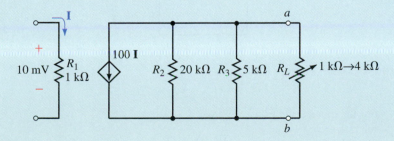

FIGURE 20–37

Solution

Step 1: Removing the load resistor from the circuit and labelling the remaining terminals *a* and *b*, we have the circuit shown in Figure 20–38.

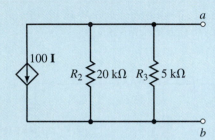

FIGURE 20–38

Steps 2 and 3: The Thévenin resistance is found by open circuiting the current source and calculating the impedance observed between the terminals *a* and *b*. Since the circuit is purely resistive, we have

$$R_{\text{Th}} = 20 \text{ k}\Omega \| 5 \text{ k}\Omega$$
$$= \frac{(20 \text{ k}\Omega)(5 \text{ k}\Omega)}{20 \text{ k}\Omega + 5 \text{ k}\Omega}$$
$$= 4 \text{ k}\Omega$$

Step 4: The open-circuit voltage between the terminals is found to be

$$\mathbf{V}_{ab} = -(100\mathbf{I})(4 \text{ k}\Omega)$$
$$= -(4 \times 10^5 \ \Omega)\mathbf{I}$$

As expected, the Thévenin voltage source is dependent upon the current **I**.

Step 5: Because the Thévenin voltage is a dependent voltage source we use the appropriate symbol when sketching the equivalent circuit, as shown in Figure 20–39.

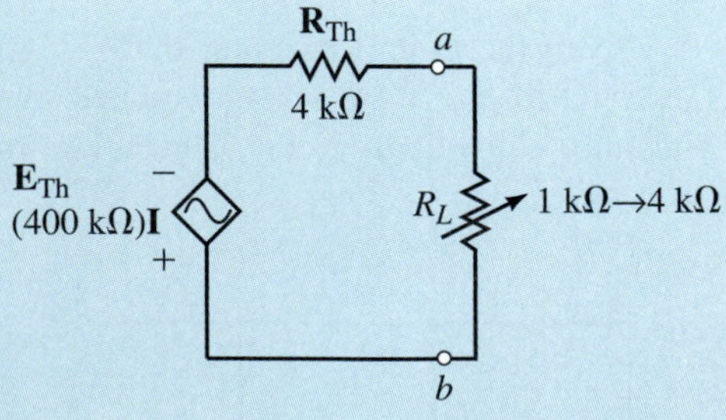

FIGURE 20–39

For the given conditions, we have

$$I = \frac{10 \text{ mV}}{1 \text{ k}\Omega} = 10 \ \mu\text{A}$$

The voltage across the load is now determined as follows:

$R_L = \mathbf{1\ k\Omega}:$ $V_{ab} = -\dfrac{1\ k\Omega}{1\ k\Omega + 4\ k\Omega}(4 \times 10^5\ \Omega)(10\ \mu A)$

$\qquad\qquad = -0.8\ V$

$R_L = \mathbf{4\ k\Omega}:$ $V_{ab} = -\dfrac{4\ k\Omega}{4\ k\Omega + 4\ k\Omega}(4 \times 10^5\ \Omega)(10\ \mu A)$

$\qquad\qquad = -2.0\ V$

For an applied voltage of 10 mV, the voltage across the load resistance will vary between 0.8 V and 2.0 V as R_L is adjusted between 1 kΩ and 4 kΩ.

If a circuit contains one or more dependent sources which are controlled by an element in the circuit being analyzed, all previous methods fail to provide equivalent circuits which correctly model the circuit's behavior. In order to determine the Thévenin or Norton equivalent circuit of a circuit having a dependent source controlled by a local voltage or current, the following steps must be taken:

1. Remove the branch across which the Norton equivalent circuit is to be found. Label the resulting two terminals *a* and *b*.

2. Calculate the open-circuit voltage (Thévenin voltage) across the two terminals *a* and *b*. Because the circuit contains a dependent source controlled by an element in the circuit, the dependent source may not be set to zero. Its effects must be considered together with the effects of any independent source(s).

3. Determine the short-circuit current (Norton current) that would occur between the terminals. Once again, the dependent source may not be set to zero, but rather must have its effects considered concurrently with the effects of any independent source(s).

4. Determine the Thévenin or Norton impedance by applying Equations 20–1 and 20–2 as follows:

$$\mathbf{Z}_N = \mathbf{Z}_{Th} = \frac{\mathbf{E}_{Th}}{\mathbf{I}_N} \qquad (20\text{–}3)$$

5. Sketch the Thévenin or Norton equivalent circuit, as shown previously in Figure 20–26. Ensure that the portion of the network that was removed in Step 1 is reinserted as part of the equivalent circuit.

EXAMPLE 20–10   For the circuit of Figure 20–40, find the Norton equivalent circuit external to the load resistor, R_L.

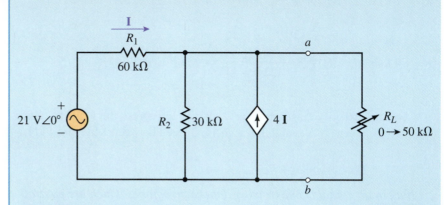

FIGURE 20–40

Solution

Step 1: After removing the load resistor from the circuit, we have the network shown in Figure 20–41.

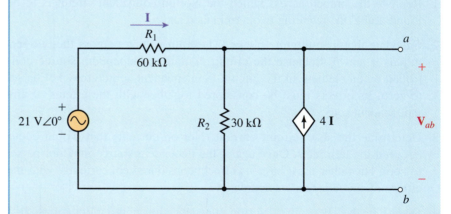

FIGURE 20–41

Step 2: At first glance, we might look into the open terminals and say that the Norton (or Thévenin) impedance appears to be 60 kΩ||30 kΩ = 20 kΩ. However, we will find that this result is incorrect. The presence of the locally controlled dependent current source makes the analysis of this circuit slightly more complicated than a circuit that contains only an independent source. We know, however, that the basic laws of circuit analysis must apply to all circuits, regardless of the complexity. Applying Kirchhoff's current law at node a gives the current through R_2 as

$$\mathbf{I}_{R_2} = \mathbf{I} + 4\mathbf{I} = 5\mathbf{I}$$

Now, applying Kirchhoff's voltage law around the closed loop containing the voltage source and the two resistors, gives

$$21 \text{ V}\angle 0° = (60 \text{ k}\Omega)\mathbf{I} + (30 \text{ k}\Omega)(5\mathbf{I}) = 210 \text{ k}\Omega\mathbf{I}$$

which allows us to solve for the current **I** as

$$\mathbf{I} = \frac{21 \text{ V}\angle0°}{210 \text{ k}\Omega} = 0.100 \text{ mA}\angle0°$$

Since the open-circuit voltage, $\mathbf{V}_{ab}$ is the same as the voltage across R_2, we have

$$\mathbf{E}_{\text{Th}} = \mathbf{V}_{ab} = (60 \text{ k}\Omega)(5)(0.1 \text{ mA}\angle0°) = 30 \text{ V}\angle0°$$

Step 3: The Norton current source is determined by placing a short-circuit between terminals a and b as shown in Figure 20–42.

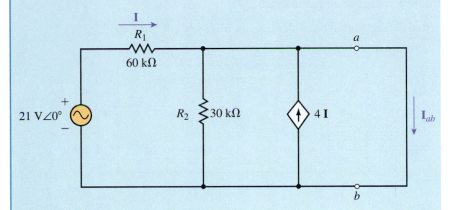

FIGURE 20–42

Upon further inspection of this circuit we see that resistor R_2 is short-circuited. The simplified circuit is shown in Figure 20–43.

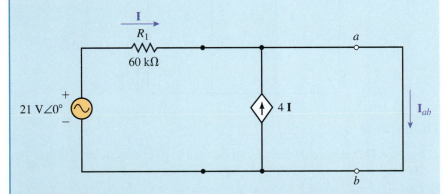

FIGURE 20–43

The short-circuit current $\mathbf{I}_{ab}$ is now easily determined by using Kirchhoff's current law at node a, and so we have

$$\mathbf{I}_{\text{N}} = \mathbf{I}_{ab} = 5\mathbf{I}$$

From Ohm's law, we have

$$\mathbf{I} = \frac{21 \text{ V}\angle0°}{60 \text{ k}\Omega} = 0.35 \text{ mA}\angle0°$$

and so

$$\mathbf{I}_N = 5(0.35 \text{ mA}\angle 0°) = 1.75 \text{ mA}\angle 0°$$

Step 4: The Norton (or Thévenin) impedance is now determined from Ohm's law as

$$\mathbf{Z}_N = \frac{\mathbf{E}_{Th}}{\mathbf{I}_N} = \frac{30 \text{ V}\angle 0°}{1.75 \text{ mA}\angle 0°} = 17.14 \text{ k}\Omega$$

Notice that this impedance is different from the originally assumed 20 kΩ. In general, this condition will occur for most circuits that contain a locally controlled voltage or current source.

Step 5: The Norton equivalent circuit is shown in Figure 20–44.

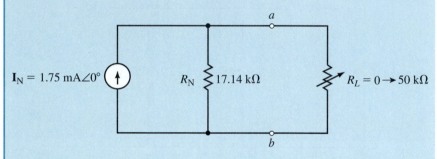

FIGURE 20–44

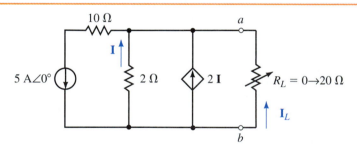

FIGURE 20–45

a. Find the Thévenin equivalent circuit external to R_L in the circuit of Figure 20–45.

b. Determine the current I_L when $R_L = 0$ and when $R_L = 20 \ \Omega$.

Answers:
a. $\mathbf{E}_{Th} = \mathbf{V}_{ab} = -3.33 \text{ V}$, $\mathbf{Z}_{Th} = 0.667 \ \Omega$
b. For $R_L = 0$: $\mathbf{I}_L = 5.00 \text{ A}$ (upward); for $R_L = 20 \ \Omega$: $\mathbf{I}_L = 0.161 \text{ A}$ (upward)

If a circuit has more than one independent source, it is necessary to determine the open-circuit voltage and short-circuit current due to each independent source while simultaneously considering the effects of the dependent source. The following example illustrates the principle.

EXAMPLE 20–11   Find the Thévenin and Norton equivalent circuits external to the load resistor in the circuit of Figure 20–46.

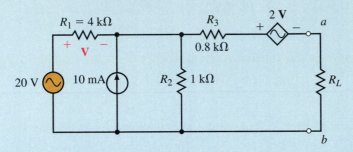

FIGURE 20–46

Solution   There are several methods of solving this circuit. The following approach uses the fewest number of steps.

Step 1: Removing the load resistor, we have the circuit shown in Figure 20–47.

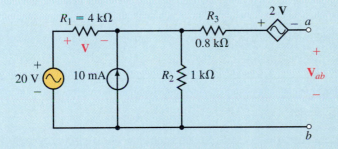

FIGURE 20–47

Step 2: In order to find the open-circuit voltage, V_{ab} of Figure 20–47, we may isolate the effects due to each independent source and then apply superposition to determine the combined result. However, by converting the current source into an equivalent voltage source, we can determine the open-circuit voltage in one step. Figure 20–48 shows the circuit that results when the current source is converted into an equivalent voltage source.

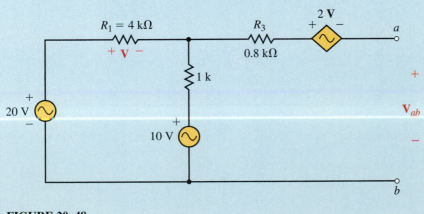

FIGURE 20–48

The controlling element (R_1) has a voltage, **V**, determined as

$$\mathbf{V} = \left(\frac{4 \text{ k}\Omega}{4 \text{ k}\Omega + 1 \text{ k}\Omega}\right)(20 \text{ V} - 10 \text{ V})$$

$$= 8 \text{ V}$$

which gives a Thévenin (open-circuit) voltage of

$$\mathbf{E}_{\text{Th}} = \mathbf{V}_{ab} = -2(8 \text{ V}) + 0 \text{ V} - 8 \text{ V} + 20 \text{ V}$$

$$= -4.0 \text{ V}$$

Step 3: The short-circuit current is determined by examining the circuit shown in Figure 20–49.

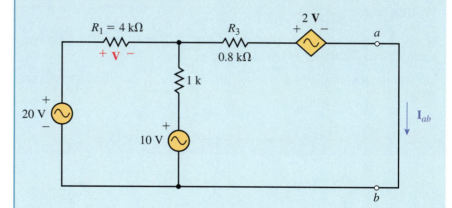

FIGURE 20–49

Once again, it is possible to determine the short-circuit current by using superposition. However, upon further reflection, we see that the circuit is easily analyzed using Mesh analysis. Loop currents $\mathbf{I}_1$ and $\mathbf{I}_2$ are assigned in clockwise directions as shown in Figure 20–50.

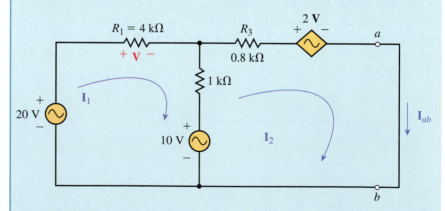

FIGURE 20–50

The loop equations are as follows:

Loop 1: $\qquad\qquad (5 \text{ k}\Omega) \mathbf{I}_1 - (1 \text{ k}\Omega) \mathbf{I}_2 = 10 \text{ V}$

Loop 2: $\qquad\qquad -(1 \text{ k}\Omega) \mathbf{I}_1 + (1.8 \text{ k}\Omega) \mathbf{I}_2 = 10 \text{ V} - 2 \text{ V}$

Notice that the voltage in the second loop equation is expressed in terms of the controlling voltage across R_1. We will not worry about this right now. We may simplify our calculations by ignoring the units in the above equations. It is obvious that if all impedances are expressed in $k\Omega$ and all voltages are in volts, then the currents $\mathbf{I}_1$ and $\mathbf{I}_2$ must be in mA. The determinant for the denominator is found to be

$$\mathbf{D} = \begin{vmatrix} 5 & -1 \\ -1 & 1.8 \end{vmatrix} = 9 - 1 = 8$$

The current $\mathbf{I}_1$ is solved by using determinants as follows:

$$\mathbf{I}_1 = \frac{\begin{vmatrix} 10 & -1 \\ 10 - 2\,\mathbf{V} & 1.8 \end{vmatrix}}{\mathbf{D}} = \frac{18 - (-1)(10 - 2\,\mathbf{V})}{8}$$

$$= 3.5 - 0.25\,\mathbf{V}$$

The above result illustrates that the current $\mathbf{I}_1$ is dependent on the controlling voltage. However, by examining the circuit of Figure 20–50, we see that the controlling voltage depends on the current $\mathbf{I}_1$, and is determined from Ohm's law as

$$\mathbf{V} = (4\ k\Omega)\mathbf{I}_1$$

or more simply as

$$\mathbf{V} = 4\mathbf{I}_1$$

Now, the current $\mathbf{I}_1$ is found as

$$\mathbf{I}_1 = 3.5 - 0.25(4\mathbf{I}_1)$$
$$2\mathbf{I}_1 = 3.5$$
$$\mathbf{I}_1 = 1.75\ \text{mA}$$

which gives $\mathbf{V} = 7.0\ \text{V}$.

Finally, the short circuit current (which in the circuit of Figure 20–50 is represented by $\mathbf{I}_2$) is found as

$$\mathbf{I}_2 = \frac{\begin{vmatrix} 5 & 10 \\ -1 & 10 - 2\,\mathbf{V} \end{vmatrix}}{\mathbf{D}} = \frac{50 - 10\,\mathbf{V} - (-10)}{8}$$

$$= 7.5 - 1.25\,\mathbf{V}$$
$$= 7.5 - 1.25(7.0\ \text{V})$$
$$= -1.25\ \text{mA}$$

This gives us the Norton current source as

$$I_{\text{N}} = I_{ab} = -1.25\ \text{mA}$$

Step 4: The Thévenin (or Norton) impedance is determined using Ohm's Law.

$$\mathbf{Z}_{\text{Th}} = \mathbf{Z}_{\text{N}} = \frac{\mathbf{E}_{\text{Th}}}{\mathbf{I}_{\text{N}}} = \frac{-4.0\ \text{V}}{-1.25\ \text{mA}} = 3.2\ k\Omega$$

The resulting Thévenin equivalent circuit is shown in Figure 20–51 and the Norton equivalent circuit is shown in Figure 20–52.

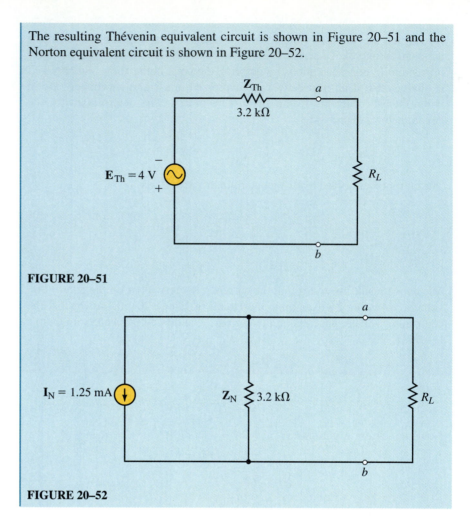

FIGURE 20–51

FIGURE 20–52

PRACTICE PROBLEMS 6

Find the Thévenin and Norton equivalent circuits external to the load resistor in the circuit of Figure 20–53.

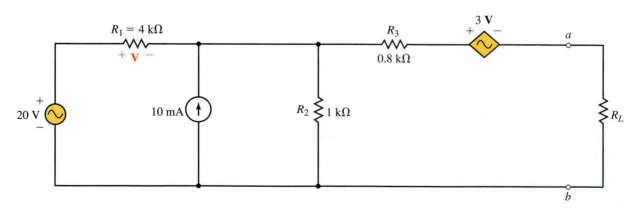

FIGURE 20–53

Answers:

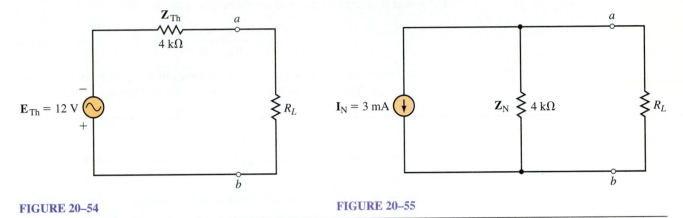

FIGURE 20–54   FIGURE 20–55

20.6   Maximum Power Transfer Theorem

The maximum power transfer theorem is used to determine the value of load impedance required so that the load receives the maximum amount of power from the circuit. Consider the Thévenin equivalent circuit shown in Figure 20–56.

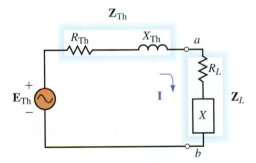

FIGURE 20–56

For any load impedance $\mathbf{Z}_L$ consisting of a resistance and a reactance such that $\mathbf{Z}_L = R_L \pm jX$, the power dissipated by the load will be determined as follows:

$$P_L = I^2 R_L$$

$$I = \frac{E_{Th}}{\sqrt{(R_{Th} + R_L)^2 + (X_{Th} \pm X)^2}}$$

$$P_L = \frac{E_{Th}{}^2 R_L}{(R_{Th} + R_L)^2 + (X_{Th} \pm X)^2}$$

Consider only the reactance portion, X, of the load impedance for the moment and neglect the effect of the load resistance. We see that the power dissipated by the load will be maximum when the denominator is kept to a minimum. If the load were to have an impedance such that $jX = -jX_{\text{Th}}$, then the power delivered to the load would be given as

$$P_L = \frac{E_{\text{Th}}^{\,2} R_L}{(R_{\text{Th}} + R_L)^2} \qquad (20\text{–}4)$$

We recognize that this is the same expression for power as that determined for the Thévenin equivalent of dc circuits in Chapter 9. Recall that maximum power was delivered to the load when

$$R_L = R_{\text{Th}}$$

For ac circuits, the maximum power transfer theorem states the following:

Maximum power will be delivered to a load whenever the load has an impedance which is equal to the complex conjugate of the Thévenin (or Norton) impedance of the equivalent circuit.

A detailed derivation of the maximum power transfer theorem is provided in Appendix C. The maximum power delivered to the load may be calculated by using Equation 20–4, which is simplified as follows:

$$P_{\text{max}} = \frac{E_{\text{Th}}^{\,2}}{4R_{\text{Th}}} \qquad (20\text{–}5)$$

For a Norton equivalent circuit, the maximum power delivered to a load is determined by substituting $E_{\text{Th}} = I_{\text{N}} Z_{\text{N}}$ into the above expression as follows:

$$P_{\text{max}} = \frac{I_{\text{N}}^{\,2} Z_{\text{N}}^{\,2}}{4R_{\text{N}}} \qquad (20\text{–}6)$$

EXAMPLE 20–12   Determine the load impedance $\mathbf{Z}_L$ which will allow maximum power to be delivered to the load in the circuit of Figure 20–57. Find the maximum power.

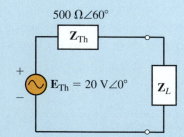

500 Ω∠60°

$\mathbf{Z}_{\text{Th}}$

$\mathbf{E}_{\text{Th}} = 20\ \text{V}\angle 0°$

$\mathbf{Z}_L$

FIGURE 20–57

Solution   Expressing the Thévenin impedance in its rectangular form, we have

$$\mathbf{Z}_{\text{Th}} = 500\ \Omega\angle 60° = 250\ \Omega + j433\ \Omega$$

In order to deliver maximum power to the load, the load impedance must be the complex conjugate of the Thévenin impedance. Hence,

$$\mathbf{Z}_L = 250\ \Omega - j433\ \Omega = 500\ \Omega\angle{-60°}$$

The power delivered to the load is now easily determined by applying Equation 20–5:

$$P_{max} = \frac{(20\ \text{V})^2}{4(250\ \Omega)} = 400\ \text{mW}$$

Given the circuit of Figure 20–57, determine the power dissipated by the load if the load impedance is equal to the Thévenin impedance, $\mathbf{Z}_L = 500\ \Omega\angle{60°}$. Compare your answer to that obtained in Example 20–12.

PRACTICE PROBLEMS 7

Answer: $P = 100$ mW, which is less than P_{max}.

Occasionally it is not possible to adjust the reactance portion of a load. In such cases, a **relative maximum power** will be delivered to the load when the load resistance has a value determined as

$$R_L = \sqrt{R_{Th}^2 + (X \pm X_{Th})^2} \qquad (20\text{–}7)$$

If the reactance of the Thévenin impedance is of the same type (both capacitive or both inductive) as the reactance in the load, then the reactances are added.

If one reactance is capacitive and the other is inductive, however, then the reactances are subtracted.

To determine the power delivered to the load in such cases, the power will need to be calculated by finding either the voltage across the load or the current through the load. Equations 20–5 and 20–6 will no longer apply, since these equations were based on the premise that the load impedance is the complex conjugate of the Thévenin impedance.

EXAMPLE 20–13 For the circuit of Figure 20–58, determine the value of the load resistor, R_L, such that maximum power will be delivered to the load.

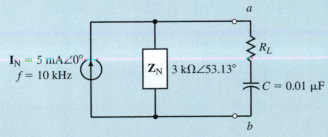

FIGURE 20–58

Solution Notice that the load impedance consists of a resistor in series with a capacitance of 0.010 μF. Since the capacitive reactance is determined by the frequency, it is quite likely that the maximum power for this circuit may only be a relative maximum, rather than the absolute maximum. For the **absolute maximum power** to be delivered to the load, the load impedance would need to be

$$\mathbf{Z}_L = 3 \text{ k}\Omega\angle-53.13° = 1.80 \text{ k}\Omega - j2.40 \text{ k}\Omega$$

The reactance of the capacitor at a frequency of 10 kHz is determined to be

$$X_C = \frac{1}{2\pi(10 \text{ kHz})(0.010 \text{ } \mu\text{F})} = 1.592 \text{ k}\Omega$$

Because the capacitive reactance is not equal to the inductive reactance of the Norton impedance, the circuit will not deliver the absolute maximum power to the load. However, relative maximum power will be delivered to the load when

$$R_L = \sqrt{R_{\text{Th}}^2 + (X^2 - X_{\text{Th}})}$$
$$= \sqrt{(1.800 \text{ k}\Omega)^2 + (1.592 \text{ k}\Omega - 2.4 \text{ k}\Omega)^2}$$
$$= 1.973 \text{ k}\Omega$$

Figure 20–59 shows the circuit with all impedance values.

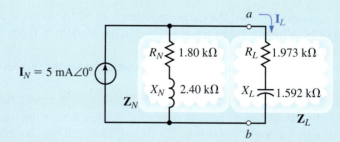

FIGURE 20–59

The load current will be

$$\mathbf{I}_L = \frac{\mathbf{Z}_\text{N}}{\mathbf{Z}_\text{N} + \mathbf{Z}_L}\mathbf{I}_\text{N}$$

$$= \frac{1.80 \text{ k}\Omega + j2.40 \text{ k}\Omega}{(1.80 \text{ k}\Omega + j2.40 \text{ k}\Omega) + (1.973 \text{ k}\Omega - j1.592 \text{ k}\Omega)}(5 \text{ mA}\angle0°)$$

$$= \frac{3 \text{ k}\Omega\angle53.13°}{3.773 \text{ k}\Omega + j0.808 \text{ k}\Omega}(5 \text{ mA}\angle0°)$$

$$= \frac{15.0 \text{ V}\angle53.13°}{3.859 \text{ k}\Omega\angle12.09°} = 3.887 \text{ mA}\angle41.04°$$

We now determine the power delivered to the load for the given conditions as

$$P_L = I_L^2 R_L$$
$$= (3.887 \text{ mA})^2(1.973 \text{ k}\Omega) = 29.82 \text{ mW}$$

If we had applied Equation 20–6, we would have found the absolute maximum power to be

$$P_{max} = \frac{(5 \text{ mA})^2 (3.0 \text{ k}\Omega)^2}{4(1.8 \text{ k}\Omega)} = 31.25 \text{ mW}$$

Refer to the Norton equivalent circuit of Figure 20–60:

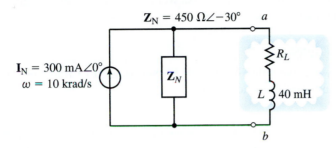

FIGURE 20–60

a. Find the value of load resistance, R_L, such that the load receives maximum power.

b. Determine the maximum power received by the load for the given conditions.

Answers: a. $R_L = 427 \, \Omega$, b. $P_L = 11.2$ W

Refer to the Thévenin equivalent circuit of Figure 20–61.

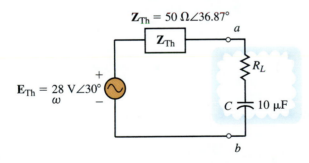

FIGURE 20–61

a. Determine the value of the unknown load resistance, R_L, which will result in a relative maximum power at an angular frequency of 1 krad/s.

b. Solve for the power dissipated by the load at $\omega_1 = 1$ krad/s.

c. Assuming that the Thévenin impedance remains constant at all frequencies, at what angular frequency, ω_2, will the circuit provide absolute maximum power?

d. Solve for the power dissipated by the load at ω_2.

(Answers are at the end of the chapter.)

ELECTRONICS WORKBENCH

PSpice

20.7 Circuit Analysis Using Computers

As demonstrated in Chapter 9, circuit analysis programs are very useful in determining the equivalent circuit between specified terminals of a dc circuit. We will use similar methods to obtain the Thévenin and Norton equivalents of ac circuits. Both Electronics Workbench and PSpice are useful in analyzing circuits with dependent sources. As we have already seen, the work in analyzing such a circuit manually is lengthy and very time consuming. In this section, you will learn how to use PSpice to find the Thévenin equivalent of a simple ac circuit. As well, we will use both programs to analyze circuits with dependent sources.

OrCAD PSpice

The following example shows how PSpice is used to find the Thévenin or Norton equivalent of an ac circuit.

EXAMPLE 20–14 Use PSpice to determine the Thévenin equivalent of the circuit in Figure 20–18. Assume that the circuit operates at a frequency $\omega = 200$ rad/s ($f = 31.83$ Hz). Compare the result to the solution of Example 20–6.

Solution We begin by using OrCAD Capture to input the circuit as shown in Figure 20–62.

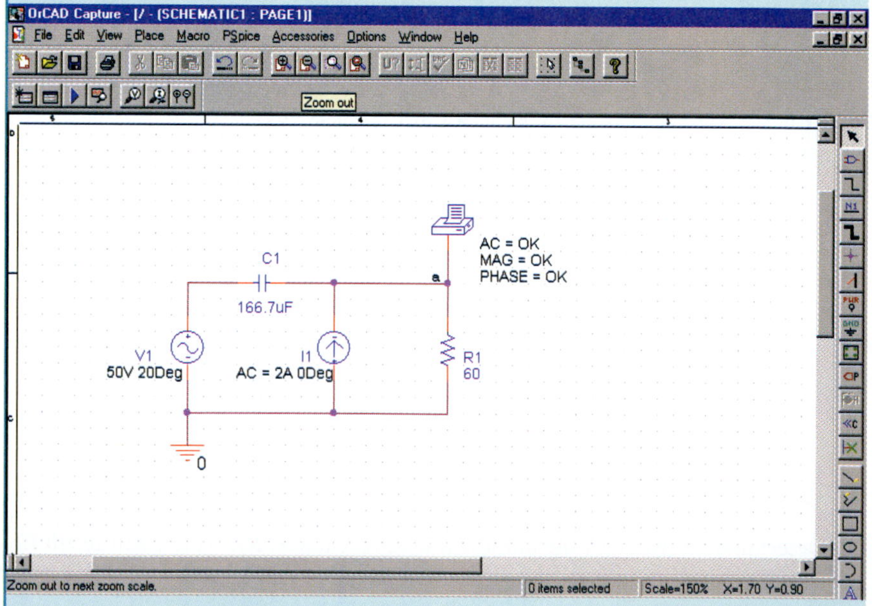

FIGURE 20–62

Notice that the load impedance has been removed and the value of capacitance is shown as

$$C = \frac{1}{\omega X_C} = \frac{1}{(200 \text{ rad/s})(30 \text{ }\Omega)} = 166.7 \text{ }\mu\text{F}$$

In order to adjust for the correct source voltage and current, we select VAC and ISRC from the source library SOURCE.slb of PSpice. The values are set for AC=**50V 20Deg** and AC=**2A 0Deg** respectively. The open-circuit output voltage is displayed using VPRINT1, which can be set to measure magnitude and phase of an ac voltage as follows. Change the properties of VPRINT1 by double clicking on the part. Use the horizontal scroll bar to find the AC, MAG, and PHASE cells. Once you have entered **OK** in each of the cells, click on Apply. Next, click on Display and select Name and Value from the display properties.

Once the circuit is entered, click on the New Simulation Profile tool and set the analysis for AC Sweep/Noise with a linear sweep beginning and ending at a frequency of **31.83Hz** (1 point). As before, it is convenient to disable the Probe postprocessor from the Probe Window tab in the simulation settings box.

As before, it is convenient to disable the Probe postprocessor prior to simulating the design. After simulating the design, the open-circuit voltage is determined by examining the output file of PSpice. The pertinent data from the output file is given as

```
FREQ           VM(a)           VP(a)
 3.183E+01     5.690E+01      -1.583E+01
```

The above result gives $\mathbf{E}_{Th} = 56.90 \text{ V}\angle -15.83°$. This is the same as the value determined in Example 20–6. Recall that one way of determining the Thévenin (or Norton) impedance is to use Ohm's law, namely

$$\mathbf{Z}_{Th} = \mathbf{Z}_N = \frac{\mathbf{E}_{Th}}{\mathbf{I}_N}$$

The Norton current is found by removing the VPRINT1 device from the circuit of Figure 20–62 and inserting an IPRINT device (ammeter) between terminal a and ground. The result is shown in Figure 20–63.

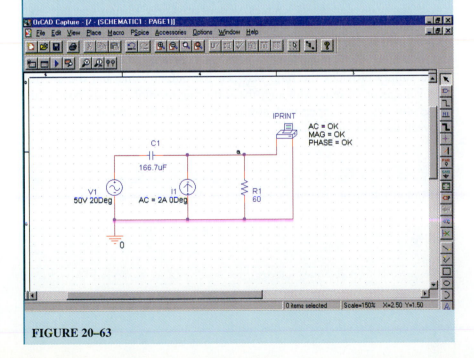

FIGURE 20–63

After simulating the design, the short-circuit current is determined by examining the output file of PSpice. The pertinent data from the output file is given as

```
FREQ          IM(V_PRINT2)IP(V_PRINT2)
3.183E+01    2.121E+00    4.761E+01
```

The above result gives $\mathbf{I}_N = 21.21 \text{ A}\angle 47.61°$ and so we calculate the Thévenin impedance as

$$\mathbf{Z}_{Th} = \frac{56.90 \text{ V}\angle -15.83°}{2.121 \text{ A}\angle 47.61°} = 26.83\Omega\angle -63.44°$$

which is the same value as that obtained in Example 20–6.

In the previous example, it was necessary to determine the Thévenin impedance in two steps, by first solving for the Thévenin voltage and then solving for the Norton current. It is possible to determine the value in one step by using an additional step in the analysis. The following example shows how to determine the Thévenin impedance for a circuit having a dependent source. The same step may also be used for a circuit having independent sources.

The following parts in OrCAD Capture are used to represent dependent sources:

Voltage-controlled voltage source: **E**

Current-controlled current source: **F**

Voltage-controlled current source: **G**

Current-controlled voltage source: **H**

When using dependent sources, it is necessary to ensure that any voltage-controlled source is placed *across* the controlling voltage and any current-controlled source is placed *in series with* the controlling current. Additionally, each dependent source must have a specified **gain.** This value simply provides the ratio between the output value and the controlling voltage or current. Although the following example shows how to use only one type of dependent source, you will find many similarities between the various sources.

EXAMPLE 20–15   Use PSpice to find the Thévenin equivalent of the circuit shown in Figure 20–46.

Solution   OrCAD Capture is used to input the circuit shown in Figure 20–64.

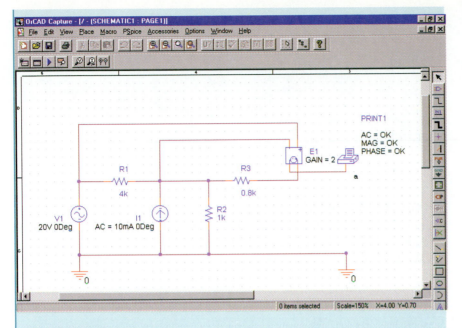

FIGURE 20–64

The voltage-controlled voltage source is obtained by clicking on the Place part tool and selecting E from the ANALOG library. Notice the placement of the source. Since R_1 is the controlling element, the controlling terminals are placed across this resistor. To adjust the gain of the voltage source, double click on the symbol. Use the scroll bar to find the GAIN cell and type **2**. In order to display this value on the schematic, you will need to click on Display and select Name and Value from the display properties.

As in the previous circuit, you will need to first use a VPRINT1 part to measure the open circuit voltage at terminal *a*. The properties of VPRINT1 are changed by typing **OK** in the AC, MAG, and PHASE cells. Click on Display and select Name and Value for each of the cells. Give the simulation profile a name and run the simulation. The pertinent data in the PSpice output file provides the open-circuit (Thévenin) voltage as follows.

```
FREQ        VM(N00431)  VP(N00431)
1.000E+03   4.000E+00   1.800E+02
```

The VPRINT1 part is then replaced with IPRINT, which is connected between terminal *a* and ground to provide the short-circuit current. Remember to change the appropriate cells using the properties editor. The PSpice output file gives the short-circuit (Norton) current as:

```
FREQ        IM(V_PRINT2)IP(V_PRINT2)
1.000E+03   1.250E-03   1.800E+02
```

The Thévenin impedance is now easily determined as

$$\mathbf{Z}_{\text{Th}} = \frac{4 \text{ V}}{1.25 \text{ mA}} = 3.2 \text{ k}\Omega$$

and the resulting circuit is shown in Figure 20–65.

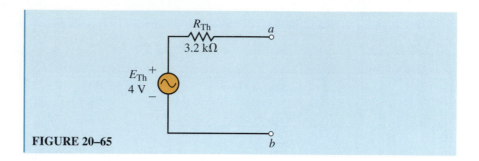

FIGURE 20–65

Electronics Workbench

Electronics Workbench has many similarities to PSpice in its analysis of ac circuits. The following example shows that the results obtained by Electronics Workbench are precisely the same as those obtained from PSpice.

EXAMPLE 20–16   Use Electronics Workbench to find the Thévenin equivalent of the circuit shown in Figure 20–47. Compare the results to those obtained in Example 20–15.

Solution   The circuit is entered as shown in Figure 20–66.

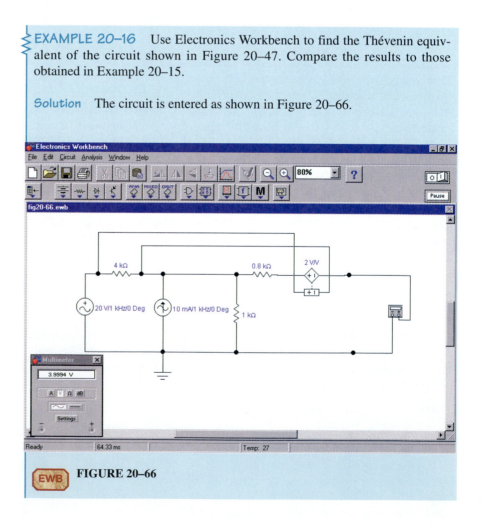

FIGURE 20–66

We use 1 kHz as the frequency of operation, although any frequency may be used. The Voltage-controlled voltage source is selected from the Sources parts bin and the gain is adjusted by double clicking on the symbol. The Value Tab is selected and the Voltage gain (E): is set to **2 V/V.** The multimeter must be set to measure ac volts. As expected, the reading on the multimeter is 4 V. Current is easily measured by setting the multimeter onto its ac ammeter range as shown in Figure 20–67. The current reading is 1.75 mA. A limitation of Electronics Workbench modeling is that the model does not indicate the phase angle of the voltage or current.

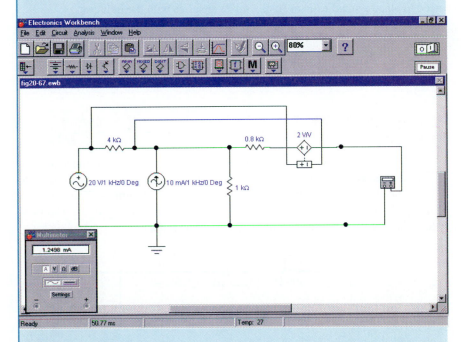

FIGURE 20–67

Now, the Thévenin impedance of the circuit is determined using Ohm's law, namely

$$\mathbf{Z}_{Th} = \frac{4 \text{ V}}{1.75 \text{ mA}} = 3.2 \text{ k}\Omega$$

These results are consistent with the calculations of Example 20–11 and the PSpice results of Example 20–15.

Use PSpice to find the Thévenin equivalent of the circuit shown in Figure 20–45. Compare your answer to that obtained in Practice Problems 5.

PRACTICE PROBLEMS 9

Answer: $\mathbf{E}_{Th} = \mathbf{V}_{ab} = -3.33 \text{ V}$, $\mathbf{Z}_{Th} = 0.667 \ \Omega$

**PRACTICE
PROBLEMS 10**

Use Electronics Workbench to find the Thévenin equivalent of the circuit shown in Figure 20–45. Compare your answer to that obtained in Practice Problems 5 and Practice Problems 9.

Answer: $\mathbf{E}_{Th} = \mathbf{V}_{ab} = -3.33$ V , $\mathbf{Z}_{Th} = 0.667\ \Omega$

**PRACTICE
PROBLEMS 11**

Use PSpice to find the Norton equivalent circuit external to the load resistor in the circuit of Figure 20–31. Assume that the circuit operates at a frequency of 20 kHz. Compare your results to those obtained in Example 20–8. Hint: You will need to place a small resistor (e.g., 1 mΩ) in series with the inductor.

Answers: $\mathbf{E}_{Th} = 7.75$ V$\angle -176.5°$ $\mathbf{I}_N = 0.1335$ A$\angle 138.5°$ $\mathbf{Z}_N = 566\Omega\angle 45°$
The results are consistent.

PUTTING IT INTO PRACTICE

I n this chapter, you learned how to solve for the required load impedance to enable maximum power transfer to the load. In all cases, you worked with load impedances that were in series with the output terminals. This will not always be the case. The circuit shown in the accompanying figure shows a load that consists of a resistor in parallel with an inductor.

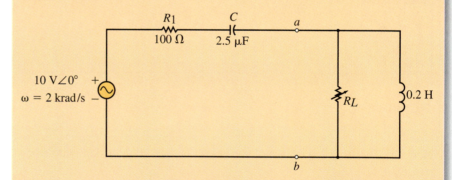

Determine the value of the resistor R_L needed to result in maximum power delivered to the load. Although several methods are possible, you may find that this example lends itself to being solved by using calculus.

PROBLEMS

20.1   Superposition Theorem—Independent Sources

1. Use superposition to determine the current in the indicated branch of the circuit in Figure 20–68.

2. Repeat Problem 1 for the circuit of Figure 20–69.

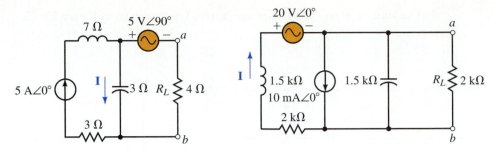

FIGURE 20–68   **FIGURE 20–69**

3. Use superposition to determine the voltage V_{ab} for the circuit of Figure 20–68.

4. Repeat Problem 3 for the circuit of Figure 20–69.

5. Consider the circuit of Figure 20–70.

  a. Use superposition to determine the indicated voltage, **V.**

  b. Show that the power dissipated by the indicated resistor cannot be determined by superposition.

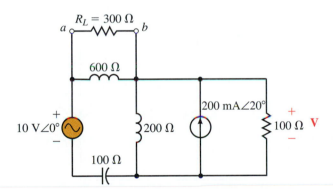

FIGURE 20–70

6. Repeat Problem 5 for the circuit of Figure 20–71.

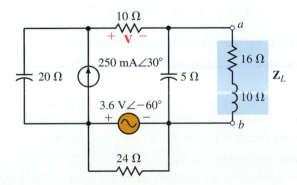

FIGURE 20–71

7. Use superposition to determine the current **I** in the circuit of Figure 20–72.

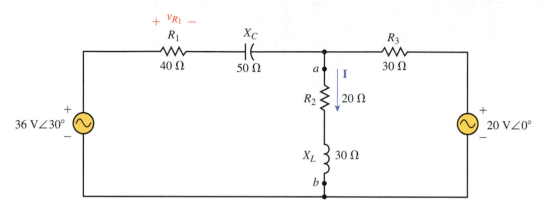

FIGURE 20–72

8. Repeat Problem 7 for the circuit of Figure 20–73.

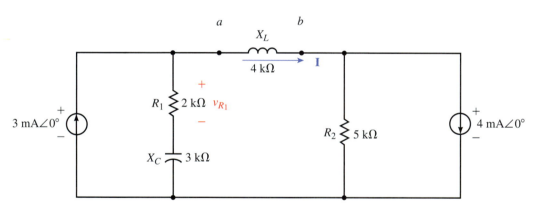

FIGURE 20–73

9. Use superposition to determine the sinusoidal voltage, v_{R_1} for the circuit of Figure 20–72.

10. Repeat Problem 9 for the circuit of Figure 20–73.

20.2   Superposition Theorem—Dependent Sources

11. Refer to the circuit of Figure 20–74.

   a. Use superposition to find $\mathbf{V}_L$.

   b. If the magnitude of the applied voltage **V** is increased to 200 mV, solve for the resulting $\mathbf{V}_L$.

12. Consider the circuit of Figure 20–75.

   a. Use superposition to find $\mathbf{V}_L$.

   b. If the magnitude of the applied current **I** is decreased to 2 mA, solve for the resulting $\mathbf{V}_L$.

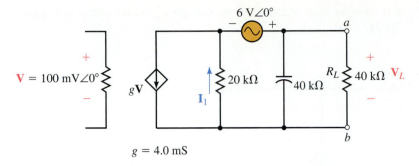

FIGURE 20–74

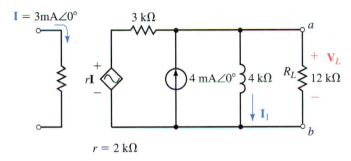

FIGURE 20–75

13. Use superposition to find the current $\mathbf{I}_1$ in the circuit of Figure 20–74.

14. Repeat Problem 13 for the circuit of Figure 20–75.

15. Use superposition to find $\mathbf{V}_L$ in the circuit of Figure 20–76.

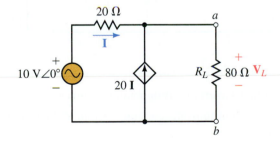

FIGURE 20–76

16. Use superposition to find $\mathbf{V}_L$ in the circuit of Figure 20–77.

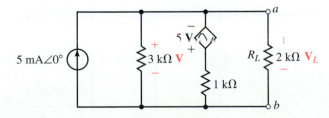

 FIGURE 20–77

17. Use superposition to determine the voltage $\mathbf{V}_{ab}$ for the circuit of Figure 20–78.

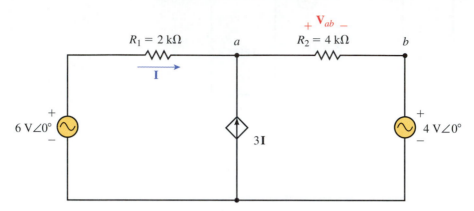

FIGURE 20–78

18. Use superposition to determine the current **I** for the circuit of Figure 20–79.

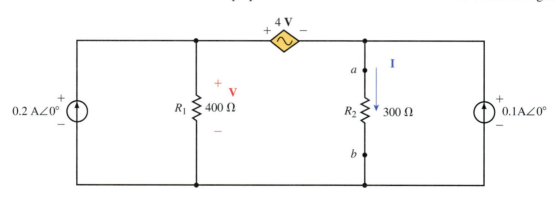

EWB **FIGURE 20–79**

20.3   Thévenin's Theorem—Independent Sources

19. Find the Thévenin equivalent circuit external to the load impedance of Figure 20–68.

20. Refer to the circuit of Figure 20–80.

   a. Find the Thévenin equivalent circuit external to the indicated load.

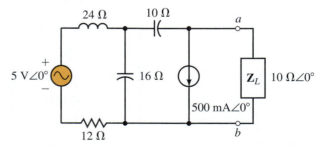

FIGURE 20–80

b. Determine the power dissipated by the load.

21. Refer to the circuit of Figure 20–81.

    a. Find the Thévenin equivalent circuit external to the indicated load at a frequency of 5 kHz.

    b. Determine the power dissipated by the load if $\mathbf{Z}_L = 100\ \Omega\angle 30°$.

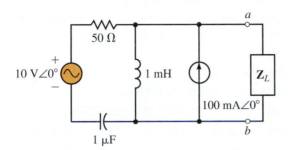

FIGURE 20–81

22. Repeat Problem 21 for a frequency of 1 kHz.

23. Find the Thévenin equivalent circuit external to R_L in the circuit of Figure 20–72.

24. Repeat Problem 23 for the circuit of Figure 20–69.

25. Repeat Problem 23 for the circuit of Figure 20–70.

26. Find the Thévenin equivalent circuit external to Z_L in the circuit of Figure 20–71.

27. Consider the circuit of Figure 20–82.

    a. Find the Thévenin equivalent circuit external to the indicated load.

    b. Determine the power dissipated by the load if $\mathbf{Z}_L = 20\ \Omega\angle -60°$.

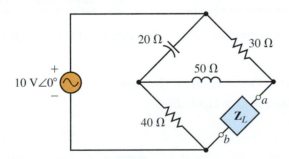

FIGURE 20–82

28. Repeat Problem 27 if a 10-Ω resistor is placed in series with the voltage source.

20.4   Norton's Theorem—Independent Sources

29. Find the Norton equivalent circuit external to the load impedance of Figure 20–68.

30. Repeat Problem 29 for the circuit of Figure 20–69.

31. a. Using the outlined procedure, find the Norton equivalent circuit external to terminals *a* and *b* in Figure 20–72.

    b. Determine the current through the indicated load.

    c. Find the power dissipated by the load.

32. Repeat Problem 31 for the circuit of Figure 20–73.

33. a. Using the outlined procedure, find the Norton equivalent circuit external to the indicated load impedance (located between terminals *a* and *b*) in Figure 20–70.

    b. Determine the current through the indicated load.

    c. Find the power dissipated by the load.

34. Repeat Problem 33 for the circuit of Figure 20–71.

35. Suppose that the circuit of Figure 20–81 operates at a frequency of 2 kHz.

    a. Find the Norton equivalent circuit external to the load impedance.

    b. If a 30-Ω load resistor is connected between terminals *a* and *b*, find the current through the load.

36. Repeat Problem 35 for a frequency of 8 kHz.

20.5   Thévenin's and Norton's Theorem for Dependent Sources

37. a. Find the Thévenin equivalent circuit external to the load impedance in Figure 20–74.

    b. Solve for the current through R_L.

    c. Determine the power dissipated by R_L.

38. a. Find the Norton equivalent circuit external to the load impedance in Figure 20–75.

    b. Solve for the current through R_L.

    c. Determine the power dissipated by R_L.

39. Find the Thévenin and Norton equivalent circuits external to the load impedance of Figure 20–76.

40. Find the Thévenin equivalent circuit external to the load impedance of Figure 20–77.

20.6   Maximum Power Transfer Theorem

41. Refer to the circuit of Figure 20–83.

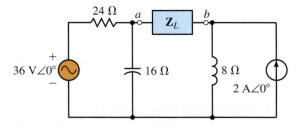

FIGURE 20–83

a. Determine the load impedance, $\mathbf{Z}_L$, needed to ensure that the load receives maximum power.

b. Find the maximum power to the load.

42. Repeat Problem 41 for the circuit of Figure 20–84.

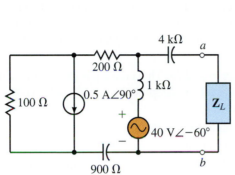

FIGURE 20–84

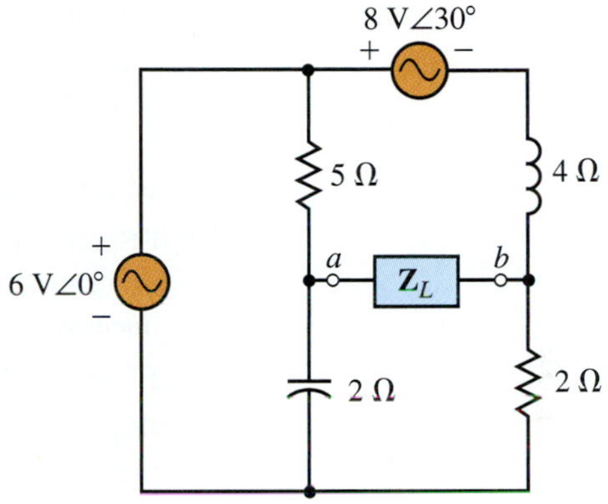

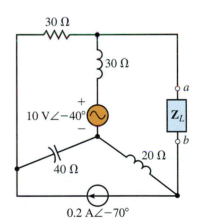

FIGURE 20–85

43. Repeat Problem 41 for the circuit of Figure 20–85.

44. Repeat Problem 41 for the circuit of Figure 20–86.

45. What load impedance is required for the circuit of Figure 20–71 to ensure that the load receives maximum power from the circuit?

46. Determine the load impedance required for the circuit of Figure 20–82 to ensure that the load receives maximum power from the circuit.

47. a. Determine the required load impedance, $\mathbf{Z}_L$, for the circuit of Figure 20–81 to deliver maximum power to the load at a frequency of 5 kHz.

b. If the load impedance contains a resistor and a 1-μF capacitor, determine the value of the resistor to result in a relative maximum power transfer.

c. Solve for the power delivered to the load in (b).

48. a. Determine the required load impedance, $\mathbf{Z}_L$, for the circuit of Figure 20–81 to deliver maximum power to the load at a frequency of 1 kHz.

b. If the load impedance contains a resistor and a 1-μF capacitor, determine the value of the resistor to result in a relative maximum power transfer.

c. Solve for the power delivered to the load in (b).

FIGURE 20–86

20.7 Circuit Analysis Using Computers

49. **PSpice** Use PSpice to find the Thévenin equivalent circuit external to R_L in the circuit of Figure 20–68. Assume that the circuit operates at a frequency of $\omega = 2000$ rad/s.

Note: PSpice does not permit a voltage source to have floating terminals. Therefore, a large resistance (e.g., 10 GΩ) must be placed across the output.

50. **PSpice** Repeat Problem 49 for the circuit of Figure 20–69.

51. **PSpice** Use PSpice to find the Norton equivalent circuit external to R_L in the circuit of Figure 20–70. Assume that the circuit operates at a frequency of $\omega = 5000$ rad/s.
   Note: PSpice cannot analyze a circuit with a short-circuited inductor. Consequently, it is necessary to place a small resistance (e.g., 1 nΩ) in series with an inductor.

52. **PSpice** Repeat Problem 51 for the circuit of Figure 20–71.
   Note: PSpice cannot analyze a circuit with an open-circuited capacitor. Consequently, it is necessary to place a large resistance (e.g., 10 GΩ) in parallel with a capacitor.

53. **PSpice** Use PSpice to find the Thévenin equivalent circuit external to R_L in the circuit of Figure 20–76. Assume that the circuit operates at a frequency of $f = 1000$ Hz.

54. **PSpice** Repeat Problem 53 for the circuit of Figure 20–77.

55. **PSpice** Use PSpice to find the Norton equivalent circuit external to $\mathbf{V}_{ab}$ in the circuit of Figure 20–78. Assume that the circuit operates at a frequency of $f = 1000$ Hz.

56. **PSpice** Repeat Problem 55 for the circuit of Figure 20–79.

57. **EWB** Use Electronics Workbench to find the Thévenin equivalent circuit external to R_L in the circuit of Figure 20–76. Assume that the circuit operates at a frequency of $f = 1000$ Hz.

58. **EWB** Repeat Problem 53 for the circuit of Figure 20–77.

59. **EWB** Use Electronics Workbench to find the Norton equivalent circuit external to $\mathbf{V}_{ab}$ in the circuit of Figure 20–78. Assume that the circuit operates at a frequency of $f = 1000$ Hz.

60. **EWB** Repeat Problem 55 for the circuit of Figure 20–79.

ANSWERS TO IN-PROCESS LEARNING CHECKS

In-Process Learning Check 1

a. $\mathbf{I} = 3.39$ A$\angle 25.34°$

b. $P_T = 230.4$ W

c. $P_1 + P_2 + P_3 = 145$ W $\neq P_T = 230.4$ W   Superposition does not apply for power.

In-Process Learning Check 2

1. Remove the inductor from the circuit. Label the remaining terminals as a and b.

2. Set the voltage source to zero by removing it from the circuit and replacing it with a short circuit.

3. Determine the values of the impedance using the given frequency. Calculate the Thévenin impedance between terminals a and b.

4. Convert the voltage source into its equivalent phasor form. Solve for the open-circuit voltage between terminals a and b.

5. Sketch the resulting Thévenin equivalent circuit.

In-Process Learning Check 3

1. Remove the resistor from the circuit. Label the remaining terminals as *a* and *b*.

2. Set the voltage source to zero by removing it from the circuit and replacing it with a short circuit.

3. Calculate the Norton impedance between terminals *a* and *b*.

4. Solve for the short-circuit current between terminals *a* and *b*.

5. Sketch the resulting Norton equivalent circuit.

In-Process Learning Check 4

a. $R_L = 80.6 \ \Omega$

b. $P_L = 3.25 \ \text{W}$

c. $\omega = 3333 \ \text{rad/s}$

d. $P_L = 4.90 \ \text{W}$

21

Resonance

OBJECTIVES

After studying this chapter, you will be able to

- determine the resonant frequency and bandwidth of a simple series or parallel circuit,
- determine the voltages, currents, and power of elements in a resonant circuit,
- sketch the impedance, current, and power response curves of a series resonant circuit,
- find the quality factor, Q, of a resonant circuit and use Q to determine the bandwidth for a given set of conditions,
- explain the dependence of bandwidth on the L/C ratio and on R for both a series and a parallel resonant circuit,
- design a resonant circuit for a given set of parameters,
- convert a series RL network into an equivalent parallel network for a given frequency.

KEY TERMS

Bandwidth
Damped Oscillations
Half-Power Frequencies
Quality Factor
Parallel Resonance
Selectivity Curve
Series Resonance

OUTLINE

Series Resonance
Quality Factor, Q
Impedance of a Series Resonant Circuit
Power, Bandwidth, and Selectivity of a Series Resonant Circuit
Series-to-Parallel RL and RC Conversion
Parallel Resonance
Circuit Analysis Using Computers

In this chapter, we build upon the knowledge obtained in previous chapters to observe how resonant circuits are able to pass a desired range of frequencies from a signal source to a load. In its most simple form, the **resonant circuit** consists of an inductor and a capacitor together with a voltage or current source. Although the circuit is simple, it is one of the most important circuits used in electronics. As an example, the resonant circuit, in one of its many forms, allows us to select a desired radio or television signal from the vast number of signals that are around us at any time.

Whereas there are various configurations of resonant circuits, they all have several common characteristics. Resonant electronic circuits contain at least one inductor and one capacitor and have a bell-shaped response curve centered at some resonant frequency, f_r, as illustrated in Figure 21–1.

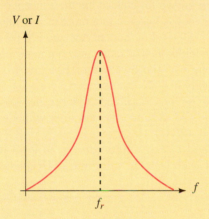

FIGURE 21–1   Response curve of a resonant circuit.

The response curve of Figure 21–1 indicates that current or voltage will be at a maximum at the resonant frequency, f_r. Varying the frequency in either direction results in a reduction of the voltage or current.

If we were to apply variable-frequency sinusoidal signals to a circuit consisting of an inductor and capacitor, we would find that maximum energy will transfer back and forth between the two elements at the resonant frequency. In an ideal LC circuit (one containing no resistance), these oscillations would continue unabated even if the signal source were turned off. However, in the practical situation, all circuits have some resistance. As a result, the stored energy will eventually be dissipated by the resistance, resulting in **damped oscillations.** In a manner similar to pushing a child on a swing, the oscillations will continue indefinitely if a small amount of energy is applied to the circuit at exactly the right moment. This phenomenon illustrates the basis of how oscillator circuits operate and therefore provides us with another application of the resonant circuit.

In this chapter, we examine in detail the two main types of resonant circuits: the **series resonant circuit** and the **parallel resonant circuit.**

Edwin Howard Armstrong—Radio Reception

EDWIN ARMSTRONG WAS BORN in New York City on December 18, 1890. As a young man, he was keenly interested in experiments involving radio transmission and reception.

After earning a degree in electrical engineering at Columbia University, Armstrong used his theoretical background to explain and improve the operation of the triode vacuum tube, which had been invented by Lee de Forest. Edwin Armstrong was able to improve the sensitivity of receivers by using feedback to amplify a signal many times. By increasing the amount of signal feedback, Armstrong also designed and patented a circuit which used the vacuum tube as an oscillator.

Armstrong is best known for conceiving the concept of superheterodyning, in which a high frequency is lowered to a more usable intermediate frequency. Superheterodyning is still used in modern AM and FM receivers and in numerous other electronic circuits such as radar and communication equipment.

Edwin Armstrong was the inventor of FM transmission, which led to greatly improved fidelity in radio transmission.

Although Armstrong was a brilliant engineer, he was an uncompromising person who was involved in numerous lawsuits with Lee de Forest and the communications giant, RCA.

After spending nearly two million dollars in legal battles, Edwin Armstrong jumped to his death from his thirteenth-floor apartment window on January 31, 1954.

21.1 Series Resonance

A simple series resonant circuit is constructed by combining an ac source with an inductor, a capacitor, and optionally, a resistor as shown in Figure 21–2a. By combining the generator resistance, R_G, with the series resistance, R_S, and the resistance of the inductor coil, R_{coil}, the circuit may be simplified as illustrated in Figure 21–2b.

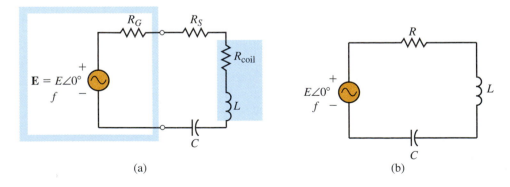

(a)   (b)

FIGURE 21–2

In this circuit, the total resistance is expressed as

$$R = R_G + R_S + R_{coil}$$

Because the circuit of Figure 21–2 is a series circuit, we calculate the total impedance as follows:

$$\mathbf{Z}_T = R + jX_L - jX_C$$
$$= R + j(X_L - X_C) \qquad (21\text{–}1)$$

Resonance occurs when the reactance of the circuit is effectively eliminated, resulting in a total impedance that is purely resistive. We know that the reactances of the inductor and capacitor are given as follows:

$$X_L = \omega L = \pi f L \qquad (21\text{–}2)$$

$$X_C = \frac{1}{\omega C} = \frac{1}{2\pi f C} \qquad (21\text{–}3)$$

Examining Equation 21–1, we see that by setting the reactances of the capacitor and inductor equal to one another, the total impedance, $\mathbf{Z}_T$, is purely resistive since the inductive reactance which is on the positive j axis cancels the capacitive reactance on the negative j axis. The total impedance of the series circuit at resonance is equal to the total circuit resistance, R. Hence, at resonance,

$$Z_T = R \qquad (21\text{–}4)$$

By letting the reactances be equal we are able to determine the series resonance frequency, ω_s (in radians per second) as follows:

$$\omega L = \frac{1}{\omega C}$$

$$\omega^2 = \frac{1}{LC} \qquad (21\text{–}5)$$

$$\omega_s = \frac{1}{\sqrt{LC}} \quad (\text{rad/s})$$

Since the calculation of the angular frequency, ω, in radians per second is easier than solving for frequency, f, in hertz, we generally express our resonant frequencies in the more simple form. Further calculations of voltage and current will usually be much easier by using ω rather than f. If, however, it becomes necessary to determine a frequency in hertz, recall that the relationship between ω and f is as follows:

$$\omega = 2\pi f \quad (\text{rad/s}) \qquad (21\text{–}6)$$

Equation 21–6 is inserted into Equation 21–5 to give the resonant frequency as

$$f_s = \frac{1}{2\pi\sqrt{LC}} \quad (\text{Hz}) \qquad (21\text{–}7)$$

The subscript s in the above equations indicates that the frequency determined is the series resonant frequency.

At resonance, the total current in the circuit is determined from Ohm's law as

$$\mathbf{I} = \frac{\mathbf{E}}{\mathbf{Z}_T} = \frac{E\angle 0°}{R\angle 0°} = \frac{E}{R}\angle 0° \qquad (21\text{–}8)$$

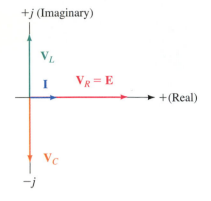

FIGURE 21–3

By again applying Ohm's law, we find the voltage across each of the elements in the circuit as follows:

$$\mathbf{V}_R = IR\angle 0° \tag{21–9}$$

$$\mathbf{V}_L = IX_L\angle 90° \tag{21–10}$$

$$\mathbf{V}_C = IX_C\angle -90° \tag{21–11}$$

The phasor form of the voltages and current is shown in Figure 21–3.

Notice that since the inductive and capacitive reactances have the same magnitude, the voltages across the elements must have the same magnitude but be 180° out of phase.

We determine the average power dissipated by the resistor and the reactive powers of the inductor and capacitor as follows:

$$P_R = I^2R \quad \text{(W)}$$
$$Q_L = I^2X_L \quad \text{(VAR)}$$
$$Q_C = I^2X_C \quad \text{(VAR)}$$

These powers are illustrated graphically in Figure 21–4.

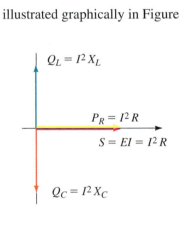

FIGURE 21–4

21.2   Quality Factor, Q

For any resonant circuit, we define the **quality factor, Q,** as the ratio of reactive power to average power, namely,

$$Q = \frac{\text{reactive power}}{\text{average power}} \tag{21–12}$$

Because the reactive power of the inductor is equal to the reactive power of the capacitor at resonance, we may express Q in terms of either reactive power. Consequently, the above expression is written as follows:

$$Q_s = \frac{I^2X_L}{I^2R}$$

and so we have

$$Q_s = \frac{X_L}{R} = \frac{\omega L}{R} \tag{21–13}$$

Quite often, the inductor of a given circuit will have a Q expressed in terms of its reactance and internal resistance, as follows:

$$Q_{coil} = \frac{X_L}{R_{coil}}$$

If an inductor with a specified Q_{coil} is included in a circuit, it is necessary to include its effects in the overall calculation of the total circuit Q.

We now examine how the Q of a circuit is used in determining other quantities of the circuit. By multiplying both the numerator and denominator of Equation 21–13 by the current, I, we have the following:

$$Q_s = \frac{IX_L}{IR} = \frac{V_L}{E} \qquad \textbf{(21–14)}$$

Now, since the magnitude of the voltage across the capacitor is equal to the magnitude of the voltage across the inductor at resonance, we see that the voltages across the inductor and capacitor are related to the Q by the following expression:

$$V_C = V_L = Q_s E \quad \text{at resonance} \qquad \textbf{(21–15)}$$

Note: Since the Q of a resonant circuit is generally significantly larger than 1, we see that the voltage across reactive elements can be many times greater than the applied source voltage. Therefore, it is always necessary to ensure that the reactive elements used in a resonant circuit are able to handle the expected voltages and currents.

EXAMPLE 21–1 Find the indicated quantities for the circuit of Figure 21–5.

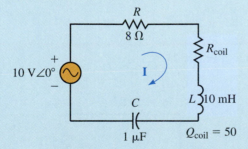

FIGURE 21–5

a. Resonant frequency expressed as ω(rad/s) and f(Hz).

b. Total impedance at resonance.

c. Current at resonance.

d. $\mathbf{V}_L$ and $\mathbf{V}_C$.

e. Reactive powers, Q_C and Q_L.

f. Quality factor of the circuit, Q_s.

Solution

a.
$$\omega_s = \frac{1}{\sqrt{LC}}$$
$$= \frac{1}{\sqrt{(10 \text{ mH})(1\,\mu\text{F})}}$$
$$= 10\ 000 \text{ rad/s}$$

$$f_s = \frac{\omega}{2\pi} = 1592 \text{ Hz}$$

b.
$$X_L = \omega L = (10\ 000 \text{ rad/s})(10 \text{ mH}) = 100\ \Omega$$
$$R_{\text{coil}} = \frac{X_L}{Q_{\text{coil}}} = \frac{100\ \Omega}{50} = 2.00\ \Omega$$
$$R_T = R + R_{\text{coil}} = 10.0\ \Omega$$
$$\mathbf{Z}_T = 10\ \Omega\angle 0°$$

c.
$$\mathbf{I} = \frac{\mathbf{E}}{\mathbf{Z}_T} = \frac{10 \text{ V}\angle 0°}{10\ \Omega\angle 0°} = 1.0 \text{ A}\angle 0°$$

d.
$$\mathbf{V}_L = (100\ \Omega\angle 90°)(1.0 \text{ A}\angle 0°) = 100 \text{ V}\angle 90°$$
$$\mathbf{V}_C = (100\ \Omega\angle -90°)(1.0 \text{ A}\angle 0°) = 100 \text{ V}\angle -90°$$

Notice that the voltage across the reactive elements is ten times greater than the applied signal voltage.

e. Although we use the symbol Q to designate both reactive power and the quality factor, the context of the question generally provides us with a clue as to which meaning to use.

$$Q_L = (1.0 \text{ A})^2(100\ \Omega) = 100 \text{ VAR}$$
$$Q_C = (1.0 \text{ A})^2(100\ \Omega) = 100 \text{ VAR}$$

f.
$$Q_s = \frac{Q_L}{P} = \frac{100 \text{ VAR}}{10 \text{ W}} = 10$$

PRACTICE PROBLEMS 1

Consider the circuit of Figure 21–6:

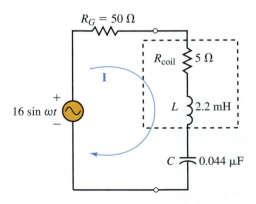

EWB   **FIGURE 21–6**

a. Find the resonant frequency expressed as ω(rad/s) and f(Hz).

b. Determine the total impedance at resonance.

c. Solve for $\mathbf{I}$, $\mathbf{V}_L$, and $\mathbf{V}_C$ at resonance.

d. Calculate reactive powers Q_C and Q_L at resonance.

e. Find the quality factor, Q_s, of the circuit.

Answers:

a. 102 krad/s, 16.2 kHz   b. 55.0 $\Omega \angle 0°$

c. 0.206 A$\angle 0°$, 46.0 V$\angle 90°$, 46.0 V$\angle -90°$ d. 9.46 VAR   e. 4.07

21.3   Impedance of a Series Resonant Circuit

In this section, we examine how the impedance of a series resonant circuit varies as a function of frequency. Because the impedances of inductors and capacitors are dependent upon frequency, the total impedance of a series resonant circuit must similarly vary with frequency. For algebraic simplicity, we use frequency expressed as ω in radians per second. If it becomes necessary to express the frequency in hertz, the conversion of Equation 21–6 is used.

The total impedance of a simple series resonant circuit is written as

$$\mathbf{Z}_T = R + j\omega L - j\frac{1}{\omega C}$$

$$= R + j\left(\frac{\omega^2 LC - 1}{\omega C}\right)$$

The magnitude and phase angle of the impedance vector, $\mathbf{Z}_T$, are expressed as follows:

$$Z_T = \sqrt{R^2 + \left(\frac{\omega^2 LC - 1}{\omega C}\right)^2} \tag{21–16}$$

$$\theta = \tan^{-1}\left(\frac{\omega^2 LC - 1}{\omega RC}\right) \tag{21–17}$$

Examining these equations for various values of frequency, we note that the following conditions will apply:

When $\omega = \omega_s$:

$$Z_T = R$$

and

$$\theta = \tan^{-1}0 = 0°$$

This result is consistent with the results obtained in the previous section.

When $\omega < \omega_s$:

As we decrease ω from resonance, Z_T will get larger until $\omega = 0$. At this point, the magnitude of the impedance will be undefined, corresponding to an open circuit. As one might expect, the large impedance occurs because the capacitor behaves like an open circuit at dc.

The angle θ will occur between of $0°$ and $-90°$ since the numerator of the argument of the arctangent function will always be negative, corresponding to an angle in the fourth quadrant. Because the angle of the impedance has a negative sign, we conclude that the impedance must appear capacitive in this region.

When $\omega > \omega_s$:

As ω is made larger than resonance, the impedance Z_T will increase due to the increasing reactance of the inductor.

For these values of ω, the angle θ will always be within $0°$ and $+90°$ because both the numerator and the denominator of the arctangent function are positive. Because the angle of $\mathbf{Z}_T$ occurs in the first quadrant, the impedance must be inductive.

Sketching the magnitude and phase angle of the impedance $\mathbf{Z}_T$ as a function of angular frequency, we have the curves shown in Figure 21–7.

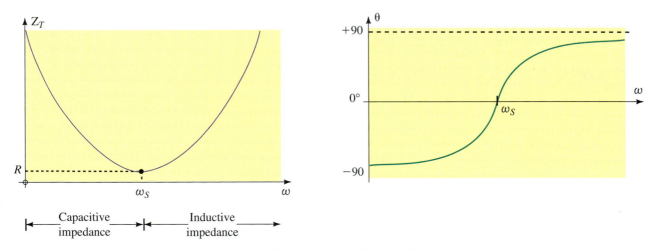

FIGURE 21–7   Impedance (magnitude and phase angle) versus angular frequency for a series resonant circuit.

21.4   Power, Bandwidth, and Selectivity of a Series Resonant Circuit

Due to the changing impedance of the circuit, we conclude that if a constant-amplitude voltage is applied to the series resonant circuit, the current and power of the circuit will not be constant at all frequencies. In this section, we examine how current and power are affected by changing the frequency of the voltage source.

Applying Ohm's law gives the magnitude of the current at resonance as follows:

$$I_{\max} = \frac{E}{R}$$ **(21–18)**

For all other frequencies, the magnitude of the current will be less than $I_{\max}$ because the impedance is greater than at resonance. Indeed, when the frequency is zero (dc), the current will be zero since the capacitor is effectively an open circuit. On the other hand, at increasingly higher frequencies, the inductor begins to approximate an open circuit, once again causing the current in the circuit to approach zero. The current response curve for a typical series resonant circuit is shown in Figure 21–8.

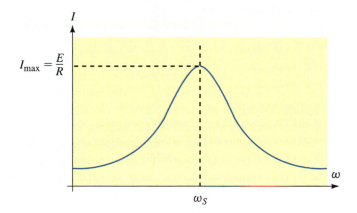

FIGURE 21–8   Current versus angular frequency for a series resonant circuit.

The total power dissipated by the circuit at any frequency is given as

$$P = I^2 R \qquad (21–19)$$

Since the current is maximum at resonance, it follows that the power must similarly be maximum at resonance. The maximum power dissipated by the series resonant circuit is therefore given as

$$P_{max} = I^2_{max} R = \frac{E^2}{R} \qquad (21–20)$$

The power response of a series resonant circuit has a bell-shaped curve called the **selectivity curve,** which is similar to the current response. Figure 21–9 illustrates the typical selectivity curve.

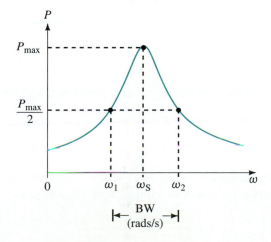

FIGURE 21–9   Selectivity curve.

Examining Figure 21–9, we see that only frequencies around ω_s will permit significant amounts of power to be dissipated by the circuit. We define the **bandwidth,** BW, of the resonant circuit to be the difference between the frequencies at which the circuit delivers half of the maximum power. The frequencies ω_1 and ω_2 are called the **half-power frequencies,** the **cutoff frequencies,** or the **band frequencies.**

If the bandwidth of a circuit is kept very narrow, the circuit is said to have a **high selectivity,** since it is highly selective to signals occuring within a very narrow range of frequencies. On the other hand, if the bandwidth of a circuit is large, the circuit is said to have a **low selectivity.**

The elements of a series resonant circuit determine not only the frequency at which the circuit is resonant, but also the shape (and hence the bandwidth) of the power response curve. Consider a circuit in which the resistance, R, and the resonant frequency, ω_s, are held constant. We find that by increasing the ratio of L/C, the sides of the power response curve become steeper. This in turn results in a decrease in the bandwidth. Inversely, decreasing the ratio of L/C causes the sides of the curve to become more gradual, resulting in an increased bandwidth. These characteristics are illustrated in Figure 21–10.

If, on the other hand, L and C are kept constant, we find that the bandwidth will decrease as R is decreased and will increase as R is increased. Figure 21–11 shows how the shape of the selectivity curve is dependent upon the value of resistance. A series circuit has the highest selectivity if the resistance of the circuit is kept to a minimum.

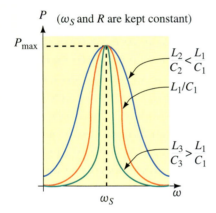

FIGURE 21–10

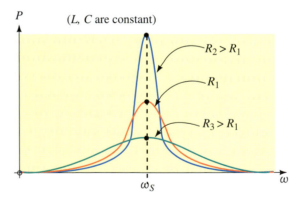

FIGURE 21–11

For the series resonant circuit the power at any frequency is determined as

$$P = I^2 R$$

$$= \left(\frac{E}{Z_T}\right)^2 R$$

By substituting Equation 21–16 into the above expression, we arrive at the general expression for power as a function of frequency, ω:

$$P = \frac{E^2 R}{R^2 + \left(\dfrac{\omega^2 LC - 1}{\omega C}\right)^2} \qquad (21\text{–}21)$$

At the half-power frequencies, the power must be

$$P_{hpf} = \frac{E^2}{2R} \qquad \text{(21–22)}$$

Since the maximum current in the circuit is given as $I_{max} = E/R$, we see that by manipulating the above expression, the magnitude of current at the half-power frequencies is

$$I_{hpf} = \sqrt{\frac{P_{hpf}}{R}} = \sqrt{\frac{E^2}{2R^2}} = \sqrt{\frac{I_{max}^2}{2}}$$

$$I_{hpf} = \frac{I_{max}}{\sqrt{2}} \qquad \text{(21–23)}$$

The cutoff frequencies are found by evaluating the frequencies at which the power dissipated by the circuit is half of the maximum power. Combining Equations 21–21 and 21–22, we have the following:

$$\frac{E^2}{2R} = \frac{E^2 R}{R^2 + \left(\dfrac{\omega^2 LC - 1}{\omega C}\right)^2}$$

$$2R^2 = R^2 + \left(\frac{\omega^2 LC - 1}{\omega C}\right)^2 \qquad \text{(21–24)}$$

$$\frac{\omega^2 LC - 1}{\omega C} = \pm R$$

$$\omega^2 LC - 1 = \pm \omega RC \quad \text{(at half-power)}$$

From the selectivity curve for a series circuit, we see that the two half-power points occur on both sides of the resonant angular frequency, ω_s.

When $\omega < \omega_s$, the term $\omega^2 LC$ must be less than 1. In this case the solution is determined as follows:

$$\omega^2 LC - 1 = -\omega RC$$

$$\omega^2 LC + \omega RC - 1 = 0$$

The solution of this quadratic equation gives the lower half-power frequency as

$$\omega_1 = \frac{-RC + \sqrt{(RC)^2 + 4LC}}{2LC}$$

or

$$\omega_1 = \frac{-R}{2L} + \sqrt{\frac{R^2}{4L^2} + \frac{1}{LC}} \qquad \text{(21–25)}$$

In a similar manner, for $\omega > \omega_s$, the upper half-power frequency is

$$\omega_2 = \frac{R}{2L} + \sqrt{\frac{R^2}{4L^2} + \frac{1}{LC}} \qquad \text{(21–26)}$$

Taking the difference between Equations 21–26 and 21–25, we find the bandwidth of the circuit as

$$BW = \omega_2 - \omega_1$$

$$= \frac{R}{2L} + \sqrt{\frac{R^2}{4L^2} + \frac{1}{LC}} - \left(-\frac{R}{2L} + \sqrt{\frac{R^2}{4L^2} + \frac{1}{LC}}\right)$$

which gives

$$\text{BW} = \frac{R}{L} \quad \text{(rad/s)} \tag{21–27}$$

If the above expression is multiplied by ω_s/ω_s we obtain

$$\text{BW} = \frac{\omega_s R}{\omega_s L}$$

and since $Q_s = \omega_s L/R$ we further simplify the bandwidth as

$$\text{BW} = \frac{\omega_s}{Q_s} \quad \text{(rad/s)} \tag{21–28}$$

Because the bandwidth may alternately be expressed in hertz, the above expression is equivalent to having

$$\text{BW} = \frac{f_s}{Q_s} \quad \text{(Hz)} \tag{21–29}$$

EXAMPLE 21–2   Refer to the circuit of Figure 21–12.

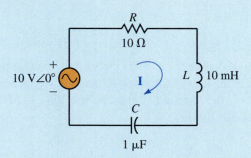

FIGURE 21–12

a. Determine the maximum power dissipated by the circuit.

b. Use the results obtained from Example 21–1 to determine the bandwidth of the resonant circuit and to arrive at the approximate half-power frequencies, ω_1 and ω_2.

c. Calculate the actual half-power frequencies, ω_1 and ω_2, from the given component values. Show two decimal places of precision.

d. Solve for the circuit current, **I**, and power dissipated at the lower half-power frequency, ω_1, found in Part (c).

Solution

a.
$$P_{\text{max}} = \frac{E^2}{R} = 10.0 \text{ W}$$

b. From Example 21–1, we had the following circuit characteristics:

$$Q_s = 10, \quad \omega_s = 10 \text{ krad/s}$$

The bandwidth of the circuit is determined to be

$$BW = \omega_s Q_s = 1.0 \text{ krad/s}$$

If the resonant frequency were centered in the bandwidth, then the half-power frequencies occur at approximately

$$\omega_1 = 9.50 \text{ krad/s}$$

and

$$\omega_2 = 10.50 \text{ krad/s}$$

c. $$\omega_1 = -\frac{R}{2L} + \sqrt{\frac{R^2}{4L^2} + \frac{1}{LC}}$$

$$= -\frac{10 \text{ }\Omega}{(2)(10 \text{ mH})} + \sqrt{\frac{(10 \text{ }\Omega)^2}{(4)(10 \text{ mH})^2} + \frac{1}{(10 \text{ mH})(1 \text{ }\mu\text{F})}}$$

$$= -500 + 10\,012.49 = 9512.49 \text{ rad/s} \quad (f_1 = 1514.0 \text{ Hz})$$

$$\omega_2 = \frac{R}{2L} + \sqrt{\frac{R^2}{4L^2} + \frac{1}{LC}}$$

$$= 500 + 10\,012.49 = 10\,512.49 \text{ rad/s} \quad (f_2 = 1673.1 \text{ Hz})$$

Notice that the actual half-power frequencies are very nearly equal to the approximate values. For this reason, if $Q \geq 10$, it is often sufficient to calculate the cutoff frequencies by using the easier approach of Part (b).

d. At $\omega_1 = 9.51249$ krad/s, the reactances are as follows:

$$X_L = \omega L = (9.51249 \text{ krad/s})(10 \text{ mH}) = 95.12 \text{ }\Omega$$

$$X_C = \frac{1}{\omega C} = \frac{1}{(9.51249 \text{ krad/s})(1 \text{ }\mu\text{F})} = 105.12 \text{ }\Omega$$

The current is now determined to be

$$\mathbf{I} = \frac{10 \text{ V}\angle 0°}{10 \text{ }\Omega + j95.12 \text{ }\Omega - j105.12 \text{ }\Omega}$$

$$= \frac{10 \text{ V}\angle 0°}{14.14 \text{ }\Omega\angle -45°}$$

$$= 0.707 \text{ A}\angle 45°$$

and the power is given as

$$P = I^2 R = (0.707 \text{ A})^2 (10 \text{ }\Omega) = 5.0 \text{ W}$$

As expected, we see that the power at the frequency ω_1 is indeed equal to half of the power dissipated by the circuit at resonance.

EXAMPLE 21–3   Refer to the circuit of Figure 21–13.

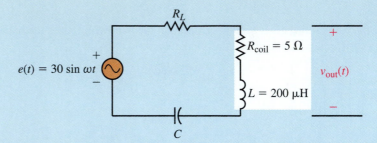

FIGURE 21–13

a. Calculate the values of R_L and C for the circuit to have a resonant frequency of 200 kHz and a bandwidth of 16 kHz.

b. Use the designed component values to determine the power dissipated by the circuit at resonance.

c. Solve for $v_{out}(t)$ at resonance.

Solution

a. Because the circuit is at resonance, we must have the following conditions:

$$Q_s = \frac{f_s}{\text{BW}}$$

$$= \frac{200 \text{ kHz}}{16 \text{ kHz}}$$

$$= 12.5$$

$$X_L = 2\pi f L$$

$$= 2\pi(200 \text{ kHz})(200 \text{ } \mu\text{H})$$

$$= 251.3 \text{ } \Omega$$

$$R = R_L + R_{coil} = \frac{X_L}{Q_s}$$

$$= 20.1 \text{ } \Omega$$

and so R_L must be

$$R_L = 20.1 \text{ } \Omega - 5 \text{ } \Omega = 15.1 \text{ } \Omega$$

Since $X_C = X_L$, we determine the capacitance as

$$C = \frac{1}{2\pi f X_C}$$

$$= \frac{1}{2\pi(200 \text{ kHz})(251.3 \text{ } \Omega)}$$

$$= 3.17 \text{ nF} \text{ } (\equiv 0.00317 \text{ } \mu\text{F})$$

b. The power at resonance is found from Equation 21–20 as

$$P_{max} = \frac{E^2}{R} = \frac{\left(\dfrac{30 \text{ V}}{\sqrt{2}}\right)^2}{20.1 \text{ } \Omega}$$

$$= 22.4 \text{ W}$$

c. We see from the circuit of Figure 21–13 that the voltage $v_{out}(t)$ may be determined by applying the voltage divider rule to the circuit. However, we must first convert the source voltage from time domain into phasor domain as follows:

$$e(t) = 30 \sin \omega t \Leftrightarrow \mathbf{E} = 21.21 \text{ V}\angle 0°$$

Now, applying the voltage divider rule to the circuit, we have

$$\mathbf{V}_{out} = \frac{(R_1 + j\omega L)}{R}\mathbf{E}$$

$$= \frac{(5\ \Omega + j251.3\ \Omega)}{20.1\ \Omega} 21.21 \text{ V}\angle 0°$$

$$= (251.4\ \Omega\angle 88.86°)(1.056 \text{ A}\angle 0°)$$

$$= 265.5 \text{ V}\angle 88.86°$$

which in time domain is given as

$$v_{out}(t) = 375 \sin(\omega t + 88.86°)$$

Refer to the circuit of Figure 21–14.

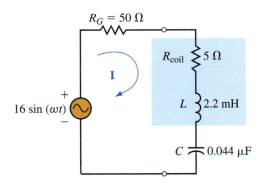

$R_G = 50\ \Omega$

R_{coil} $5\ \Omega$

$\mathbf{I}$

$16 \sin(\omega t)$ +

L 2.2 mH

C $0.044\ \mu$F

FIGURE 21–14

a. Determine the maximum power dissipated by the circuit.

b. Use the results obtained from Practice Problem 1 to determine the bandwidth of the resonant circuit. Solve for the approximate values of the half-power frequencies, ω_1 and ω_2.

c. Calculate the actual half-power frequencies, ω_1 and ω_2, from the given component values. Compare your results to those obtained in Part (b). Briefly explain why there is a discrepancy between the results.

d. Solve for the circuit current, $\mathbf{I}$, and power dissipated at the lower half-power frequency, ω_1, found in Part (c).

Answers:

a. 2.33 W

b. BW = 25.0 krad/s (3.98 kHz), $\omega_1 \cong 89.1$ krad/s, $\omega_2 \cong 114.1$ krad/s

c. $\omega_1 = 89.9$ krad/s, $\omega_2 \cong 114.9$ krad/s. The approximation assumes that the power-frequency curve is symmetrical around ω_s, which is not quite true.

d. $\mathbf{I} = 0.145$ A$\angle 45°$, $P = 1.16$ W

IN-PROCESS
LEARNING
CHECK 1

Refer to the series resonant circuit of Figure 21–15.

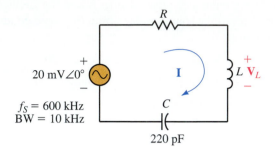

FIGURE 21–15

Suppose the circuit has a resonant frequency of 600 kHz and a bandwidth of 10 kHz:

a. Determine the value of inductor L in henries.

b. Calculate the value of resistor R in ohms.

c. Find $\mathbf{I}$, $\mathbf{V}_L$, and power, P, at resonance.

d. Find the approximate values of the half-power frequencies, f_1 and f_2.

e. Using the results of Part (d), determine the current in the circuit at the lower half-power frequency, f_1, and show that the power dissipated by the resistor at this frequency is half the power dissipated at the resonant frequency.

(Answers are at the end of the chapter.)

IN-PROCESS
LEARNING
CHECK 2

Consider the series resonant circuit of Figure 21–16:

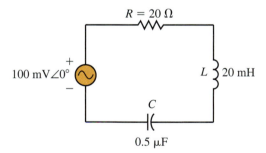

FIGURE 21–16

a. Solve for the resonant frequency of the circuit, ω_s, and calculate the power dissipated by the circuit at resonance.

b. Determine Q, BW, and the half-power frequencies, ω_1 and ω_2, in radians per second.

c. Sketch the selectivity curve of the circuit, showing P (in watts) versus ω (in radians per second).

d. Repeat Parts (a) through (c) if the value of resistance is reduced to 10 Ω.

e. Explain briefly how selectivity depends upon the value of resistance in a series resonant circuit.

(Answers are at the end of the chapter.)

21.5 Series-to-Parallel *RL* and *RC* Conversion

As we have already seen, an inductor will always have some series resistance due to the length of wire used in the coil winding. Even though the resistance of the wire is generally small in comparison with the reactances in the circuit, this resistance may occasionally contribute tremendously to the overall circuit response of a parallel resonant circuit. We begin by converting the series *RL* network as shown in Figure 21–17 into an equivalent parallel *RL* network. It must be emphasized, however, that *the equivalence is only valid at a single frequency, ω.*

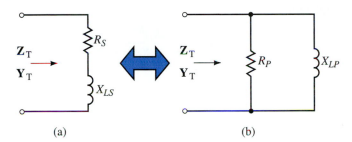

(a)   (b)

FIGURE 21–17

The networks of Figure 21–17 can be equivalent only if they each have the same input impedance, $\mathbf{Z}_T$ (and also the same input admittance, $\mathbf{Y}_T$).

The input impedance of the series network of Figure 21–17(a) is given as

$$\mathbf{Z}_T = R_S + jX_{LS}$$

which gives the input admittance as

$$\mathbf{Y}_T = \frac{1}{\mathbf{Z}_T} = \frac{1}{R_S + jX_{LS}}$$

Multiplying numerator and denominator by the complex conjugate, we have

$$\mathbf{Y}_T = \frac{R_S - jX_{LS}}{(R_S + jX_{LS})(R_S - jX_{LS})}$$

$$= \frac{R_S - jX_{LS}}{R_S^2 + X_{LS}^2} \tag{21–30}$$

$$= \frac{R_S}{R_S^2 + X_{LS}^2} - j\frac{X_{LS}}{R_S^2 + X_{LS}^2}$$

From Figure 21–17(b), we see that the input admittance of the parallel network must be

$$\mathbf{Y}_T = G_P - jB_{LP}$$

which may also be written as

$$\mathbf{Y}_T = \frac{1}{R_P} - j\frac{1}{X_{LP}} \tag{21–31}$$

The admittances of Equations 21–30 and 21–31 can only be equal if the real and the imaginary components are equal. As a result, we see that for a given frequency, the following equations enable us to convert a series *RL* network into its equivalent parallel network:

$$R_P = \frac{R_S^2 + X_{LS}^2}{R_S} \tag{21-32}$$

$$X_{LP} = \frac{R_S^2 + X_{LS}^2}{X_{LS}} \tag{21-33}$$

If we were given a parallel *RL* network, it is possible to show that the conversion to an equivalent series network is accomplished by applying the following equations:

$$R_S = \frac{R_P X_{LP}^2}{R_P^2 + X_{LP}^2} \tag{21-34}$$

$$X_{LS} = \frac{R_P^2 X_{LP}}{R_P^2 + X_{LP}^2} \tag{21-35}$$

The derivation of the above equations is left as an exercise for the student.

Equations 21–32 to 21–35 may be simplified by using the quality factor of the coil. Multiplying Equation 21–32 by R_S/R_S and then using Equation 21–13, we have

$$R_P = R_S \frac{R_S^2 + X_{LS}^2}{R_S^2}$$
$$R_P = R_S(1 + Q^2) \tag{21-36}$$

Similarly, Equation 21–33 is simplified as

$$X_{LP} = X_{LS} \frac{R_S^2 + X_{LS}^2}{X_{LS}^2}$$
$$X_{LP} = X_{LS}\left(1 + \frac{1}{Q^2}\right) \tag{21-37}$$

The quality factor of the resulting parallel network must be the same as for the original series network because the reactive and the average powers must be the same. Using the parallel elements, the quality factor is expressed as

$$Q = \frac{X_{LS}}{R_S}$$
$$= \frac{\left(\dfrac{R_P^2 X_{LP}}{R_P^2 + X_{LP}^2}\right)}{\left(\dfrac{R_P X_{LP}^2}{R_P^2 + X_{LP}^2}\right)} \tag{21-38}$$
$$= \frac{R_P^2 X_{LP}}{R_P X_{LP}^2}$$
$$Q = \frac{R_P}{X_{LP}}$$

EXAMPLE 21–4   For the series network of Figure 21–18, find the *Q* of the coil at $\omega = 1000$ rad/s and convert the series *RL* network into its equivalent parallel network. Repeat the above steps for $\omega = 10$ krad/s.

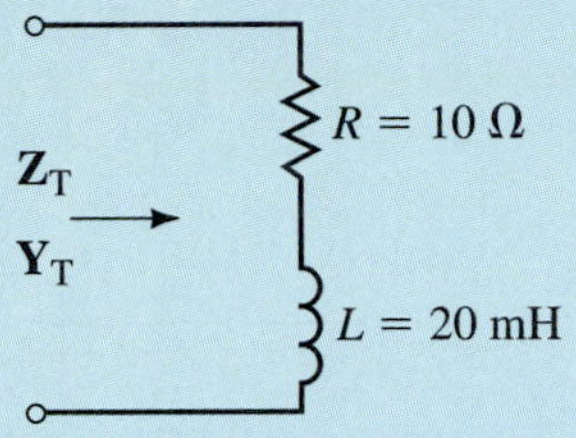

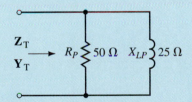

Z_T $\longrightarrow$
Y_T

$R = 10\ \Omega$

$L = 20\ \text{mH}$

FIGURE 21–18

Solution

For $\omega = 1000$ rad/s,

$$X_L = \omega L = 20\ \Omega$$

$$Q = \frac{X_{LS}}{R_S} = 2.0$$

$$R_P = R_S(1 + Q^2) = 50\ \Omega$$

$$X_{LP} = X_{LS}\left(1 + \frac{1}{Q^2}\right) = 25\ \Omega$$

The resulting parallel network for $\omega = 1000$ rad/s is shown in Figure 21–19.

Z_T $\longrightarrow$
Y_T

$R_P \gtrless 50\ \Omega$ $X_{LP} \gtrless 25\ \Omega$

FIGURE 21–19

For $\omega = 10$ krad/s,

$$X_L = \omega L = 200\ \Omega$$

$$Q = \frac{X_{LS}}{R_S} = 20$$

$$R_P = R_S(1 + Q^2) = 4010\ \Omega$$

$$X_{LP} = X_{LS}\left(1 + \frac{1}{Q^2}\right) = 200.5\ \Omega$$

The resulting parallel network for $\omega = 10$ krad/s is shown in Figure 21–20.

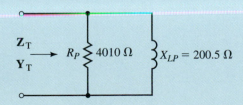

Z_T $\longrightarrow$
Y_T

$R_P \gtrless 4010\ \Omega$ $X_{LP} = 200.5\ \Omega$

FIGURE 21–20

EXAMPLE 21–5 Find the Q of each of the networks of Figure 21–21 and determine the series equivalent for each.

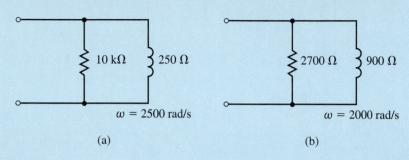

FIGURE 21–21

Solution For the network of Figure 21–21(a),

$$Q = \frac{R_P}{X_{LP}} = \frac{10 \text{ k}\Omega}{250 \text{ }\Omega} = 40$$

$$R_S = \frac{R_P}{1 + Q^2} = \frac{10 \text{ k}\Omega}{1 + 40^2} = 6.25 \text{ }\Omega$$

$$X_{LS} = QR_S = (40)(6.25 \text{ }\Omega) = 250 \text{ }\Omega$$

$$L = \frac{X_L}{\omega} = \frac{250 \text{ }\Omega}{2500 \text{ rad/s}} = 0.1 \text{ H}$$

For the network of Figure 21–21(b),

$$Q = \frac{R_P}{X_{LP}} = \frac{2700 \text{ }\Omega}{900 \text{ }\Omega} = 3$$

$$R_S = \frac{R_P}{1 + Q^2} = \frac{2700 \text{ }\Omega}{1 + 3^2} = 270 \text{ }\Omega$$

$$X_{LS} = QR_S = (3)(270 \text{ }\Omega) = 810 \text{ }\Omega$$

$$L = \frac{X_L}{\omega} = \frac{810 \text{ }\Omega}{2000 \text{ rad/s}} = 0.405 \text{ H}$$

The resulting equivalent series networks are shown in Figure 21–22.

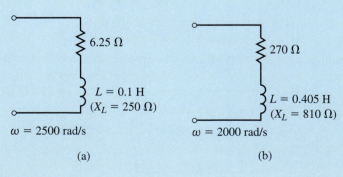

FIGURE 21–22

Refer to the networks of Figure 21–23.

100 Ω 50 mH	2 kΩ 0.4 H
(a)	(b)

FIGURE 21–23

a. Find the quality factors, Q, of the networks at $\omega_1 = 5$ krad/s.

b. Use the Q to find the equivalent parallel networks (resistance and reactance) at an angular frequency of $\omega_1 = 5$ krad/s.

c. Repeat Parts (a) and (b) for an angular frequency of $\omega_2 = 25$ krad/s.

Answers:

a. $Q_a = 2.5$ $Q_b = 1.0$

b. Network a: $R_P = 725\ \Omega$ $X_{LP} = 290\ \Omega$
    Network b: $R_P = 4\ k\Omega$ $X_{LP} = 4\ k\Omega$

c. Network a: $Q_a = 12.5$ $R_P = 15.725\ k\Omega$ $X_{LP} = 1.258\ k\Omega$
    Network b: $Q_a = 5$ $R_P = 52\ k\Omega$ $X_{LP} = 10.4\ k\Omega$

The previous examples illustrate two important points which are valid if the Q of the network is large $(Q \geq 10)$.

1. The resistance of the parallel network is approximately Q^2 larger than the resistance of the series network.

2. The inductive reactances of the series and parallel networks are approximately equal. Hence

$$R_P \cong Q^2 R_S \qquad (Q \geq 10) \qquad\qquad \textbf{(21–39)}$$

$$X_{LP} \cong X_{LS} \qquad (Q \geq 10) \qquad\qquad \textbf{(21–40)}$$

Although we have performed conversions between series and parallel *RL* circuits, it is easily shown that if the reactive element is a capacitor, the conversions apply equally well. In all cases, the equations are simply changed by replacing the terms X_{LS} and X_{LP} with X_{CS} and X_{CP} respectively. The Q of the network is determined by the ratios

$$Q = \frac{X_{CS}}{R_S} = \frac{R_P}{X_{CP}} \qquad\qquad \textbf{(21–41)}$$

**PRACTICE
PROBLEMS 4**

Consider the networks of Figure 21–24:

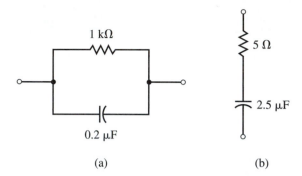

(a)   (b)

FIGURE 21–24

a. Find the Q of each network at a frequency of $f_1 = 1$ kHz.

b. Determine the series equivalent of the network in Figure 21–24(a) and the parallel equivalent of the network in Figure 21–24(b).

c. Repeat Parts (a) and (b) for a frequency of $f_2 = 200$ kHz.

Answers:

a. $Q_a = 1.26$ $Q_b = 12.7$

b. Network a: $R_S = 388\ \Omega$ $X_{CS} = 487\ \Omega$
    Network b: $R_P = 816\ \Omega$ $X_{CP} = 64.1\ \Omega$

c. Network a: $Q_a = 251$ $R_S = 0.0158\ \Omega$ $X_{CS} = 3.98\ \Omega$
    Network b: $Q_a = 0.0637$ $R_P = 5.02\ \Omega$ $X_{CP} = 78.9\ \Omega$

**IN-PROCESS
LEARNING
CHECK 3**

Refer to the networks of Figure 21–25:

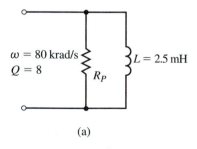   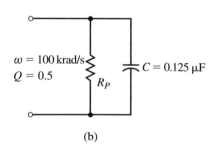

(a)   (b)

FIGURE 21–25

a. Determine the resistance, R_P, for each network.

b. Find the equivalent series network by using the quality factor for the given networks.

(Answers are at the end of the chapter.)

21.6   Parallel Resonance

A simple parallel resonant circuit is illustrated in Figure 21–26. The parallel resonant circuit is best analyzed using a constant-current source, unlike the series resonant circuit which used a constant-voltage source.

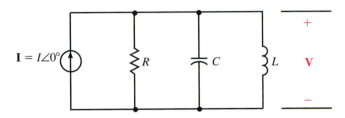

FIGURE 21–26   Simple parallel resonant circuit.

Consider the *LC* "tank" circuit shown in Figure 21–27. The tank circuit consists of a capacitor in parallel with an inductor. Due to its high *Q* and frequency response, the tank circuit is used extensively in communications equipment such as AM, FM, and television transmitters and receivers.

The circuit of Figure 21–27 is not exactly a parallel resonant circuit, since the resistance of the coil is in series with the inductance. In order to determine the frequency at which the circuit is purely resistive, we must first convert the series combination of resistance and inductance into an equivalent parallel network. The resulting circuit is shown in Figure 21–28.

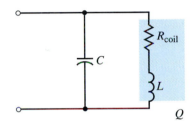

FIGURE 21–27

At resonance, the capacitive and inductive reactances of the circuit of Figure 21–28 are equal. As we have observed previously, placing equal inductive and capacitive reactances in parallel effectively results in an open circuit at the given frequency. The input impedance of this network at resonance is therefore purely resistive and given as $Z_T = R_P$. We determine the resonant frequency of a tank circuit by first letting the reactances of the equivalent parallel circuit be equal:

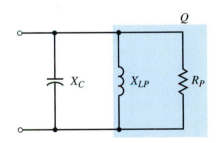

FIGURE 21–28

$$X_C = X_{LP}$$

Now, using the component values of the tank circuit, we have

$$X_C = \frac{(R_{coil})^2 + X_L^{\ 2}}{X_{LS}}$$

$$\frac{1}{\omega C} = \frac{(R_{coil})^2 + (\omega L)^2}{\omega L}$$

$$\frac{L}{C} = (R_{coil})^2 + (\omega L)^2$$

which may be further reduced to

$$\omega = \sqrt{\frac{1}{LC} - \frac{R^2}{L^2}}$$

Factoring $\sqrt{LC}$ from the denominator, we express the parallel resonant frequency as

$$\omega_P = \frac{1}{\sqrt{LC}} \sqrt{1 - \frac{(R_{coil})^2 C}{L}} \qquad\qquad \textbf{(21–42)}$$

Notice that if $R_{\text{coil}}^2 \ll L/C$, then the term under the radical is approximately equal to 1.

Consequently, if $L/C \geq 100R_{\text{coil}}$, the parallel resonant frequency may be simplified as

$$\omega_p \cong \frac{1}{\sqrt{LC}} \qquad \text{(for } L/C \geq 100R_{\text{coil}}\text{)} \qquad \textbf{(21–43)}$$

Recall that the quality factor, Q, of a circuit is defined as the ratio of reactive power to average power for a circuit at resonance. If we consider the parallel resonant circuit of Figure 21–29, we make several important observations.

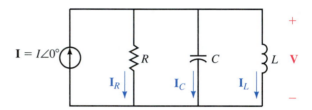

FIGURE 21–29

The inductor and capacitor reactances cancel, resulting in a circuit voltage simply determined by Ohm's law as

$$\mathbf{V} = \mathbf{IR} = IR\angle 0°$$

The frequency response of the impedance of the parallel circuit is shown in Figure 21–30.

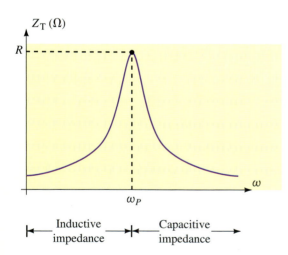

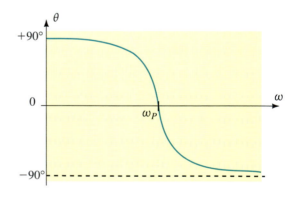

FIGURE 21–30   Impedance (magnitude and phase angle) versus angular frequency for a parallel resonant circuit.

Notice that the impedance of the entire circuit is maximum at resonance and minimum at the boundary conditions ($\omega = 0$ rad/s and $\omega \to \infty$). This result is exactly opposite to that observed in series resonant circuits which have minimum impedance at resonance. We also see that for parallel circuits, the impedance will appear inductive for frequencies less than the resonant frequency, ω_P. Inversely, the impedance is capacitive for frequencies greater than ω_P.

The Q of the parallel circuit is determined from the definition as

$$Q_P = \frac{\text{reactive power}}{\text{average power}}$$

$$= \frac{V^2/X_L}{V^2/R} \qquad (21\text{–}44)$$

$$Q_P = \frac{R_L}{X_L} = \frac{R}{X_C}$$

This is precisely the same result as that obtained when we converted an *RL* series network into its equivalent parallel network. If the resistance of the coil is the only resistance within a circuit, then the circuit Q will be equal to the Q of the coil. However, if the circuit has other sources of resistance, then the additional resistance will reduce the circuit Q.

For a parallel *RLC* resonant circuit, the currents in the various elements are found from Ohm's law as follows:

$$\mathbf{I}_R = \frac{\mathbf{V}}{\mathbf{R}} = \mathbf{I} \qquad (21\text{–}45)$$

$$\mathbf{I}_L = \frac{\mathbf{V}}{X_L \angle 90°}$$

$$= \frac{V}{R/Q_P} \angle -90° \qquad (21\text{–}46)$$

$$= Q_P I \angle -90°$$

$$\mathbf{I}_C = \frac{\mathbf{V}}{X_C \angle -90°}$$

$$= \frac{V}{R/Q_P} \angle 90° \qquad (21\text{–}47)$$

$$= Q_P I \angle 90°$$

At resonance, the currents through the inductor and the capacitor have the same magnitudes but are 180° out of phase. Notice that the magnitude of current in the reactive elements at resonance is Q times greater than the applied source current. Because the Q of a parallel circuit may be very large, we see the importance of choosing elements that are able to handle the expected currents.

In a manner similar to that used in determining the bandwidth of a series resonant circuit, it may be shown that the half-power frequencies of a parallel resonant circuit are

$$\omega_1 = \frac{1}{2RC} - \sqrt{\frac{1}{4R^2C^2} + \frac{1}{LC}} \quad \text{(rad/s)} \qquad (21\text{–}48)$$

$$\omega_2 = \frac{1}{2RC} + \sqrt{\frac{1}{4R^2C^2} + \frac{1}{LC}} \quad \text{(rad/s)} \qquad (21\text{–}49)$$

The bandwidth is therefore

$$BW = \omega_2 - \omega_1 = \frac{1}{RC} \quad \text{(rad/s)} \qquad (21\text{--}50)$$

If $Q \geq 10$, then the selectivity curve is very nearly symmetrical around ω_P, resulting in half-power frequencies which are located at $\omega_P \pm BW/2$.

Multiplying Equation 21–50 by ω_P/ω_P results in the following:

$$BW = \frac{\omega_P}{R(\omega_P C)} = \frac{X_C}{R}\omega_P$$

$$BW = \frac{\omega_P}{Q_P} \quad \text{(rad/s)} \qquad (21\text{--}51)$$

Notice that Equation 21–51 is the same for both series and parallel resonant circuits.

EXAMPLE 21–6   Consider the circuit shown in Figure 21–31.

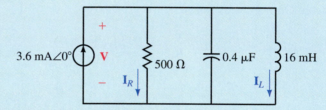

EWB   **FIGURE 21–31**

a. Determine the resonant frequencies, ω_r(rad/s) and f_r(Hz) of the tank circuit.

b. Find the Q of the circuit at resonance.

c. Calculate the voltage across the circuit at resonance.

d. Solve for currents through the inductor and the resistor at resonance.

e. Determine the bandwidth of the circuit in both radians per second and hertz.

f. Sketch the voltage response of the circuit, showing the voltage at the half-power frequencies.

g. Sketch the selectivity curve of the circuit showing P(watts) versus ω(rad/s).

Solution

a. $$\omega_P = \frac{1}{\sqrt{LC}} = \frac{1}{\sqrt{(16 \text{ mH})\,(0.4 \text{ }\mu\text{F})}} = 12.5 \text{ krad/s}$$

$$f_P = \frac{\omega}{2\pi} = \frac{12.5 \text{ krad/s}}{2\pi} = 1989 \text{ Hz}$$

b. $$Q_P = \frac{R_P}{\omega L} = \frac{500 \text{ }\Omega}{(12.5 \text{ krad/s})\,(16 \text{ mH})} = \frac{500 \text{ }\Omega}{200 \text{ }\Omega} = 2.5$$

c. At resonance, $\mathbf{V}_C = \mathbf{V}_L = \mathbf{V}_R$, and so

$$\mathbf{V} = \mathbf{IR} = (3.6 \text{ mA}\angle 0°)(500 \text{ } \Omega\angle 0°) = 1.8 \text{ V}\angle 0°$$

d.
$$\mathbf{I}_L = \frac{\mathbf{V}_L}{\mathbf{Z}_L} = \frac{1.8 \text{ V}\angle 0°}{200 \text{ } \Omega\angle 90°} = 9.0 \text{ mA}\angle -90°$$

$$\mathbf{I}_R = \mathbf{I} = 3.6 \text{ mA}\angle 0°$$

e.
$$\text{BW(rad/s)} = \frac{\omega_P}{Q_P} = \frac{12.5 \text{ krad/s}}{2.5} = 5 \text{ krad/s}$$

$$\text{BW(Hz)} = \frac{\text{BW(rad/s)}}{2\pi} = \frac{5 \text{ krad/s}}{2\pi} = 795.8 \text{ Hz}$$

f. The half-power frequencies are calculated from Equations 21–48 and 21–49 since the Q of the circuit is less than 10.

$$\omega_1 = -\frac{1}{2RC} + \sqrt{\frac{1}{4R^2C^2} + \frac{1}{LC}}$$

$$= -\frac{1}{0.0004} + \frac{1}{1.6 \times 10^{-7}} + \frac{1}{6.4 \times 10^{-9}}$$

$$= -2500 + 12\,748$$

$$= 10\,248 \text{ rad/s}$$

$$\omega_2 = \frac{1}{2RC} + \sqrt{\frac{1}{4R^2C^2} + \frac{1}{LC}}$$

$$= \frac{1}{1.0004} + \sqrt{\frac{1}{1.6 \times 10^{-7}} + \frac{1}{6.4 \times 10^{-9}}}$$

$$= 2500 + 12\,748$$

$$= 15\,248 \text{ rad/s}$$

The resulting voltage response curve is illustrated in Figure 21–32.

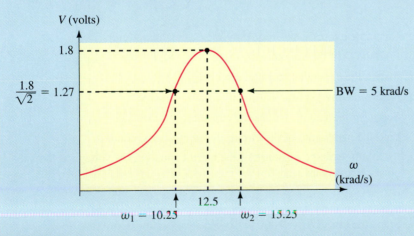

FIGURE 21–32

g. The power dissipated by the circuit at resonance is

$$P = \frac{V^2}{R} = \frac{(1.8 \text{ V})^2}{500 \text{ } \Omega} = 6.48 \text{ mW}$$

The selectivity curve is now easily sketched as shown in Figure 21–33.

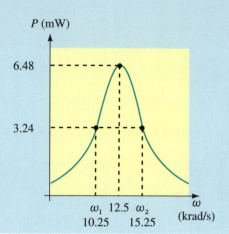

FIGURE 21–33

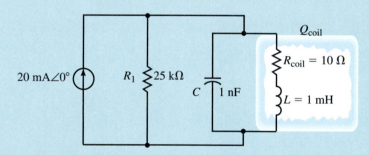

EXAMPLE 21–7   Consider the circuit of Figure 21–34.

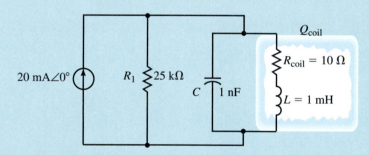

EWB   **FIGURE 21–34**

a. Calculate the resonant frequency, ω_r, of the tank circuit.

b. Find the Q of the coil at resonance.

c. Sketch the equivalent parallel circuit.

d. Determine the Q of the entire circuit at resonance.

e. Solve for the voltage across the capacitor at resonance.

f. Find the bandwidth of the circuit in radians per second.

g. Sketch the voltage response of the circuit showing the voltage at the half-power frequencies.

Solution

a. Since the ratio $L/C = 1000 \geq 100R_{coil}$, we use the approximation:

$$\omega_r = \frac{1}{\sqrt{LC}} = \frac{1}{\sqrt{(1 \text{ mH})(1 \text{ nF})}} = 1 \text{ Mrad/s}$$

b. $$Q_{coil} = \frac{\omega L}{R_{coil}} = \frac{(1 \text{ Mrad/s})(1 \text{ mH})}{10 \text{ }\Omega} = 100$$

c.
$$R_P \cong Q^2 R_{coil} = (100)^2 (10 \ \Omega) = 100 \ k\Omega$$

$$X_{LP} \cong X_{LS} = \omega L = (1 \ \text{Mrad/s}) (1 \ \text{mH}) = 1 \ k\Omega$$

The circuit of Figure 21–35 shows the circuit with the parallel equivalent of the inductor.

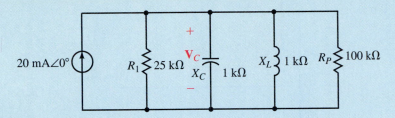

FIGURE 21–35

We see that the previous circuit may be further simplified by combining the parallel resistances:

$$R_{eq} = R_1 \| R_P = \frac{(25 \ k\Omega)(100 \ k\Omega)}{25 \ k\Omega + 100 \ k\Omega} = 20 \ k\Omega$$

The simplified equivalent circuit is shown in Figure 21–36.

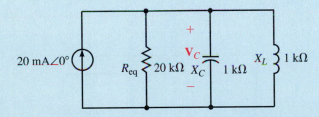

FIGURE 21–36

d.
$$Q = \frac{R_{eq}}{X_L} = \frac{20 \ k\Omega}{1 \ k\Omega} = 20$$

e. At resonance,

$$\mathbf{V}_C = \mathbf{I} R_{eq} = (20 \ \text{mA} \angle 0°) (20 \ k\Omega) = 400 \ \text{V} \angle 0°$$

f.
$$\text{BW} = \frac{\omega_r}{Q} = \frac{1 \ \text{Mrad/s}}{20} = 50 \ \text{krad/s}$$

g. The voltage response curve is shown in Figure 21–37. Since $Q \geq 10$, the half-power frequencies will occur at the following angular frequencies:

$$\omega_1 \cong \omega_r - \frac{\text{BW}}{2} = 1.0 \ \text{Mrad/s} - \frac{50 \ \text{krad/s}}{2} = 0.975 \ \text{Mrad/s}$$

and

$$\omega_2 \cong \omega_r + \frac{\text{BW}}{2} = 1.0 \ \text{Mrad/s} + \frac{50 \ \text{krad/s}}{2} = 1.025 \ \text{Mrad/s}$$

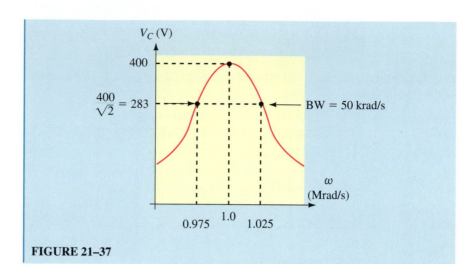

FIGURE 21–37

EXAMPLE 21–8   Determine the values of R_1 and C for the resonant tank circuit of Figure 21–38 so that the given conditions are met.

$L = 10$ mH, $R_{coil} = 30\ \Omega$

$f_P = 58$ kHz

BW = 1 kHz

Solve for the current, $\mathbf{I}_L$, through the inductor.

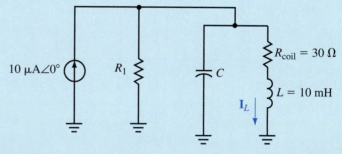

FIGURE 21–38

Solution

$$Q = \frac{f_P}{\text{BW(Hz)}} = \frac{580\text{ kHz}}{10\text{ kHz}} = 58$$

Now, because the frequency expressed in radians per second is more useful than hertz, we convert f_P to ω_P:

$$\omega_P = 2\pi f_P = (2\pi)(58\text{ kHz}) = 364.4\text{ krad/s}$$

The capacitance is determined from Equation 21–43 as

$$C = \frac{1}{\omega_P^2 L} = \frac{1}{(364.4\text{ krad/s})^2(10\text{ mH})} = 753\text{ pF}$$

Solving for the Q of the coil permits us to easily convert the series RL network into its equivalent parallel network.

$$Q_{coil} = \frac{\omega_L}{R_{coil}}$$

$$= \frac{(364.4 \text{ krad/s})(10 \text{ mH})}{30 \text{ }\Omega}$$

$$= \frac{3.644 \text{ k}\Omega}{30 \text{ }\Omega} = 121.5$$

$$R_P \cong Q_{coil}^2 R_S = (121.5)^2(30 \text{ }\Omega) = 443 \text{ k}\Omega$$

$$X_{LP} \cong X_{LS} = 3644 \text{ }\Omega$$

The resulting equivalent parallel circuit is shown in Figure 21–39.

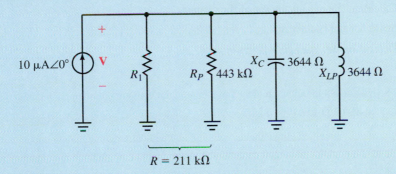

FIGURE 21–39

The quality factor, Q, is used to determine the total resistance of the circuit as

$$R = QX_C = (58)(3.644 \text{ k}\Omega) = 211 \text{ k}\Omega$$

But

$$\frac{1}{R} = \frac{1}{R_1} + \frac{1}{R_P}$$

$$\frac{1}{R_1} = \frac{1}{R} - \frac{1}{R_P} = \frac{1}{211 \text{ k}\Omega} - \frac{1}{443 \text{ k}\Omega} = 2.47 \text{ }\mu S$$

And so

$$R_1 = 405 \text{ k}\Omega$$

The voltage across the circuit is determined to be

$$\mathbf{V} = \mathbf{I}R = (10 \text{ }\mu A\angle 0°)(211 \text{ k}\Omega) = 2.11 \text{ V}\angle 0°$$

and the current through the inductor is

$$\mathbf{I}_L = \frac{\mathbf{V}}{R_{coil} + jX_L}$$

$$= \frac{2.11 \text{ V}\angle 0°}{30 + j3644 \text{ }\Omega} = \frac{2.11 \text{ V}\angle 0°}{3644 \text{ }\Omega\angle 89.95°} = 579 \text{ }\mu A\angle -89.95°$$

PRACTICE PROBLEMS 5

Refer to the circuit of Figure 21–40:

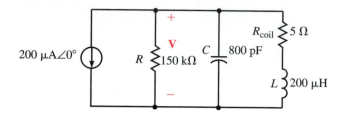

EWB **FIGURE 21–40**

a. Determine the resonant frequency and express it in radians per second and in hertz.

b. Calculate the quality factor of the circuit.

c. Solve for the bandwidth.

d. Determine the voltage **V** at resonance.

Answers:
a. 2.5 Mrad/s (398 kHz)   b. 75
c. 33.3 krad/s (5.31 kHz)   d. 7.5 V∠180°

IN-PROCESS LEARNING CHECK 4

Refer to the parallel resonant circuit of Figure 21–41:

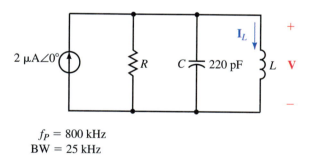

$f_P = 800$ kHz
BW $= 25$ kHz

FIGURE 21–41

Suppose the circuit has a resonant frequency of 800 kHz and a bandwidth of 25 kHz.

a. Determine the value of the inductor, *L,* in henries.

b. Calculate the value of the resistance, *R,* in ohms.

c. Find **V**, $\mathbf{I}_L$, and power, *P,* at resonance.

d. Find the appoximate values of the half-power frequencies, f_1 and f_2.

e. Determine the voltage across the circuit at the lower half-power frequency, f_1, and show that the power dissipated by the resistor at this frequency is half the power dissipated at the resonant frequency.

(Answers are at the end of the chapter.)

21.7   Circuit Analysis Using Computers

PSpice is particularly useful in examining the operation of resonant circuits. The ability of the software to provide a visual display of the frequency response is used to evaluate the resonant frequency, maximum current, and bandwidth of a circuit. The Q of the given circuit is then easily determined.

OrCAD PSpice

EXAMPLE 21–9   Use OrCAD PSpice to obtain the frequency response for current in the circuit of Figure 21–12. Use cursors to find the resonant frequency and the bandwidth of the circuit from the observed response. Compare the results to those obtained in Example 21–2.

Solution
OrCAD Capture CIS is used to input the circuit as shown in Figure 21–42. For this example, the project is titled **EXAMPLE 21-9.** The voltage source used in this example is VAC and the value is changed to AC=**10V 0Deg.** In order to obtain a plot of the circuit current, use the Current Into Pin tool as shown.

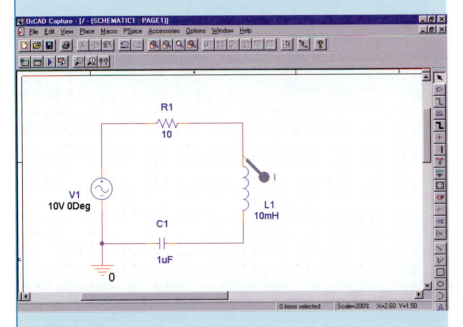

FIGURE 21–42

Next, we change the simulation settings by clicking on the New Simulation Profile tool. Give the simulation a name such as **Series Resonance** and click on Create. Once you are in the simulation settings box, click on the Analysis tab and select AC Sweep/Noise as the analysis type. The frequency can be swept either linearly or logarithmically (decade or octave). In this example we select a logarithmic sweep through a decade. In the box titled AC Sweep Type, click on ⊙Logarithmic and select Decade. Type the following values as the settings. $\underline{S}$tart Frequency: **1kHz,** $\underline{E}$nd Frequency: **10kHz,** and Points/$\underline{D}$ecade: **10001.** Click OK.

Click on the Run tool. If there are no errors, the PROBE postprocessor will run automatically and display I(L1) as a function of frequency. You will notice that the selectivity curve is largely contained within a narrow range of frequencies. We may zoom into this region as follows. Select the $\underline{P}$lot menu, and click on Axis $\underline{S}$ettings menu item. Click on the X Axis tab and select $\underline{U}$ser Defined Data Range. Change the values to **1kHz** to **3kHz.** Click OK. The resulting display is shown in Figure 21–43.

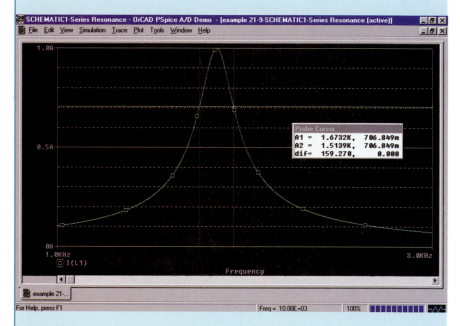

FIGURE 21–43

Finally, we may use cursors to provide us with the actual resonant frequency, the maximum current and the half-power frequencies. Cursors are obtained as follows. Click on $\underline{T}$race, $\underline{C}$ursor, and $\underline{D}$isplay. The positions of the cursors are adjusted by using either the mouse or the arrow and <Shift> keys. The current at the maximum point of the curve is obtained by clicking on $\underline{T}$race, $\underline{C}$ursor, and Ma$\underline{x}$. The dialog box provides the values of both the frequency and the value of current. The bandwidth is determined by determining the frequencies at the half-power points (when the current is 0.707 of the maximum value). We obtain the following results using the cursors:

$$I_{max} = 1.00\,\text{A}, f_S = 1.591\,\text{kHz}, f_1 = 1.514\,\text{kHz}, f_2 = 1.673\,\text{kHz}, \text{BW} = 0.159\,\text{kHz}.$$

These values correspond very closely to those calculated in Example 21–2.

EXAMPLE 21–10 Use OrCAD PSpice to obtain the frequency response for the voltage across the parallel resonant circuit of Figure 21–34. Use the PROBE postprocessor to find the resonant frequency, the maximum voltage (at resonance), and the bandwidth of the circuit. Compare the results to those obtained in Example 21–7.

Solution
This example is similar to the previous example, with minor exceptions. The OrCAD Capture program is used to enter the circuit as shown in Figure 21–44. The ac current source is found in the SOURCE library as IAC. The value of the ac current source is changed AC=**20mA 0Deg.** The Voltage Level tool is used to provide the voltage simulation for the circuit.

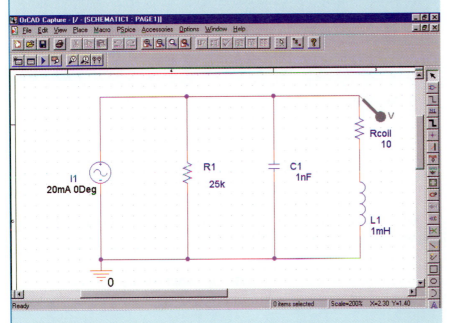

FIGURE 21–44

Use the New Simulation Profile tool to set the simulation for a logarithmic ac sweep from 100kHz to 300kHz with a total of 10001 points per decade. Click on the <u>P</u>lot menu to select the Axis <u>S</u>ettings. Change the x-axis to indicate a <u>U</u>ser Defined range from **100kHz** to **300kHz** and change the y-axis to indicate a <u>U</u>ser Defined range from **0V** to **400V.** The resulting display is shown in Figure 21–45.

As in the previous example, we use cursors to observe that the maximum circuit voltage, $V_{max} = 400$ V, occurs at the resonance frequency as $f_P = 159.2$ kHz (1.00 Mrad/s). The half-power frequencies are determined when the output voltage is at 0.707 of the maximum value, namely at $f_1 = 155.26$ kHz (0.796 Mrad/s) and $f_2 = 163.22$ kHz (1.026 Mrad/s). These frequencies give a bandwidth of BW = 7.96 kHz (50.0 krad/s). The above results are the same as those found in Example 21–7.

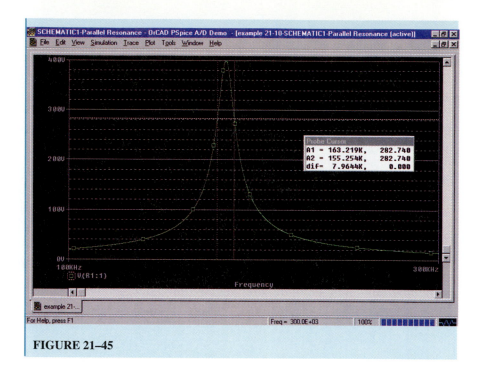

FIGURE 21–45

**PRACTICE
PROBLEMS 6**

Use OrCAD PSpice to obtain the frequency response of voltage, V versus f, for the circuit of Figure 21–40. Use cursors to determine the approximate values of the half-power frequencies and the bandwidth of the circuit. Compare the results to those obtained in Practice Problem 5.

Answers:
$V_{max} = 7.50$ V, $f_P = 398$ kHz, $f_1 = 395.3$ kHz , $f_2 = 400.7$ kHz, BW $= 5.31$ kHz

PUTTING IT INTO PRACTICE

You are the transmitter specialist at an AM commercial radio station, which transmits at a frequency of 990 kHz and an average power of 10 kW. As is the case for all commercial AM stations, the bandwidth for your station is 10 kHz. Your transmitter will radiate the power using a 50-Ω antenna. The accompanying figure shows a simplified block diagram of the output stage of the transmitter. The antenna behaves exactly like a 50-Ω resistor connected between the output of the amplifier and ground.

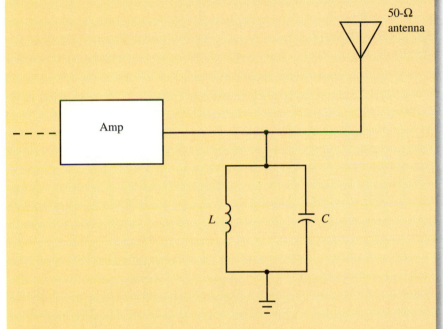

Transmitter stage of a commercial AM radio station.

You have been asked to determine the values of *L* and *C* so that the transmitter operates with the given specifications. As part of the calculations, determine the peak current that the inductor must handle and solve for the peak voltage across the capacitor. For your calculations, assume that the transmitted signal is a sinusoidal.

21.1   Series Resonance

PROBLEMS

1. Consider the circuit of Figure 21–46:

   a. Determine the resonant frequency of the circuit in both radians per second and hertz.

   b. Calculate the current, **I**, at resonance.

   c. Solve for the voltages $\mathbf{V}_R$, $\mathbf{V}_L$, and $\mathbf{V}_C$. (Notice that the voltage $\mathbf{V}_L$ includes the voltage dropped across the internal resistance of the coil.)

   d. Determine the power (in watts) dissipated by the inductor. (Hint: The power will not be zero.)

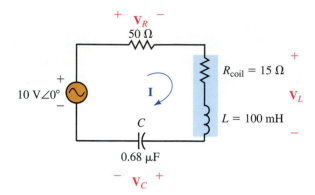

FIGURE 21–46

2. Refer to the circuit of Figure 21–47:

   a. Determine the resonant frequency of the circuit in both radians per second and hertz.

   b. Calculate the phasor current, **I.**

   c. Determine the power dissipated by the circuit at resonance.

   d. Calculate the phasor voltages, $\mathbf{V}_L$ and $\mathbf{V}_R$.

   e. Write the sinusoidal form of the voltages v_L and v_R.

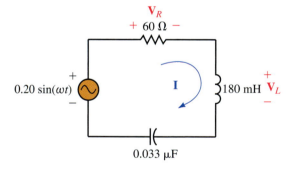

FIGURE 21–47

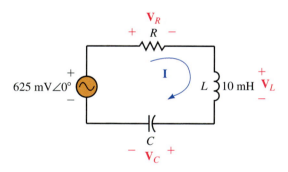

FIGURE 21–48

3. Consider the circuit of Figure 21–48:

   a. Determine the values of R and C such that the circuit has a resonant frequency of 25 kHz and an rms current of 25 mA at resonance.

   b. Calculate the power dissipated by the circuit at resonance.

   c. Determine the phasor voltages, $\mathbf{V}_C$, $\mathbf{V}_L$, and $\mathbf{V}_R$.

   d. Write the sinusoidal expressions for the voltages v_C, v_L, and v_R.

4. Refer to the circuit of Figure 21–49.

   a. Determine the capacitance required so that the circuit has a resonant frequency of 100 kHz.

   b. Solve for the phasor quantities $\mathbf{I}$, $\mathbf{V}_L$, and $\mathbf{V}_R$.

   c. Find the sinusoidal expressions for i, v_L, and v_R.

   d. Determine the power dissipated by each element in the circuit.

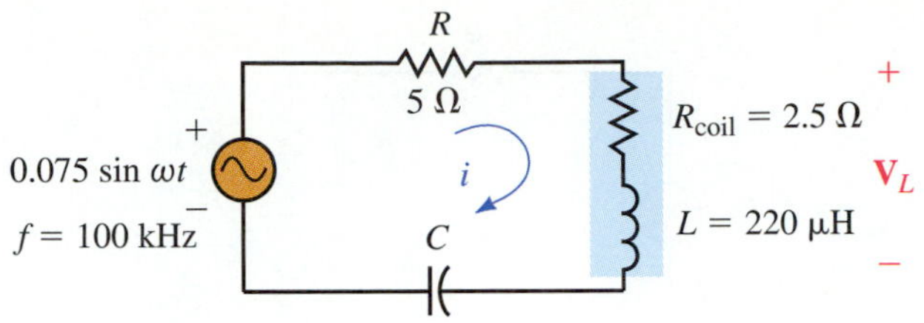

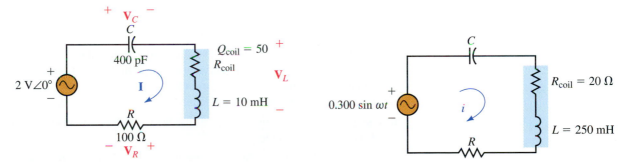

FIGURE 21–49

21.2   Quality Factor, Q

5. Refer to the circuit of Figure 21–50:

a. Determine the resonant frequency expressed as ω(rad/s) and f(Hz).

b. Calculate the total impedance, $\mathbf{Z}_T$, at resonance.

c. Solve for current $\mathbf{I}$ at resonance.

d. Solve for $\mathbf{V}_R$, $\mathbf{V}_L$, and $\mathbf{V}_C$ at resonance.

e. Calculate the power dissipated by the circuit and evaluate the reactive powers, Q_C and Q_L.

f. Find the quality factor, Q_S, of the circuit.

FIGURE 21–50   FIGURE 21–51

6. Suppose that the circuit of Figure 21–51 has a resonant frequency of $f_S = 2.5$ kHz and a quality factor of $Q_S = 10$:

a. Determine the values of R and C.

b. Solve for the quality factor of inductor, Q_{coil}.

c. Find $\mathbf{Z}_T$, $\mathbf{I}$, $\mathbf{V}_C$, and $\mathbf{V}_R$ at resonance.

d. Solve for the sinusoidal expression of current i at resonance.

e. Calcuate the sinusoidal expressions v_C and v_R at resonance.

f. Calculate the power dissipated by the circuit and determine the reactive powers, Q_C and Q_L.

7. Refer to the circuit of Figure 21–52:

a. Design the circuit to have a resonant frequency of $\omega = 50$ krad/s and a quality factor $Q_S = 25$.

b. Calculate the power dissipated by the circuit at the resonant frequency.

c. Determine the voltage, $\mathbf{V}_L$, across the inductor at resonance.

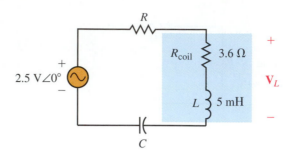

FIGURE 21–52

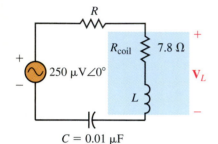

FIGURE 21–53

8. Consider the circuit of Figure 21–53:

   a. Design the circuit to have a resonant frequency of $\omega = 400$ krad/s and a quality factor $Q_S = 10$.

   b. Calculate the power dissipated by the circuit at the resonant frequency.

   c. Determine the voltage, $\mathbf{V}_L$, across the inductor at resonance.

21.3   Impedance of a Series Resonant Circuit

9. Refer to the series resonant circuit of Figure 21–54.

   a. Determine the resonant frequency, ω_S.

   b. Solve for the input impedance, $\mathbf{Z}_T = Z\angle\theta$, of the circuit at frequencies of $0.1\omega_S$, $0.2\omega_S$, $0.5\omega_S$, ω_S, $2\omega_S$, $5\omega_S$, and $10\omega_S$.

   c. Using the results from (b), sketch a graph of Z (magnitude in ohms) versus ω (in radians per second) and a graph of θ (in degrees) versus ω (in radians per second). If possible, use log-log graph paper for the former and semilog graph paper for the latter.

   d. Using your results from (b), determine the magnitude of current at each of the given frequencies.

   e. Use the results from (d) to plot a graph of I (magnitude in amps) versus ω (in radians per second) on log-log graph paper.

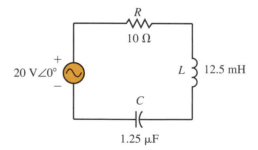

FIGURE 21–54

10. Repeat Problem 9 if the 10-Ω resistor is replaced with a 50-Ω resistor.

21.4   Power, Bandwidth, and Selectivity of a Series Resonant Circuit

11. Refer to the circuit of Figure 21–55.

   a. Find ω_S, Q, and BW (in radians per second).

   b. Calculate the maximum power dissipated by the circuit.

c. From the results obtained in (a) solve for the approximate half-power frequencies, ω_1 and ω_2.

d. Calculate the actual half-power frequencies, ω_1 and ω_2 using the component values and the appropriate equations.

e. Are the results obtained in (c) and (d) comparable? Explain.

f. Solve for the circuit current, **I**, and power dissipated at the lower half-
   power frequency, ω_1, determined in (d).

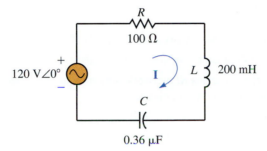

FIGURE 21–55

12. Repeat Problem 11 for the circuit of Figure 21–56.

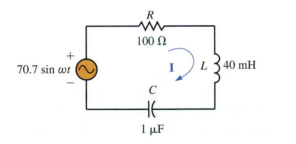

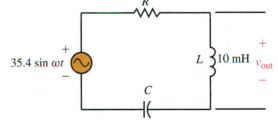

FIGURE 21–56   FIGURE 21–57

13. Consider the circuit of Figure 21–57.

   a. Calculate the values of R and C for the circuit to have a resonant fre-
      quency of 200 kHz and a bandwidth of 16 kHz.

   b. Use the designed component values to determine the power dissipated by
      the circuit at resonance.

   c. Solve for v_{out} at resonance.

14. Repeat Problem 13 if the resonant frequency is to be 580 kHz and the band-
    width is 10 kHz.

21.5   Series-to-Parallel *RL* and *RC* Conversion

15. Refer to the series networks of Figure 21–58.

   a. Find the Q of each network at $\omega = 1000$ rad/s.

   b. Convert each series *RL* network into an equivalent parallel network, hav-
      ing R_P and X_{LP} in ohms.

   c. Repeat (a) and (b) for $\omega = 10$ krad/s.

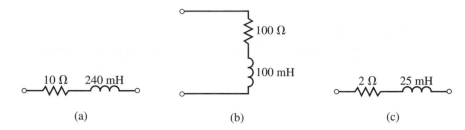

FIGURE 21–58

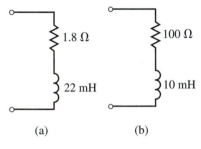

FIGURE 21–59

16. Consider the series networks of Figure 21–59.
    a. Find the Q of each coil at $\omega = 20$ krad/s.
    b. Convert each series RL network into an equivalent parallel network consisting of R_P and X_{LP} in ohms.
    c. Repeat (a) and (b) for $\omega = 100$ krad/s.

17. For the series networks of Figure 21–60, find the Q and convert each network into its parallel equivalent.

300 Ω 20 Ω	90 Ω 45 Ω	2500 Ω 500 Ω
$\omega = 9$ krad/s	$\omega = 100$ rad/s	$\omega = 377$ rad/s
(a)	(b)	(c)

FIGURE 21–60

18. Derive Equations 21–34 and 21–35, which enable us to convert a parallel RL network into its series equivalent. (Hint: Begin by determining the expression for the input impedance of the parallel network.)

19. Find the Q of each of the networks of Figure 21–61 and determine the series equivalent of each. Express all component values in ohms.

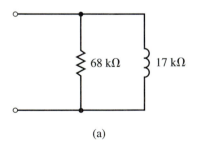

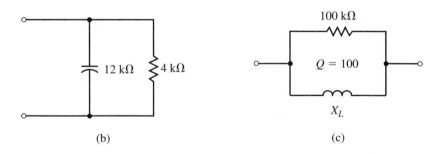

FIGURE 21–61

20. Repeat Problem 19 for the networks of Figure 21–62.

21. Determine the values of L_S and L_P in henries, given that the networks of Figure 21–63 are equivalent at a frequency of 250 krad/s.

22. Determine the values of C_S and C_P in farads, given that the networks of Figure 21–64 are equivalent at a frequency of 48 krad/s.

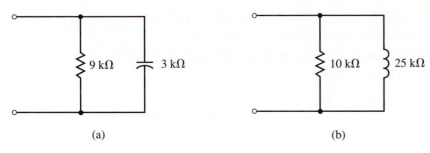

(a)

(b)

(c)

FIGURE 21–62

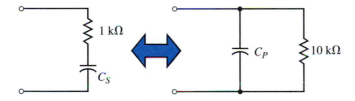

FIGURE 21–63

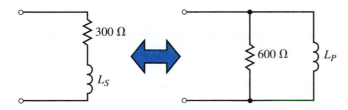

FIGURE 21–64

21.6   Parallel Resonance

23. Consider the circuit of Figure 21–65.

    a. Determine the resonant frequency, ω_P, in radians per second.

    b. Solve for the input impedance, $\mathbf{Z}_T = Z\angle\theta$, of the circuit at frequencies of $0.1\omega_P$, $0.2\omega_P$, $0.5\omega_P$, ω_P, $2\omega_P$, $5\omega_P$, and $10\omega_P$.

    c. Using the results obtained in (b), sketch graphs of Z (magnitude in ohms) versus ω (in radians per second) and θ (in degrees) versus ω. If possible, use log-log graph paper for the former and semilog for the latter.

    d. Using the results from (b), determine the voltage $\mathbf{V}$ at each of the indicated frequencies.

    e. Sketch a graph of magnitude V versus ω on log-log graph paper.

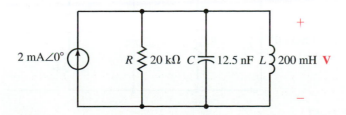

FIGURE 21–65

24. Repeat Problem 23 if the 20-kΩ resistor is replaced with a 40-kΩ resistor.

25. Refer to the circuit shown in Figure 21–66.

   a. Determine the resonant frequencies, ω_P(rad/s) and f_P(Hz).

   b. Find the Q of the circuit.

   c. Calculate $\mathbf{V}$, $\mathbf{I}_R$, $\mathbf{I}_L$, and $\mathbf{I}_C$ at resonance.

   d. Determine the power dissipated by the circuit at resonance.

   e. Solve for the bandwidth of the circuit in both radians per second and hertz.

   f. Sketch the voltage response of the circuit, showing the voltage at the half-power frequencies.

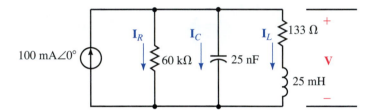

FIGURE 21–66

26. Repeat Problem 25 for the circuit of Figure 21–67.

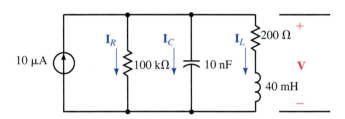

FIGURE 21–67

27. Determine the values of R_1 and C for the resonant tank circuit of Figure 21–68 so that the given conditions are met. Solve for current $\mathbf{I}_L$ through the inductor.

$L = 25$ mH, $R_{\text{coil}} = 100$ Ω

$f_P = 50$ kHz

BW $= 10$ kHz

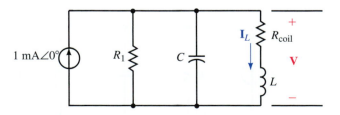

FIGURE 21–68

28. Determine the values of R_1 and C for the resonant circuit of Figure 21–68 so that the given conditions are met. Solve for the voltage, **V**, across the circuit.

    $L = 50$ mH, $R_{coil} = 50\ \Omega$

    $\omega_P = 100$ krad/s

    BW = 10 krad/s

29. Refer to the circuit of Figure 21–69.

    a. Determine the value of X_L for resonance.

    b. Solve for the Q of the circuit.

    c. If the circuit has a resonant frequency of 2000 rad/s, what is the bandwidth of the circuit?

    d. What must be the values of C and L for the circuit to be resonant at 2000 rad/s?

    e. Calculate the voltage $\mathbf{V}_C$ at resonance.

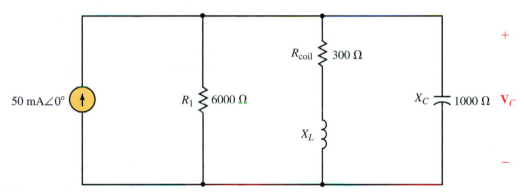

FIGURE 21–69

30. Repeat Problem 29 for the circuit of Figure 21–70.

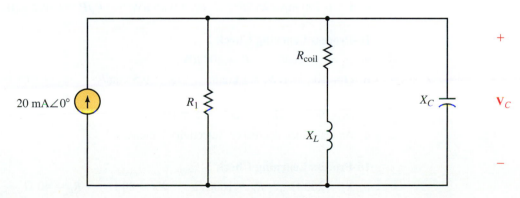

FIGURE 21–70

21.7   Circuit Analysis Using Computers

31. **PSpice**   Use OrCAD PSpice to input the circuit of Figure 21–55. Use the Probe postprocessor to display the response of the inductor voltage as a function of frequency. From the display, determine the maximum rms voltage, the resonant frequency, the half-power frequencies, and the bandwidth. Use the results to determine the quality factor of the circuit.

32. **PSpice**   Repeat Problem 31 for the circuit of Figure 21–56.

33. **PSpice**   Use OrCAD PSpice to input the circuit of Figure 21–66. Use the Probe postprocessor to display the response of the capacitor voltage as a function of frequency. From the display, determine the maximum rms voltage, the resonant frequency, the half-power frequencies, and the bandwidth. Use the results to determine the quality factor of the circuit.

34. **PSpice**   Repeat Problem 33 for the circuit of Figure 21–67.

35. **PSpice**   Use the values of C and L determined in Problem 29 to input the circuit of Figure 21–69. Use the Probe postprocessor to display the response of the capacitor voltage as a function of frequency. From the display, determine the maximum rms voltage, the resonant frequency, the half-power frequencies, and the bandwidth. Use the results to determine the quality factor of the circuit.

36. **PSpice**   Use the values of C and L determined in Problem 30 to input the circuit of Figure 21–70. Repeat the measurements of Problem 35.

ANSWERS TO IN-PROCESS LEARNING CHECKS

In-Process Learning Check 1

a. $L = 320 \ \mu H$

b. $R = 20.1 \ \Omega$

c. $\mathbf{I} = 0.995 \ mA\angle0°$ $\mathbf{V}_L = 1.20 \ V\angle90°$ $P = 20.0 \ \mu W$

d. $f_1 = 595 \ kHz$ $f_2 = 605 \ kHz$

e. $\mathbf{I} = 0.700 \ mA\angle45.28°$ $P_1 = 9.85 \ \mu W$ $P_1/P = 0.492 \cong 0.5$

In-Process Learning Check 2

a. $\omega_S = 10 \ krad/s$ $P = 500 \ \mu W$

b. $Q = 10$ $BW = 1 \ krad/s$ $\omega_1 = 9.5 \ krad/s$ $\omega_2 = 10.5 \ krad/s$

d. $\omega_S = 10 \ krad/s$ $P = 1000 \ \mu W$ $Q = 20$ $BW = 0.5 \ krad/s$
   $\omega_1 = 9.75 \ krad/s$ $\omega_2 = 10.25 \ krad/s$

e. As resistance decreases, selectivity increases.

In-Process Learning Check 3

a. $X_L = 200 \ \Omega$ $R_P = 1600 \ \Omega$ $X_C = 80 \ \Omega$ $R_P = 40 \ \Omega$

b. $X_{LS} = 197 \ \Omega$ $R_S = 24.6 \ \Omega$ $X_{CS} = 16 \ \Omega$ $R_S = 32 \ \Omega$

In-Process Learning Check 4

a. $L = 180 \ \mu H$

b. $R_p = 28.9 \ k\Omega$

c. $\mathbf{V} = 57.9 \ mV\angle 0°$ $\mathbf{I}_L = 64.0 \ \mu A\angle -90°$ $P = 115 \ nW$

d. $f_1 = 788 \ kHz$ $f_2 = 813 \ kHz$

e. $\mathbf{V} = 41.1 \ mV\angle 44.72°$ $P = 58 \ nW$

22

Filters and the Bode Plot

OBJECTIVES

After studying this chapter you will be able to

- evaluate the power gain and voltage gain of a given system,
- express power gain and voltage gain in decibels,
- express power levels in dBm and voltage levels in dBV and use these levels to determine power gain and voltage gain,
- identify and design simple (first-order) *RL* and *RC* low-pass and high-pass filters and explain the principles of operation of each type of filter,
- write the standard form of a transfer function for a given filter. The circuits which are studied will include band-pass and band-stop as well as low- and high-pass circuits,
- compute τ_c and use the time constant to determine the cutoff frequency(ies) in both radians per second and hertz for the transfer function of any first-order filter,
- sketch the Bode plot showing the frequency response of voltage gain and phase shift of any first-order filter,
- use PSpice to verify the operation of any first-order filter circuit.

KEY TERMS

Amplifier
Attenuator
Bode Plots
Cutoff Frequency
Decibels
Filters
Transfer Functions

OUTLINE

The Decibel
Multistage Systems
Simple *RC* and *RL* Transfer Functions
The Low-Pass Filter
The High-Pass Filter
The Band-Pass Filter
The Band-Reject Filter
Circuit Analysis Using Computers

In the previous chapter we examined how LRC resonant circuits react to changes in frequency. In this chapter we will continue to study how changes in frequency affect the behavior of other simple circuits. We will analyze simple low-pass, high-pass, band-pass, and band-reject filter circuits. The analysis will compare the amplitude and phase shift of the output signal with respect to the input signal.

As their names imply, low-pass and high-pass filter circuits are able to pass low frequencies and high frequencies while blocking other frequency components. A good understanding of these filters provides a basis for understanding why circuits such as amplifiers and oscilloscopes are not able to pass all signals from their input to their output.

Band-reject filters and band-pass filters are circuits which are similar to resonant circuits with the exception that these circuits are able to reject or pass certain frequencies without the need for LC combinations.

The analysis of all filters may be simplified by plotting the output/input voltage relationship on a semilogarithmic graph called a *Bode plot*.

Alexander Graham Bell

ALEXANDER GRAHAM BELL was born in Edinburgh, Scotland on March 3, 1847. As a young man, Bell followed his father and grandfather in research dealing with the deaf.

In 1873, Bell was appointed professor of vocal physiology at Boston University. His research was primarily involved in converting sound waves into electrical fluctuations. With the encouragement of Joseph Henry, who had done a great deal of work with inductors, Bell eventually developed the telephone.

In his now-famous accident in which Bell spilled acid on himself, Bell uttered the words "Watson, please come here. I want you." Watson, who was on another floor, ran to Bell's assistance.

Although others had worked on the principle of the telephone, Alexander Graham Bell was awarded the patent for the telephone in 1876. The telephone he constructed was a simple device which passed a current through carbon powder. The density of the carbon powder was determined by air fluctuations due to the sound of a person's voice. When the carbon was compressed, resistivity would decrease, allowing more current.

Bell's name has been adopted for the decibel, which is the unit used to describe sound intensities and power gain.

Although the invention of the telephone made Bell wealthy, he continued experimenting in electronics, air conditioning, and animal breeding. Bell died at the age of 75 in Baddeck, Nova Scotia on August 2, 1922.

PUTTING IT IN PERSPECTIVE

22.1   The Decibel

In electronics, we often wish to consider the effects of a circuit without examining the actual operation of the circuit itself. This **black-box** approach is a common technique used to simplify transistor circuits and for depicting integrated circuits which may contain hundreds or even thousands of elements. Consider the system shown in Figure 22–1.

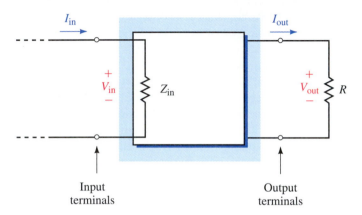

FIGURE 22–1

Although the circuit within the box may contain many elements, any source connected to the input terminals will effectively see only the input impedance, Z_{in}. Similarly, any load impedance, R_L, connected to the output terminals will have voltage and current determined by certain parameters in the circuit. These parameters usually result in an output at the load which is easily predicted for certain conditions.

We now define several terms which are used to analyze any system having two input terminals and two output terminals.

The **power gain,** A_P, is defined as the ratio of output signal power to the input signal power:

$$A_P = \frac{P_{out}}{P_{in}} \tag{22–1}$$

We must emphasize that the total output power delivered to any load can never exceed the total input power to a circuit. When we refer to the power gain of a system, we are interested in only the power contained in the ac signal, so we neglect any power due to dc. In many circuits, the ac power will be significantly less than the dc power. However, ac power gains in the order of tens of thousands are quite possible.

The **voltage gain,** A_v, is defined as the ratio of output signal voltage to the input signal voltage:

$$A_v = \frac{V_{out}}{V_{in}} \tag{22–2}$$

As mentioned, the power gain of a system may be a very large. For other applications the output power may be much smaller than the input power, resulting in a loss, or **attenuation.** Any circuit in which the output signal power is greater than the input signal power is referred to as an **amplifier.**

Conversely, any circuit in which the output signal power is less than the input signal power is referred to as an **attenuator.**

The ratios expressing power gain or voltage gain may be either very large or very small, making it inconvenient to express the power gain as a simple ratio of two numbers. The **bel,** which is a logarithmic unit named after Alexander Graham Bell, was selected to represent a ten-fold increase or decrease in power. Stated mathematically, the power gain in bels is given as

$$A_{P(bels)} = \log_{10} \frac{P_{out}}{P_{in}}$$

Because the bel is an awkwardly large unit, the **decibel** (dB), which is one tenth of a bel, has been adopted as a more acceptable unit for describing the logarithmic change in power levels. One bel contains 10 decibels and so the power gain in decibels is given as

$$A_{P(dB)} = 10 \log_{10} \frac{P_{out}}{P_{in}} \qquad\qquad (22\text{–}3)$$

If the power level in the system increases from the input to the output, then the gain in dB will be positive. If the power at the output is less than the power at the input, then the power gain will be negative. Notice that if the input and output have the same power levels, then the power gain will be 0 dB, since log 1 = 0.

EXAMPLE 22–1   An amplifier has the indicated input and output power levels. Determine the power gain both as a ratio and in dB for each of the conditions:

a. $P_{in} = 1$ mW, $P_{out} = 100$ W.

b. $P_{in} = 4\ \mu W$, $P_{out} = 2\ \mu W$.

c. $P_{in} = 6$ mW, $P_{out} = 12$ mW.

d. $P_{in} = 25$ mW, $P_{out} = 2.5$ mW.

Solution

a.
$$A_P = \frac{P_{out}}{P_{in}} = \frac{100\ W}{1\ mW} = 100\ 000$$
$$A_{P(dB)} = 10 \log_{10}(100\ 000) = (10)(5) = 50\ dB$$

b.
$$A_P = \frac{P_{out}}{P_{in}} = \frac{2\ \mu W}{4\ \mu W} = 0.5$$
$$A_{P(dB)} = 10 \log_{10}(0.5) = (10)(-0.30) = -3.0\ dB$$

c.
$$A_P = \frac{P_{out}}{P_{in}} = \frac{12\ mW}{6\ mW} = 2$$
$$A_{P(dB)} = 10 \log_{10}2 = (10)(.30) = 3.0\ dB$$

d.
$$A_P = \frac{P_{out}}{P_{in}} = \frac{2.5\ mW}{25\ mW} = 0.10$$
$$A_{P(dB)} = 10 \log_{10}0.10 = (10)(-1) = -10\ dB$$

The previous example illustrates that if the power is increased or decreased by a factor of two, the resultant power gain will be either $+3$ dB or -3 dB, respectively. You will recall that in the previous chapter a similar reference was made when the half-power frequencies of a resonant circuit were referred to as the 3-dB down frequencies.

The voltage gain of a system may also be expressed in dB. In order to derive the expression for voltage gain, we first assume that the input resistance and the load resistance are the same value. Then, using the definition of power gain as given by Equation 22–3, we have the following:

$$A_{P(\text{dB})} = 10 \log_{10} \frac{P_{\text{out}}}{P_{\text{in}}}$$

$$= 10 \log_{10} \frac{V_{\text{out}}^2/R}{V_{\text{in}}^2/R}$$

$$= 10 \log_{10} \left(\frac{V_{\text{out}}}{V_{\text{in}}} \right)^2$$

which gives

$$A_{P(\text{dB})} = 20 \log_{10} \frac{V_{\text{out}}}{V_{\text{in}}}$$

Because the above expression represents a decibel equivalent of the voltage gain, we simply write the voltage gain in dB as follows:

$$A_{v(\text{dB})} = 20 \log_{10} \frac{V_{\text{out}}}{V_{\text{in}}} \qquad\qquad (22\text{–}4)$$

EXAMPLE 22–2   The amplifier circuit of Figure 22–2 has the given conditions. Calculate the voltage gain and power gain in dB.

$Z_{\text{in}} = 10\text{ k}\Omega$

$R_L = 600\text{ }\Omega$

$V_{\text{in}} = 20\text{ mV}_{\text{rms}}$

$V_{\text{out}} = 500\text{ mV}_{\text{rms}}$

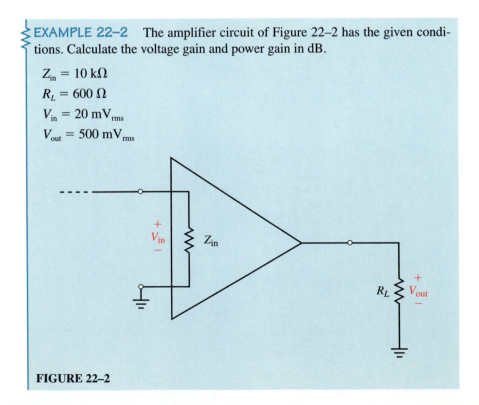

FIGURE 22–2

Solution   The voltage gain of the amplifier is

$$A_v = 20 \log_{10} \frac{V_{out}}{V_{in}}$$

$$= 20 \log_{10} \frac{500 \text{ mV}}{20 \text{ mV}} = 20 \log_{10} 25 = 28.0 \text{ dB}$$

The signal power available at the input of the amplifier is

$$P_{in} = \frac{V_{in}^2}{Z_{in}} = \frac{(20 \text{ mV})^2}{10 \text{ k}\Omega} = 0.040 \ \mu\text{W}$$

The signal at the output of the amplifier has a power of

$$P_{out} = \frac{V_{out}^2}{R_L} = \frac{(500 \text{ mV})^2}{600 \ \Omega} = 416.7 \ \mu\text{W}$$

The power gain of the amplifier is

$$A_P = 10 \log_{10} \frac{P_{out}}{P_{in}}$$

$$= 10 \log_{10} \frac{416.7 \ \mu\text{W}}{0.040 \ \mu\text{W}}$$

$$= 10 \log_{10}(10\ 417) = 40.2 \text{ dB}$$

Calculate the voltage gain and power gain (in dB) for the amplifier of Figure 22–2, given the following conditions:

PRACTICE PROBLEMS 1

$$Z_{in} = 2 \text{ k}\Omega$$
$$R_L = 5 \ \Omega$$
$$V_{in} = 16 \ \mu\text{V}_{rms}$$
$$V_{out} = 32 \text{ mV}_{rms}$$

Answers: 66.0 dB, 92.0 dB

In order to convert a gain from decibels into a simple ratio of power or voltage, it is necessary to perform the inverse operation of the logarithm; namely, solve the unknown quantity by using the exponential. Recall that the following logarithmic and exponential operations are equivalent:

$$y = \log_b x$$
$$x = b^y$$

Using the above expressions, Equations 22–3 and 22–4 may be used to determine expressions for power and voltage gains as follows:

$$\frac{A_{P(dB)}}{10} = \log_{10} \frac{P_{out}}{P_{in}}$$

$$\frac{P_{out}}{P_{in}} = 10^{A_{P(dB)}/10} \qquad\qquad (22\text{–}5)$$

$$\frac{A_{v(dB)}}{20} = \log_{10}\frac{V_{out}}{V_{in}}$$

$$\frac{V_{out}}{V_{in}} = 10^{A_{v(dB)}/20} \qquad\qquad \textbf{(22–6)}$$

EXAMPLE 22–3   Convert the following from decibels to ratios:

a. $A_P = 25$ dB.

b. $A_P = -6$ dB.

c. $A_v = 10$ dB.

d. $A_v = -6$ dB.

Solution

a.
$$A_P = \frac{P_{out}}{P_{in}} = 10^{A_{P(dB)}/10}$$
$$= 10^{25/10} = 316$$

b.
$$A_P = 10^{A_{P(dB)}/10}$$
$$= 10^{-6/10} = 0.251$$

c.
$$A_v = \frac{V_{out}}{V_{in}} = 10^{A_{v(dB)}/20}$$
$$= 10^{10/20} = 3.16$$

d.
$$A_v = 10^{A_{v(dB)}/20}$$
$$= 10^{-6/20} = 0.501$$

Applications of Decibels

Decibels were originally intended as a measure of changes in acoustical levels. The human ear is not a linear instrument; rather, it responds to sounds in a logarithmic fashion. Because of this peculiar phenomenon, a ten-fold increase in sound intensity results in a perceived doubling of sound. This means that if we wish to double the sound heard from a 10-W power amplifier, we must increase the output power to 100 W.

The minimum sound level which may be detected by the human ear is called the **threshold of hearing** and is usually taken to be $I_0 = 1 \times 10^{-12}$ W/m^2. Table 22–1 lists approximate sound intensities of several common sounds. The decibel levels are determined from the expression

$$\beta(dB) = 10 \log_{10}\frac{I}{I_0}$$

Some electronic circuits operate with very small power levels. These power levels may be referenced to some arbitrary level and then expressed in decibels in a manner similar to how sound intensities are represented. For

TABLE 22–1   Intensity Levels of Common Sounds

Sound	Intensity Level (dB)	Intensity (W/m²)
threshold of hearing, I_0	0	10^{-12}
virtual silence	10	10^{-11}
quiet room	20	10^{-10}
watch ticking at 1 m	30	10^{-9}
quiet street	40	10^{-8}
quiet conversation	50	10^{-7}
quiet motor at 1 m	60	10^{-6}
busy traffic	70	10^{-5}
door slamming	80	10^{-4}
busy office room	90	10^{-3}
jackhammer	100	10^{-2}
motorcycle	110	10^{-1}
loud indoor rock concert	120	1
threshold of pain	130	10

instance, power levels may be referenced to a standard power of 1 mW. In such cases the power level is expressed in dBm and is determined as

$$P_{dBm} = 10 \log_{10}\frac{P}{1\ mW} \qquad (22\text{–}7)$$

If a power level is referenced to a standard of 1 W, then we have

$$P_{dBW} = 10 \log_{10}\frac{P}{1\ W} \qquad (22\text{–}8)$$

EXAMPLE 22–4   Express the following powers in dBm and in dBW.

a. $P_1 = 0.35\ \mu W$.

b. $P_2 = 20\ mW$.

c. $P_3 = 1000\ W$.

d. $P_4 = 1\ pW$.

Solution

a.
$$P_{1(dBm)} = 10 \log_{10}\frac{0.35\ \mu W}{1\ mW} = -34.6\ dBm$$

$$P_{1(dBW)} = 10 \log_{10}\frac{0.35\ \mu W}{1\ W} = -64.6\ dBW$$

b.
$$P_{2(dBm)} = 10 \log_{10}\frac{20\ mW}{1\ mW} = 13.0\ dBm$$

$$P_{2(dBW)} = 10 \log_{10}\frac{20\ mW}{1\ W} = -17.0\ dBW$$

c.
$$P_{3(dBm)} = 10 \log_{10}\frac{1000\text{ W}}{1\text{ mW}} = 60 \text{ dBm}$$

$$P_{3(dBW)} = 10 \log_{10}\frac{1000\text{ W}}{1\text{ W}} = 30 \text{ dBW}$$

d.
$$P_{4(dBm)} = 10 \log_{10}\frac{1\text{ pW}}{1\text{ mW}} = -90 \text{ dBm}$$

$$P_{4(dBW)} = 10 \log_{10}\frac{1\text{ pW}}{1\text{ W}} = -120 \text{ dBW}$$

Many voltmeters have a separate scale calibrated in decibels. In such cases, the voltage expressed in dBV uses 1 V_{rms} as the reference voltage. In general, any voltage reading may be expressed in dBV as follows:

$$V_{dBV} = 20 \log_{10}\frac{V_{out}}{1\text{ V}}$$

(22–9)

PRACTICE
PROBLEMS 2

Consider the resistors of Figure 22–3:

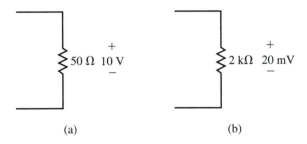

(a)   (b)

FIGURE 22–3

a. Determine the power levels in dBm and in dBW.

b. Express the voltages in dBV.

Answers:
a. 33.0 dBm (3.0 dBW), −37.0 dBm (−67.0 dBW)

b. 20.0 dBV, −34.0 dBV

22.2   Multistage Systems

Quite often a system consists of several stages. In order to find the total voltage gain or power gain of the system, we would need to solve for the product of the individual gains. The use of decibels makes the solution of a multistage system easy to find. If the gain of each stage is given in decibels, then the resultant gain is simply determined as the summation of the individual gains.

Consider the system of Figure 22–4, which represents a three-stage system.

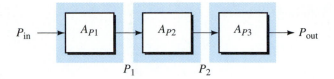

FIGURE 22–4

The power at the output of each stage is determined as follows:

$$P_1 = A_{P1} P_{in}$$
$$P_2 = A_{P2} P_1$$
$$P_{out} = A_{P3} P_2$$

The total power gain of the system is found as

$$\begin{aligned}
A_{PT} &= \frac{P_{out}}{P_{in}} = \frac{A_{P3} P_2}{P_{in}} \\
&= \frac{A_{P3} (A_{P2} P_1)}{P_{in}} \\
&= \frac{A_{P3} A_{P2} (A_{P1} P_i)}{P_{in}} \\
&= A_{P1} A_{P2} A_{P3}
\end{aligned}$$

In general, for *n* stages the total power gain is found as the product:

$$A_{PT} = A_{P1} A_{P2} \cdots A_{Pn} \qquad \textbf{(22–10)}$$

However, if we use logarithms to solve for the gain in decibels, we have the following:

$$\begin{aligned}
A_{PT}(dB) &= 10 \log_{10} A_{PT} \\
&= 10 \log_{10} (A_{P1} A_{P2} \cdots A_{Pn}) \\
A_{PT}(dB) &= 10 \log_{10} A_{P1} + 10 \log_{10} A_{P2} + \cdots + 10 \log_{10} A_{Pn}
\end{aligned}$$

The total power gain in decibels for *n* stages is determined as the summation of the individual decibel power gains:

$$A_{P(dB)} = A_{P1(dB)} + A_{P2(dB)} + \cdots + A_{Pn(dB)} \qquad \textbf{(22–11)}$$

The advantage of using decibels in solving for power gains and power levels is illustrated in the following example.

EXAMPLE 22–5 The circuit of Figure 22–5 represents the first three stages of a typical AM or FM receiver.

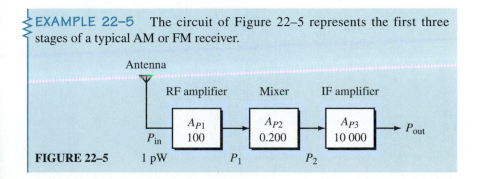

FIGURE 22–5

Find the following quantities:

a. $A_{P1(dB)}$, $A_{P2(dB)}$, and $A_{P3(dB)}$.

b. $A_{PT(dB)}$.

c. P_1, P_2, and P_{out}.

d. $P_{in(dBm)}$, $P_{1(dBm)}$, $P_{2(dBm)}$, and $P_{out(dBm)}$.

Solution

a. $A_{P1(dB)} = 10 \log_{10} A_{P1} = 10 \log_{10} 100 = 20 \text{ dB}$

$A_{P2(dB)} = \log_{10} A_{P2} = 10 \log_{10} 0.2 = -7.0 \text{ dB}$

$A_{P3(dB)} = 10 \log_{10} A_{P3} = 10 \log_{10}(10\ 000) = 40 \text{ dB}$

b. $A_{PT(dB)} = A_{P1(dB)} + A_{P2(dB)} + A_{P3(dB)}$

$= 20 \text{ dB} - 7.0 \text{ dB} + 40 \text{ dB}$

$= 53.0 \text{ dB}$

c. $P_1 = A_{P1}P_{in} = (100)(1 \text{ pW}) = 100 \text{ pW}$

$P_2 = A_{P2}P_1 = (100 \text{ pW})(0.2) = 20 \text{ pW}$

$P_{out} = A_{P3}P_2 = (10\ 000)(20 \text{ pW}) = 0.20\ \mu\text{W}$

d. $P_{in(dBm)} = 10 \log_{10}\dfrac{P_{in}}{1 \text{ mW}} = 10 \log_{10}\dfrac{1 \text{ pW}}{1 \text{ mW}} = -90 \text{ dBm}$

$P_{1(dBm)} = 10 \log_{10}\dfrac{P_1}{1 \text{ mW}} = 10 \log_{10}\dfrac{100 \text{ pW}}{1 \text{ mW}} = -70 \text{ dBm}$

$P_{2(dBm)} = 10 \log_{10}\dfrac{P_2}{1 \text{ mW}} = 10 \log_{10}\dfrac{20 \text{ pW}}{1 \text{ mW}} = -77.0 \text{ dBm}$

$P_{out(dBm)} = 10 \log_{10}\dfrac{P_{out}}{1 \text{ mW}} = 10 \log_{10}\dfrac{0.20\ \mu\text{W}}{1 \text{ mW}} = -37.0 \text{ dBm}$

Notice that the power level (in dBm) at the output of any stage is easily determined as the sum of the input power level (in dBm) and the gain of the stage (in dB). It is for this reason that many communication circuits express power levels in decibels rather than in watts.

PRACTICE PROBLEMS 3

Calculate the power level at the output of each of the stages in Figure 22–6.

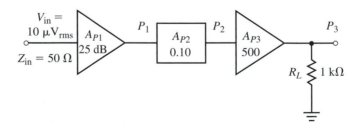

FIGURE 22–6

Answers: $P_1 = -62 \text{ dBm}$, $P_2 = -72 \text{ dBm}$, $P_3 = -45 \text{ dBm}$

1. Given that an amplifier has a power gain of 25 dB, calculate the output power level (in dBm) for the following input characteristics:

   a. $V_{in} = 10$ mV$_{rms}$, $Z_{in} = 50$ Ω.

   b. $V_{in} = 10$ mV$_{rms}$, $Z_{in} = 1$ kΩ.

   c. $V_{in} = 400$ μV$_{rms}$, $Z_{in} = 200$ Ω.

2. Given amplifiers with the following output characteristics, determine the output voltage (in volts rms).

   a. $P_{out} = 8.0$ dBm, $R_L = 50$ Ω.

   b. $P_{out} = -16.0$ dBm, $R_L = 2$ kΩ.

   c. $P_{out} = -16.0$ dBm, $R_L = 5$ kΩ.

IN-PROCESS
LEARNING
CHECK 1

(Answers are at the end of the chapter.)

22.3 Simple *RC* and *RL* Transfer Functions

Electronic circuits usually operate in a highly predictable fashion. If a certain signal is applied at the input of a system, the output will be determined by the physical characteristics of the circuit. The frequency of the incoming signal is one of the many physical conditions which determine the relationship between a given input signal and the resulting output. Although outside the scope of this text, other conditions that may determine the relationship between the input and output signals of a given circuit are temperature, light, radiation, etc.

For any system subjected to a sinusoidal input voltage, as shown in Figure 22–7, we define the **transfer function** as the ratio of the output voltage phasor to the input voltage phasor for any frequency ω (in radians per second).

$$\mathbf{TF}(\omega) = \frac{\mathbf{V}_{out}}{\mathbf{V}_{in}} = A_v \angle \theta \qquad (22\text{–}12)$$

Notice that the definition of transfer function is almost the same as voltage gain. The difference is that the transfer function takes into account both amplitude and phase shift of the voltages, whereas voltage gain is a comparison of only the amplitudes.

From Equation 22–12 we see that the amplitude of the transfer function is in fact the voltage gain. The phase angle, θ, represents the phase shift between the input and output voltage phasors. The angle θ will be positive if the output leads the input and negative if the output lags the input waveform.

If the elements within the block of Figure 22–7 are resistors, then the output and the input voltages will always be in phase. Also, since resistors have the same value at all frequencies, the voltage gain will remain constant at all frequencies. (This text does not take into account resistance variations due to very high frequencies.) The resulting circuit is called an attenuator since resistance within the block will dissipate some power, thereby reducing (or attenuating) the signal as it passes through the circuit.

If the elements within the block are combinations of resistors, inductors, and capacitors, then the output voltage and phase will depend on frequency

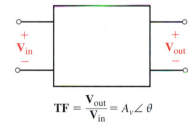

$$\mathbf{TF} = \frac{\mathbf{V}_{out}}{\mathbf{V}_{in}} = A_v \angle \theta$$

FIGURE 22–7

since the impedances of inductors and capacitors are frequency dependent. Figure 22–8 illustrates an example of how voltage gain and phase shift of a circuit may change as a function of frequency.

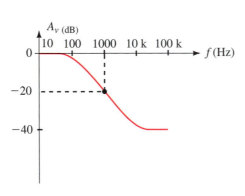

(a) Voltage gain as a function of frequency.

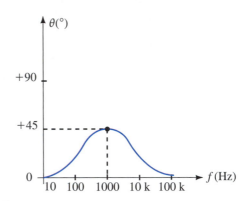

(b) Phase shift as a function of frequency.

FIGURE 22–8 Frequency response of a circuit.

In order to examine the operation of the circuit over a wide range of frequencies, the abscissa (horizontal axis) is usually shown as a logarithmic scale. The ordinate (vertical axis) is usually shown as a linear scale in decibels or degrees. Such graphs are said to be semilogarithmic since one scale is logarithmic and the other is linear.

By examining the frequency response of a circuit, we are able to determine at a glance the voltage gain (in dB) and phase shift (in degrees) for any sinusoidal input at a given frequency. For example, at a frequency of 1000 Hz, the voltage gain is −20 dB and the phase shift is 45°. This means that the output signal is one tenth as large as the input signal and the output leads the input by 45°.

From the frequency response of Figure 22–8, we see that the circuit which corresponds to this response is able to pass low-frequency signals, while at the same time partially attenuating high-frequency signals. Any circuit which passes a particular range of frequencies while blocking others is referred to as a **filter** circuit. Filter circuits are usually named according to their function, although certain filters are named after their inventors. The filter having the response of Figure 22–8 is referred to as a stepfilter, since the voltage gain occurs between two limits (steps). Other types of filters which are used regularly in electrical and electronic circuits include low-pass, high-pass, band-pass, and band-reject filters.

Although the design of filters is a topic in itself, we will examine a few of the more common types of filters which are used extensively.

Sketching Transfer Functions

Transfer functions are easily sketched on semilogarithmic graph paper by following a few basic steps. Consider a transfer function given as

$$\mathbf{TF} = \frac{1}{1 + j0.01\omega}$$

Several important characteristics are evident by examining the transfer function above. The magnitude of the transfer function (voltage gain) will be less than 1 for all frequencies other than zero hertz (dc). As the frequency increases, the voltage gain will decrease. For this reason, the circuit which has the above transfer function is called a **low-pass filter.** Finally, we see that the output voltage will lag the input voltage at all frequencies other than at zero hertz.

We define the **cutoff frequency,** ω_c (also called the **break frequency**), as the frequency at which the imaginary component of $1 + j0.01\omega$ is equal to the real component. This results in a voltage gain which is equal to

$$A_v = \frac{1}{\sqrt{2}} = 0.7071$$

From the transfer function, we see that the cutoff frequency occurs when the angular frequency $\omega = \omega_c = 1/0.01 = 100$ rad/s. At the cutoff frequency, the voltage gain (in dB) is $A_v = -3.0$ dB and the phase shift is $\theta = -45°$. To determine the cutoff frequency in hertz, we simply perform the following operation:

$$f_c = \frac{\omega_c}{2\pi}$$

The voltage gain (in decibels) at any angular frequency is found from the transfer function as

$$A_v(\text{dB}) = -20 \log(\sqrt{1 + (0.01\omega)^2})$$

or

$$A_v(\text{dB}) = -10 \log[1 + (0.01\omega)^2]$$

and the phase shift is given as

$$\theta = -\arctan(0.01\omega)$$

The frequency response of the transfer function may be sketched by determining the voltage gain and phase shift at various angular frequencies around the cutoff frequency. Table 22–2 provides the voltage gain and phase shift for frequencies between $\omega = 0.01\omega_c$ and $\omega = 100\omega_c$.

TABLE 22–2   Frequency Response of a Low-Pass Filter

ω (rad/s)	TF(ω)	A_v (V/V)	A_v (db)	θ (°)
1	1.00∠−0.6°	1.000	0.00	−0.6°
5	0.999∠−2.9°	0.999	0.00	−2.9°
10	0.995∠−5.7°	0.995	−0.04	−5.7°
20	0.981∠−11.3°	0.981	−0.17	−11.3°
50	0.894∠−26.6°	0.894	−0.97	−26.6°
100	0.707∠−45°	0.707	−3.01	−45.0°
200	0.447∠−63.4°	0.447	−6.99	−63.4°
500	0.196∠−78.9°	0.196	−14.2	−78.9°
1000	0.100∠−84.3°	0.100	−20.0	−84.3°
2000	0.050∠−87.1°	0.050	−26.0	−87.1°
10k	0.010∠−89.4°	0.010	−40.0	−89.4°

The data of Table 22–2 are plotted in Figure 22–9.

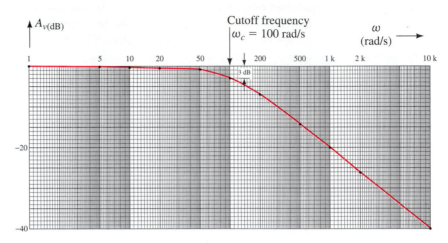

(a) Voltage gain as a function of frequency.

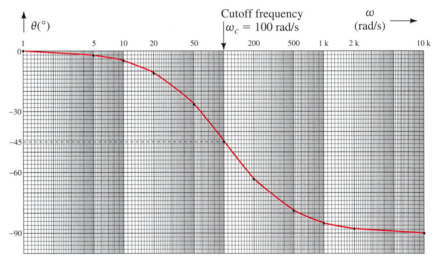

(b) Phase shift as a function of frequency.

FIGURE 22–9 Frequency response of a low-pass filter circuit.

While the previous method can be used to provide a frequency response for any circuit, it is tedious and very time-consuming. A better method is to approximate the results as follows:

For $\omega \ll \omega_c$ ($\omega \le 0.1\omega_c$),

$$A_v \simeq 0 \text{ dB}$$
$$\theta \simeq 0$$

For $\omega \gg \omega_c$ ($\omega \ge 10\omega_c$),

$$A_v \simeq -20 \log(0.01\omega)$$
$$\theta \simeq -90°$$

The above information is plotted on semilogarithmic graphs using straight lines to approximate the actual frequency response. Such graphs are called **Bode plots.** The actual response will approach the straight lines for frequencies much less than and much greater than the cutoff frequency. For this reason, the straight lines are also called the **asymptotes** of the transfer function. Figure 22–10 shows the Bode plot for a low-pass filter.

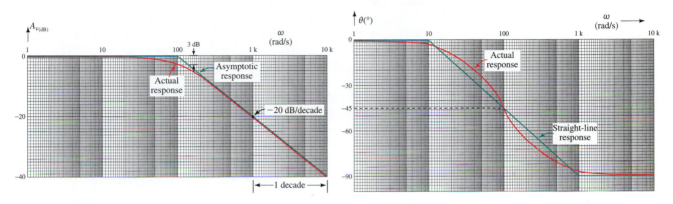

FIGURE 22–10   Frequency response of a low-pass filter.

The Bode plot illustrates several very important points. Because the abscissa of the Bode plot is not measured linearly, the slope of the asymptotes cannot be expressed in the usual terms of rise over run. A **decade** is defined as the ten-fold increase or decrease in frequency. For instance, increasing a frequency from 10 Hz to 100 Hz represents a decade. Similarly, a ten-fold change in angular frequency is also a decade.

An alternate way of describing a frequency change is by using the term **octave.** An octave is defined as a twofold increase or decrease in frequency. Therefore, a frequency of 400 Hz is one octave higher than a frequency of 200 Hz, while 800 Hz is two octaves higher than 200 Hz.

Both voltage gain and the phase shift have changes which are easily expressed in terms of a decade. The asymptotic response for voltage gain shows that the voltage gain will be approximately 0 dB for frequencies less than ω_c. For frequencies above ω_c, the voltage gain changes at a rate of -20 dB for each decade that the frequency increases (-20 dB/decade). A change of -20 dB/decade is equivalent to having a change of -6 dB/octave.

The straight-line response for phase shift illustrates that for frequencies less than $0.1\omega_c$ the phase shift will be approximately $0°$. For frequencies above $10\omega_c$ the phase shift will be essentially constant at $-90°$. In the region between $0.1\omega_c$ and $10\omega_c$, the phase shift may be approximated to be $-45°$/decade although it actually follows a curved line in this region.

All transfer functions of first-order filters (the only types of filters covered in this text) can be written as

$$\mathbf{TF} = \frac{(j\omega\tau_{Z_1})(1 + j\omega\tau_{Z_2}) \cdots (1 + j\omega\tau_{Z_n})}{(j\omega\tau_{P_1})(1 + j\omega\tau_{P_2}) \cdots (1 + j\omega\tau_{P_m})}$$

where each of the cutoff frequencies (in radians per second) is found as

$$\omega = \frac{1}{\tau}$$

Since voltage gain is expressed in dB, the Bode plot of any transfer function is determined from the summation of the effects due to the various terms in the transfer function.

EXAMPLE 22–6 Sketch the straight-line approximation of the following transfer function:

$$\mathbf{TF} = \frac{j0.01\omega}{1 + j0.1\omega}$$

Solution The voltage gain in dB is given as

$$A_{v(dB)} = 20\log(0.01\omega) - 20\log(\sqrt{1 + (0.1\omega)^2})$$

and the phase shift is given as

$$\theta = 90° - \arctan(0.1\omega)$$

The individual terms of the voltage gain are shown in Figure 22–11(a) and the individual terms of the phase shift are shown in Figure 22–11(b).

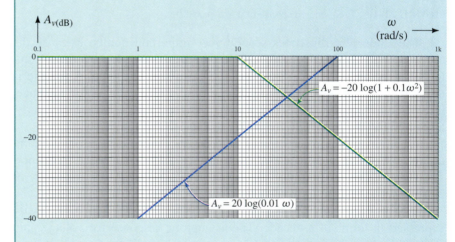

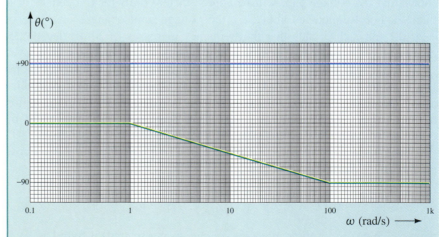

FIGURE 22–11

The combined results are shown in Figure 22–12. The total voltage gain at each frequency is determined by summing the gain due to each term of the transfer function. Similarly, the total phase shift at each frequency is the summation of phase shifts due to each of the two terms.

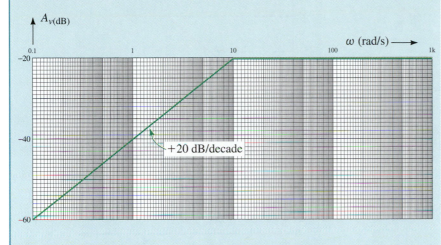

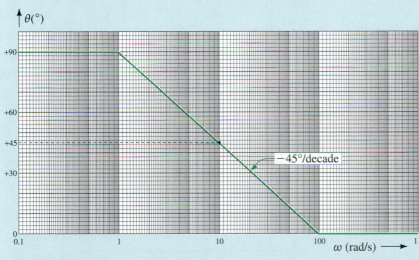

FIGURE 22–12

From the Bode plot of Figure 22–12, we conclude that the filter having this response is a high-pass filter with a cutoff frequency of 10 rad/s. The high-frequency gain of this filter is -20 dB.

Sketch the straight-line approximation for each of the transfer functions given.

a. $\mathbf{TF} = 1 + j0.02\omega$

b. $\mathbf{TF} = \dfrac{100}{1 + j0.005\omega}$

PRACTICE PROBLEMS 4

$$\text{c. } \mathbf{TF} = \frac{1 + j0.02\omega}{20(1 + j0.002\omega)}$$

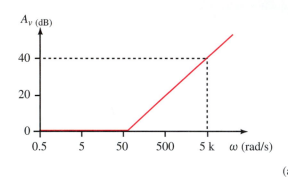
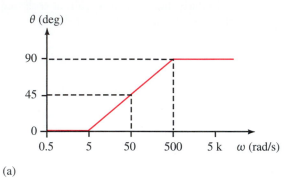

(a)

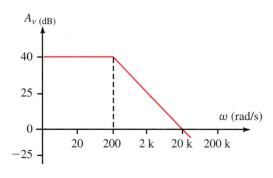
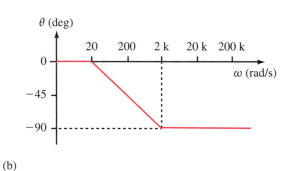

(b)

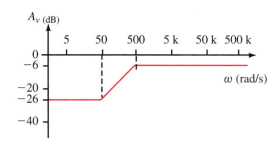
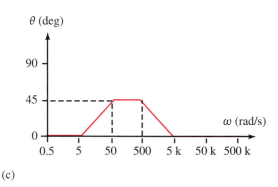

(c)

FIGURE 22–13

Writing Transfer Functions

The transfer function of any circuit is found by following a few simple steps. As we have already seen, a properly written transfer function allows us to easily calculate the cutoff frequencies and quickly sketch the corresponding Bode plot for the circuit. The steps are as follows:

1. Determine the boundary conditions for the given circuit by solving for the voltage gain when the frequency is zero (dc) and when the frequency

approaches infinity. The boundary conditions are found by using the following approximations:

At $\omega = 0$,　　inductors are short circuits,
　　　　　　　capacitors are open circuits.

As $\omega \to \infty$,　inductors are open circuits,
　　　　　　　capacitors are short circuits.

　By using the above approximations, all capacitors and inductors are easily removed from the circuit. The resulting voltage gain is determined for each boundary condition by simply applying the voltage divider rule.

2. Use the voltage divider rule to write the general expression for the transfer function in terms of the frequency, ω. In order to simplify the algebra, all capacitive and inductive reactance vectors are written as follows:

$$\mathbf{Z}_C = \frac{1}{j\omega C}$$

and

$$\mathbf{Z}_L = j\omega L$$

3. Simplify the resulting transfer function so that it is in the following format:

$$\mathbf{TF} = \frac{(j\omega\tau_{Z_1})(1 + j\omega\tau_{Z_2}) \cdots (1 + j\omega\tau_{Z_n})}{(j\omega\tau_{P_1})(1 + j\omega\tau_{P_2}) \cdots (1 + j\omega\tau_{P_m})}$$

Once the function is in this format, it is good practice to verify the boundary conditions found in Step 1. The boundary conditions are determined algebraically by first letting $\omega = 0$ and then solving for the resulting dc voltage gain. Next, we let $\omega \to \infty$. The various $(1 + j\omega\tau)$ terms of the transfer function may now be approximated as simply $j\omega\tau$, since the imaginary terms will be much larger (≥ 10) than the real components. The resultant gain will give the high-frequency gain.

4. Determine the break frequency(ies) at $\omega = 1/\tau$ (in radians per second) where the time constants will be expressed as either $\tau = RC$ or $\tau = L/R$.

5. Sketch the straight-line approximation by separately considering the effects of each term in the transfer function.

6. Sketch the actual response of the circuit from the approximation. The actual voltage gain response will be a smooth, continuous curve which follows the asymptotic curve but which usually has a 3-dB difference at the cutoff frequency(ies). This approximation will not apply if two cutoff frequencies are separated by less than a decade. The actual phase shift response will have the same value as the straight-line approximation at the cutoff frequency. At frequencies one decade above and one decade below the cutoff frequency, the actual phase shift will be 5.71° from the straight-line approximation.

　These steps will now be used in analyzing several important types of filters.

22.4 The Low-Pass Filter

The *RC* Low-Pass Filter

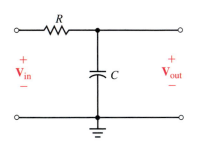

FIGURE 22–14 *RC* low-pass filter.

The circuit of Figure 22–14 is referred to as a low-pass *RC* filter circuit since it permits low-frequency signals to pass from the input to the output while attenuating high frequency signals.

At low frequencies, the capacitor has a very large reactance. Consequently, at low frequencies the capacitor is essentially an open circuit resulting in the voltage across the capacitor, $\mathbf{V}_{out}$, to be essentially equal to the applied voltage, $\mathbf{V}_{in}$.

At high frequencies, the capacitor has a very small reactance, which essentially short circuits the output terminals. The voltage at the output will therefore approach zero as the frequency increases. Although we are able to easily predict what happens at the two extremes of frequency, called the **boundary conditions,** we do not yet know what occurs between the two extremes.

The circuit of Figure 22–14 is easily analyzed by applying the voltage divider rule. Namely,

$$\mathbf{V}_{out} = \frac{\mathbf{Z}_C}{\mathbf{R} + \mathbf{Z}_C}\mathbf{V}_{in}$$

In order to simplify the algebra, the reactance of a capacitor is expressed as follows:

$$\mathbf{Z}_C = -j\frac{1}{\omega C} = -j\frac{j}{j\omega C} = \frac{1}{j\omega C} \tag{22–13}$$

The transfer function for the circuit of Figure 22–13 is now evaluated as follows:

$$\mathbf{TF}(\omega) = \frac{\mathbf{V}_{out}}{\mathbf{V}_{in}} = \frac{\dfrac{1}{j\omega C}}{R + \dfrac{1}{j\omega C}} = \frac{\dfrac{1}{j\omega C}}{\dfrac{1 + j\omega RC}{j\omega C}}$$

$$= \frac{1}{1 + j\omega RC}$$

We define the cutoff frequency, ω_c, as the frequency at which the output power is equal to half of the maximum output power (3 dB down from the maximum). This frequency occurs when the output voltage has an amplitude which is 0.7071 of the input voltage. For the *RC* circuit, the cutoff frequency occurs at

$$\omega_c = \frac{1}{\tau} = \frac{1}{RC} \tag{22–14}$$

Then the transfer function is written as

$$\mathbf{TF}(\omega) = \frac{1}{1 + j\dfrac{\omega}{\omega_c}} \tag{22–15}$$

The above transfer function results in the Bode plot shown in Figure 22–15.

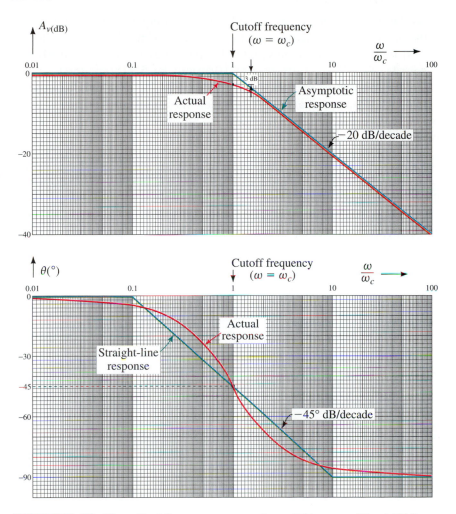

FIGURE 22–15   Normalized frequency response for an *RC* low-pass filter. (a) Voltage gain response. (b) Phase shift response.

Notice that the abscissas (horizontal axes) of the graphs in Figure 22–15 are shown as a ratio of ω/ω_c. Such a graph is called a **normalized plot** and eliminates the need to determine the actual cutoff frequency, ω_c. The normalized plot will have the same values for all low-pass *RC* filters. The actual frequency response of the *RC* low-pass filter can be approximated from the straight-line approximation by using the following guidelines:

1. At low frequencies ($\omega/\omega_c \le 0.1$) the voltage gain is approximately 0 dB with a phase shift of about 0°. This means that the output signal of the filter is very nearly equal to the input signal. The phase shift at $\omega = 0.1\omega_c$ will be 5.71° less than the straight-line approximation.

2. At the cutoff frequency, $\omega_c = 1/RC$ ($f_c = 1/2\pi RC$), the gain of the filter is −3 dB. This means that at the cutoff frequency, the circuit will deliver

half the power that it would deliver at very low frequencies. At the cutoff frequency, the output voltage will lag the input voltage by 45°.

3. As the frequency increases beyond the cutoff frequency, the amplitude of the output signal decreases by a factor of approximately ten for each ten-fold increase in frequency; namely the voltage gain is -20 dB per decade. The phase shift at $\omega = 10\omega_c$ will be 5.71° greater than the straight-line approximation, namely at $\theta = -84.29°$. For high frequencies ($\omega/\omega_c \geq 10$), the phase shift between the input and output voltage approaches $-90°$.

The *RL* Low-Pass Filter

A low-pass filter circuit may be made up of a resistor and an inductor as illustrated in Figure 22–16.

In a manner similar to that used for the *RC* low-pass filter, we may write the transfer function for the circuit of Figure 22–16 as follows:

$$\mathbf{TF} = \frac{\mathbf{V}_{out}}{\mathbf{V}_{in}}$$

$$= \frac{\mathbf{R}}{\mathbf{R} + \mathbf{Z}_L} = \frac{R}{R + j\omega L}$$

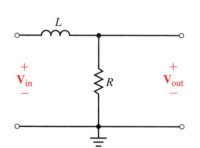

FIGURE 22–16   *RL* low-pass filter.

Now, dividing the numerator and the denominator by *R,* we have the transfer function expressed as

$$\mathbf{TF} = \frac{1}{1 + j\omega\dfrac{L}{R}}$$

Since the cutoff frequency is found as $\omega_c = 1/\tau$, we have

$$\omega_c = \frac{1}{\tau} = \frac{1}{\dfrac{L}{R}} = \frac{R}{L}$$

and so

$$\mathbf{TF} = \frac{1}{1 + j\dfrac{\omega}{\omega_c}} \tag{22–16}$$

Notice that the transfer function for the *RL* low-pass circuit in Equation 22–16 is identical to the transfer function of an *RC* circuit in Equation 22–15. In each case, the cutoff frequency is determined as the reciprocal of the time constant.

EXAMPLE 22–7   Sketch the Bode plot showing both the straight-line approximation and the actual response curves for the circuit of Figure 22–17. Show frequencies in hertz.

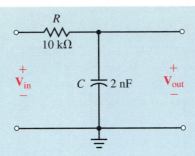

FIGURE 22–17

Solution   The cutoff frequency (in radians per second) for the circuit occurs at

$$\omega_c = \frac{1}{\tau} = \frac{1}{RC}$$

$$= \frac{1}{(10 \text{ k}\Omega)(2 \text{ nF})} = 50 \text{ krad/s}$$

which gives

$$f_c = \frac{\omega_c}{2\pi} = \frac{50 \text{ krad/s}}{2\pi} = 7.96 \text{ kHz}$$

In order to sketch the Bode plot, we begin with the asymptotes for the voltage gain response. The circuit will have a flat response until $f_c = 7.96$ kHz. Then the gain will drop at a rate of 20 dB for each decade increase in frequency. Therefore, the voltage gain at 79.6 kHz will be -20 dB, and at 796 kHz the voltage gain will be -40 dB. At the cutoff frequency for the filter, the actual voltage gain response will pass through a point which is 3 dB down from the intersection of the two asymptotes. The frequency response of the voltage gain is shown in Figure 22–18(a).

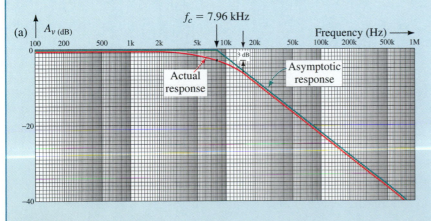

FIGURE 22–18   *(Continues)*

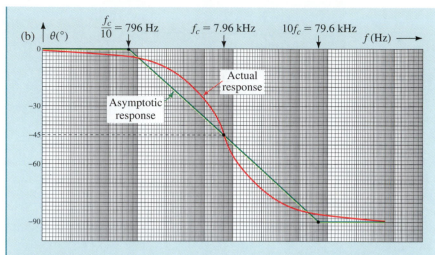

FIGURE 22–18 *(Continued)*

Next, we sketch the approximate phase shift response. The phase shift at 7.96 kHz will be $-45°$. At a frequency one decade below the cutoff frequency (at 796 Hz) the phase shift will be approximately equal to zero, while at a frequency one decade above the cutoff frequency (at 79.6 kHz) the phase shift will be near the maximum of $-90°$. The actual phase shift response will be a curve which varies slightly from the asymptotic response, as shown in Figure 22–18(b).

EXAMPLE 22–8 Consider the low-pass circuit of Figure 22–19:

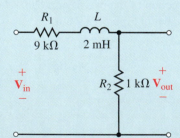

 FIGURE 22–19

a. Write the transfer function for the circuit.
b. Sketch the frequency response.

Solution

a. The transfer function of the circuit is found as

$$\mathbf{TF} = \frac{\mathbf{V}_{out}}{\mathbf{V}_{in}} = \frac{R_2}{R_2 + R_1 + j\omega L}$$

which becomes

$$\mathbf{TF} = \frac{R_2}{R_1 + R_2}\left(\frac{1}{1 + j\omega\dfrac{L}{R_1 + R_2}}\right)$$

b. From the transfer function of Part a), we see that the dc gain will no longer be 1 (0 dB) but rather is found as

$$A_{v(\text{dc})} = 20\log\left(\frac{R_2}{R_1 + R_2}\right)$$

$$= 20\log\left(\frac{1}{10}\right)$$

$$= -20\text{ dB}$$

The cutoff frequency occurs at

$$\omega_c = \frac{1}{\tau} = \frac{1}{\dfrac{L}{R_1 + R_2}}$$

$$\omega_c = \frac{R_1 + R_2}{L} = \frac{10\text{ k}\Omega}{2\text{ mH}}$$

$$= 5.0\text{ Mrad/s}$$

The resulting Bode plot is shown in Figure 22–20. Notice that the frequency response of the phase shift is precisely the same as for other low-pass filters. However, the response of the voltage gain now starts at -20 dB and then drops at a rate of -20 dB/decade above the cutoff frequency, $\omega_c = 5$ Mrad/s.

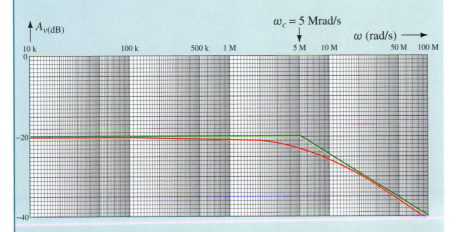

FIGURE 22–20 *(Continues)*

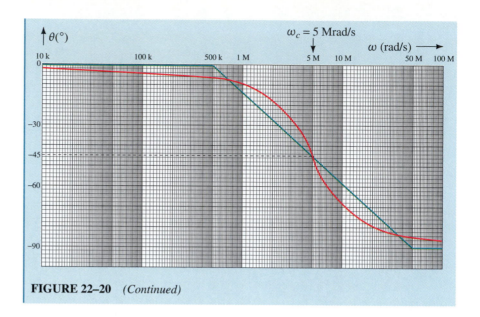

FIGURE 22–20   *(Continued)*

PRACTICE
PROBLEMS 5

Refer to the low-pass circuit of Figure 22–21:

EWB   **FIGURE 22–21**

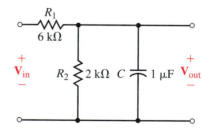

a. Write the transfer function for the circuit.

b. Sketch the frequency response.

Answers:

a. $\mathbf{TF}(\omega) = \dfrac{0.25}{1 + j\omega 0.0015}$

b.

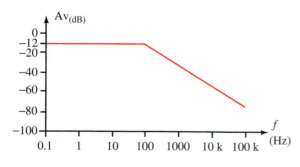
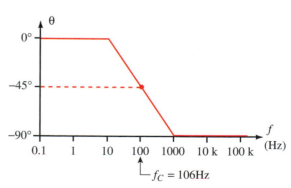

FIGURE 22–22

a. Design a low-pass *RC* filter to have a cutoff frequency of 30 krad/s. Use a 0.01-μF capacitor.

b. Design a low-pass *RL* filter to have a cutoff frequency of 20 kHz and a dc gain of −6 dB. Use a 10-mH inductor. (Assume that the inductor has no internal resistance.)

IN-PROCESS
LEARNING
CHECK 2

(Answers are at the end of the chapter.)

22.5 The High-Pass Filter

The *RC* High-Pass Filter

As the name implies, the **high-pass filter** is a circuit which allows high-frequency signals to pass from the input to the output of the circuit while attenuating low-frequency signals. A simple *RC* high-pass filter circuit is illustrated in Figure 22–23.

At low frequencies, the reactance of the capacitor will be very large, effectively preventing any input signal from passing through to the output. At high frequencies, the capacitive reactance will approach a short-circuit condition, providing a very low impedance path for the signal from the input to the output.

The transfer function of the low-pass filter is determined as follows:

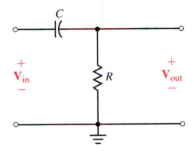

FIGURE 22–23 *RC* high-pass filter.

$$\text{TF} = \frac{\mathbf{V}_{\text{out}}}{\mathbf{V}_{\text{in}}} = \frac{\mathbf{R}}{\mathbf{R} + \mathbf{Z}_C}$$

$$= \frac{R}{R + \dfrac{1}{j\omega C}} = \frac{R}{\dfrac{j\omega RC + 1}{j\omega C}} = \frac{j\omega RC}{1 + j\omega RC}$$

Now, if we let $\omega_c = 1/\tau = 1/RC$, we have

$$\text{TF} = \frac{j\dfrac{\omega}{\omega_c}}{1 + j\dfrac{\omega}{\omega_c}} \qquad\qquad (22\text{–}17)$$

Notice that the expression of Equation 22–17 is very similar to the expression for a low-pass filter, with the exception that there is an additional term in the numerator. Since the transfer function is a complex number which is dependent upon frequency, we may once again find the general expressions for voltage gain and phase shift as functions of frequency, ω.

The voltage gain is found as

$$A_v = \frac{\dfrac{\omega}{\omega_c}}{\sqrt{1 + \left(\dfrac{\omega}{\omega_c}\right)^2}}$$

which, when expressed in decibels becomes

$$A_{v(dB)} = 20 \log \frac{\omega}{\omega_c} - 10 \log\left[1 + \left(\frac{\omega}{\omega_c}\right)^2\right] \qquad (22\text{–}18)$$

The phase shift of the numerator will be a constant 90°, since the term has only an imaginary component. The overall phase shift of the transfer function is then found as follows:

$$\theta = 90° - \arctan\frac{\omega}{\omega_c} \qquad (22\text{–}19)$$

In order to sketch the asymptotic response of the voltage gain we need to examine the effect of Equation 22–18 on frequencies around the cutoff frequency, ω_c.

For frequencies $\omega \leq 0.1\omega_c$, the second term of the expression will be essentially equal to zero, and so the voltage gain at low frequencies is approximated as

$$A_{v(dB)} \cong 20 \log \frac{\omega}{\omega_c}$$

If we substitute some arbitrary values of ω into the above approximation, we arrive at a general statement. For example, by letting $\omega = 0.01\omega_c$, we have the voltage gain as

$$A_v = 20 \log(0.01) = -40 \text{ dB}$$

and by letting $\omega = 0.1\omega_c$, we have

$$A_v = 20 \log(0.1) = -20 \text{ dB}$$

In general, we see that the expression $A_v = 20 \log (\omega/\omega_c)$ may be represented as a straight line on a semilogarithmic graph. The straight line will intersect the 0-dB axis at the cutoff frequency, ω_c, and have a slope of +20 dB/decade.

For frequencies $\omega \gg \omega_c$, Equation 22–18 may be expressed as

$$A_{v(dB)} \cong 20 \log \frac{\omega}{\omega_c} - 10 \log\left[\left(\frac{\omega}{\omega_c}\right)^2\right]$$

$$= 20 \log \frac{\omega}{\omega_c} - 20 \log \frac{\omega}{\omega_c}$$

$$= 0 \text{ dB}$$

For the particular case when $\omega = \omega_c$, we have

$$A_{v(dB)} = 20 \log 1 - 10 \log 2 = -3.0 \text{ dB}$$

which is exactly the same result that we would expect, since the actual response will be 3 dB down from the asymptotic response.

Examining Equation 22–19 for frequencies $\omega \leq 0.1\omega_c$, we see that the phase shift for the transfer function will be essentially constant at 90°, while for frequencies $\omega \geq 10\omega_c$ the phase shift will be approximately constant at 0°. At $\omega = \omega_c$ we have $\theta = 90° - 45° = 45°$.

Figure 22–24 shows the normalized Bode plot of the high-pass circuit of Figure 22–23.

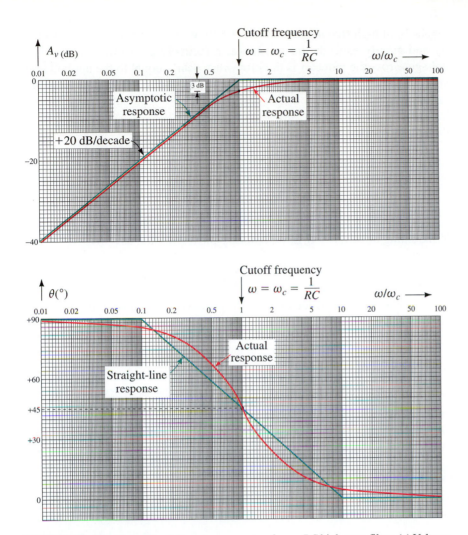

FIGURE 22–24   Normalized frequency response for an *RC* high-pass filter. (a) Voltage gain response. (b) Phase shift response.

The *RL* High-Pass Filter

A typical *RL* high-pass filter circuit is shown in Figure 22–25.

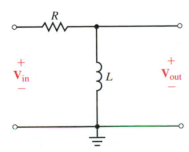

FIGURE 22–25   *RL* high-pass filter.

At low frequencies, the inductor is effectively a short circuit, which means that the output of the circuit is essentially zero at low frequencies.

Inversely, at high frequencies, the reactance of the inductor approaches infinity and greatly exceeds the resistance, effectively preventing current. The voltage across the inductor is therefore very nearly equal to the applied input voltage signal. The transfer function for the high-pass *RL* circuit is derived as follows:

$$\text{TF} = \frac{\mathbf{Z}_L}{\mathbf{R} + \mathbf{Z}_L}$$

$$= \frac{j\omega L}{R + j\omega L} = \frac{j\omega \dfrac{L}{R}}{L + j\omega \dfrac{L}{R}}$$

Now, letting $\omega_c = 1/\tau = R/L$, we simplify the expression as

$$\text{TF} = \frac{j\omega\tau}{1 + j\omega\tau}$$

The above expression is identical to the transfer function for a high-pass *RC* filter, with the exception that in this case we have $\tau = L/R$.

EXAMPLE 22–9 Design the *RL* high-pass filter circuit of Figure 22–26 to have a cutoff frequency of 40 kHz. (Assume that the inductor has no internal resistance.) Sketch the frequency response of the circuit expressing the frequencies in kilohertz.

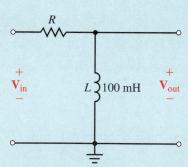

FIGURE 22–26

Solution The cutoff frequency, ω_c, in radians per second is

$$\omega_c = 2\pi f_c = 2\pi(40 \text{ kHz}) = 251.33 \text{ krad/s}$$

Now, since $\omega_c = R/L$, we have

$$R = \omega_c L = (251.33 \text{ krad/s}(100 \text{ mH}) = 25.133 \text{ k}\Omega$$

The resulting Bode plot is shown in Figure 22–27.

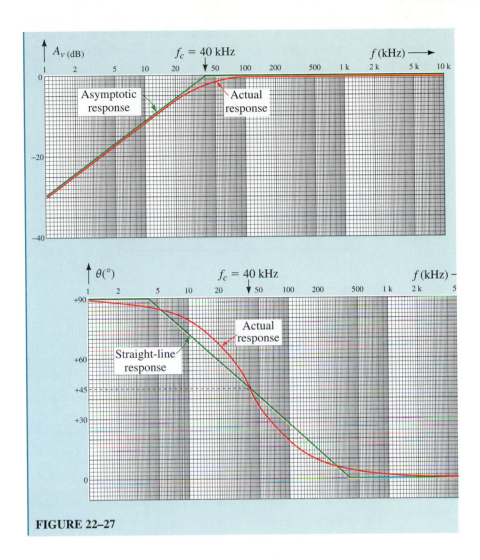

FIGURE 22–27

Consider the high-pass circuit of Figure 22–28:

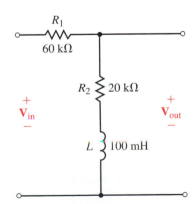

FIGURE 22–28

a. Write the transfer function of the filter.

b. Sketch the frequency response of the filter. Show the frequency in radians per second. (Hint: The frequency response of the filter is a step response.)

Answers:

a. $\mathbf{TF}(\omega) = \left(\dfrac{R_2}{R_1 + R_2}\right)\left(\dfrac{1 + j\omega\dfrac{L}{R_2}}{1 + j\omega\dfrac{L}{R_1 + R_2}}\right)$

$f_1 = 31.8$ kHz (200 krad/s), $f_2 = 127$ kHz (800 krad/s)

b.

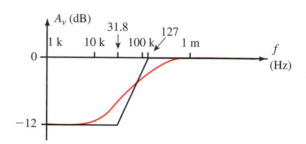

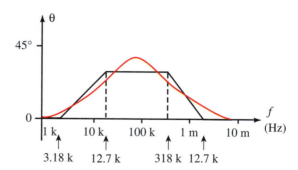

FIGURE 22–29

IN-PROCESS LEARNING CHECK 3

1. Use a 0.05-μF capacitor to design a high-pass filter having a cutoff frequency of 25 kHz. Sketch the frequency response of the filter.

2. Use a 25-mH inductor to design a high-pass filter circuit having a cutoff frequency of 80 krad/s and a high-frequency gain of -12 dB. Sketch the frequency response of the filter.

(Answers are at the end of the chapter.)

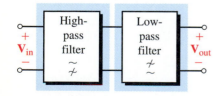

FIGURE 22–30   Block diagram of a band-pass filter.

22.6   The Band-Pass Filter

A **band-pass filter** will permit frequencies within a certain range to pass from the input of a circuit to the output. All frequencies which fall outside the desired range will be attenuated and so will not appear with appreciable power at the output. Such a filter circuit is easily constructed by using a low-pass filter cascaded with a high-pass circuit as illustrated in Figure 22–30.

Although the low-pass and high-pass blocks may consist of various combinations of elements, one possibility is to construct the entire filter network from resistors and capacitors as shown in Figure 22–31.

The bandwidth of the resulting band-pass filter will be approximately equal to the difference between the two cutoff frequencies, namely,

$$\text{BW} \cong \omega_2 - \omega_1 \quad (\text{rad/s}) \qquad (22\text{–}20)$$

The above approximation will be most valid if the cutoff frequencies of the individual stages are separated by at least one decade.

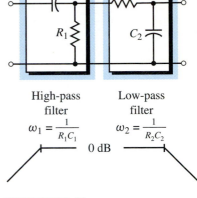

High-pass   Low-pass
filter   filter
$$\omega_1 = \frac{1}{R_1 C_1} \qquad \omega_2 = \frac{1}{R_2 C_2}$$
0 dB

FIGURE 22–31

EXAMPLE 22–10   Write the transfer function for the circuit of Figure 22–32. Sketch the resulting Bode plot and determine the expected bandwidth for the band-pass filter.

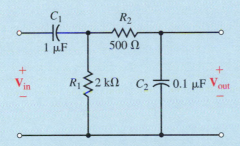

FIGURE 22–32

Solution   While the transfer function for the circuit may be written by using circuit theory, it is easier to recognize that the circuit consists of two stages: one a low-pass stage and the other a high-pass stage. If the cutoff frequencies of each stage are separated by more than one decade, then we may assume that the impedance of one stage will not adversely affect the operation of the other stage. (If this is not the case, the analysis is complicated and is outside the scope of this textbook.) Based on the previous assumption, the transfer function of the first stage is determined as

$$\text{TF}_1 = \frac{\mathbf{V}_1}{\mathbf{V}_{\text{in}}} = \frac{j\omega R_1 C_1}{1 + j\omega R_1 C_1}$$

and for the second stage as

$$\text{TF}_2 = \frac{\mathbf{V}_{\text{out}}}{\mathbf{V}_1} = \frac{1}{1 + j\omega R_2 C_2}$$

Combining the above results, we have

$$\text{TF} = \frac{\mathbf{V}_{\text{out}}}{\mathbf{V}_{\text{in}}} = \frac{(\text{TF}_2)(\mathbf{V}_1)}{\dfrac{\mathbf{V}_1}{\text{TF}_1}} = \text{TF}_1\text{TF}_2$$

which, when simplified, becomes

$$\text{TF} = \frac{\mathbf{V}_{\text{out}}}{\mathbf{V}_{\text{in}}} = \frac{j\omega\tau_1}{(1 + j\omega\tau_1)(1 + j\omega\tau_2)} \qquad (22\text{–}21)$$

where $\tau_1 = R_1C_1 = 2.0$ ms and $\tau_2 = R_2C_2 = 50$ μs. The corresponding cutoff frequencies are $\omega_1 = 500$ rad/s and $\omega_2 = 20$ krad/s. The transfer function of Equation 22–21 has three separate terms which, when taken separately, result in the approximate responses illustrated in Figure 22–33.

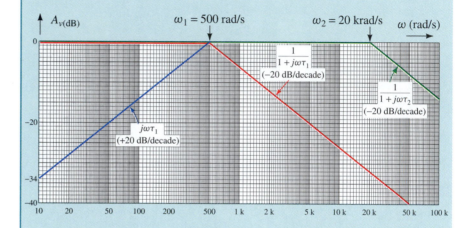

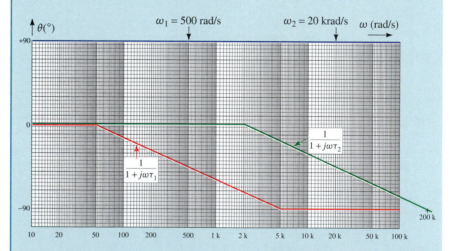

FIGURE 22–33

The resulting frequency response is determined by the summation of the individual responses as shown in Figure 22–34.

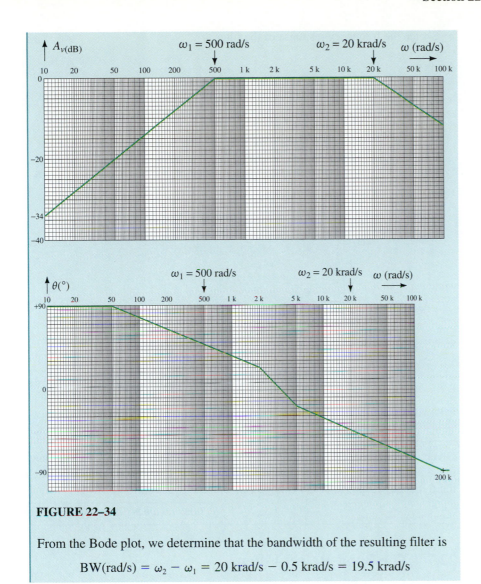

FIGURE 22–34

From the Bode plot, we determine that the bandwidth of the resulting filter is

$$\text{BW(rad/s)} = \omega_2 - \omega_1 = 20 \text{ krad/s} - 0.5 \text{ krad/s} = 19.5 \text{ krad/s}$$

Refer to the band-pass filter of Figure 22–35:

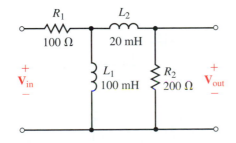

  FIGURE 22–35

a. Calculate the cutoff frequencies in rad/s and the approximate bandwidth.

b. Sketch the frequency response of the filter.

Answers:

a. $\omega_1 = 1.00$ krad/s, $\omega_2 = 10.0$ krad/s, BW $= 9.00$ krad/s

b.

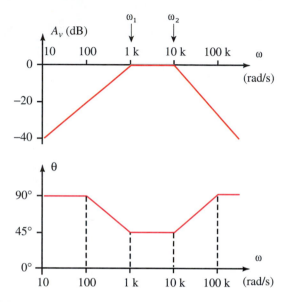

FIGURE 22–36

IN-PROCESS
LEARNING
CHECK 4

Given a 0.1-μF capacitor and a 0.04-μF capacitor, design a band-pass filter having a bandwidth of 30 krad/s and a lower cutoff frequency of 5 krad/s. Sketch the frequency response of the voltage gain and the phase shift.

(Answers are at the end of the chapter.)

22.7   The Band-Reject Filter

The **band-reject filter** has a response which is opposite to that of the band-pass filter. This filter passes all frequencies with the exception of a narrow band which is greatly attenuated. A band-reject filter constructed of a resistor, inductor, and capacitor is shown in Figure 22–37.

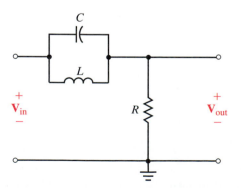

FIGURE 22–37   Notch filter.

Notice that the circuit uses a resonant tank circuit as part of the overall design. As we saw in the previous chapter, the combination of the inductor and the capacitor results in a very high tank impedance at the resonant frequency. Therefore, for any signals occurring at the resonant frequency, the output voltage is effectively zero. Because the filter circuit effectively removes any signal occurring at the resonant frequency, the circuit is often referred to as a **notch filter.** The voltage gain response of the notch filter is shown in Figure 22–38.

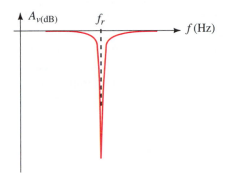

FIGURE 22–38   Frequency response of voltage gain for a notch filter.

For low-frequency signals, the inductor provides a low-impedance path from the input to the output, allowing these signals to pass from the input and appear across the resistor with minimal attenuation. Conversely, at high frequencies the capacitor provides a low-impedance path from the input to the output. Although the complete analysis of the notch filter is outside the scope of this textbook, the transfer function of the filter is determined by employing the same techniques as those previously developed.

$$\mathbf{TF} = \frac{\mathbf{R}}{\mathbf{R} + \mathbf{Z}_L \| \mathbf{Z}_C}$$

$$= \frac{R}{R + \dfrac{(j\omega L)\left(\dfrac{1}{j\omega C}\right)}{j\omega L + \dfrac{1}{j\omega C}}}$$

$$= \frac{R}{R + \dfrac{j\omega L}{1 - \omega^2 LC}}$$

$$= \frac{R(1 - \omega^2 LC)}{R - \omega^2 RLC + j\omega L}$$

$$\mathbf{TF} = \frac{1 - \omega^2 LC}{1 - \omega^2 LC + j\omega \dfrac{L}{R}} \qquad\qquad (22\text{–}22)$$

Notice that the transfer function for the notch filter is significantly more complicated than for the previous filter circuits. Due to the presence of the complex quadratic in the denominator of the transfer function, this type of filter circuit is called a **second-order filter.** The design of such filters is a separate field in electronics engineering.

Actual filter design often involves using operational amplifiers to provide significant voltage gain in the passband of the filter. In addition, such active filters have the advantage of providing very high input impedance to prevent loading effects. Many excellent textbooks are available for assistance in filter design.

22.8   Circuit Analysis Using Computers

ELECTRONICS
WORKBENCH

PSpice

Despite the complexity of designing a filter for a particular application, the analysis of the filter is a relatively simple process when done by computer. We have already seen the ease with which PSpice may be used to examine the frequency response of resonant circuits. In this chapter we will again use the Probe postprocessor of PSpice to plot the frequency characteristics of a particular circuit. We find that with some minor adjustments, the program is able to simultaneously plot both the voltage gain (in decibels) and the phase shift (in degrees) of any filter circuit.

Electronics Workbench provides displays that are similar to those obtained in PSpice. The method, though, is somewhat different. Since Electronics Workbench simulates actual lab measurements, an instrument called a Bode plotter is used by the software. As one might expect, the Bode plotter provides a graph of the frequency response (showing both the gain and phase shift) of a circuit even though no such instrument is found in a real electronics lab.

OrCAD PSpice

EXAMPLE 22–11   Use the PROBE postprocessor of PSpice to view the frequency response from 1 Hz to 100 kHz for the circuit of Figure 22–32. Determine the cutoff frequencies and the bandwidth of the circuit. Compare the results to those obtained in Example 22–10.

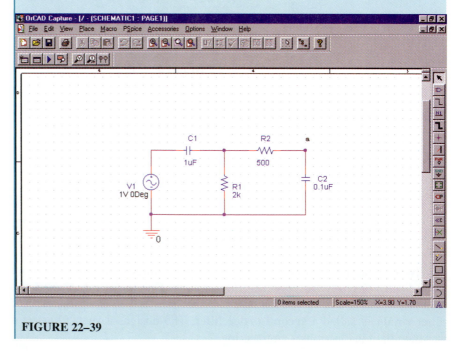

FIGURE 22–39

Solution   The OrCAD Capture program is used to enter the circuit as shown in Figure 22–39. The analysis is set up to perform an ac sweep from 1 Hz to 100 kHz using 1001 points per decade. The signal generator is set for a magnitude of 1 V and a phase shift of 0°.

In this example, we show both the voltage gain and the phase shift on the same display. Once the PROBE screen is activated, two simultaneous displays are obtained by clicking on Plot/Add Plot to Window. We will use the top display to show voltage gain and the bottom to show the phase shift.

PSpice does not actually calculate the voltage gain in decibels, but rather determines the output voltage level in dBV, referenced to 1 V_{rms}. (It is for this reason that we used a supply voltage of 1 V.) The voltage at point a of the circuit is obtained by clicking on Trace/Add Trace and then selecting **DB(V(C2:1))** as the Trace Expression. Cursors are obtained by clicking on Tools/Cursor/Display. The maximum of the function is found by clicking on Tools/Cursor/Max. The cursor indicates that the maximum gain for the circuit is −1.02 dB. We find the bandwidth of the circuit (at the −3 dB frequencies) by moving the cursors (arrow keys and <Ctrl> arrow keys) to the frequencies at which the output of the circuit is at −4.02 dB. We determine $f_1 = 0.069$ kHz, $f_2 = 3.67$ kHz, and BW = 3.61 kHz. These results are consistent with those found in Example 22–10.

Finally, we obtain a trace of the phase shift for the circuit as follows. Click anywhere on the bottom plot. Click on Trace/Add Trace and then select **P(V(C2:1))** as the Trace Expression. The range of the ordinate is changed by clicking on the axis, selecting the Y Axis tab, and setting the User defined range for a value of **−90d** to **90d.** The resulting display is shown in Figure 22–40.

FIGURE 22–40

a. Use OrCAD Capture to input the circuit of Figure 22–35.

b. Use the Probe postprocessor to observe the frequency response from 1 Hz to 100 kHz.

c. From the display, determine the cutoff frequencies and use the cursors to determine the bandwidth.

d. Compare the results to those obtained in Practice Problem 7.

EXAMPLE 22–12   Use Electronics Workbench to obtain the frequency response for the circuit of Figure 22–32. Compare the results to those obtained in Example 22–11.

Solution   In order to perform the required measurements, we need to use the function generator and the Bode plotter, both located in the Instruments parts bin. The circuit is constructed as shown in Figure 22–41.

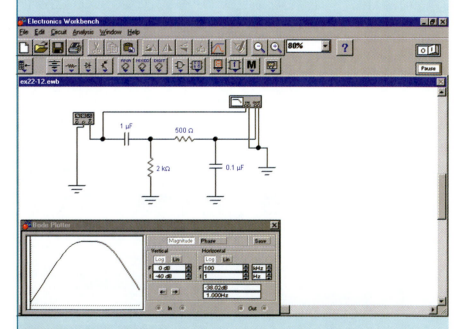

EWB   **FIGURE 22–41**

The Bode plotter is adjusted to provide the desired frequency response by first double clicking on the instrument. Next, we click on the Magnitude button. The Vertical scale is set to log with values between −40 dB and 0 dB. The Horizontal scale is set to log with values between 1 Hz and 100 kHz. Similarly, the Phase is set to have a Vertical range of −90° to 90°. After clicking on the run button, the Bode plotter provides a display of either the voltage gain response or the phase response. However, both displays are shown simultaneously by clicking on the Display Graphs icon. By using the cursor feature, we obtain the same results as those found in Example 22–11. Figure 22–42 shows the frequency response as viewed using the Display Graphs feature.

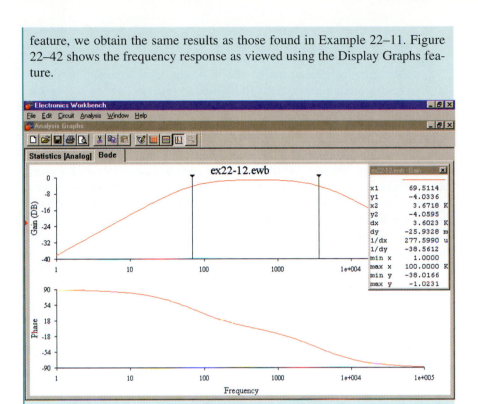

FIGURE 22–42

Use Electronics Workbench to obtain the frequency response for the circuit of Figure 22–21. Compare the results to those obtained in Practice Problem 5.

PRACTICE PROBLEMS 9

PUTTING IT INTO PRACTICE

As a designer for a sound studio, you have been asked to design bandpass filters for a color organ that will be used to provide lighting for a rock concert. The color organ will provide stage lighting that will correspond to the sound level and frequency of the music. You remember from your physics class that the human ear can perceive sounds from 20 Hz to 20 kHz.

The specifications say that the audio frequency spectrum is to be divided into three ranges. Passive *RC* filters will be used to isolate signals for each range. These signals will then be amplified and used to control lights of a particular color. The low-frequency components (20 to 200 Hz) will control blue lights, the mid-frequency components (200 Hz to 2 kHz) will control green lights, and the high-frequency components (2 kHz to 20 kHz) will control red lights.

Although the specifications call for 3 band-pass filters, you realize that you can simplify the design by using a low-pass filter with a break frequency of 200 Hz for the low frequencies and a high-pass filter with a break frequency of 2 kHz for the high frequencies. To simplify your work, you decide to use only 0.5-μF capacitors for all filters.

Show the design for each of the filters.

PROBLEMS

22.1   The Decibel

1. Refer to the amplifier shown in Figure 22–43. Determine the power gain both as a ratio and in decibels for the following power values.

   a. $P_{in} = 1.2$ mW, $P_{out} = 2.4$ W
   b. $P_{in} = 3.5$ μW, $P_{out} = 700$ mW
   c. $P_{in} = 6.0$ pW, $P_{out} = 12$ μW
   d. $P_{in} = 2.5$ mW, $P_{out} = 1.0$ W

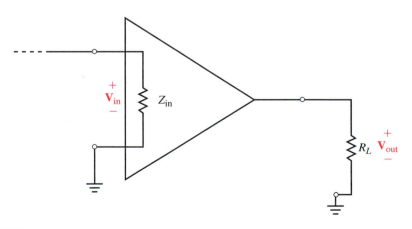

FIGURE 22–43

2. If the amplifier of Figure 22–43 has $Z_{in} = 600\ \Omega$ and $R_L = 2\ k\Omega$, find P_{in}, P_{out}, and $A_P(dB)$ for the following voltage levels:

   a. $V_{in} = 20\ mV$, $\quad\quad V_{out} = 100\ mV$

   b. $V_{in} = 100\ \mu V$, $\quad\quad V_{out} = 400\ \mu V$

   c. $V_{in} = 320\ mV$, $\quad\quad V_{out} = 600\ mV$

   d. $V_{in} = 2\ \mu V$, $\quad\quad\quad V_{out} = 8\ V$

3. The amplifier of Figure 22–43 has $Z_{in} = 2\ k\Omega$ and $R_L = 10\ \Omega$. Find the voltage gain and power gain both as a ratio and in dB for the following conditions:

   a. $V_{in} = 2\ mV$, $\quad\quad\quad P_{out} = 100\ mW$

   b. $P_{in} = 16\ \mu W$, $\quad\quad\ V_{out} = 40\ mV$

   c. $V_{in} = 3\ mV$, $\quad\quad\quad P_{out} = 60\ mW$

   d. $P_{in} = 2\ pW$, $\quad\quad\quad V_{out} = 80\ mV$

4. The amplifier of Figure 22–43 has an input voltage of $V_{in} = 2\ mV$ and an output power of $P_{out} = 200\ mW$. Find the voltage gain and power gain both as a ratio and in dB for the following conditions:

   a. $Z_{in} = 5\ k\Omega$, $\quad\quad\quad R_L = 2\ k\Omega$

   b. $Z_{in} = 2\ k\Omega$, $\quad\quad\quad R_L = 10\ k\Omega$

   c. $Z_{in} = 300\ k\Omega$, $\quad\quad R_L = 1\ k\Omega$

   d. $Z_{in} = 1\ k\Omega$ $\quad\quad\quad R_L = 1\ k\Omega$

5. The amplifier of Figure 22–43 has an input impedance of $5\ k\Omega$ and a load resistance of $250\ \Omega$. If the power gain of the amplifier is 35 dB, and the input voltage is 250 mV, find P_{in}, P_{out}, V_{out}, A_v, and $A_v(dB)$.

6. Repeat Problem 5 if the input impedance is increased to $10\ k\Omega$. (All other quantities remain unchanged.)

7. Express the following powers in dBm and in dBW:

   a. $P = 50\ mW$

   b. $P = 1\ W$

   c. $P = 400\ nW$

   d. $P = 250\ pW$

8. Express the following powers in dBm and in dBW.

   a. $P = 250\ W$

   b. $P = 250\ kW$

   c. $P = 540\ nW$

   d. $P = 27\ mW$

9. Convert the following power levels into watts:

   a. $P = 23.5\ dBm$

   b. $P = -45.2\ dBW$

   c. $P = -83\ dBm$

   d. $P = 33\ dBW$

10. Convert the following power levels into watts:

    a. $P = 16\ dBm$

    b. $P = -43\ dBW$

c. $P = -47.3$ dBm

d. $P = 29$ dBW

11. Express the following rms voltages as voltage levels (in dBV):

   a. 2.00 V

   b. 34.0 mV

   c. 24.0 V

   d. 58.2 μV

12. Express the following rms voltages as voltage levels (in dBV):

   a. 25 μV

   b. 90 V

   c. 72.5 mV

   d. 0.84 V

13. Convert the following voltage levels from dBV to rms voltages:

   a. -2.5 dBV

   b. 6.0 dBV

   c. -22.4 dBV

   d. 10.0 dBV

14. Convert the following voltage levels from dBV to rms voltages:

   a. 20.0 dBV

   b. -42.0 dBV

   c. -6.0 dBV

   d. 3.0 dBV

15. A sinusoidal waveform is measured as 30.0 $V_{p\text{-}p}$ with an oscilloscope. If this waveform were applied to a voltmeter calibrated to express readings in dBV, what would the voltmeter indicate?

16. A voltmeter shows a reading of 9.20 dBV. What peak-to-peak voltage would be observed on an oscilloscope?

22.2   Multistage Systems

17. Calculate the power levels (in dBm) at the output of each of the stages of the system shown in Figure 22–44. Solve for the output power (in watts).

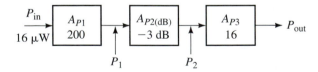

FIGURE 22–44

18. Calculate the power levels (in dBm) at the indicated locations of the system shown in Figure 22–45. Solve for the input and output powers (in watts).

19. Given that power $P_2 = 140$ mW as shown in Figure 22–46. Calculate the power levels (in dBm) at each of the indicated locations. Solve for the voltage across the load resistor, R_L.

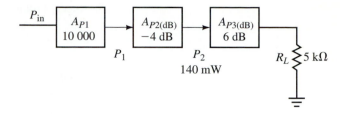

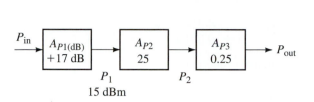

FIGURE 22–45

FIGURE 22–46

20. Suppose that the system of Figure 22–47 has an output voltage of 2 V:

    a. Determine the power (in watts) at each of the indicated locations.

    b. Solve for the voltage, V_{in}, if the input impedance of the first stage is 1.5 kΩ.

    c. Convert V_{in} and V_L into voltage levels (in dBV).

    d. Solve for the voltage gain, A_v (in dB).

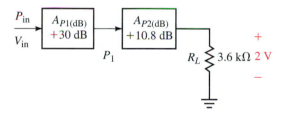

FIGURE 22–47

21. A power amplifier (P.A.) with a power gain of 250 has an input impedance of 2.0 kΩ and is used to drive a stereo speaker (output impedance of 8.0 Ω). If the output power is 100 W, determine the following:

    a. Output power level (dBm), input power level (dBm)

    b. Output voltage (rms), input voltage (rms)

    c. Output voltage level (dBV), input voltage level (dBV)

    d. Voltage gain in dB

22. Repeat Problem 21 if the amplifier has a power gain of 400 and Z_{in} = 1.0 kΩ. The power delivered to the 8.0-Ω speaker is 200 W.

22.3   Simple *RC* and *RL* Transfer Functions

23. Given the transfer function

$$\text{TF} = \frac{200}{1 + j0.001\omega}$$

    a. Determine the cutoff frequency in radians per second and in hertz.

    b. Sketch the frequency response of the voltage gain and the phase shift responses. Label the abscissa in radians per second.

24. Repeat Problem 23 for the transfer function

$$\text{TF} = \frac{1 + j0.001\omega}{200}$$

25. Repeat Problem 23 for the transfer function

$$\mathbf{TF} = \frac{1 + j0.02\omega}{1 + j0.001\omega}$$

26. Repeat Problem 23 for the transfer function

$$\mathbf{TF} = \frac{1 + j0.04\omega}{(1 + j0.004\omega)(1 + j0.001\omega)}$$

27. Repeat Problem 23 for the transfer function

$$\mathbf{TF}(\omega) = \frac{j0.02\omega}{1 + j0.02\omega}$$

28. Repeat Problem 23 for the transfer function

$$\mathbf{TF}(\omega) = \frac{j0.01\omega}{1 + j0.005\omega}$$

22.4   The Low-Pass Filter

29. Use a 4.0-μF capacitor to design a low-pass filter circuit having a cutoff frequency of 5 krad/s. Draw a schematic of your design and sketch the frequency response of the voltage gain and the phase shift.

30. Use a 1.0-μF capacitor to design a low-pass filter with a cutoff frequency of 2500 Hz. Draw a schematic of your design and sketch the frequency response of the voltage gain and the phase shift.

31. Use a 25-mH inductor to design a low-pass filter with a cutoff frequency of 50 krad/s. Draw a schematic of your design and sketch the frequency response of the voltage gain and the phase shift.

32. Use a 100-mH inductor (assume $R_{coil} = 0\ \Omega$) to design a low-pass filter circuit having a cutoff frequency of 15 kHz. Draw a schematic of your design and sketch the frequency response of the voltage gain and the phase shift.

33. Use a 36-mH inductor to design a low-pass filter having a cutoff frequency of 36 kHz. Draw a schematic of your design and sketch the frequency response of the voltage gain and the phase shift.

34. Use a 5-μF capacitor to design a low-pass filter circuit having a cutoff frequency of 100 krad/s. Draw a schematic of your design and sketch the frequency response of the voltage gain and the phase shift.

35. Refer to the low-pass circuit of Figure 22–48:

    a. Write the transfer function for the circuit.

    b. Sketch the frequency response of the voltage gain and phase shift.

36. Repeat Problem 35 for the circuit of Figure 22–49.

22.5   The High-Pass Filter

37. Use a 0.05-μF capacitor to design a high-pass filter to have cutoff frequency of 100 krad/s. Draw a schematic of your design and sketch the frequency response of the voltage gain and the phase shift.

38. Use a 2.2-nF capacitor to design a high-pass filter to have a cutoff frequency of 5 kHz. Draw a schematic of your design and sketch the frequency response of the voltage gain and phase shift.

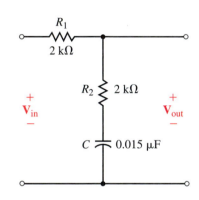

FIGURE 22–48

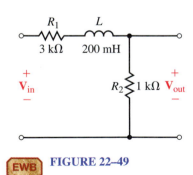

EWB **FIGURE 22–49**

39. Use a 2-mH inductor to design a high-pass filter to have a cutoff frequency of 36 krad/s. Draw a schematic of your design and sketch the frequency response of the voltage gain and phase shift.

40. Use a 16-mH inductor to design a high-pass filter circuit having a cutoff frequency of 250 kHz. Draw a schematic of your design and sketch the frequency response of the voltage gain and the phase shift.

41. Refer to the high-pass circuit of Figure 22–50.

    a. Write the transfer function for the circuit.

    b. Sketch the frequency response of the voltage gain and phase shift.

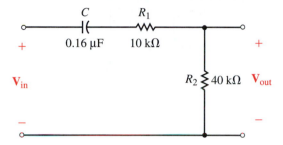

FIGURE 22–50

42. Repeat Problem 41 for the high-pass circuit of Figure 22–51.

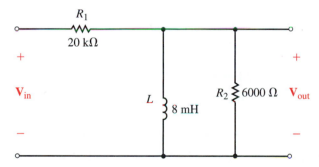

FIGURE 22–51

22.6   The Band-Pass Filter

43. Refer to the filter of Figure 22–52.

    a. Determine the approximate cutoff frequencies and bandwidth of the filter. (Assume that the two stages of the filter operate independently.)

    b. Sketch the frequency response of the voltage gain and the phase shift.

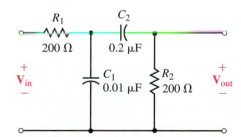

FIGURE 22–52

44. Repeat Problem 43 for the circuit of Figure 22–53.

45. a. Use two 0.01-μF capacitors to design a band-pass filter to have cutoff frequencies of 2 krad/s and 20 krad/s.

    b. Draw your schematic and sketch the frequency response of the voltage gain and the phase shift.

    c. Do you expect that the actual cutoff frequencies will occur at the designed cutoff frequencies? Explain.

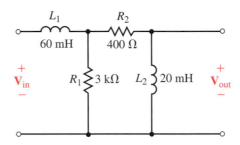

FIGURE 22–53

46. a. Use two 10-mH inductors to design a band-pass filter to have cutoff frequencies of 25 krad/s and 40 krad/s.

    b. Draw your schematic and sketch the frequency response of the voltage gain and the phase shift.

    c. Do you expect that the actual cutoff frequencies will occur at the designed cutoff frequencies? Explain.

22.7   The Band-Reject Filter

47. Given the filter circuit of Figure 22–54:

    a. Determine the "notch" frequency.

    b. Calculate the Q of the circuit.

    c. Solve for the bandwidth and determine the half-power frequencies.

    d. Sketch the voltage gain response of the circuit, showing the level (in dB) at the "notch" frequency.

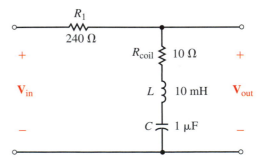

FIGURE 22–54

48. Repeat Problem 47 for the circuit of Figure 22–55.

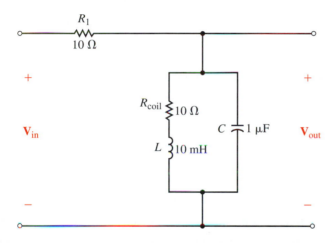

FIGURE 22–55

49. Repeat Problem 47 for the circuit of Figure 22–56.

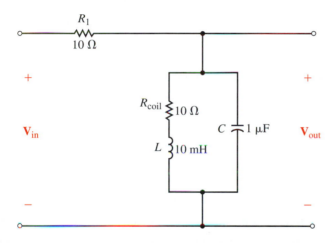

FIGURE 22–56

50. Repeat Problem 47 for the circuit of Figure 22–57.

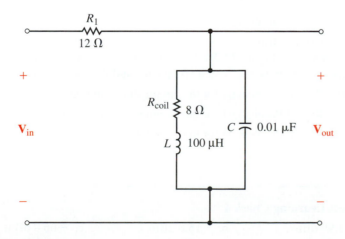

FIGURE 22–57

22.8   Circuit Analysis Using Computers

51. **PSpice**   Use OrCAD Capture to input the circuit of Figure 22–58. Let the circuit sweep through frequencies of 100 Hz to 1 MHz. Use the Probe post-processor to display the frequency response of voltage gain (in dBV) and phase shift of the circuit.

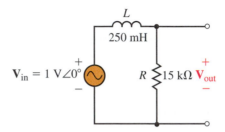

 FIGURE 22–58

52. **PSpice**   Repeat Problem 51 for the circuit shown in Figure 22–49.

53. **PSpice**   Use OrCAD Capture to input the circuit of Figure 22–52. Use the Probe postprocessor to display the frequency response of voltage gain (in dBV) and phase shift of the circuit. Select a suitable range for the frequency sweep and use the cursors to determine the half-power frequencies and the bandwidth of the circuit.

54. **PSpice**   Repeat Problem 53 for the circuit shown in Figure 22–53.

55. **PSpice**   Repeat Problem 53 for the circuit shown in Figure 22–54.

56. **PSpice**   Repeat Problem 53 for the circuit shown in Figure 22–55.

57. **PSpice**   Repeat Problem 53 for the circuit shown in Figure 22–56.

58. **PSpice**   Repeat Problem 53 for the circuit shown in Figure 22–57.

59. **EWB**   Use Electronics Workbench to obtain the frequency response for the circuit of Figure 22–58. Let the circuit sweep through frequencies of 100 Hz to 1 MHz.

60. **EWB**   Repeat Problem 59 for the circuit of Figure 22–49.

61. **EWB**   Use Electronics Workbench to obtain the frequency response for the circuit shown in Figure 22–54. Select a suitable frequency range and use cursors to determine the "notch" frequency and the bandwidth of the circuit.

62. **EWB**   Repeat Problem 61 for the circuit shown in Figure 22–55.

63. **EWB**   Repeat Problem 61 for the circuit shown in Figure 22–56.

64. **EWB**   Repeat Problem 61 for the circuit shown in Figure 22–57.

ANSWERS TO IN-PROCESS LEARNING CHECKS

In-Process Learning Check 1

1. a. -1.99 dBm   b. -15.0 dBm   c. -66.0 dBm

2. a. 0.561 V_{rms} b. 0.224 V_{rms} c. 0.354 V_{rms}

In-Process Learning Check 2

a. $R = 3333\ \Omega$ in series with $C = 0.01\ \mu F$ (output across C)

b. $R_1 = 630\ \Omega$ in series with $L = 10$ mH and $R_2 = 630\ \Omega$ (output across R_2)

In-Process Learning Check 3

1. $\mathbf{TF} = \dfrac{j\omega(6.36 \times 10^{-6})}{1 + j\omega(6.36 \times 10^{-6})}$

   $C = 0.05\ \mu F$ is in series with $R = 127.3\ \Omega$ (output across R)

2. $\mathbf{TF} = \dfrac{j\omega(3.125 \times 10^{-6})}{1 + j\omega(12.5 \times 10^{-6})}$

   $R_1 = 8\ k\Omega$ is in series with $L = 25\ \text{mH} \| R_2 = 2.67\ k\Omega$ (output across the parallel combination)

In-Process Learning Check 4

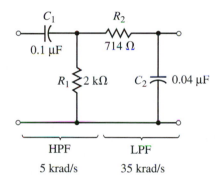

FIGURE 22–59

23

Three-Phase Systems

OBJECTIVES

After studying this chapter, you will be able to

- describe three-phase voltage generation,
- represent three-phase voltages and currents in phasor form,
- describe standard three-phase load connections,
- analyze balanced three-phase circuits,
- compute active power, reactive power, and apparent power in a three-phase system,
- measure power using the two-wattmeter method and the three-wattmeter method,
- analyze simple, unbalanced three-phase circuits,
- apply Electronics Workbench and PSpice to three-phase problems.

KEY TERMS

Balanced Systems
Floating
Line Current
Line Voltage
Neutral
Phase Current
Phase Sequence
Phase Voltage
Single-Phase Equivalent
Two-Wattmeter Method
Unbalanced Systems
Watts Ratio Curve

OUTLINE

Three-Phase Voltage Generation
Basic Three-Phase Circuit Connections
Basic Three-Phase Relationships
Examples
Power in a Balanced System
Measuring Power in Three-Phase Circuits
Unbalanced Loads
Power System Loads
Circuit Analysis Using Computers

So far, we have looked only at single-phase systems. In this chapter, we consider three-phase systems. (Three-phase systems differ from single-phase systems in that they use a set of three voltages instead of one.) Three-phase systems are used for the generation and transmission of bulk electrical power. All commercial ac power systems, for example, are three-phase systems. Not all loads connected to a three-phase system need be three-phase, however—for example, the electric lights and appliances used in our homes require only single-phase ac. To get single-phase ac from a three-phase system, we simply tap off one of its phases.

Three-phase systems may be **balanced** or **unbalanced.** If a system is balanced, it can be analyzed by considering just one of its phases. (This is because, once you know the solution for one phase, you can write down the solutions to the other two phases with no further computation other than the addition or subtraction of an angle.) This is significant because it makes the analysis of balanced systems only slightly more complex than the analysis of single-phase systems. Since most systems operate close to balance, many practical problems can be dealt with by assuming balance. This is the approach used in practice.

Three-phase systems possess economic and operating advantages over single-phase systems. For example, for the same power output, three-phase generators cost less than single-phase generators, produce uniform power rather than pulsating power, and operate with less vibration and noise.

We begin the chapter with a look at three-phase voltage generation.

Nikola Tesla

AS NOTED IN CHAPTER 15, the advent of the commercial electrical power age began with a fierce battle between Thomas A. Edison and George Westinghouse over the use of dc versus ac for the infant electrical power industry. Edison vigorously promoted dc while Westinghouse promoted ac. Tesla settled the argument in favor of ac with his development of the three-phase power system, the induction motor, and other ac devices. Coupled with the creation of a practical power transformer, these developments made long-distance transmission of electrical energy possible and ac became the clear winner.

Tesla was born in Smiljan, Croatia in 1856 and emigrated to the United States in 1884. During part of his career, he was associated with Edison, but the two had a falling out and became bitter rivals. Tesla made many important contributions in the fields of electricity and magnetism (he held over 700 patents), and the SI unit of magnetic flux density (the "tesla") is named after him. Tesla was also primarily responsible for the selection of 60 Hz as the standard power system frequency in North America and much of the world.

23.1   Three-Phase Voltage Generation

Three-phase generators have three sets of windings and thus produce three ac voltages instead of one. To get at the idea, consider first the elementary single-phase generator of Figure 23–1. As coil AA' rotates, it produces a sinusoidal waveform $e_{AA'}$ as indicated in (b). This voltage can be represented by phasor $\mathbf{E}_{AA'}$ as shown in (c).

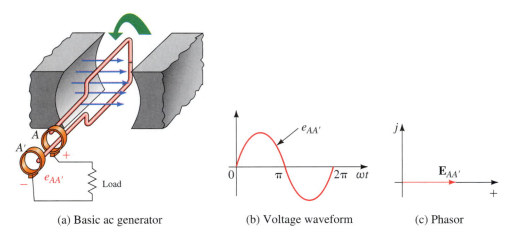

(a) Basic ac generator   (b) Voltage waveform   (c) Phasor

FIGURE 23–1   A basic single-phase generator.

If two more windings are added as in Figure 23–2, two additional voltages are generated. Since these windings are identical with AA' (except for their position on the rotor), they produce identical voltages. However, since coil BB' is placed 120° behind coil AA', voltage $e_{BB'}$ lags $e_{AA'}$ by 120°; similarly, coil CC', which is placed ahead of coil AA' by 120°, produces voltage $e_{CC'}$ that leads by 120°. Waveforms are shown in (b) and phasors in (c). As indicated, the generated voltages are equal in magnitude and phase displaced

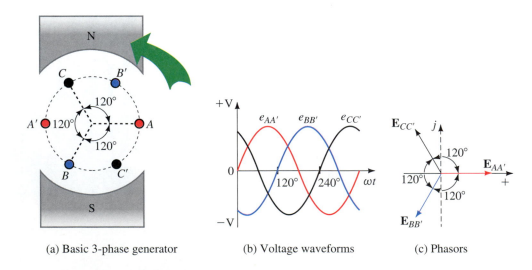

(a) Basic 3-phase generator   (b) Voltage waveforms   (c) Phasors

FIGURE 23–2   Generating three-phase voltages. Three sets of coils are used to produce three balanced voltages.

by 120°. Thus, if $\mathbf{E}_{AA'}$ is at 0°, then $\mathbf{E}_{BB'}$ will be at −120° and $E_{CC'}$ will be at +120°. Assuming an rms value of 120 V and a reference position of 0° for phasor $\mathbf{E}_{AA'}$ for example, yields $\mathbf{E}_{AA'} = 120 \text{ V}\angle 0°$, $\mathbf{E}_{BB'} = 120 \text{ V}\angle -120°$ and $\mathbf{E}_{CC'} = 120 \text{ V}\angle 120°$. Such a set of voltages is said to be balanced. Because of this fixed relationship between balanced voltages, you can, if you know one voltage, easily determine the other two.

PRACTICE PROBLEMS 1

a. If $\mathbf{E}_{AA'} = 277 \text{ V}\angle 0°$, what are $\mathbf{E}_{BB'}$ and $\mathbf{E}_{CC'}$?

b. If $\mathbf{E}_{BB'} = 347 \text{ V}\angle -120°$, what are $\mathbf{E}_{AA'}$ and $\mathbf{E}_{CC'}$?

c. If $\mathbf{E}_{CC'} = 120 \text{ V}\angle 150°$, what are $\mathbf{E}_{AA'}$ and $\mathbf{E}_{BB'}$?

Sketch the phasors for each set.

Answers:

a. $\mathbf{E}_{BB'} = 277 \text{ V}\angle -120°$; $\mathbf{E}_{CC'} = 277 \text{ V}\angle 120°$

b. $\mathbf{E}_{AA'} = 347 \text{ V}\angle 0°$; $\mathbf{E}_{CC'} = 347 \text{ V}\angle 120°$

c. $\mathbf{E}_{AA'} = 120 \text{ V}\angle 30°$; $\mathbf{E}_{BB'} = 120 \text{ V}\angle -90°$

23.2   Basic Three-Phase Circuit Connections

The generator of Figure 23–2 has three independent windings: *AA'*, *BB'*, and *CC'*. As a first thought, you might try connecting loads using six wires as in Figure 23–3(a). This will work, although it is not a scheme that is used in practice. Nonetheless, some useful insights can be gained from it. To illustrate, assume a voltage of 120 V for each coil and a 12-ohm resistive load. With $\mathbf{E}_{AA'}$ as reference, Ohm's law applied to each circuit yields

$$\mathbf{I}_A = \mathbf{E}_{AA'}/R = 120 \text{ V}\angle 0°/12 \ \Omega = 10 \text{ A}\angle 0°$$

$$\mathbf{I}_B = \mathbf{E}_{BB'}/R = 120 \text{ V}\angle -120°/12 \ \Omega = 10 \text{ A}\angle -120°$$

$$\mathbf{I}_C = \mathbf{E}_{CC'}/R = 120 \text{ V}\angle 120°/12 \ \Omega = 10 \text{ A}\angle 120°$$

These currents form a balanced set, as shown in Figure 23–3(b).

Four-Wire and Three-Wire Systems

Each load in Figure 23–3(a) has its own return wire. What if you replace them with a single wire as in (c)? By Kirchhoff's current law, the current in this wire (which we call the **neutral**) is the phasor sum of $\mathbf{I}_A$, $\mathbf{I}_B$, and $\mathbf{I}_C$. For the balanced 12-ohm load,

$$\mathbf{I}_N = \mathbf{I}_A + \mathbf{I}_B + \mathbf{I}_C = 10 \text{ A}\angle 0° + 10 \text{ A}\angle -120° + 10 \text{ A}\angle 120°$$

$$= (10 \text{ A} + j0) + (-5 \text{ A} - j8.66 \text{ A}) + (-5 \text{ A} + j8.66 \text{ A}) = 0 \text{ amps}$$

Thus, the return wire carries no current at all! (This result is always true regardless of load impedance, provided the load is balanced, i.e., all phase impedances are the same.) In practice, power systems are normally operated close to balance. Thus, the return current, while not necessarily zero, will be quite small, and the neutral wire can be made smaller than the other three conductors. This configuration is called a **four-wire system** and is one of the systems used in practice.

NOTES...

A Comment on Generator Construction

Except for small generators, most three-phase generators do not actually use the construction of Figure 23–2. Instead, they use a fixed set of windings and a rotating magnetic field. The design of Figure 23–2 was chosen to illustrate three-phase voltage generation because it is easier to visualize.

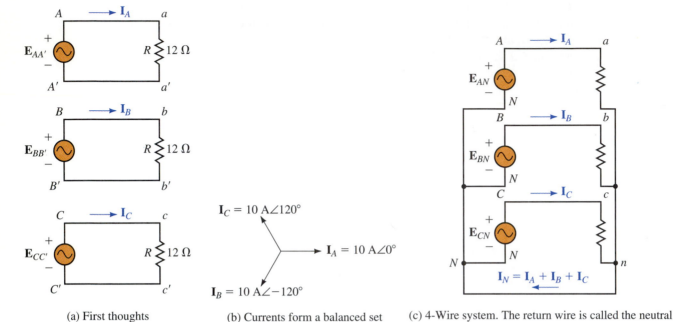

(a) First thoughts   (b) Currents form a balanced set   (c) 4-Wire system. The return wire is called the neutral

EWB   **FIGURE 23–3**   Evolution of three-phase connections.

The outgoing lines of Figure 23–3(c) are called **line** or **phase conductors.** They are the conductors that you see suspended by insulators on transmission line towers.

Symbology

Having joined points A', B', and C' in Figure 23–3(c), we now drop the A', B', and C' notation and simply call the common point N. The voltages are then renamed $\mathbf{E}_{AN}$, $\mathbf{E}_{BN}$, and $\mathbf{E}_{CN}$. They are known as **line-to-neutral voltages.**

Standard Representation

Three-phase circuits are not usually drawn as in Figure 23–3. Rather, they are usually drawn as in Figure 23–4. (Figure 23–4(a), for example, shows Figure 23–3(c) redrawn in standard form.) Note that coil symbols are used to represent generator windings rather than the circle symbol that we use for single phase.

As Figure 23–4(a) shows, the circuit that we have been looking at is a **four-wire, wye-wye (Y-Y) circuit.** A variation, the **three-wire wye-wye circuit,** is shown in (b). Three-wire wye-wye circuits may be used if the load can be guaranteed to remain balanced, since under balanced conditions the neutral conductor carries no current. However, for practical reasons (discussed in Section 23.7), most wye-wye systems use four wires.

Delta-Connected Generators

Now consider Δ connection of the generator windings. Theoretically, this is possible as indicated in Figure 23–5. However, there are practical difficulties. For example, when generators are loaded, distortions occur in the coil

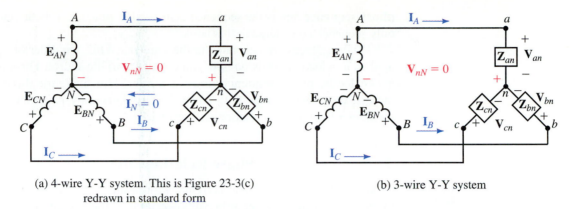

(a) 4-wire Y-Y system. This is Figure 23-3(c) redrawn in standard form

(b) 3-wire Y-Y system

FIGURE 23–4   Conventional representation of three-phase circuits. Both are Y-Y systems.

voltages due to magnetic fluxes produced by load currents. In Y-connected generators, these distortions cancel, but in Δ-connected generators, they do not. These distortions create a third harmonic current that circulates within the windings of the Δ-connected generator, lowering its efficiency. (You will learn about third harmonics in Chapter 25). For this and other reasons, Δ-connected generators are seldom used and will not be discussed in this book.

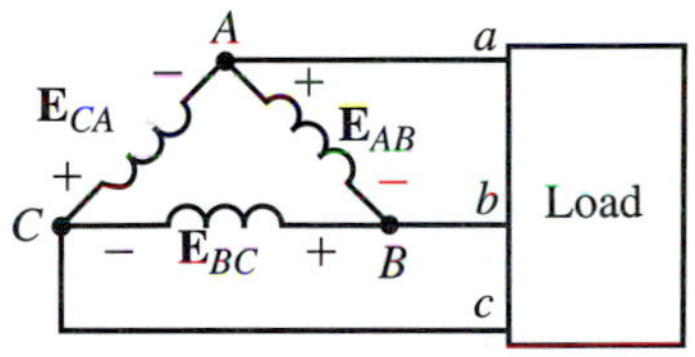

FIGURE 23–5   A delta-connected generator. For practical reasons, delta generators are seldom used.

Neutral-Neutral Voltage in a Wye-Wye Circuit

In a balanced Y-Y system, neutral current is zero because line currents sum to zero. As a consequence, the voltage between neutral points is zero. To see why, consider again Figure 23–4(a). Assume that the wire joining points n and N has impedance $\mathbf{Z}_{nN}$. This yields voltage $\mathbf{V}_{nN} = \mathbf{I}_N \times \mathbf{Z}_{nN}$. But since $\mathbf{I}_N = 0$, $\mathbf{V}_{nN} = 0$, regardless of the value of $\mathbf{Z}_{nN}$. Even if the neutral conductor is absent as in (b), $\mathbf{V}_{nN}$ is still zero. Thus, *in a balanced Y-Y system, the voltage between neutral points is zero.*

EXAMPLE 23–1   Assume the circuits of Figure 23–4(a) and (b) are balanced. If $\mathbf{E}_{AN} = 247\ \text{V}\angle 0°$, what are $\mathbf{V}_{an}$, $\mathbf{V}_{bn}$, and $\mathbf{V}_{cn}$?

Solution   In both cases, the voltage $\mathbf{V}_{nN}$ between neutral points is zero. Thus, by KVL, $\mathbf{V}_{an} = \mathbf{E}_{AN} = 247\ \text{V}\angle 0°$. Since the system is balanced, $\mathbf{V}_{bn} = 247\ \text{V}\angle -120$ and $\mathbf{V}_{cn} = 247\ \text{V}\angle 120°$.

Phase Sequence

Phase sequence refers to the order in which three-phase voltages are generated. Consider again Figure 23–2. As the rotor turns in the counterclockwise direction, voltages are generated in the sequence $e_{AA'}$, $e_{BB'}$, and $e_{CC'}$ as indicated by waveforms (b) and the phasor set (c) and the system is said to have an **ABC phase sequence.** On the other hand, if the direction of rotation were reversed, the sequence would be *ACB.* Sequence *ABC* is called the **positive**

phase sequence and is the sequence generated in practice. It is therefore the only sequence considered in this book.

While voltages are generated in the sequence *ABC,* the order of voltages applied to a load depends on how you connect it to the source. For most balanced loads, phase sequence doesn't matter. However, for three-phase motors, the order is important, since if you reverse any pair of wires, the direction of the motor's rotation will reverse.

23.3   Basic Three-Phase Relationships

To keep track of voltages and currents, we use the symbols and notations of Figure 23–6. Capital letter subscripts are used at the source and lowercase letters at the load. As usual, *E* is used for source voltage and *V* for voltage drops.

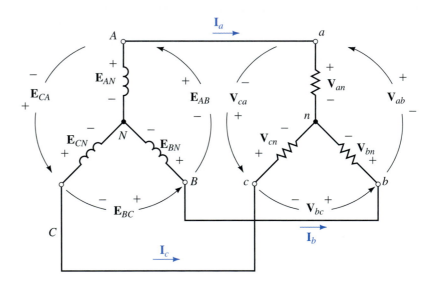

(a) For a Y, phases are defined from line to neutral

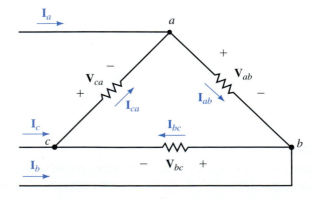

(b) For a Δ, phases are defined from line to line

FIGURE 23–6   Symbols and notation for 3-phase voltages and currents.

Definitions

Line (also called **line-to-line**) voltages are voltages between lines. Thus, $\mathbf{E}_{AB}$, $\mathbf{E}_{BC}$, and $\mathbf{E}_{CA}$ are line-to-line voltages at the generator, while $\mathbf{V}_{ab}$, $\mathbf{V}_{bc}$, and $\mathbf{V}_{ca}$ are line-to-line voltages at the load.

Phase voltages are voltages across phases. For a Y load, phases are defined from line to neutral as indicated in (a); thus, $\mathbf{V}_{an}$, $\mathbf{V}_{bn}$, and $\mathbf{V}_{cn}$ are phase voltages for a Y load. For a Δ load, phases are defined from line to line as shown in (b); thus, $\mathbf{V}_{ab}$, $\mathbf{V}_{bc}$, and $\mathbf{V}_{ca}$ are phase voltages for a Δ. As you can see, for a Δ load, phase voltages and line voltages are the same thing. For the generator, $\mathbf{E}_{AN}$, $\mathbf{E}_{BN}$, and $\mathbf{E}_{CN}$ are phase voltages.

Line currents are the currents in the line conductors. Only a single subscript is needed. You can use either $\mathbf{I}_a$, $\mathbf{I}_b$, and $\mathbf{I}_c$ as in Figure 23–6 or $\mathbf{I}_A$, $\mathbf{I}_B$, and $\mathbf{I}_C$ as in Figure 23–4. (Some authors use double subscripts such as $\mathbf{I}_{Aa}$, but this is unnecessary.)

Phase currents are currents through phases. For the Y load Figure 23–6(a), $\mathbf{I}_a$, $\mathbf{I}_b$, and $\mathbf{I}_c$ pass through phase impedances and are therefore phase currents. For the Δ load (b), $\mathbf{I}_{ab}$, $\mathbf{I}_{bc}$, and $\mathbf{I}_{ca}$ are phase currents. As you can see, for a Y load, phase currents and line currents are the same thing.

Phase impedances for a Y load are the impedances from a-n, b-n, and c-n [Figure 23–6(a)] and are denoted by the symbols $\mathbf{Z}_{an}$, $\mathbf{Z}_{bn}$, and $\mathbf{Z}_{cn}$. For a Δ load (b), phase impedances are $\mathbf{Z}_{ab}$, $\mathbf{Z}_{bc}$, and $\mathbf{Z}_{ca}$. In a balanced load, impedances for all phases are the same, i.e., $\mathbf{Z}_{an} = \mathbf{Z}_{bn} = \mathbf{Z}_{cn}$, etc.

Line and Phase Voltages for a Wye Circuit

We now need the relationship between line and phase voltages for a Y circuit. Consider Figure 23–7. By KVL, $\mathbf{V}_{ab} - \mathbf{V}_{an} + \mathbf{V}_{bn} = 0$. Thus,

$$\mathbf{V}_{ab} = \mathbf{V}_{an} - \mathbf{V}_{bn} \qquad (23\text{–}1)$$

Now, assume a magnitude V for each phase voltage and take $\mathbf{V}_{an}$ as reference. Thus, $\mathbf{V}_{an} = V\angle 0°$ and $\mathbf{V}_{bn} = V\angle -120°$. Substitute these two into Equation 23–1:

$$\mathbf{V}_{ab} = V\angle 0° - V\angle -120° = V(1 + j0) - V(-0.5 - j0.866)$$
$$= V(1.5 + j0.866) = 1.732\, V\angle 30° = \sqrt{3}V\angle 30°$$

But $\mathbf{V}_{an} = V\angle 0°$. Thus,

$$\mathbf{V}_{ab} = \sqrt{3}\mathbf{V}_{an}\angle 30° \qquad (23\text{–}2)$$

Equation 23–2 shows that the magnitude of $\mathbf{V}_{ab}$ is $\sqrt{3}$ times the magnitude of $\mathbf{V}_{an}$ and that $\mathbf{V}_{ab}$ leads $\mathbf{V}_{an}$ by 30°. This is shown in phasor diagram form in Figure 23–8(a). Similar relationships hold for the other two phases. This is shown in (b). Thus, *for a balanced Y system, the magnitude of line-to-line voltage is $\sqrt{3}$ times the magnitude of the phase voltage and each line-to-line voltage leads its corresponding phase voltage by 30°. From this you can see that the line-to-line voltages also form a balanced set.* (Although we developed these relationships with $\mathbf{V}_{an}$ in the 0° reference position, they are true regardless of the choice of reference.) They also hold at the source. Thus,

$$\mathbf{E}_{AB} = \sqrt{3}\mathbf{E}_{AN}\angle 30° \qquad (23\text{–}3)$$

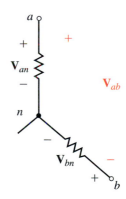

FIGURE 23–7

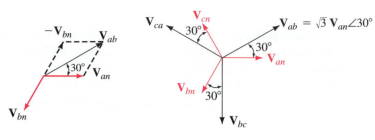

(a) $\mathbf{V}_{ab} = \mathbf{V}_{an} - \mathbf{V}_{bn}$ (b) Line and phase voltages each
                                                                form a balanced set

FIGURE 23–8   Voltages for a balanced Y load. If you know one voltage, you can determine the other five by inspection.

EXAMPLE 23–2

a. Given $\mathbf{V}_{an} = 120\ V\angle-45°$. Determine $\mathbf{V}_{ab}$ using Equation 23–2.

b. Verify $\mathbf{V}_{ab}$ by direct substitution of $\mathbf{V}_{an}$ and $\mathbf{V}_{bn}$ into equation 23–1.

Solution

a. $\mathbf{V}_{ab} = \sqrt{3}\mathbf{V}_{an}\angle30° = \sqrt{3}(120\ V\angle-45°)(1\angle30°) = 207.8\angle-15°.$

b. $\mathbf{V}_{an} = 120\ V\angle-45°$. Thus, $\mathbf{V}_{bn} = 120\ V\angle-165°$.

   $\mathbf{V}_{ab} = \mathbf{V}_{an} - \mathbf{V}_{bn} = (120\ V\angle-45°) - (120\ V\angle-165°)$

   $\qquad = 207.8\ V\angle-15°$ as before.

Nominal Voltages

While Example 23–2 yields 207.8 V for line-to-line voltage, we generally round this to 208 V and refer to the system as a 120/208-V system. These are nominal values. Other sets of nominal voltages used in practice are 277/480-V and 347/600-V.

EXAMPLE 23–3   For the circuits of Figure 23–4, suppose $\mathbf{E}_{AN} = 120\ V\angle0°$.

a. Determine the phase voltages at the load.

b. Determine the line voltages at the load.

c. Show all voltages on a phasor diagram.

Solution

a. $\mathbf{V}_{an} = \mathbf{E}_{AN}$. Thus, $\mathbf{V}_{an} = 120\ V\angle0°$. Since the system is balanced, $\mathbf{V}_{bn} = 120\ V\angle-120°$ and $\mathbf{V}_{cn} = 120\ V\angle120°$.

b. $\mathbf{V}_{ab} = \sqrt{3}\mathbf{V}_{an}\angle30° = \sqrt{3} \times 120\ V\angle(0° + 30°) = 208\ V\angle30°$. Since line voltages form a balanced set, $\mathbf{V}_{bc} = 208\ V\angle-90°$ and $\mathbf{V}_{ca} = 208\ V\angle150°$.

c. The phasors are shown in Figure 23–9.

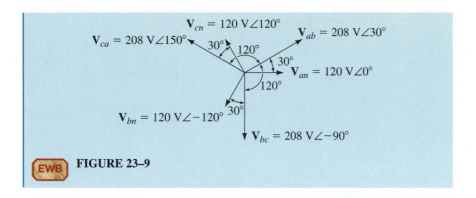

FIGURE 23–9

Equations 23–2 and 23–3 permit you to calculate line voltages from phase voltages. Rearranging them yields equation 23–4 which permits you to calculate phase voltage from line voltage.

$$\mathbf{V}_{an} = \frac{\mathbf{V}_{ab}}{\sqrt{3}\angle 30°} \quad \mathbf{E}_{AN} = \frac{\mathbf{E}_{AB}}{\sqrt{3}\angle 30°} \qquad \text{(23–4)}$$

For example, if $\mathbf{E}_{AB} = 480 \text{ V}\angle 45°$, then

$$\mathbf{E}_{AN} = \frac{\mathbf{E}_{AB}}{\sqrt{3}\angle 30°} = \frac{480 \text{ V}\angle 45°}{\sqrt{3}\angle 30°} = 277 \text{ V}\angle 15°$$

A Milestone

You have now reached an important milestone. *Given any voltage at a point in a balanced, three-phase Y system, you can, with the aid of Equation 23–2 or 23–4, determine the remaining five voltages by inspection,* i.e., by simply shifting their angles and multiplying or dividing magnitude by $\sqrt{3}$ as appropriate.

For a balanced Y generator, $\mathbf{E}_{AB} = 480 \text{ V}\angle 20°$.

a. Determine the other two generator line voltages.

b. Determine the generator phase voltages.

c. Sketch the phasors.

PRACTICE
PROBLEMS 2

Answers:

a. $\mathbf{E}_{BC} = 480 \text{ V}\angle -100°$; $\mathbf{E}_{CA} = 480 \text{ V}\angle 140°$

b. $\mathbf{E}_{AN} = 277 \text{ V}\angle -10°$; $\mathbf{E}_{BN} = 277 \text{ V}\angle -130°$; $\mathbf{E}_{CN} = 277 \text{ V}\angle 110°$

Currents for a Wye Circuit

As you saw earlier, for a Y load, line currents are the same as phase currents. Consider Figure 23–10. As indicated in (b),

$$\mathbf{I}_a = \mathbf{V}_{an}/\mathbf{Z}_{an} \qquad \text{(23–5)}$$

Similarly for $\mathbf{I}_b$ and $\mathbf{I}_c$. Since $\mathbf{V}_{an}$, $\mathbf{V}_{bn}$, and $\mathbf{V}_{cn}$ form a balanced set, *line currents $\mathbf{I}_a$, $\mathbf{I}_b$, and $\mathbf{I}_c$ also form a balanced set.* Thus, if you know one, you can determine the other two by inspection.

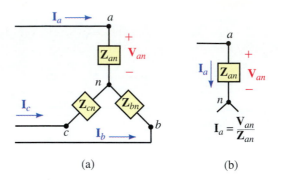

(a)   (b)

FIGURE 23–10   Determining currents for a Y load.

EXAMPLE 23–4   For Figure 23–11, suppose $\mathbf{V}_{an}$ = 120 V∠0°.

a. Compute $\mathbf{I}_a$, then determine $\mathbf{I}_b$ and $\mathbf{I}_c$ by inspection.

b. Verify by direct computation.

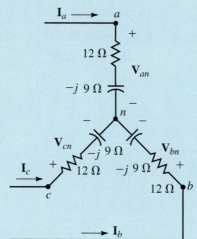

EWB   **FIGURE 23–11**

Solution

a. $\mathbf{I}_a = \dfrac{\mathbf{V}_{an}}{\mathbf{Z}_{an}} = \dfrac{120\angle 0°}{12 - j9} = \dfrac{120\angle 0°}{15\angle -36.87°} = 8.0\ \text{A}\angle 36.87°$

$\mathbf{I}_b$ lags $\mathbf{I}_a$ by 120°. Thus, $\mathbf{I}_b = 8\ \text{A}\angle -83.13°$.
$\mathbf{I}_c$ leads $\mathbf{I}_a$ by 120°. Thus, $\mathbf{I}_c = 8\ \text{A}\angle 156.87°$.

b. Since $\mathbf{V}_{an}$ = 120 V∠0°, $\mathbf{V}_{bn}$ = 120 V∠−120°, and $\mathbf{V}_{cn}$ = 120 V∠120°.
Thus,

$$\mathbf{I}_b = \frac{\mathbf{V}_{bn}}{\mathbf{Z}_{bn}} = \frac{120\angle -120°}{15\angle -36.87°} = 8.0\ \text{A}\angle -83.13°$$

$$\mathbf{I}_c = \frac{\mathbf{V}_{cn}}{\mathbf{Z}_{cn}} = \frac{120\angle 120°}{15\angle -36.87°} = 8.0\ \text{A}\angle 156.87°$$

These agree with the results obtained in (a).

1. If $\mathbf{V}_{ab} = 600\ \text{V}\angle 0°$ for the circuit of Figure 23–11, what are $\mathbf{I}_a$, $\mathbf{I}_b$, and $\mathbf{I}_c$?

2. If $\mathbf{V}_{bc} = 600\ \text{V}\angle -90°$ for the circuit of Figure 23–11, what are $\mathbf{I}_a$, $\mathbf{I}_b$, and $\mathbf{I}_c$?

Answers:
1. $\mathbf{I}_a = 23.1\ \text{A}\angle 6.9°$; $\mathbf{I}_b = 23.1\ \text{A}\angle -113.1°$; $\mathbf{I}_c = 23.1\ \text{A}\angle 126.9°$

2. $\mathbf{I}_a = 23.1\ \text{A}\angle 36.9°$; $\mathbf{I}_b = 23.1\ \text{A}\angle -83.1°$; $\mathbf{I}_c = 23.1\ \text{A}\angle 156.9°$

Line and Phase Currents for a Delta Load

Consider the delta load of Figure 23–12. Phase current $\mathbf{I}_{ab}$ can be found as in (b).

$$\mathbf{I}_{ab} = \mathbf{V}_{ab}/\mathbf{Z}_{ab} \qquad (23\text{–}6)$$

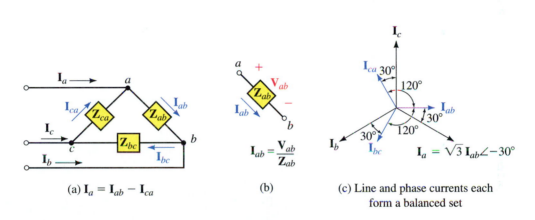

(a) $\mathbf{I}_a = \mathbf{I}_{ab} - \mathbf{I}_{ca}$

(b) $\mathbf{I}_{ab} = \dfrac{\mathbf{V}_{ab}}{\mathbf{Z}_{ab}}$

(c) Line and phase currents each form a balanced set

FIGURE 23–12   Currents for a balanced Δ load. If you know one current, you can determine the other five by inspection.

Similar relationships hold for $\mathbf{I}_{bc}$ and $\mathbf{I}_{ca}$. Since line voltages are balanced, phase currents are also balanced. Now consider again Figure 23–12(a). KCL at node *a* yields

$$\mathbf{I}_a = \mathbf{I}_{ab} - \mathbf{I}_{ca} \qquad (23\text{–}7)$$

After some manipulation, this reduces to

$$\mathbf{I}_a = \sqrt{3}\mathbf{I}_{ab}\angle -30° \qquad (23\text{–}8)$$

Thus, the magnitude of $\mathbf{I}_a$ is $\sqrt{3}$ times the magnitude of $\mathbf{I}_{ab}$, and $\mathbf{I}_a$ lags $\mathbf{I}_{ab}$ by 30°. Similarly, for the other two phases. Thus, *in a balanced Δ, the magnitude of the line current is $\sqrt{3}$ times the magnitude of the phase current, and each line current lags its corresponding phase current by 30°.* Since phase currents are balanced, line currents are also balanced. This is shown in (c). To find phase currents from line currents, use

$$\mathbf{I}_{ab} = \frac{\mathbf{I}_a}{\sqrt{3}\angle -30°} \qquad (23\text{–}9)$$

A Second Milestone

You have reached a second milestone. *Given any current in a balanced, three-phase Δ load, you can, with the aid of Equations 23–8 or 23–9, determine all remaining currents by inspection.*

EXAMPLE 23–5 Suppose $\mathbf{V}_{ab} = 240\ \mathrm{V}\angle 15°$ for the circuit of Figure 23–13.

a. Determine the phase currents.

b. Determine the line currents.

c. Sketch the phasor diagram.

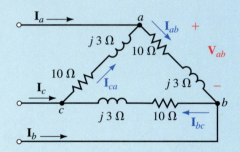

FIGURE 23–13

Solution

a. $\mathbf{I}_{ab} = \dfrac{\mathbf{V}_{ab}}{\mathbf{Z}_{ab}} = \dfrac{240\angle 15°}{10 + j3} = 23.0\ \mathrm{A}\angle -1.70°$

Thus,

$$\mathbf{I}_{bc} = 23.0\ \mathrm{A}\angle -121.7° \quad \text{and} \quad \mathbf{I}_{ca} = 23.0\ \mathrm{A}\angle 118.3°$$

b. $\mathbf{I}_a = \sqrt{3}\mathbf{I}_{ab}\angle -30° = 39.8\ \mathrm{A}\angle -31.7°$

Thus,

$$\mathbf{I}_b = 39.8\ \mathrm{A}\angle -151.7° \quad \text{and} \quad \mathbf{I}_c = 39.8\ \mathrm{A}\angle 88.3°$$

c. Phasors are shown in Figure 23–14.

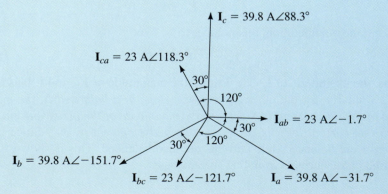

FIGURE 23–14

1. For the circuit of Figure 23–13, if $\mathbf{I}_a = 17.32$ A$\angle 20°$, determine
    a. $\mathbf{I}_{ab}$
    b. $\mathbf{V}_{ab}$
2. For the circuit of Figure 23–13, if $\mathbf{I}_{bc} = 5$ A$\angle -140°$, what is $\mathbf{V}_{ab}$?

Answers:
1. a. 10 A$\angle 50°$
    b. 104 V$\angle 66.7°$
2. 52.2 V$\angle -3.30°$

The Single-Phase Equivalent

By now it should be apparent that if you know the solution for one phase of
a balanced system, you effectively know the solution for all three phases. We
will now formalize this viewpoint by developing the **single-phase equivalent** approach to solving balanced systems. Consider a Y-Y system with line
impedance. The system may be either a three-wire system or a four-wire system with neutral conductor impedance. In either case, since the voltage
between neutral points is zero, you can join points n and N with a **zero-
impedance conductor** without disturbing voltages or currents elsewhere in
the circuit. This is illustrated in Figure 23–15(a). Phase a can now be isolated as in (b). Since $V_{nN} \equiv 0$ as before, the equation describing phase a in
circuit (b) is the same as that describing phase a in the original circuit; thus,
circuit (b) can be used to solve the original problem. If there are Δ loads
present, convert them to Y loads using the Δ-Y conversion formula for balanced loads: $\mathbf{Z}_Y = \mathbf{Z}_\Delta/3$, from Chapter 19. This procedure is valid regardless
of the configuration or complexity of the circuit. We will look at its use in
Section 23.4.

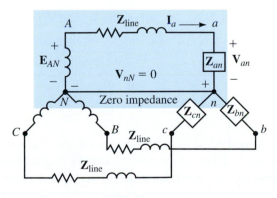

(a) Original circuit: $\mathbf{E}_{AN} = \mathbf{I}_a \mathbf{Z}_{line} + \mathbf{V}_{an}$

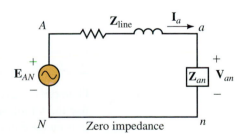

(b) Single phase equivalent: $\mathbf{E}_{AN} = \mathbf{I}_a \mathbf{Z}_{line} + \mathbf{V}_{an}$

FIGURE 23–15 Reducing a circuit to its single phase equivalent.

Selecting a Reference

Before you solve a three-phase problem, you need to select a reference. For
Y circuits, we normally choose $\mathbf{E}_{AN}$ or $\mathbf{V}_{an}$; for Δ circuits, we normally
choose $\mathbf{E}_{AB}$ or $\mathbf{V}_{ab}$.

Summary of Basic Three-Phase Relationships

Table 23–1 summarizes the relationships developed so far. Note that in balanced systems (Y or Δ), *all* voltages and *all* currents are balanced.

TABLE 23–1   Summary of Relationships (Balanced System). All Voltages and Currents Are Balanced

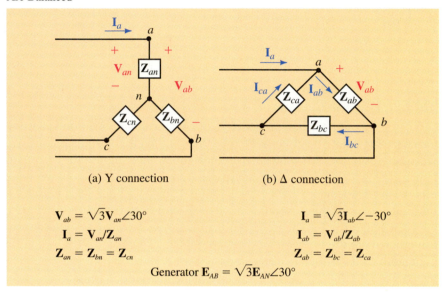

(a) Y connection   (b) Δ connection

$$\mathbf{V}_{ab} = \sqrt{3}\mathbf{V}_{an}\angle 30° \qquad\qquad \mathbf{I}_{a} = \sqrt{3}\mathbf{I}_{ab}\angle{-30°}$$

$$\mathbf{I}_{a} = \mathbf{V}_{an}/\mathbf{Z}_{an} \qquad\qquad \mathbf{I}_{ab} = \mathbf{V}_{ab}/\mathbf{Z}_{ab}$$

$$\mathbf{Z}_{an} = \mathbf{Z}_{bn} = \mathbf{Z}_{cn} \qquad\qquad \mathbf{Z}_{ab} = \mathbf{Z}_{bc} = \mathbf{Z}_{ca}$$

$$\text{Generator } \mathbf{E}_{AB} = \sqrt{3}\mathbf{E}_{AN}\angle 30°$$

IN-PROCESS
LEARNING
CHECK 1

1. For Figure 23–4(a), if $\mathbf{E}_{AN} = 277\ \text{V}\angle{-20°}$, determine all line and phase voltages, source and load.

2. For Figure 23–4(a), if $\mathbf{V}_{bc} = 208\ \text{V}\angle{-40°}$, determine all line and phase voltages, source and load.

3. For Figure 23–11, if $\mathbf{I}_{a} = 8.25\ \text{A}\angle 35°$, determine $\mathbf{V}_{an}$ and $\mathbf{V}_{ab}$.

4. For Figure 23–13, if $\mathbf{I}_{b} = 17.32\ \text{A}\angle{-85°}$, determine all voltages.

(Answers are at the end of the chapter.)

23.4   Examples

Generally, there are several ways to solve most problems. We usually try to use the simplest approach. Thus, sometimes we use the single-phase equivalent method and sometimes we solve the problem in its three-phase configuration. Generally, if a circuit has line impedance, we use the single-phase equivalent method; otherwise, we may solve it directly.

EXAMPLE 23–6   For Figure 23–16, $\mathbf{E}_{AN} = 120\ \text{V}\angle 0°$.

a. Solve for the line currents.

b. Solve for the phase voltages at the load.

c. Solve for the line voltages at the load.

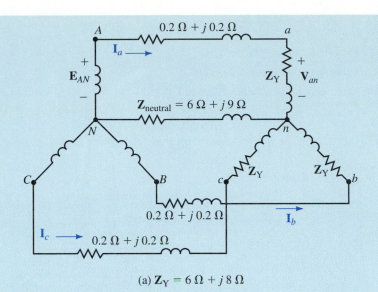

(a) $\mathbf{Z}_Y = 6\ \Omega + j\ 8\ \Omega$

(b) Single phase equivalent. Since the neutral conductor in (a)
   carries no current, its impedance has no effect on the solution

 FIGURE 23–16   A Y-Y problem.

Solution

a. Reduce the circuit to its single-phase equivalent as shown in (b).

$$\mathbf{I}_a = \frac{\mathbf{E}_{AN}}{\mathbf{Z}_T} = \frac{120\angle 0°}{(0.2 + j0.2) + (6 + j8)} = 11.7\ \text{A}\angle -52.9°$$

Therefore,

$$\mathbf{I}_b = 11.7\ \text{A}\angle -172.9° \quad \text{and} \quad \mathbf{I}_c = 11.7\ \text{A}\angle 67.1°$$

b. $\mathbf{V}_{an} = \mathbf{I}_a \times \mathbf{Z}_{an} = (11.7\angle -52.9°)(6 + j8) = 117\ \text{V}\angle 0.23°$

Thus,

$$\mathbf{V}_{bn} = 117\ \text{V}\angle -119.77° \quad \text{and} \quad \mathbf{V}_{cn} = 117\ \text{V}\angle 120.23°$$

c. $\mathbf{V}_{ab} = \sqrt{3}\mathbf{V}_{an}\angle 30° = \sqrt{3} \times 117\angle(0.23° + 30°) = 202.6\ \text{V}\angle 30.23°$

Thus,

$$\mathbf{V}_{bc} = 202.6\ \text{V}\angle -89.77° \quad \text{and} \quad \mathbf{V}_{ca} = 202.6\ \text{V}\angle 150.23°$$

Note the phase shift and voltage drop across the line impedance. Note also
that the impedance of the neutral conductor plays no part in the solution,
since no current passes through it because the system is balanced.

For the circuit of Figure 23–17, $\mathbf{V}_{an} = 120\ \mathrm{V}\angle0°$.

a. Find the line currents.

b. Verify that the neutral current is zero.

c. Determine generator voltages $\mathbf{E}_{AN}$ and $\mathbf{E}_{AB}$.

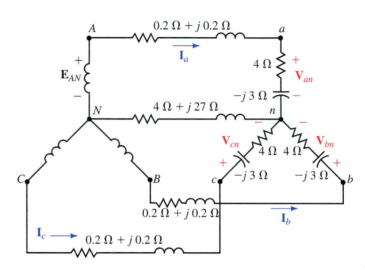

FIGURE 23–17

Answers:

a. $\mathbf{I}_a = 24\ \mathrm{A}\angle36.9°$; $\mathbf{I}_b = 24\ \mathrm{A}\angle-83.1°$; $\mathbf{I}_c = 24\ \mathrm{A}\angle156.9°$

b. $24\ \mathrm{A}\angle36.9° + 24\ \mathrm{A}\angle-83.1° + 24\ \mathrm{A}\angle156.9° = 0$

c. $\mathbf{E}_{AN} = 121\ \mathrm{V}\angle3.18°$; $\mathbf{E}_{AB} = 210\ \mathrm{V}\angle33.18°$

EXAMPLE 23–7   For the circuit of Figure 23–18, $\mathbf{E}_{AB} = 208\ \mathrm{V}\angle30°$.

a. Determine the phase currents.

b. Determine the line currents.

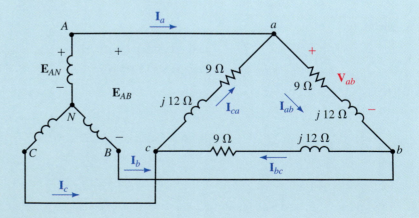

FIGURE 23–18   A Y-Δ problem.

Solution

a. Since this circuit has no line impedance, the load connects directly to the source and $\mathbf{V}_{ab} = \mathbf{E}_{AB} = 208\,\text{V}\angle 30°$. Current $\mathbf{I}_{ab}$ can be found as

$$\mathbf{I}_{ab} = \frac{\mathbf{V}_{ab}}{\mathbf{Z}_{ab}} = \frac{208\angle 30°}{9 + j12} = \frac{208\angle 30°}{15\angle 53.1°} = 13.9\,\text{A}\angle -23.13°$$

Thus,

$$\mathbf{I}_{bc} = 13.9\,\text{A}\angle -143.13° \text{ and } \mathbf{I}_{ca} = 13.9\,\text{A}\angle 96.87°$$

b. $\mathbf{I}_a = \sqrt{3}\mathbf{I}_{ab}\angle -30° = \sqrt{3}(13.9)\angle(-30° -23.13°) = 24\,\text{A}\angle -53.13°$

Thus,

$$\mathbf{I}_b = 24\,\text{A}\angle -173.13° \quad \text{and} \quad \mathbf{I}_c = 24\,\text{A}\angle 66.87°$$

EXAMPLE 23–8 For the circuit of Figure 23–19(a), the magnitude of the line voltage at the generator is 208 volts. Solve for the line voltage $\mathbf{V}_{ab}$ at the load.

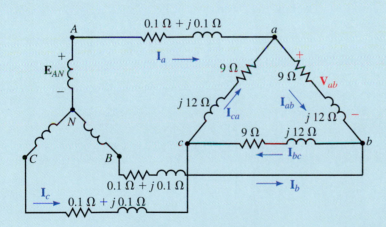

(a) $\mathbf{Z}_Y = \dfrac{\mathbf{Z}_\Delta}{3} = 3\,\Omega + j\,4\,\Omega$

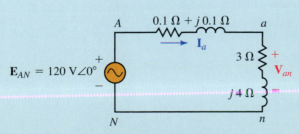

(b) Single phase equivalent

FIGURE 23–19 A circuit with line impedances.

Solution Since points *A-a* and *B-b* are not directly joined, $\mathbf{V}_{ab} \neq \mathbf{E}_{AB}$ and we cannot solve the circuit as in Example 23–7. Use the single-phase equivalent.

Phase voltage at the source is $208/\sqrt{3} = 120$ V. Choose $\mathbf{E}_{AN}$ as reference: $\mathbf{E}_{AN} = 120\ \text{V}\angle 0°$.

$$\mathbf{Z}_Y = \mathbf{Z}_\Delta/3 = (9 + j12)/3 = 3\ \Omega + j4\ \Omega.$$

The single-phase equivalent is shown in (b). Now use the voltage divider rule to find $\mathbf{V}_{an}$:

$$\mathbf{V}_{an} = \left(\frac{3 + j4}{3.1 + j4.1}\right) \times 120\angle 0° = 117\ \text{V}\angle 0.22°$$

Thus,

$$\mathbf{V}_{ab} = \sqrt{3}\mathbf{V}_{an}\angle 30° = \sqrt{3}(117\ \text{V})\angle 30.22° = 203\ \text{V}\angle 30.22°$$

EXAMPLE 23–9 For the circuit of Figure 23–19(a), the generator phase voltage is 120 volts. Find the Δ currents.

Solution Since the source voltage here is the same as in Example 23–8, load voltage $\mathbf{V}_{ab}$ will also be the same. Thus,

$$\mathbf{I}_{ab} = \frac{\mathbf{V}_{ab}}{\mathbf{Z}_{ab}} = \frac{203\ \text{V}\angle 30.22°}{(9 + j12)\Omega} = 13.5\ \text{A}\angle -22.9°$$

and,

$$\mathbf{I}_{bc} = 13.5\ \text{A}\angle -142.9° \quad \text{and} \quad \mathbf{I}_{ca} = 13.5\ \text{A}\angle 97.1°$$

EXAMPLE 23–10 A Y load and a Δ load are connected in parallel as in Figure 23–20(a). Line voltage magnitude at the generator is 208 V.

a. Find the phase voltages at the loads.

b. Find the line voltages at the loads.

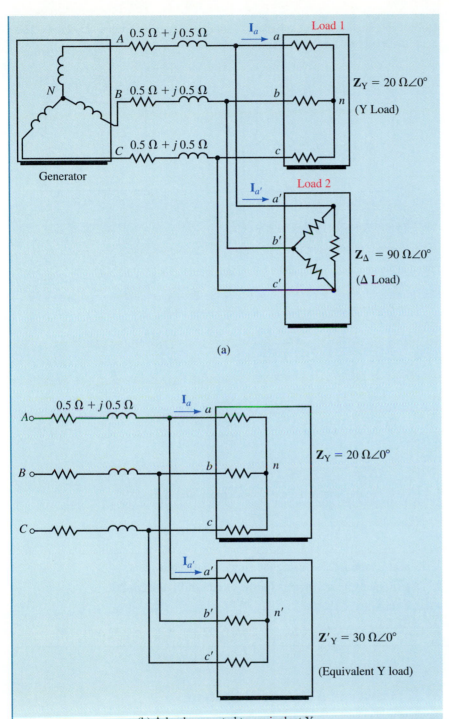

Load 1

$0.5\ \Omega + j\ 0.5\ \Omega$

$\mathbf{I}_a$

$\mathbf{Z}_Y = 20\ \Omega\angle 0°$

(Y Load)

Generator

Load 2

$\mathbf{I}_{a'}$

$\mathbf{Z}_\Delta = 90\ \Omega\angle 0°$

(Δ Load)

(a)

$0.5\ \Omega + j\ 0.5\ \Omega$

$\mathbf{I}_a$

$\mathbf{Z}_Y = 20\ \Omega\angle 0°$

$\mathbf{I}_{a'}$

$\mathbf{Z}'_Y = 30\ \Omega\angle 0°$

(Equivalent Y load)

(b) Δ load converted to equivalent Y

EWB   FIGURE 23–20

Solution   Convert the Δ load to a Y load. Thus, $\mathbf{Z}'_Y = \frac{1}{3}\mathbf{Z}_\Delta = 30\ \Omega\angle 0°$ as in (b). Now join the neutral points N, n, and n' by a zero-impedance conductor to get the single-phase equivalent, which is shown in Figure 23–21(a). The parallel load resistors can be combined as in (b).

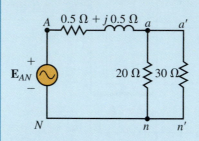

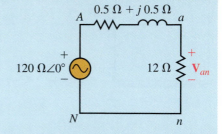

(a) Single phase equivalent of Figure 23-20   (b) Reduced circuit

FIGURE 23–21

a. Phase voltage is 208 V/$\sqrt{3}$ = 120 V. Select $\mathbf{E}_{AN}$ as reference; $\mathbf{E}_{AN}$ = 120 V∠0°. Using the voltage divider rule yields

$$\mathbf{V}_{an} = \left(\frac{12}{12.5 + j0.5}\right) \times 120\angle 0° = 115.1 \text{ V}\angle -2.29°$$

Thus,

$$\mathbf{V}_{bn} = 115.1 \text{ V}\angle -122.29° \quad \text{and} \quad \mathbf{V}_{cn} = 115.1 \text{ V}\angle 117.71°$$

b. $\mathbf{V}_{ab} = \sqrt{3}\mathbf{V}_{an}\angle 30° = \sqrt{3}(115.1 \text{ V})\angle(-2.29° + 30°) = 199 \text{ V}\angle 27.71°$

Thus,

$$\mathbf{V}_{bc} = 199 \text{ V}\angle -92.29° \quad \text{and} \quad \mathbf{V}_{ca} = 199 \text{ V}\angle 147.71°$$

These are the line voltages for both the Y and Δ loads.

PRACTICE PROBLEMS 6

1. Repeat Example 23–7 using the single-phase equivalent.
2. Determine Δ phase currents for the circuit of Figure 23–20(a).

Answers:
2. $\mathbf{I}_{a'b'}$ = 2.22 A∠27.7°; $\mathbf{I}_{b'c'}$ = 2.22 A∠−92.3°; $\mathbf{I}_{c'a'}$ = 2.22 A∠147.7°

23.5   Power in a Balanced System

To find total power in a balanced system, determine power to one phase, then multiply by three. Per phase quantities can be found using the formulas of Chapter 17. Since only magnitudes are involved in many power formulas and calculations and since magnitudes are the same for all three phases, we can use a simplified notation. We will use V_ϕ for magnitude of phase voltage, I_ϕ for phase current, V_L and I_L for line voltage and line current respectively, and Z_ϕ for phase impedance.

Active Power to a Balanced Wye Load

First, consider a Y load (Figure 23–22). The power to any phase as indicated in (b) is the product of the magnitude of the phase voltage V_ϕ times the magnitude of the phase current I_ϕ times the cosine of the angle θ_ϕ between them. Since the angle between voltage and current is always the angle of the load impedance, the power per phase is

$$P_\phi = V_\phi I_\phi \cos \theta_\phi \quad \text{(W)} \tag{23–10}$$

where θ_ϕ is the angle of $\mathbf{Z}_\phi$. Total power is

$$P_T = 3P_\phi = 3V_\phi I_\phi \cos \theta_\phi \quad \text{(W)} \tag{23–11}$$

It is also handy to have a formula for power in terms of line quantities. For a Y load, $I_\phi = I_L$ and $V_\phi = V_L/\sqrt{3}$, where I_L is the magnitude of the line current and V_L is the magnitude of the line-to-line voltage. Substituting these relations into Equation 23–11 and noting that $3/\sqrt{3} = \sqrt{3}$ yields

$$P_T = \sqrt{3}V_L I_L \cos \theta_\phi \quad \text{(W)} \tag{23–12}$$

This is a very important formula and one that is widely used. Note carefully, however, that θ_ϕ is the angle of the load impedance and not the angle between V_L and I_L.

Power per phase can also be expressed as

$$P_\phi = I_\phi^2 R_\phi = V_R^2/R_\phi \quad \text{(W)} \tag{23–13}$$

where R_ϕ is the resistive component of the phase impedance and V_R is the voltage across it. Total power is thus

$$P_T = 3I_\phi^2 R_\phi = 3V_R^2/R_\phi \quad \text{(W)} \tag{23–14}$$

Reactive Power to a Balanced Wye Load

Equivalent expressions for reactive power are

$$Q_\phi = V_\phi I_\phi \sin \theta_\phi \quad \text{(VAR)} \tag{23–15}$$

$$= I_\phi^2 X_\phi = V_X^2/X_\phi \quad \text{(VAR)} \tag{23–16}$$

$$Q_T = \sqrt{3}V_L I_L \sin \theta_\phi \quad \text{(VAR)} \tag{23–17}$$

where X_ϕ is the reactive component of $\mathbf{Z}_\phi$ and V_X is the voltage across it.

Apparent Power

$$S_\phi = V_\phi I_\phi = I_\phi^2 Z_\phi = \frac{V_\phi^2}{Z_\phi} \quad \text{(VA)} \tag{23–18}$$

$$S_T = \sqrt{3}V_L I_L \quad \text{(VA)} \tag{23–19}$$

Power Factor

$$F_p = \cos \theta_\phi = P_T/S_T = P_\phi/S_\phi \tag{23–20}$$

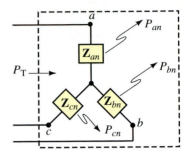

(a) $P_T = P_{an} + P_{bn} + P_{cn} = 3 P_\phi$

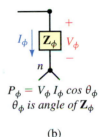

$$P_\phi = V_\phi I_\phi \cos \theta_\phi$$
θ_ϕ is angle of $\mathbf{Z}_\phi$

(b)

FIGURE 23–22 For a balanced Y, $P_\phi = P_{an} = P_{bn} = P_{cn}$.

EXAMPLE 23–11   For Figure 23–23, the phase voltage is 120 V.

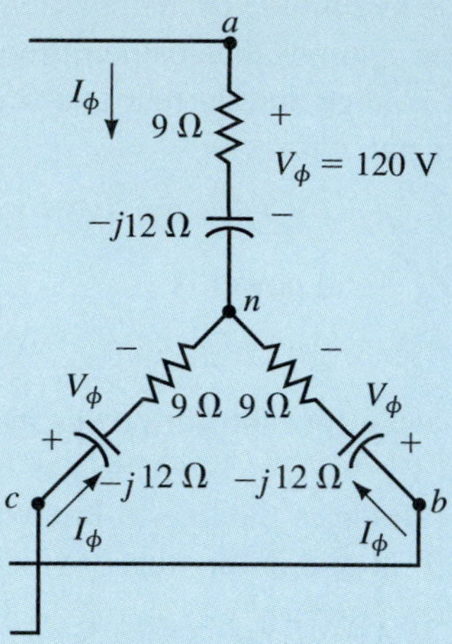

FIGURE 23–23

a. Compute active power to each phase and total power using each equation of this section.
b. Repeat (a) for reactive power.
c. Repeat (a) for apparent power.
d. Find the power factor.

Solution   Since we want to compare answers for the various methods, we will use 207.8 V for the line voltage rather than the nominal value of 208 V to avoid truncation error in our computations.

$$\mathbf{Z}_\phi = 9 - j12 = 15\ \Omega\angle -53.13°. \text{ Thus, } \theta_\phi = -53.13°.$$
$$V_\phi = 120 \text{ V and } I_\phi = V_\phi/Z_\phi = 120 \text{ V}/15\ \Omega = 8.0 \text{ A}.$$
$$V_R = (8 \text{ A})(9\ \Omega) = 72 \text{ V} \quad \text{and} \quad V_X = (8 \text{ A})(12\ \Omega) = 96 \text{ V}$$

a. $P_\phi = V_\phi I_\phi \cos\theta_\phi = (120)(8)\cos(-53.13°) = 576 \text{ W}$

$P_\phi = I_\phi^2 R_\phi = (8^2)(9) = 576 \text{ W}$

$P_\phi = V_R^2/R_\phi = (72)^2/9 = 576 \text{ W}$

$P_T = 3P_\phi = 3(576) = 1728 \text{ W}$

$P_T = \sqrt{3}V_L I_L \cos\theta_\phi = \sqrt{3}(207.8)(8)\cos(-53.13°) = 1728 \text{ W}$

b. $Q_\phi = V_\phi I_\phi \sin\theta_\phi = (120)(8)\sin(-53.13°) = -768 \text{ VAR}$

   $= 768 \text{ VAR (cap.)}$

$Q_\phi = I_\phi^2 X_\phi = (8)^2(12) = 768 \text{ VAR (cap.)}$

$Q_\phi = V_X^2/X_\phi = (96)^2/12 = 768 \text{ VAR (cap.)}$

$Q_T = 3Q_\phi = 3(768) = 2304 \text{ VAR (cap.)}$

$Q_T = \sqrt{3}V_L I_L \sin\theta_\phi = \sqrt{3}(207.8)(8)\sin(-53.13°) = -2304 \text{ VAR}$

   $= 2304 \text{ VAR (cap.)}$

c. $S_\phi = V_\phi I_\phi = (120)(8) = 960$ VA

$\quad S_T = 3S_\phi = 3(960) = 2880$ VA

$\quad S_T = \sqrt{3}V_L I_L = \sqrt{3}(207.8)(8) = 2880$ VA

Thus, all approaches yield the same answers.

d. The power factor is $F_p = \cos\theta_\phi = \cos 53.13° = 0.6$.

Power to a Balanced Delta Load

For a Δ load [Figure 23–24(a)],

$$P_\phi = V_\phi I_\phi \cos\theta_\phi \quad \text{(W)} \qquad\qquad \text{(23–21)}$$

where θ_ϕ is the angle of the Δ impedance. Note that this formula is identical with Equation 23–10 for the Y load. Similarly for reactive power, apparent power, and power factor. Thus, all power formulas are the same. Results are tabulated in Table 23–2. *Note: In all of these formulas, θ_ϕ is the angle of the load impedance, i.e., the angle of $\mathbf{Z}_{an}$ for Y loads and $\mathbf{Z}_{ab}$ for Δ loads.*

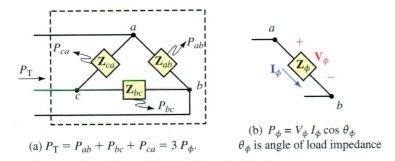

(a) $P_T = P_{ab} + P_{bc} + P_{ca} = 3\,P_\phi$.

(b) $P_\phi = V_\phi I_\phi \cos\theta_\phi$

θ_ϕ is angle of load impedance

FIGURE 23–24   For a balanced Δ, $P_\phi = P_{ab} = P_{bc} = P_{ca}$.

TABLE 23–2   Power Formulas for Balanced Wye and Delta Circuits

Active power	$P_\phi = V_\phi I_\phi \cos\theta_\phi = I_\phi^2 R_\phi = \dfrac{V_R^2}{R_\phi}$
	$P_T = \sqrt{3}V_L I_L \cos\theta_\phi$
Reactive power	$Q_\phi = V_\phi I_\phi \sin\theta_\phi = I_\phi^2 X_\phi = \dfrac{V_x^2}{X_\phi}$
	$Q_T = \sqrt{3}V_L I_L \sin\theta_\phi$
Apparent power	$S_\phi = V_\phi I_\phi = I_\phi^2 Z_\phi = \dfrac{V_\phi^2}{Z_\phi}$
	$S_T = \sqrt{3}V_L I_L$
Power factor	$F_p = \cos\theta_\phi = \dfrac{P_T}{S_T} = \dfrac{P_\phi}{S_\phi}$
Power triangle	$\mathbf{S}_T = P_T + jQ_T$

EXAMPLE 23–12 Determine per phase and total power (active, reactive, and apparent) for Figure 23–25. Use $V_\phi = 207.8$ V in order to compare results.

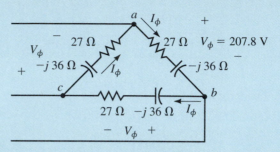

FIGURE 23–25

Solution

$$\mathbf{Z}_\phi = 27 - j36 = 45\ \Omega\angle{-53.13°}, \quad \text{so} \quad \theta_\phi = -53.13°$$

$$V_\phi = 207.8\ \text{V} \quad \text{and} \quad I_\phi = V_\phi/Z_\phi = 207.8\ \text{V}/45\ \Omega = 4.62\ \text{A}$$

$$P_\phi = V_\phi I_\phi \cos\theta_\phi = (207.8)(4.62)\cos(-53.13°) = 576\ \text{W}$$

$$Q_\phi = V_\phi I_\phi \sin\theta_\phi = (207.8)(4.62)\sin(-53.13°) = -768\ \text{VAR}$$

$$= 768\ \text{VAR (cap.)}$$

$$S_\phi = V_\phi I_\phi = (207.8)(4.62) = 960\ \text{VA}$$

$$P_T = 3P_\phi = 3(576) = 1728\ \text{W}$$

$$Q_T = 3Q_\phi = 3(768) = 2304\ \text{VAR (cap.)}$$

$$S_T = 3S_\phi = 3(960) = 2880\ \text{VA}$$

Note that the results here are the same as for Example 23–11. This is to be expected since the load of Figure 23–23 is the Y equivalent of the Δ load of Figure 23–25.

PRACTICE PROBLEMS 7

Find total active, reactive, and apparent power for the circuit of Example 23–12 using the formulas for P_T, Q_T, and S_T from Table 23–2.

Power and the Single-Phase Equivalent

You can also use the single-phase equivalent in power calculations. Here, all single-phase active, reactive, and apparent power formulas apply. The power is of course just the power for one phase.

EXAMPLE 23–13 The total power to the balanced load of Figure 23–26 is 6912 W. The phase voltage at the load is 120 V. Compute the generator voltage $\mathbf{E}_{AB}$, magnitude and angle.

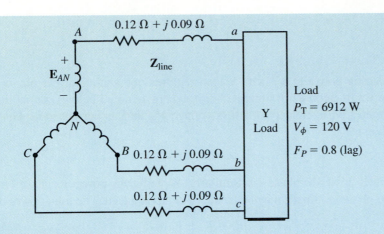

FIGURE 23–26

Solution Consider the single-phase equivalent in Figure 23–27.

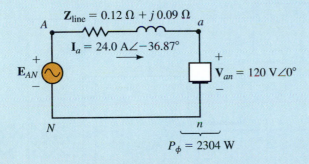

FIGURE 23–27

$$P_{an} = P_T/3 = \frac{1}{3}(6912) = 2304 \text{ W}$$

$$V_{an} = 120 \text{ V}$$

$$\theta_{an} = \cos^{-1}(0.8) = 36.87°$$

$$P_{an} = V_{an}I_a\cos\theta_{an}$$

Therefore,

$$I_a = \frac{P_{an}}{V_{an}\cos\theta_{an}} = \frac{2304}{(120)(0.8)} = 24.0 \text{ A}$$

Select $\mathbf{V}_{an}$ as reference. $\mathbf{V}_{an} = 120 \text{ V}\angle 0°$. Thus, $\mathbf{I}_a = 24 \text{ A}\angle{-36.87°}$ (since power factor was given as lagging).

$$\mathbf{E}_{AN} = \mathbf{I}_a \times \mathbf{Z}_{line} + \mathbf{V}_{an}$$
$$= (24\angle{-36.87})(0.12 + j0.09) + 120\angle 0° = 123.6 \text{ V}\angle 0°$$
$$\mathbf{E}_{AB} = \sqrt{3}\mathbf{E}_{AN}\angle 30° = 214.1 \text{ V}\angle 30°$$

23.6   Measuring Power in Three-Phase Circuits

The Three-Wattmeter Method

Measuring power to a 4-wire Y load requires one wattmeter per phase (i.e., three wattmeters) as in Figure 23–28 (except as noted below). As indicated, wattmeter W_1 is connected across voltage $\mathbf{V}_{an}$ and its current is $\mathbf{I}_a$. Thus, its reading is

$$P_1 = V_{an}I_a\cos\theta_{an}$$

which is power to phase *an*. Similarly, W_2 indicates power to phase *bn* and W_3 to phase *cn*. Loads may be balanced or unbalanced. The total power is

$$P_T = P_1 + P_2 + P_3 \qquad\qquad \textbf{(23–22)}$$

If the load of Figure 23–28 could be guaranteed to always be balanced, only one wattmeter would be needed. P_T would be 3 times its reading.

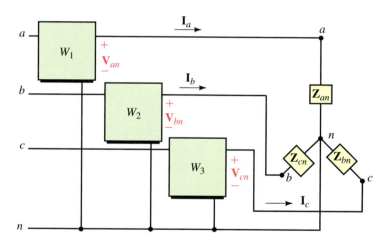

FIGURE 23–28   Three-wattmeter connection for a 4-wire load.

The Two-Wattmeter Method

While three wattmeters are required for a four-wire system, for a three-wire system, only two are needed. The connection is shown in Figure 23–29. Loads may be Y or Δ, balanced or unbalanced. The meters may be connected in any pair of lines with the voltage terminals connected to the third line. The total power is the algebraic sum of the meter readings.

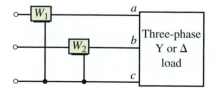

FIGURE 23–29   Two-wattmeter connection. Load may be balanced or unbalanced.

Determining Wattmeter Readings

Recall from Chapter 17, the reading of a wattmeter is equal to the product of the magnitude of its voltage, the magnitude of its current, and the cosine of the angle between them. For each meter, you must carefully determine what this angle is. This is illustrated next.

EXAMPLE 23–14 For Figure 23–30, $\mathbf{V}_{an} = 120\ \text{V}\angle 0°$. Compute the readings of each meter, then sum to determine total power. Compare P_T to the P_T found in Example 23–11.

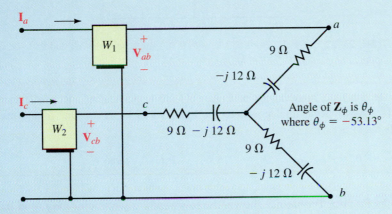

FIGURE 23–30

Solution $\mathbf{V}_{an} = 120\ \text{V}\angle 0°$. Thus, $\mathbf{V}_{ab} = 208\ \text{V}\angle 30°$ and $\mathbf{V}_{bc} = 208\ \text{V}\angle -90°$.
  $\mathbf{I}_a = \mathbf{V}_{an}/\mathbf{Z}_{an} = 120\ \text{V}\angle 0°/(9 - j12)\ \Omega = 8\ \text{A}\angle 53.13°$. Thus, $\mathbf{I}_c = 8\ \text{A}\angle 173.13°$.
  First consider wattmeter 1, Figure 23–31. Note that W_1 is connected to terminals *a-b;* thus it has voltage $\mathbf{V}_{ab}$ across it and current $\mathbf{I}_a$ through it. Its reading is therefore $P_1 = V_{ab}I_a \cos\theta_1$, where θ_1 is the angle between $\mathbf{V}_{ab}$ and $\mathbf{I}_a$. $\mathbf{V}_{ab}$ has an angle of 30° and $\mathbf{I}_a$ has an angle of 53.13°. Thus, $\theta_1 = 53.13° - 30° = 23.13°$ and $P_1 = (208)(8)\cos 23.13° = 1530\ \text{W}$.

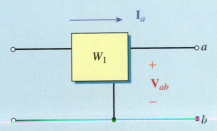

FIGURE 23–31 $P_1 = V_{ab}\,I_a \cos\theta_1$ where θ_1 is the angle between $\mathbf{V}_{ab}$ and $\mathbf{I}_a$.

  Now consider wattmeter 2, Figure 23–32. Since W_2 is connected to terminals *c-b,* the voltage across it is $\mathbf{V}_{cb}$ and the current through it is $\mathbf{I}_c$. But $\mathbf{V}_{cb} = -\mathbf{V}_{bc} = 208\ \text{V}\angle 90°$ and $\mathbf{I}_c = 8\ \text{A}\angle 173.13°$. The angle between $\mathbf{V}_{cb}$ and

$\mathbf{I}_c$ is thus $173.13° - 90° = 83.13°$. Therefore, $P_2 = V_{cb}I_c\cos\theta_2 = (208)(8)\cos$ $83.13° = 199$ W and $P_T = P_1 + P_2 = 1530 + 199 = 1729$ W. (This agrees well with the answer of 1728 W that we obtained in Example 23–11.) Note that one of the wattmeters reads lower than the other. (This is generally the case for the two-wattmeter method.)

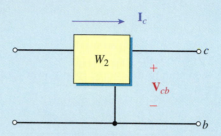

FIGURE 23–32 $P_2 = V_{cb}I_c\cos\theta_2$ where θ_2 is the angle between $\mathbf{V}_{cb}$ and $\mathbf{I}_c$.

PRACTICE PROBLEMS 8

Change the load impedances of Figure 23–30 to $15\ \Omega\angle70°$. Repeat Example 23–14. (As a check, total power to the load is 985 W. *Hint:* One of the meters reads backwards.)

Answers: $P_1 = -289$ W; $P_2 = 1275$ W; $P_1 + P_2 = 986$ W

NOTES...

To understand why a wattmeter may read backward, recall that it indicates the product of the magnitude of its voltage times the magnitude of its current times the cosine of the angle between them. This angle is not the angle θ_ϕ of the load impedance; Rather, it can be shown that, for a balanced load, one meter will indicate $V_\phi I_\phi \cos(\theta_\phi - 30°)$ while the other will indicate $V_\phi I_\phi \cos(\theta_\phi + 30°)$. If the magnitude of $(\theta_\phi + 30°)$ or $(\theta_\phi - 30°)$ exceeds $90°$, its cosine will be negative and the corresponding meter will read backward.

As Practice Problem 8 shows, the low-reading wattmeter may read backward. (If this happens, reverse either its voltage or current connection to make it read upscale.) Its reading must then be subtracted. Thus, if P_h and P_ℓ are the high-reading and low-reading meters respectively,

$$P_T = P_h - P_\ell \qquad (23-23)$$

Watts Ratio Curve

The power factor for a balanced load can be obtained from the wattmeter readings using a simple curve called the **watts ratio curve,** shown in Figure 23–33.

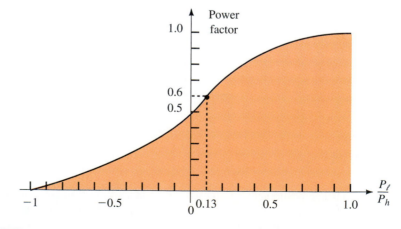

FIGURE 23–33   Watts ratio curve. Valid for balanced loads only.

EXAMPLE 23–15   Consider again Figure 23–30.

a. Determine the power factor from the load impedance.

b. Using the meter readings of Example 23–14, determine the power factor from the watts ratio curve.

Solution

a. $F_p = \cos \theta_\phi = \cos 53.13° = 0.6$.

b. $P_\ell = 199$ W and $P_h = 1530$ W. Therefore, $P_\ell/P_h = 0.13$. From Figure 23–33, $F_p = 0.6$.

The problem with the watts ratio curve is that values are difficult to determine accurately from the graph. However, it can be shown that

$$\tan \theta_\phi = \sqrt{3}\left(\frac{P_h - P_\ell}{P_h + P_\ell}\right) \qquad (23\text{–}24)$$

From this, you can determine θ_ϕ and then you can compute the power factor from $F_p = \cos \theta_\phi$.

23.7   Unbalanced Loads

For unbalanced loads, none of the balanced circuit relationships apply. Each problem must therefore be treated as a three-phase problem. We will look at a few examples that can be handled by fundamental circuit techniques such as Kirchhoff's laws and mesh analysis. Source voltages are always balanced.

Unbalanced Wye Loads

Unbalanced four-wire Y systems without line impedance are easily handled using Ohm's law. However, for three-wire systems or four-wire systems with line and neutral impedance, you generally have to use mesh equations or computer methods. One of the problems with unbalanced three-wire Y systems is that you get different voltages across each phase of the load and a voltage between neutral points. This is illustrated next.

EXAMPLE 23–16   For Figure 23–34(a), the generator is balanced with line-to-line voltage of 208 V. Select $\mathbf{E}_{AB}$ as reference and determine line currents and load voltages.

SAFETY NOTE...

Voltage at a neutral point can be dangerous. For example, in Figure 23–34, if the neutral is grounded at the source, the voltage at the load neutral is *floating* at some potential relative to ground. Since we are accustomed to thinking that neutrals are at ground potential and hence safe to touch, there is a potential safety hazard here.

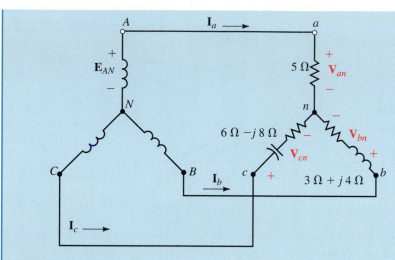

EWB   **FIGURE 23–34**

Solution   Redraw the circuit as shown in Figure 23–35, then use mesh analysis. $\mathbf{E}_{AB} = 208 \text{ V}\angle 0°$ and $\mathbf{E}_{BC} = 208 \text{ V}\angle -120°$.

$$\textit{Loop 1: } (8 + j4)\mathbf{I}_1 - (3 + j4)\mathbf{I}_2 = 208 \text{ V}\angle 0°$$
$$\textit{Loop 2: } -(3 + j4)\mathbf{I}_1 + (9 - j4)\mathbf{I}_2 = 208 \text{ V}\angle -120°$$

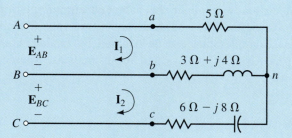

FIGURE 23–35

These equations can be solved using standard techniques such as determinants or MathCAD. The solutions are

$$\mathbf{I}_1 = 29.9 \text{ A}\angle -26.2° \quad \text{and} \quad \mathbf{I}_2 = 11.8 \text{ A}\angle -51.5°$$

$$\text{KCL: } \mathbf{I}_a = \mathbf{I}_1 = 29.9 \text{ A}\angle -26.2°$$
$$\mathbf{I}_b = \mathbf{I}_2 - \mathbf{I}_1 = 19.9 \text{ A}\angle 168.5°$$
$$\mathbf{I}_c = -\mathbf{I}_2 = 11.8 \text{ A}\angle 128.5°$$

$$\mathbf{V}_{an} = \mathbf{I}_a \mathbf{Z}_{an} = (29.9 \text{A}\angle -26.2°)\,(5) = 149.5 \text{ V}\angle -26.2°$$
$$\mathbf{V}_{bn} = \mathbf{I}_b \mathbf{Z}_{bn} \quad \text{and} \quad \mathbf{V}_{cn} = \mathbf{I}_c \mathbf{Z}_{cn}.$$

Thus,

$$\mathbf{V}_{bn} = 99 \text{ V}\angle -138.4° \quad \text{and} \quad \mathbf{V}_{cn} = 118 \text{ V}\angle 75.4°$$

1. Using KVL and the results from Example 23–16, compute the voltage between neutral points n and N of Figure 23–34.

2. For the circuit of Figure 23–36, $\mathbf{E}_{AN} = 120$ V∠0°. Compute the currents, power to each phase, and total power. Hint: This is actually quite a simple problem. Because there is a neutral, source voltages are applied directly to the load and $\mathbf{V}_{an} = \mathbf{E}_{AN}$, etc.

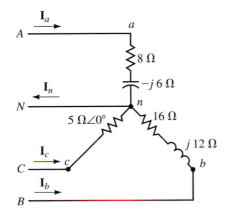

FIGURE 23–36

Answers:

1. $\mathbf{V}_{nN} = 30.8$ V∠168.8°

2. $\mathbf{I}_a = 12$ A∠36.9°; $\mathbf{I}_b = 6$ A∠−156.9°; $\mathbf{I}_c = 24$ A∠120°; $\mathbf{I}_n = 26.8$ A∠107.2°
   $P_a = 1152$ W; $P_b = 576$ W; $P_c = 2880$ W; $P_T = 4608$ W

Unbalanced Delta Loads

Systems without line impedance are easily handled since the source voltage is applied directly to the load. However, for systems with line impedance, use mesh equations.

EXAMPLE 23–17 For the circuit of Figure 23–37, the line voltage is 240 V. Take $\mathbf{V}_{ab}$ as reference and do the following:

a. Determine the phase currents and sketch their phasor diagram.

b. Determine the line currents.

c. Determine the total power to the load.

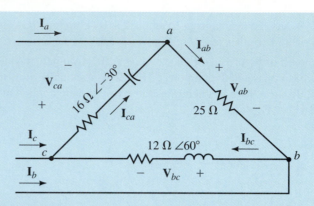

FIGURE 23–37

Solution

a. $\mathbf{I}_{ab} = \mathbf{V}_{ab}/\mathbf{Z}_{ab} = (240\ \text{V}\angle 0°)/25\ \Omega = 9.6\ \text{A}\angle 0°$

$\mathbf{I}_{bc} = \mathbf{V}_{bc}/\mathbf{Z}_{bc} = (240\ \text{V}\angle -120°)/(12\ \Omega\angle 60°) = 20\ \text{A}\angle -180°$

$\mathbf{I}_{ca} = \mathbf{V}_{ca}/\mathbf{Z}_{ca} = (240\ \text{V}\angle 120°)/(16\ \Omega\angle -30°) = 15\ \text{A}\angle 150°$

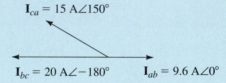

FIGURE 23–38   Phase currents for the circuit of Figure 23–37.

b. $\mathbf{I}_a = \mathbf{I}_{ab} - \mathbf{I}_{ca} = 9.6\ \text{A}\angle 0° - 15\ \text{A}\angle 150° = 23.8\ \text{A}\angle -18.4°$

$\mathbf{I}_b = \mathbf{I}_{bc} - \mathbf{I}_{ab} = 20\ \text{A}\angle -180° - 9.6\ \text{A}\angle 0° = 29.6\ \text{A}\angle 180°$

$\mathbf{I}_c = \mathbf{I}_{ca} - \mathbf{I}_{bc} = 15\ \text{A}\angle 150° - 20\ \text{A}\angle -180° = 10.3\ \text{A}\angle 46.9°$

c. $P_{ab} = V_{ab}I_{ab}\cos\theta_{ab} = (240)(9.6)\cos 0° = 2304\ \text{W}$

$P_{bc} = V_{bc}I_{bc}\cos\theta_{bc} = (240)(20)\cos 60° = 2400\ \text{W}$

$P_{ca} = V_{ca}I_{ca}\cos\theta_{ca} = (240)(15)\cos 30° = 3118\ \text{W}$

$P_T = P_{ab} + P_{bc} + P_{ca} = 7822\ \text{W}$

EXAMPLE 23–18   A pair of wattmeters are added to the circuit of Figure 23–37 as illustrated in Figure 23–39. Determine wattmeter readings and compare them to the total power calculated in Example 23–17.

Solution $P_1 = V_{ac}I_a\cos\theta_1$, where θ_1 is the angle between $\mathbf{V}_{ac}$ and $\mathbf{I}_a$. From Example 23–17, $\mathbf{I}_a = 23.8\ \text{A}\angle -18.4°$ and $\mathbf{V}_{ac} = -\mathbf{V}_{ca} = 240\ \text{V}\angle -60°$. Thus, $\theta_1 = 60° - 18.4° = 41.6°$. Therefore,

$$P_1 = (240)(23.8)\cos 41.6° = 4271\ \text{W}$$

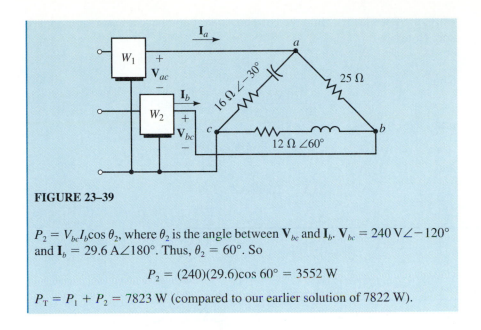

FIGURE 23–39

$P_2 = V_{bc}I_b\cos\theta_2$, where θ_2 is the angle between $\mathbf{V}_{bc}$ and $\mathbf{I}_b$. $\mathbf{V}_{bc} = 240\,\text{V}\angle -120°$
and $\mathbf{I}_b = 29.6\,\text{A}\angle 180°$. Thus, $\theta_2 = 60°$. So

$$P_2 = (240)(29.6)\cos 60° = 3552\,\text{W}$$

$P_T = P_1 + P_2 = 7823\,\text{W}$ (compared to our earlier solution of 7822 W).

23.8   Power System Loads

Before we leave this chapter, we look briefly at how both single-phase and three-phase loads may be connected to a three-phase system. (This is necessary because residential and business customers require only single- phase power, while industrial customers sometimes require both single-phase and three-phase power.) Figure 23–40 shows how this may be done. (Even this is simplified, as real systems contain transformers. However, the basic principles are correct.) Two points should be noted here.

1. In order to approximately balance the system, the utility tries to connect one-third of its single phase loads to each phase. Three-phase loads generally are balanced.

2. Real loads are seldom expressed in terms of resistance, capacitance, and inductance. Instead, they are described in terms of power, power factor, and so on. This is because most loads consist of electric lights, motors, and the like which are never described in terms of impedance. (For example, you purchase light bulbs as 60-W bulbs, 100-W bulbs, etc., and electric motors as ½ horsepower, etc. You never ask for a 240-ohm light!)

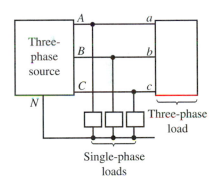

FIGURE 23–40 Single-phase loads are tapped off the three-phase lines.

23.9   Circuit Analysis Using Computers

PSpice and Electronics Workbench may be used to analyze three-phase systems (balanced or unbalanced, Y or Δ connected). As usual, PSpice yields full phasor solutions, but as of this writing, Workbench yields magnitudes only. Because neither software package allows component placement at an angle, Y and Δ circuits must be drawn with components placed either horizontally or vertically as in Figures 23–42 and 23–45, rather than in their traditional three-phase form. To begin, consider the balanced 4-wire Y circuit of Figure 23–41. First, find the currents so that you have a basis for comparison. Note

$$X_C = 53.05 \ \Omega$$

Thus

$$\mathbf{I}_a = \frac{120 \ \text{V} \angle 0°}{(30 - j53.05) \ \Omega} = 1.969 \ \text{A} \angle 60.51°$$

and

$$\mathbf{I}_b = 1.969 \ \text{A} \angle -59.49°$$

and

$$\mathbf{I}_c = 1.969 \ \text{A} \angle -179.49°.$$

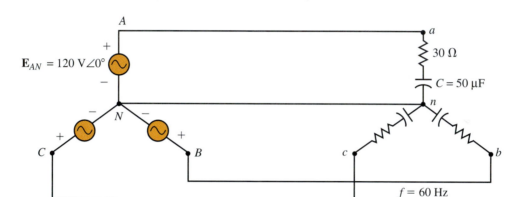

FIGURE 23–41 Balanced system for computer analysis.

Electronics Workbench

Draw the circuit on the screen as in Figure 23–42. Make sure the voltage sources are oriented with their + ends as shown and that their phase angles are set appropriately. (Workbench does not accept negative angles, thus use 240° instead of −120° for $\mathbf{E}_{BN}$.) Double click the ammeters and set them to

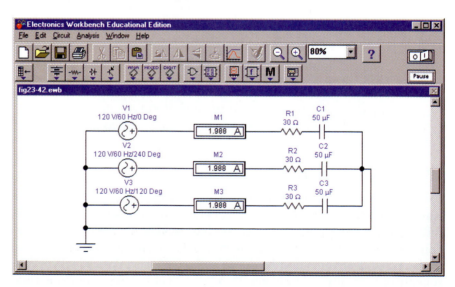

FIGURE 23–42 Solution of the circuit of Figure 23–41 by Electronics Workbench.

AC. Activate the circuit by clicking the power switch. Workbench computes answers of 1.989 for the currents (instead of 1.969 as we calculated above), an error of 0.02 A (approximately 1%).

OrCAD PSpice

Draw the circuit on the screen as in Figure 23–43 with source + terminals oriented as shown. Double click each source in turn and, in its Property Editor, set its magnitude to 120V with phase angle as indicated. Similarly, double click each IPRINT device and set MAG to **yes,** PHASE to **yes,** and AC to **yes.** Via the New Profile icon, select AC Sweep/Noise, set the Start and End frequencies to **60Hz** and the number of computational points to **1.** Run the simulation, open the Output File (found under View in the results window), then scroll down until you find the answers. (Figure 23–44 shows the answer for current $\mathbf{I}_a$.) Examining the results, you will see that they agree exactly with what we calculated earlier.

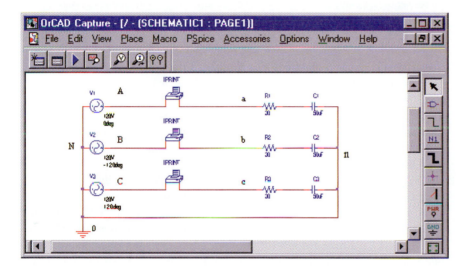

FIGURE 23–43 Solution of the circuit of Figure 23–41 by PSpice.

FREQ	IM(V_PRINT1)	IP(V_PRINT1)
6.000E+01	1.969E+00	6.051E+01

FIGURE 23–44 PSpice yields $\mathbf{I}_a = 1.969\ \text{A}\angle 60.51°$ for the circuit of Figure 23–41.

Since the circuit is balanced, you should be able to remove the neutral conductor from between *N-n.* If you do however, you will get errors. This is because PSpice requires a dc path from every node back to the reference, but because capacitors look like open circuits to dc, it sees node *n* as floating. A simple solution is to put a very high value resistor (say 100 kΩ) in the path between *N-n.* Its value is not critical; it just has to be large enough to look like an open circuit. Try it and note that you get the same answers as before.

FINAL NOTES...

1. Electronics Workbench and PSpice make no distinction between balanced circuits and unbalanced circuits; to the computer, they are simply circuits. Thus, no special considerations are necessary for unbalanced loads.
2. Delta loads may be drawn as in Figure 23–45.
3. To annotate the diagram as in Figure 23–43 (i.e., with points A, B, C, N, etc. identified), click the Place text icon on the tool palette, key in the text, and place as desired.

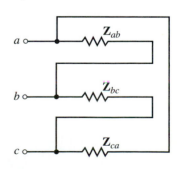

FIGURE 23–45 Representing a Δ load.

PUTTING IT INTO PRACTICE

You have been sent to a job site to supervise the installation of a 208-V, three-phase motor. The motor drives a machine, and it is essential that the motor rotate in the correct direction (in this case, clockwise), else the machine may be damaged. You have a drawing that tells you to connect Line *a* of the motor to Line *A* of the three-phase system, Line *b* to Line *B,* etc. However, you find that the three-phase lines have no markings, and you don't know which line is which. Unfortunately, you cannot simply connect the motor and determine which direction it turns for fear of damage.

You think about this for awhile, then you come up with a plan. You know that the direction of rotation of a three-phase motor depends on the phase sequence of the applied voltage, so you make a sketch (sketch A). As indicated in part (a) of that drawing, the motor rotates in the correct direction when *a* is connected to *A, b* to *B,* etc. You reason that the phase sequence you need is ...*A-B-C-A-B-C*... (since the direction of rotation depends only on phase sequence) and that it really doesn't matter to which line *a* is connected as long as the other two are connected such that you provide sequencing in this order to the motor. To convince yourself, you make several more sketches (parts b and c). As indicated, in (b) the sequence is ...*B-C-A-B-C-A*... (which fits the above pattern) and the motor rotates in the correct direction, but for (c) the motor rotates in the reverse direction. (Show why.)

You also remember reading about a device called a *phase sequence indicator* that allows you to determine phase sequence. It uses lights and a capacitor as in sketch B. To use the device, connect terminal *a* to the three-phase line that you have designated *A* and connect terminals *b* and *c* to the other two lines. The lamp that lights brightly is the one attached to line *B.* After a few calculations, you ask the plant electrician for some standard 120-V 60-W light bulbs and sockets. From your tool kit, you retrieve a 3.9-μF capacitor (one that is rated for ac operation). You make a slight modification to the schematic of sketch B, solder the parts together, wrap exposed wires with electrical tape (for safety), then connect the device and identify the three-phase lines. You then connect the motor and it rotates in the correct direction. Prepare an analysis to show why the lamp connected to *B* is much brighter than the lamp connected to line *C.* (Note: Don't forget that you made a slight change to the schematic. Hint: If you use a single 60-W lamp in each leg as shown, they will burn out.)

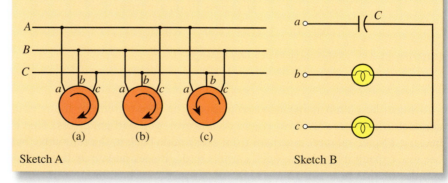

Sketch A   Sketch B

23.2   Basic Three-Phase Circuit Connections

1. As long as the loads and voltages of Figure 23–3 are balanced (regardless of their actual value), currents $\mathbf{I}_A$, $\mathbf{I}_B$, and $\mathbf{I}_C$ will sum to zero. To illustrate, change the load impedance from 12 Ω to 15 Ω∠30° and for $\mathbf{E}_{AA'} = 120\,\text{V}\angle0°$, do the following:

   a. Compute currents $\mathbf{I}_A$, $\mathbf{I}_B$, and $\mathbf{I}_C$.

   b. Sum the currents. Does $\mathbf{I}_A + \mathbf{I}_B + \mathbf{I}_C = 0$?

2. For Figure 23–3(c), $\mathbf{E}_{AN} = 277\,\text{V}\angle-15°$.

   a. What are $\mathbf{E}_{BN}$ and $\mathbf{E}_{CN}$?

   b. If each resistance is 5.54 Ω, compute $\mathbf{I}_A$, $\mathbf{I}_B$, and $\mathbf{I}_C$.

   c. Show that $\mathbf{I}_N = 0$.

3. If the generator of Figure 23–2 is rotated in the clockwise direction, what will be the phase sequence of the generated voltages?

23.3   Basic Three-Phase Relationships

4. For the generators of Figure 23–4, $\mathbf{E}_{AN} = 7620\,\text{V}\angle-18°$.

   a. What are the phase voltages $\mathbf{E}_{BN}$ and $\mathbf{E}_{CN}$?

   b. Determine the line-to-line voltages.

   c. Sketch their phasor diagram.

5. For the loads of Figure 23–4, $\mathbf{V}_{bc} = 208\,\text{V}\angle-75°$.

   a. Determine the line-to-line voltages $\mathbf{V}_{ab}$ and $\mathbf{V}_{ca}$.

   b. Determine the phase voltages.

   c. Sketch their phasor diagram.

6. Repeat Problem 5 if $\mathbf{V}_{ca} = 208\,\text{V}\angle90°$.

7. For the load of Figure 23–46, $\mathbf{V}_{an} = 347\,\text{V}\angle15°$. Determine all line currents. Sketch their phasor diagram.

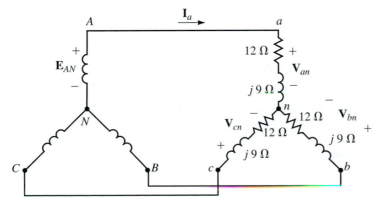

FIGURE 23–46

8. For the load of Figure 23–46, if $\mathbf{I}_a = 7.8\,\text{A}\angle-10°$, determine the phase voltages and line voltages. Sketch their phasor diagram.

9. A balanced Y load has impedance $\mathbf{Z}_{an} = 14.7\,\Omega\angle16°$. If $\mathbf{V}_{cn} = 120\,\text{V}\angle160°$, determine all line currents.

10. For a balanced delta load, $\mathbf{I}_{ab} = 29.3\,\text{A}\angle 43°$. What is $\mathbf{I}_a$?

11. For the circuit of Figure 23–47, $\mathbf{V}_{ab} = 480\,\text{V}\angle 0°$. Find the phase and line currents.

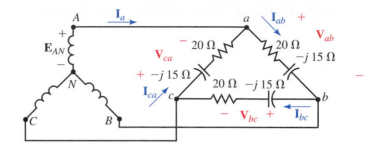

FIGURE 23–47

12. For the circuit of Figure 23–47, if $\mathbf{I}_a = 41.0\,\text{A}\angle -46.7°$, find all the phase currents.

13. For the circuit of Figure 23–47, if $\mathbf{I}_{ab} = 10\,\text{A}\angle -21°$, determine all the line voltages.

14. For the circuit of Figure 23–47, if line current $\mathbf{I}_a = 11.0\,\text{A}\angle 30°$, find all the phase voltages.

15. A balanced Y load has phase impedance of $24\,\Omega\angle 33°$ and line-to-line voltage of 600 V. Take $\mathbf{V}_{an}$ as reference and determine all phase currents.

16. A balanced Δ load has phase impedance of $27\,\Omega\angle -57°$ and phase voltage of 208 V. Take $\mathbf{V}_{ab}$ as reference and determine

  a. Phase currents   b. Line currents

17. a. For a certain balanced Y load, $\mathbf{V}_{ab} = 208\,\text{V}\angle 30°$, $\mathbf{I}_a = 24\,\text{A}\angle 40°$ and $f = 60$ Hz. Determine the load (R and L or C).

  b. Repeat (a) if $\mathbf{V}_{bc} = 208\,\text{V}\angle -30°$ and $\mathbf{I}_c = 12\,\text{A}\angle 140°$.

18. Consider Figure 23–12(a). Show that $\mathbf{I}_a = \sqrt{3}\,\mathbf{I}_{ab}\angle -30°$.

19. At 60 Hz, a balanced Δ load has current $\mathbf{I}_{bc} = 4.5\,\text{A}\angle -85°$. The line voltage is 240 volts and $\mathbf{V}_{ab}$ is taken as reference.

  a. Find the other phase currents.

  b. Find the line currents.

  c. Find the resistance R and capacitance C of the load.

20. A Y generator with $\mathbf{E}_{AN} = 120\,\text{V}\angle 0°$ drives a balanced Δ load. If $\mathbf{I}_a = 43.6\,\text{A}\angle -37.5°$, what are the load impedances?

23.4 Examples

21. For Figure 23–48, $\mathbf{V}_{an} = 120\,\text{V}\angle 0°$. Draw the single-phase equivalent and:

  a. Find phase voltage $\mathbf{E}_{AN}$, magnitude and angle.

  b. Find line voltage $\mathbf{E}_{AB}$, magnitude and angle.

22. For Figure 23–48, $\mathbf{E}_{AN} = 120\,\text{V}\angle 20°$. Draw the single-phase equivalent and:

  a. Find phase voltage $\mathbf{V}_{an}$, magnitude and angle.

  b. Find line voltage $\mathbf{V}_{ab}$, magnitude and angle.

23. For Figure 23–47, $\mathbf{E}_{AN} = 120\,\text{V}\angle -10°$. Find the line currents using the single-phase equivalent method.

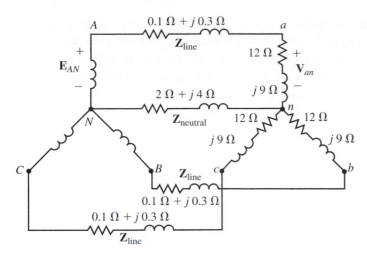

FIGURE 23–48

24. Repeat Problem 23 if $\mathbf{E}_{BN} = 120\ \text{V}\angle-100°$.

25. For Figure 23–47, assume the lines have impedance $\mathbf{Z}_{\text{line}}$ of $0.15\ \Omega + j0.25\ \Omega$ and $\mathbf{E}_{AN} = 120\ \text{V}\angle 0°$. Convert the Δ load to a Y and use the single-phase equivalent to find the line currents.

26. For Problem 25, find the phase currents in the Δ network.

27. For the circuit of Figure 23–48, assume $\mathbf{Z}_{\text{line}} = 0.15\ \Omega + j0.25\ \Omega$ and $\mathbf{V}_{ab} = 600\ \text{V}\angle 30°$. Determine $\mathbf{E}_{AB}$.

28. For Figure 23–49, $\mathbf{Z}_Y = 12\ \Omega + j9\ \Omega$ and $\mathbf{Z}_\Delta = 27\ \Omega + j36\ \Omega$. At the Y load, $\mathbf{V}_{an} = 120\ \text{V}\angle 0°$.

    a. Draw the single-phase equivalent.

    b. Find generator voltage $\mathbf{E}_{AN}$.

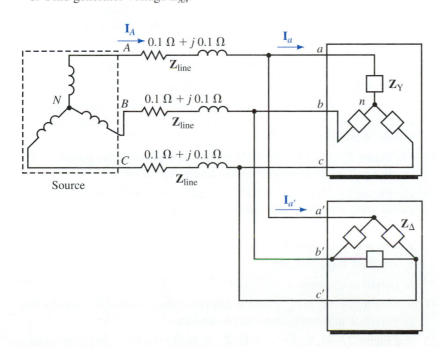

FIGURE 23–49

29. Same as Problem 28 except that the phase voltage at the Δ load is $V_{a'b'} = 480\ V\angle30°$. Find generator voltage $\mathbf{E}_{AB}$, magnitude and angle.

30. For Figure 23–49, $\mathbf{Z}_Y = 12\ \Omega + j9\ \Omega$ and $\mathbf{Z}_\Delta = 36\ \Omega + j27\ \Omega$. Line current $\mathbf{I}_A$ is $46.2\ A\angle-36.87°$. Find the phase currents for both loads.

31. For Figure 23–49, $\mathbf{Z}_Y = 15\ \Omega + j20\ \Omega$, $\mathbf{Z}_\Delta = 9\ \Omega - j12\ \Omega$ and $\mathbf{I}_{a'b'} = 40\ A\angle73.13°$. Find Y phase voltage $\mathbf{V}_{an}$, magnitude and angle.

23.5 Power in a Balanced System

32. For the balanced load of Figure 23–50, $V_{ab} = 600\ V$. Determine per phase and total active, reactive, and apparent power.

33. Repeat Problem 32 for the balanced load of Figure 23–51, given $E_{AN} = 120\ V$.

34. For Figure 23–46, $E_{AN} = 120$ volts.

    a. Determine per phase real, reactive, and apparent power.

    b. Multiply per phase quantities by 3 to get total quantities.

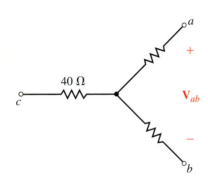

FIGURE 23–50

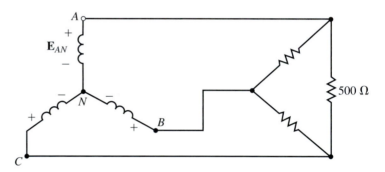

FIGURE 23–51

35. For Figure 23–46, compute the total real, reactive, and apparent power using the P_T, Q_T, and S_T formulas of Table 23–2. (Use $V_L = 207.8\ V$ rather than the nominal value of 208 V here.) Compare these to the results of Problem 34.

36. For Figure 23–47, $E_{AB} = 208$ volts.

    a. Determine per phase real, reactive, and apparent power.

    b. Multiply per phase quantities by 3 to get total quantities.

37. For Figure 23–47, $E_{AB} = 208\ V$. Compute the total real, reactive, and apparent power using the P_T, Q_T, and S_T formulas of Table 23–2. Compare to the results of Problem 36.

38. For Figure 23–11, if $V_{an} = 277\ V$, determine the total active power, total reactive power, total apparent power, and power factor.

39. For Figure 23–13, if $V_{ab} = 600\ V$, determine the total power, total reactive power, total apparent power, and power factor.

40. For Figure 23–17, if $V_{an} = 120\ V$, determine the total power, total reactive power, and total apparent power,

    a. supplied to the load.

    b. output by the source.

41. For Figure 23–18, if $V_{ab} = 480\ V$, determine the total power, total reactive power, total apparent power, and power factor.

42. For Figure 23–49, let $\mathbf{Z}_{\text{line}} = 0\ \Omega$, $\mathbf{Z}_Y = 20\ \Omega\angle0°$, $\mathbf{Z}_\Delta = 30\ \Omega\angle10°$, and $E_{AN} = 120\ V$.

a. Find the total real power, reactive power, and apparent power to the Y load.

b. Repeat (a) for the Δ load.

c. Determine total watts, VARs, and VA using the results of (a) and (b).

43. V_{ab} = 208 V for a balanced Y load, P_T = 1200 W, and Q_T = 750 VAR (ind.). Choose $\mathbf{V}_{an}$ as reference and determine $\mathbf{I}_a$. (Use power triangle.)

44. A motor (delivering 100 hp to a load) and a bank of power factor capacitors are connected as in Figure 23–52. The capacitors are rated Q_C = 45 kVAR. Reduce the problem to its single-phase equivalent, then compute the resultant power factor of the system.

45. The capacitors of Figure 23–52 are connected in Y and each has the value C = 120 μF. Compute the resultant power factor. Frequency is 60 Hz.

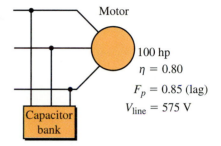

FIGURE 23–52

23.6 Measuring Power in Three-Phase Circuits

46. For Figure 23–53:

a. Determine the wattmeter reading.

b. If the load is balanced, what is P_T?

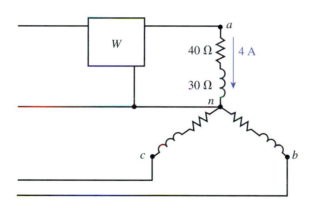

FIGURE 23–53

47. For Figure 23–46, the generator phase voltage is 120 volts.

a. Sketch three wattmeters correctly into the circuit.

b. Compute the reading of each wattmeter.

c. Sum the readings and compare this to the result of 2304 W obtained in Problem 34.

48. Figures 23–29 and 23–30 show two ways that two wattmeters may be connected to measure power in a three-phase, three-wire circuit. There is one more way. Sketch it.

49. For the circuit of Figure 23–54, $\mathbf{V}_{ab}$ = 208 V∠30°.

a. Determine the magnitude and angle of the currents.

b. Determine power per phase and total power, P_T.

c. Compute the reading of each wattmeter.

d. Sum the meter readings and compare the result to P_T of (b).

50. Two wattmeters measure power to a balanced load. Readings are P_h = 1000 W and P_ℓ = −400 W. Determine the power factor of the load from Equation 23–24 and from Figure 23–33. How do they compare?

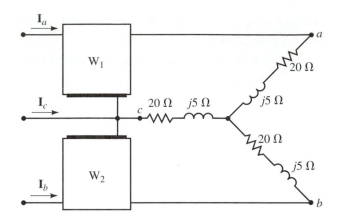

FIGURE 23–54

51. Consider the circuit of Figure 23–54.

    a. Compute the power factor from the angle of the phase impedances.

    b. In Problem 49, wattmeter readings $P_h = 1164$ W and $P_\ell = 870$ W were determined. Substitute these values into Equation 23–24 and compute the power factor of the load. Compare the results to (a).

52. For the balanced load of Figure 23–55, $\mathbf{V}_{ab} = 208$ V∠0°.

    a. Compute phase currents and line currents.

    b. Determine power per phase and total power P_T.

    c. Compute the reading of each wattmeter, then sum the wattmeter readings and compare to P_T from (b).

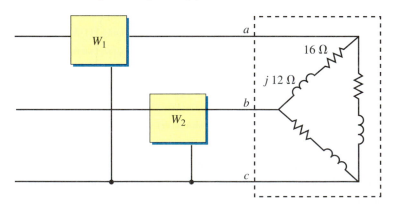

FIGURE 23–55

23.7   Unbalanced Systems

53. For Figure 23–56, $R_{ab} = 60$ Ω, $\mathbf{Z}_{bc} = 80$ Ω $+ j60$ Ω. Compute

    a. Phase and line currents.

    b. Power to each phase and total power.

54. Repeat Problem 53 if $P_{ab} = 2400$ W and $\mathbf{Z}_{bc} = 50$ Ω∠40°.

55. For Figure 23–57, compute the following:

    a. Line currents, magnitudes and angles.

    b. The neutral current.

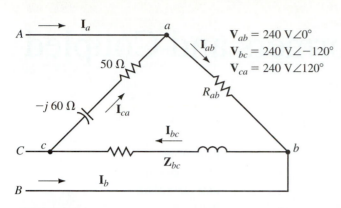

$$\mathbf{V}_{ab} = 240 \text{ V}\angle0°$$
$$\mathbf{V}_{bc} = 240 \text{ V}\angle-120°$$
$$\mathbf{V}_{ca} = 240 \text{ V}\angle120°$$

FIGURE 23–56

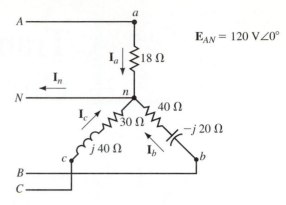

$$\mathbf{E}_{AN} = 120 \text{ V}\angle0°$$

FIGURE 23–57

   c. Power to each phase.

   d. Total power to the load.

56. Remove the neutral conductor from the circuit of Figure 23–57 and compute the line currents. *Hint:* Use mesh equations.

57. From Problem 56, $\mathbf{I}_a = 1.94 \text{ A}\angle-0.737°$, $\mathbf{I}_b = 4.0 \text{ A}\angle-117.7°$ and $\mathbf{I}_c = 3.57 \text{ A}\angle91.4°$. Compute the following:

   a. Voltages across each phase of the load.

   b. Voltage between the neutral of the load and the neutral of the generator.

23.9 Circuit Analysis Using Computers

For the following, use either Electronics Workbench or PSpice. With Workbench, you get only magnitude; with PSpice, solve for magnitude and angle. Caution: For PSpice, insert the IPRINT devices so that current enters the positive terminal. Otherwise, the phase angle will be in error by 180°. (See Figure 23–43.)

58. **EWB** or **PSpice** For the balanced system of Figure 23–46, let $\mathbf{E}_{AN} = 347 \text{ V}\angle15°$, $L = 8.95$ mH, and $f = 160$ Hz. Solve for line currents.

59. **EWB** or **PSpice** For the balanced system of Figure 23–47, let $\mathbf{E}_{AN} = 277 \text{ V}\angle-30°$, $C = 50 \text{ }\mu\text{F}$, and $f = 212$ Hz. Solve for phase currents and line currents.

60. **EWB** or **PSpice** Repeat Problem 59 with C replaced with $L = 11.26$ mH.

61. **EWB** or **PSpice** For Figure 23–57, let $L = 40$ mH, $C = 50 \text{ }\mu\text{F}$, and $\omega = 1000$ rad/s. Solve for line and neutral currents.

In-Process Learning Check 1

1. $\mathbf{V}_{an} = \mathbf{E}_{AN} = 277 \text{ V}\angle-20°$; $\mathbf{V}_{bn} = \mathbf{E}_{BN} = 277 \text{ V}\angle-140°$;
   $\mathbf{V}_{cn} = \mathbf{E}_{CN} = 277 \text{ V}\angle100°$; $\mathbf{V}_{ab} = \mathbf{E}_{AB} = 480 \text{ V}\angle10°$;
   $\mathbf{V}_{bc} = \mathbf{E}_{BC} = 480 \text{ V}\angle-110°$; $\mathbf{V}_{ca} = \mathbf{E}_{CA} = 480 \text{ V}\angle130°$

2. $\mathbf{V}_{an} = \mathbf{E}_{AN} = 120 \text{ V}\angle50°$; $\mathbf{V}_{bn} = \mathbf{E}_{BN} = 120 \text{ V}\angle-70°$;
   $\mathbf{V}_{cn} = \mathbf{E}_{CN} = 120 \text{ V}\angle170°$; $\mathbf{V}_{ab} = \mathbf{E}_{AB} = 208 \text{ V}\angle80°$;
   $\mathbf{V}_{bc} = \mathbf{E}_{BC} = 208 \text{ V}\angle-40°$; $\mathbf{V}_{ca} = \mathbf{E}_{CA} = 208 \text{ V}\angle-160°$

3. $\mathbf{V}_{an} = 124 \text{ V}\angle-1.87°$; $\mathbf{V}_{ab} = 214 \text{ V}\angle28.13°$

4. $\mathbf{V}_{ab} = 104 \text{ V}\angle81.7°$; $\mathbf{V}_{bc} = 104 \text{ V}\angle-38.3°$; $\mathbf{V}_{ca} = 104 \text{ V}\angle-158.3°$

24

Transformers and Coupled Circuits

OBJECTIVES

After studying this chapter, you will be able to

- describe how a transformer couples energy from its primary to its secondary via a changing magnetic field,
- describe basic transformer construction,
- use the dot convention to determine transformer phasing,
- determine voltage and current ratios from the turns ratio for iron-core transformers,
- compute voltage and currents in circuits containing iron-core and air-core transformers,
- use transformers to impedance match loads,
- describe some basic transformer applications,
- determine transformer equivalent circuits,
- compute iron-core transformer efficiency,
- use Electronics Workbench and PSpice to solve circuits with transformers and coupled circuits.

KEY TERMS

Air-Core Transformer
Autotransformer
Coefficient of Coupling
Copper Loss
Core Loss
Coupled Circuit
Current Ratio
Dot Convention
Ferrite Core
Ideal Transformer
Impedance Matching
Iron Core
Leakage Flux
Loosely Coupled
Magnetizing Current
Mutual Inductance
Open-Circuit Test
Primary
Reflected Impedance
Secondary
Short-Circuit Test
Step-Up/Step-Down
Tightly Coupled
Transformer
Turns Ratio
Voltage Ratio

OUTLINE

Introduction
Iron-Core Transformers: The Ideal Model
Reflected Impedance
Transformer Ratings
Transformer Applications
Practical Iron-Core Transformers
Transformer Tests
Voltage and Frequency Effects
Loosely Coupled Circuits
Magnetically Coupled Circuits with Sinusoidal Excitation
Coupled Impedance
Circuit Analysis Using Computers

In our study of induced voltage in Chapter 13 we found that the changing magnetic field produced by current in one coil induced a voltage in a second coil wound on the same core. A device built to utilize this effect is the **transformer.**

Transformers have many applications. They are used in electrical power systems to step up voltage for long-distance transmission and then to step it down again to a safe level for use in our homes and offices. They are used in electronic equipment power supplies to raise or lower voltages, in audio systems to match speaker loads to amplifiers, in telephone, radio, and TV systems to couple signals, and so on.

In this chapter, we look at transformer fundamentals and the analysis of circuits containing transformers. We discuss transformer action, types of transformers, voltage and current ratios, applications, and so on. Both iron-core and air-core transformers are covered.

George Westinghouse

ONE OF THE DEVICES that made possible the commercial ac power system as we know it today is the transformer. Although Westinghouse did not invent the transformer, his acquisition of the transformer patent rights and his manufacturing business helped make him an important player in the battle of dc versus ac in the emerging electrical power industry (see Chapter 15). In concert with Tesla (see Chapter 23), Westinghouse fought vigorously for ac against Edison, who favored dc. In 1893, Westinghouse's company built the Niagara Falls power system using ac, and the battle was over with ac the clear winner. (Ironically, in recent years dc has been resurrected for use in commercial electrical power systems because it is able to transmit power over longer distances than ac. However, this was not possible in Edison's day, and ac was and still is the correct choice for the commercial electrical power system.)

George Westinghouse was born in 1846 in Central Bridge, New York. He made his fortune with the invention of the railway air brake system. He died in 1914 and was elected to the Hall of Fame for Great Americans in 1955.

24.1   Introduction

A transformer is a magnetically **coupled circuit,** i.e., a circuit in which the magnetic field produced by time-varying current in one circuit induces voltage in another. To illustrate, a basic iron-core transformer is shown in Figure 24–1. It consists of two coils wound on a common core. Alternating current in one winding establishes a flux which links the other winding and induces a voltage in it. Power thus flows from one circuit to the other via the medium of the magnetic field, with no electrical connection between the two sides. The winding to which we supply power is called the **primary,** while the winding from which we take power is called the **secondary.** Power can flow in either direction, as either winding can be used as the primary or the secondary.

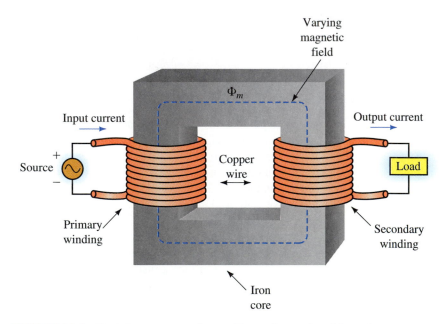

FIGURE 24–1   Basic iron-core transformer. Energy is transferred from the source to the load via the transformer's magnetic field with no electrical connection between the two sides.

Transformer Construction

Transformers fall into two broad categories, iron-core and air-core. We begin with **iron-core** types. Iron core transformers are generally used for low frequency applications such as audio- and power-frequency applications. Figures 24–2 and 24–3 show a few examples of iron-core transformers.

Iron (actually a special steel called transformer steel) is used for cores because it increases the coupling between coils by providing an easy path for magnetic flux. Two basic types of iron-core construction are used, the **core type** and the **shell type** (Figure 24–4). In both cases, cores are made from laminations of sheet steel, insulated from each other by thin coatings of ceramic or other material to help minimize eddy current losses (Section 24.6).

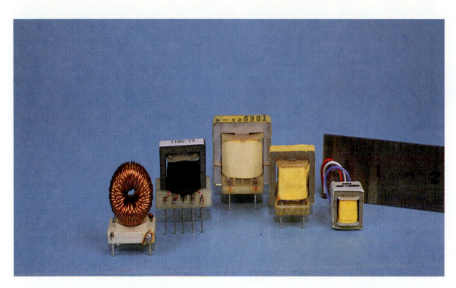

FIGURE 24–2   Iron-core transformers of the type used in electronic equipment. *(Courtesy Transformer Manufacturers Inc.)*

FIGURE 24–3   Distribution transformer (cutaway view) of the type used by electric utilities to distribute power to residential and commercial users. The tank is filled with oil to improve insulation and to remove heat from the core and windings. *(Courtesy Carte International Inc.)*

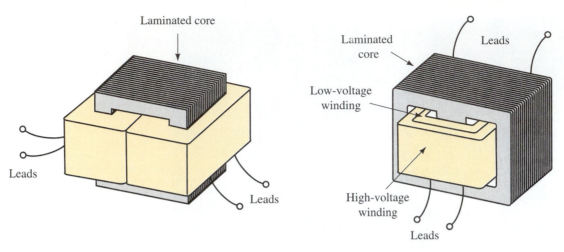

FIGURE 24–4 For the core type (left), windings are on separate legs, while for the shell type both windings are on the same leg. (Adapted with permission from Perozzo, *Practical Electronics Troubleshooting,* © 1985, Delmar Publishers Inc.)

Iron, however, has considerable power loss due to hysteresis and eddy currents at high frequencies, and is thus not useful as a core material above about 50 kHz. For high-frequency applications (such as in radio circuits), **air-core** and **ferrite-core** types are used. Figure 24–5 shows a ferrite-core device. Ferrite (a magnetic material made from powdered iron oxide) greatly increases coupling between coils (compared with air) while maintaining low losses. Circuit symbols for transformers are shown in Figure 24–6.

(a) Iron-core

(b) Air-core

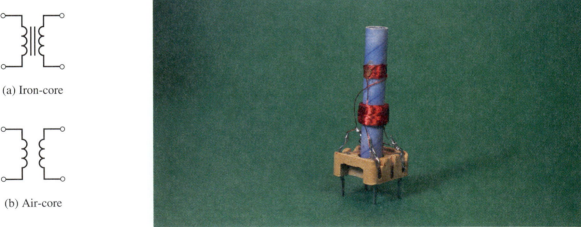

FIGURE 24–5 A slug-tuned ferrite-core radio frequency transformer. Coupling between coils is varied by a ferrite slug positioned by an adjusting screw.

(c) Ferrite-core

FIGURE 24–6 Transformer schematic symbols.

Winding Directions

One of the advantages of a transformer is that it may be used to change the polarity of an ac voltage. This is illustrated in Figure 24–7 for a pair of iron-core transformers. For the transformer of (a) the primary and secondary voltages are in phase (for reasons to be discussed later), while for (b) they are 180° out of phase.

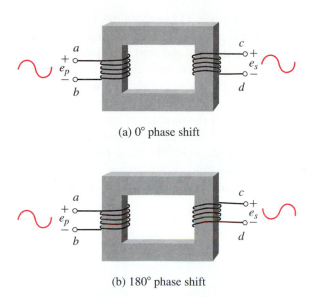

(a) 0° phase shift

(b) 180° phase shift

FIGURE 24–7   The relative direction of the windings determines the phase shift.

Tightly Coupled and Loosely Coupled Circuits

If most of the flux produced by one coil links the other, the coils are said to be **tightly coupled.** Thus, iron-core transformers are tightly coupled (since close to 100% of the flux is confined to the core and thus links both windings). For air- and ferrite-core transformers, however, much less than 100% of the flux links both windings. They are therefore **loosely coupled.** Since air-core and ferrite-core devices are loosely coupled, the same principles of analysis apply to both and we treat them together in Section 24.9.

Faraday's Law

All transformer operation is described by Faraday's law. Faraday's law (in SI units) states that the voltage induced in a circuit by a changing magnetic field is equal to the rate at which the flux linking the circuit is changing. When Faraday's law is applied to iron-core and air-core transformers, however, the results that emerge are quite different: Iron-core transformers are found to be characterized by their turns ratios, while air-core transformers are characterized by self- and mutual inductances. We begin with iron-core transformers.

24.2 Iron-Core Transformers: The Ideal Model

At first glance, iron-core transformers appear quite difficult to analyze because they have several characteristics such as winding resistance, core loss, and leakage flux that appear difficult to handle. Fortunately, these effects are small and often can be neglected. The result is the **ideal transformer.** Once you know how to analyze an ideal transformer, however, it is relatively easy to go back and add in the nonideal effects. This is the approach we use here.

To idealize a transformer, (1) neglect the resistance of its coils, (2) neglect its core loss, (3) assume all flux is confined to its core, and (4) assume that negligible current is required to establish its core flux. (Well-designed iron-core transformers are very close to this ideal.)

We now apply Faraday's law to the ideal transformer. Before we do this, however, we need to determine flux linkages. The flux linking a winding (as determined in Chapter 13) is the product of the flux that passes through the winding times the number of turns through which it passes. For flux Φ passing through N turns, flux linkage is $N\Phi$. Thus, for the ideal transformer (Figure 24–8), the primary flux linkage is $N_p\Phi_m$, while the secondary flux linkage is $N_s\Phi_m$, where the subscript "m" indicates mutual flux, i.e., flux that links both windings.

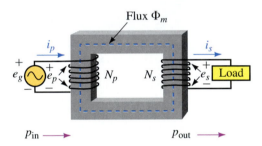

FIGURE 24–8   Ideal transformer. All flux is confined to the core and links both windings. This is a "tightly-coupled" transformer.

Voltage Ratio

Now apply Faraday's law. Since the flux linkage equals $N\Phi$ and since N is constant, the induced voltage is equal to N times the rate of change of Φ, i.e., $e = N d\Phi/dt$. Thus, for the primary,

$$e_p = N_p \frac{d\Phi_m}{dt} \qquad (24\text{–}1)$$

while for the secondary

$$e_s = N_s \frac{d\Phi_m}{dt} \qquad (24\text{–}2)$$

Dividing Equation 24–1 by Equation 24–2 and cancelling $d\Phi_m/dt$ yields

$$\frac{e_p}{e_s} = \frac{N_p}{N_s} \qquad (24\text{–}3)$$

Equation 24–3 states that *the ratio of primary voltage to secondary voltage is equal to the ratio of primary turns to secondary turns.* This ratio is called the **transformation ratio** (or **turns ratio**) and is given the symbol a. Thus,

$$a = N_p/N_s \qquad (24\text{–}4)$$

and

$$e_p/e_s = a \qquad (24\text{–}5)$$

For example, a transformer with 1000 turns on its primary and 250 turns on its secondary has a turns ratio of 1000/250 = 4. This is referred to as a 4:1 ratio.

Since the ratio of two instantaneous sinusoidal voltages is the same as the ratio of their effective values, Equation 24–5 can also be written as

$$E_p/E_s = a \qquad (24\text{–}6)$$

As noted earlier, e_p and e_s are either in phase or 180° out of phase, depending on the relative direction of coil windings. We can therefore also express the ratio of voltages in terms of phasors as

$$\mathbf{E}_p/\mathbf{E}_s = a \qquad (24\text{–}7)$$

where the relative polarity (in phase or 180° out of phase) is determined by the direction of the coil windings (Figure 24–7).

Step-Up and Step-Down Transformers

A **step-up** transformer is one in which the secondary voltage is higher than the primary voltage, while a **step-down** transformer is one in which the secondary voltage is lower. Since $a = E_p/E_s$, a step-up transformer has $a < 1$, while for a step-down transformer, $a > 1$. If $a = 1$, the transformer's turns ratio is **unity** and the secondary voltage is equal to the primary voltage.

EXAMPLE 24–1   Suppose the transformer of Figure 24–7(a) has 500 turns on its primary and 1000 turns on its secondary.

a. Determine its turns ratio. Is it step-up or step-down?

b. If its primary voltage is $e_p = 25 \sin \omega t$ V, what is its secondary voltage?

c. Sketch the waveforms.

Solution

a. The turns ratio is $a = N_p/N_s = 500/1000 = 0.5$. This is a step-up transformer.

b. From Equation 24–5, $e_s = e_p/a = (25 \sin \omega t)/0.5 = 50 \sin \omega t$ V.

c. Primary and secondary voltages are in phase as noted earlier. Figure 24–9 shows the waveforms.

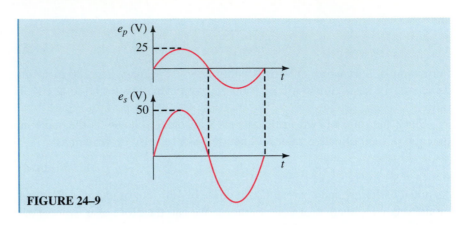

FIGURE 24–9

≷ EXAMPLE 24–2   If the transformers of Figure 24–7 have 600 turns on their primaries and 120 turns on their secondaries, and $\mathbf{E}_p = 120 \text{ V} \angle 0°$, what is $\mathbf{E}_s$ for each case?

Solution   The turns ratio is $a = 600/120 = 5$. For transformer (a), $\mathbf{E}_s$ is in phase with $\mathbf{E}_p$. Therefore, $\mathbf{E}_s = \mathbf{E}_p/5 = (120 \text{ V} \angle 0°)/5 = 24 \text{ V} \angle 0°$. For transformer (b), $\mathbf{E}_s$ is 180° out of phase with $\mathbf{E}_p$. Therefore, $\mathbf{E}_s = 24 \text{ V} \angle 180°$.

**PRACTICE
PROBLEMS 1**

Repeat Example 24–1 for the circuit of Figure 24–7(b) if $N_p = 1200$ turns and $N_s = 200$ turns.

Answer: $e_s = 4.17 \sin(\omega t + 180°)$

Current Ratio

Because an ideal transformer has no power loss, its efficiency is 100% and thus power in equals power out. Consider again Figure 24–8. At any instant, $p_{in} = e_p i_p$ and $p_{out} = e_s i_s$. Thus,

$$e_p i_p = e_s i_s \tag{24–8}$$

and

$$\frac{i_p}{i_s} = \frac{e_s}{e_p} = \frac{1}{a} \tag{24–9}$$

since $e_s/e_p = 1/a$. (This means that if voltage is stepped up, current is stepped down, and vice versa.) In terms of current phasors and current magnitudes, Equation 24–9 can be written as

$$\frac{\mathbf{I}_p}{\mathbf{I}_s} = \frac{I_p}{I_s} = \frac{1}{a} \tag{24–10}$$

For example, for a transformer with $a = 4$, and $\mathbf{I}_p = 2 \text{ A} \angle -20°$, $\mathbf{I}_s = a\mathbf{I}_p = 4(2 \text{ A} \angle -20°) = 8 \text{ A} \angle -20°$, Figure 24–10.

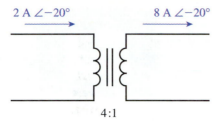

FIGURE 24–10 Currents are inverse to the turns ratio.

Polarity of Induced Voltage: The Dot Convention

As noted earlier, an iron core transformer's secondary voltage is either in phase with its primary voltage or 180° out of phase, depending on the relative direction of its windings. We will now demonstrate why.

A simple test, called the **kick test** (sometimes used by electrical workers to determine transformer polarity), can help establish the idea. The basic circuit is shown in Figure 24–11. A switch is used to make and break the circuit (since voltage is induced only while flux is changing).

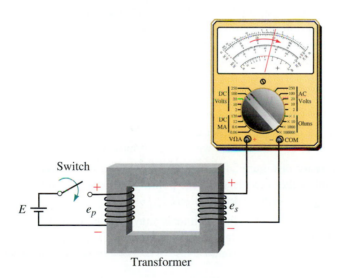

FIGURE 24–11 The "kick" test. For the winding directions shown, the meter kicks upscale at the instant the switch is closed. (This is the transformer of Figure 24–7a.)

For the winding directions shown, at the instant the switch is closed, the voltmeter needle "kicks" upscale, then settles back to zero. To understand why, we need to consider magnetic fields. Before we start, however, let us place a dot on one of the primary terminals; in this case, we arbitrarily choose the top terminal. Let us also replace the voltmeter by its equivalent resistance (Figure 24–12).

At the instant the switch is closed, the polarity of the dotted primary terminal is positive with respect to the undotted primary terminal (because the + end of the source is directly connected to it). As current in the primary builds, it creates a flux in the upward direction as indicated by the blue arrow (recall the right-hand rule). According to Lenz's law, the effect that results

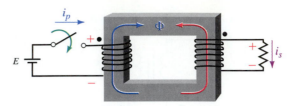

FIGURE 24–12   Determining dot positions.

must oppose the cause that produced it. The effect is a voltage induced in the secondary winding. The resulting current in the secondary produces a flux which, according to Lenz's law, must *oppose the buildup of the original flux,* i.e., it must be in the direction of the red arrow. Applying the right-hand rule, we see that secondary current must be in the direction indicated by i_s. Placing a plus sign at the tail of this arrow shows that the top end of the resistor is positive. This means that the top end of the secondary winding is also positive. Place a dot here. Dotted terminals are called **corresponding** terminals.

As you can see, corresponding terminals are positive (with respect to their companion undotted terminals) at the instant the switch is closed. If you perform a similar analysis at the instant the switch is opened, you will find that both dotted terminals are negative. Thus, *dotted terminals have the same polarity at all instants of time.* What we have developed here is known as the **dot convention for coupled circuits.**

PRACTICAL NOTES...

1. While we developed the dot convention using a switched dc source, it is valid for ac as well. In fact, we will use it mostly for ac.

2. The resistor of Figure 24–12 is required only to work through the physics to establish the secondary voltage polarity. You can now remove it without affecting the resulting dot position.

3. In practice, corresponding terminals may be marked with dots, by color coded wires, or by special letter designations.

EXAMPLE 24–3   Determine the waveform for e_s in the circuit of Figure 24–13(a).

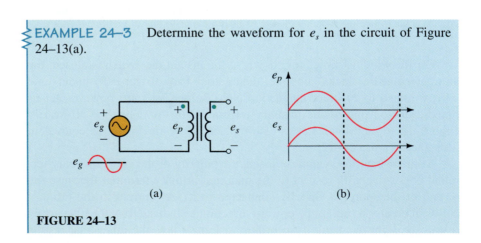

FIGURE 24–13

Solution Dotted terminals have the same polarity (with respect to their undotted terminals) at all instants. During the first half-cycle, the dotted end of the primary coil is positive. Therefore, the dotted end of the secondary is also positive. During the second half-cycle, both are negative. The polarity marks on e_s mean that we are looking at the polarity of the top end of the secondary coil with respect to its bottom end. Thus, e_s is positive during the first half-cycle and negative during the second half cycle. It is therefore in phase with e_p as indicated in (b). Thus, if $e_p = E_{m_p} \sin \omega t$, then $e_s = E_{m_s} \sin \omega t$.

1. Determine the equation for e_s in the circuit of Figure 24–14(b).

PRACTICE
PROBLEMS 2

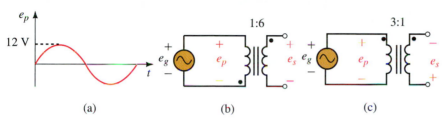

| (a) | (b) | (c) |

FIGURE 24–14

2. Repeat Problem 1 for Figure 24–14(c).
3. If $\mathbf{E}_g = 120 \text{ V} \angle 30°$, determine $\mathbf{E}_s$ for each transformer of Figure 24–14.
4. Where do the dots go on the transformers of Figure 24–7?

Answers:
1. $e_s = 72 \sin(\omega t + 180°) \text{ V}$
2. $e_s = 4 \sin(\omega t + 180°) \text{ V}$
3. For (b), $\mathbf{E}_s = 720 \text{ V} \angle 180°$; for (c), $\mathbf{E}_s = 40 \text{ V} \angle 180°$.
4. For (a), place dots at a and c. For (b), place dots at a and d.

Analysis of Simple Transformer Circuits

Simple transformer circuits may be analyzed using the relationships described so far, namely $\mathbf{E}_p = a\mathbf{E}_s$, $\mathbf{I}_p = \mathbf{I}_s/a$, and $P_{in} = P_{out}$. This is illustrated in Example 24–4. More complex problems require some additional ideas.

EXAMPLE 24–4 For Figure 24–15(a), $\mathbf{E}_g = 120 \text{ V} \angle 0°$, the turns ratio is 5:1, and $\mathbf{Z}_L = 4 \, \Omega \angle 30°$. Find

a. the load voltage,
b. the load current,
c. the generator current,
d. the power to the load,
e. the power output by the generator.

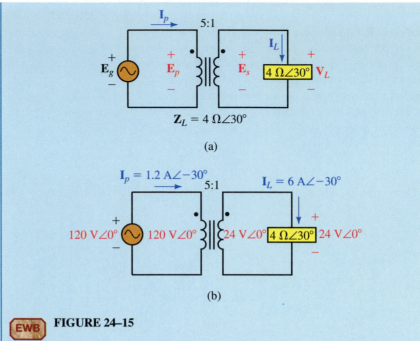

FIGURE 24–15

Solution

a. $\mathbf{E}_p = \mathbf{E}_g = 120\text{ V}\angle 0°$

    $\mathbf{V}_L = \mathbf{E}_s = \mathbf{E}_p/a = (120\text{ V}\angle 0°)/5 = 24\text{ V}\angle 0°$

b. $\mathbf{I}_L = \mathbf{V}_L/\mathbf{Z}_L = (24\text{ V}\angle 0°)/(4\ \Omega\angle 30°) = 6\text{ A}\angle -30°$

c. $\mathbf{I}_g = \mathbf{I}_p$. But $\mathbf{I}_p = \mathbf{I}_s/a = \mathbf{I}_L/a$. Thus,

    $\mathbf{I}_g = (6\text{ A}\angle -30°)/5 = 1.2\text{ A}\angle -30°$

    Values are shown on Figure 24–15(b).

d. $P_L = V_L I_L \cos\theta_L = (24)(6)\cos 30° = 124.7\text{ W}.$

e. $P_g = E_g I_g \cos\theta_g$, where θ_g is the angle between $\mathbf{E}_g$ and $\mathbf{I}_g$. $\theta_g = 30°$. Thus, $P_g = (120)(1.2)\cos 30° = 124.7\text{ W}$, which agrees with (d) as it must since the transformer is lossless.

IN-PROCESS LEARNING CHECK 1

1. A transformer has a turns ratio of $1:8$. Is it step-up or step-down? If $E_p = 25$ V, what is E_s?

2. For the transformers of Figure 24–7, if $a = 0.2$ and $\mathbf{E}_s = 600\text{ V}\angle -30°$, what is $\mathbf{E}_p$ for each case?

3. For each of the transformers of Figure 24–16, sketch the secondary voltage, showing both phase and amplitude.

4. For Figure 24–17, determine the position of the missing dot.

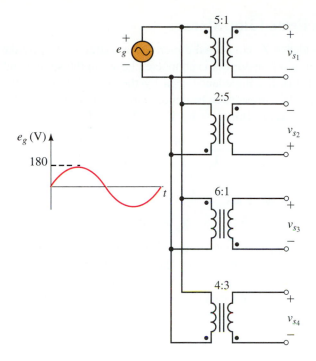

FIGURE 24–16

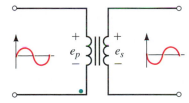

FIGURE 24–17

5. Figure 24–18 shows another way to determine dotted terminals. First, arbitrarily mark one of the primary terminals with a dot. Next, connect a jumper wire and voltmeter as indicated. From the voltmeter readings, you can determine which of the secondary terminals should be dotted. For the two cases indicated, where should the secondary dot be? (*Hint:* Use KVL.)

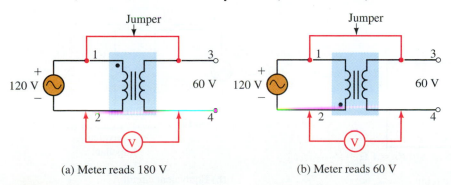

(a) Meter reads 180 V   (b) Meter reads 60 V

EWB   **FIGURE 24–18**   Each transformer is rated 120V/60V.

(Answers are at the end of the chapter.)

24.3 Reflected Impedance

A load impedance $\mathbf{Z}_L$ connected directly to a source is seen by the source as $\mathbf{Z}_L$. However, if a transformer is connected between the source and load as in Figure 24–19(a), the impedance seen by the source is now quite different. Let this equivalent impedance be denoted as $\mathbf{Z}_p$. Then, $\mathbf{I}_p = \mathbf{E}_g/\mathbf{Z}_p$. Rearranging yields $\mathbf{Z}_p = \mathbf{E}_g/\mathbf{I}_p$. But $\mathbf{E}_g = \mathbf{E}_p$, $\mathbf{E}_p = a\mathbf{E}_s$, and $\mathbf{I}_p = \mathbf{I}_s/a$. Thus,

$$\mathbf{Z}_p = \frac{\mathbf{E}_p}{\mathbf{I}_p} = \frac{a\mathbf{E}_s}{\left(\dfrac{\mathbf{I}_s}{a}\right)} = a^2\frac{\mathbf{E}_s}{\mathbf{I}_s} = a^2\frac{\mathbf{V}_L}{\mathbf{I}_L}$$

However $\mathbf{V}_L/\mathbf{I}_L = \mathbf{Z}_L$. Thus,

$$\mathbf{Z}_p = a^2\mathbf{Z}_L \qquad\qquad (24\text{–}11)$$

This means that $\mathbf{Z}_L$ now looks to the source like the transformer's turns ratio squared times the load impedance. The term $a^2\mathbf{Z}_L$ is referred to as the load's **reflected impedance.** Note that it retains the load's characteristics, that is, a capacitive load still looks capacitive, an inductive load still looks inductive, and so on.

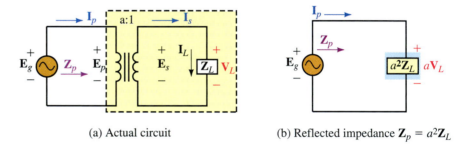

(a) Actual circuit (b) Reflected impedance $\mathbf{Z}_p = a^2\mathbf{Z}_L$

FIGURE 24–19 Concept of reflected impedance. From the primary terminals, $\mathbf{Z}_L$ looks like an impedance of $a^2\mathbf{Z}_L$ with voltage $a\mathbf{V}_L$ across it and current $\mathbf{I}_L/a$ through it.

Equation 24–11 shows that a transformer can make a load look larger or smaller, depending on its turns ratio. To illustrate, consider Figure 24–20. If a 1-Ω resistor were connected directly to the source, it would look like a 1-Ω

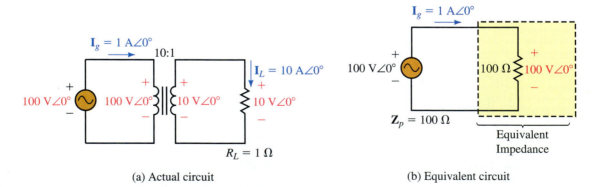

(a) Actual circuit (b) Equivalent circuit

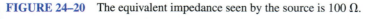

FIGURE 24–20 The equivalent impedance seen by the source is 100 Ω.

resistor and the generator current would be 100A∠0°. However, when connected to a 10:1 transformer, it looks like a $(10)^2(1\,\Omega) = 100\text{-}\Omega$ resistor and the generator current is only 1 A∠0°.

The concept of reflected impedance is useful in a number of ways. It permits us to match loads to sources (such as amplifiers) as well as provides a better way to solve complex transformer problems.

EXAMPLE 24–5 Use the reflected impedance idea to solve for primary and secondary currents and the load voltage for the circuit of Figure 24–21(a). (Note: This is the same circuit we solved earlier in Example 24–4. While there is no advantage to the reflected impedance idea over our previous solution for a problem as simple as this, the advantages are considerable for complex problems.)

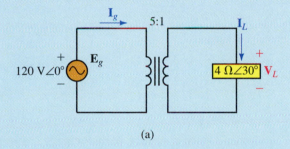

(a)

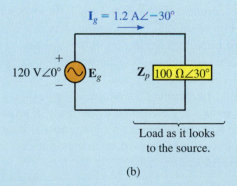

Load as it looks
to the source.

(b)

FIGURE 24–21

Solution

$$\mathbf{Z}_p = a^2\mathbf{Z}_L = (5)^2(4\,\Omega\angle 30°) = 100\,\Omega\angle 30°.$$

The equivalent circuit is shown in (b).

$$\mathbf{I}_g = \mathbf{E}_g/\mathbf{Z}_p = (120\,\text{V}\angle 0°) / (100\,\Omega\angle 30°) = 1.2\,\text{A}\angle -30°$$
$$\mathbf{I}_L = a\mathbf{I}_p = a\mathbf{I}_g = 5(1.2\,\text{A}\angle -30°) = 6\,\text{A}\angle -30$$
$$\mathbf{V}_L = \mathbf{I}_L\mathbf{Z}_L = (6\,\text{A}\angle -30°)(4\,\Omega\angle 30°) = 24\,\text{V}\angle 0°$$

The answers are the same as in Example 24–4.

24.4   Transformer Ratings

Transformers are rated in terms of voltage and apparent power (for reasons discussed in Chapter 17). Rated current can be determined from these ratings. Thus a transformer rated 2400/120 volt, 48 kVA, has a current rating of 48 000 VA/2400 V = 20 A on its 2400-V side and 48 000 VA/120 V = 400 A on its 120-V side (Figure 24–22). This transformer can handle a 48-kVA load, regardless of power factor.

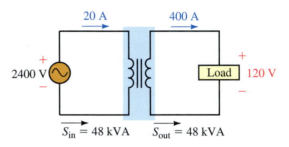

FIGURE 24–22   Transformers are rated by the amount of apparent power and the voltages that they are designed to handle.

24.5   Transformer Applications

Power Supply Transformers

On electronic equipment, **power supply transformers** are used to convert the incoming 120 Vac to the voltage levels required for internal circuit operation. A variety of commercial transformers are made for this purpose. The transformer of Figure 24–23, for example, has a multi-tapped secondary winding, each tap providing a different output voltage. It is intended for laboratory supplies, test equipment, or experimental power supplies.

Figure 24–24 illustrates a typical use of a power supply transformer. First, the incoming line voltage is stepped down, then a rectifier circuit (a circuit that uses diodes to convert ac to dc using a process called rectification) converts the ac to pulsating dc, a filter smooths it, and finally, a voltage regulator (an electronic device used to maintain a constant output voltage) regulates it to the required dc value.

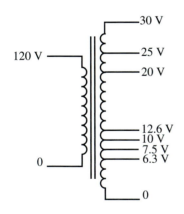

FIGURE 24–23   A multi-tapped power supply transformer. The secondary is tapped at various voltages.

Transformers in Power Systems

Transformers are one of the key elements that have made commercial ac power systems possible. Transformers are used at generating stations to raise voltage for long-distance transmission. This lowers the transmitted current and hence the I^2R power losses in the transmission line. At the user end, transformers reduce the voltage to a safe level for everyday use. A typical residential connection is shown in Figure 24–25. The taps on the primary permit the electric utility company to compensate for line voltage drops. Transformers located far from substations, for example, have lower input voltages (by a few percent) than those close to substations due to voltage drops on distribution lines. Taps permit the turns ratio to be changed to com-

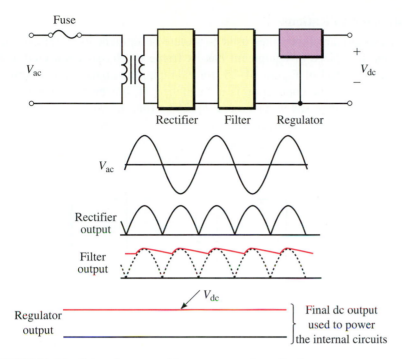

FIGURE 24–24   A transformer used in a power supply application.

pensate. Note also the split secondary. It permits 120-V and 240-V loads to be supplied from the same transformer.

The transformer of Figure 24–25 is a single-phase unit (since residential customers require only single phase). By connecting its primary from line to neutral (or line to line), the required single-phase input is obtained from a three-phase line.

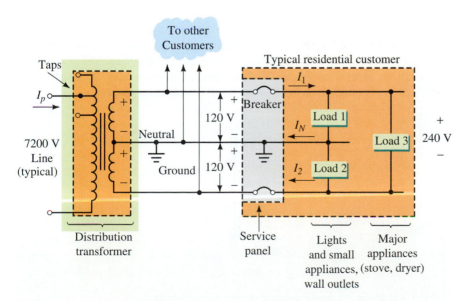

FIGURE 24–25   Typical distribution transformer supplying residential loads.

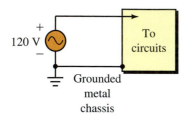

(a) Chassis is safe

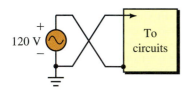

(b) Chassis is at 120 V

FIGURE 24–26   If the connections are inadvertently reversed as in (b), you will get a shock if you are grounded and you touch the chassis.

Isolation Applications

Transformers are sometimes used to isolate equipment for safety or other reasons. If a piece of equipment has its frame or chassis connected to the grounded neutral of Figure 24–25, for example, the connection is perfectly safe as long as the connection is not changed. If, however, connections are inadvertently reversed as in Figure 24–26(b) (due to faulty installation for example), a dangerous situation results. A transformer used as in Figure 24–27 eliminates this danger by ensuring that the chassis is never directly connected to the "hot" wire. Isolation transformers are made for this purpose.

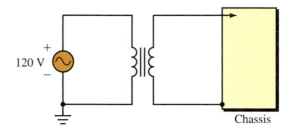

FIGURE 24–27   Using a transformer for isolation.

Impedance Matching

As you learned earlier, load impedance $\mathbf{Z}_L$, when viewed through a transformer, has a reflected value of $\mathbf{Z}_p = a^2\mathbf{Z}_L$. This means that a transformer can be used to raise or lower the apparent impedance of a load by choice of turns ratio. This is referred to as **impedance matching.** Impedance matching is sometimes used to match loads to amplifiers to achieve maximum power transfer. If the load and source are not matched, a transformer can be inserted between them as illustrated next.

EXAMPLE 24–6   Figure 24–28(a) shows the schematic of a multi-tap sound distribution transformer with various taps that permit the matching of speakers to amplifiers. Over their design range, speakers are basically resistive. If the speaker of Figure 24–29(a) has a resistance of 4 Ω, what transformer ratio should be chosen? What is the power to the speaker?

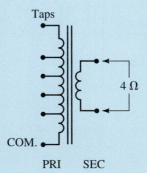

FIGURE 24–28   A tapped sound distribution transformer.

Solution   Make the reflected resistance of the speaker equal to the internal (Thévenin) resistance of the amplifier. Thus, $Z_p = 400 \ \Omega = a^2 Z_L = a^2(4 \ \Omega)$. Solving for a yields

$$a = \sqrt{\frac{Z_p}{Z_L}} = \sqrt{\frac{400 \ \Omega}{4 \ \Omega}} = \sqrt{100} = 10$$

Now consider power. Since $Z_p = 400 \ \Omega$, Figure 24–29(b), half the source voltage appears across it. Thus power to Z_p is $(40 \ \text{V})^2/(400 \ \Omega) = 4 \ \text{W}$. Since the transformer is considered lossless, all power is transferred to the speaker. Thus, $P_{\text{speaker}} = 4 \ \text{W}$.

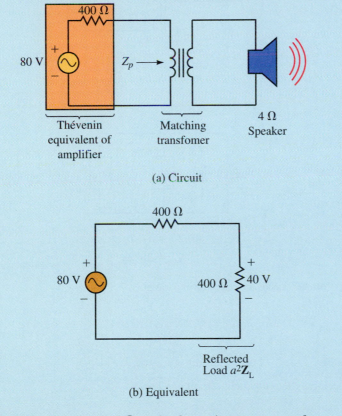

(a) Circuit

(b) Equivalent

FIGURE 24–29   Matching the 4-Ω speaker for maximum power transfer.

Determine the power to the speaker of Figure 24–29 if the transformer is not present (i.e., the speaker is directly connected to the amplifier). Compare to Example 24–6.

PRACTICE PROBLEMS 3

Answer: 0.157 W (dramatically lower)

Transformers with Multiple Secondaries

For a transformer with multiple secondaries (Figure 24–30), each secondary voltage is governed by the appropriate turns ratio, that is, $\mathbf{E}_1/\mathbf{E}_2 = N_1/N_2$ and $\mathbf{E}_1/\mathbf{E}_3 = N_1/N_3$. Loads are reflected in parallel. That is, $\mathbf{Z}'_2 = a_2{}^2\mathbf{Z}_2$ and $\mathbf{Z}'_3 = a_3{}^2\mathbf{Z}_3$ appear in parallel in the equivalent circuit, (b).

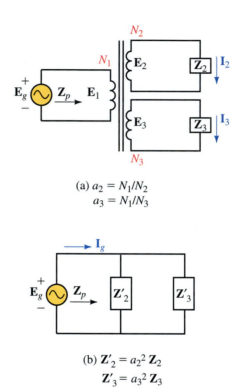

(a) $a_2 = N_1/N_2$
$a_3 = N_1/N_3$

(b) $\mathbf{Z}'_2 = a_2{}^2\,\mathbf{Z}_2$
$\mathbf{Z}'_3 = a_3{}^2\,\mathbf{Z}_3$

FIGURE 24–30   Loads are reflected in parallel.

EXAMPLE 24–7   For the circuit of Figure 24–31(a),

a. determine the equivalent circuit,

b. determine the generator current,

c. show that apparent power in equals apparent power out.

Solution

a. See Figure 24–31(b).

b. $\mathbf{I}_g = \dfrac{\mathbf{E}_g}{\mathbf{Z}'_2} + \dfrac{\mathbf{E}_g}{\mathbf{Z}'_3} = \dfrac{100\angle 0°}{10} + \dfrac{100\angle 0°}{-j10} = 10 + j10 = 14.14\ \text{A}\angle 45°$

c. Input: $S_{\text{in}} = E_g I_g = (100\ \text{V})(14.14\ \text{A}) = 1414\ \text{VA}$
   Output: From Figure 24–31(b), $P_{\text{out}} = (100\ \text{V})^2/(10\ \Omega) = 1000\ \text{W}$ and
   $Q_{\text{out}} = (100\ \text{V})^2/(10\ \Omega) = 1000\ \text{VAR}$. Thus $S_{\text{out}} = \sqrt{P_{\text{out}}{}^2 + Q_{\text{out}}{}^2} = 1414\ \text{VA}$
   which is the same as S_{in}.

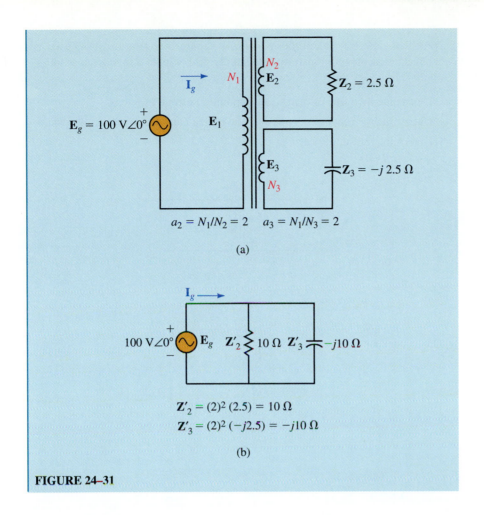

$$\mathbf{Z'}_2 = (2)^2\,(2.5) = 10\ \Omega$$
$$\mathbf{Z'}_3 = (2)^2\,(-j2.5) = -j10\ \Omega$$

(b)

FIGURE 24–31

Autotransformers

An important variation of the transformer is the **autotransformer** (Figure 24–32). Autotransformers are unusual in that their primary circuit is not electrically isolated from their secondary. However, they are smaller and cheaper than conventional transformers for the same load kVA since only part of the load power is transferred inductively. (The remainder is transferred by direct conduction.) Figure 24–32 shows several variations. The transformer of (c) is variable by means of a slider, typically, from 0% to 110%.

For analysis, an autotransformer may be viewed as a standard two-winding transformer reconnected as in Figure 24–33(b). Voltage and current relationships between windings hold as they do for the standard connection. Thus, if you apply rated voltage to the primary winding, you will obtain rated voltage across the secondary winding. Finally, since we are assuming an ideal transformer, apparent power out equals apparent power in.

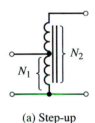

(a) Step-up

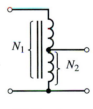

(b) Step-down

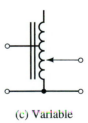

(c) Variable

FIGURE 24–32 Autotransformers.

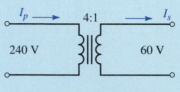

EXAMPLE 24–8   A 240/60-V, 3-kVA transformer [Figure 24–33(a)] is reconnected as an autotransformer to supply 300 volts to a load from a 240-V supply [Figure 24–33(b)].

a. Determine the rated primary and secondary currents.
b. Determine the maximum apparent power that can be delivered to the load.
c. Determine the supply current.

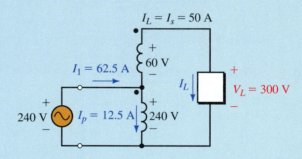

(a) 3 kVA transformer

(b) Used as an autotransformer

EWB   **FIGURE 24–33**

Solution

a. Rated current = rated kVA/rated voltage. Thus,

$$I_p = 3\,\text{kVA}/240\,\text{V} = 12.5\,\text{A} \quad \text{and} \quad I_s = 3\,\text{kVA}/60\,\text{V} = 50\,\text{A}$$

b. Since the 60-V winding is rated at 50 A, the transformer can deliver 50 A to the load [Figure 24–33(b)]. The load voltage is 300 V. Thus,

$$S_L = V_L I_L = (300\,\text{V})(50\,\text{A}) = 15\,\text{kVA}$$

This is five times the rated kVA of the transformer.

c. Apparent power in = apparent power out:

$$240 I_1 = 15\,\text{kVA}.$$

Thus, $I_1 = 15\,\text{kVA}/240\,\text{V} = 62.5\,\text{A}$. Current directions are as shown.

As a check, KCL at the junction of the two coils yields

$$I_1 = I_p + I_L = 12.5 + 50 = 62.5\,\text{A}$$

1. For each of the circuits of Figure 24–34, determine the required answer.

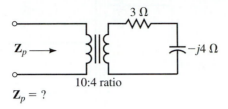

Z_p ⟶

10:4 ratio

Z_p = ?

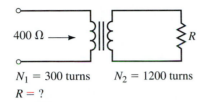

400 Ω ⟶

N_1 = 300 turns N_2 = 1200 turns

R = ?

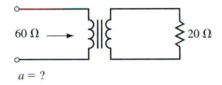

60 Ω ⟶   20 Ω

a = ?

FIGURE 24–34

2. For Figure 24–35, if $a = 5$ and $\mathbf{I}_p = 5\,\text{A}\angle{-60°}$, what is $\mathbf{E}_g$?

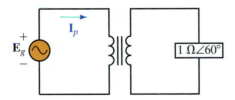

$\mathbf{E}_g$ $\mathbf{I}_p$ $1\,\Omega\angle 60°$

FIGURE 24–35

3. For Figure 24–35, if $\mathbf{I}_p = 30\,\text{mA}\angle{-40°}$ and $\mathbf{E}_g = 240\,\text{V}\angle 20°$, what is a?

4. a. How many amps can a 24-kVA, 7200/120-V transformer supply to a 120-V unity power factor load? To a 0.75 power factor load?

   b. How many watts can it supply to each load?

5. For the transformer of Figure 24–36, between position 2 and 0 there are 2000 turns. Between taps 1 and 2 there are 200 turns, and between taps 1 and 3 there are 300 turns. What will the output voltage be when the supply is connected to tap 1? To tap 2? To tap 3?

6. For the circuit of Figure 24–37, what is the power delivered to a 4-ohm speaker? What is the power delivered if an 8-ohm speaker is used instead? Why is the power to the 4-ohm speaker larger?

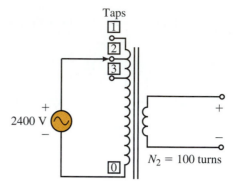

FIGURE 24–36

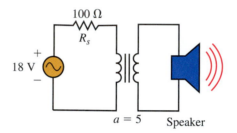

FIGURE 24–37

7. The autotransformer of Figure 24–38 has a 58% tap. The apparent power of the load is 7.2 kVA. Calculate the following:

a. The load voltage and current.

b. The source current.

c. The current in each winding and its direction.

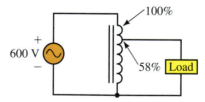

FIGURE 24–38

(Answers are at the end of the chapter.)

24.6   Practical Iron-Core Transformers

In Section 24.2, we idealized the transformer. We now add back the effects that we ignored.

Leakage Flux

While most flux is confined to the core, a small amount (called **leakage flux**) passes outside the core and through air at each winding as in Figure 24–39(a). The effect of this leakage can be modeled by inductances L_p and L_s

as indicated in (b). The remaining flux, the **mutual flux** Φ_m, links both windings and is accounted for by the ideal transformer as previously.

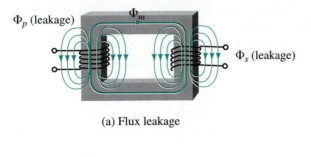

(a) Flux leakage

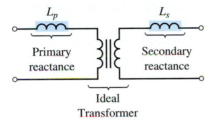

(b) Equivalent

FIGURE 24–39   Leakage flux can be modeled by small inductances.

Winding Resistance

The effect of coil resistance can be approximated by resistances R_p and R_s as shown in Figure 24–40. The effect of these resistances is to cause a slight power loss and hence a reduction in efficiency as well as a small voltage drop. (The power loss associated with coil resistance is called **copper loss** and varies as the square of the load current.)

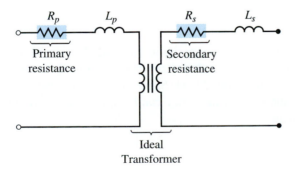

FIGURE 24–40   Adding the winding resistance to the model.

Core Loss

Losses occur in the core because of **eddy currents** and **hysteresis.** First, consider eddy currents. Since iron is a conductor, voltage is induced in the core as flux varies. This voltage creates currents that circulate as "eddies" within the core itself. One way to reduce these currents is to break their path of circulation by constructing the core from thin laminations of steel rather than using a solid block of iron. Laminations are insulated from each other by a coat of ceramic, varnish, or similar insulating material. (Although this does not eliminate eddy currents, it greatly reduces them.) Power and audio transformers are built this way (Figure 24–4). Another way to reduce eddy currents is to use powdered iron held together by an insulating binder. Ferrite cores are made like this.

Now consider hysteresis. Because flux constantly reverses, magnetic domains in the core steel constantly reverse as well. This takes energy. However, this energy is minimized by using special grain-oriented transformer steel.

The sum of hysteresis and eddy current loss is called **core loss** or **iron loss.** In a well-designed transformer, it is small, typically one to two percent of the transformer rating. The effect of core loss can be modeled as a resistor, R_c in Figure 24–41. Core losses vary approximately as the square of applied voltage. As long as voltage is constant (which it normally is), core losses remain constant.

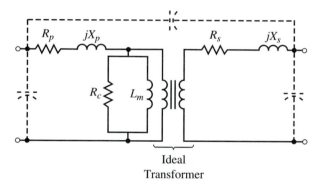

FIGURE 24–41   Final iron-core transformer equivalent circuit.

Other Effects

We have also neglected **magnetizing current.** In a real transformer, however, some current is required to magnetize the core. To account for this, add path L_m as shown in Figure 24–41. Stray capacitances also exist between various parts of the transformer. They can be approximated by lumped capacitances as indicated.

The Full Equivalent

Figure 24–41 shows the final equivalent with all effects incorporated. How good is it? Calculations based on this model agree exceptionally well with measurements made on real transformers. However, the circuit is complex and awkward to use. In practice, therefore, only those elements of the model that affect a given application are actually retained. For example, at power frequencies, the effect of capacitance is negligible and the capacitors are omitted for power frequency analysis.

Voltage Regulation

Because of the internal impedance of a transformer (Figure 24–41) voltage drops occur inside the transformer. Therefore, the output voltage of a transformer under load is different from its voltage under no load. This change in voltage (expressed as a percentage of full-load voltage) is termed **regulation.** For regulation analysis, parallel branches R_c and L_m and stray capacitance have negligible effect and can be neglected. This yields the simplified circuit of Figure 24–42(a). Even greater simplification is achieved by reflecting secondary impedances into the primary. This yields the circuit of (b). The reflected load voltage is $a\mathbf{V}_L$ and the reflected load current is $\mathbf{I}_L/a$. Regulation calculations are performed using this simplified circuit.

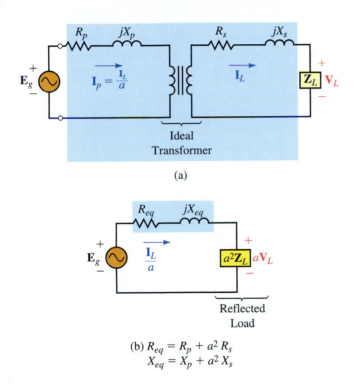

(a)

(b) $R_{eq} = R_p + a^2 R_s$
$X_{eq} = X_p + a^2 X_s$

FIGURE 24–42 Simplifying the equivalent.

EXAMPLE 24–9 A 10:1 transformer has primary and secondary resistance and reactance of 4 Ω + j4 Ω and 0.04 Ω + j0.04 Ω respectively as in Figure 24–43.

a. Determine its equivalent circuit.
b. If $\mathbf{V}_L = 120\text{ V}\angle0°$ and $\mathbf{I}_L = 20\text{ A}\angle-30°$, what is the supply voltage, $\mathbf{E}_g$?
c. Determine the regulation.

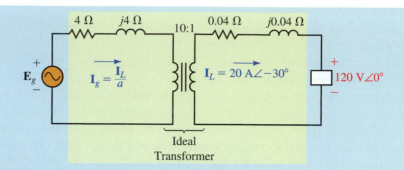

EWB   **FIGURE 24–43**

Solution

a. $R_{eq} = R_p + a^2 R_s = 4\ \Omega + (10)^2(0.04\ \Omega) = 8\ \Omega$

  $X_{eq} = X_p + a^2 X_s = 4\ \Omega + (10)^2(0.04\ \Omega) = 8\ \Omega$

  Thus, $\mathbf{Z}_{eq} = 8\ \Omega + j8\ \Omega$ as shown in Figure 24–44.

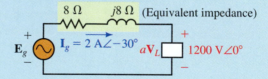

FIGURE 24–44

b. $a\mathbf{V}_L = (10)(120\ \text{V}\angle 0°) = 1200\ \text{V}\angle 0°$ and $\mathbf{I}_L/a = (20\ \text{A}\angle -30°)/10 = 2\ \text{A}\angle -30°$. From KVL, $\mathbf{E}_g = (2\ \text{A}\angle -30°)(8\ \Omega + j8) + 1200\ \text{V}\angle 0° = 1222\ \text{V}\angle 0.275°$.

  Thus, there is a phase shift of 0.275° across the transformer's internal impedance and a drop of 22 V, requiring that the primary be operated slightly above its rated voltage. (This is normal.)

c. Now consider the no-load condition (Figure 24–45). Let V_{NL} be the no-load voltage. As indicated, $a V_{NL} = 1222$ V. Thus, $V_{NL} = 1222/a = 1222/10 = 122.2$ volts and

$$\text{regulation} = \frac{V_{NL} - V_{FL}}{V_{FL}} \times 100 = \frac{122.2 - 120}{120} \times 100 = 1.83\%$$

  Note that only magnitudes are used in determining regulation.

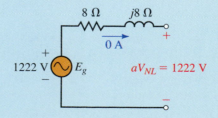

FIGURE 24–45   No-load equivalent: $a V_{NL} = E_g$.

PRACTICAL NOTES...

1. From Figure 24–45, $a = E_g/V_{NL}$. This means that the turns ratio is the ratio of input voltage to output voltage at no load.

2. The voltage rating of a transformer (such as 1200/120 V) is referred to as its *nominal rating*. The ratio of nominal voltages is the same as the turns ratio. Thus, for a nonloaded transformer, if nominal voltage is applied to the primary, nominal voltage will appear at the secondary.

3. Transformers are normally operated close to their nominal voltages. However, depending on operating conditions, they may be a few percent above or below rated voltage at any given time.

PRACTICE
PROBLEMS 4

A transformer used in an electronic power supply has a nominal rating of 120/12 volts and is connected to a 120-Vac source. Its equivalent impedance as seen from the primary is $10\ \Omega + j10\ \Omega$. What is the magnitude of the load voltage if the load is 5 ohms resistive? Determine the regulation.

Answer: 11.8 V; 2.04%

Transformer Efficiency

Efficiency is the ratio of output power to input power.

$$\eta = \frac{P_{out}}{P_{in}} \times 100\% \qquad \text{(24–12)}$$

But $P_{in} = P_{out} + P_{loss}$. For a transformer, losses are due to I^2R losses in the windings (called copper losses) and losses in the core (called core losses). Thus,

$$\eta = \frac{P_{out}}{P_{out} + P_{loss}} \times 100\% = \frac{P_{out}}{P_{out} + P_{copper} + P_{core}} \times 100\% \qquad \text{(24–13)}$$

Large power transformers are exceptionally efficient, of the order of 98 to 99 percent. The efficiencies of smaller transformers are around 95 percent or better.

24.7   Transformer Tests

Losses may be determined experimentally using the **short-circuit test** and the **open-circuit test.** (These tests are used mainly with power transformers.) They provide the data needed to determine a transformer's equivalent circuit and to compute its efficiency.

The Short-Circuit Test

Figure 24–46 shows the test setup for the short-circuit test. Starting at 0 V, gradually increase E_g until the ammeter indicates the rated current. (This occurs at about 5 percent rated input voltage.) Since core losses are proportional to the square of voltage, at 5 percent rated voltage, core losses are negligible. The losses that you measure are therefore only copper losses.

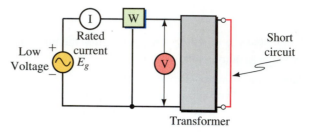

FIGURE 24–46 Short-circuit test.

EXAMPLE 24–10 Measurements on the high side of a 240/120-volt, 4.8-kVA transformer yield $E_g = 11.5$ V and $W = 172$ W at the rated current of $I = 4.8$ kVA/240 = 20 A. Determine $\mathbf{Z}_{eq}$.

Solution See Figure 24–47. Since $Z_L = 0$, the only impedance in the circuit is $\mathbf{Z}_{eq}$. Thus, $Z_{eq} = E_g/I = 11.5$ V/20 A = 0.575 Ω. Further, $R_{eq} = W/I^2 = 172$ W/$(20$ A$)^2 = 0.43$ Ω. Therefore,

$$X_{eq} = \sqrt{Z_{eq}^2 - R_{eq}^2} = \sqrt{(0.575)^2 - (0.43)^2} = 0.382 \ \Omega$$

and $\mathbf{Z}_{eq} = R_{eq} + jX_{eq} = 0.43 \ \Omega + j0.382 \ \Omega$ as shown in (b).

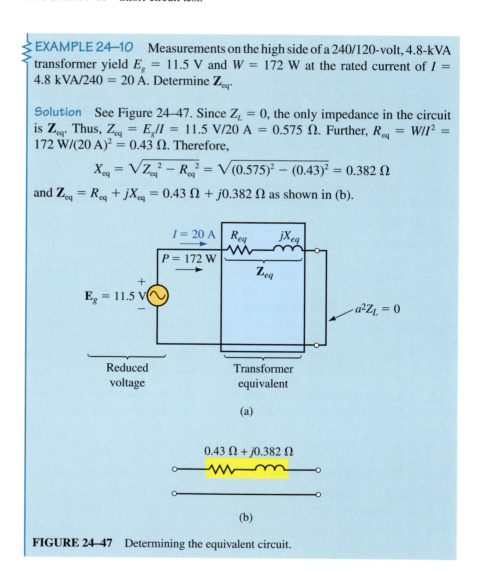

FIGURE 24–47 Determining the equivalent circuit.

The Open-Circuit Test

The setup for the open-circuit test is shown in Figure 24–48. Apply the full rated voltage. Since the load current is zero, only exciting current results. Since the exciting current is small, power loss in the winding resistance is negligible, and the power that you measure is just core loss.

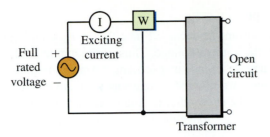

FIGURE 24–48 Open-circuit test.

EXAMPLE 24–11 An open-circuit test on the transformer of Example 24–10 yields a core loss of 106 W. Determine this transformer's efficiency when supplying the full, rated VA to a load at unity power factor.

Solution Since the transformer is supplying the rated VA, its current is the full, rated current. From the short-circuit test, the copper loss at full rated current is 172 W. Thus,

$$\text{copper loss} = 172 \text{ W}$$
$$\text{core loss} = 106 \text{ W (Measured above)}$$
$$\text{output} = 4800 \text{ W (Rated)}$$
$$\text{input} = \text{output} + \text{losses} = 5078 \text{ W}$$

Thus

$$\eta = P_{\text{out}}/P_{\text{in}} = (4800 \text{ W}/5078 \text{ W}) \times 100 = 94.5\%$$

Copper loss varies as the square of load current. Thus, at half the rated current, the copper loss is $(\frac{1}{2})^2 = \frac{1}{4}$ of its value at full rated current. Core loss remains constant since applied voltage remains constant.

For the transformer of Example 24–11, determine power input and efficiency at half the rated VA output, unity power factor.

PRACTICE PROBLEMS 5

Answer: 2549 W;   94.2%

24.8   Voltage and Frequency Effects

Iron-core transformer characteristics vary with frequency and voltage. To determine why, we start with Faraday's law, $e = N d\Phi/dt$. Specializing this to the sinusoidal ac case, it can be shown that

$$E_p = 4.44 f N_p \Phi_m \qquad\qquad \textbf{(24–14)}$$

where Φ_m is the mutual core flux.

Effect of Voltage

First, assume constant frequency. Since $\Phi_m = E_p/4.44fN_p$, the core flux is proportional to the applied voltage. Thus, if the applied voltage is increased, the core flux increases. Since magnetizing current is required to produce this flux, magnetizing current must increase as well. An examination of the *B-H* curves of Chapter 12 shows that magnetizing current increases dramatically when flux density rises above the knee of the curve; in fact, the effect is so pronounced that the magnetizing current is no longer negligible, but may exceed the load current! For this reason, power transformers should be operated only at or near their rated voltage.

Effect of Frequency

Audio transformers must operate over a range of frequencies. Consider again $\Phi_m = E_p/4.44fN_p$. As this indicates, decreasing the frequency increases the core flux and hence the magnetizing current. At low frequencies, this larger current increases internal voltage drops and hence decreases the output voltage as indicated in Figure 24–49. Now consider increasing the frequency. As frequency increases, leakage inductance and shunt capacitance cause the voltage to fall off. To compensate for this, audio transformers are sometimes designed so that their internal capacitances resonate with their inductances to extend the operating range. This is what causes the peaking at the high-frequency end of the curve.

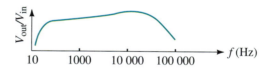

FIGURE 24–49   Frequency response curve, audio transformer.

A transformer with a nominal rating of 240/120 V, 60 Hz, has its load on the 120-V side. Suppose $R_p = 0.4\ \Omega$, $L_p = 1.061$ mH, $R_s = 0.1\ \Omega$, and $L_s = 0.2653$ mH.

a. Determine its equivalent circuit as per Figure 24–42(b).

b. If $\mathbf{E}_g = 240\ \text{V}\angle 0°$ and $\mathbf{Z}_L = 3 + j4\ \Omega$, what is $\mathbf{V}_L$?

c. Compute the regulation.

(Answers are at the end of the chapter.)

24.9   Loosely Coupled Circuits

We now turn our attention to coupled circuits that do not have iron cores. For such circuits, only a portion of the flux produced by one coil links another and the coils are said to be **loosely coupled.** Loosely coupled circuits cannot be characterized by turns ratios; rather, as you will see, they are characterized by self- and mutual inductances. Air-core transformers, ferrite-core transformers, and general inductive circuit coupling fall into this category. In this section, we develop the main ideas.

Voltages in Air-Core Coils

To begin, consider the isolated (noncoupled) coil of Figure 24–50. As shown in Chapter 13, the voltage across this coil is given by $v_L = L di/dt$, where i is the current through the coil and L is its inductance. Note carefully the polarity of the voltage; the plus sign goes at the tail of the current arrow. Because the coil's voltage is created by its own current, it is called a **self-induced voltage.**

Now consider a pair of coupled coils (Figure 24–51). When coil 1 alone is energized as in (a), it looks just like the isolated coil of Figure 24–50; thus its voltage is

$$v_{11} = L_1 di_1/dt \quad \text{(self-induced in coil 1)}$$

where L_1 is the self-inductance of coil 1 and the subscripts indicate that v_{11} is the voltage across coil 1 due to its own current. Similarly, when coil 2 alone is energized as in (b), its self-induced voltage is

$$v_{22} = L_2 di_2/dt \quad \text{(self-induced in coil 2)}$$

For both of these self-voltages, note that the plus sign goes at the tail of their respective current arrows.

Mutual Voltages

Consider again Figure 24–51(a). When coil 1 is energized, some of the flux that it produces links coil 2, inducing voltage v_{21} in coil 2. Since the flux here is due to i_1 alone, v_{21} is proportional to the rate of change of i_1. Let the constant of proportionality be M. Then,

$$v_{21} = M di_1/dt \quad \text{(mutually induced in coil 2)}$$

v_{21} is the **mutually induced voltage** in coil 2 and M is the **mutual inductance** between the coils. It has units of henries. Similarly, when coil 2 alone is energized as in (b), the voltage induced in coil 1 is

$$v_{12} = M di_2/dt \quad \text{(mutually induced in coil 1)}$$

When both coils are energized, the voltage of each coil can be found by superposition; *in each coil, the induced voltage is the sum of its self-voltage plus the voltage mutually induced due to the current in the other coil.* The sign of the self term for each coil is straightforward: It is determined by placing a plus sign at the tail of the current arrow for the coil as shown in Figures 24–51(a) and (b). The polarity of the mutual term, however, depends on whether the mutual voltage is additive or subtractive.

Additive and Subtractive Voltages

Whether self- and mutual voltages add or subtract depends on the direction of currents through the coils relative to their winding directions. This is best described in terms of the dot convention. Consider Figure 24–52(a). Comparing the coils here to Figure 24–12, you can see that their top ends correspond and thus can be marked with dots. Now let currents enter both coils at dotted ends. Using the right-hand rule, you can see that their fluxes add. The total flux linking coil 1 is therefore the *sum* of that produced by i_1 and i_2;

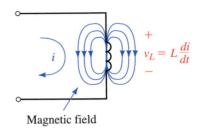

FIGURE 24–50   Place the plus sign for the self-induced voltage at the tail of the current arrow.

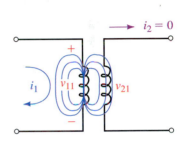

(a) v_{11} is the self-induced voltage in coil 1; v_{21} is the mutual voltage in coil 2

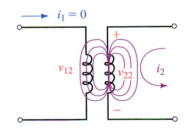

(b) v_{22} is the self-induced voltage in coil 2; v_{12} is the mutual voltage in coil 1

FIGURE 24–51   Self and mutual voltages.

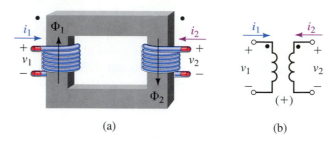

(a)   (b)

FIGURE 24–52   When both currents enter dotted terminals, use the + sign for the mutual term in Equation 24–15.

therefore, the voltage across coil 1 is the sum of that produced by i_1 and i_2. That is,

$$v_1 = L_1 \frac{di_1}{dt} + M \frac{di_2}{dt} \qquad \textbf{(24–15a)}$$

Similarly, for coil 2,

$$v_2 = M \frac{di_1}{dt} + L_2 \frac{di_2}{dt} \qquad \textbf{(24–15b)}$$

Now consider Figure 24–53. Here, the fluxes oppose and the flux linking each coil is the *difference* between that produced by its own current and that produced by the current of the other coil. Thus, the sign in front of the mutual voltage terms will be negative.

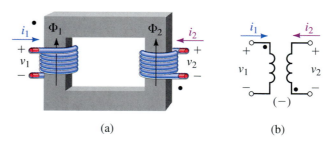

(a)   (b)

FIGURE 24–53   When one current enters a dotted terminal and the other enters an undotted terminal, use the − sign for the mutual term in Equation 24–15.

The Dot Rule

As you can see, the signs of the mutual voltage terms in Equations 24–15 are positive when both currents enter dotted terminals, but negative when one current enters a dotted terminal and the other enters an undotted terminal. Stated another way, *the sign of the mutual voltage is the same as the sign of its self-voltage when both currents enter dotted (or undotted) terminals, but is opposite when one current enters a dotted terminal and the other enters an undotted terminal.* This observation provides us with a procedure for determining voltage polarities in coupled circuits.

1. Assign a direction for currents i_1 and i_2.

2. Place a plus sign at the tail of the current arrow for each coil to denote the polarity of its self-induced voltage.

3. If both currents enter (or both leave) dotted terminals, make the sign of the mutually induced voltage the same as the sign of the self-induced voltage when you write the equation.

4. If one current enters a dotted terminal and the other leaves, make the sign of the mutually induced voltage opposite to the sign of the self-induced voltage.

EXAMPLE 24–12   Write equations for v_1 and v_2 of Figure 24–54(a).

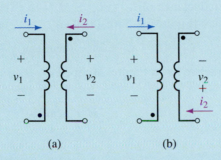

(a)   (b)

FIGURE 24–54

Solution   Since one current enters an undotted terminal and the other enters a dotted terminal, place a minus sign in front of M. Thus,

$$v_1 = L_1 \frac{di_1}{dt} - M \frac{di_2}{dt}$$

$$v_2 = -M \frac{di_1}{dt} + L_2 \frac{di_2}{dt}$$

Write equations for v_1 and v_2 of Figure 24–54(b).

Answer: Same as Equation 24–15.

PRACTICE
PROBLEMS 6

Coefficient of Coupling

For loosely coupled coils, not all of the flux produced by one coil links the other. To describe the degree of coupling between coils, we introduce a **coefficient of coupling,** *k*. Mathematically, *k* is defined as the ratio of the flux that links the companion coil to the total flux produced by the energized coil. For iron-core transformers, almost all the flux is confined to the core and links both coils; thus, *k* is very close to 1. At the other extreme (i.e., isolated coils where no flux linkage occurs), $k = 0$. Thus, $0 \le k < 1$. Mutual inductance depends on *k*. It can be shown that mutual inductance, self-inductances, and the coefficient of coupling are related by the equation

$$M = k\sqrt{L_1 L_2} \qquad (24\text{–}16)$$

Thus, the larger the coefficient of coupling, the larger the mutual inductance.

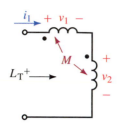

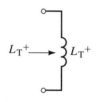

(a) $L_1' = L_1 + M$; $L_2' = L_2 + M$

$L_T^+ \longrightarrow$ L_T^+

(b) $L_T^+ = L_1 + L_2 + 2M$

FIGURE 24–55 Coils in series with additive mutual coupling.

Inductors with Mutual Coupling

If a pair of coils are in close proximity, the field of each coil couples the other, resulting in a change in the apparent inductance of each coil. To illustrate, consider Figure 24–55(a), which shows a pair of inductors with self-inductances L_1 and L_2. If coupling occurs, the effective coil inductances will no longer be L_1 and L_2. To see why, consider the voltage induced in each winding—it is the sum of the coil's own self-voltage plus the voltage mutually induced from the other coil. Since current is the same for both coils, $v_1 = L_1 di/dt + M di/dt = (L_1 + M)di/dt$, which means that coil 1 has an effective inductance of $L_1' = L_1 + M$. Similarly, $v_2 = (L_2 + M)di/dt$, giving coil 2 an effective inductance of $L_2' = L_2 + M$. The effective inductance of the series combination [Figure 24–55(b)] is then

$$L_T^+ = L_1 + L_2 + 2M \quad \text{(henries)} \tag{24–17}$$

If coupling is subtractive as in Figure 24–56, $L_1' = L_1 - M$, $L_2' = L_2 - M$, and

$$L_T^- = L_1 + L_2 - 2M \quad \text{(henries)} \tag{24–18}$$

EXAMPLE 24–13   Three inductors are connected in series (Figure 24–56). Coils 1 and 2 interact, but coil 3 does not.

FIGURE 24–56

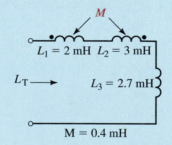

M

$L_1 = 2$ mH $L_2 = 3$ mH

$L_T \longrightarrow$ $L_3 = 2.7$ mH

$M = 0.4$ mH

a. Determine the effective inductance of each coil.

b. Determine the total inductance of the series connection.

Solution

a. $L_1' = L_1 - M = 2$ mH $- 0.4$ mH $= 1.6$ mH

  $L_2' = L_2 - M = 3$ mH $- 0.4$ mH $= 2.6$ mH

  L_1' and L_2' are in series with L_3. Thus,

b. $L_T = 1.6$ mH $+ 2.6$ mH $+ 2.7$ mH $= 6.9$ mH

The same principles apply when more than two coils are coupled. Thus, for the circuit of Figure 24–57, $L_1' = L_1 - M_{12} - M_{31}$, etc.

For two parallel inductors with mutual coupling, the equivalent inductance is

$$L_{eq} = \frac{L_1 L_2 - M^2}{L_1 + L_2 \pm 2M} \tag{24–19}$$

If the dots are at the same ends of the coils, use the + sign. For example, if $L_1 = 20$ mH, $L_2 = 5$ mH, and $M = 2$ mH, then $L_{eq} = 4.57$ mH if the dots are both at the same ends of the coils, and $L_{eq} = 3.31$ mH when the dots are at opposite ends.

For the circuit of Figure 24–57, different "dot" symbols are used to represent coupling between sets of coils.

PRACTICE
PROBLEMS 7

a. Determine the effective inductance of each coil.

b. Determine the total inductance of the series connection.

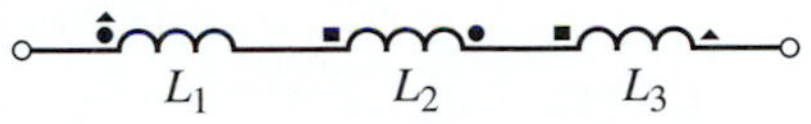

L_1 L_2 L_3

$L_1 = 10$ mH $M_{12} = 2$ mH (Mutual inductance between coils 1 and 2) (●)
$L_2 = 40$ mH $M_{23} = 1$ mH (Mutual inductance between coils 2 and 3) (■)
$L_3 = 20$ mH $M_{31} = 0.6$ mH (Mutual inductance between coils 3 and 1) (▲)

FIGURE 24–57

Answers:
a. $L_1' = 7.4$ mH; $L_2' = 39$ mH; $L_3' = 20.4$ mH

b. 66.8 mH

The effect of unwanted mutual inductance can be minimized by physically separating coils or by orienting their axes at right angles. The latter technique is used where space is limited and coils cannot be spaced widely. While it does not eliminate coupling, it can help minimize its effects.

24.10 Magnetically Coupled Circuits with Sinusoidal Excitation

When coupling occurs between various parts of a circuit (whether wanted or not), the foregoing principles apply. However, since it is difficult to continue the analysis in general, we will change over to steady state ac. This will permit us to look for the main ideas. We will use the mesh approach. To use the mesh approach, (1) write mesh equations using KVL, (2) use the dot convention to determine the signs of the induced voltage components, and (3) solve the resulting equations in the usual manner using determinants or a computer program such as MATHCAD.

To specialize to the sinusoidal ac case, convert voltages and currents to phasor form. To do this, recall from Chapter 16 that inductor voltage in phasor form is $\mathbf{V}_L = j\omega L\mathbf{I}$. (This is the phasor equivalent of $v_L = L \, di/dt$, Figure 24–50.) This means that $L \, di/dt$ becomes $j\omega L\mathbf{I}$; in a similar fashion, $M \, di_1/dt \Rightarrow j\omega M\mathbf{I}_1$ and $M \, di_2/dt \Rightarrow j\omega M\mathbf{I}_2$. Thus, in phasor form Equations 24–15 become

$$\mathbf{V}_1 = j\omega L_1\mathbf{I}_1 + j\omega M\mathbf{I}_2$$
$$\mathbf{V}_2 = j\omega M\mathbf{I}_1 + j\omega L_2\mathbf{I}_2$$

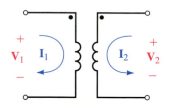

FIGURE 24–58 Coupled coils with sinusoidal ac excitation.

These equations describe the circuit of Figure 24–58 as you can see if you write KVL for each loop. (Verify this.)

EXAMPLE 24–14 For Figure 24–59, write the mesh equations and solve for $\mathbf{I}_1$ and $\mathbf{I}_2$. Let $\omega = 100$ rad/s, $L_1 = 0.1$ H, $L_2 = 0.2$ H, $M = 0.08$ H, $R_1 = 15\ \Omega$, and $R_2 = 20\ \Omega$.

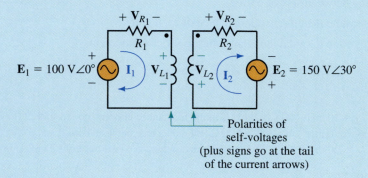

FIGURE 24–59 Air-core transformer example.

Solution $\omega L_1 = (100)(0.1) = 10\ \Omega$, $\omega L_2 = (100)(0.2) = 20\ \Omega$, and $\omega M = (100)(0.08) = 8\ \Omega$. Since one current enters a dotted terminal and the other leaves, the sign of the mutual term is opposite to the sign of the self term. KVL yields

Loop 1: $\mathbf{E}_1 - R_1\mathbf{I}_1 - j\omega L_1\mathbf{I}_1 + j\omega M\mathbf{I}_2 = 0$

Loop 2: $\mathbf{E}_2 - j\omega L_2\mathbf{I}_2 + j\omega M\mathbf{I}_1 - R_2\mathbf{I}_2 = 0$

Thus,

$$(15 + j10)\mathbf{I}_1 - j8\mathbf{I}_2 = 100\angle 0°$$

$$-j8\mathbf{I}_1 + (20 + j20)\mathbf{I}_2 = 150\angle 30°$$

These can be solved by the usual methods such as determinants or by computer. The answers are $\mathbf{I}_1 = 6.36\angle -6.57°$ and $\mathbf{I}_2 = 6.54\angle -2.24°$.

EXAMPLE 24–15 For the circuit of Figure 24–60, determine $\mathbf{I}_1$ and $\mathbf{I}_2$.

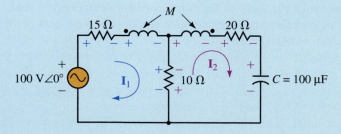

$L_1 = 0.1$ H, $L_2 = 0.2$ H, $M = 80$ mH, $\omega = 100$ rad/s

FIGURE 24–60

Solution $\omega L_1 = 10\ \Omega$, $\omega L_2 = 20\ \Omega$, $\omega M = 8\ \Omega$, and $X_C = 100\ \Omega$.

Loop 1: $100\angle 0° - 15\mathbf{I}_1 - j10\mathbf{I}_1 + j8\mathbf{I}_2 - 10\mathbf{I}_1 + 10\mathbf{I}_2 = 0$

Loop 2: $-10\mathbf{I}_2 + 10\mathbf{I}_1 - j20\mathbf{I}_2 + j8\mathbf{I}_1 - 20\mathbf{I}_2 - (-j100)\mathbf{I}_2 = 0$

Thus:

$$(25 + j10)\mathbf{I}_1 - (10 + j8)\mathbf{I}_2 = 100\angle 0°$$
$$-(10 + j8)\mathbf{I}_1 + (30 - j80)\mathbf{I}_2 = 0$$

Solution yields $\mathbf{I}_1 = 3.56\ \text{A}\angle -18.6°$ and $\mathbf{I}_2 = 0.534\ \text{A}\angle 89.5°$

Refer to the circuit of Figure 24–61.

PRACTICE PROBLEMS 8

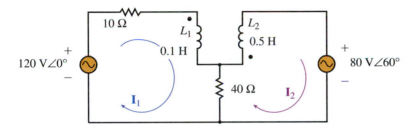

FIGURE 24–61 $M = 0.12\ \text{H}$, $\omega = 100\ \text{rad/s}$.

a. Determine the mesh equations.

b. Solve for currents $\mathbf{I}_1$ and $\mathbf{I}_2$.

Answer:

a. $(50 + j10)\mathbf{I}_1 - (40 - j12)\mathbf{I}_2 = 120\angle 0°$
$-(40 - j12)\mathbf{I}_1 + (40 + j50)\mathbf{I}_2 = -80\angle 60°$

b. $\mathbf{I}_1 = 1.14\ \text{A}\angle -31.9°$ and $\mathbf{I}_2 = 1.65\ \text{A}\angle -146°$

24.11   Coupled Impedance

Earlier we found that an impedance $\mathbf{Z}_L$ on the secondary side of an iron-core transformer is reflected into the primary side as $a^2\mathbf{Z}_L$. A somewhat similar situation occurs in loosely coupled circuits. In this case however, the impedance that you see reflected to the primary side is referred to as **coupled impedance.** To get at the idea, consider Figure 24–62. Writing KVL for each loop yields

Loop 1: $\mathbf{E}_g - \mathbf{Z}_1\mathbf{I}_1 - j\omega L_1\mathbf{I}_1 - j\omega M\mathbf{I}_2 = 0$

Loop 2: $-j\omega L_2\mathbf{I}_2 - j\omega M\mathbf{I}_1 - \mathbf{Z}_2\mathbf{I}_2 - \mathbf{Z}_L\mathbf{L}_2 = 0$

which reduces to

$$\mathbf{E}_g = \mathbf{Z}_p\mathbf{I}_1 + j\omega M\mathbf{I}_2 \qquad\qquad \textbf{(24–20a)}$$

$$0 = j\omega M\mathbf{I}_1 + (\mathbf{Z}_s + \mathbf{Z}_L)\mathbf{I}_2 \qquad\qquad \textbf{(24–20b)}$$

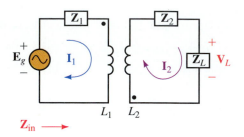

FIGURE 24–62

where $\mathbf{Z}_p = \mathbf{Z}_1 + j\omega L_1$ and $\mathbf{Z}_s = \mathbf{Z}_2 + j\omega L_2$. Solving Equation 24–20b for $\mathbf{I}_2$ and substituting this into Equation 24–20a yields, after some manipulation,

$$\mathbf{E}_g = \mathbf{Z}_p\mathbf{I}_1 + \frac{(\omega M)^2}{\mathbf{Z}_s + \mathbf{Z}_L}\mathbf{I}_1$$

Now, divide both sides by $\mathbf{I}_1$, and define $\mathbf{Z}_{in} = \mathbf{E}_g/\mathbf{I}_1$. Thus,

$$\mathbf{Z}_{in} = \mathbf{Z}_p + \frac{(\omega M)^2}{\mathbf{Z}_s + \mathbf{Z}_L} \qquad (24\text{–}21)$$

The term $(\omega M)^2/(\mathbf{Z}_s + \mathbf{Z}_L)$, which reflects the secondary impedances into the primary, is the coupled impedance for the circuit. Note that since secondary impedances appear in the denominator, they reflect into the primary with reversed reactive parts. Thus, capacitance in the secondary circuit looks inductive to the source and vice versa.

EXAMPLE 24–16   For Figure 24–62, let $L_1 = L_2 = 10$ mH, $M = 9$ mH, $\omega = 1000$ rad/s, $\mathbf{Z}_1 = R_1 = 5\ \Omega$, $\mathbf{Z}_2 = 1\ \Omega - j5\ \Omega$, $\mathbf{Z}_L = 1\ \Omega + j20\ \Omega$ and $\mathbf{E}_g = 100$ V$\angle 0°$. Determine $\mathbf{Z}_{in}$ and $\mathbf{I}_1$.

Solution

$\omega L_1 = 10\ \Omega$. Thus, $\mathbf{Z}_p = R_1 + j\omega L_1 = 5\ \Omega + j10\ \Omega$.

$\omega L_2 = 10\ \Omega$. Thus, $\mathbf{Z}_s = \mathbf{Z}_2 + j\omega L_2 = (1\ \Omega - j5\ \Omega) + j10\ \Omega = 1\ \Omega + j5\ \Omega$.

$\omega M = 9\ \Omega$ and $\mathbf{Z}_L = 1\ \Omega + j20\ \Omega$. Thus,

$$\mathbf{Z}_{in} = \mathbf{Z}_p + \frac{(\omega M)^2}{\mathbf{Z}_s + \mathbf{Z}_L} = (5 + j10) + \frac{(9)^2}{(1 + j5) + (1 + j20)}$$

$$= 8.58\ \Omega\angle 52.2°$$

$$\mathbf{I}_1 = \mathbf{E}_g/\mathbf{Z}_{in} = (100\angle 0°)/(8.58\angle 52.2°) = 11.7\ \text{A}\angle -52.2°$$

The equivalent circuit is shown in Figure 24–63.

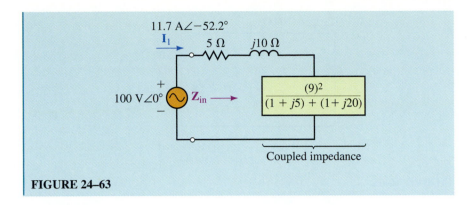

FIGURE 24–63

For Example 24–16, let $R_1 = 10 \ \Omega$, $M = 8$ mH, and $\mathbf{Z}_L = (3 - j8) \ \Omega$. Determine $\mathbf{Z}_{in}$ and $\mathbf{I}_1$.

PRACTICE
PROBLEMS 9

Answers: $28.9 \ \Omega\angle 41.1°$; 3.72 A$\angle -41.1°$

24.12 Circuit Analysis Using Computers

Electronics Workbench and PSpice may be used to solve coupled circuits. (PSpice handles both loosely coupled circuits and tightly coupled (iron-core) transformers, but as of this writing, Workbench handles only iron-core devices.) As a first example, let us solve for generator and load currents and the load voltage for the circuit of Figure 24–64. First, manually determine the answers to provide a basis for comparison. Reflecting the load impedance using $a^2\mathbf{Z}_L$ yields the equivalent circuit of Figure 24–65. From this,

$$\mathbf{I}_g = \frac{100 \ \text{V}\angle 0°}{200 \ \Omega + (200 \ \Omega - j265.3 \ \Omega)} = 208.4 \ \text{mA}\angle 33.5°$$

Thus,

$$\mathbf{I}_L = a\mathbf{I}_g = 416.8 \ \text{mA}\angle 33.5°$$

and

$$\mathbf{V}_L = \mathbf{I}_L\mathbf{Z}_L = 34.6 \ \text{V}\angle -19.4°.$$

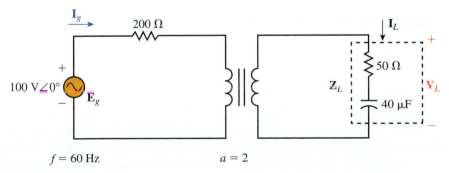

FIGURE 24–64   Iron-core transformer circuit for first Electronics Workbench and PSpice example.

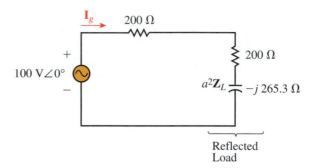

FIGURE 24–65

Electronics Workbench

Draw the circuit as in Figure 24–66. (The transformer is found in the Basic parts bin. It models the transformer in a straightforward manner using its turns ratio.) Double click the transformer symbol, select Models, default, ideal, click Edit and set Magnetizing Inductance (LM) to a very high value, e.g., **10000 H.** (This is L_m of Figure 24–41, which theoretically for an ideal transformer is infinity.) Set the turns ratio to **2,** leave everything else alone, then click OK, OK. Set all the meters to AC, then activate the circuit. Note the magnitudes of the currents and the load voltage. They agree quite well with the answers computed above.

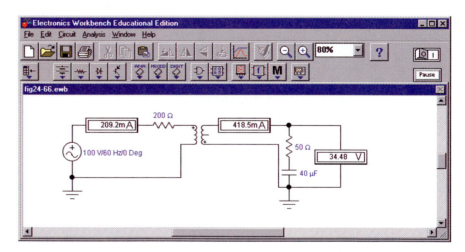

FIGURE 24–66 Electronics Workbench solution for the circuit of Figure 24–64.

OrCADSpice

First, read Item #2 from "PSpice and EWB Notes for Coupled Circuits" (see page 1022). As indicated, transformer element XFRM_LINEAR may be used to model iron-core transformers based solely on their turns ratios. To do this, set coupling $k = 1$, choose an arbitrarily large value for L_1, then let $L_2 = L_1/a^2$ where a is the turns ratio. (The actual values for L_1 and L_2 are not critical, they simply must be very large.) For example, arbitrarily choose

$L_1 = 100000$ H, then compute $L_2 = 100000/(2^2) = 25000$ H. This sets $a = 2$. Now proceed as follows. Create the circuit of Figure 24–64 on the screen as Figure 24–67. Use source VAC, set up as shown. Double click VPRINT1 and then set AC, MAG, and PHASE to **yes** in the Properties Editor. Repeat for the IPRINT devices. Double click the transformer and set COUPLING to **1**, L1 to **100000H,** and L2 to **25000H.** Select AC Sweep and set <u>S</u>tart and <u>E</u>nd frequencies to **60HZ**. Complete the remainder of the setup, run the simulation, then scroll through the Output File. You should find $\mathbf{I}_g = 208.3$ mA∠33.6°, $\mathbf{I}_L = 416.7$ mA∠33.6° and $\mathbf{V}_L = 34.6$ V∠-19.4°. Note that these agree almost exactly with the computed results.

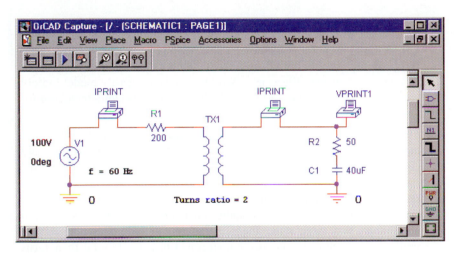

FIGURE 24–67   PSpice solution for the circuit of Figure 24–64.

As a final PSpice example, consider the loosely coupled circuit of Figure 24–59, drawn on the screen as Figure 24–68. Use VAC for the sources

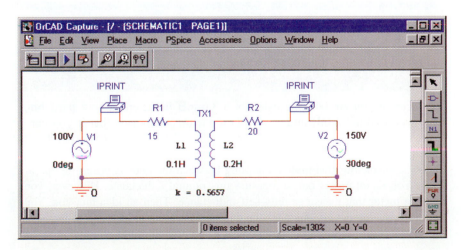

FIGURE 24–68   PSpice solution for Example 24–14.

and XFRM_LINEAR for the transformer. (Be sure to orient Source 2 as shown.) Compute $k = \dfrac{M}{\sqrt{L_1 L_2}} = 0.5657$. Now double click the transformer symbol and set $L_1 = \mathbf{0.1H}$, $L_2 = \mathbf{0.2H}$, and $k = \mathbf{0.5657}$. Set $f = \mathbf{15.9155Hz}$. (Remember, PSpice uses f, not ω.). Complete the remainder of the setup, then run the simulation. When you scroll the Output File, you will find $\mathbf{I}_1 = 6.36\ \mathrm{A}\angle{-6.57°}$ and $\mathbf{I}_2 = 6.54\ \mathrm{A}\angle{-2.23°}$ as determined earlier in Example 24–14.

PSPICE AND EWB NOTES FOR COUPLED CIRCUITS

1. Electronics Workbench uses the ideal transformer equations $\mathbf{E}_p/\mathbf{E}_s = a$ and $\mathbf{I}_p/\mathbf{I}_s = 1/a$ to model transformers. It also permits you to set winding resistance, leakage flux, and exciting current effects as per Figure 24–41. As of this writing, however, Electronics Workbench cannot easily model loosely coupled circuits. (See the Delmar web site at **www.electronictech.com** for a method.)

2. The PSpice transformer model XFRM_LINEAR is based on self-inductances and the coefficient of coupling and is thus able to handle loosely coupled circuits directly. It is also able to model tightly coupled circuits (such as iron-core transformers). To see how, note that basic theory shows that for an ideal iron-core transformer, $k = 1$ and L_1 and L_2 are infinite, but their ratio is $L_1/L_2 = a^2$. Thus, to approximate the transformer, just set L_1 to an arbitrary very large value, then compute $L_2 = L_1/a^2$. This fixes a, permitting you to model iron-core transformers based solely on their turns ratio.

3. The sign of the coefficient of coupling to use with PSpice depends on the dot locations. For example, if dots are on adjacent coil ends (as in Figure 24–59), make k positive; if dots are on opposite ends (Figure 24–62), make k negative.

4. PSpice requires grounds on both sides of a transformer.

PUTTING IT INTO PRACTICE

A circuit you are building calls for a 3.6-mH inductor. In your parts bin, you find a 1.2-mH and a 2.4-mH inductor. You reason that if you connect them in series, the total inductance will be 3.6 mH. After you build and test the circuit, you find that it is out of spec. After careful reasoning, you become suspicious that mutual coupling between the coils is upsetting operation. You therefore set out to measure this mutual inductance. However, you have a meter that measures only self-inductance. Then an idea hits you. You de-energize the circuit, unsolder the end of one of the inductors and measure total inductance. You get 6.32 mH. What is the mutual inductance?

24.1   Introduction

1. For the transformers of Figure 24–69, sketch the missing waveforms.

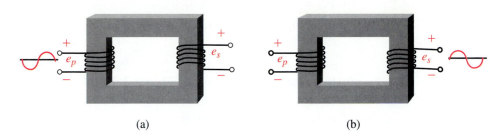

(a)   (b)

FIGURE 24–69

24.2   Iron-Core Transformers: The Ideal Model

2. List the four things that you neglect when you idealize an iron-core transformer.

3. An ideal transformer has $N_p = 1000$ turns and $N_s = 4000$ turns.
   a. Is it step-up or step-down voltage?
   b. If $e_s = 100 \sin \omega t$, what is e_p when wound as in Figure 24–7(a)?
   c. If $E_s = 24$ volts, what is E_p?
   d. If $\mathbf{E}_p = 24 \text{ V}\angle 0°$, what is $\mathbf{E}_s$ when wound as in Figure 24–7(a)?
   e. If $\mathbf{E}_p = 800 \text{ V}\angle 0°$, what is $\mathbf{E}_s$ when wound as in Figure 24–7(b)?

4. A 3:1 step-down voltage transformer has a secondary current of 6 A. What is its primary current?

5. For Figure 24–70, determine the phase relationship for v_1, v_2, and v_3. Determine the expression for each.

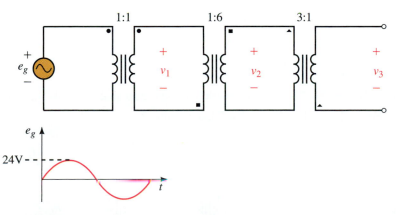

FIGURE 24–70

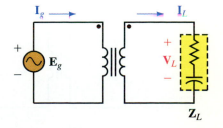

6. If, for Figure 24–71, $\mathbf{E}_g = 240 \text{ V}\angle 0°$, $a = 2$, and $\mathbf{Z}_L = 8 \, \Omega - j6 \, \Omega$, determine the following:
   a. $\mathbf{V}_L$ b. $\mathbf{I}_L$ c. $\mathbf{I}_g$

FIGURE 24–71

7. If, for Figure 24–71, $\mathbf{E}_g = 240\text{ V}\angle0°$, $a = 0.5$, and $\mathbf{I}_g = 2\text{ A}\angle20°$, determine the following:

   a. $\mathbf{I}_L$ b. $\mathbf{V}_L$ c. $\mathbf{Z}_L$

8. If, for Fig. 24–71, $a = 2$, $\mathbf{V}_L = 40\text{ V}\angle0°$, and $\mathbf{I}_g = 0.5\text{ A}\angle10°$, determine $\mathbf{Z}_L$.

9. If, for Fig. 24–71, $a = 4$, $\mathbf{I}_g = 4\text{ A}\angle30°$, and $\mathbf{Z}_L = 6\ \Omega\ -j8\ \Omega$, determine the following:

   a. $\mathbf{V}_L$ b. $\mathbf{E}_g$

10. If, for the circuit of Figure 24–71, $a = 3$, $\mathbf{I}_L = 4\text{ A}\angle25°$, and $\mathbf{Z}_L = 10\ \Omega\angle-5°$, determine the following:

    a. Generator current and voltage.

    b. Power to the load.

    c. Power output by the generator.

    d. Does $P_{\text{out}} = P_{\text{in}}$?

24.3 Reflected Impedance

11. For each circuit of Figure 24–72, determine $\mathbf{Z}_p$.

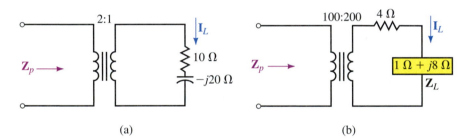

(a)   (b)

FIGURE 24–72

12. For each circuit of Figure 24–72, if $\mathbf{E}_g = 120\text{ V}\angle40°$ is applied, determine the following, using the reflected impedance of Problem 11.

    a. $\mathbf{I}_g$ b. $\mathbf{I}_L$ c. $\mathbf{V}_L$

13. For Figure 24–72(a), what turns ratio is required to make $\mathbf{Z}_p = (62.5 - j125)\ \Omega$?

14. For Figure 24–72(b), what turns ratio is required to make $\mathbf{Z}_p = 84.9\angle58.0°\ \Omega$?

15. For each circuit of Figure 24–73, determine $\mathbf{Z}_T$.

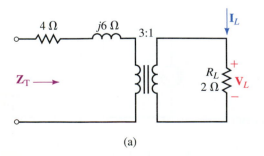

(a)

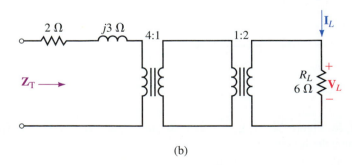

(b)

FIGURE 24–73

16. For each circuit of Figure 24–73, if a generator with $\mathbf{E}_g = 120\,\text{V}\angle-40°$ is applied, determine the following:

    a. $\mathbf{I}_g$ b. $\mathbf{I}_L$ c. $\mathbf{V}_L$

24.4 Transformer Power Ratings

17. A transformer has a rated primary voltage of 7.2 kV, $a = 0.2$, and a secondary rated current of 3 A. What is its kVA rating?

18. Consider a 48 kVA, 1200/120-V transformer.

    a. What is the maximum kVA load that it can handle at $F_p = 0.8$?

    b. What is the maximum power that it can supply to a 0.75 power factor load?

    c. If the transformer supplies 45 kW to a load at 0.6 power factor, is it overloaded? Justify your answer.

24.5 Transformer Applications

19. The transformer of Figure 24–25 has a 7200-V primary and center-tapped 240-V secondary. If Load 1 consists of twelve 100-W lamps, Load 2 is a 1500-W heater, and Load 3 is a 2400-W stove with $F_p = 1.0$, determine

    a. I_1 b. I_2 c. I_N d. I_p

20. An amplifier with a Thévenin voltage of 10 V and Thévenin resistance of 128 Ω is connected to an 8-Ω speaker through a 4 : 1 transformer. Is the load matched? How much power is delivered to the speaker?

21. An amplifier with a Thévenin equivalent of 10 V and R_{Th} of 25 Ω drives a 4-Ω speaker through a transformer with a turns ratio of $a = 5$. How much power is delivered to the speaker?

22. For Problem 21, what turns ratio do you need to get an output of 1 W to the speaker?

23. For Figure 24–30(a), $a_2 = 2$ and $a_3 = 5$, $\mathbf{Z}_2 = 20\,\Omega\angle50°$, $\mathbf{Z}_3 = (12 + j4)\,\Omega$ and $\mathbf{E}_g = 120\,\text{V}\angle0°$. Find each load current and the generator current.

24. It is required to connect a 5-kVA, 120/240-V transformer as an autotransformer to a 120-V source to supply 360 V to a load.

    a. Draw the circuit.

    b. What is the maximum current that the load can draw?

    c. What is the maximum load kVA that can be supplied?

    d. How much current is drawn from the source?

24.6 Practical Iron-Core Transformers

25. For Figure 24–74, $\mathbf{E}_g = 1220\,\text{V}\angle0°$.

    a. Draw the equivalent circuit,

    b. Determine $\mathbf{I}_g$, $\mathbf{I}_L$, and $\mathbf{V}_L$.

26. For Figure 24–74, if $\mathbf{V}_L = 118\,\text{V}\angle0°$, draw the equivalent circuit and determine

    a. $\mathbf{I}_L$ b. $\mathbf{I}_g$ c. $\mathbf{E}_g$

    d. no-load voltage   e. regulation

27. A transformer delivering $P_{\text{out}} = 48$ kW has a core loss of 280 W and a copper loss of 450 W. What is its efficiency at this load?

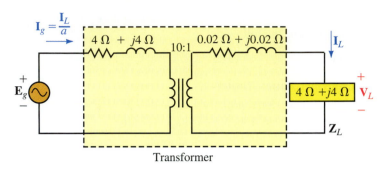

FIGURE 24–74

FIGURE 24–74

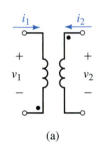

(a)

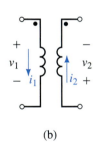

(b)

24.7 Transformer Tests

28. A short-circuit test (Figure 24–46) at rated current yields a wattmeter reading of 96 W, and an open-circuit test (Figure 24–48) yields a core loss of 24 W.

   a. What is the transformer's efficiency when delivering the full, rated output of 5 kVA at unity F_p?

   b. What is its efficiency when delivering one quarter the rated kVA at 0.8 F_p?

24.9 Loosely Coupled Circuits

29. For Figure 24–75,

$$v_1 = L_1 \frac{di_1}{dt} \pm M \frac{di_2}{dt}, \qquad v_2 = \pm \frac{M di_1}{dt} + L_2 \frac{di_2}{dt}$$

   For each circuit, indicate whether the sign to use with M is plus or minus.

30. For a set of coils, $L_1 = 250$ mH, $L_2 = 0.4$ H, and $k = 0.85$. What is M?

31. For a set of coupled coils, $L_1 = 2$ H, $M = 0.8$ H and the coefficient of coupling is 0.6. Determine L_2.

32. For Figure 24–51(a), $L_1 = 25$ mH, $L_2 = 4$ mH, and $M = 0.8$ mH. If i_1 changes at a rate of 1200 A/s, what are the primary and secondary induced voltages?

33. Everything the same as Problem 32 except that $i_1 = 10 \, e^{-500t}$ A. Find the equations for the primary and secondary voltages. Compute them at $t = 1$ ms.

34. For each circuit of Figure 24–76, determine L_T.

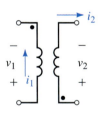

(c)

FIGURE 24–75

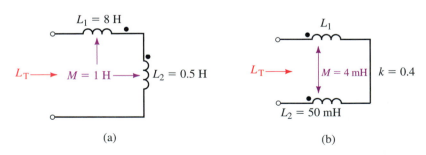

(a)   (b)

FIGURE 24–76

35. For Figure 24–77, determine L_T.

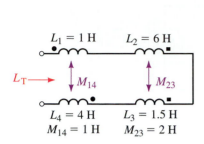

$L_1 = 1$ H $L_2 = 6$ H

$L_T \longrightarrow$ M_{14} M_{23}

$L_4 = 4$ H $L_3 = 1.5$ H
$M_{14} = 1$ H $M_{23} = 2$ H

FIGURE 24–77

$L_1 = 1.0$ H $k = 0.8$
$L_2 = 4.0$ H $f = 60$ Hz

FIGURE 24–78

36. For the circuit of Figure 24–78, determine **I**.

37. The inductors of Figure 24–79 are mutually coupled. What is their equivalent inductance? If $f = 60$ Hz, what is the source current?

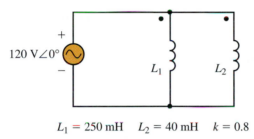

$L_1 = 250$ mH $L_2 = 40$ mH $k = 0.8$

FIGURE 24–79   Coupled parallel inductors.

24.10   Magnetically Coupled Circuits with AC Excitation

38. For Figure 24–59, $R_1 = 10$ Ω, $R_2 = 30$ Ω, $L_1 = 100$ mH, $L_2 = 200$ mH, $M = 25$ mH, and $f = 31.83$ Hz. Write the mesh equations.

39. For the circuit of Figure 24–80, write mesh equations.

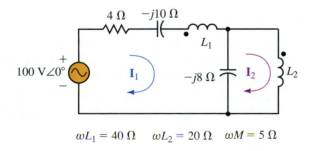

$\omega L_1 = 40$ Ω $\omega L_2 = 20$ Ω $\omega M = 5$ Ω

FIGURE 24–80

40. Write mesh equations for the circuit of Figure 24–81.

41. Write mesh equations for the circuit of Figure 24–82. (This is a very challenging problem.)

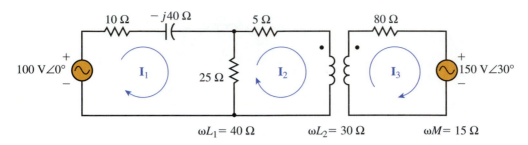

FIGURE 24–81

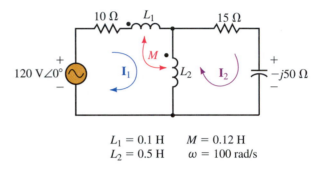

$L_1 = 0.1$ H $M = 0.12$ H
$L_2 = 0.5$ H $\omega = 100$ rad/s

FIGURE 24–82

24.11   Coupled Impedance

42. For the circuit of Figure 24–83,

  a. determine $\mathbf{Z}_{in}$

  b. determine $\mathbf{I}_g$.

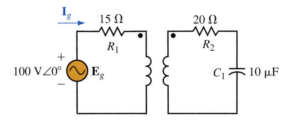

FIGURE 24–83

$L_1 = 0.1$ H; $L_2 = 0.2$ H; $M = 0.08$ H; $f = 60$ Hz

24.12   Circuit Analysis Using Computers

Notes: (1) At the time of writing, Electronics Workbench solves for magnitude only. (2) With PSpice, orient the IPRINT devices so that current enters the positive terminal. Otherwise the phase angles will be in error by 180°.

42. **EWB** or **PSpice** An iron-core transformer with a 4:1 turns ratio has a load consisting of a 12-Ω resistor in series with a 250-µF capacitor. The transformer is driven from a 120-V∠0°, 60-Hz source. Use Electronics Workbench or PSpice to determine the source and load currents. Verify the answers by manual computation.

43. **EWB** or **PSpice** Using Workbench or PSpice, solve for the primary and secondary currents and the load voltage for Figure 24–84.

44. **PSpice** Using PSpice, solve for the source current for the coupled parallel inductors of Figure 24–79. Hint: Use XFRM_LINEAR to model the two inductors. You will need two very low-value resistors to avoid creating a source-inductor loop.

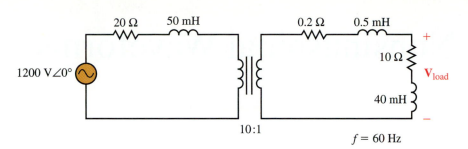

FIGURE 24–84

45. Solve for the currents of Figure 24–60 using PSpice. Compare these to the answers of Example 24–15.

46. Solve for the currents of Figure 24–61 using PSpice. Compare these to the answers of Practice Problem 8.

47. Solve Example 24–16 for current $\mathbf{I}_1$ using PSpice. Compare answers. Hint: If values are given as X_L and X_C, you must convert them to L and C.

In-Process Learning Check 1

1. Step-up; 200 V

2. a. 120 V∠−30°   b. 120 V∠150°

3.

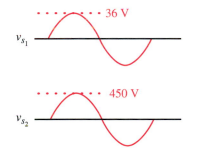

4. Secondary, upper terminal.

5. a. Terminal 4   b. Terminal 4

In-Process Learning Check 2

1. $\mathbf{Z}_p = 18.75\ \Omega - j25\ \Omega$; $R = 6400\ \Omega$; $a = 1.73$

2. 125 V∠0°

3. 89.4

4. a. 200 A; 200 A   b. 24 kW; 18 kW

5. Tap 1: 109.1 V;   Tap 2: 120 V;   Tap 3: 126.3 V

6. 0.81 W; 0.72 W;   Maximum power is delivered when $R_s = a^2 R_L$.

7. a. 348 V;   20.7 A   b. 12 A   c. 12 A↓   8.69 A ↑

In-Process Learning Check 3

1. a. $\mathbf{Z}_{eq} = 0.8\ \Omega + j0.8\ \Omega$

   b. 113.6 V∠0.434°

   c. 5.63%

25

Nonsinusoidal Waveforms

OBJECTIVES

After studying this chapter, you will be able to

- solve for the coefficients of the Fourier series of a simple periodic waveform, using integration,
- use tables to write the Fourier equivalent of any simple periodic waveform,
- sketch the frequency spectrum of a periodic waveform, giving the amplitudes of various harmonics in either volts, watts, or dBm,
- calculate the power dissipated when a complex waveform is applied to a resistive load,
- determine the output of a filter circuit given the frequency spectrum of the input signal and the frequency response of the filter,
- use PSpice to observe the actual response of a filter circuit to a nonsinusoidal input signal.

KEY TERMS

Fourier Series
Frequency Spectrum
Fundamental Frequency
Distortion
Harmonic Frequency
Spectrum Analyzer

OUTLINE

Composite Waveforms
Fourier Series
Fourier Series of Common Waveforms
Frequency Spectrum
Circuit Response to a Nonsinusoidal Waveform
Circuit Analysis Using Computers

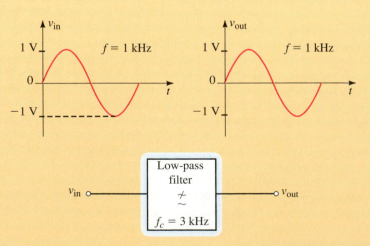

FIGURE 25–1

In our analysis of ac circuits, we have dealt primarily with sinusoidal waveforms. Although the sinusoidal wave is the most common waveform in electronic circuits, it is by no means the only type of signal used in electronics. In previous chapters we observed how sinusoidal signals were affected by the characteristics of components within a circuit. For instance, if a 1-kHz sinusoid is applied to a low-pass filter circuit having a cutoff frequency of 3 kHz, we know that the signal appearing at the output of the filter will be essentially the same as the signal applied to the input. This effect is illustrated in Figure 25–1.

One would naturally expect that the 1-kHz low-pass filter would allow any other 1-kHz signal to pass from input to the output without being distorted. Unfortunately, this is not the case.

In this chapter we will find that any periodic waveform is composed of numerous sinusoidal waveforms, each of which has a unique amplitude and frequency. As we have already seen, circuits such as the low-pass filter and the resonant tank circuit do not allow all sinusoidal frequencies to pass from the input to the output in the same manner. As a result, the output signal may be dramatically different from the signal applied at the input. For example, if we were to apply a 1-kHz square wave to a low-pass filter having a 3-kHz cutoff, the output would appear as shown in Figure 25–2.

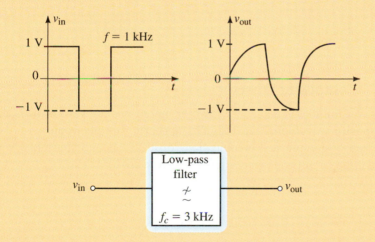

FIGURE 25–2

Although the frequency of the square wave is less than the 3-kHz cutoff frequency of the filter, we will find that the square wave has many high-frequency components which are well above the cutoff frequency. It is these components which are affected in the filter and this results in the distortion of the output waveform.

Once a periodic waveform is reduced to the summation of sinusoidal waveforms, it is a fairly simple matter to determine how various frequency components of the original signal will be affected by the circuit. The overall response of the circuit to a particular waveform can then be found.

Jean Baptiste Joseph Fourier

FOURIER WAS BORN IN AUXERRE, Yonne, France on March 21, 1768. As a youth, Fourier reluctantly studied for the priesthood in the monastery of Saint-Benoit-sur Loire. However, his interest was in mathematics. In 1798, Fourier accompanied Napoleon to Egypt, where he was made a governor. After returning to France, Fourier was particularly interested in the study of heat transfer between two points of different temperature. He was appointed joint secretary of the Academy of Sciences in 1822.

In 1807, Fourier announced the discovery of a theorem which made him famous. Fourier's theorem states that any periodic waveform can be written as the summation of a series of simple sinusoidal functions.

Using his theorem, Fourier was able to develop important theories in heat transfer, which were published in 1822 in a book titled *Analytic Theory of Heat.*

Although still used to describe heat transfer, Fourier's theorem is used today to predict how filters and various other electronic circuits operate when subjected to a nonsinusoidal periodic function.

Fourier died in Paris on May 16, 1830 as the result of a fall down the stairs.

25.1   Composite Waveforms

Any waveform that is made up of two or more separate waveforms is called a **composite waveform.** Most signals appearing in electronic circuits are comprised of complicated combinations of dc levels and sinusoidal waves. Consider the circuit and signal shown in Figure 25–3.

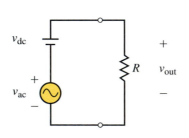
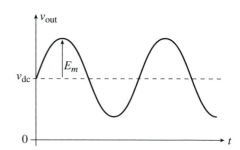

FIGURE 25–3

The voltage appearing across the load is determined by superposition as the combination of the ac source in series with a dc source. The result is a sine wave with a dc offset. As one might expect, when a composite wave is applied to a load resistor, the resulting power is determined by considering the effects of both signals. The rms voltage of the composite waveform is determined as

$$V_{rms} = \sqrt{V_{dc}^2 + V_{ac}^2} \qquad (25-1)$$

where V_{ac} is the rms value of the ac component of waveform and is found as
$V_{ac} = \dfrac{E_m}{\sqrt{2}}$. The power delivered to a load will be determined simply as

$$P_{load} = \dfrac{V_{rms}^2}{R_{load}}$$

The following example illustrates this principle.

EXAMPLE 25–1   Determine the power delivered to the load if the waveform of Figure 25–4 is applied to a 500-Ω resistor.

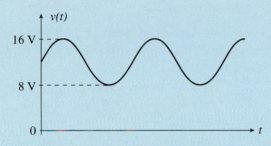

FIGURE 25–4

Solution   By examining the waveform, we see that the average value is $V_{dc} = 12$ V and the peak value of the sinusoidal is $V_m = 16\ V - 12\ V = 4$ V. The rms value of the sinusoidal waveform is determined to be $V_{ac} = (0.707)(4\ V) = 2.83$ V. Now we find the rms value of the composite wave as

$$\begin{aligned}
V_{rms} &= \sqrt{(12\ V)^2 + (2.83\ V)^2}\\
&= \sqrt{152\ V^2}\\
&= 12.3\ V
\end{aligned}$$

and so the power delivered to the load is

$$P_{load} = \dfrac{(12.3\ V)^2}{500\ \Omega} = 0.304\ W$$

Determine the power delivered to the load if the waveform of Figure 25–5 is applied to a 200-Ω resistor.

PRACTICE PROBLEMS 1

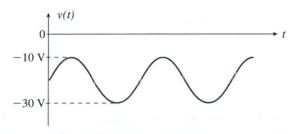

FIGURE 25–5

Answer: 2.25 W

25.2 Fourier Series

In 1826, Baron Jean Baptiste Joseph Fourier developed a branch of mathematics which is used to express any periodic waveform as an infinite series of sinusoidal waveforms. Although it seems as though we are turning a simple waveform into a more complicated form, you will find that the resulting expression actually simplifies the analysis of many circuits which respond differently to signals of various frequencies. Using Fourier analysis, any periodic waveform can be written as a summation of sinusoidal waveforms as follows:

$$f(t) = a_0 + a_1 \cos \omega t + a_2 \cos 2\omega t + \cdots + a_n \cos n\omega t + \cdots \\ + b_1 \sin \omega t + b_2 \sin 2\omega t + \cdots + b_n \sin n\omega t + \cdots \qquad \text{(25–2)}$$

The coefficients of the individual terms of the Fourier series are found by integrating the original function over one complete period. The coefficients are determined as

$$a_0 = \frac{1}{T} \int_{t_1}^{t_1 + T} f(t)\, dt \qquad \text{(25–3)}$$

$$a_n = \frac{2}{T} \int_{t_1}^{t_1 + T} f(t)\cos n\omega t\, dt \qquad \text{(25–4)}$$

$$b_n = \frac{2}{T} \int_{t_1}^{t_1 + T} f(t)\sin n\omega t\, dt \qquad \text{(25–5)}$$

Notice that Equation 25–2 indicates that the Fourier series of a periodic function may contain both a sine and a cosine component at each frequency. These individual components can be combined to give a single sinusoidal expression as follows:

$$a_n \cos nx + b_n\sin nx = a_n\sin(nx + 90°) + b_n\sin nx \\ = c_n\sin (nx + \theta)$$

where

$$c_n = \sqrt{a_n^2 + b_n^2} \qquad \text{(25–6)}$$

and

$$\theta = \tan^{-1}\left(\frac{a_n}{b_n}\right) \qquad \text{(25–7)}$$

Therefore, the Fourier equivalent of any periodic waveform may be simplified as follows:

$$f(t) = a_0 + c_1\sin(\omega t + \theta_1) + c_2\sin(2\omega t + \theta_2) + \cdots$$

The a_0 term is a constant which corresponds to the average value of the periodic waveform and the c_n coefficients give the amplitudes of the various sinusoidal terms. Notice that the first sinusoidal term ($n = 1$) has the same frequency as the original waveform. This component is referred to as the **fundamental frequency** of the given waveform. All other frequencies are integer multiples of the fundamental frequency and are called the **harmonic**

frequencies. When $n = 2$, the resulting term is called the second harmonic; when $n = 3$, we have the third harmonic, etc. Using Equations 25–3 to 25–7, it is possible to derive the Fourier series for any periodic function.

EXAMPLE 25–2 Write the Fourier series for the pulse waveform shown in Figure 25–6.

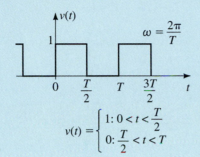

$$v(t) = \begin{cases} 1: 0 < t < \dfrac{T}{2} \\ 0: \dfrac{T}{2} < t < T \end{cases}$$

FIGURE 25–6

Solution The various coefficients are calculated by integrating as follows:

$$a_0 = \frac{1}{T}\int_0^{T/2}(1)\,dt + \frac{1}{T}\int_{T/2}^{T}(0)\,dt = \frac{1}{2}$$

$$a_n = \frac{2}{T}\int_0^{T/2}(1)\cos n\omega t\,dt + \frac{2}{T}\int_{T/2}^{T}(0)\,dt$$

$$= \frac{2}{T}\left[\left(\frac{1}{n\omega}\right)\sin n\omega t\right]_0^{T/2}$$

$$= \frac{1}{n\pi}\sin n\pi = 0$$

Notice that all $a_n = 0$ since $\sin n\pi = 0$ for all n.

$$b_1 = \frac{2}{T}\int_0^{T/2}(1)\sin \omega t\,dt + \frac{2}{T}\int_{T/2}^{T}(0)\,dt$$

$$= \frac{2}{T}\left[-\left(\frac{1}{\omega}\right)\cos \omega t\right]_0^{T/2}$$

$$= -\frac{1}{\pi}\left[\cos\left(\frac{2\pi t}{T}\right)\right]_0^{T/2}$$

$$= -\frac{1}{\pi}[(-1) - (1)] = \frac{2}{\pi}$$

$$b_2 = \frac{2}{T}\int_0^{T/2}(1)\sin 2\omega t\,dt + \frac{2}{T}\int_{T/2}^{T}(0)\,dt$$

$$= \frac{2}{T}\left[-\left(\frac{1}{2\omega}\right)\cos 2\omega t\right]_0^{T/2}$$

$$= -\frac{1}{2\pi}\left[\cos\left(\frac{4\pi t}{T}\right)\right]_0^{T/2}$$

$$= -\frac{1}{2\pi}[(1) - (1)] = 0$$

$$b_3 = \frac{2}{T}\int_0^{T/2} (1) \sin 3\omega t \, dt + \frac{2}{T}\int_{T/2}^{T} (0) \, dt$$

$$= \frac{2}{T}\left[-\left(\frac{1}{3\omega}\right)\cos 3\omega t \right]_0^{T/2}$$

$$= -\frac{1}{3\pi}\left[\cos\left(\frac{6\pi t}{T}\right) \right]_0^{T/2}$$

$$= -\frac{1}{3\pi}[(-1) - (1)] = \frac{2}{3\pi}$$

For all odd values of n, we have $b_n = 2/n\pi$ since $\cos n\pi = -1$. Even values of n give $b_n = 0$ since $\cos n\pi = 1$.

The general expression of the Fourier series for the given pulse wave is therefore written as

$$v(t) = \frac{1}{2} + \frac{2}{\pi}\sum_1^{\infty}\frac{\sin n\omega t}{n} \qquad n = 1, 3, 5, \dots \qquad \text{(25–8)}$$

By examining the general expression for the pulse wave of Figure 25–6, several important characteristics are observed. The Fourier series confirms that the wave shown has an average value of $a_0 = 0.5$. In addition, the pulse wave has only odd harmonics. In other words, a pulse wave having a frequency of 1 kHz would have harmonic components occurring at 3 kHz, 5 kHz, etc. Although the given wave consists of an infinite number of sinusoidal components, the amplitudes of successive terms decrease as n increases.

If we were to consider only the first four nonzero frequency components of the pulse wave, we would have the following expression:

$$v(t) = 0.5 + \frac{2}{\pi}\sin \omega t + \frac{2}{3\pi}\sin 3\omega t + \frac{2}{5\pi}\sin 5\omega t$$

$$+ \frac{2}{7\pi}\sin 7\omega t \qquad \text{(25–9)}$$

The graphical representation of the above expression is shown in Figure 25–7.

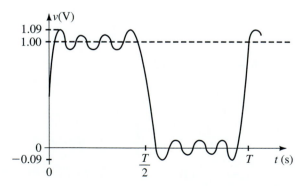

FIGURE 25–7

Although this waveform is not identical with the given pulse wave, we see that the first four nonzero harmonics provide a reasonable approximation of the original waveform.

Derivations of Fourier series for certain waveforms are simplified due to symmetry that occurs in the wave form. We will examine three types of symmetry; even, odd, and half-wave symmetry. Each type of symmetry results in consistent patterns in the Fourier series. The waveforms of Figure 25–8 are symmetrical around the vertical axis and are said to have **even symmetry** (or **cosine symmetry**).

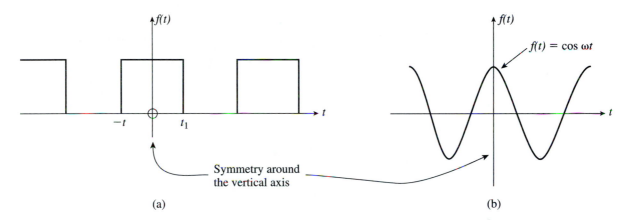

(a)   (b)

FIGURE 25–8 Even symmetry (cosine symmetry).

Waveforms having even symmetry will always have the form

$$f(-t) = f(t) \qquad \text{(Even symmetry)} \qquad (25\text{–}10)$$

When the Fourier series of an even symmetry waveform is written, it will only contain cosine (a_n) terms and possibly an a_0 term. All sine (b_n) terms will be zero.

If the portion of the waveform to the right of the vertical axis in each signal of Figure 25–9 is rotated 180°, we find that it will exactly overlap the portion of the waveform to the left of the axis. Such waveforms are said to have **odd symmetry.**

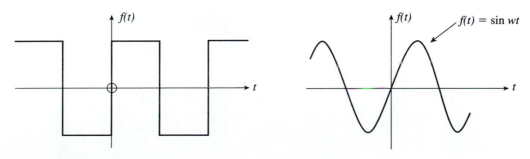

FIGURE 25–9 Odd symmetry (sine symmetry).

Waveforms having odd symmetry will always have the form

$$f(-t) = -f(t) \qquad \text{(Odd symmetry)} \tag{25-11}$$

When the Fourier series of an odd symmetry waveform is written, it will only contain sine (b_n) terms and possibly an a_0 term. All cosine (a_n) terms will be zero.

If the portion of the waveform below the horizontal axis in Figure 25–10 is the mirror image of the portion above the axis, the waveform is said to **half-wave symmetry.**

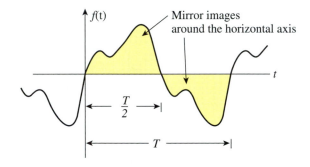

FIGURE 25–10   Half-wave symmetry.

Waveforms having half-wave symmetry will always have the form

$$f(t + T) = -f(t) \qquad \text{(Half-wave symmetry)} \tag{25-12}$$

When the Fourier series of a half-wave symmetry waveform is written, it will have only odd harmonics and possibly an a_0 term. All even harmonic terms will be zero.

If we refer back to the waveform of Figure 25–6, we see that it has both odd symmetry and half-wave symmetry. Using the above rules, we would expect to find only sine terms and odd harmonics. Indeed, we see that Equation 25–9 has these conditions.

PRACTICE PROBLEMS 2

Consider the ramp function shown in Figure 25–11.

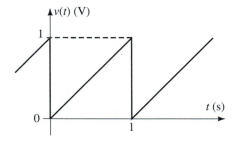

FIGURE 25–11

a. Does this wave form show symmetry?

b. Use calculus to determine the Fourier expression for $v(t)$.

c. Verify that the a_0 term of the Fourier series is equal to the average value of the waveform.

d. From the Fourier expression, is the ramp function made up of odd harmonics, even harmonics, or all harmonic components? Briefly justify your answer.

Answers:

a. Odd symmetry

b. $v(t) = 0.5 - \dfrac{1}{\pi}\sin(2\pi t) - \dfrac{1}{2\pi}\sin(4\pi t) - \dfrac{1}{3\pi}\sin(6\pi t) \cdots$

c. $a_0 = 0.5$ V

d. All harmonic components are present since the function does not have half-wave symmetry.

Without using calculus, determine a method of rewriting the expression of Equation 25–9 to represent a square wave having an amplitude of 1 V as illustrated in Figure 25–12. *Hint:* Notice that the square wave is similar to the pulse waveform with the exception that its average value is zero and that the peak-to-peak value is twice that of the pulse wave.

IN-PROCESS
LEARNING
CHECK 1

FIGURE 25–12

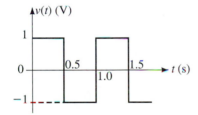

(Answers are at the end of the chapter.)

25.3   Fourier Series of Common Waveforms

All periodic waveforms can be converted into their Fourier equivalent series by using integration as shown in Section 25.2. Integration of common waveforms is time-consuming and prone to error. A more simple approach is to use tables such as Table 25–1, which gives the Fourier series of several common waveforms encountered in electrical circuits.

TABLE 25–1 Fourier Equivalents of Common Waveforms ($\omega = 2\pi/T$)

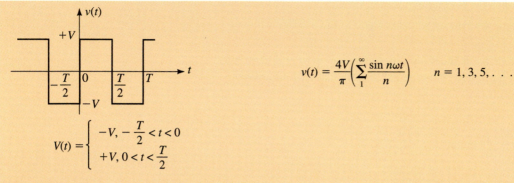

$$v(t) = \frac{4V}{\pi}\left(\sum_{1}^{\infty}\frac{\sin n\omega t}{n}\right) \qquad n = 1, 3, 5, \ldots$$

$$V(t) = \begin{cases} -V, -\dfrac{T}{2} < t < 0 \\ +V, 0 < t < \dfrac{T}{2} \end{cases}$$

FIGURE 25–13

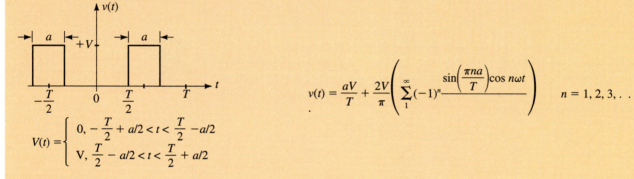

$$v(t) = \frac{aV}{T} + \frac{2V}{\pi}\left(\sum_{1}^{\infty}(-1)^n\frac{\sin\left(\dfrac{\pi na}{T}\right)\cos n\omega t}{n}\right) \qquad n = 1, 2, 3, \ldots$$

$$V(t) = \begin{cases} 0, -\dfrac{T}{2} + a/2 < t < \dfrac{T}{2} - a/2 \\ V, \dfrac{T}{2} - a/2 < t < \dfrac{T}{2} + a/2 \end{cases}$$

FIGURE 25–14

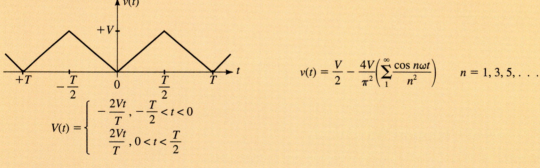

$$v(t) = \frac{V}{2} - \frac{4V}{\pi^2}\left(\sum_{1}^{\infty}\frac{\cos n\omega t}{n^2}\right) \qquad n = 1, 3, 5, \ldots$$

$$V(t) = \begin{cases} -\dfrac{2Vt}{T}, -\dfrac{T}{2} < t < 0 \\ \dfrac{2Vt}{T}, 0 < t < \dfrac{T}{2} \end{cases}$$

FIGURE 25–15

TABLE 25–1   Fourier Equivalents of Common Waveforms ($\omega = 2\pi/T$) *(continued)*

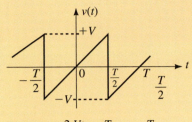

$$V(t) = \frac{2\,Vt}{T}, \; -\frac{T}{2} < t < \frac{T}{2}$$

FIGURE 25–16

$$v(t) = -\frac{2V}{\pi}\left(\sum_1^\infty (-1)^n \frac{\sin n\omega t}{n}\right) \qquad n = 1, 2, 3, \ldots$$

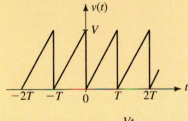

$$V(t) = \frac{Vt}{T}, \; 0 < t < T$$

FIGURE 25–17

$$v(t) = \frac{V}{2} - \frac{V}{\pi}\left(\sum_1^\infty \frac{\sin n\omega t}{n}\right) \qquad n = 1, 2, 3, \ldots$$

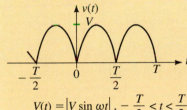

$$V(t) = \left|V \sin \omega t\right|, \; -\frac{T}{2} < t < \frac{T}{2}$$

FIGURE 25–18

$$v(t) = \frac{2V}{\pi} - \frac{4V}{\pi}\left(\frac{\cos 2\omega t}{1 \cdot 3} + \frac{\cos 4\omega t}{3 \cdot 5} + \frac{\cos 6\omega t}{5 \cdot 7} + \cdots\right)$$

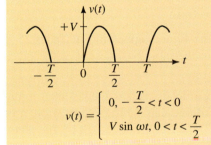

$$v(t) = \begin{cases} 0, \; -\dfrac{T}{2} < t < 0 \\[2mm] V \sin \omega t, \; 0 < t < \dfrac{T}{2} \end{cases}$$

$$v(t) = \frac{V}{\pi} + \frac{V}{2}\sin \omega t - \frac{2V}{\pi}\left(\frac{\cos 2\omega t}{1 \cdot 3} + \frac{\cos 4\omega t}{3 \cdot 5} + \frac{\cos 6\omega t}{5 \cdot 7} + \cdots\right)$$

FIGURE 25–19

The following example illustrates how a given waveform is converted into its Fourier series equivalent.

EXAMPLE 25–3   Use Table 25–1 to determine the Fourier series for the ramp function of Figure 25–20.

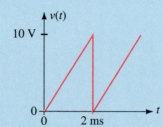

FIGURE 25–20

Solution   The amplitude of the waveform is 10 V and the angular frequency of the fundamental is $\omega = 2\pi/(2 \text{ ms}) = 1000\pi$ rad/s. The resulting series is determined from Table 25–1 as

$$v(t) = \frac{10}{2} - \frac{10}{\pi}\sin 1000\pi t - \frac{10}{2\pi}\sin 2000\pi t$$

$$-\frac{10}{3\pi}\sin 3000\pi t - \frac{10}{4\pi}\sin 4000\pi t - \cdots$$

$$= 5 - 3.18 \sin 1000\pi t - 1.59 \sin 2000\pi t$$

$$-1.06 \sin 3000\pi t - 0.80 \sin 4000\pi t - \cdots$$

If a given waveform is similar to one of the types shown in Table 25–1 but is shifted along the time axis, it is necessary to include a phase shift with each of the sinusoidal terms. The phase shift is determined as follows:

1. Determine the period of the given waveform.

2. Compare the given waveform with the figures appearing in Table 25–1 and select which of the waveforms in the table best describes the given wave.

3. Determine whether the given waveform leads or lags the selected figure of Table 25–1. Calculate the amount of the phase shift, as a fraction, t, of the total period. Since one complete cycle is equivalent to 360°, the phase shift is determined as

$$\phi = \frac{t}{T} \times 360°$$

4. Write the resulting Fourier expression for the given waveform. If the given waveform leads the selected figure of Table 25–1, then add the angle ϕ to each term. If the given waveform lags the selected figure, then subtract the angle ϕ from each term.

EXAMPLE 25–4   Write the Fourier expression for the first four nonzero sinusoidal terms of the waveform shown in Figure 25–21.

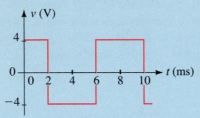

FIGURE 25–21

Solution   **Step 1:** The period of the given waveform is $T = 8.0$ ms, which gives a frequency of $f = 125$ Hz or an angular frequency of $\omega = 250\pi$ rad/s.

Step 2: From Table 25–1, we see that the given waveform is similar to the square wave of Figure 25–13.

Step 3: The waveform of Figure 25–21 leads the square wave of Figure 25–13 by an amount equivalent to $t = 2$ ms. This corresponds to a phase shift of

$$\phi = \frac{2 \text{ ms}}{8 \text{ ms}} \times 360° = 90°$$

Step 4: The Fourier expression for the first four terms of the waveform of Figure 25–21 is now written as follows:

$$v(t) = \frac{4(4)}{\pi}\sin(250\pi t + 90°) + \frac{4(4)}{3\pi}\sin[3(250\pi t + 90°)]$$

$$+ \frac{4(4)}{5\pi}\sin[5(250\pi t + 90°)] + \frac{4(4)}{7\pi}\sin[7(250\pi t + 90°)]$$

The above expression may be left as the summation of sine waves. However, since the cosine wave leads the sine wave by 90°, the expression may be simplified as a summation of cosine waves without any phase shift. This results in the following:

$$v(t) = 5.09 \cos 250\pi t - 1.70 \cos 750\pi t + 1.02 \cos 1250\pi t - 0.73 \cos 1750 \pi t$$

Write the Fourier expression for the first four nonzero sinusoidal terms of the waveform shown in Figure 25–22. Express each term as a sine wave rather than as a cosine wave.

PRACTICE
PROBLEMS 3

FIGURE 25–22

Answer: $v(t) = \dfrac{48}{\pi^2}\sin(250\pi t - 135°) + \dfrac{48}{3^2\pi^2}\sin(750\pi t + 135°)$

$$+ \frac{48}{5^2\pi^2}\sin(1250\pi t + 45°) + \frac{48}{7^2\pi^2}\sin(1750\pi t - 45°)$$

The waveforms of Table 25–1 provide most of the commonly observed waveforms. Occasionally, however, a particular waveform consists of a combination of several simple waveforms. In such a case, it is generally easiest if we first redraw the original waveform as the summation of two or more recognizable waveforms. Then the Fourier series of each of the individual component waves is determined. Finally, the resultant is expressed as the summation of the two series.

EXAMPLE 25–5   Write the first four nonzero sinusoidal terms of the Fourier series for the waveform of Figure 25–23.

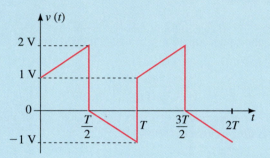

FIGURE 25–23

Solution   The waveform of Figure 25–23 is made up of a combination of waves as illustrated in Figure 25–24.

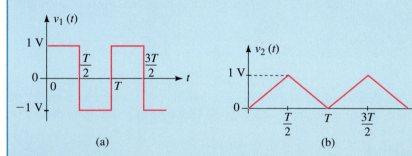

FIGURE 25–24

The Fourier series of each of the waveforms is determined from Table 25–1 as

$$v_1(t) = \frac{4}{\pi} \sin \omega t + \frac{4}{3\pi} \sin 3\omega t + \frac{4}{5\pi} \sin 5\omega t + \cdots$$

and

$$v_2(t) = \frac{1}{2} - \frac{4}{\pi^2} \cos \omega t - \frac{4}{3^2 \pi^2} \cos 3\omega t - \frac{4}{5^2 \pi^2} \cos 5\omega t - \cdots$$

When these series are added algebraically, we get

$$v(t) = v_1(t) + v_2(t)$$
$$= 0.5 + 1.27 \sin \omega t - 0.41 \cos \omega t$$
$$+ 0.42 \sin 3\omega t - 0.05 \cos 3\omega t$$
$$+ 0.25 \sin 5\omega t - 0.02 \cos 5\omega t$$
$$+ 0.18 \sin 7\omega t - 0.01 \cos 7\omega t$$
$$\cdot$$
$$\cdot$$
$$\cdot$$

The above series may be further simplified by using Equations 25–6 and 25–7 to provide a single coefficient and phase shift for each frequency. The resultant waveform is written as

$$v(t) = 0.5 + 1.34 \sin(\omega t - 17.7°) + 0.43 \sin(3\omega t - 6.1°)$$
$$+ 0.26 \sin(5\omega t - 3.6°) + 0.18 \sin(7\omega t - 2.6°)$$

A composite waveform is made up of the summation of the waveforms illustrated in Figure 25–25.

PRACTICE PROBLEMS 4

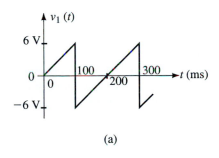

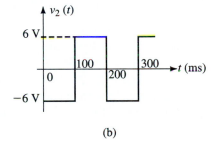

(a)   (b)

FIGURE 25–25

a. Sketch the composite waveform showing all voltage levels and time values.

b. Write the Fourier expression for the resultant waveform, $v(t)$.

Answers:

a.

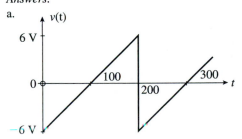

FIGURE 25–26

b. $v(t) = \dfrac{12}{\pi} \sin(10\pi t) - \dfrac{36}{2\pi} \sin(20\pi t) + \dfrac{12}{3\pi} \sin(30\pi t) - \cdots$

Consider the composite waveform of Figure 25–27:

FIGURE 25–27

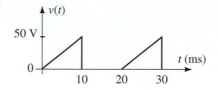

a. Separate the given waveform into two recognizable waveforms appearing in Table 25–1.

b. Use Table 25–1 to write the Fourier series for each of the component waveforms of (a).

c. Combine the results to determine the Fourier series for $v(t)$.

(Answers are at the end of the chapter.)

25.4   Frequency Spectrum

Most waveforms that we have observed were generally shown as a function of time. However, they may also be shown as a function of frequency. In such cases, the amplitude of each harmonic is indicated at the appropriate frequency. Figure 25–28 shows the display of a 1-kHz sine wave in both the time domain and the frequency domain, while Figure 25–29 shows the corresponding displays for a pulse waveform.

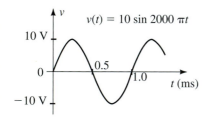

(a) Time-domain display of a 1-kHz sine wave.

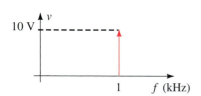

(b) Frequency-domain display of a 1-kHz sine wave.

FIGURE 25–28

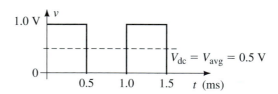

(a) Time domain display of a 1-kHz pulse wave.

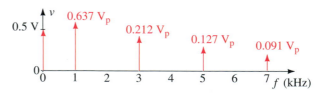

(b) Frequency domain display of a 1-kHz pulse wave.

FIGURE 25–29

The frequency spectrum of the pulse wave shows the average value (or dc value) of the wave at a frequency of 0 kHz and illustrates the absence of even harmonics. Notice that the amplitude of successive harmonic components decreases fairly quickly.

The rms voltage of the composite waveform of Figure 25–29 is determined by considering the rms value of each frequency. The resultant rms voltage is found as

$$V_{rms} = \sqrt{V_{dc}^2 + V_1^2 + V_2^2 + V_3^2 + \cdots}$$

where each voltage, V_1, V_2, etc., represents the rms value of the corresponding harmonic component. Using the first five nonzero terms, we would find the rms value of the pulse-wave of Figure 25–29 to be

$$V_{rms} = \sqrt{(0.500)^2 + \left(\frac{0.637}{\sqrt{2}}\right)^2 + \left(\frac{0.212}{\sqrt{2}}\right)^2 + \left(\frac{0.127}{\sqrt{2}}\right)^2 + \left(\frac{0.091}{\sqrt{2}}\right)^2}$$

$$= 0.698 \text{ V}$$

This value is only slightly smaller than the actual value of $V_{rms} = 0.707$ V.

If the pulse wave were applied to a resistive element, power would be dissipated as if each frequency component had been applied independently. The total power can be determined by the summation of the individual contributions of each frequency. In order to calculate the power dissipated at each sinusoidal frequency, we need to first convert the voltages into rms values. The frequency spectrum may then be represented in terms of power rather than as voltage.

EXAMPLE 25–6 Determine the total power dissipated by a 50-Ω resistor if the pulse waveform of Figure 25–29 is applied to the resistor. Consider the dc component and the first four nonzero harmonics. Indicate the power levels (in watts) on a frequency distribution curve.

Solution The power dissipated by the dc component is determined as

$$P_0 = \frac{V_0^2}{R_L} = \frac{(0.5 \text{ V})^2}{50 \text{ Ω}} = 5.0 \text{ mW}$$

The power dissipated by any resistor subjected to a sinusoidal frequency is determined as

$$P = \frac{V_{rms}^2}{R_L} = \frac{\left(\frac{V_P}{\sqrt{2}}\right)^2}{R_L} = \frac{V_P^2}{2R_L}$$

For the pulse wave of Figure 25–29, the power due to each of the first four nonzero sinusoidal components is found as follows:

$$P_1 = \frac{\left(\frac{2\text{ V}}{\pi}\right)^2}{(2)(50\text{ Ω})} = 4.05 \text{ mW}$$

$$P_3 = \frac{\left(\dfrac{2\,V}{3\pi}\right)^2}{(2)(50\,\Omega)} = 0.45 \text{ mW}$$

$$P_5 = \frac{\left(\dfrac{2\,V}{5\pi}\right)^2}{(2)(50\,\Omega)} = 0.16 \text{ mW}$$

$$P_7 = \frac{\left(\dfrac{2\,V}{7\pi}\right)^2}{(2)(50\,\Omega)} = 0.08 \text{ mW}$$

Figure 25–30 shows the power levels (in milliwatts) as a function of frequency.

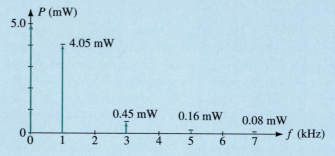

FIGURE 25–30

Using only the dc component and the first four nonzero harmonics, the total power dissipated by the resistor is $P_T = 9.74$ mW. From Chapter 15, the actual rms voltage of the pulse waveform is found to be

$$V_{\text{rms}} = \sqrt{\frac{(1\,V)^2(0.5\text{ ms})}{1.0\text{ ms}}} = 0.707 \text{ V}$$

Therefore, using the rms voltage, the power dissipated by the resistor is found as

$$P = \frac{(0.707\,V)^2}{50\,\Omega} = 10.0 \text{ mW}$$

Although the pulse waveform has power contained in components with frequencies above the seventh harmonic, we see that more than 97% of the total power of a pulse waveform is contained in only the first seven harmonics.

Power levels and frequencies of the various harmonics of a periodic waveform may be measured with an instrument called a **spectrum analyzer,** shown in Figure 25–31.

Some spectrum analyzers are able to display either voltage levels or power levels in the frequency domain, while most others display only power levels (in dBm). When displaying power levels, the spectrum analyzer usually uses a reference 50-Ω load. Figure 25–32 shows the display of a 1.0-V pulse waveform as it would appear on a typical spectrum analyzer.

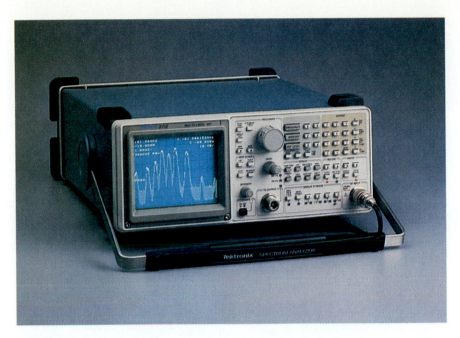

FIGURE 25–31 Spectrum analyzer (*Courtesy of Tektronix Inc.*)

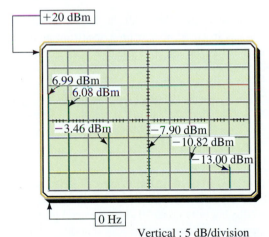

Vertical : 5 dB/division
Horizontal : 1 kHz/division
Reference : +20 dBm

FIGURE 25–32 Spectrum analyzer display of a 1-kHz pulse wave having a peak of 1 V.

Notice that the spectrum analyzer has a reference of +20 dBm and that this reference is shown at the top of the display rather than at the bottom. The vertical scale of the instrument is measured in decibels, where each vertical division corresponds to 5 dB. The horizontal axis for the spectrum analyzer is scaled in hertz, where each division in Figure 25–32 corresponds to 1 kHz.

PRACTICAL NOTES...

Spectrum analyzers are very sensitive instuments. As a result, great care must be taken to ensure that the input power never exceeds the rated maximum. When in doubt, it is best to insert an extra attenuator to lower the amount of power entering the spectrum analyzer.

EXAMPLE 25–7   A spectrum analyzer with a 50-Ω input is used to display the power levels in dBm of the Fourier series components of the ramp waveform shown in Figure 25–33.

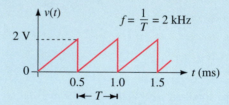

FIGURE 25–33

Determine the voltage and power levels of the various components and sketch the resultant display as it would appear on a spectrum analyzer. Assume that the spectrum analyzer has the same vertical and horizontal settings as those shown in Figure 25–32.

Solution   The Fourier series of the given waveform is determined from Table 25–1 as

$$v(t) = \frac{2}{2} - \frac{2}{\pi}\sin \omega t - \frac{2}{2\pi}\sin 2\omega t - \frac{2}{3\pi}\sin 3\omega t - \cdots$$

Because the fundamental frequency occurs at $f = 2$ kHz, we see that harmonic frequencies will occur at 4 kHz, 6 kHz, etc. However, because the spectrum analyzer is able to display only up to 10 kHz, we need go no further.

The dc component will have an average value of $v_0 = 1.0$ V, as expected. The rms values of the harmonic sinusoidal waveforms are determined as

$$V_{\text{rms}} = \frac{V_p}{\sqrt{2}}$$

which gives the following:

$$V_{1(\text{rms})} = \frac{2}{\pi\sqrt{2}} = 0.450 \text{ V}$$

$$V_{2(\text{rms})} = \frac{2}{2\pi\sqrt{2}} = 0.225 \text{ V}$$

$$V_{3(\text{rms})} = \frac{2}{3\pi\sqrt{2}} = 0.150 \text{ V}$$

$$V_{4(rms)} = \frac{2}{4\pi\sqrt{2}} = 0.113 \text{ V}$$

$$V_{5(rms)} = \frac{2}{5\pi\sqrt{2}} = 0.090 \text{ V}$$

The above rms voltages are used to calculate the powers (and power levels in dBm) of the various harmonic components.

$$P_0 = \frac{(1.0 \text{ V})^2}{50 \text{ }\Omega} = 20.0 \text{ mW} \equiv 10 \log \frac{20 \text{ mW}}{1 \text{ mW}} = 13.0 \text{ dBm}$$

$$P_1 = \frac{(0.450 \text{ V})^2}{50 \text{ }\Omega} = 4.05 \text{ mW} \equiv 10 \log \frac{4.05 \text{ mW}}{1 \text{ mW}} = 6.08 \text{ dBm}$$

$$P_2 = \frac{(0.225 \text{ V})^2}{50 \text{ }\Omega} = 1.01 \text{ mW} \equiv 10 \log \frac{1.01 \text{ mW}}{1 \text{ mW}} = 0.04 \text{ dBm}$$

$$P_3 = \frac{(0.150 \text{ V})^2}{50 \text{ }\Omega} = 0.45 \text{ mW} \equiv 10 \log \frac{0.45 \text{ mW}}{1 \text{ mW}} = -3.5 \text{ dBm}$$

$$P_4 = \frac{(0.113 \text{ V})^2}{50 \text{ }\Omega} = 0.25 \text{ mW} \equiv 10 \log \frac{0.25 \text{ mW}}{1 \text{ mW}} = -6.0 \text{ dBm}$$

$$P_5 = \frac{(0.090 \text{ V})^2}{50 \text{ }\Omega} = 0.16 \text{ mW} \equiv 10 \log \frac{0.16 \text{ mW}}{1 \text{ mW}} = -7.9 \text{ dBm}$$

A spectrum analyzer would indicate a display similar to that shown in Figure 25–34.

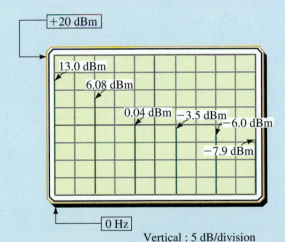

Vertical : 5 dB/division
Horizontal : 1 kHz/division
Reference : +20 dBm

FIGURE 25–34

The sawtooth waveform of Figure 25–35 is applied to a 50-Ω spectrum analyzer. Sketch the display that would be observed, assuming that the spectrum analyzer has the same vertical and horizontal settings as those shown in Figure 25–32.

PRACTICE PROBLEMS 5

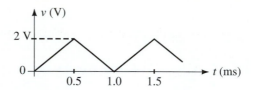

FIGURE 25–35

Answers: P_{dc} = +13.0 dBm, P_{1kHz} = +8.2 dBm, P_{3kHz} = −10.9 dBm, P_{5kHz} = −19.8 dBm. (All other components are less than −20 dBm and so will not appear.)

25.5   Circuit Response to a Nonsinusoidal Waveform

We have determined that all periodic, nonsinusoidal waveforms are comprised of numerous sinusoidal components together with a dc component. In Chapter 22 we observed how various frequencies were affected when they were applied to a given filter. We will now examine how the frequency components of a waveform will be modified when applied to the input of a given filter.

Consider what happens when a pulse waveform is applied to a bandpass filter tuned to the third harmonic, as shown in Figure 25–36.

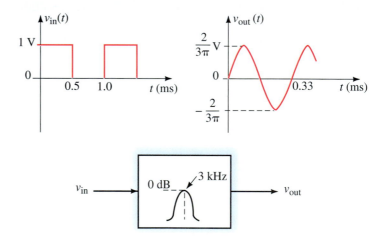

FIGURE 25–36

Since the bandpass filter is tuned to the third harmonic, only the frequency component corresponding to this harmonic will be passed from the input of the filter to the output. Further, because the filter has a voltage gain of 0 dB at the center frequency, the amplitude of the resulting sinusoidal output will be the same voltage level as the amplitude of the original third harmonic. All other frequencies, including the dc component, will be attenuated by the filter, and so are effectively eliminated from the output.

This method is used extensively in electronic circuits to provide frequency multiplication, since any distorted waveform will be rich in harmonics. The desired frequency component is easily extracted by using a tuned filter circuit. Although any integer multiplication is theoretically possible, most frequency multiplier circuits are either frequency doublers or frequency triplers since higher-order harmonics have much lower amplitudes.

In order to determine the resulting waveform after it passes through any other filter, it is necessary to determine the amplitude and phase shift of numerous harmonic components.

EXAMPLE 25–8 The circuit of Figure 25–37 has the frequency response shown in Figure 25–38.

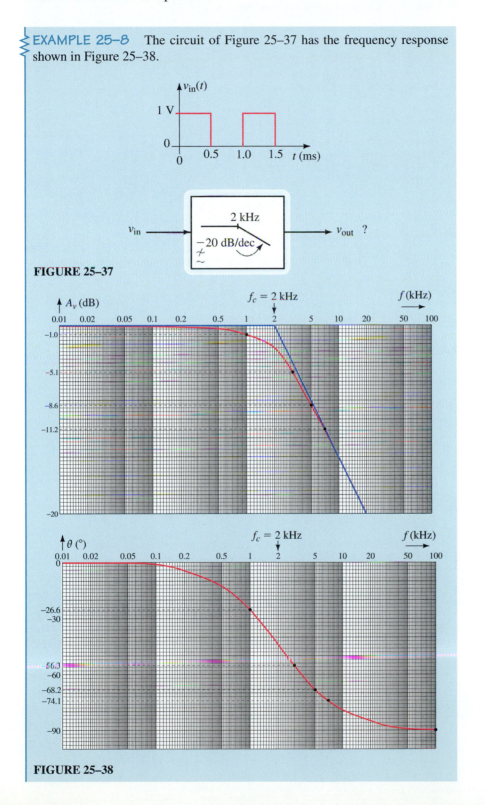

FIGURE 25–37

FIGURE 25–38

a. Determine the dc component at the output of the low-pass filter.

b. Calculate the amplitude and corresponding phase shift of the first four non-zero sinusoidal output components.

Solution

a. From previous examples, we have determined that the given waveform is expressed by the following Fourier series:

$$v(t) = 0.5 + \frac{2}{\pi} \sin \omega t + \frac{2}{3\pi} \sin 3\omega t + \frac{2}{5\pi} \sin 5\omega t + \frac{2}{7\pi} \sin 7\omega t$$

Since the circuit is a low-pass filter, we know that the dc component will pass from the input to the output without being attenuated. Therefore

$$V_{0(out)} = V_{0(in)} = 0.5 \text{ Vdc}$$

b. By examining the frequency response of Figure 25–36, we see that all sinusoidal components will be attenuated and phase shifted. From the graphs we have the following:

1 kHz:	$A_{V1} = -1.0$ dB,	$\Delta\theta_1 = -26.6°$
3 kHz:	$A_{V3} = -5.1$ dB,	$\Delta\theta_3 = -56.3°$
5 kHz:	$A_{V5} = -8.6$ dB,	$\Delta\theta_5 = -68.2°$
7 kHz:	$A_{V7} = -11.2$ dB,	$\Delta\theta_7 = -74.1°$

The amplitudes of the various harmonics at the output of the filter are determined as follows:

$$V_{1(out)} = \left(\frac{2}{\pi}\right) 10^{-1.0/20} = 0.567 \text{ V}_p$$

$$V_{3(out)} = \left(\frac{2}{3\pi}\right) 10^{-5.1/20} = 0.118 \text{ V}_p$$

$$V_{5(out)} = \left(\frac{2}{5\pi}\right) 10^{-8.6/20} = 0.047 \text{ V}_p$$

$$V_{7(out)} = \left(\frac{2}{7\pi}\right) 10^{-11.2/20} = 0.025 \text{ V}_p$$

The Fourier series of the output waveform is now approximated as

$$v(t) = 0.5 + 0.567 \sin (\omega t - 26.6°) + 0.118 \sin (3\,\omega t - 56.3°)$$
$$+ 0.047 \sin (5\,\omega t - 68.2°) + 0.025 \sin (7\,\omega t - 74.1°)$$

Various computer-aided design (CAD) and mathematical applications programs are able to generate a time-domain display of a waveform from a mathematical expression. When the above waveform is plotted in the time domain, it appears as illustrated in Figure 25–39.

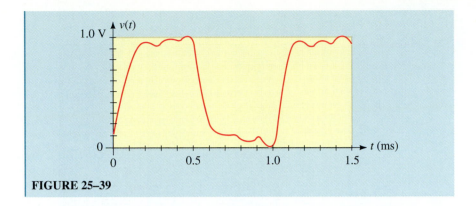

FIGURE 25–39

The voltage waveform of Figure 25–40 is applied to a high-pass filter circuit having the frequency response shown in Figure 25–41.

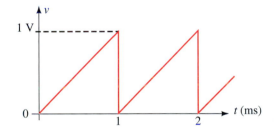

FIGURE 25–40

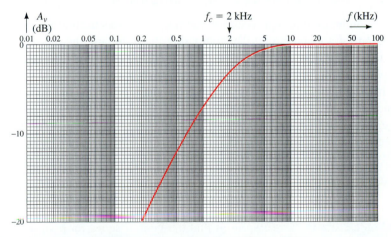

(a) Voltage gain response for a high-pass filter

FIGURE 25–41 *(Continues)*

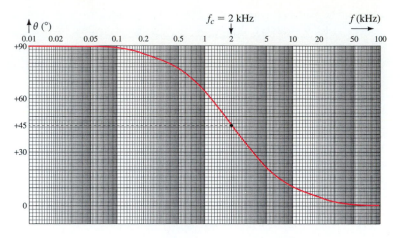

(b) Phase shift response for a high-pass filter

FIGURE 25–41   *(Continued)*

a. Determine the dc component at the output of the high-pass filter.

b. Calculate the amplitude and corresponding phase shift of the first four nonzero sinusoidal output components.

Answers:
a. zero

b. 1 kHz: 0.142 V_P $-116°$
    2 kHz: 0.113 V_P $-135°$
    3 kHz: 0.088 V_P $-147°$
    4 kHz: 0.071 V_P $-154°$

25.6   Circuit Analysis Using Computers

PSpice can be used to help visualize the frequency spectrum at the input and the output of a given circuit. By comparing the input and the output, we are able to observe how a given circuit distorts the waveform due to attenuation and phase shift of the various frequency components.

In the following example, we use a low-pass filter that has a cutoff frequency of 3 kHz. We will observe the effects of the filter on a 1-V pulse wave. In order to complete the required analysis, it is necessary to set up PSpice correctly.

EXAMPLE 25–9   Use PSpice to find the Fourier series for both the input and the output waveforms for the circuit shown in Figure 25–42. Use the Probe postprocessor to obtain both a time-domain and a frequency-domain display of the input and output waveforms.

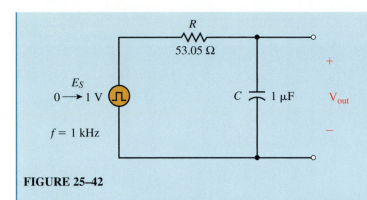

FIGURE 25–42

Solution   The circuit is input as shown in Figure 25–43.

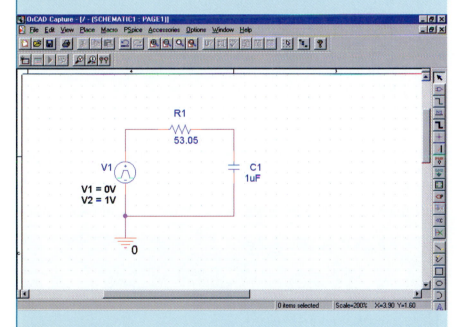

FIGURE 25–43

The voltage source is a pulse generator and is obtained from the SOURCE library by calling for VPULSE. The properties for the pulse generator are set as follows. V1=**0V**, V2=**1V**, TD=**0**, TR=**0.01us**, TF=**0.01us**, PW=**0.5ms**, PER=**1.0ms**.

   We begin the analysis by first setting the simulation settings for Time Domain (Transient) analysis. Run to time is set for **2.0ms** and the Maximum step size is set for **2us.** Now the analysis can be run.

   Once in the Probe window, we will simultaneously display both the input and the output waveforms. Click on Trace and Add Trace. Enter **V(V1:+)**, **V(C1:1)** in the Trace Expression dialog box. You will see the time-domain output as shown in Figure 25–44.

FIGURE 25–44

In order to obtain the frequency-domain display, simply click on Trace and Fourier. You will need to adjust the range of the abscissa by clicking on Plot and Axis Settings. Click on the X Axis tab and change the range from 0Hz to 10kHz. The display will appear as shown in Figure 25–45.

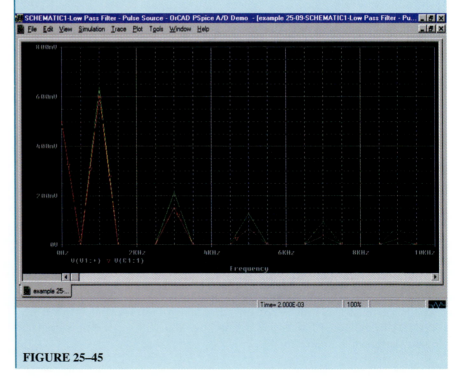

FIGURE 25–45

Notice that the output voltage of the third harmonic (3 kHz) is approximately 0.15 V, while the input voltage of the same harmonic is approximately 0.21 V. As expected, this represents approximately 3 dB of attenuation between the input and the output at the cutoff frequency.

PUTTING IT INTO PRACTICE

One method of building a frequency multiplier circuit is to generate a signal which is "rich" in harmonics. A full-wave rectifier is a circuit that converts a sine wave (which consists of only one frequency component) into one which appears as shown in Figure 25–17. As we see, the waveform at the output of the full-wave rectifier is composed of an infinite number of harmonic components. By applying this signal to a narrow–band-pass filter, it is possible to select any one of the components. The resulting output will be a pure sine wave at the desired frequency.

If a sine wave with an amplitude of 10 V is applied to a full-wave rectifier, what will be the amplitude and frequency at the output of a passive filter tuned to the third harmonic? Assume that there are no losses in the full-wave rectifier or in the filter circuit.

25.1   Composite Waveforms

PROBLEMS

1. a. Determine the rms voltage of the waveform shown in Figure 25–46.

   b. If this waveform is applied to a 50-Ω resistor, how much power will be dissipated by the resistor?

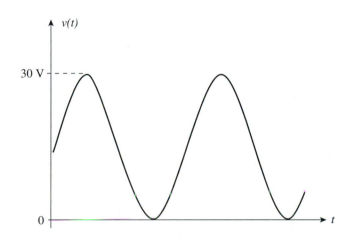

FIGURE 25–46

2. Repeat Problem 1 if the waveform of Figure 25–47 is applied to a 250-Ω resistor.

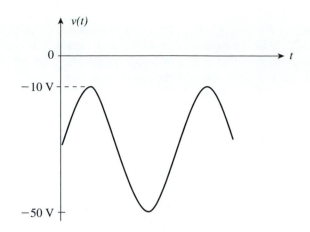

FIGURE 25–47

3. Repeat Problem 1 if the waveform of Figure 25–48 is applied to a 2.5-kΩ resistor.

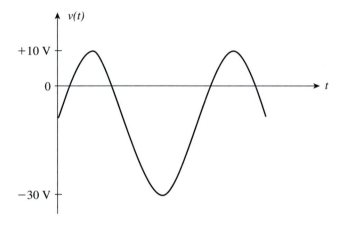

FIGURE 25–48

4. Repeat Problem 1 if the waveform of Figure 25–49 is applied to a 10-kΩ resistor.

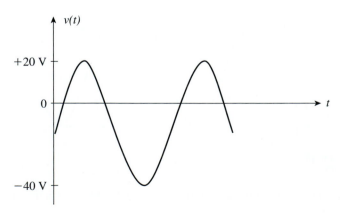

FIGURE 25–49

25.2 Fourier Series

5. Use calculus to derive the Fourier series for the waveform shown in Figure 25–50.

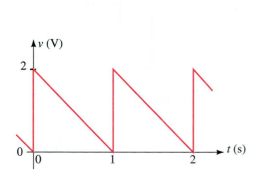

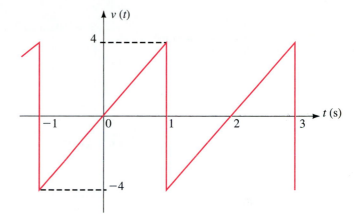

FIGURE 25–50

FIGURE 25–51

6. Repeat Problem 5 for the waveform shown in Figure 25–51.

25.3 Fourier Series of Common Waveforms

7. Use Table 25–1 to determine the Fourier series for the waveform of Figure 25–52.

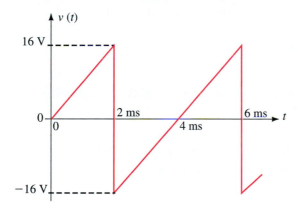

FIGURE 25–52

8. Repeat Problem 7 for the waveform of Figure 25–53.

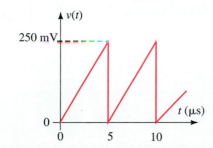

FIGURE 25–53

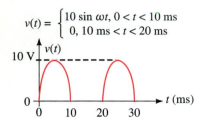

$$v(t) = \begin{cases} 10 \sin \omega t, & 0 < t < 10 \text{ ms} \\ 0, & 10 \text{ ms} < t < 20 \text{ ms} \end{cases}$$

FIGURE 25–54

9. Repeat Problem 7 for the waveform of Figure 25–54.
10. Repeat Problem 7 for the waveform of Figure 25–55.

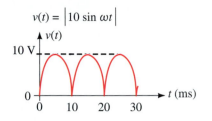

$$v(t) = |10 \sin \omega t|$$

FIGURE 25–55

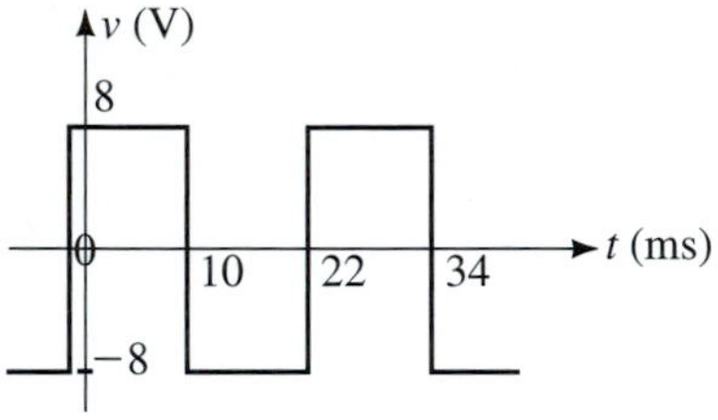

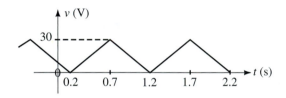

FIGURE 25–56

11. Write the expression including the first four sinusoidal terms of the Fourier series for the waveform of Figure 25–56.
12. Repeat Problem 11 for the waveform of Figure 25–57.

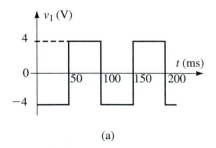

FIGURE 25–57

13. A composite waveform is made up of the two periodic waves shown in Figure 25–58.

    a. Sketch the resulting waveform.

    b. Write the Fourier series of the given waveforms.

    c. Determine the Fourier series of the resultant.

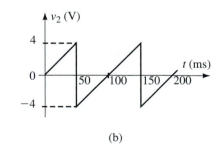

(a)   (b)

FIGURE 25–58

14. Repeat Problem 13 for the periodic waveforms shown in Figure 25–59.

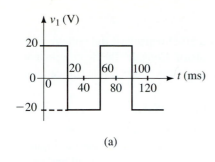

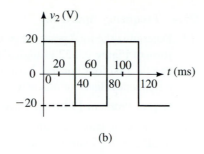

(a)   (b)

FIGURE 25–59

15. A composite waveform is made up of the two periodic waves shown in Figure 25–60.

   a. Sketch the resulting waveform.

   b. Solve for the dc value of the resultant.

   c. Write the Fourier series of the given waveforms.

   d. Determine the Fourier series of the resultant.

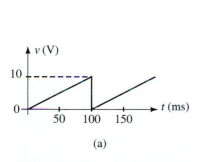

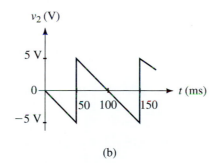

(a)   (b)

FIGURE 25–60

16. Repeat Problem 15 for the waveforms shown in Figure 25–61.

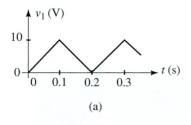

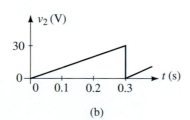

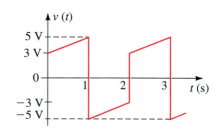

FIGURE 25–62

(a)   (b)

FIGURE 25–61

17. The waveform of Figure 25–62 is made up of two fundamental waveforms from Table 25–1. Sketch the two waveforms and determine the Fourier series of the composite wave.

18. The waveform of Figure 25–63 is made up of a dc voltage combined with two fundamental waveforms from Table 25–1. Determine the dc voltage and sketch the two waveforms. Determine the Fourier series of the composite wave.

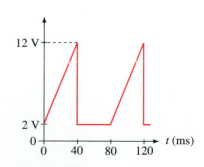

FIGURE 25–63

25.4   Frequency Spectrum

19. Determine the total power dissipated by a 50-Ω resistor, if the voltage waveform of Figure 25–52 is applied to the resistor. Consider the dc component and the first four nonzero harmonics. Indicate the power levels (in watts) on a frequency distribution curve.

20. Repeat Problem 19 for the waveform of Figure 25–53.

21. A spectrum analyzer with a 50-Ω input is used to measure the power levels in dBm of the Fourier series components of the waveform shown in Figure 25–54. Determine the power levels (in dBm) of the dc component and the first four nonzero harmonic components. Sketch the resultant display as it would appear on a spectrum analyzer.

22. Repeat Problem 21 if the waveform of Figure 25–55 is applied to the input of the spectrum analyzer.

25.5   Circuit Response to a Nonsinusoidal Waveform

23. The circuit of Figure 25–64 has the frequency response shown in Figure 25–65.

    a. Determine the dc component at the output of the filter.

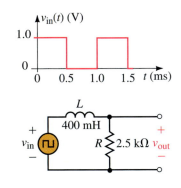

FIGURE 25–64

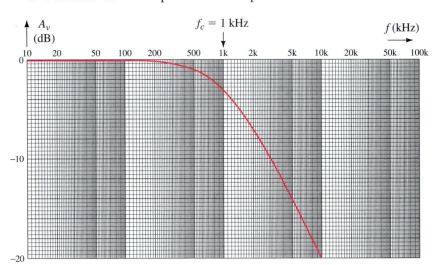

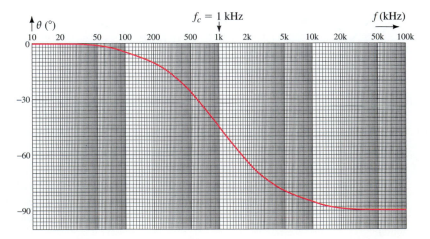

FIGURE 25–65

b. Calculate the amplitude and corresponding phase shift of the first four nonzero sinusoidal output components.

24. Repeat Problem 33 for Figures 25–66 and 25–67.

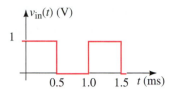

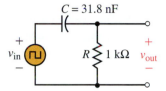

FIGURE 25–66

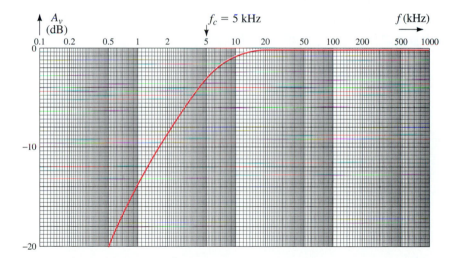

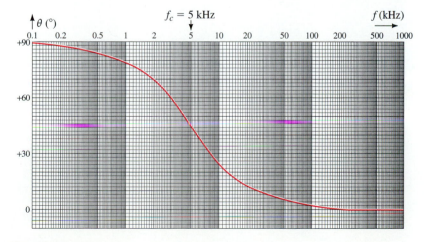

FIGURE 25–67

25.6   Circuit Analysis Using Computers

25. **PSpice**   Use PSpice to find the Fourier series for both the input and output waveforms for the circuit of Figure 25–64. Use the Probe postprocessor to obtain both a time-domain and frequency-domain display of the output waveform. Compare your results to those obtained in Problem 23.

26. **PSpice**   Repeat Problem 25 for the circuit of Figure 25–66. Compare your results to those obtained in Problem 24.

ANSWERS TO IN-PROCESS LEARNING CHECKS

In-Process Learning Check 1

The waveform of Figure 25–12 has an average value of zero. Therefore, $a_0 = 0$. The waveform also has a peak-to-peak value which is double that of Figure 25–6, which means that the amplitude of each harmonic must be doubled.

$$v(t) = \frac{4}{\pi}\sin \omega t + \frac{4}{3\pi}\sin 3\omega t + \frac{4}{5\pi}\sin 5\omega t + \cdots$$

In-Process Learning Check 2

a. Figure 25–15 with $V_m = 25$ V and $T = 20$ ms

b. Figure 25–16 with $V_m = 25$ V and $T = 20$ ms

c. $v(t) = 12.5 + 18.9 \sin(100\pi t - 32.48°) - 7.96 \sin(200\pi t) + 5.42 \sin(300\pi t - 11.98°) - 3.98 \sin(400\pi t)$

OrCAD–PSpice A/D

APPENDIX
A

OrCAD PSpice (formerly MicroSim PSpice) is a circuit simulation program designed to analyze electrical and electronic problems using a graphical representation of the circuit. Anyone who has worked with previous versions of the PSpice software will find many similarities between those versions and the latest release. The applications used in this textbook will help make the transition a little easier. For those who have never worked with PSpice, this appendix provides the required steps to guide you through the sometimes-complicated world of software application. A detailed description of the software is provided in the *OrCAD Capture User's Guide*. For an evaluation CD of OrCAD, contact OrCAD at their web site, www.orcad.com. The software consists of several parts, of which we use only the schematic capture and PSpice simulation systems.

A.1 Getting Started

You will need to install the software from the CD-ROM. The instructions provided by OrCAD will guide you through this process. Since the installation cannot proceed successfully with virus detection software running, it is necessary to disable this software during the installation.

After you have installed the OrCAD software, the Capture program can be started as follows:

1. Click the Start menu on the Windows desktop,
2. Select the OrCAD Demo sub-folder from the Program item,
3. Click on Capture CIS Demo. The Capture *session frame* will open.

Your schematic design and processing will all be done in this frame. To start a new project in the OrCAD Capture session frame, click on the File menu, New menu item, and Project item. A New Project box will appear as shown in Figure A–1.

You will need to give your project a name such as **Project 1** and indicate the location (in which subdirectory) the project is to be saved. In order to perform a PSpice simulation, it is necessary to select the Analog or Mixed-Signal Circuit Wizard in the same box. Click OK.

A.2 Entering a Simple Circuit with OrCAD Capture

Although each circuit simulation uses different components, the illustration that follows shows some of the common commands that are used to construct a schematic, assign values to the parts and run the PSpice simulation. The PSpice examples throughout the textbook show various applications of

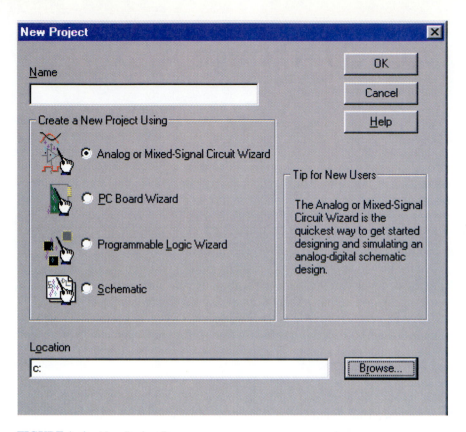

FIGURE A–1   New Project Box

the software. The *OrCAD Capture User's Guide* and the *OrCAD PSpice A/D User's Guide* provide additional information on using OrCAD software.

1. Once the project has been named, OrCAD prompts you with the libraries that you would like to use. In most cases, the default libraries are sufficient. For this project, we will add two additional libraries; BREAKOUT and EVAL. Click on **breakout.olb,** then on **Add>>**. Similarly, click on **eval.olb,** then on **Add>>**. All libraries that are available for the project are shown in the right-hand box. If you wish to remove a library from a project, the steps are similar. Click on **Finish.** You will now be in the *schematic page editor.*

2. Click anywhere in the grid of the schematic page editor. You will observe that a *tool palette* has been provided as part of the editor. The tool palette allows you to select various tools with which to create a schematic page. This palette is easily moved and resized by placing the cursor at the top of the palette and dragging it (using the left mouse button) to a new location. Figure A–2 shows the tool palette and the corresponding button functions as well as other tools that are used to simplify simulation and viewing of the circuit characteristics.

3. The **Place part** tool is used to construct the circuit shown in Figure A–3.

4. We enter a voltage source by first selecting the SOURCE library. Next, we may either scroll through the various parts contained in this library

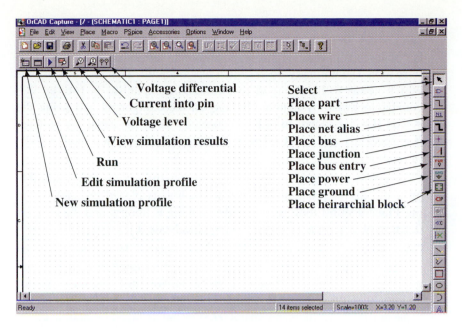

FIGURE A–2   Tool Palette

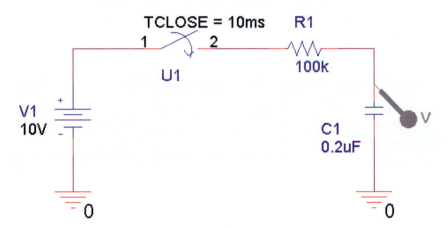

FIGURE A–3

or more easily, move the cursor into the <u>P</u>art box and type **VDC** followed by clicking on OK. The dc source can now be placed anywhere on the schematic by clicking in the appropriate position. If we had to place more than one source, we would do so by moving the cursor and repeating the operation. In this example, we need only one source. The placement tool is disabled by first right clicking and then selecting **End Mode** or by pressing the **Esc** key.

Notice that the default value of the voltage source is 0V. The value is easily changed by first double clicking on the default value and then entering a new value in the V<u>a</u>lue box. In this case we enter **10V,** followed by clicking on OK. PSpice does not require that we show the units. However if units are shown, there must not be any spaces between the magnitude and the units.

5. Other components are selected using the **Place part** tool. The capacitor and resistor are both obtained from the ANALOG library and the switch is obtained from the EVAL library and then connected using the **Place wire** tool. The normally-open switch has a default, TClose value of 0ms which needs to be changed to **10ms.** The capacitor is rotated 90° counterclockwise by double clicking on the component and using the **<Ctrl R>** command. In order to correctly show the capacitor voltage, it is necessary to use **<Ctrl R>** three times to result in a total of 270° of rotation.

6. PSpice will not simulate the design unless the circuit has a suitable ground. The reference node (ground symbol) is placed by clicking on the **Place ground** tool, selecting the SOURCE library, and clicking on 0. (In PSpice, ground is always assigned as Node 0.)

7. Once the circuit is completed, it can be simulated. However, if the circuit were simulated at this point, we would find that the results would not accurately represent the actual operation. We must give the capacitor an initial condition. Since it is initially uncharged, the voltage across the capacitor is 0V. Component characteristics are changed by double clicking on the part, which opens the *Property editor* of Capture. Next we click on the **Parts** tab (bottom left) of the Property editor. In order to set the initial condition of the capacitor to 0V, we enter **0** into the *IC* (initial condition) box of the editor. You may find it is necessary to scroll across to locate the appropriate property. The result is shown in Figure A–4.

FIGURE A–4   Property Editor

The changes are made to the component properties by clicking on Apply. The Property editor is closed by clicking on X.

8. Using a similar approach, the Property editor is used to change the default resistance of the switch in its open condition from 1Meg (1 MΩ) to **1G** (1 GΩ).

9. Place a voltage probe as shown in the circuit of Figure A–3 by clicking on the **Voltage Level** tool and dragging it to the location shown.

10. The design is now complete and is saved by clicking on the **Save document** tool. You will need to save all schematic changes.

A.3   Simulating a Circuit in OrCAD PSpice

1. The Capture window will display the *Project manager* as shown in Figure A–5. Once in the project manager, click on the **New Simulation Profile** tool. Enter a name for the profile (e.g., **RC Timer**) in the Name field.

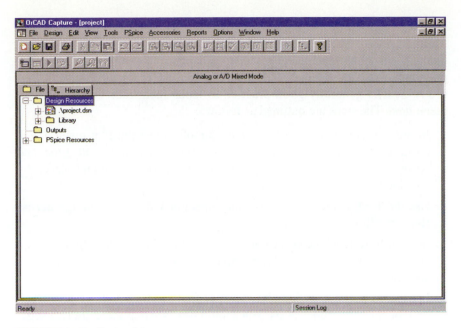

FIGURE A–5 Project Manager

2. The *Simulation Settings* are used to control how the circuit is analyzed. For the given design, click on the <u>A</u>nalysis tab. In <u>A</u>nalysis Type: select Time Domain (Transient). In <u>R</u>un to time: (TSTOP) enter **100ms.** In Transient options select <u>M</u>aximum step size select enter **20us.** Finally, click on OK to return to the Project Manager. Click on the Run tool.

3. If there are no errors in the design, the *Probe* window will open.

4. Once in the Probe window, you will see the capacitor voltage as shown in Figure A–6.

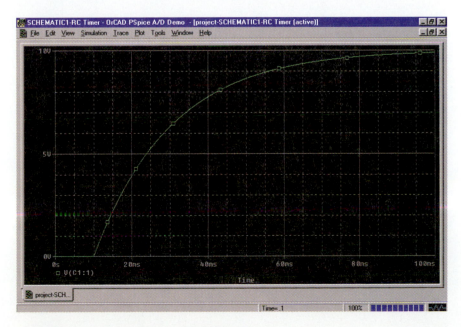

FIGURE A–6 Probe Display—Single Trace

Displaying and Labeling Multiple Traces

Multiple traces may be graphed either on the same vertical axis or using independent y-axis scales. For instance, it is possible to simultaneously display both capacitor voltage V(C1:1) and current I(C1) for the circuit of Figure A–3. It is generally good practice to label each trace using the label menu item. The steps are outlined as follows:

1. In order to add a label to the display, simply select the Plot menu. Click on Label. You may select from several options, including Text, Line, and Arrow. The various labels are positioned using the mouse and placed by clicking in the correct location.

2. Use the Text menu item to label the capacitor voltage as **Voltage across the capacitor.**

3. In order to re-label the vertical axis, select Plot, click on Axis Settings. . . and select the Y Axis tab. Type **Capacitor voltage** in the Axis Title box and click on OK.

4. In order to obtain a second plot, we must first generate another axis. Click on Plot and then Add Y Axis. You should now see that there are two separate vertical axes.

5. Add a current trace by clicking on the Trace menu and selecting the Add Trace menu item. Click on I(C1) to obtain a plot of capacitor current. Click OK.

6. The plot is labeled by selecting Plot, Label, and then Text, Line, or Arrow.

7. Finally, the current axis is labeled by first selecting the appropriate axis with the mouse. Notice that as you select an axis, >> appears adjacent to the axis. Select Plot, Axis Settings. . . and click on the Y Axis tab. Type **Capacitor current** in the Axis Title box and click on OK. The display will appear as shown in Figure A–7.

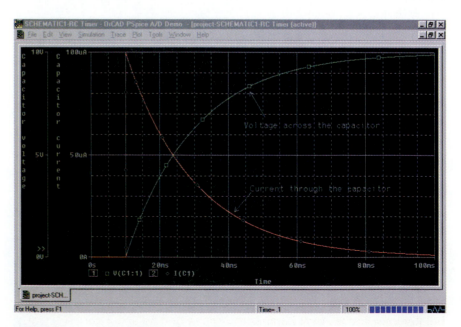

FIGURE A–7   Probe Display—Two Traces

A.4 Syntax and Other Things

In designing circuits with OrCAD Capture, you can use either scientific notation or standard engineering prefixes. For example, you may specify a 12 000-ohm resistor as 12000, 1.2E04, 12k, or 0.012Meg. The previous illustration shows that PSpice does not require units. However, if units are used, there must not be a space between the magnitude and the units. In other words, you may enter 12kohms but not 12k ohms. Table A–1 shows the prefixes used by PSpice. Notice that M cannot be used to represent mega (10^6), since this prefix is used to represent milli (10^{-3}).

TABLE A–1 Standard Prefixes used by PSpice

Symbol	Scale	Name
T or t	10^{12}	tera
G or g	10^{9}	giga
MEG or meg	10^{6}	mega
K or k	10^{3}	kilo
M or m	10^{-3}	milli
U or u	10^{-6}	micro
N or n	10^{-9}	nano
P or p	10^{-12}	pico

Inductors and Voltage Sources

PSpice does not permit voltage sources and inductors to be connected in loops. To get around this shortcoming, we simply connect a very small resistor (e.g. 1E-06) in series with the inductor.

Capacitors and Voltage Sources

PSpice requires a dc path from all nodes to the reference node (ground). If no dc path exists (because of capacitors for example), we may simply insert a high-value resistor from the node to the reference node.

Component Orientation

Whenever a component is placed into a circuit, it has a default position as determined by the library from which the component was taken. Any component can be rotated from its default orientation by entering **<Ctrl R>,** which rotates the component 90° counterclockwise. In order to ensure that PSpice provides the results that we expect, we must remember that current through the device is calculated as going from terminal 1 to terminal 2. Various component orientations are shown in Figure A–8. Clearly, if the component is connected incorrectly, it is likely that the current will be negative.

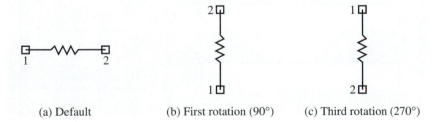

| (a) Default | (b) First rotation (90°) | (c) Third rotation (270°) |

FIGURE A–8

A.5 Translating Schematics from MicroSim into OrCAD Capture

The Schematics-to-Capture translator converts MicroSim schematics, symbol libraries, and package libraries to Capture designs and libraries. It is possible to translate designs generated in MicroSim Schematics from versions 5.4 to 8.0.

The following steps guide you through the process of translating existing PSpice schematics into Capture schematics. You will find that the appearance of the circuit is much the same as it was in older versions of PSpice.

1. Follow the steps outlined in section A.1 Getting Started. As always, you will need to give your project a name.

2. In order to open an existing schematic, click on the File menu and select Import Design. The Import Design dialog box shown in Figure A-9 will open.

3. Click on the PSpice tab.

4. In the Open text box, enter the location and file name of the schematic that you wish to simulate. You may use the Browse button to find the file in your computer.

5. In the Save As text box, enter the location and file name of the schematic that you wish to simulate. The default location is the same as the location of the original schematic file. The default name is the same as the original file name with a .OPJ extension. You may use the Browse button to find the file in your computer.

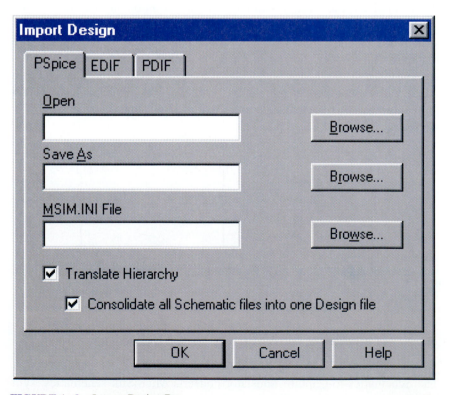

FIGURE A–9 Import Design Box

6. In the MSIM.INI File text box, enter the location of the MSIM.INI file. It will generally be located in the Windows directory of your computer. If you were using a demonstration version of the MicroSim PSpice program, the file will likely be named **msim_evl.ini.** This means that you would need to enter **C:\Windows\msim_evl.ini** in the text box.

7. Click OK. You may now edit or simulate the circuit as described previously.

Solution of Simultaneous Linear Equations

Simultaneous linear equations appear often in the solution of problems in electrical/electronics technology. The solutions of these equations has been simplified by a branch of mathematics called linear algebra. Although the actual theorems and proofs are well outside the scope of this textbook, we will use some of the principles of linear algebra to solve simple linear equations. The following is a set of n simultaneous linear equations in n unknowns:

$$a_{11}x_1 + a_{12}x_2 + \cdots + a_{1n}x_n = b_1$$
$$a_{21}x_1 + a_{22}x_2 + \cdots + a_{2n}x_n = b_2$$
$$\cdot$$
$$\cdot$$
$$\cdot$$
$$a_{n1}x_1 + a_{n2}x_2 + \cdots + a_{nn}x_n = b_n$$

The above equations may also be expressed in matrix form as

$$\mathbf{AX = B}$$

where

$$\mathbf{A} = \begin{bmatrix} a_{11}a_{12} \cdots a_{1n} \\ a_{21}a_{22} \cdots a_{2n} \\ \cdot \quad \cdot \quad\quad \cdot \\ \cdot \quad \cdot \quad\quad \cdot \\ \cdot \quad \cdot \quad \cdots \\ a_{n1}a_{n2} \cdots a_{nn} \end{bmatrix}, \quad \mathbf{X} = \begin{bmatrix} x_1 \\ x_2 \\ \cdot \\ \cdot \\ \cdot \\ x_n \end{bmatrix}, \quad \mathbf{B} = \begin{bmatrix} b_1 \\ b_2 \\ \cdot \\ \cdot \\ \cdot \\ b_n \end{bmatrix}$$

Substitution

Although simultaneous linear equations may be expressed in several unknowns, we begin with the most simple, namely two simultaneous linear equations in two unknowns. Consider the equations below:

$$a_{11}x_1 + a_{12}x_2 = b_1 \qquad\qquad \textbf{(B–1)}$$

$$a_{21}x_1 + a_{22}x_2 = b_2 \qquad\qquad \textbf{(B–2)}$$

If we multiply Equation B–1 by a_{22} and Equation B–2 by a_{12}, we have

$$a_{11}a_{22}x_1 + a_{12}a_{22}x_2 = a_{22}b_1$$
$$a_{12}a_{21}x_1 + a_{12}a_{22}x_2 = a_{12}b_2$$

Subtracting, we obtain

$$a_{11}a_{22}x_1 - a_{12}a_{21}x_1 = a_{22}b_1 - a_{12}b_2$$

which gives

$$x_1 = \frac{a_{22}b_1 - a_{12}b_2}{a_{11}a_{22} - a_{12}a_{21}}$$

Similarly, we solve for the unknown x_2 as

$$x_2 = \frac{a_{11}b_2 - a_{21}b_1}{a_{11}a_{22} - a_{12}a_{21}}$$

EXAMPLE B–1   Use substitution to find the solutions for the following following linear equations:

$$2x_1 + 8x_2 = -2$$
$$x_1 + 2x_2 = 5$$

Solution   Rewriting the first equation, we have

$$2x_1 = -2 - 8x_2$$
$$x_1 = -1 - 4x_2$$

Now, substituting the above expression into the second equation, we have

$$(-1 - 4x_2) + 2x_2 = 5$$
$$-2x_2 = 6$$
$$x_2 = -3$$

Finally, we have

$$x_1 = -1 - 4(-3) = 11$$

Determinants

While substitution may be used for solving simultaneous linear equations in two variables, it is lengthy and particularly complicated when solving for more than two unknowns. An easier method used for solving simultaneous linear equations involves using *determinants*. We begin by expressing the simultaneous linear equations (B–1) and (B–2) as a product of matrices:

Column 1   Column 2   Column 3

$$\begin{bmatrix} a_{11} & a_{12} \\ a_{21} & a_{22} \end{bmatrix} \begin{bmatrix} x_1 \\ x_2 \end{bmatrix} = \begin{bmatrix} b_1 \\ b_2 \end{bmatrix} \qquad \textbf{(B–3)}$$

A determinant is a set of coefficients which has the same number of rows and columns and which may be expressed as a single value. The number of rows (or columns) defines the *order* of a determinant. The second-order determinant corresponding to the coefficients of the matrix equation (B–3) consists of the elements in columns 1 and 2 and is expressed as

$$D = \begin{vmatrix} a_{11} a_{12} \\ a_{21} a_{22} \end{vmatrix}$$

The value of the second-order determinant is found by taking the product of the upper left term and the lower right term (elements of the principal diagonal) and then subtracting the product of the lower left term and the upper right term (elements of the secondary diagonal). The result is given as

$$D = a_{11} a_{22} - a_{12} a_{21}$$

The unknowns of the simultaneous linear equations are found by using a technique called *Cramer's rule.* In applying this rule, we need to solve the following determinants:

$$x_1 = \frac{\begin{vmatrix} b_1 a_{12} \\ b_2 a_{22} \end{vmatrix}}{\begin{vmatrix} a_{11} a_{12} \\ a_{21} a_{22} \end{vmatrix}} \quad \frac{a_{22} b_1 - a_{12} b_2}{a_{11} a_{22} - a_{21} a_{12}}$$

and

$$x_2 = \frac{\begin{vmatrix} a_{11} b_1 \\ a_{21} b_2 \end{vmatrix}}{\begin{vmatrix} a_{11} a_{12} \\ a_{21} a_{22} \end{vmatrix}} \quad \frac{a_{11} b_2 - a_{21} b_1}{a_{11} a_{22} - a_{21} a_{12}}$$

The application of Cramer's rule gives the solution for each unknown by first placing the determinant of the coefficient matrix in the denominator. The numerator is then developed by using the same determinant with the exception that the coefficients of the variable to be found are replaced by the coefficients of the solution matrix. The resulting solutions are precisely those found when we used substitution.

EXAMPLE B–2   Use determinants to find solutions for the following linear equations:

$$2x_1 + 8x_2 = -2$$
$$x_1 + 2x_2 = 5$$

Solution   The determinant of the denominator is found as

$$D = \begin{vmatrix} 2 & 8 \\ 1 & 2 \end{vmatrix} = (2)(2) - (1)(8) = -4$$

The variables are now calculated as

$$x_1 = \frac{\begin{vmatrix} -2 & 8 \\ 5 & 2 \end{vmatrix}}{-4} = \frac{(-2)(2) - (5)(8)}{-4} = \frac{-44}{-4} = 11$$

and

$$x_2 = \frac{\begin{vmatrix} 2 & -2 \\ 1 & 5 \end{vmatrix}}{-4} = \frac{(2)(5) - (1)(-2)}{-4} = \frac{12}{-4} = -3$$

The solution of third-order simultaneous linear equations is similar to the method used for solving second-order equations. Consider the following third-order simultaneous linear equation:

$$a_{11}x_1 + a_{12}x_2 + a_{13}x_3 = b_1$$
$$a_{21}x_1 + a_{22}x_2 + a_{23}x_3 = b_2$$
$$a_{31}x_1 + a_{32}x_2 + a_{33}x_3 = b_3$$

The corresponding matrix equation is shown as follows:

$$\begin{bmatrix} a_{11} a_{12} a_{13} \\ a_{21} a_{22} a_{23} \\ a_{31} a_{32} a_{33} \end{bmatrix} \begin{bmatrix} x_1 \\ x_2 \\ x_3 \end{bmatrix} = \begin{bmatrix} b_1 \\ b_2 \\ b_3 \end{bmatrix}$$

The value of the third-order determinant may be found in one of several ways. The first method works for only third-order determinants, while the second method is a more general approach which evaluates any order of determinant.

Method I. This method works only for third-order determinants:

1. Begin by writing the original columns of the third-order determinant.

2. Copy the first two columns, placing them to the right of the original determinant.

3. Add the product of the elements of the principal diagonal to the products of the adjacent two parallel diagonals to the right of the principal diagonal.

4. Subtract the product of the elements of the secondary diagonal and also subtract the products of the elements along the two other parallel diagonals.

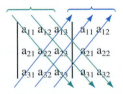

The resultant determinant is written as

$$D = a_{11}a_{22}a_{33} + a_{12}a_{23}a_{31} + a_{13}a_{21}a_{32} - a_{31}a_{22}a_{13} - a_{32}a_{23}a_{11} - a_{33}a_{21}a_{12}$$

EXAMPLE B–3 Evaluate the following determinant:

$$D = \begin{vmatrix} 3 & 1 & -2 \\ 1 & -2 & 3 \\ 2 & 3 & 2 \end{vmatrix}$$

Solution We begin by rewriting the first two columns as follows:

$$D = \begin{vmatrix} 3 & 1 & -2 \\ 1 & -2 & 3 \\ 2 & 3 & 2 \end{vmatrix} \begin{matrix} 3 & 1 \\ 1 & -2 \\ 2 & 3 \end{matrix}$$

Now, adding the products of the principal diagonal and adjacent diagonals and subtracting the products of the secondary diagonal and adjacent diagonals, we have

$$D = (3)(-2)(2) + (1)(3)(2) + (-2)(1)(3)$$
$$- (2)(-2)(-2) - (3)(3)(3) - (2)(1)(1)$$
$$= -49$$

METHOD II. This evaluation of determinants is achieved by expansion by minors. The *minor* of an element is the determinant which remains after deleting the row and the column in which the element lies. The value of any nth-order determinant is found as follows:

1. For any row or column, find the product of each element and the determinant of its minor.

2. A product is given a positive sign if the sum of the row and the column of the element is even. The product is given a negative sign if the sum is odd.

3. The value of the determinant is the sum of the resulting terms.

As before, Cramer's rule is used to solve for the unknowns, x_1, x_2, and x_3, by using determinants and replacing the appropriate terms of the numerator with the terms of the solution matrix. The resulting determinants and solutions are given as follows:

$$x_1 = \frac{\begin{vmatrix} b_1 & a_{12} & a_{13} \\ b_2 & a_{22} & a_{23} \\ b_3 & a_{32} & a_{33} \end{vmatrix}}{\begin{vmatrix} a_{11} & a_{12} & a_{13} \\ a_{21} & a_{22} & a_{23} \\ a_{31} & a_{32} & a_{33} \end{vmatrix}}$$

By expansion of minors, the determinant of the denominator is found as

$$D = + a_{11} \begin{vmatrix} a_{22} & a_{23} \\ a_{32} & a_{33} \end{vmatrix} - a_{21} \begin{vmatrix} a_{12} & a_{13} \\ a_{32} & a_{33} \end{vmatrix} + a_{31} \begin{vmatrix} a_{12} & a_{13} \\ a_{22} & a_{23} \end{vmatrix}$$
$$= a_{11}(a_{22}a_{33} - a_{23}a_{32}) - a_{21}(a_{12}a_{33} - a_{13}a_{32}) + a_{31}(a_{12}a_{23} - a_{13}a_{22})$$

The solution for x_1 is now found to be

$$x_1 = \frac{+ b_1 \begin{vmatrix} a_{22} & a_{23} \\ a_{32} & a_{33} \end{vmatrix} - b_2 \begin{vmatrix} a_{12} & a_{13} \\ a_{32} & a_{33} \end{vmatrix} + b_3 \begin{vmatrix} a_{12} & a_{13} \\ a_{22} & a_{23} \end{vmatrix}}{D}$$
$$= \frac{b_1(a_{22}a_{33} - a_{23}a_{32}) - b_2(a_{12}a_{33} - a_{13}a_{32}) + b_3(a_{12}a_{23} - a_{13}a_{22})}{}$$

Similarly, for x_2, we get

$$x_2 = \frac{\begin{vmatrix} a_{11} & b_1 & a_{13} \\ a_{21} & b_2 & a_{23} \\ a_{31} & b_3 & a_{33} \end{vmatrix}}{\begin{vmatrix} a_{11} & a_{12} & a_{13} \\ a_{21} & a_{22} & a_{23} \\ a_{31} & a_{32} & a_{33} \end{vmatrix}}$$

$$= \frac{-b_1(a_{21}a_{33} - a_{23}a_{31}) + b_2(a_{11}a_{33} - a_{13}a_{31}) + b_3(a_{11}a_{23} - a_{13}a_{21})}{}$$

and for x_3 we have

$$x_3 = \frac{\begin{vmatrix} a_{11} & a_{12} & b_1 \\ a_{21} & a_{22} & b_2 \\ a_{31} & a_{32} & b_3 \end{vmatrix}}{\begin{vmatrix} a_{11} & a_{12} & a_{13} \\ a_{21} & a_{22} & a_{23} \\ a_{31} & a_{32} & a_{33} \end{vmatrix}}$$

$$= \frac{b_1(a_{21}a_{32} - a_{22}a_{31}) - b_2(a_{11}a_{32} - a_{12}a_{31}) + b_3(a_{11}a_{22} - a_{12}a_{21})}{}$$

EXAMPLE B–4 Solve for x_1 in the following system of linear equations using minors.

$$3x_1 + x_2 - 2x_3 = 1$$
$$x_1 - 2x_2 + 3x_3 = 11$$
$$2x_1 + 3x_2 + 2x_3 = -3$$

Solution The determinant of the denominator is evaluated as follows:

$$D = \begin{vmatrix} 3 & 1 & -2 \\ 1 & -2 & 3 \\ 2 & 3 & 2 \end{vmatrix}$$

$$= +(3)\begin{vmatrix} -2 & 3 \\ 3 & 2 \end{vmatrix} - (1)\begin{vmatrix} 1 & -2 \\ 3 & 2 \end{vmatrix} + (2)\begin{vmatrix} 1 & -2 \\ -2 & 3 \end{vmatrix}$$

$$= (3)(-4 - 9) - (2 + 6) + (2)(3 - 4)$$

$$= -49$$

and so the unknown x_1 is calculated to be

$$
x_1 = \frac{\begin{vmatrix} 1 & 1 & -2 \\ 11 & -2 & 3 \\ -3 & 3 & 2 \end{vmatrix}}{-49}
$$

$$
= \frac{+(1)\begin{vmatrix} -2 & 3 \\ 3 & 2 \end{vmatrix} - (11)\begin{vmatrix} 1 & -2 \\ 3 & 2 \end{vmatrix} + (-3)\begin{vmatrix} 1 & -2 \\ -2 & 3 \end{vmatrix}}{-49}
$$

$$
= \frac{(-4 - 9) - (11)(2 + 6) - (3)(3 - 4)}{-49}
$$

$$
= 2
$$

PRACTICE PROBLEM

Use expansion by minors to solve for x_2 and x_3 in Example B–4.

Answers: $x_2 = -3$, $x_3 = 1$

Maximum Power Transfer Theorem

Figure C–1 shows the Thévenin equivalent of a dc circuit.

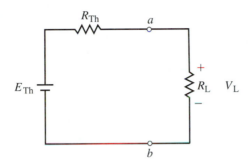

FIGURE C–1

For the above circuit, the values of E_{Th} and R_{Th} are constant. Therefore, power delivered to the load is determined as a function of the load resistance and is given as

$$P_L = \frac{V_L^2}{R_L} = \frac{\left(\dfrac{R_L E_{Th}}{R_L + R_{Th}}\right)^2}{R_L} = \frac{E_{Th}^2 R_L}{(R_L + R_{Th})^2} \qquad \textbf{(C–1)}$$

Maximum power will be delivered to R_L when the first derivative, $\dfrac{dP_L}{dR_L} = 0$.

Applying the quotient rule, $\dfrac{d}{dx}\left(\dfrac{u}{v}\right) = \dfrac{v\dfrac{du}{dx} - u\dfrac{dv}{dx}}{v^2}$, we find the derivative of power with respect to load resistance as

$$\begin{aligned}
\frac{dP_L}{dR_L} &= \frac{(R_L + R_{Th})^2(E_{Th})^2 - (E_{Th}^2 R_L)(2)(R_L + R_{Th})}{(R_L + R_{Th})^4} \\
&= \frac{E_{Th}^2[(R_L + R_{Th})^2 - 2R_L(R_L + R_{Th})]}{(R_L + R_{Th})^4}
\end{aligned} \qquad \textbf{(C–2)}$$

Now, since the first derivative can only be zero if the numerator of the above expression is zero, and since E_{Th} is a constant, we have

$$(R_L + R_{Th})^2 - 2R_L(R_L + R_{Th}) = 0 \qquad \textbf{(C–3)}$$

And so,

$$\begin{aligned}
R_L^2 + 2R_L R_{Th} + R_{Th}^2 - 2R_L R_L - 2R_L R_{Th} &= 0 \\
R_{Th}^2 - R_L^2 &= 0 \qquad \textbf{(C–4)} \\
R_L &= R_{Th}
\end{aligned}$$

Figure C–2 shows the Thévenin equivalent of an ac circuit.

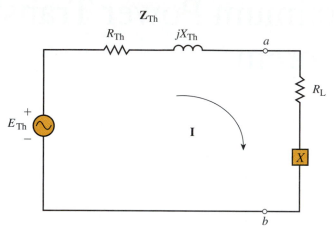

FIGURE C–2

For the above circuit, the values of E_{Th}, R_{Th}, and X_{Th} are constant. Although X_{Th} is shown as an inductor, it could just as easily be a capacitor. The power delivered to the load is determined as a function of the load impedance as

$$P_L = I^2 R_L = \frac{E_{Th}^2 R_L}{(R_L + R_{Th})^2 + (X + X_{Th})^2} \tag{C–5}$$

Maximum power is dependent on two variables, R_L and X. Therefore, we will need to solve for partial derivatives. Maximum power will be transferred to the load when $\frac{\partial P_L}{\partial R_L} = 0$ and $\frac{\partial P_L}{\partial X} = 0$.

We begin by finding $\frac{\partial P_L}{\partial X}$. Using $\frac{d}{dx}\left(\frac{1}{v}\right) = -\frac{1}{v^2}\frac{dv}{dx}$, we get

$$\frac{\partial P_L}{\partial X} = -\frac{(E_{Th}^2 R_L)(2)(X + X_{Th})}{[(R_L + R_{Th})^2 + (X + X_{Th})^2]^2} \tag{C–6}$$

Now, since the partial derivative can only be zero if the numerator of the above expression is zero, and since E_{Th} and R_L are treated as constants, we have

$$X + X_{Th} = 0$$

or

$$X = -X_{Th} \tag{C–7}$$

This result implies that if the Thévenin impedance contains an inductor of magnitude X, the load must contain a capacitor with the same magnitude. (Conversely, if the Thévenin impedance were to contain a capacitor, then the load impedance would need to have an inductor of the same magnitude.)

Next, we determine the partial derivative, $\frac{\partial P_L}{\partial R_L}$ of equation (C–5). Applying the quotient rule,

$$\frac{d}{dx}\left(\frac{u}{v}\right) = \frac{v\frac{du}{dx} - u\frac{dv}{dx}}{v^2}$$

we get

$$\frac{\partial P_L}{\partial R_L} = \frac{[(R_L + R_{Th})^2 + (X + X_{Th})^2](E_{Th}^2) - (E_{Th}^2 R_L)[(2)(R_L + R_{Th})]}{[(R_L + R_{Th})^2 + (X + X_{Th})^2]^2} \quad \text{(C–8)}$$

Now, since the partial derivative can only be zero if the numerator of the above expression is zero, and since E_{Th} and X are treated as constants, we have

$$(R_L + R_{Th})^2 + (X + X_{Th})^2(E_{Th}^2) - (E_{Th}^2 R_L)(2)(R_L + R_{Th}) = 0$$
$$(R_L + R_{Th})^2 + (X + X_{Th})^2 - (R_L)(2)(R_L + R_{Th}) = 0$$
$$R_L^2 + 2R_L R_{Th} + R_{Th}^2 + (X + X_{Th})^2 - 2R_L^2 - 2R_L R_{Th} = 0$$
$$R_{Th}^2 - R_L^2 + (X + X_{Th})^2 = 0$$

In general, the above equation determines the value of load resistance regardless of the load reactance. Therefore, we have

$$R_L = \sqrt{R_{Th}^2 + (X + X_{Th})^2} \quad \text{(C–9)}$$

This result shows that the load resistance is dependent upon the load reactance, X. If the reactance of the load is the same type as the Thévenin reactive component (both inductive or both capacitive), then the reactances are added. If the load reactance is the opposite type of the Thévenin reactive component, then the reactances are subtracted. If the reactance of the load can be adjusted to result in maximum power transfer ($X = -X_{Th}$), then equation (C–9) is simplified to the expected result, namely

$$R_L = R_{Th} \quad \text{(C–10)}$$

Answers to Selected Odd-Numbered Problems

CHAPTER 1

1. a. 1620 s   b. 2880 s   c. 7427 s
   d. 26 110 W   e. 2.45 hp   f. 8280°
3. a. 0.84 m^2 b. 0.0625 m^2
   c. 0.02 m^3 d. 0.0686 m^3
5. 4500 parts/h
7. 11.5 km/l
9. 150 rpm
11. 8.33 mi
13. 0.508 m/s
15. Machine 1: \$25.80/h;   Machine 2: \$25.00/h;   Machine 2
17. a. 8.675×10^3 b. 8.72×10^{-3} c. 1.24×10^3
    d. 3.72×10^{-1} e. 3.48×10^2 f. 2.15×10^{-7}
    g. 1.47×10^1
19. a. 1.25×10^{-1} b. 8×10^7
    c. 2.0×10^{-2} d. 2.05×10^4
21. a. 10   b. 10   c. 3.6×10^3
    d. 15×10^4 e. -12.0
23. 1.179; 4.450; Direct computation is less work for these examples.
25. 6.24×10^{18}
27. 62.6×10^{21}
29. 1.16 s
31. 13.4×10^{10} l/h
33. a. kilo, k   b. mega, M
    c. giga, G   d. micro, μ
    e. milli, m   f. pico, p
35. a. 1.5 ms   b. 27 μs   c. 350 ns
37. a. 150,   0.15   b. 0.33,   33
39. a. 680 V   b. 162.7 W
41. 1.5 kW
43. Radio signal, 16.68 ms; Telephone signal, 33.33 ms; The radio signal by 16.65 ms.
45. 2.35
47. a. 1.44   b. 1.46   c. 1.45

CHAPTER 2

1. a. 10^{29} b. 10.4×10^{23}
3. Increases by a factor of 24.
5. a. Material with many free electrons (i.e., material with 1 electron in the valence shell).
   b. Inexpensive and easily formed into wires.
   c. Full valence shell. Therefore, no free electrons.
   d. The large electrical force tears electrons out of orbit.
7. It has an excess or deficiency of electrons.
9. 2 μC; (Attraction)
11. 0.333 μC, 1.67 μC; both $(+)$ or both $(-)$
13. 30.4 μC
15. 27.7 μC $(+)$
17. 24 V
19. 2400 V
21. 4.25 mJ
23. 4.75 C
25. 50 mA
27. 334 μC
29. 3 mA
31. 80 A
33. 18 V,   0.966 A
35. 1.54 V
37. 50 h
39. 11.7 h
41. 267 h
43. (c) Both
45. The voltmeter and ammeter are interchanged.
47. If you exceed a fuse's voltage rating, it may arc over when it "blows."

CHAPTER 3

1. a. 3.6 Ω b. 0.90 Ω
   c. 36.0 kΩ d. 36.0 mΩ
3. 0.407 inch
5. 300 m = 986 feet
7. 982×10^{-8} $\Omega \cdot$ m (Resistivity is less than for carbon.)

9. $2.26 \times 10^{-8} \, \Omega \cdot m$ (This alloy is not as good a conductor as copper.)

11. AWG 22: 4.86 Ω

    AWG 19: 2.42 Ω

    Diameter of AWG 19 is 1.42 times the diameter of AWG 22. The resistance of AWG 19 is half the resistance of an equal length of AWG 22.

13. AWG 19 should be able to handle 4 A.

    AWG 30 can handle about 0.30 A.

15. 405 meters

17. a. 256 CM   b. 6200 CM   c. 1910 MCM

19. a. 16.2 Ω b. 0.668 Ω c. $2.17 \times 10^{-3} \, \Omega$

21. a. 4148 CM = 3260 sq mil b. 0.0644 inch

23. a. 1600 CM = 1260 sq mil b. 1930 feet

25. $R_{-30°C} = 40.2 \, \Omega$ $R_{0°C} = 46.1 \, \Omega$ $R_{200°C} = 85.2 \, \Omega$

27. a. Positive temperature coefficient

    b. $0.00385 \, (°C)^{-1}$

    c. $R_{0°C} = 18.5 \, \Omega$ $R_{100°C} = 26.2 \, \Omega$

29. 16.8 Ω

31. $T = -260°C$

33. a. $R_{ab} = 10 \, k\Omega$ $R_{bc} = 0 \, \Omega$

    b. $R_{ab} = 8 \, k\Omega$ $R_{bc} = 2 \, k\Omega$

    c. $R_{ab} = 2 \, k\Omega$ $R_{bc} = 8 \, k\Omega$

    d. $R_{ab} = 0 \, k\Omega$ $R_{bc} = 10 \, k\Omega$

35. a. 150 k$\Omega \pm 10\%$

    b. 2.8 $\Omega \pm 5\%$ with a reliability of 0.001%

    c. 47 M$\Omega \pm 5\%$

    d. 39 $\Omega \pm 5\%$ with a reliability of 0.1%

37. Connect the ohmmeter between the two terminals of the light bulb. If the ohmmeter indicates an open circuit, the light bulb is burned out.

39. AWG 24 has a resistance of 25.7 Ω/1000 ft. Measure the resistance between the two ends and calculate the length as

$$\ell = \frac{R}{0.0257 \, \Omega/ft}$$

41. a. 380 Ω

    b. 180 Ω

    c. Negative temperature coefficient. Resistance decreases as temperature increases.

43. a. 4.0 S   b. 2.0 mS

    c. 4.0 μS   d. 0.08 μS

45. 2.93 mS

CHAPTER 4

1. a. 2 A   b. 7.0 A

   c. 5 mA   d. 4 μA

   e. 3 mA   f. 6 mA

3. a. 40 V   b. 0.3 V

   c. 400 V   d. 0.36 V

5. 96 Ω

7. 28 V

9. 6 A

11. Red, Red, Red

13. 22 V

15. a. 2.31 A   b. 2.14 A

17. 2.88 V

19. 4 Ω

21. 400 V

23. 3.78 mA

25. a. + 45 V −   b. 4 A ($\rightarrow$)

    c. − 90 V +   d. 7 A ($\leftarrow$)

27. 3.19 J/s; 3.19 W

29. 36 W

31. 14.1 A

33. 47.5 V

35. 50 V,   5 mA

37. 37.9 A

39. 2656 W

41. 23.2 V; 86.1 mA

43. 361 W $\rightarrow$ 441 W

45. a. 48 W ($\rightarrow$)   b. 30 W ($\leftarrow$)

    c. 128 W ($\leftarrow$)   d. 240 W ($\rightarrow$)

47. a. 1.296×10^6 J   b. 360 Wh   c. 2.88 cents

49. 26 cents

51. $5256

53. 5 cents

55. 51.5 kW

57. 82.7%

59. 2.15 hp

61. 8.8 hp

63. 1.97 hp

65. $137.45

67. a. 10 Ω b. 13.3 Ω

CHAPTER 5

1. a. +30 V   b. −90 V

3. a. +45 V   b. −60 V   c. +90 V   d. −105 V

5. a. 7 V   b. $V_2 = 4$ V $V_1 = 4$ V

7. $V_3 = 12$ V $V_4 = 2$ V

9. a. 10 kΩ   b. 2.94 MΩ   c. 23.4 kΩ

11. Circuit 1: 1650 Ω, 6.06 mA   Circuit 2: 18.15 kΩ, 16.5 mA

13. a. 10 mA   b. 13 kΩ   c. 5 kΩ

 d. $V_{3\text{-}k\Omega} = 30$ V $V_{4\text{-}k\Omega} = 40$ V $V_{1\text{-}k\Omega} = 10$ V $V_R = 50$ V

 e. $P_{1\text{-}k\Omega} = 100$ mW $P_{3\text{-}k\Omega} = 300$ mW $P_{4\text{-}k\Omega} = 400$ mW
 $P_R = 500$ mW

15. a. 40 mA   b. $V_{R_1} = 12$ V $V_{R_3} = 10$ V   c. 26 V

17. a. $V_{R_2} = 4.81$ V $V_{R_3} = 3.69$ V

 b. 1.02 mA   c. 7.32 kΩ

19. a. 457 Ω   b. 78.8 mA

 c. $V_1 = 9.45$ V $V_2 = 3.07$ V $V_3 = 6.14$ V $V_4 = 17.33$ V

 d. $V_T = 36$ V

 e. $P_1 = 0.745$ W $P_2 = 0.242$ W $P_3 = 0.484$ W
 $P_4 = 1.365$ W

 f. R_1: 1 W R_2: 1/4 W R_3: 1/2 W R_4: 2 W

 g. 2.836 W

21. a. 0.15 A   b. 0.115 mA

23. Circuit 1: $V_{6\text{-}\Omega} = 6$ V $V_{3\text{-}\Omega} = 3$ V $V_{5\text{-}\Omega} = 5$ V $V_{8\text{-}\Omega} = 8$ V
 $V_{2\text{-}\Omega} = 2$ V $V_T = 24$ V

 Circuit 2: $V_{4.3\text{-}k\Omega} = 21.6$ V $V_{2.7\text{-}k\Omega} = 13.6$ V
 $V_{7.8\text{-}k\Omega} = 39.2$ kΩ $V_{9.1\text{-}k\Omega} = 45.7$ V $V_T = 120$ V

25. Circuit 1:

 a. $R_1 = 0.104$ kΩ $R_2 = 0.365$ kΩ $R_3 = 0.730$ kΩ

 b. $V_1 = 2.09$ V $V_2 = 7.30$ V $V_3 = 14.61$ V

 c. $P_1 = 41.7$ mW $P_2 = 146.1$ mW $P_3 = 292.2$ mW

 Circuit 2:

 a. $R_1 = 977$ Ω $R_2 = 244$ Ω $R_3 = 732$ Ω

 b. $V_1 = 25.0$ V $V_2 = 6.25$ V $V_3 = 18.75$ V

 c. $P_1 = 640$ mW $P_2 = 160$ mW $P_3 = 480$ mW

27. a. 0.2 A   b. 5.0 V   c. 1 W

 d. $R_T = 550$ Ω $I = 0.218$ A $V = 5.45$ V $P = 1.19$ W

 e. Life expectancy decreases.

29. Circuit 1: $V_{ab} = 9.39$ V $V_{bc} = 14.61$ V

 Circuit 2: $V_{ab} = +25.0$ V $V_{bc} = +6.25$ V

31. Circuit 1:
 $V_{3k\text{-}\Omega} = 9$ V $V_{9k\text{-}\Omega} = 27$ V $V_{6k\text{-}\Omega} = 18$ V $V_a = 45$ V

 Circuit 2: $V_{330\text{-}\Omega} = 2.97$ V $V_{670\text{-}\Omega} = 6.03$ V $V_a = 3.03$ V

33. a. 109 Ω   b. 9.20 V

35. a. 9 kΩ   b. 49 kΩ   c. 249 Ω

 d. A 1.0-V range is not possible since the meter movement requires a minimum of 2 V to give full scale deflection.

37. $R_s = 3.5$ kΩ
 25% deflection: $R_x = 13.5$ kΩ
 50% deflection: $R_x = 4.5$ kΩ
 75% deflection: $R_x = 1.5$ kΩ
 100% deflection: $R_x = 0$

39. Circuit 1:
 $I_{actual} = 0.375$ mA $I_{measured} = 0.3745$ mA
 loading error = 0.125%

 Circuit 2:
 $I_{actual} = 0.375$ mA $I_{measured} = 0.3333$ mA
 loading error = 11.1%

41. Circuit 1:

 a. 1 A

 b. $V_{6\text{-}\Omega} = 6$ V $V_{3\text{-}\Omega} = 3$ V $V_{5\text{-}\Omega} = 5$ V $V_{8\text{-}\Omega} = 8$ V
 $V_{2\text{-}\Omega} = 2$ V

 Circuit 2:

 a. 5.06 mA

 b. $V_{4.3\text{-}k\Omega} = 21.7$ V $V_{2.7\text{-}k\Omega} = 13.6$ V $V_{7.8\text{-}k\Omega} = 39.1$ V
 $V_{9.1\text{-}k\Omega} = 45.6$ V

43. a. 78.8 mA

 b. $V_1 = 9.45$ V $V_2 = 3.07$ V $V_3 = 6.14$ V $V_4 = 17.33$ V

CHAPTER 6

1. a. A and B are in series; D and E are in series; C and F are parallel

 b. B, C, and D are parallel

 c. A and B are parallel; D and F are parallel; C and E are in series

 d. A, B, C, and D are parallel

5. a. $I_1 = 3$ A $I_2 = -1$ A

 b. $I_1 = 7$ A $I_2 = 2$ A $I_3 = -7$ A

 c. $I_1 = 4$ mA $I_2 = 20$ mA

7. a. $I_1 = 1.25$ A $I_2 = 0.0833$ A $I_3 = 1.167$ A $I_4 = 1.25$ A

 b. $R_3 = 4.29$ Ω

9. a. $I_1 = 200$ mA $I_2 = 500$ mA $I_3 = 150$ mA
 $I_4 = 200$ mA

 b. 2.5 V

 c. $R_1 = 12.5$ Ω $R_3 = 16.7$ Ω $R_4 = 50$ Ω

11. a. $R_T = 2.4$ Ω $G_T = 0.417$ s

 b. $R_T = 32$ kΩ $G_T = 31.25$ μs

 c. $R_T = 4.04$ kΩ $G_T = 247.6$ μs

13. a. 2.0 MΩ   b. 450 Ω

15. a. $R_1 = 1250$ Ω $R_2 = 5$ kΩ $R_3 = 250$ Ω

 b. $I_{R_1} = 0.40$ A $I_{R_2} = 0.10$ A

 c. 2.5 A

17. a. 900 mV   b. 4.5 mA

19. a. 240 Ω   b. 9.392 kΩ   c. 1.2 kΩ

21. $I_1 = 0.235$ mA $I_2 = 0.706$ mA $I_3 = 1.059$ mA
    $R_1 = 136$ kΩ $R_2 = 45.3$ kΩ $R_3 = 30.2$ kΩ

23. a. 12.5 kΩ b. 0 c. $75 \, \Omega$

25. $R_T \cong 15 \, \Omega$

27. $I = 0.2$ A $I_1 = 0.1$ A $= I_2$

29. a. $I_1 = 2$ A $I_2 = 8$ A   b. $I_1 = 4$ mA $I_2 = 12$ mA

31. a. $I_1 = 6.48$ mA $I_2 = 9.23$ mA $I_3 = 30.45$ mA
       $I_4 = 13.84$ mA

    b. $I_1 = 60$ mA $I_2 = 30$ mA $I_3 = 20$ mA $I_4 = 40$ mA
       $I_5 = 110$ mA

33. $12 \, \Omega$

35. a. $8 \, \Omega$

    b. 1.50 A

    c. $I_1 = 0.50$ A $I_2 = 0.25$ A $I_3 = 0.75$ A

    d. $\Sigma I_{in} = \Sigma_{out} = 1.50$ A

37. a. $25 \, \Omega$ $I = 9.60$ A

    b. $I_1 = 4.0$ A $I_2 = 2.40$ A $I_3 = 3.20$ A
       $I_4 = 5.60$ A

    c. $\Sigma I_{in} = \Sigma I_{out} = 9.60$ A

    d. $P_1 = 960$ W $P_2 = 576$ W $P_3 = 768$ W
       $P_T = 2304$ W $= P_1 + P_2 + P_3$

39. a. $I_1 = 1.00$ A $I_2 = 2.00$ A $I_3 = 5.00$ A $I_4 = 4.00$ A

    b. 12.00 A

    c. $P_1 = 20$ W $P_2 = 40$ W $P_3 = 100$ W $P_4 = 80$ W

41. a. $R_1 = 2$ kΩ $R_2 = 8$ kΩ $R_3 = 4$ kΩ $R_4 = 6$ kΩ

    b. $I_{R_1} = 24$ mA $I_{R_2} = 6$ mA $I_{R_4} = 8$ mA

    c. $I_1 = 20$ mA $I_2 = 50$ mA

    d. $P_2 = 288$ mW $P_3 = 576$ mW $P_4 = 384$ mW

43. $I_1 = 8.33$ A $I_2 = 5.00$ A $I_3 = 2.50$ A $I_4 = 7.50$ A
    $I_T = 15.83$ A
    The rated current of the fuse will be exceeded; the fuse will
    "blow."

45. $25.1 \, \Omega$

47. $I_{range} = 10$ A $I = 6.2$ A

49. a. $V_{measured} = 20$ V   b. loading effect $= 33.3\%$

51. 25.2 V

53. $I_1 = 4.0$ A $I_2 = 2.4$ A $I_3 = 3.2$ A

55. 20 V

57. $I_1 = 4.0$ A $I_2 = 2.4$ A $I_3 = 3.2$ A

CHAPTER 7

1. a. $R_T = R_1 + R_5 + [(R_2 + R_3)\|R_4]$

   b. $R_T = (R_1\|R_2) + (R_3\|R_4)$

3. a. $R_{T_1} = R_1 + [(R_3 + R_4)\|R_2] + R_5$ $R_{T_2} = R_5$

   b. $R_{T_1} = R_1 + (R_2\|R_3\|R_5)$ $R_{T_2} = R_5\|R_3\|R_2$

7. a. $1500 \, \Omega$ b. 2.33 kΩ

9. $R_{ab} = 140 \, \Omega$ $R_{cd} = 8.89 \, \Omega$

11. a. $R_T = 314 \, \Omega$

    b. $I_T = 63.7$ mA $I_1 = 19.2$ mA $I_2 = 44.5$ mA
       $I_3 = 34.1$ mA $I_4 = 10.4$ mA

    c. $V_{ab} = 13.6$ V $V_{bc} = -2.9$ V

13. a. $I_1 = 5.19$ mA $I_2 = 2.70$ mA $I_3 = 1.081$ mA
       $I_4 = 2.49$ mA $I_5 = 1.621$ mA $I_6 = 2.70$ mA

    b. $V_{ab} = 12.43$ V $V_{cd} = 9.73$ V

    c. $P_T = 145.3$ mW $P_1 = 26.9$ mW $P_2 = 7.3$ mW
       $P_3 = 3.5$ mW $P_4 = 30.9$ mW $P_5 = 15.8$ mW
       $P_6 = 7.0$ mW $P_7 = 53.9$ mW

15. Circuit (a):

    a. $I_1 = 4.5$ mA $I_2 = 4.5$ mA $I_3 = 1.5$ mA

    b. $V_{ab} = -9.0$ V

    c. $P_T = 162$ mW $P_{6\text{-k}\Omega} = 13.5$ mW $P_{3\text{-k}\Omega} = 27.0$ mW
       $P_{2\text{-k}\Omega} = 40.5$ mW $P_{4\text{-k}\Omega} = 81.0$ mW

    Circuit (b):

    a. $I_1 = 0.571$ A $I_2 = 0.365$ A $I_3 = 0.122$ A $I_4 = 0.449$ A

    b. $V_{ab} = -1.827$ V

    c. $P_T = 5.14$ W $P_{10\text{-}\Omega} = 3.26$ W $P_{16\text{-}\Omega} = 0.68$ W
       $P_{5\text{-}\Omega} = 0.67$ W $P_{6\text{-}\Omega} = 0.36$ W $P_{8\text{-}\Omega} = 0.12$ W
       $P_{4\text{-}\Omega} = 0.06$ W

17. $I_1 = 93.3$ mA $I_2 = 52.9$ mA $I_Z = 40.4$ mA $V_1 = 14$ V
    $V_2 = 2.06$ V $V_3 = 7.94$ V $P_T = 2240$ mW $P_1 = 1307$ mW
    $P_2 = 109$ mW $P_3 = 420$ mW $P_Z = 404$ mW

19. $R = 31.1 \, \Omega \rightarrow 3900 \, \Omega$

21. $I_C = 1.70$ mA $V_B = -1.97$ V $V_{CE} = -8.10$ V

23. a. $I_D = 3.6$ mA   b. $R_S = 556 \, \Omega$ c. $V_{DS} = 7.6$ V

25. $I_C \cong 3.25$ mA $V_{CE} \cong -8.90$ mA

27. a. $V_L = 0 \rightarrow 7.2$ V   b. $V_L = 2.44$ V   c. $V_{ab} = 9.0$ V

29. $V_{bc} = 7.45$ V $V_{ab} = 16.55$ V

31. a. $V_{out(min)} = 0$ V $V_{out(max)} = 40$ V

    b. $R_2 = 3.82$ kΩ

33. 0 V, 8.33 V, 9.09 V

35. a. 11.33 V   b. 8.95 V   c. 44.1% d. 1.333 V

37. a. Break the circuit between the 5.6-Ω resistor and the volt-
       age source. Insert the ammeter at the break, connecting
       the red ($+$) lead of the ammeter to the positive terminal
       of the voltage source and the black ($-$) lead to the
       5.6-Ω resistor.

    b. $I_{1(loaded)} = 19.84$ mA $I_{2(loaded)} = 7.40$ mA
       $I_{3(loaded)} = 12.22$ mA

    c. loading effect $(I_1) = 19.9\%$
       loading effect $(I_2) = 18.0\%$
       loading effect $(I_3) = 22.3\%$

39. 12.0 V, 30.0 V, 5.00 A, 3.00 A, 2.00 A

41. 14.1 V

43. 12.0 V, 30.0 V, 5.00 A, 3.00 A, 2.00 A

CHAPTER 8

1. 38 V

3. a. 12 mA   b. $V_S = 4.4$ V $V_1 = 2.0$ V

5. $I_1 = 400 \mu$A $I_2 = 500 \mu$A

7. $P_T = 7.5$ mW $P_{50\text{-}k\Omega} = 4.5$ mW $P_{150\text{-}k\Omega} = 1.5$ mW
   $P_{\text{current source}} = 1.5$ mW

   Note: The current source is absorbing energy from the circuit rather than providing energy.

9. Circuit (a):
   0.25 A-source in parallel with a 20-Ω resistor

   Circuit (b):
   12.5-mA source in parallel with a 2-kΩ resistor

11. a. 7.2 A

    b. $E = 3600$ V $I_L = 7.2$ A

13. a. 21.45 V   b. 6.06 mA   c. 0.606 V

15. $V_2 = -80$ V $I_1 = -26.7$ mA

17. $V_{ab} = -7.52$ V $I_3 = 0.133$ mA

19. $I_1 = 0.467$ A $I_2 = 0.167$ A $I_3 = 0.300$ A

21. $I_2 = -0.931$ A

23. a. $(8\ \Omega)I_1 + 0\ I_2 - (10\ \Omega)I_3 = 24$ V
       $0\ I_1 + (4\ \Omega)I_2 + (10\ \Omega)I_3 = 16$ V
       $I_1 - I_2 + I_3 = 0$

    b. $I = 3.26$ A

    c. $V_{ab} = -13.89$ V

25. $I_1 = 0.467$ A $I_2 = 0.300$ A

27. $I_2 = -0.931$ A

29. $I_1 = -19.23$ mA $V_{ab} = 2.77$ V

31. $I_1 = 0.495$ A $I_2 = 1.879$ A $I_3 = 1.512$ A

33. $V_1 = -6.73$ V $V_2 = 1.45$ V

35. $V_1 = -6$ V $V_2 = 20$ V

37. $V_{6\Omega} = 17.0$ V

39. Network (a): $R_1 = 6.92\ \Omega$ $R_2 = 0.77\ \Omega$ $R_3 = 62.33\ \Omega$
    Network (b): $R_1 = 1.45\ k\Omega$ $R_2 = 2.41\ k\Omega$ $R_3 = 2.03\ k\Omega$

41. Network (a): $R_A = 110\ \Omega$ $R_B = 36.7\ \Omega$ $R_C = 55\ \Omega$
    Network (b): $R_A = 793\ k\Omega$ $R_B = 1693\ k\Omega$ $R_C = 955\ k\Omega$

43. $I = 6.67$ mA

45. $I = 0.149$ A

47. a. The bridge is not balanced.

    b. $(18\ \Omega)I_1 - (12\ \Omega)I_2 - (6\ \Omega)I_3 = 15$ V
       $-(12\ \Omega)I_1 + (54\ \Omega)I_2 - (24\ \Omega)I_3 = 0$
       $-(6\ \Omega)I_1 - (24\ \Omega)I_2 + (36\ \Omega)I_3 = 0$

    c. $I = 38.5$ mA

    d. $V_{R_5} = 0.923$ V

49. $I_{R_5} = 0$ $I_{RS} = 60$ mA $I_{R_1} = I_{R_3} = 45$ mA
    $I_{R_2} = I_{R_4} = 15$ mA

51. $I_{R_1} = 0.495$ A $I_{R_2} = 1.384$ A $I_{R_3} = 1.879$ A
    $I_{R_4} = 1.017$ A $I_{R_5} = 0.367$ A

53. $I_{R_1} = 6.67$ mA $I_{R_2} = 0$ $I_{R_3} = 6.67$ mA
    $I_{R_4} = 6.67$ mA $I_{R_5} = 6.67$ mA $I_{R_6} = 13.33$ mA

CHAPTER 9

1. $I_{R_1} = 75$ mA (up) $I_{R_2} = 75$ mA (to the right)
   $I_{R_3} = 87.5$ mA (down) $I_{R_4} = 12.5$ mA (to the right)

3. $V_a = -3.11$ V $I_1 = 0.1889$ A

5. $E = 30$ V $I_L(1) = 2.18$ mA $I_L(2) = 2.82$ mA

7. $R_{Th} = 20\ \Omega$ $E_{Th} = 10$ V $V_{ab} = 6.0$ V

9. $R_{Th} = 2.02\ k\Omega$ $E_{Th} = 1.20$ V $V_{ab} = -0.511$ V

11. a. $R_{Th} = 16\ \Omega$ $E_{Th} = 5.6$ V

    b. When $R_L = 20\ \Omega$: $V_{ab} = 3.11$ V
       When $R_L = 50\ \Omega$: $V_{ab} = 4.24$ V

13. a. $E_{Th} = 75$ V $R_{Th} = 50\ \Omega$

    b. $I = 0.75$ A

15. a. $E_{Th} = 50$ V $R_{Th} = 3.8\ k\Omega$

    b. $I = 13.21$ mA

17. a. $R_{Th} = 60\ k\Omega$ $E_{Th} = 25$ V

    b. $R_L = 0$: $I = -0.417$ mA
       $R_L = 10\ k\Omega$: $I = -0.357$ mA
       $R_L = 50\ k\Omega$: $I = -0.227$ mA

19. a. $E_{Th} = 28.8$ V, $R_{Th} = 16\ k\Omega$

    b. $R_L = 0$: $I = 1.800$ mA
       $R_L = 10\ k\Omega$: $I = 1.108$ mA
       $R_L = 50\ k\Omega$: $I = 0.436$ mA

21. $E_{Th} = 4.56$ V $R_{Th} = 7.2\ \Omega$

23. a. $E_{Th} = 20$ V $R_{Th} = 200\ \Omega$

    b. $I = 55.6$ mA (upward)

25. $I_N = 0.5$ A, $R_N = 20\ \Omega$, $I_L = 0.2$ A

27. $I_N = 0.594$ mA, $R_N = 2.02\ k\Omega$, $I_L = 0.341$ mA

29. a. $I_N = 0.35$ A, $R_N = 16\ \Omega$

    b. $R_L = 20\ \Omega$: $I_L = 0.156$ A
       $R_L = 50\ \Omega$: $I_L = 0.085$ A

31. a. $I_N = 1.50$ A, $R_N = 50\ \Omega$

    b. $I_N = 1.50$ A, $R_N = 50\ \Omega$

33. a. $I_N = 0.417$ mA, $R_N = 60\ k\Omega$

    b. $I_N = 0.417$ mA, $R_N = 60\ k\Omega$

35. a. $I_N = 0.633$ A, $R_N = 7.2\ \Omega$

    b. $I_N = 0.633$ A, $R_N = 7.2\ \Omega$

37. a. 60 kΩ   b. 2.60 mW

39. a. 31.58 Ω

    b. 7.81 mW

41. a. $R_1 = 40\ k\Omega$ or $160\ k\Omega$, $R_2 = 160\ k\Omega$ or $40\ k\Omega$

    b. 3.125 W

43. $E = 1.5625$ V

45. $I = 0.054$ A, $P_L = 0.073$ W

47. $I = 0.284$ mA, $P_L = 0.807$ W

49. a. $I = 0.24$ A

 b. $I = 0.24$ A

 c. Reciprocity does apply.

51. a. $V = 22.5$ V

 b. Reciprocity does apply.

53. $E_{Th} = 10$ V, $R_{Th} = 20 \ \Omega$

 $I_N = 0.5$ A, $R_N = 20 \ \Omega$

55. $R_L = 2.02$ kΩ for maximum power.

CHAPTER 10

1. a. 800 μC b. 2 μF c. 100 μC

 d. 30 V e. 150 V f. 1.5 μF

3. 200 V

5. 420 μC

7. 73 pF

9. 5.65×10^{-4} m^2

11. 117 V

13. a. 2.25×10^{12} N/C b. 0.562×10^{12} N/C

15. 4.5 kV

17. 3.33 kV

19. a. points b. spheres c. points

21. 24.8 μF

23. 77 μF

25. 3.86 μF

27. a. 9.6 μF b. 13 μF c. 3.6 μF d. 0.5 μF

29. 9 μF

31. 60 μF; 30 μF

33. 81.2 μF; 1.61 μF

35. The 10-μF capacitor is in parallel with the series combination of the 1-μF and 1.5-μF capacitors.

37. a. $V_1 = 60$ V; $V_2 = V_3 = 40$ V

 b. $V_1 = 50$ V; $V_2 = V_3 = 25$ V; $V_4 = 25$ V; $V_5 = 8.3$ V; $V_6 = 16.7$ V

39. 14.4 V; 36 V; 9.6 V

41. 800 μF

43. -50 mA from 0 to 1 ms; 50 mA from 1 ms to 4 ms; 0 mA from 4 ms to 6 ms; 50 mA from 6 ms to 7 ms; -75 mA from 7 ms to 9 ms.

45. $-23.5 \ e^{-0.05t}$ μA

47. 0 mJ, 0.25 mJ, 1.0 mJ, 1.0 mJ, 2.25 mJ, 0 mJ

CHAPTER 11

1. a. 0 V; 5 A b. 20 V; 0 A

3. a. Short circuit b. Voltage source c. Open circuit

 d. $i(0^-) = $ current just before $t = 0$ s; $i(0^+) = $ current just after $t = 0$ s

5. 15.1 V

7. a. $45(1 - e^{-80t})$ V b. $90e^{-80t}$ mA

 c.
t (ms)	v_C (V)	i_C (mA)
0	0	90
20	35.9	18.2
40	43.2	3.67
60	44.6	0.741
80	44.93	0.150
100	44.98	0.030

9. $40(1 - e^{-t/39 \text{ ms}})$ V $10.3e^{-t/39 \text{ ms}}$ mA 28.9 V 2.86 mA

11. 40 μs; 200 μs

13. v_C: 0, 12.6, 17.3, 19.0, 19.6, 19.9 (all V)

 i_C: 5, 1.84, 0.675, 0.249, 0.092, 0.034 (all A)

15. 25 kΩ; 8 μF

17. 45 V; 4.5 kΩ; 0.222 μF

19. 2.5 A

21. a. $20 + 10e^{-25\,000t}$ V b. $-2.5e^{-25\,000t}$ A

 c. v_C starts at 30 V and decays exponentially to 20 V in 200 μs. i_C is 0 A at $t = 0^-$, -2.5 A at $t = 0^+$, and decays exponentially to zero in 200 μs.

23. a. $50e^{-2t}$ V b. $-2e^{-2t}$ mA c. 0.5 s

 d. v_C: 50 V, 18.4 V, 6.77 V, 2.49 V, 0.916 V, 0.337 V

 i_C: -2 mA, -0.736 mA, -0.271 mA, -0.0996 mA, -0.0366 mA, -0.0135 mA

25. 14.4 V

27. a. 200 V; -12.5 mA b. 8 ms

 c. $200e^{-125t}$ V, $-12.5e^{-125t}$ mA

29. $45(1 - e^{-t/0.1857})$ V, 28.4 V (same)

31. a. $60(1 - e^{-500t})$ V b. $1.5e^{-500t}$ A

33. 90 V; 15 kΩ; 100 μF

35. $V_{C_1} = 65$ V; $V_{C_2} = 10$ V; $V_{C_3} = 55$ V; $I_T = 0.5$ A

37. 14.0 μF

39. a. 5 μs b. 40% c. 200 000 pulses/s

41. 0.8 μs; 0.8 μs; 4 μs

43. 6.6 ns

45. -17.8 V (theoretical)

47. 8.65 V (theoretical)

49. Verification point: At $t = 20$ ms, -17.8 V and -0.179 A

51. 29.3 V, 0.227 mA

CHAPTER 12

1. a. A_1 b. 1.4 T
3. 0.50 T
5. 1.23×10^{-3} Wb
7. 0.45 T;   0.9 T
9. 0.625 T
11. $B = \dfrac{\Phi}{A} = \dfrac{\mathscr{F}/\mathscr{R}}{A} = \dfrac{\mathscr{F}}{\mathscr{R}A} = \dfrac{NI}{\left(\dfrac{\ell}{\mu A}\right)A} = \mu\left(\dfrac{NI}{\ell}\right) = \mu H$
13. $N_1 I_1 = H_1\ell_1 + H_2\ell_2$; $N_2 I_2 = H_2\ell_2 - H_3\ell_3$
15. 0.47 A
17. 0.88 A
19. 0.58 A
21. 0.53 A
23. 0.86 A
25. 3.7 A
27. 4.4×10^{-4} Wb
29. 1.06×10^{-4} Wb

CHAPTER 13

1. 225 V
3. 6.0 V
5. 150 mV
7. 0.111 s
9. 79.0 μH
11. $L = \dfrac{N\Phi}{I} = \dfrac{N(B_g A_g)}{I} = \dfrac{N(\mu_0 H_g)A_g}{I}$

$= \dfrac{N\mu_0\left(\dfrac{NI}{\ell_g}\right)A_g}{I} = \dfrac{\mu_0 N^2 A_g}{\ell_g}$
13. 4 H
15. a. 4 H   c. 5 A
17. 84 mH
19. 4.39 mH
21. a. 21 H   b. 2 H   c. 20 H
    d. 4 H   e. 4 mH
23. 9 μH
25. Circuit (a): 6 H; 1.5 H
    Circuit (b): 2 H; 8 H
27. 1.6 H, in series with 6 H∥4 H
29. 1.2 H in series with 8 H∥12 H
31. a. 1 H in series with 3 μF   b. 2 H in series with 10 μF
    c. 10 Ω, 10 H, and 25 μF in series
    d. 10 Ω in series with 40 Ω∥(50 H in series with 20 μF)

33. 0.32 J
35. The path containing L_1 and L_2 is open.

CHAPTER 14

1. a. open circuit
   b. Circuit (a): 1.6 A
      Circuit (b): 6 A; 60 V
      Circuit (c): 0 A; E
      Circuit (d): 2 A; 30 V
3. $v_{R_1} = 180$ V; $v_{R_2} = 120$ V; $v_{R_3} = 60$ V; $v_{R_4} = 32$ V
   $v_{R_5} = 28$ V; $v_{R_6} = 0$ V; $i_T = 21$ A; $i_1 = 18$ A
   $i_2 = 3$ A; $i_3 = 1$ A; $i_4 = i_5 = 2$A; $i_6 = 0$ A
5. a. 50 ms   b. 250 ms   c. $3(1 - e^{-20t})$ A; $180e^{-20t}$ V

   d.

t	i_L (A)	v_L (V)
0	0	180
τ	1.90	66.2
2τ	2.59	24.4
3τ	2.85	8.96
4τ	2.95	3.30
5τ	2.98	1.21

7. a. 0.2 s   b. 1 s   c. $20\, e^{-5t}$ V; $(1 - e^{-5t})$ A
   d. v_L: 20, 7.36, 2.71, 0.996, 0.366, 0.135 (all V)
      i_L: 0, 0.632, 0.865, 0.950, 0.982, 0.993 (all A)
9. $-182 (1 - e^{-393t})$ mA; $-40e^{-393t}$ V; -134 mA; -10.5 V
11. 80 V;   20Ω;   2 H
13. 40 V;   4 kΩ;   2 H
15. a. 2 ms   b. $5e^{-500t}$ A; $-1250e^{-500t}$ V
    c.

t	i_L (A)	v_L (V)
0	5	-1250
τ	1.84	-460
2τ	0.677	-169
3τ	0.249	-62.2
4τ	0.092	-22.9
5τ	0.034	-8.42

17. -365 V
19. $R_1 = 20\ \Omega$; $R_2 = 30\ \Omega$
21. 5.19 A
23. a. 8.89 μs   b. $-203e^{-t/8.89\mu s}$ V; $5e^{-t/8.89\mu s}$ mA
    c. -27.3 V; 0.675 mA
25. a. 10 ms   b. $90 (1 - e^{-t/10\, ms})$ mA; $36e^{-t/10\, ms}$ V
    c. 2.96 V; 82.6 mA
27. 103.3 mA;   3.69 V
29. 33.1 V
33. $i_L(25$ ms$) = 126$ mA; $i_L(50$ ms$) = 173$ mA

CHAPTER 15

1. AC voltage is voltage whose polarity cycles periodically between positive and negative. AC current is current whose direction cycles periodically.

3. a. The magnitude of a waveform (such as a voltage or current) at any instant of time.

   b. 0, 10, 20, 20, 20, 0, −20, −20, 0 (all V)

5. 0 mA, 2.5 mA, 5 mA, 5 mA, 5 mA, 0 mA, −5 mA, −5 mA, 0 mA

7. a. 2 Hz   b. 10 Hz   c. 62.5 kHz

9. 7 ms; 142.9 Hz

11. 15 V; 6 mA

13. 149 919 cycles

15. 100 μs; 400 μs; 2500 Hz

17. 144.4 V

19. 100 V

21. a. 0.1745   b. 0.4363   c. 1.3963

   d. 2.618   e. 6.1087   f. 10.821

23. 43.3 A;   25 A;   −49.2 A;   −50 A

25. V_m = 85.1 V. Waveform is like Figure 15–25 except T = 50 μs.

27. a. 62.83×10^6 rad/s   b. 188.5 rad/s   c. 157.1 rad/s

   d. 314.2 rad/s   e. 1571 rad/s

29. a. $v = 170 \sin 377t$ V   b. $i = 40 \sin 628t$ μA

   c. $v = 17 \sin 52.4 \times 10^3 t$ V

31. $v = 20 \sin 125.7t$ V

33. 0, 28.8, −11.4, −22 (all mA)

35. a. $5 \sin(1000 t + 36°)$ mA

   b. $10 \sin(40\pi t + 120°)$ A

   c. $4 \sin(1800\pi t − 45°)$ V

37. 4.46 V; −3.54 V; 0.782 V

39. $v = 100 \sin(3491t + 36°)$ V

41. 6.25 ms; 13.2 ms; 38.2 ms

43. a. 20°; i leads   b. in phase

   c. 50°; i_1 leads   d. 60°; i leads

45. a. A leads by 90°   b. A leads by 150°

47. Zero for each

49. a. 1.1 A   b. −5 V   c. 1.36 A

51. a. Similar to Figure 15–61(b), except positive peak is 40 V, negative peak is −10 V, and V_{avg} = 15 V. T = 120 ms.

   b. 40 V;   −10 V

   c. 27.5 V;   36.7 V;   2.5 V;   −6.65 V

   d. 15 V

53. 2.80 V

55. a. 12 V   b. 17.0 mA

   c. 19.7 V   d. 48.9 V

57. a. 5 mA   b. 23.2 V

59. a. 8.94 A   b. 16.8 A

61. 24 V; Its magnitude is always 24 V; therefore, it produces the same average power to a resistor as a 24 V battery.

63. 26.5 mA

65. a. 108 V   b. 120 V

   c. 14.6 V   d. 422 V

67. 6.25 ms

CHAPTER 16

1. a. $13\angle 67.4°$ b. $10.8\angle −33.7°$

   c. $17\angle 118.1°$ d. $10.8\angle −158.2°$

5. a. $7 + j6$ b. $1 + j10$

   c. $15.4 − j3.50$ d. $−4.64 + j1.86$

   e. $−7.2 − j12.1$

7. a. $14.2\angle −23.8°$ b. $1.35\angle −69.5°$ c. $5.31\angle 167.7°$

9. a. $10 \sin(\omega t + 30°)$ V   b. $15 \sin(\omega t − 10°)$ V

11. a. 10 V$\angle 30°$; 15 V$\angle −20°$ b. 11.5 V$\angle 118.2°$

   c. $11.5 \sin(\omega t + 118.2°)$ V

13. a. 17.7 mA $\angle 36°$; 28.3 mA $\angle 80°$; 42.8 mA $\angle 63.3°$

   b. $60.5 \sin(\omega t + 63.3°)$ mA

15. a. 4.95 mA $\angle 0°$; 2.83 mA $\angle −90°$; 4.2 mA $\angle 90°$; 5.15 mA $\angle 15.9°$

   b. $7.28 \sin(\omega t + 15.9°)$ mA

17. a. $10 \sin \omega t$ A   b. $10 \sin(\omega t + 27°)$ A

   c. $204 \sin(\omega t − 56°)$ mV   d. $204 \sin(\omega t − 157°)$ μV

19. a. $147 \sin(\omega t + 39°)$ V; $183.8 \sin(\omega t + 39°)$ V

   b. $330.8 \sin(\omega t + 39°)$ V; Identical

21. a. $5 \sin(\omega t − 30°)$ A   b. $5 \sin(\omega t − 105°)$ A

   c. $10 \sin(\omega t + 120°)$ V   d. $10 \sin(\omega t + 100°)$ V

23. a. $0.531 \sin(377t − 90°)$ A

   b. $31.8 \sin(6283t − 90°)$ mA

   c. $0.4 \sin(500t − 90°)$ A

25. a. 8.74 kHz;   b. 50.9 mH; 130°

27. a. 530.5 Ω b. 31.83 Ω c. 400 Ω

29. a. $1.89 \sin(377t + 90°)$ A

   b. $79.6 \sin(2\pi \times 400t − 150°)$ mV

31. a. 48 $\Omega\angle 0°$ b. $j37.7$ Ω c. $−j50$ Ω

33. a. $\mathbf{V}_R$ = 25 V$\angle 0°$; $\mathbf{V}_L$ = 12.5 V$\angle 90°$; $\mathbf{V}_C$ = 5 V$\angle −90°$

   b. $v_R = 35.4 \sin\omega t$ V; $v_L = 17.7 \sin(\omega t + 90°)$ V;

        $v_C = 7.07 \sin(\omega t − 90°)$ V

35. a. 39.8 Hz   b. 6.37 μF

37. Theoretical: 45.2 mA

39. 1.59 A$\angle -90°$

41. 7.96 A

CHAPTER 17

1. When p is +, power flows from source to load. When p is −, power flows out of load. Out of the load.

3. a. 1000 W and 0 VAR; 500 W and 0 VAR

   b. 1500 W and 0 VAR

5. 151 VAR (cap.)

7. a. 10 Ω   b. 6 Ω   c. 265 μF

9. 30 Ω

11. 160 W; 400 VAR (ind.)

13. 900 W; 300 VAR (ind.)

15. a. 20 Ω   b. 6 Ω   c. 8 Ω   d. 1.2 H

17. 125 VA

19. 1150 W; 70 VAR (cap.); 1152 VA; $\theta = -3.48°$

21. 2.36 Ω

23. 120 Ω

25. a. 721 W;   82.3 VAR (cap.);   726 VA

   b. $I = 6.05$ A;   No

27. a. Across the load   b. 73.9 μF

29. 57.3 kW

31. 2598 W

33. 160 Ω

35. Same as Fig. 17–5 with peak value of $p = 3.14$ W.

37. A sawtooth wave oscillating between 8 W and −8 W

CHAPTER 18

1. a. 0.125 $\sin \omega t$

3. a. $1.87 \times 10^{-3} \sin(\omega t + 30°)$

5. a. $1.36 \sin(\omega t - 90°)$

7. a. $1333 \sin(2000\pi t + 30°)$

9. a. $62.5 \sin(10000t - 90°)$

11. a. $67.5 \sin(20000t - 160°)$

13. Network (a): 31.6 Ω$\angle 18.43°$

   Network (b): 8.29 kΩ$\angle -29.66°$

15. a. 42.0 Ω$\angle 19.47° = 39.6$ Ω $+ j14.0$ Ω

17. $R = 1.93$ kΩ, $L = 4.58$ mH

19. $R = 15$ Ω, $C = 1.93$ μF

21. a. $\mathbf{Z}_T = 50$ Ω$\angle -36.87°$, $\mathbf{I} = 2.4$ A$\angle 36.87°$,
   $\mathbf{V}_R = 96$ V$\angle 36.87°$, $\mathbf{V}_L = 48$ V$\angle 126.87°$,
   $\mathbf{V}_C = 120$ V$\angle -53.13°$

c. 230.4 W

d. 230.4 W

23. a. $\mathbf{Z}_T = 45$ Ω$\angle -36.87°$

   b. $i = 0.533 \sin(\omega t + 36.87°), v_R = 19.20 \sin(\omega t + 36.87°)$,
   $v_C = 25.1 \sin(\omega t - 53.13°)$, $v_L = 10.7 \sin(\omega t - 126.87°)$

   e. 5.12 W

   f. 5.12 W

25. a. $\mathbf{V}_R = 9.49$ V$\angle -18.43°$, $\mathbf{V}_L = 11.07$ V$\angle 71.57°$,
   $\mathbf{V}_C = 7.91$ V$\angle -108.43°$

   b. $\Sigma \mathbf{V} = 10.00$ V$\angle 0°$

27. a. $\mathbf{V}_C = 317$ V$\angle -30°$, $\mathbf{V}_L = 99.8$ V$\angle 150°$

   b. 25 Ω

29. a. $\mathbf{V}_C = 6.0$ V$\angle -110°$

   b. $\mathbf{V}_Z = 13.87$ V$\angle 59.92°$

   c. 69.4 Ω$\angle 79.92°$

   d. 1.286 W

31. Network (a): 199.9 Ω$\angle -1.99°$,
   Network (b): 485 Ω$\angle -14.04°$

33. a. $\mathbf{Z}_T = 3.92$ kΩ$\angle -78.79°$, $\mathbf{I}_T = 2.55$ mA$\angle 78.69°$,
   $\mathbf{I}_1 = 0.5$ mA$\angle 0°$, $\mathbf{I}_2 = 10.0$ mA$\angle -90°$,
   $\mathbf{I}_3 = 12.5$ mA$\angle 90°$

   d. 5.00 mW

35. a. 5.08 kΩ$\angle 23.96°$

   b. 152.3 V$\angle 23.96°$

37. 2.55 Ω$\angle 81.80°$

39. Network (a): $\mathbf{I}_R = 10.00$ mA$\angle -31.99°$,
   $\mathbf{I}_L = 4.00$ mA$\angle -121.99°$, $\mathbf{I}_C = 4.35$ mA$\angle 58.01°$
   Network (b): $\mathbf{I}_R = 9.70$ mA$\angle -44.04°$,
   $\mathbf{I}_{C_1} = 1.62$ mA$\angle 45.96°$, $\mathbf{I}_{C_2} = 0.81$ mA$\angle 45.96°$

41. $\mathbf{I}_L = 2.83$ mA$\angle -135°$, $\mathbf{I}_C = 3.54$ mA$\angle 45°$,
   $\mathbf{I}_R = 0.71$ mA$\angle -45°$, $\Sigma \mathbf{I}_{out} = \Sigma \mathbf{I}_{in} = 1.00$ mA$\angle 0°$

43. a. 6.25 A$\angle 90°$

   b. 72.1 Ω

   c. 8.00 A$\angle 51.32°$

45. a. $\mathbf{Z}_T = 31.38$ Ω$\angle 67.52°$, $\mathbf{I}_L = 3.82$ A$\angle -67.52°$,
   $\mathbf{I}_C = 3.42$ A$\angle -40.96°$, $\mathbf{I}_R = 1.71$ A$\angle -130.96°$

   c. $P_R = 175.4$ W

   d. $P_T = 175.4$ W

47. a. $\mathbf{Z}_T = 10.53$ Ω$\angle 10.95°$, $\mathbf{I}_T = 1.90$ A$\angle -10.95°$,
   $\mathbf{I}_1 = 2.28$ A$\angle -67.26°$, $\mathbf{I}_2 = 2.00$ A$\angle 60.61°$

   b. $\mathbf{V}_{ab} = 8.87$ V$\angle 169.06°$

49. a. $\mathbf{Z}_T = 7.5$ kΩ$\angle 0°$, $\mathbf{I}_1 = 0.75$ mA$\angle 0°$, $\mathbf{I}_2 = 0.75$ mA$\angle 90°$,
   $\mathbf{I}_3 = 0.79$ A$\angle -71.57°$

   b. $\mathbf{V}_{ab} = 7.12$ V$\angle 18.43°$

51. $\omega_C = 2000$ rad/s

53. $f_C = 3.39$ Hz

55. Network (a): 5.5-kΩ resistor in series with a 9.0-kΩ inductive reactance

Network (b): 207.7-Ω resistor in series with a 138.5-Ω inductive reactance

57. $\omega = 1$ krad/s: $Y_T = 0.01$ S $+ j0$, $Z_T = 100$ Ω

$\omega = 10$ krad/s: $Y_T = 0.01$ S $+ j0$, $Z_T = 100$ Ω

CHAPTER 19

1. a. 5.00 V$\angle 180°$

   b. 12.50 V$\angle 0°$

   c. 15.00 V$\angle -120°$

3. a. 3.20 mV$\angle 180°$

   b. 8.00 mV$\angle 180°$

   c. 9.60 mV$\angle -120°$

5. 7.80 V$\angle -120°$

7. Circuit (a): $E = 54$ V$\angle 0°$, $V_L = 13.5$ V$\angle 0°$

   Circuit (b): $E = 450$ mV$\angle -60°$, $V_L = 439$ mV$\angle -47.32°$

9. a. 4.69 V$\angle 180°$

   b. $E = (7.5$ MΩ$)I$, $V = 4.69$ V$\angle 180°$

11. a. $(4\,\Omega + j2\,\Omega)I_1 - (4\,\Omega)I_2 = 20$ V$\angle 0°$

    $-(4\,\Omega)I_1 + (6\,\Omega + j4\,\Omega)I_2 = 48.4$ V$\angle -161.93°$

    b. $I_1 = 2.39$ A$\angle 72.63°$ $I_2 = 6.04$ A$\angle 154.06°$

    c. $I = 6.15$ A$\angle -3.33°$

13. a. $(12\,\Omega - j16\,\Omega)I_1 + (j15\,\Omega)I_2 = 13.23$ V$\angle -79.11°$

    $(j15\,\Omega)I_1 + 0I_2 = 10.27$ V$\angle -43.06°$

    b. $I_1 = 0.684$ A$\angle -133.06°$ $I_2 = 1.443$ A$\angle -131.93°$

    c. $V = 11.39$ V$\angle -40.91°$

15. 27.8 V$\angle 6.79°$ $I = 6.95$ mA$\angle 6.79°$

17. a. $(0.417$ S$\angle 36.87°)V_1 - (0.25$ S$\angle 90°)V_2 = 3.61$ A$\angle 56.31°$

    $-(0.25$ S$\angle 90°)V_1 + (0.083$ S$\angle 90°)V_2 = 7.00$ A$\angle 90°$

    b. $V_1 = 30.1$ V$\angle 139.97°$ $V_2 = 60.0$ V$\angle 75.75°$

    c. $I = 13.5$ V$\angle -44.31°$

19. a. $(0.0893$ S$\angle 22.08°)V_1 - (0.04$ S$\angle 90°)V_2 = 0.570$ A$\angle 93.86°$

    $-(0.04$ S$\angle 90°)V_1 + (0.06$ S$\angle 90°)V_2 = 2.00$ A$\angle 180°$

    b. $V_1 = 17.03$ V$\angle 18.95°$ $V_2 = 31.5$ V$\angle 109.91°$

    c. $V = 11.39$ V$\angle -40.91°$

21. $(0.372\ \mu$S$\angle -5.40°)V = 10.33$ mA$\angle 1.39°$

    27.8 V$\angle 6.79°$ $I = 6.95$ mA$\angle 6.79°$

    As expected, the answers are the same as those in Problem 15.

23. Network (a):

    $Z_1 = 284.4\,\Omega\angle -20.56°$ $Z_2 = 94.8\,\Omega\angle 69.44°$
    $Z_3 = 31.6\,\Omega\angle 159.44°$

Network (b):

$Z_1 = 11.84$ kΩ$\angle 9.46°$ $Z_2 = 5.92$ kΩ$\angle -80.54°$
$Z_3 = 2.96$ kΩ$\angle -80.54°$

25. $I_T = 0.337$ A$\angle -2.82°$

27. a. $Z_T = 3.03\,\Omega\angle -76.02°$

    b. $I = 5.28$ A$\angle 76.02°$ $I_1 = 0.887$ A$\angle -15.42°$

29. a. $Z_2 = 1\,\Omega - j7\,\Omega = 7.07\,\Omega\angle -81.87°$

    b. $I = 142.5$ mA$\angle 52.13°$

31. $Z_1Z_4 = Z_2Z_3$ as required.

35. $R_3 = 50.01\,\Omega$, $R_1 = 253.3\,\Omega$

39. Same as Figure 19–21

41. Same as Problem 26

CHAPTER 20

1. $I = 4.12$ A$\angle 50.91°$

3. 16 V$\angle -53.13°$

5. a. $V = 15.77$ V$\angle 36.52°$

   b. $P_{(1)} + P_{(2)} = 1.826$ W $\neq P_{100\text{-}\Omega} = 2.49$ W

7. 0.436 A$\angle -9.27°$

9. $19.0 \sin(\omega t + 68.96°)$

11. a. $V_L = 1.26$ V$\angle 161.57°$ b. $V_L = 6.32$ V$\angle 161.57°$

13. 0.361 mA$\angle -3.18°$

15. $V_L = 9.88$ V$\angle 0°$

17. 1.78 V

19. $Z_{Th} = 3\,\Omega\angle -90°$ $E_{Th} = 20$ V$\angle -90°$

21. a. $Z_{Th} = 37.2\,\Omega\angle 57.99°$ $E_{Th} = 9.63$ V$\angle 78.49°$

    b. 0.447 W

23. $Z_{Th} = 22.3\,\Omega\angle -15.80°$ $E_{Th} = 20.9$ V$\angle 20.69°$

25. $Z_{Th} = 109.9\,\Omega\angle -28.44°$ $E_{Th} = 14.5$ V$\angle -91.61°$

27. a. $Z_{Th} = 20.6\,\Omega\angle 34.94°$ $E_{Th} = 10.99$ V$\angle 13.36°$

    b. $P_L = 1.61$ W

29. $Z_N = -j3\,\Omega$ $I_N = 6.67$ A$\angle 0°$

31. a. $Z_N = 22.3\,\Omega\angle -15.80°$ $I_N = 0.935$ A$\angle 36.49°$

    b. 0.436 A$\angle -9.27°$

    c. 3.80 W

33. a. $Z_N = 109.9\,\Omega\angle -28.44°$ $I_N = 0.131$ A$\angle -63.17°$

    b. 0.0362 A$\angle -84.09°$

    c. 0.394 W

35. a. $Z_N = 14.1\,\Omega\angle 85.41°$ $I_N = 0.181$ A$\angle 29.91°$

    b. 0.0747 A$\angle 90.99°$

37. a. $Z_{Th} = 17.9\,\Omega\angle -26.56°$ $E_{Th} = 1.79$ V$\angle 153.43°$

    b. $0.0316\ \mu$A$\angle 161.56°$

    c. $40.0\ \mu$W

39. $\mathbf{E}_{Th} = 10\ V\angle 0°$ $\mathbf{I}_N = 10.5\ A\angle 0°$ $\mathbf{Z}_{Th} = 0.952\ \Omega\angle 0°$

41. a. $\mathbf{Z}_L = 8\ \Omega\angle 22.62°$ b. $40.2\ W$

43. a. $\mathbf{Z}_L = 2.47\ \Omega\angle 21.98°$ b. $1.04\ W$

45. $4.15\ \Omega\angle 85.24°$

47. a. $\mathbf{Z}_L = 37.2\ \Omega\angle -57.99°$
    b. $19.74\ \Omega$
    c. $1.18\ W$

49. $\mathbf{Z}_{Th} = 3\ \Omega\angle -90°$ $\mathbf{E}_{Th} = 20\ V\angle -90°$

51. $\mathbf{Z}_{Th} = 109.9\ \Omega\angle -28.44°$ $\mathbf{E}_{Th} = 14.5\ V\angle -91.61°$

53. $\mathbf{E}_{Th} = 10\ V\angle 0°$ $\mathbf{I}_N = 10.5\ A\angle 0°$ $\mathbf{Z}_{Th} = 0.952\ \Omega\angle 0°$

55. $\mathbf{Z}_N = 0.5\ k\Omega\angle 0°$ $\mathbf{I}_N = 4.0\ mA\angle 0°$

57. $\mathbf{E}_{Th} = 10\ V\angle 0°$ $\mathbf{I}_N = 10.5\ A\angle 0°$ $\mathbf{Z}_N = 0.952\ \Omega\angle 0°$

59. $\mathbf{Z}_N = 0.5\ k\Omega\angle 0°$ $\mathbf{I}_N = 4.0\ mA\angle 0°$

CHAPTER 21

1. a. $\omega_s = 3835\ rad/s$ $f_s = 610.3\ Hz$
   b. $\mathbf{I} = 153.8\ mA\angle 0°$
   c. $\mathbf{V}_C = 59.0\ V\angle -90°$ $\mathbf{V}_L = 59.03\ V\angle 87.76°$
      $\mathbf{V}_R = 7.69\ V\angle 0°$
   d. $P_L = 0.355\ W$

3. a. $R = 25.0\ \Omega$ $C = 4.05\ nF$
   b. $P = 15.6\ mW$
   c. $X_C = 1.57\ k\Omega$ $\mathbf{V}_C = 39.3\ V\angle -90°$
      $\mathbf{V}_L = 39.3\ V\angle 90°$
      $\mathbf{V}_R = \mathbf{E} = 0.625\ V\angle -90°$
   d. $v_C = 55.5\ \sin(50,000\pi t - 90°)$
      $v_L = 55.5\ \sin(50,000\pi t + 90°)$
      $v_R = 0.884\ \sin(50,000\pi t)$

5. a. $\omega_s = 500\ rad/s$ $f_s = 79.6\ kHz$
   b. $\mathbf{Z}_T = 200\ \Omega\angle 0°$
   c. $\mathbf{I} = 10\ mA\angle 0°$
   d. $\mathbf{V}_R = 1\ V\angle 0°$ $\mathbf{V}_L = 50.01\ V\angle 88.85°$
      $\mathbf{V}_C = 50\ V\angle -90°$
   e. $P_T = 20\ mW$ $Q_C = 0.5\ VAR\ (cap.)$
      $Q_L = 0.5\ VAR\ (ind.)$
   f. $Q_s = 25$

7. a. $C = 0.08\ \mu F$ $R = 6.4\ \Omega$
   b. $P_T = 0.625\ W$
   c. $\mathbf{V}_L = 62.5\ V\angle 90°$

9. $\omega_s = 8000\ rad/s$

11. a. $\omega_s = 3727\ rad/s$ $Q = 7.45$ $BW = 500\ rad/s$
    b. $P_{max} = 144\ W$
    c. $\omega_1 \approx 3477\ rad/s$ $\omega_2 \approx 3977\ rad/s$
    d. $\omega_1 = 3485.16\ rad/s$ $\omega_2 = 3985.16\ rad/s$

e. The results are close, although the approximation will yield some error if used in further calculations. The error would be less if Q were larger.

13. a. $R = 1005\ \Omega$ $C = 63.325\ pF$
    b. $P = 0.625\ W$ $\mathbf{V}_L = 312.5\ V\angle 90°$
    c. $v_{out} = 442\ \sin(400\pi \times 10^3 t + 90°)$

15. Network (a):
    a. $Q = 24$
    b. $R_P = 5770\ \Omega$ $X_{LP} = 240\ \Omega$
    c. $Q = 240$ $R_P = 576\ \Omega$
       $X_{LP} = 2400\ \Omega$
    Network (b):
    a. $Q = 1$
    b. $R_P = 200\ \Omega$ $X_{LP} = 200\ \Omega$
    c. $Q = 10$ $R_P = 10.1\ k\Omega$
       $X_{LP} = 1.01\ k\Omega$
    Network (c):
    a. $Q = 12.5$
    b. $R_P = 314.5\ \Omega$ $X_{LP} = 25.16\ \Omega$
    c. $Q = 125$ $R_P = 31.25\ \Omega$ $X_{LP} = 250\ \Omega$

17. Network (a):
    $Q = 15$ $R_P = 4500\ \Omega$ $X_{CP} = 300\ \Omega$
    Network (b):
    $Q = 2$ $R_P = 225\ \Omega$ $X_{CP} = 112.5\ \Omega$
    Network (c):
    $Q = 5$ $R_P = 13\ k\Omega$ $X_{CP} = 2600\ \Omega$

19. Network (a):
    $Q = 4$ $R_s = 4\ k\Omega$ $X_{LS} = 16\ k\Omega$
    Network (b):
    $Q = 0.333$ $R_s = 3.6\ k\Omega$ $X_{CS} = 1.2\ k\Omega$
    Network (c):
    $Q = 100$ $R_s = 10\ \Omega$ $X_{LS} = 1\ k\Omega$

21. $L_s = 1.2\ mH$ $L_P = 2.4\ mH$

23. a. $\omega_P = 20\ krad/s$

25. a. $\omega_P = 39.6\ krad/s$ $f_P = 6310\ Hz$
    b. $Q = 6.622$
    c. $\mathbf{V} = 668.2\ V\angle 0°$ $\mathbf{I}_R = 11.14\ mA\angle 0°$
       $\mathbf{I}_L = 668.2\ mA\angle -82.36°$
       $\mathbf{I}_C = 662.2\ mA\angle 90°$
    d. $P_T = 66.82\ W$
    e. $BW = 5.98\ krad/s$ $BW = 952\ Hz$

27. $R_1 = 4194\ k\Omega$ $C = 405\ pF$ $\mathbf{I}_L = 5.0\ mA\angle -89.27°$

29. a. $900\ \Omega$
    b. 1.862

c. 1074 rad/s

d. $C = 500$ nF $L = 0.450$ H

e. 93.1 V$\angle 0°$

33. $V = 668.2$ V, $f_P = 6310$ Hz,   BW $= 952$ Hz, $Q = 6.622$

35. $V = 93.1$ V, $f_P = 318.3$ Hz,   BW $= 170.9$ Hz, $Q = 1.86$

CHAPTER 22

1. a. 2000 (33.0 dB)   b. 200,000 (53.0 dB)

   c. 2×10^6 (63.0 dB)   d. 400 (26.0 dB)

3. a. $A_P = 50 \times 10^6$ (77.0 dB) $A_V = 500$ (54.0 dB)

   b. $A_P = 10$ (10.0 dB) $A_V = 0.224$ $(-13.0$ dB)

   c. $A_P = 13.3 \times 10^6$ (71.3 dB) $A_V = 258$ (48.2 dB)

   d. $A_P = 320 \times 10^6$ (85.1 dB) $A_V = 1265$ (62.0 dB)

5. $P_{in} = 12.5\ \mu$W $P_{out} = 39.5$ mW
   $V_{out} = 3.14$ V $A_V = 12.6$ $[A_V]_{dB} = 22.0$ dB

7. a. 17.0 dBm $(-13.0$ dBW)

   b. 30.0 dBm (0 dBW)

   c. -34.0 dBm $(-64.0$ dBW)

   d. -66.0 dBm $(-96.0$ dBW)

9. a. 0.224 W   b. 30.2 μW   c. 5.01 pW   d. 1995 W

17. $P_1 = 5.05$ dBm $P_2 = 2.05$ dBm $P_3 = 14.09$ dBm
    $P_0 = 25.6$ dBm

19. $P_1 = 25.5$ dBm $P_{in} = -14.5$ dBm $V_{out} = 52.8$ V

23. a. $\omega_C = 1000$ rad/s $f_C = 159.2$ Hz

25. a. $\omega_1 = 50$ rad/s $f_C = 1000$ rad/s

35. a. T.F. $= \dfrac{1 + j0.00003\omega}{1 + 0.00006\omega}$

43. a. Low-pass filter: $\omega_C = 500$ rad/s
    High-pass filter: $\omega_C = 25$ krad/s
    BW $= 475$ krad/s

   c. The actual cutoff frequencies will be close to the designed values since the break frequencies are separated by more than one decade.

45. a. $R_1 = 5$ kΩ $R_2 = 50$ kΩ

   c. The actual frequencies will not occur at the designed values since they are 1 decade apart.

47. a. 10 krad/s

   b. 10

   c. BW $= 1$ krad/s $\omega_1 = 9.5$ krad/s $\omega_2 = 10.5$ krad/s

   d. At resonance, $[A_V]_{dB} = -28.0$ dB

49. a. 10 krad/s

   b. 10

   c. BW $= 1$ krad/s $\omega_1 = 9.5$ krad/s $\omega_2 = 10.5$ krad/s

   d. At resonance, $[A_V]_{dB} = -0.09$ dB

CHAPTER 23

1. a. 8 A$\angle -30°$;   8 A$\angle -150°$;   8 A$\angle 90°$ b. yes

3. ACB.

5. a. $V_{ab} = 208$ V$\angle 45°$; $V_{ca} = 208$ V$\angle 165°$

   b. $V_{an} = 120$ V$\angle 15°$; $V_{bn} = 120$ V$\angle -105°$;
      $V_{cn} = 120$ V$\angle 135°$

7. $I_a = 23.1$ A$\angle -21.9°$; $I_b = 23.1$ A$\angle -141.9°$;
   $I_c = 23.1$ A$\angle 98.1°$

9. 8.16 A$\angle 24°$;   8.16 A$\angle -96°$;   8.16 A$\angle 144°$

11. $I_{ab} = 19.2$ A$\angle 36.9°$; $I_{bc} = 19.2$ A$\angle -83.1°$;
    $I_{ca} = 19.2$ A$\angle 156.9°$; $I_a = 33.3$ A$\angle 6.9°$;
    $I_b = 33.3$ A$\angle -113.1°$; $I_c = 33.3$ A$\angle 126.9°$

13. $V_{ab} = 250$ V$\angle -57.9°$; $V_{bc} = 250$ V$\angle -177.9°$;
    $V_{ca} = 250$ V$\angle 62.1°$

15. 14.4 A$\angle -33°$;   14.4 A$\angle -153°$;   14.4 A$\angle 87°$

17. a. $R = 3.83$ Ω; $C = 826\ \mu$F

    b. $R = 7.66$ Ω; $L = 17.1$ mH

19. a. $I_{ab} = 4.5$ A$\angle 35°$; $I_{ca} = 4.5$ A$\angle 155°$

    b. $I_a = 7.79$ A$\angle 5°$; $I_b = 7.79$ A$\angle -115°$;
       $I_c = 7.79$ A$\angle 125°$

    c. $R = 43.7$ Ω; $C = 86.7\ \mu$F

21. a. 122.1 V$\angle 0.676°$

    b. 212 V$\angle 30.676°$

23. $I_a = 14.4$ A$\angle 26.9°$; $I_b = 14.4$ A$\angle -93.1°$;
    $I_c = 14.4$ A$\angle 146.9°$

25. $I_a = 14.4$ A$\angle 34.9°$

27. 611 V$\angle 30.4°$

29. b. 489 V$\angle 30°$

31. 346 V$\angle -10°$

33. $P_\phi = 86.4$ W; $Q_\phi = 0$ VAR; $S_\phi = 86.4$ VA; For totals, multiply by 3.

35. 2303 W;   1728 VAR (ind.);   2879 VA

37. 4153 W;   3115 VAR (cap.);   5191 VA

39. 99.1 kW;   29.7 kVAR (ind.);   103 kVA;   0.958

41. 27.6 kW;   36.9 kVAR (ind.);   46.1 kVA; 0.60

43. 3.93 A$\angle -32°$

45. 0.909

47. a. Same as Figure 23–28   b. 768 W   c. 2304 W

49. a. $I_a = 5.82$ A$\angle -14.0°$; $I_b = 5.82$ A$\angle -134.0°$;
       $I_c = 5.82$ A$\angle 106°$

    b. $P_\phi = 678$ W; $P_T = 2034$ W

    c. $W_1 = 1164$ W; $W_2 = 870$ W

    d. 2034 W

51. a. 0.970   b. 0.970

53. a. $\mathbf{I}_{ab} = 4\,A\angle 0°$; $\mathbf{I}_{bc} = 2.4\,A\angle -156.9°$;
 $\mathbf{I}_{ca} = 3.07\,A\angle 170.2°$; $\mathbf{I}_a = 7.04\,A\angle -4.25°$;
 $\mathbf{I}_b = 6.28\,A\angle -171.4°$; $\mathbf{I}_c = 1.68\,A\angle 119.2°$

 b. $P_{ab} = 960\,W$; $P_{bc} = 461\,W$; $P_{ca} = 472\,W$; 1893 W

55. a. $\mathbf{I}_a = 6.67\,A\angle 0°$; $\mathbf{I}_b = 2.68\,A\angle -93.4°$;
 $\mathbf{I}_c = 2.4\,A\angle 66.9°$

 b. $7.47\,A\angle -3.62°$

 c. $P_{an} = 800\,W$; $P_{bn} = 288\,W$; $P_{cn} = 173\,W$

 d. 1261 W

57. a. $\mathbf{V}_{an} = 34.9\,V\angle -0.737°$; $\mathbf{V}_{bn} = 179\,V\angle -144°$;
 $\mathbf{V}_{cn} = 178\,V\angle 145°$

 b. $\mathbf{V}_{nN} = 85.0\,V\angle 0.302°$

59. $\mathbf{I}_{ab} = 19.2\,A\angle 36.87°$; $\mathbf{I}_a = 33.2\,A\angle 6.87°$

61. $\mathbf{I}_a = 6.67\,A\angle 0°$; $\mathbf{I}_b = 2.68\,A\angle -93.4°$;
 $\mathbf{I}_c = 2.40\,A\angle 66.9°$; $\mathbf{I}_N = 7.64\,A\angle -3.62°$

CHAPTER 24

1. a. e_s is in phase with e_p.

 b. e_p is 180° out of phase with e_s.

3. a. Step-up b. $25\,\sin\omega t$ V

 c. 6 V d. $96\,V\angle 0°$

 e. $3200\,V\angle 180°$

5. $v_1 = 24\,\sin\omega t$ V; $v_2 = 144\,\sin(\omega t + 180°)$ V;
 $v_3 = 48\,\sin\omega t$ V

7. a. $1\,A\angle 20°$ b. $480\,V\angle 0°$ c. $480\,\Omega\angle -20°$

9. a. $160\,V\angle -23.1°$ b. $640\,V\angle -23.1°$

11. a. $40\,\Omega - j80\,\Omega$ b. $1.25\,\Omega + j2\,\Omega$

13. 2.5

15. a. $22\,\Omega + j6\,\Omega$ b. $26\,\Omega + j3\,\Omega$

17. 108 kVA

19. a. 20 A b. 22.5 A c. 2.5 A d. 0.708 A

21. 0.64 W

23. $3\,A\angle -50°$; $1.90\,A\angle -18.4°$; $1.83\,A\angle -43.8°$

25. b. $2.12\,A\angle -45°$; $21.2\,A\angle -45°$; $120.2\,V\angle 0°$

27. 98.5%

29. All are minus.

31. 0.889 H

33. $-125\,e^{-500t}$ V; $-4\,e^{-500t}$ V; -75.8 V; -2.43 V

35. 10.5 H

37. 27.69 mH; $11.5\,A\angle -90°$

39. $(4 + j22)\,\mathbf{I}_1 + j13\,\mathbf{I}_2 = 100\angle 0°$
 $j13\,\mathbf{I}_1 + j12\,\mathbf{I}_2 = 0$

41. $(10 + j84)\,\mathbf{I}_1 - j62\,\mathbf{I}_2 = 120\,V\angle 0°$
 $-j62\,\mathbf{I}_1 + 15\,\mathbf{I}_2 = 0$

43. $0.644\,A\angle -56.1°$; $6.44\,A\angle -56.1°$; $117\,V\angle 0.385°$

CHAPTER 25

1. a. 18.37 V

 b. 6.75 W

3. a. 17.32 V

 b. 0.12 W

5. $v(t) = 1 + \dfrac{2}{\pi}\sin\omega t + \dfrac{2}{2\pi}\sin(2\,\omega t) + \dfrac{2}{3\pi}\sin(3\omega t) + \ldots$

7. $v(t) = \dfrac{32}{\pi}\sin 500\,\pi t - \dfrac{32}{2\pi}\sin(1000\pi t) + \dfrac{32}{3\pi}\sin(1500\pi t) - \ldots$

9. $v(t) = \dfrac{10}{\pi} + 5\sin\omega t - \dfrac{20}{\pi}\left[\dfrac{\cos(2\omega t)}{3} + \dfrac{\cos(4\omega t)}{15} \ldots\right]$

11. $v(t) = \dfrac{32}{\pi}\sin(\omega t + 30°) + \dfrac{32}{3\pi}\sin[3(\omega t + 30°)] + \dfrac{32}{5\pi}\sin$
 $[5(\omega t + 30°)] + \dfrac{32}{7\pi}\sin[7(\omega t + 30°)]$

13. b. $v_1 = -\dfrac{16}{\pi}\sin\omega t - \dfrac{16}{3\pi}\sin 3\omega t - \dfrac{16}{5\pi}\sin 5\omega t - \ldots$

 $v_2 = \dfrac{8}{\pi}\sin\omega t - \dfrac{8}{2\pi}\sin 2\omega t + \dfrac{8}{3\pi}\sin 3\omega t - \ldots$

 c. $v = -\dfrac{8}{\pi}\sin\omega t - \dfrac{8}{2\pi}\sin 2\omega t - \dfrac{8}{3\pi}\sin 3\omega t - \ldots$

15. b. $V_{\text{avg}} = 5$ V

 c. $v_1 = 5 - \dfrac{10}{\pi}\sin\omega t - \dfrac{10}{2\pi}\sin 2\omega t - \dfrac{10}{3\pi}\sin 3\,\omega t - \ldots$

 $v_2 = -\dfrac{10}{\pi}\sin\omega t + \dfrac{10}{2\pi}\sin 2\omega t - \dfrac{10}{3\pi}\sin 3\omega t + \ldots$

 d. $v_1 + v_2 = 5 - \dfrac{20}{\pi}\sin\omega t - \dfrac{20}{3\pi}\sin 3\omega t - \dfrac{10}{5\pi}\sin 5\omega t - \ldots$

17. $v_1 + v_2 = \dfrac{16}{\pi}\sin\omega t - \dfrac{4}{2\pi}\sin 2\omega t + \dfrac{16}{3\pi}\sin 3\,\omega t - \ldots$

19. $P = 1.477$ W

21. $P_0 = 23.1$ dBm $P_1 = 24.0$ dBm $P_2 = 26.1$ dBm
 $P_3 = 2.56$ dBm $P_4 = -4.80$ dBm

23. a. $V_0 = 0.5$ V

 b. $V_1 = 0.90\,V_P$ $\theta_1 = -45°$
 $V_3 = 0.14\,V_P$ $\theta_2 = -63°$
 $V_5 = 0.05\,V_P$ $\theta_5 = -79°$
 $V_7 = 0.03\,V_P$ $\theta_7 = -82°$

Glossary

ac   Abbreviation for alternating current; used to denote periodically varying quantities such as ac current, ac voltage, and so on.

admittance (Y)   A vector quantity (measured in siemens, s) which is the reciprocal of impedance. **Y = 1/Z.**

alternating current   Current that periodically reverses in direction, commonly called an ac current.

alternating voltage   Voltage that periodically changes in polarity, commonly called an ac voltage. The most common ac voltage is the sine wave.

American Wire Gauge (AWG)   An American standard for classifying wire and cable.

ammeter   An instrument that measures current.

ampere (A or amp)   The SI unit of electrical current, equal to a rate of flow of one coulomb of charge per second.

ampere-hour (Ah)   A measure of the storage capacity of a battery.

angular frequency (ω)   Frequency of an ac waveform in radians/s. $\omega = 2\pi f$ where f is frequency in Hz.

apparent power (S)   The power that apparently flows in an ac circuit. It has components of real power and reactive power, related by the power triangle. The magnitude of apparent power is equal to the product of effective voltage times effective current. Its unit is the VA (volt-amp).

atom   The basic building block of matter. In the Bohr model, an atom consists of a nucleus of positively charged protons and uncharged neutrons, surrounded by negatively charged orbiting electrons. An atom normally consists of equal numbers of electrons and protons and is thus uncharged.

attenuation   The amount that a signal decreases as it passes through a system. The attenuation is usually measured in decibels, dB.

audio frequency   A frequency in the range of human hearing, which is typically from about 15 Hz. to 20 kHz.

autotransformer   A type of transformer with a partially common primary and secondary winding. Part of its energy is transferred magnetically and part conductively.

average of a waveform   The mean value of a waveform, obtained by algebraically summing the areas above and below the zero axis of the waveform, divided by the cycle length of the waveform. It is equal to the dc value of the waveform as measured by an ammeter or a voltmeter.

balanced   (1) For a bridge circuit, the voltage between midpoints on its arms is zero. (2) In three-phase systems, a load that is identical for all three phases.

band-pass filter   A circuit that permits signals within a range of frequencies to pass through a circuit. Signals of all other frequencies are prevented from passing through the circuit.

band-stop filter (or notch filter)   A circuit designed to prevent signals within a range of frequencies from passing through a circuit. Signals of all other frequencies freely pass through the circuit.

bandwidth (BW)   The difference between the half-power frequencies for any resonant, band-pass, or band-stop filter. The bandwidth may be expressed in either hertz or radians per second.

Bode plot   A straight line approximation that shows how the voltage gain of a circuit changes with frequency.

branch   A portion of a circuit that occurs between two nodes (or terminals).

branch current   The current through a branch of a circuit.

capacitance   A measure of charge storage capacity, for example, of a capacitor. A circuit with capacitance opposes a change in voltage. Unit is the farad (F).

capacitor   A device that stores electrical charges on conductive "plates" separated by an insulating material called a dielectric.

cascade   Two stages of a circuit are said to be in a cascade connection when the output of one stage is connected to the input of the next stage.

CGS system   A system of units based on centimeters, grams, and seconds.

charge   (1) The electrical property of electrons and protons that causes a force to exist between them. Electrons are negatively charged while protons are positively charged. Charge is denoted by Q and is defined by Coulomb's law. (2) An excess or deficiency of electrons on a body. (3) To store electricity as in to charge a capacitor or charge a battery.

choke   Another name for an inductor.

circuit   A system of interconnected components such as resistors, capacitors, inductors, voltages sources, and so on.

circuit breaker   A resettable circuit protection device that trips a set of contacts to open the circuit when current reaches a preset value.

circuit common   The reference point in a electrical circuit from which voltages are measured.

circular mil (CM)   A unit used to specify the cross-section area of a cable or wire. The circular mil is defined as the area contained in a circle having a diameter of 1 mil (0.001 inch).

coefficient of coupling (k)   A measure of the flux linkage between circuits such as coils. If $k = 0$, there is no linkage; if $k = 1$, all of the flux produced by one coil links another. The mutual inductance M between coils is related to k by the relationship $M = k\sqrt{L_1L_2}$, where L_1 and L_2 are the self-inductances of the coils.

coil   A term commonly used to denote inductors or windings on transformers.

conductance (G)   The reciprocal of resistance. Unit is the siemens (S).

conductor   A material through which charges move easily. Copper is the most common metallic conductor.

copper loss   The I^2R power loss in a conductor due to its resistance, for example the power loss in the windings of a transformer.

core   The form or structure around which an inductor or the coils of a transformer are wound. The core material affects the magnetic properties of the device.

core loss   Power loss in the core of a transformer or inductor due to hysteresis and eddy currents.

coulomb (C)   The SI unit of electrical charge, equal to the charge carried by 6.24×10^{18} electrons.

Coulomb's law   An experimental law which states that the force (in Newtons) between charged particles is $F = Q_1Q_2/4\pi \in r^2$, where Q_1 and Q_2 are the charges (in coulombs), r is the distance between their centers in meters, and $\in$ is the permittivity of the medium. For air, $\in = 8.854 \times 10^{-12} F/m$.

critical temperature   The temperature below which a material becomes a superconductor.

current (I or i)   The rate of flow of electrical charges in a circuit, measured in amperes.

current source   A practical current source can be modeled as an ideal current source in parallel with an internal impedance.

cutoff frequency, f_c or ω_c   The frequency at which the output power of a circuit is reduced to half of the maximum output power. The cutoff frequency may be measured in either hertz, (Hz) or radians per second, (rad/s).

cycle   One complete variation of an ac waveform.

decade   A tenfold change in frequency.

decibel (dB)   A logarithmic unit used to represent an increase (or decrease) in power levels or sound intensity.

delta (Δ)   A small change (increment or decrement) in a variable. For example, if current changes a small amount from i_1 to i_2, its increment is $\Delta i = i_2 - i_1$, while if time changes a small amount from t_1 to t_2, its increment is $\Delta t = t_2 - t_1$.

delta load   A configuration of circuit components connected in the shape of a Δ (Greek letter delta). Sometimes called a pi (π) load.

derivative   The instantaneous rate of change of a function. It is the slope of the tangent to the curve at the point of interest.

dielectric   An insulating material. The term is commonly used with reference to the insulating material between the plates of a capacitor.

dielectric constant ($\in$)   A common name for permittivity.

differentiator   A circuit whose output is proportional to the derivative of its input.

diode   A two-terminal component made of semiconductor material, which permits current in one direction while preventing current in the opposite direction.

direct current (dc)   Unidirectional current such as that from a battery.

DMM   A digital multimeter that displays results on a numeric readout. In addition to voltage, current, and resistance, some dmms measure other quantities such as frequency and capacitance.

duty cycle   The ratio of on time to the duration of a pulse waveform, expressed in percent.

eddy current   A small circulating current. Usually refers to the unwanted current that is induced in the core of an inductor or transformer by changing core flux.

effective resistance   Resistance defined by $R = P/I^2$. Effective resistance is larger than dc resistance due to skin effect and other effects such as power losses.

effective value   An equivalent dc value of a time varying waveform, hence, that value of dc that has the same heating effect as the given waveform. Also called *rms* (root mean square) value. For sinusoidal current, $I_{eff} = 0.707 I_m$, where I_m is the amplitude of the ac waveform.

efficiency (η)   The ratio of output power to input power, usually expressed as a percentage. $\eta = P_{out}/P_{in} \times 100\%$.

electron   A negatively charged atomic particle. *See* atom.

energy (W)   The ability to do work. Its SI unit is the joule; electrical energy is also measured in kilowatt-hours (kWh).

fall time (t_f)   The time it takes for a pulse or step to change from its 90% value to its 10% value.

farad (F)   The SI unit of capacitance, named in honor of Michael Faraday.

ferrite   A magnetic material made from powdered iron oxide. Provides a good path for magnetic flux and has low enough eddy current losses that it is used as a core material for high frequency inductors and transformers.

field   A region in space where a force is felt, hence a force field. For example, magnetic fields exist around magnets and electric fields exist around electric charges.

field intensity   The strength of a field.

filter   A circuit that passes certain frequencies while rejecting all other frequencies.

flux A way of representing and visualizing force fields by drawing lines that show the strength and direction of a field at all points in space. Commonly used to depict electric or magnetic fields.

free electron An electron that is weakly bound to its parent atom and is thus easily broken free. For materials like copper, there are billions of free electrons per cubic centimeter at room temperature. Since these electrons can break free and wander from atom to atom, they form the basis of an electric current.

frequency (f) The number of times that a cycle repeats itself each second. Its SI unit is the hertz (Hz).

gain The ratio of output voltage, current, or power to the input. Power gain for an amplifier is defined as the ratio of ac output power to ac input power, $A_p = P_{out}/P_{in}$. Gain may also be expressed in decibels. In the case of power gain, $A_p(dB) = 10 \log P_{out}/P_{in}$.

gauss The unit of magnetic flux density in the CGS system of units.

giga (G) A prefix with a value of 10^9.

ground (1) An electrical connection to earth. (2) A circuit common. (*See* circuit common.) (3) A short to ground, such as a ground fault.

harmonics Integer multiples of a frequency.

henry (H) The SI unit of inductance, named in honor of Joseph Henry.

hertz (Hz) The SI unit of frequency, named in honor of Heinrich Hertz. One Hz equals one cycle per second.

high-pass filter A circuit which readily permits frequencies above the cutoff frequency to pass from the input to the output of the circuit, while attenuating frequencies below the cutoff frequency. (*See* cutoff frequency).

hysteresis loss Power loss in a ferromagnetic material caused by the reversal of magnetic domains in a time varying magnetic field.

ideal current source A current source having an infinite shunt (parallel) impedance. An ideal current source is able to provide the same current to all loads (except an open circuit). The voltage across the current source is determined by the value of the load impedance.

ideal transformer A transformer having no losses and characterized by its turns ratio $a = N_p/N_s$. For voltage, $\mathbf{E}_p/\mathbf{E}_s = a$, while for current $\mathbf{I}_p/\mathbf{I}_s = 1/a$.

ideal voltage source A voltage source having zero series impedance. An ideal voltage source is able to provide the same voltage across all loads (except a short circuit). The current through the voltage source is determined by the value of the load impedance.

impedance (Z) Total opposition that a circuit element presents to sinusoidal ac in the phasor domain. $\mathbf{Z} = \mathbf{V}/\mathbf{I}$ ohms, where $\mathbf{V}$ and $\mathbf{I}$ are voltage and current phasors respectively. Impedance is a complex quantity with magnitude and angle.

induced voltage Voltage produced by changing magnetic flux linkages.

inductance (L) That property of a coil (or other current-carrying conductor) that opposes a change in current. The SI unit of inductance is the henry.

inductor A circuit element designed to posses inductance, e.g., a coil of wire wound to increase its inductance.

instantaneous value The value of a quantity (such as voltage or current) at some instant of time.

insulator A material such as glass, rubber, bakelite, and so on, that does not conduct electricity.

integrator A circuit whose output is proportional to the integral of its input.

internal impedance The impedance that exists internally in a device such as a voltage source.

ion An atom that has become charged. If it has an excess of electrons, it is a negative ion, while if it has a deficiency, it is a positive ion.

joule (J) The SI unit of energy, equal to one newton-meter.

kilo A prefix with the value of 10^3.

kilowatt-hour (kWh) A unit of energy equal to 1000 W times one hour and commonly used by electrical utilities.

Kirchhoff's current law An experimental law which states that the sum of the currents entering a junction is equal to the sum leaving.

Kirchhoff's voltage law An experimental law that states that the algebraic sum of voltages around a closed path in a circuit is zero.

lagging load A load in which current lags voltage (e.g., an inductive load).

leading load A load in which current leads voltage (e.g., a capacitive load).

linear circuit A circuit in which relationships are proportional. In a linear circuit, current is proportional to voltage.

load (1) The device that is being driven by a circuit. Thus, the lamp in a flashlight is the load. (2) The current drawn by a load.

low-pass filter A circuit that permits frequencies below the cutoff frequency to pass through from the input to the output of the circuit, while attenuating frequencies above the cutoff frequency. (*See* cutoff frequency.)

magnetic flux density (B) The number of magnetic flux lines per unit area, measured in the SI system in tesla (T), where one T = one Wb/m^2.

magnetomotive force (mmf) The flux producing ability of a coil. In the SI system, the mmf of a coil of N turns with current I is NI ampere-turns.

maxwell (Mx) The CGS unit of magnetic flux Φ.

mega (M) A prefix with the value of 10^6.

micro (μ) A prefix with the value of 10^{-6}.

milli (m) A prefix with the value of 10^{-3}.

multimeter   A multifunction meter used to measure a variety of electrical quantities such as voltage, current, and resistance. Its function and range is selected by a switch. (*See also* DMM.)

mutual inductance (M)   The inductance between circuits (such as coils) measured in henries. The voltage induced in one circuit by changing current in another circuit is equal to M times the rate of change of current in the first circuit.

nano (n)   A prefix with the value of 10^{-9}.

neutron   An atomic particle with no charge. (*See* atom.)

node   A junction where two or more components connect in an electric circuit.

ohm (Ω)   The SI unit of resistance. Also used as the unit for reactance and impedance.

ohmmeter   An instrument for measuring resistance.

open circuit   A discontinuous circuit, hence one that does not provide a complete path for current.

oscilloscope   An instrument that electronically displays voltage waveforms on a screen. The screen is ruled with a scaled grid to permit measurement of the waveform's characteristics.

parallel   Elements or branches are said to be in a parallel connection when they have exactly two nodes in common. The voltage across all parallel elements or branches is exactly the same.

peak   The maximum instantaneous value (positive or negative) of a waveform.

peak-to-peak   The magnitude of the difference between a waveform's maximum and minimum values.

period (T)   The time for a waveform to go through one cycle. $T = 1/f$ where f is frequency in Hz.

periodic   Repeating at regular intervals.

permeability (μ)   A measure of how easy it is to magnetize a material. $B = \mu H$, where B is the resulting flux density and H is the magnetizing force that creates the flux.

permittivity ($\in$)   A measure of how easy it is to establish electric flux in a material. (*See also* relative dielectric constant and Coulomb's law.)

phase shift   The angular difference by which one waveform leads or lags another, hence the relative displacement between time varying waveforms.

phasor   A way of representing the magnitude and angle of a sine wave graphically or by a complex number. The magnitude of the phasor represents the rms value of the ac quantity and its angle represents the waveform's phase.

pico (p)   A prefix with the value of 10^{-12}.

potentiometer   A three-terminal resistor consisting of a fixed resistance between two end terminals and a third terminal that is connected to a movable wiper arm. When the end terminals are connected to a voltage source, the voltage between the wiper and either of the other terminals is adjustable.

power (P, p)   The rate of doing work, with units of watts, where one watt equals one joule per second. Also called real or active power.

power factor   The ratio of active power to apparent power, equal to cos θ, where θ is the angle between the voltage and the current.

power triangle   A way to represent the relationship between real power, reactive power, and apparent power using a triangle.

primary   The winding of a transformer to which we connect the source.

proton   A positively charged atomic particle. (*See* atom).

pulse   A short duration voltage or current that abruptly changes from one value to another, then back again.

pulse width   The duration of a pulse. For non-ideal pulses, it is measured at the 50% amplitude point.

quality factor (Q)   (1) A figure of merit. Q for a coil is the ratio of its reactive power to its real power. The higher the Q, the more closely the coil approaches the ideal. (2) A measure of the selectivity of a resonant circuit. The higher the Q, the narrower the bandwidth.

reactance (X)   The opposition that a reactive element (capacitance or inductance) presents to sinusoidal ac, measured in ohms.

reactive power   A component of power that alternately flows into then out of a reactive element, measured in VARs (volt-amps reactive). Reactive power has an average value of zero and is sometimes called "wattless" power.

rectifier   A circuit, generally consisting of a least one diode, which permits current in only one direction.

regulation   The change in voltage from no-load to full-load expressed as a percentage of full load voltage.

relative dielectric constant ($\in_r$)   The ratio of the dielectric constant of a material to that of a vacuum.

relay   A switching device that is opened or closed by an electrical signal. May be electromechanical or electronic.

reluctance   The opposition of a magnetic circuit to the establishment of flux.

resistance (R)   The opposition to current that results in power dissipation. Thus, $R = P/I^2$ ohms. For a dc circuit, $R = V/I$, while for an ac circuit containing reactive elements, $R = V_R/I$, where V_R is the component of voltage across the resistive part of the circuit.

resistor   A circuit component designed to posses resistance.

resonance, resonant frequency   The frequency at which the output power of an L-R-C circuit is at a maximum. $f = 1/(2\pi\sqrt{LC})$

rheostat   A variable resistor connected so that current through the circuit is controlled by the position of the wiper.

rise time (t_r)   The time that it takes for a pulse or step to change from its 10% value to its 90% value.

rms value   The root-mean-square value of a time varying waveform. (*See* effective value.)

saturation   The condition of a ferromagnetic material where it is fully magnetized. Thus, if the magnetizing force (current in a coil for example) is increased, no significant increase in flux results.

schematic diagram   A circuit diagram that uses symbols to represent physical components.

secondary winding   The output winding of a transformer.

selectivity   A measure of the ability of a resonant circuit to select a very narrow band of frequencies and reject all others. The higher the Q, the narrower the bandwidth and hence, the greater the selectivity.

semiconductor   A material such as silicon from which transistors, diodes, and the like are made.

series circuit   A closed loop of elements where two elements have no more than one common terminal. In a series circuit, there is only one current path and all series elements have the same current.

short circuit   A short circuit occurs when two terminals of an element or branch are connected together by a low-resistance conductor. When a short circuit occurs, very large currents may result in sparks or a fire, particularly when the circuit is not protected by a fuse or circuit breaker.

SI System   The international system of units used in science and engineering. It is a metric system and includes the standard units for length, mass, and time (e.g., meters, kilograms, and seconds), as well as the electrical units (e.g., volts, amperes, ohms, and so on).

siemens (S)   A unit of measure for conductance, admittance, and susceptance. The siemens is the reciprocal of ohm.

sine wave   A periodic waveform that is described by the trigonometric sine function. It is the principle waveform used in ac systems.

skin effect   At high frequencies, the tendency of current to travel in a thin layer near the surface of a conductor.

steady state   The condition of operation of a circuit after transients have subsided.

step   An abrupt change in voltage or current, as for example when a switch is closed to connect a battery to a resistor.

superconductor   A conductor that has no internal resistance. Current will continue unimpeded through a superconductor even though there is no externally applied voltage or current source.

susceptance   The reciprocal of reactance. Unit is the siemens.

tank circuit   A circuit consisting of an inductor and capacitor connected in parallel. Such an *L-C* circuit is used in oscillators and receivers to provide maximum signal at the resonant frequency. (*See* selectivity.)

temperature coefficient   (1) The rate at which resistance changes as the temperature changes. A material has a positive temperature coefficient if the resistance increases with an increase in temperature. Conversely, a negative temperature coefficient means that resistance decreases as temperature is increased. (2) Similarly for capacitance. The change in capacitance is due to changes in the characteristics of its dielectric with temperature.

tesla (T)   The SI unit of magnetic flux density. One T = one Wb/m².

time constant (τ)   A measure of how long a transient lasts. For example, during charging, capacitor voltage changes by 63.2% in one time constant, and for all practical purposes, charges fully in five time constants, For an *RC* circuit, $\tau = RC$ seconds and for an *RL* circuit, $\tau = L/R$ seconds.

transformer   A device with two or more coils in which energy is transferred from one winding to the other by electromagnetic action.

transient   A temporary or transitional voltage or current.

turns ratio (a)   The ratio of primary turns to secondary turns; $a = N_p/N_s$.

valence shell   The outermost shell of an atom.

volt   The unit of voltage in the SI system.

voltage (V, v, E, e)   Potential difference created when charges are separated, as for example by chemical means in a battery. If one joule of work is required to move a charge of one coulomb from one point to another, the potential difference between the points is one volt.

voltage source   A practical voltage source can be modeled as an ideal voltage source in a series with an internal impedance.

watt (W)   The SI unit of active power. Power is the rate at which work is done; one watt equals one joule/s.

watthour (Wh)   A unit of energy, equal to one watt times one hour. One Wh = 3600 joules.

waveform   The variation versus time of a time varying signal, hence, the shape of a signal.

weber (Wb)   The SI unit of magnetic flux.

work (W)   The product of force times distance, measured in joules in the SI system, where one joule equals one newton-meter.

wye load   A configuration of circuit components connected in the shape of a Y. Sometimes called a star or T load.

Index

Note: Page numbers in **bold type** reference non-text material.

A

ABC phase sequence, 939
Absolute dielectric constant, 389–90
Absolute zero, defined, 70
AC (Alternating current), 40–41
    amplitude, 559
    direct current superimposed, 584
    effective values and, 585–89
    energy relationships for, 663–64
    frequency, 557
    measurement, 590–92
    peak value, 559–61
    peak-to-peak value, 559
    period, 557–59
    phasors and, 570–79
        phase difference, 574–78
    RMS values and, 588–89
    sine waves, angular and graphical relationships for, 561–65
    sinusoidal, 551
    source, symbol for, 550
    symbol for, 130
    voltage, generating, 551–54
    voltage and current,
        conventions of, 554–57
        as functions of time, 565–70
    waveforms, average value and, 579–85
Ac circuits
    computer circuit analysis, Thévenin or Norton equivalent, 818–24
    computer circuit analysis of, 592–95, 630–34, 664–65, 768–71
    measuring power in, 659–61
    mesh (loop) analysis of, 744–51
    nodal analysis of, 751–58
    Ohm's law for, 674–81
        capacitors and, 679–81
        inductors and, 677–78
    Ohm's law for resistors and, 674–77
    parallel. *See* Ac parallel circuits
    series. *See* Ac series circuits

Ac parallel circuits, 693–98
Ac resistance, 114
Ac series circuits, 681–89
Ac voltage
    generating, 551–54
    measurement, 590–92
        average responding meter and, 590
        representing by complex numbers, 611–12
    summing, 612–16
Active power, 642–43
    equations, 654–55
Addition, using powers of ten, 11
Admittance diagram, 693
Air gaps, magnetic circuits and, 468
Air-core
    coils, voltages in, 1011
    inductors, 497–98
    resistors, symbol for, **505**
    transformer, 982
        symbol for, **982**
Alkaline batteries, 41
Alloy resistors, 73–74
Alternating current (Ac). *See* Ac (Alternating current)
American Wire Gauge (AWG), 62
Ammeters, 46
    design, 198–200
    loading effects, 158–59
    symbols for, **50**
Ampere, 39–40
    hour rating, 43
Ampère, André Marie, 327
Ampere's circuital law, 474–75
Amplifier, defined, 884
Amplitude, waveform, 559
Analog
    multimeter, reading, 49
    ohmmeter, 75, **76**
Angular velocity, sine wave, 562–63
Apparent power, 650–51
    power triangle and, 651–55
Area, effect of, on capacitance, 387
Armstrong, Edwin Howard, 836
Asymptotes, defined, 897

Atom, 31
    polarized, 393
    representation of, **31**
Atomic theory, review of, 30–35
Attenuation, 884
Attenuator, defined, 885
Autopolarity, 48
Autotransformers, 999–1002
Average power, 644
Average responding meter, 590
Average values
    Ac waveforms, 579–80
        area under a curve, 580–82
        numerical method of approximating, 582–84
        sine wave average, 582
AWG (American Wire Gauge), 62

B

Back voltage, 496–97
Balanced bridge, defined, 303–304
Balanced delta load, power to a, 957–58
Balanced system, power in, 954–59
Balanced wye load, active power to, 955–57
Band frequencies, defined, 844
Band-pass filter, 914–18
Band-reject filter, 918–20
Bandwidth
    defined, 844
    series resonant circuit and, 842–50
Bardeen, John, 783
Base, defined, 10
BASIC programming language, 20
Batteries
    capacity, defined, 43
    described, 41
    primary, 41
    secondary, 41
    in series/parallel, 44
    symbol for, **38**
    types/applications of, 41–43
Bel, defined, 885
Bell, Alexander Graham, 883

B–H
  curves, 472–73
  relationship between, 472
Bilateral
  defined, 265
  network, defined, 333
Block diagrams, 16
Bode plots, defined, 897
Boundary conditions, defined, 902
Branch
  defined, 222
  superposition theorem and, 328–32
Branch-circuit analysis, Kirchhoff's laws
    and, 275–79
Brattain, Walter, 783
Break frequency, defined, 895
Bridge networks, 303–12, 762–67
  balanced, 303–304
  Hay bridge, 766
  Maxwell bridge, 764–66
  Schering bridge, 766–67
  unbalanced, 305–12
Bunsen, R.W., 129

C

C programming language, 20
Capacitance, 386–87, 393, 629
  of a capacitor, 387
  defined, 385, 386–87
  energy and power for alternating cur-
      rent, 664
  factors affecting, 387–90
    area, 387
    dielectric, 388–89
    dielectric constant, 389–90
    space, 388
  parasitic, 506
  sinusoidal alternating current and,
      623–27
  stray, 506
  see also Capacitor(s)
Capacitance circuit, phase load in,
    623–24
Capacitive
  load, power to, 646–48
  loading, 444
  reactance,
    defined, 624–25
    variation of with frequency, 626
Capacitor(s)
  capacitance of, 387
  charging, 418–19
    equations, 422–27, 430–38
    current/voltage and, 404–407

defined, 385
  dielectric absorption and, 395
  discharge, equations, 429–30
  discharging, 419–21
  energy stored by, 407
  equivalent series resistance (ESR) and,
    395
  failures, 408–409
  with an initial voltage, 427–28
  leakage current and, 394
  nonideal effects and, 394–95
  Ohm's law and, 679–81
  in parallel, 400–401
  parallel-plate,
    capacitance of, 389
    field of, 392–93
  in series, 401–403
    voltage divider rule for, 403–404
  steady state conditions, 419
  temperature coefficient and, 395
  testers, 409
  troubleshooting, 408–409
  types of, 395–99
    ceramic, 396
    electrolytic, 397–98
    mica, 397
    plastic film, 396–97
    surface mount, 398
    variable, 398
  v-i relationship, 405–407
  voltage rating, 394
  see also Capacitance
Capacity, battery, defined, 43
Carbon film resistor, 73
Carbon-Zinc batteries, 42
  capacity-current drain of selected, **43**
Cavendish, Henry, 641
CDR (Current divider rule)
  Kirchhoff's current law and, 698–701
  parallel circuits and, 190–95
Cell, symbol for, **38**
Ceramic capacitors, 396
Charged, defined, 386
Charge(s)
  conductors and, 34
  Coulomb unit and, 35
  Coulomb's law and, 33
  electrical, 32
  insulators and, 34
  semiconductors and, 34
  symbol for, 386
Charging
  capacitor(s), 418–19
    equations, 422–27

Chassis ground, symbol for, **142**
Circuit analysis
  beginnings of, 60
  computer-aided, 115–20
    ac circuits, 592–95, 630–34,
      664–65, 768–71
    coupled circuits, 1019–22
    filter design and, 920–24
    frequency spectrum, 1056–58
    multiloop circuits, 312–15
    network theorem verification and,
      364–70
    parallel circuits, 203–207
    resonant circuits, 867–71
    series circuits, 160–63
    series-parallel circuits, 714–20
    three-phase systems, 967–70
  linear bilateral networks, Norton's the-
      orem and, 341–51
  series-parallel circuits, 247–54
  using computers, 18–20
Circuit breakers, 51, **52**
Circuit connections, three-phase, 937–40
Circuit diagrams, 16–18
  block, 16
  pictorial, 17
  schematic, 17–18
Circuit ground, 142–43
  symbol for, **142**
Circuit Maker, 19
Circuit response, to nonsinusoidal wave-
    forms, 1052–56
Circuit simulation software, 18–20
Circuit theory, equations of, **30**
Circuit voltage, during energization,
    524
Circuits
  Ac parallel. *See* Ac parallel circuits
  Ac series. *See* Ac series circuits
  inductive, interrupting current in,
    527–29
  initial condition, 521–22
  single source, reciprocity theorem and,
    361–63
  time constant of, 525–27
Circular mil, **66**
  defined, 65
  wire resistance and, 65–68
Closed loop, defined, 132
Coefficient of coupling, 1013
Coil
  flux linkage of, 497
  resistor, resistance of, 505–506
  self-inductance of, 498

Common waveforms, Fourier series and, 1039–46
Complex numbers
    in Ac analysis, 611–17
        calculators for, 610
    addition/subtraction of, 608
    conjugates of, 610
    conversion between forms, 606–608
    defined, 606
    geometrical representation of, 606
    multiplication/division of, 609
Components, series, interchanging, 138
Composite waveforms, 1032–33
Computer-aided circuit analysis, 18–20, 115–20
    ac circuits, 592–95, 630–34, 664–65, 768–71
        Thévenin or Norton equivalent, 818–24
    coupled circuits, 1019–22
    filter design and, 920–24
    frequency spectrum, 1056–58
    multiloop circuits, 312–15
    network theorem verification and, 364–70
    parallel circuits, 203–207
    resonant circuits, 867–71
    series-parallel circuits, 247–50, 714–17, 714–20
    solving *RL* transients, 537–41
    three-phase systems, 967–70
    transient analysis with, 444–50
Conductance, 85–86
    defined, 85
    symbol for, 85
Conductor(s), 34
    with circular cross-section, **61**
    with rectangular cross-section, **61**
    resistivity of, **68**
Conservation of energy, law of, 110–11
Constant current source, 266–68
Constant dc current, symbol for, **507**
Constantum resistors, 74
Continuity of current for inductance, 520
Convention for coupled circuits, 988
Conventional current direction, 40
Conventional flow, **130**
Conversion factors, defined, 9
Copper wire, solid, table, 63
Corresponding terminals, 988
Cosine symmetry, 1037
Coulomb, defined, 35
Coulomb's law, 33
    constant in, 391
Counter EMF, 496–97

Coupled circuits, computer-aided circuit analysis, 1019–22
Coupled impedance, 1017–19
Critical temperature
    superconductors and, 87
    symbol for, 87
Current divider rule (CDR), 190–95
    Kirchhoff's current law and, 698–701
Current gun, 591
Current ratio, iron-core transformers, 986–91
Current source(s)
    constant, 266–68
    ideal,
        defined, 268
        symbol for, 266
    in parallel, 272–75
    in series, 272–75
Current(s), 38–41
    ampere and, 39–40
    buildup transients, 523–27
    capacitors and, 404–407
    continuity of, 520
    direction, 40, 101–104
    discontinuous, 418
    during energization, 523
    as functions of time, 565–70
    interrupting, in inductive circuits, 527–29
    measuring, 46–50
        with multimeter, 49
    with phase shift, 568–70
    representing by complex numbers, 611–12
    sources of, reciprocity theorem and, 362–63
    summing, 612–16
    superposition theorem and, 328–32
    time-varying, symbol for, 405
    value/direction, indicating, 103–104
Current-to-voltage converter, 591
Cutoff frequencies, defined, 844, 895
Cycle, defined, 550

D
DC (Direct current)
    alternating current superimposed, 584
    constant, symbol for, **507**
    generators, 46
    sources of, 41–46
    symbol for, 38, 130
de Forest, Lee, 836
Decade, defined, 897
Decibel (dB), 884–90
    application of, 888–90

defined, 883
De-engerizing transients, 529–31
Defined formulas, **30**
Delta-connected generators, 938–39
Delta load
    line and phase currents for, 945
    power to a balanced, 957–58
    unbalanced, 965–67
Delta-to-wye conversions, 296–303, 758–62
Demagnetization process, 486
Dependent sources, 738–39
    Norton's theorem, 803–13
    superposition theorem, 789–91
    Thévenin's theorem, 803–13
Derived formulas, **30**
Diagrams, circuit, 16–18
Dielectric absorption, capacitors and, 395
Dielectric(s), 393–94
    breakdown, 393
    constant, 389–90
    defined, 386
    effect of, on capacitance, 388–89
    strength of, 393
Digital multimeter (DMM), voltage/current measurement with, 47–48
Digital ohmmeter, 75, **76**
Diodes, 83–84
    defined, 83
Direct current (DC)
    generators, 46
    sources of, 41–46
    symbol for, 38, 130
Discharge, described, 386
Discharging
    capacitor(s), 419–21
    equations, 429–30
Division, using powers of ten, 11
DMM (Digital multimeter), voltage/current measurement with, 47–48
Dot rule, 1012–13
Double subscripts, 143–46
Dynamic resistance, 114–15

E
Earth ground, defined, 143
Eddy currents, 662
    core loss and, 1003–1004
Edison, Thomas Alva, 549, 935
Effective resistance, 662–63
Effective values
    alternating current, 585–89
    defined, 585
    general equations for, 587–88
    for sine waves, 585–88

Efficiency, 111–13
Electric cell, 95
Electric circuit, defined, 130
Electric current, defined, 39
Electric field, 385
    intensity, 391
Electric fields, 390–93
    defined, 390
Electric flux, 390, 392
    density, 391
        symbol for, 391
    lines, symbol for, 391
Electric generator, 265
Electrical charge. *See* Charges
Electrical circuit analysis. *See* Circuit
        analysis
Electrical systems, power in, 105–106
Electrolytic capacitors, 397–98
Electromagnetic induction, described,
        494–96
Electromagnetics
    André Marie Ampère and, 327
    force due to, 484–85
Electromagnetism, 464
Electron flow direction, 40
Electronic power supplies, 44–45
Electronic signal generators, 554
Electronic systems, power in, 105–106
Electronics Workbench (EWB), 19,
        115–17
    ac circuits, 592–93, 630–31, 768
        Thévenin or Norton equivalent of,
            822–24
    filter design and, 920–24
    multiloop circuits, 312–14
    parallel circuits, 203–205
    series circuits, 160
    series-parallel circuits, 247–50,
        714–17
    solving *RL* transients with, 537–39
    three-phase systems, 968–69
    transient analysis using, 444–47
Electrons, 31
    free, 33
Energy
    formula for, 109
    stored by inductance, 509–10
Engineering notation, 13–14
English system of units, 7–8
    converting, 9–10
    wire tables and, 62
Equipment, reasons why rated in VA, 656
Equivalent series resistance (ESR) and,
        capacitors and, 395
ESR (Equivalent series resistance) and,
        capacitors and, 395

Even symmetry, 1037
EWB (Electronics Workbench). *See*
        Electronics Workbench (EWB)
Experimental formulas, **30**
Exponent, defined, 10
Exponential functions, 421

F

Fall times, 441
Farad
    Michael Faraday and, 385
    symbol for, 387
Faraday, Michael, 385, 461
    electromagnetic induction and, 494
    field concept of, 385, 391
Faraday's law, 495
    transformers and, 983
Farads, per meter, symbol for, 389
Ferrite-core
    inductors, symbol for, **505**
    transformer, 982
        symbol for, **982**
Ferromagnetic materials, 462–64
Fessenden, Reginald, 673
Field, concept of, 385, 391
Field intensity, electric, 391
Film resistors, 73
Film/foil capacitors, 396–97
Fixed resistors, 72–75
    defined, 72
Flux, electric, symbol for lines of, 391
Flux
    density, 465–66
    electric, 390, 392
        density, 391
    lines, 385, 462
        symbol for, 391
    linkage, of a coil, 497
Force, lines of, 385
FORTRAN programming language, 20
Forward biased diode, 84
Forward region, diodes and, 84
Fourier, Jean Baptiste Joseph, 1032
Fourier series, 1034–39
    of common waveforms, 1039–46
Fourier's theorem, 1032
Four-wire
    systems, 937–38
    wye-wye (Y-Y) circuit, 938
Franklin, Benjamin, 221
Free electrons, 33
Frequency
    defined, 557
    effects of, 704–10
        RC circuits and, 705–707
        RL circuits, 707–708

RLC circuits, 708–10
    on transformers, 1009–10
    inductive reactance variation with,
        622
    spectrum, 1046–52
Fringing, 468
Fundamental frequency, defined, 1034
Fuses, 51, **52**

G

Galvani, Luigi, 177
Galvanometer, 177
Generators
    delta-connected, 938–39
    direct current (DC), 46
    electric, 265
    signal electronic, 554
Ground, defined, 142

H

Half-power frequencies, defined, 844
Half-wave symmetry, 1038
Hall, E.H., 487
Hall effect, 487
    gaussmeters, 487
Harmonic frequency, defined, 1034–35
Hay bridge, 766
Heat transfer, Fourier and, 1032
Henry, defined, 493
Hertz, Heinrich Rudolph, 673
Hertz (Hz), defined, 557
High selectivity circuit, defined, 844
High-pass filter, 909–14
Hysteresis, 662
    core loss and, 1003–1004

I

Ideal current source
    defined, 268
    symbol for, **266**
Impedance
    coupled, 1017–19
    defined, 627, 681
    diagram, defined, 681
    matching, 996–97
    reflected, 992–93
    of series resonant circuit, 841–42
    three in parallel, 696–98
In phase, defined, 574
Inch, defined, 8
Independent sources, 738
    Norton's theorem, 797–803
    Thévenin's theorem, 791–96
Induced voltage, 496–98
    computing, 501–503
    polarity of, 987–89

Inductance
  concept of, 627–30
  defined, 627
  effect of, 519
  energy and power for alternating current, 663–64
  energy stored by, 509–10
  inductor,
    core type and, 504–505
    variable and, 505
  open-circuit equivalent of, 521
  in series and parallel, 503–504
  and sinusoidal Ac, 619–22
  steady state DC and, 507–509
  stray, 506
Induction, 496–98
Inductive circuit
  interrupting current in, 527–29
  phase lag in, 619–20
Inductive kick, 528
Inductive load, power to, 644–46
Inductive reactance, 620–22
  variations of, with frequency, 622
Inductor voltage, 521
Inductors
  air-core, 497–98
  circuit symbols, 505
  coil resistance, 505–506
  core types, inductance and, 504–505
  defined, 493
  equivalent at switching, 529
  iron-core, 497–98
  with mutual couplings, 1014–15
  Ohm's law and, 677–78
  troubleshooting hints, 510–11
Inferred absolute temperature, defined, 70
Initial condition circuits, 521–22
Instantaneous power, defined, 642
Instantaneous value, defined, 553
Instruments, loading effects of, 241–47
Insulators, 34
Integrated circuit packages, 73
Integrated resistor network, **74**
Internal resistance, voltage sources, 149–55
Ions, 33
Iron-core inductors, 497–98
  symbol for, **505**
Iron-core transformers, 980, 983–91
  core loss, 1003–1004
  current ratio, 986–91
  leakage flux and, 1002–1003
  symbol for, **982**
  voltage ratio, 984–85
  voltage regulation, 1005–1007

winding resistance, 1003
  see also Transformers

J
Joule, defined, 8

K
Kick test, 987
Kilogram, defined, 8
Kilowatthours, symbol for, 109
Kirchhoff, Gustav Robert, 129
Kirchhoff's circuit law
  multiple source circuits, 275–79
  nodal analysis and, 288–96
Kirchhoff's current law
  current divider rule and, 698–701
  parallel circuits and, 179–83
  shunt resistor and, 198
Kirchhoff's voltage law, 132–33
  mesh (loop) analysis and, 280–87
  multiple source circuits, 275–79
  resistors in parallel and, 183–88
  symbol for, 132

L
Lag, phasors and, 574
Lagging power factor, 655, 687
Lead, phasors and, 574
Lead-Acid batteries, 43
Leading power factor, 655, 687
Leakage flux, iron-core transformers, 1002–1003
Lenz's law, 496
Leyden jar, 221
Line, defined, 938, 941
Line currents
  defined, 941
  for a delta load, 945
Line voltages, for a wye circuit, 941–42
Linear
  defined, 265
  resistance, 114
Linear bilateral network
  analysis, 280–87
    Norton's theorem and, 341–51
    Thévenin's theorem and, 332–41
Linear network, defined, 333
Lines of force, 385, 462
Line-to-line, defined, 941
Line-to-neutral voltages, 938
Lithium batteries, 42–43
Load
  power to,
    capacitive, 646–48
    inductive, 644–46
    resistive load, 643–44

Loaded output, 239
Loading effects
  defined, 241
  of instruments, 241–47
  voltmeter, 200–203
Loads, symbols for, 100
Loop analysis. *See* Mesh (loop) analysis
Loosely coupled circuits, 983, 1010–15
Low-pass filter, 902–909
  defined, 895
  frequency response of, **895**
Low power factor wattmeter, 661
Low selectivity circuit, defined, 844

M
Magnet, poles of, 462
Magnetic circuits, 467–68
  air gaps and, 468
  Ampere's circuital law and, 474–75
  with DC excitation, 470–71
  fringing and, 468
  laminated cores and, 469
  Ohm's law for, 471
  parallel, 470
  series, 469
    given NI, find ϕ, 482–84
    given ϕ, find NI, 475–80
  series-parallel, 480–81
Magnetic fields
  density of, 471–74
  described, 462–64
  measuring, 487
Magnetic flux, 462, 465–66
  opposition to, 470
  source of, 470
Magnetic materials, 462–64
  properties of, 485–86
Magnetically coupled circuits, sinusoidal excitation with, 1015–17
Magnetically coupled device, defined, 980
Magnetization curves, 471–74
Magnetizing
  current, 1004
  force, 471–74
Magnetomotive force (mmf), defined, 470
Make-before-break switch, 200
Manganin resistors, 74
Marconi, Guglielmo, 673
Math, prepackaged software, 20
Maximum power transfer theorem, 352–57, 813–17
Maxwell, James Clerk, 461, 641
Maxwell bridge, 764–66

Megajoule, symbol for, 110
Meissner, Walther, 87
Meissner effect, 87
Mesh (Loop) analysis, 280–87
    ac analysis and, 744–51
    format approach for, 284–87
Metal film resistors, 73
Metal oxide resistors, 73
Metallized-film capacitors, 396
Meter, defined, 7–8
Meter loading, defined, 158
Metric system of units, 7–8
    unit sizes, **7**, 8
    wire tables and, 62
Mica capacitors, 397
Millman's theorem, 359–61
mmf (Magnetomotive force), 470
Molded carbon composition resistor,
    72
Multimeter, 46
    analog, reading, 49
    current measurement, 47–48
    current select, 48
    voltage measurement, 47–48
    voltage select, 48
Multimeters, symbols, 50
Multiplication, using powers of ten,
    11
Multistage systems, 890–93
Mutual
    conductance, 292–93
    couplings, inductors with, 1014–15
    distance, 1011
    voltage, 1011
Mutually induced voltage, 1011

N

n equal resistors in parallel, 185
Negative charged, defined, 386
Negative temperature coefficient
    capacitors and, 395
    defined, 69
Network analysis, linear bilateral net-
    work, Thévenin's theorem and,
    332–41
Network theorems
    maximum power transfer theorem,
        352–57
    Millman's theorem, 359–61
    Norton's theorem, 341–51
    reciprocity theorem, 361–63
    substitution theorem, 357–58
    superposition theorem, 328–32
    Thévenin's theorem, 332–41
Network(s)

internal resistor, **74**
    potential difference analysis in, 288–96
        format approach, 292–96
Neutral, defined, 937
Neutral-Neutral voltage wye-wye circuit,
    939
Newton, defined, 8
Nickel chromium resistors, 73–74
Nickel-cadmium batteries, 43
Nodal analysis
    of ac circuits, 751–58
    Kirchhoff's circuit law and, 288–96
        format approach, 292–96
Nodes, defined, 178
Nominal voltages, 942–43
Nonlinear resistance, 83–84, 114–15
Nonohmic devices, defined, 83
Nonsinusoidal waveforms
    circuit response to, 1052–56
    see also Waveforms
Normal magnetization curve, 486
Normalized plot, defined, 903
Norton resistance, maximum power
        transfer theorem and, 352
Norton's theorem, 341–51
    computer-aided, 818–24
    dependent sources, 803–13
    independent sources, 797–803
Notation, power of ten, 10–12
Nucleus, atom, 31
Numerical accuracy, significant digits
        and, 14–16

O

Octave, defined, 897
Odd symmetry, 1037
Oersted, Hans Christian, 461
Ohm, Georg Simon, 60, 95
Ohmic device, defined, 83
Ohmic resistance, 114
Ohmmeter, 78–81
    defined, 78
    design of, 155–57
    testing, capacitors with,
        408–409
Ohms, symbol for, 60
Ohm's law, 95, 96–100
    Ac circuits and, 674–81
        capacitors and, 679–81
        inductors and, 677–78
        resistors and, 674–77
    in graphical form, 100
    magnetic circuits and, 471
    plotting, PSpice, 119
    source conversion and, 740

Open circuit(s), 100
    defined, 81
    equivalent of an inductance,
        521
    illustrated, **80**
Open coil, troubleshooting, 510
Open-circuit test, transformer,
        1008–1009
OrCAD PSpice, 117–20
    Ac circuits, 592–93, 631–34, 769–71
    filter design and, 920–24
    frequency spectrum, 1056–58
    multiloop circuits, 313–15
    network theorem verification and,
        368–70
    parallel circuits, 205–207
    resonant circuits, 867–71
    series circuits, 161–63
    series-parallel circuits, 251–54,
        714–20
    solving RL transients with, 539–41
    Thévenin or Norton equivalent of,
        818–22
    three-phase system, 969
    transient analysis using, 447–50
Oscilloscopes, alternating current mea-
        surement, 591
Out of phase, defined, 574

P

Padder capacitors, 398
Parallel cells, 44
Parallel circuits, 177, 178–79
    analysis of, 195–97
    computer analysis of, 203–207
Parallel elements, 176
Parallel inductance, 503–504
Parallel magnetic circuits, 470
Parallel networks, current divider rule
        and, 190–95
Parallel resonance, 857–66
Parallel voltage sources, Millman's theo-
        rem and, 359–61
Parallel-plate capacitor
    capacitance of, 389
    field of, 392–93
Parasitic capacitance, 506
Peak value, waveform, 559–61
Peak-to-peak
    value, waveform, 559
    voltage, 559
Period, defined, 440
Periodic waveform, defined, 550
Permeability, property of core material,
        470

Permanent-magnet, moving coil (PMMC), 198
Permittivity, defined, 389
Phase conductors, defined, 938
Phase currents
  defined, 941
  for a delta load, 945
Phase difference, 574–78
  defined, 574
Phase impedance, defined, 941
Phase lag, in inductive circuits, 619–20
Phase load, in capacitance circuits, 623–24
Phase sequence, 939–40
Phase voltages
  defined, 941
  for a wye circuit, 941–42
Phasors, 570–79
  defined, 570
  as shifted sine waves, 572–73
Photocells, 82
  defined, 178
  symbol for, **83**
Photoconductive cells, 82, **83**
Pictorial diagrams, 17
Pi-tee conversion, 296–303
Plastic film capacitors, 396–97
PMMC (Permanent–magnet, moving coil), 198
Point sources, voltage, 148
Polarity
  alternating current, 550–51, 554
  of induced voltage, 987–89
  voltage, 101, 131
Polarized atoms, 393
Positive charged, defined, 386
Positive phase sequence, 939–40
Positive temperature coefficient
  capacitors and, 395
  defined, 69
Potential difference analysis, in networks, 288–96
Potential energy, 37
  see also Voltage
Potentiometers, 239–41
  defined, 75
Power, 104–107
  ac circuits, measuring power in, 659–61
  active, equations, 654–55
  active/real, 642–43
  to an inductive load, 644–46
  apparent, 650–51
    power triangle and, 651–55
  average, 644
  to a capacitive load, 646–48

in complex circuits, 648–50
defined, 104
direction convention, 107–108
  symbol for positive, 107
instantaneous, 642
measuring, 107
reactive, 643
  defined, 645
  equations, 654–55
  power triangle and, 651–55
real, power triangle and, 651–55
to a resistive load, 643–44
series resonant circuit and, 842–50
Power factor
  angle, 655
  correction, 656–59
  defined, 655
Power gain, defined, 884
Power resistors, **74**
Power sources, electronic, 44–45
Power supply transformers, 994
Power system
  loads, 967
  transformers in, 994–95
Power of ten notation, 10–12
Power triangle, 651–55
  defined, 652
Powers, 12
Prefixes, 13–14
PRF. *See* Pulse repetition frequency
Primary winding, 980
Programming languages, 2
PRR. *See* Pulse repetition rate
PSpice, 19–20, 117–20
  ac circuits, 592–93, 631–34, 769–71
  frequency spectrum, 1056–58
  multiloop circuits, 313–15
  network theorem verification and, 368–70
  Ohm's law, plotting, 119–20
  parallel circuits, 205–207
  resonant circuits, 867–71
  series circuits, 161–63
  series-parallel circuits, 251–54, 714–20
  Thévenin or Norton equivalent of, 818–22
  three-phase system, 969
  transient analysis using, 447–50
Pulse, defined, 440
Pulse repetition frequency (PRF), defined, 440
Pulse repetition rate (PRR), defined, 440
Pulse train, defined, 440
Pulse width, 441
  effects of, 441–43

Q
Quality factor (Q), resonant circuit, 838–41

R
Radian measure, of a sine wave, 563–64
RC circuit(s), 418
  frequency effects on, 705–707
  pulse response of, 440–44
  series-to-parallel conversion, 851–56
  in steady state DC, 436–38
  timing application, 438
  transfer functions, 893–901
RC high-pass filter, 909–14
RC low-pass filter, 902–904
Reactive power, 643
  defined, 645
  equations, 654–55
  power triangle and, 651–55
Real power, 642–43
  power triangle and, 651–55
Reciprocity theorem, 361–63
Reference ground, symbol for, **142**
Reflected impedance, 992–93
Relative dielectric constant, defined, 388–89
Relative maximum power, 815
Relative permittivity, defined, 388–89
Reluctance, defined, 470
Resistance, 17, 628
  defined, 60
  dynamic, 114–15
  effective, 662–63
  internal, voltage sources, 149–55
  linear, 114
  measuring, 78–81
  nonlinear, 83–84, 114–15
  ohmic, 114
  and sinusoidal Ac, 617–19
  symbol for, 60
  temperature effects on, 69–71
Resistive circuit
  basic, **60**
  defined, 683
Resistive load, power to, 643–44
Resistors
  color coding of, 76–77
  defined, 60
  internal, arrangement, **74**
  Ohm's law and, 674–77
  in parallel, 183–88
    n equal, 185
    three, 188
    two, 186–87
  power rating of, 107
  in series, **130**, 134–36

superposition theorem and, 328–32
types of, 72–75
Resonance, parallel, 857–66
Resonant circuit
  computer-aided Three-phase, 867–71
  defined, 835
  quality factor (Q), 838–41
  series. *See* Series resonant circuit
Reverse
  biased diode, 84
  region, diodes and, 84
Rheostats, defined, 75
Right-hand rule, electromagnetism and, 464
Rise times, 441
RL circuits
  frequency effects on, 707–708
  series-to-parallel conversion, 851–56
  transfer functions, 893–901
RL low-pass filter, 904–909
RL transients, solving with computers, 537–41
RLC circuits
  frequency effects on, 708–710
  with sinusoidal excitation, 617
RMS
  true measurement, 591
  values, alternating current and, 588–89
Root mean square. *See* RMS

S
Schematic diagrams, 17–18
Schering bridge, 766–67
Scientific notation, 13–14
Second, defined, 8
Secondary winding, 980
Selectivity, series resonant circuit and, 842–50
Selectivity curve, series resonant circuit and, 843
Self-healing capacitors, 396–97
Self-induced voltage, 1011
Self-inductance, 499–501
  of a coil, 498
Semiconductors, 34
Series cells, 44
Series circuits, 130–31
  ac. *See* Ac series circuits
  defined, 130
  voltage divider rule and, 190
Series components, interchanging, 138
Series connection, defined, 130
Series inductance, 503–504
Series magnetic circuits, 469
  given NI, find φ, 482–84
  given φ, find NI, 475–80

Series resonance, 836–38
Series resonant circuit, 836
  bandwidth and, 842–50
  impedance of, 841–42
  power and, 842–50
  selectivity and, 842–50
Series-parallel
  combinations, **178**
  magnetic circuits, 480–81
Series-parallel circuits, 701–704
  analysis of, 223–30
  applications of, 231–39, 710–14
    computer-aided, 714–20
Series-parallel network, 222–23
Series-to-parallel
  RC conversion, 851–56
  RL conversion, 851–56
Shelf life, defined, 397
Shells, atomic, 31
Shockley, William Bradford, 783
Short circuit
  defined, 81
  illustrated, **80**
  transformer test, 1007–1008
Shorts, troubleshooting, 510–11
Shunt circuit. *See* Parallel circuit
Shunt resistor, Kirchhoff's current law and, 198
SI system of units, 7–8
  converting, 9–10
  unit sizes, **7**, 8
  wire tables and, 62
Signal generators, electronic, 554
Significant digits
  defined, 14
  numerical accuracy and, 14–16
Simulation Program with Integrated Circuit Emphasis (SPICE). *See* SPICE
Sine wave
  angular velocity, 562–63
  basic equation for, 561
  defined, 550
  effective values for, 585–88
  graphing, 564
  radian measure, 563–64
  rate of change of, 590
  shifted, phasors representing, 572–73
Single-phase equivalent, 947
  power and, 958–59
Single source circuits, reciprocity theorem and, 361–63
Single subscripts, 147
Sinusoidal Ac
  capacitance and, 623–27

inductance and, 619–22
resistance and, 617–19
voltage, 550, 551
Sinusoidal excitation
  magnetically coupled circuits with, 1015–17
  with R, L, and C circuits, 617
Skin effect, 663
SMDs (Surface mount devices), 398
Software
  circuit simulation, 18–20
  prepackaged math, 20
Solar cells, 46
Source conversion(s), 268–72, 739–43
Sources, voltage, symbol for, 100
Spacing, effect of, on capacitance, 388
Specimen, magnetizing a, 485–86
Spectrum analyzer, 1048–49
SPICE, 19
Square mil, **66**
  defined, 65
Square wave, defined, 440
Steady state conditions, capacitor(s), 419
Steady state DC, inductance and, 507–509
Steinmetz, Charles Proteus, 605
Step-down, symbol for, 999
Step-up, symbol for, 999
Stranded wire, 63–64
Stray capacitance, 506
Stray inductance, 506
Subscripts, voltage, 143–48
Substitution theorem, 357–58
Subtraction, using powers of ten, 11
Superconductors, 86–88
Superposition theorem, 328–32
  dependent sources, 789–91
  independent sources, 784–88
Surface mount capacitors, 398
Surface mount devices (SMDs), 398
Susceptance of the component, defined, 693
Switches, 51
  make-before-break, 200

T
Telegraph, 265
Temperature coefficient
  capacitors and, 395
  defined, 69
  symbol for, 69
Temperature intercept, defined, 70
Temperature, resistance and, 69–71
Terminal voltage, defined, 149
Tesla, defined, 465

Tesla, Nikola, 465, 935
Thermistors, 81–82
    defined, 81
    symbol for, **82**
Thévenin resistance, maximum power
        transfer theorem and, 352
Thévenin's theorem, 332–41
    computer-aided, 818–24
    dependent sources, 803–13
    independent sources, 791–96
Three-phase circuits, measuring power
        in, 960–63
Three-phase system
    balanced, power in, 954–59
    circuit connections, 937–40
    problem solving examples,
        948–54
    relationships, 940–48
        currents for a wye circuit, 943–45
        line and phase currents for a delta
            load, 945
        nominal voltages, 942–43
        single-phase equivalent, 947
    voltage generation, 936–37
Three-wire
    systems, 937–38
    wye-wye circuit, 938
Threshold of hearing, defined, 888
Tightly coupled circuits, 983
Time constant, 425
    of circuits, 525–27
Time scale, 533–36
    Ac cycle and, 553
Transducer, defined, 81
Transfer function
    defined, 893
    RC and *RL* circuits, 893–901
    sketching, 894–900
    writing, 900–901
Transformers
    additive voltages, 1011–12
    applications of, 994–1002
    coefficient of coupling, 1013
    construction of, 980–83
    defined, 980
    dot rule and, 1012–13
    efficiency of, 1007
    Faraday's law and, 983
    frequency effects, 1009–10
    impedance matching, 996–97
    iron-core. *See* Iron-core transformers
    with multiple secondaries, 998
    mutual voltages and, 1011
    in power systems, 994–95
    ratings of, 994

reflected impedance and, 992–93
    subtractive voltages, 1011–12
    tests of, 1007–1009
    voltage effects, 1009–10
    see also Iron-core transformers
Transient analysis
    time in, 421
    using computers, 444–45
Transient(s)
    current buildup, 523–27
    de-energizing, 529–31
    defined, 520
    duration of a, 425–27
    RL, solving with computers, 537–41
Transistor
    defined, 233
    development of, 783
Trimmer capacitors, 398
True RMS measurement, 591
Two-wattmeter method, 960

U

Unbalanced bridge, 305–12
    defined, 305
Unbalanced loads, 963–67
Units
    converting, 9
    SI system of, 7–8
    sizes of, **7**, 8
Unity power factor, 655
    correction, 656
Universal bias circuit, 236
Unloaded output, 239
US Customary system of units, 7–8
    converting, 9–10
    wire tables and, 62

V

Valence
    electron, 31
    shell, 31
Value, determining voltage/current, 567
Variable capacitors, 398
Variable inductors, inductance and, 505
Variable resistors, 75
    symbol for, **505**
Variable, symbol for, 999
Varistors, 85
VDR (Voltage divider rule), 190
    for series capacitors, 403–404
Volt, defined, 37
Volta, Alessandro, 95
Voltage
    Ac,
        measurement, 590–92

representing by complex numbers,
        611–12
    effects of on transformers, 1009–10
Voltage divider rule (VDR), 190
    for series capacitors, 403–404
Voltage gain, defined, 884
Voltage generation, Three-phase, 836–37
Voltage ratio, iron-core transformers,
        984–85
Voltage regulation, transformers,
        1005–1007
Voltage regulator, 237
Voltage(s), 36–38
    Ac, generating, 551–54
    across resistors, symbol for, 102
    breakdown, 393–94
    capacitors and, 404–407
    circuit, during energization, 524
    defined, 37–38
    divider rule, 139–42
    as functions of time, 565–70
    induced, 496–98
        computing, 501–503
    inductor, 521
    measuring, 46–50
        with multimeter, 48
    parallel sources of, Millman's theorem
        and, 359–61
    peak-to-peak, 559
    with phase shift, 568–70
    point sources, 148
    polarity, 101, 131
    potential energy and, 37
    practical sources of, 36–37
    sinusoidal Ac, 550, 551
    sources,
        in parallel, 189–90
        in series, 137
        internal resistance of, 149–55
        reciprocity theorem and, 361
    subscripts, 143–48
    superposition theorem and, 328–32
    symbols, 100
    time-varying, symbol for, 405
Voltmeters, 46
    loading effects, 200–203
    schematic symbols for, **50**
VOM (Volt-ohm-milliammeter), 46
von Helmholtz, Hermann Ludwig Ferdi-
        nand, 673, 737

W

Watt
    defined, 8
    symbol for, 104

Watt, James, 8
Watthour meter, 110
Wattmeter
    determining readings of, 961–62
    low power factor, 661
Watts ratio curve, 962–63
Waveform(s)
    alternating current, average value and,
        579–85
    amplitude, 559
    common, Fourier series and, 1039–46
    composite, 1032–33
    frequency of, 557
    frequency spectrum and, 1046–52
    nonsinusoidal. *See* Nonsinusoidal
        waveforms
    peak value, 559–61

peak-to-peak value, 559
period, 557–59
periodic, defined, 550
symmetry of, 1037–38
Westinghouse, George, 549, 935, 979
Wheatstone, Charles, 265, 303
Wheatstone bridge, 265
Winding directions, transformer, 982–83
Winding resistance, iron-core transform-
    ers, 1003
Wire
    resistance of, circular mils and, 65–68
    solid copper, table, 63
    stranded, 63–64
    tables,
        electrical, 62–65
        solid copper, 63

Wire-wound resistors, 73, 74
Wye circuit
    currents for, 943–45
    line and phase voltages for, 941–42
Wye load
    active power to balanced, 955–57
    unbalanced, 963–65
Wye-to-Delta conversions, 758–62

Z
Zener diode, 237
Zero impedance conductor, 947
Zero temperature coefficient, capacitors
    and, 395